T0418219

ELECTRIC, ELECTRONIC AND CONTROL ENGINEERING

PROCEEDINGS OF THE 2015 INTERNATIONAL CONFERENCE ON ELECTRIC, ELECTRONIC AND CONTROL ENGINEERING (ICEECE 2015), PHUKET ISLAND, THAILAND, 5–6 MARCH 2015

Electric, Electronic and Control Engineering

Editors

Fun Shao
Digital Library Department, Library of Huazhong University of Science and Technology, China

Wise Shu
Art College, Hubei Open University, China

Tracy Tian
Bos'n Academic Service Centre, China

CRC Press is an imprint of the
Taylor & Francis Group, an **informa** business

A BALKEMA BOOK

CRC Press/Balkema is an imprint of the Taylor & Francis Group, an informa business

Typeset by diacriTech, Chennai, India
Printed and bound in China by CTPS DIGIPRINTS

Published by: CRC Press/Balkema
P.O. Box 11320, 2301 EH Leiden, The Netherlands
e-mail: Pub.NL@taylorandfrancis.com
www.crcpress.com – www.taylorandfrancis.com

ISBN: 978-1-138-02842-5 (Hardback)
ISBN: 978-1-315-67504-6 (eBook PDF)

Electric, Electronic and Control Engineering – Shao, Shu & Tian (Eds)
© 2015 Taylor & Francis Group, London, ISBN 978-1-138-02842-5

Table of contents

Electric, Electronic and Control Engineering – Shao, Shu & Tian (Eds)

Preface

Electric, Electronic and Control Engineering contains the contributions present at the 2015 International Conference on Electric, Electronic and Control Engineering (ICEECE 2015, Phuket Island, Thailand, 5–6 March 2015). The book is divided into four main topics:

- Electric and Electronic Engineering
- Mechanic and Control Engineering
- Information and Communication Technology
- Environmental and Industrial Technology

Considerable attention is also paid to education science, chemical engineering, hydraulic engineering and civil engineering, etc. The book will be useful and invaluable to professionals and academics in electric & electronic, mechanic & control engineering and information & Communication and environmental & industrial technology.

Electric, Electronic and Control Engineering – Shao, Shu & Tian (Eds)
© 2015 Taylor & Francis Group, London, ISBN 978-1-138-02842-5

Acknowledgement

The authors, who contributed their papers to the conference, should be thanked first for their trust and supports. It's also grateful for e-science website of Chinese Academy of Science, which provided a platform to release information of conference, process the submitted papers online and communicate with authors directly. Of course, the efforts of all members in organizing committees can't be ignored. At last, thanks to all editors at Bos'n Academic Service Centre and CRC Press.

Electric, Electronic and Control Engineering – Shao, Shu & Tian (Eds)
© 2015 Taylor & Francis Group, London, ISBN 978-1-138-02842-5

Organizing committee

Program Chairs

- Dr. Max Lee, *Union University, United States of America*
- Dr. Shee Wung, *The Chinese University of Hong Kong*

General Chairs

- Prof. HongT Yu, *Huazhong University of Science and Technology, China*
- Dr. Fun Shao, *Huazhong University of Science and Technology, China*

Technical Committees

- Prof. Efu Liu, *China Pharmaceutical University, China*
- Dr. Long Tan, *Xiangya School of Medicine, Central South University, China*
- PhD. Chen Chen, *Huazhong University of Science and Technology, China*
- PhD. Fang Xiao, *Wuhan National Laboratory For Optoelectronics, China*
- Prof. Zuo Tang, *Central South University, China*
- PhD. Yhong Zhao, *Huazhong University of Science and Technology, China*
- PhD. Yuan Liu, *Technology Center of Foton Motor Co., Ltd., China*
- Dr. Bifa Zhu, *Hubei University of traditional Chinese Medicine, China*
- PhD. Fanze Hua, *Huazhong University of Science and Technology, China*
- PhD. Lifeng Wong, *Beijing University of Posts & Telecommunications, China*

Secretary General

- Doreen Dong, *Bos'n Academic Service Centre, China*
- Rita Sun, *Bos'n Academic Service Centre, China*

Electric, Electronic and Control Engineering – Shao, Shu & Tian (Eds)
© 2015 Taylor & Francis Group, London, ISBN: 978-1-138-02842-5

Adaptive algorithm of parallel genetic optimization based on orthogonal wavelet of space diversity

Xiaogen Pei
Communication Command Department of the Armored Force Institute, Bengbu, China
School of Information Science and Engineering, Shandong University, Jinan, China

Shaohua Zheng
Communication Command Department of the Armored Force Institute, Bengbu, China

ABSTRACT: Diversity technology can effectively resist channel multipath fading, and balanced technology can effectively inhibit the inter-symbol interference, so the diversity technology and balanced technology combined can effectively improve the quality of communication. Simultaneously, this paper uses parallel genetic algorithm to optimize the space diversity orthogonal wavelet adaptive algorithm, taking the space points on each branch equalizer weight vector as the son species of parallel genetic algorithms for selection, crossover, and mutation; between each species to each other and regularly send the best individual fitness; eliminate the worst individual fitness; and take diversity branch output signal and input orthogonal wavelet adaptive device combined. The computer simulation results show that the fast algorithm convergence speed and small steady-state errors can achieve the global optimal solution.

KEYWORDS: Parallel genetic algorithm; Space diversity; Wavelet transform; Adaptive.

1 INTRODUCTION

In a communication system, the channel's multipath fading and the transmission time delay phenomenon seriously affect the reliability of digital communication between transmitter and receiver. An effective method that is used for eliminating the channel fading is adopting diversity technology, which includes frequency diversity, time diversity, and space diversity, compared with the spatial diversity, frequency diversity, and time diversity that will take up too much bandwidth. And the adaptive technology is one of the most effective ways to overcome the inter-symbol interference, so the combination of space diversity and adaptive technique can effectively overcome channel fading and inter-symbol interference, thus improving the performance of the communication system [1]. Through orthogonal wavelet transform, orthogonal wavelet adaptive algorithm can reduce the correlation between signal and noise, thus accelerating the convergence speed [2]. However, the initialization of an adaptive algorithm for weight vector is sensitive, which makes the algorithm easily fall into the local minimum value and even divergence [3].

Genetic Algorithm (GA) provides a generic framework for solving complex system optimization problems, and it does not depend on problems of specific areas; problems of species have a strong robustness. However, the traditional genetic algorithm efficiency is not high and is easy in premature convergence. The parallel genetic algorithm using a traditional genetic algorithm intrinsic parallel mechanism aims at improving the accuracy of the algorithm efficiency and the precision of the solution, avoiding premature convergence, and accelerating convergence speed [4].

On the basis of the earlier analysis, this paper will have reference to the parallel genetic algorithm based on the orthogonal wavelet adaptive algorithm of space diversity; by using the parallel genetic algorithm for the spatial diversity equalizer, the weight vector of each branch is optimized, which is the diversity branch output signal input orthogonal wavelet adaptive device after the merger. The computer simulation shows that the fast algorithm convergence speed and small steady-state errors can get the global optimal solution.

2 NORM OF THE ADAPTIVE ALGORITHM BASED ON ORTHOGONAL WAVELET TRANSFORM

The norm of the adaptive algorithm based on orthogonal wavelet (WTCMA) principle is shown in figure 1.

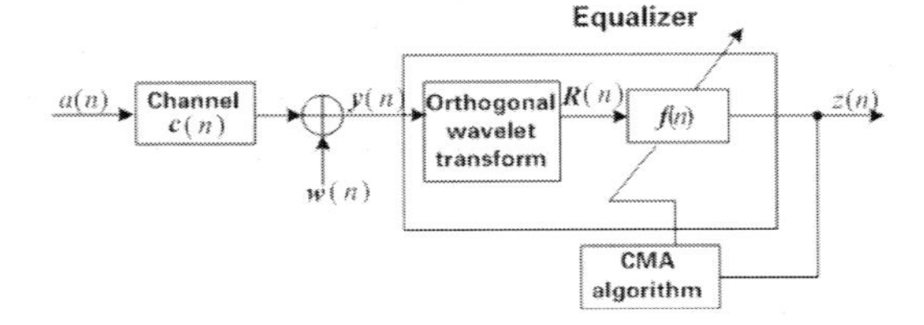

Figure 1. Adaptive algorithm based on orthogonal wavelet transform.

In figure 1, n shows the time sequence, $\boldsymbol{a}(n)$ shows the launch signal, $\boldsymbol{c}(n)$ shows the channel impulse response, $\boldsymbol{w}(n)$ shows the channel output of additive white Gaussian noise, $\boldsymbol{y}(n)$ shows the equalizer input signal, $\boldsymbol{R}(n)$ shows the signal after orthogonal wavelet transform, $\boldsymbol{f}(n)$ shows the equalizer weight vector, and $\boldsymbol{z}(n)$ shows the signal after equilibrium.

The norm of the adaptive algorithm of orthogonal wavelet transform for the cost function is as follows:

$$J(\boldsymbol{f}) = E\{[R_2 - |z(n)|^2]^2\} \tag{1}$$

$$R_2 = \frac{E(|a(n)|^4)}{E(|a(n)|^2)} \tag{2}$$

$$z(n) = \boldsymbol{f}^H(n)\boldsymbol{R}(n) \tag{3}$$

$$\boldsymbol{R}(n) = \boldsymbol{Q}\boldsymbol{y}(n) \tag{4}$$

$$\boldsymbol{f}(n+1) = \boldsymbol{f}(n) + \mu\hat{\boldsymbol{R}}^{-1}(n)\boldsymbol{R}(n)e(n)z^*(n) \tag{5}$$

$$e(n) = z(n)[|z(n)|^2 - R_2] \tag{6}$$

In the formula, $\boldsymbol{Q}$ shows the orthogonal wavelet transform matrix, superscript H shows conjugate transpose, $e(n)$ shows the error function, μ shows the step length, $\hat{\boldsymbol{R}}^{-1}(n) = \text{diag}[\sigma^2_{j,0}(n), \sigma^2_{j,1}(n), \ldots, \sigma^2_{j,k_j}(n), \sigma^2_{J+1,0}(n), \ldots, \sigma^2_{J,k_j}(n)]$, among them $\sigma^2_{j,k_j}(n)$, $\sigma^2_{J+1,k_j}(n)$, respectively, for $r_{j,k}(n)$, $s_{J,k}(n)$ of the average power estimation. By the following recursive formula, we get:

$$\begin{cases} \hat{\sigma}^2_{j,k}(n+1) = \beta\hat{\sigma}^2_{j,k}(n) + (1-\beta)|r_{j,k}(n)|^2 \\ \hat{\sigma}^2_{J+1,k}(n+1) = \beta\hat{\sigma}^2_{J+1,k}(n) + (1-\beta)|s_{J,k}(n)|^2 \end{cases} \tag{7}$$

However, the WTCMA algorithm has easy convergence to the local minimum points, and the performance of channel fading resistance is poor. In order to overcome the WTCMA performance defects, in combination with the parallel genetic algorithm, the space diversity, and orthogonal wavelet adaptive algorithm, the adaptive performance should be improved.

3 BALANCE TECHNOLOGY BASED ON SPACE DIVERSITY

3.1 *Spatial diversity equalizer*

Space diversity also calls the antenna diversity and uses more of the diversity form in the communication, in simple terms, uses multiple receiving antennas to receive signals [5]. Because the decline of each primitive received signal can be regarded as independent of each other, and the fading probability of all channel colleagues is very low, so the space diversity technology is one of the effective ways to eliminate the decline. As a result, the space diversity technology is applied to the adaptive equalizer and can effectively overcome the channel fading and inter-symbol interference. Spatial diversity equalizer (SDE) principle is shown in figure 2.

In figure 2, $\boldsymbol{a}(n)$ is independent with the distribution of the emission signal sequence $\boldsymbol{c}_l(n)$ is the NO. l road channel of impulse response vector:

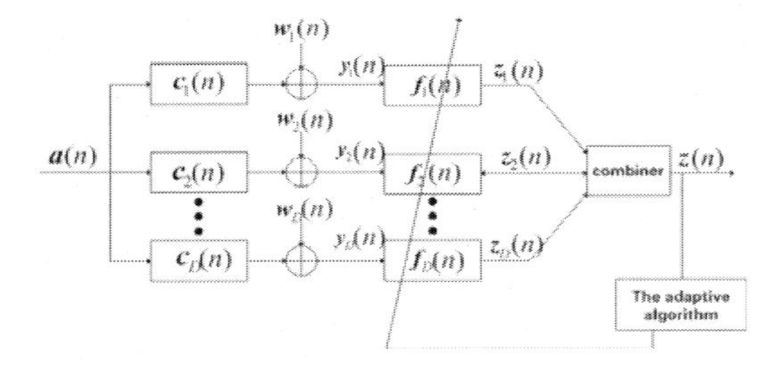

Figure 2. Spatial diversity equalizer structure.

$\boldsymbol{w}_l(n)$ is the NO. l road of Gaussian white noise sequence; $\boldsymbol{y}_l(n)$ shows the NO. l road of input signal vector; $\boldsymbol{f}_l(n)$ is the NO. l road of the equalizer's weight vector; among them, $\boldsymbol{M}_f$ shows the length of the equalizer; $\boldsymbol{z}_l(n)$ is the NO. l road of the output of the equalizer; and among them, $\boldsymbol{z}(n)$ shows the combined output signal.

Figure 2 shows that the adaptive spatial diversity in each branch is made up of sub-channels and the sub-adaptive device. Signals are received through the sub-channel and the sub-equalizer to merge processing; merging processing methods have a choice to merge, with maximal ratio combination and an equal-gain merger. Among them, the equal-gain merger is the most easy to implement, so this article will use an equal-gain merger method.

3.2 *Adaptive algorithm of space diversity based on equal-gain merger*

Two paths of the equal-gain merger of space diversity adaptive device structure are shown in figure 3.

According to the figure, two paths of the equal-gain spatial diversity equalizer contain two prior to filters (each path set a filter) $f_F^{(i)}(n)$ and a rear filter $f_B(n)$. Weight vector of three filters is united by the adaptive algorithm based on decision feedback adjustment, then

$$z(n)=\sum_{i=1}^{2} y^{(i)}(n)[f_F^{(i)}(n)]^T - \hat{a}(n-1)f_B(n) \quad (8)$$

$$f_F^{(1)}(n+1)=f_F^{(1)}(n)-\mu_F z(n)[|z(n)|^2-R^2]\,[y^{(1)}(n)]^* \quad (9)$$

$$f_F^{(2)}(n+1)=f_F^{(2)}(n)-\mu_F z(n)[|z(n)|^2-R^2]\,[y^{(2)}(n)]^* \quad (10)$$

$$f_B(n+1)=f_B(n)-\mu_B z(n)[|z(n)|^2-R^2]\hat{a}^*\,(n-1) \quad (11)$$

Taking the just cited equal-gain merger of space diversity adaptive device apparatus for EG – SDE,

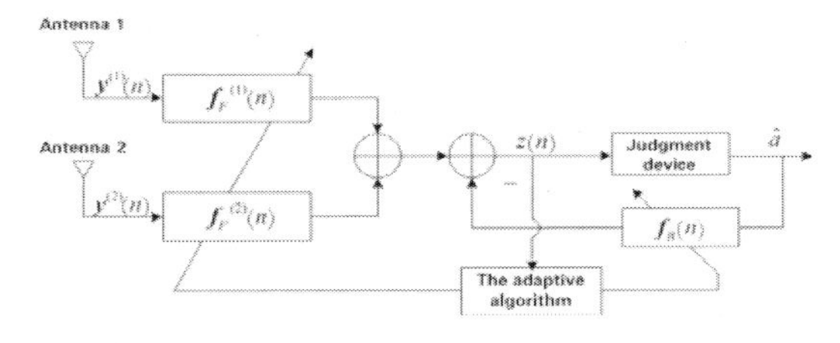

Figure 3. Equal-gain merger of space diversity adaptive device structure.

4 INTRODUCING PARALLEL GENETIC OPTIMIZATION OF SPACE DIVERSITY WAVELET ADAPTIVE ALGORITHM

4.1 *Parallel genetic algorithm*

Parallel Genetic Algorithm (PGA) is a suitable kind for complex constrained optimization problem of global optimization ability. Parallel genetic algorithm mainly hasthreecategories:(1)themaster–slaveparallelgenetic algorithm; (2) coarse-grained degree parallel genetic algorithm; and (3) fine-grained degree parallel genetic algorithm[6]. Among them, the coarse-grained parallel genetic algorithm distributes several sub-populations to their corresponding processor; each processor has not only independent calculation fitness but also independent selection, restructure crossover, and mutation operation, and regularly sends each other the best individual fitness, thus speeding up to meet the requirements of the termination conditions. Currently, it is the most widely used parallel genetic algorithm [7].

Migration is a parallel genetic algorithm introduced into a new operator that points to the course of evolution of neutron population exchange between individual processes; the general migration method sends the best individual in the group to other subgroups. Through migration, it can accelerate better individuals in the group communication, improve convergence rate and precision of the solution[8]. Compared with the single population, only a small number of individual evaluation calculation workload is needed. As a result, even with a single processor, in a serial way (pseudo parallel), a computer could implement the parallel algorithm, thus producing good results [9]. So, the use of a migration operator makes the parallel algorithm more suitable for global optimization with a small amount of calculation.

The most basic migration model is the ring topology model, and individual transfer occurs only in the adjacent sub-population. In the adjacent transfer model, the transfer occurs only in the close neighbor set. The models are shown in figure 4.

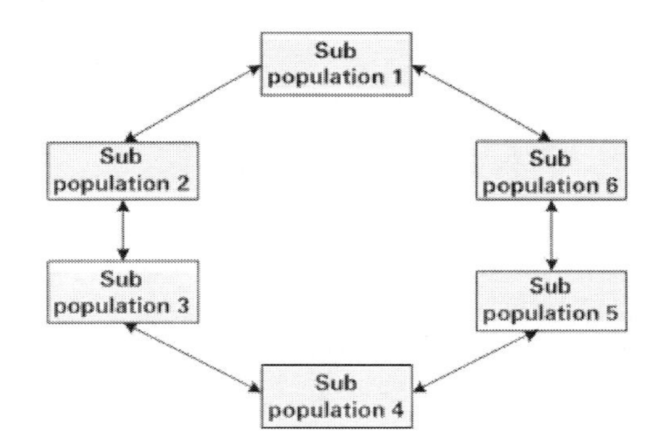

Figure 4. Ring topology model figure.

This article considers the combination of the coarse-grained degree parallel genetic algorithm based on migration of ring topologies, space diversity, and orthogonal wavelet transform and adaptive algorithm; the parallel genetic algorithm has fast convergence speed, global optimal solutions, and orthogonal wavelet transform that reduce the correlation of signal and noise to the space diversity adaptive algorithm. These are optimized by an equal-gain merger, and we get the space diversity orthogonal wavelet adaptive algorithm based on parallel genetic optimization.

4.2 *Space diversity orthogonal wavelet adaptive algorithm based on parallel genetic optimization*

By reference to the parallel genetic algorithm and the orthogonal wavelet transform based on the equal-gain merger of adaptive algorithms, we get the spatial diversity based on parallel genetic optimization; this algorithm has a high convergence rate, precision of solution, and we can get the global optimal solution. Three paths of space diversity orthogonal wavelet adaptive algorithm based on parallel genetic optimization principles are shown in figure 5.

In the figure, n shows the time sequence, $\boldsymbol{a}^{(i)}(n)$ shows the NO. i road launch signal, $\boldsymbol{c}^{(i)}(n)$ shows the NO. i road channel impulse response, $\boldsymbol{w}^{(i)}(n)$ shows the NO. i road channel output end additive Gaussian white noise, $\boldsymbol{f}_F^{(i)}(n)$ shows the NO. i road pre-equalizer weight vector, $\boldsymbol{y}^{(i)}(n)$ shows the NO. i road output signal, $\boldsymbol{y}(n)$ shows the signal after the equal-gain merger, $z(n)$ shows the output signal, $\boldsymbol{f}_B(n)$ shows the orthogonal wavelet adaptive device weight vector, and $\boldsymbol{J}^{(i)}(f_m)$shows the NO. i road parallel genetic algorithm's son population of cost function.

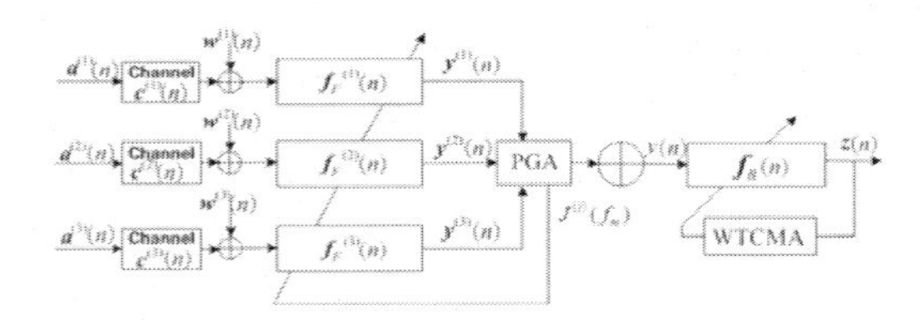

Figure 5. Block diagram of space diversity principle orthogonal wavelet adaptive algorithm based on parallel genetic optimization.

In this algorithm, three forward equalizers are contained (each path set an equalizer), with each branch of the equalizer weight vector as the decision variable of parallel genetic algorithms, the design initial population $Chrom^{(i)} = [f_1, f_2, \cdots, f_M]$, of which M is individual number. In the population of every individual, $f_m (1 \leq m \leq M)$ corresponds to a balancer weight vector. This paper constructs the initial population based on the characteristics of parallel genetic algorithms, namely there are three son populations, with each son population corresponding to a path. Considering the characteristics of the adaptive, the weight vector of each module value will be less than 1, so will the search as the [0, 1].

Assuming the received signal sequence of length for N, using the time average instead of the statistical average, formula (1) often shows that the constant model algorithm of the cost function can be calculated by the following formula:

$$J^{(i)}(f_m) = \frac{\sum_{j=1}^{N} (R - |y_m^{(i)}(j)|^2)^2}{N} \tag{12}$$

In the formula, m shows the equalizer weight vector of individual serial number, N shows the received signal sequence length for each generation, i shows the son population number, and $y_m^{(i)}(j)$ shows each equalizer weight vector of the output signal of the individual by each path. Using the earlier formulas as the objective function of parallel genetic algorithms, and solving its minimum value, we obtain the best individual as the optimal weights coefficient of the adaptive algorithm. Due to the $J^{(i)}(f_m) > 0$, therefore, the individual fitness functions are as follows:

$$Fit^{(i)}(m) = \frac{1}{J^{(i)}(f_m)} \tag{13}$$

This paper uses the coarse-grained degree parallel genetic algorithm based on migration of ring topology, with each branch of the equalizer weight vector as a sub-population of the parallel genetic algorithm. Entering the parallel genetic algorithm, each sub-population is assigned to each processor's independent fitness calculation, selection, crossover, and mutation operations, and every generation of a ring topology operation on migration. Each sub-population has copies with the highest fitness individuals to adjacent sub-populations and replaces its lowest fitness of individuals; this can avoid algorithm premature convergence and can improve the convergence speed. Meeting termination conditions of the genetic algorithm, each branch of output signal after the equal-gain merger, we should input the rear equalizer, the equalizer weight vector by the adaptive algorithm adjustment based on orthogonal wavelet transform. This has played an important role in small range search and search speed. The orthogonal wavelet adaptive algorithm based on parallel genetic optimization and space diversity makes full use of the parallel genetic algorithm, space diversity, and orthogonal wavelet adaptive algorithms along with their respective advantages, and a better adaptive effect is obtained.

5 COMPUTER SIMULATION EXPERIMENT

In order to test the performance of the PGA-SD-WTCMA algorithm, we should compare it with the WTCMA and the EG-SDE serving as the objects, having the computer simulation experiments. In this paper, the genetic algorithm program uses a genetic algorithm toolbox (GATBX) based on the Matlab. Channel using mixed-phase underwater acoustic channels c_1 = [0.3132, -0.104, 0.8908, 0.3132], minimum-phase underwater acoustic channels c_2 = [0.8264, 0.1653, 0.1653], and two-size underwater acoustic channel c_3 = [-0.35, 0, 0, 1]. Emission signal is 16QAM, signal-to-noise ratio is 20DB, and the length of the equalizer is 16.

The algorithm parameter selection is as follows:

WTCMA used the channel c_1, the initial weight vector of the NO.4 tap took 1, the other took 0, and the step length took 0.000015.

EG-SDE used the channel c_1 and c_2; among them, CMA's initial weight vector of the NO.4 tap took 1, the other took 0, and the step length took 0.000005.

PGA-SD-WTCMA used the channels c_1, c_2, and c_3; in the WTCMA algorithm, the parts initial weight vector of the NO.5 tap took 1, the other took 0, and the step length took 0.00003. The sub-population size took 20, crossover probability took 0.7, the mutation probability took 1 out of 16 and every generation migration, termination conditions for evolution to the tenth generation.

The orthogonal wavelet change adopted Db4 wavelet in the WTCMA and the PGA-SD-WTCMA; the decomposition layer number is 2; β took a value of 0.99; and the power initialization value is 10.

The simulation results are shown in figure 6.

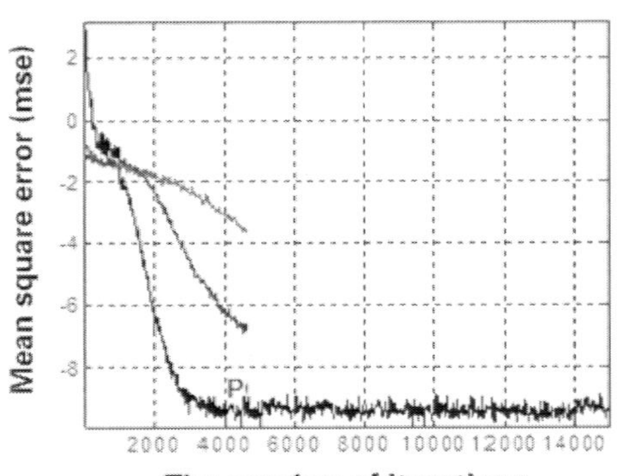

(A) The mean square error of curve

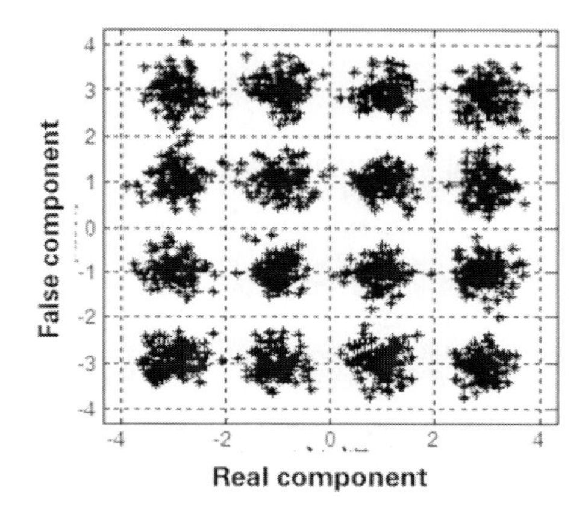

(B) WTCMA output signal

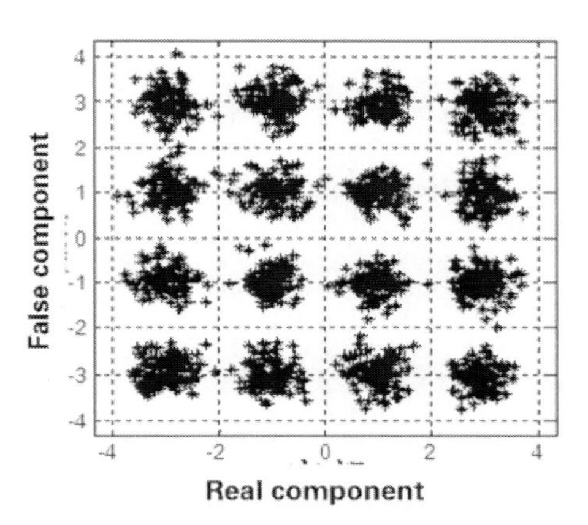

(C) EG-SDE output signal

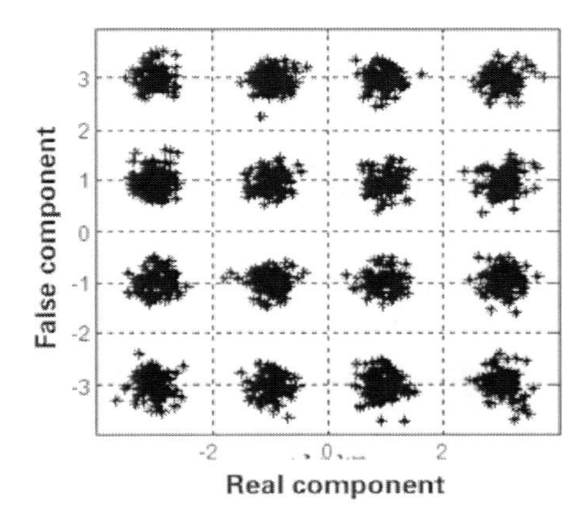

(D) PGA-SD-WTCMA output signal

Figure 6. Computer simulation results.

Figure (A) shows that the convergence rate of the PGA-SD-WTCMA than the WTCMA and the EG-SDE increased nearly 7000 times and 2500 times, respectively, and the mean square error decreased nearly 3 dB and 2 dB, respectively.

Based on figure (B)-(D), after the PGA-SD-WTCMA optimization of equalizer output data error than the WTCMA and the EG-SDE decreased significantly, the constellation diagram is clearer.

These results show that the orthogonal wavelet adaptive algorithm based on spatial diversity by optimization of the parallel genetic algorithm can effectively accelerate the convergence speed and reduce the mean square error (MSE); performance is obviously improved.

6 CONCLUSION

In order to overcome the influence of the multipath effect and channel fading, the combination of the space diversity, parallel genetic algorithm, orthogonal wavelet transform, and adaptive algorithm proposed the orthogonal wavelet adaptive algorithm based on parallel genetic optimization and space diversity. The algorithm makes full use of space diversity and decreases the influence of the multipath effect, using the parallel genetic algorithm for the optimization of each branch of the space diversity of the equalizer weight vector, accelerating the algorithm convergence speed and global convergence, and using the auto-correlation based on orthogonal wavelet transform to reduce the signal, so that the algorithm has good performance. The computer simulation results verify the effectiveness of the algorithm.

REFERENCES

[1] Alain Y K, Gerard F. (2009) Blind equalization of nonlinear channels using a tensor decomposition with code/space/time diversities. *Signal Processing*, 89(2):133–143.

[2] Han Y G. (2007) *The Research of Adaptive Device Design and Algorithm Simulation Based on Wavelet Transform*, Anhui: Anhui University of Science and Technology.

[3] Li Y, Li K, Lu X. (2009) The Neural network adaptive algorithm based on genetic optimization, *Journal of China's North University*, 30(2):137–142.

[4] Wang X P, Cao L M. (2003) *Genetic Algorithm Theory, Application and Software Implementation*. Xi'an: Xi'an Jiaotong University Press.

[5] Din X J. (2010) *The Orthogonal Wavelet Adaptive Algorithm Based on Diversity Technology*, Nanjing: Nanjing Information Engineering University.

[6] Chen Y F, Tian Y X. An improved parallel genetic algorithm to solve TSP.

[7] Xue S J, Guo S Y, Bai D L. (2008) The analysis and research of parallel genetic algorithm. *Wireless Communications, Networking and Mobile Computing, 2008. WiCOM'08. 4th International Conference on*, pp:1–4.

[8] Zdenek K F. (2004) Parallel genetic algorithm: advances, computing trends, applications and perspectives. *Parallel and Distributed Processing Symposium*, pp:162–169.

[9] Rebaudengo, Sonza R D. (1993) An experiment analysis of the effects of migration in Parallel Genetic Algorithms. *Parallel and Distributed Processing*, pp:232–238.

Electric, Electronic and Control Engineering – Shao, Shu & Tian (Eds)
© 2015 Taylor & Francis Group, London, ISBN: 978-1-138-02842-5

Research and application of comprehensive evaluation and detection analysis platform of transformer bushing state

Anquan Cai, Qilin Zhang, Yawei Li & Qiang Zeng
Bazhong Power Supply Company, State Grid Sichuan Electric Power Company, Bazhong Sichuan, China

ABSTRACT: Substation transformer is an important part of power system and transformer bushing is also one of the important electric transmission and transformation equipment. According to the main problem of insulation destruction of transformer bushing, the author starts from causes of accelerating the insulating material aging of bushing and proposes the construction of comprehensive evaluation and detection analysis platform of transformer bushing state. From aspects of both hardware and software, the author realizes on-line monitoring on the insulation state of main transformer bushing with such technologies as digital sensor, signal de-noising processing based on wavelet transform, bushing data acquisition, and accurate influence of dynamic change of busbar voltage frequency on end screen current.

KEYWORDS: dielectric loss; end screen; state monitoring; comprehensive evaluation.

1 SUBJECT OVERVIEW

Substation transformer is an important part of power system and transformer bushing is also one of the important electric transmission and transformation equipment. Massive power outage brought by various insulation destructions of bushing will cause a huge loss to the whole national economy and affect the safe operation of the power grid as well as the safety of the power station staff. Therefore, the insulation state of the main transformer in operation is of great significance for the safety of the power system. Operation and maintenance staff should have an obligation to strengthen monitoring and diagnosis on electrical equipment insulation so as to find hidden troubles timely and guarantee the safe operation of electrical equipment.

Under the comprehensive effect of electricity, magnet, external force and environment, the insulation state of bushing will become gradually aged, the performance will gradually decrease, and more severely, there will be no function of insulation. This is the aging of insulation. However, the aging of insulation is a process with slow changes. In many cases, problem and degree of insulation aging cannot be directly found. There might be severe failures of bushing and even insulation breakdown if insulation aging is in the critical stage and maintenance measures cannot be taken timely, causing equipment damage and even explosion and fire so as to influence the normal operation of a transformer. There might also possibilities of emergent power accidents, causing casualties, massive power outage and huge direct and indirect economic losses.

2 RESEARCH IDEA OF THE SUBJECT

After the electrification of bushing, the phase angle of leakage current and voltage is 90° in advance. Therefore, if the insulation performance decreases, there will be impedance component of the same phase of voltage in leakage current. Any change of capacitance and dielectric loss (tanδ) would lead to the change of leakage current (change of capacitance and impedance component). The on-line monitoring system of dielectric loss and capacitance of bushing detects leakage current of a set of three-phase bushing through the end screen sensor installed at the test tap of bushing, then represents the change of capacitance and dielectric loss through vector analysis of leakage current, and obtains the on-line trend display of leakage current through the comparison with system initial values, which can be acquired by factory test or on-site off-line test of bushing.

One of the processing technologies of leakage current is the vector current superposition, the principle of which is that the change of any component of leakage current would have impact on the vector sum of leakage current. Because the impedance component of leakage current with single bushing is related to dielectric loss of insulation and the change of the impedance component also has small impact on the vector sum of leakage current, small changes of dielectric loss can be detected with this technology, so that small changes of single leakage current can be also detected through current vector. The change of an insulation bushing can be inferred through phase changes of current and vector.

In this subject, through detecting the leakage current situation of high-voltage bushing end screen of transformer, the author calculates dielectric loss and capacitance of transformer bushing, establishes the comprehensive monitoring and expert diagnosis system with multiple functions and parameters, carries out concentrated supervision and diagnosis on electrical equipment of substation, improves reliability and sensitivity of monitoring system, fully grasps states and defects of high-voltage bushing of transformer, develops artificial intelligence technology and increases the level of intelligence, so as to realize unattended operation and remote control, maintenance and management of transformer, reduce labor intensity, increase working efficiency and quality, and realize greater efficiency by reducing staff. Meanwhile, it can also reduce costs and improve efficiency of investment. Timely and effective maintenance can prolong the service life of equipment, which improves efficiency of power generation by establishing resource saving and environmentally friendly substations.

3 DESIGN RESEARCH OF THE PLATFORM

The design idea of the system can be generally divided into two parts, namely the hardware part and the backstage system part. Hardware monitoring sensor is installed on the attachment of the transformer bushing end screen to carry out on-line monitoring on the leakage current of bushing and transmit monitoring data to a data repeater. Data repeater transmits data back to backstage host station system after package. The design idea of software consists of two parts, basic data and functional module. The part of basic data is used for storing equipment account data of various substations, transformers and bushings. Monitoring data are received, modeled, calculated and analyzed through functional module, which finally displays the results of analysis and calculation and gives alarms to cross-border data.

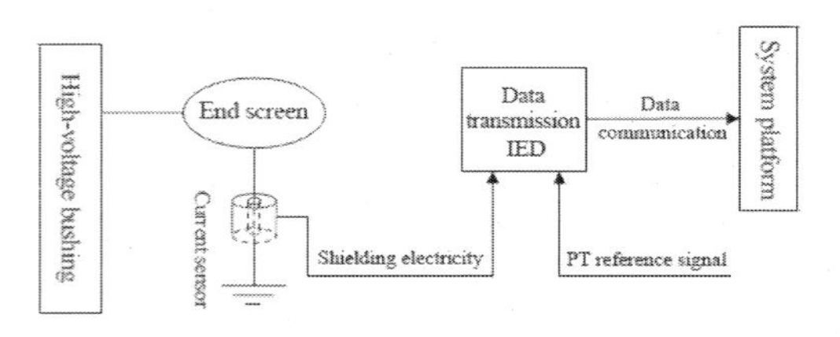

Figure 1. The design idea of the system.

4 KEY TECHNOLOGIES

- Development of digital sensor

Digital sensor is the product of microelectronic technique, computer technology and automatic test technology, the inner part of which contains sensor, A/D convertor, signal processor, memorizer (or register) and interface circuit. It replaces the original 4-20mA standard analog signals with digital signals so as to guarantee the accuracy of data acquisition. Data is transmitted in the way of digital coding to ensure data correctness.

Digital sensor represents the development trend of sensor. It is able to realize self-management through standard digital bus interface. After exchange of detected signals, information communications with upper equipment can be conducted in a standard protocol through a digital interface.

In field operation, there is a great deal of random noises in signals captured by the front perception part of the digital sensor, which inevitably influences the data acquisition process with noises and clutters. As for front-end noises, low-frequency and high-frequency noises are mainly filtered by low-pass and high-pass filters. But useful signals are also filtered in this way.

- De-noising processing of signals based on wavelet transform

By adding threshold de-noising, relevance de-noising and the maximum de-noising algorithms to the DSP processor of digital sensor, the de-noising algorithm based on wavelet transform equipped with digital sensor eliminates noises of signal and makes acquired data more close to real data.

- Bushing data acquisition

A great deal of monitoring data of transformer bushing and maintenance data of electric power can be acquired through cooperation with transformer manufacturers and power departments, which are mined with classifications and stored in expert database for analytical judgment of a later period.

- Accurate influence of dynamic change of busbar voltage frequency on end screen current

The system uses waveform of periodic continuous voltage and current for excessive acquisition. Through frequency calculation on voltage waveform from continuous excessive acquisition, the ZCP phase angle can be determined after the determination of voltage waveform frequency so as to make sure that correct impedance current value of dielectric loss calculation can be obtained in current waveform of end screen and guarantee the correctness of the system in terms of data.

5 SYSTEM FUNCTION DESIGN

The system mainly includes two modules, hardware detection equipment and software data processing.

Data acquisition of bushing state is completed by the hardware detection module, which is then provided to backstage analysis software.

The backstage analysis software includes the following parts:

5.1 *Data acquisition module*

Data acquisition module is an important link of the system software. After acquisition by sensor detection equipment with high precision, the required state for detection and data parameters are sent to data concentration units through wire transmission and transmitted to backstage host through optical fiber, 485 bus or wireless private network. The data acquisition module of backstage host starts operation. Various data are stored in database in different classification for the convenience of data recall of other modules. Through advanced data fault-tolerant technology, the system is able to analyze the accuracy of acquired data. When it is judged that there might be fault of data, data acquisition module automatically recalls data from data concentration units and makes comparison with data of the last time after an accuracy analysis on recalled data so as to guarantee the accuracy of data and avoid calculation error of data processing module caused by data transmission error.

5.2 *Data processing module*

Data processing module is also very important in the whole system. After acquiring and storing data in database accurately, data processing module receives start-up information and starts processing on data. The international advanced algorithm is adopted by the system, the highly accurate calculation and results of which can truly reflect the current state of equipment. It is convenient for other modules to conduct processing on calculation results and for detection staff to determine equipment state in line with calculation results.

5.3 *Off-limit alarm module*

The system is able to set a variety of critical limits as needed. Off-limit alarm module will alarm when test data and calculation data are in excess of critical value. The alarm includes two levels, warning and alarming. Warning means that management staff will be reminded to focus on equipment state when test data and calculation results are in excess of lower critical value. Alarming means that the equipment state is in emergency and management staff will be required to focus on equipment state immediately and adjust operating state of equipment when test data and calculation results are in excess of higher critical value. Off-limit alarm module is able to send messages to relevant management staff according to the needs of users so as to make sure that relevant staff can know hazardous situations of equipment at any time.

5.4 *Data modeling module*

Data modeling is an abstract organization of all kinds of data in real world, which determines the scope of database and organizational form of data until a practical database is established. In this system, data modeling module, data acquisition module and data processing module are mutually relied on, being able to convert analog data of bushing into digital data that are stored in database, and then convert physical model into mathematical model to guarantee the calculability of test results so as to guarantee working stability of data acquisition module and data processing module and provide a strong support for the display of test data and calculation results.

5.5 *Module of Chinese and foreign expert database*

The system gathers various information into a huge expert database, including relevant testing guidelines of State Grid, operational test results of similar equipment at home and abroad, equipment state determined by the current system test results and later various accidents. The expert module is the most important module of the whole system. Judgment and comparison of state diagnosis module and risk evaluation module on equipment test results and functions like risk evaluation on current and future condition of equipment both rely on true and accurate data of this module. In a manner of speaking, equipment state cannot be evaluated, hidden troubles of equipment cannot be inferred before failure and potential safety hazards cannot be eliminated in advance without expert database module.

5.6 *State diagnosis module*

The system will make a comparison between calculation results and data of expert database state of various periods immediately and show the comparison results of expert database to users. Users can directly know the current state of equipment from diagnosis results, the stage of current state in the whole service life, accidents of current state, and the percentage of accidents in expert database of the same level, which are for the convenience of management staff to fully understand equipment state and make comprehensive conclusion on equipment state.

5.7 *Risk evaluation module*

Risk evaluation module also needs to rely on authoritative expert database module. Through comparison between current monitoring data and calculation results and mass data in expert database, the module analyzes the possibility of failures in the state and the time period between the current and future failure and alarm point. For example, possible failures or alarms in the near future will increase the risk evaluation level and remind management staff so as to make sure that management staff can know equipment state and potential risk from all aspects and support management staff to formulate a maintenance schedule in future.

5.8 *Trend judgment module*

Trend judgment module can conclude the trend of future operation state with trend inference method through the analysis on historical test data and calculation results of equipment. It can also carry out state diagnosis on arbitrary state point of equipment state curve can make comparison between real data and expert database so as to conclude accidents of the state, estimate the alarm time of equipment state, help management staff to estimate time of failure and alarm and formulate better maintenance schedule.

6 CONCLUSION

This platform is installed in 110kV Zhaipo Substation of Bazhong Power Supply Company in Sichuan province for on-line monitoring test, successfully realizing on-line monitoring on the insulation state of three-phase bushing A, B and C of the main transformer. Hundreds of groups of data tested in one week are analyzed with the software, results of which show that the on-line monitoring system is in good operational condition. End screen grounding current I and equivalent capacitance C are relatively stable while dielectric loss ($\tan\delta$) has slight fluctuation. But they are in the normal range. Data changes have certain cycle, reflecting the influence of external factors on insulating properties like ambient temperature and humidity, busbar voltage and insulation grade.

The backstage expert diagnosis and evaluation system shows the results of measurement and calculation. Statistical chart of discrete distribution data is established through periodic monitoring data and historical data for data classification. Estimation on service life of bushing is then carried out through fuzzy statistical probability. With the constant accumulation of monitoring data, the artificial intelligence algorithm library gradually adjusts the fuzzy statistical probability factor κ so that it is more faithful to the precise critical value and the service life of bushing can be estimated more accurately.

REFERENCES

[1] Sun, J.X., Wu. G.Y., Zhou, L.J., Chen, L., Wen, D.B., Du, P.D. & Liu, Y.C. (2012) Research on the improvement of dielectric loss factor measurement of transformer bushing under the working condition of intermittent heavy load. *China Railway Science*, 33(5).

[2] Yin, S.L., Xia, L. & Wang, K.F. (2012) Diagnosis and analysis on a defect of transformer bushing. *Journal of Anhui Technical College of Mechanical and Electrical Engineering*, 17(1).

[3] Bi, J. (2002) Analysis on high-voltage bushing accidents of 110kV transformer. *Transformer*, 39(3).

[4] Cao, F.X. & Li, Y.Z. (2012) Analysis and processing on heating problem of bushing joint of 220kV transformer. *Transformer*, 49(2).

Electric, Electronic and Control Engineering – Shao, Shu & Tian (Eds)
 ISBN: 978-1-138-02842-5

Designing a state transition circuit with Creator2.0 and its PSoC realization

Yiping Liu, Wei Wang, Jiadong Huang, Xing He, Tao Huang & Baokai Liu
College of Engineering, Shanghai Second Polytechnic University, Shanghai, China

ABSTRACT: Designing a state transition circuit is one of the principal tasks of digital electronics. In our design, two methods are used to realize the state transition required. One is the traditional way that uses D flip-flops and gate circuit for design realization and realizes hardware-based state transition by designing, compiling, debugging and downloading to PSoC using schematic method; the other uses a counter and a lookup table for state transition which also downloads to PSoC and achieves hardware realization. The result indicates that both methods are able to achieve sequential logics on single-chip PSoC.

KEYWORDS: Creator2.0; PSoC; lookup table; state transition circuit.

1 INTRODUCTION

Most of the electronic systems in use nowadays are a mixture of digital signals with analog signals. Semiconductor devices used in designing an electronic system are generally minor integrated circuits, discrete components and programmable devices, whereas the majority of the programmable devices are either digital or analog ones alone, which necessitates the connection of other external devices when building up a system and thus complicates the system design. In such cases, the essential requirements for system design are diluted since a lot of effort is spared for building systems and removing troubles.

The Programmable System on Chip (PSoC) introduced by Cypress Semiconductor is a combination of a microcontroller, a programmable digit array and a programmable analog array designed for "in-system programmability", which provides for both the resources of a normal electronic system and the development trend of modern electronic design methodology. Use of the resulting electronic system is not only convenient, but also gives the best play to people's creativity.

In this paper, two methods are used to realize state transition circuit on new software and new hardware as a reform from the traditional way of thinking.

2 ABOUT CREATOR2.0 SOFTWARE AND PSOC HARDWARE

PSoC Creator2.0 is a full-function graphical hardware/software design and programming environment with innovative graphic design interfaces that allow hardware design, software design and debugging, project compilation and downloading with PSoC3 or PSoC5 chips. Its graphical design entry simplifies the task of configuring a particular component so that the designer can select the functions he needs from the component library and place it into his design. All parameterized components have an EDITOR dialog box that allows the designer to configure the functions as needed.

The PSoC Creator2.0 software platform automatically configures the clock pulse and wiring I/O to the selected pins and generates application program interface function API for a given application to control the hardware. It is very easy to modify the PSoC configuration, such as add a component, set its parameters and rebuild a project. At any stage of the development, the designer may modify the hardware configuration or even the target processor. He may also modify C compiler and perform performance evaluation.

3 DESIGN OF STATE TRANSITION CIRCUIT

If we have to realize state transition as shown in Fig 1 as practically needed, two methods may be used to design the circuit we need.

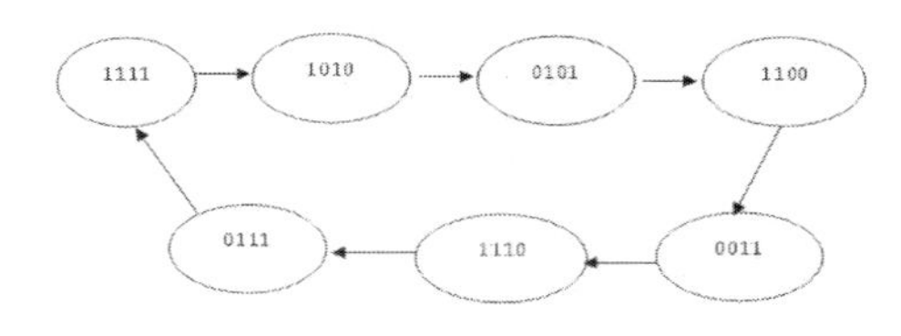

Figure 1. State diagram.

3.1 *Method 1*

Design principle: the normal method for sequential logic circuit by using D flip-flops and gate circuit.

3.1.1 *Logic design process*

1 Excitation table

This is realized with four D flip-flops. The state is $Q_3^n Q_2^n Q_1^n Q_0^n$. From the state diagram and the properties of the D flip-flops we get an excitation table as shown in Table 1.

Table 1. State transition circuit excitation table.

Q_3^n	Q_2^n	Q_1^n	Q_0^n	Q_3^{n+1}	Q_2^{n+1}	Q_1^{n+1}	Q_0^{n+1}	D_3	D_2	D_1	D_0
0	0	0	0	X	X	X	X	X	X	X	X
0	0	0	1	X	X	X	X	X	X	X	X
0	0	1	0	X	X	X	X	X	X	X	X
0	0	1	1	1	1	1	0	1	1	1	0
0	1	0	0	X	X	X	X	X	X	X	X
0	1	0	1	1	1	0	0	1	1	0	0
0	1	1	0	X	X	X	X	X	X	X	X
0	1	1	1	1	1	1	1	1	1	1	1
1	0	0	0	X	X	X	X	X	X	X	X
1	0	0	1	X	X	X	X	X	X	X	X
1	0	1	0	0	1	0	1	0	1	0	1
1	0	1	1	X	X	X	X	X	X	X	X
1	1	0	0	0	0	1	1	0	0	1	1
1	1	0	1	X	X	X	X	X	X	X	X
1	1	1	0	0	1	1	1	0	1	1	1
1	1	1	1	1	0	1	0	1	0	1	0

2 Driving equation by K-map simplication, we get:

$$D_3 = Q_0^n$$

$$D_2 = \overline{Q_3^n} + Q_1^n \overline{Q_0^n}$$

$$D_1 = Q_1^n Q_0^n + \overline{Q_0^n}$$

$$D_0 = \overline{Q_0^n} + \overline{Q_3^n} Q_2^n Q_1^n$$

3 Check if self-starting is possible and work out the respective state tables as shown in Table 2 from the resulting driving equation described above.

Table 2. State transition table.

Q_3^n	Q_2^n	Q_1^n	Q_0^n	Q_3^{n+1}	Q_2^{n+1}	Q_1^{n+1}	Q_0^{n+1}	D_3	D_2	D_1	D_0
0	0	0	0	0	1	0	1	0	1	0	1
0	0	0	1	1	1	0	0	1	1	0	0
0	0	1	0	0	1	0	1	0	1	0	1
0	0	1	1	1	1	1	0	1	1	1	0
0	1	0	0	0	1	1	1	0	1	1	1
0	1	0	1	1	1	0	0	1	1	0	0
0	1	1	0	0	1	1	1	0	1	1	1
0	1	1	1	1	1	1	1	1	1	1	1
1	0	0	0	0	0	0	1	0	0	0	1
1	0	0	1	1	0	0	0	1	0	0	0
1	0	1	0	0	1	0	1	0	1	0	1
1	0	1	1	1	0	1	0	1	0	1	0
1	1	0	0	0	0	1	1	0	0	1	1
1	1	0	1	1	0	0	0	1	0	0	0
1	1	1	0	0	1	1	1	0	1	1	1
1	1	1	1	1	0	1	0	1	0	1	0

Draw a state diagram as shown in Fig 2.

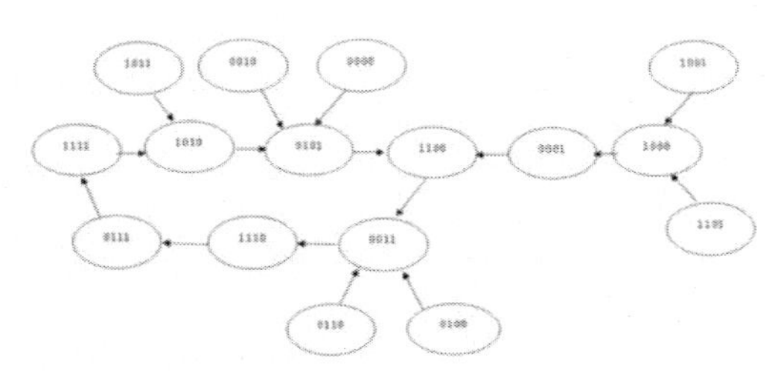

Figure 2. Check the self-starting state diagram.

We can see that all the invalid states will come back to the valid state after several CPs and therefore can realize self-starting. This design is feasible.

3.2 *Realization on Creator 2.0 software*

Fig 3 shows a schematic of the state transition circuit designed with Creator2.0.

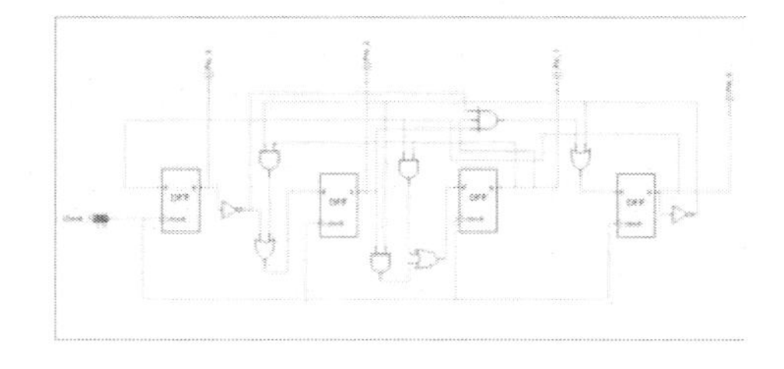

Figure 3. Schematic showing the state transition circuit designed on Creator2.0.

Fig 4 shows the designed pin assignment.

Alias	Name	Pin		Lock
	Pin_2	P0[2] OpAmp+	▾	☑
	Pin_3	P0[3] OpAmp-, DSM:ExtVref	▾	☑
	Pin_1	P0[1] OpAmp:out	▾	☑
	Pin_0	P0[0] OpAmp:out	▾	☑

Figure 4. Designed pin assignment.

Finally, the Pin_0, Pin_1, Pin_2 and Pin_3 outputs will be connected to the LED indicator.

3.2.1 *Method 2*

Cypress's PSoC provides lookup table resources containing a maximum of five inputs and eight outputs to execute any logic function. Fig 5 shows the symbols of a lookup table.

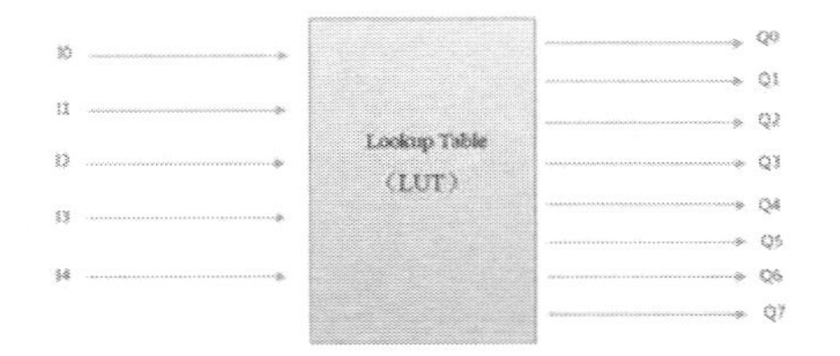

Figure 5. Lookup table symbols.

The principle of a lookup table can be expressed by the equation below:

$$Q_0 = f_1(I_0, I_1, ..., I_4);$$

$$Q_1 = f_2(I_0, I_1, ..., I_4);$$

$$Q_7 = f_8(I_0, I_1, ..., I_4);$$

Here: f1(), f2(), f3(), f4(), f5(), f6(), f7() and f8() are expressions of logic logarithms. A lookup table realizes combinational logic in a digital logic. To realize combinational logic with LUT, complex combinational logic relations can be realized as long as the correlation between I0, ...I4 and Q0, ..., Q7 is known.

Fig 6 shows the circuit diagram of a three binary up counter designed with D flip-flops and a lookup table. By looking up the coding rule from the truth table, and finding out the input-output relation using K-map simplification, we can realize the core functions of a three binary up counter based on D flip-flops and a lookup table. The clock pulse LUK is a 1Hz count pulse. The three outputs Q2, Q1 and Q0 are controlled by the clock pulse. The state changes in sequence from 000 to 111.

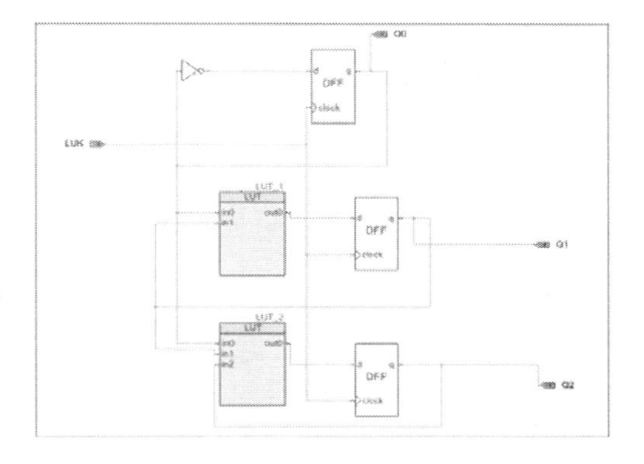

Figure 6. Schematic of three binary up counter.

Table 3. Lookup table LUT_1.

in1	In0	out0
0	0	0
0	1	1
1	0	1
1	1	0

Table 4. Lookup table LUT_2.

in2	in1	in0	out0
0	0	0	0
0	0	1	0
0	1	0	0
0	1	1	1
1	0	0	1
1	0	1	1
1	1	0	1
1	1	1	0

Fig 7 shows the schematic of a state transition circuit. The control principle is to connect a lookup table via a three binary up counter and then realize state transition by corresponding the lookup table to different states.

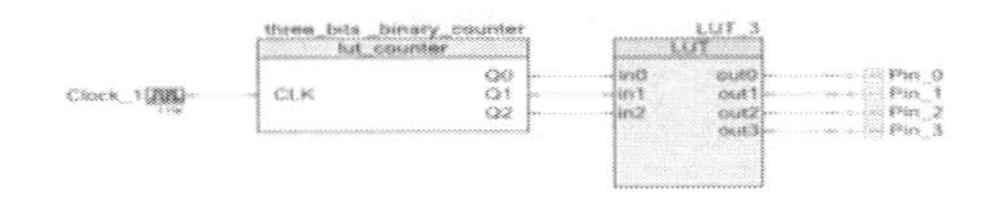

Figure 7. Schematic showing how state transition is realized by using a counter-controlled lookup table.

We realized output state control by using a lookup table. We connected the three inputs of lookup table LUT_3 to the output of the three binary up counter, and then designed the output state of the lookup table according to our specific requirement as shown in Table 3. Next, we connected the four outputs of the lookup table to the LED indicator. For the purpose of our design, we define 1 as ON and 0 as OFF. While counting, we had the outputs change in sequence according to the state shown in Fig 1, i.e. (1010) →(0101) →(1100) →(0011) →(1110) →(0111) →(1111) and then repeat.

Table 5. Lookup table LUT_3.

in2	in1	in0	out3	out2	out1	out0
0	0	0	1	0	1	0
0	0	1	0	1	0	1
0	1	0	1	1	0	0
0	1	1	0	0	1	1
1	0	0	1	1	1	0
1	0	1	0	1	1	1
1	1	0	1	1	1	1
1	1	1	1	0	1	0

4 REALIZATION ON PSOC HARDWARE

Figs 3 and 7 are designed, compiled and downloaded to the PSoC chips. When the LED indicator is connected according to the designed pins, the LED indicator will light up according to the state shown in Fig 1. As shown in Fig 8, both methods were able to achieve the required state transition.

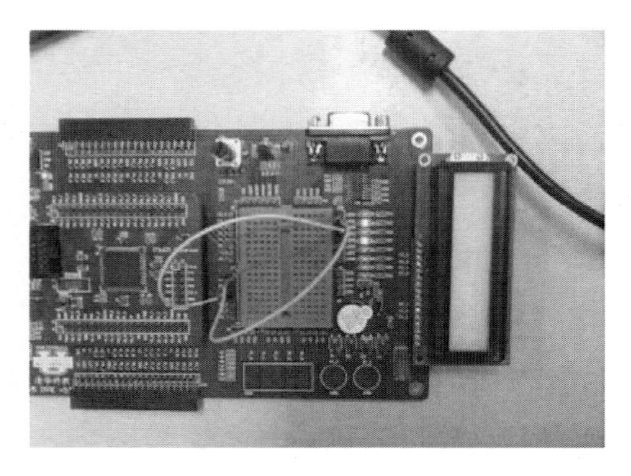

Figure 8. Hardware realization of state transition circuit.

5 CONCLUSIONS

By using Creator2.0, we can either design a state transition circuit by using the traditional method for digital circuit design or designing it with a counter-controlled lookup table designed with D flip-flops, compiling it and downloading it to PSoC hardware, which allows easy realization of a state transition circuit on single chips. State transition circuit is widely applied to control circuit with good referential and practical value.

ACKNOWLEDGEMENT

This paper is supported by Discipline Construction Foundation of Shanghai Second Polytechnic University (GN: XXKYS1402); School Foundation of Shanghai Second Polytechnic University in 2013 (GN: EGD13XQD20); Students Science & Technology Project of Shanghai Second Polytechnic University in 2014 (GN: 2014-xjkj-033).

REFERENCES

[1] Zhu, M.C. & Li, X.B. (2008) *PSoC Principle and Application Design*. Beijing: China Machine Press, (3).
[2] He, B. (2012) *The Design Guide of PSoC Analog and Digital Circuit*. Beijing: Chemical Industry Press, (6).
[3] He, B. (2011) *Design Guide of Programmable System on chip*. Beijing: Chemical Industry Press, (8).
[4] Lin, Z.Q. & Jiang, H.P. (2012) *Signal generation circuit: Principle and practical design*. Beijing: Posts & Telecom Press, (2).
[5] Ding, W.R. & Xing, B.B. (2013) On the practical teaching of PSoC-based technology. *Experimental Technology & Management*, 30(6):105–107.
[6] Han, X.X., Gao, X.D. & Zhang, C.Y. (2011) General counter design based on PSOC. *Journal of Heilongjiang Institute of Technology*, 25(3):49.
[7] Zhang, Z.W. & Hu, Z.J. (2011) PSoC based counting system for output pulse of multichannel fiberoptic gyroscope. *Journal of Xi'an Technological University*, 31(3):281–282.
[8] Lu, H.F. & Han, B. (2003) The study on the functions of a programmable system on chip (PSoC) and its application in motor soft start. *Development & Application*, 19(12):29–30.
[9] Ashby, Robert. (2005) *Designer's Guide to the Cypress PSoC*. Burlington, MA. Elsevier Newnes.

Electric, Electronic and Control Engineering – Shao, Shu & Tian (Eds)
 ISBN: 978-1-138-02842-5

An obstacle avoidance scheme for autonomous robot based on PCNN

Tao Xu
College of Electronic Information & Control Engineering, Beijing University of Technology, Beijing, China
School of Mechanical and Electrical Engineering, Henan Institute of Science and Technology, Xinxiang Henan, China

Songmin Jia, Zhengyin Dong & Xiuzhi Li
College of Electronic Information & Control Engineering, Beijing University of Technology, Beijing, China

ABSTRACT: The obstacle recognition and segmentation in image sequences has become one of the key technologies of the robot obstacle avoidance. In this paper, a new method to obstacle regions extraction from images for mobile robots is proposed. In the proposed method a Pulse Coupled Neural Network (PCNN) and an improved Chan–Vese (C-V) level set algorithm are applied for obstacle recognition and classification through a robotic vision system. Then A* search algorithm is used to achieve path planning and graph traversals. The result shows that the method can efficiently extract obstacles region in the field of view. Furthermore, the validity and practicability of the proposed approach was validated by a lot of experiments on the mobile robot Pioneer3-DX.

KEYWORDS: *Obstacle avoidance; PCNN; C-V level set model; Path planning; Autonomous Robot.*

1 INTRODUCTION

The obstacle avoidance scheme and the shortest path from the starting point to the finishing point is a common problem for autonomous Robot. This problem needs to be done in real time which becomes a challenge for large maps with complex environment as path planning is computationally very expensive [1]. One of the earliest and most well-known problems for such a system, especially for indoor domains, is the generation of collision free global path for the robot to move to a given point in a dynamic environment [2]. Conventional path planning approaches employed for such motion planning can be categorized into two types including global methods and local methods. The global methods such as road map [3], cell decomposition [4] and distance transform [5] which were able to search for possible paths in the whole workspace. The local methods have been proposed to provide effective path searching, such as the Potential Field method [6, 7] and the grid-based algorithms [8]. However, most of these algorithms suffer from time inefficiency in their computation and are not designed for use in real time path planning. Also, Potential Field methods are known to suffer unwanted local minima in which the robot gets stuck in a U-shaped obstacle [9].

In recently years, many new models were used for path planning that includes fluid model [10], dynamic waveexpansionmodel[11]andneuralnetworks[12,13]. PCNN is a result of research effort on the development of artificial neuron model that was capable of emulating the behavior of cortical neurons observed in the visual cortices of animal [14, 15]. Recent research shows that the spatio-temporal dynamics of PCNNs provide good computational capability for solving a number of optimization problems [1]. Hong [16] presents a modified model of PCNNs, multi-output model of pulse coupled neural networks (MPCNNs), to solve the shortest path problem. The modified model requires fewer neurons than the other models, guaranteeing the shortest path and a solution that is independent of the complexity of the search space. However, since the model employs an unconstrained autowave, it searches the whole space, irrespective of where the target is located, hence this unconstrained search leads to time inefficiency [2].

In order to facilitate an informed search for autonomous robot, a vision-based obstacle avoidance system needs to acquire the environment information, separate obstacles and describe them mathematically to give out the obstacle avoidance strategy. So the obstacle recognition and segmentation in image sequences has become one of the key technologies of the robot obstacle avoidance. In this paper, we present a novel obstacle avoidance scheme. First, we use PCNN to achieve the binarization region abstraction of the images. In order to make the obstacle region segmentation more exact, the improved C-V model [17] was used in each divided area. Furthermore, based on A* search algorithm, we use the position of target neuron to focus a controlled search in its direction. Also, it employs a variable threshold unique to each neuron. These modifications enable the proposed scheme to significantly improve the optimal path query times.

Rest of this paper is organized as follows. In section II covers some related research on PCNN method. In section III we present the improved C-V model and A* search algorithm. The details of our method will be given in Section IV. In Section V, experimental results are given and discussed. Finally conclusions are given in Section VI.

2 PCNN MODEL

A PCNN neuron shown in Fig. 1 contains two main compartments: the Feeding and Linking compartments [18]. Each of these communicates with neighbouring neurons through the synaptic weights M and W respectively. Each one retains its previous state but with a decay factor. Only the Feeding compartment receives the input stimulus, S. The values of these two compartments are determined by:

$$F_{ij}[n] = e^{\alpha_F \delta_n} F_{ij}[n-1] + S_{ij} + V_F \sum_{kl} M_{ijkl} Y_{kl}[n-1] \quad (1)$$

$$L_{ij}[n] = e^{\alpha_L \delta_n} L_{ij}(n-1) + V_L \sum_{kl} W_{ijkl} Y_{kl}[n-1] \quad (2)$$

Where the (i, j) pair is the position of neuron. F and L are feeding inputs and linking inputs, respectively. Y is the pulse output; α_F and α_L are time constants for feeding and linking; V_F and V_L are normalizing constants. If the receptive fields of M and W change then these constants are used to scale the resultant correlation to prevent saturation.

The modulation fields generate the internal activity of each neuron, is modeled as follows:

$$U_{ij}[n] = F_{ij}[n]\{1 + \beta L_{ij}[n]\} \quad (3)$$

Where β is strength of the linking.

The internal state of the neuron is compared to a dynamic threshold Θ, to produce the output Y, by

$$Y_{ij}[n] = \begin{cases} 1 & U_{ij}[n] > \Theta_{ij}[n] \\ 0 & Otherwise \end{cases} \quad (4)$$

$$\Theta_{ij}[n] = e^{\alpha_\Theta \delta_n} \Theta_{ij}[n-1] + V_\Theta Y_{ij}[n] \quad (5)$$

α_Θ and V_Θ are the time constant and the normalization constant, respectively. If U_{ij} is greater than the threshold, the output of neuron (i, j) turns into 1, neuron (i, j) fires, then Y_{ij} feedbacks to make Θ_{ij} rise over U_{ij} immediately, then the output of neuron (i, j) turns into 0, which produces a pulse output. It is clear that the pulse generator is responsible for the modeling of the refractory period of spiking.

Some related research on PCNN method for mobile robot path planning is studied. The search proceeds in the form of a pulse through which neighboring neurons are coupled. When the internal activity of a particular neuron exceeds the threshold level, it fires. In this process, internal activity of the obstacle neurons is kept zero, hence avoiding their firing. Path is traced backwards from robot to target through a sequence of parents of the neuron.

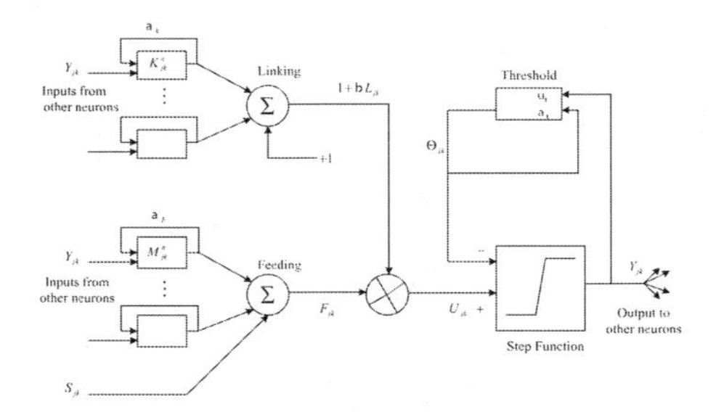

Figure 1. Schematic representation of a PCNN model.

3 IMPROVED C-V MODEL AND A* SEARCH ALGORITHM

3.1 *Improved C-V model*

Chan and Vese [17] proposed an algorithm for decomposing the image into two regions with piecewise constant approximations by minimizing the energy of the Mumford and Shah functional

$$\varepsilon^{CV}(c_1, c_2, C) = \mu Length(C) + \lambda_1 \int_{inside(C)} |f_o(x) - c_1|^2 dx + \lambda_2 \int_{outside(C)} |f_o(x) - c_2|^2 dx \quad (6)$$

Where μ, λ_1, and λ_2 are positive parameters. The constants c_1 and c_2 are the averages of f_o inside and outside of C, respectively. Chan and Vese replaced the unknown curve C by the level-set function ϕ(x). Then the energy functional $\varepsilon^{CV}(c_1, c_2, C)$ can be rewritten as:

$$\varepsilon^{CV}(c_1, c_2, \phi) = \mu \int_\Omega \delta_\varepsilon(\phi(x)) |\nabla \phi(x)| dx + \lambda_1 \int_\Omega |f_o(x) - c_1|^2 H_\varepsilon(\phi(x)) dx + \lambda_2 \int_\Omega |f_o(x) - c_2|^2 (1 - H_\varepsilon(\phi(x))) dx \quad (7)$$

By applying the gradient descent method, we obtain the following equation:

$$\frac{\partial \phi}{\partial t} = \delta_{\varepsilon}(\phi)\left[\mu \nabla \cdot (\frac{\nabla \phi}{|\nabla \phi|}) - \lambda_1 (f_o(x) - c_1)^2 + \lambda_2 (f_o(x) - c_2)^2\right] \quad (8)$$

The level set based algorithm of Chan and Vese can be used to process the image with a large amount of noise and detect objects whose boundaries cannot be defined by gradient. So with PCNN and C-V model we can find unknown obstacles through a robotic vision system.

3.2 *A* Search algorithm*

A* [19] is a search algorithm which has been extensively employed to the problems of path planning and graph traversals. It determines the path with the smallest cost (represented by f(x)) in a graph or a network based upon the length of path already covered (given by g(x)) and a admissible heuristics (h(x)) representing an approximation of the path to be covered. Cost of each path is given as follows:

$$f(x) = g(x) + \omega h(x) \quad (9)$$

where ω represents the weight function.

A* search algorithm exploration is carried out based upon a cost function and then shortest path is extracted through parent–child relations. Fig. 2 shows the path determined by A*.

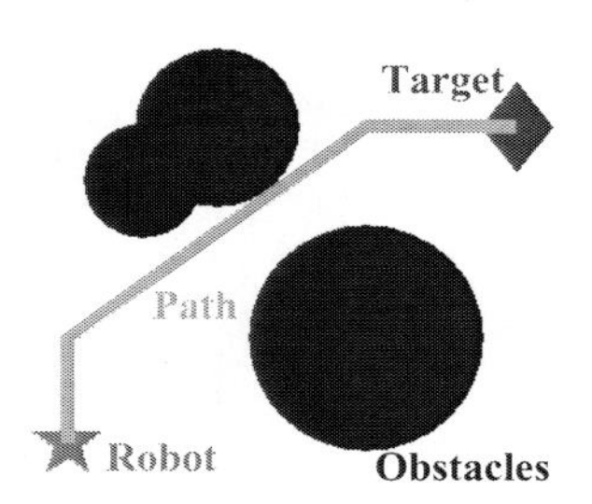

Figure 2. A* search algorithm for robot path planning.

From Fig. 2 we can find that the path determined by A* search algorithm can be used for known obstacles avoidance scheme.

4 OUR MODEL

Frame diagram of our proposed method is showed in Fig. 3. The PCNN and an improved C-V level set algorithm are applied for obstacle recognition and classification through a robotic vision system. Then A* search algorithm is used to achieve path planning and graph traversals.

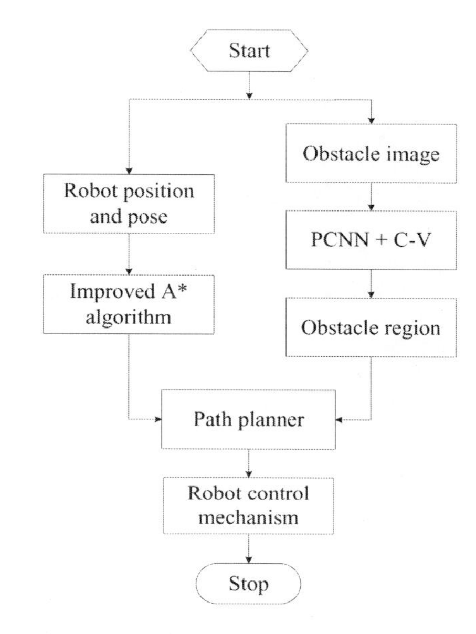

Figure 3. Frame diagram of our model.

An important characteristic of the human biological visual process is that visual attention changes as the change of scene. In order to simulate the process accurately, we have improved traditional PCNN by C-V level set algorithm. Then we can get the localization information of unknown obstacles through a robotic vision system.

Based on A* search algorithm, PCNN method was used for autonomous robot path planning. The search proceeds in the form of a pulse through which neighboring neurons are coupled. When the internal activity of a particular neuron exceeds the threshold level, it fires. In this process, internal activity of the obstacle neurons is kept zero, hence avoiding their firing. Path is traced backwards from robot to target through a sequence of parents of the neuron.

5 EXPERIMENTAL RESULTS AND DISCUSSION

5.1 *Obstacles extraction at actual scene*

To verify the effectiveness of our algorithm, we use this algorithm to extract obstacles region through the robotic vision system. The result was shown in Fig. 4.

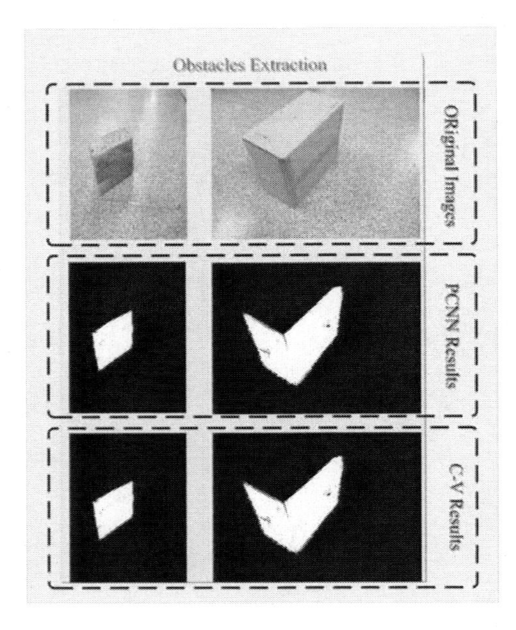

Figure 4. Experimental results of obstacles extraction.

The first line is the original images get form the robotic vision system. The second line is the obstacles extraction results by PCNN. There are some obstacles regions not ignited. So the C-V model was used to achieve effectively segment. The results were showed in the third line.

5.2 *Path planning*

Indoor experiments are conducted to verify the validity and accuracy of the algorithm. The Pioneer3-DX is chosen as experimental platform, which is 44cm × 38cm × 22cm aluminium body with 19cm diameter drive wheels and weighs about 9kg. It is depicted in Fig. 3. This platform is qualified for indoor research activities like mapping, obstacle avoidance, path planning, etc. In our research, the URG scanning laser rangefinder (LRF) offers both serial (RS-232) and USB interfaces to provide laser scans and we equipped a Point Grey Flea2 camera system which is fixed on the mobile robot. The camera is connected via 1394B to a 3.1GHz Intel i3 computer with 4G RAM and an NVIDIA GT 520 GPU.

Based on A* search algorithm, PCNN method was used for autonomous robot path planning. When the internal activity of a particular neuron exceeds the threshold level, it fires. Initially the target neuron's internal activity is set greater than the threshold level to start the firing process. The fired neuron prepares its neighbors to fire at a later time thus becoming their parent. The result was shown in Fig. 5.

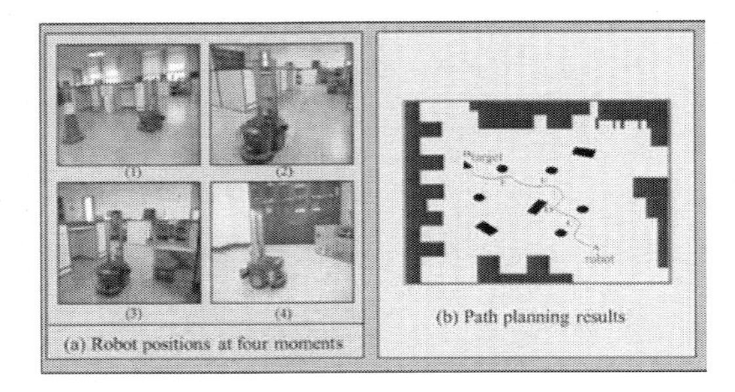

Figure 5. Path planning result of our model.

Fig. 5 (a) shows the four moments of robot movements and Fig. 5 (b) shows the experiment and the path of robot. From the results, we can see that the robot can avoid obstacles and move in an optimal path using our path planning strategy based on PCNN model.

6 CONCLUSIONS

A novel approach based on a hybrid model for obstacle avoidance and path planning of autonomous robots is presented. We use PCNN to achieve the binarization region abstraction of the images. In order to make the obstacle region segmentation more exact, the improved C-V model was used in each divided area. Then we can get unknown obstacles through a robotic vision system. Furthermore, based on A* search algorithm, we use the position of target neuron to focus a controlled search in its direction. Also, it employs a variable threshold unique to each neuron. These modifications enable the proposed scheme to significantly improve the optimal path query times. The experimental results show that the method can efficiently extract obstacles region in the field of view. Furthermore, the validity and practicability of the proposed approach was validated by a lot of experiments on the mobile robot Pioneer3-DX.

In a subsequent study, dynamics constraints shall be introduced to the detected paths, thereby facilitating the motion of mobile robot. And further research includes the depth estimation and simultaneous localization and mapping.

ACKNOWLEDGMENT

The research work is supported by National Natural Science Foundation (61175087, 61105033).

REFERENCES

[1] S. U. Ahmed, U. A. Malik, and K. F. Iqbal, *et al.* "Sparsed Potential-PCNN for Real Time Path Planning and Indoor Navigation Scheme for Mobile Robots," International Conference on Mechatronics and Automation, 2011,pp. 1729–1734.

[2] A. S. Usman, K. Faraz, and I. Mazhar, "Guided Autowave Pulse Coupled Neural Network (GAPCNN) based real time path planning and an obstacle avoidance scheme for mobile robots," Robotics and Autonomous Systems, vol. 62, no. 3, pp. 474–486, 2014.

[3] O. Takahashi and R. J. Schilling, "Motion planning in a plane using generalized Voronoi diagrams, " Robotics and Automation, IEEE Transactions on, vol.5, no.2, pp.143–150, Apr 1989.

[4] F. Lingelbach, "Path planning for mobile manipulation using probabilistic celldecomposition," in: Proceedings, 2004 IEEE/RSJ International Conference on Intelligent Robots and Systems, IROS 2004, vol. 3, 2004, pp. 2807–2812.

[5] J. C. Elizondo-Leal and T. G. Ramírez, "An Exact Euclidean Distance Transform for Universal Path Planning," Electronics, Robotics and Automotive Mechanics Conference (CERMA), 2010, vol., no., pp.62–67, Sept. 28 2010-Oct. 1 2010.

[6] S.S. Ge and Y.J. Cui, "Dynamic motion planning for mobile robots using potential field method," Auton. Robots 13 (2002) pp. 07–222.

[7] Q. D. Zhu, Y. J. Yan, Z. Y. Xing, "Robot Path Planning Based on Artificial Potential Field Approach with Simulated Annealing," Intelligent Systems Design and Applications, 2006. ISDA'06. Sixth International Conference on, vol.2, no., pp.622–627, 16–18 Oct. 2006.

[8] Y. R. Hu; S. X. Yang, "A knowledge based genetic algorithm for path planning of a mobile robot," Robotics and Automation, 2004. Proceedings. ICRA '04. 2004 IEEE International Conference on, vol.5, no., pp. 4350–4355 Vol.5, 26 April-1 May 2004.

[9] K. Sertac and F. Emili, "Sampling-based algorithms for optimal motion planning," Int. J. Robot. Res. 30 (7) (2011) 846–894.

[10] D. Gingras, E. Dupuis, G. Payre, *et al.* "Path planning based on fluid mechanics for mobile robots using unstructured terrain models," Robotics and Automation (ICRA), 2010 IEEE International Conference on, vol., no., pp.1978–1984, 3–7 May 2010.

[11] V. Lebedev, J Steil, J. Ritter, "The dynamic wave expansion neural network model for robot motion planning in time-varying environments," Neural Netw. vol 18, pp.267–285,2005.

[12] B. Hao and X. F. Dai, "The collision-free motion of robot with Fuzzy neural network," Industrial and Information Systems (IIS), 2010 2nd International Conference on, vol.2, no., pp.219–222, 10-11 July 2010.

[13] Q. Hong, S.X. Yang, A.R.Willms, *et al.* "Real-time robot path planning based on a modified pulse-coupled neural network model," IEEE Trans. Neural Netw. 20 (11) (2009) 1724–1739.

[14] H. J. Caulfield, J. M. Kinser, "Finding the shortest path in the shortest time using PCNN's," Neural Networks, IEEE Transactions on , vol.10, no.3, pp.604–606, May 1999.

[15] R. Eckhorn, H. J. Reitboeck, M. Arndt, *et al.* "A neural network for feature linking via synchronous activity: results from cat visual cortex and from simulations," Models Brain Funct 1989:255–72

[16] Q. Hong and Y. Zhang, "A new algorithm for finding the shortest paths using PCNNs," Chaos, Solitons and Fractals 33(2007), pp.1220–1229

[17] Y. B. Li and K. Junseok, "An unconditionally stable hybrid method for image segmentation," Applied Numerical Mathematics, 82(2014), pp.32–43

[18] T. Lindblad and J. M. Kinser, Image Processing Using Pulse-Coupled Neural Networks. New York: Springer-Verlag, 2013.

[19] P. E. Hart, N. J. Nilsson, B. Raphael, "A formal basis for the heuristic determination of minimum cost paths, " IEEE Trans. Syst. Sci. Cybern. 4 (2) (1968) 100–107.

Electric, Electronic and Control Engineering – Shao, Shu & Tian (Eds)
 ISBN: 978-1-138-02842-5

A study on thematic map of power grid operation based on GIS

Yi Liu, Zhaohui Cui, Dongya Wu & Junli Zhao
State Grid Henan Electric Power Company, Zhengzhou Henan, China

ABSTRACT: Electric power demand in China is growing rapidly with the development of economy. Electric power information construction is also speeding up. Drawings of power business have developed from paper drawings to electronic drawings. GIS is one of the most advanced technologies to realize electronic drawings. In this paper, the author briefly introduces GIS as well as the development and application status of GIS, and analyzes in detail the current application of GIS of power thematic map in power enterprise, aiming at providing a practical reference for current studies in related fields of electric power information.

KEYWORDS: GIS; single line diagram; internal wiring diagram; system diagram of major network; diagram of contaminated area; thunder and lightning distribution diagram.

1 INTRODUCTION

Geographic Information System (GIS) is an emerging interdisciplinary subject, integrating computer science, geography, topography, environmental science, economics, space science, and information and management science. Based on geographic space database and with the support of computer hardware and software, GIS collects, manages, operates, analyzes, simulates and displays spatial data and provides spatial and dynamic geographic information in time.

GIS has been used in different aspects of power system related to spatial information on account of its powerful function of data analysis and spatial analysis. In this paper, the author mainly introduces the development and application status of GIS and electric power GIS, analyzes the variety of currently used power thematic map and points out major functions of power thematic map based on GIS, providing an important reference and a necessary way of thinking for the current construction of related fields of electric power information.

2 DEVELOPMENT SITUATION OF GIS IN POWER ENTERPRISE

On the basis of the fact that GIS system can vividly reflect equipment and operational condition of the power supply system, the application of GIS system is becoming wider and wider in power supply department. It is highly necessary and helpful to establish an electric power GIS system with spatial location as well as attribute information and make effective connections with other power supply module.

A feature of electric power GIS is that power grid devices are labeled on an electronic map in the form of graphic symbols. Attribute information of devices is stored in a database. A one-to-one mapping relation between graphic symbol and attribute information is established, being able to realize the function of mutual query and simple analysis.

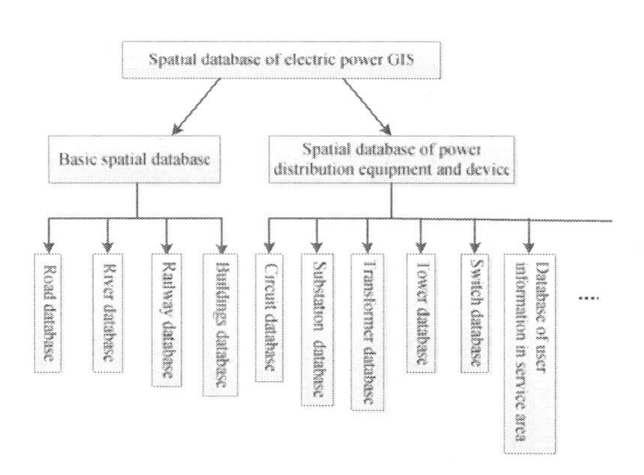

Figure 1. The mapping relation between graphic symbol and attribute information.

Geographic information system not only realizes hierarchical display and independent management of geographic graphs, power transmission and distribution line and electrical equipment, but also provides various kinds of convenient graphic drawing functions, such as power thematic maps like circuit alignment chart, primary wiring diagram of substation (distribution station), cable cross-section diagram, circuit wiring diagram, network wiring diagram, circuit vertical section diagram, and other line chart, pie chart, bar chart, etc. Besides, all these graphs are automatically consistent with data for the convenience of maintenance. Automatic or semi-automatic mapping can save a great deal of drawing costs for enterprises.

3 RESEARCH AND APPLICATION OF GIS IN ELECTRIC POWER THEMATIC MAP

3.1 *Geographic layout*

All electric power circuits, power equipment distribution and communication circuit diagrams are hierarchically displayed with the geographic background as the base map.

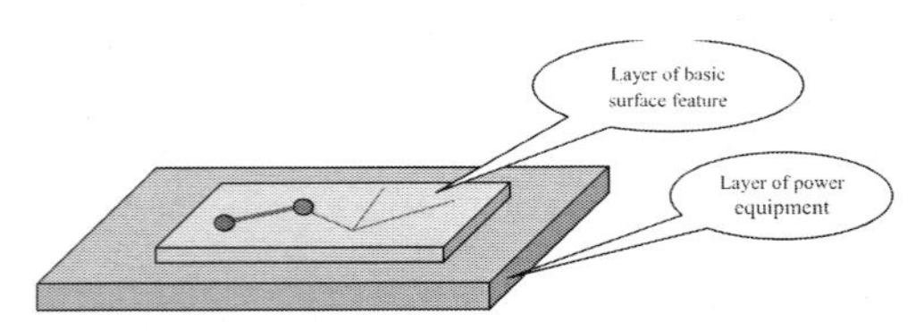

Figure 2. Diagram of geographic layout.

3.1.1 *Geographic background image*

Geographic background images include road map, buildings distribution diagram, administrative map and topographic map.

The layer of geographic background image of the system supports electronic maps of a variety of scale accuracies. Topographic maps with mapping accuracies of 1:10000, 1:2000, 1:1000, and 1:500 in different coordinate systems can be selected. The topographic map with scale accuracy of 1:10000 should be provided in power grid coverage. Major cities and towns in downtown area and suburb area should satisfy the topographic map with scale accuracy of 1:2000.

3.1.2 *Hierarchical information of power grid*

According to different categories of power facilities and different voltage grades, hierarchical information of power grid is convenient for the establishment of power grid topological relation and reasonable power grid GIS layer structure to meet the requirements of production and management. It can be divided into power transmission layer set, 10kV power distribution network layer set, 0.4kV low voltage layer set, and other layers like primary wiring layer of substation, primary wiring layer of open and close station (distribution room, disconnecting link room), contaminated area, thunder area, layer of ice coverage area, crossing layer and various thematic map layer of system.

With modeling tools of geographic location, it is able to carry out unified map drawings of power station, substation, power transmission line and power distribution line (10kV, 380V, 220V) and conduct management on layers of power equipment. Lines and equipment of different voltages (220kV, 110kV, 66kV, 10kV, 380V, and 220V) can be also displayed in layers.

Various graphic operation functions like zooming in, zooming out and roaming and various graphic selection functions like rectangular selection, circular selection and polygon selection are provided in this system. In the process of modeling, functions of smart capture and backspacing are also provided for the rapid construction of power grid model.

Meanwhile, the system provides modeling tools for internal wiring and conducts drawing of internal wiring relation of substation, switching station and loop net box.

Figure 3. Geographic information graph of 10kv line of power supply bureau in saertu.

3.1.3 *Attribute setting of power transmission and distribution equipment*

Attributes of all power transmission and distribution equipment and customer equipment are managed by database software. Management content includes editing, recording and managing on basic attributes data of equipment, such as substation (switching station, power distribution station), power transmission and distribution line (overhead line, cable), electric pole, small substation, distribution transformer, high-voltage customer, switch, lightning arrester, etc.

3.1.4 *Statistic analysis of power grid*

GIS provides powerful data statistics functions as well as a basis for scientific decision-making. Statistical functions include:

1 Provide direct statistics of equipment number, customer number, circuit capacity and load on circuit, cable, substation and accessory equipment;

2 Calculate the distance of circuit length of two arbitrary points on circuit;
3 Provide statistics of equipment number, customer number, circuit capacity and load in accordance with the administrative region, small region or region defined by customer;
4 Provide respective statistics of distribution and amount of the same equipment state;
5 Provide statistics of change amount of equipment resource (increment, decrement, etc.) in line with a given time period;
6 Conduct spatial statistics and analysis on load density of power grid, including regional load density, load intensity of circuits and equipment, load factor of circuits and equipment, etc.
7 Generate thematic map of statistical results, including pie chart, bar chart, density chart, etc.
8 Power supply and sale volume, theoretical line loss, practical line loss, and load record of power transmission and distribution line.

3.1.5 *Line operation and analysis*

Carry out operations on various devices of power transmission and distribution line, power distribution station and wiring chart of switching station, and display data of switches, PT and CT in power transmission and distribution net.

1 Analysis on switch/switch blade operations and power supply and outage range. It is able to carry out breaking and closing operations on arbitrary switch and disconnecting link of each line and substation, provide real-time display of states of line and power supply and outage after operations, and conduct analysis on power supply and outage range.
2 Electric power supply analysis. Choose a line randomly on power grid chart and find the power supply source of this line through network topology.
3 Line impedance analysis. It carries out automatic calculation and analysis on line impedance of power grid.
4 Real-time remote signaling deflection analysis. With real-time remote signaling deflection signals of SCADA, it conducts automatic deflection of switches in power distribution station and automatic analysis on power supply and outage range of power grid according to the state after deflection.
5 Real-time display of remote signals: After the networking with SCADA system, real-time information of remote signals load and current can be displayed on diagram of power distribution station so as to realize the function of real-time monitoring.

3.1.6 *Customer failure alarm*

Establish knowledge base and model base of customer telephone alarm in the system. When there are more than two failure alarm calls in one region or information of alarm phone number, customer number, power outage phenomenon and failure phenomenon are manually input by operator on duty, distribution and transformation failures or failure of a certain line can be automatically determined through real-time information provided by SCADA system and acquired from power business system on the equipment management chart with the background of geographic information provided by GIS. It displays failure address and power outage state on power transmission and distribution network diagram, and prints out information of affected region and customer.

3.1.7 *Power outage management*

1 Scheduled power outage
In the stage of scheduled power outage, scheduled information that should be input includes power outage range, start time, power outage duration, power outage line, power outage equipment, etc. System operational process of power outage date, as well as system recovery notice after completion, is automatically generated.
2 Power outage operation
In the operation power outage, information of power outage plan will be reflected on the monitored object system to carry out outage operation. Power outage operation must be conducted under the condition of completely no error in real-time monitoring objects and planned operation process.

3.1.8 *Decision support of power transmission and distribution grid*

Provide decision support for power transmission and distribution grid with a variety of auxiliary decision-making models of the system.

1 Decision of power outage isolation point: When alarm signals of failure and outage are received or certain equipment or certain line needs to be inspected and maintained, decision-making model can be used to determine the optimal outage isolation points within the minimum outage range.
2 Load transfer decision: When there is failure outage or inspection outage, part of loads is transferred to other substations or other lines of the substation. The optimal load transfer scheme can be calculated with load transfer model.
3 Tour inspection decision: According to the established topological model base and tour inspection method base, the system can automatically determine the optimal inspection decision scheme

from one point to multiple points or from multiple points to one point and print out inspection route map and schedule table.

4 Reconstruction optimization decision: The system automatically provides the optimal network reconstruction scheme for power supply with network analysis models established in the system, so as to achieve the highest optimization goal, such as uniform power load of the whole region, the most reasonable load and no-load factors of lines and equipment, and large quantity of power supply and sale.

3.2 *Single line diagram*

3.2.1 *Function objectives*

Convert the single line (feeder line) of wiring chart of geographic power grid into a single line diagram horizontally and vertically. Functions of extracting equipment type, layout spacing, direction, station building style, and simplified station wiring can be automatically defined by users. The single line diagram mentioned in this paper refers to the single line diagram of distribution grid.

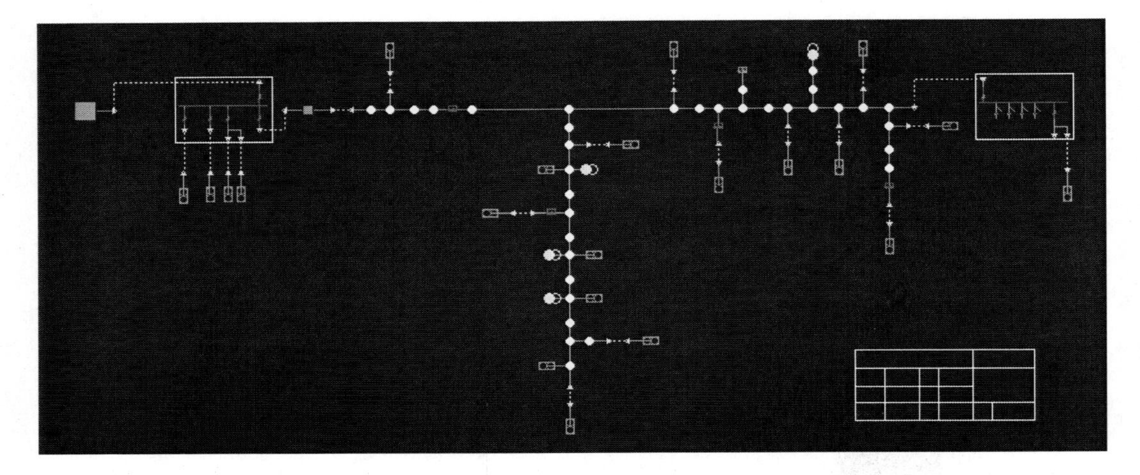

Figure 4. A horizontal and vertical single line diagram.

3.2.2 *Technological route*

1 Data sources. Equipment connection relation of topology, correlation and subordination of geographic chart and internal station chart and spatial data are data sources.
2 Definition of extraction type
There are two types of extraction. One is a crucial type and the other is an optional type. The necessary type is the equipment type that must exist in a single line diagram; the optional type is a dispensable equipment type in the single line diagram.
3 Extraction type of feeder line
Find the outgoing switch according to the given feeder line ID and carry out topological correlation analysis on geographic chart and internal station chart from the start point.
4 Filter type
Carry out filtering in accordance with layout parameters set by the system and users (such as equipment type selection, unit interval, layout direction, simplified internal wiring of station, etc.).
5 Layout adjustment of single line diagram
Adjust the layout of the single line diagram with the function of general editing.
6 Differential analysis
The function of differential analysis is used for displaying inconsistencies of line and geographic chart as well as providing the function of modification and regeneration.
7 Generation of single line diagram
The process of generating a single line diagram is shown as below, mainly consisting of four parts that are extraction, filter, layout and storage.

3.3 *In-station wiring chart*

Station building devices in geographic chart, such as substation, power distribution station and box substation, have the access to in-station chart, which is displayed both horizontally and vertically. Different voltage grades are differentiated with different colors. Mark names of feeder line at terminals of outlet line of substation. In-station wiring chart can realize functions of in-station

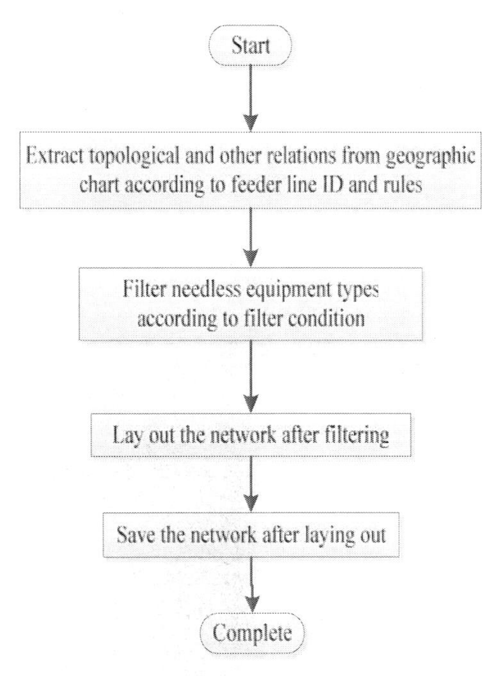

Figure 5. The process of generating a single line diagram.

chart browsing, in-station wiring, and geographic location query of station building and wiring copy of equipment terminal box. Functional diagram is presented below:

3.4 *System diagram of major network*

3.4.1 *Functional objectives*

1 Only station building, switch, pole and tower, wire, cable and terminals can be retained in this diagram.
2 Substation diagram is generated in box form. Different voltage grades are displayed with different colors.
3 Switches of outgoing line are displayed in different colors according to different voltage grades.
4 Delete relevant switches of outgoing line and wiring when deleting station buildings. Delete wiring when deleting switches.

3.4.2 *Technological route*

1 Data sources.
Equipment connection relation of topology, correlation and subordination of geographic chart and internal station chart and spatial data are data sources.

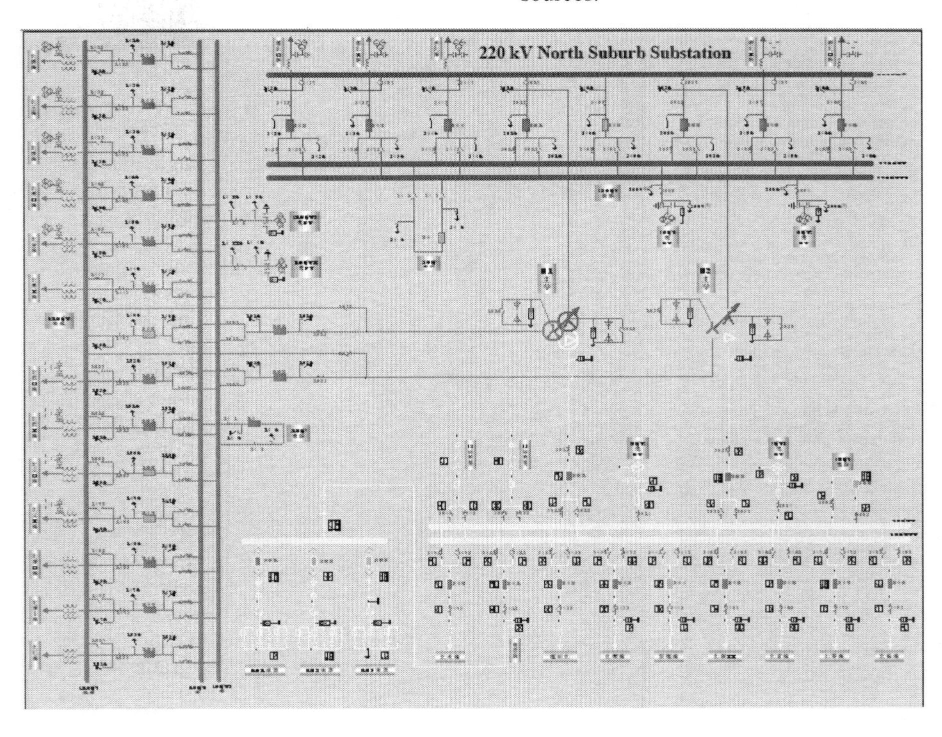

Figure 6. Functional diagram.

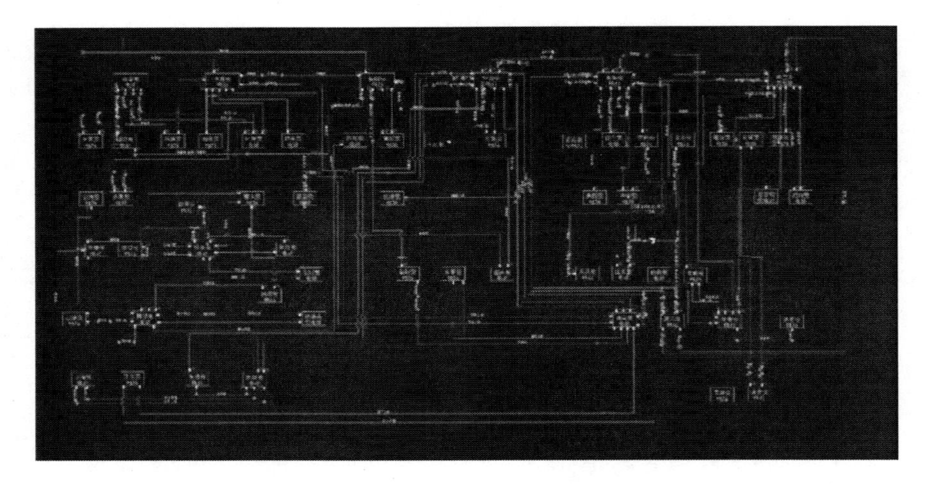

Figure 7. Substation diagram.

2 Definition of extraction type

There are two types of extraction. One is a necessary type and the other is an optional type. The necessary type is the equipment type that must exist in a single line diagram, such as generatrix, pole and tower, terminals or equipment (line equipment) relied by other devices like wire and in-station connecting line. The optional type is a dispensable equipment type in a single line diagram, like lightning arrester, etc. Optional types can be defined by users.

3 Filter type

Carry out filtering in accordance with layout parameters set by the system and users (such as equipment type selection, unit interval, layout direction, simplified internal wiring of station, etc.).

4 Layout way

Lay out the network after filtering horizontally and vertically.

5 Layout adjustment of system diagram

Adjust the layout of the system diagram with the function of general editing.

6 Differential analysis

Make comparison of devices between system diagram and geographic chart. Delete devices of system diagram that are more than geographic chart and add more devices of system diagram that are less than geographic chart.

3.5 *Diagram of contaminated area*

It provides power distribution grid with the function of displaying in contamination grade. The contaminated range is displayed in various regions within the province and power transmission lines of the necessary grade are displayed as well. It also displays affected power transmission equipment in appointed contaminated range (lines and towers).

3.5.1 *Functional objectives*

1 Diagram of contaminated area of the whole province is displayed in line with voltage grade of power transmission line, including 110, 220 and 500. It is able to display lines of any voltage grade or multiple voltage grades.
2 Local diagram of contaminated area is displayed in accordance with regions (Prefectural electric power bureau).
3 Situation query of affected equipment through diagram of contaminated area. Make query about the equipment list of affected electric transmission line according to the diagram of contaminated area.

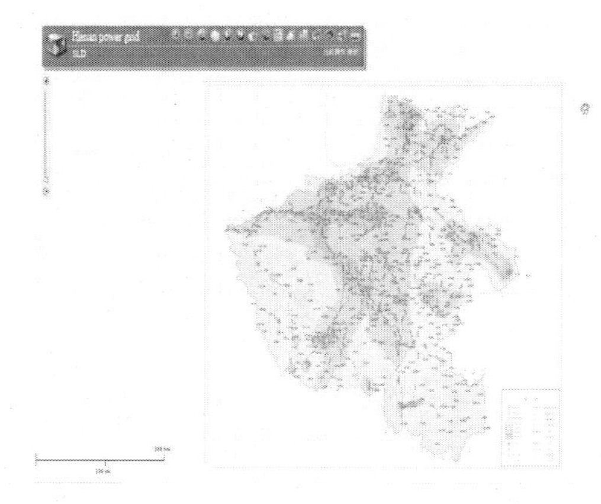

Figure 8. Single line diagram of Henan power grid.

3.5.2 *Technological route*

1 Acquisition of basic data
Integrate data of contaminated area of prefectural bureau in one layer according to the current data of diagram of contaminated area, and add relevant unit code and marks of contamination level, which is labeled according to current classifications (a\b\c\d\e\f\g).

2 Complete diagram of contaminated area
Overlay the data layer of contaminated area and geographic chart to form the complete diagram.

3.6 *Lightning distribution diagram*

3.6.1 *Functional objective*

It displays the situation of lightning distribution on spatial geographic chart according to the acquired statistical analysis data of lightning.

3.6.2 *Technological route*

1 Acquisition of basic data
Apply colors to the current data of lightning. Use dark colors for areas of large lightning density and light colors for areas of small lightning density. A rendered diagram of lightning distribution can be generated.

2 Complete diagram of lightning distribution
Overlay the diagram of lightning distribution and geographic chart to form the complete diagram.

3.6.3 *Functional description*

1 Dynamic generation of lightning distribution diagram
After obtaining the information of thunderbolt point of a certain time period, the power grid GIS platform is able to generate real-time lightning distribution diagram according to start time and end time.

2 Editing and maintenance of lightning distribution diagram
It provides the function of graphic drawing and attributes editing for lightning distribution diagram.

3 Lightning information inquiry
In geographic chart, choose a point to set buffer radius and inquiry time period. Make an inquiry about lighting information of appointed time period in buffer range and display the inquiry results in the table.

4 PROSPECT AND OUTLOOK OF POWER THEMATIC MAPS BASED ON GIS

Development and application of power thematic maps based on GIS will significantly enhance safety, economy and reliability of power transmission and distribution equipment operation, realizing the organic combination of massive information of power transmission and distribution grid and geographic chart information. It directly reflects the practical operation of power grid and brings huge convenience for different levels of management staff, operation staff and field operation staff. Especially for drawings modification and management of drawings review process, it realizes paperless operation. The establishment of power thematic maps database helps working staff to make scientific and reasonable decisions as soon as possible. All these not only greatly improve working efficiency and management quality but also make for close cooperation and mutual coordination among different departments, so as to improve the social image of power supply enterprise, bring huge social benefit and produce considerable economic benefits for power supply enterprise, such as:

1 Save a great deal of human, physical and financial resources;
2 Avoid all kinds of losses caused by decision-making mistakes;
3 Reduce information circulation

With the development of GIS, power thematic maps based on GIS will provide safer and more reliable service for the electric power industry and bring larger social benefits and economic benefits for power supply enterprises.

5 CONCLUSION

To sum up, GIS can be applied to various kinds of power thematic maps so as to meet different requirements of business. Power thematic maps based on GIS can fully improve working efficiency, save energy and reduce emission, strengthen drawing management process, bring great convenience for working staff, and produce enormous social benefits and economic benefits as well. It will inevitably become the new development trend of power thematic maps.

REFERENCES

[1] Liu, G. (2000) *Geographic Information System—Basic part*. Beijing: China Electric Power Press.
[2] Li, X.X. (2003) *Application of GIS in Environmental Science and Engineering*. Beijing: Electronic Industry Press.
[3] Guo, X.H. (2001) A brief discussion on geographic information system and its application. *Inner Mongolia Science & Technology and Economy*, 2.

Electric, Electronic and Control Engineering – Shao, Shu & Tian (Eds)
 ISBN: 978-1-138-02842-5

Capacity ratio relation of energy storage and intermittent DG and practical estimation method

Yue Zhang, Fenyan Yang & Jie Zeng
Electric Power Research Institute, Guangdong Power Grid Corporation, Guangzhou Guangdong, China

Guangyu Zou & Lei Dong
Tianjin Tianda Qiushi New Technology of Electric Power Co. Ltd., Tianjin, China

ABSTRACT: Energy storage is the key of solving the connection problem of intermittent distributed generation. According to the planning technology of energy storage, the capacity ratio relation of energy storage and distributed generation is still not determined. In this paper, the author uses data analog simulation to analyze the relation between the maximum fluctuating value of DG and the optimal energy storage power capacity as well as the relation between energy storage power capacity and energy capacity. Because it is very difficult to provide maximum fluctuation values of wind and light in practical project, the relation between rated power and the maximum fluctuating value of DG is also analyzed so as to get the ratio law of energy storage and DG, through which the approximate optimal energy storage power capacity and energy capacity can be acquired by simple manual calculation instead of the optimization algorithm of energy storage.

KEYWORDS: distributed generation; energy storage; planning; capacity ratio.

1 INTRODUCTION

As clean and renewable energies, wind and light get rapid development both at home and abroad. But the speed of development is severely impeded due to the intermittent nature. At present, more and more living examples are proving that energy storage is an important solution to this problem [1]. However, current energy storage is expensive and the problem of energy storage capacity allocation has become a hot topic today. Aiming at the minimum economic cost or the optimal technical indicators, the existing optimization method of energy storage capacity establishes corresponding model and objective function to obtain conclusive results with a certain algorithm. In this paper, based on the current mainstream optimization method of energy storage capacity, the author carries out a study on relevant problems and conducts optimization calculation of energy storage capacity on a great deal of instance data. Then the author summarizes the law to acquire the capacity ratio relation between energy storage and intermittent DG and proposes a simple and practical energy capacity determination method based on this.

2 CAPACITY RATIO RELATION BETWEEN ENERGY STORAGE AND DG AND PRACTICAL ESTIMATION METHOD

2.1 *The relation between the maximum fluctuation value of DG power and energy storage power capacity*

2.1.1 *Data of capacity optimization*

Results of optimal calculation of energy storage capacity on intermittent DG power data are shown in Table 1.

Power fluctuation data of a certain DG project in typical days are presented in Table 1, of which Example 1 to Example 7 are data of the same project in different days and Example 8 to Example 12 are data of other five projects.

Time windows are within the whole data length, representing continuous periods of time that are extracted for inspecting fluctuation features in certain time scale. Power sampling times of six projects are different, so time windows of projects are also different. In this paper, sampling time of 2 times is taken as the time window. The maximum fluctuation value becomes larger with time window because the range

Table 1. Relation of DG power fluctuation and energy storage capacity.

Examples	Sampling time(/s)	Time window (/s)	Maximum fluctuation (/kW)	Power capacity (/kW)	Calculated value (/kW)
Example 1	600	1200	0.87	0.69	0.62
Example 2	600	1200	5.99	4.79	4.25
Example 3	600	1200	0.88	0.49	0.62
Example 4	600	1200	4.54	3.46	3.23
Example 5	600	1200	2.41	1.98	1.71
Example 6	600	1200	0.88	0.52	0.62
Example 7	600	1200	5.99	4.79	4.25
Example 8	30	60	17.6	13.11	12.51
Example 9	120	240	65.6	47.35	46.62
Example 10	600	1200	86.6	65.76	61.55
Example 11	60	120	314.8	298.87	223.74
Example 12	1800	3600	358.5	194.36	254.8-

of data acquisition becomes larger. Time window of sampling time of 2 times is relatively reasonable.

2.1.2 *Conclusion and analysis of law*

From optimization results in Table 1, the formula of relation between rated power of energy storage and the maximum fluctuation value of DG can be expressed as follows:

$$P = Flu_{\max} \times Q \div \sqrt{\eta} \tag{1}$$

In this formula, P is the calculated value of power and $Flu_{\max}$ is the maximum fluctuation value, the unit of which is kW. Q stands for the maximum fluctuation cut value, ranging between 0 and 1. η stands for overall efficiency. In this paper, the overall efficiency is 88%, which means that charge efficiency and discharge efficiency are equal. If the two are not equal in practical application, they can be substituted into respectively so as to select the maximum value.

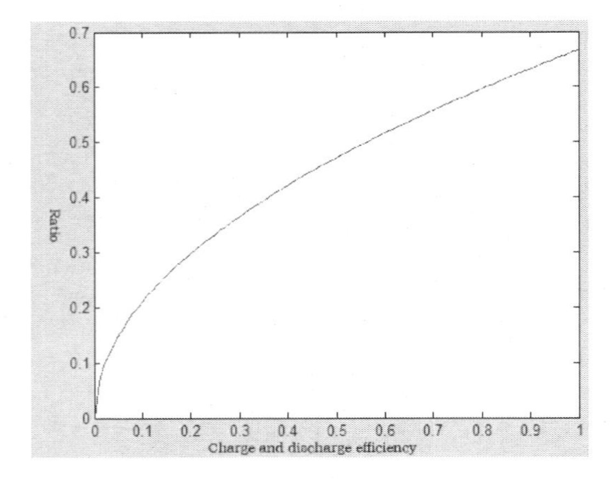

Figure 1. Relation between power fluctuation ratio and η.

Figure 1 is inferred from Formula (1). It can be concluded that the overall efficiency of charge and discharge η is in square relation with power fluctuation ratio. Starting from η, the ratio changes obviously with η. It means that there is a large increase of energy storage power capacity P on the condition of the overall charge and discharge efficiency with equal interval. The ratio tends to be gentle with the increase of η.

In Formula (1), Q refers to the ratio of the larger part of the total fluctuation power after the maximum fluctuation power of DG is cut into two parts by the smooth curve. Explain the meaning of photovoltaic sampling data of a specific area on March 29th.

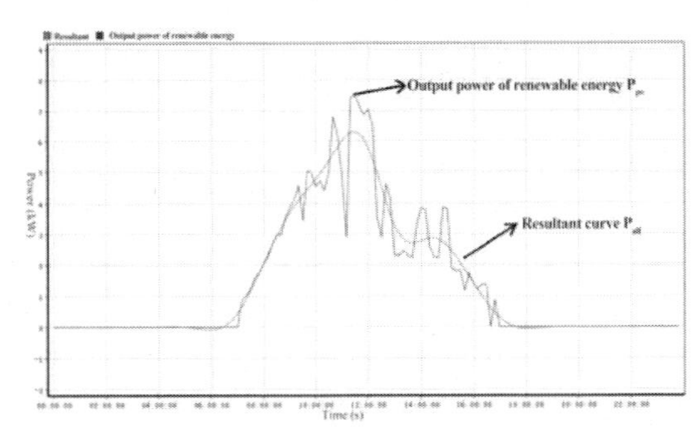

Figure 2. Sampling value of photovoltaic and smoothing effect.

As shown in Figure 2, P_{pv} is the actual photovoltaic power amplitude and P_{all} is the smoothing effect of energy storage power. It can be known from P_{pv} that the largest photovoltaic fluctuation in one day appears around 11 o'clock. When P_{pv} is higher than P_{all}, redundant photovoltaic electricity needs to be absorbed for energy storage charging. When P_{pv} is lower than P_{all}, insufficient electricity needs to be discharged for energy storage discharging. If P_{all} passes through 2/3 of the largest fluctuation of P_{pv}, about 2/3 of electric power need to be discharged by energy storage. This is the largest power of one-time charge and discharge of energy storage in one day. Then the energy storage power capacity can be calculated by dividing the efficiency of discharging η_c.

It can be seen that the largest fluctuation cut value fluctuates around 2/3 within a narrow range. 2/3 can be taken in practical calculation.

2.2 *The relation of DG rated power and the maximum fluctuation value*

Resources of wind and light are different across the country, so maximum power percentages of rated power in different areas are also different, which can be used to define power generation coefficient of wind and light η_{ge}. The practical maximum output power of DG at the installation site is obtained by multiplying rated power of DG P_{RP} with power generation coefficient of wind and light η_{ge}. The ratio of the maximum fluctuation power of DG and the maximum power of practical generation is defined as η_p, namely fluctuation ratio for short. The satisfaction relation is expressed with the following formula:

$$Flu_{max} <= P_{RP} \times \eta_{ge} \times \eta_p \quad (2)$$

In this formula, P_{RP} is the rated power of wind and light, and η_{ge} is the power generation coefficient of wind and light determined by wind speed, light and temperature of local area. According to a great deal of simulation results, the maximum fluctuation value would not be larger than 0.6 of the practical maximum output power of DG. In practical calculation, the value can be taken as 0.6.

2.3 *Ratio relationship between energy storage capacity and power capacity*

A lot of researches have been made on energy storage capacity at home and abroad. In this paper, the author proposes the concept of the optimal time duration to explore the relation between power capacity and energy capacity. Time duration refers to charge/discharge time of energy storage on the condition of maintaining the rated power.

2.3.1 *Data analysis*

Compare energy storage power of optimization cases with energy capacity. Results are shown in Table 2.

The last column stands for time duration. Data larger than 0.45 can be approximately taken as 0.5 and 1.05 is approximately taken as 1. Thus in 13 groups of data, 92.3% time duration ranges between 0.5 and 1. It means that only one group of data is not within the range. In this paper, the author selects 0.5-1h as the optimal time duration. After the determination of power capacity, energy capacity can be selected according to the criterion of 0.5-1h.

Table 2. Fluctuation parameters of energy.

Examples	Power capacity (/kW)	Energy capacity (kWh)	Time duration (/h)
Example 1	0.69	0.39	0.57
Example 2	4.79	1.54	0.32
Example 3	0.49	0.31	0.63
Example 4	3.46	1.7	0.49
Example 5	1.98	1.18	0.60
Example 6	0.52	0.44	0.85
Example 7	4.79	2.2	0.46
Example 8	13.11	6.19	0.47
Example 9	47.35	24.61	0.52
Example 10	65.76	41.57	0.63
Example 11	298.87	314.00	1.05
Example 12	194.36	161.98	0.83

2.3.2 *Conclusion and analysis of law*

It can be known from the computational formula of energy storage capacity that the calculation of energy storage capacity is not only related to the maximum value and the minimum value of energy storage in the process of charging and discharging but also related to difference values of the initial value and bound of SOC. The optimal initial SOC is selected in this paper, being able to guarantee that the maximum value and the minimum value of energy storage in the process of charging and discharging respectively locate between SOC_{up} and SOC_{low}. Besides, SOC_{up}=100% and SOC_{low}=40%, namely SOC_{up}-SOC_{low}=60%. This means that if the initial SOC is the optimal, the best time duration is 0.5-1h when the difference value of SOC is 60%. In practical application, the initial value of SOC can be optimized in line with the principle of this paper. Difference values of top and bottom limitations might vary with photovoltaic value and fan model. Best lasting times of different D-values of SOC are shown in Table 3:

Table 3. Relation between D-value of SOC and the best lasting time.

D-value of SOC (%)	45~55	55~65	65~75	75~85	85~95
Best lasting time (h)	0.6~1.2	0.5~1.0	0.45~0.9	0.4~0.8	0.35~0.7

It can be seen from Table 3 that best lasting times provided in this paper are range values. Intermediate value in range can be selected in practical applications.

For example, the value is selected as 0.9 when the D-value of SOC ranges between 45% and 55%.

2.4 *Practical calculation method of energy storage capacity based on ratio relation*

According to simultaneous formulas of (1) and (2) in 2.1, the simplified calculation formula of energy storage capacity can be expressed as:

$$P = P_{RP} \times \eta_{ge} \times \eta_{p} \times Q \div \sqrt{\eta} \qquad (3)$$

In this formula, P_{RP} is the rated power of wind and light, η_{ge} is the power generation coefficient of wind and light, η_p is the fluctuation ratio, Q is the cut value of the maximum fluctuation, and η is the overall efficiency.

Here, P_{RP} and η_{ge} need to be determined in line with practical situation. It can be known from above that, in practical applications, η_p is 0.6 and Q is 2/3. η is determined by properties of energy storage, the range of which is around 88%. Substitute the above-mentioned data into Formula (3), so that:

$$P = P_{RP} \times \eta_{ge} \times 0.43 \qquad (4)$$

It can be seen from Formula (4) that the ratio relation between energy storage capacity and the practical maximum power of DG is 0.43.

According to 2.4, the calculation formula of energy storage capacity can be concluded as:

$$E = P \times \mathrm{n} \qquad (5)$$

In Formula (5), n is the optimal hour. Specific results can be acquired from Table 3.

To sum up, the power capacity of energy storage can be obtained according to DG rated power, power generation coefficient of wind and light, as well as overall efficiency of energy storage. Energy capacity of energy storage can be obtained on the basis of power capacity of energy storage and the known D-value of SOC.

It can be concluded from analysis of the three sections that energy storage capacity can be acquired on the condition of knowing the DG rate power, power generation coefficient of wind and light related to regional environmental factors and relevant parameters of energy storage (overall efficiency of energy storage and D-value of SOC). According to analysis and conclusion of experimental data in this paper, the required energy storage capacity can be calculated by the above-mentioned four data without the method of capacity optimization.

3 ANALYSIS OF CALCULATION CASE

3.1 *Basic conditions of calculation case*

In this paper, an area with the rated power of 1500 kW photovoltaic value is taken as the example. The power generation coefficient of wind and light can be determined as 1 according to local weather and illumination. This means that the fluctuation of photovoltaic value of the area reaches 60% of the rate power. The overall efficiency of energy storage is 88% and the up and bottom limitations of SOC are respectively 100% and 40%. PV power of charge and discharge in one day is shown in Figure 3.

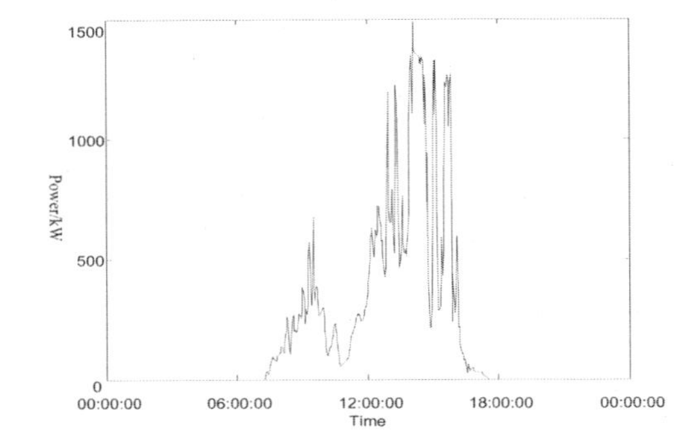

Figure 3. PV power in one day.

3.2 *Calculation of energy storage capacity in line with the method of this paper*

According to Formula (3), it can be obtained that the maximum fluctuation value of photovoltaic power generation is 900kW. The power capacity of energy storage is 639.6kW according to Formula (1). From the law analysis of Table 3, 0.5-1h is selected as the optimal time duration of energy storage. Thus, the energy storage capacity ranges between 319.8kWh to 639.6kWh. In this paper, the intermediate value is proposed to be selected, which is 479.7kWh.

3.3 *Comparison with results of literature [6]*

Carry out optimization on energy storage capacity with the method in Literature [6], which is also introduced in Chapter 1 of this paper. The power curve of photovoltaic power generation is shown in Figure 4.

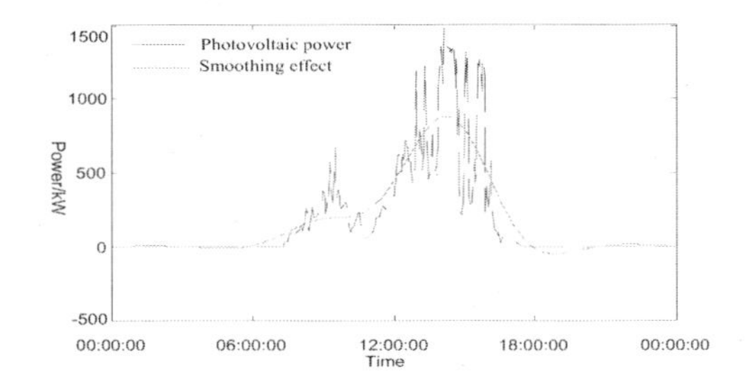

Figure 4. PV sampling value and smoothing effect.

Compare the optimization results of this paper and the results of Literature [6] after the optimization calculation. Results of comparison are given in Table 4.

Table 4. The comparison result.

Methods	Power capacity of energy storage / kW	Energy capacity of energy storage kWh
Paper	639.6	479.7
Literature [6]	670.98	393.73

Smoothing effect on photovoltaic power of the method in Literature [6] is presented in Figure 4. Results of comparison between the paper and Literature [6] are shown in Table 4. It can be seen from Table 4 that there is a small difference between methods of this paper and Literature [6] in terms of power capacity of energy storage. As for energy capacity, the method in Literature [6] is within the scope of the method in this paper. The practicability of the estimation method of this paper is thus verified.

4 CONCLUSION

Starting from the maximum generated power of distributed generation, the author obtains the relation between the rated power and the energy of energy storage, and carries out in-depth analysis on the law. Generally, it is difficult to obtain data of the maximum fluctuation of distributed generation in actual projects. Therefore, the author proceeds with an analysis on the maximum power fluctuation value and the rated power of distributed generation at the end of this paper. It can be concluded from the analysis that the most appropriate ratio of power capacity of energy storage and the rated power of DG is 0.43, while the ratio of energy capacity and power capacity needs to be determined by D-value of SOC, which is provided in Table 3. With the method in this paper instead of complex optimization method of energy storage capacity, the value of energy storage capacity can be estimated on the condition of knowing the rated power of distributed generation, power generation coefficient of wind and light, charge and discharge efficiency of energy storage and SOC limitations of energy storage.

REFERENCES

[1] Hessami, M.A. & Bowly, D. R. (2011) Economic feasibility and optimization of an energy storage system for Portland wind farm. *Applied Energy*, 88:2755–2763.

[2] Rahul, W., Jay, A. & Rick, M. (2007) Economics of electric energy storage for energy arbitrage and regulation in New York. *Energy Policy*, 35(4):2558–2568.

[3] He, X., Lecomte, R. & Nekrassov, A. et al. (2011) *Compressed air energy storage multi-stream value assessment on the French energy market*. IEEE Trondheim PES Power Tech, NTNU. Trondheim, Norway, 2011: 1–6.

[4] Yusuke Hida, Yuki Ito & Ryuichi Yokoyama, et al. (2010) A study of optimal capacity of PV and batter energy storage system distributed in demand side. *45th International Universities Power Engineering Conference*, Cardiff, Wales, UK: 2010: 1–5.

[5] Liang, L., Li, J.L. & Hui, D. (2011) Optimal configuration of capacity of energy storage devices in large-scale wind power plant. *High Voltage Technology*, 37(4):930–936.

[6] Wang, C.S., Yu, B. & Xiao, J. et al. (2012) Energy storage capacity optimization method of power generation system output fluctuation of smooth and renewable energy sources. *Proceedings of the CSEE*, 32(16):1–8.

Electric, Electronic and Control Engineering – Shao, Shu & Tian (Eds)
© 2015 Taylor & Francis Group, London, ISBN: 978-1-138-02842-5

Design of English educational games and SPSS analysis on application effect

Yanghai Sun
Foreign Affairs Office of Scientific Research, Anyang Vocational and Technical College, Anyang Henan, China

ABSTRACT: Educational game is now a new teaching model of English teaching activities. As a new teaching model assisting English teaching, educational game plays an immeasurable role in English teaching. Based on the theoretical foundation of combining educational games and English teaching, the author concludes that the combination of educational games and English teaching accords with the objective law of teaching. The effect of combining the two can be anticipated. Then, the author gives the design process of English educational games and designs an English educational game of finding the post office of another city in line with the subject content of freshman. At last, with the method of teaching experiment, the author sets an experimental class to which English educational games are applied and a control class that is traditionally taught for two months, and carries out achievement test for their learning conditions, so as to verify the superiority of English proficiency in experimental class that is compared to the control class. Results indicate that English educational games designed in this paper play a significant and positive role in the teaching process, which can be popularized tentatively.

KEYWORDS: English teaching; educational games; waterfall model; SPSS19.0; significant difference.

1 INTRODUCTION

Juan Ang (2013) points out that it will have a significant impact on the reform of educational and learning methods that educational games coincide with the new curriculum concept with the ideology of "teaching through lively activities" [1]. The educational field of this paper is English teaching. However, there is a great variety of English educational games on the market, effects of which have certain defects. Therefore, it is quite necessary to master design principles and methods of educational games and integrate educational games with practical teaching.

Efforts on the attempt and exploration of combining educational games and English teaching have been made by many people, who made the development and application of educational games wider and wider by promoting the process of integrating educational games and educational teaching. Based on the theory of constructivism, Feifeng Wang (2014) designs an English game learning platform according to problems existing in current higher vocational English teaching and features of higher vocational students. A semester of teaching practice is conducted through the platform, effects of which have been initially proved that the real combination of knowledge and games not only meets interests and hobbies of students but also helps students learn English easily [2]. Liping Wang (2014) indicates that the current studies on educational games mainly involve theoretical aspect while quite less in English aspect. However, educational games provide a new possibility for English learners in terms of independent learning. With the example of "English learning in Dreamland", the possibility of assisting English independent learning by educational games is also explored [3]. Wen Wang (2014) points out that traditional teaching methods are unable to meet the requirements of modern teaching after the release of new curriculum standard. As for English teaching in higher vocational college, educational games break through limitations of traditional teaching modes, playing an important role in improving teaching efficiency. It is concluded that educational games effectively improve learning enthusiasm of students and propose higher requirements on comprehensive quality of English teachers [4].

Based on previous studies, the author designs English educational games for English courses of the first semester at college and verifies the superiority of English educational games designed in this paper in term of improving English proficiency according to the differences of two months of periodic examination performance between the experimental class and the control class.

2 THEORETICAL FOUNDATION OF COMBINING EDUCATIONAL GAMES AND ENGLISH TEACHING

"People who know it are no better than those who love it; those who love it are no better than those who love to know it." The ancients suggested that interest has a significant impact on learning effects. English is quite essential in examination, work and communication. Today, English is particularly important in a world of globalization. In order to improve English learning efficiency and learning interests of students, the author designs educational games combined with English teaching so as to provide a theoretical basis for the development of English teaching and lifelong English learning of students.

2.1 *Definition and features of educational games*

Educational game is a secondary thing, which becomes popular with the rapid development of video games. Games are conducted on the premise of resources. The purpose of educational game design is to experience happiness during games and develop some kind of ability while accomplishing games. The research lab of Chinese distance education market (2004) indicates that software of computer games that cultivate knowledge, skill, intelligence, emotion, attitude, and value of game users is called educational game [5]. Zhiting Zhu, et al (2005) defines educational games as entertainment technology, holding that entertainment technology is the theory as well as the practice of promoting the combination of learner's life experience and educational purpose and means by constructing, using and managing proper technological process and resources on the basis of respecting learners' current life value [6]. Aikui Tian (2007) points out that, with the aid of digital media like computer, network and multimedia, computer games with certain educational significance that combines advanced teaching concepts and learning methods with game scenario and carries out teaching through games can be called educational games [7]. The author believes that the definition of educational games provided by Aikui Tian is more appropriate.

According to conventional criteria, educational games can be divided into Role-playing Game, Action Game, Simulation Game, Adventure Game and Strategy Game. According to supporting platform, educational games can be divided into standalone educational games and on-line educational games. According to educational objectives, educational games can be divided into cognitive games, emotional games and motor skill games. Functions of educational games include constructing learning environment, purifying mind of students, creating learning situation, inspiring learning interests, activating classroom atmosphere, fulfilling course content, overcoming psychological barriers and achieving self-worth. Therefore, educational games have such features as entertainment, instructive nature, interactivity and challenges.

The purpose of this paper is to provide a scientific and reasonable design scheme of educational games that are combined with English teaching. So, game design concept should also instruct the design of educational games. Andrew Rollings, et al. (2004) suggests that game design is divided into three special areas, namely core mechanisms, storytelling and narrative, and interactivity. Here, core mechanisms refer to game rules; storytelling and narrative refer to game background; interactivity refer to game operating methods of students [8]. The model is presented in Figure 1:

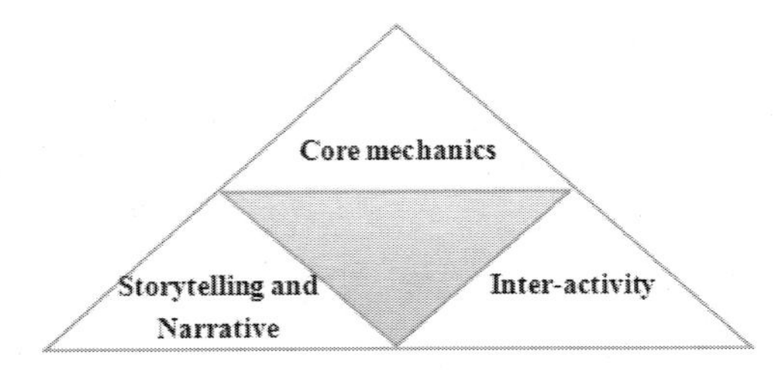

Figure 1. Model of designing special areas of educational games.

2.2 *Combination of educational games and English teaching*

Weixu Zhang (2011) points out that, with the further development of curriculum reform, it can be proved by a great deal of teaching activities of first-rate teachers that teaching effects are significantly improved in courses with game teaching. The emergence of educational games is the development and enrichment of game teaching theory. The application of educational games in teaching not only activates classroom atmosphere but also inspires learning interests of students [9]. According to *English Curriculum Standard of Ordinary Senior High School of Full-time Compulsory Education* formulated by the Ministry of Education in 2001, it is necessary to pay attention to emotions of students, create relaxed and harmonious teaching atmosphere, and broaden the channel of learning and using English with modern educational technology. Teachers should take full advantage of modern educational technology to explore English teaching resources, broaden learning channel of students and improve learning ways of students so as to improve teaching performance [10]. Therefore, from properties of educational resources and features of educational games, it can be affirmed that educational game belongs to educational resources, playing the same role of teaching assistance as multimedia course material, video data and recording. As an important manifestation of mind emancipation, development and utilization of curriculum resources provide a practical basis for schools and teachers in exploring teaching resources and accord with the ideology of curriculum reform. This is the significance of educational games in the establishment English curriculum resources.

Yalin Ying (2005) indicates that educational games applied in teaching can be divided into explorative games, diagnostic games, cooperative games and virtual games [11]. Different educational games are able to improve different knowledge and skills of students. There is a wide range of English educational games, which can be generally divided into games of keyboard input cultivation, games of word memory ability improvement, games of English application capability cultivation and games of surviving in English context. Typical games are shown below:

1 A-Skill operation mode: Typing games are developed by Kingsoft. The background of the game "Speed" is a traditional scene of the police and thief. Students can choose to be the role of a policeman as well as a thief in stand-alone mode or multi-player mode of LAN. This game helps the improvement of word spelling ability to some extent.
2 B-Knowledge memory mode: "Babel elf" is an application of Renren, a famous dating site in China. Combining the playing method of the most popular SNS game, it has beautiful pictures and plentiful scenes, which help to increase vocabulary of students to some extent.
3 C-Problem exploration mode: "Wulong Academy", developed by Shanghai Network Technology Development co., LTD, aims at learning English with the English learning tool of 2D turn-based MMORPG network game form. Students are required to do a great deal of English exercises during the game process, including listening comprehension and oral practices, so that the English proficiency can be increased unconsciously.
4 D-Virtual simulation mode: The open and interactive virtual environment provides a great possibility for the realization of "learner-centered" concept of Second Life, which should be realized in each kind of teaching activity through active interactions. It not only requires active participation of learners but also provides autonomous control and selection for learners. Students can not only harvest their outlook of life and values but also carry out social activities with English and improve English proficiency of different aspects in the game.

Parts of interfaces of the above-mentioned four typical games are presented in Figure 2:

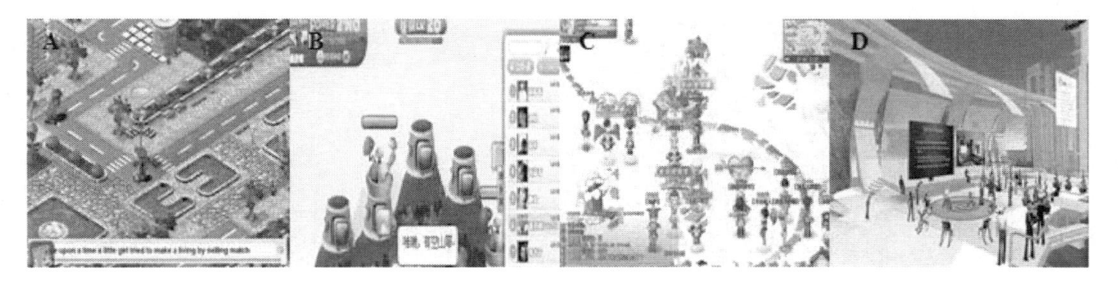

Figure 2. Interfaces of four typical English educational games.

3 DEVELOPMENT TECHNOLOGIES AND DESIGN PROCESS OF ENGLISH EDUCATIONAL GAMES

There are various English educational games on the market, users of which are the majority of people. Firstly, it is necessary to analyze the technological principle of game development in order to design English educational games with targets. Only in this way games can be tailored for relevant people.

Development technologies of English educational games mainly include artificial intelligence technology, network communication technology and tools for educational games development. Artificial intelligence technology is an important branch of computer science technology, the main content of which is to provide computers with thoughts that are similar to human. This can be achieved in two ways. One way relies on traditional programming and the other relies on genetic algorithm and artificial neural network. Artificial intelligence technologies realized in the two ways provide the same feeling for people. Network communication technology covers such aspects as computer, database, information resource, network, sensor, etc. and has such functions as resource sharing, information transmission and centralized processing, integrated information service, etc. Video games develop from stand-alone operation to internet with the development of this technology. Although there is no web-based educational game at present, the successful operation of web-based video games has reflected the prospect of web-based educational game. Tools for educational games development include programming language, animation software and professional game development software. Representative programming languages are C, C++ and java high-level languages. Representative animation software includes Flash and internally installed Action Script. Professional game development software includes RPG maker and Game maker.

The design of English educational game in this paper can be divided into plot design, game objectives setting and game rules setting. Zhongjian Bai (2009) points out that development mode of waterfall model includes S1-Feasibility analysis, S2-Demand analysis, S3-Profile design, S4-Detailed design, S5-Coding, S6-Testing and S7-Run and maintenance. It has an effective promoting function on eliminating unstructured features and reducing software complexity [12].

The objective of feasibility analysis stage is to analyze the feasibility of developing an edutainment games with the combination of classroom English teaching. Two tasks are expected to be accomplished in demand analysis stage. One task is to carry out search with advanced network on similar educational games. The other task is to conduct a detailed analysis on each function of software on the condition of determining the feasibility of developing software. Coding stage aims at coding for a certain event of a game. Different results are triggered by correct and wrong answers. This requires a complete game framework in designing stage, which exists in the whole process of game development. Programming of each event requires immediate test to see if development effect is consistent with the expected effect. In addition, it is better to find out problems of third party testing games after development. Therefore, the design process of English educational games of the waterfall model is shown in Figure 3.

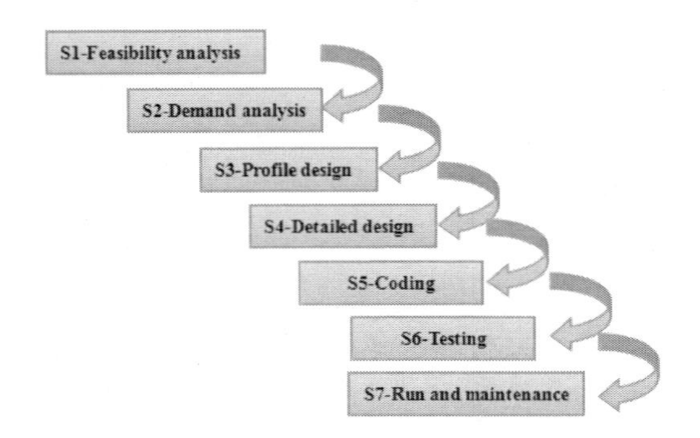

Figure 3. Design process of English educational games of waterfall model.

4 DESIGN AND IMPLEMENTATION OF ENGLISH EDUCATIONAL GAME

4.1 *Design scheme*

English educational games designed in this paper are tailored for freshmen of higher vocational college. The design process contains plot design, game objectives setting and game rules setting. The above-mentioned three modules are designed as follows:

Plot design: The clue of this game is to find the post office of another city. There are four scenes. The first scene is set at school and classroom, where teachers should test the listening ability of students. If the game is completed perfectly, an important task will be assigned to Jack from teacher, which is to deliver an important letter to a neighboring city. The second scene is set outside school, where Jack is required to finish the task with his classmate Mali. But Mali is asked to check the learning condition of Jack. Only when Jack finishes his learning perfectly can Mali agree to join the task of finding the post office and delivering the letter. The learning condition of vocabulary of Jack is examined in this scene. The third scene is set at a forest park. There are several important levels on the road, which can be only passed by mastering enough vocabulary. This aims at examining the total learning condition of vocabulary. The fourth scene is set at the city where the post office is situated. The precise location of the post office can be acquired by having conversations with different people. This aims mainly at examining the learning condition of sentence patterns of asking directions and the ability of grasping keywords in listening comprehension. Classic fighting scenes of RPG games are also added in this game in order to increase the entertainment of the game.

Game objectives setting: RPG Maker XP software is applied in this paper to design a game of 15 minutes, where students are required to comprehend the meaning of the short passage according to English recordings in the game, extract the main point of the short passage according to the scene, comprehend the general idea with new words in the context, and choose right answers according to the listening information. English should be used in the game to assign tasks. Students are required to get meanings of new words according to the context, master the logical relationship among sentences and obtain important information of the text so as to continue the game. Finding the post office is the main clue of the game and explorative scenes are also set in the game, where students are able to acquire knowledge, find problems, analyze problems, comprehend problems, design schemes for solving problems and solve problems ultimately. In this game, intelligence of students can be enhanced and thoughts of students can be expanded. Functions of RPG Maker XP software, such as display options, condition differences, loop, break loop, label, and skip label, are applied in this game to design logic events, so that different structures can be acquired from different choices. As for some levels, students can only get to a new question by answering the current one correctly. Otherwise, questions should be answered repeatedly.

There are five points of the main content of the game rule:

1 Keys of up, down, right and left on the keyboard are used to control the moving direction of figures. Use the correct key to choose an answer when there are options.
2 Priorities of map elements in this game are set so that figure can only move on permitted elements.
3 Homeless dog or monsters can choose to fight or run away.
4 Only by meeting the requirements of the game rule at answering level can the next level be accessed by figure. Otherwise, the next link cannot be accessed.
5 Progressing file can be recorded so that game players are able to start the rest of the game from the record.

4.2 *Scheme implementation*

The implementation of English educational game mainly includes four aspects, namely interface design, game framework design, game event design and interactive design. Interface design contains start interface, campus scene interface, classroom scene interface, small town interface, wild hotel interface, etc. Unnecessary details are not provided here. Scheme implementations of other three aspects are emphasized here.

The English educational game designed in this paper is unfolded according to the plot. So, the linear structure of the game framework is shown in Figure 4:

The whole plot of the English educational game is completed through the design of events. Game events design refers to the happening of events in the process of the game. Tasks of the figure and structures of the figure and other events are completed through events design. The game development software RPG Maker XP provides strong ability of event processing. Small squares would appear at game editing interface when event layer is clicked. There are new events on each square when the right-hand button is clicked or the mouse is double-clicked. Events can be edited through event windows. The so-called events are different results of contact between the figure and objects. Event management can be applied to relationships among characters and any events.

Interactive design refers to interpersonal interaction and human-computer interaction. The English educational game designed in this paper is a standalone game, which mainly conducted through human-computer interaction. Interactive modes of this game are keys and keyboard control and interaction of animation and sound effects. The specific condition is presented as below:

1 Keys and keyboard control: Students use keys of up, down, right and left on the keyboard to control the moving direction of figure, which can be controlled according to the wishes of students. There are three different options on the start interface, which can be selected by students with keys of up, down, right and left. Use "ENTER" on the keyboard to determine the option. The step of using "ENTER" provides students their own choices so as to interact with computer time. The plot of the game is carried on through selections at different time. Some of the options are automatically executed. For example, dialogues among characters can be taken as human-computer interaction.
2 Animation and sound effects: Animation and sound effects also provide an experience of game interaction. The link of asking directions is realized by sound effects. Interactive experience on hearing of students can be enhanced by adding conversations to sound library. Appropriate option settings and effective animation and sound design can reflect game interaction, making game players fully devote themselves into the game and acquire knowledge from the game.

The key technology and design skill of English educational game include the introduction of homemade figure elements, introduction of homemade sound, increase and decrease of money, automatic display of objects and reading of scripting language in homemade files. The game should be popularized after development, so packaging and encryption are important components of game development. There are two methods through which the developing method of a complete game can be seen by other people:

Method 1. The RPG Maker XP development environment should be equipped by players. Game.exe would be automatically generated in disk directory

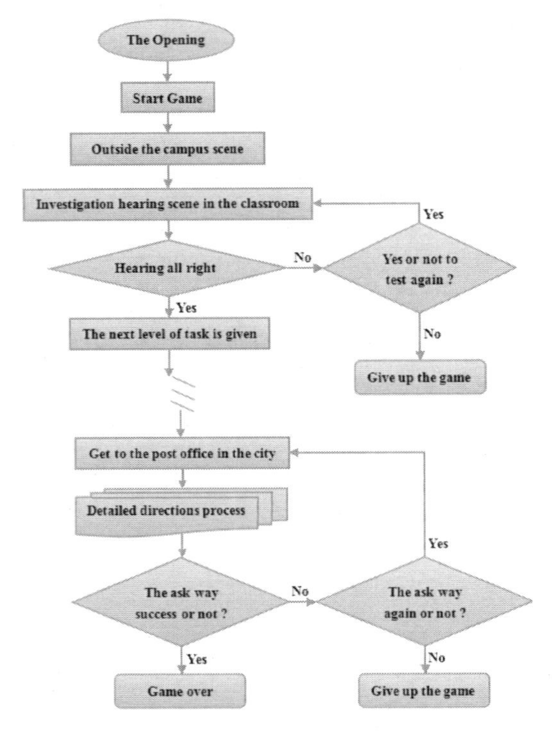

Figure 4. Linear structured framework of English educational game.

where the game is recorded at game testing. Players are able to play the game only by copying the file folder of game developing to their own computer.

Method 2. Formal release, steps of which are presented as follows:

STEP 1. "Game" precedes "Choose RTP" in project selection menu. Three settings are selected as "None" and then save the project.

STEP 2. Copy all materials directly from installation folder of RPG Maker XP to game directory.

STEP 3. Install RPG Maker XP in the corresponding directory.

STEP 4. In project menu, select "File" that precedes "Game data compression", generate check box of encrypted file, and finally generate executable file "post office.EXE".

5 EFFECT ANALYSIS OF THE COMBINATION OF EDUCATIONAL GAMES AND ENGLISH TEACHING

The method of performance comparison of teaching experiment test is applied to verifying the remarkable result of English learning brought by English educational games.

Teaching experiment process: Select 32 students of Class A and 33 students of Class B from XX College

Table 1. Results of overall analysis on scores of the experimental class and the control class.

			Frequency		Percentage		Effective percentage		Accumulative percentage	
			Experiment	Control	Experiment	Control	Experiment	Control	Experiment	Control
Effective	/	5.5	/	1	/	3.0	/	3.0	/	3.0
	/	6.0	/	2	/	6.1	/	6.1	/	9.1
	6.5	/	1	/	3.0	/	3.1	/	3.1	/
	7.0	7.0	5	8	15.2	24.2	15.6	24.2	18.8	33.3
	7.5	7.5	1	5	3.0	6.1	3.1	6.1	21.9	39.4
	8.0	8.0	5	8	15.2	24.2	15.6	24.2	37.5	63.6
	8.5	8.5	3	3	9.1	9.1	9.4	9.1	46.9	72.7
	9.0	9.0	7	5	21.2	15.2	21.9	15.2	68.8	87.9
	9.5	9.5	4	2	12.1	6.1	12.5	6.1	81.3	93.9
	10.0	10.0	6	2	18.2	6.1	18.8	6.1	100.0	100.0
	Total		32	33	97.0	100.0	100.0	100.0		
Deficient	System		1	0	3.0	0				
	Total		33	33	100.0	100.0				

	Sample number	Minimum	Maximum	Average	Standard deviation
Experiment	32	6.50	10.00	8.6094	1.08311
Control	33	5.50	10.00	7.9394	1.13028
Effective	32				

Table 2. Analysis results of the independent-samples T test.

Class	Experiment class	Control class				Equal variances assumed	Unequal variances assumed
Sample number	32	33	Levene test of variance equation		F	.023	
					Sig.	.880	
Average	8.6094	7.9394	T test of mean equation		t	2.439	2.440
					df	36	62.992
Standard difference	1.08311	1.13025			Sig. Two sides	.018	.017
					Mean difference	.66998	.66998
Standard error of average	.19147	.19675			Standard error	.27472	.27454
				Confidence interval of 95% difference	Lower limit	.12099	.12136
					Upper limit	1.21897	1.21860

as the objects. Class A is the experimental class, to which the edutainment teaching method of English educational game is applied; Class B is the control class, where the traditional English teaching method is applied. The learning period lasts for two months and a test in line with examination syllabus is carried out at the end of the second month.

According to the comparison of test performance of the experimental class and the control class, use the statistical description function of software SPSS19.0 to carry out overall analysis on scores of the two classes. Analysis results are summarized in Table 1.

It can be known from results in Table 1 that the average score of the experimental class is higher than that of the control class. And it can be also known from the comparative analysis that the proportion of students with full marks and high scores in the experimental class is obviously higher than that of the control class. In order to further make sure whether there are obvious differences between the two classes in terms of score, the method of the independent-samples T test is applied in this paper for verification, results of which are shown in Table 2. It can be known from Table 2 that the corresponding probability of F statistics in the verification of two population variances is 0.023. Differences of scores in this analysis are not quite obvious, so the significance level α is 0.05. Because P (0,880) is larger than 0.05, it can be known according to the principle of independent-samples T test that there are non-obvious differences between population variances of the two classes. Therefore, the results of the first line of T test are applied to inspect the ensemble average. The observed value of T statistics is 2.439 and the corresponding probability is 0.018, which is 0.05 less than the significance level. Thus it can be inferred that there are significant differences between ensemble averages of the two classes.

6 CONCLUSION

1 Through the study on features of English teaching and the theoretical foundation of combining education with games, it can be concluded that learning enthusiasm and learning effect of students can be promoted by the combination of the two.
2 Emphasize the design process and rules of educational game, and designs a game of finding the post office of another city in line with the new curriculum content of the first semester at college. It can be known from the design process and scheme implementation that the educational game designed in this paper is featured by its advantage of easy access, which provides a basis for the propagation.
3 In order to verify the advantage of combining educational games with English teaching, the author carries out two months of teaching experiment on Class A and Class B of college with the educational game designed in this paper and concludes the significant advantages of combining educational games with English teaching through monthly examination scores.

REFERENCES

[1] Ang, J. (2013) Design and development of Fruit Matching in primary English teaching based on Flash. *Journal of Guizhou Normal University (Natural Science Edition)*, 31(6):100–105.
[2] Wang, F.F. (2014) Research on design and practice of game learning platform of higher vocational English. *Journal of Hubei Economy Academy (Humanities and Social Science Edition)*, 11(3):210–211.

[3] Wang, L.P. (2014) An explorative analysis on independent English learning assisted by educational games——Take "English learning in Dreamland" as the example. *Journal of Mudanjiang Institute of Education*, (4):76–77.
[4] Wang, W. (2014) Applications of educational games in primary English teaching. *Elementary and Secondary Education*, (3): 241–242.
[5] Research Lab of China Distance Education Market. (2004) A research report on educational game industry. *China Distance Education*, 15(03):32–33.
[6] Zhu, A.T. *et al.* (2005) Edutainment technology: A new field of educational technology. *China Educational Technology*, 25(14): 15–18.
[7] Tian, A.K. (2007) *A Study on Digital Educational Games that Support Independent Learning*. Shanghai: Huadong Normal University.
[8] Andrew Rollings and Emest Adams. (2004) *Game Design Technology* (translated by Jin, M & Zhang, C.F.). Beijing: China Environmental Science Press.
[9] Zhang, W.X. (2011) *Design and research of educational games of English teaching in middle school*. Shenyang: Shenyang Normal University.
[10] Ministry of Education. (2001) *English Curriculum Standard of Ordinary Senior High School of Full-time Compulsory Education*, Beijing: Beijing Normal University Press, pp:28–31.
[11] Yin, Y.L. (2005) *Worth Exploring and Model Construction of the Integration of Video Games and Courses*. Wuhan: Huazhong Normal University.
[12] Bai, Z.J. *et al.* (2009) *Software Engineering: Theories and Practices*. Beijing: Higher Education Press, 8–9.

Electric, Electronic and Control Engineering – Shao, Shu & Tian (Eds)
© 2015 Taylor & Francis Group, London, ISBN: 978-1-138-02842-5

Design and FEA of a light truck two-leveled leaf spring

Huajie Wu & Xiaofeng Dai
Yangzhou Polytechnic Institute, Yangzhou Jiangsu, China
Yangzhou Tong An Automobile Performance Test Co., Ltd., Yangzhou Jiangsu, China

Zaixiang Zheng
Yangzhou University, Yangzhou Jiangsu, China

ABSTRACT: The objective of this article is the two-leveled leaf spring, which is composed of main leap spring and auxiliary spring. This article mainly described the calculation process of the leaf spring's parameter, checked the leaf spring model's stress, and conducted finite-element analysis on its body structure. This paper calculated the leaf spring stiffness through the geometric mean method. The paper also calculated the width, thickness, number, and length of each piece spring; the spring assembly's arc height and radius of curvature in a free state. According to the corresponding parameter, the author established the finite-element model and carried out the finite-element static analysis. The simulation result is in agreement with the leaf spring's design parameter and is reasonable.

KEYWORDS: leaf spring; design parameter; stiffness; FEA.

1 INTRODUCTION

Two characteristics of the leap spring are exhibited when it is compared with other elastic elements. It has a guiding and shock absorption role. Whether the light truck in no-load or full-load condition, two-leveled leaf spring with nonlinear characteristics reveals good frequency. The author designed two-leveled leaf spring's parameter for a light truck under load, to further understand how the leaf spring meets the static strength. Simultaneously, the article provides a theoretical basis for smooth leaf spring design.

2 DESIGN OF LIGHT TRUCK LEAF SPRING SUSPENSION

According to the structural characteristics of the light truck chassis frame, the leaf spring suspension structure's type is vertical, and the leaf spring has a symmetrical arrangement. According to the requirements of GB/T1222-1984, 60Si2M is selected as the material of the leaf spring. The diameter of the spring is 30mm. The light truck's design carriage requirement for quality of rear axle is 2000Kg when the truck is no-load, and the value is 3500Kg when the truck is full-load. This means that every pair of the leaf spring suspension load is 7663.6 N when the truck is no-load, and the value is 15013.6 N when the truck is full-load.

2.1 *Main parameter*

2.1.1 *Stiffness*

The static deflection of automobile suspension affects the vehicle's comfort. The closer the static deflection and vibration frequency of front suspension and rear suspension to each other, the smaller the car assembly of resonance will be, and the more comfortable it will be. Formula 1 involves the relationship between frequency (n) and deflection (f_c).

$$n = 5 \Big/ \sqrt{f_c} \tag{1}$$

Usually, the static deflection range is from 5 to 9 cm; the author estimates the static deflection is 7.72 cm. General experience shows the suspension dynamic deflection range as follows: 6–8cm, and the author estimates the dynamic deflection is 6.2 cm.

Assuming that the leaf spring is a linear characteristic, the stiffness of the main spring is f_k when the auxiliary spring contacts the frame. P_0 is the letter appellation of burden when the truck is no-load, and

P_m is the letter appellation of burden when the truck is full-load. Natural frequencies of the main spring can be calculated by formula 2.

$$N = \frac{300}{\sqrt{P/C}} \tag{2}$$

The design requires good comfort, which means requiring small natural frequency changes when the car actually runs. There are two kinds of situations: One is in the whole load range, where the frequency changes are minimal; the other is a frequency change in the process of auxiliary spring contacts where the frame is minimal. But these two solutions are contradictory. These are extended out of the two calculation methods: geometric mean method and average load method. The difference between them is that the former is suitable for heavy load condition, which requires comfort, whereas the latter is often applicable to a light load state. The designed truck will be used in the heavy load condition, so the geometric mean method is applied to calculate the parameters. The precondition of calculation is natural frequency that will not change during the auxiliary spring contacts frame no matter whether the truck is in no-load or full-load condition.

The calculation process is as follows:

$$\begin{cases} f_m = f_k \\ f_0 = f_a \end{cases} \tag{3}$$

$$f_m = \frac{P_m}{C_1 + C_2},\ f_k = \frac{P_k}{C_1},\ f_0 = \frac{P_0}{C_1},\ f_a = \frac{P_k}{C_1 + C_2}$$

Through formula 3, the stiffness of the main spring can be got, $C_1 = 138.95\ N/mm$; the stiffness of the auxiliary spring can be got, $C_2 = 55.53\ N/mm$; and the stiffness of the two-leveled leaf spring can be got, $C_c = 195.4\ N/mm$. The critical load of the work for auxiliary is $P_k = \sqrt{P_0 P_m} = 10726.52(N)$.

2.1.2 *Width, thickness, and piece number*

The light truck's leaf spring is symmetric. The formula of the section beam is used to calculate stiffness and strength, but it needs to introduce coefficient δ to correct results. The author calculated the leaf spring's total inertia moment J_0 according to the correct coefficient.

$$c = \frac{1}{\delta}\frac{48EJ_0}{(L-ks)^3} = \frac{1}{\delta}\frac{48E\sum_{k=1}^{n} J_k}{(L-ks)^3} = \frac{1}{\delta}\frac{48EnJ_1}{(L-ks)^3} \tag{4}$$

$$\frac{n}{\delta} - \frac{(L-ks)^3}{48EJ_1} = 0 \tag{5}$$

In the formula, S - U bolt center distance; k - invalid length factor when U bolt assembly is completed (Rigid clamping: $k = 0.5$; flexible clamping: $k = 0$); c - leaf spring vertical stiffness; δ - disturbance enlargement coefficient; E - material elastic modulus; and J_k - inertia moment.

The formula of the total cross-section leaf spring coefficient is as follows:

$$W_0 \geq F_W (L-ks)/(4[\sigma_w]) \tag{6}$$

For the leap spring's piece number, the less number of the spring's piece will be beneficial to manufacturing and assembly; it can also reduce the friction between the pieces and improve the vehicle ride comfort. Weakness in the leaf spring mathematical model is not accurate, and the reason is the difference between the leaf spring and the equal-strength beam. In addition, the utilization rate of the material is low in the situation. The spring pieces' number, width, and thickness affect the total penetration resistance moment and spring vertical stiffness. If width and thickness are increased, the pieces' number must be reduced. In addition, steel section size and size should comply with the Chinese standard profile size. The multi-plate leaf spring's general number selection range is 6–14 pieces. The pieces' number of the auxiliary spring is often less than 6.

The last expression $\frac{n}{\delta} - \frac{(L-ks)^3}{48EJ_1}$ can be used to describe the specific thickness and the number of pieces. Because the pieces' number should be an integer, the formula of the final result cannot be zero; when the result (Δ) is close to zero, the result is reasonable.

$$\Delta = \frac{n(1-\eta)^3}{3[0.5 - 2\eta + (1.5 - \ln\eta)\eta^2]} - \frac{(L-ks)^3 c}{48EJ_1} \tag{7}$$

The results show that if the main spring pieces' number is 11, width is 75mm, thickness is 8mm, and disturbance enlargement coefficient is 1.3991, and Δ is the most close to zero in all the data, which is also that it is the design parameter for the main spring. The results also show that if the auxiliary spring pieces' number is 4, width is 75mm, thickness is 6mm, disturbance enlargement coefficient is 1.2828, and Δ is the most close to zero in all the data, which is the design parameter for the auxiliary spring. According to the physical model of design parameters, the length of the auxiliary spring is rather shorter than that of the main spring, and the addition of thin plates is not suitable, so it is difficult to accord it with the ideal of the geometric mean method completely. From what has been discussed earlier, the author believes the stiffness is bigger than the actual result.

2.1.3 *Length*

The length of each piece directly affects quality and service life of the leaf spring. If parameters such as length and curvature of each piece are determined, each stress condition of the spring is also determined, so pieces' stress should be reasonably distributed. Two methods are used to determine the length, that is mapping method and calculation method. The author chooses the former method. The specific process was performed by using AutoCAD software. It is based on this principle that the shape of the leap spring unfolds along the thickness closest to the trapezoidal shape. This is done in a particular way: First, draw the point whose value is the piece's thickness cubic value along the y coordinate; then, in the X coordinate, draw the point whose value is half master L and u-bolt S. Connecting these two points, there is a triangular plate spring expansion drawing. The pieces' length of the main spring is 1270,1170,1070,960,860,760,660, 550,450,350,250; the pieces' length of the auxiliary spring is 970,770,560,350. Their unit is mm.

3 ARC HEIGHT AND RADIUS

The free state of the leaf spring is shown in figure 1:

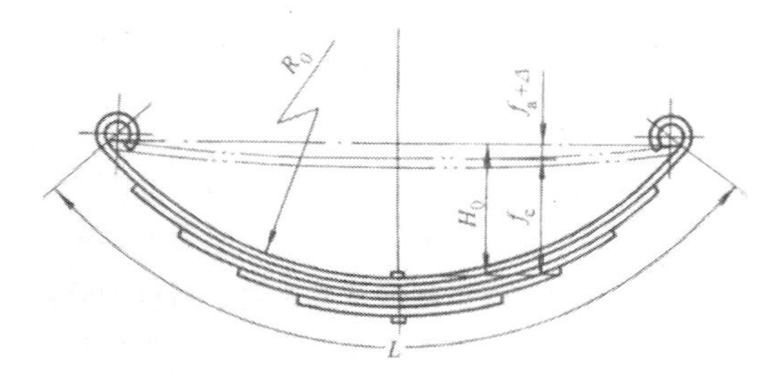

Figure 1. Leaf spring's arc height in a free state.

The following calculation formula is used for the leap spring's arc height in a free state:

$$H_0 = f_a + f_c + \Delta \quad (8)$$

The letters in the formula: f_a - arc height in full-load; f_c- static deflection in full-load. The author ignores leaf spring plastic deformation in the process of vibration. The frame height is high enough, and it ensures that the leaf spring will be able to get enough dynamic deflection without interference. Usually, the scope of value of f_a is 10–20mm, and the author believes this value is 10mm.

$$\Delta = \frac{s(3L - s)(f_a + f_c)}{2L^2} \quad (8)$$

Through this formula 8, the author calculated arc height of the leaf spring's main spring in free state: $\Delta = 13.83(mm)$, $H_{01} = 100.67(mm)$; arc height of the leaf spring's auxiliary spring in free state: $\Delta = 17.9(mm)$, $H_{01} = 104.7(mm)$ is also calculated.

Radius (R_0) of curvature for spring assembly:

$$R_0 = \frac{H_0^2 + L^2/4}{2H_0} \approx \frac{L^2}{8H_0} \quad (9)$$

This value must be an integer, the radius of curvature for the main spring is 2000mm, and the radius of curvature for the auxiliary spring is 1120mm.

4 STRENGTH CHECK

After calculating the static deflection of the main spring and the auxiliary spring, the load distribution of both can be got in no-load and full-load situations. The author deduced the load distribution formula. According to the working process of the leaf spring,

$$\text{When truck is no-load: } \begin{cases} P_{01} = f_{01}C' \\ P_{02} = 0 \end{cases} \quad (10)$$

$$\text{When truck is full-load: } \begin{cases} P_1 = f_1C' \\ P_2 = f_2C' \end{cases} \quad (11)$$

P_{01},P_{02} is the letter appellation of burden of the main spring and auxiliary spring, respectively, when truck is no-load, and P_1,P_2 is the letter appellation of burden of the main spring and auxiliary spring, respectively, when truck is full-load.

The author obtained the following result:

$P_{01} = 7664.04(N);P_{02} = 0;P_1 = 13185.22(N);$
$P_2 = 1828.4(N)$

The average static stress:

$$\begin{cases} \sigma_1 = \overline{\sigma_1} \times f \\ \sigma_2 = \overline{\sigma_2} \times f_2 \end{cases} \quad (12)$$

Limit stress:

$$\begin{cases} \sigma_{1\max} = \overline{\sigma_1} \times (f_1 + f_d) \\ \sigma_{2\max} = \overline{\sigma_2} \times (f_2 + f_d) \end{cases} \quad (13)$$

The letters in the formula: σ_1,σ_2- the average static stress of the main spring and auxiliary spring, respectively; $\sigma_{1\max}$,$\sigma_{2\max}$- the limit stress of the main spring and auxiliary spring, respectively. $f_d = d\sqrt{f_{c1}}$ (The value scope of f_d is 2.5 to 3.5 when the truck is running.

The result of the strength check based on the earlier formula: static stress of the main spring, $\sigma_1 = 463.75(MPa)$; static stress of the auxiliary spring $\sigma_2 = 233.84(MPa)$. Allowable stress range is 450–550 MPa for the main spring and 200–250 MPa for the auxiliary spring. The result meets design requirements. Limit stress of the main spring, $\sigma_{1\max} = 970.27(MPa)$; stress of the auxiliary spring $\sigma_{2\max} = 970.40(MPa)$. Allowable limit stress range is 900–1000 MPa for the main spring; the result also meets design requirements.

5 FINITE-ELEMENT ANALYSIS

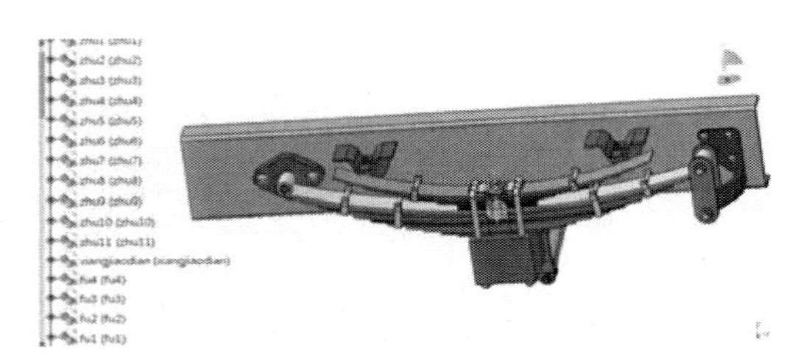

Figure 2. Spring assembly physics model.

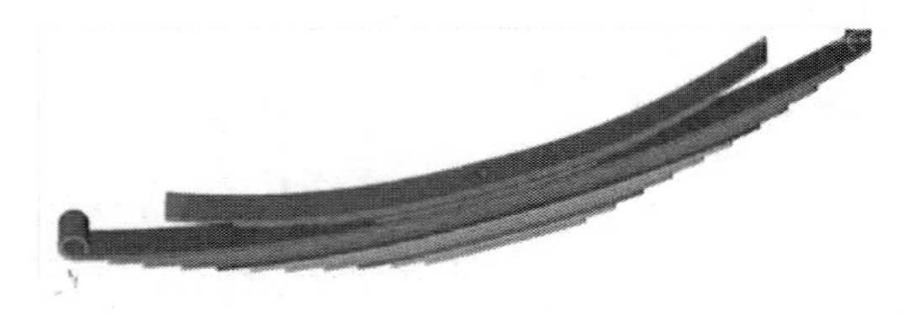

Figure 3. Spring assembly finite-element model.

The spring assembly physics model was established in CATIA, shown as figure 2. The finite-element model was established in Hypermesh, shown as figure 3. The author applied force on the finite-element model of the main spring and auxiliary spring during pre-processing, and the value of force is 8576N and 1920N. The deflection result of simulation is shown as follows:

The author also applied force in the leap spring assembly, and the value of force is 15013N. The deflection of the leaf spring assembly is shown as follows:

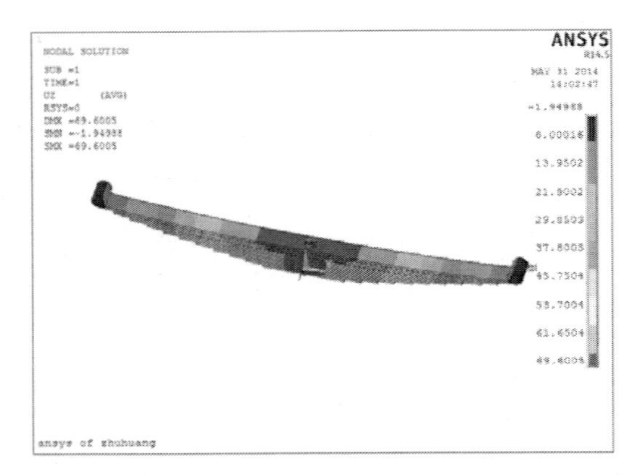

Figure 4. The main spring deflection distribution.

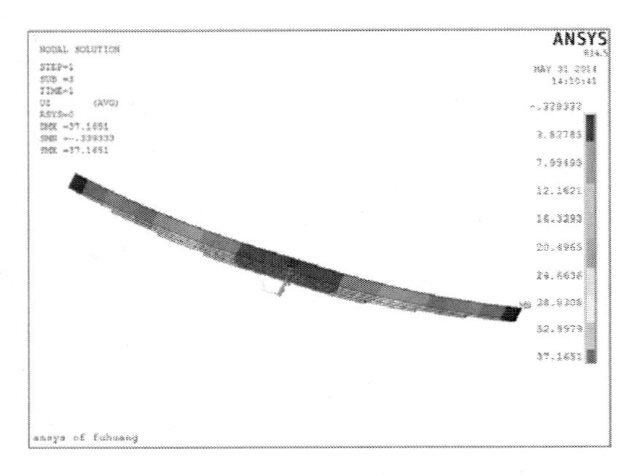

Figure 5. The auxiliary spring deflection distribution.

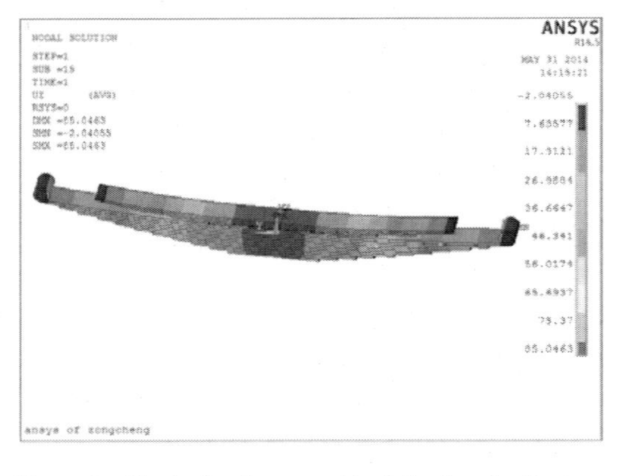

Figure 6. The leaf spring assembly deflection distribution.

The stiffness result of the main spring is 123.21 N/mm based on simulation, compared with the previous calculation results - 139.6 N/mm; the value difference is about 10%. For the auxiliary spring, the simulation result is 51.66N/mm, the calculation result is 55.8N/mm, and the value difference is about 8%. For the auxiliary spring, the simulation result is

176.63N/mm, the calculation result is 195.4N/mm, and the value difference is about 9.6%. Results in conformity are reasonable. The friction coefficient has a great influence on the finite-element model. The large friction coefficient will prevent deformation of the leaf spring, which can decrease the deflection and increase the stiffness. The error of integer for the length of spring's piece also affects the simulation result, because in the calculation model length is not an integer.

6 CONCLUSION

The main content of this paper is to design the leaf spring's parameter for light truck, including width, thickness, length, and pieces' number. First, the author chose working frequency of the leaf spring and then calculated its stiffness. Then, the author got the design parameter based on the geometric mean method, and a check was carried out to ensure stress safety. Compared with the result of the finite-element analysis, the design parameter is reasonable.

REFERENCES

[1] Chen Jiarui. (2009) *Automobile Structure*, Beijing: China Machine Press.

[2] Zhang Bingfang. (1996) Design calculation of the two-stage steel plate compound spring. *Automobile Technology*, 10:49.

[3] Ekici B. (2005) Multi-response optimisation in a three-link leaf-spring mode. *Vehicle Design*, 38(4):326.

[4] Li Lingyang. (2013) A research on the modeling methods of leaf spring. *Automotive Engineering*, 35(7):660.

[5] Kong, Y.S., Omar, M.Z. (2013) Explicit nonlinear finite element geometric analysis of parabolic leaf spring under various loads. *The Scientific World Journal*, 1:11.

Electric, Electronic and Control Engineering – Shao, Shu & Tian (Eds)
 ISBN: 978-1-138-02842-5

A brief discussion on research and application of intelligent management and control alarm platform of unattended transformer

Fuchang Li, Ke Hu, Hang Zhang, Bing Song & Zongjun Hu
Beibei Power Supply Company, State Grid Chongqing Electric Power Company, Chongqing, China

ABSTRACT: With the development of production technology, requirements of different industries pertaining to power supply are becoming higher and higher. Therefore, in order to guarantee stable and reliable power supply, not only power supply units in the power grid need to be inspected and maintained carefully but also the substation environment of power supply units needs to be monitored effectively. The substation environment also has an impact on the performance of the power supply unit and the stability of power supply.

KEYWORDS: substation; unattended operation; alarm platform.

1 INTRODUCTION

With the overall popularization of the unattended transformer management mode, the monitoring system of the unattended transformer is currently being widely used. The basic principle of the monitoring system of the unattended transformer is similar to this: Conduct real-time monitoring according to operation state data of the unattended transformer in the power grid (including operation state of equipment in master control room, real-time image of operation state of main transformer, breaker and disconnecting link, alarm information of intelligent devices for fire alarm, and burglary prevention) and transmit real-time operation state data to the monitoring center, so that management staff of the monitoring center can carry out control and analysis on the operation of each unattended transformer in accordance with real-time operation state data. The monitoring system of the unattended transformer increases the security and reliability of operation and maintenance of the unattended transformer, realizes visual monitoring and scheduling of the power grid, and makes regulation and operation of the power grid safer and more reliable. The traditional monitoring system of the unattended transformer is usually implemented through analog video monitoring technology, which is able to carry out on-site video surveillance and simple alarm information processing and transmission. However, limited by structure and principle, the traditional monitoring system of the unattended transformer is unable to realize remote and real-time integrated control so that management staff of the monitoring center cannot understand and determine the operation state of the unattended transformer in a timely and accurate manner. For this reason, it is necessary to develop a new monitoring system of the unattended transformer so as to solve the problem that remote and real-time integrated control cannot be realized by the traditional monitoring system of the unattended transformer.

Transformer monitoring focuses mainly on environmental monitoring and equipment monitoring. Traditional environmental monitoring is generally carried out through human inspection, which, however, can only conduct monitoring regularly or irregularly within a certain time interval. Changes of environmental parameters of transformers in this interval cannot be effectively monitored, so it is very likely to have missing inspection or inaccurate inspection results and changes of environmental parameters might also cause power supply accidents. Intelligent monitoring systems are nowadays being gradually developed in order to increase the accuracy of monitoring results and conduct all-weather monitoring. And current intelligent monitoring systems such as camera monitoring are also developed. But there are still problems as follows: Monitoring results need to be screened from mass testing data by monitoring staff for comparison and abnormal data are screened by human power, so data are frequently missed without timely discovery. Besides, once an accident happens in a certain substation, such as fire and water immersion, the loss of an accident will be magnified if the substation cannot be regulated in a timely manner and can be only processed by emergency staff on site.

2 BACKGROUND OF INTELLIGENT MANAGEMENT AND CONTROL ALARM PLATFORM OF UNATTENDED TRANSFORMER

In current technologies, two methods are adopted to avoid access of illegal person to the substation. One method involves carrying out guard or walk-around inspection by working staff. But this method causes a waste of human resources and is unable to play the role of effective security and protection. There is no guarantee that each time illegal break-in can be found. The other method is realized by an access control system. Illegal people can be eliminated through the badge number of working staff (namely secret code), working card, or fingerprint. However, there are still deficiencies of these methods. For example, an illegal person can easily steal badge number or working card for illegal break-in. Compared with the former two verification approaches, fingerprint is safer. But there is also the possibility of it being stolen. Even with the access control system, it is still required to have guarding staff. Besides, it is difficult to feed on-site situations back to the monitoring center in time.

There is another common problem of the access control system or human walk-around inspection. If it is unable to make an alarm in time when someone breaks into the substation and takes illegal actions against power supply equipment, abnormal situations happen to the substation environment or equipment, maintenance workers perform wrong operations on the equipment, or illegal people steal equipment, power supply failures will be caused with a severe economic loss.

Therefore, it is necessary to develop a comprehensive monitoring system for the unattended transformer, which is not only able to conduct all-weather monitoring on security and protection, environment, and equipment operation state of substation but also able to take cooperative operations according to security and protection, environment, and equipment real-time monitoring, so as to guarantee the stability of the substation and power supply with no manual intervention and less human cost.

3 DESIGN STRUCTURE OF INTELLIGENT MANAGEMENT AND CONTROL ALARM PLATFORM OF UNATTENDED TRANSFORMER

The intelligent management and control alarm platform of the unattended transformer is a distributed structure with levels and partitions. It consists of three levels, including the master station system of substation video and environment monitoring of provincial level, the master station system of substation video and environment monitoring of prefecture level, and the end system of substation video and environment monitoring, which are shown in figure 1.

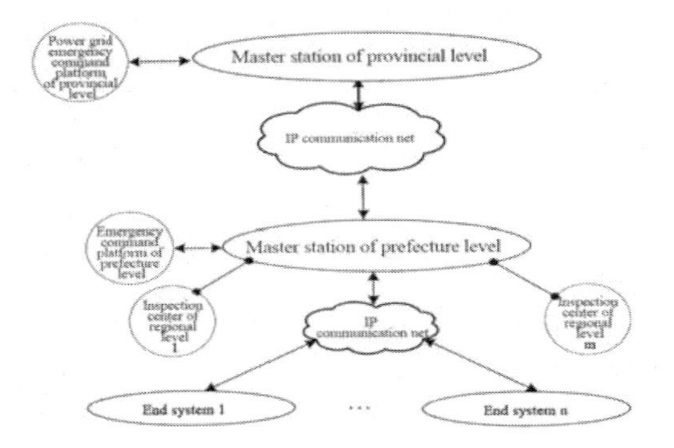

Figure 1. Schematic diagram of remote centralized monitoring system.

Intelligent management and control alarm platform of the unattended transformer supports multi-stage networking structure for the convenience of dilatation.

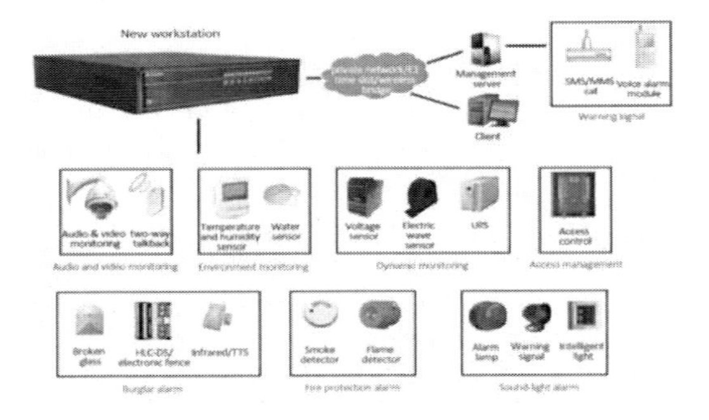

Figure 2. Comprehensive chart of intelligent management and control alarm platform of unattended transformer.

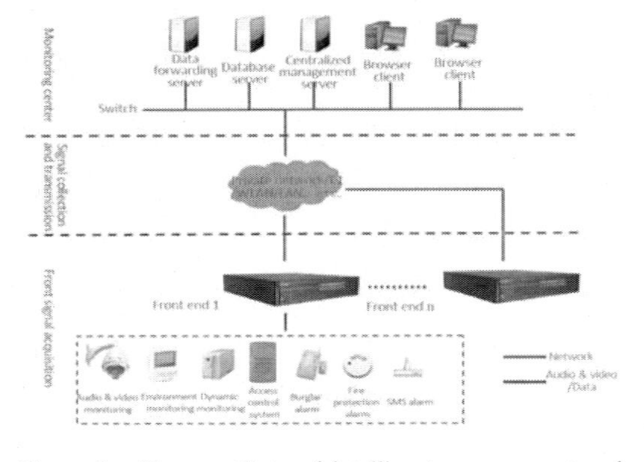

Figure 3. The constitute of intelligent management and control alarm platform of unattended transformer.

The intelligent management and control alarm platform of the unattended transformer consists of three parts, namely front-end signal acquisition and processing part, signal collection and transmission part, and backstage control part (monitoring center).

4 FEATURES OF INTELLIGENT MANAGEMENT AND CONTROL ALARM PLATFORM OF UNATTENDED TRANSFORMER

The intelligent management and control alarm platform of the unattended transformer includes security monitoring subsystem, environment monitoring subsystem, equipment monitoring subsystem, monitoring center, and output unit. The security monitoring subsystem, environment monitoring subsystem, and equipment monitoring subsystem mentioned earlier are connected to the monitoring center for data exchange. The monitoring center is connected to the output unit for information output.

The security monitoring subsystem includes access control module, outside image detection equipment, authentication server and equipment, inside image detection equipment, outside authentication and request equipment, and inside authentication and request equipment. The outside image detection equipment is installed at the export of substation, being connected with an authentication server for image information output. Information interactions between the authentication server and authentication equipment are conducted in a wireless way, whereas data interactions between the authentication server and the monitoring center are conducted in a wired and/or wireless way. The inside image detection equipment is installed for power supply equipment of the substation, the output end of which is connected with an authentication server. Both the outside authentication equipment and the output end of the inside authentication equipment are connected with the authentication server.

The environment monitoring subsystem includes the detection module of environmental parameters of substation, information acquisition and processing module, and environmental output module. The output end of the detection module is connected to the input end of the information acquisition module. The output end of the information acquisition module is connected with the processing module. The output end of the processing module is connected with the environmental output module mentioned earlier, and the processing module is connected with the monitoring center for data interaction. The detection module includes one or two combinations of smoke sensor, temperature sensor, humidity sensor, and water sensor. Output ends of sensors in detection modules are connected with the input end of the information acquisition module.

The third generation of network monitoring technology is adopted for the intelligent management and control alarm platform of the unattended transformer. It is able to conduct independent monitoring management, real-time monitoring, and remote centralized management on single or multiple substations, not only realizing cross-platform operation but also monitoring the working condition of substation equipment through a network.

5 CONCLUSION

The intelligent management and control alarm platform of the unattended transformer is able to conduct all-weather unattended monitoring on security protection, environment, and equipment. The dual authentication function of the security monitoring subsystem is able to prevent the no-operation phenomenon when there are failures of substation environment and equipment, so as to avoid further deterioration of power supply failure as well as a great loss.

REFERENCES

[1] Zheng, D.X. (2007) Security and environmental monitoring alarm system of unattended substation. *Electric Power Information Technology*, 09.

[2] Teng, X.H. (2010) Analysis and processing of remote failures of unattended substation. *Rural Electrification*, 09.

[3] Chen, M.B. (2009) A research on automation system design of unattended substation. *Management & Technology of SME*, (10).

[4] Su, Z.J. (2013) Discussion on the technology of unattended 110kV intelligent substation. *China Electric Power Education*, (30).

[5] Spagnolo, P., Orazio, T.D., Leo, M. & Distante, A. (2006) Moving object segmentation by background subtraction and temporal analysis. *Image and Vision Computing*.

[6] Wu, Q.Z., Cheng, H.Y. & Fan, K.C. (2004) Motion detection based on two-piece linear approximation for cumulative histo2grams of ratio images in intelligent transportation system. *Proceeding of the 2004 IEEE International Conference on Networking, Sensing & Control*.

[7] Ruth Aguilar Ponce, Ashok Kumar, J. (2007) A network of sensor-based framework for automated visual surveillance. *Journal of Network and Computer Applications*.

Electric, Electronic and Control Engineering – Shao, Shu & Tian (Eds)
 ISBN: 978-1-138-02842-5

Research and implementation of yoga system based on virtual reality

Shuangjun Li
Physical Education Institute, Hubei Institute of Science and Technology, Xianning Hubei, China

Zhengda Xiong
School of Computing, Huazhong University of Science and Technology, Wuhan Hubei, China

ABSTRACT: Virtual reality is a human-computer interface that realistically simulates human behaviors such as seeing, listening, and moving in a natural environment. It has a very significant effect in guiding exercises. Hence, virtual reality has good prospects in applications of teaching and learning of yoga. Starting from the significance of the relationship between the two, the author of this paper designs a yoga system based on virtual reality and implements the system through simulation. The following aspects are mainly investigated in this paper: explanation of the significance of combining immersive, interactive, and imaginative features of virtual reality technology with teaching and learning of yoga; proposing the key technology of realizing the yoga system by virtual reality-progressive mesh generation algorithm of real-time graph formation, so as to provide a theoretical basis for the effective interaction of the human–machine interface and the implementation of the system; expounding on the key point of each link and functional expectations of the yoga system designed in this paper from three parts that constitute the yoga system simulation based on virtual reality technology, namely an input system, a virtual environment generator, and an output system; conducting functional design in line with the core content of the yoga system based on virtual reality of graphics, behavior, interaction, and communication; proposing the development process from high-level abstraction to low-level detailed design realized by the CFM model and the "top-down" method and providing technical support for the simulation implementation of the yoga system; 3D modeling of figures in the yoga system with software such as Maya, 3dmax, and AutoCAD; and the implementation of system simulation.

KEYWORDS: virtual reality; yoga system; functional design; 3D modeling; system simulation.

1 INTRODUCTION

Bijuan Deng et al. (2010) pointed out that yoga exercises alleviate mental stress and tension of people, stabilize the autonomic nerve, and achieve psychological stability mainly through basic asana, thoracic respiration, abdominal respiration, and yoga relaxation. It is of great significance for adjusting psychological balance, maintaining mental health, [1] and shaping a beautiful figure. The development of new ways of teaching and learning has become inevitable due to the significance of yoga exercises. Meanwhile, the technological progress of virtual reality has become gradually matured and will play an important role in the future art field. As a new medium, virtual reality can not only convert static art into dynamic art but also serve as an important link connecting creators and appreciators of art. Existing virtual reality systems related to art include Videoplace, Mandale, Virtual Actors, Virtual Museum, Virtual Music, and so on. A yoga system based on virtual reality is designed in this paper in order to expand approaches of teaching and learning of yoga and provide theoretical guidance for the development of yoga exercises and the application of virtual reality technology.

Many people have made efforts on the application of virtual reality technology, providing not only support for the expansion of application fields of the technology but also a theoretical basis for the design and implementation of the yoga system. Xin Cai (2010) pointed out that training tasks can be accomplished by the cooperation of auditory sense, vision, and touch in a virtual environment through interactive devices with the support of a network environment and multiple users' viewing interaction. Structural design, technical framework, network communication, and application examples of a multiple users' coordination training system based on virtual reality were also introduced. Besides, this system has a remarkable practicability [2]. Xin Gao (2004) successfully designed and realized a 3D landscape stereoscopic roam system of the bicycle based on a microcomputer platform with the following devices: a computer configured by a CPU of Pentium IIII 2.4GHz, an internal storage of 2GB, a hard disk of 80GB and a stereoscopic display card of Quadro4 750XGL 128MB, an exercise bicycle equipped with large anti-vibration flywheels, a 21-inch flat ViewSonic display with a 120Hz refresh rate, a pair of active stereoscopic glasses equipped with an infrared emitter, an operating system of Windows 2000, stereo image generation software and stereo display interface software programmed by Visual C++6.0, Open GL 3D graphics standard, and 3Dmax and ArcInfo8.0.2 3D modeling tools. Besides, it is perfectly able to simulate the real feeling of cycling [3]. Hua Chen et al. (2001) indicated that VRML constructs a 3D scene of real-time rendering in flexible and effective ways and realizes interactions with the scene in an event-driven way. All of these provide a great convenience for the establishment of a virtual reality system mainly featured by real-time 3D interaction. A scheme of constructing a virtual reality system based on VRML is also proposed, which is regarded as a quick and effective software method of realizing the virtual reality system [4].

The author analyzes features of yoga and the principle of designing a virtual reality system based on previous studies and carries out simulation realization on the yoga system on the basis of expounding on the graphic real-time generation algorithm of the yoga system based on virtual reality technology, system simulation components, and functional design.

2 FEATURES OF YOGA EXERCISES AND THE SIGNIFICANCE OF COMBINING YOGA WITH VIRTUAL REALITY TECHNOLOGY

Yoga maintains a stable state both psychologically and mentally through the regulation of breathing and adjusts body shapes by imitating postures of animals for the purpose of strengthening body function. Ying Zhu et al. (2004) indicated that modern yoga training not only pays more attention to the coordination of strength, breathing, and movement but also stresses the control of moving and motionless processes so as to increase muscle mass and body flexibility of exercisers. Movements of concentration and endurance are simple and practical [5].

Another feature of modern yoga is the accurately targeted movement design as well as the humanity. With proper strength of movement, exercisers move their bodies to the music and slow down their breathing naturally and smoothly. According to medical explanations, the success of yoga lies in the balance of the human nervous system and endocrine system. It has influences on other systems of a human body to maintain the entire balance. It is also quite helpful for body building and stress releasing. This exercise is extensively popular due to its features. Women have a particularly great interest in yoga.

Yoga is a sport technique with strong practicability. There are also certain differences among learners. Thus, it is quite necessary to establish a targeted learning system. The application of virtual reality technology can provide great help for yoga. Lian Liu et al. (2014) pointed out that, as an emerging human–computer interaction means of the computer field, virtual reality technology provides users with a 3D virtual world that is similar to the real world and featured by immersive, interactive, and imaginative natures. The virtual reality system receives users' interactions and other instructions through input devices, such as helmet and gloves, which are then processed in the system and sense organs of sight, hearing, and touch are influenced by some output devices, so that people have the same feeling of behaving in the real world [6]. Jin Sun (2014) pointed out that users are able to carry out real-time interactions with the computer system in the virtual environment. These kinds of interactions are conducted through body and language, so that a multidimensional anthropomorphic space can be constructed [7]. The principle of virtual reality technology is to generate behavior data after the processing of original data collected by sensors. Behavior data are then input into a virtual environment for interactions with users. Information of the virtual environment is finally sent back to sensors. Features of the virtual reality technology are shown in figure 1:

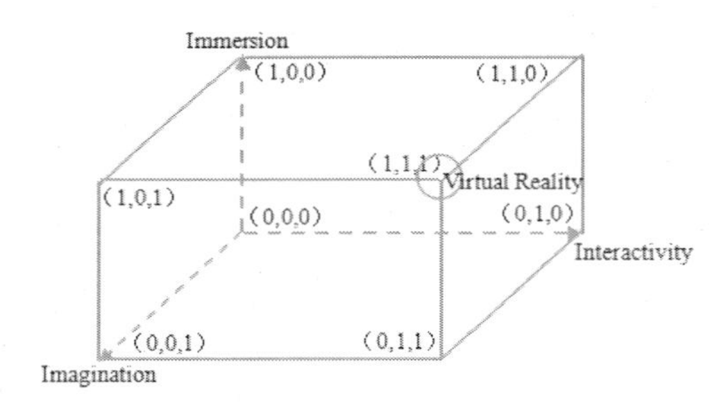

Figure 1. Features of the virtual reality technology.

To sum up, the combination of virtual reality technology and yoga training is quite preferable, being able to popularize this exercise and promote the improvement of yoga exercisers in terms of sport technique. The application of virtual reality technology plays the role of stimulating the learning ability of yoga exercisers and develops a new approach for the training of modern sport technique. The study in this paper involves the design and implementation of the yoga system based on virtual reality technology, aiming at providing a broader platform for the application of science and technology and providing a new mode for the scientific exercise of yoga.

3 ACCELERATION MODEL OF GRAPH FORMATION

The key technology of realizing the yoga system by virtual reality technology is the real-time graph formation. Thus, it is essential to choose or design a scientific and reasonable acceleration model of graph formation. A typical model is the LOD model, namely the level of detail. As shown in Table 1, A - global/local error measurement, B - topological characteristics, C - compact multi-resolution, D - applicable model type, E - feature edges, F - transition of LOD, and G - bounded/boundless error.

Multiple different simplified models frequently need to be generated when the LOD [7] model is constructed with the general mesh simplification algorithm. In addition, a large number of levels are usually needed in order to ensure the continuous transition between LOD models and to avoid the leaping sense in level switching. However, it is very difficult for a set of multiple LOD models to be effectively stored, transmitted, and compressed in mesh. For this reason, the progressive mesh generation algorithm is adopted in this paper. The progressive mesh is set up on the basis of the mesh simplification algorithm, the simplified operation of which is the edge collapse. The process is to transform a triangular mesh $M_n \in M$ into a relatively rough and simplified mesh $M_0 \in M$ through a series of edge collapses. The feature of this method is that local mesh elements changed by simplified iterations are identical: a vertex, an edge, and two triangles. Thus, it is able to generate a structured information sequence. The operation of edge collapse and vertex split is shown in figure 2:

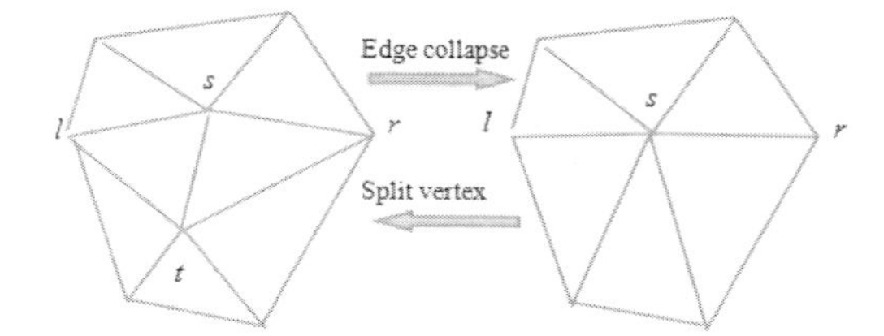

Figure 2. Operation of edge collapse and vertex split.

It can be observed from figure 2 that the edge collapse operation and the related vertex split operation is a pair of a dual operation. A set of n edge collapse operation sequences correspondingly generates a set of continuous and approximate mesh sequences:

$$M_n \xrightarrow{edgecol_{n-1}} M_{n-1} \xrightarrow{edgecol_{n-2}} \cdots M_1 \xrightarrow{edgecol_0} M_0$$

Similarly, the transformation can be also conducted through vertex split. A subsequent mesh of a high level of detail can be obtained from the most simplified meshM_0. The two operations are reversible. The process is shown next:

$$M_0 \xrightarrow{vsplpt_0} M_1 \xrightarrow{vsplpt_1} \cdots M_{n-1} \xrightarrow{vsplpt_{n-1}} M_n$$

Because the operation of edge collapse is a reversible process, the binary relation of progressive mesh can be defined as follows: $\left(M_0, \{vsplpt_0, vsplpt_1, \cdots, vsplpt_{n-1}\}\right)$

Table 1. Characteristic contrast of typical LOD-generating algorithm.

Compared algorithms	Self-adaptive subdivision algorithm	Vertex clustering algorithm	Vertex deletion algorithm	Envelop grid algorithm	Plane merger algorithm	Progressive mesh algorithm	Second-order error algorithm	Wavelet multi-resolution analysis
Types	Self-adaptive subdivision	Sampling	Vertex deletion	Vertex deletion	Plane merger	Edge collapse	Edge collapse	Wavelet
A	A1	A2	A2	A1	A1	A1	A1	A1
B	B1	B1	B1	B1	B1	B1	B2	B1
C	C2	C2	C2	C2	C2	C1	C1	C1
D	D1	D2	D3	D3	D2	D4	D2	D3
E	E1	E2	E1	E1	E1	E1	E2	E2
F	F1	F2	F2	F2	F2	F1	F1	F1
G	G1	G2	G2	G1	G1	G2	G2	G1

Note: A1-global, A2-local; B1-maintained, B2-not maintained; C1-support, C2-nonsupport; D1-regular quadrilateral mesh, D2-arbitrary mesh model, D3-manifold surface, D4-2d manifold; E1-maintained, E2-not maintained; F1-continuous, F2-discontinuous; G1-bounded, G2-boundless.

4 YOGA SYSTEM SIMULATION BASED ON VIRTUAL REALITY TECHNOLOGY

4.1 *Components of the yoga system simulation*

The yoga system simulation based on virtual reality technology consists of an input system, a virtual environment's generator, and an output system. The structure of the system is given in figure 6.

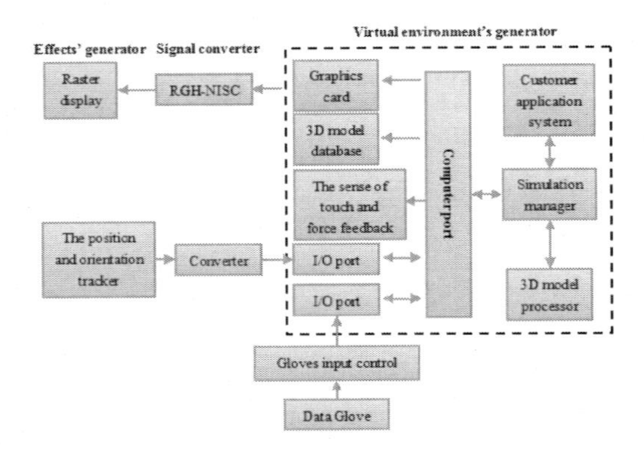

Figure 6. Structure chart of the yoga system simulation based on virtual reality technology.

It can be concluded from figure 6 that the input system consists of the position and orientation tracker, data glove, gloves input controller, and converter. The virtual environment's generator consists of customer application system, 3D model processor, 3D model database, simulation manager, high-performance computer, I/O port, graphics card, feedback device of touch and force, and so on. The output system is composed of a signal convertor and an effects generator. In the output system, users are able to have a stereo sense of vision and hearing and make natural interactions with the virtual environment. Yoga exercisers can have the feeling of being completely immersed in the virtual environment [8].

The system obtains original data through the input system, data gloves, and other devices. Original data are input into the virtual environment's generator and analyzed by computer software. Virtual effects can be finally achieved after data information passing through the signal converter and the raster display. Model driver software mainly provides an interface for the access to the 3D model and controls the 3D model through interface invocation so as to complete all functions. As for visual tracking and viewpoint sensing, a video camera can be used in X-Y planar array. The position and orientation of tracking objects can be calculated through light projection at different time points and different positions on the graphic plane. Viewpoint sensing and reality technology must be combined so that the sight at some point can be determined by multiple positioning methods. The 3D harmonic technology mainly focuses on providing users with a lifelike auditory feeling. Effects of the technology are not the same when the user's position is changed.

The yoga system designed in this paper should have functions such as organizing training scenarios and equipment, capturing movement data, collecting physiological and psychological data, repeating and displaying movements, and analyzing training effects and scientific selection of materials.

4.2 *Functional design*

It could be observed that the virtual reality system mainly consists of three parts, namely an application module, a system module, and a dialog box module. The core content of the functional design of the yoga system based on virtual reality includes graph, behavior, interaction, and communication of objects. The structure chart of the CFM [9] model with multiple components is shown in figure 7:

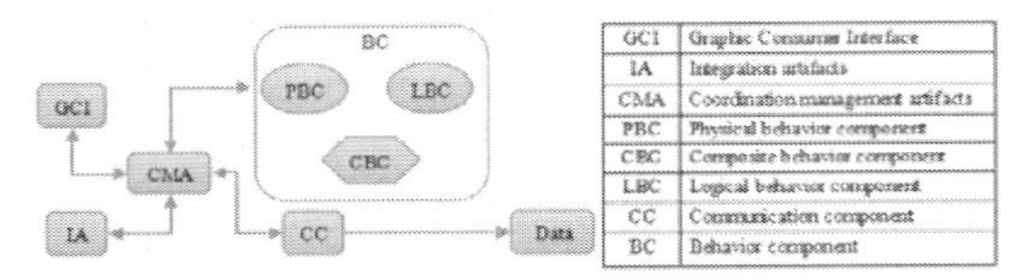

Figure 7. Structure chart of CFM model.

The graphic consumer interface in figure 7 refers to the graphic representation of objects in the system. The behavior component is the description of all object behaviors in the system. The interaction component is used for indicating the input and output orientation of objects and how to change behaviors of objects. The coordination management component describes the control and coordination mechanism of communication between objects in the system. The communication component realizes external communication between objects and communication between data elements or component applications [10].

The design process of establishing a virtual reality system with the CFM model is divided into a top-level design and a down-level design. A solution for design is formed after the completion of the top-level design phase. This solution is an abstraction with the high level. An integrated framework model structure with multiple components is used for conceptual guidance. The input of the top-level design phase is the functional description of the virtual reality system. The output is the high-level expression of data elements and objects in the system. Properties of graph, behavior, interaction, and internal/external communication of objects are defined on the basis of high-level abstraction. The output level is the input level of the down-level design. Detailed expression and formal realization of the top-level design are completed in the top-level design phase. The development process of detailed design from high-level abstraction to the bottom level is implemented in this paper through the "top-down" method in literature [9]. Implementation steps are shown next:

Top-level design phase

TL_STEP1: Determine data elements [Data elements refer to data that realize the external entity and the internal communication of the virtual reality system].

TL_STEP2: Determine objects [Make clear potential objects in the system; determine legal objects; distinguish between virtual objects and physical objects].

TL_STEP3: Object model [3.1 Graph (high-level description of virtual object graph); 3.2 Behavior (clearly indicate all kinds of behaviors of objects; classify behaviors into physical behaviors, logical behaviors, and composite behaviors; enable designers to understand behaviors visually by organizing a behavior description table); 3.3 Interaction (assign input of objects; conclude the reason of behavior changes caused by input; make clear the information sent by changed behavior components to other components; use state transition diagram to design the interaction part); 3.4 Internal communication (inspect all behaviors of objects; determine potential conflict behaviors; determine communication requests that might lead to potential conflicts; determine the priority of communication); 3.5 External communication (relevant control and coordination requirements of establishing external communication of objects)].

TL_STEP4: Determine the coordination and management mechanism [The coordination and management components design model [11] is used for realizing the control function of the coordination and management components and for coordinating all communication between objects in order to reduce conflicts between object behaviors].

Down-level design phase

DL_STEP1: Build up graphic standard [The down-level graphic design aims at establishing the entity of a graphic model and at connecting the graphic model with object behavior].

DL_STEP2: Build up behavior standard [PMIW language is used for the down-level design of behavior [12]. The implementation process of PMIW determines discrete components and continuous components of behaviors. Continuous behaviors are expressed with a data flow diagram, and discrete behaviors are expressed with a static diagram and a state transition diagram [13]].

DL_STEP3: Build up interaction standard [The interaction process of the down-level design is specifically expressed in TL_STEP3_3.3. The process is realized by PMIW language].

DL_STEP4: Build up internal communication standard [The internal communication standard of the down-level design is the settlement mechanism of the timing sequence conflict of a communication request. Timing sequence arrangement can be completed by competitive conflict settlement of all behaviors].

DL_STEP5: Build up external communication standard [The external communication standard of the down-level design indicates how the information transmission mechanism is realized by communication components. Synchronous, asynchronous, timeout, and batch information transmission mechanisms can be adopted. The timeout mechanism is used in this paper to build up external communication standard].

4.3 *Implementation of the yoga system simulation based on virtual reality technology*

The 3D models of yoga learners (referred to as yoga exercisers) and exercise scenarios need to be constructed in the yoga system simulation based on virtual reality technology. The 3D models studied in this paper are 3D human models. Modeling software applied in the construction of these models includes Maya, 3dmax, AutoCAD, and so on. The process of 3D modeling [14] is shown in figure 3:

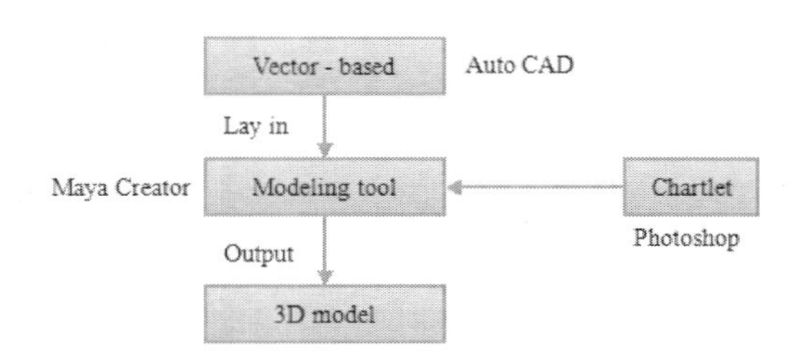

Figure 3. Process of 3D modeling.

In this paper, hierarchical presentation is used for the modeling of virtual yoga exercisers that consists of skin layer, bone layer, and muscle layer. VRML and MPEG-4 are two important criteria for the structure of body bone. Both of them support the presentation of the virtual human body. H-Anim, a sub-criterion of VRML, is used for describing virtual human body models, the expression form of which is the tree form. With the gravity center of the human body as the root node, joints as child nodes, and bones as connecting lines, the tree form attaches the geometry model of each body to relevant bones. The hierarchical structure model of the human body tree is shown in figure 4:

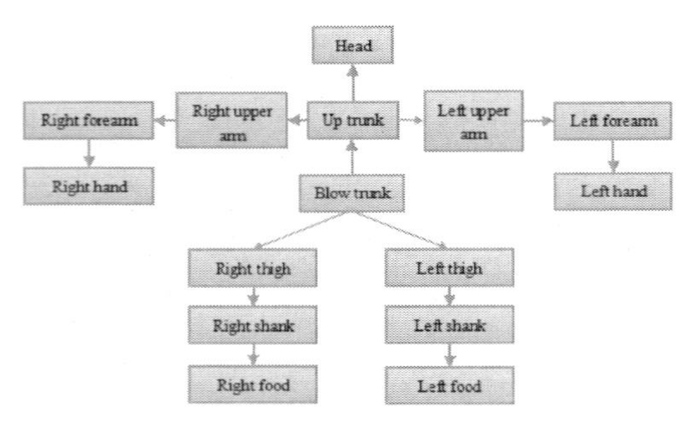

Figure 4. Hierarchical structure model of human body tree.

MPEG-4 defines the facial motion parameter FAP and the facial model definition parameter FDP of a virtual human body. The progressive process of three parts, namely the bone structure of the virtual human body, the virtual human body model fixed with bones and skin, and the colored model of the virtual human body, is shown in figure 5. In the bone structure diagram of the virtual human body, bones are expressed with 3D rigid bodies and connected through joints. There is no relative shift but only spin between the father node and child nodes in the bone layer. The virtual human body model fixed with bones and skin is modeled by the modeling software 3Dmax. The colored virtual human body is modeled by the software Maya, which designs the hair and clothing of the virtual human body with powerful simulation technology. Clothing colors can be realized through a UV coordinate.

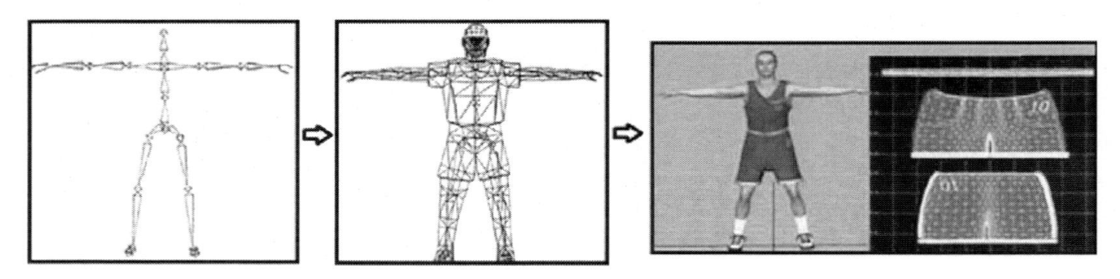

Figure 5. Modeling process of a colored virtual human body.

5 CONCLUSION

Based on the overview of values and the development path of yoga, it is concluded in this paper that virtual reality technology can effectively improve skills of yoga learners. Thus, a study on the yoga system based on virtual reality is carried out in this paper. The following conclusions can be drawn through discussions in this paper:

1 The key technology of realizing the yoga system by virtual reality technology is the real-time graph formation. Thus, it is essential to choose or design a scientific and reasonable acceleration model of pattern formation. The progressive mesh generation algorithm is proposed in this paper on account of defects of the classical LOD model. Besides, the algorithm has certain advantages in storage, transmission, and mesh compression.
2 The functional design of the virtual reality system is composed of three parts, namely the application module, the system module, and the dialog box module. The design of the CFM model applied in this paper can be divided into a top-level design and a down-level design. Both the design processes are quite effective and provide a basis for the implementation of the yoga system simulation.
3 The hierarchical structure model of the human body tree designed in this paper, namely bone structure of virtual human body–>virtual human body model fixed with bones and skin–>colored model of virtual human body, has functions such as organizing training scenarios and equipment, capturing movement data, collecting physiological and psychological data, repeating and displaying movements, and analyzing training effects and scientific selection of materials.

REFERENCES

[1] Deng, B.J. *et al.* (2010) Indicators' comparison of mental health of PE majored students before and after yoga class. *Fujian Sports Science and Technology*, 29(1):57–59.

[2] Cai, X. *et al.* (2010) Research and implementation of the collaborative training system based on virtual reality. *Modern Educational Technology*, 20(4):120–123.

[3] Gao, X. *et al.* (2004) Research and implementation of the bicycle roaming system based on virtual reality technology. *Application of Electronic Technique*, (9):18–20.

[4] Chen, H. *et al.* (2001) Study on virtual reality system based on VRML. *Computer Engineering*, 27(7):83–85.

[5] Zhu, Y. *et al.* (2004) Study on the influence of yoga exercising on physiological indexes of female college students. *Journal of Guangzhou Sport University*, 24(2):61–64.

[6] Liu, L. *el al.* (2014) Application status and design requirements of virtual reality technology in dace teaching. *China Educational Technology*, (329):85–88.

[7] Hao, Q.H. (2006) *A Study on LOD Models of Complex Shapes in Virtual Reality*, Zhengzhou: Zhengzhou University.

[8] Sun, J. (2014) A discussion on the application of virtual reality technology in aerobics training. *Electronic Test*, (19):76–78.

[9] Zhang, X.S. *et al.* (1999) *Virtual Reality Technology and Programming Skills*. Changsha: Press of National University of Defense Technology.

[10] Zhang, W.B. *et al.* (2005) Modeling and systematic design of the integrated framework of virtual reality system. *Design and Research*, (3):3–5.

[11] Jacob R JK, et al. (1999) A software model and specification language for non-WIMP user interfaces. *ACM Transactions on Computer-Human Interaction*, 6(1):1–46.

[12] Harel D, A Naamad. (1996) The statement semantics of state charts. *ACM Transactions on Software Engineering and Methodology*, 5(4):293–333.

[13] Gamma E, R Helm, et al. (1995) *Design Patterns, Elements of Reusable Object-Oriented Software*. Reading. MA: Addison-Wesley.

[14] Wang, S. (2012) Application of virtual reality technology in PE education and training. *Shanxi Education*, (3):116–117.

[15] Liu, G.W. (2014) An exploration on physical education modes based on virtual reality technology. *Electronic Test*, (18):63–35.

Electric, Electronic and Control Engineering – Shao, Shu & Tian (Eds)
© 2015 Taylor & Francis Group, London, ISBN: 978-1-138-02842-5

On fusing substation video surveillance with visual analysis under integrated dispatch and control

Junliu Zhang, Peilin Chen, Lijun Feng, Mingdi Li, Xingquan Zhao & Yang Liu
Dispatch & Control Center, State Grid Shanxi Electric Power Company, Taiyuan Shanxi, China

ABSTRACT: To meet the demand for applying the central control of a video surveillance system, a grid smart monitoring system fusing remote video surveillance with intelligent video analysis under integrated dispatch and control is introduced. A description is included on the current conditions of existing video surveillance systems, the essentials and key expertise of intelligent visual analysis, and its seamless fusion with the existing substation video processing equipment. By processing video data via a network and using image processing and artificial intelligence methods such as feature extraction, target tracking, and SVM classifier, it is able to identify any exposure to intrusion, flame, equipment hazard, or bad weather in live areas within the boundary of a substation within the first instance and to interact with the technical support system to further improve the ability to analyze and handle failures or troubles.

KEYWORDS: video surveillance; image processing; image identification; background subtraction; SVM classifier.

1 INTRODUCTION

With the full-span construction of the "big run" system and the continued deepening of the "integrated dispatch and control" initiative of Shanxi power grid, the unattended substation central monitoring and control of Shanxi Power Dispatch and Control Center has also stepped into a quality-efficiency upgrade phase. Electrical information monitoring systems already built and activated by the center include grid weather alarm system, lightning positioning system, video surveillance system, [1] and smart grid dispatch and control support system, which have the capability of acquiring and summarizing scene video and the regional environmental parameters of the controlled substations via power communication network and the virtual realization of real-time accurate delivery of information. However, such information as required for monitoring the substation site is delivered to the dispatch surveillance terminal only in the form of a simple data packet, and data are run on their own without information sharing or exchange. More importantly, substation video surveillance systems serve as a "video camera" in most cases, that is, the video of the period is recorded by DVR and verified after an abnormality or contingency has already taken place. As consequential damage or impact already exists and is irretrievable, the surveillance means nothing but a passive action equaling to "locking the stable after the horse is stolen." Hence, under the present control mode, as more substations are involved for surveillance, the monitoring personnel's ability to effectively review video images of the control substations and quality of patrol will be low and may become even lower due to information overload.

As the smart grid continues to grow, this monotonous, passive surveillance mode has to be reformed without delay. An intelligent video analysis and alarm system is needed to give the monitoring personnel a full view of the ambient conditions, equipment operation, security, and fire protection of the controlled substation; intensify the monitoring of major threats to the production safety of the substation, including bad weather, fire, equipment explosion, equipment failure, facility theft, and vandalism; and perform crucial, accurate detection and alarm on-site abnormalities so as to allow quick and wise decisions and immediate responses to events and effectively improve the real-time monitoring and safety level of unmanned substations.

Intelligent visual analysis relies on its powerful data processing ability and allows quick analysis of video data by using various algorithms [2], filters information of no interest to users, and highlights valuable key information interesting to the monitoring personnel through active push and multi-alarm, which not only relieves the video monitoring personnel from heavy labor and minimizes false alarms or alarm failures but also improves the handling efficiency of alarm events. In addition, the key information extracted, after being fused with other grid monitoring systems, will also become a high-value application [3].

2 APPLICATION DEMAND OF INTELLIGENT VISUAL ANALYSIS IN POWER GRID

So far, video surveillance systems already built in the power system have been virtually "monitoring" other than "controlling." Intelligent visual analysis, on the contrary, relies on its own capabilities of giving alarms in advance and performing active preventions so that all events are controllable, can be controlled, and are brought under control, and the overall safety control level of the power system is improved.

The demand of fusing intelligent visual analysis with integrated dispatch and control is reflected in the following aspects: a) moving target detection: An alert boundary can be virtually drawn on the video screen to detect objects that go beyond this boundary. This can be used for the operational safety control of a substation; b) flame detection: Flames or smoke are detected fully based on visual light image analysis and comprehensively evaluated through distributed video data analysis to allow a more accurate early alarm; c) leakage or seepage detection of oil-filled equipment [4], which allows different priority alarms on any significant oil leakage, dripping, or seepage in oil-filled equipment within the area of surveillance; d) state of knife switch [5], which allows real-time discrimination of state changes of the knife switch by video record monitoring; e) impurity detection, which allows analysis of remaining or hanging impurities or in the equipment area or on the bus or jumper through camera video analysis; and f) interaction with the technical support system, which allows interaction with equipment failure or defect information, automatic positioning and pop-out of video monitoring pictures, and related information of the faulted equipment and related facilities when information related to emergent switch tripping or equipment failure or abnormalities appears in the technical support system.

3 VIDEO SURVEILLANCE INTELLIGENT ANALYSIS AND ALARM SYSTEM UNDER INTEGRATED DISPATCH AND CONTROL

After intensive analysis of the application demand of intelligent video surveillance in its power grid system, Shanxi Power Corporation Dispatch and Control Center started investigating the application of video surveillance, alarm, and analysis under integrated dispatch and control by making the best use of the video network and data in the existing video surveillance information base platform.

The system is a C/S framework involving station-end front analysis and rear-end central management, where the station-end intelligent analysis and processing terminal is connected to the DVR video acquisition device of the video surveillance system through Ethernet, which allows immediate obtaining DVR channel video information and camera memory information as well as real-time analysis of the monitored objects and scenes. In real system communication, the client is able to communicate with the server on a real-time basis after certification. Once there is an alarm information, it is delivered via the network to the center's analysis and processing management server, which then classifies the information as predefined and pushes it to the authorized client according to the specified authority. Figure 1 shows a sketch of the system framework.

Seamless fusion with the existing video surveillance and technical support systems is permitted by the Ethernet. Figure 2 shows the system network topology.

The merits of this framework and arrangement include flexible arrangement, central management, easy maintenance, and compatibility with the multilevel administration of the power grid that allows application to management hierarchy.

4 DETECTION ALGORITHM

As for the application to intelligent video analysis in the typical complex scene of a substation, visual-based target feature detection and modal classification are generally used.

1 **Scene background modeling**
Pictures and videos collected at the substation worksite are used for static scene modeling, and the model thus built is updated from time to time with newly collected images. In our study, mixture Gaussian method is used to build a model, which needs a number of probability distributions to describe the scene and allow background separation:

$$\omega_{k,t} = (1-\alpha)\omega_{k,t} - 1 + \alpha(M_{k,t}) \tag{1}$$

$$\mu_t = (1-\beta)\mu_{t-1} + \beta X_t \tag{2}$$

$$\sigma_t^2 = (1-\beta)\sigma_{t-1}^2 + \beta(X_t - \mu_t)I(X_t - \mu_t) \tag{3}$$

$$M_{k,t} = \begin{cases} 1, \text{Matched} \\ 0, \text{Dismatched} \end{cases} \tag{4}$$

$$\beta = \alpha\eta(X|\mu_k, \sigma_k) \tag{5}$$

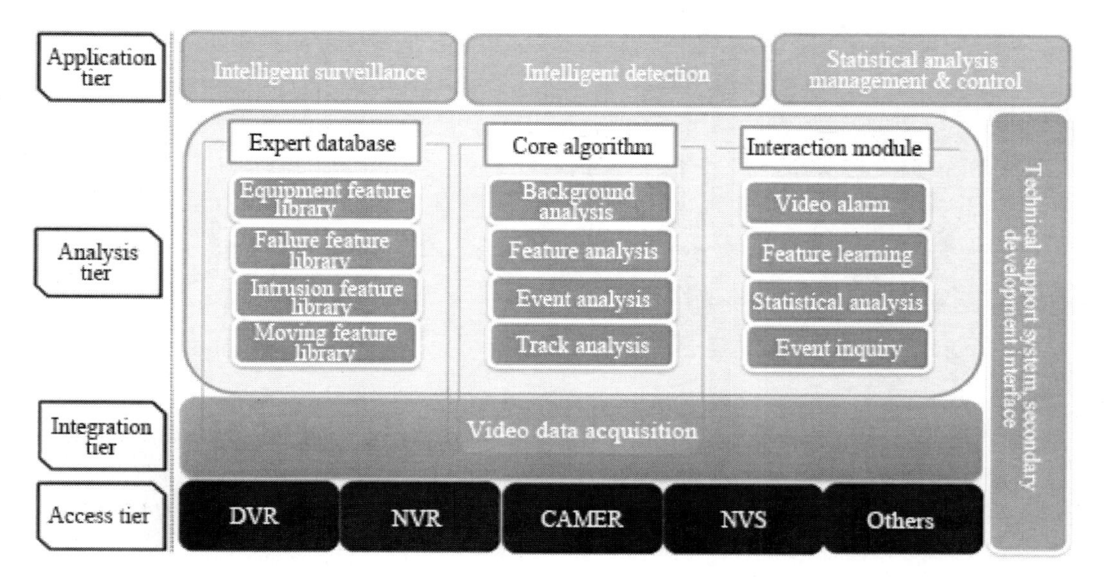

Figure 1. Sketch showing the system framework.

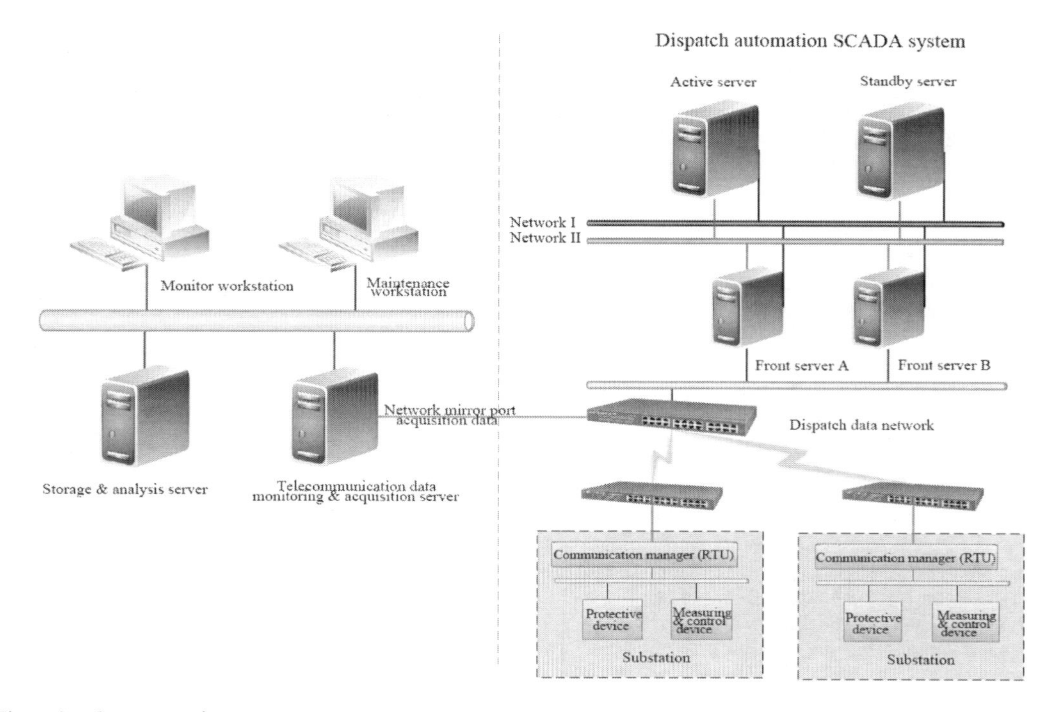

Figure 2. System topology.

Here, α is a constant term that represents the update rate of this mixture Gaussian model; 1/ α is the parameter variation rate of this model; $M_{k,t}$ is a constant term as defined by formula 4; and β is an auto-adjusting term that represents the parameter learning rate of this mixture model as defined by formula 5. It is noted that formulas 2 and 3 only provide updates for Gaussian distributions that match them and keep the parameters unchanged for other Gaussian distributions that do not match them.

2 **Target detection**
After background information of the scene is obtained, newly obtained images are detected from time to time by background subtraction to exclude entry of any suspicious target. Any suspicious target is locked for subsequent tracking and behavior analysis.
The operating process is given in figure 3.

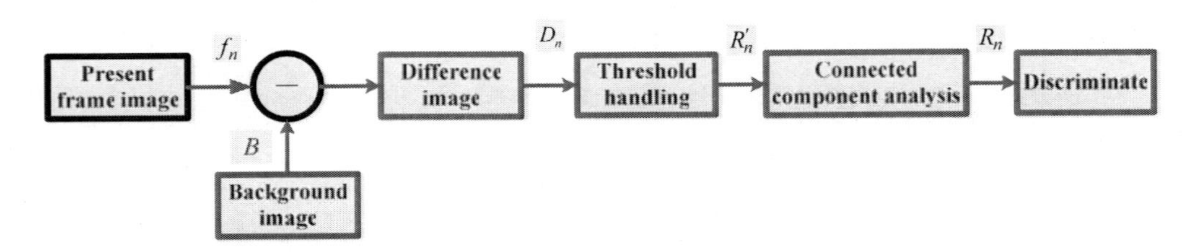

Figure 3. Computational process.

$$D_n(x,y) = |f_n(x,y) - B(x,y)| \tag{6}$$

Here, f_n represents the present frame image; B is the background image; and $B(x,y)$ and $f_n(x,y)$ are the gray scales at (x,y) of the background image and the present frame image, respectively, which are differentially operated according to the earlier formula before the absolute value is assumed to generate a differential image Dn.

Given the threshold T and binarizing the differential image using formula 6, values larger than this threshold are assigned as 1, which represents the movement target point. Those smaller than this threshold are assigned as 0, which represents the background point. In this manner, we get a binarized image R_n in which the target is separate from the background.

3 **Feature extraction**
For feature detection of transformers or disconnectors in the substation scene, the histogram of oriented gradient (HOG) feature algorithm is used. HOG features, when used with SVM classifier, allow an accurate detection of equipment state changes.
The proposed HOG feature extraction algorithm first grays the detection target (regards the image as an x, y, z (gray-scale) 3D image), standardizes (normalizes) the color space of the inputted image using Gamma correction to minimize the effect of local shadows and illumination changes of the image, and inhibits noise disturbance. It calculates the gradient of each pixel of the image (including the size and direction) mainly to capture contour information and further weaken illumination disturbance. The image is divided into small cells (e.g. 6*6 pixels/cells). By calculating the gradient histogram of each cell (number of different gradients), we are able to generate a descriptor for each cell. Every few cells are grouped into a block (e.g. 3*3 cells/blocks), and the feature descriptors of all cells within one such block are connected in series to the HOG descriptor of such an image (the target to be detected); this is the ultimate feature vector available for classification.

4 **Modal classification**
After building a data sample library and extracting features, all these features are inputted into SVM and trained. To ensure quick speed, linear SVM is used for the training. In our study, SVM classifier is used to classify the modes of the detection targets. Support vector machines (SVM) are used as two basic classifiers. SVMs with unsatisfactory classification accuracy are weighted and voted using AdaBoost algorithm to increase their classification accuracy. Figure 4 shows the algorithm training.

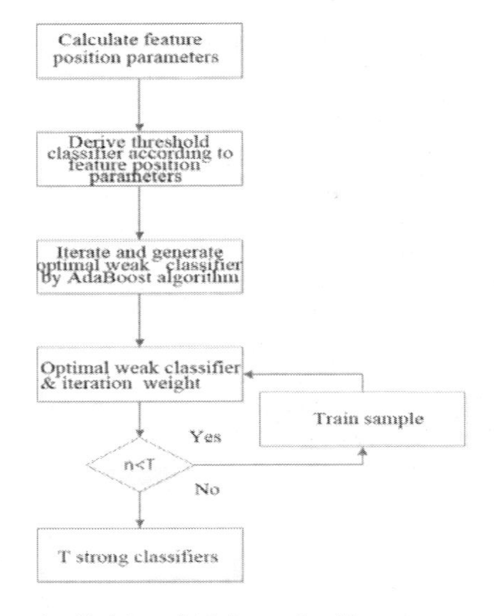

Figure 4. Training of AdaBoost algorithm.

5 CONCLUSIONS

An overall application solution of how the existing video surveillance system can be effectively and seamlessly fused with intelligent visual analysis under integrated dispatch and control is introduced and verified by an experiment. The proposed solution and method allows effective intrusion detection, impurity detection, flame detection, and oil leakage/seepage detection. The algorithm is robust, and the system is well structured for real-time detection of video information. It also allows interaction with the technical support system with respect to analysis application. For dispatch and control professionals, this is a means that greatly improves the monitoring efficiency and remote failure identification. As such, it is imperative to further expand the application of the smart grid to powers system and start fusing the existing video surveillance system with intelligent video surveillance to give better play to video surveillance systems in the production safety of the power grid.

REFERENCES

[1] Qin Jian, Zeng Xiaoping, Li Yongming, et al. (2008) Intelligent surveillance system based on video analysis. *Computer Engineering & Applications*.

[2] Zheng Shibao. (2009) Intelligent video monitoring technique and application. *Television Technology*, 01.

[3] Sun Fengjie, Cui Weixin, Zhang Jinbao, et al. (2005) Application of remote digital video monitoring and image recognition technology in power system. *Power System Technology*, 29(5).

[4] Hou Sizu, Tian Xincheng, Lu Xu. (2007) Design and research of video supervisory control system of substation. *Relay*, 35.

[5] Zhang Hao, Wang Wei, Xu Lijie. (2010) Application of image recognition technology in electrical equipment online monitoring. *Power System Protection and Control*, 38(6).

Electric, Electronic and Control Engineering – Shao, Shu & Tian (Eds)
© 2015 Taylor & Francis Group, London, ISBN: 978-1-138-02842-5

A study on the system of teaching quality based on classification rule

Fan Cui
Pingdingshan Institute of Education, Pingdingshan Henan, China

ABSTRACT: With the continuous expansion of the college enrollment, lots of problems have arisen in teaching. With explicit significance of the digital technique used for teaching research in colleges, this study analyzes the current situation of college education and proposes a new data mining classification technique in order to improve the teaching quality in colleges. Studies of data mining techniques have gained lots of achievements both at home and abroad as well as applications in almost all fields but education.

KEYWORDS: digital technique, data mining, college education.

1 INTRODUCTION

The digital technique has been generally applicable in many fields such as commerce, finance, and marketing management. On the contrary, it is rarely used in education. Treatment of students' data such as information and scores in colleges still remains a simple query and backup. Expansion of college enrollment that has kept growing for decades and causes a sharp increase in college students gives rise to lots of teaching problems, resulting in traditional methods of teaching management not being able to adapt to the development of society. A large amount of data that are accumulated in the process of teaching cannot play its due role because of the inadequate processing of data. Therefore, how to explore and take full advantage of data is becoming a serious question that needs to earn the attention of every teacher in colleges and universities. For example, teachers just keep an account of excellent students, good students, qualified students, and unqualified students as the analysis on students' scores without the knowledge of why they got such a score. If teachers could find reasons for influencing students' performance, this would play a positive role to a certain extent.

2 ANALYSIS ON PRESENT SITUATION OF COLLEGE TEACHING

At present, it is generally agreed that such problems exist in the teaching of colleges, especially teaching of basic courses.

1 Shortage of teachers. Because of the expansion of college enrollment and a sharp increase in college students, the number of faculty is far behind the growth in the number of students. It makes teachers have no access to getting and tracking the students' learning achievements. In addition, it is growing weaker on the guidance of students. And moreover, it will cause a reduction in teaching quality over time.
2 Teaching hours of the majority of basic courses are less than the required hours. It is generally accepted that more knowledge is necessary for today's competitive society. Many colleges tend to compress the basic course hours, for more learning content can be taught during class hours. Therefore, how to complete the task of teaching in a limited time and make students acquire more knowledge is a question that concerns all students and teachers.
3 Low foundation of the students. It provides teachers with big challenges in completing teaching tasks that college students graduate from different regions that have varied standards of elementary education. For example, teachers who conduct basic computer courses may face problems whereby students have a dramatic difference in the acquaintance of computers.

To fix the problem, most teachers focus on improving teaching quality in the analysis of current teaching conditions. Generally speaking, teachers will harvest a huge accumulation of data; however, most of the data cannot play a due role owing to the poor treatment of data.

3 DATA MINING IN DIGITAL TECHNIQUE

Data mining refers to a process of decision support via analysis of deep data. Using data mining technique, we can get a comprehensive analysis of the internal relationships between final results and correlative factors that is helpful in making teaching evaluation. For example, data mining tools could solve problems such as finding out factors that may influence students' learning results by analysis of the database system about students' scores, that is, what traditional evaluation cannot be achieved. The results of evaluation using data mining analysis will result in unexpected consequences.

Category algorithm based on data mining manages lots of data information by transferring it into classification rules for getting useful messages from the same kind of data instead of the traditional database query method.

4 STUDY ON SCHEMES

4.1 *Establish objects and goals*

We collect data from 210 students by asking them to fill the form on the online survey system for the study, and attempt to analyze which element has the best effect on learning computer network courses. What is more, the results can be used to analyze other professional courses offered by schools for the sake of conducting teaching affairs.

4.2 *Data collection*

First, a table of students' information, which can be acquired from the student management system, includes students' basic information such as name, student id, gender, major, and class.

Second, it is a questionnaire of students' information that is mainly derived from filling in the questionnaire. In order to minimize the repeated questionnaires from one person, students must type the accurate name and student number once they log on to the system and just have one chance to submit. Considering that there is the possibility for wrong information to be filled in the questionnaire, we add students' self-evaluation and suggestions about teaching methods to increase the accuracy of the information. On comparing the investigation data of random sampling questions, we can ensure the authenticity and reliability of the essential information as well as the credibility of further data mining results.

Third, score information, which has two parts of content. One is the score of the network security course. The other is the summary score in advanced courses of network security, especially basic courses, including the basis of computer engineering and network infrastructure; that is to say, we calculate the average score of the two courses as the final basic score in this study. Both the parts are acquired from the teaching management system.

4.3 *Data process*

4.3.1 *Data integration*

"Data integration" means a combination of multiple data sources. In this study, using database technology, the basic database of students' achievement analysis integrates multiple database files gathered through data collection.

4.3.2 *Data cleaning process*

Students who have been absent from the test or have a delayed test will get a "null" on the score list. That is to say, we cannot get a corresponding analysis from these data. After cleaning up the database by deleting the redundant records, 200 records remain, which is 96.2% of the respondents.

4.3.3 *Data transformation*

Using the concept of layering technology to transform the continuous attribute values into discrete attribute values is called "discretization." Histogram analysis is a relatively simple method of discretization, including equal width binning and equal depth binning. Width binning attributes values to an equal loss of parts or evaluates the scores of network security courses such as Y (YES) for scores above 75 and N (NO) for scores less than 75.

4.3.4 *Data reduction*

The purpose of data reduction is reducing the size of data mining without an impact on the final result. The use of the subtractive dimension method searching for real useful attributes from initial attributes helps reduce the number of attributes and variables.

4.3.5 *Classification of data mining*

The objective of data mining classification is to construct the treelike model of score analysis and decision. Considering that the training set is not very large, we compare ID3 with C4.5 algorithm and choose the ID3 algorithm to classify the data in this study. In simple terms, the ID3 algorithm selects the attributes based on the maximization of information gain contained in nodes of the generated decision tree. Information gain that values the entropy of difference between the sample size information and test attributes expectation is actually the core of the ID3 algorithm.

Under the assumption of general set $P(X_1,X_2,......X_P)$, we conclude the sample data as follows:

$X_i= [(X_{i1},X_{i2},......X_{in})$, i=1,2,......,n, name the two training subsets PX and NX separately. P is the set of positive examples. N is the set of counter-examples. So, the entropy of sample size information is as follows:

$$I(P,N)=-\frac{P}{P+N}\log(\frac{P}{P+N})-\frac{N}{N+P}\log(\frac{N}{N+P})$$

Assuming random variable X as the root test attribute of decision tree, X_i has k different discrete values such as V_1, V_2,......, V_k. Dividing X into k subsets and assuming that P_j is the number of positive examples and n_j is the number of counter-examples in the subset j, the entropy of this subset can be expressed as I (P_j, n_j). The information of expectation attributes tested by X is as follows:

$$E(X_i)=\sum_{j=1}^{k}\frac{Pj}{Pj+Nj}I(Pj,Nj)$$

So, the information gain of root node X is

$$Gain(X_i)=I(P,N)-E(X_i)$$

The ID3 attribute selection strategy chooses the maximum attribute information gain as the test attribute. Taking the training set shown in Table 1 as an example, we illustrate the generated decision tree model of good or bad scores.

Table 1. Set of training data.

No.	Interest	Learning effects	Scores of basic courses	Teaching evaluation	Score evaluation of network security
1	little	average	Good	good	N
2	very	average	good	bad	N
3	little	average	good	average	N
4	little	excellent	good	good	Y
5	very	excellent	good	average	Y
6	very	bad	bad	good	N
...	...	...	...	...	...

According to the principle of the ID3 algorithm, we classify the root nodes in the first place, and then construct the decision tree by the following three steps:

First of all, calculate the information entropy required to classify the given sample. In Table 1, there are a total of 200 samples containing 90 Y together with 110 N.

$$I(S_1,S_2)=I(90,100)=-\frac{90}{200}\log_2\frac{90}{200}-\frac{110}{200}\log_2\frac{110}{200}=0.993$$

Second, calculate the information gain ratio of every attribute. We need to calculate the information entropy of every subset divided by attribute values. In the case that attribute "interest" values "vary," there are 80 samples in class Y and 40 samples in class N.

$$I(S_1,S_2)=I(80,40)=-\frac{80}{120}\log_2\frac{80}{120}-\frac{40}{120}\log_2\frac{40}{120}=0.918$$

And there are 30 samples in class Y, 50 samples in class N when attribute "interest" values are "little."

$$I(S_1,S_2)=I(30,50)=-\frac{30}{80}\log_2\frac{30}{80}-\frac{50}{80}\log_2\frac{50}{80}=0.912$$

Calculate the information gain of attribute "interest":

$$Gain(\text{interest})=I(S1,S2)\text{-}E(\text{interest})=0.077$$

Third, determine the test attribute. Thus, we can see that the E (learning effects) is the minimum; that is to say, the learning effects with the greatest contribution to classification provide the maximum amount of information, namely the Gain (learning effects) is the biggest.

This is an iterative process until we finally get the decision tree as shown next:

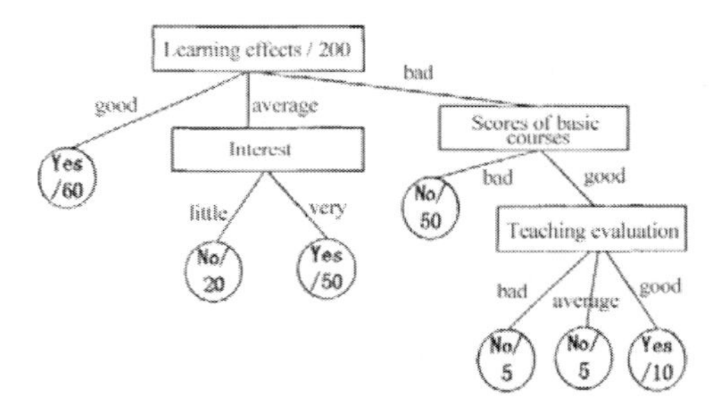

Figure 1. Final decision tree based on ID3 algorithm.

5 GENERATED CLASSIFICATION RULES

The advantage of the decision tree is taking the classification rules directly. The main purpose of this study is finding out elements that affect students' scores of network security courses. The generated classification rules are as shown in table 2:

6 PROSPECT

In this study, we conduct a preliminary inquiry into data mining technology and its application in college educational systems. Besides, there are still few questions that need to be improved:

First, there are some unreasonable parts of the algorithm used in this study, such as data need to be discrete continuously in the transformation stage. In addition, the incomplete data that are always deleted

Table 2. Generated classification rules.

IF Learning effects	Connected relation	State of affairs	Connected relation	State of affairs	Results
good	then	IF scores are good	-	-	good
average	and	interest: very	Then	IF scores are good	good
average	and	interest: little	And	IF scores are good	bad
bad	and	scores of basic courses: bad	Then	IF scores are good	bad
bad	and	scores of basic courses: good	And	IF scores are good	good
bad	and	scores of basic courses: average; teaching method: average	Then	IF scores are good	bad
bad	and	scores of basic courses: average; teaching method: bad	Then	IF scores are good	bad

or re-entered by hand deserve to be handled with more proper measures. On account of time limit, we demand programming instead of manual work to ensure the completion of data management.

Second, data selection is not adequate in this study. The data that are chosen are limited. In students' scores analysis, only computer professional students and a portion of teachers' teaching data are included, so that the mining results will inevitably have their limitations and one-sided certainty.

Third, the mining algorithm should be more efficient. Because data size of the data algorithm is very large, only when the mining algorithm becomes very efficient and falls within a user-acceptable scope can accurate mining results be obtained.

Fourth, the absence of data mining function in the current college–student management system not only results in great challenges for our study but also determine directions for our future work, which is a combination with the college management system, data mining, and analysis to conduct our teaching work properly

Although data mining technology has been widely used and has made some achievements in the field of insurance, telephone, and so on, its application in the field of education still remains original. What is more, products of data mining technology are basically in the theoretical study stage, namely there are hardly any forming products of data mining. With the maturation of the data mining theory and the development of education, data mining technology will be widely used in education to play a promoting role in educational work.

7 CONCLUSION

Data mining is the decision tree classification algorithm based on the ID3 algorithm, which is a kind of commonly used algorithm. We attempt to apply this kind of algorithm to the teaching in colleges, in order to discover useful knowledge. It is not only conducive to the reform of college teaching and education but can also reflect certain problems existing in education. It is very meaningful for college teaching management decisions. Compared with other research conducted on the same topic, the combination of data mining and school performance management system with a teaching evaluation system for docking is very rare, so further analysis and research is necessary.

REFERENCES

[1] C Romero, S Ventura. (2007) Educational data mining: A survey from 1995 to 2005. *Expert Systems with Applications*, (1):135–146.

[2] Alfred E. Hartemink, Megan R. Balks, Zuengsang Chen *et al.* (2014) The joy of teaching soil science. *Geoderma*, (11):217–218.

[3] Wu Yan (2011) Application research of data mining technology about teaching quality assessment in colleges and universities. *Procedia Engineering*, 15:4241–4245.

[4] Diana C. Woodhouse (2009) *Rationalism and the Professional Development of Graduate Teaching Associates*. California: San Jose State University.

[5] Wang Yang, Yu Jun, Li Zhenhua (2012) Research on the planning and construction of platform of

comprehensive informatization service in colleges and universities, *Procedia Engineering*, (19):69–73.
[6] Li Gang, Li Yan. (2014) Digital information technology and university teaching. Shanxi Science and Technology, (3):89–90.
[7] Cui Huaguo. (2013) Research on the layout design of course teaching in Colleges and Universities Digital Media Technology. *Journal of Chuzhou University*, (4):122–124.
[8] Zhou Yuan. (2010) The investigation analysis and strategy research of the effect of multimedia teaching in colleges and universities. *China University Teaching*. (3):86–88.
[9] Zhou Kejiang, Huang Yue, Jiang Hua, Luo Qin. (2011) The application framework of grid technology in the digital university under the mechanism of construction and sharing of regional distance education teaching resource. *Technology and application of digital, Chinese University Review*, (4):26–27.
[10] Ren Xi, Hou Jianjun, Li Zhaohong, Lu Yong, Zeng Tao. (2010) Research on the teaching of 'digital electronic technology' course in the exploration. *Journal of Electrical and Electronic Engineering Education*, (4):28–29.

Electric, Electronic and Control Engineering – Shao, Shu & Tian (Eds)
© 2015 Taylor & Francis Group, London, ISBN: 978-1-138-02842-5

Design of a high efficiency 2.45-GHz rectifier for low input power energy harvesting

Quanqi Zhang, Huachao Deng & Hongzhou Tan
Department of Electronics and Communication Engineering, Sun Yat-sen University, Guangzhou Guangdong, China
SYSU-CMU Shunde International Joint Research Institute, Shunde Guangdong, China

ABSTRACT: This paper presents a high-efficiency 2.45-GHz rectifying circuit with microstrip line for harvesting low input RF power effectively. The presented rectifier is designed and printed on a FR4 substrate. The rectifying circuit which mainly composed of voltage double diode BAT15-04W, open stub matching circuit and harmonic suppression circuit. The voltage double diode BAT15-04W is a low-cost surface mounted Schottky diode which is suitable for low power harvesting. The open stub matching circuit is used to match the rectifier to 50Ω. The harmonic suppression circuit is designed for conversion efficiency optimization. The measured results of rectifying circuit showed that the RF-to-dc conversion efficiency can reach 75.3% and the DC power is higher than 3.45V for 5dBm input power at 5KΩ DC load.

KEYWORDS: rectifier; low input RF power; high efficiency; harmonic suppression.

1 INTRODUCTION

In recent years, with the increasing development of wireless communication devices, the wireless power transmission system has grown up to be an interesting topic for an energy transmission[1-3]. WPT has been implemented in numerous fields, such as RFID and telemetry or in some areas that are dangerous or difficult to reach[4-7]. In this paper, we pay our attention to the frequency of 2.400-2.484 GHz (IEEE 802.11 b/g) which is located at the center of un-licensed ISM (Industrial-Scientific-Medical) band.

The most important component of wireless power transmission system is the rectenna. As shown in Figure 1, it mainly contains a receiving antenna which collects microwave incident power and a rectifying circuit that can transforms the microwave power into useful DC power. The rectifying circuit plays a crucial role in this system. In general, a rectifier is composed of the capacitor coupling, a matching circuit, several Schottky diodes and a low pass filter (LPF). The low pass filter (LPF) is often used to reduce the ripple DC power and reject harmonics created by the diode. The matching circuit is employed for matching between the antenna and rectifying circuit.

Several rectification structures including single diode or dual diode circuits have been presented in many works[8-10]. For a given input power level, the efficiency of the rectifier mainly depends on the suppression of harmonic components generated by diode. In addition, the level of the load is also an important factor that cannot be overlooked. In this work, we will utilize the combination of radial stub and linear open stub to suppress the fundamental frequency, second and third harmonic component of output power, thus improve the efficiency of rectification. The rectifying circuitry is designed and printed on a FR4 substrate. The dielectric constant (ε) and loss tangent (tan δ) of the substrate of the rectifying circuit are 2.55 and 0.0015, respectively. The thickness and the height of the substrate are 35μm and 0.8mm, respectively.

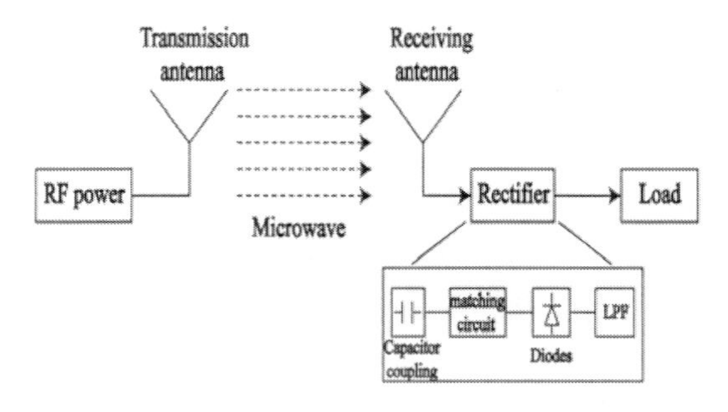

Figure 1. Wireless power transmission system.

2 RECTIFIER DESIGN

The rectifying circuit that we used in this paper consists of a voltage doubler Schottky diode, a dc-pass filter and a resistive load. Figure 2 shows circuit schematic of the proposed rectifier by coupling Momentum electromagnetic simulator and Harmonic Balance of Agilent ADS (Advanced Design system).

The Schottky diode is located between the matching circuit and the dc-pass filter. In this paper, the Schottky diode part number is chosen BAT15-04W which produced by the Infineon Technology Company. The used Schottky diodes are characterized by a low threshold voltage V_j (224mV), a low parasitic capacitance C_{j0} (138.5fF) and a low serial resistor R_s (5Ω). So it is a low-cost surface mounted Schottky diode that can convert the low microwave power into higher DC power. The model of BAT15-04W including the parasitic parameters of the plastic package is shown in Figure 3.

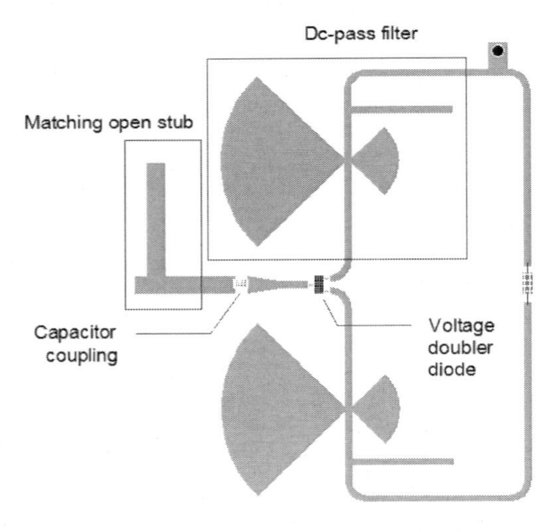

Figure 2. Schematic diagram of rectifying circuit.

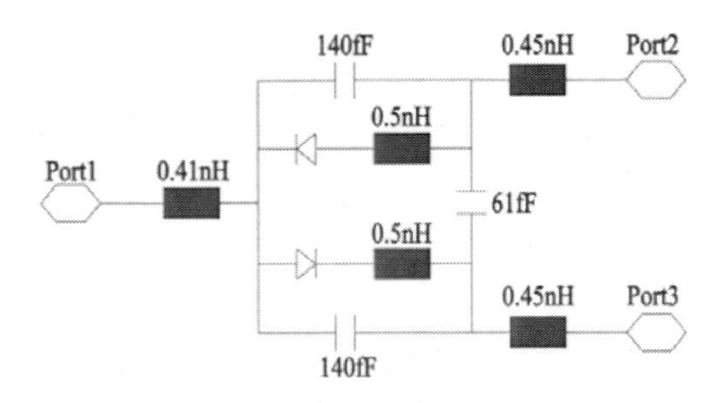

Figure 3. Package model of the Schottky diode BAT15-04W.

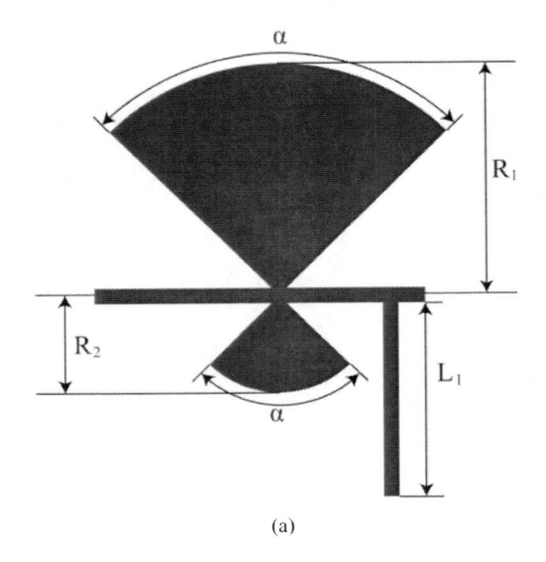

(a)

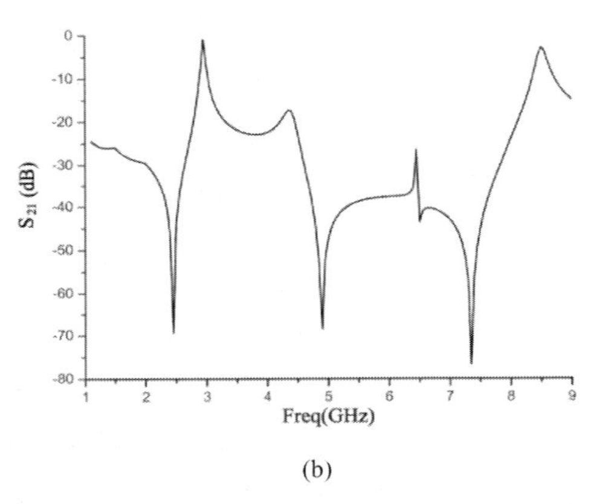

(b)

Figure 4. (a) The circuit the proposed dc-pass filter. (b) The harmonic suppression effect of the proposed dc-pass filter. The parameters are: R_1=13.3mm, R_2=5.2mm, L_1=10.9mm,α=90°.

Some high-order harmonics will be generated due to the nonlinear characteristics of the diode. In order to block these harmonics and improve the efficiency of this rectifier, a dc-pass filter is commonly inserted between the diode and the load. In this design, we use two radial stubs to suppress the fundamental frequency and the third harmonic respectively. Meanwhile, the linear open stub was used to suppress the second harmonic. Those higher harmonic components are very small, so we refuse to conceive here. The circuit and the harmonic suppression effect of the proposed dc-pass filter are shown in Figure 4.As we can see, the 2.45-GHz frequency component, the second harmonic and the third harmonic are significantly reduced (<-60 dB).

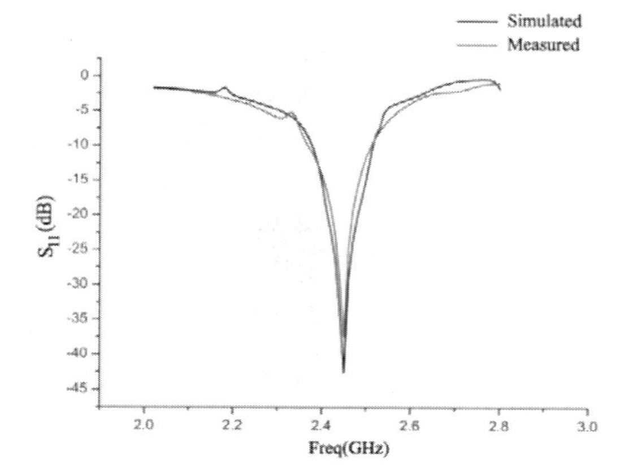

Figure 5. Simulated and measured S11 characteristics of the tuned rectifier.

The simulated and measured |S11| of the proposed rectifying circuit are shown in Figure 5. As can be seen, a good impedance matching is obtained in 2.45GHz both in simulation or measurement. The measured |S11| is -37dB at 2.45 GHz and lower than -10dB covers the frequency range from 2.37GHz to 2.51GHz, so it reveals a good performance of the transition.

3 SIMULATION AND MEASUREMENT RESULTS

The photograph of the proposed rectifier which is composed of a matching circuit, a voltage doubler Schottky diode, a dc-pass filter and a resistive load are presented in Figure 6.

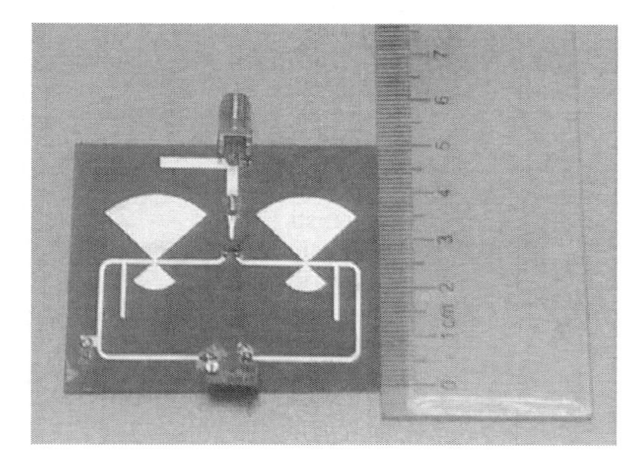

Figure 6. Photograph of the rectifier.

It has been measured by a signal generator and a conventional voltmeter. The signal generator was used to supply a 2.45GHz RF power. The conventional voltmeter was used for measuring the voltage of the load. The RF-to-dc efficiency (η) of the rectifier is defined as follows:

$$\eta = \frac{P_{DC}}{P_{inc}} \times 100\% = \frac{V_L^2}{R_{Load} P_{inc}} \times 100\% \tag{1}$$

Where P_{inc} is the incident RF power, P_{DC} is the DC output power and, R_{Load} is the DC load and V_L is the output DC voltage.

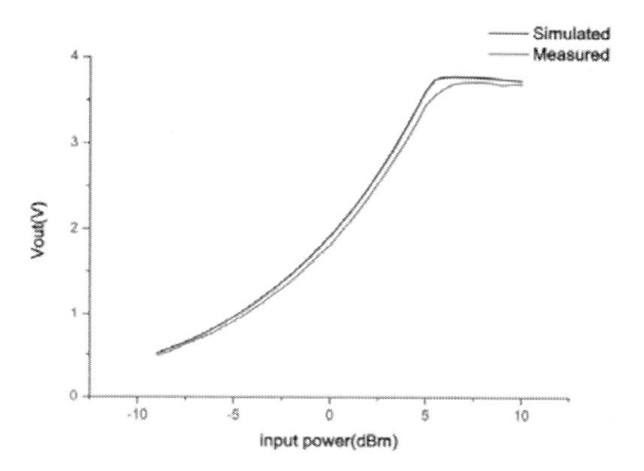

Figure 7. Output dc voltage against input RF power.

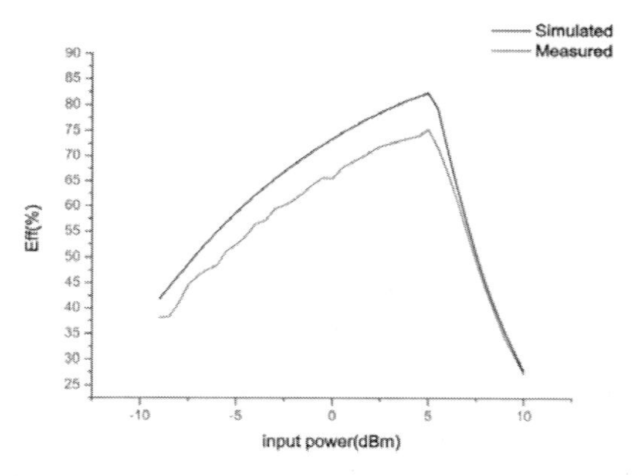

Figure 8. Efficiency of the rectifier against input RF power.

The simulated and measured output dc voltages against the input power from -10 to 10dBm are shown in Figure 7. An optimal resistive load (5kΩ) is used on both simulation and measurement. Obviously, the measured rectifier can obtain 3.45V output dc voltage at 5dBm input power. The simulated and measured rectifier efficiencies as a function of input power with a 5kΩ resistor at 2.45 GHz are shown in Figure 8. It can be observed that when the input power is 5dBm, the measured efficiency achieves its peak value of 75.3%. When the input power is between -2.5 to 6.5dBm, the efficiency can remain higher than 60%.

4 CONCLUSIONS

This paper has pointed out a rectifying circuit based on the low-cost surface mounted Schottky diode BAT15-04W and the measurement of the rectifier revealed that it can harvest low input RF power effectively. The measurement result has revealed that the rectifier exhibits an efficiency of 75.3% at 5dBm input power level. In the future work, in order to construct a complete microwave rectenna, we will concern on adding an end-fire antenna with high gain and directional between the transmitting antenna and rectifier.

REFERENCES

[1] W.C. Brown. (1984) The history of power transmission by radio wave. *IEEE Transactions on Microwave Theory and Techniques*, MTT-32(9):1230–1242.

[2] Naoki Shinohara. (2011) Power without wires, *IEEE Microwave Magazine*, 12(7):S64–S73.

[3] Franceschetti G. and Gervasio V. (2012) Wireless Power Transmission. A new science is borne, 2012 IEEE MTT-S International Microwave Workshop Series on Innovative Wireless Power Transmission: Technologies, Systems, and Applications (IMWS), pp:21–23+10–11.

[4] Casares M, Vuran M.C. and Velipasalar S. (2008) Design of a wireless vision sensor for object tracking in wireless vision sensor networks. Second ACM/IEEE International Conference on Distributed Smart Cameras, pp:1–9+7–11.

[5] Yuan Gao, Yuanjin Zheng, Shengxi Diao , Wei-Da Toh *et al.* (2011) Low-power ultrawideband wireless telemetry transceiver for medical sensor applications. *IEEE Transactions on Biomedical Engineering*, pp:768–772.

[6] Lingfei Mo, Shaopeng Liu, Gao R.X, John D *et al.* (2012) Wireless design of a multisensor system for physical activity monitoring. *IEEE Transactions on Biomedical Engineering*, 59(11):3230–3237.

[7] Boaventura A.J.S and Carvalho N.B. (2013) A batteryless RFID remote control system. *IEEE Transactions on Microwave Theory and Techniques*, pp:2727–2736.

[8] A. Douyere, J.D. Lan Sun Luk and F. Alicalapa. (2008) High efficiency microwave rectenna circuit: modeling and design. *Electronics Letters*, 44(24).

[9] H. Takhedmit, B. Merabet, L. Cirio, B. Allard, F. costa, C. Vollaire and O. Picon (2010) A 2.45-GHz low cost and efficient Rectenna, *Proceedings of the Fourth European Conference on Antennas and Propagation*, pp:1–5+12–16.

[10] Franciscatto B.R., Freitas V., Duchamp, J.M, Defay C. *et al*, (2013) High-efficiency rectifier circuit at 2.45 GHz for low-input-power RF energy harvesting, 2013 European Microwave Conference (EuMC), pp:507–510+6–10.

Electric, Electronic and Control Engineering – Shao, Shu & Tian (Eds)
© 2015 Taylor & Francis Group, London, ISBN: 978-1-138-02842-5

Investment portfolio model design based on multi-objective fuzzy comprehensive evaluation method

Zhaoxia Shang
Shandong Institute for Product Quality Inspection, Jinan Shandong, China

Yi Wang
Shandong External Science and Technology Exchange Center, Jinan Shandong, China

Xuan Liu
College of Extended Education, Chengdu University of Technology, Chengdu Sichuan, China

ABSTRACT: Considering the high risk, fuzziness and scientific quantization deficiency of investment portfolio, to establish a scientific and easy-to-use evaluation system is the main approach to solve the issues. A fuzzy comprehensive evaluation method is introduced in this paper to quantize each aspect of the investment portfolio and unify the dimensions. It combines qualitative factors as well as quantitative factors and can make equal, fair and effective assessment for the investment portfolio project by utilizing all information.

KEYWORDS: multi-objective, fuzzy comprehensive evaluation, algorithm model.

1 INTRODUCTION

Many real-life optimization problems can be solved by the multi-objective model. Normally valuation is not restricted only to yes or no. Objects connect with each other, restrict each other, and even conflict with each other. This kind of complex system is very difficult to evaluate by quantitative analysis. At that time we can turn to multi-objective fuzzy evaluation system for help. Common multi-objective evolution algorithms include multi-sexual genetic algorithm [1], VEGA algorithm [2], MOGA algorithm [3], NS—GA algorithm [4], etc. The multi-objective fuzzy comprehensive evaluation method proposed in this paper is also an evolution multi-objective algorithm. It considers multiple factors comprehensively based on fuzzy mathematics principle and conducts evaluation using fuzzy set principle. This method was proposed in 1965 by Zadeh L A [5], the automatic control expert from University of California. It has been evolved in practice by a number of researchers after that and now it has developed to a set of evaluation methods in various types. Its research objects have the characteristic of certain intension and uncertain extension. Fuzzy mathematics makes treatment by using membership function [6]. Fuzzy mathematics is a method that deals with fuzzy phenomena using mathematics. Classical set theories only confined its expressive force to the concept with certain extension. It clearly defined: every set must be composed of the clear element, the membership relationship from element to set must be clear. It either belongs or not. The either-or relationship limits its application in some area. So the mainstream of fuzzy mathematics development focuses on application. Multi-objective fuzzy comprehensive evaluation method is an effective multi-objective decision method that can make comprehensive evaluation on objects influenced by many factors.

2 MODEL DESIGN

Generally speaking, both uncertainties and certainties can affect the portfolio evaluation results while the uncertainties can only be evaluated by fuzzy evaluation method. An innovative fuzzy evaluation method is proposed in this section. Every objective is evaluated separately. Membership function and fuzzy matrix are set consecutively. After fuzzy calculation, quantitative evaluation results and the results from evaluation function are integrated. Then the final comprehensive value is given. The starting point of this algorithm design is to achieve balance in subjective evaluation and objective evaluation. The main procedures are described in detail as following:

Step 1: The factors, namely the contents which are to be evaluated, are listed according to their attributes, such as performance, standard, circumstances, etc. Secondary or third-level evaluation factors can also be set if necessary. They are ladder-related. Define the evaluation factors set as A.

$$A=\left(A_1, A_2, \ldots A_n\right), \tag{1}$$

Step 2: Define the evaluation level set as B.

$$B=\left(B_1, B_2, \ldots B_m\right), \quad (2)$$

Step 3: Determine the weight (w) by importance of every level's factor. The rationality of weight allocation determines the rationality of result.

$$W=\left(\mu_1, \mu_2, \cdots, \mu_m\right) \quad (3)$$

Step 4: Let experts grade the factors then integrate the results. Minimum number of experts is 5 and all of them must have rich experience in related industry.

$$a=\left(a_1, a_2, \ldots a_n\right) \quad (4)$$

Step 5: Set fuzzy evaluation matrix. Sometimes the matrix is incomplete because certain experts might feel unconfident about some factors and choose not to grade them. Normally the missing factor r_{ij} can be got by equation (5), (6). Considering the application, we cannot give up the whole evaluation result just because of few missing factors. Therefore equation (7) was designed to make up the matrix. The final fuzzy evaluation matrix is shown as (8).

$$r_{ij} = r_{ik} r_{kj}, \quad (5)$$

$$r_{ij} = r_{ik} r_{kk} \ldots r_{kj}, \quad (6)$$

$$r_{ij} = \frac{r_{i1} + \cdots + r_{i(n-1)}}{n-1}, \quad (7)$$

$$R=\begin{pmatrix} r_{11} & r_{12} & \cdots & r_{1n} \\ r_{21} & r_{22} & \cdots & r_{2n} \\ \vdots & \vdots & \vdots & \vdots \\ r_{m1} & r_{m2} & \cdots & r_{mn} \end{pmatrix}, \quad (8)$$

Step 6: Conduct fuzzy calculation by using operator $M(\wedge, \vee)$. Every object is treated by using normalization treatment.

$$s_k = \bigvee_{j=1}^{m}(\mu_j \wedge r_{jk}) = \max_{1 \le j \le m}\left\{\min\left(\mu_j, r_{jk}\right)\right\}, \quad k=1,2,\cdots,n, \quad (9)$$

$$S = W \circ R = \left(\mu_1, \mu_2, \cdots, \mu_m\right) \circ \begin{pmatrix} r_{11} & r_{12} & \cdots & r_{1n} \\ r_{21} & r_{22} & \cdots & r_{2n} \\ \vdots & \vdots & \vdots & \vdots \\ r_{m1} & r_{m2} & \cdots & r_{mn} \end{pmatrix} = \left(s_1, s_2, \cdots, s_n\right), \quad (10)$$

$$\underset{\infty}{S} = \left(\underset{\infty}{s_1}, \underset{\infty}{s_2}, \cdots, \underset{\infty}{s_n}\right), \quad (11)$$

Step 7: Quantize the evaluation level and calculate the fuzzy evaluation value shown as (12) and (13).

$$U=\left(\eta_1, \eta_2, \cdots, \eta_m\right), \quad (12)$$

$$Z = \eta_1 * \underset{\infty}{s_1} + \eta_2 * \underset{\infty}{s_2} + \ldots + \eta_m * \underset{\infty}{s_n}, \quad (13)$$

Step 8: Allocate weight to game evaluation function and fuzzy evaluation value; then calculate the final evaluation value as follows:

$$Spor = V_1 M_{eas} + V_2 Z, \quad (14)$$

It shall satisfy the following condition:

$$V_1 + V_2 = 1, \quad (15)$$

Step 9: Conduct comparison and screening of the portfolio solutions according to the evaluation value.

3 NUMERICAL MODELING

Based on the above algorism, portfolio model described in table 1 is analyzed using fuzzy evaluation value. Evaluation factors are original cost, operating cost, expected return and risk index respectively. Evaluation grade is excellent, good, medium or poor. Suppose that an evaluation team comprised of 10 experts assessed the portfolio solutions and voted. Table 1 shows the number of experts who voted for each evaluation degree.

Table 1. Investment project fuzzy evaluation table.

	Excellent	Good	Medium	Poor
Original cost/ ten thousand dollars B_1	6	3	1	0
Operating cost/ ten thousand dollars B_2	7	1	2	0
Expected return/ ten thousand dollars B_3	5	2	3	0
Risk index/ degree B_4	7	3	0	0

Set fuzzy evaluation matrix R. It is the multi-factor evaluation matrix after grading by the experts and integration. During the integration the number of experts shall be transferred to percentage.

$$R=\begin{bmatrix} R_1 \\ R_2 \\ R_3 \\ R_4 \end{bmatrix}=\begin{bmatrix} 0.6 & 0.3 & 0.1 & 0 \\ 0.7 & 0.1 & 0.2 & 0 \\ 0.5 & 0.2 & 0.3 & 0 \\ 0.7 & 0.3 & 0 & 0 \end{bmatrix}$$

Give weight to four factors according to their influence degree and the importance among evaluation factors.

$$W=\begin{pmatrix} 0.2 & 0.1 & 0.3 & 0.4 \end{pmatrix}$$

Conduct fuzzy calculation by using operator $M(\wedge,\vee)$. The fuzzy evaluation set S can be calculated with the maximum membership principle:

$$S=W\circ R=(0.2,0.1,0.3,0.4)\begin{bmatrix} 0.6 & 0.3 & 0.1 & 0 \\ 0.7 & 0.1 & 0.2 & 0 \\ 0.5 & 0.2 & 0.3 & 0 \\ 0.7 & 0.3 & 0 & 0 \end{bmatrix}$$

$$=\begin{bmatrix} 0.4 & 0.3 & 0.3 & 0 \end{bmatrix}$$

"$\circ$" is not simple matrix multiplication. Take the case of first column, the calculation principle is shown below:

$b_1=(0.2\wedge 0.6)\vee(0.1\wedge 0.7)\vee(0.3\wedge 0.5)\vee(0.4\wedge 0.7)=0.2\vee 0.1\vee 0.3\vee 0.4=0.4$

$\underset{\infty}{S}$ can be get by using normalization treatment:

$$\underset{\infty}{S}=\begin{pmatrix} \frac{0.4}{0.4+0.3+0.3} & \frac{0.3}{0.4+0.3+0.3} & \frac{0.3}{0.4+0.3+0.3} & \frac{0}{0.4+0.3+0.3} \end{pmatrix}$$

$$=\begin{pmatrix} 0.4 & 0.3 & 0.3 & 0 \end{pmatrix}$$

The result after quantification of the evaluation level is: U= (100, 80, 50, 20)

The overall fuzzy grade of the scheme is: Z=0.4*100+0.3*80+0.3*50+0*20=79

4 CONCLUSIONS

The fuzzy grading can be used not only as the assessment standard among different schemes in the same industry, but also as the comparison reference in different industries. It is noteworthy that the fuzzy data used in fuzzy comprehensive evaluation method do not mean "random" data; the rationality of data selection can affect the result significantly. Only when the index level is divided clearly and the weight is given rationally, can the method help the decision-makers.

The method gives scores according to the experiences of expert system based on multi-objective optimization. It borrows expert opinions and carries on the quantification to the indexes with the membership principle of fuzzy mathematic. Combining by the characteristic of investment portfolio, it comprehensively evaluates the fuzzy grade of objectives being evaluated by multiple indexes. The paper designs a new method that can eliminate deviation aiming at the complementation problem of incomplete matrix and does not affect overall evaluation.

The example shows the Multi-objective fuzzy comprehensive evaluation innovative method has some advantages in diversified investment portfolio area. In the assessment process the experiences of expert is functioning adequately. This method is more scientific than other export assessment methods. This algorithm is helpful to the selection the best from two or more alternatives.

REFERENCES

[1] J. Lis and A. E. Eiben, T. Fukuda, T. Furuhashi. (1996) A Multi-sexual Genetic Algorithm for Multi-objective Optimization. *Proceedings of the 1996 International Conference on Evolutionary Computation*, Nagoya, Japan.

[2] J. D. Schaffer. (1984) *Multiple Objective Optimization with Vector Evaluated Genetic Algorithms*. Nashville, TN (USA): Vanderbilt University.

[3] C. M. Fonse & P. J. Fleming. (1995) An Overview of Evolutionary Algorithms in Multi-Objective Optimization. *Evolutionary computation*, 3(1):1–16.

[4] N. Stinivas & D. Kalyanmoy. (1994) Multi-objective optimization using non-dominated sorting in genetic algorithms. *Evolutionary Computation*, 2(3):221–248.

[5] L. A. Zadeh. (1965) Fuzzy sets. *Information and Control*, 8(2):338–353.

[6] Xi Chen. (2010) Multi-objective Fuzzy Comprehensive Evaluation Method applied in measurement management project research. *Information Technology*, 27(27):50.

Electric, Electronic and Control Engineering – Shao, Shu & Tian (Eds)
© 2015 Taylor & Francis Group, London, ISBN: 978-1-138-02842-5

Design and implementation of the system of impact location based on acoustics detecting technique

Daoheng Fang
School of Communication and Information Engineering, University of Electronic Science and Technology of China, Chengdu, China

Haitao Jia
Research Institute of Electronic Science and Technology, University of Electronic Science and Technology of China, Chengdu, China

ABSTRACT: Acoustic positioning technique receives sound through acoustic sensing devices, and processes the received signals to locate the sound source. In this paper, we propose a dual-acoustic-sensor-array based on acoustic positioning technique to precisely locate the sound source in a large area. The geometric relationship of each array acoustic wave arrival angle is exploited to achieve the sound source location. Simulation experiments are conducted on a sound source with 5 km×5 km area, with the two acoustic sensor arrays being at a distance of 100m, and the array radius being 0.5m. Experimental results show that the measurement error in the perspective of the wave path difference time is only in the microsecond (us) order of magnitude.

KEYWORDS: acoustic measurement; positioning; lateral.

1 INTRODUCTION

Acoustic positioning technique is used extensively in areas such as underwater acoustic positioning and bomb site positioning. The principle of an acoustic positioning system is to calculate the coordinate of the sound generation point with the absolute time sensed by arrays of sound sensors, while taking the influence of environmental factors into consideration [1]. Measurement accuracy is crucial for an acoustic positioning system. However, conventional acoustic positioning systems is limited by the versatile acoustic environment in presence of varying temperature, wind speed, multipath interferences, and so on.

In this paper, a dual-acoustic-sensor-array based on acoustic positioning technique (DASA) is proposed to precisely locate the sound source in a large area.

2 DASA SYSTEM

In the DASA system, two sensor arrayes are utilized for acoustics detection. Each of the two sensor array is composed of eight receiving units evenly placed in a circular. The radius of the circular array is one meter, and the distance between the two circular arrays is 100 meters. The layout of the DASA system is shown in Fig. 1.

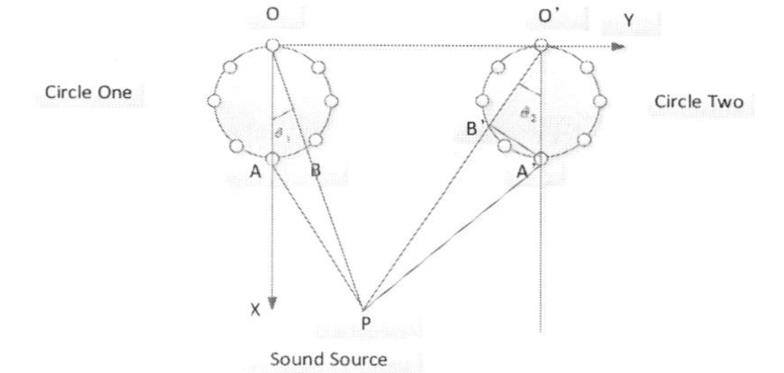

Figure 1. Layout of the DASA system.

The sixteen sensor devices in the two sensor arrays work in a timing synchronization mode. Therefore, after the sound source sends out a sound, the DASA system obtains fifteen independent delay values. In this work, the sensor receiving time is represented as t_{ij}, (i=1, 2), (j=1, 2, …, 8), where i represents a sensor array, j represents the sensor in an array. So t_{1j} is the receiving time of the jth sensor in the first sensor array, while t_{2j} is the receiving time of the jth sensor in the second sensor array. Let $c = 334m/s$ and $t_{1\min}$ represents the maximum and the minimum value in t_{1j}, respectively, while $t_{2\max}$ and $t_{2\min}$ represent the maximum and the minimum value in t_{2j}. In Fig.1, the four sensor devices which have the receiving time $c = 334m/s$, $t_{1\min}$ $t_{2\max}$ and $t_{2\min}$, respectively, are labeled with O, A, O′, and A′. It is straightforward to see that, OA and O 'A' are the diameter of the two sensor arrays, while P is the sound source location. The OP line crosses the first sensor array circle at the point B, and the O′P′ line crosses the second sensor array circle at the point B'.

Because the distance from P to the two sensor arrays is relatively large, i.e. in the thousands of meters (Km) order of magnitude, the sensor array circle diameter becomes relatively small. Therefore, the length of the two line pairs PA and PB, PA′ and PB′ can be considered as the same [4],

$$PA=PB,\ PA' = PB' \tag{1}$$

The direction angle θ_1, θ_2 respectively reflects the an angle made by the line OP and the X axis, the line PO 'and the line O′ A′. With the circle and triangle theory, we can get

$$\begin{cases} \cos\theta_1 = c(t_{1\max} - t_{1\min})/(2r) = (c\tau_1)/(2r) \\ \cos\theta_2 = c(t_{2\max} - t_{2\min})/(2r) = (c\tau_2)/(2r) \end{cases} \tag{2}$$

Where r is the radius of the sensor array circle, C is the speed of sound. Since we use one parameter to represent the radius of the two sensor array circles, we imply the two sensor array have the same radius.

Then we can determine the △POO′ of the two corners and edges as

$$\begin{cases} \angle POO' = \pi/2 \pm \theta_1 \\ \angle PO'O = \pi/2 \pm \theta_2 \\ OO' = D \end{cases} \tag{3}$$

where D represents the center distance of the two sensor array circles.

To determine the position of the point P, we need to calculate the direction of θ_1 and θ_2 as well as the size of the line OP distance [5].

We mainly use the four sensors which has the receiving time $c = 334m/s$, $t_{1\min}$ $t_{2\max}$ and $t_{2\min}$, to determine the △POO′. Because the sound source is located in the extension of the line diameter OA and O 'A′ , the geometric relationship of θ_1, θ_2, ∠POO′ and ∠PO′O is exploited to calculate the sound source location. As shown in Fig. 2, three scenarios can happen:

1 if the point P is in the middle of the extension of the line OA and the line O 'A′ , then we have

$$\angle POO' = \pi/2 - \theta_1, \angle PO'O = \pi/2 - \theta_2, \angle OPO' = \theta_1 + \theta_2 \tag{4}$$

2 if P is in the left side of the extension of the line OA, then we have

$$\angle POO' = \pi/2 + \theta_1, \angle PO'O = \pi/2 - \theta_2, \angle OPO' = \theta_2 - \theta_1 \tag{5}$$

3 if the point P is in the right side of the extension of the line O 'A′ , then we have.

$$\angle POO' = \pi/2 - \theta_1, \angle PO'O = \pi/2 + \theta_2, \angle OPO' = \theta_1 - \theta_2 \tag{6}$$

Therefore, we can get the following equation with the sine theorem

$$\frac{OO'}{\sin\angle OPO'} = \frac{OP}{\sin\angle OO'P} \tag{7}$$

$$OP = \frac{OO' \times \sin\angle OO'P}{\sin\angle OPO'} \tag{8}$$

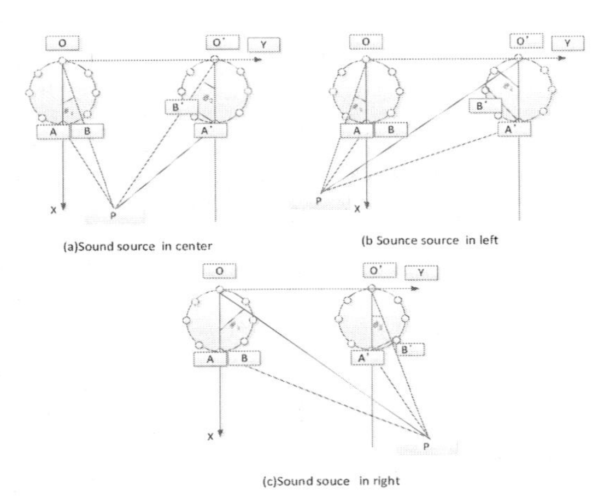

Figure 2. Three kinds of location under the different conditions of model.

Therefore, for the three scenarios, OP is given by

$$OP = \frac{D\cos\theta_2}{\sin(\theta_1 + \theta_2)},$$

$$OP = \frac{D\cos\theta_2}{\sin(\theta_2 - \theta_1)},$$

$$OP = \frac{D\cos\theta_2}{\sin(\theta_1 - \theta_2)}$$

Similarly, for the size of the line OP distance, we also have three scenarios:

1 When the middle position of the sound source is located in the extension line OA and O 'A′, which corresponds to the first scenario:

$$x = OP\cos\theta_1 = \frac{D\cos\theta_2}{\sin(\theta_1+\theta_2)}\cos\theta_1$$

$$x = OP\sin\theta_1 = \frac{D\cos\theta_2}{\sin(\theta_1+\theta_2)}\sin\theta_1$$

2 When the sound source P is located on the X axis on the left, which corresponds to the second scenario:

$$x = OP\cos\theta_1 = \frac{D\cos\theta_2}{\sin(\theta_2-\theta_1)}\cos\theta_1$$

$$x = OP\sin\theta_1 = \frac{D\cos\theta_2}{\sin(\theta_2-\theta_1)}\sin\theta_1$$

3 When the sound source P is located at the right to the extension line of O 'A' , which corresponds to third scenario:

$$x = OP\cos\theta_1 = \frac{D\cos\theta_2}{\sin(\theta_1-\theta_2)}\cos\theta_1$$

$$x = OP\sin\theta_1 = \frac{D\cos\theta_2}{\sin(\theta_1-\theta_2)}\sin\theta_1$$

3 ANALYSIS OF MEASUREMENT ERROR

Since the direction angle θ_1, θ_2 and the time delay error δ_t are the main sources of the measurement error , we will analyze the impact of these factors to the accuracy of the proposed DASA technique. In addition, sound propagation, temperature and wind speed variation in the air, and the sensor installation accuracy will also impact the positioning accuracy.

The direction angle θ_1, θ_2 error is mainly due to the approximation of PA=PB, PA′=*PB′*. We use the Monte Carlo simulation method to evaluate the error caused by the approximation of the direction angle.

Firstly, considering the changes of direction angle δ_θ with the changes of D. In the Monte Carlo simulation, we assume r=1 m, c=340 m/s, the sound source area of 5 Km×5 Km. As shown in Fig. 3, δ_θ is less than 0.3 in most cases when changing D.

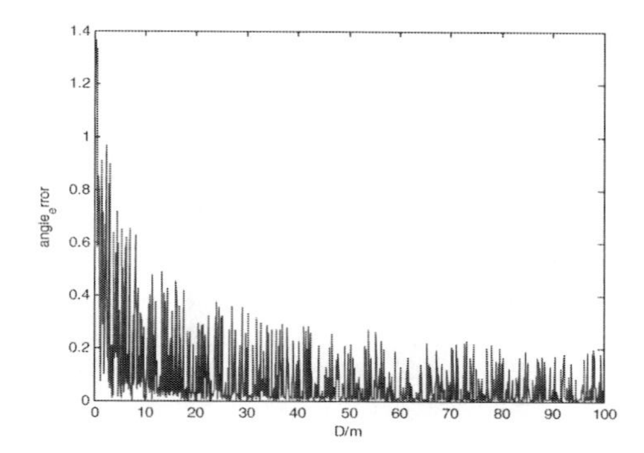

Figure 3. The relationship between δ_θ and D.

Secondly, we evaluate the changes with the radius of r considering the deviation of the direction angle. Here we set D=100 m, c=340 m/s, the sound source area of 5 Km×5 Km. From Fig 4, we can see that the deviation of the direction angle $\delta_\theta$increases with the increase of r, so we need to keep the sensor array circle small.

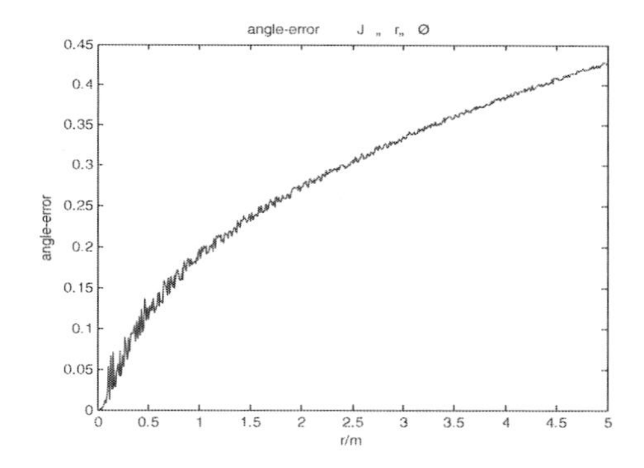

Figure 4. The deviation direction angle varies with the radius r.

Therefore, when the sound source area is large, the desirable distance between the two sensor arrays is D =100 m, and the radius r of the sensor array is 0.5 m, then the measurement error induced by the direction angle approximation is δ_θ = 0.098. We can see, the approximation of PA=PB, PA′=*PB′* only induces a very small direction angle error.

Now we analyze the measurement error in the perspective of the wave path difference time [6]. As shown in Fig 5, the sensor array is composed of eight sensors, and the circular radius r =1m.

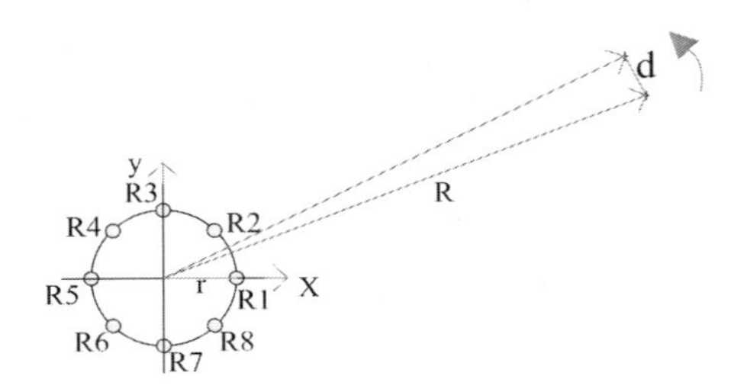

Figure 5. Sensor array consisting of eight sensors.

Assuming the sensor is evenly deployed in the circular array, then the sensor position is given by

$$R1 = (1,0)$$
$$R2 = (\sqrt{2}/2, \sqrt{2}/2)$$
$$R3 = (0,1)$$
$$R4 = (-\sqrt{2}/2, \sqrt{2}/2)$$
$$R5 = (-1,0)$$
$$R6 = (-\sqrt{2}/2, -\sqrt{2}/2)$$
$$R7 = (0,-1)$$
$$R8 = (\sqrt{2}/2, -\sqrt{2}/2)$$

We assume the sound source is located at a distance R=5000m to the sensor array. When the sound source is on a counter clockwise direction 22.5° to the X axis, to achieve a minimum distance error for each of the sensor in the array, the "initial position of the sound source" is $Rd = (R*\cos(22.5°), R*\sin(22.5°))$, then we can have the distance for each sensor as follows,

d1=4.999076135114856e+03
d2=4.999076135114856e+03
d3=4.999617401929506e+03
d4=5.000382768781171e+03
d5=5.000923894174467e+03
d6=5.000923894174467e+03
d7=5.000382768781171e+03
d8=4.999617401929506e+03

If the sound source is counter clockwise away of d=80m, then sound source angle rotates for $\theta = 2*\arcsin((d/2)/R)$, and

$$Rd\_new = (R*\cos(22.5° + \theta), R*\sin(22.5° + \theta)),$$

then the distance between the new position of the sound source to each of the sensor is given by

D1=4.999082377259900e+03
D2=4.999070129519177e+03
D3=4.999602668164370e+03
D4=5.000367939311467e+03
D5=5.000917654328004e+03
D6=5.000929897543968e+03
D7=5.000397500247727e+03
D8=4.999632233625389e+03

So we can calculate the new rotation position and the initial position of the distance difference:

D_d1=0.006242145044780
D_d2=-0.006005595678289
D_d3=-0.014733765136953
D_d4=-0.014829469704637
D_d5=-0.006239846463359
D_d6=0.006003369500831
D_d7=0.014731466556441
D_d8=0.014831695883004

Obviously, the smallest wave path difference is D_d6, while the largest is D_d8. When the sound propagation speed in the air is $c = 334m/s$, we can get the minimum and the maximum wave path difference time as follows

T1=D_d6/c= 1.797416018213014e-05s

T2=D_d8/c= 4.440627509881348e-05s

We can see the measurement error in the perspective of the wave path difference time is only in the microsecond (us) order of magnitude.

4 DISCUSSIONS

This section discusses some parameters values we utilized in the work. Sound propagation velocity in air is 340m/s. The auditory system of normal human audible frequency range is 20Hz~20KHz, because the point of the sound source is in the audible frequency range, so the sampling frequency in 96KHz, which is more than four times of the highest sound source frequency, satisfying the Nyquist frequency rule [2].

The targeted sound frequency is 20Hz~60Hz, which means its corresponding wavelength is 17.20m~5.73m. When using the acoustic array direction finding, generally requires the array spacing is less than the wavelength of the signal. Therefore, we set the radius of the sensor array 1m is rational.

On the other hand, the intensity of sound varies with distance law [3]:

$$I = \frac{1}{d^2} \tag{9}$$

where I is the sound intensity, d is the distance. Therefore, for the sensor in the distance of d receiving the sound, the signal attenuation is about 20log$d$ dB. For example, within the distance of 2km, the signal attenuation is about 66dB. If considering the sound source and the distance of the array in 2km with the 12 bit ADC, with dynamic range being about 72dB, then we need to use at least a 12 ADC to meet the requirements of accracy.

For the dual sensor arrays, the accuracy of the time synchronization affects the key parameters of the system. If the time cannot be synchronized well, then the proposed technique is not suitable for positioning. In this case, if the two sensor arrays locate 100m away from each other, then GPS timing synchronization should be employed.

5 CONCLUSIONS

In this paper, we propose a dual-acoustic-sensor-array based acoustic positioning technique to precisely locate the sound source in a large area. The geometric relationship of each array acoustic wave arrival angle is exploited to achieve the sound source location. The proposed technique can precisely locate the sound source.

REFERENCES

[1] Chen Weixing, Zhang Chuanyi. (2009) The impact location system based on acoustic detection technology. *Automation*, (04):36–38+42.

[2] Wang Xiliang. (2004) *Leak Detection and Location Technology Research of Long Pipeline Based on Network Control*. Daqing: Daqing Petroleum Institute.

[3] Jiang Haiwei. (2008) Development of the System of Automatic Detection of Impact. Nanjing: Nanjing University of Science and Technology.

[4] Zhou Wei. (2012) *Research on Algorithms of Moving Object Discovering and Tracking for Video Surveillance*. Hefei: University of Science and Technology of China.

[5] Fu Bo. (2013) *Several Key Problems Research on Intelligent Monitoring Analysis System*. Changchun: Jilin University.

[6] Fu Zhizhong, Liu Lingqiao, Xian Haiying. (2010) Human computer interaction research and realization based on leg movement analysis. *ICACIA 2010*.

[7] Zhizhong Fu, Haiying Xian, Jin Xu, Xiaoqi Ge. (2010) Evaluation of motion blur parameter based on cepstrum domain of the intentional restored image. *ICCP2010*.

Electric, Electronic and Control Engineering – Shao, Shu & Tian (Eds)
© 2015 Taylor & Francis Group, London, ISBN: 978-1-138-02842-5

Application in robot of the three-dimensional force tactile sensor research based on PVDF

Qi Pan, Zhou Wan & Shilin Yi
School of Information Engineering and Automation, Kunming University of Science and Technology, Kunming, China

ABSTRACT: Detecting the force information of an object matters regarding whether a robot hand may smoothly grasp the object or not, and the force information may be fully reflected in the three-dimensional direction force. There are still some defects in current tactile sensors for detecting three-dimensional force during grasping. Thus, in this paper, an attempt was made to design a PVDF-based tactile sensor for three-dimensional force. The structural design of the sensor was presented, and a signal conditioning circuit was designed. Meanwhile, a mathematical model was established for the piezoelectric film and the structure of the sensor head. Besides, the sensor was tested and validated, and the results show that the sensor was effective for detecting three-dimensional force in the process of grasping an object with a robot hand.

KEYWORDS: tactile sensor; PVDF; three-dimensiona force.

1 INTRODUCTION

The more wide use of the robot was promoted by the continuous improvement of automation and intelligence in modern industries, thus contributing to the study of robot sensor technology, which has become very important. The research conducted on robot sensors mainly includes integration and awareness composite and the direction of the mini, so the tactile detection is an important part of robot sensing function. Research of tactile sensors began in the 1970s in foreign countries: In Japan, Saga University used soft good toughness of a pressure-sensitive conductive rubber-developed tactile sensing system for detecting temperature, hardness, and thermal conductivity. In China, Luo Zhizeng by using an anisotropic piezo-resistive material CSA developed a high-resolution flexible tactile sensor, but it was not flexible and the cost was high. A PVDF-based tactile sensor for three-dimensional force was presented in this paper; it is used for rectifying the defects of the current robot tactile sensor for three-dimensional force information detection in the grab process and is combined with the piezoelectric properties and related physical parameters of PVDF to design a high practical value robot tactile sensor with the outstanding features of small volume, simple and reasonable structure, high sensitivity, and one that can effectively detect the force information in the three-dimensional direction of an object. Compared with the traditional tactile sensor, the detected three-dimensional force information in this design can fully respond to the force information of an object in the grasping process.

2 MEASURING PRINCIPLE OF THE SENSOR

PVDF is an abbreviation of polyvinylidene fluoride and also a new type of piezoelectric polymers. According to the characteristics of the PVDF piezoelectric film, the polarization plane will exhibit a certain amount of charge when the polarized film suffers a pressure deformation in certain directions. By exporting the charge through a relevant circuit and converting it into an electrical signal, we can measure the related surface's specific pressure deformation information of the piezoelectric film by detecting a change in the electrical signal at this time. We can select a four-prism and paste the PVDF piezoelectric film on its four sides, respectively. The PVDF piezoelectric film pasted on each side of the four-prism will be subject to different pressures when a force F is exerted on the four-prism contact face, and there will be a charge proportional to the force. The strength of the electrical signal of each film that is generated will not be the same because of the different pressure on each film. Thus, we can calculate the component of each side of the four-prism by detecting the electrical signal, so the force information in the three-dimensional direction of the object at this time can be pulled out.

3 STRUCTURE DESIGN OF THE SENSOR

The sensing head is the core part of this design; in order to detect the three-dimensional force information of the contact surface effectively and according to the analysis of the three-dimensional force measurement principle, we use figure 1 as the structure of a sensing head.

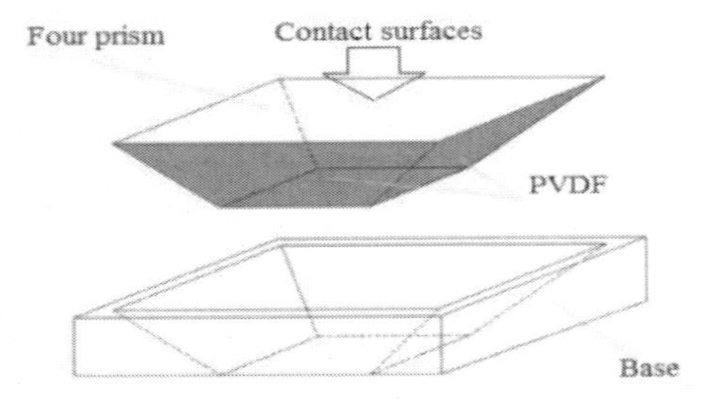

Figure 1. The sensor head structure design.

Figure 1 shows that the sensor head is mainly composed of a base, a four-prism, and PVDF piezoelectric film composition. The base as a support and each surface of the four-prism is pasted with a PVDF piezoelectric thin film. The PVDF piezoelectric film that is pasted on each surface of the four-prism will have a charge generated because of the contact surface stressed by force. We can analyze the component that each surface suffered by detecting the different charge of each surface generated, which can effectively pull out the three-dimensional force information related to the contact surface of the sensor with the object.

4 MODEL AND CHARACTERISTIC ANALYSIS OF THE SENSOR

4.1 *PVDF sensing model and characteristic analysis*

A PVDF piezoelectric film micro unit forces information, and the representation of the piezoelectric element coordinate system is shown in figure 2.

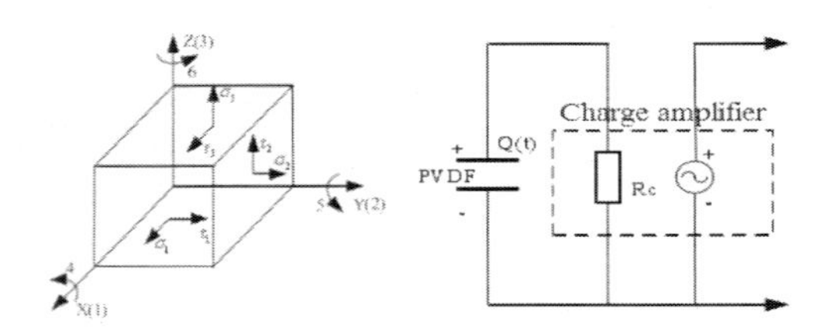

Figure 2. PVDF force information and equivalent model.

Here, we use numerical subscript to indicate the plane of the piezoelectric element and the associated force directions, X, Y, and Z, three axial, respectively, by 1, 2, and 3 to represent, around X, Y, and Z, three-axial tangential effect, by 4, 5, and 6 are, respectively, expressed. Generally speaking, X-axis represents the tensile direction, which is due to the PVDF piezoelectric film having a large piezoelectric constant in the direction of extension; Z-axis direction that is perpendicular to not only the membrane surface, expressed as the polarization direction, but also the force direction, which is usually due to the piezoelectric constant, is the maximum polarization. When the sensing element is the PVDF piezoelectric film, its piezoelectric equation is expressed by equation 1:

$$\delta_i = \sum_{i,j} d_{ij}\delta_j \tag{1}$$

Among them: The charge density is the piezoelectric coefficient, is the stress, $i = 1-3, j = 1-6$

If the force on the PVDF sensor unit has changed, the charge Δq that is produced by the piezoelectric film can be simultaneously obtained.

$$\Delta q = \sum_{j=1}^{3} d_{3j}\Delta\sigma_j \tag{2}$$

In the earlier formula, Δq represents changes in charge amount per unit area; $d_{3j}$is on behalf of the direction of the piezoelectric constant; and $\Delta\sigma_j$ indicates the amount of changes of the direction of stress.

Along with the applied stress and the revocation of the stress, the basic shape of the response is shown in Figure 3 (a), 3 (b).

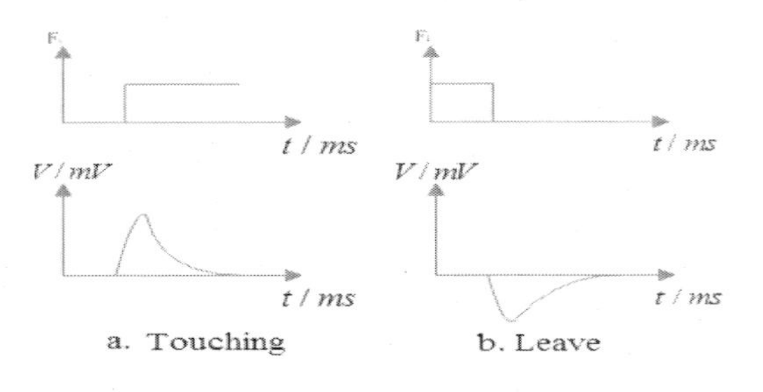

Figure 3. PVDF piezoelectric sensor step response curve.

4.2 *Sensor head structure model and characterization*

A part of the stress analysis diagram of the sensor head is shown in figure 4.

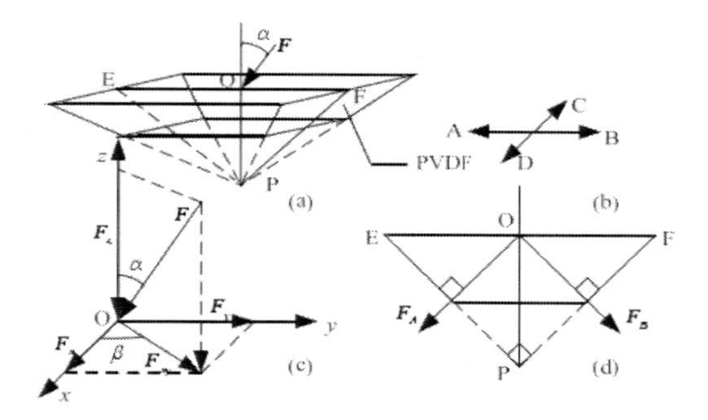

Figure 4. Force analysis schematic of sensor head.

Figure 4 (b), 4 (c) shows $F_x = (F_C, F_D)$, $F_y = (F_A, F_B)$; figure 4 (d) shows a sectional view of a tactile sensor head. If we extend both sides, it will be an isosceles triangle profile; this leads to withstanding the force points O, A, and B on both sides of the vertical line. Then, EP and FP are the pedals located above the midpoint to arrive at F_A, F_B direction, as shown in figure 4 (d). Similarly seen are *FC, FD* direction.

In the external excitations, the PVDF piezoelectric film produced is proportional to the amount of charge Q with the external forces, through the signal conditioning circuit, the charge of Q into the amount of voltage V. According to the film, A, B, C,and D generate the voltage value of the F_A, F_B, F_C, F_D component of V_A, V_B, V_C, V_D and can be obtained, so that we can clearly distinguish the three directions x, y, and z.

Excitation voltage PVDF piezoelectric film is produced in line with the following formula:

$$V \propto SF \tag{3}$$

In this formula, S represents the ratio of the piezoelectric coefficient; F represents pressure on size.

5 CONDITIONING CIRCUIT

The output charge amount of the PVDF-based robot tactile sensor for three-dimensional force is extremely weak, while the sensor has high internal resistance and high-frequency interference exists in the resulting signal. Hence, it needs a conditioning circuit for the charge signal PVDF produced post-optimization. This paper adopts three-stage amplifier circuits to solve the problem.

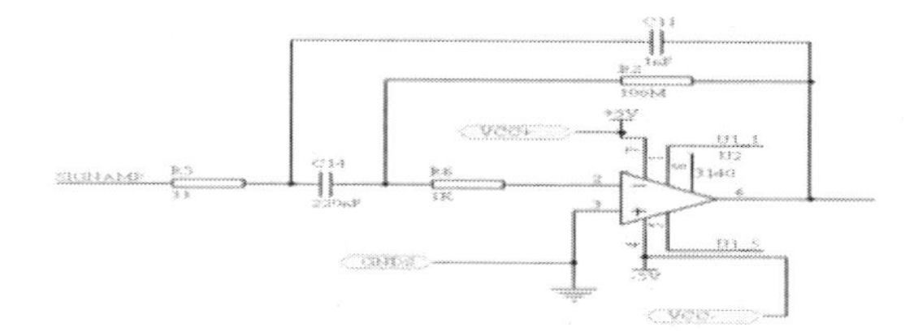

Figure 4. First-stage amplifier circuit (signal amplification and conversion).

The first stage uses the preamplifier circuit, and the main function is to amplify the signal and the input charge into a voltage output.

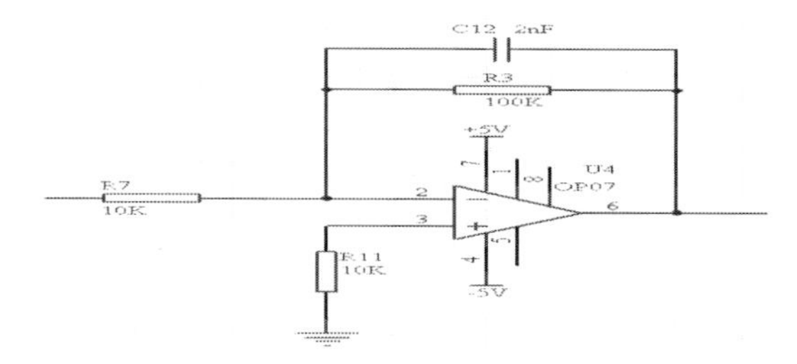

Figure 5. Second-stage amplifier circuit (low-pass filter).

The second level is the use of a low-pass filter with gain, for filtering the high-frequency interference in tactile signals.

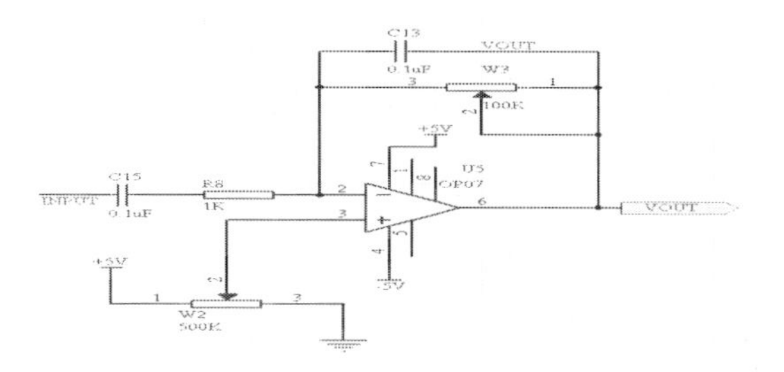

Figure 6. Third-stage amplifier circuit (band-pass filter).

The third stage uses the band-pass filter as the last-stage amplifier circuit, and the signal is generated by the PVDF three-dimensional force sensor to optimize the processing of the last step.

6 TEST AND DATA ANALYSIS

In order to detect the three-dimensional force information that an object suffered in a different grasp state more realistically, in accordance with the clamping direction and the horizontal plane forming an angle that is not the same for extracting the information, respectively, we conduct specific experiments based on the relevant physical mechanics theory. For convenience, we choose three sets of experimental data with the typical characteristic angle (0 degrees, 45 degrees, 90 degrees) of the clamping force to analyze the three-dimensional force information; after the calculation, we can obtain the following tactile response curve:

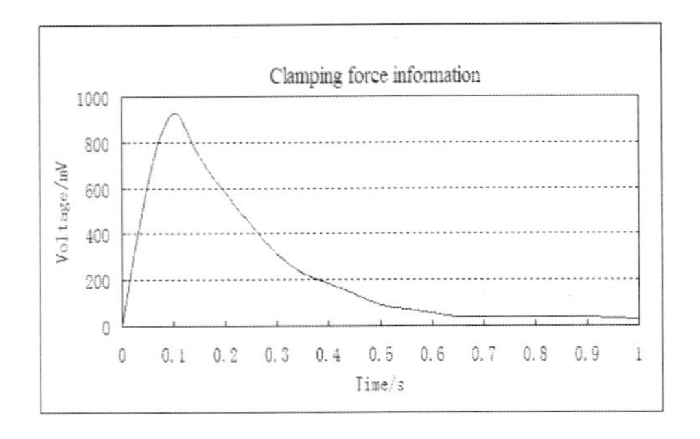

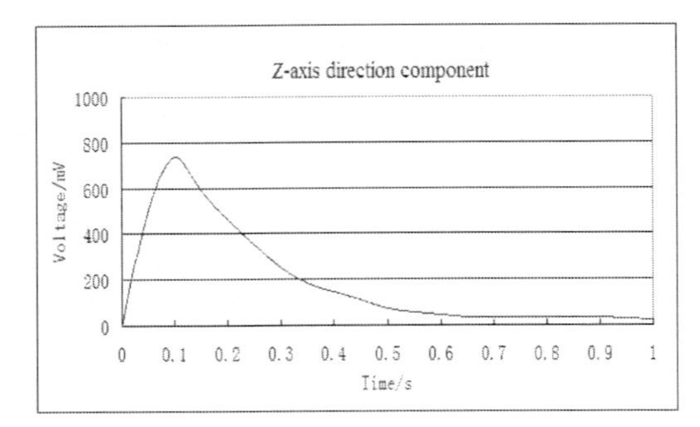

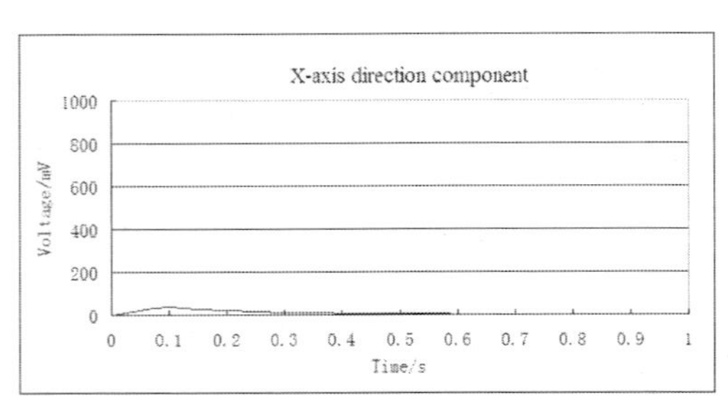

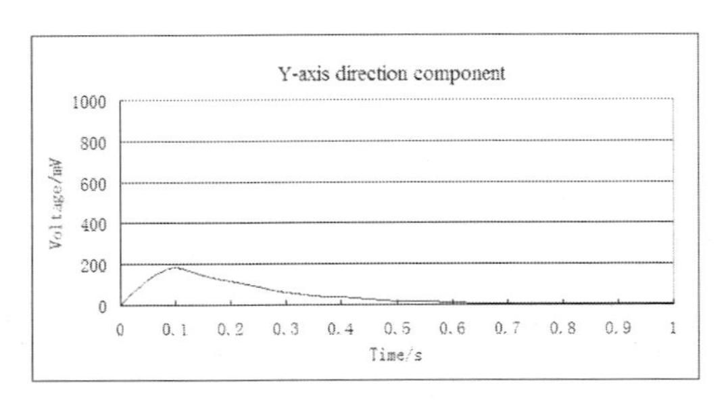

Figure 7. Clamping direction 0 degrees to the horizontal.

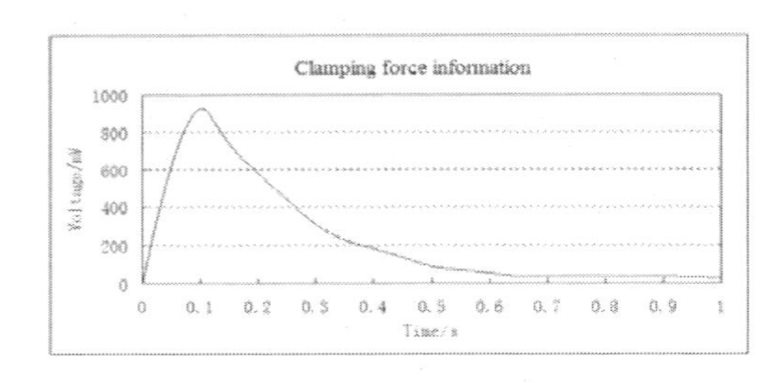

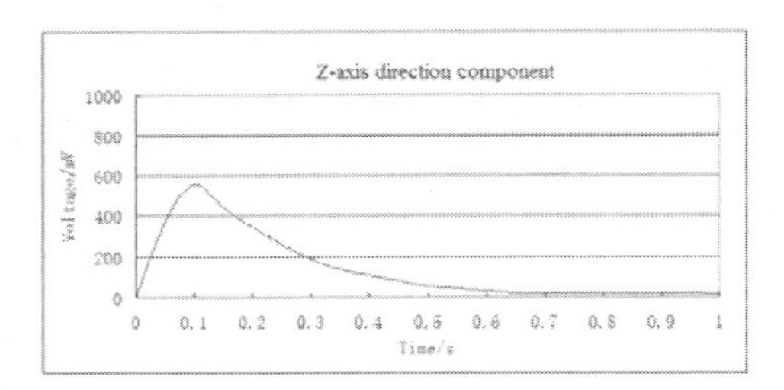

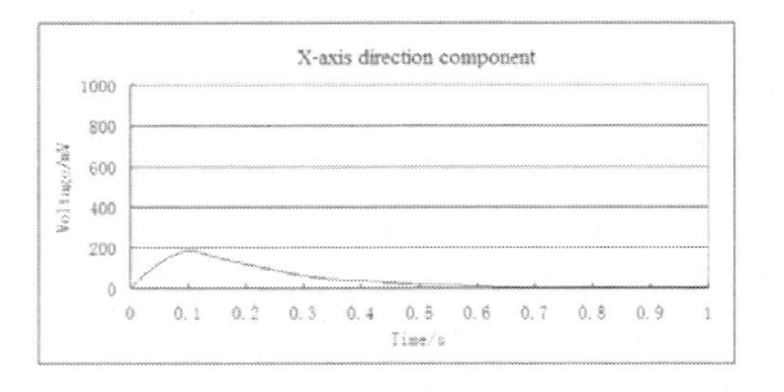

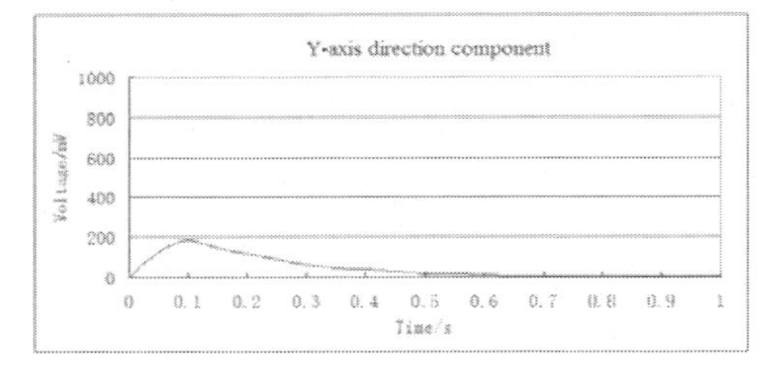

Figure 8. Clamping direction 45 degrees to the horizontal.

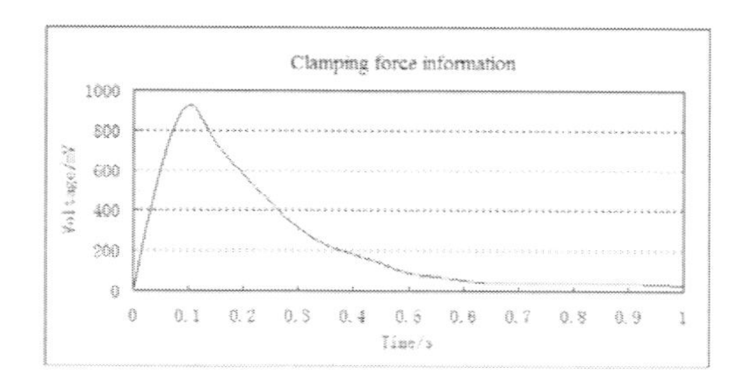

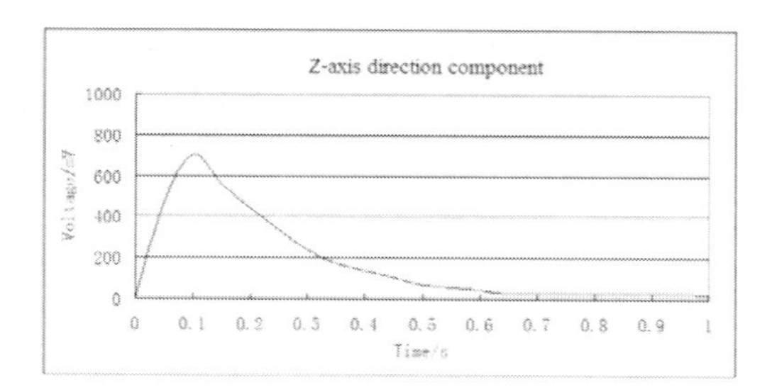

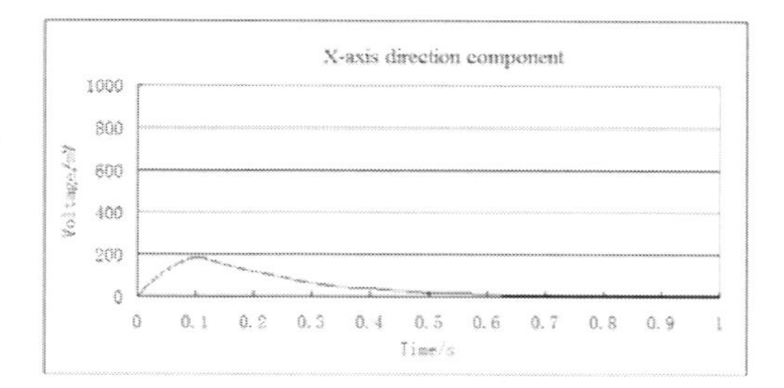

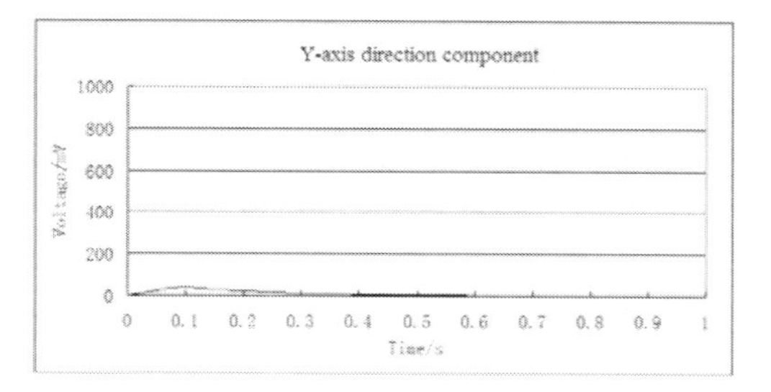

Figure 9. Clamping direction 90 degrees to the horizontal.

The earlier tactile response curves can show us that, when the clamping direction changes, the robot grabs the same object at different angles, and the component size of each contact surface of the sensor will be changed. This indicates that the detected three-dimensional force information that the contact surface of the sensor with the object suffered involves real-time changes with the force case that the contact surface suffered. It is suggested that this design can detect the three-dimensional force information that the object suffered when the robot and objects effectively come into contact, and it has high sensitivity.

7 CONCLUSION

This paper presented a kind of tactile sensor based on the PVDF piezoelectric film for the robot. Compared with the traditional robot tactile sensor, the sensor that we can get in this design can effectively detect the three-dimensional force information that the object suffered when the robot hand came into contact with the object after some practical tests, and with the outstanding features of high sensitivity, simple and reasonable structure, flexibility, and a high practicability and promotional value.

REFERENCES

[1] Luo Zhizeng. (2006) *Multisensory Robot Sensor System and Multi-Information Fusion Technology*. pp:12–13.

[2] Chen Dongyue. (2007) *Study has a Number of Perceptual and Cognitive Abilities of Intelligent Robots Problem*. (10):3–18.

[3] Wen Dianying. (1998) Polyvinylidene fluoride (PVDF) film piezoelectric response to shock loading. *Functional Polymer Materials*, pp:1505–1508.

[4] Liu Shaoqiang, Huang weiyi, Wang Aimin. (2002) History status and trends of robot tactile sensing technology research and development. Robots, 24 (4): 362–366.

[5] Wang Yanshen. (2010) PVDF piezoelectric film based force measuring system, *Research Journal of Applied Sciences, Engineering and Technology*, 4(16):2857–2861.

[6] Huang Ying, Ming Xiaohui, Xiang Bei. (2008) A new type of robot and other three-dimensional force flexible tactile sensor design. *Sensing Technology*, 21(10):1695–1699.

[7] Liu Jingcheng, Liu, Min. (2005) Design of new three-dimensional force sensor based on piezoelectric. *Piezoelectric Sound and Light*, 27 (6):643–645.

Electric, Electronic and Control Engineering – Shao, Shu & Tian (Eds)
© 2015 Taylor & Francis Group, London, ISBN: 978-1-138-02842-5

Application of auto-control in car repair

Xinjian Yuan, Zhaona Lu
School of Automotive Engineering, Nantong Polytechnic College, Nantong Jiangsu, China

ABSTRACT: Auto-control car repair has proved to be an effective solution for the low accuracy and slow speed faced by manual diagnosis, and it provides higher operational reliability and lower repair difficulty, especially in some special car repair processes such as the testing of chassis output power and the diagnosis of the internal-combustion engine. The operation of auto-control, however, requires an acceptable environment as well as modern electronic testing apparatuses. In this paper, the conditions of or solutions to auto-control car repair are investigated by relying on scientific evaluation methods.

KEYWORDS: car repair; auto-control; fuzzy evaluation.

1 INTRODUCTION

Car development is already integrating and further enhancing high technologies such as e-information and electrical automation. This development, however, calls for good adaptability of car-related professionals to processes in the car industry, especially in car repair service. Changes in car repair involve knowledge related to car repair, updates of the repair apparatuses, as well as substantial ability for car repair operations. As such, it is highly necessary that car repair stays in pace with the development of the car industry, and skilled workers selected, equipment updated, and repair techniques improved on the basis of car development.

Technical diagnosis in car repair aims at testing or evaluating the behavior of some part of the car with the appropriate instrumentation in a scientific manner, locating the fault, and repairing it. As manual testing requires high experience and ability or complicated disassembly of the car itself, which is low in accuracy and slow in speed and sometimes may even fail to solve any problem effectively, we need to find some more scientific means. As technology advances, scientific instruments are more extensively used and have made great progress in solving car problems. These instruments, characterized by high accuracy and, most importantly, quick diagnosis, provide higher operational reliability and lower repair difficulty, as is the case of the many electronic apparatuses used these days in auto-control car repair.

2 MODELING

Auto-control is typically used in the testing and diagnosis of the internal-combustion engine, the repair of the gas engine, the testing and analysis of oil samples, the testing and repair of the chassis output power, and the testing of the car braking behavior; it is also used to repair the car in an acceptable operating environment by using modern electronic testing techniques. Table 1 gives the scheme of auto-control car repair.

Car repair is highly important. In Japan, for example, an Automobile Service Promotion Association established in Japan releases regular basis information on the performance of some new cars and car service data. In the 1980s, a repair training center was founded in Japan's car repair industry to train car repair professionals. With the emergence of the car industry, a survey was able to obtain data related to the number and items of repair for the variety of cars used.

As shown in Figure 1, we have a brief view of the auto-control car repair process. In this paper, auto-control car repair as a part of this process is evaluated.

2.1 *Building a fuzzy evaluation model*

As the operation of auto-control requires an acceptable environment as well as modern electronic testing apparatuses, it has some demerits, including site requirements, despite its efficient and quick testing

Table 1. Auto-control testing and repair methods.

Testing Aspects	Testing Process	Auto-control Equipment Used
Gas engine repair	Tests single-cylinder power equilibrium, vacuum degree behind air damper, mixture concentration, engine cylinder air tightness	Integrated performance analyzer for gas engine
Abnormal sound testing & diagnosis in internal-combustion engine	Tests abnormal sound by amplifying it 5- to 8-fold and by frequency spectral analysis of images	Frequency spectral analyzer Fast stethoscope High-sense detectoscope
Chassis output power testing	Tests output power of car chassis by simulating road operation. The dynamometer drum produces eddy current, and the success rate is displayed on the instrument panel. The engine power is estimated on the basis of the chassis output power as a means to evaluate the dynamic behavior of car engine and power consumption from chassis transmission.	Chassis dynamometer
Oil sample testing & analysis	As metal particles resulting from friction come into the lube and hydraulic oil, it is possible to analyze the composition of the metal particles by sampling the oil and separate high-strength and gradient magnetic fields.	Oil spectral analyzer
Car braking behavior testing	The braking force is displayed on the console instrument panel, and the total braking force and braking distance are calculated.	Inertia brake tester Reaction brake tester

Table 2. Car repair status in Japan.

Applied Type	Check Period/ month	Number of Check Items
Private care	6	16
	12	60
	24	102
Private wagon	6	41
	12	120
Commercial car (passenger, tax) etc	1	42
	3	94
	12	149

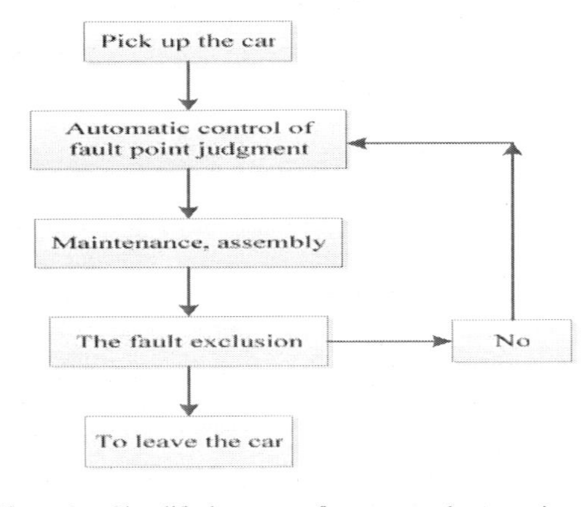

Figure 1. Simplified process of auto-control car repair.

means. Figure 2 shows a simplified model of a multimodal system based on practical fuzzy evaluation.

2.2 *Evaluating auto-control status in car repair in the light of the fuzzy evaluation model*

First of all, the auto-control status in car repair is preliminarily evaluated with respect to the conditions for auto-control repair, the extent of auto-control repair, and the repair status, as shown in Tables 2–5.

On the basis of fuzzy evaluation modeling, we establish a factor set U, where $U=(U_1\ \ U_2\ \ U_3\ \ U_4)$.

Based on the four important factors, that is, the extent of auto-control repair U_1, the conditions for auto-control repair U_1, the repair status U_3, and the miscellaneous U_4, we get Table 3. Small factor sets are established among these four important factors.

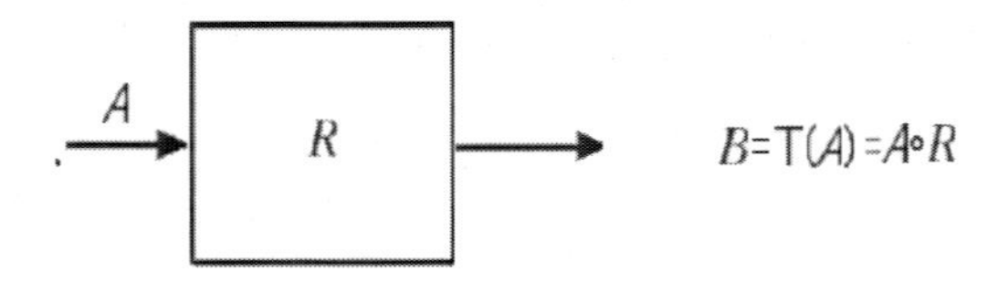

Figure 2. Simplified model.

Table 3. Evaluation index system for auto-control in car repair.

Extent of auto-control repair U_1	Conditions for auto-control repair U_2	Repair status U_3	Miscellaneous U_4
Fault diagnosis u_{11}	Repair job class u_{21}	Software development and design aspects u_{31}	Repair accessory operation u_{41}
Fault identification u_{12}	Social dispersibility u_{22}	Repair cost u_{32}	Supervision from repair and car administrations u_{42}
Fault recovery u_{13}	Repair-related economic system u_{23}	Operational consumption saved u_{33}	Management mode u_{43}
Assemble, replace u_{14}	Requirements for auto-control technician u_{24}	Miscellaneous u_{34}	

We get evaluation sets based on the factors presented in Table 3.

$$U_1 = \{u_{11}, u_{12}, u_{13}, u_{14}\}$$

$$U_2 = \{u_{21}, u_{22}, u_{23}, u_{24}\}$$

$$U_3 = \{u_{31}, u_{32}, u_{33}, u_{34}\}$$

$$U_4 = \{u_{41}, u_{42}, u_{43}\}$$

A fuzzy evaluation model applied to fuzzy computation involves a number of uncertain factors. Here, the following operations are performed in the light of fuzzy evaluation:

1 We set up a factor set U,

$$U = \begin{pmatrix} U_1 & U_2 & \cdots & U_k \end{pmatrix}$$

2 And an evaluation set V(evaluation set),

$$V = \begin{pmatrix} V_1 & V_2 & \cdots & V_n \end{pmatrix}$$

3 Then, we determine the priority domain according to the general ranking system:

$$V = \{V_1, V_2, V_3, V_4\}$$

$$= \{\text{very good, good, average, bad}\}$$

We establish the fuzzy mapping of the evaluation matrix from U to V and obtain a fuzzy relation as shown in the matrix next.

$$R = \begin{bmatrix} r_{11} & r_{12} & \cdots & r_{1n} \\ r_{21} & r_{22} & \cdots & r_{2n} \\ \vdots & \vdots & & \vdots \\ r_{m1} & r_{m2} & \cdots & r_{mn} \end{bmatrix}$$

After a comprehensive fuzzy evaluation of the training status of the PE training postgraduates, we set up a weight set, $A = (a_1, a_2, \cdots, a_n)$, which satisfies

$$\sum_{i=1}^{n} a_i = 1 \quad a_i \geq 0$$

Each row of the fuzzy relation R represents the degree of influence of the influencing factors on the object judgment, whereas each column of R represents the degree of influence of the influencing factors in this column on the object judgment.

$$\sum_{i=1}^{n} r_{ij} \quad j = 1, 2, 3, \cdots, m$$

After collecting and analyzing data, we get the priority of the four factors by the level of importance as presented in Table 4.

Table 4. Summary priority of four factors by level of importance.

Class	Priority 1	Priority 2	Priority 3	Priority 4
Extent of auto-control repair U_1	24	6	5	1
Conditions for auto-control repair U_2	0	3	19	12
Repair status U_3	0	9	15	12
Miscellaneous U_4	4	19	10	2

From Table 4, we get the priority matrix of the four factors, that is, the extent of auto-control repair U_1, the conditions for auto-control repair U_2, the repair status U_3, and miscellaneous U_4.

$$U_1 = \{24, 6, 5, 1\}$$

$$U_2 = \{0, 3, 19, 12\}$$

$$U_3 = \{0, 9, 15, 12\}$$

$$U_4 = \{4, 19, 10, 2\}$$

From priority 1 to priority 2, we get the weighted vector

$$\beta = \{\beta_1, \beta_2, \beta_3, \beta_4\} = \{0.4, 0.3, 0.2, 0.1\}$$

$$U_i^* = U_i \cdot \beta^T$$

$$U_1^* = 13, U_2^* = 9.3, \mathrm{U_2} U_3^* = 3.9, U_4^* = 6.6$$

By normalizing it,

$$U_1^* = 0.34, U_2^* = 0.31, U_3^* = 0.19, U_4^* = 0.16$$

We have

$$\bar{A} = (\ 0.34 \quad 0.31 \quad 0.19 \quad 0.16\ )$$

We establish the membership degree of the evaluation markers as shown in Table 5.

By this means, we get the evaluations of the indexes for the extent of auto-control repair U_1, the conditions for auto-control repair U_2, the repair status U_3, and miscellaneous U_4, as shown in Table 6.

We get fuzzy sets for the single-tier weight factors, such as

$$U_1^* = \{U_{11}, U_{12}, U_{13}, U_{14}\} = \{0.24, 0.36, 0.26, 0.14,\}$$

$$U_2^* = \{U_{21}, U_{22}, U_{23}, U_{24}\} = \{0.50, 0.1, 0.25, 0.14\}$$

$$U_1^* = \{U_{31}, U_{32}, U_{33}, U_{34}\} = \{0.4, 0.3, 0.1, 0.2\}$$

$$U_1^* = \{U_{41}, U_{42}, U_{43}\} = \{0.3, 0.4, 0.3\}$$

We establish the conversion functions for the priority evaluation of the appropriate factors as shown next:

$$u_{v1}(u_1) = \begin{cases} 0.5(1 + \dfrac{u_i - k_1}{u_i - k_2}), & u_i \geq k_1 \\ 0.5(1 - \dfrac{k_1 - u_i}{k_1 - k_2}), & k_2 \leq u_i < k_1 \\ 0 \quad , & u_i < k_2 \end{cases} \tag{1}$$

Table 5. Membership degree of evaluation markers.

Evaluation method	Assigned fraction zone			
	0-60	60-80	80-90	90-100
Very good	0	0	0.05	0.95
Good	0	0.05	0.9	0.05
Average	0.05	0.9	0.05	0
Bad	0.95	0.05	0	0

Table 6. Evaluation values of indexes for auto-control in car repair.

Tier Index	Evaluation Value	Tier Index	Evaluation Value
Fault diagnosis u_{11}	Very good	Software development and design aspects u_{31}	Very good
Fault identification u_{12}	Average	Repair cost u_{32}	Good
Fault recovery u_{13}	Very good	Operational consumption saved u_{33}	Good
Assemble, replace u_{14}	Very good	Miscellaneous u_{34}	Average
Repair job class u_{21}	Average	Repair accessory operation u_{41}	Good
Social dispersibility u_{22}	Very good	Supervision from repair and car management authorities u_{42}	Average
Repair-related economic system u_{23}	Good	Management mode	Good
Requirements for auto-control technician u_{24}	Good		

$$u_{v2}(u_1)=\begin{cases}0.5(1-\dfrac{u_i-k_1}{u_i-k_2}), & u_i \geq k_1\\ 0.5(1+\dfrac{k_1-u_i}{k_1-k_2}), & k_2 \leq u_i < k_1\\ 0.5(1-\dfrac{u_i-k_3}{k_2-k_3}), & k_3 \leq u_i < k_2\\ 0.5(1-\dfrac{k_3-u_i}{k_2-u_i}), & u_i < k_3\end{cases} \quad (2)$$

$$u_{v1}(u_1)=\begin{cases}0, & u_i \geq k_2\\ 0.5(1-\dfrac{k_1-u_i}{k_2-k_3}), & k_3 \leq u_i < k_2\\ 0.5(1+\dfrac{k_3-u_i}{k_2-u_i}), & u_i < k_3\end{cases} \quad (3)$$

According to the evaluations in Table 5, and in the light of the membership degree of the evaluation markers, we get the evaluation sets of all aspects of the extent of auto-control repair U_1, the conditions for auto-control repair U_2, the repair status U_3, and the miscellaneous U_4:

Extent of auto-control repair

$$U_1=\begin{pmatrix}0 & 0 & 0.05 & 0.95\\ 0 & 0.05 & 0.9 & 0.05\\ 0 & 0 & 0.05 & 0.95\\ 0 & 0.05 & 0.9 & 0.05\end{pmatrix}$$

Conditions for auto-control repair

$$U_2=\begin{pmatrix}0 & 0 & 0.05 & 0.95\\ 0 & 0.05 & 0.09 & 0.05\\ 0 & 0 & 0.05 & 0.95\\ 0 & 0.05 & 0.9 & 0.05\end{pmatrix}$$

$$\text{Repair status } U_3=\begin{pmatrix}0 & 0.05 & 0.9 & 0.05\\ 0 & 0.05 & 0.9 & 0.05\\ 0 & 0 & 0.95 & 0.05\\ 0 & 0.05 & 0.9 & 0.05\end{pmatrix}$$

$$\text{Miscellaneous } U_4=\begin{pmatrix}0 & 0 & 0.05 & 0.95\\ 0 & 0.05 & 0.9 & 0.05\\ 0 & 0 & 0.05 & 0.95\end{pmatrix}$$

Next, we perform the following operation according to the fuzzy evaluation:

$$B=A\cdot R$$

$$=(a_1,a_2,a_3,\cdots,a_n)\cdot\begin{bmatrix}r_{11} & r_{12} & \cdots & r_{1n}\\ r_{21} & r_{22} & \cdots & r_{2n}\\ \vdots & \vdots & & \vdots\\ r_{m1} & r_{m2} & \cdots & r_{mn}\end{bmatrix}$$

$$=(b_1,b_2,b_3,\cdots,b_n)$$

The fuzzy set on V is the evaluation set B.

We perform the following operation according to the fuzzy evaluations:

$$B=A\cdot R$$

$$=(a_1,a_2,a_3,\cdots,a_n)\cdot\begin{bmatrix}r_{11} & r_{12} & \cdots & r_{1n}\\ r_{21} & r_{22} & \cdots & r_{2n}\\ \vdots & \vdots & & \vdots\\ r_{m1} & r_{m2} & \cdots & r_{mn}\end{bmatrix}$$

$$=(b_1,b_2,b_3,\cdots,b_n)$$

The fuzzy set on V is the evaluation set B. The four evaluation sets, that is, the extent of auto-control repair U_1, the conditions for auto-control repair U_2, the repair status U_3, and the miscellaneous U_4, are calculated next:

$$B_i=A_i\cdot R_i$$

By normalizing the resulting B_i, we obtain the following fuzzy evaluation matrix:

$$\bar{B}=\begin{pmatrix}B_1\\ B_2\\ B_3\\ B_4\end{pmatrix}=\begin{pmatrix}0.07 & 0.17 & 0.25 & 0.52\\ 0.11 & 0.26 & 0.43 & 0.62\\ 0.14 & 0.23 & 0.12 & 0.24\\ 0.13 & 0.23 & 0.28 & 0.38\end{pmatrix}$$

and, eventually, the synthetic evaluation value:

$$Z=U^*\cdot B=\begin{pmatrix}0.22 & 0.33 & 0.24 & 0.21\end{pmatrix}$$

3 CONCLUSIONS

Auto-control is typically used in the testing and diagnosis of the internal-combustion engine, the repair of the gas engine, the testing and analysis of oil samples, the testing and repair of the chassis output power, and the testing of the car braking behavior, and it has proved to be particularly effective in repairing precision parts. In this paper, the application of auto-control in car repair is investigated and analyzed with respect to the extent of auto-control repair, the conditions for auto-control repair, the repair status, and so on. The resulting evaluation value of auto-control in car repair is "good." However, despite the high accuracy, fast speed, high operational reliability, and low repair difficulty, auto-control has some demerits depending on the conditions for auto-control repair as well as the nature and size of the one-dimensional repair service. Nevertheless, for major firms, auto-control is still an excellent solution to car repair.

REFERENCES

[1] Yao, H., Yao, G.P. (2005) *Structure and Repair of Automobile Electric Control System*. Beijing: Beijing Institute of Technology Press.

[2] Yao, T.F., Zhao, J. (2010) Application of computer technology in car repair. *Communications Science & Technology Heilongjiang*, (9):22–23.

[3] Pang, L.X. (2014) Car repair and maintenance: Current situation and countermeasures. *Net Friends*, (06).

[4] Liu, J.X. (2006) Discussion on the development tendency of automobile maintenance industry in China. *Journal of Chongqing Jiaotong University*, 25(S1).

[5] Japan Automobile Standards Internationalization Center. (1991) Safety and pollution control system for motor vehicles [R]. Edited and compiled by Chinese Automobile Technology & Research Center. *Country Report of Japan*.

[6] Zhang, Y., Wang, Y. (2013) Optimal models and scheduling algorithm of auto dealership. *Public Communication of Science & Technology*, (23).

[7] Yao, T.F., Zhao, J. (2010) Application of computer technology in car repair. *Communications Science & Technology Heilongjiang*, (09):159–160.

[8] Ling, P.X., Lin, W.Q. (2006) Application of automatic control in car. *China Electric Power Education*, S3:109–111.

Electric, Electronic and Control Engineering – Shao, Shu & Tian (Eds)
© 2015 Taylor & Francis Group, London, ISBN: 978-1-138-02842-5

The development history of engineering cost consulting industry in mainland China

Kecheng Li
Chongqing Construction Vocational College, Chongqing, China

Jiahui Yan
Chongqing Construction Cost Association, Chongqing, China

Jie Ding
Chongqing City Construction Cost Office Co. Ltd., Chongqing, China

ABSTRACT: The market-oriented engineering cost consulting service in mainland China dates from the early 1990s and undergoes its embryonic stage, disconnecting and reconstructing stage and standardized management stage. Improved systems including professional qualification management, engineering cost consulting business management and industrial self-regulation management were established by 2006. Since its foundation in 1990, China Engineering Cost Association has been entrusted with daily management work of the engineering cost consulting industry by the construction administrative department of the State Council of Central People's Government in order to enhance the sound development of the engineering cost consulting industry in mainland China. At present, the engineering cost consulting industry is an important industry of the tertiary industry.

KEYWORDS: engineering cost; consulting industry; development; review.

1 INTRODUCTION

The engineering cost consulting industry belongs to the tertiary industry, providing the construction industry of the secondary industry with specialized consulting services like construction project investment and engineering cost determination and control. In mainland China, the market-oriented engineering cost consulting service in mainland China dates from the early 1990s. The engineering cost consulting industry has played an irreplaceable role in the construction industry and construction project transaction over the two decades. So far, it has become one of the indispensable market entities of the construction industry and the construction project transaction market.

According to the working and managerial experience of the engineering cost consulting industry in mainland China, the author attempts to make a brief review on the development history of the engineering cost consulting industry in mainland China.

2 THE BUDDING ENGINEERING COST CONSULTING INDUSTRY

Before 1980s, the planned economic system was implemented in mainland China. There was completely no need for the market-oriented engineering cost consulting industry in the society. Since the reform and opening-up in the early 1980s, the planned economic system has gradually transited to the market economic system with Chinese characteristics. With the deepening of the reform and opening-up, the construction industry and the construction project transaction market have had demand for the market-oriented engineering cost consulting service by the early 1990s. At that time, norm management department of the government, audit and appraisal department of the government, investment audit department of CCB, technical and economic department of the designing institute, large and medium accounting firms and infrastructure management department of state-owned large factories and mines took the lead in providing the society with market-oriented engineering cost consulting services so as to meet the demand of the society for engineering cost consulting services. Some of these departments are subordinate or affiliated to functional departments of the government while some are subordinate or affiliated to state-owned large factories and mines. They do not exist in the form of independent market-oriented engineering cost consulting industry. In spite of this, market-oriented engineering cost consulting services provided by these departments mark the emerging of the engineering cost consulting industry in mainland China.

It is stipulated in the file of No.19 [1997] issued by the Central Committee of CPC and the State Council that institutions affiliated to functional departments of the government, including investment consultancy, assets evaluation, accounting firm and law office, should be transformed into independent social intermediary service agencies. Disconnection and reconstruction of economic verification service agencies were started thusly. In 1998, independent market-oriented engineering cost consulting business emerged at the right moment. There are quite a number of first emerged consulting businesses of engineering cost existing in the form of collective ownership instead of partnership or corporate system. With the emerging of these independent market-oriented engineering cost consulting businesses, the industry of engineering cost consulting in mainland China has started regardless of business forms.

3 DISCONNECTION AND RECONSTRUCTION OF ENGINEERING COST CONSULTING BUSINESS

In October 1999, the State Council of CPC set up a leading group to clear and reorganize social intermediary agencies of economic verification and issued the file of No.92 [1999] *A Notification of Clearing and Reorganizing Social Intermediary Agencies of Economic Verification from the General Office of the State Council*, stipulating that all social intermediary agencies of economic verification should be disconnected and reconstructed after clearing and reorganizing and any government departments shall not establish social intermediary agencies of economic verification. In July 2000, the General Office of the State Council of CPC forwarded *Suggestions on the Disconnection and Reconstruction of Economic Verification Service Agencies and Government Departments* in the file of No.51 [2000] issued by the leading group of clearing and reorganizing social intermediary agencies of economic verification.

Disconnection means that social intermediary agencies of economic verification completely stop the subordinate or affiliated relationship with government departments and state-owned enterprises and institutions in terms of personnel, finance (including capital, entity, property right, etc.), business and name. Reconstruction means that social intermediary agencies of economic verification are established by people with professional qualification in organizational forms stipulated by relevant laws and regulations.

In September 2000, the former Ministry of Construction of CPC stipulated in *A Notification of the Disconnection and Reconstruction of Engineering Cost Consulting Agencies with Government Departments* (No.208 [2000] of the building standard) that engineering cost consulting agencies belonging to the range of disconnection and reconstruction should be completely disconnected with attached institutions in terms of personnel, finance, business and name. Engineering cost consulting agencies after disconnection should be reconstructed into social intermediary agencies in partnership or system with limited liability established by cost engineers in organizational forms stipulated by relevant laws and regulations.

In December 2000, the former General office of the Ministry of Construction of CPC issued *Several Opinions on Implementing 'A Notification of the Disconnection and Reconstruction of Engineering Cost Consulting Agencies with Government Departments'* (No.50 [2000] of the building standard). It is stipulated that the name of engineering cost consulting agencies of partnership after disconnection and reconstruction should be: the name of territory + trade number + words reflecting properties of engineering cost business + firm. The name of engineering cost consulting agencies of the system with limited liability after disconnection and reconstruction should be: the name of territory + trade number + words reflecting properties of engineering cost business + firm Co., LTD.

In June 2002, the leading group of clearing and reorganizing social intermediary agencies of economic verification the State Council of CPC issued *A Notification of Standardizing Management on Engineering Cost Consulting Industry* (No.6 [2002] of clean-up and rectification of the State Council), stipulating that all engineering cost consulting agencies engaged in intermediary service activities, especially agencies specialized in engineering cost consultation, should be disconnected and reconstructed into independent social intermediary agencies, and measures should be taken to break industry monopoly and departmental blockade of the engineering cost consulting market.

In July 2002, the former Ministry of Construction of CPC forwarded *A Notification of Standardizing Management on Engineering Cost Consulting Industry of the Leading Group of Clearing and Reorganizing Social Intermediary Agencies of Economic Verification the State Council of CPC* in the file of No.51 [2000], stipulating that all regions and departments should conduct a spot check to engineering cost consulting agencies of level A and B within the jurisdiction (or affiliated) according to the requirement in Article 3 of the *Notification*. Results should be reported to the standard norm department of the Ministry of Construction by the end of October 2002. All regions and departments should take active measures to modify relevant regulations impeding market opening and fair competitions in line with requirements of the *Notification*, and results should be reported to the standard norm department of the Ministry of Construction by the end of December 2002.

Engineering cost consulting businesses in mainland China have completed the disconnection in 2003. Almost all businesses were reconstructed in the form of companies with limited liability.

4 PERSONNEL QUALIFICATION OF ENGINEERING COST CONSULTING INDUSTRY

One of the most important conditions of industrial development is the quality improvement of industrial personnel. In order to guarantee the quality of engineering cost management and maintain the public interest of the nation and the society, the former Ministry of Personnel and the Ministry of Construction of CPC issued *Temporary Provisions on Qualification System of Cost Engineers* (No.77 [1996]) in 1996, stipulating that the qualification system of cost engineers should be implemented in the field of engineering cost and posts like valuation, assessment, investigation(audit), control and management should be equipped with professionals with the qualification of cost engineer. Qualification tests for cost engineer follow the unified outline, propositions and organization, which are held once a year in principle. The Ministry of Construction is in charge of the preparation of the outline, compiling of training materials and propositions, and unified planning and training before examination. The training is conducted according to the principle of being separated from test and participating voluntarily. The Ministry of Personnel is in charge of examining the outline, subjects and propositions as well as organizing and authorizing various test related affairs. Supervision, inspection and guidance are conducted together with the Ministry of Construction to determine the qualification. Qualified examinees can complete registration formalities at local cost engineer registration and management institutions of provincial or ministerial level within three months after getting the certificate.

The qualification test of cost engineer was first held by the former Ministry of Personnel and the former Ministry of Construction of CPC in September 1998. A qualification test of cost engineer was held every year since 2000 except 1999. In January 2000, the former Ministry of Construction of CPC issued *Regulations on Cost Engineer Registration* in the form of the 75th order. In December 2006, the former Ministry of Construction of CPC issued *Regulations on Certified Cost Engineers* in the form of the 105th order, which was implemented since March 1, 2007. *Regulations on Cost Engineer Registration* (the 75th order of the Ministry of Construction) issued on January 21, 2000 was abolished at the same time.

According to the spirit of the file, *A Notification on Self-regulation of Preliminary Budget Personnel under the Centralization of CECA* (the No.69 [2005] of building standards), China Engineering Cost Association formulated and issued *Interim Measures of National Management on Cost Engineers* in June 2006.

So far, most professionals of engineering cost consulting industry in mainland China have passed the qualification test and acquired the qualification of certified cost engineers or the qualification of members of the national construction cost.

5 THE GRADE OF QUALIFICATION OF ENGINEERING COST CONSULTING BUSINESS

As early as the budding stage of engineering cost consulting industry in mainland China, the former Ministry of Construction of CPC issued *Regulations on Qualification of Engineering Cost Consulting Units (Trial implementation)* in the form of No.133 [1996] in March 1996 and *Implementing Rules of 'Regulations on Qualification of Engineering Cost Consulting Units (Trial implementation)'* in the form of No.316 [1996] in May 1996 so as to normalize the development of engineering cost consulting industry. The grade of engineering cost consulting units can be classified into Class A, Class B and Class C.

In January 2000, the former Ministry of Construction of CPC issued *Regulations on Engineering Cost Consulting Units* in the form of 74th order, stipulating that the grade of engineering cost consulting units is divided into Class A and Class B. The former *Regulations on Qualification of Engineering Cost Consulting Units (Trial implementation)* and *Implementing Rules of 'Regulations on Qualification of Engineering Cost Consulting Units (Trial implementation)'* were abolished at the same time.

In March 2006, the former Ministry of Construction of CPC issued *Regulations on Engineering Cost Consulting Businesses* in the form of 149th order, stipulating that the grade of engineering cost consulting businesses is divided into Class A and Class B. *Regulations on Engineering Cost Consulting Units* (the 74th order) issued by the Ministry of Construction on January 25, 2000 was abolished at the same time.

6 SELF-REGULATION MANAGEMENT OF ENGINEERING COST CONSULTING INDUSTRY

It is stipulated in *Regulations on Engineering Cost Consulting Businesses* issued by the former Ministry of Construction of CPC in the form of 149th order that engineering cost consulting businesses should be encouraged to join the organization of engineering cost consulting industry. The organization of engineering cost consulting industry should reinforce the management of industrial self-regulation.

The organization of engineering cost consulting industry is China Engineering Cost Association. It was founded in July 1990 and called CECA for short. It is a national non-profit industrial social organization formed voluntarily by engineering cost consulting businesses, engineering cost administrative units, certified cost engineers and senior experts and scholars of engineering cost.

China Engineering Cost Association receives operational guidance and supervision from the Civil Administration Department and the Construction Administration Department of the State Council of CPC. Entrusted and approved by the Construction Administration Department of the State Council, CECA is responsible for such specific affairs as daily management of engineering cost consulting industry, test for certified cost engineers, registration and continuing education, and team construction of cost engineers. It organizes industrial training, carries out business communication and popularizes advanced experience of engineering cost consultation and management. Activities like the selection and recommendation of outstanding units, individuals and engineering cost consulting achievements can be carried out in line with relevant regulations after approval. China Engineering Cost Association builds a relationship with international organizations and counterparts worldwide on behalf of the engineering cost consulting industry of China and certified cost engineers of China, fulfilling responsibilities and obligations as a member of international organizations and providing services for international exchange and cooperation.

China Engineering Cost Association has played a significant role in promoting the establishment and normalization of self-regulation management as well as the healthy development of the market-oriented engineering cost consulting industry in mainland China since the establishment.

7 CONCLUSION

The market-oriented engineering cost consulting service in mainland China dates from the early 1990s and undergoes its embryonic stage, disconnecting and reconstructing stage and standardized management stage. Improved systems including professional qualification management, engineering cost consulting business management and industrial self-regulation management were established by 2006. So far, the engineering cost consulting industry has become an important industry of the tertiary industry, playing an irreplaceable role in maintaining the normal economic order of the construction industry and the construction engineering transaction market as well as legitimate interests of construction project investors and contractors. From now on, the engineering cost consulting industry in mainland China will inevitably continue to develop and make larger contributions to the economic development of mainland China.

REFERENCES

[1] The Central Committee of CPC, the State Council. *A Notification of Deepening Financial Reform, Rectifying Financial Order and Preventing Financial Risk* (No.19 [1997] of CPC), 1997-12-6/2014-08-22.

[2] The State Council. *A Notification of Clearing and Reorganizing Social Intermediary Agencies of Economic Verification from the General Office of the State Council* (No.92 [1999] of the State Council), 1999-10-31/2014-08-22.

[3] The Ministry of Construction. *A Notification of the Disconnection and Reconstruction of Engineering Cost Consulting Agencies with Government Departments* (No.208 [2000] of the building standard), 2000-9-22/2014-08-22.

[4] The Ministry of Construction, the Ministry of Personnel. *Temporary Provisions on Qualification System of Cost Engineers* (No.77 [1996] of the Ministry of Personnel), 1996-08-26/2014-08-22.

[5] The Ministry of Construction. *Regulations on Certified Cost Engineers* (the 105th order), 2006-12-25/2014-08-22.

[6] The Ministry of Construction. *Regulations on Qualification of Engineering Cost Consulting Units (Trial implementation)* (No.133 [1996] of the building standard), 1996-03-06/2014-08-22.

[7] The Ministry of Construction. *Regulations on Engineering Cost Consulting Businesses* (the 149th order), 2006-03-22/2014-08-22.

Electric, Electronic and Control Engineering – Shao, Shu & Tian (Eds)
 ISBN: 978-1-138-02842-5

Insulator ESDD prediction based on least-squares support vector machines

Hongling Li, Xishan Wen & Naiqiu Shu
School of Electrical Engineering, Wuhan University, Wuhan, China

ABSTRACT: There is an instantaneous release of acoustic energy when a partial discharge occurs in polluted insulators. By detecting the sound signals, the pollution severities of insulators can be monitored online. The Equal Salt Deposit Density (ESDD) is the source of confirming polluted degree and mapping polluted areas. Artificial contaminations tests proved that there is a complicated nonlinear relationship between the Acoustic Emission (AE) signals emitted by the polluted insulators and the development of contamination discharge. Based on the testing data, the forecast model of ESDD of the insulator is built by using Least-Squares Support Vector Machines (LS-SVM). In the predicting model, the multiple variables come from the AE signals that are chosen as the input variables and then the ESDD is chosen as the output variable. The model parameters are tuned with the cross-validation method, and the feasibility of the model is proved by some data in the laboratory simulation. Simulation results show that the LS-SVM predicting model is capable of learning quite well from the raw data samples while processing good forecast and generalization ability.

KEYWORDS: insulator; AE; automation monitoring; LS-SVM; ESDD predicting.

1 INTRODUCTION

The surface of the outdoor insulators is covered by airborne pollutants due to natural or industrial pollution. In recent years, the contamination extent of insulators gets worse more quickly than earlier because of heavy air pollution. As the surface becomes moist because of rain, fog, or dew, the pollution layer becomes conductive. Then, the leakage current flows through the conducting surface film, and partial discharges may occur. The repetition rate and amplitude of partial discharges tends to increase with increasing levels of pollution and may, eventually, result in complete flashover. It greatly threatens the safety of the power grid. It is statistic that pollution flashover is the second main reason for the great number of power system accidents; moreover, the damage caused from pollution insulator flashover is 10 times as that from thunder-stroke accidents. [1] Large-scale pollution flashover happened in lots of countries in the world and resulted in great economic loss.

To prevent polluted insulator flashover accidents, lots of detection methods are used to monitor the pollution status of insulators. These are equivalent salt-attached density method, surface pollution conductivity method, leakage current method, pulse current method, and so on. The first two methods are off-line detection methods, which cannot provide precise monitoring results. As the conventional electrical measurements of polluted insulator partial discharge, leakage current method and pulse current method have certain inherent limitations. So, AE technology is presented, which is a non-electrical technique. When a partial discharge occurs in polluted insulators, there is an instantaneous release of acoustic energy. Therefore, using the acoustic transducer to detect the sound signals emitted by the polluted insulators, and dealing with the signals, the pollution severities of insulators can be monitored and accordingly threaten the insulation of power grid that is evaluated online. The AE method is immune to electromagnetic noise and can be used for different structural insulators with simple instrumentation. But it is easy to be influenced by the surrounding noise. Hence, the valid data processing method should be adopted in AE detection of insulator discharge.

As a new classification and regression tool, support vector machines improve the generalization performance through structural risk minimization. And then, it can perfectly solve the practical problem of small sample, nonlinear, high dimension, and local minimum. In least-support vector machines (LS-SVM), analytical solutions can be obtained by solving linear equations instead of a quadratic programming problem. [2, 3] Thus, the calculating speed is greatly improved; simultaneously, the good resolving performance is maintained. Owing to its simplification and validity, the LS-SVM is applied in many fields. [4, 5] In this paper, the LS-SVM is presented in the data processing of AE signals emitted by polluted insulator discharge. Using the analysis and processing of the LS-SVM method, the insulator ESDD forecasting model is built based on the artificial contaminations testing data. Then, the insulator ESDD can be predicted, which is very important for flashover prediction and condition-based maintenance.

2 PRINCIPLE OF LS-SVM REGRESSION ARITHMETIC

Considering a given regression sample data set $\{x_k, y_k\}_{k=1}^{N}$, where N is the total number of training data pairs, input vector $x_k \in R^n$, and output class $y_k \in R$. According to the support vector machines theory, the input space is mapped into a feature space F with the nonlinear function $\Phi(\cdot)$ being the corresponding mapping function. In the feature space, optimal linear regression function is constructed:

$$f(x) = \omega^T \cdot \Phi(x) + b \quad (1),$$

where ω is the weight vector of the hyperplane, and scalar b is the offset.

Simultaneously, a distant concept is introduced by using the structural risk minimization principle, and the dot product operation of high-dimension space is skilfully replaced by kernel function of the original space. Then, the nonlinear estimation function is translated into the linear estimation function of high-dimension space. It avoids the complex calculate.

According to the structural risk minimization principle, searching ω and b is to minimize the following function:

$$R = \frac{1}{2} \cdot \|\omega\|^2 + \frac{1}{2} \gamma \cdot R_{emp} \quad (2),$$

where $\|\omega\|^2$ controls the model complexity; γ is the normalization parameter, and it controls the punitive degree. R_{emp} is the error control function. Choosing a different loss function can construct different SVM.

The 2-norm of error is selected as the loss function in optimizing objects in LS-SVM. Then, the optimization problem is defined as follows:

$$\min J(\omega, \xi) = \min\left(\frac{1}{2} \cdot \omega^T \omega + \frac{1}{2} \gamma \cdot \sum_{k=1}^{N} \xi_k^2\right)$$
$$s.t. y_k = \omega^T \cdot \Phi(x_k) + b + \xi_k, k = 1, \cdots, N \quad (3),$$

where vector ξ_k is the error between the actual output data and the predictive output data. It is very difficult to solve the optimization problem of (3). So, the optimization problem is transformed into its dual space, and the Lagrange function is defined as follows:

$$L = J(\omega, \xi) - \sum_{k=1}^{N} a_k \left[\omega^T \cdot \Phi(x_k) + b + \xi_k - y_k\right] \quad (4),$$

where α_k is Lagrange multiplier. According to the KKT condition, the following equation and constraint condition can be obtained:

$$\begin{cases} \frac{\partial L}{\partial \omega} = 0 \to \omega = \sum_{k=1}^{N} a_k \Phi(x_k) \\ \frac{\partial L}{\partial b} = 0 \to \sum_{k=1}^{N} a_k = 0 \\ \frac{\partial L}{\partial \zeta_k} = 0 \to a_k = \gamma \xi_k \\ \frac{\partial L}{\partial a_k} = 0 \to \omega^T \cdot \Phi(x_k) + b + \xi_k - y_k = 0 \end{cases} \quad (5)$$

After elimination of ω and ξ in (5), the following linear equations are given:

$$\begin{bmatrix} 0 & 1^T \\ 1 & K + \gamma^{-1} I \end{bmatrix} \begin{bmatrix} b \\ a \end{bmatrix} = \begin{bmatrix} 0 \\ Y \end{bmatrix} \quad (6)$$

With $1 = [1, \cdots, 1]^T$, $Y = [y_1, \cdots, y_N]^T$, $a = [a_1, \cdots, a_N]$ and $K(\cdot, \cdot)$ is the kernel function. Then, analytical solutions of a and b can be obtained, and the predicting output is obtained:

$$y(x) = \sum_{k=1}^{N} a_k \Phi(x)^T \Phi(x_k) + b = \sum_{k=1}^{N} a_k K(x, x_k) + b \quad (7)$$

So, the nonlinear predicting model is defined as follows:

$$y(x) = \sum_{k=1}^{N} a_k K(x, x_k) + b \quad (8)$$

3 PREDICTING MODEL OF INSULATOR ESDD

ESDD represents the equivalent Nacl amount in every area of insulators (mg/cm^2). It is only related to the pollution severity, the component and the nature of the contaminations. The ESDD parameter is widely used to describe the state of contaminated insulators. In the 1950s, the ESDD was first used as an important characteristic parameter to scale the pollution severity in Japan. In the 1960s, the measurement of ESDD was first carried out in Northeast of China and it was widely accepted all over China by 1977. Nowadays, ESDD has become a formal contamination criterion for grading the pollution severity. However, easements of the

ESDD parameter are time consuming and difficult to automate. [6]

The AE technology is a valuable non-destructive detection method. On condition that the AE signals generated from the polluted insulators discharge are detected by the acoustic transducer, it is possible to monitor the discharger process and even the polluted insulators are still working online. That is, the acoustic signals can reflect the contamination severities of insulators. But there is a complicated non-linear relationship between the discharge acoustic signals and the contamination severities of the insulator surface. Traditional signal processing approaches are complicated comparatively and have low precision. Owing to its simplification and validity, the LS-SVM method can solve the small sample and non-linear problem. Using the LS-SVM method to process the acoustical signals effectively, ESDD can be inferred through the signals, for flashover prediction and condition-based maintenance.

4 ESDD AND AE SIGNALS

It is generally considered that insulator contamination flashover is a continuous development process of the partial arc, that is, the appearance of contamination flashover is a gradational process. So, it is feasible to continually online monitor the discharge ultrasonic signals emitted by polluted insulators, for flashover prediction and condition-based maintenance.

From the point of view that discharge is the release of energy, acoustic, heated, electromagnetic, and optical energy are the portion of the released energy. It is considered that it is proportionate between the acoustic energy and the general released energy in [7]. However, practically the proportion is uncertain, because the discharge is affected by several factors. The proportion relationship is still affirmative from the statistic point of view.

In order to check the validity of the AE monitoring method through clarifying the fundamental characteristics of AE signals emitted by the polluted insulator, the artificial pollution test is done in the laboratory.

The testing voltage is increased gradually until a flashover occurs. The AE signals from the insulators string were detected with the AE monitoring system in the whole flashover process. Figure 1 shows the waveform of AE signals of insulator samples at 0.03 mg.cm^{-2} ESDD level, which expresses a light degree of contamination. The testing voltage is increasing gradually. And the transition includes the waveforms of the AE signal at different stages in the process of flashover.

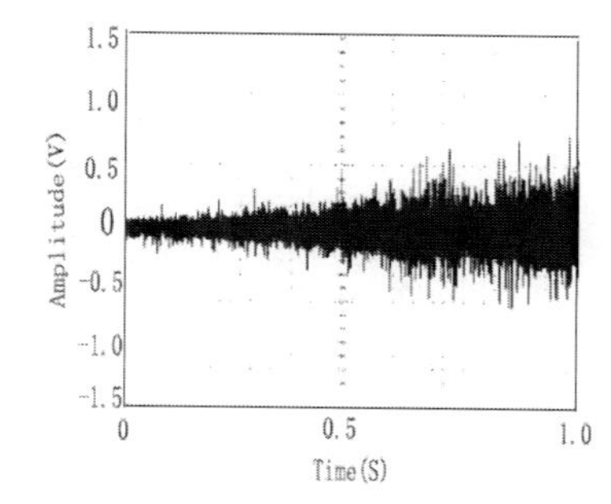

Figure 1. Waveform of AE signals of insulators at 0.03 mg.cm^{-2} ESDD level.

As shown in figure 1, the voltage amplitude of the AE signals increases with the increase of contaminator discharge energy. At the beginning, the amplitude of AE signals is small. With the increase of the testing voltage, discharges occurred locally and developed gradually. Accordingly, the amplitude and frequency of the AE signal are increased as shown in figure 1.

The same artificial contamination tests with insulators at a different ESDD level are performed. Figure 2 shows the waveform of AE signals of samples at a different ESDD level. The testing voltage is 10KV.

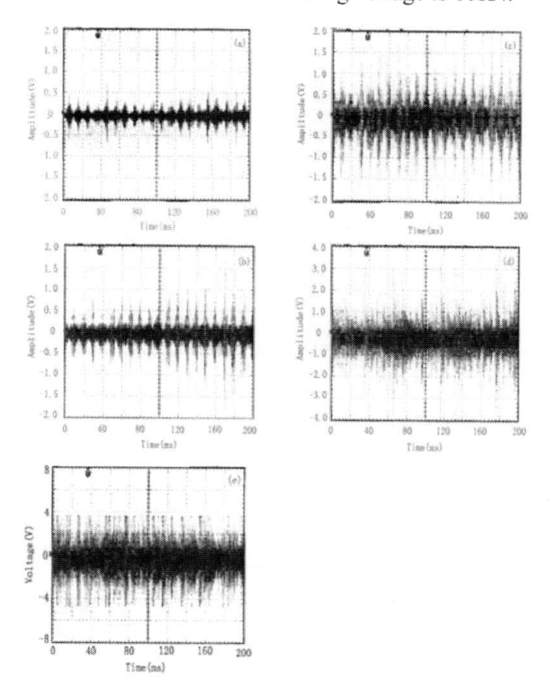

Figure 2. Waveform of AE signals of insulators at a different ESDD level.

At a light degree of contamination, the amplitude of the amplitude of AE in (a) was extremely small.

With the increase of ESDD level, the amplitude and frequency of AE signal are discontinuously augmented as shown in (b) and (c). When discharges occurred more and more, the AE signal becomes intense as shown in (d) and (e).

All testing results showed a close relationship between AE signals variations and imminence of flashover. With the increase of polluted degree, the AE signals accordingly increased. The amplitude of the signals became greater, and the wave cycle became a dense pulse. Therefore, it is valid to judge the intensity of contamination that caused flashover through detecting the AE signal. But it is not the simple linear relationship between the AE signals and the insulator polluted degree (or ESDD). They have a complicated nonlinear relationship. Figure 3 shows the two variable values of AE signals at four different ESDD levels (0.03, 0.07, 0.1, and 0.2 mg.cm^{-2}), where Umax is the maxim amplitude of signals amplitude and Smax is the maxim surface, that is, the maxim surrounding areas of the signal encircle line are in a power frequency period.

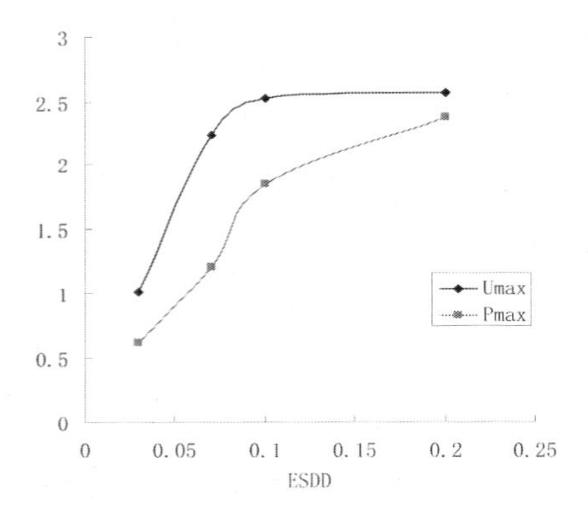

Figure 3. Characteristic values of AE signals at different ESDD.

5 PREDICTION MODEL OF INSULATOR ESDD

1 *Select the characteristic variations of the predicting model*: As shown in the testing results, the intension of insulator partial discharge is related to the amplitude and frequency of the AE signals. So, the following variables that are derived from the AE signals are chosen as the input variables of the predicting model, maxim amplitude, and maxim surface. Therefore, maxim amplitude is the peak value of signal amplitude under certain voltage and contamination severities. Maxim surface is the maxim surrounding areas of the signal encircle line in a power frequency period. And then, the insulator ESDD is chosen as the output variable.
2 *Pretreatment of data*: Normalize the input data using the pre-process function.
3 *Select the kernel function:* Choose Basis Function (RBF) Kernel as the kernel function.

$$K(x, x_k) = \exp\left\{-\frac{\|x - x_k\|^2}{2\sigma^2}\right\} \quad (9),$$

where σ is the kernel parameter.

4 *Confirm the model parameters:* The main parameters of the LS-SVM regre ssion model are the regularization parameter γ and the kernel parameter σ. The model parameters are tuned with the cross-validation method, and the feasibility of the model is proved by some data in the laboratory simulation. The simulation software is LS-SVMlab1.5.

First, a set of parameters is given, which are shown as follows:

$$\begin{aligned} S_\gamma &= \{800,500,200,10,5,1,0.5,0.2,\ 0.1,0.01\} \\ S_\sigma &= \{100,10,5,2,1,0.5,\ 0.2,0.1,0.05,0.01\} \end{aligned} \quad (10)$$

Then, we need to train the data with every giving γ_i and σ_j, and calculate the corresponding training relative error.

$$\varepsilon = \frac{1}{N}\sum_{i=1}^{N}\frac{|Q(i) - T(i)|}{T(i)} \times 100\% \quad (11)$$

Choose the parameters with the least relative error as the new reference parameter value and then give a new set of parameters.

Repeat the earlier work and get the optical model parameters. The eventual model parameters are chosen as $\gamma = 140$ and $\sigma = 1.1$, and the corresponding average training relative error is $e = 3.64\%$.

5 *Assessment results:* The feasibility of the model is proved by some data in the laboratory simulation. Table 1 shows the prediction of ESDD results of partial samples, and the average testing relative error is $e = 4.68\%$. Compared with the average training relative error, it can be seen that the LS-SVM predicting model is capable of learning quite well from the raw data samples while good prediction and generalization ability.

It can be seen from the predicting results that the relative error of some ESSD predicting data is large, such as in samples 2, 15, and 16. But our major concern is the polluted degree, so the predicting model can meet the practical monitoring request of polluted insulators.

Table 1. Testing results of the insulator ESDD with LS-SVM model.

Testing sample	Umax V	Smax V.mS	Measuring ESSD mg*cm^{-2}	Predicting ESSD mg*cm^{-2}	Relative error %
1	1.02	7.72	0.03	0.0309	3.00
2	1.84	20.30	0.15	0.1378	8.13
3	1.51	14.15	0.07	0.0713	1.86
4	2.14	23.82	0.20	0.1987	0.65
5	0.98	6.08	0.03	0.0310	3.33
6	1.24	12.10	0.07	0.0652	6.86
7	0.99	6.28	0.03	0.0305	1.67
8	1.14	11.80	0.06	0.0598	0.33
9	1.01	7.53	0.03	0.0295	1.67
10	1.70	18.33	0.10	0.1051	5.10
11	0.91	6.53	0.03	0.0294	2.00
12	1.50	16.33	0.08	0.0801	0.13
13	1.51	14.15	0.07	0.0706	0.86
14	2.18	23.65	0.20	0.1990	0.5
15	1.61	15.85	0.08	0.0732	8.50
16	2.04	22.82	0.18	0.1995	10.83
17	1.91	20.50	0.15	0.1396	6.93
18	1.73	16.76	0.08	0.0784	2.00
19	0.86	6.42	0.03	0.0238	2.67
20	2.06	21.59	0.20	0.1931	3.45

The optimized problem is translated into the solution of a linear equation group using the LS-SVM method. It is simple and suitable to select the model parameters through the cross-validation method. The gained optimal parameters make the LS-SVM realize the compromise between the model complication and the minimum error. Therefore, the prediction model based on the LS-SVM has good generalization ability.

6 CONCLUSIONS

To apply the AE technology in the monitoring of polluted insulator flashover, seeking a more effective processing method of AE signals is very important. Nowadays, the principal forecast methods of the insulator contamination degree are based on traditional statistics and neural network. Using the traditional statistics method, perfect results cannot be obtained when dealing with the small sample. And there is the problem of slow converge, structure selection, and local minimum in the neural network. The disadvantages mentioned earlier are avoided by using LS-SVM. Using the analysing and processing method of the LS-SVM, the ESDD of the polluted insulator can be predicted with high precision. Then, the degree of contamination can be assessed according to the insulator ESDD, for flashover prediction and condition-based maintenance. Testing results show that LS-SVM can realize a good forecast of polluted insulator ESDD, under finite sample condition. It produced a simple and valid method to monitor the insulator ESDD in line and had a certain practical application.

ACKNOWLEDGMENT

The authors would like to thank Wuhan High-voltage Research Institute of State Grid Corporation China for the use of their high-voltage facilities to conduct the testing presented in this paper, Z.D. Yang and M. Li for their parts in the testing.

REFERENCES

[1] He Wei, Chen Tao, Liu Xiaoming. (2000) Online monitoring system of faulty insulator based on non-touching UV pulse method. *Automation of Electric Power Systems*, 30(10):69–74.

[2] Fan Ju, Liang Xidong, Yin Yu, Wang Chengsheng, and Chen Ling. (2000) Application of acoustic emission technology on structure design and quality control of composite insulators. *Proceedings of the 6th International Conference on Properties and Applications of Dielectric Materials*, pp:358–361.

[3] J.A.K. Suykens, L. Lukas, J. Wandewalle. (2000) Sparse approximation using least squares support vector machines, *In Proceeding of the IEEE International Symposium on Circuits and Systems, ISCAS 2000*, pp:757–760.

[4] J.A.K. Suykens, T.V. Gestel, J.D. Brabanter, *et al.* (2002) *Least Squares Support Vector Machines*. Singapore: World Scientific.

[5] S.S. Mohamed, M.M.A. Salama, M. Kamel, *et al.* (2004) Region of interest based prostate tissue characterization using least square support vector machine LS-SVM. *Lecture Notes in Computer Science*, pp:51–58.

[6] Andrea Cavallini, S. Chandrasekar, Gian Carlo Montanari, Francesco Puletti. (2007) Inferring ceramic insulator pollution by an innovative approach resorting to PD detection. *IEEE Transactions on Dielectrics and Electrical Insulation*, 14(1):23–29.

[7] L.E. Lundgaard. (1992) Partial discharge XIV: Acoustic Partial discharge detection-Practical application. *IEEE Electrical Insulation Magazine*, 8(5):34–43.

Electric, Electronic and Control Engineering – Shao, Shu & Tian (Eds)
 ISBN: 978-1-138-02842-5

A Reliable Privacy-preserving Attribute-based Encryption

Tianfang Wang
Management Team of Postgraduate Students, Engineering University of CAPF, Xi'an Shanxi, China

Longjun Zhang
Department of Information engineering, Engineering University of CAPF, Xi'an Shanxi, China

Cheng Guo
Management Team of Postgraduate Students, Engineering University of CAPF, Xi'an Shanxi, China

ABSTRACT: This paper points out the existing security flaws in the attribute-based encryption: The legitimate users' identity privacy can easily be stolen by attackers through the DDH test. To solve this defect, a new Reliable Privacy-Preserving Attribute-Based Encryption (RPP-ABE) with enhanced security has been designed. After the analysis of the RPP-ABE, it's conceivable that this scheme can not only ensure the user's identity information not to be obtained by attackers via DDH test, but also improve the security flaws in the existing schemes, and the reliable access control is thus guaranteed. The experiments and analysis on RPP-ABE show that the memory overhead and computation overhead are relatively ideal.

KEYWORDS: Cloud computing, privacy preservation, ABE, encryption.

1 INTRODUCTION

Currently, the attribute-based encryption has already realized the data sharing in distrustful environment, and can effectively ensure the users' access control to the data within the scope of authority. In these schemes, the identity-based encryption designed by Waters in literature [1] has higher efficiency, but the identity privacy of legitimate users is easily stolen by attackers. Given this loophole, this paper explores and designs a Reliable Privacy-preserving Attribute-based Encryption (hereinafter referred to as RPP-ABE), which can protect the identity privacy of legitimate users from being stolen by attackers. Besides, the reliability analysis of this scheme shows that it is more reliable and securer on the access control of files compared with the scheme in the literature [2]. As to the memory overhead, the RPP-ABE scheme is consistent with the scheme in the literature [2] with fairly higher efficiency.

2 PREVIOUS SCHEMES AND ANALYSIS

The RPP-ABE designed in this chapter is improved based on the previous schemes which are briefly introduced and analyzed as follows.

2.1 *Waters' scheme*

Waters designs an efficient identity-based encryption scheme which can effectively ensure users' access control to files in the literature [1], but it may lead to the leakage of legitimate users' identity information.

There are three entities in this scheme: the Authority, the file owner and the user, wherein the Authority is the safety parameter of system and carries out management on the secret key. The identity information of the user is described by a bit string V with the length of n, wherein v_i represents the ith bit in the bit string, and the identity V of certain user is defined as $V = \{i \mid v_i = 1\}$, $V \subseteq \{1, 2, ..., n\}$.

This scheme is divided into 4 steps: Setup, KeyGen, Encrypy, and Decrypt. The specific algorithm is as follows.

1 *Setup*(1^λ): The Authority generates the system parameter through the security parameter 1^λ.

G and G_T are both the multiplication cyclic group with the order of prime P, wherein g is the generator of the group G and the bilinear mapping $e: G \times G \rightarrow G_T$. The Authority randomly selects an $\alpha \in Z_p$, supposes $g_1 = g^\alpha$, and randomly selects a $g_2 \in G$. In addition, the Authority randomly selects a $u' \in G$ and a vector $U = (u_i)$ with the length of n, wherein $u_i \in G, 1 \le i \le n$. In this way, the algorithm of public key is $PK =< g, g_1, g_2, u', U >$, and the algorithm of master key is $MK =< g_2^\alpha >$.

2 *KeyGen*(v, MK, PK): the generation of user private key.

The Authority randomly selects an $r \in Z_p$, then the private key generated for the user with the identity V is: $d_v =< d_1 = g_2^{\alpha}(u'\prod_{i\in V} u_i)^r, d_2 = g^r >$

3 *Encrypt*(PK, M, v): Realize the encryption of message M.

The encryptor randomly selects $t \in Z_p$. The cipher text obtained is: $C =< C_1 = e(g_1, g_2)^t M, C_2 = g^t, C_3 = (u'\prod_{i\in V} u_i)^t >$.

4 *Decrypt*(C, d_v): Decrypt the cipher text C.

Use the secret key $d_v =< d_1, d_2 >$ to recover the plaintext M from the cipher text C. The process is:

$$C_1 \frac{e(d_2, C_3)}{e(d_1, C_2)} = (e(g_1, g_2)^t M) \frac{e(g^r, (u'\prod_{i\in V} u_i)^t)}{e(g_2^{\alpha}(u'\prod_{i\in V} u_i)^r, g^t)}$$

$=$M

2.2 *Analysis of Waters' scheme*

In Waters' scheme, the user identity is represented by the bit string with the length of n, and the cipher text with the user identity takes on a one-to-one relationship. Due to the fixed length of the private key and cipher text, the matching operation during the encryption process is fixed for two times, so the scheme has fairly high efficiency.

Although the cipher text does not explicitly demonstrate the user identity that can be decrypted, for certain cipher text C, $(C_2, C_3) = (g^t, (u'\prod_{i\in V} u_i)^t)$ constitutes a DDH tuple. Due to its nature, the user identity V can be obtained from the cipher text. For example, for a user with the identity of v', there can be a similarly defined V', and the attacker can achieve his goal through the DDH tuple test, namely, through judging whether $e(C_2, u'\prod_{i\in V'} u_i) = e(C_3, g)$ can be established, to determine whether V' has been used in the cipher text, that is, to determine whether the cipher text can be decrypted by the user with the identity of v'. Because the number of the possibility for the user identity can be 2^n, the identity of the legitimate user that can decrypt the cipher text can be known through at least 2^n times of DDH tuple test. Therefore, the identity privacy of legitimate users in this scheme can be easily obtained by attackers.

3 DESIGN OF RPP-ABE

From the analysis we can see that although the Waters' scheme [1] can ensure the legitimate access control of users to files, it cannot effectively protect the identity privacy of legitimate users. This section, based on previous schemes, designs the RPP-ABE which greatly enhances the security of the system and improves the reliability of the scheme on the premise of protecting the identity privacy of legitimate users from being obtained by attackers.

3.1 *Relevant definitions*

In order to achieve the protection of the identity privacy of legitimate users, this scheme in this paper adopts the bilinear matching under the order group of composite number, whose definition is a little different from the common prime order group. The bilinear matching under the order group of composite number has been defined in the literature [3] for the first time.

To prove the security, the scheme in this paper has been reduced to the DBDH problem. The following will offer their definitions and explanations on the user attribute and file access structure involved in this scheme.

Definition 3.1 (bilinear matching under the order group of composite number)
Suppose p, r are two different primes, G, G_T are two multiplication cycle groups with the order of the composite number $N = p \cdot r$, and e is a bilinear mapping $e: G \times G \rightarrow G_T$, which has the following properties:

1 Bilinear: $\forall g, h \in G, a, b \in Z_N^*$, then $e(g^a, h^b) = e(g, h)^{ab}$
2 Non-degeneracy: $\exists g, h \in G$, and makes the order of e(g, h) is N in the group G_T;
3 Computability: Mapping can be computed efficiently in the polynomial time.

Use G_p, G_r to represent subgroups in the group G, and the order is p, r respectively. Suppose $G = G_p \times G_r$, and there is such property: if $h_p \in G_p$ and $h_r \in G_r$, then $e(h_p, h_r) = 1$

The proof is as follows: Suppose g, g_p, g_r are the generators of groups G, G_p, G_r respectively, then g^r can generate the group G_p, g^p can generate the group G_r, then, for certain $\partial_1, \partial_2 \in Z_N^*$, there are $h_p = (g^r)^{\partial_1}$ and $h_r = (g^p)^{\partial_2}$, then $e(h_p, h_r) = e((g^r)^{\partial_1}, (g^p)^{\partial_2}) = e(g^{\partial_1}, g^{\partial_2})^{pr} = 1$

Definition 3.2 (DBDH problem under the order group of composite number)
Suppose bilinear mapping $e: G \times G \rightarrow G_T$, G, G_T are the two multiplication cycle groups with the order of composite number $N = pr$, and p, r are two different primes, G_p, G_r are subgroups of the group of G. The orders are respectively p, r, marked as $G = G_p \times G_r$. Select a generator $g_p \in G_p, g_r \in G_r$, and the random number $a, b, c \in Z_N^*$. Judge a six tuple $(g_p, g_r, g_p^a, g_p^b, g_p^c, Z)$ as the input, and see whether $Z = e(g_p, g_p)^{abc}$ is established.

To define an algorithm $\Re$, the advantage σ can be used to solve the DBDH problem. If

$Adv(\Re) = |\Pr[\Re(g_p, g_r, g_p^a, g_p^b, g_p^c, e(g_p, g_p)^{abc}) = 0] -$
$\Pr[\Re(g_p, g_r, g_p^a, g_p^b, g_p^c, e(g_p, g_p)^z) = 0]| \geq \sigma$

Wherein $z \in Z_N^*$. If you can't find an algorithm of polynomial time to solve the DBDH problem through at least the advantage σ, then the DBDH problem is thorny.

3.2 *Basic framework design of the scheme*

The entities involved in this scheme can be divided into four parties, respectively the Trusted Authority (TA), Data Provider (DP), Data Demander (DD) and Storage Center (SC).

As a trusted third party, TA plays a key role in the scheme. It is responsible for the initialization of the system, the management of system properties and user property, and the issue of corresponding private key for users according to the user's attribute set.

As the owner of data, DP encrypts the data and then transfers them to the SC. DD is the user who applies to access the data; TA issues the user's private key; DD decrypts and then obtains the data through using the private key after acquiring the data cipher text from the SC. DD and DP are collectively referred to as the user, and a DP often can also become a DD, and vice versa.

SC is responsible for the storage of mass data information. Due to the weak safety protection of SC, in many cases, it cannot be trusted. Therefore, in this scheme, what SC stores is the encrypted data cipher text. Even if it is attacked, the attacker cannot obtain the data plaintext.

The basic framework of the RPP-ABE is shown in the Figure 1.

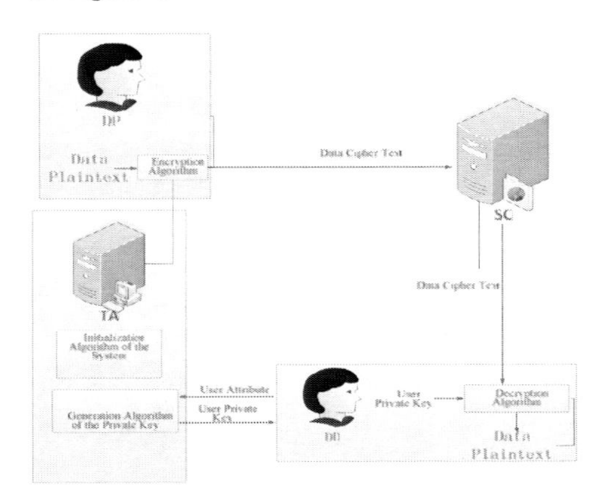

Figure 1. The basic framework of the RPP-ABE.

The basic process of the RPP-ABE is as follows:

1 Setup: It is executed by TA which generates the group and bilinear equivalent algebraic structure that the scheme needs according to the security parameter. Meanwhile, TA generates the public key (PK) and master key (MK) of system, wherein PK is open to all users and MK, as shall be kept secretely by TA.
2 Key Generation: It is executed by TA which uses PK and MK to generate private key for all users according to the attribute uploaded by the user.
3 Encryption: It is locally executed by DP which uses public key *PK* and the access structure of files to realize the encryption of data. Then DP sends the cipher text of files to SC.
4 Decryption: It is executed by DD which obtains the required data cipher text from the SC. DD uses the decryption cipher text of user private key obtained from TA to get the data plaintext.

The basic process of this scheme is shown as Figure 2.

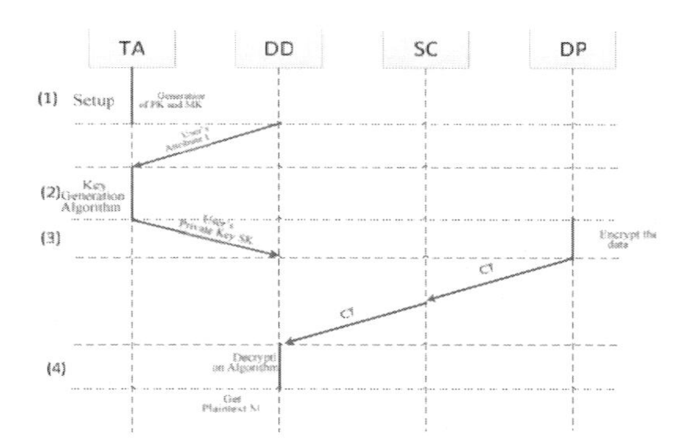

Figure 2. Scheme flow.

3.2.1 *Setup*

Setup algorithm is executed by TA, which selects the group that the scheme needs, random number and bilinear equivalent algebraic structure, to generate the public key (PK) and master key (MK) of system. PK is open to all users, while MK, as secret parameters, shall be kept solely of by TA.

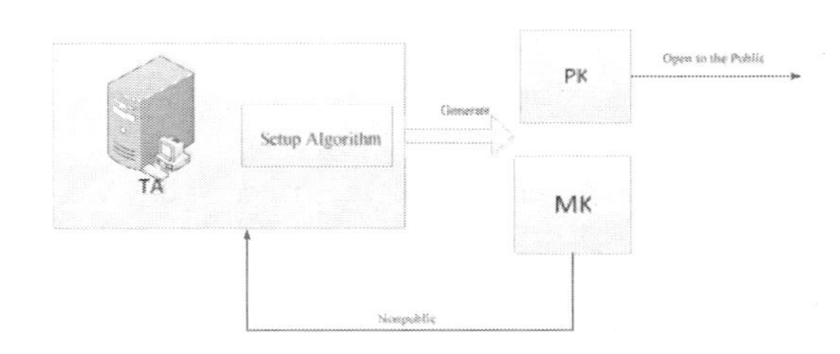

Figure 3. Setup.

The specific algorithm for $Setup() \rightarrow \{PK, MK\}$ is as follows,

1 Construct a group G with an order of composite number N, wherein $N = p \cdot r, G = G_p \times G_r$ and G_p, G_r are respectively the multiplication cyclic groups of prime order P and r. The generators of subgroups G_p and G_r are respectively marked as g_p and g_r.
2 Randomly selects, $t \in Z_N^*$, and calculate $\partial = s \cdot t, \partial \in Z_N^*$, and the result will be marked as $g_1 = g_p^{\partial}$. Randomly select $g_2 \in G_p, g_3 \in G_r$ $h \in Z_N$, mark them as $g' = g_p^h$, and calculate $X = e(g_1, g')$. Uniformly select $a_i \in Z_N^* (1 \le i \le n)$ for each attribute in the system, where n is the number of attributes in the system, and then calculate $A_i = g_2^{a_i + a_i^2}$, wherein $1 \le i \le n$.
3 Generate system public key $PK =< g_p, g_r, s, t, g_3, X, \{A_i\}_{1 \le i \le n} >$, and master key $MK =< g_2, h, \{a_i\}_{1 \le i \le n} >$.

3.2.2 *Private key generation*
Private key generation is executed by TA. After a user sends his attribute list L to TA, TA uses PK and MK to generate private key SK, and sends it to the user.

The following is the specific algorithm $KeyGen(PK, MK, L) \rightarrow SK$ for private key generation,

1 Set users' attributes as $L = \{v_1, \cdots v_n\}$, and then TA calculates $\mu = \sum (a_i + a_i^2)$ for a user's attribute L.
2 TA randomly selects $t \in Z_N^*$, and calculates $D_0 = g_p^t$ and $D_1 = g_p^h \cdot g_2^{\mu}$. The user's corresponding private key is $SK =< D_0, D_1 >$.
3 TA shall send the private key SK to the corresponding user through a secure channel.

3.2.3 *Encryption*
Encryption of file is executed by DP, and DP uses public key PK and the access structure of file W to realize the encryption of the plaintext of file. Then DP sends the cipher text of file that it gets to SC. The specific algorithm for key encryption algorithm is as follows:

1 The access structure of file is $W = \{w_1, \cdots w_n\}$, in which $w_i \in S_i$.
2 The plaintext that needs to be encrypted is M. Combined with the access structure of plaintext M, calculate $C_0 = g_p^{\partial}$ and $C' = M \cdot e(g_1, g')$, obtain the corresponding part from the public key, and calculate $C_1 = (\prod_{v_i \in W} A_i)^s \cdot g_3$.
3 After each plaintext is encrypted, the cipher text obtained is $CT =< C', C_0, C_1 >$, and then DP will send the cipher text CT to SC.

3.2.4 *Decryption*
When the data demander (DD) needs to access a file, firstly he should obtain cipher text CT of the file from SC, use the private key obtained from TA to decrypt the cipher text, and in the course of decryption, the bilinear operation is used. Whether the user can use his private key to decrypt the cipher text of file successfully, lies in whether the user's attribute can meet the access structure of file.

The decryption algorithm $Decrypt(CT, SK) \rightarrow M$ is as follows:

$$M = C' \cdot \frac{e(D_0, C_1)}{e(D_1, C_0)}$$

Any user, who has the decryption ability, namely user's attribute meets the requirements to visit the access structure of file, can use the decryption algorithm, and use his private key to decrypt the cipher text of file successfully.

4 ANONYMITY ANALYSIS OF SCHEME

In section 1, it is elaborated that waters' scheme fails to achieve anonymity, which means that the attacker can obtain valid users' identity information by DDH test.

In the scheme designed in this paper, due to the fact that an additional multiplication with the elements in group G_r is added in the cipher-text component (C_0, C_1), the cipher text is $C_0 = g_p^{\partial}, C_1 = (\prod_{v_i \in W} A_i)^s \cdot g_3$, instead of $C_0 = g_p^{\partial}, C_1 = (\prod_{v_i \in W} A_i)^s$. If the latter is the cipher text, through selecting a certain access structure W' and judging whether $e(C_0, \prod_{v_i \in W'} A_i) = e(C_1, g_p)$ is true, the attacker can verify whether W' and W are consistent, and through continuous tests, the attacker can eventually obtain the identity information of user's who can visit the file validly; But if the cipher text is $C_0 = g_p^{\partial}, C_1 = (\prod_{v_i \in W} A_i)^s \cdot g_3$, then DDH test attacks are invalid, because even the right access structure W is used for test, the DDH test $e(C_0, \prod_{v_i \in W} A_i) = e(C_1, g_p)$ is false.

Therefore, this scheme can resist the attackers' DDH test attacks, so as to hide the access structure of file effectively, and ensure that the valid users' identity privacy will not be obtained by attackers.

5 EXPERIMENT SIMULATION

Through experiment simulation, the RPP-ABE scheme and the existing ABE scheme are compared in computation overhead and memory overhead respectively, to verify the performance of the RPP-ABE scheme.

5.1 *Experimental environment*

The experimental devices are Intel (R) Core (TM) 2 DOU processor, 2.80GHz, 4.00GB memory, and 2TB HP 7200r/s SATA hard disk, with the kernel version of 2.6.31Ubuntu10.04 operating system. The experiment is based on the open source code repositories PBC, and uses the toolkit of CPABE-0.10 in PBC to write the program.

5.2 *Experimental results and analysis*

In the existing and typical CP-ABE scheme, the memory overheads in literature [4] and [5] are ideal, as its user's private key and master key are short in length; the computing overheads in literature [2] and [5] are ideal, as the encryption time and decryption time are short. So compare the RPP-ABE scheme with literature [4] and [5] respectively in the lengths of private key and master key to measure the scheme's memory overhead; compare the RPP-ABE scheme with literature [2] and [5] respectively in encryption time and decryption time to measure the scheme's computation overhead. In the experiment, the number of values for each attribute is uniformly set as 4.

1 The comparative analysis on memory overhead

Compare the RPP-ABE scheme with schemes designed in literature [4] and [5] separately in the length of private key, and with the increase in the number of system attributes, the comparison in the length of private key for three schemes is shown in Figure 4.

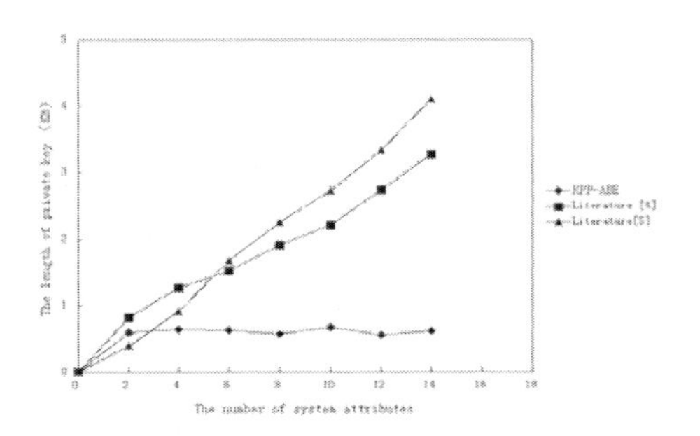

Figure 4. The comparison in the length of private key for three schemes.

The length of private key in RPP-ABE scheme basically remains unchanged with the increase of the number of system attributes, so it is constant, which is independent from the number of system attributes; both the lengths of private key in schemes of literature [4] and [5] grow linearly with the increase of system attributes. Therefore, the scheme in this paper is superior to those in literature [4] and [5] in the length of private key.

Compare the length of master key in RPP-ABE scheme and CP-ABE scheme designed in literature [2] and [5] respectively, and according to the different numbers of system attributes, the comparison in the length of master key for the three schemes is shown in Figure 5.

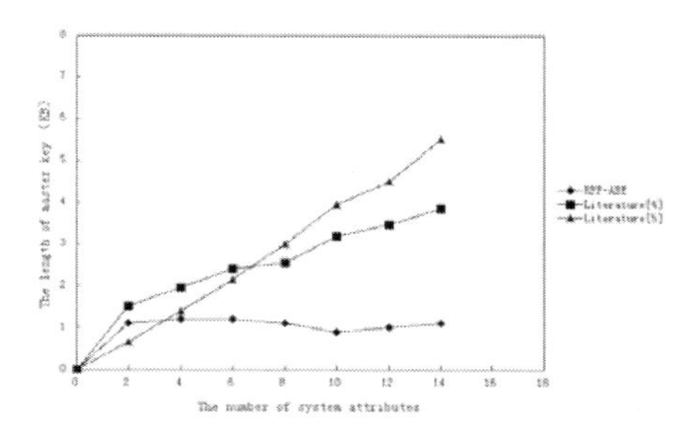

Figure 5. The comparison in the length of master key for three schemes.

As seen from Figure 5, the length of master key in RPP-ABE scheme basically remains unchanged with the increase of the number of system attributes, so it is constant; both the lengths of master key in schemes of literature [4] and [5] grow linearly with the increase of the number of system attributes. Therefore, the scheme in this paper is superior to those of literature [4] and [5] in the length of master key.

Therefore both the lengths of private key and master key in RPP-ABE scheme are superior to those in schemes of literature [4] and [5].

2 The Comparative Analysis on Computation Overhead

Respectively use the encryption algorithm in RPP-ABE scheme and the encryption algorithm in scheme of literature [2] and [5] to encrypt the data file of 128KB, and the relation between the required encryption time and the number of system attributes is shown in Figure 6.

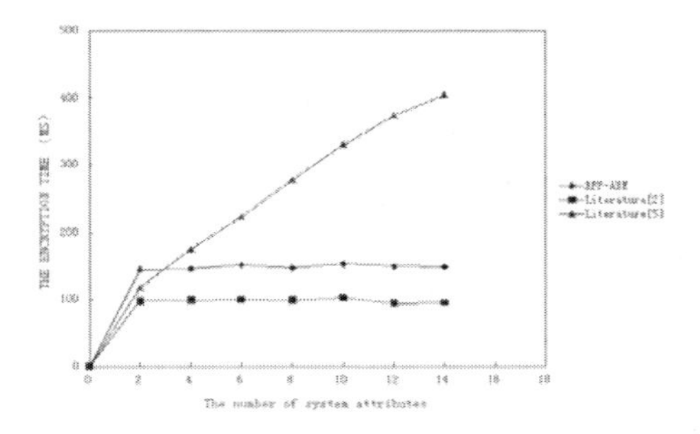

Figure 6. The comparison in encryption time for different schemes.

As seen from Figure 6, the encryption time in RPP-ABE scheme and the scheme of literature [2] has no relation with the number of system attributes, so it is constant; the encryption time in the scheme of literature [5] grows linearly with the increase of the number of system attributes and its encryption time is obviously longer than the above mentioned two. Although the encryption time in RPP-ABE scheme is a little longer than that in the scheme of literature [2], due to the fact that both of them are in the constant-level, the difference between them in cloud computing environment can be ignored.

Respectively use the decryption algorithm in RPP-ABE scheme and the encryption algorithm in scheme of literature [2] and [5] to decrypt the cipher text of 64KB, and the relation between the required decryption time and the number of system attributes is shown in Figure 7.

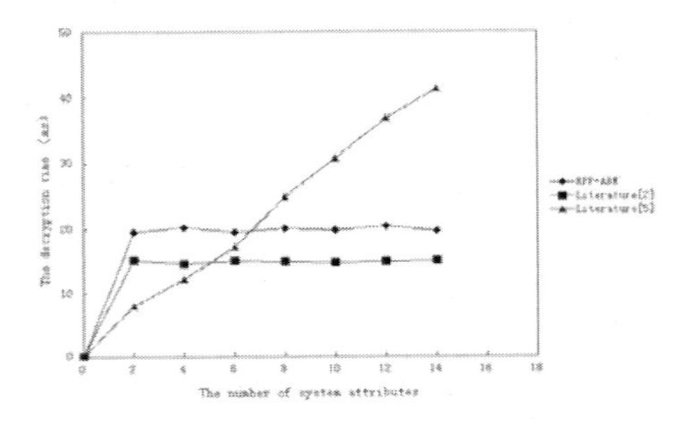

Figure 7. The comparison in decryption time for different schemes.

As seen from Figure 7, the required time for decryption of ciphertext in RPP-ABE scheme and the scheme in literature [2] has nothing to do with the number of system attributes, so it is a constant; while the required time for decryption of ciphertext in literature [5] grows linearly with the increase of the number of system attributes; when the number of system attributes is within 6, the decryption time of literature [5] is ideal, but when the number of system attributes is more, the decryption time of literature [5] is much longer than the time in the RPP-ABE scheme and the scheme of literature [2].

Through the experiment, in which the scheme in this paper and the rest CP-ABE schemes are compared in the performance, it is can be seen that on memory overhead, the scheme in this paper is superior to the other schemes; on computation overhead, the schemes in this paper and literature [2] are superior to the scheme in literature [5]. Although the computation overhead of the scheme in this paper is slightly larger than that of the literature [2], both of them are in constant-level, so the difference is little.

6 CONCLUSIONS

Security issues of cloud storage influence the development of cloud storage applications, and the reasonable and effective encryption scheme not only can guarantee the security of data stored in the cloud, but also should ensure the anonymity of the data owners and users, thereby improving the cloud storage service users' trust in the cloud storage service. With the introduction of encryption scheme based on CP - ABE algorithm, this paper ensures the confidentiality of user data, at the same time makes some attempts to realize the protection of user's private data, and the experimental results show that, the mechanism can ensure the security of data, and at the same time guarantee user's private data from leakage. Compared with the existing schemes, the various performance of this scheme is more ideal. In the follow-up work, the method will be improved to support more flexible access control policy.

REFERENCES

[1] Goyal V, Pandey A, Sahai A, Waters B. (2006) Attribute-Based encryption for fine-grained access control of encrypted data. In: Juels A, Wright RN, Vimercati SDC, eds. *Proc. of the 13th ACM Conf. on Computer and Communications Security, CCS 2006*. Alexandria: ACM Press, pp:89–98.

[2] Xiaohui Li, Dawu Gu, Yanli Ren, Ning Ding, Kan Yuan. Efficient ciphertext-policy attribute based encryption with hidden policy. In: Y. Xiang et al. (eds.) *Proc. of 5th Internet and Distributed Computing Systems (IDCS), 2012, LNCS*, (7647):146–159.

[3] Sims K. (2007) IBM introduces ready-to-use cloud computing collaboration services get clients started with cloud computing. Available from: http://www-03.ibm.com/press/us/en/pressrelease/22613.

[4] Cheung L, Newport C. (2007) Provably secure ciphertext policy ABE. *Proceedings of the 14th ACM Conference on Computer and Communication Security*, pp:456–465.

[5] Nishide T, Yoneyama K, Ohta K. (2008) Attribute-Based encryption with partially hidden encryptor-specified access structures. In: Bellovin SM, Gennaro R, Keromytis A, Yung M, (eds). *Proc. of the Applied Cryptography and Network Security*. Berlin, Heidelberg: Springer-Verlag, pp:111–129.

Electric, Electronic and Control Engineering – Shao, Shu & Tian (Eds)
© 2015 Taylor & Francis Group, London, ISBN: 978-1-138-02842-5

Research on the agility of C2 organization

Shuai Chen
Graduate Management Unit No.14, CAPF Engineering University, Xi'an, China

Xuejun Ren
Department of Information Engineering, CAPF Engineering University, Xi'an, China

Wenjing Shao
Graduate Management Unit No.13, CAPF Engineering University, Xi'an, China

ABSTRACT: This paper researched on agility of C2 organization, imagined a battle case and assign tasks between platforms by improved MPLDS algorithm, proving the reliability of the result, changing the task demand resources randomly, through the analysis of the case it can be concluded that agility and task finishing rate have a certain positive correlation.

KEYWORDS: Command and control; C2 organization; agility; adaptability; positive correlation.

1 INTRODUCTION

The concept of Agile C2 Organization was first brought about by Alberts & Hayes [1] in 2003. Adam Forsyth et al. [2] dug into the issue of information refinery for a C2 organization featured with its agility. Lenard Simpson et al. [3] analyzed the features of an agility-based organization network and issues like dynamic troop distribution, etc. Reiner K. et al.[4] performed qualitative analysis on the improvements in the agility of a C2 organization in terms of the role and influence of individuals in an organization. However, the construction of an agile C2 organization and the process of such construction are still at the early conceptual and experimental stages. This paper is presented in an attempt to provide a new approach of agile organization design on basis of conceptual and case analyses.

2 C2 ORGANIZATION

2.1 *Concept*

A C2 organization represents a command and control architecture designed for a particular battlefield environment, completed by the chain of command and control between commanding nodes (i.e. platforms, resources and decision-making entities) and the whole operation process to complete the mission. [5]

2.2 *Agility of C2 organization*

The modern C2 combat is a complex and dynamic process full of uncertainty. Military organizations must improvise new methods and measures, and put them into work immediately for the achievement of their goals. The Canadian armed forces believe a more agile C2 organizing system should be established in cope with diversified military operations.

Agility represents an organization's ability to maintain its efficiency to an acceptable level in the changed environment. Agility is defined by NATO SAS-085 in its research works as "the ability to influence, deal with and/or take advance of varied situations". Albert and Hayes in their *Power to the Edge* believed that agility was achieved by the coordination of 6 properties: Robustness, Elasticity, Responsiveness, Flexibility, Innovativeness and Adaptability.

Robustness: "An ability to maintain efficiency in diversified missions and under diversified situations [6] ", described as:

$$O_{rob} = \frac{\sum_{i}\left(N_{r_i} / N_{\max_{r_i}}\right)}{n} \quad (1)$$

Suppose an organization O_f is in possess of a resource set $\{(r_i,N_{ri}),\cdots(r_n,N_{rn})\}$, representing the types and capability of each resource. The task resource demand and resource capability are described as $\{(r_i,Nmax_{ri}),\cdots(r_n,Nmax_{rn})\}$. So we have the ability of O_f to successfully deal with changes in missions in progress as O_{rob} (i.e. Robustness), which determines the ability of a C2 organization to complete the proposed mission and to cope with emergency situations key to the accomplishment of missions in the environment of uncertainty. It should be pointed out that robustness is defined by the ability taking effect in a certain scope, instead of a certain mission. An non-robust organization relies on the known and estimated mission environment parameters, which in most cases are time variant and highly likely to result in the structural failure of its functionality and the suspension and even total failure of ongoing missions[7]-[10].

Adaptability: "An ability to change the working process or organizational structure", described as:

$$O_{ada} = A_{exch}\left\langle W_{str}\ or\ W_{pro}\right\rangle \tag{2}$$

Pulakos et al. described adaptability as the ability to take effective action in necessary cases in respond to the unforeseeable events of situations, in order to make changes of the organizational structure or the operation schemes correspondingly. In the last 15 years, theoretical and empirical researches have been conducted on issues of adaptability. Stagl et al. pointed out that the development and changes in an organization's structure, ability, behavior and/cognitive activities were proofs of an organization's adaptability; Lepin carried on from Stagl's research and observed the adaptability of roles in an organization, suggesting the "efficiency of an organization can be improved by adjusting the roles of its members responsively and irregularly". His research suggested that the connection between the composition of a team (e.g. cognitive ability and openness) and its efficiency was strengthened by the adaptability of its role structure when faced with unexpected situations in its missions.

Responsiveness: "An ability to react to the changed environment", described as:

$$O_{resp} = \left\langle A_{res}; A_{subj}\right\rangle \tag{3}$$

A_{res} represents the existing resource capability, and A_{subj} the objective capability of the decision-making entity. A decision-making entity makes objective judgments on battlefield situations, mobilizes the troops at hand for tasks, and regulates its tasking process and personnel structure immediately when the battlefield environment changes, in order to maintain the combat capability of the troops and the continuity of its missions. Responsiveness indirectly reflects the resource of an organization to carry out missions and the ability of a decision-making entity to make decisions.

Flexibility: "The ability to achieve a goal through different approaches and to switch seamlessly between different ways", described as:

$$O_{flex} = A_{exch}\left\langle M_{con}; M_{recon}; M_{syn}; M_{coll}; M_{bor}\right\rangle \tag{4}$$

NATO SAS-065 in their maturity model specified five C2 methods, i.e. Conflicting C2 (M_{con}) (a C2 without a group, or the member entity of the C2 behaves independently from the group), Reconciling C2 (M_{recon}) (AC2 group with each of its member entities actively avoiding conflicts in their intents, plans or actions, in order to avoid the negative effect among the entities, for which they constantly take efforts to identify the potential conflicts. The proper allocation of missions, time, space and/or resources may be a viable way to avoid conflicts), Synchronized C2 (M_{syn}) (A C2 organization behaves collectively under an shared intent by synchronizing the various intents of each member entity, in order to achieve the uniformity in its actions, for which the member entities must take the overall efficiency of the organization into account), Collaborating C2 (M_{coll}) (A C2 organization with all member entities sharing a plan and a unified intent, seeking the maximization of the organization's overall efficiency), and Border C2 (M_{bor}) (A C2 organization with its functionality relying on its self-synchronization and the self-organization of its member entities toward a unified intent, with all its entities exerting themselves for the efficiency of the organization), among which M_{bor} requires the highest level of organization and infrastructure standards, and also provides the highest mission capability.

Elasticity: "An ability to operate in a slightly downgraded manner, and to recover its functionality after being damaged", described as:

$$O_{ela} = \left\langle A_{de}\ and\ A_{re}\right\rangle \tag{5}$$

Especially

$$\begin{cases} A_{de_i} < A_{re_j}\ \ when\ E_{plat_i} > E_{plat_j} \\ or \end{cases} \tag{6}$$

The decision-making entity and platforms of a traditional organization may change overtime as the mission progresses, resulting in the degradation or structural incompleteness of the organization, compromising the organization's ability to carry on its mission, or the total loss of functionality due to enemy attack, which is exactly a C2 organization is trying to avoid, which is able to reshape itself immediately in situations like organizational degradation or damage for the requirements of new missions it undertakes. Meanwhile, the organizational elements (i.e. decision-making entities, platforms and tasks) in the volatile battlefield environment may change dynamically overtime, with the status and structure of the organization changed along with for the optimal organization-mission match[11]-[12].

Innovativeness: "An ability to do new things, or that of doing experienced th`vings in a new way", described as:

$$O_{inno} = \left\langle A_{subj}; A_{res}\right\rangle \tag{7}$$

Innovativeness represents the willingness of the commander or the commanding element to seek ways to accomplish its mission in existing mission environments, which relies on the initiative of the commander or commanding element, as those with higher initiative are more innovative, and their initiative will make difference in organizational structure and procedures, bringing their activities with totally different results.

The agility of a C2 organization is defined by all these features as:

$$O_{agi} = \langle O_{rob}, O_{ada}, O_{resp}, O_{flex}, O_{ela}, O_{inno} \rangle \tag{8}$$

And the agile ability is defined as:

$$A_{ali} = W_{abi} \langle T; D; (\sum R * W_R) \rangle \tag{9}$$

In which W_{abi} represents the weight of ability, T the ability of the organization, D the (objective) decision-making ability, the R resource ability, and W_R the weight of resources. The ability of a C2 organization to undertake diversified tasks, and restructure itself when facing the changed battlefield environment for the accomplishment of its missions, is determined precisely by its agility.

3 CASE ANALYSIS

This case designs a contingent of 14 communication teams handling 12 task missions, and there are five decision-making entities, which assign tasks between platforms by improved MPLDS algorithm, the final task - contingent allocation results are shown in figure 1.

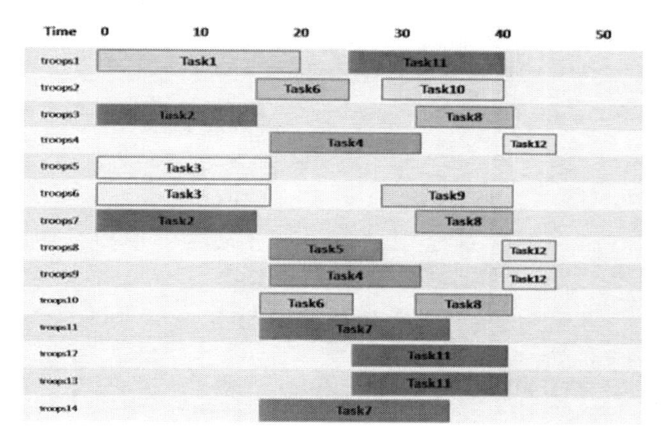

Figure 1. Final task - contingent allocation results.

Changing the task demand resources randomly, we get 4038 task variant samples, considering the team collaboration, we produce 61 new team establishments, respectively for the 75 communication units to carry out, the team can perform tasks with samples of all tasks to do, it is concluded that the task completion, stack, and agility, the results are shown in figure 2:

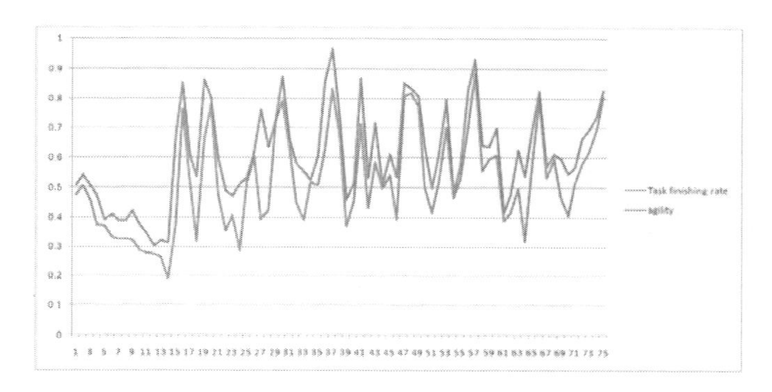

Figure 2. Task finishing rate-agility.

By the graph, you can see that the distribution of task completion and agility show the same trend, having a positive relationship, at the same time, for relatively abundant resources division, sample task completion rate and agility of the sample distribution is more concentrated, we can see that the organizational agility metrics reflects task progress and result forces organizations' duration to the changes the tasks have.

4 CONCLUSIONS

The C2 organization with high agility is capable of performing diversified tasks in volatile battlefield environments, and is able to restructure itself to the requirements of particular battlefield conditions and combat missions, optimizing the task force (including the resource allocation, force deployment and targets to strike), in order to deal the deadliest attack to the enemy at minimum cost, which is believed to be the major combat organization pattern in future military conflicts where the C2 organization will play an irreplaceable role.

REFERENCES

[1] S. Alberts, R. E. Hayes. (2003) *Power to the Edge: Command and Control in the Information Age* (Information Age Transformation Series), Washington, D.C.: CCRP Publications.

[2] Ford, David N., Voyer, John J., Wilkinson, Janet M. Could. (2000) Building learning organizations in engineering cultures: case study. *Journal of Management in Engineering*, 16(4):72–83.

[3] Dr Anthony H. Dekker. (2012) Analyzing team C2 behavior using games and agent. *CCRP17th*.

[4] Reiner K. Huber. (2012) Achieving agile C2 by adopting higher levels of C2 maturity. CCRP17th.

[5] Baoxin Xiu, Bin Li, Weiming Zhang, Zhong Liu, Dongsheng Yang, Jincai Huang. (2011) Robust design of C2 organizational structure. *Fire Control and Command Control*, (7).

[6] David S. Alberts. (2011) *The Agility Advantage: A Survival Guide for Complex Enterprises and Endeavors*, Washington, D.C.: CCRP Publications.

[7] Levchuk G M, et al. (2003) Congruence of human organizations and missions; Theory versus data. In: *International Command and Control Research and Technology Symposium. Washington, DC. June. 2003*, pp:368.

[8] Entin E. et al. (2003) When do organizations need to change-Part two: Incongruence in action. In: *International Command and Control Research and Technology Symposium, Washington, DC, June, 2003*, pp:585.

[9] Diedrich F, et al. (2003) When do organizations need to change-Part one: Coping with organizational incongruence. In: *International Command and Control Research and Technology Symposium, Washington, DC, June, 2003*, pp:125.

[10] Kleinman D L, et al. (2003) Scenario design for the empirical testing of organizational congruence. In: *International Command and Control Research and Technology Symposium, Washington, DC, June, 2003*, pp:324.

[11] Haixiao Liu. (2001) *Design and Adjust Methodology of Command and Control Organization*. Changsha: National University of Defense Technology.

[12] Qiang Sun, Dongsheng Yang, Weiming Zhang. (2011) Understanding and comprehend about command and control and related terms. *Fire Control and Command Control*, (7).

Electric, Electronic and Control Engineering – Shao, Shu & Tian (Eds)
 ISBN: 978-1-138-02842-5

Experimental research and analysis on acoustic emission from polluted insulator discharge

Hongling Li, Naiqiu Shu & Xishan Wen
School of Electrical Engineering, Wuhan University, Wuhan, China

ABSTRACT: When a partial discharge occurs in a pollution insulator, there is an instantaneous release of acoustic energy. By detecting the sound signals emitted by polluted insulator discharge, the pollution severities of insulators in-service can be monitored and accordingly threaten the insulation of the power grid, which is evaluated. In order to verify the feasibility of the Acoustic Emission (AE) method, numerous polluted insulators discharge is researched systematically. The detection principle and the apparatus of the AE method are introduced. Several contamination experiments of two-unit insulator strings and long insulator strings were performed, and real discharge AE signals were collected. Then, the common characteristics of AE wave were summarized. The AE waves have the same developing trend. With the increase of the testing voltage, the amplitude and frequency of the AE signals are increased correspondingly. It is proved that there is a corresponding relationship between the AE signals emitted by the polluted insulators and the development of contamination discharge. Experimental and analyzing results indicated that the polluted insulator strings flashover process can be monitored by acoustic detection apparatus, for flashover prediction.

KEYWORDS: *Acoustic emission (AE); acoustic transducer; monitoring; polluted insulator; partial discharge.*

1 INTRODUCTION

To prevent polluted insulator flashover accidents, lots of monitoring methods have been adopted, such as the equivalent salt deposits density (ESDD) method, the surface conductance, the leakage current, and so on. The ESDD is the source of defining pollution classes and mapping pollution areas. However, it is time consuming and difficult to automate [1]. The surface conductance can generally reflect the pollution and wetting degree of the insulator surface. But it is difficult for conducting measurement. Leakage current measurement is also popular. It is shown that the leakage current surge count for an outdoor insulator can provide indications about the amount of conductive pollution. However, the leakage current is easily affected by the environment. The application of the leakage current measurement also usually requires reconstructing insulators in order to install highly specialized current transducers, which causes difficulty in installation and extends the apparatus purchase. [2]

With the rapid development of online monitoring technology, existing methods have been consummated continuously, and new methods, such as acoustic ultrasonic detection [3], ultra-violet pulse method [4], and laser-induced fluorescence (LIF) spectroscopy [5], have also been introduced continuously. All these methods have their advantages and inherent shortcomings.

In this paper, the acoustic emission technology (AE technology) is presented with regard to polluted insulator discharge detection. The AE method is one of the non-destructive diagnostic methods used to measure, detect, and localize partial discharge in insulation systems of power facilities. [6, 7] The AE method can be made immune to electromagnetic interference. The acoustic partial discharge detection has been applied to SF6 insulated switchgears (GIS) [8], capacitors [9], transformers, [10] and so on.

In recent years, several studies have been performed for the application of the AE method in insulator monitoring [11, 12]. But the characteristics of AE signals generated from the polluted insulators are still largely unexplored. Therefore, an in-depth research of the relevant acoustic theory is required.

The monitoring principle and measuring apparatus of the AE method are introduced in this paper. The main experimental results of AE signals from polluted insulator discharge are also given. Finally, an analysis and discussion of these results are presented.

2 DETECTION PRINCIPLE AND APPARATUS OF AE METHOD

2.1 *Detection principle*

It is generally considered that the process of insulator pollution flashover can be divided into the following four phases macroscopically: deposition of airborne particles on insulator surface, wetting of the insulator surface, appearance of dry bands and partial discharge, and development of partial discharge and complete of flashover. Therefore, deposition of contamination and wetting can be considered the causation of contamination flashover. The appearance of partial discharge and development is the process of discharge. It can be seen that contamination flashover is a continuous development process of the partial arc. That is, the appearance of contamination flashover is a gradational process. So, it is feasible to continually online monitor the discharge ultrasonic signals emitted by polluted insulators, for flashover prediction and condition-based maintenance.

From the point of view that discharge is the release of energy, acoustic, heated, electromagnetic, and optical energy are the portion of the released energy. It is considered that it is proportionate between the acoustic energy and the general released energy in [7]. However, practically the proportion is uncertain, because the discharge is affected by several factors. The proportion relationship is still affirmative from the statistic point of view.

2.2 *Detection apparatus unit*

The AE detection apparatus is shown in figure 1. It consists of a broad bandwidth acoustic transducer, the signal processing unit, and the oscillograph. The AE signals from the discharge source are converted into electrical signals by the transducer. By using the broadband acoustic transducer, the spectrum analyse of the AE signal can be carried through, which is redound to an in-depth understanding of the characteristics of AE signals and the process of discharge. The converted electrical signals are amplified and processed to the usable voltage levels by the signal processing unit. Then, the processed signals are processed and displayed in the oscillograph.

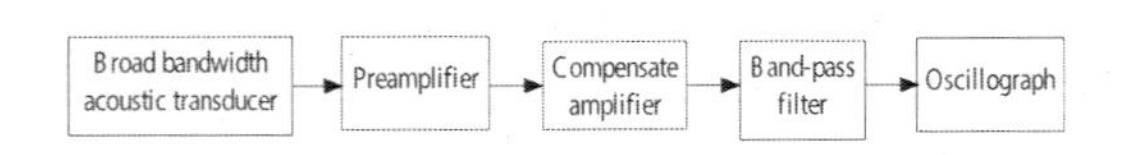

Figure 1. Sketch of AE detection apparatus.

It is well known that acoustic waves radiate in all directions from the discharging source, and acoustic losses result in exponential decreases of intensity with distance along the ultrasonic wave propagation path as it moves away from the point source. In order to fit the signal level of the AE detection apparatus, the studied acoustic transducer is combined by a piezoelectric sensor and focused paraboloid. As shown in figure 2, the Piezoelectricity ceramic wafer is used as the sensor of the studied transducer because of its merits of small volume, good performance, and low cost. And the focused paraboloid can flock together sound waves to strengthen sound intensity from a long distance at focus. So the studied acoustic transducer has a high sensitivity.

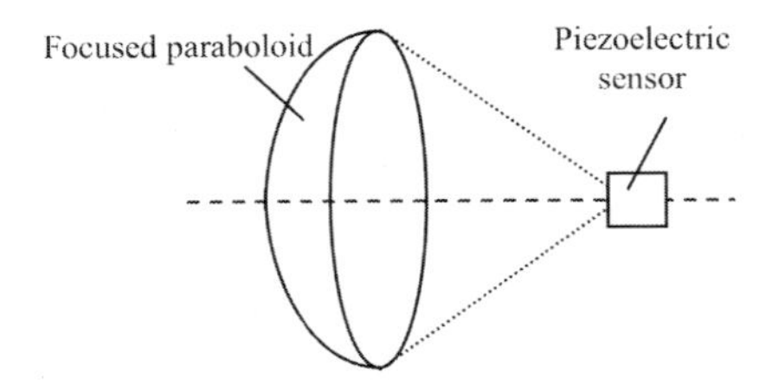

Figure 2. Sketch of acoustic transducer.

3 AE EXPERIMENTS AND ANALYSIS OF POLLUTED INSULATOR DISCHARGE

3.1 *Experiments of the two-unit insulator string*

In order to check the validity of the AE monitoring method through clarifying the fundamental characteristics of AE signals emitted by the polluted insulator, the pollution discharge experiments of a suspension insulator string of two units were done in the laboratory. The solid pollution layer method is used in artificial pollution experiments, using NaCl to simulate the conductive substance, and kaolin for infusible substance. Testing work was carried out in a fixed fog chamber (volume 2.5×1.7×1.8m), testing the insulator string (insulator type XP3-160, shed spacing 155 mm, shed diameter 254 mm and creepage distance 350mm) at different wetting conditions in order to monitor basic parameters from the acoustic signals in real time for variations over time. The testing apparatus arrangement is shown in figure 3.

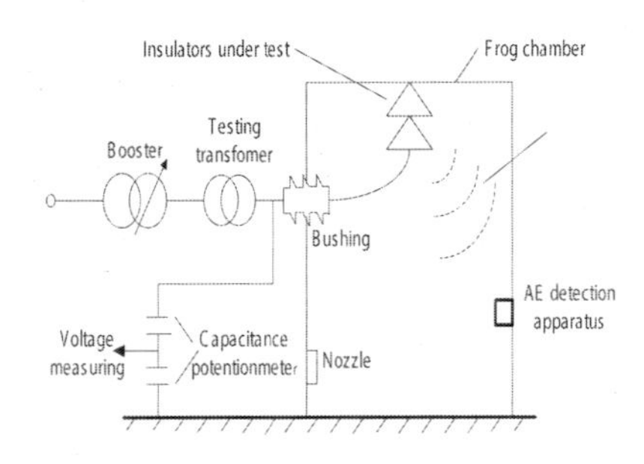

Figure 3. Sketch of testing apparatus arrangement.

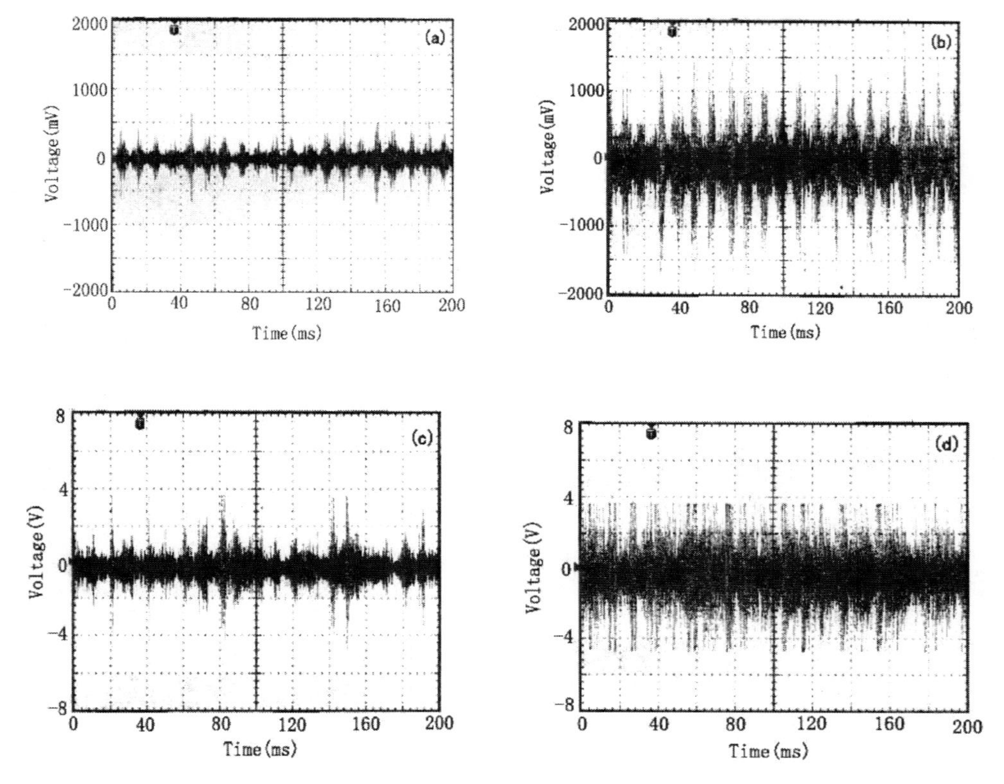

Figure 4. Acoustic emission signals of two-unit insulator string.

The two aged ceramic units of suspension insulators made a mist-shaped pollution liquid stick. Test voltage is educed through the capacitance for measurement. The artificial fog chamber is equipped with steam that is generated by boiling water, coming into the chamber through the nozzle. The steam input rate is the IEC specified value ($50 g.m^{-3}.h$).

Before testing, the insulator string was carefully and totally washed off. Then, the dosage of NaCl and clay (kaolin) was calculated according to the request of salt density. The value can be converted to the ESDD, and by dividing the equivalent salt quantity in the solution (mg) by the specified surface area of the insulator (cm^2). Then, the sample dried was contaminated by using the collocated slurry. After the contamination layer was dry, the sample was wet with steam. The test voltage was increased gradually until a flashover occurred. The AE signals from the insulator were detected with the AE detection apparatus in the whole flashover process. Figure 4 shows the waveform of AE signals of samples at 0.03 $mg.cm^{-2}$ ESDD level, which expresses a light degree of contamination. The testing voltage is increasing from (a) to (d).

As shown in figure 4, the intensity of the AE signals was increased with the increasing of testing voltage. That is, the voltage amplitude and frequency of AE waves were increased with the increase of the sound press of discharge. The transition can be divided into four stages in the process of flashover. At the first stage of discharge, the leakage current was small, and the acoustic wave was shown in (a). The AE signals reached a peak value corresponding to the positive and negative peak value of power frequency voltage, and it reached the least value at the zero point of supply voltage. It proved that the discharge was the strongest when testing voltage reached its peak value, and there was an obvious extinguishing of the discharge arc at the zero point. The partial discharges developed continuously in (b) and (c), and the discharge sound was larger and larger. There was no obvious extinguishing of the arc in (d). Because the discharge was intense in the situation, the energy storing in the electric arc was very large. When the discharge was continuous, the pulse of AE signals was dense.

Based on the experiments, the common characteristics of the AE wave from polluted insulator discharge were summarized. Then, three characteristic values (maxim amplitude, maxim surface, and amount of continuous periods) were selected to express the AE signals. As shown in Table 1. maxim amplitude (MA) is the peak value of signals amplitude under certain voltage and contamination severities. Maxim surface (MS) is the maxim surrounding areas of the signal encircle line in a power frequency period. The amount of continuous periods (AP) is the amount of periods whose surrounding area exceeds enactment value. The three characteristic values are relative to the amplitude and frequency of the AE signals. It is seen that the three characteristic values are increased with the increase of testing voltage.

Table 1. Characteristic values of AE signals at 0.1 $Mg.Cm^{-2}$ esdd level under different voltages.

U(KV)	5	10	15	20
MA(mV)	512	2520	3760	4200
MS(mV.ms)	5660	18384	34060	38060
AP	0	0	4	18

The same artificial contamination experiments were performed with the insulator string at different ESDD levels.

Table 2 consists of the characteristic values of AE signals at different ESDD levels under a 10KV voltage. It is seen that the three characteristic values are also increased with the advancing of contamination.

Table 2. Characteristic values of AE signals at different esdd levels under 10KV voltage.

ESDD($mg.cm^{-2}$)	0.03	0.07	0.1	0.2
MA(mV)	1008	2240	2520	2560
MS(mV.ms)	6090	12100	18384	23806
AP	0	1	2	5

3.2 *Long insulator string experiments*

The flashover path of long insulator stings is different from the single insulator. So, the flashover voltage is not the summation of a single insulator. In order to verify the validity of the AE method, polluted insulator discharge experiments of long high-voltage insulator strings are necessary.

Experimental work of long insulator strings was carried out in a big fixed fog chamber (volume 24×24×26m). The testing insulator strings (insulator type FC12P/146, 30 units) are at 0.06 $mg.cm^{-2}$ ESDD level. The testing method and AE detection apparatus are the same as that of the two-unit insulator testing. Figure 5 shows the AE waveform from the sample. The testing voltage increased from (a) to (d).

As shown in figure 5, the AE waves of long insulator strings have the same developing trend as that of the two-unit insulator string. The transition also can be divided into four stages in the process of flashover. At the beginning, discharges occurred rarely on the surface of samples. Thus, the amplitude of the AE signal in (a) was also extremely small. With the humidity of the fog chamber being increased, the discharges occurred continuously and developed stage by stage. Accordingly, the amplitude and frequency of the AE signal were discontinuously augmented, as shown in (b) and (c). When discharges occurred more and more on the surface of samples, the AE signal continuously occurred in (d). The persistent waveform in (d) indicated flashover just approaching. It should be noted that the more discharge approaches flashover, the more the AE signal becomes intense.

The same contamination experiments with different types of insulator strings at different ESDD levels were performed.

All experimental results show a close relationship between AE signals variations and imminence of flashover. The intensity of the polluted discharge is relative to the amplitude and the frequency of the AE waves. With the increase of the testing voltage, the intensity of polluted discharge is enhanced correspondingly. That is, the amplitude is greater and greater; the signal pulse is denser and denser. Thus, it is valid to judge the intensity of contamination that caused flashover through detecting the AE signal.

4 CONCLUSIONS

The AE method is an effective non-destructive detection method, and it can be introduced into monitoring the flashover process of the polluted insulator. While assisting with pollution discharge experiments, the characteristics of the AE signals are explored in-depth. It is proved that there is a corresponding relationship between the AE signals emitted by the insulators under test and the development of contamination discharge. Thus, AE monitoring and analysis offer a new method for flashover prediction and condition-based maintenance.

To apply the AE technology to monitoring of polluted insulator flashover, there is much work to be done as follows. Analyzing flashover mechanism of polluted insulators; conducting tests on various types of samples made of different materials; seeking more effective processing methods of AE signals; and researching possible influences on

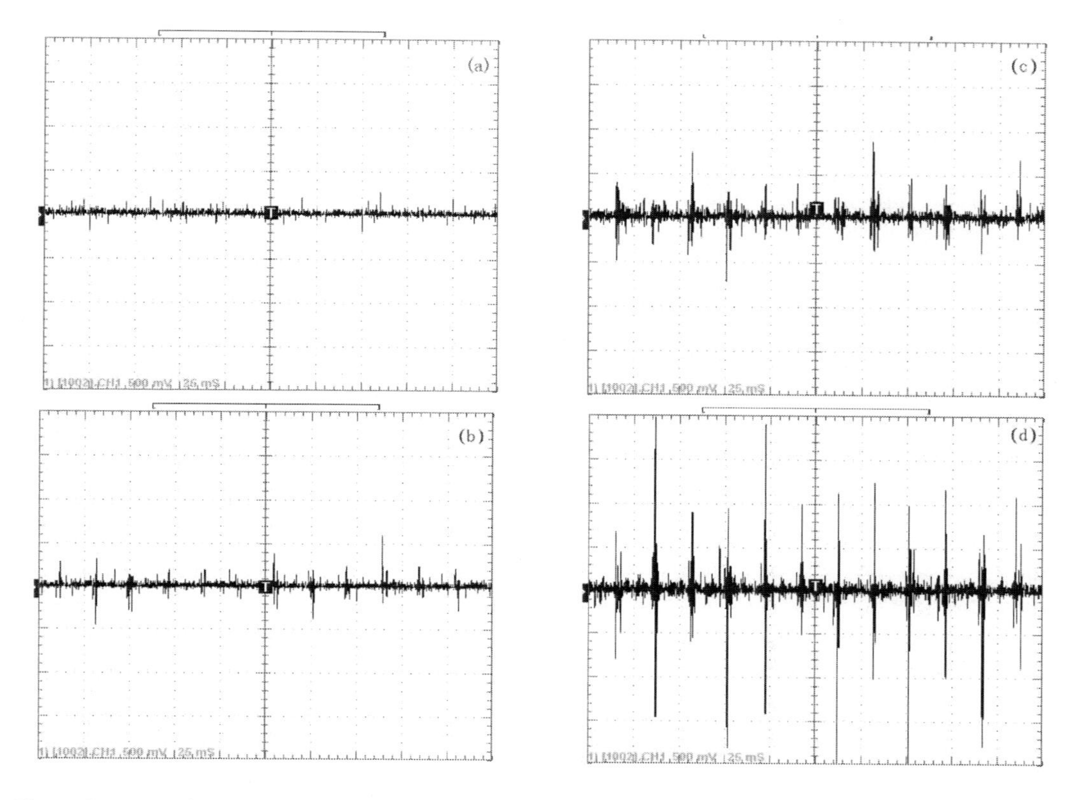

Figure 5. Waveform of AE signals of long insulator strings.

the AE measurements associated with the environment to make the AE monitoring method better and practical.

ACKNOWLEDGMENTS

The authors would like to thank Wuhan High-voltage Research Institute of State Grid Corporation China for the use of their high-voltage facilities to conduct the testing presented in this paper, Z.D. Yang and D. Wang for their parts in the testing.

REFERENCES

[1] Andrea Cavallini, S. Chandrasekar, Gian Carlo Montanari, Francesco Puletti. (2007) Inferring ceramic insulator pollution by an innovative approach resorting to PD detection. *IEEE Transactions on Dielectrics and Electrical Insulation*, 14(1):23–29.

[2] Bo Li, Xiujie Wang, Liu Nian. (2006) Remote online monitoring system for suspension insulator strings. *IEEE ISIE 2006*, July 9-12, 2006, Montreal, Quebec, Canada, pp:2738–2742.

[3] Toshinori Tsuji, Shingo Matsumoto, Takatoshi Sakata, Shingo Nakayama, Masahisa Otsubo. (2005) Basic study on acoustic noise of polluted insulator and waveform analysis method. *Proceedings of 2005 International Symposium on Electrical Insulating Materials*, June 5–9, 2005, Kitakyushu. Japan, pp:694–697.

[4] Yang Ji, Xu Tao, Tang Jianjun. (2007) Online detection system for contaminated insulators based on ultra-violet pulse method. *2007 Annual Report Conference on Electrical Insulation and Dielectric Phenomena*, pp:558–561.

[5] A. Larssont, A. Roslund, A. D. Dernfalk ,S. M. Gubanski. (2001) Remote and non-intrusive diagnostics of high-voltage insulation materials using laser-induced fluorescence spectroscopy. *2001 Annual Report Conference on Electrical Insulation and Dielectric Phenomena*, pp:428–431.

[6] L.E.Lundgard. (1992) Practical discharge Part XIII: Acoustic partial discharge detection-fundamental considerations. *IEEE ElectricalInsulation Magazine*, 8(4): 25–31.

[7] L.E.Lundgaard. (1992) Partial discharge XIV: Acoustic partial discharge detection-practical application. *IEEE Electrical Insulation Magazine*, 8(5):34–43.

[8] L.E.Lundgard, M.Runde, B.Skyberg. (1990) Acoustic diagnoses of gas insulated substations, *IEEE PD-5*, (4):1751–1759.

[9] R.T.Harrold, T.W.Bakin. (1979) Ultrasonic sensing of partial discharges within microfarad value AC capacitors. *IEEE PAS-98*, (2):444–448.
[10] Y.Lu, X.Tan, X.Hu. (2000) PD detection and localization by acoustic measurements in an oilfilled transformer. *IEE Proc. Sci. Meas.Technol.*, 147(2):81–85.
[11] Pei, C.M., Shu, N.Q., Li, L., Li, Z.P., Peng, H. (2008) On-line monitoring of insulator contamination causing flashover based on acoustic emission. *DRPT 2008*, April 2008, pp:1667–1671.
[12] Fan Ju, Liang Xidong, Yin Yu, Wang Chengsheng, and Chen Ling. (2000) Application of acoustic emission technology on structure design and quality control of composite insulators. *Proceedings of The 6th International Conference on Properties and Applications of Dielectric Materials*, pp:358–361.

Electric, Electronic and Control Engineering – Shao, Shu & Tian (Eds)
© 2015 Taylor & Francis Group, London, ISBN: 978-1-138-02842-5

A fault location method for micro-grid based on distributed decision

Fei Tang,
NARI Technology Development Co., Ltd., Nanjing, China

Wei Gao
NARI-TECH Nanjing Control Systems Ltd, Nanjing, China

ABSTRACT: Considering that a Micro-Grid's (MG's) operating mode is very complicated, the power flow changes frequently and short current is limited, so this paper proposes a new fault location method based on distributed decision. First, the associated region is determined on the basis of network segmentation using a tree structure. Then, the node information is stored based on the chain table model. Finally, based on the node connectivity and fault information of the fault-related region, fault location with a high precision is realized. For using static region division, which can be dynamically addressed, in this algorithm, it reduces restriction for real-time performance and has certain adaptability for changes in MG network structure and scarcity of protection information. In addition, the fault current criterion proposed in this paper can greatly improve sensitivity, and the results of simulation verify it.

KEYWORDS: Micro-grid (MG); fault location; associated region division; fault criterion.

1 INTRODUCTION

In recent years, distributed generation with its small investment, generating flexibility, compatible with the environment and other characteristics have been developing rapidly. However, without a constraint to accessing distributed power will bring many problems for the distribution network [3]. Therefore, experts and scholars put forward the concept of a micro-grid, and hope through the effective integration of distributed power, load, and energy storage devices within a certain area, together with the corresponding control and protection methods, to solve problems caused by the distributed power generation and network.

However, the control principle and protection method of the micro-grid is quite complex, mainly reflected in: each distributed power generation with a flexible approach in the micro-grid and with the micro-grid trend changing; each distributed power in a micro-grid is based on power electronics interface control, provided the fault current does not exceed twice the rated current [6]. The micro-grid has two modes named "parallel operation" and "off-grid run," and protection strategies of the two modes should have consistency. As a result, the traditional protection of the distribution network will no longer apply to the micro-grid and will need to seek new protection and control principle.

Domestic and foreign scholars have carried out research on the protection of the micro-grid; the results can be divided into two categories: (i) without the aid of means of communication, optimize and modify the traditional protection; (ii) by the means of communication, build a digital network-style protection. With DG permeability increasing, however, the protection device based on single-point electrical information obviously cannot satisfy the request of the micro-grid. Therefore, building a micro-level communications network, using multiple electrical information comprehensive judgment of the micro-grid, should help solve the problem of the protection of the micro-grid, providing a new way of thinking.

However, in the digital network protection, the protection algorithm of this study is mostly based on centralized decision-making wide-area protection mode [3, 6, 9], through the comprehensive judgment of the faults of all the protection devices in the micro-grid to locate the fault. The protection device, however, accepts that the scope of the fault information is not the bigger the better, mainly for two reasons: (i) Receiving and processing a wide range of information is not easy to satisfy the quickness requirements of the protection; (ii) faults may only have an impact on the local area; the scope of the fault information is essential for fault diagnosis, and the information outside the scope is out of matter. Therefore, this article introduces the wide-area protection mode of distribution decisions to the micro-grid system; in view of the actual micro-grid model, introduces a classification method of the associated regional protection device based on the tree structure and list method; and only combines various regions within the region associated with protection device fault information. In this way, you can determine faults. Meanwhile, considering the fault current supplied through the power of distributed power electronic interface is smaller and not detected easily; fault current criterion is given based on the gain factor; and the instance simulation verifies the sensitivity of the criterion.

2 MICRO-GRID WIDE-AREA PROTECTION SYSTEM

2.1 *System framework of micro-grid*

To meet the needs of different functions, the micro-grid can have a variety of structures [4]. Studying a kind of complex micro-grid structure and extending it to a simple micro-grid or even a complex micro-grid structure with a multiple simple micro-grid is more universal. Figure 1 depicts a more complex micro-grid structure.

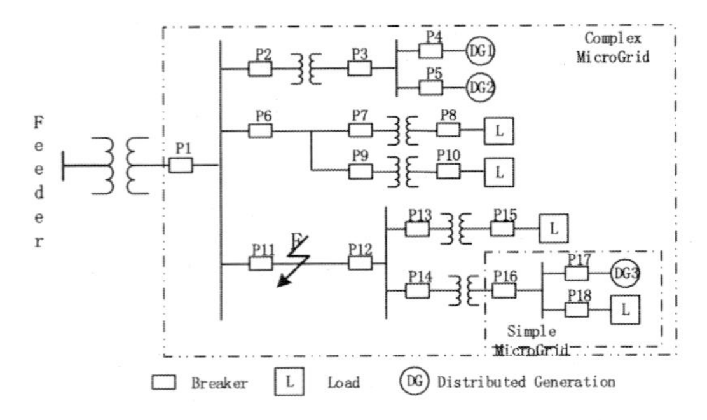

Figure 1. Framework of micro-grid.

Simple micro-grids generally contain only one class of distributed power supply; its function and design is relatively simple, as only in order to realize the combination supply of cooling, heating, and power (CCHP) or ensure the power supply of the critical load. However, the complex micro network not only contains a kind of distributed power supply but also may consist of a number of different simple piconets or is made up of the distributed power supply with a variety of complementary coordinated operations [8].

2.2 *System framework of micro-grid wide-area protection*

At present, the wide-area protection system mainly adopts two kinds of structure forms: centralized decision-making type and distribution of decision-making type [9]. The centralized decision-making structure consists of protection master station and protection device, which are installed dispersedly. Then, a protection master station is used to collect the information of each protection device, perform main comprehensive judgment, and locate the fault. In the distributed decision-making structure, the function of the main weakened is protected; protection work is performed by dispersed installation of the protective device. The protection is responsible not only for the installation point of information collection, operation, and transfer but also for complete fault location and judgment.

Since the centralized decision base has high dependence on the master station of the protection and communication system, this paper studies the wide-area protection system that adopts a distributed decision base and cancels the protected master station; retains only the protection device and communication interface of the substation main unit (SMU) to upload information, update tasks and data. The protection device and micro-network circuit breaker have one-to-one correspondence; we need to collect electrical information of various branches, then calculate and compare it, and at last complete protection tasks. The concrete structure is shown in figure 2.

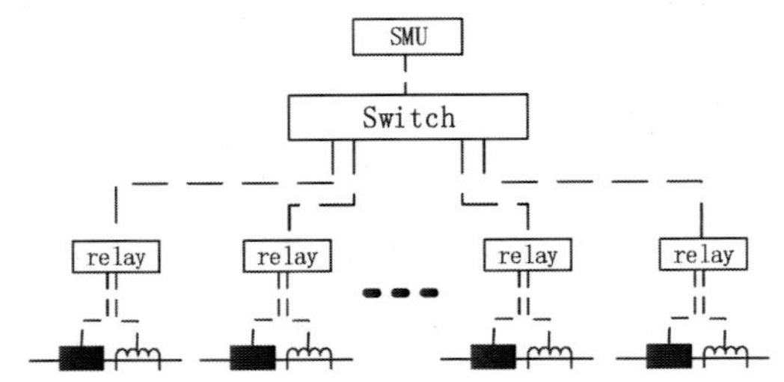

Figure 2. Framework of distributed decision-based protection system.

3 RELATED DIVISION OF PROTECTION DEVICE

The protection devices of the distributed decision-based wide-area protection system, in theory, can obtain the information at any measuring point of the system. However, in practical application, exchanging information blindly in a wide range will undoubtedly increase the burden on the communication system; reduce the reliability of the communication system, also being meaningful for the protection system [10]. To make various protection devices work quickly and reliably, we must define the scope of its information exchange, through failure of the exchange of information within a certain area to locate faults. This section will conduct research of the division of protection device association area (hereinafter referred to as correlation area).

3.1 *Network topology description*

Micro-grids are generally designed based on low-voltage distribution network or formed directly on the basis of the distribution network; therefore, their structure can be equivalent to the separate open-loop tree-type network [12]. The tree-type network consists of nodes and the branches that connect the nodes. Each node in the tree-type network is derived by a guide; this boot node is the root node of the network.

For two different nodes Pi and Pj in the tree network: If Pi connects with Pj directly and is closer to the root node, then it is the father node of Pj; if Pi connects with Pj directly and is more away from the root node, then it is a child node of Pj; if Pi and Pj have the same father node, they are sibling nodes for each other; and if Pi has no children node, it is called the "leaf node." Figure 3 is the tree-type network topology of the micro-grid as shown in figure 1; each node in the figure and the protection device and circuit breaker in the micro-grid have one-to-one correspondence.

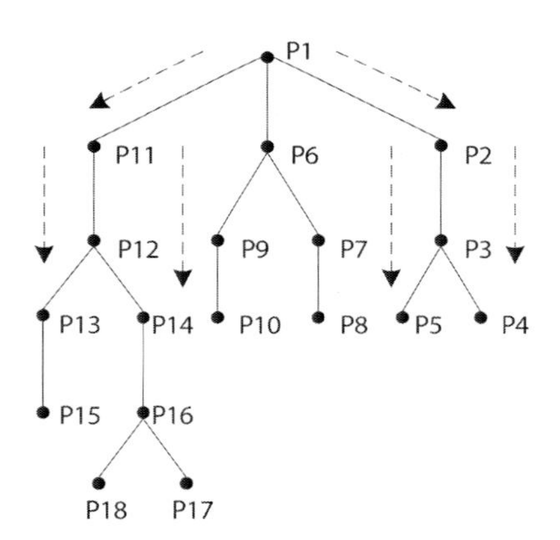

Figure 3. Topology of tree-type micro-grid network.

Figure 3 shows that P1 is the root node of the tree and also the father node for P2, P6, and P11. The P7 and P9 are siblings and both are the child nodes of P6. Such as P4, P5 and P8 are leaf nodes.

Due to any node in the tree-type network to the path of the root node has uniqueness, then positive direction of the network current can be set according to the network topology. Considering that the structure of the micro-grid network is more than a power source, feeder-positive direction can be assumed as the flow direction of the power of the entire network when the network is supplied by only a power [12]. The power source can choose any power source in the network, and this paper considers system-side power as the only power source. Then, in the actual design, the positive direction of the current is from the father node to its child node, shown by the arrows in figure 3.

After the positive direction of the network current had been determined, fault information function of the protection device was created. The fault information function is established in this paper:

$$f(i)=\begin{cases} 1 & \text{node } i \text{ detects fault current and the direction is same with the positive direction} \\ -1 & \text{node } i \text{ detects fault current and the direction opposite to the positive direction} \\ 0 & \text{node } i \text{ does not detect the fault current} \end{cases}$$

3.2 *Related division*

Known by the general principles of fault isolation, we can realize the isolation of minimum area is the area network determined by two or more adjacent breakers. But for the peripheral network, the minimum area is downstream area divided by the peripheral feeder circuit breaker. In the framework of the micro-grid, the minimum area is a segment of the feeder, distribution transformer, bus bar, T-contact region, or peripheral feeder established by two or more circuit breakers. In the tree-type structure, it is any node in addition to the leaf nodes surrounded by all adjacent areas of child nodes or the downstream area of the leaf node form alone. Then, let us consider this minimum region as the related area of protection devices, the maximum limit to reduce the fault isolation area. Simultaneously, we can also narrow the scope of protection of information exchange, as well as reduce the burden of network communication.

Tree-type network topology indicates the relationship between the nodes. Due to the fault location needing to make the related area as the basic unit of analysis, this section will search and divide the tree-type network based on the linked list method, form the set of nodes associated with the region, and express its storage methods. It should be pointed out that search and classification should be combined with information of the node, and the storage of the node information has a variety of ways, such as storing the node information while providing a pointer to the location of the parent node, or a pointer to the location of the child and sibling node, and so on. And also these forms of information storage can only describe a tree-type framework, but this article does not specify it. Specific steps of division of the related area are as follows:

i. Cite queue structure, press all the nodes information into the queue and give numbers, and establish a pointer P to point to the first node of the queue. Go to step (2).

ii. Establish a sub-list; keep the address of the node that the pointer P points to into the header. Go to step (3).
iii. Judge the number of sub nodes which are adjacent to the current node. If the number is 0 then go to step (5); If the number is not 0; then go to step (4).
iv. Adopt sequence traversal method, give the adjacent child nodes of the current node number, and deposit into the list one by one.
v. Determine whether the pointer P points to the end of the queue: If it does, end the process of dividing; if not, make the pointer P point to the next node of the queue. Go to step (2).

It should be pointed out that, in step (1), the order of the node information pressed into the queue does not impact the result of division. While adding new nodes, press them into the bottom of the queue directly. The division flowchart of the micro-grid shown in figure 1 can be represented as in figure 4. Each sub-list terminates with the sign Λ, and the dots it contains is the node set of the associated region.

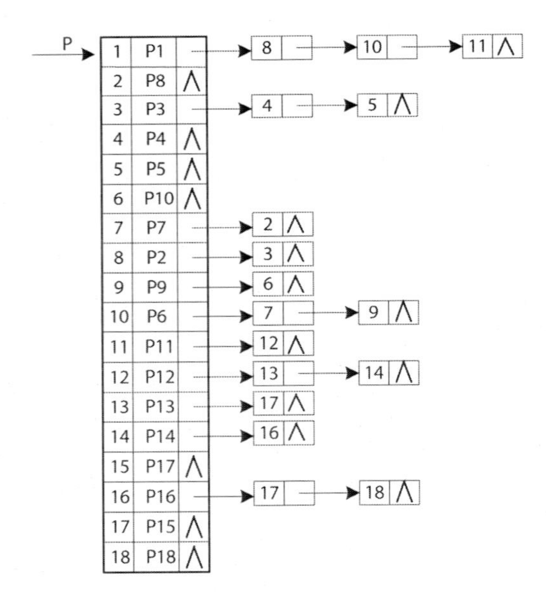

Figure 4. Division flowchart of micro-grid shown in Figure 1.

3.3 *Dressing of associated region in several circumstances*

Micro-grid power supply is flexible, and the operating environment of the protection system is complex. This paper will discuss several scenarios that affect results of the division of the associated region:

1) Distributed power supply/return

Known from the structure of the micro-grid, regardless of the distributed power accesses from which voltage level, it is always located in the peripheral of the network, namely the leaf node of the tree structure. When there are distributed power inputs, they are simply numbered and deposited in the corresponding list by their father node. When there are distributed power cuts, simply remove them from the corresponding list of their father node.

2) Information of protection miss / protection exit

When information of protection is missed or protection exits, the original associated region will be a lack of fault information to accomplish fault diagnosis. By expanding correlation area, combine it with the method of the parent and child nodes information of the missing information to complete fault judgment. The specific trimming method enables the corresponding list of missing information node to merge into the corresponding list of its father node and removes the dot of the missing information node.

3) The back-up protection

Back-up protection is about to extend the scope of protection to downstream of the network lines, bus, or distribution transformer. Scilicet extends the associated region to the downstream of the adjacent related area corresponding to the tree structure. Specific trimming method enables the corresponding list of all the child nodes in the related area to merge into the corresponding list of their father nodes and removes the dot of the missing information node. The synthetic chain indicates the associated regions of the backup protection.

Based on this situation, the trim of the associated region only involves the corresponding list operation of the individual node, without re-dividing the associated area.

4 FAULT DIAGNOSIS

The division of the associated regional specifies the scope of information exchange of each protection device. When a failure occurs, each protection device compares and calculates the information coming from other protection devices in the associated region to determine the fault section. However, due to the special nature of the micro-grid fault current, the realization of fault location also needs the help of a certain criterion and methods. This section introduces an applicable failure criterion for the micro-grid and a method for determining the fault section.

4.1 *Criterion of the fault current based on the gain factor*

In the low-voltage distribution network of our country, three-step current protection is usually used as its main protection principle, generally using current instantaneous trip protection combined with definite time of over-current protection. However, distributed power in the micro-grid has access to the micro-grid based on power electronic devices, whose inverter device usually takes a different control mode, and no matter what control mode is adopted, for the protection of power electronics, all its control loops need to add a current limit unit to limit the short-circuit current that is contributed by the distributed power while system failure is occurring, which is generally not more than twice the rated current. Simultaneously, both randomness of the output power of distributed power and change of the micro-grid operation mode can make the normal current protection lack sufficient sensitivity [13].

When micro-grid failure occurs, the distributed power will increase the output power in order to maintain the stability of the system voltage and frequency. When the fault point is close to the distributed power, it will lead the output power of the power supply to reach the limit, and the system voltage will not be able to continue to maintain the normal level. The closer the point of failure to the installation point of protection, the voltage drops more seriously. Type (1) is the per-unit value of voltage at protection installation when three-phase short circuit of the system occurs; type (2) is the per-unit value of voltage when interphase short circuit occurs; and type (3) is the per-unit value of voltage when the single-phase short circuit occurs in the neutral point grounding system.

$$u^{(3)} = \frac{I^{(3)} \cdot Z_L}{U_S} \quad (1)$$

$$u^{(2)} = \frac{2I^{(2)} \cdot Z_L}{\sqrt{3}U_S} \quad (2)$$

$$u^{(3)} = \frac{I^{(3)} \cdot Z_L}{U_S} \quad (3),$$

where U_S is the reference value of the phase, and Z_L is the distance from the installation of the protection to the point of failure.

Obviously, u can accurately reflect the distance from the point of failure to the installation of the protection; the closer the point of failure, the lower the voltage. Based on these characteristics, this paper presents a criterion of fault current using the amount of voltage as a fault current gain factor, such as type (4).

$u = {U_P}/{U_S}$,U_P is the circuit phase voltage of protection measure, and u is the gain factor. Conditions for over-current protection action are as follows:

$$I_P/u > I_{set} \quad (4),$$

where I_P is the current of the protection measure. It should be noted that I_{set} will be set, respectively, according to the different current direction, which means the over-current protection setting values of forward/reverse circulation.

4.2 *Determination of fault region*

The fault current criterion that is based on gain factor increases the sensitivity of the protection. When the micro-grid failure occurs, the protection devices in each associated region can interact through fault information and determine the fault region with the associated region as the basic unit of analysis.

Set L_N to be a list for storing the node set of associated area and L_N has also trimmed the associated area under the circumstances mentioned in section 2.3; its storage nodes from header to footer of the table are sequentially $i_0, i_1 \cdots i_n$.When failure occurs, the corresponding fault information function is $f(i_0), f(i_1) \cdots f(i_n)$. General principles of the fault area judgment show that when the fault current gets through the associated regions, namely the fault current inflows and outflows the associated area, this association region is the non-fault region; When a fault current flows into the associated area and does not outflow , the associated region is the fault region.

Thus, we define $F(L_N)$ as a fault diagnosis function of the associated region.

$$F(L_N) = \begin{cases} 1 & L_N \textit{ shows that} \text{ association region is fault region.} \\ 0 & L_N \textit{ shows that} \text{ association region is non - fault region.} \end{cases}$$

Then, the criterion of the associated regions in failure as L_N shown is represented as follows:

① $f(i_0) = 1 \quad f(i_j) \neq 1, j = 1, 2 \quad n;$

② $f(i_0) = 0 \quad f(i_j) \neq 1, j = 1, 2 \cdots n;$

$\exists f(i_k) = -1, k \in (1, n)$

If any one of the conditions ① and ② is satisfied, the associated region is considered a fault area, that is, $F(L_N) = 1$.

5 SIMULATION CALCULATION

The simulation model of the micro-grid structure as shown in figure 1 is established. The distributed power is by a supply of 400 V access; the micro-grid system consists of the neutral grounding operation mode. Line lengths are 0.1 KM, line order parameter: X1=X2=0.0723Ω/km, R1=R2=0.253Ω/km, X0=0.289Ω/km, R0=1.01Ω/km, each load is 0.5MVA, and all the load power factor is 0.9.DG1; DG2 adopts PQ control mode, and its maximum output power is 1MVA. Distributed power DG3 uses the VF control mode (provide voltage support), and the maximum output power is 1MVA. The maximum output current of all the distributed power is 1.5 times of its rated current.

When the micro-grid is connecting to the main network, because of the fault current that comes from the main network, this can help increase the fault current, and the protection sensitivity can be guaranteed. So this paper focuses on sensitivity of the fault current criteria based on the gain factor under the micro-grid off-grid operation. Let us assume that F of the micro-grid as shown in figure 1, respectively, occurs as a single-phase short circuit, phase short circuit, three-phase short-circuit fault, and research on the sensitivity of protection P11.

Figure 5 (a), 5 (b), and 5 (c) are the comparison of the protection starting current and fault current based on the gain factor when the system is, respectively, under single-phase, phase, and three-phase short circuit fault (top of the curve is the protection starting current based on gain factor). Let us assume that a fault occurs in the 2s and lasts 500ms.

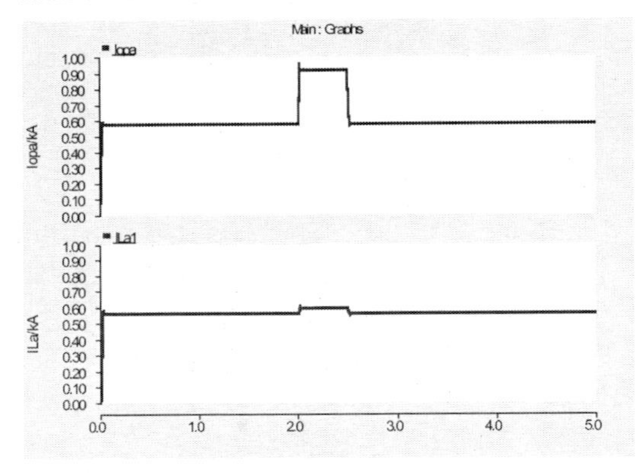

Figure 5(a). Phase grounding fault.

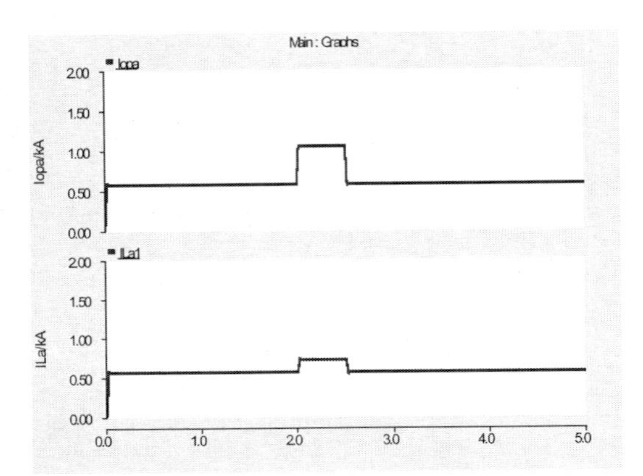

Figure 5(b). AB phase to phase fault.

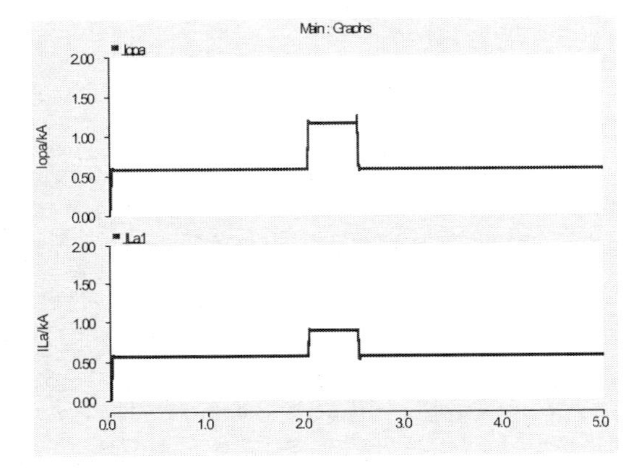

Figure 5(c). Three phases grounding fault.

Under various fault types, action situation of the protection P11 is as shown in table 1:

Table 1 shows that the fault current criteria based on gain factor used in this article can improve the sensitivity of protection for different fault types. The protection sensitivity increases most obviously under the single-phase fault.

Table1. Analysis of action situation of the protection P11.

Fault types	Load current before failure (KA)	Fault current Value (KA)	Protection starting current based on the gain factor (KA)	Improve sensitivity value (%)
Single-phase short circuit	0.563	0.591	0.916	54.99%
Interphase short circuit	0.563	0.725	1.052	45.10%
Three phase short circuit	0.563	0.888	1.151	29.62%

6 CONCLUSION

At present, the study of the micro-power grid protection system is still in the exploratory stage. This paper is based on the structure and characteristics of the micro-power grid and proposes a new method for fault location based on related division. Through the fault information exchange of associated areas of the protection device, this method combines with the logic computation of fault information function f(i) , which can determine the fault state of the associated region. This paper elaborates associated region division and the dressing method by using a tree structure and chain storage, proposes a criterion of fault current based on the gain facto, defines fault diagnosis logic of the associated region, and also carries out the simulation of the fault current criterion. The fault location method avoids the fact that protection gathers a large amount of information. It has a very good adaptability of being flexible toward the micro-grid structure and a lower level of short-circuit current. Simulation proves the validity of the criterion.

REFERENCES

Nikkhajoei H & Lasseter R H. (2009) Distributed generation interface to the CERTS microgrid. *IEEE Trans. on Power Delivery*, 24(3):1598–1608.

Salomonsson D & Sannino A. (2007) Low voltage DC distribution system for commercial power systems with sensitive electronic loads. *IEEE Trans. on Power Delivery*, 22(3):16202–1627.

Zhang Zongbao, Yuan Rongxiang & Zhang Shuhua, *et al.* (2010) Research on micro grid protection a new power system. *Power System Protection and Control*, 38(18):204–209.

Driesen J & Katiraei F. (2008) Design for distributed energy resources. *IEEE Power and Energy Magazine*, 6 (3):30240.

Zhao Shanglin, Wu Zaijun & Hu Minqiang, *et al.* (2010) Thought about protection of distributed generation and micro grid. *Automation of Electric Power Systems*, 34(1):73–77.

Dong Peng. (2009) *Research on Micro Grid Control and Protection Strategy*. Beijing: North China electric power university.

WANG Chengshan & LI Peng. (2010) Development and challenges of distributed generation, the micro-grid and smart distribution system. *Automation of Electric Power Systems*, 34(2):10–14.

Ding Lei & Pan Zhencun. (2007) *Hierarchical Islanding Protection and Control of Distribution Network with Multi-micro Grids*. Jinan: Shandong University.

Cong Wei, Pan Zhencun & Zhao Jianguo. (2006) A wide-area relaying on protection algorithm based on longitudinal comparison principle. *Proceedings of the CSEE 2006*, 26(21):8–14.

TANG Fei & LU Yuping. (2010) New fault location algorithm for distributed generation system. *Power System Protection and Control*, 38(20): 62–68.

Weng Lantian, Liu Kaipei & Liu Xiaoli, *et al.* (2009) Chain table algorithm for fault location of complicated distribution network. *Transactions of China Electrotechnical Society*, 24(5):190–196.

IEEE Std. 1547–2003 IEEE Standard for Interconnecting Distributed Resources with Electric[S]. New York: The Institute of Electrical and Electronics Engineers Inc, 2003.

Electric, Electronic and Control Engineering – Shao, Shu & Tian (Eds)
© 2015 Taylor & Francis Group, London, ISBN: 978-1-138-02842-5

Application of computer automatic control technology in the industrial production site

Zhou Lv
Research Institute of Mechanical Electronic Engineering, Taiyuan University of Technology, Taiyuan, China

ABSTRACT: In recent years, with the continuous development and innovation of science and technology, computer technology has been applied to all walks of life. Among them, the industrial production as the main backing for China's economic development has also been affected in the process of sustainable development in information technology. Nowadays, the computer automatic control technology has been effectively developed and used in the field of industrial production in China, which leads to the economic benefits and profits of China fundamentally. In the 80s of the last century, to promote the development of enterprises, many large enterprises in our country as well as plants actively introduced foreign production equipment and production lines, which promotes production and realized automated production though, also raises new requirements for production data acquisition. Therefore, in the case of the status quo of corporate development, it has currently become one of the most important issues to infuse the computer automatic control technology with such development. This article aims to introduce the main methods of computer automatic control technology to related personnel together with the application phenomenon, and provide the theoretical basis for their work in the future, thus contributing to the sustainability of China's industrial production.

KEYWORDS: computer automatic control technology, industrial production, application; motor test.

1 INTRODUCTION

With the rapid development of technology, many of large and medium enterprises in China adopt computer control technology in the production process. By using this technique, it is possible to effectively develop industrial production in detection, control, optimization, management and other aspects, thus contribute to the high quality and high efficiency of the profits of industrial production, and improve the goals of safe and stable production of enterprises fundamentally. [1] Where, under the applications of control theory, instrument and apparatus and various information technologies, the computer automatic control technology has been effectively developed, and in the effective integration of automatic control theory and computer technology, our industrial production moves towards a new level.

The computer automatic control technology has been widely used in a variety of industrial sites, and computer, network communication and industrial control technologies basically cover all workshops and production lines. [2] Where the more familiar ones, such as engine manufacturing plants, automobile factories and steel car plants, bill printing factories have all implemented computer control technology. However, the industrial site is different from other places, whose environment is not only bad, but also with strong magnetic field, high voltage, dust and noise, etc. For this phenomenon, the computer control system in industrial site is required to be with a sound performance and able to perform interference in all kinds of magnetic fields. In addition, there will be many devices in the industrial site, but such equipment are not only expensive but also difficult to be operated, which also requires the computer control system to have high performance with trouble-free in continuous operation, otherwise, it will bring economic losses to the enterprise.

Integration of industrial control technology and industrial computer automatic control technology enables industrial site to successfully complete data acquisition and processing during the construction, and promotes the computer automatic control system to be properly used in the industrial production at high reliability, interference resistance and precise process. [3]

In this article, part of control programs is taken as examples to carry out analysis of and research on the application of computer automatic control technology in industrial sites.

2 RESEARCH ON COMPUTER AUTOMATIC CONTROL IN ENGINE TESTING

The high-speed engine manufacturing industry is an extremely important part of our industry. Products of which are involved in the development have not only high permeability, but also a good experimental availability. However, one of the most important parts in the industrial site is to design a diesel engine consistent with relevant standards and with high performance. In the development of technology, the test controller designed by a computer automatic control and monitoring system is the main control unit developed based on the development status. Mount the controller in the area of heat engine test or test network, and then it will supervise the work on the and then calculate and analyze the data required for tests. In addition, in all steps of computer controlled test, the test data and method shall proceed automatically or manually, wherein the main form of manual is: adjust the data displayed in the screen with the computer system, and analyze them in strict accordance with relevant instructions and signals, and process and display the performance value of each item in adjustment. See Fig.1.

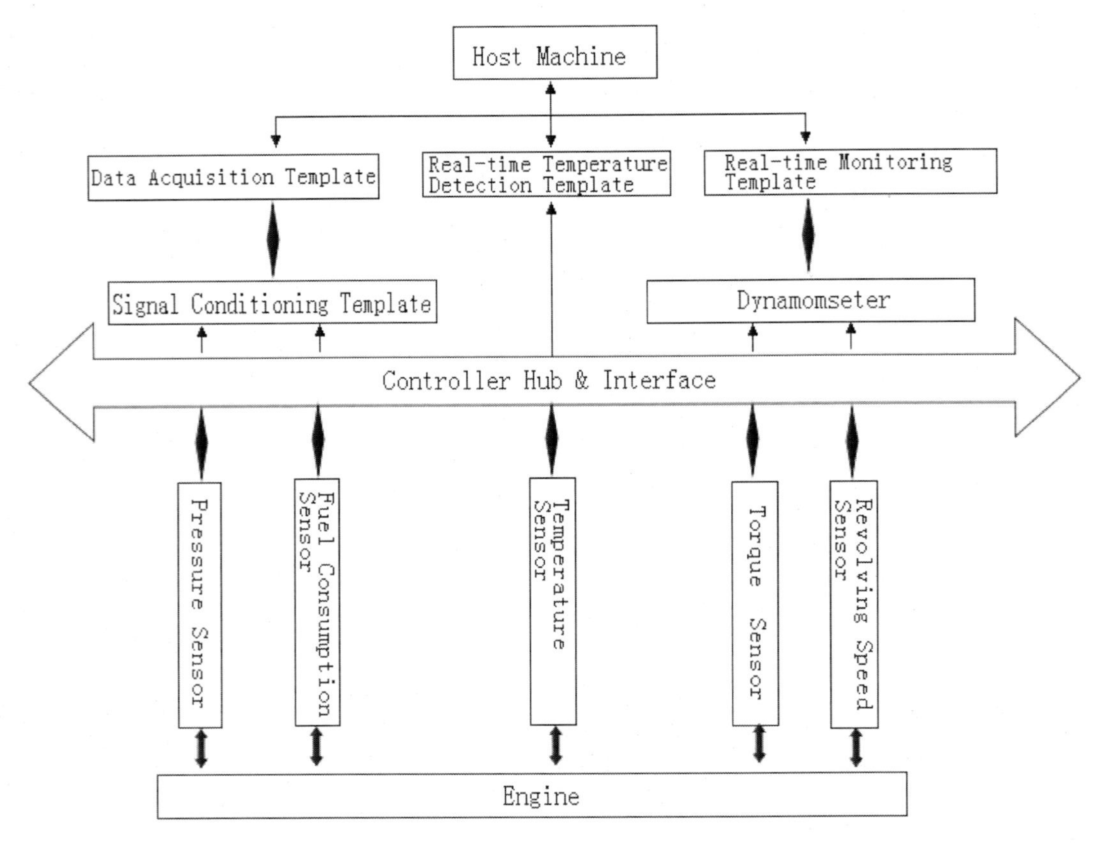

Figure 1. Design framework of computer automatic control in engine testing.

heat engine. [4] Further, in the process of analyzing and researching several engines of different types, the engine can be connected to other computers, so as to obtain the primary data on the engine in a short time.

The main component of the test controller is a computer, which can collect and control signals when it is connected with several other PCB circuits, and thus form a connected mode with several machines. When the programming of main program is done by the computer system, the engine would have to analyze all tests, and control each step in a strict manner,

The main function of the controller in a computer monitoring test in operation is to sort out the control parts in the text system. When the computer system is started, the engine shall be moved to on the test bed, and connected interfaces well with related devices. In addition, the drive shaft and the dynamometer shall also be accurately calibrated. After successful connection between them, the basic efficiency of test can be improved fundamentally and the major time of trial occurred can reduce. The auto-calibration capability of test controller can measure the basic performance of engine on the recombination test bed, and then develop unified standards.

When the test controller is applied to the computer control system, it will play a role in many aspects, which mainly depends on the basic characteristics of the engine, and the engine will be analyzed and researched in the process of testing in strict accordance with the test standards. Part of the test in progress needs to be completed manually by the operator, however, all instructions in the computer system are designed in advance because it has no timeliness, therefore, in the computer system programming, the test results are shown on the display so as to help the tester accelerate the test. When the engine rotates at different loads, the controller is capable of recording all tasks, and in the process of computer operation, electrical signals of all sensors can complete corresponding data, and take record. [5] Therefore, in terms of basic function, the computer controller system is able to determine the time required for the engine, when the stability condition of the engine in operation is also in connection with the basic performance of the engine. Computer monitor can be used to shorten the time and reduce the basic time required for an engine to achieve the quality requirement.

Saving the engine performance data obtained throughout the test is an extremely important part of the whole procedure, which requires a computer to complete it, and the data will perform as the main basis for future maintenance. [6] When the base station computer is combined with the test bed, a common feature is that the test beds are all equipped with the same computers, and each computer can contact with the central computer, if the test controller is affected by certain factors, it will produce failure, and the test bed will be affected. However, as the control and supervision on the test bed is not completed by the computer alone, the test controller can be updated along with the development of the times and possible to meet the basic requirements of users.

The computer automatic control system in the test bed is the main application form in computer industry, which has been widely used in China, in short, only if the degree of automatic control is strengthened, can the productivity of industrial equipment be brought into full play.

3 ANALYSIS OF THE AUTOMATIC CONTROL TECHNOLOGY IN HYDRAULIC EXCAVATOR

Applying the computer automatic control system to the hydraulic excavator can reduce fuel consumption and lower operating costs. When the excavator can not move, the engine can be used to slow down the system, thus saving fuel. At the same time, when the lever is in the neutral position, the entire computer control system can make decisions according to relevant signals, and then slow down the excavator. When the signal is received by the solenoid valve, the hydraulic cylinder will change and gradually extend. Although at this moment the fuel excavator handle does not withdraw, the engine has shown a state of deceleration, and in such development situation, a loss of fuel can be avoided arising from the unloading of the excavator. In addition, the automatic control technology can also be used in the bucket to control it automatically. For the excavation personnel without many years of experience, the use of computer automatic control technology enables them to dig on the flat ground, but they cannot accurately control the tilt angle of the bucket. [7] However, in the use of computer automatic control, the bucket can be pre-set according to the system indicates and the tilt angle can be set, which not only protect the safety of operators, but also improves the work efficiency in the overall construction. For example, the value of the throttle control system can be set in advance by using the computer control system, the speed of the engine is possible to change through controller, actuator and throttle lever, and it is possible to make the displacement sensor change by quickly pulling and measuring, and thus has the throttle controlled. See Fig.2 for details.

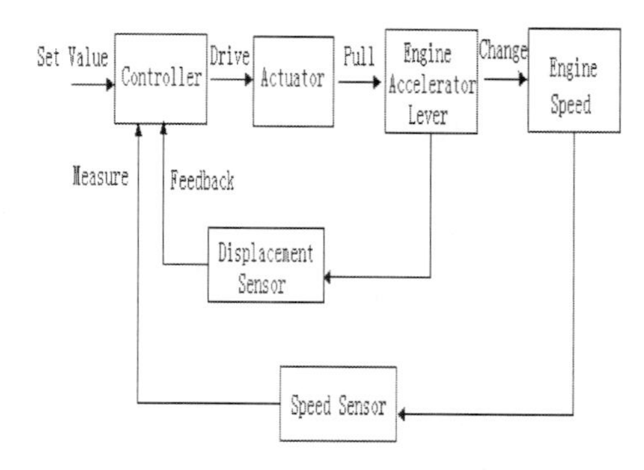

Figure 2. Design framework of excavator accelerator automatic control.

In the operation of the excavator, the use of automatic control technology can improve the reliability and operating comfort of the excavator, e.g., as for the common sensors, the control system can monitor the major temperature of the hydraulic oil, and change the flow rate and pressure parameters accordingly, in the change and adjustment according to the parameters, the state of excavator can also change accordingly, and the faults can also be predicted in advance. The automatic adjustment of cab temperature and remote control technology can ensure the comfort of the operators, thus improving the efficiency and quality of work.

4 HRL MICROCOMPUTER CONTROL WEIGHING SYSTEM IN NUCLEAR FUEL PLANTS

HRL microcomputer control system is a relatively complete computer automatic detecting and control system. By using this system, it is possible to achieve data acquisition, data processing, document processing and other functions, and capable of weighing in the industrial site, filling pellets of nuclear fuel in rods and so forth assisted by this system.

HRL microcomputer control system includes two layers of network structures, where the bottom layer is the field bus capable of data acquisition, device control and connecting the device layer with the upper layer so as to connect the controller with the master control room, and further complete information management and information services.

Therein, the main procedures of using HRL microcomputer control system to fill pellets of nuclear fuel in rods and acquiring data are:

To begin with, the operator should start the system, turn on the host machine and boost relevant software. The operator must set the punch number in the place of information input, input the maximum and minimum numbers in the corresponding cladding tubes, enter the real-time control window during the operation, and implement weighing when the weighing image appears on the screen.

Then, when the tubes sequentially enter the corresponding position, the system must read the barcodes in the empty tubes, and weight these tubes. In the case this work cannot proceed due to some failure, the system can repeat operation, and send instructions to the weighing platform. At the same time, the maintenance personnel should handle it promptly and troubleshoot the fault, if it is removed, then the command to go to the weighing platform is available to be sent.

After that, wait until such tubes come to the No.2 rod filling table, then open the control column in the computer, and read the barcodes of the tubes on the table. In normal circumstance, a batch includes 18 rods, after the operation on this batch is completed, the computer will send a command of closure, when the operator inputs the batch number of pellets filled in the tubes into it, testing and weighing can be carried out to the next batch of pellets. [8]

This system can be used not only for weighing control, but also for data processing, data backup and report output. In addition, in the whole process of pellet filling, the computer will form a operation condition diagram in accordance with the current situation, by which the weighing can be described specifically and intuitively.

As the system shows, its man-machine interface runs well and stable during the operation, and the system has a very high reliability, whose accuracy can reach 0.1%, and the accuracy of the sensor can reach 0.05 %. This system is the most important part of the construction in the current industrial site.

5 AUTOMATIC CONTROL TECHNOLOGY IN TEXTILE MACHINERY

In the production of textile machinery, many rings have applied computer automatic control technology. Among them, the carding machine is relatively important. During the operation of carding machine, the air pressure of vacuuming is relatively large, and the air quantity will as well have an important impact on the quality of sliver processed. The use of automatic control technology can optimize the air pressure and change the air quantity fundamentally, and adjust the air quantity in accordance with the actual situation so as to ensure the overall quality of carding. In addition, the latter part of fabric quality inspection generally adopts the artificial inspection. Such inspection method takes a long time, and the fabric defects detected are relatively small and difficult to be detected. The computer automatic control technology can systemize the operation on the fabric, and at the same time in accordance with its own rules, find out the existing defects of the fabric.

6 CONCLUSION

The computer automatic control system consists essentially of a host machine, terminal control equipment and sensors. In accordance with different control requirements in production sites, the designers can programme corresponding signals for detection, conduct data processing, control the entire process, and store it in the computer. In practice, the computer system can control the electrical signals according to the software, and promote the operation of electrical equipment in accordance with the instructions of the host machine so as to meet the control requirements.

In the development of technology, the computer technology as an advanced technology can provide the people with more reliable data acquisition and automatic control systems, and make the work in the industrial sites more convenient, fast and accurate. In addition, the network technology developed on this basis also contributes to the development of industrial automation in a certain extent, and the combination of the both in the industrial site can make the production efficiency of enterprises develop actively.

REFERENCES

[1] Zhang Tielin, Application of Computer Automatic Control Technology in Industrial Production Field, Science and Technology Innovation Herald, 19, pp. 44, 2014.

[2] Li Min & Sun Fenglai, Analysis of Application Computer Technology in Industrial Automatic Control System, Computer CD Software and Applications, 11, pp. 125–126, 2014.

[3] Zhou Ping, Application of Computer Automatic Control Technology in Industrial Production Field, China Market, 35, pp. 38–39, 2007.

[4] Wang Wei & Chai Tianyou, New Development Trend of Industrial Control System, Control Engineering, 1, pp. 1–5, 2006.

[5] Wang Ji & Zhang Pin, Analysis of Application of Computer in Automatic Control Practices, Digital Technology and Application, 6, pp. 28, 2013.

[6] Zhao Li, Application of Computer Automatic Control Technology in Silicon Reduction and Iron Extraction of Concentrate, Express Information of Mining Industry, 9, pp. 98–99+104, 2008.

[7] Zhang Yuchi, Application of Field Control Technology in Industrial Enterprises, Journal of Hunan Industry Polytechnic, 1, pp. 24–26, 2005.

[8] Jia Juhong, Study on Application of Computer in Automatic Control, Computer CD Software and Applications, 3, pp. 146–148, 2014.

Electric, Electronic and Control Engineering – Shao, Shu & Tian (Eds)
© 2015 Taylor & Francis Group, London, ISBN: 978-1-138-02842-5

Risk evaluation for wind power project based on grey hierarchy method

Cunbin Li, Kun Zhao & Chenhao Ma
School of Economics and Management, North China Electric Power University, Beijing, China

ABSTRACT: As one of the most valuable new energy, wind power has broad prospects accompanied with many risks. The risks faced by the wind power projects can be summarized as five categories, namely wind resource risk, engineering risk, wind power technology risk, market risk and operational risk. This paper comprehensively use grey system theory and the analytic hierarchy process to establish a multi-level grey evaluation model for wind power project quantitative risk assessment. The results show that this method can effectively identify and assess the potential risks of wind power projects, and provide a scientific theory support for wind power investors to grasp and avoid risks.

KEYWORDS: Grey Hierarchy Evaluation; wind power projects; risk assessment.

1 INTRODUCTION

Long-term excessive consumption of fossil fuels has led to the environmental pollution and energy supply pressure increasing. As a clean renewable energy, wind energy attracts more and more widespread concern around the world. As China has enormous wind energy reserves, coupled with the impact from oil energy crisis and coal pollution, the wind power industry will usher in an unprecedented surge of development. However, wind power has also revealed a number of risks at the same time.

In recent years, both domestic and foreign scholars have conducted corresponding research on wind power risk. Literature [4] identified the risks in the conduction process of wind power project and proposed corresponding risk prevention countermeasures; literature [5] studied the risks caused by wind power integration on a large scale; literature [6] treated Fujian Province as the research object and expounded the investment risks of wind power project systematically, with a focus on the economic risks. As can be seen, the current research primarily focuses on risk analysis and description, the quantitative evaluation hidden in wind power is lacking. Based on current research, this paper makes quantitative evaluation on China's wind power risks by use of grey hierarchy model.

2 RISK IDENTIFICATION AND CLASSIFICATION OF WIND POWER PROJECT

Combined with past project and literature review, the risks involved in wind power projects can be divided into five categories: wind resource risk, engineering risk, wind power generation technology risk, market risk, and operation and manage risk.

2.1 *Wind resource risk*

Wind resource condition is the most direct factor affecting the investment returns. Wind resource assessment result is directly related to the project investment. This risk is mainly reflected in three aspects: regional climate change, environmental risks surrounding wind farm, and urban planning effect.

2.2 *Engineering risk*

The economics of wind power project largely depends on the scientificity and controllability of the wind farm selection and construction phase. At first, whether the wind farm site is reasonable directly determines the result of the wind power project. After entering the construction phase, the construction cost and the progressiveness of construction technology will affect desired objectives.

2.3 *Wind power generation technology risk*

Considering previous experience, the wind turbine replacement and the management and survivor capacity of equipment manufacturer also contribute to the risks. Wind power generation technology risk is mainly reflected in three aspects: wind power equipment manufacturing capacity, technology maturity, management and survivor capacity of equipment manufacturer.

2.4 *Market risk*

Intermittence and volatility of wind power have influence on stable operation of power grid, so power grid companies may have resentment based on the consideration of business interests. Besides, lagging energy storage technologies and renewable energy grid-connection technologies cannot ensure wind power integration in full, which means the reduction of wind power integration rate. In addition, the pricing policies also influence the economics of wind power project.

2.5 *Operation and manage risk*

Operation and manage risk refers to the loss uncertainty caused by the investors' blunders in the process of actual argument and organizing the implementation. It mainly includes: the implementation risks caused by errors in the previous feasibility study and measures, construction organization, supervision and management, and external coordination; the investment risk caused by the investment decision errors of wind farm construction; the accuracy of predicting the possible losses in construction and operation; the real-time tracking, reason finding and promptly resolving possible risk. Therefore, this kind of risk depends on investor's management experience, management team quality and external related resources.

According to the preceding analysis, an evaluation index system can be given and seen in Table 1.

Table1. Risk evaluation index system of wind power project.

Risk factors	Risk sub-factors
Wind resource risk U1	①regional climate change U11; ②environmental risks surrounding wind farm U12; ③urban planning effect U13
Engineering risk U2	①wind farm selection U21; ②construction cost U22; ③"three control" U23; ④construction technology U24
Wind power generation technology risk U3	①wind power equipment manufacturing capacity U31; ② technology maturity U32; ③management and survivor capacity of equipment manufacturer U33
Market risk U4	①wind power integration rate U41; ②electricity price policy U42; ③profit margin U43
Operation and manage risk U5	①investment management experience U51; ②team quality U52; ③external resources U53

3 GRAY HIERARCHY EVATLUATION MODEL OF WIND POWER PROJECT RISK

3.1 *Index weight determination of grey hierarchy evaluation*

This paper obtains expert weighting sets by the use of Group AHP and gets the mean after excluding the isolated views that are away from the center mode through pattern recognition so that the result is closer to the objective reality. At last, the first level indicator weight vector $W_U = (\omega_{U_1}, \omega_{U_2}, ..., \omega_{U_m})$ and second level indicator weight vector $W_{U_i} = (\omega_{i1}, \omega_{i2}, ..., \omega_{in_i})$ are obtained, in which $\omega_{ij} (i = 1, 2, ..., m; j = 1, 2, ..., n_i)$ represents the third level indicator weight vector.

3.2 *Scoring criteria and sample matrix of evaluation index*

Taking the maximum distinguishing ability of human thinking and the higher risk of investment project into account, the risk level is divided into 5 grades which are assigned with 1, 3, 5, 7 and 9. If the level is distributed among them, it can be assigned with 2, 4, 6 or 8.

Select t experts to score each indicator. Assuming a_{ijk} denotes the scores of U_{ij} from Expert $k\,(k=1,...,t)$, and then the scores of U_{ij} from all experts are $a_{ij1}, a_{ij2}, ..., a_{ijt}$. On this basis, an evaluation sample matrix $D = (a_{ij1}, a_{ij2}, ..., a_{ijt})$ of a particular investment project can be obtained.

3.3 *Determination of evaluation gray classes*

Set evaluation gray class number to be h, $h = 1, 2 \cdots, n$, which means that there are n classes. The grey classes mean different risk levels: lower risk, low risk, moderate risk, high risk and higher risk, namely $n = 5$. In order to describe these gray classes, the whitening weight functions of them are needed to be determined [8].

As to the 1st gray class-lower risk ($h = 1$), set gray number $\otimes_1 \in [0, 1, 9]$, and whitening weight function f_1 is as follows:

$$f_1(a_{ijk}) = \begin{cases} 1, a_{ijk} \in [0,1] \\ \dfrac{9 - a_{ijk}}{8}, a_{ijk} \in [1,9] \end{cases} \tag{1}$$

Similarly, the following can be obtained:

$$f_2(a_{ijk}) = \begin{cases} \dfrac{a_{ijk}-1}{2}, a_{ijk} \in [1,3] \\ \dfrac{9-a_{ijk}}{6}, a_{ijk} \in [3,9] \end{cases} \tag{2}$$

$$f_3(a_{ijk}) = \begin{cases} \dfrac{a_{ijk}-1}{4}, a_{ijk} \in [1,5] \\ \dfrac{9-a_{ijk}}{4}, a_{ijk} \in [5,9] \end{cases} \tag{3}$$

$$f_4(a_{ijk}) = \begin{cases} \dfrac{a_{ijk}-1}{6}, a_{ijk} \in [1,7] \\ \dfrac{9-a_{ijk}}{2}, a_{ijk} \in [7,9] \end{cases} \tag{4}$$

$$f_5(a_{ijk}) = \begin{cases} \dfrac{a_{ijk}}{9}, a_{ijk} \in [0,9] \\ 1, a_{ijk} \in [9,+\infty) \end{cases} \tag{5}$$

3.4 *Gray evaluation coefficient calculation*

Assuming the whitening weight functions of U_{ij} belonging to the h^{st} gray class are $f_h(a_{ij1}), f_h(a_{ij2}), \dots, f_h(a_{ijt})$. Thus, the total whitening weight function of U_{ij} belonging to the h^{st} gray class is $\sum_{k=1}^{t} f_h(a_{ijk})$, and its corresponding total whitening weight is $\sum_{h=1}^{n}\sum_{k=1}^{t} f_h(a_{ijk})$. So the gray class coefficient is $r_{ijh} = \frac{\sum_{k=1}^{t} f_h(a_{ijk})}{\sum_{h=1}^{n}\sum_{k=1}^{t} f_h(a_{ijk})}$, which indicates the membership degree of U_{ij} at h^{st} gray class.

3.5 *Comprehensive evaluation value calculation and correction*

Set gray evaluation vector to be $r_{ij} = (r_{ij1}, r_{ij2}, \dots, r_{ijp})$. Based on second level indicator weight vector $W_{U_i} = (u_{i1}, u_{i2}, \dots, u_{in_i})$, carry out second level comprehensive evaluation of U_i by use of fuzzy operator, and get the gray evaluation vector $U_i = W_{U_i} \circ R_i = (\tau_{i1}, \tau_{i2}, \dots, \tau_{ip})$, wherein $\tau_{ij} = \vee\{(w_{is} \wedge r_{sj}) | 1 \le s \le p\}$[9]. Among this, gray evaluation matrix R_i can be denoted by:

$$R_i = \begin{pmatrix} r_{i1} \\ r_{i2} \\ M \\ r_{in_i} \end{pmatrix} = \begin{pmatrix} r_{i11} & r_{i12} & M & r_{i1p} \\ r_{i21} & r_{i22} & M & r_{i2p} \\ M & M & M & M \\ r_{in_i1} & r_{in_i2} & M & r_{in_ip} \end{pmatrix}. \tag{6}$$

Carry out first level comprehensive evaluation of the total object U, and get the comprehensive evaluation vector $U = W_U \circ R = (\tau_1, \tau_2, \dots, \tau_p)$ of U, wherein $R = (U_1, U_2, \dots, U_m)^T$.

Based on the total object evaluation vector $U = (\tau_1, \tau_2, \dots, \tau_p)$, in accordance with the maximum membership principle, we can get $\tau^* = \max(\tau_1, \tau_2, \dots, \tau_p)$. If take the gray class corresponding to τ^* as risk level, there exists a certain probability of failure, so it needs to be corrected. This paper adopts the weighted average method for correction. Carry out normalization process to $U = (\tau_1, \tau_2, \dots, \tau_p)$. Take τ_i as the weight, and carry out weighted average for each evaluation level $V_K (\kappa = 1, 2, \dots, p)$. According to above assignments of the gray classes, let $\omega = (1, 3, 5, 7, 9)^T$, $W = U \times \omega$, and corresponding gray class of this corrected value is the corrected risk level.

4 NUMERICAL EXAMPLES

A wind farm is located in Hami, which is with 99MW of planning capacity and 66 wind turbines of Gold wind GW77/1500 type whose stand-alone capacity is 1500kW. This wind farm officially started construction on April 12, 2012 and its utilization hour is expected to be 2330h.

Using Group AHP, get the weight sets of U_i : $W_{U_1} = (0.433, 0.309, 0.258)$, $W_{U_2} = (0.266, 0.222, 0.373, 0.139)$, $W_{U_3} = (0.302, 0.216, 0.482)$, $W_{U_4} = (0.495, 0.23, 0.275)$, $W_{U_5} = (0.45, 0.229, 0.321)$. These sets all satisfy the consistency test.

Invite 5 experts to compose the assessment team. These experts respectively score the risk degrees based on three level indicators. The ample evaluation matrix is obtained:

$$D = \begin{bmatrix} 7 & 5 & 5 & 7 & 6 & 7 & 3 & 7 & 6 & 8 & 8 & 5 & 6 & 7 & 4 & 6 \\ 7 & 4 & 3 & 6 & 6 & 6 & 2 & 6 & 5 & 8 & 8 & 3 & 6 & 7 & 4 & 5 \\ 6 & 5 & 5 & 6 & 5 & 6 & 1 & 6 & 6 & 7 & 7 & 4 & 5 & 7 & 3 & 5 \\ 7 & 5 & 3 & 5 & 5 & 7 & 1 & 7 & 5 & 8 & 8 & 5 & 5 & 6 & 2 & 6 \\ 6 & 4 & 4 & 6 & 5 & 8 & 2 & 6 & 6 & 8 & 7 & 4 & 6 & 8 & 3 & 7 \end{bmatrix}^T \tag{7}$$

The gray weight matrixes are:

$$R_1 = \begin{pmatrix} 0.093 & 0.124 & 0.186 & 0.371 & 0.227 \\ 0.167 & 0.223 & 0.273 & 0.182 & 0.155 \\ 0.198 & 0.266 & 0.238 & 0.158 & 0.141 \end{pmatrix}$$

$$R_2 = \begin{pmatrix} 0.095 & 0.143 & 0.215 & 0.379 & 0.168 \\ 0.103 & 0.138 & 0.207 & 0.414 & 0.138 \\ 0.09 & 0.12 & 0.18 & 0.361 & 0.248 \\ 0.491 & 0.218 & 0.109 & 0.073 & 0.109 \end{pmatrix}$$

$$R_3 = \begin{pmatrix} 0.095 & 0.127 & 0.19 & 0.38 & 0.208 \\ 0.102 & 0.136 & 0.204 & 0.408 & 0.149 \\ 0.071 & 0.094 & 0.142 & 0.283 & 0.409 \end{pmatrix}$$

$$R_4 = \begin{pmatrix} 0.076 & 0.101 & 0.152 & 0.304 & 0.367 \\ 0.187 & 0.25 & 0.25 & 0.167 & 0.146 \\ 0.131 & 0.175 & 0.263 & 0.237 & 0.193 \end{pmatrix}$$

Carry out second level risk comprehensive evaluation and then first level comprehensive evaluation

$$R = \begin{bmatrix} W_{U_1} \circ R_1 \\ W_{U_2} \circ R_2 \\ W_{U_3} \circ R_3 \\ W_{U_4} \circ R_4 \\ W_{U_5} \circ R_5 \end{bmatrix} = \begin{bmatrix} 0.198 & 0.266 & 0.273 & 0.371 & 0.227 \\ 0.139 & 0.143 & 0.215 & 0.361 & 0.248 \\ 0.102 & 0.136 & 0.204 & 0.302 & 0.409 \\ 0.187 & 0.229 & 0.252 & 0.33 & 0.296 \\ 0.229 & 0.227 & 0.261 & 0.326 & 0.295 \end{bmatrix}$$, and

then first level comprehensive evaluation $U = W_U \circ R =$ $(0.187, 0.229, 0.263, 0.298, 0.298)$. After the normalization process, we can get $U' = (0.147, 0.18, 0.205, 0.234, 0.234)$. So we can obtain the corrected value $W = U \times \omega = 5.456$. According to the gray class corresponding to this corrected value, the risk level of this wind power project is moderate risk.

5 CONCLUSIONS

This paper makes quantitative research on China's wind power project risk:

1 According to the gray hierarchy model, the comprehensive evaluation value of this project is 5.456, and the conclusion is moderate risk. The result contributes to controlling and avoiding risks for investors.
2 Compared with other evaluation methods, this method maximizes the use of evaluation information of all kinds of gray class degrees, so that it avoids result failure and adapts to the project risk characteristics. Thus, this method is suitable for risk evaluation of wind power project.
3 It is helpful to comprehensively and intuitively understand wind power project risk and to take measures to minimize the negative impacts of quantitative mathematical evaluation model, which will effectively promote the healthy and rapid development of China's wind power industry.

ACKNOWLEDGEMENT

Fund Project: National Natural Science Foundation of China (71071054, 71271084); technology project of State Grid Corporation of China (521820140017)

REFERENCES

[1] Qiu Weidong, Wang Zhidong, Li Jun. (2008) Analysis on issues related to wind power development in China. *Electric Power Technologic Economics*, 20(1):17–19.

[2] Shen Youxing, Guo Lingli, Zeng Ming. (2008) Wind power development experience of Denmark and its reference to China. *East China Electric Power*, 36(11):153–158.

[3] Fang Chuanglin. (2007) Analysis and prospect of wind power development goals in China. *Energy of China*, 29(12):30–33.

[4] Chen Zhong, Kang Ning. (2003) Identification and analysis of risk factors for wind power projects. *Heilongjiang Electric Power*, 25(6):415–417.

[5] Ren Boqiang, Jiang Chuanwen. (2009) A review on the economic dispatch and risk management considering wind power in the power market. *Renewable and Sustainable Energy Reviews*, 13(8):2169–2174.

[6] Ouyang Wane. (2008) Risk analysis and management strategies of wind power investment in Fujian. *Development Research*, 12(7):37–38.

[7] Zhao Shanshan, Zhang Dongxia, Yin Yonghua, *et al.* (2011) Pricing policy and risk management strategy for wind power considering wind integration. *Power System Technology*, 35(5):142–145.

[8] Dong Fenyi, Xiao Meidan, Liu Bin, *et al.* (2010) Construction method of whitening weight function in grey system teaching. *Journal of North China Institute of Water Conservancy and Hydroelectric Power*, 31(3):97–99.

[9] Li Angui, Zhang Zhihong, Meng Yan, *et al.* (2009) *Fuzzy Mathematics and Its Applications*. Beijing: Metallurgical Industry Press.

Electric, Electronic and Control Engineering – Shao, Shu & Tian (Eds)
© 2015 Taylor & Francis Group, London, ISBN: 978-1-138-02842-5

Automatic control system of automotive light based on image recognition

Zhaona Lu & Xinjian Yuan
School of Automotive Engineering, Nantong Institute of Technology, Nantong Jiangsu, China

ABSTRACT: This paper discusses a new automatic control system of automotive light based on image recognition from the perspective of computer, proposes the analytical method of light image recognition, elaborates on deficiencies of the current automatic control system of automotive light based on image recognition, and studies the feasibility algorithm of image recognition in terms of automotive light. Results show that the algorithm features high recognition rate and short processing time so that the system can be adapted for real-time applications. It is finally verified through experiments that the algorithm of image recognition features relatively higher accuracy compared to manual measurement result.

KEYWORDS: image recognition; automotive light; gray level.

1 INTRODUCTION

Automotive lighting system is a major factor for automobile safety. Light condition not only concerns vehicle operation but also serves as an important guarantee of traffic accidents prevention and people's travel.

Power measurement and component sensing are adopted in domestic studies on automatic control system of lighting. But there are certain deficiencies in power measurement and position determination. There have been lots of efforts made on light recognition. For example, in the study of light monitoring system of image recognition, Shucai Zhang proposed a detection method of light detection equipment with computer's image recognition and designed a corresponding module, providing a practicable reference for the study on the algorithm of recognizing light through images. In regard to automatic image acquisition of machine vision, Xianjun Qiao proposed an automatic image recognition technology of liquid preparation of light detection, providing a reference for the study on light recognition through images.

Based on previous studies, this paper not only suggests that the model of the automatic control system of automotive light can be constructed with the new approach of image recognition, but also makes an argumentation with the comparison method that the new approach is reasonable and effective.

2 LIGHT PROCESSING MODEL OF IMAGE RECOGNITION

In order to realize reliable monitoring on the equipment state of automotive light, it needs to avoid interferences through software recognition under the condition of relatively normal information collection and to collect data under the normal condition of automotive operation.

Image recognition of automotive light needs to be processed in accordance with different lighting conditions. Preprocessing on light extraction is required in order to improve the recognition performance. Different lights need to be labeled in the form of image to enhance the control performance, including image conversion and binary processing.

2.1 *Color conversion of automotive light image*

Use a digital camera to extract three colors, namely blue, green and red. All colors in nature can be obtained by combining the three colors in different proportions. Monochromatic lights consist of 256 levels. The total number of colors is larger than the cube of 256.

In automotive light processing, relevant information of color would be contained in colored images, so images can be processed accordingly.

Images of gray level include 256 levels. Zero stands for black and white. Gray is interpenetrated between black and white. The gray-level image is labeled in this way.

The major basis of recognizing lighting system with the gray-level image is that the recognition of the gray-level image is quick and convenient and RGB gray-level images of corresponding color palette are the same. In addition, image data and the reality are corresponding. However, colored images of RGB are not the same. New RGB images can be obtained after the processing. Each pixel can be expressed with only one corresponding byte. Colored images of RGB need to be processed in different classifications so as to prolong the image processing time. All the information is preserved after the processing of images. Thus, images processed in this way have great advantages.

There are mainly three processing methods of gray-level image:

1 The method of maximum value

Take the maximum value in order to obtain a relatively high brightness of gray-level images.

$$R=G=B=max(R,G,B) \tag{1}$$

2 The method of average value

Take the average value of lights in three colors in order to obtain a relatively gentle brightness of images:

$$R=G=B=(R+G+B)/3 \tag{2}$$

3 The method of weighted average

Obtain indicators after weighting by assigning different values to three primary colors. Different weights have different effects.

$$R=G=B=(WrR+WgG+WbB)/(Wr+Wg+Wb) \tag{3}$$

The gray level transformation equation of relevant colored images is given below:

$$Y=0.299R+0.587G+0.114B \tag{4}$$

2.2 *Gray level image transformation*

The target light area of pixel needs to be strengthened in order to reduce the interference of background in the search of light images. The binary process of image is the conversion of two relatively obvious gray level images, namely black and white. Use 0 and 1 to represent the two colors and use two bytes to represent two relevant colors. Images created in this way are binary images. Pixels of binary image are normally selected between 0 and 1 in a relatively ideal circumstance.

If a gray level image is divided into N columns and M lines, the pixel of gray level is expressed with y. The pixel after processing is expressed with g. So, the corresponding function can be expressed as:

$$g(x,y)=\begin{cases} 0, f(x,y)<T \\ 1, f(x,y)\geq T \end{cases} \tag{5}$$

In the above formula, T stands for the threshold value. The gray level image becomes two distinct colors after the binary processing, namely black and white. Different threshold values can be selected by separating different results.

2.3 *Selection of threshold values in binary processing*

Gray-level values of image are the statistical results of the same image of all gray levels. Take the value of gray level between 0 and 255. The maximum value is expressed in white while the minimum value is expressed in black. The corresponding probability is expressed with p and the function is expressed as:

$$P(g_k)=\frac{n_k}{n}, \quad 0\leq g_k\leq 255 \tag{6}$$

The peak point and the valley point in the image need to be calculated for the selection of threshold value so as to find the minimum peak value. Corresponding codes are given below:

Input: gray level image

```
% the maximum entropy
clc
clear all
tic
b=imread('d:\image\tn_im7b.jpg','jpg');
subplot(1,2,1);
imshow(b);
ng=zeros(1,256);
gray=255;
[m,l]=size(b);
for i=1:1:m
    for j=1:1:l
       temp=double(b(i,j));
       if temp<=gray
           gray=temp;
       end
       ng(temp+1)=ng(temp+1)+1;
    end
end
for i=1:1:256
    hg(i)=ng(i)/(m*l);
end
for t=gray:1:255 % t is the gray level
    E0=0;
    E1=0;
    P0=0;
    for i=0:1:t
        P0=P0+hg(i+1);
    end
    P1=1-P0;
       if P0~=0
          for i=0:1:t
```

```
            R0=hg(i+1)/P0;
             if R0~=0
                E0=E0-R0*log(R0);
             end
        end
      end
      if P1~=0
        for i=t+1:1:255
            R1=hg(i+1)/P1;
             if R1~=0
                E1=E1-R1*log(R1);
             end
        end
      end
    n(t+1)=E0+E1;
    if t==gray
       Nm=n(t+1);
    end
    if n(t+1)>=Nm
      Nm=n(t+1);
      Topt=t;
    end
end
Topt
image1=ones(m,l);
for i=1:m
   for j=1:l
     if b(i,j)<=Topt
         image1(i,j)=0;
     end
   end
end
subplot(1,2,2);
imshow(image1);
toc
```

3 SBEL GRADIENT ALGORITHM OF IMAGE

Pixel values of image are different because intensities of automotive light are different. It is necessary to conduct gradient processing on images. Establish the gradient equation according to the rotational invariance of the partial differential equation:

$$G_{rad} f(x,y) = \begin{bmatrix} \frac{\partial f}{\partial x} \\ \frac{\partial f}{\partial y} \end{bmatrix} \tag{7}$$

The gradient modulus of the point is:

$$GM(x,y) = |G_{rad} f(x,y)| = \sqrt{(\frac{\partial f}{\partial x})^2 + (\frac{\partial f}{\partial y})^2} \tag{8}$$

Relatively clear images can be obtained through the above-mentioned process, the step of which is:

$$S_x(i) = f_i(x+1,y-1) + f_i(x+1,y+1) - f_i(x-1,y-1)$$
$$f_i(x-1,y+1) + 2f_i(x+1,y) - 2f_i(x-1,y)$$

$$S_y(i) = f_i(x-1,y+1) + f_i(x+1,y+1) - f_i(x-1,y-1)$$
$$- f_i(x+1,y-1) + 2f_i(x,y+1) - 2f_i(x,y-1)$$

$$G_s(i) = \sum_{(x,y)} (\sqrt{(S_x^2(i) + S_y^2(i))}) \tag{9}$$

Relevant threshold values can be obtained by comparing the gray value of each gradient in images. The gray value of pixel point is the gradient value expressed with large threshold value.

3.1 *Image segmentation*

Steps of solving appropriate threshold values are:

Measure the image grey level histogram and then select the initial threshold of relevant gray level range;

Calculate the initial threshold

$$T_0 = \frac{Z_1 + Z_L}{2} \tag{10}$$

Z stands for the maximum and the minimum gray levels.

Segment an image into a background image and a target image according to the above formula, and calculate the corresponding average gray value z, so:

$$Z_0 = \frac{\sum_{Z(i,j) < T_k} Z(i,j)^* N(i,j)}{\sum_{Z(i,j) < T_k} N(i,j)} \tag{11}$$

$$Z_B = \frac{\sum_{Z(i,j) < T_k} Z(i,j)^* N(i,j)}{\sum_{Z(i,j) < T_k} N(i,j)} \tag{12}$$

In the above formula, the gray value of (i, j) is expressed with z (i, j). N is the weight coefficient of relevant gray value, which is normally 0 and 1.

$$T_{k+1} = k(Z_0 + Z_B) \tag{13}$$

The iteration stops when it reaches the step k+1=k. Otherwise, the iteration goes back to the previous step.

The proportion of light images with byte pixel and background pixel is:

$$S = \frac{\sum_{Z(i,j)} < T_k^{Z(i,j)}}{\sum_{Z(i,j)} < T_k^{Z(i,j)}} \tag{14}$$

The proportion of byte pixel and background pixel is expressed with S, which also suggests whether the selection of threshold value is correct or not. Thus, there is a certain connection between S and k.

Namely:

When k=0.65, S>2.5

When k=0.55, S<1.8

S is constant when k is constant

The above-mentioned iterative coefficient represents an automatic regulation method. It is highly adaptable to automotive light recognition of image. Image thresholds are relatively accurate.

The study on the positioning method based on image

The histogram projection of vertical direction and horizontal direction is adopted under the condition that bytes of light and background are relatively larger. The number between the white pixel and the gray level value can be obtained through the projection of automotive light in vertical and horizontal direction and the corresponding image can be drawn as well. The gray level only contains black and white, so the wave trough and the wave peak can be obtained by segmenting the image horizontally. Then the edge is cut off after searching. The corresponding formula is:

$$\text{White(row)}=\sum_{col=1}^{n}\text{image(row,col)}\,row\in[1,m],\ col\in[1,n] \quad (15)$$

The vertical coordinate of each point in relevant images can be acquired through the above formula.

The corresponding horizontal coordinate can be acquired in a similar way. Images are obtained in line with the number of white pixel points.

Steps are given below:

Firstly, calculate

$$G(i,j)=\begin{cases}1,|F(i,j)-F(i,j+1)|!=0,|F(i,j)-F(i+1,j)|!=0\\0,\text{Others}\end{cases} \quad (16)$$

Secondly, calculate the horizontal projection, which is scanned on the condition that m plus 1and the gray level transition of black and white is satisfied. The value of m stands for the number of gray level transition. The number of transitions in each line is:

$$m=\sum_{i=2}^{n}|G(i)-G(i-1)| \quad (17)$$

Use $L(t>=T)\geq a$ for constraint. t is the number of transitions. L is the line number of continuous scanning that satisfying the condition. a is the corresponding experience number. Two edges of the left and the right can be obtained by projection.

Namely:

$$b(x,y)=f(top:bottom,left:right) \quad (18)$$

4 ALGORITHM VERIFICATION

The image processing algorithm of the automatic control system of automotive light recognition mainly involves processing at the early stage, complex edge detection at the later stage and image drawing. The preprocessing of image is actually the extraction of image filtering features and the transition of gray level. Features can be reflected through the background.

The data volume of image preprocessing is relatively large. FPGA is adopted in this paper for analysis and processing. The efficiency of the algorithm is improved by segmentation and recognition. An experiment is carried out in this paper in order to verify the features.

Images can be obtained according to the above-mentioned coordinate of each pixel point. Conduct systematic detection of the obtained images after manual measurement and make mutual comparisons. The table is given below:

Table 1. Manual experimental measurement and image recognition.

Number	Manual measurement/ mm	Image recognition system/mm	Relative error value
1	83.9	83.3	0.7
2	78.8	79.4	0.25
3	64.8	64.2	0.78
4	50.3	51.5	1.0
5	34.0	33.6	1.2

It can be concluded from the table that the relative error between the manual experiment results and image recognition is quite large. Meanwhile, image recognition features higher accuracy, the efficiency of which is higher than manual measurement.

5 CONCLUSION

In this paper, a series of studies are carried out on the automatic control system of automotive light recognition and corresponding images are drawn according to light features. It can be concluded through experimental comparison that image recognition is prompt and accurate, featuring good reliability and feasibility. It functions well in adapting different lights and is convenient for control and management.

REFERENCES

[1] Cheng, L.L. & Chen, W.H. (2010) Study on fast image matching algorithm of SMD defect detection. *Computer Applications and Software*, 27(11):108–131.

[2] Yanamura Y., Goto M., Nishiyama D., Soga M., Nakatani H., Saji H. (2008) Extraction and tracking of the license plate using Hough transform and voted block matching. *IEEE Intelligent Vehicles Symposium*, 5(2):232–242.
[3] Emiris D.M., Koulouriotis D.E. (2001) Automated optic recognition of alphanumeric content in car license plates in a semi-structured environment. *IEEE Transaction on Pattern Analysis and Machine Intelligence*, 7(8):120–127.
[4] Chang, S.L., Chen, L.S. & Chung, Y.C. *et al.* (2004) Automatic license plate recognition. *IEEE Transactions on Intelligent Transportation Systems*, 5(1):41–52.
[5] Lowe D.G. (1999) Object recognition from local scale-invariant features. *Proceedings of the 7th IEEE International Conference on Computer Vision*, Greece, pp:1149–1152.
[6] Lu, X.B., Zhang, G.H. *et al.* An accurate license plate location method based on binary image. *Journal of Southeast University (Natural science edition)*, 35(6):970–978.
[7] Hu, Y.C. & Yang, S.M. (2009) A study on image matching algorithm. Electronic Instrumentation Customer, (2):15–18.
[8] Dai, S.J., Huang, H., Gao, Z.Y., *et al.* Vehicle-logo recognition method based on Tchebichef moment invariant and SVM. *Proceedings of the 9th WRI World Congress on Software Engineering*, China, 2009, 21–22.
[9] Li, R. (2008) *A Study on Video-based Vehicle Detection and Tracking Technology*. Shanghai: Shanghai Jiaotong University, pp:24–28.
[10] Psyllos A., Anagnostopoulos C.N. & Kayafas E. (2010) Vehicle logo recognition using a SIFT-based enhanced matching scheme. *IEEE Transactions on Intelligent Transportation Systems*, 11(2):322–328.
[11] Wang, X.M., Gu, L.J. & Ren, R.Z. (2007) Target image detection based on improved template matching algorithm. *Journal of Jilin University (information science edition)*, 25(1): 32–38.
[12] M.H. terBrugge, J.H. Stevens, J.A.G. Nijhuisicense. (2008) Plate recognition of DTC-NNS. *Proceedings of the IEEE Inter-national Work shop on Cellular Neural Networks and their Applications*, 12(4):13–24.
[13] Guo, J. & Shi, P.F. (2002). License plate location method based on color and texture analysis. *Journal of Image and Graphics*, 7(5): 471–477.
[14] Wang, Y.N. et al. (2002) *Technology of Computer Image Processing and Recognition*. Beijing: Higher Education Press.

Electric, Electronic and Control Engineering – Shao, Shu & Tian (Eds)
© 2015 Taylor & Francis Group, London, ISBN: 978-1-138-02842-5

The research and application of in vitro diagnosis using smart grid dispatch data

Yuxiang Liao, Hongbing Li, Huaxing Yu, Feng Li, Zilong Mu
State Grid Chongqing Jiangbei Power Supply Subsidiary Company, Chongqing, China

ABSTRACT: In the face of the challenges in State Grid Chongqing Electric Power Company, Chongqing grid dispatch data monitoring and troubleshooting, particularly the incorrect remote data monitoring, jumping, telesignal point-list inconsistency and non-responsiveness to remote commands, as a result of the large volumes of data involved, a concept of in vitro diagnosis using smart grid dispatch data is introduced. The rationality of this solution is secured through key-point research of vitro diagnosis and this solution has already been successfully applied to a Chongqing electric power company.

KEYWORDS: Dispatch data; in vitro diagnosis; monitoring; telediagnosis; ethernet

1 BACKGROUND

Grid dispatch data involves all aspects of the electric power system and monitoring grid dispatch data is extremely significant on determining the normal functioning of the power grid. Identifying and diagnosing abnormal data present in a database is very critical for avoiding equipment failures or accidents. However, remote monitoring and diagnosis of electric power equipment is no longer impossible in the presence of powerful network technology and the colossal data system. Nevertheless, as the dispatch data system is fairly sizable and part of it has to be separately monitored and diagnosed, in vitro diagnosis has come into being to answer this call.

In vitro diagnosis, as opposed to in vivo diagnosis, monitors related data acquired by software from equipment and monitoring data and makes professional diagnosis on samples obtained from the matrix as commonly practiced in the medical field.

For the purpose of this paper, in vitro diagnosis monitors the service data of secondary grid equipment from a dispatch database and diagnoses any abnormality when needed. It is a professional monitoring and diagnosis technology designed for grid O&M data with more focus on timely data monitoring and professional diagnosis.

2 NECESSITY

While great importance is attached to the monitoring of secondary equipment, the monitoring and diagnosis of secondary equipment is still beyond the control of O&M departments due to the involvement of a colossal dispatch network. This is typically represented by:

1 The mass monitoring data. Dispatch data covers the whole spectrum of the electric power system from power generation to delivery, transformation, distribution through consumption. Monitoring therefore implies an arduous task. There are frequently cases when abnormalities in the monitoring data or the system are not discovered soon enough, when effective information is unavailable to address such abnormalities, and when potential risks in the monitoring system is not identified in time.
2 Impossibility to obtain useful insights into protocols. Widely acceptable O&M monitoring systems like HTTP, DNS, SMTP and POP3 are rarely used in an industrial control environment while it is impossible to obtain useful insights into those more commonly used in such environment.
3 Impossibility to reproduce the site. As remote data acquisition is performed by coordination among a number of systems, in the event of incorrect remote data monitoring, jumping, telesignal point-list inconsistency and non-responsiveness to remote commands, the station end is not able to restore historical information. This further adds to the difficulty of diagnosis and troubleshooting.

3 REALIZATION OF IN VITRO DIAGNOSIS

In our solution, Ethernet, a widely used LAN technology nowadays, is used to monitor online all data involved in a dispatch automation system to perform real-time data parsing, detection, analysis and judgment according to the respective telecontrol protocols, generate abnormality or failure alarm information, and inform and dispatch O&M personnel of the automation system via SMS platform to allow active alarm and quick treatment of remote failures.

3.1 *Network architecture of in vitro diagnosis*

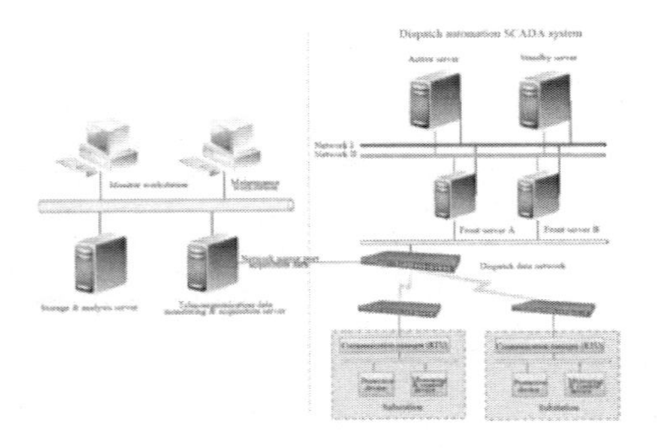

Figure 1. Network architecture of in vitro diagnosis

3.2 *Software framework of in vitro diagnosis*

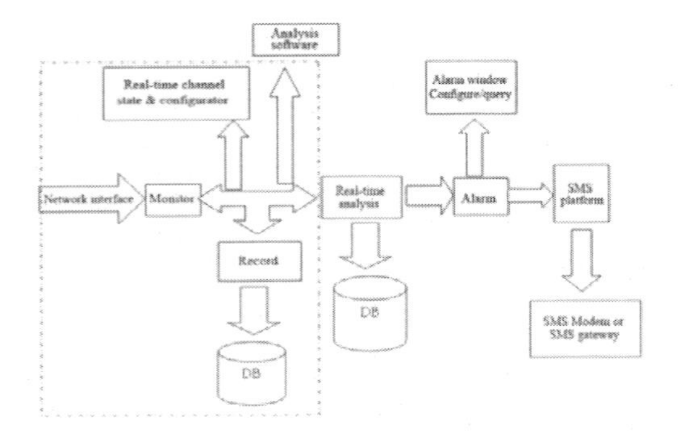

Figure 2. Software framework of in vitro diagnosis

3.3 *Key-point research of in vitro diagnosis*

1 Realization of Ethernet remote monitoring

Ethernet uses CSMA/CD (carrier sense multiple access and collision detection) protocol, the most widely used communication protocol standard among present LANs. It allows online monitoring and complete recording of all types of remote channels in a system, and also supports conventional digital channels. All message data on these channels are obtained from a mirror port without any influence on the communication of the dispatch automation system, thereby guaranteeing the data security of the monitored channels.

This technology allows continuous monitoring of data at all stations and through all communication channels, and review of all commands and uploaded information of any period anytime.

2 Real-time storage and parsing of batch remote data

This technique automatically and quickly stores remote channel data from online monitoring according to channel parameter configuration information and provides important raw communication data for historical analysis and failure analysis of remote data. In terms of real-time remote data analysis, frame details tree, cross-link list and frame data bits analysis to ensure accurate data analysis.

Frame details tree abstracts online monitoring data processing processes according to the tree theory following the principle of hierarchical parsing technique, parses monitoring data hierarchically by reference to related data parameters, and links the decomposed parameters into a link tree that covers all nodes. It also updates parameters locally and tree-parses or restructures them as necessary, decouples monitoring data to share things in common and adds or modifies the parsing or parameter processing methods anytime.

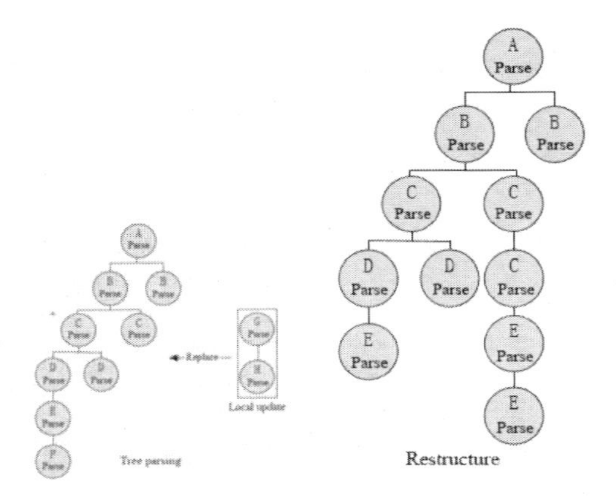

Figure 3. Sketch showing frame details tree data analysis technique

3 Real-time monitoring and alarming of remote data

This technique not only measures remote channel operation and alarms and informs by SMS any abnormality, but also monitors all downlink order information, including manually issued instructions, identifies causes of any failure by screening related data and informs related persons on such failure.

4 TECHNICAL INNOVATIONS

Technical innovations in this solution mainly include:

1. Ethernet allows online monitoring of all different types of remote channels in a system within the extensive LAN as well as online monitoring and recording of all data for analysis.
2. Remote data mass memory allows automatic and quick storage of remote channel data from online monitoring according to channel parameter configuration information for the first time.
3. By applying frame details technique, this solution ensures significantly higher intelligence and accuracy of in vitro diagnosis by means of tree parsing, local updates and restructuring.
4. Remote communication online detection and failure alarm technique detects all dispatch data in the system and the "four-tele" data of telemetering, telesignaling, telecontrol and teleadjusting, and identifies, alarms and deals with failures immediately, which greatly improve the automation and intelligence level of the system.

5 CONCLUSIONS

In vitro diagnosis is used to monitor secondary electric equipment data as a solution to the long-existing problem with remote dispatch including incorrect data monitoring, jumping, telesignal point-list inconsistency and non-responsiveness to remote commands. Application to a Chongqing power company proves greatly higher data monitoring accuracy and failure treatment timeliness for the smart grid dispatch.

REFERENCES

[1] Xing Ningzhe. (2012) Analysis and research of several problems with the North China Power Dispatch Data Network. *North China Electric Power*, 12(03):46–48.

[2] Liu Lirong, Liu Zongqi, Wang Yudong. (2010) Discussion on the Second Plane Construction Scheme of State Grid Dispatching Digital Network. *Sci-Tech Information Development & Economy*, 12(12): 99–101.

[3] Ni Jianhong, Lu Guanghong. (2009) Research on VPN-based different implementations. *Application Research of Computers*, 22(7):257–260.

[4] Liu Dongmei. (2011) Comparison of BGP/MPLSVPN and IPSecVPN. *China Data Communication Network*, 12(5):78–80.

[5] Guo Shigang. (2011) MPLS VPN technology of IP bearer network. *Modern Science & Technology of Telecommunications*, (9):2–3.

[6] M. Boucadair, P. Morand, I. Borges, et al. (2008) Enhancing the service ability of IMS-based multimedia services: preventing core service failures. *The Second International Conference on Next Generation Mobile Applications, Services and Technologies*, 9:444–449.

[7] Christakidis, A., Efthymiopoulos, N., Fiedler, (2011) J.VITAL++, a new communication paradigm: embedding P2P technology in next generation networks. *IEEE Communications Magazine*, 49(1):84–91.

[8] Li Haihua. (2011) Analysis of Data Transmit Process in BGP MPLS VPN. *Computer Technology and Development*, 21(6):5–Í6.

Electric, Electronic and Control Engineering – Shao, Shu & Tian (Eds)
 ISBN: 978-1-138-02842-5

Design and implementation of a novel parallel FFT algorithm based on SIMD-BF model

Shiceng Zhang*, Yunhua Li, Yi Li & Bin Tian
State Key Laboratory of Integrated Services Networks, Xidian University, Xi'an, China

ABSTRACT: The Fast Fourier Transform (FFT) algorithm has promoted the digital development of information systems fields immensely. Among the available algorithms, the operations including anti, left-shift and zero-padding, are required to obtain the W index (W=$e^{-j2\pi/N}$, where N denotes the length of the data used for the FFT operation). In order to reduce the complexity of those algorithms, this paper proposes a novel parallel FFT algorithm which has been utilized in the single instruction multiple data streams–butterfly (SIMD-BF) network. This algorithm acquires the law of W index easily to conduct recursive operations by making good use of the Decimation-In-Frequency (DIF) method. Analysis results demonstrate that the clearer and simpler law of the W index can be obtained, and that the number of complex multiplication and the quantities of processors and storage units which are required by this algorithm are halved in the SIMD-BF model. What's more, each level of the butterfly network has the same architecture, which leads to the lower requirement of the hardware. And at the same time, this algorithm is also applicable to the Single Instruction Multiple Data Streams-Perfect Shuffle (SIMD-PS) model.

KEYWORDS: SIMD-BF; FFT; Parallel Algorithm; Decimation-in-Frequency; Decimation-in-Time.

1 INTRODUCTION

Thanks to the speed limits of current devices, the parallel computing algorithms for high-performance computers have received a lot of attention [1]-[4] home and abroad. To date, parallel algorithms [5]-[8] and the Fast Fourier Transform (FFT) parallel algorithm [2]-[5] have been developed greatly. In the SIMD model, one of common parallel computing models, the same instructions are executed in a synchronous manner at the same time, which is particularly suitable for data intensive calculations such as many multimedia applications.

Therefore, the research on the parallel FFT algorithm based on the SIMD-BF model is of great significance. In this paper, a novel radix-2 DIF parallel FFT algorithm is proposed. This algorithm is based on the homeostatic geometrical structure of the flow diagram in the SIMD-BF model. In this model, the data sequence is inputted in the bit-reversed manner first of all. Then, the output data are the discrete Fourier transform (DFT) values of $x_N(n)$ after the input data are calculated with the specific law of the W index in the BF network. In addition, this algorithm with the simpler characteristics of the W index is also suitable for the implementation in the SIMD-PS model.

This paper is structured as follows. Section II briefly introduces the current FFT algorithms based on the SIMD-BF model. A novel parallel FFT algorithm based on the SIMD-BF model is proposed in Section III, and its basic principles and implementation are illustrated in detail. In addition, the advantages of this proposed algorithm over the existing parallel algorithms are also verified by taking the DFT operation of a data sequence with the length of $N=8$ for an example in this section. Section IV concludes this paper.

2 THE CURRENT FFT PARALLEL ALGORITHMS BASED ON THE SIMD-BF MODEL

2.1 *The basic principles of current FFT parallel algorithms [8]*

The existing algorithms are the implementations of the radix-2 FFT based flow diagram with the decimation-in-time (DIT) method in the BF network. Assume that there is a data sequence of length N. It is denoted by $x_N(n)$, $n=0,1,2,\ldots,N-1$. The basic principles of these algorithms are as follows.

*Corresponding author: 1187582207@qq.com, yunhuali0816@126.com, 2785554985@qq.com, btian@xidian.edu.cn

In the BF network, there are $(N+1)\log_2 N$ processors, arranged in $1+\log_2 N$ rows. Each row has a lot of the processors, and the processor on column i and row r is represented by $P_{r,i}$. $W^{\exp(r,i)}$ is distributed in each processor, where $\exp(r,i)$ is the element of the W index. Here, if the binary representation of i is $i=(a_0,a_1,\cdots,a_{N-1})_2$, then the index $\exp(r,i)=(a_r,a_{r-1},\cdots,a_2,a_1,0,0,\cdots,0,)_2$, i.e., inversing the first r bits of i, and then replacing the $(k-r)$ bits with 0. When the calculation begins, the original sequence $x_N(n)$ is placed into the memory of the processors on the row 0. The processors on the row 1 will calculate according to the corresponding inputs, including the complex additions and the complex multiplications. The intermediate calculation results are transmitted to the memory of the next row of the processors, until the entire recursive process is completed. The data in the memory of the last row of processors needs to undergo the bit-reversed address. Then the DFT values of the sequence $x_N(n)$ are obtained.

Fig. 1 shows the DIF-FFT flow diagram of a data sequence of the length $N=8$. In the figure, the "O" represents the nodes where the calculation takes place, while the "●" stands for the signal nodes.

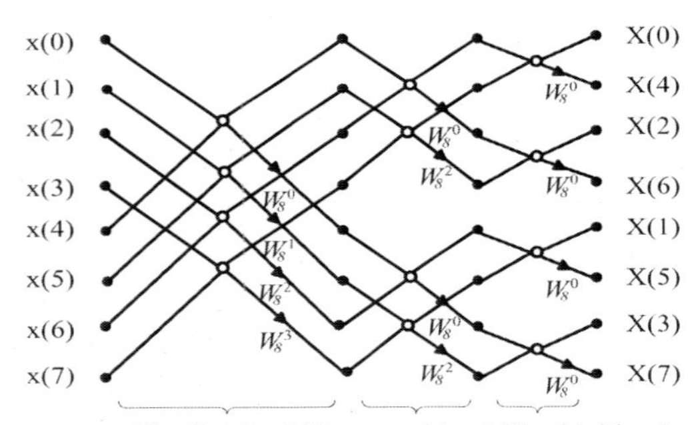

Figure 1. The dif-fft flow diagram of a data sequence of the length $N=8$.

2.2 *The implementation of the existing algorithms based on the SIMD-BF model[8]*

The implementation structure of the FFT operation on a data sequence of the length $N=8$ in the BF network is provided in Fig. 2.

In Fig. 2, the distribution of the elements of the W index on each row of the BF network is given as follows:

Row1(the first level)	W_8^0	W_8^0	W_8^0	W_8^0	W_8^4	W_8^4	W_8^4	W_8^4
Row2(the second level)	W_8^0	W_8^0	W_8^4	W_8^4	W_8^2	W_8^2	W_8^6	W_8^6
Row3(the third level)	W_8^0	W_8^2	W_8^4	W_8^6	W_8^1	W_8^3	W_8^5	W_8^7

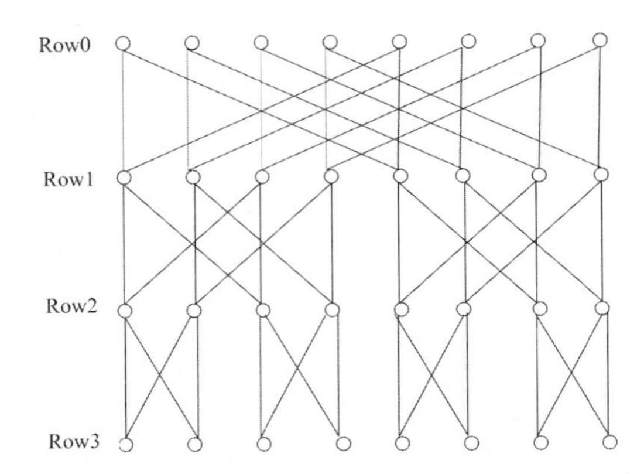

Figure 2. The implementation structure of the fft algorithm on a data sequence with the length $N=8$.

3 A NOVEL PARALLEL FFT ALGORITHM BASED ON THE SIMD-BF MODEL

3.1 *The fundamentals of the novel parallel FFT algorithm*

According to the analysis of the existing algorithms, this paper proposes a novel parallel FFT algorithm based on the SIMD-BF model. This algorithm forms a new flow diagram with a homeostatic geometrical structure by changing the flow diagram of the radix-2 DIF FFT in the current algorithms.

In this algorithm, the calculation nodes are kept to connect with the fixed memory units and to store the fixed W index all the time. No matter how the calculation nodes are transformed, all of the flow diagrams are always equivalent and the correct output results will be obtained. However, the data sequence is extracted from memory units and is stored into memory units in a different order.

On the basis of the analysis above, Fig. 3 presents the BF flow diagram with a homeostatic geometrical structure when the length of the data sequence is $N=8$. From the figure, it is very clear that the top half of the transmission coefficients in each level is all 1. Enter a data sequence $x_N(n)$ as the input data after the bit-reversed addressing. And then the outputs are the DFT values of $x_N(n)$ in the right order.

3.2 *The implementation of this algorithm*

In the BF network, the processors of the first $N/2$ columns can be replaced by the adders. The adder or the processor on row r and column i is denoted by $P_{r,i}$. The subscript (r, i) refers to the coordinates of row r and column i, where $0\le i\le N-1$ and $1\le r\le k$. Here, the row r corresponds to the above mentioned calculation level in the calculating process. According to

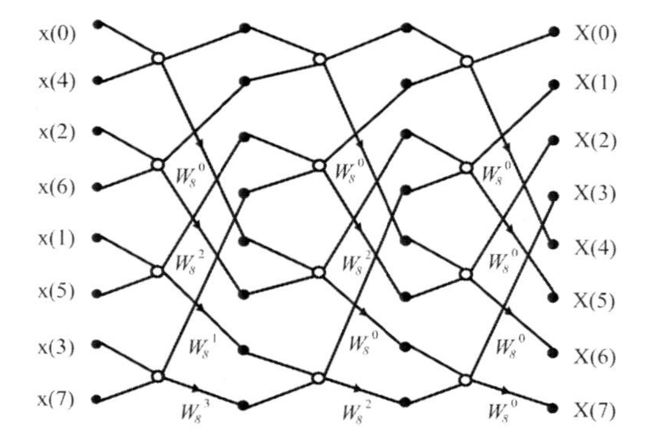

Figure 3. The BF flow diagram with a homeostatic structure when $N = 8$.

the solution for the W index shown above, the power of the element in the W index at coordinates (r, i) in the BF network is defined as $\exp(r, i)$. If $i < N/2 - 1$, $\exp(r, i)$ is equal to 0. When the column i is equal to or more than $N/2$, we set up $j = (i - N/2)_{10} = (m_0, m_1, m_3, \ldots, m_{k-1})_2$, and $j = (m_{k-1}, m_{k-2}, \ldots, m_0)_2$. Then we obtain the index of the row r and the column i $\exp(r, i) = (m_{k-1}, m_{k-2}, \ldots, m_0, 0, 0, \ldots 0)_2$, i.e., the last $(r - 1)$ bits of j are 0.

When the calculation begins, $x_N(n)$ undergoes bit-reversed addressing to become the new sequence $x_N(n)$. $x_N(n)$ is placed in parallel in the memory of the processors on row 0. The processors on row 1 will perform the calculations about their corresponding input data. The intermediate calculation results are transmitted to the memory units of the next row of processors, until the whole recursive process is completed. The values stored in the last row of processors are the DFT values of the sequence $x_N(n)$.

Detailed description of this algorithm is presented as follows:

input: $x_N(n)$ //$x_N(n)$ is a random input sequence
output: $X(i) = d_{k,i}$ // $X(i)$ is an output sequence
begin
(1) bitrev $(x_N(n), N)$ // $x_N(n)$ is inversed
for $n = 0$ to $N - 1$ parallel do
$d_{o,n} \leftarrow x_N(n)$
end for
(2) for $r = 1$ to k do
for $j = 0$ to $N/2 - 1$ parallel do
$d_{r,j} \leftarrow d_{r-1,2j} + d_{r-1,2j+1}$
$d_{r,j+N/2} \leftarrow (d_{r-1,2j} - d_{r-1,2j+1}) \times W^{\exp(r,j)}$
end for
end for
end

Fig. 4 provides the implementation structure of the FFT algorithm on a sequence with the length of $N = 8$ in the SIMD-BF model.

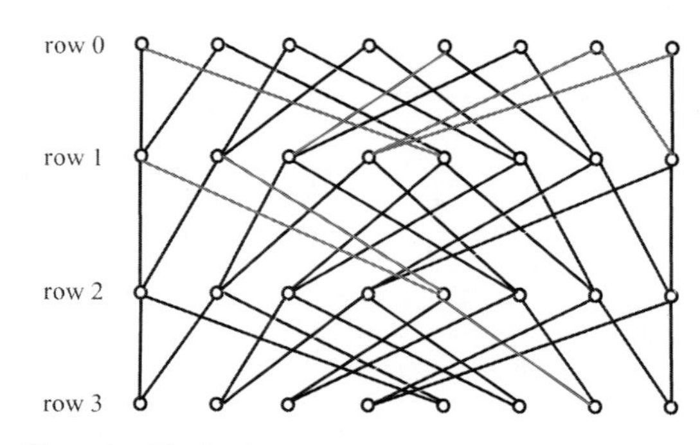

Figure 4. The implementation structure of the FFT algorithm on a sequence with the length $N = 8$.

In Fig. 4, the distribution of the elements of the W index on each row of the BF network is as follows:

Row1(the first level)	W_8^0	W_8^0	W_8^0	W_8^0	W_8^0	W_8^2	W_8^1	W_8^3
Row2(the second level)	W_8^0	W_8^0	W_8^0	W_8^0	W_8^0	W_8^2	W_8^0	W_8^2
Row3(the third level)	W_8^0	W_8^0	W_8^0	W_8^0	W_8^0	W_8^0	W_8^0	W_8^0

Hence, the analysis of this algorithm is summarized as follows:

- The connection method of each stage of the BF network is the same, i.e., the pathways of the signals are the same, which is also suitable to the characteristic of SIMD-PS model.
- The elements of the W index in the processors of the first N/2 columns on each row have a power of 0. Thus, there is no need for the storage units but only the additions need to be performed. The adders can take the place of these processors. As a result, the number of the processors and storage units is reduced greatly.
- The number of the complex multiplications is reduced by half.

Table 1. Complexity comparisons between the existing alogrithms and the proposed algorithm.

	The existing algorithms	The proposed algorithm
Number of Processors	$(\log_2 N + 1) \times N$	$[(\log_2 N + 1) \times N]/2$
Number of Complex Multiplications	$(\log_2 N \times N)$	$(\log_2 N \times N)/2$

CONCLUSIONS

Because of the superiority of this parallel FFT algorithm, we will continue to do some researches on the parallel algorithm based on the SIMD-PS model [6]-[8]. As for the different variations of the FFT flow diagram, as long as the positions of the calculation nodes and their elements in the W index remain constant, the results of the FFT operation are unchanged. The only difference lies in the extraction and memory of the data sequence. The algorithm proposed in the paper results in the 0 values of the powers of the elements to locate at the top N/2 columns of W index matrix. Through the analysis, we obtain the simpler characteristics of the W index, and that the FFT operations are implemented on the SIMD-BF model on the basis of such characteristics conveniently and flexibly. The implementation results show that this proposed algorithm not only lowers the requirements for the hardware devices, but also reduces the calculation complexity of FFT operations, as shown in Table 1.

ACKNOWLEDGMENT

This work is supported by Transponding Satellite Navigation System.

REFERENCES

[1] Schatzman, James C. (1996) Accuracy of the discrete Fourier transform and the fast Fourier transform. *SIAM Journal on Scientific Computing*, (17)5:1150–1166.

[2] Hwang, Kai, and Zhiwei Xu. (1998) *Scalable Parallel Computing: Technology, Architecture, Programming*. McGraw-Hill, Inc.

[3] Jones, Douglas L. (2006) Decimation-in-frequency (DIF) radix-2 FFT. Washington, USA.

[4] He, Shousheng, and Mats Torkelson. (1998) Design and implementation of a 1024-point pipeline FFT processor. Custom Integrated Circuits Conference, *Proceedings of the IEEE 1998*.

[5] Jamieson, Leah H., Philip T. Mueller Jr, and Howard Jay Siegel. (1986) FFT algorithms for SIMD parallel processing systems. *Journal of Parallel and Distributed Computing*, 3(1):48–71.

[6] Kalé, L., Skeel, R., Bhandarkar, M., Brunner, R., Gursoy, A., Krawetz, N. & Schulten, K. (1999) NAMD2: greater scalability for parallel molecular dynamics. *Journal of Computational Physics*, 151(1):283–312.

[7] Ullman, Jeffrey D. (1984) Computational aspects of VLSI. Computer Science Press.

[8] Stone, Harold S. (1971) Parallel processing with the perfect shuffle. *IEEE Transactions on Computers*, 20(2):153–161.

Electric, Electronic and Control Engineering – Shao, Shu & Tian (Eds)
 ISBN: 978-1-138-02842-5

Research and implementation of CellSense biosensor based on environmental engineering

Qilin Liu
Department of Tea and Food, Wuyi University, Wuyishan Fujian, China

ABSTRACT: With biological components as functional identification components, biosensors convert non-electric signals of perceived materials into measurable electric information. It is of great significance for the current environmental pollution abatement in our country. In order to construct a new type of biosensor pointedly, this paper takes the returned sludge of the first sewage treatment plant of Jingtang Company, Shougang Group as the bacterial source, screens Psychrobacter (PS) and Bacillus Subtilis (BS) of the highest strain matching rate with RT1 and RT2 through biochemical properties and physiological tests of bacterial and the similarity identification of 16SrDNA, and constructs two types of new CellSense biosensors with the two bacilli as indicator microorganisms. It can be concluded from experimental tests that: bacillus subtilis and psychrobacter perform well in toxicity analysis; as indicator microorganisms, bacterial strains at the late growth stage make the toxicity test of the two biosensors more accurate; the influence on the output current of the two types of new biosensors is relatively small when the salinity of the base solution for analysis is lower than 3%; the influence on the output current of the two types of new biosensors is also small when the pH value of the base solution for analysis ranges between 6 and 8.

KEYWORDS: Biological component; culture cycle; salinity; pH value; inhibition ratio; output current

1 INTRODUCTION

At present, human life has been severely influenced by the exhaustion of ecological resources and the aggravation of environmental pollution. Therefore, people have an even stronger sense of survival crisis for their dependable environment. It is an important thing to conduct accurate detection in order to improve the current ecological environment. People are in urgent need of means and methods for sustained, rapid and online environmental quality monitoring. Biosensor [1] has drawn extensive attention because of its advantages like high sensitivity, low cost, simple and convenient, rapid, good reproducibility and online inspection. Practical results and broad application prospects of the environmental detection have won the consensus of counterparts at home and abroad [2].

Main applications of biosensor in environmental monitoring include aquatic monitoring, pesticide control and atmospheric monitoring. Hongbin Luo (2014) pointed out that biochemical oxygen demand (BOD) is the most common indicator for the monitoring of the organic matter pollution degree in water. Water eutrophication is an important factor for water pollution. The total number of bacteria is an important monitoring index of water quality and the concentration of heavy metal ions in waste water is also an important monitoring index [3]. S Jouanneau et al. (2014) summarized various methods of determining BOD and believed that biosensor determination is a method consuming the least time with less market use [4]. Shi H et al. (2013) developed an automatic online optical biosensor system (AOBS), which is able to carry out sustained and real-time monitoring on micro-cystic toxins LR (MC-LR) with high sensitivity and specificity [5]. Qianyu Zhang et al. (2013) designed an electric potential sensor based on chemoautotrophic ammonia oxidizing bacteria that can be used for the detection of pollutants in water [6]. Roberto et al. (2002) developed a biosensor of bacterial fluorescein with green fluorescent protein (GFP), which determines arsenite and arsenate to the level of sub-microgram with good effect and less cost. It is able to conduct long-term online monitoring on arsenic contaminated regions [7]. Qimei Luo et al. (2007) believed that enzyme biosensors have advantages in terms of pesticide leftover detection that are incomparable by traditional detection methods [8]. Hua Zhou et al. (2008) fixed bovine serum albumin and acetylcholine esterase on the surface of carbon nanotube electrodes on graphite electrode substrate. This new type of biosensor can be used in the detection of organophosphorus pesticides, which performs well in the detection of parathion-methyl, dimethoate and DDVP [9]. Jianping Li et al. (2003) developed an electrochemical biosensor, which detects herbicides on site rapidly [10]. Zhongming Lu et al. (2003) developed a CO_2 fiber optical chemical sensor, which features small volume and low energy consumption and is suitable for long-term on-site automatic monitoring. The optimal accuracy is +1.21ppm and the response time (99%) is about 2 minutes when the concentration range of CO_2 is 0-194ppm [11]. Charlesp et al. (2000) composed a microbial sensor with immobilized nitrifying bacteria, porous permeable membrane and oxygen electrode. It is able to determine the concentration of NO_x in the air indirectly, the detection limit of which is 1×10^{-8}mol/L [12]. Carballor et al. (2003) developed a sensor by combining the flow injection analysis with modified glassy carbon of poly-porphyrin and nickel compound that is simply manufactured with low cost. It detects the content of SO_2 rapidly and effectively, the detection limit of which reaches 0.15mg/L [13]. Yantao Zhu et al. (2008) analyzed comprehensively the research status of formaldehyde gas sensors at home and abroad, summarized properties of metallic oxide sensors, surface acoustic wave sensors, electrochemical sensors and chemical luminescence sensors, pointed out problems of formaldehyde gas sensors, and predicted the development trend [14].

Based on previous studies, this paper carries out a study on the construction of new CellSense biosensors with psychrobacter (PS) and bacillus subtilis (BS) as indicator microorganisms and conducts relevant tests for the performance so as to provide a theoretical basis as well as guidance for the widespread use of biosensor in daily life.

2 OPERATING PRINCIPLE OF BIOSENSOR

Hongbin Luo (2013) pointed out that biosensor is a kind of chemical sensor that detects on the basis of biological reaction. With bioactive and immobilized biological sensitive materials like microorganism, enzyme, antigen and antibody as functional identity components, biosensor converts external influences on physicochemical properties into electric signals through signal conversion components, which are then amplified through signal amplification devices, so as to realize quantitative determination of materials [15]. The basic principle of biosensor is to produce response signals through interactions between detected materials and biological sensing components, which are received, processed, converted and output through signal processing components, so as to realize analysis and detection on materials [3]. The schematic diagram of biosensor is shown in Figure 1 [3].

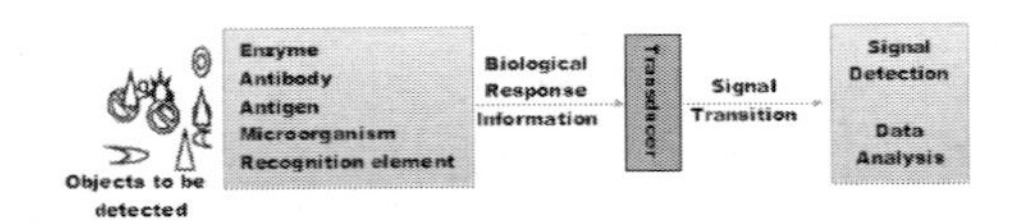

Figure 1. Schematic diagram of biosensor.

Biosensor can be generally classified from aspects of biological recognition components (enzyme sensor, microbial sensor, tissue and organ sensor, organelle sensor, immune-sensor, DNA sensor, etc.), determination indexes (sensor of electrode active material determination and sensor of respiratory activity determination), signal converters (electrochemical biosensor, semi-conduct biosensor, olfactory biosensor, calorimetric biosensor, photometric biosensor, acoustic determination biosensor, etc.), generation methods of output signal (biologically affinitive, metabotropic or catalytic biosensor), objects of converter (pH converter, O_2 converter, CO_2 converter, NH_3 converter, etc.) and measuring signals (electric potential sensor, Ampere sensor, impedance sensor, sonic sensor, etc.) [16]. Based on biological recognition components, this paper selects typical heavy metal ions, organic compounds, surface active agents and antibiotics as the main research objects according to features of combined pollution of heavy metals and organic matters and screens psychrobacter (PS) and bacillus subtilis (BS) as indicator microorganisms so as to construct a new type of CellSense biosensor. It also studies the application performance of this biosensor in toxicity analysis of binary mixtures and single toxins like heavy metals, organic matters, surface active agents and antibiotics.

3 THE CONSTRUCTION OF CELLSENSE BIOSENSOR BASED ON PS AND BS

CellSense biosensor is a kind of whole-cell current-mode microbial sensor based on oxidation-reduction mediums. The operating principle is: through oxidation-reduction mediums, electrons generated by the respiration of microorganisms fixed on electrode surface are transmitted between microbial cells and electrode surface and biological current signals are detected by the system. If there are toxic pollutants, respiration of microbial cells are reflected in the change of current intensity due to the inhabitation. The toxicity can be determined in accordance with the influence degree of pollutants on bioelectric current. The evaluation index is 50% effective concentration (EC_{50}). Major components of the sensor are Module and Data collecting and processes. The analysis system and micro electrodes of CellSense biosensor are shown in Figure 2 [17]. This system has such advantages as wide detection range, fast detection speed, flexibility, high sensitivity, good repeatability and good exploitability [18].

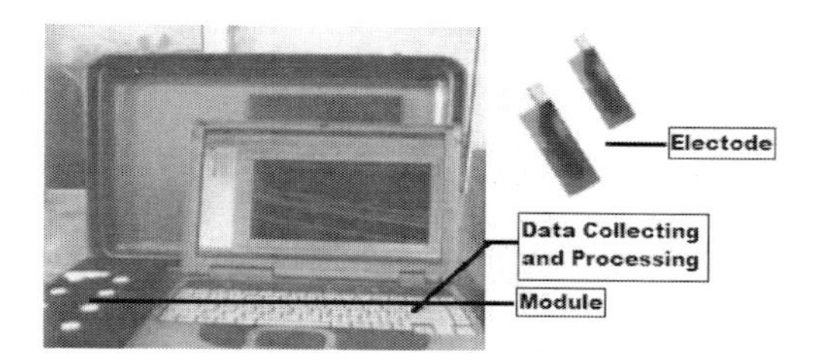

Figure 2. Diagram of the analysis system and micro electrodes of the biosensor.

The length and the thickness of the major structure of the micro electrode of CellSense biosensor are respectively 3cm and 0.5mm. The material is polycarbonate (studies show that [19] Nuclepore Track-Etced Membranes, Whatman with a bore diameter of 0.4μm is the best). Both sides are printed with conductive graphite electrode and Ag/AgCl solid reference electrode, shown in Figure 2. Specific steps of preparing microbial electrode with the direct fixing method are given below:

STEP1. Take cultivated PS and BS in fixed quantities for a centrifugal treatment of 5000rpm for 30 minutes. Pour some clear solution and transfer the wet bacteria to a centrifugal tube of 1.5mL. Wash the bacterial with 0.85% NaCl solution for three times. Pour out the solution after a centrifugal treatment of 10000rpm for 2.5 minutes and remove the liquid attached to the surface of the wet bacterial with a pipette.

STEP2. Take 0.5g wet bacterial and put it into 200μl 0.85% NaCl solution. Mix up the bacterial and the solution with a sterilized toothpick. Draw 7μl of the solution with a pipette and apply on the electrode surface evenly. After drying at room temperature, seal polycarbonate with a double faced adhesive tape in the surface area of the bacterial and preserve in a refrigerator of 4°C

Operating steps of CellSense biosensor are as follows:

STEP1. Put the biological electrode in a respiratory substrate of 0.85% NaCl solution for 30 minutes.

STEP2. Fixed the biological electrode at the end with microbial cells and immerge it in flat glass bottle with 10mL respiratory substrate.

STEP3. Click "start test" [20] and then add oxidation-reduction mediums after stabilization for 5 to 10 minutes. Add tested samples for toxicity test after the stabilization of electric signal (for about 15 minutes).

Output current signals of all biosensor electrodes still have differences in the same blank state due to the differences of fixed quantity and living condition of microbial cells. But the trend of change has a certain consistency. Therefore, in the calculation, it is necessary to carry out normalization processing on the output current of different microbial electrodes in the first place. And then calculate inhabitation rate of different time nodes in line with normalized current. The formula is shown below:

$$\text{Inhibition}(\%) = 1 - \left(I_{m,n} \times I_{ctrl,0}\right) / \left(I_{m,0} \times I_{ctrl,n}\right) \quad (1)$$

In the formula, $I_{m,n}$ is the nth current value recorded by the mth biosensor after the beginning of the test. $I_{m,0}$ is the current value recorded by the mth biosensor after the beginning of the test. $I_{ctrl,n}$ stands for the nth current value recorded by the biosensor used as a comparison after the beginning of the test. $I_{ctrl,0}$ stands for the current value recorded by the biosensor used as a comparison at the beginning of the test. The beginning of the test refers to the moment when tested samples are added in.

4 EXPERIMENTAL ANALYSIS

4.1 *Identification of sensitive microbial strains and growth curve drawing*

Culture mediums used in the experiment include broth peptone medium (RT), Czapek medium (CS), potato culture medium (TD), Gause No.1 medium (GS) and LB culture medium. Refer to literature [17] for specific preparation.

The returned sludge of the first sewage treatment plant of Jingtang Company Shougang Group in Tangshan Caofeidian is taken as the bacterial source. Two strains of bacteria are respectively named as RT1 and RT2.

Physiological characteristics of bacterial can be determined by A- gram staining, B- capsule staining, C- spore staining, D- catalase experiment, E- amylolysis, F- methyl red test and G-gelatin experiment. Refer to literature [21] for specific experimental steps. Cultivate RT1 and RT2 separately with scoring method so as to isolate single bacterial colonies. Observe morphological characteristics of colonies on LB culture medium+ agar. There are significant differences between the two bacterial strains in morphological characteristics. The bacterial colony of RT1 is round and wet with neat edges, smooth surfaces, swelling planes, high viscosity and no pigment. The bacterial colony RT2 is milk white and round with swelling surfaces, folded surfaces, irregular edges and micro jags. It can be known that the two bacterial strains are rod-shaped bacteria by observing morphological characteristics with an ordinary optics microscope. Results of physiological tests of the two bacterial strains in literature [21] are shown in Table 1:

Table.1 Testing results of major physiological indicators of RT1 and RT2

Types	A	B	C	D	E	F	G
RT1	Positive	With a capsule	No spore	Positive	Negative	Positive	Positive
RT2	Positive	With a capsule	With a spore	Positive	Positive	Negative	Positive

Reaction systems and reaction conditions of 16SrDNA sequence similarity analysis of bacterial strains are shown in Table 2:

Table.2 Reaction systems and reaction conditions

Reaction systems	Volume of sterilized water	Reaction conditions		
		Reaction temperature	Reaction time	Time of cycle
Degeneration reaction liquid	1 μl	94°C	5 min	1 cycle
PCR Premix	25 μl	94°C	1min	
Forward primer(20pmol/μl)	0.5 μl	55°C	1min	30 cycle
Reverse primer2(20pmol/μl)	0.5 μl	72°C	1.5min	
16S-free H_2O	23 μl	72°C	5min	1 cycle
Total	50 μl			

Experimental methods are shown below:

1 Degeneration: inoculate LB solid medium with bacterial strains for cultivation of 24 hours. Pick a small amount of bacterial from the culture medium and put into 10μl sterilized water for centrifugation. Take the supernatant as the template.
2 PCR amplification: use TaKaRa 16Sr DNA Bacterial Identification PCR Kit (Code No.D310) to amplify target fragment with PCR.
3 Sequencing: the forward primer (Seq forward) and the reverse primer (Seq reverse) of 16SrDNA similarity analysis are respectively 5'-GAGCGGATAACAATTTCACACGG-3' and 5'-CGCCAGGGTTTTCCCAGTCGAC-3'. 16SrDNA makes homologous comparison between sequencing results and 16SrDNA sequence listed in GenBank with BLAST software. Find homologous sequences in GenBank database with Blast tool and software like Cluastx and PhyloDraw provided by NCBI according to sequencing results. Extract results of DNA and amplification products of PCR are tested with 0.8% agarose gel electrophoresis. The electrophoresis buffer solution is 0.5×TBE with a voltage of 150V and electrophoresis time of 20 minutes. The concentration of ethidium bromide (EB) is 1μg/mL.

Results of 16SrDNA similarity analysis of RT1 and RT2 are as follows: 16SrDNA basic groups of RT1 and RT2 are respectively 1453bp and 1464bp; PS has relatively higher homology with RT1, the matching ratio of which is above 98%; BS has relatively higher homology with RT2, the matching ratio of which reaches 100%.

Inoculate a certain amount of microorganisms from a slant to a 150mL conical flask filled with 5mL LB culture medium with an inoculating loop. The temperature is 30 and the rotating speed of the shaker is 150rpm. After 24 hours' cultivation, inoculate the culture solution again to the LB culture medium of 50mL in an inoculation concentration of 0.2% for bacterial amplification. According to the growth rate of bacterial, take a sample for the determination of OD_{600} (Optical density at 600 nm) at intervals of appropriate time [22]. Growth curves of PS and BS can be determined by the method of turbidimetry. Results are shown in Figure 3. It can be known from Figure 3: the growth of PS in 4-20h is an increase of geometric progression, which is in the logarithmic phase of a bacterial growth cycle; the variation of OD_{600} is trivial in 24-40h and the growth rate of PS is quite slow, which is in the late phase logarithmic growth and a stable phase; the value of OD_{600} presents a downward trend after 40h and the growth of PS is in a decreasing phase. The growth of BS in 0-16h is an increase of geometric progression, which is in the logarithmic phase of a bacterial growth cycle; the variation of OD_{600} is trivial in 16-36h and the growth rate of PS is quite slow, which is in the platform stage of logarithmic growth; the value of OD_{600} presents a downward trend after 36h and the growth of PS is in a decreasing phase.

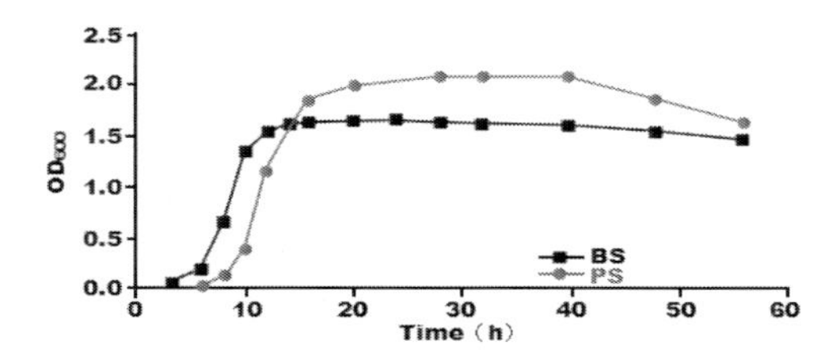

Figure 3. Growth curves of PS and BS.

4.2 *Toxicity test of CellSense biosensor based on PS and BS*

This section carries out an analysis on the influence of three factors, namely microbial growth cycle, salinity and pH value, on toxicity analysis of CellSense biosensor based on PS and BS so as to provide guidance for practical applications of new type of biosensor.

Microorganisms in different growth stages have different metabolic characteristics [17]. For this reason, analysis capabilities of biosensors prepared by microorganisms of different cultivating stages are also different. According to the influence of microbial growth cycle on toxicity analysis performance of biosensor, this paper takes wet bacteria of the late stage, stationary stage and decreasing stage of logarithmic growth of BS and PS and prepares BS and PS microbial electrodes with the direct fixing method of polycarbonate membrane, which are respectively used for the toxicity analysis of Cd^{2+}. Inhibition curve of Cd^{2+} on BS and PS are shown in Figure 4:

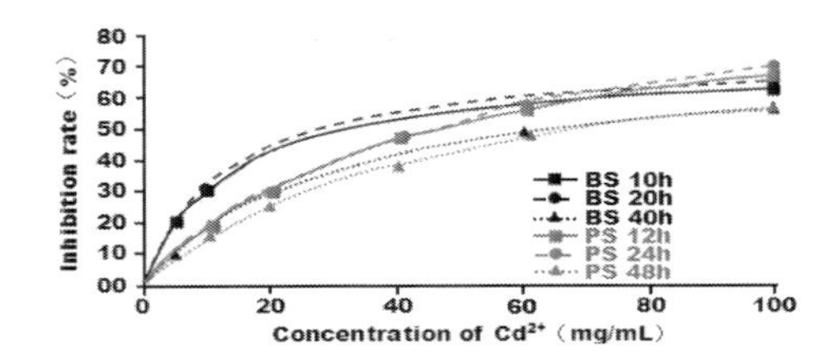

Figure 4. The influence of BS and PS on toxicity analysis of biosensor at different cultivating stages.

It can be concluded from Figure 4: biosensors prepared at 10h and 20h of the BS cultivating cycle are more sensitive to the acute toxicity of Cd^{2+}. Inhabitation rates of the two are similar. The sensitivity of toxicity analysis of the biosensor prepared at 40h of the cultivating cycle declines dramatically. So, BS in the late stage and the stable stage of logarithmic growth is more sensitive to the toxicity of metal ions. But the sensitivity on toxicity pollutants declines with the decrease of cell activity in the decreasing stage. Biosensors constructed based on PS have the same sensitivity of toxicity of metal ions with biosensors constructed based on BS in terms of trend. In conclusion: the bacteria number of microorganisms in the logarithmic phase is smaller than that in the stable phase. Therefore, it is more accurate to conduct toxicity test of heavy metal ions and organic matters with biosensors prepared by BS of 20 hours' cultivating cycle and PS of 24 hours' cultivating cycle.

Add NaCl solution of different concentrations (0.5%-10%) to respiratory substrate after the stabilization of current signals of electron transfer medium added in CellSense biosensor. Test the influence of salinity on current signals of CellSense biosensors based on BS and PS. Test results are shown in Figure 5:

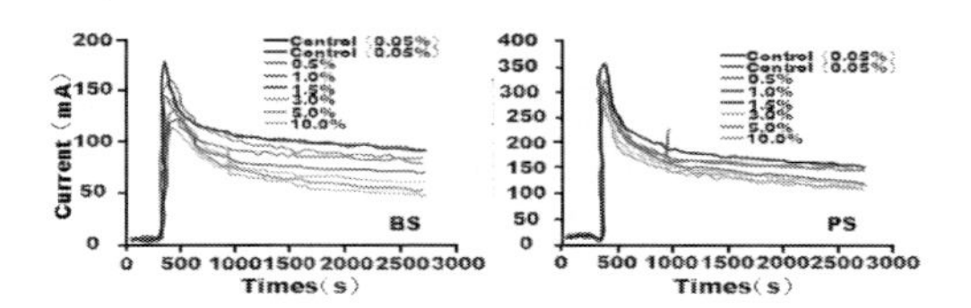

Figure 5. Influence of salinity on current signals of CellSense biosensors based on BS and PS.

It can be known from Figure 5: current difference between BS and PS biosensors are trivial when the salinity of analysis solution is lower than 3%. But current signals of CellSense biosensor are reduced obviously when the salinity is higher than 5%. Therefore, both PS and BS are in normal active state when the salinity is lo wer than 3% and the activity is inhibited significantly with the rising of salinity. It would be better to control the salinity of analysis solution under 3% when CellSense biosensors of BS and PS are used for toxicity analysis.

Regulate the pH value of respiratory substrate with 10% HCl and NaOH solutions after the stabilization of current signals of electron transfer medium added in CellSense biosensor so as to test the influence of pH value on current signals of two CellSense biosensors based on BS and PS.

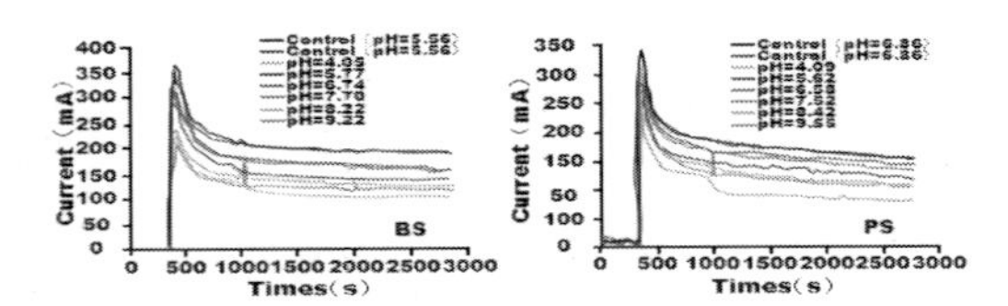

Figure 6. Influence of pH value on current signals of CellSense biosensors based on BS and PS.

From pH values in Figure 6: the current of biosensor decreases when pH<6; the smaller the pH value is, the more obvious the current of biosensor decreases; there is a fluctuation of current signal at the instant of adding samples when pH>9. And then the current recovers with no changes. So, it would be better to control the salinity of analysis solution within 6-8 when CellSense biosensors of BS and PS are used for toxicity analysis.

5 CONCLUSION

Based on the analysis on operating principle of biosensor and the identification of RT1 and RT2 bacterial strains, this paper constructs two types of new CellSense biosensors of bacillus subtilis and psychrobacter and tests the influence of different cultivating cycles, salinities and pH values on two types of biosensors. The following conclusions can be drawn according to the discussion and experimental results of this paper:

1 Bacillus subtilis and psychrobacter perform well in toxicity analysis.
2 The growth of bacillus subtilis and psychrobacter is divided into logarithmic phase, late growth phase, stable phase and decreasing phase. Bacterial strains of different cultivating stages have significant differences in toxicity analysis. It is more accurate to conduct toxicity test of heavy metal ions and organic matters with biosensors constructed by bacterial strains of late growth phase.
3 The influence on output current of two CellSense biosensors is relatively small when the salinity of analysis solution is lower than 3% while the influence becomes large when the salinity is higher than 5%.
4 The influence on output current of two CellSense biosensors constructed by bacillus subtilis and psychrobacter is relatively large in acidic condition while the influence is small in alkaline condition.

References

[1] Kuswandi B, et al. (2001) Optical fiber biosensors based on immobilized enzymes. *Analyst*, 126(8): 1469–1491.

[2] Wu, B.C. et al. (1999) *Modern Environmental Monitoring Technology (1st Edition)*, Beijing: China Environmental Science Press.

[3] Luo, H.B. (2014) Applications of biosensor in environmental monitoring, *Journal of Dongguan Institute of Technology*, 21(5):78–82.

[4] S. Jouanneau, et al. (2014) Thouand. Methods for assessing biochemical oxygen demand (BOD): A review. *Water Research*, 49(1):62–82.

[5] Shi. H., et al. (2013) Automated online optical biosensor system for continuous real-time determination of microcystin-LR with high sensitivity and specificity: Early warning for cyan-toxin risk in drinking water sources. *Environmental Science Technology*, 47:4434–4441.

[6] Zhang, Q.Y., et al. (2013) A potentiometric flow biosensor based on ammonia-oxidizing bacteria for the detection of toxicity in water. *Sensors*, 13(6):6936–6945.

[7] Roberto F F., et al. (2002) Evaluation of a GFP reporter gene construct for environmental arsenic detection Talanta, 58(1):181–188.

[8] Luo, Q.M., et al. (2007) Research progress of enzyme fixing technology of pesticide residue monitoring biosensor, *Chemical Sensors*, 27(3):17–22.

[9] Zhou, H., et al. (2008) Application progress of CNT modification enzyme biosensor, *Sensor World*, (4): 6–10.

[10] Li, J.P., et al. (2003) A study on thylakoid membrane biosensor based on herbicide detection by antagonism, *Chemical Journal of Chinese Universities*, 24(3):404–409.

[11] Lu, Z.M., et al. (2003) The development of fiber optical chemical CO_2 sensor, *High Technology Letters*, (11):38–43.

[12] Charles P T, et al. (2000) On-site immune-analysis of nitrate and nitro-aromatic compounds in groundwater. *Environmental Science Technology*, 34(21):464–465.

[13] Carballor R, et al. (2003) Sensors and Actuators B. *Chemical*, (88):155–161.

[14] Zhu, Y.T., et al. (2008) China measurement technology, *Measurement Technology*, 34(1):100–104.

[15] Luo, H.B. (2013) Application of modern molecular biology techniques in environmental monitoring, *Journal of Dongguan Institute of Technology*, 20(3):73–77.

[16] Luo, Z.Y., et al. (2005) Biosensors used for environmental monitoring, *Biotechnology*, (4):95–98.

[17] Zhao, H.N. (2009) *A Study on the Toxicity Analysis Performance of CELLSENSE Biosensor Based on New Toxic Sensitive Strains*, Shanghai: Tongji University.

[18] Farte M., et al. (2001) Toxicity assessment of organic pollution in waste waters using a bacterial biosensor. *Analytical Chemistry*, 426(2):155–165.

[19] Gu, M.B., et al. (2001) Soil biosensor for the detection of PAH toxicity using an immobilized recombinant bacterium and a bio-surfactant. *Biosensors & Bioelectronics*, 16 (9):667–674.

[20] Bhatia R., et al. (2003) Combined physical-chemical and biological sensing in environmental monitoring. *Biosensors & Bioelectronics*, 18(6):669–672.

[21] Huang, X.L. (2000) *Microbiological experiment instructions*, Beijing: Higher Education Press.

[22] Hou, L.P., et al. (2002) A study on acute toxicity and joint toxicity of cadmium and zinc on grass carp, *Freshwater Fisheries*, 32(3):44–46.

Electric, Electronic and Control Engineering – Shao, Shu & Tian (Eds)
© 2015 Taylor & Francis Group, London, ISBN: 978-1-138-02842-5

New construction methods for the optimum Grassmannian sequences in CDMA

Lixin Wang, Huan Hu*, Junbin Yang
College of Communication Engineering, Hangzhou Dianzi University, Hangzhou Zhejiang, China

ABSTRACT: CDMA code sequences constructed with Equiangular Tight Frames (ETFs) are called the optimum Grassmannian sequences which, when used in the CDMA system, minimize inter-user interferences and, when there's a change in the activated users, correlate these interferences to the number of code sequences only. The challenge is, however, that ETFs are not so easy to construct. This paper proposes a new method on the basis of best antipodal spherical coding and improved genetic algorithm and presents how a (10, 16) real-valued ETFs and a (11, 12) binary ETF are realized through numerical experiments and the existing Conference Matrix, Alternating Projection Algorithm and Genetic Algorithm become invalid under the same dimensions. In terms of construction a binary ETF, this new method provides higher convergence rate than some normal genetic algorithms.

KEYWORDS: CDMA; code sequence; Grassmannian frames; ETF; MWBE

1 INTRODUCTION

A CDMA system distinguishes signals from different users with the code sequence of these users and different codebooks could make much difference to the system performance. Within the frame theory, a Grassmannian frame is one that minimizes the maximum correlation among all frame elements, and a frame is called the optimum Grassmannian frame or an equiangular tight frame (ETF) when this correlation meets the Welch bound (WB) [1]. ETFs are widely used in compressive sampling [2,3] and MIMO [4] since the correlation among their frame elements is the lowest and that between any two frame elements is the same. In the same way, ETFs are very suitable when used to construct frequency spread sequences of a CDMA system, which are widely concerned as the optimum Grassmannian sequences or MWBE (maximum Welch bound equality) sequences [5-7]. However, how to construct an ETF is a widely known challenge [8] and an ETF exists only when the dimension m and the number of column vectors are certain given values [9,10]. Each of the few existing construction methods has its own limitations. Conference Matrix, for example, only constructs ETFs at $N = 2m$ provided that $N = p^a + 1$ for a real-valued frame or $N = 2^{a+1}$ for a Complex-valued frame, where p is an odd prime number and a is a nonnegative integer [11], while Alternating Project Algorithm [12,13] and Genetic Algorithm (GA) [14,15] are positioned to construct low-dimension Grassmannian frames only. In tthis paper, a real-valued ETF construction method based on best antipodal spherical codes (BASC) and a binary ETF construction method based on simulated annealing genetic algorithm (SAGA) are proposed. Numerical simulation indicates that, compared with some of the existing construction methods, andboth the two new methods proposed have better dimensional fitness and higher convergence rate.

*Corresponding author: *hh02188171@126.com*

2 OPTIMUM GRASSMANNIAN SEQUENCE

Let us suppose that N sequences are contained in the code sequence set of a synchronous CDMA system, w_k is the code sequence of user k sized $m \times 1$, and $\|w_k\| = 1$, $k = 1, 2, \cdots, N$. We define matrix $W = [w_1, \cdots, w_N]$, then the relation between the total squared correlation (TSC) of W and the maximum correlation functtion $I_{max}(W)$ is [16]:

$$\mathrm{TSC}(W) = \sum_{k=1}^{N}\sum_{l=1}^{N} \left|w_k^H w_l\right|^2 \geq \frac{N^2}{m} \tag{1}$$

$$I_{max}(W) = \max_{1 \leq k < l \leq N} \left|w_k^H w_l\right| \geq \sqrt{\frac{N-m}{(N-1)m}} \tag{2}$$

The right-side parts of (1) and (2) are the WB. The equality of (1) holds up that is, when TSC is the minimum, W is called a WBE (Welch Bound Equality) sequence under which the total system capacity is the largest [17]. In the frame theory, a WBE sequence is actually a uniform tight frame [18]. When (2) equalizes, W is called the optimum Grassmannian sequence or a MWBE sequence. The condition under which (2) equalizes is:

$$\left|\boldsymbol{w}_k^{\mathrm{H}}\boldsymbol{w}_l\right| = \sqrt{\frac{N-m}{(N-1)m}}, \forall k \neq l \tag{3}$$

That is, the correlation between any two sequences is the same. As such, the optimum Grassmannian sequence is actually an ETF.

The best merit of a system using WBE code sequences is that, when all users are activated, each user is subject to the same inter-user interferences [5]. This, however, is generally not maintained by the new activated codebook once there is a change in the activated users when a user leaves the system. For example, a WBE sequence codebook contains N sequences of which only $N \leq N$ sequences are activated. When $N = N$, each user is subject to the same inter-user interferences, but when $m \leq N < N$, the non-MWBE codebook will no longer bear this feature. The "equiangular" nature of a MWBE sequence correlates the inter-user interference of a system to the number of activated users only and not to the exact sequences involved. Under a fixed number of activated users, the inter-user interference is able to stay unchanged.

3 BASC-BASED REAL-VALUED ETF CONSTRUCTION

Let us use $\Omega_m(0,1)$ to indicate the unit sphere in m-dimensional Euclidean space $\boldsymbol{R}^m$ whose center of circle is the origin of coordinates. A finite vector set $\{\boldsymbol{s}_k\}_{k=1}^{N}$ comprising N points on $\Omega_m(0,1)$ is called a spherical code recorded as $\boldsymbol{C}_S(m,N)$. A best spherical code (BSC) is a spherical code when all points on the sphere are most "uniformly" distributed, recorded as $\boldsymbol{C}_{BS}(m,N)$. When a spherical code is a BSC, either the smallest angle between vectors is the maximum, or the maximum inner product between vectors is the smallest. That is:

$$\boldsymbol{C}_{BS}(m,N) = \min_{C_S}\left\{\max_{\substack{1\leq i,j\leq N\\ i\neq j}}\langle \boldsymbol{s}_i, \boldsymbol{s}_j\rangle\right\} \tag{4}$$

To find a BSC, we can think of N points on $\Omega_m(0,1)$ as N particles that move freely on the sphere and these particles have certain repulsion against each other, the magnitude of which repulsion is decided only by the distance between the particles so that these particles start from an initial state and gradually move to a steady state, maximizing the minimum distance between the particles. This balanced state can be expressed as [19]:

$$\left\{\boldsymbol{s}_k = \frac{\sum_{l\neq k}\left[(\boldsymbol{s}_k - \boldsymbol{s}_l)/\left|\boldsymbol{s}_k - \boldsymbol{s}_l\right|^{v}\right]}{\left|\sum_{l\neq k}\left[(\boldsymbol{s}_k - \boldsymbol{s}_l)/\left|\boldsymbol{s}_k - \boldsymbol{s}_l\right|^{v}\right]\right|}\right\}_{k=1}^{N} \tag{5}$$

Here: v is a natural number larger than or equal to 2. Equation (5) can be understood as under the balanced state, the direction in which particle $\boldsymbol{s}_k$ received the composite of forces from other particles is collinear with the direction of $\boldsymbol{s}_k$. To make it easier to write, for any vector u, we assume $\underline{\boldsymbol{u}} = \frac{\boldsymbol{u}}{|\boldsymbol{u}|}$, then we get the relation below out of (5):

$$\left\{\underline{\boldsymbol{s}}_k = \underline{\sum_{l\neq k}\frac{\boldsymbol{s}_k - \boldsymbol{s}_l}{\left|\boldsymbol{s}_k - \boldsymbol{s}_l\right|^{v}}}\right\}_{k=1}^{N} \tag{6}$$

We define the mapping below:

$$\Phi\left[\boldsymbol{C}_S(m,N)\right] = \left\{\underline{\boldsymbol{s}_k + \alpha\underline{\boldsymbol{h}}_k}\right\}_{k=1}^{N} \tag{7}$$

Here: α is the attenuation coefficient; $\underline{\boldsymbol{h}}_k$ is the right-side part of (6) and can be regarded as the composite of forces received by particle k from other particles; $\alpha\underline{\boldsymbol{h}}_k$ is the displacement of $\boldsymbol{s}_k$ under these forces. Given the right magnitude of α, the iteration process

$$\boldsymbol{C}_S(m,N)^{(l+1)} = \Phi\left[\boldsymbol{C}_S(m,N)^{(l)}\right], \quad l = 0,1,\ldots \tag{8}$$

will converge. Using the iteration process of (8), as long as v is large enough, the resulting spherical code will be very close to a BSC.

An antipodal code, recorded as $\boldsymbol{C}_{AS}(m,2N)$, can be obtained from any non-antipodal spherical code, i.e. $\boldsymbol{C}_{AS}(m,2N) = \left[\boldsymbol{C}_S(m,N) \quad -\boldsymbol{C}_S(m,N)\right]$ so that the antipodal point of any codeword is still a codeword. As described for BSC above, the BASC $C_{BAS}(m, 2N)$ can also be obtained from the equation below:

$$C_{BAS}(m, 2N) = \min_{C_{AS}} \left\{ \max_{\substack{1 \le i, j \le 2N \\ i \ne j}} \langle s_i, s_j \rangle \right\} \quad (9)$$

Here: $\max_{\substack{1 \le i, j \le 2N \\ i \ne j}} \langle s_i, s_j \rangle$ can be decomposed below according to the distribution of s_i and s_j in the set $\{s_k\}_{k=1}^{2N}$:

$$\max_{\substack{1 \le i, j \le 2N \\ i \ne j}} \langle s_i, s_j \rangle = \max \left\{ \max_{\substack{1 \le i, j \le N \\ i \ne j}} \langle s_i, s_j \rangle, \max_{1 \le i < N+1 \le j \le 2N} \langle s_i, s_j \rangle, \max_{\substack{N+1 \le i, j \le 2N \\ i \ne j}} \langle s_i, s_j \rangle \right\} \quad (10)$$

Order $X = \max_{\substack{1 \le i, j \le N \\ i \ne j}} \langle s_i, s_j \rangle$, $Y = \max_{1 \le i < N+1 \le j \le 2N} \langle s_i, s_j \rangle$, $Z = \max_{\substack{N+1 \le i, j \le 2N \\ i \ne j}} \langle s_i, s_j \rangle$.

As for Z, according to the symmetric nature of the codebook, we can assume $s_i = -s_{i_A}$, $s_j = -s_{j_A}$, where $1 \le i_A, j_A \le N$ and $i_A \ne j_A$, then we have

$$Z = \max_{\substack{1 \le i_A, j_A \le N \\ i_A \ne j_A}} \langle -s_{i_A}, -s_{j_A} \rangle = \max_{\substack{1 \le i_A, j_A \le N \\ i_A \ne j_A}} \langle s_{i_A}, s_{j_A} \rangle = X \quad (11)$$

In the same way, for Y, we may assume $s_j = -s_{j_A}$, where $1 \le j_A \le N$. If we want the maximum of $\langle s_i, s_j \rangle$, we have to satisfy $s_i \ne -s_j$. So $j_A \ne i$, then we have

$$Y = \max_{\substack{1 \le i, j_A \le N \\ i \ne j_A}} \langle s_i, -s_{j_A} \rangle = \max_{\substack{1 \le i, j_A \le N \\ i \ne j_A}} -\langle s_i, s_{j_A} \rangle = \max_{\substack{1 \le i', j' \le N \\ i' \ne j'}} -\langle s_{i'}, s_{j'} \rangle \quad (12)$$

According to (11) and (12), we get:

$$\max_{\substack{1 \le i, j \le 2N \\ i \ne j}} \langle s_i, s_j \rangle = \max \left\{ \max_{\substack{1 \le i, j \le N \\ i \ne j}} \langle s_i, s_j \rangle, \max_{\substack{1 \le i', j' \le N \\ i' \ne j'}} -\langle s_{i'}, s_{j'} \rangle \right\} = \max_{\substack{1 \le i, j \le N \\ i \ne j}} \{ |\langle s_i, s_j \rangle| \} \quad (13)$$

and then

$$C_{BAS}(m, 2N) = \min_{C_{AS}} \left\{ \max_{\substack{1 \le i, j \le N \\ i \ne j}} |\langle s_i, s_j \rangle| \right\} \quad (14)$$

Hence, when an antipodal spherical code is a BASC, the maximum correlation among the first N codewords is the minimum or, in other words, $\{s_k\}_{k=1}^{N}$ constitutes a Grassmannian frame.

When construction a BASC, according to (6), we can simplify $\underline{h}_k$ into

$$\left\{ \underline{h}_k = \frac{\sum_{l \ne k} \left(\frac{s_k - s_l}{|s_k - s_l|^v} + \frac{s_k + s_l}{|s_k + s_l|^v} \right)}{} \right\}_{k=1}^{N} \quad (15)$$

The iteration process then is

$$C_S(m, 2N)^{(l+1)} = \Phi\left[C_S(m, 2N)^{(l)}\right], \quad l = 0, 1, \cdots \quad (16)$$

The Grassmannian frames algorithm constructed based on BASC can be summarized below:

1 Initialize the dimension m, the number of column vectors N, the attenuation coefficient α=0.9 and v=2;

2 A random spherical code $C_S(m, N)$ is produced, out of which an antipodal spherical code $C_{AS} = [C_S \quad -C_S]$ is derived;

3 Calculate $\underline{h}_k = \frac{\sum_{l \ne k} \left(\frac{s_k - s_l}{|s_k - s_l|^v} + \frac{s_k + s_l}{|s_k + s_l|^v} \right)}{}$, $k = 1, \cdots, N$;

4 We get a new spherical code $C_S(m, N) = \{\underline{s_k + \alpha \underline{h}_k}\}_{k=1}^{N}$ and a new antipodal spherical code $C_{AS} = [C_S \quad -C_S]$;

5 Repeat steps 3) and 4) until we satisfy $|\underline{h}_k - s_k| < \varepsilon$, $k = 1, \cdots, N$, where ε is the threshold which is a very small constant;

6 Order $v = 2v$, $\alpha = \frac{1}{v-1}$;

7 Return to step 3) until v comes to a large enough value; and

8 Obtain the Grassmannian frame $\{f_k\}_{k=1}^{N} = \{s_k\}_{k=1}^{N}$ and end the computation.

The simulation below assumes $m = 7$, $8 \le N \le 40$. The maximum correlation of the resulting frame is given in Figure 1 below. When N is 8, 14 and 28, the maximum correlation of the frame meets the WB, or the resulting frame is an ETF coincident with the conditions for the existence of an ETF discussed in reference [7].

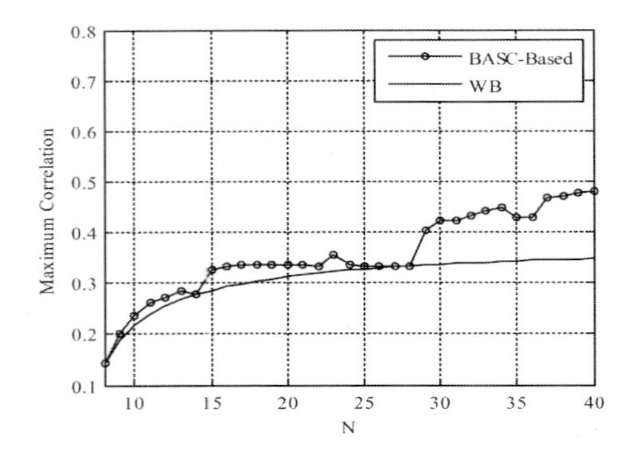

Figure 1. Maximum correlation of a Grassmannian frame at $m = 7$ constructed under BASC construction

Below is the realization result of a (10, 16) real-valued ETF:

$$\left[\begin{matrix}
0.0671 & 0.1121 & 0.8087 & -0.4008 & -0.0911 & 0.0057 & -0.1970 & 0.4798 \\
-0.2121 & 0.0154 & -0.0217 & -0.4923 & 0.1257 & -0.2889 & -0.6752 & -0.1907 \\
-0.3660 & -0.0960 & 0.3056 & -0.3003 & 0.3807 & -0.3563 & 0.3217 & -0.3738 \\
-0.4689 & 0.2024 & 0.1793 & 0.2951 & -0.4754 & 0.1375 & 0.0338 & -0.1212 \\
0.2035 & 0.1800 & 0.0911 & -0.3227 & -0.3307 & 0.2129 & -0.1818 & -0.5332 \\
-0.4218 & 0.3076 & -0.1873 & 0.1885 & 0.1555 & 0.5288 & -0.0039 & 0.0579 \\
0.4707 & -0.1611 & 0.3384 & 0.4157 & -0.5015 & -0.0524 & -0.3725 & -0.1624 \\
0.3228 & 0.4573 & 0.1511 & 0.2517 & 0.2196 & -0.5868 & 0.2535 & 0.5130 \\
-0.1788 & -0.4512 & -0.1600 & -0.0012 & 0.1217 & -0.0604 & 0.1624 & 0.0146 \\
0.1401 & 0.6095 & 0.1174 & 0.2096 & 0.3961 & 0.3080 & -0.3718 & -0.0399
\end{matrix}\right.$$

$$\left.\begin{matrix}
-0.1763 & -0.1438 & -0.3178 & -0.5018 & -0.0538 & 0.1225 & 0.0811 & -0.4820 \\
-0.3753 & -0.3720 & 0.0503 & 0.4292 & -0.0175 & -0.3908 & -0.2709 & -0.0564 \\
-0.1038 & 0.5534 & -0.1476 & 0.2215 & -0.1962 & 0.0008 & 0.5390 & 0.0670 \\
-0.5001 & -0.1525 & 0.0078 & 0.1974 & -0.5912 & 0.5903 & -0.0847 & -0.0058 \\
0.2054 & -0.2651 & 0.2642 & -0.4498 & -0.5559 & -0.2581 & -0.0032 & 0.3815 \\
-0.1224 & -0.4191 & -0.1981 & -0.0220 & 0.2581 & -0.2931 & 0.7203 & 0.1729 \\
-0.3346 & 0.3411 & 0.0220 & 0.1004 & 0.3245 & -0.2550 & 0.2983 & 0.1614 \\
-0.2869 & -0.1498 & 0.1364 & 0.0222 & -0.3177 & -0.2372 & 0.0873 & 0.4711 \\
-0.5602 & 0.0486 & 0.5183 & -0.4712 & -0.1444 & -0.4295 & -0.0347 & -0.4385 \\
0.0432 & 0.3525 & 0.6913 & 0.2012 & 0.0774 & 0.1599 & 0.0755 & -0.3793
\end{matrix}\right]$$

4 SAGA-BASED BINARY ETF CONSTRUCTION

In a real CDMA system, for easier realization, the code sequences are generally finite elements as exemplified by the commonly seen binary sequences (the sequence value is ± 1) and quaternary sequences (the sequence value is $\pm j$ or $\pm j$). Hence it is necessary to explore for a construction algorithm for binary ETFs. For this purpose, we improved the normal genetic algorithm (GA) into a simulated annealing genetic algorithm (SAGA) and used this algorithm to realize the construction of a binary ETF.

Genetic algorithm is a global optimization search algorithm that follows laws based on laws similar to the Darwinian Theory and natural inheritance. When genetic algorithm is used to construct a Grassmannian frame, each individual represents a frame. Under the action of the fitness function, the population continues to evolve before a frame under which the maximum correlation is the lowest, i.e. a Grassmannian frame, is eventually obtained. The overall process of SAGA is summarized below:

1 Randomly initialize 200 $m \times N$ binary frames as a population;
2 Calculate the fitness function of each individual in the population and prioritize the individuals by fitness;
3 Evaluate the best individual. Stop if the requirements are met and proceed to the next step if not;
4 Select, cross over, mutate, migrate; and
5 Simulate annealing to produce a new population. Return to step 2).

Here, a fitness function is defined as:

$$F = \sum_{k,l}^{N} \left(\theta_{k,l} - \theta_{op} \right)^2 \tag{17}$$

Here: $\theta_{k,l} = \cos^{-1}\left(\left| \langle f_k, f_l \rangle \right| \right)$, $1 \le k, l \le N$ and $k \ne l$; θ_{op} is the angle between the frame elements when an ETF is satisfied, i.e.:

$$\theta_{op} = \cos^{-1}\left(\sqrt{\frac{N-m}{m(N-1)}} \right) \tag{18}$$

Hence, for the purpose of this algor ithm, the smaller the fitness function, the better the individual, and the entire search process is looking for the smallest F .

During the evolution, the first 20 individuals by priority are crossed over with the remaining individuals to produce a new population at the crossover probability of 0.8. The crossover is a uniform two-point crossover, i.e. two "loci" are randomly selected out of each pair of crossover individuals. The genes are divided into three parts, one of which is randomly exchanged between two individuals. During mutation, a frame element is replaced by a random $m \times 1$ binary vector at the mutation probability of $0.9/t$, where t *is* the present evolutional generation. This way, as the evolution generations increase, the mutation probability reduces. In the course of evolution, a migration takes place every 20 generations. That is, the last %20 individuals by priority are replaced by new random binary frames at the maximum evolutional generation of 500.

Despite the high global search ability, genetic algorithm is prone to "prematurity" and poor search efficiency in the last stage of evolution. Simulated annealing algorithm is a global optimization algorithm proposed by reference to the cooling process of solid matters in thermal equilibrium. It is good local search ability and protects the search processing from coming into the local best solution. By combining genetic algorithm with simulated annealing method we are able to draw the strengths and avoid the weaknesses of these algorithms and ensure that the new algorithm has better global search ability [20]. In our study, we use the evolutional generation t of genetic algorithm as the annealing time of simulated annealing method and assume that the initial annealing temperature $T_0 = 10000$. The annealing process is described below.

1 Calculate the present temperature $T = T_0 \times 0.99^t$;
2 For the individual i $(1 \le i \le 200)$ of the present population of genetic method, calculate the fitness function F_i, and the total correlation between frame element $\boldsymbol{f}_k$ and the other frame elements, i.e. $\mu(\boldsymbol{f}_k) = \sum_{1 \le l \le N, l \ne k} |\langle \boldsymbol{f}_k, \boldsymbol{f}_l \rangle|$, $1 \le k \le N$;
3 Replace the frame element that makes $\mu(\boldsymbol{f}_k)$ the maximum with a random binary column vector $\boldsymbol{v}$, and assume $\upsilon \ne \pm \boldsymbol{f}_k$, $1 \le k \le N$, to get a new individual $\hat{i}$; and
4 Calculate the fitness $F_{\hat{i}}$ of the new individual. Accept the new individual if $F_{\hat{i}} < F_i$, otherwise accept the new accept at the probability $\exp\left(-\frac{F_{\hat{i}} - F_i}{T}\right)$.

Figure 2 shows each generation of the best individual fitness function under normal genetic algorithm and SAGA constructions at (6, 16)-ETF. From this chart, SAGA is faster and more effective in search and needs only half of the generations for convergence used by normal genetic algorithm.

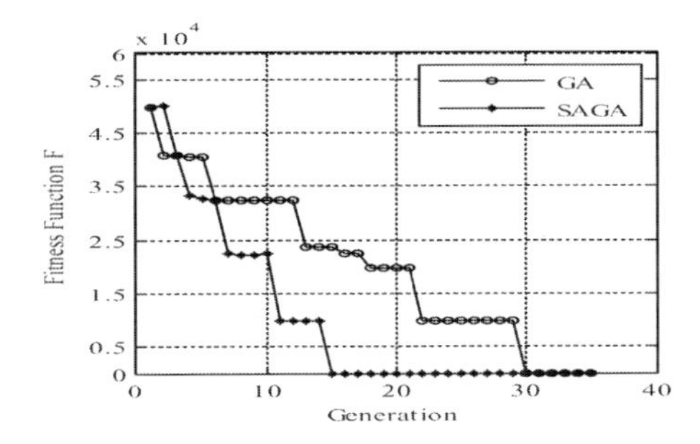

Figure 2. (6, 16)-ETF convergence process under GA and SAGA constructions

Below is the realization result of a (11, 12) binary ETF:

$$\begin{bmatrix} - & - & - & - & + & + & + & + & + & + & + & + \\ + & + & - & + & - & + & - & + & + & + & - & + \\ - & - & - & + & - & + & - & + & - & - & + & - \\ - & + & - & + & - & - & + & - & + & - & + & + \\ + & + & - & - & + & - & - & + & - & - & + & + \\ + & - & - & + & + & - & - & - & + & + & + & - \\ - & + & - & + & + & - & + & + & - & + & - & - \\ + & - & - & - & - & - & + & + & + & - & - & - \\ - & - & + & + & + & - & - & + & + & - & - & + \\ + & + & + & + & + & + & + & + & + & - & + & - \\ + & - & - & + & + & + & + & - & - & - & - & + \end{bmatrix}$$

Here: "+" means $\sqrt{\frac{1}{11}}$ and "_" means $-\sqrt{\frac{1}{11}}$.

5 CONCLUSIONS

A BASC-based real-valued ETF construction method and a SAGA-based binary ETF construction method are proposed. The CDMA frequency spread codebook constructed with ETFs has the lowest correlation and features equal correlation between sequences. However, in practical application, the frequency spread sequences of CDMA have to come to a substantial code length, demanding better dimensional fitness from the ETF construction method. Subsequent efforts may include further exploration for construction methods of higher-dimensional ETFs as well as a multiuser detection method based on the optimal Grassmannian sequences to make better use of the features of ETFs.

REFERENCES

[1] Strohmer T and Heath Jr R W. (2003) Grassmannian frames with applications to coding and communication. *Applied and Computational Harmonic Analysis*, 14(3):257-275.

[2] Tsiligianni E, Kondi L and Katsaggelos A. (2014) Construction of incoherent unit norm tight frames with application to compressed sensing. *IEEE Transactions on Information Theory*, 60(4):2319-2330.

[3] Yan W, Wang Q, and Shen Y. (2014) Shrinkage-Based Alternating Projection Algorithm for Efficient Measurement Matrix Construction in Compressive Sensing. *IEEE Transactions on Instrumentation and Measurement*, 63(5):1073-1084.

[4] Medra A and Davidson T. (2014) Flexible codebook design for limited feedback systems via sequential smooth optimization on the Grassmannian manifold. *IEEE Transactions on Signal Processing*, 62(5):1305-1318.

[5] Heath Jr R W and Strohmer T, Paulraj Arogyaswami J. (2003) Grassmannian signatures for CDMA systems. *Global Telecommunications Conference (GLOBECOM'03), San Francisco, USA*, pp:1553-1557.
[6] Heath Jr R W, Strohmer T, and Paulraj A J. (2006) On quasi-orthogonal signatures for CDMA systems. *IEEE Transactions on Information Theory*, 52(3):1217-1226.
[7] Vanhaverbeke F and Moeneclaey M. (2006) Binary signature sets for increased user capacity on the downlink of CDMA systems. *IEEE Transactions on Wireless Communications*, 5(7):1795-1804.
[8] Soltanalian M, Naghsh M and Stoica P. (2014) On Meeting the Peak Correlation Bounds. *IEEE Transactions on Signal Processing*, 62(5):1210-1220.
[9] Sustik M A, Tropp J A, Dhillon I S, *et al.* (2007) On the existence of equiangular tight frames. *Linear Algebra and its Applications*, 426(2-3):619-635.
[10] Redmond D J. (2009) *Existence and Construction of Real-Valued Equiangular Tight Frames*. Kansas: University of Missouri-Columbia.
[11] Strohmer T. (2008) A note on equiangular tight frames. *Linear Algebra and its Applications*, 429(1):326-330.
[12] Tropp J A, Dhillon I S, Heath Jr R W, *et al.* (2005) Designing constructed tight frames via an alternating projection method. *IEEE Transactions on Information Theory*, 51(1):188-209.
[13] Heath R W, Tropp J A, Dhillon I S, *et al.* (2004) Construction of equiangular signatures for synchronous CDMA systems. *2004 IEEE Eighth International Symposium on Spread Spectrum Techniques and Applications, Sydney Australia*, pp:708-712.
[14] Isaacs J C and Roberts R. (2009) Constructions of equiangular tight frames with genetic algorithms. *IEEE International Conference on Systems, Man and Cybernetics, San Antonio, USA*, pp:595-598.
[15] Isaacs J C. (2011) Stochastic constructions of equiangular tight frames. *Digital Signal Processing Workshop and IEEE Signal Processing Education Workshop (DSP/SPE), Sedona, USA*, pp:66-71.
[16] Hu H and Wu J. (2014) New constructions of codebooks nearly meeting the welch bound with equality. *IEEE Transactions on Information Theory*, 60(2):1348-1355.
[17] Rose C, Ulukus S, and Yates R D. (2002) Wireless systems and interference avoidance. *IEEE Transactions on Wireless Communications*, 1(3):415-428.
[18] Waldron S. (2003) Generalized Welch bound equality sequences are tight frames. *IEEE Transactions on Information Theory*, 49(9):2307-2309.
[19] Lazich D E, Zörlein H, and Bossert M. (2013) Low coherence sensing matrices based on best spherical codes. *Proceedings of 2013 9th International ITG Conference on Systems, Communication and Coding (SCC), Munich, Germany*, pp:1-6.
[20] Lv Guangming, Sun Xiaomeng, and Wang Jian. (2011) A simulated annealing-new genetic algorithm and its application. *2011 International Conference on Electronics and Optoelectronics (ICEOE), Dalian, China*, pp:246-249.

Electric, Electronic and Control Engineering – Shao, Shu & Tian (Eds)
© 2015 Taylor & Francis Group, London, ISBN: 978-1-138-02842-5

Research and application of data mining technology and operational analysis of the integration of power grid regulation and control

Dongya Wu, Junli Zhao, Zhaohui Cui & Yi Liu
State Grid Henan Electric Power Company, Zhengzhou Henan, China

ABSTRACT: The so-called integration of regulation and control is a system platform combining two relatively independent work types: regulation and control. This system saves not only labor costs but also site costs. Regulation and control are operated separately in daily management of the transformer substation, so the two types of work are completely disconnected and data are completely distributed. Thus, this often leads to unclear tasks, complex procedures, and a relatively long period in system maintenance.

KEYWORDS: Integration of power grid regulation and control; data mining

1 RESEARCH BACKGROUND

The so-called integration of regulation and control is a system platform combining two relatively independent types of work: regulation and control. This system saves not only labor costs but also site costs. Regulation and control are operated separately in daily management of the transformer substation, so the two types of work are completely disconnected and data are completely distributed. Thus, this often leads to unclear tasks, complex procedures, and a relatively long period in system maintenance. Besides, data of the operating state of each grid device cannot provide management staff with effective operational information. This is neither for the good of management staff in judging the state of substation equipment nor for the good of the operation of the whole power grid system. The solution proposed in this paper focuses mainly on two aspects, namely the study on the system platform of integrated management of regulation and control and the construction of a deep data mining platform.

2 TECHNICAL SCHEME OF DATA MINING TECHNOLOGY AND THE OPERATIONAL ANALYSIS OF INTEGRATION OF POWER GRID REGULATION AND CONTROL

"Regulation" and "control" are conducted separately in a traditional power system. So, the overall operating cost is relatively high and the management is disjointed. The cross-platform polymorphic integration system adopted in this paper is a hybrid technology platform combining software and hardware based on IEC61970 and IEC61968. The information model based on CIM design adopts the currently advanced CORBA component technology internationally, which is able to realize integration and information exchange with other applications, carry out deep data mining on operational data of all equipment, and conduct centralized monitoring and unified regulation on all substations. All integrated systems adopt the layout of redundant backup so as to enhance the system reliability and security. The overall structure of the system is given next.

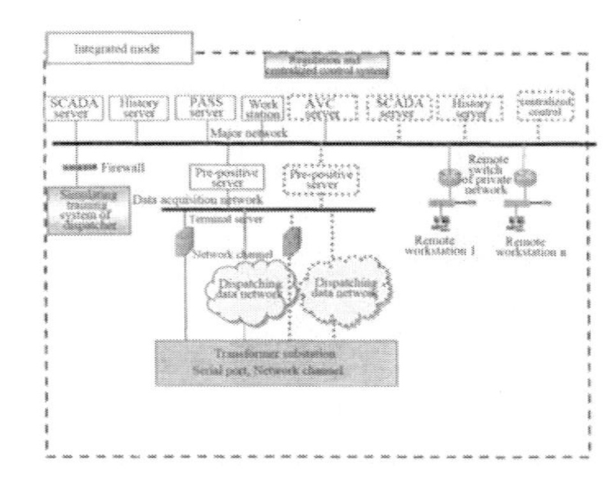

Figure 1. Overall structure of the system.

The platform of data mining technology and the operational analysis of integration of power grid regulation and control can be mainly divided into eight layers. The first layer is the hardware support layer, which constructs a hardware platform for future data storage and data mining as the basic equipment layer of the system. The second layer is the operating system layer. It exists with the equipment layer complementarily. The first layer and the second layer are the base layer of the software and hardware of integrated regulation and control platform and data mining platform. The third layer is the base communication layer, which is constructed based on TCP/IP protocol or electric power communication protocol. It provides a platform for data transmission. The fourth layer is the database service layer, providing a powerful storage platform for all kinds of data as well as a basic platform for refined data mining. The fifth layer is the RTE internal soft bus. The sixth layer is the public service layer, mainly including a series of public affairs such as file service, management service, data transmission service, real-time data service, event service, and so on. The seventh layer is the functional layer. It provides the integrated control system with functional components, including information distribution of data acquisition, SCADA regulation, SCADA monitoring, intelligent classified alarm, authority management of integrated control, software against incorrect operation, intelligent device alarm, maintenance and modeling of unified data, and monitoring information management. The eighth layer is the human-computer interaction layer, which provides various kinds of visual and effective images for integrated control staff, system maintenance staff, and management staff through the front-end display interface of different components.

3 FUNCTIONS OF THE PLATFORM OF DATA MINING TECHNOLOGY AND THE OPERATIONAL ANALYSIS OF INTEGRATION OF POWER GRID REGULATION AND CONTROL

1 It is able to achieve fine management in the power grid and effectively improves the operating speed of grid equipment through this platform. With this mode, monitoring staff can know the operating state of grid equipment at the first time. Once the abnormal operation is found, the situation can be reported to the dispatching center at the first time so that dispatching staff can directly make judgments on properties of malfunctions. The correctness of malfunction conclusion can be improved in this manner. Dispatching staff on duty can directly give orders to monitoring staff in emergency so that monitoring staff can isolate malfunction sites in a timely manner through remote operations and avoid unnecessary loss caused by equipment faults.
2 It is able to realize intensive operation through this platform so as to effectively avoid the waste of human resources. The integration mode of regulation and control can not only effectively reduce the number of staff on duty but also reduce investment on monitoring equipment and sites. This platform can realize the optimized layout of human resources, equipment resources, and site resources, thus achieving the goal of optimal configuration effectively.
3 The construction of the platform of data mining technology and the operational analysis of integration of power grid regulation and control improves the power grid equipment. The technology of power grid construction and maintenance can be enhanced more effectively. The new networking technology brings a technological reform to power grid system. The new technology after the reform promotes management staff to complete power grid construction and maintenance with high efficiency and less operating costs.
4 With this platform, all kinds of data can be incorporated in a database. The OLAP multidimensional data analysis platform can be then used in the multidimensional fine mining of grid operating data to find data that are beneficial to the judgments on the operating state of the integration of power grid regulation and control.

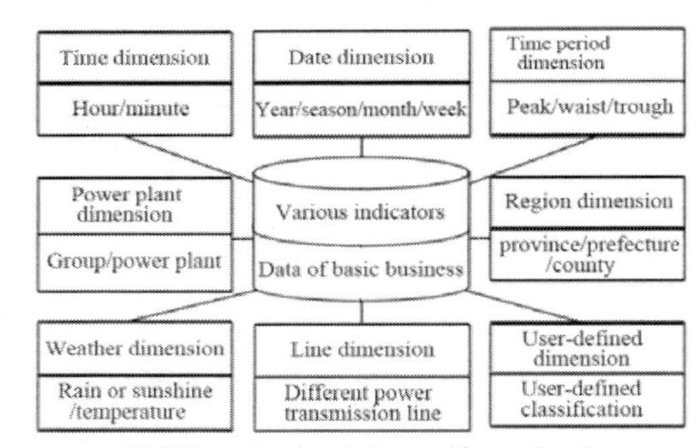

Figure 2. Multidimensional analysis on grid operating data.

In fact, electric power companies of various regions are not only constantly strengthening talents reserve and equipment renovation but also conducting upgrading and transformation on various kinds of old transformer substations with the aim of successfully transiting from the traditional management mode to the management mode of integrated regulation and control. This system carries out unified monitoring on substations and unified dispatching on staff through the image video monitoring mode and the intelligent network anti-error mode. The function of intelligent signal push separates information into layers, levels, and streams so as to realize the centralization and unification of substations as well as the function of integrated monitoring.

4 OPERATIONAL FEATURES OF THE SYSTEM

1 The dispatching system and the monitoring system are operated on a hardware platform. Staff of remote centralized control can know the operating state of equipment at any time through the monitoring platform. This system is able to realize real-time dispatching, operation, and maintenance in different layers and regions. The system structure enhances the reliability and security of the platform through a redundancy backup system so as to realize highly efficient management of regulators, who are able to master operational faults of equipment quickly and eliminate equipment faults in a timely manner.
2 Mutual backup of the channel level and the pre-positive level in the entire process of the platform construction: The system can be normally accessed by any plant as long as there is one normal channel so as to realize the channel-level mutual backup; the system can operate normally as long as there is one normal pre-positive server so as to realize the pre-positive level backup.
3 A particularly large amount of data would be generated in grid operation by some server systems such as SCADA, EMS, and CPS, which are usually assumed as daily work summary and report without fine and deep mining. Data generated by these systems always contain the operational law of the whole power grid and information of safe production. So, the system is more to the benefit of effective classification and fine extraction of historical data. Resources of historical data can be applied more efficiently after the correlation analysis according to the result of extraction.
4 Data management can be enhanced by this platform so that data can be maintained uniformly. In addition, relevant data of each sector can be directly extracted through the platform login system and data of the redundancy backup system requires only automatic backup through system settings instead of manual backup. The problem of data synchronism can be solved in this manner. It not only helps management staff carry out fine and multidimensional data mining further but also is convenient for management staff to conduct effective management and scientific analysis on the system and discover the law of power utilization.

Although the system provides regulators with convenience in power system management, still numerous problems need to be modified. Take the Electric Power Company of the State Grid in Henan and its subordinate electric power companies for instance. Although the integrated system of regulation and control brings convenience for the province's power transmission, power distribution, and user management, the storage and classification of a great deal of data and the correlation analysis create a huge amount of workload for regulators. Besides, fault location and fault solution of the system are extremely accurate and enormous data storage maintenance increases the overall operating cost.

5 CONCLUSION

This paper mainly introduces not only the research status of the platform of data mining technology and the operational analysis of integration of power grid regulation and control but also the research and application of the system construction scheme. Regulation and centralized control are integrated in one platform with an organic combination of data mining technology. This platform is more to the benefit of regulators in mastering the operating state of the power grid. Besides, researchers are able to conduct deep mining on historical data so as to master the correlation factors between devices and between the power transmission grid and the power distribution grid.

REFERENCES

[1] Han, B.M., Liu, Y., Li, T., *et al.* (2012) Construction practice of the integration system of power grid regulation and control in Tangshan. *Value Engineering.*

[2] Shi, Z.L. & Xu, Z.L. (2011) Formal starting of the integration of power grid regulation and control in Jinhua. *China Electric Power News*, 2011-12-27.

[3] Wang, R.X. & Jiang, M.J. (2011) New stage of the integration of power grid regulation and control in companies of Beijing. *State Grid News*, 2011-01-14.

[4] Xu, X.H., Chen, L.J., Zhong, X.Q. *et al.* (2009) Introduction of Intelligent Power Grid. Beijing: *China Electric Power Press*, pp:2-7.

[5] Ozansoy C.R., Zayegh A., Kalam A. (2008) Time synchronization in an IEC 61850 based substation automation system. *Power Engineering Conference, IEEE*, 3

[6] Yu, F. (2009) Development and implementation of intelligent power grid. *Electric Power Technology*, (09):1-6.

[7] Boudaoud K. (2000) Network Security Management with Intelligent Agents. *Network Operations and Management Symposium.*

Electric, Electronic and Control Engineering – Shao, Shu & Tian (Eds)
© 2015 Taylor & Francis Group, London, ISBN: 978-1-138-02842-5

A study on the city coordination scheme of industrial integration

Yaguang Qi
College of Economics, Zhejiang University, Hangzhou Zhejiang, China

ABSTRACT: In this paper, a study is carried out on the industry agglomeration and the integrated and coordinated development of city industry from the perspective of city labor division. First, relevant theories of industry agglomeration and metropolitan circle are elaborated. Second, the industry agglomeration of city labor division is analyzed from the mechanism. It is concluded that different economic entities are integrated as a network organization to reduce market transaction costs and promote labor division when the market is regarded as a method of coordinated labor division. But this network organization features optimal agglomeration, because it is related to soft and hard environments of areas. Third, an empirical analysis is conducted on the agglomeration effect and the transactional efficiency with the data of the Yangtze River Delta region from 2002 to 2012 as an example, concluding that the development of industry-oriented city integration depends on whether the industry agglomeration effect is expanded continuously. The formation of industry agglomeration is the product of specialized labor division as well as an economic space form for people to reduce transaction costs of specialized labor division and obtain increasing returns of labor division. Finally, a coordination scheme for the problem of labor division revenue decrease is proposed based on a comparative analysis on the agglomeration effect and the transactional efficiency of the Yangtze River Delta region from 2002 to 2012 in the same coordinate system, aiming at providing suggestions for the coordinated development of industrial integration in our country.

KEYWORDS: The perspective of city labor division; industry agglomeration; utility function; agglomeration effect; transactional efficiency.

1 INTRODUCTION

Jun Liu et al. (2010) pointed out that industry agglomeration, called the "development impetus" of the regional economy, is able to promote economic growth. However, industry agglomeration causes unbalanced development of regional economy and widens the gap of regional development to some extent while simultaneously promoting economic growth [1]. Therefore, many people have made studies on characteristics of industry agglomeration for the healthy development of city industries. Qiang Li et al. (2014) carried out an empirical study on the agglomeration degree of 37 industries in nine cities of the Wan-jiang urban belt with location entropy indexes, proposing that it is necessary for the Wan-jiang urban belt to increase undertaking efforts of industries with a high agglomeration degree, undertake industrial professions in line with advantages, and enhance the industrial competitiveness of core cities in the process of industrial transfer [2]. Husen An et al. (2010) indicated that it is obliged to break the circulating cumulative causal mechanism that constantly strengthens regional advantages of developed areas in order to achieve regional coordinated development. The degree of integration between developed areas and underdeveloped areas must be reduced so as to break this circulating cumulative mechanism. The proposition of "policy gradient" has to be put forward between developed areas and underdeveloped areas [3]. Guangming Li et al. (2014) established a co-integration VAR model by calculating the comprehensive evaluation index of industry agglomeration, the comprehensive evaluation index of logistics industry, and the comprehensive evaluation index of economic development level, and they analyzed the dynamic relationship among industry agglomeration, logistics industry, and economic development of Urumchi, Xinjiang from 1995 to 2012 with co-integration analysis and impulse response function [4].

Based on previous studies, this paper studies the industry agglomeration and the integrated and coordinated development from the perspective of city labor division in two aspects, namely the mechanism and the empirical study. And the city coordination scheme of industrial integration is proposed as well on the basis of the mechanism research and the empirical analysis.

2 RELEVANT THEORIES OF INDUSTRY AGGLOMERATION AND METROPOLITAN CIRCLE

Industry agglomeration means a large number of related enterprises concentrate in a specific region to form an industrial community, which is similar to a biological organism. It has the law of evolution and development of its own [5]. The coordinated development of a city requires the analysis on characteristics of relevant industry agglomeration. Industry agglomeration has certain relevance with regard to industrial chain, industrial cluster, and national policy.

The extension and integration of the industrial chain forms an industrial cluster, which has certain requirements of the industrial chain [6]. Products or services in a region have a long enough production chain and can be decomposed. Enterprises in a region should be able to carry out reasonable specialized labor division and cooperation in terms of the industrial chain. Industry agglomeration changes with the economic development and it has its own life cycle. The industrial cluster is formed by industry agglomeration in a mature stage. It is an important organizational form of industries clustering in space. The state and local governments usually implement cluster initiative in order to promote the development of clusters [7]. There is an interactive relationship between the government and cluster enterprises. The government's efforts on the industrial cluster mainly include the supply of public facilities and services [8], the supply of institution [9], the maintenance of market environment, [10] and economic guidance.

There are mainly three models of industrial cluster, namely X1–market-oriented model, X2–government-supported model, and X3–model with plans. The three models of industrial cluster are compared from five angles: Y1–features of market model, Y2–the government's role, Y3–major problems, Y4–types of industrial clusters, and Y5–typical nations and regions. Results are shown in Table 1:

Table 1. Results of the comparison among various industrial cluster models.

Y X	X1	X2	X3
Y1	Improved	Immature	Completely ignored
Y2	Indirectness, auxiliary	Directness	Completely dependent
Y3	Obvious gap of competitive advantage	Uncoordinated role of government and market	Deviates from actual economic construction
Y4	Covers almost all types	Key and dominant industrial cluster for promoting national and regional competitive advantages	Industrial cluster areas with policies and strategies
Y5	Countries with developed market economy in Europe and America, such as the United States, Germany, Italy, etc.	Countries and regions with post-developed economy, such as Japan, Taiwan, Korea, India, etc.	Countries with the planned economic system, such as the former Soviet Union, China before the reform and opening up, etc.

The essence of a metropolitan circle is a spatial concept based on "core-edge," which emphasizes the origin destination and the area size. It is related to transportation and communication ability. Specifically, a metropolitan circle should have a relatively large central city and adjacent regions with transportation and communication ability that can be affected by the central city. The center and peripheral regions of a metropolitan circle are connected through developed infrastructures such as transportation and communication. Featuring high spatial integration, the two couple, interact mutually, and connect closely through a frequent stream of people, capital flow, and information flow. A metropolitan circle is an effective spatial organization structure in developed countries. It has an intrinsic regularity of distribution. The spatial structure diagram of a metropolitan circle in the United States is shown in figure 1:

3 ANALYSIS ON THE MECHANISM OF INDUSTRY AGGLOMERATION FROM THE PERSPECTIVE OF CITY LABOR DIVISION

Songlin Zhang et al. (2010) indicated that spatial agglomeration is the result of the interaction of micro subjects in the process of location selection. However, labor division of micro subjects is one of the basic factors for promoting the interaction [11]. With the aim of studying industry agglomeration and providing reasonable suggestions for the coordinated development of city integration, this paper carries out an

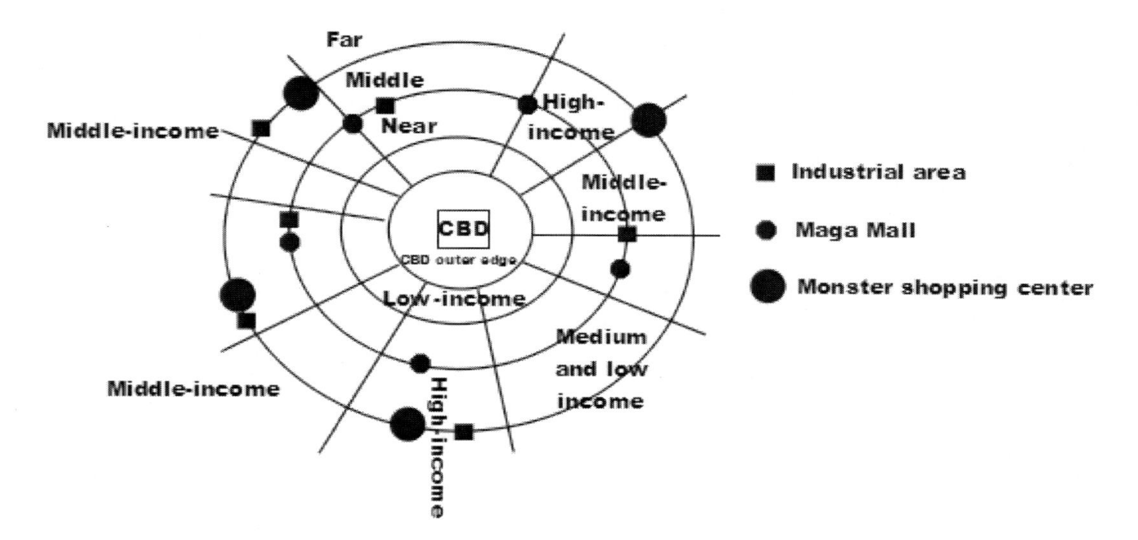

Figure 1. Spatial structure diagram of a metropolitan area in the United States.

analysis on the mechanism of spatial agglomeration from the perspective of city labor division so as to lay a theoretical basis for the analysis on industry agglomeration.

Krugman (1991) pointed out that a self-sufficient society cannot play the role of scale economy. In this case, there is only a world composed of self-sufficient family farms [12]. Xiaokai Yang (2003) indicated further that each economic entity in a self-sufficient society has to produce agricultural products that belong to land-intensive products. For this reason, economic entities in a self-sufficient society have to live in a vast space dispersedly so that the agglomeration cannot be formed [13]. As for enterprises and the market, there are two kinds of labor divisions, namely enterprise labor division and social labor division. Qi Liang (2004) pointed out that enterprises are able to establish a network relationship through agglomeration and reduce transaction costs by close spatial distance. Besides, common industrial culture, regional culture, and value are advantageous to enterprises to establish an enterprise network and a social network based on reputation so that the two sides can deal easily without breaking a contract. In addition, time and expenses on searching market information can be also saved [14]. Therefore, different economic entities are integrated as a network organization to reduce market transaction costs and promote labor division when the market is regarded as a method of coordinated labor division. But the formation of this network organization is closely related to soft and hard environments of areas. In order to elaborate the mechanism of industry agglomeration from the perspective of labor division more precisely, it is presumed in this paper that there are two economic entities in an economic system and each economic entity exists as the role of producer and consumer. A producer can make a choice between products x and y, whereas a consumer should consume two products simultaneously.

If the two are divided in line with labor, they have to pay certain transaction costs when buying products from the other side. Suppose the transaction cost coefficient of purchasing unit product is $I-k$. Here, $k$means that the transaction cost coefficient ranges between 0 and 1. Economic entities can conduct production and transaction dispersedly or intensively. In order to analyze the area of agglomeration, it is assumed that there are two areas, namely A and B. The two areas are different in quality, that is, area B is superior to area A in terms of soft and hard environment. The transactional efficiency of area A is lower than that of area B. Suppose transactional efficiencies of decentralized production and centralized production in area A and B are, respectively, k'',k',k^*, so $k''<k'<k^*$. Utility functions of economic entities are presented in Formula (1) to (5):

$$\max \quad u=\left(x+k_r x^d\right)\left(y+k_i y^d\right) \tag{1}$$

$$s.t \quad x^{t^D} = x+x^t = L_x^a \tag{2}$$

$$y^{t^D} = y+y^t = L_y^a \tag{3}$$

$$L_x + L_y = 1 \tag{4}$$

$$P_x\left(x^t-x^d\right)+P_y\left(y^t-y^d\right)=0 \tag{5}$$

Variables in Formulas (1) to (5) are nonnegative numbers. Formula (1) is a special Cobb–Douglas utility function. Formulas (2) and (3) represent production functions of x and y, respectively. Formula (4) stands for the resource constraint, which means that the total volume of resources put on production is 1. Formula (5) stands for the budget constraint, which means that the total amount of sale is equal to that of purchase. Here, x and y stand for the self-sufficient amount of two products, x^d and y^d are the purchase amount of two products, and x^t and y^t are the sale amount of products x and y. Labor inputs of producing x and y are, respectively, represented by L_x and L_y. Prices of x and y are, respectively, represented by P_x and P_y. The index of individual specialization degree is represented by a.

The model jus described can be classified into two kinds, namely the self-sufficient structure and the complete labor division structure. In the self-sufficient structure, both the economic entities choose the self-sufficient model. Problems of the two economic entities shown in Formula (6) can be concluded according to the basic model:

$$\max \quad u_1 = xy$$
$$s.t \quad \begin{cases} x = L_x^a \\ y = L_y^a \\ L_x + L_y = 1 \end{cases} \tag{6}$$

It can be calculated through the optimization equation shown in Formula (6) that the equilibrium utility of each economic entity is $u_1^* = 2^{-2a}$.

In the complete labor division structure, two economic entities are divided in line with labor. There are two labor division modes, (x/y) and (y/x). According to the calculation method of the self-sufficient mode, equilibrium utilities of two labor division modes shown in Formula (7) can be calculated as u_2^*, u_3^*.

$$\left.\begin{aligned} u_2^* &= \frac{kP_x}{4P_y} \\ u_3^* &= \frac{kP_y}{4P_x} \\ u_2^* &= u_3^* \end{aligned}\right\} \Rightarrow P_x = P_y \} \Rightarrow u_2^* = u_3^* = \frac{k}{4} \tag{7}$$

In Formula (7), $u_2^* = u_3^*$ is the condition for equal utility.

When $u_1^* > u_2^* = u_3^*$, both the economic entities would choose the self-sufficient mode. The self-sufficient structure is a general equilibrium structure. When $u_2^* = u_3^* > u_1^*$, both the economic entities would choose the complete labor division mode. The complete labor division structure is a general equilibrium structure. Thus, the corner equilibrium shown in Table 2 can be concluded.

Table 2. Corner equilibrium.

Value range of k	General equilibrium str ucture
$0 < k < 2^{2-2a}$	Self-sufficient mode
$2^{2-2a} < k < 1$	Complete labor division mode

The following conclusions can be drawn in line with the analysis on the modes just described:

1 Only agglomeration can increase the transactional efficiency to a certain threshold value and promote labor division. Spatial agglomeration would then appear.
2 If economic entities are able to realize labor division and cooperation in two areas, the area with a better soft and hard environment would be selected for agglomeration.
3 Labor division is the essential condition for spatial agglomeration instead of the sufficient condition.

The conclusions mentioned earlier suggest that: the spatial agglomeration of economic entities is not a simple concentration. It forms inseparable connections in the process of clustering. This is consistent with the viewpoint proposed in literature [15] that agglomeration is the structure of spatial interactions of economic entities through location selection. Economic entities do not transform from the dispersed self-sufficient state to the centralized self-sufficient state. They cluster together so as to evolve from the self-sufficient state to the labor division state. A connection of labor division and cooperation is formed among economic entities due to the state transformation. On the contrary, it is the spatial interaction caused by labor division that promotes the emergence of agglomeration. The purpose of labor division of economic entities lies in the pursuit of more economies of labor division and less transactional costs. But economies of labor division acquired in two areas with the same labor division level are equal. So, entities would choose to cluster at the area with less transactional costs. According to this paper, the area where the agglomeration first appears is decided by the rational choice of economic entities. It means that the deepening of labor division is promoted so that higher transactional efficiency can be obtained. Whether economic entities conduct labor division and cooperation is determined by weighing the conflict between the labor division economy and the transactional cost. If the transaction cost is low enough, economic entities can still make up for the transactional cost by the labor division economy and get extra labor division economy under the condition of dispersed location. They still choose labor division and cooperation in this way.

4 EMPIRICAL ANALYSIS

The development of industry-oriented city integration depends on whether the industry agglomeration effect is expanded continuously. The formation of industry agglomeration is the product of specialized labor division as well as an economic space form for people to reduce transaction costs of specialized labor division and obtain increasing returns of labor division. For this reason, the paper carries out an empirical analysis on agglomeration effect and transactional efficiency so as to verify the mechanism of industry agglomeration from the perspective of city labor division and provide a basis for the formulation of the coordination scheme.

4.1 *Verification of agglomeration effect*

In this paper, the parameter of returns to scale h is adopted as the agglomeration effect indicator of industry agglomeration. It is calculated in accordance with industrial output values, profits, and fixed assets. The following variables are needed in the calculation of h: β-profit elasticity; γ-profit elasticity of fixed assets; P-actual profit of industry; Q-total industrial output value; and K-net value of fixed assets. The calculation method is shown in Formula (8):

$$\left.\begin{array}{c} P = AQ^{\beta}K^{\gamma} \\ \Updownarrow \\ \ln P = \ln A + \beta \ln Q + \gamma \ln K \end{array}\right\} \Rightarrow h = 1 + \frac{\gamma}{1-\beta} \qquad (8)$$

The industrial cluster has an agglomeration effect when the parameter of returns to scale $h \geq 1$. The higher the value is, the greater the industry agglomeration effect appears. There is no agglomeration effect in the macro economy or industries when the parameter of returns to scale $h < 1$.

In this paper, the total profit P, the industrial output Q, and the net value of fixed assets K of the Yangtze River Delta region in the past 18 years are taken as the empirical data for the moving and retrogression process. Parameters of returns to scale from 2002 to 2012 are given in Table 3:

The trend of industry agglomeration effect of the Yangtze River Delta region is shown in figure 2:

Table 3. Industry agglomeration effect of the Yangtze River Delta region from 2002 to 2012.

Year	2002	2003	2004	2005	2006	2007	2008	2009	2010	2011	2012
h	3.88	4.68	6.35	3.05	2.15	1.27	1.28	1.35	2.33	3.72	4.97

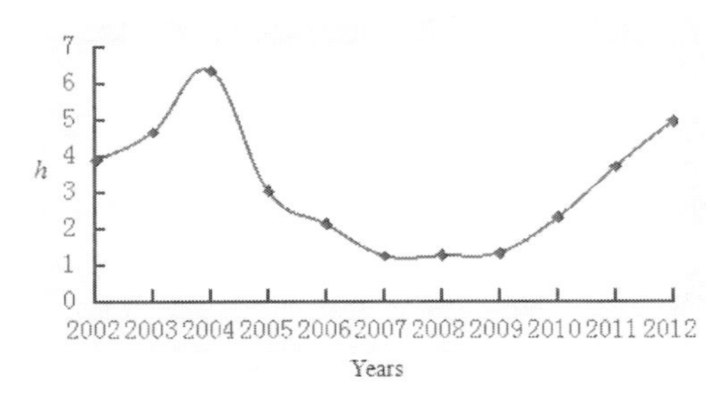

Figure 2. Trend of industry agglomeration effect of the Yangtze River Delta region.

It can be observed from figure 2 that the industry agglomeration effect of the Yangtze River Delta region of 2004 reaches its peak. The major reason is the development of the export-oriented economy in the Yangtze River Delta region. The growth of export in 2004 is extremely strong due to the impact of the world economic recovery and the adjustment of export tax rebate policy. The total self-support export of the whole year reaches $208.3 billion, the growth rate of which is as high as 47.3%. The proportion of the export amount of the nation reaches 35.1%, which is 2.9% higher than that in 2003. The gross value of production takes 14% of the whole nation. The degree of dependence on export reaches 59.9%, which is 10.7% higher than that in 2003. The industry agglomeration effect of the Yangtze River Delta region from 2007 to 2009 reaches the minimum. The major reason is that it is influenced by the financial bubble crisis caused by the subprime crisis. Prices in China soared substantially. But the effect was changed obviously due to the state macro-control from 2010 to 2012. This is also the main reason that the industry agglomeration effect of the Yangtze River Delta region increased continuously from 2010 to 2012. From the perspective of specialized labor division, the formation of industry agglomeration is the product of specialized labor division as well as an economic space form for people to reduce transaction costs of specialized labor division and obtain increasing returns of labor division. Therefore, the following empirical analysis on trading efficiency is carried out for the study on the industry agglomeration effect of the Yangtze River Delta region.

4.2 *Verification of transactional efficiency based on factor analysis*

Transaction cost is explained in the new institutional economics as follows [16]:

1 The government, government management, or relevant laws and regulations can reduce transaction costs and influence transactional efficiency.
2 Communication technology and electronic commerce can reduce transaction costs significantly and increase the total transactional efficiency of economic entities.
3 Educational level, education degree, and literacy rate can greatly reduce transaction costs and increase transactional efficiency.

According to the explanation of transaction cost in the new institutional economics, the indicator system of transactional efficiency is formulated in Table 4:

In this paper, relevant data of 32 provinces in 2012 are collected to solve weight coefficients of 8 factors, which are then used in the calculation of factor scores of the Yangtze River Delta region from 2002 to 2012. The data process is shown next:

STEP1. Conduct processing on original data in the same direction.

STEP2. Conduct normalized and standardized processing on data with the same trend.

STEP3. Obtain correlation matrix, characteristic value, contribution rate, and accumulative contribution rate with SPSS software.

STEP4. Extract main factors.
The principle of factor extraction is that the characteristic value of the main factor should be larger than 1 and the accumulative contribution rate of variance reaches 80% of the main factor. Results from SPSS are given in Table 5.

The correlation matrix **R** (8×3) is shown in Formula (8).

Table 4. Indictors of transactional efficiency.

Indicator level		Indicator symbol
Institutional aspect	Ratio of dependence on foreign trade	X1
	The proportion of fixed assets of non-state-owned economy in fixed assets investment of the whole society	X2
Government aspect	The proportion of potential digging and reform capital of enterprise in fiscal expenditure	X3
	The proportion of government expenditure	X4
	The ratio between the number of granted patents and the number of scientific and technical personnel	X5
	Educational level: the number of people with higher education	X6
Traffic aspect	The sum of road and railway per square kilometer	X7
	Volume of telecommunication business	X8

Table 5. Results from SPSS.

	Initial Eigenvalue			Extraction Sums of Squared Loadings		
Component	Total	% of Variance	Cumulative %	Total	% of Variance	Cumulative %
1	4.087	51.089	51.089	4.087	51.089	51.089
2	1.419	17.735	68.824	4.419	17.735	68.824
3	1.037	12.957	81.781	1.037	12.957	81.781
4	.618	7.728	89.509			
5	.352	4.402	93.912			
6	.223	2.789	96.700			
7	.203	2.536	99.236			
8	.061	.764	100.000			

It can be concluded from the results of SPSS that the first main factors with higher scores are Z1, Z2, Z5, Z6, Z7, and Z8; the second main factors with higher scores are Z3 and Z4; and the third main factors with higher scores are Z2 and Z4. The comprehensive evaluation scoring model of transactional efficiency of industry agglomeration can be constructed if the proportion of the variance contribution rate of each main factor in the total variance is taken as the weight coefficient. The comprehensive scoring model is shown in Formula (10).

$$\mathbf{R} = \begin{bmatrix} & Z1 & Z2 & Z3 & Z4 & Z5 & Z6 & Z7 & Z8 \\ 1 & 0.879 & 0.647 & 0.176 & -0.552 & 0.932 & 0.731 & 0.779 & 0.741 \\ 2 & 0.340 & -0.357 & 0.925 & 0.296 & 0.129 & 0.282 & -0.235 & -0.281 \\ 3 & -0.179 & 0.587 & -0.071 & 0.639 & -0.057 & 0.392 & 0.136 & -0.266 \end{bmatrix}^T \tag{9}$$

$$F = 62\% \times F1 + 22\% \times F2 + 16\% \times F3 \tag{10}$$

It can be observed from Formula (10) that the weight of the first main factors is the largest, the weight of the second main factors takes the second place, and the weight of the third main factors is the smallest. The transactional efficiency of industry agglomeration of the Yangtze River Delta region from 2002 to 2012 is shown in figure 3:

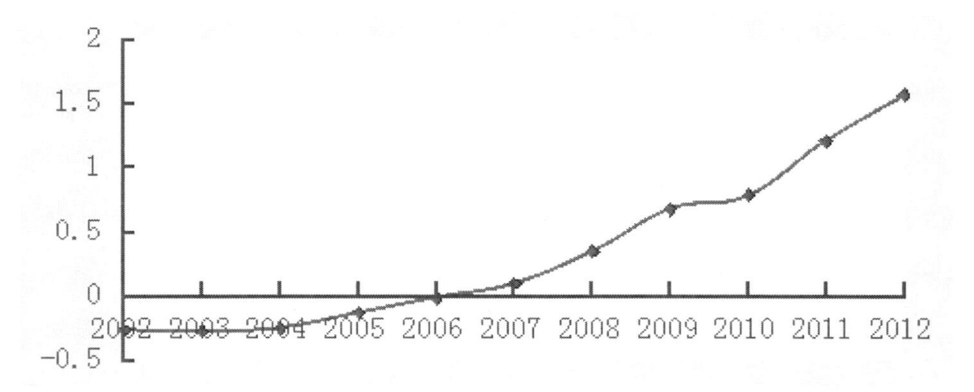

Figure 3. Transactional efficiency of industry agglomeration of the Yangtze River Delta region from 2002 to 2012.

It can be observed from figure 3 that transactional efficiency of the Yangtze River Delta region from 2002 to 2012 presents a rising tendency. The score of the transactional efficiency of the Yangtze River Delta region before 2006 is a negative. This is because the industrial integration process of the region still remains on the surface, which is caused by the slow progress. The network in the industry agglomeration system composed by enterprises, research institutions, intermediary institutions, and the government is initially reasonable, being able to obtain an agglomerate economy. With the development of agglomeration, rigid systems would lead to extra costs, namely the increase of costs for constructing and maintaining the network structure. However, the further integration of industries in the Yangtze River Delta region led to a significant growth of the transactional efficiency of industry agglomeration in this region. As for the Yangtze River Delta region, the decline of transactional costs of the industry agglomeration is now playing a positive role.

4.3 *Comparative analysis of agglomeration effect and transactional cost and the design of feasible schemes*

Figures 2 and 3 are painted in the same coordinate system for comparative analysis so that the industry agglomeration features from the perspective of city labor division can be analyzed more comprehensively.

It can be concluded from the results in figure 4 that the transactional efficiency increased with the growth of the industry agglomeration effect before 2004. It means that the decline of transactional cost

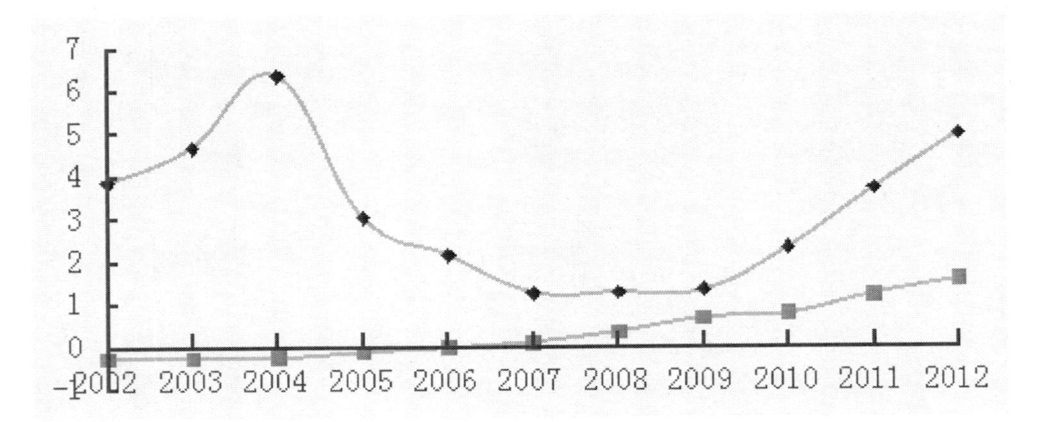

Figure 4. Industry agglomeration effect and transactional efficiency of the Yangtze River Delta region.

promoted the increase of the industry agglomeration effect. However, the industry agglomeration effect has declined since 2004 instead of exhibiting continuous increase with the growth of transactional efficiency. Empirical results indicate that the development of industry agglomeration in the Yangtze River Delta region is faced with the problem of agglomeration benefits decline. It can be observed from the comparison of industry agglomeration effect and transactional efficiency that the stagnation problem of the expansion of agglomeration effect lies in the decline of labor division benefits not of transaction cost. For this reason, the following schemes are proposed in this paper:

1 Constituent parts of the region conduct the cooperation of labor division through reasonable institutional arrangement and form a reasonable regional labor division system so as to realize the regional coordinated development and promote the sustainable development of regional industry agglomeration.
2 Enhance the regional labor division effect based on modularized industry agglomeration and lay out all links in a large region in the way of modularized industry agglomeration with the bond of the regional industrial chain.
3 The process of promoting the integration of the Yangtze River Delta region involves deepening the industrial labor division. But the system layout of labor division in the Yangtze River Delta region should be started globally by constructing a regional industrial layout. Shanghai should focus on a minority of industrial productions that are technology intensive, port type, and metropolitan type and have capitals with a competitive advantage. It is an important step for Zhejiang province to actively participate in specialized production and labor division of the Yangtze River Delta region that labor-intensive industries transferred by Shanghai and Jiangsu province should be actively absorbed. However, industry agglomeration in Zhejiang province should focus on labor-intensive industries so as to increase its manufacturing share in the whole Yangtze River Delta region. Jiangsu province should absorb some labor-intensive and capital-intensive industries transferred by Shanghai.

5 CONCLUSIONS

Based on the elaboration on relevant theories of industry agglomeration and metropolitan circle, this paper not only carries out an analysis on the industry agglomeration mechanism, agglomeration effect, and transactional cost from the perspective of city labor division but also proposes city coordination schemes for the industrial integration of the Yangtze River Delta region based on the previous analysis.

1 Spatial agglomeration of economic entities is not a simple concentration. Only agglomeration can increase the transactional efficiency to a certain threshold value and promote labor division. Spatial agglomeration would then appear.
2 The purpose of labor division of economic entities lies in the pursuit of more economies of labor division and less transactional costs. But economies of labor division acquired in two areas with the same labor division level are equal. So, entities would choose to cluster at the area with a better soft and hard environment.
3 Whether economic entities conduct labor division and cooperation is determined by weighing the conflict between the labor division economy and

the transactional cost. If the transaction cost is low enough, economic entities can still make up for the transactional cost by the labor division economy and get extra labor division economy under the condition of dispersed location. They still choose labor division and cooperation in this way.

4 Industrial cluster has an agglomeration effect when the parameter of returns to scale $h \geq 1$. The higher the value is, the greater the industry agglomeration effect appears. There is no agglomeration effect in the macro economy or industries when the parameter of returns to scale $h < 1$.
5 The development of industry agglomeration in the Yangtze River Delta region is faced with the problem of agglomeration benefits decline. It can be known from the comparison of industry agglomeration effect and transactional efficiency that the stagnation problem of the expansion of agglomeration effect lies in the decline of labor division benefits and not of transaction cost.
6 In the Yangtze River Delta region, the deepening of industrial labor division can promote the integration process. It is the integration process of regional industrial structures that industries are integrated spatially to form a regional industrial system with close relations and distinct labor division. Resource efficiency would be improved with the industrial optimal configuration of regional resources so that industry agglomeration can deepen industrial labor division constantly along with sustained, rapid, and stable growth.

REFERENCES

[1] Liu, J. *et al.* (2010) Reasonable distribution of industry agglomeration in a region and the promoting strategy, *Reform of the Economic System*, (2):112–117.

[2] Li, Q. *et al.* (2014) A study on the issue of carrying on industry transfer of Wanjiang urban belt from the perspective of industry agglomeration, *Journal of UEST of China (Social science edition)*, 16(1):45–50.

[3] An, H.S. *et al.* (2010) The endogenous mechanism of the spatial agglomeration of economic activities and regional coordinated development, *Social Sciences in Nanjing*, (1):22–29.

[4] Li, G.M. *et al.* (2014) An empirical study on the dynamic relation among industry agglomeration, logistics industry and urban economic development, *Commercial Times*, (21):39–41.

[5] Song, L.Y. (2008) *A Study on the Industry Agglomeration of Harbin-Daqing-Qiqihar Industrial Corridor and Countermeasures*, Harbin: Harbin Institute of Technology.

[6] Rosenthal. (2001) Determinants of agglomeration, *Journal of Urban Economics*, 12(35):191–209.

[7] Michael E. Porer. (1998) Clusters in a global economics of competition, *Harvard Business Review*, pp:231–236.

[8] Liu, K.J. (2002) Knowledge spillover, industry agglomeration and policy choice of regional high-tech industries, *Productivity Research*, (1):97–106.

[9] Dekle. R. J. Eaton. (1999) Agglomeration and land rents, evidence from the prefectures, *Journal of Urban Economic*, (46):200–214.

[10] Wu, X.P. (2003) The theory of industry agglomeration and its development, *Journal of Sichuan Administration College*, (2):25–27.

[11] Zhang, S.L. *et al.* (2010) From location choice to spatial agglomeration: an analytical framework based on labor division. Future and Development, (7):36–39.

[12] Krugman, P. (1991) Increasing returns and economic geography, *Journal of Political Economy*, (99):483–499

[13] Yang, X.K. (2003) *Development Economics*, Beijing: Social Sciences Academic Press.

[14] Liang, Q. (2004) *Discussion on Industry Agglomeration*, Beijing: Commercial Press.

[15] Hao, S.Y. (2007) *The Principle of Regional Economy*, Shanghai: Shanghai People's Publishing House, Gezhi Press.

[16] Chen, W.X. (2009) *A Study on the Sustainable Development of Industry Agglomeration in the Yangtze River Delta Region from the Perspective of Labor Division*, Beijing: Renmin University of China.

Electric, Electronic and Control Engineering – Shao, Shu & Tian (Eds)
© 2015 Taylor & Francis Group, London, ISBN: 978-1-138-02842-5

Power system black-start recovery subsystems partition based on improved CNM community detection algorithm

Yikui Liu, Tianqi Liu, Qian Li & Xiaotong Hu
School of Electrical Engineering and Information, Sichuan University, Chengdu Sichuan, China

ABSTRACT: Partitioning power systems into several subsystems to realize parallel recovery is significant for speeding up the recovery processes. For this purpose, according to the community structure property of complex networks, a black-start subsystems partition algorithm of a power system that is based on the improved CNM community detection algorithm was proposed. Edge weight, which is based on power flow and can reflect the connection degree of nodes, was defined. Utilizing the edge weight and the improvements, this algorithm can effectively and reasonably partition power systems into several subsystems that not only connected closely inside and connected sparsely with each other but also had startup powers. This algorithm takes edge weight into consideration and uses modularity to measure the community structure. Simulations on IEEE 39-bus system and IEEE 118-bus system show the validity of the algorithm.

KEYWORDS: power system black-start; subsystem partitioning; CNM community discovery algorithm; community; complex network.

1 INTRODUCTION

With the continuous expansion of the scale of the power grid, the interaction between the various factors of the power grid becomes more and more obvious and the possibility of large area blackout caused by cascading failure is becoming bigger and bigger [1]. In recent years, large area blackouts that occurred worldwide had caused bad social effects and great economic loss[2,3]. After blackouts happen on interconnected power grids, the systems will rely on generator units, called "startup power," which have a self-starting ability to recover. So, according to certain strategies, power systems can be partitioned into several subsystems to implement parallel recovery. This is significant for accelerating the processes of a system's recovery.

Black-start subsystems partition of the power system is a complex decision-making optimization problem. In recent years, many domestic and foreign experts had published plenty of beneficial research about the black-start subsystems partition problem from different aspects and carried out some experiments in real power systems [4-6]. Reference [7] utilized fuzzy C-clustering to partition power systems into subsystems that take the core buses as centers. Reference [8] transformed the black-start problem into the Boolean decision problem and introduced the ordered binary decision diagrams (two-ordered binary decision diagram, OBDD) to model and solve the problem. However, OBDD was going to search all the solution space, which means that exponent overflow will occur when OBDD is applied on a relatively large-scale power system. Reference [9] integrated the importance indexes of recovery paths between nodes and startup powers into path weight and transformed power system networks into weighted networks. Based on the weighted networks, the taboo search algorithm was applied to partition the power systems into reasonable subsystems. Reference [10], by taking subsystem partition strategy, recovery paths of internal nodes, and recovery order into unified consideration and utilizing the shortest path algorithm and classical genetic algorithm, optimized subsystems' parallel recovery schemes. However, because genetic codes will not only account for the partition of the nodes but also consider the recovery sequences of the nodes, which means large code space and low computation efficiency, it is hard to converge to the optimal solution. Reference [11] took the uncertainty of transmission lines' restoration, recovery time after partitioning, and the sum of generators' output in limitative recovery time into consideration to build up target function and then solved it by the particle swarm algorithm. But this method is difficult to determine the coefficients of the target function when it is in the face of multiple targets. The low efficiency problem of the algorithm is also remarkable. Reference [12], based on community structure theory in complex network theory, used the spectral clustering algorithm to partition power systems into a fixed number of subsystems. Reference [13], also based on community structure theory, utilized the fast spectral segmentation method proposed by Wu and Huberman, which is on the basis of resistance networks' voltage spectrum and takes the similarity of nodes' voltage as the criterion of partitioning subsystems. In this way, power systems can be partitioned into a fixed number of subsystems. However, partition methods proposed in [12-13] found it hard to deal with the ambiguous nodes, which are the connection of two subsystems and can only partition power systems into a predetermined number of subsystems.

The power system networks are viewed as complex networks in this paper and edge weight, which reflects the degree of connection between two adjacent nodes, is defined according to power flow distribution of the system under the normal operation state. According to the community property of power system networks and pointing at black-start recovery subsystems partition, the CNM community discovery algorithm was improved. The improved algorithm accounts for edge weight and considers modularity as the partition index. Importantly, it can ensure that each subsystem has one startup power.

2 COMMUNITY STRUCTURE DETECTION OF POWER SYSTEM

2.1 *Community structure of power system*

Along with the in-depth study of a complex network's properties, many real networks had been found to have a common property, namely, community structure. The network is composed of a plurality of "communities" whose internal nodes' connection is relatively tight but where the connection between each other is relatively sparse.

Community structure theory in complex network theory had been applied in natural, social, and biological areas. Power system networks, as a real complex network, also have the community property [14]. The basic principles of black-start recovery subsystems partition are similar to the definition of community structure. That is, subsystems are closely electrically connected inside and the electrical connection between subsystems is relatively sparse. Therefore, it is feasible to apply community detection principles in partitioning recovery subsystems. So, it is unsurprising to apply community detection methods in complex network theory in partitioning black-start recovery subsystems as well. And communities detected by community detection methods in a complex network can be viewed as a subsystem in a power system.

2.2 *CNM community detection algorithm*

2.2.1 *Modularity*

Modularity is a commonly used measurement of the quality of community detection [15]. Its basic idea is to compare the partitioned network with the corresponding null model to measure the quality of community detection. Modularity is defined as follows:

$$Q=\sum_{v=1}^{n}\left[\frac{l_v}{M}-\left(\frac{d_v}{2M}\right)^2\right] \tag{1},$$

where n is the number of communities l_v is the number of edges included in community v d_v is the sum of node degree in community v and M is the total number of edges in a network.

Edge weight measures the connection degree between nodes in a network; w_{ij} represents the edge weight of the edge-connected node i and node j. (2) represents the sum of the edge weight of all edges in a network.

$$W=\frac{1}{2}\sum_{i,j}w_{ij} \tag{2}$$

The modularity-accounted edge weight is defined as follows [16]:

$$\begin{cases} Q=\sum_{v=1}^{n}\left[\frac{W_v}{W}-\left(\frac{S_v}{2W}\right)^2\right] \\ W_v=\frac{1}{2}\sum_{i,j\in v}w_{ij} \\ S_v=\sum_{i\in v}\sum_{j\neq i}w_{ij}\delta(i,j) \end{cases} \tag{3},$$

where W_v is the sum of edge weight of all edges in community v; S_v is the sum of edge weight of all edges that are connected with any internal nodes of community v; and $\delta(i,j)$ is the node connection function. When node i and j are connected, the function value is 1; otherwise, it is 0.

2.2.2 *CNM algorithm accounting edge weight*

CNM algorithm is a community detection algorithm that is based on modularity [17], and it belongs to the condensation algorithm [18]. The CNM algorithm constructs a modularity increment matrix ΔQ directly. And then it uses the greedy strategy to choose the maximum element ΔQ_{vk} every time and merges the corresponding two communities, v and k. After the merger, the elements of ΔQ will be updated to carry out the next merger until the largest element of ΔQ

changes from positive to negative. After that, the algorithm ends.

The processes of the algorithm are as follows:

1 Initialization: Assume that each node is an independent community. The initial ΔQ can be obtained by (4) and modularity $Q = 0$.

$$\begin{cases} \Delta Q_{ij} = \begin{cases} e_{ij} - a_i a_j, & \delta(i,j)=1 \\ 0, & \delta(i,j)\neq 1 \end{cases} \\ e_{ij} = \begin{cases} \frac{w_{ij}}{2W}, & \delta(i,j)=1 \\ 0, & \delta(i,j)\neq 1 \end{cases} \\ a_i = \frac{\sum_{j\neq i} w_{ij}\delta(i,j)}{2W} \end{cases} \quad (4),$$

where e_{ij} is the corresponding element in initial matrix **e**; a_i, a_j are the corresponding elements in auxiliary vector **a**.

2 Select the largest element ΔQ_{vk} in $\mathbf{\Delta Q}$ and merge the corresponding communities v and k into k after recording the merger. Recalculate the auxiliary vector **a** and update modularity increment matrix $\mathbf{\Delta Q}$ by (5).

$$\Delta Q'_{ck} = \begin{cases} \Delta Q_{vc} + \Delta Q_{kc}, \partial(v,c)=1 \cap \partial(k,c)=1 \\ \Delta Q_{vk} - 2a_c a_k, \partial(v,c)=1 \cap \partial(k,c)=0 \\ \Delta Q_{ck} - 2a_c a_v, \partial(v,c)=0 \cap \partial(k,c)=1 \end{cases} \quad (5)$$

where $\partial(v,k)$ is the community connection function; When community v and k are connected, function value is 1; otherwise, it is 0.

Remove row v , column v in $\mathbf{\Delta Q}$ and calculate modularity by (6) after the merger.

$$Q = Q + \Delta Q_{vk} \quad (6)$$

3 Repeat 2) until the maximum element in $\mathbf{\Delta Q}$ changes from positive to negative.

3 PARTITION OF BLACK-START RECOVERY SUBSYSTEMS

3.1 *Principles of partition*

Partition of black-start recovery subsystems aims at utilizing startup powers to realize parallel recovery in subsystems. After subsystems are recovered, the whole system is rebuilt though reconnecting subsystems. For respective recovery, there must be at least one startup power in each subsystem. In addition, in order to avoid long-distance power transmission, the internal electrical connection in a subsystem should be tight. Simultaneously, in order to avoid the shock when connecting subsystems in the whole-system recovery stage, the electrical connection between subsystems should be loose. Moreover, Reference [8] also proposed considering the "retrospective principle." That is, "subsystems partition" should refer to the normal operation mode of the power grid structure as far as possible.

3.2 *Power system network with edge weight*

Power systems actually are networks formed by power plants, substations, and transmission lines. Abstract the power plants and substations into nodes, transmission lines into edges to obtain the abstract network. Multi-circuit transmission lines connecting the same two nodes should be abstracted into one edge.

Edge weight in the network measures the connection degree of two adjacent nodes. In view of the thought of weight factor in [19] and the retrospective principle, active power P and reactive power Q transmitted by the transmission line under the normal operating mode were considered factors that influence the edge weight of power system networks.

The greater the active power P is, the more power is transmitted between those two nodes, which means the closer the connection between those two nodes. The greater Q is, the greater the reactive power demand of one of the two terminals of a line is under the normal operation mode, which means the relatively greater loss. So, it is less possible to restore the line loaded with great reactive power Q in black-start recovery process, in other words, connection can be considered loose. To sum up, under the normal operation mode, the active power P transmitted by a line is positively correlated with its edge weight and the reactive power Q is negatively correlated. The edge weight is defined as follows:

$$w_{ij} = \alpha P_{ij} - \beta Q_{ij} \quad (7),$$

where P_{ij} and Q_{ij} are active power and reactive power transmitted between node i and j under normal operation mode, respectively. When nodes i and j are connected by multi-circuit transmission lines, P_{ij} and Q_{ij} are the sum of the active power and reactive power transmitted by multi-circuits transmission lines respectively; $\alpha \in [0,1]$ is active power weight factor; and $\beta \in [0,1]$ is reactive power weight factor.

4 BLACK-START RECOVERY SUBSYSTEMS PARTITION ALGORITHM

4.1 *The shortages of using CNM algorithm in partition*

The CNM algorithm is based on modularity. Modularity is widely used to measure the connection degree between internal nodes of a community and the connection degree between communities, but it does not satisfy the demand in black-start that there must be at least one startup power in each subsystem. Therefore, the CNM algorithm cannot be applied to black-start recovery subsystems partition directly.

Two main situations will result in subsystems without any startup powers in direct use of the CNM algorithm to partition:

1 Imbalanced position distribution of startup powers
When the grid has strong community structure and startup powers, all of which are located in one or some communities, the community without a startup power will be considered a subsystem.

In figure 1, sub-networks A, B, and C in a power grid are closely connected within, but loosely connected between each other. The CNM algorithm will partition the grid into A, B, and C, three communities with maximum modularity.

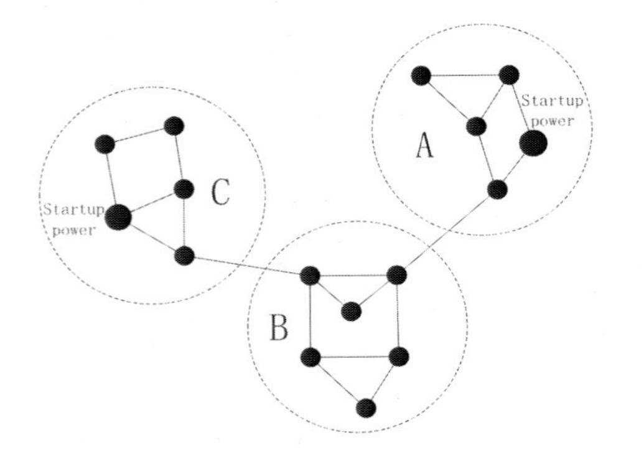

Figure 1. Imbalanced position distribution of startup powers.

2 Subsystems formed by transmission line branches
When the power grid has long transmission line branches and the branch networks formed by those branches have a strong community structure, the branch networks will be partitioned as independent communities. The branch networks that do not have startup powers will be regarded as subsystems.

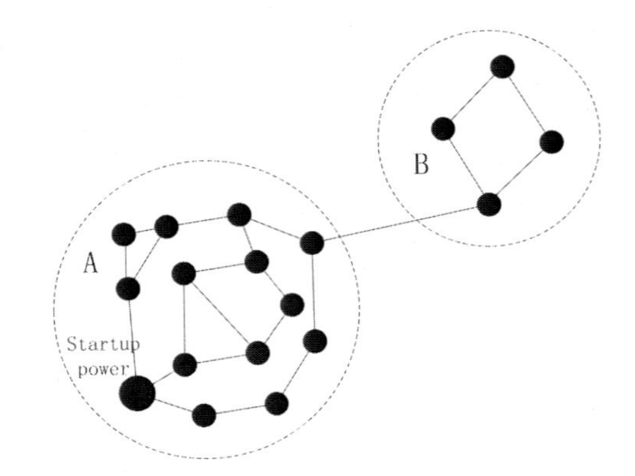

Figure 2. Subsystems composed by transmission line branches.

In figure 2, community A contained one or more startup powers. Community B actually was a branch network of community A. However, since B had a strong community structure, modularity was maximum when community B was partitioned as an independent community.

4.2 *Improved CNM algorithm*

Based on figure 3, it can be observed that the partition with maximum modularity may not ensure that all subsystems contained at least one startup power. Therefore, a compromise must be made between each subsystem containing startup powers and getting the maximum possible modularity. In other words, the maximum possible modularity was pursued under the premise that every community contained startup powers. For the earlier analysis, the following two improvements were made in the CNM algorithm:

1 Choose the maximum element ΔQ_{vk} in $\mathbf{\Delta Q}$ with the requirement that at least one community contains startup powers. That is, to choose the maximum element in rows and columns that correspond to the communities that had startup powers.

When merging two communities, always retain the label of the community with startup powers. If both the communities contain startup powers, retain one of them.

2 The algorithm does not end, even when all the elements in $\mathbf{\Delta Q}$ change from positive to negative. Continue to select the largest element (negative) and merge the corresponding communities. For a n nodes network, after $n-1$ times of mergers, all nodes will be included in a community and the modularity will be 0. Therefore, the algorithm is set to end after $n-2$times of mergers.

3 $n-2$ times of mergers corresponded with $n-2$ modularity values and $n-2$ partition modes. Order all the partition modes with modularity values in descending order and check whether each community of the partition mode contains startup powers. Because only the label of the community with startup powers was retained every time, if all retained labels were labels of the community with startup powers, the partition mode was available. Once such a partition mode is discovered, the algorithm ends.

4.3 *Algorithm flow*

1 Initialize $\mathbf{\Delta Q}$ by (4), set merger time $num = 1$, modularity $Q = 0$.
2 Select the largest element ΔQ_{vk} in rows and columns as representing the community with startup powers in $\mathbf{\Delta Q}$ and merge the corresponding communities v and k. Retain the label of the community with startup powers and remove the other label and the corresponding rows and columns in $\mathbf{\Delta Q}$. Recalculate the auxiliary vector $\mathbf{a}$ and update $\mathbf{\Delta Q}$ by (5).
3 Calculate modularity by (6) after the merger. Set $num = num + 1$.
4 If $num < n-1$, go to 2); otherwise, go to 5).
5 Order all the partition modes with modularity values in descending order and check whether each community of the partition mode contains startup powers. Once a satisfied mode is found, the algorithm ends. The algorithm flow chart is shown in figure 3.

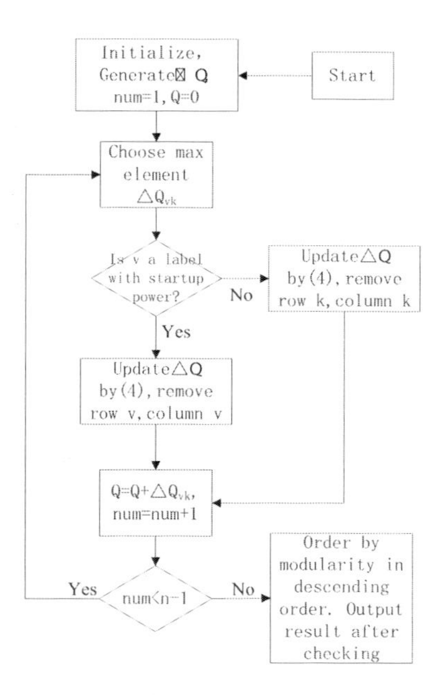

Figure 3. Algorithm flow chart.

5 PERFORMANCE EVALUATION

5.1 *IEEE 39-bus system*

As shown in figure 4, the IEEE 39-bus system was taken as an example to verify the effectiveness of the proposed algorithm. The system was partitioned with and without accounting for the edge weight, respectively. Set node 37, 32 as startup powers. Without accounting for the edge weight, all edge weights were set as 1. The partition result was shown in figure 4; modularity $Q = 0.3455$. Accounting for the edge weight, calculate power flow first and calculate edge weight; set $\alpha = 1$ and $\beta = 0$, by (7).The partitioning result is shown in figure 5; modularity $Q = 0.3401$.

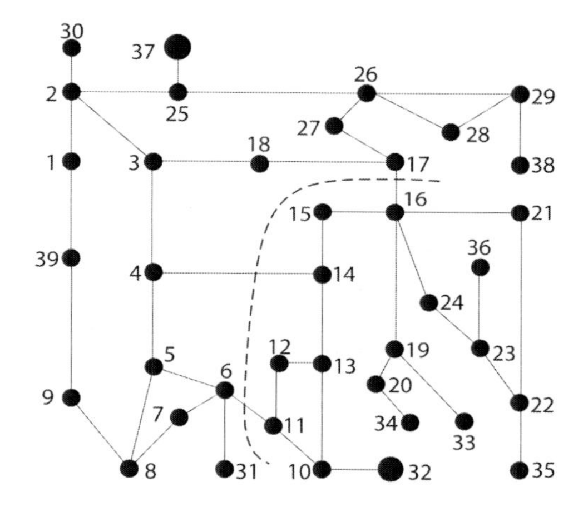

Figure 4. Partition without accounting for edge weight.

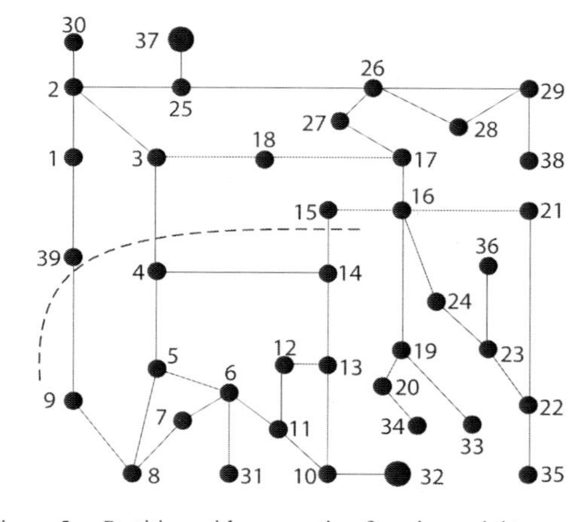

Figure 5. Partition with accounting for edge weight.

In figures 4 and 5, obvious differences are seen between the partition results with and without accounting for edge weight. Without accounting for edge weight, edges connected the nodes in the same degree, which can not reflect the degree of connection between the nodes in the system under normal operation mode. In figure 4, edge 16–17 was disconnected; whereas in figure 5, edge 16–17 connected two areas. This is because edge 16–17 transmitted a lot of power, the edge weight is large, which means nodes 16 and 17 were considered as being closely connected.

5.2 *IEEE 118-bus system*

To further verify the effectiveness of the algorithm in complex networks, the IEEE 118-bus system was partitioned. As Reference [13], set nodes 62, 100, and 113 as startup powers. Calculate power flow and set $\alpha = 1$, $\beta = 0$. Calculate edge weight by (7). The partition result is shown in figure 6; modularity $Q = 0.3760$. Simultaneously, partition the system with the CNM algorithm, which is without any improvements. At the first partition, the system was partitioned into two subsystems, as shown in figure 8 by the dotted line 1. Then, remove the left part of the dotted line and do the second partition. The partition result is shown in figure 7 by the dotted line 2.

Comparing the results of the two algorithms, only node 47 belonged to different subsystems. The load of edges 46–47, 69 – 47, and 49 – 47, including network loss, were 31.5MW, 58.7MW, and 9.6MW, respectively. It is obvious that nodes 69 and 47 were connected more closely. Since the improved CNM algorithm that uses local greedy strategy was revised to ensure that each subsystem contains startup powers, there may be a loss in modularity. But the example also showed that the improved CNM algorithm can detect community structure in complex networks, since the partition result was similar to the CNM algorithm's result.

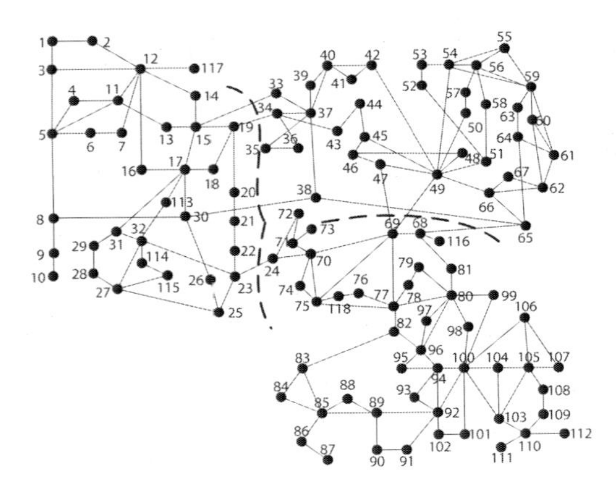

Figure 6. Partition result of 118-bus system accounting for edge weight.

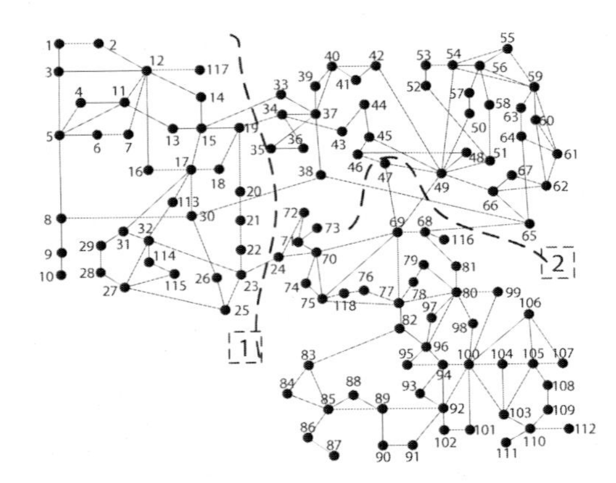

Figure 7. Partition result of 118-bus system using the CNM algorithm

6 CONCLUSIONS

This paper proposed the definition of a network edge weight based on the distribution of system power flow under the normal operation mode. This definition can reasonably and effectively reflect the connection degree between nodes of the power system network. This paper analyzed the shortages of using the CNM community detection algorithm in the black-start recovery subsystem partition directly and improved the CNM algorithm aimed at these shortages. The improved CNM algorithm, considering network edge weight measured by modularity, can effsectively partition the system into subsystems that are connected closely inside but have sparse connection with each other and certainly startup powers.

REFERENCES

[1] Shi Libao. Shi Zhongying. Yao Liangzhong. *et al.* (2010) A review of mechanism of large cascading failure blackouts of modern power system. *Power System Technology*. 34(3):49–54.

[2] Liu Qiang. Shi Libao. Zhou Ming. Li Gengyin. *et al.* (2007) Survey of power system restoration control. *Electric Power Automation Equipment*, 27(11):104–110.

[3] Adibi M M. Fink L H. (2006) Overcoming restoration challenges associated with major power system disturbances-restoration from cascading failures. *IEEE Power and Energy Magazine*. 4(5):68–77.

[4] Yang Ke. Liu Junyong. He Xingqi. Liu Bosi. (2010) Research and experiment of restoration control during Sichuan power grid black-start (1): guiding principles and voltage control. *Electric Power Automation Equipment*, 30(6):100–104.

[5] Yang Ke. Liu Junyong. He Xingqi. Liu Bosi. (2010) Research and experiment of restoration control during Sichuan power grid black-start (2): frequency. Paralleling in control and secondary equipment configuration. *Electric Power Automation Equipment*, 30(7):34–39.

[6] Yan Fengjun. Wang Ning. (2007) Power plant black-start scheme and its experiment. *Electric Power Automation Equipment*, 27(1):122–125.

[7] Kamwa I. Pradhan A K. Joos G. (2009) Fuzzy partitioning of a real power system for dynamic vulnerability assessment. *IEEE Transaction on Power Systems*, 24(3):1356–1365.

[8] Liu Yingshang. Wu Wenchuan. Feng Yongqing. *et al.* (2008) Black-start zone partitioning based on ordered binary decision diagram method. *Proceedings of the CSEE*, 28(10):26–31.

[9] Wu Ye. Fang Xinyan. Zhang Yan. *et al.* (2010) Tabu search algorithm based black-start zone partitioning. *Power System Protection and Control*, 38(10):6–11.

[10] Yu Yixin. Luan Wenpeng. (2009) Smart grid and its implementations. *Proceedings of the CSEE*, 29(10):41–46.

[11] Liang Haiping. Hao Jie. Gu Xueping. (2012) Optimization of system partitioning schemes for black-start restoration considering the successful rate of node restoration. *Transactions of China Electrotechnical Society*, 27(11):230–238.

[12] Liang Haiping. Gu Xueping. (2013) Black-start network partitioning based on spectral clustering. *Power System Technology*, 37(2):372–377.

[13] Liang Haiping. Gu Xueping. (2010) Black-start subsystem partitioning based on similarity of nodes voltage. *Power System Protection and Control*, 41(14):81–86.

[14] Lin Zhenzhi. Wen Fushuan. Zhou Hao. (2009) A new algorithm for restoration subsystem division based on community structure of complex network theory. *Automation of Electric Power Systems*, 33(12):12–16.

[15] Newman M E J. Girvan M. (2004) Finding and evaluating community structure in networks. *Phys. Rev. E.*, 69(2):026113.

[16] Wang Xiaofan. Li Xiang. Chen Guanrong. (2012) *Network Science: An Introduction*. Beijing: Higher Education Press, pp:133–134.

[17] Clauset A. Newman M E J. Moore C. (2004) Finding community structure in very large networks. *Phys. Rev. E.*, 70(6):066111.

[18] Newman M E J. (2004) Fast algorithm for detecting community structure in networks. *Phys. Rev. E.*, 69:066133.

[19] Zhao Dawei. Liu Tianqi. Tang Jian. (2012) A new method for black-start path optimization based on path and node weight factors. *Automation of Electric Power Systems*, 36(15):1–6.

Electric, Electronic and Control Engineering – Shao, Shu & Tian (Eds)
 ISBN: 978-1-138-02842-5

Discussion on the application design of vertical greening in urban public spaces

Dujiang Long & Dong Wang
Guilin Institute of Tourism, Guilin Guangxi, China

ABSTRACT: As central urban building clusters gathering the modern population, urban space development is undergoing unprecedented changes. The densely distributed skyscrapers, the ever sprawling urban construction land and the contracting greenbelts are greatly confining the sunlight, green spaces, ecology and air, significantly affecting the living quality of urban residents and arousing people's reflection on the living environment and spaces in which we exist. Against this background, the construction and development of an eco city has become one of the focuses of concern across the globe. As one of the key techniques for building an eco city, vertical greening is moving towards globalization, diversification and large-scale. In this paper, the research outputs and progress of the application design of vertical greening in urban public spaces are outlined; the technical means and design methods available for creating vertical green landscapes in urban public spaces are analyzed; the important values of vertical greening are demonstrated by experimental study. The conclusion is that vertical greening has presented pronounced values in reducing solar radiation and lowering the ambient and indoor temperatures of buildings and is therefore well worth recognition.

KEYWORDS: urban; public space; vertical greening; design; form.

Over a long time, people tended to look at the extent of industrialization as a gauge that measures the strength of a nation's economy or, in other words, they tended to "look at the GDP growth as the only criterion and driver for building an economic powder". This invariable development philosophy has led to global environmental and energy crisis and forced man to reexamine how it has lived. In this context, urban greening has aroused great concern from the public. In the face of urban land shortage, there's no wonder building vertical greening is appealing to interested parties. In China, however, vertical greening is still far from meeting the demand of urban development since the techniques used are still very simple and the species of plants are quite monotonous. From vertical planting practices outside China, vertical greening in its real sense requires high levels of techniques. Besides stable techniques, good vertical greening also relies on artistic esthetics of the plant model patterns, especially in the greening design of urban public spaces. How to work out a better design that allows higher levels of vertical green landscape application and creates more artistic and eco-friendly vertical green landscapes is a topic well worth concern. In this paper, key aspects of the application design of vertical greening in urban public spaces are examined and discussed.

1 RESEARCH PROGRESS

It is imperative that modern urban development today provide man with a living dimension closer to Nature and friendlier to the eco environment. A green city is a signature of man-nature integration and closeness. When planning and constructing a green network, if specific technical views are turned to vertical greening of building walls, needless to say our city will become more brilliant and bright. Among many other merits, vertical greening helps relieve urban green land shortage and create favorable environmental conditions for building an eco city; caters for the objective requirements of linking urban buildings to the eco environment; and allows extension of green vegetation in the longitudinal space, which increases the esthetic values of urban environment, transforms static greening to dynamic greening, covers the symbolic nature of modernist buildings with the cultural properties of Nature and return the buildings to the man-nature origin. Driven by these merits, countries and regions around the world have come to pay more attention to the vertical greening in urban public spaces, started to research into this field and made rich and diverse progresses.

1.1 *Research progress outside China*

Development and application of the urban vertical greening technique was recognized by developed European and American nations as early as in the 1970s and 80s, when the US, UK, France, Germany and Japan started to regard urban greening as an important national strategy and established a self-awareness in the social public. In Paris, for example, a greening product comprising aluminum frame structure and plastic fiber carpet was produced and applied over the walls of downtown Paris and successfully made up a "vertical garden" full of natural smell, which not only offered a decoration for the building walls, but also added to the liveliness of the urban environment. Brazil succeeded in developing a hollow brick containing soil, fertilizer and grass seeds that was directly used to build houses. In addition to esthetic values, this plant wall also helped reduce noise and air pollution. More importantly, vertical greening was also incorporated into the legal protection of many jurisdictions and played an important role in improving urban environment, realizing low carbon and mitigating emissions.

1.2 *Research progress inside China*

Against the urban greening efforts in Chinese cities that involve dismantling unauthorized buildings, setting up green spaces, importing large trees and planting lawns, vertical greening in China is still in a stagnant or slow-moving stage. As decision makers are more concerned about lawn installation and ground-level tree planting, some even purchase large numbers of gigantic trees. Bringing old or big trees into cities is indeed the best of a bad bunch: while it provides some visual effect at a time, it is not ecological sustainable and results in significant hidden hazard to the urban eco environment. Over the past few years, planting wall as a special form of vertical greening has found its place in China and can serve as a technical reference for designing China's own vertical greening. It also lays a foundation for urban vertical greening in future.

2 METHOD USED

Our study is centered round the application design of vertical greening in urban public spaces and uses a macro-to-micro scheme following four steps of "background analysis, theoretical research, case study and experimental verification". Work performed in these four steps is summarized below.

2.1 *Background analysis*

Building vertical greening both has a long history and great potential for further development. A review on the background of building vertical greening tells us what we have done, what we can do and what we will do, helps us identify the objectives of our study and shows us the direction of our study.

2.2 *Theoretical research*

This step constitutes the "macro understanding" of vertical greening in the design and application study of building spaces. The method used is to summarize the necessity and feasibility of urban three-dimensional and vertical greening by analyzing the development process of natural ecology, using eco esthetics as the philosophical basis for reviewing the research and design of urban vertical greening.

2.3 *Case study*

In this step, the angle at which vertical greening design is analyzed is shrunk from macro to micro. Taking an urban public building as an example, we discuss what vertical greening effect is expected when building vertical greening is applied with different technical means. In this way, we are able to examine the strengths and limitations of different technical conditions so as to extract and analyze and eventually provide reference for subsequent design of new vertical greening systems.

2.4 *Experimental verification*

In this step, we measure the temperatures of two buildings: a building intervened by vertical greening as the experiment subject and another building not intervened by vertical greening as the control subject, take the indoor and outdoor temperatures of different directions and different floors to collect temperature data of different time periods for comparison and thereby demonstrate how vertical greening helps reduce solar radiation and lower the indoor and outdoor temperatures.

3 RESEARCH OUTPUTS

As a greening form that occupies the minimum land area but provides the maximum green area, vertical greening has been widely recognized and studied among greening professionals. As a special industry, vertical greening does not only provide some merits incomparable for horizontal greening, it also has some complex, technical and systematic problems completely different from ground planting. As far as urban public spaces are concerned, there are typically two types of designs for realizing vertical greening. One is the traditional form of techniques and the other is the manual form of construction. Now, we are going to demonstrate what vertical greening can be presented in the application of building vertical greening by using different technical means.

3.1 *Traditional form of techniques*

Climbing or hanging plants are attached to the exterior wall of the building. Free climbing is the natural behavior of climbing plants. Free climbing wall greening is to make use of the adsorption ability and thorns of plants that have their own climbing ability and do not need artificial wall frames. As long as the wall is smooth enough, these plants can climb up the wall of the building and form a completely natural state with little or no assistance of artificial equipment. This gives the maximum play to the natural properties of the plants and makes them green plants most easily used over building walls. In designing this form of wall greening, considering the biological characteristics of the plants, the climbing wall must be something not too smooth like exposed stone, artificial stone or cement mortar wall where the planting density is more suitable and provides the landscape that decorates the building façade in the natural manner while giving corresponding eco effect. There are a good variety of climbing plants, most of which are woody vines and the rest are annual, biannual or perennial herbal plants. They can be selected according to the environment in which they are intended to be used, such as light, moisture, temperature and/or soil. Heliophiles like radix campsis and wisteria and the majority of annual herbal climbing plants like towel gourd, cypress vine and gourd can be used in a well sunlit environment. Shade plants like money plant, bindwood and kadsura are more suitable for landscaping under woods or on the shaded side of a building. But wall-attaching landscape takes a long time to form, since this form of greening typically relies on the growing of plants which itself takes a relatively long time. Also, different climbing plants vary in their growth spaces and coverage abilities, especially in the later stage of plant growth, when the trunk and branches grow heavily, possibly beyond the control of their shapes. An improper design of plant density could limit the functions of a building when the plants sprawl out of control. If the plants grow too well and are not drawn properly, some of the branches may shelter the windows. Also, if plants directly climb up the building wall by relying on their own properties, they may be rooted in the building material and, over time, can damage the wall of the building. As a result, while this technique allows the plants to show their natural state and adds interest to the building, the problems cannot be ignored. Care must be exercised when applying this technique in real practice.

3.2 *Manual form of construction*

3.2.1 *Planting container*

When this technique is used, a framework parallel to the wall is put up immediately on or 510m off the wall and assisted by a drip or spray irrigation system. Next, prebuilt planting containers are placed into the hollow spaces of the containers to achieve vertical greening. One of the most important merits of this form of vertical greening design is its high flexibility in plant selection, automatic irrigation and easy replacement of plants. It is suitable for temporary landscaping of plants and flowers. Its limitation, however, lies in that the framework has to be fixed on the wall, which frequently leads to water leakage at the fixing point and erosion on the framework that can both limit the service life of the system. Furthermore, the drip irrigation system is also exposed to blockage that eventually results in plant deaths due to lack of water.

3.2.2 *Modular planting*

When this technique is used, modular components are used to grow plants and achieve wall greening. According to module division, there are three forms of presentation. The first is the use of unitary modules where the greening plane is composed of a number of unitary modules, each of which is a complete "planting box" consisting of a box, the matrix and the plant. Each unitary module is a relatively independent unit that can be separated in whole. When separated, they are individual separate "flowerpots". When combined, they are closely connected into an integral greening plane. This both allows easy replacement and stabilizes the greening system. The second is the use of a support frame consisting of a steel structure installed on the building façade. When installing the unitary modules, the structure system is installed on the wall before the modules are installed one by one on the carrying framework to achieve the intended greening effect. The installation process is similar to that of the drying hanging process of marble in which the carrying framework is placed before the modules are installed one by one. The third is the use of an irrigation system, which irrigates the plants by means of water dripping. Besides watering the plants, the irrigation system is also responsible for supplying liquid fertilizer to the plants grown in each of the unitary modules. The irrigation system generally comprises water pumps, water pipes, a programmable controller, fertilizer feeders, solenoid valves, filters and anti-clog micro irrigation pipes. The water pipe attached to the frame is sometimes directly mounted on the structure and water the plants in a distributed manner on the building wall through a booster pump. From the main supply pipe to the startup of the irrigation system, the green surface is divided into a number of groups. The irrigation system is automatically controlled by computer programs, which allows automatic water supply circulation according to the type and growth period of the plant, the height of the building façade, the directions of the building façade, the climate conditions and the growth and nutrition requirements of the plant, demonstrating the superior adaptability of the entire technical scheme.

3.2.3 *Attached planting*

This technique applies to the greening of mass wall areas and creates garden-like greening effect on the building wall with different combinations of plants. When this technique is used, two layers of polyamide sheets are attached to PVC panel and covered with a synthetic fiber carpet, which not only allows the plant to root well, but also helps retain moisture. Then, it is fixed on a metal frame mounted over the wall. A drip irrigation system is inserted on the building surface to supply water for the plant and liquid fertilizer for the plant to grow. While attachment design provides the designer with more flexible spaces than any other form of vertical greening and can create very attractive landscape, the eventual landscape formed is fully dependent on the artistic accomplishments and even the painting expertise of the designer. For designers of this form of vertical design, therefore, it can be a great challenge.

3.2.4 *Eco co-growth pyramid*

When this technique is used, balconies and spatial platforms are used to create out-stretching green plants of various sizes that form a vertical greening system spanning the longitudinal space with the urban environment. There's no limitation on the size of the planting containers or flower pots. Sufficient spaces are provided to hold the growth matrix, which provides the existence spaces for plants that require plenty of growth matrices in order to grow. Nor is there limitation on the size of the plant. Small-sized arbors, for example, can provide lively vertical greening effect with their large crowns and beautiful shape. This form of greening, when properly used in urban public spaces according to the particular environment, can contribute to the eco-diversity recovery and creates energy-efficient vertical green landscape on the one hand, and enhance the urban greening and accelerate eco system recovery on the other hand.

4 EXPERIMENTING

4.1 *Experiment subject and instrumentation*

The general guideline of our experiment is to select a building intervened by vertical greening as the experiment subject and another building not intervened by vertical greening as the control object, take the indoor and outdoor temperatures of different directions and different floors to collect temperature data of different time periods for comparison and thereby demonstrate how vertical greening helps reduce solar radiation and lower the indoor and outdoor temperatures. Instruments used during the experiment include an electronic thermometer and a traditional thermometer. In operation, the electronic thermometer was used to measure the temperature on the east outer façade of the building and the measurement was recorded every 1.0h.

4.2 *Experimental result*

Figures 1~6 show the experiment result from the operation process described above.

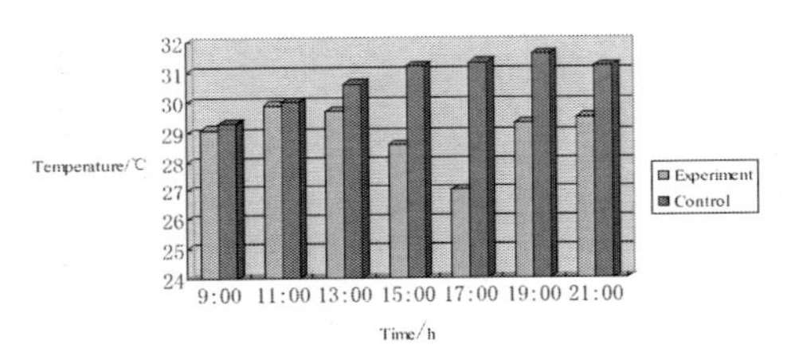

Figure 1. Comparison indoor temperature of floor 1.

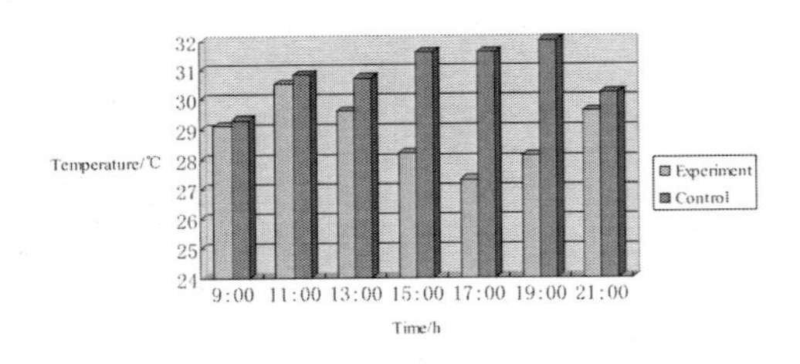

Figure 2. Comparison indoor temperature of floor 5.

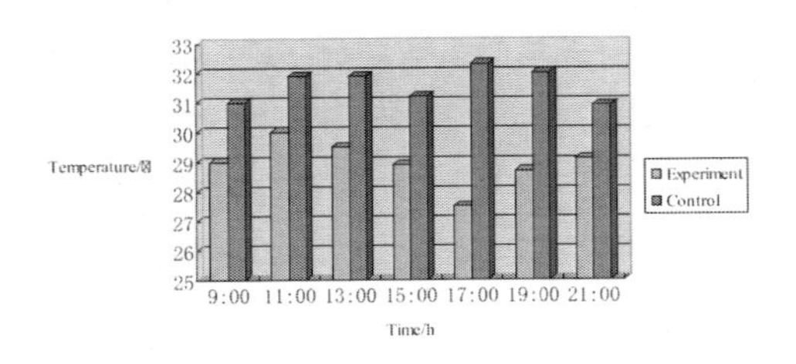

Figure 3. Comparison indoor temperature of floor 10.

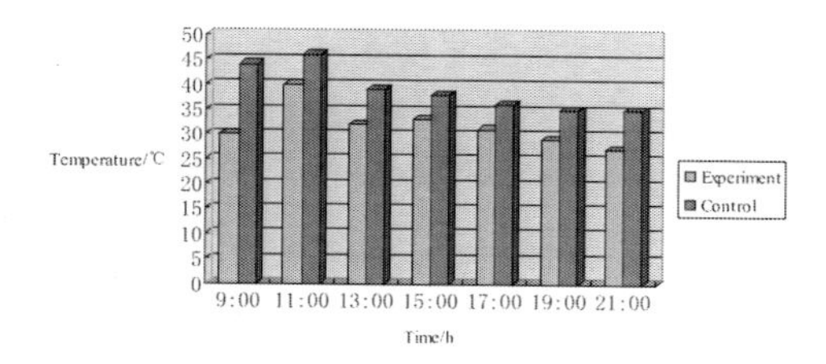

Figure 4. Comparison east façade temperature of floor 1.

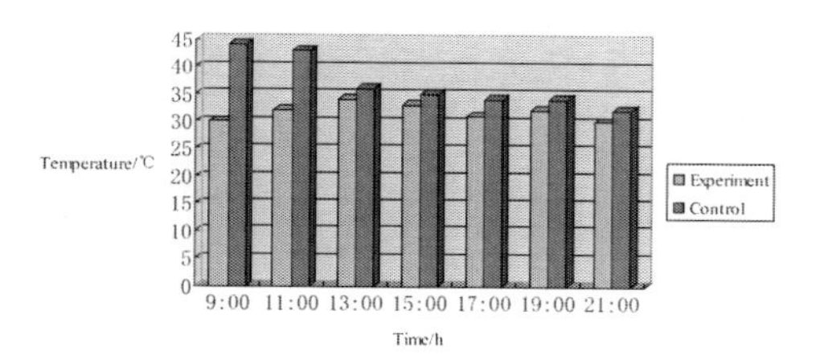

Figure 5. Comparison east façade temperature of floor 5.

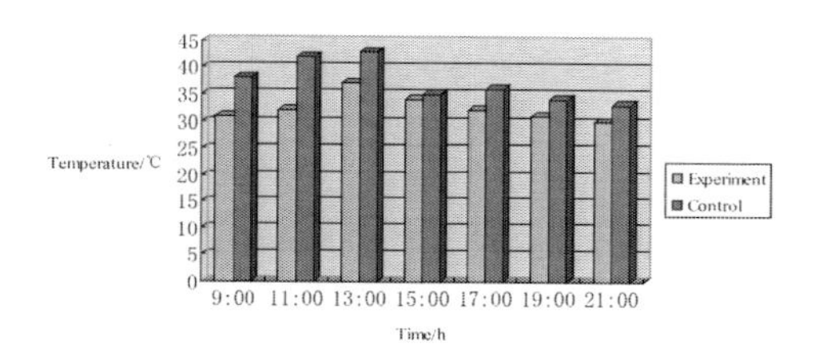

Figure 6. Comparison east façade temperature of floor 10.

Analysis of the data above shows that the temperature values, especially the indoor temperature values, of the building intervened by vertical greening is lower than the one not intervened by vertical greening in every direction and on every floor, and the higher the floor, the larger the difference. This demonstrates that vertical greening helps reduce solar radiation, lower the ambient temperature and indoor temperature of the building.

5 CONCLUSIONS

Three-dimensional greening is defined by a multi-level, multipurpose form of greening or beautification on the façade, roof, substructure and superstructure spaces of a building or a structure in order to improve the urban eco environment and beautify the urban environment. Vertical greening, as a very critical part of three-dimensional greening, is highly recognized and concerned among the various forms of three-dimensional greening as it covers a very considerable area with very modest land occupation. In our study, key aspects of the application design of vertical greening in urban public spaces are examined and discussed. The typical forms and technical means of vertical greening design are examined. Experiment is performed and data are obtained to demonstrate the important role played by vertical greening in improving the eco environment of urban spaces. It is recommended that, in subsequent construction of eco cities, controlling the vertical greening design and better applying and practicing vertical greening techniques will have very profound practical implications and values and are well worth recognition.

ACKNOWLEDGEMENT

This paper is supported by Research on the Influence and Application of Vertical Greening on the Tourist City Image of Guilin (GN: 2013YB02); Application Research on Vertical Greening of Cities in Guangxi Province (GN: YB2014249).

REFERENCES

[1] Liu, X., Li, Y.D. (2013) Applied research on vertical greening and new superior plants--Taking Rizhao City for example. *Acta Agriculturae Shanghai*, 29(3):86–88.
[2] Li, Y. (2014) Development trend and countermeasures of space vertical greening in China. *Journal of Anhui Agricultural Sciences*, (19):6302+6308.
[3] Zhu, H.X., Wang, C. (2004) Vertical greening--An effective way to extend urban space. *Chinese Garden*, 20(3):28–31.
[4] Chen, L., Zhou, J.H., Li, R.C. *et al.* (2011) Investigation on the vertical virescence status in the city proper of Quanzhou. *Journal of Anhui Agricultural Sciences*, 39(27):16893–16895+16898.
[5] Ding, S.J., Li, G.G., Li, J.L. *et al.* (2006) Study on the evergreen and flower plant disposition in the vertical greening for flyovers. *Chinese Garden*, 22(2):85–91.
[6] Shi, L.C., Wang, X.C., Huang, H.Y. *et al.* (2005) Natural resources of climber plants and cultivars selection of vertical gardening plants in Guangxi. *Guangxi Flora*, 25(2):174–178.
[7] Yangjin, Y.L., Ge, Y.Y., Tang, B. *et al.* (2014) Investigation and analysis on the current vertical greening in Hangzhou. *Journal of Zhejiang Agricultural Sciences*, (8):1187–1192.
[8] Chen, H., Chen, H., Qi, H. *et al.* (2012) Application of wild plant resources of vertical greening plants from Qin-Ba Mountain area in landscaping. *Northern Horticulture*, (9):92–95.

Electric, Electronic and Control Engineering – Shao, Shu & Tian (Eds)
 ISBN: 978-1-138-02842-5

New designation to multi-parameter measurement system based on chemical oscillation reaction

Xiaoning Chen, Hongyan He*, Tan Zhang, Hongting Dong & Dexiang Zhang
College of Electrical Engineering and Automation, Anhui University, Hefei Anhui, China

ABSTRACT: A chemical oscillation reaction has more and more applications in many fields. In this paper, we design a kind of multi-parameter measurement system based on a chemical oscillation reaction. The system can measure voltage, temperature, pH, and intensity of pressure simultaneously. Its hardware circuit adopts STM32F103CBT6 as a main controller. Instrument control and data acquisition are provided by software written in VC++. The software is able to analyze, deal with, contrast, save data, and display graphics on the computer. The experimental results show that compared with traditional methods this system can be operated conveniently and have higher precision.

KEYWORDS: chemical oscillation reaction; multi-parameter; data collection; high precision; measurement system.

1 INTRODUCTION

According to ex-chemists, some reactions that existed far from equilibrium are called "chemical oscillation reactions," in which reactants, intermediates, and products periodically suffered from repeating with time[1]. It has far-ranging applications in many fields such as agriculture, biology, pharmaceutical, life science, and so on. Thus, many scientific workers paid more and more attention to chemical oscillation reactions and conducted researched on them. However, conventional recorders existing nowadays are immature and mainly manifested in low precision, single-measure parameter, complex exercise, and imperfect analysis of data, which are bad for data collection as well as for comparison and analysis. To solve these problems, this paper designed a multi-parameter measurement system based on a chemical oscillating reaction aimed at measuring voltage, temperature, pH, and intensity of pressure signals generated during reactions[2-3]. The system consisted of upper computer, lower computer, chemical oscillation measurement system, and some peripheral equipment. The lower computer was an extended system choosing STM32F103CBT6 as the hardware control platform, supported with other operational amplifiers to magnify data. The upper computer was a computer installing a software development system, used to analyze, deal with, and contrast the data and displayed graphics. Experimental results showed that the system can measure parameters accurately and reliably, greatly reducing experimental time and enhancing the accuracy of tests.

*Corresponding author: *hhywh1205@163.com*

2 DETECTION PRINCIPLE

We present a novel multi-parameter measurement system to study chemical oscillation reactions in this paper. An oscillatory chemical reaction in the cerium that catalyzed potassium bromate-malonic acid-sulfuric system was chosen to examine the performance of this system[4-6]. Changes in potential during the oscillating reaction will occur when blending potassium bromate with malonic acid, sulfuric acid, and cerium sulfate:

$$2H^{+}+2BrO_3^{-}+2CH_2(COOH)_2 \rightarrow 2BrCH(COOH)_2+3CO_2+4H \quad (1)$$

$$4Ce^{4+}+BrCH(COOH)_2+H_2O+HOBr \rightarrow 2Br^{-}+4Ce^{3+}+3CO_2+4H^{+} \quad (2)$$

With the reaction in process, the concentration of bromide ions increases until it is higher than critical, and process B will occur.

$$BrO_3^{-}+HBrO_2+H^{+} \rightarrow BrO_2+H_2O \quad (3)$$

$$BrO_2+Ce^{3+}+H^{+} \rightarrow HBrO_2+Ce^{4+} \quad (4)$$

$$2HBrO_2 \rightarrow BrO_3^{-}+HOBr+H^{+} \quad (5)$$

Thus, the trace of vibration appears in the system between process A and B, with the concentration variable. Oscillation reaction finished when Bromide had run out. During the experiment, we can see on the micro that there is a damped oscillation in the voltage signal between calomel electrode and platinum electrode; the solution's temperature, PH, and intensity of pressure will also change. Meanwhile, the system we designed is capable of studying the influence caused by one parameter.

3 SYSTEM HARDWARE DESIGN

System hardware design mainly refers to the data collection connector instrument, shown as a measurement system in figure 1. The lower computer is an MCU-extended system made with STM32F103CBT6 as the core. The upper computer is a computer whose designation is directly related to stability and reliability. The lower computer played a leading role in measuring export controlling quantity and collecting quantity. For saving data, displaying charts, and printing results, the upper computer played a leading role. In this way, we can ensure system reliability and by making full use of a computer's strong function we can do the complex data process and chart display.

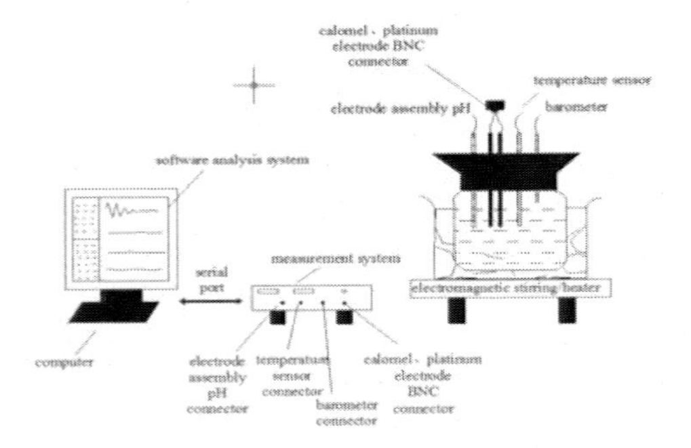

Figure 1. Chemical oscillating reaction parameter measurement system.

The hardware structure block diagram of the system is shown in figure 2, mainly consisting of the following modules: single-chip minimum system module, operational amplifier signal process module, hardware filtering module, and power supply module.

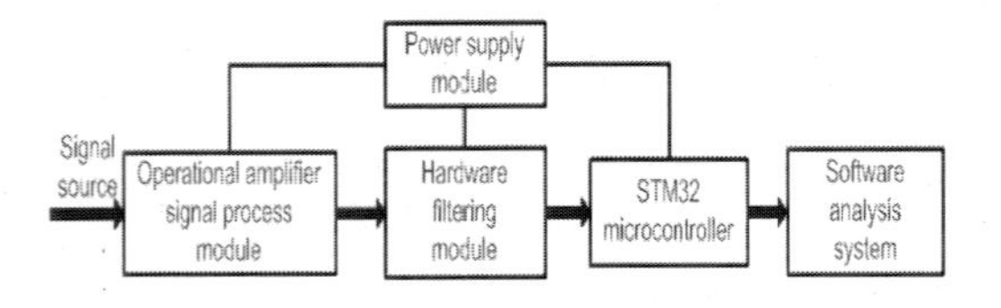

Figure 2. System hardware structure block diagram.

The single-chip minimum system needs to collect, transmit, transfer data rapidly, and communicate with the upper computer by means of the serial port. We pick STM32F103CBT6 as the micro-controller because of its high running speed, low cost, low power consumption, and rich peripheral [7]. For this system, we design a single-chip minimum system as shown in figure 3.

AD620 is a high-precision and adjustable amplifier used to magnify voltage in the paper. LTC1051 is a kind of chopper-stabilizer high-precision amplifier used as an adder. Both these amplifiers are integrated into the operational amplifier signal process module whose usage is to process voltage signals. It is uncertain whether plus or minus voltage can be produced when the chemical oscillating reaction proceeds. Thus, the adder constituted by LTC1051 mentioned earlier can be solved minus voltage signal acquisition. The system can also adjust as well as detect scale and precision by regulating external resistance easily. This module also integrates high-performance and high-input impedance TLC4502 operational amplifier using TL431 to provide the reference voltage aimed at handling the high-impedance composite electrode of the PH value of the output signal. Temperature is detected by 1-wire digital temperature sensor DS18B20 with high precision and strong anti-interference capacity. Intensity of pressure is acquired by the sensor PTZ-612 whose measure scale is 0–5 standard atmosphere pressure.

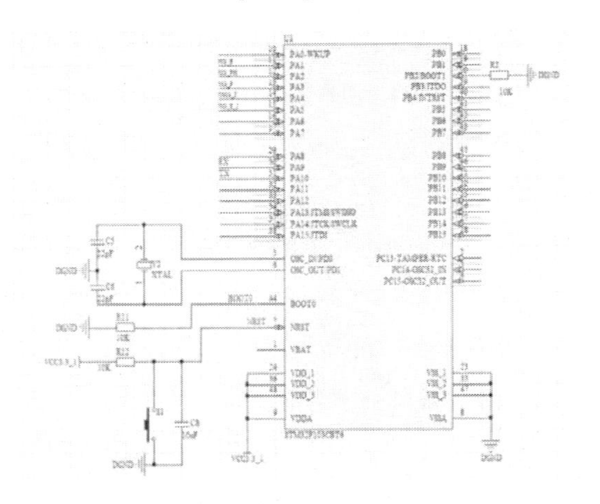

Figure 3. Single-chip minimum system.

In this paper, we adopt the fourth-order low-pass Butter-worth filter to filter noise waves in the hardware filter module in order to lead to fewer interferences.

Whether a system can operate stably greatly depends on the power supply module's stability. In this system, the power supply module provides 3.3 V, +5 V, and -5 V, for which 3.3 V is supply for the micro-controller whereas +5 V and -5 V are for various kinds of sensors and amplifiers. In order to keep stable voltage, application of three low chop-out, low in-band ripple constant voltage chip is fine and necessary.

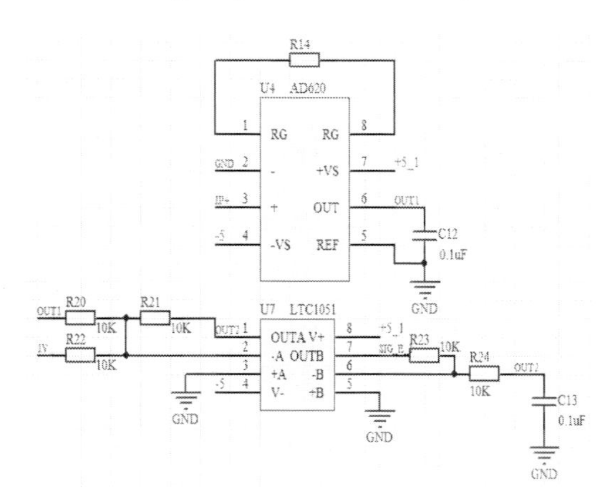

Figure 4. Voltage signal process module.

4 SYSTEM SOFTWARE DESIGN

The hardware system flow chart is given in figure 5. The description is as follows: (1) initializing module configuration; (2) AD collection, DMA transmission, and software filter; (3) entering into loop body, feeding dogs, doing data processing according to software flag bit, and interruption checking during loop body; (4) setting software flag bit to one after entering interruption, sending processed data to software analysis system for further analysis and display; (5) exiting interruption, going back to loop body.

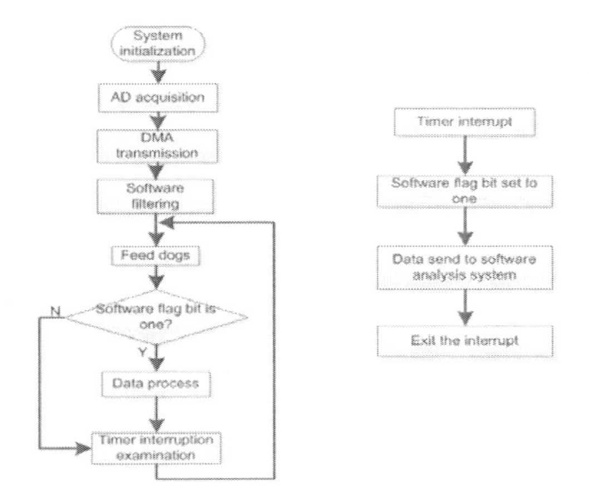

Figure 5. Hardware flow chart.

The software system flow chart is given in figure 6. Applications of the upper computer need are programmed by MSComm controls in VC++ based on VS2008 MFC [8-11].

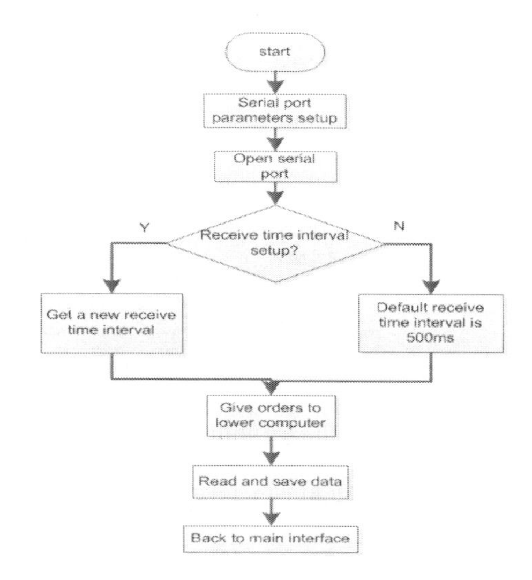

Figure 6. Software flow chart.

5 EXPERIMENTAL RESULTS

For limited conditions, we do the tests under laboratory conditions to verify the practicability and reliability of the system. Table 1 mainly shows measurements and actual values of different parameters. Voltage, temperature, and pH signals were detected in the same way. The data were gained under laboratory conditions whose process was as follows. For voltage signal, five sets of different voltage (They were 300 mV, 380 mV, 460 mV, 540 mV, and 620 mV, respectively.) exported by DC power were considered input voltage. Every set is measured five times; every thirty seconds, it records the data and averages every ten data bits as the result. The intensity of pressure is atmospheric indoor pressure due to limited lab conditions.

Table 1. Measurements and actual value of parameters.

Parameter	Actual value					Measurements				
	1	2	3	4	5	1	2	3	4	5
Voltage/mV	300	380	460	540	620	302.35	377.92	463.13	542.04	622.44
Temperature/℃	16	19	22	25	28	16.06	19.12	22.68	25.56	28.5
PH	1.54	5.24	7.13	8.05	10.42	1.49	5.28	7.21	8.01	10.37
Pressure/kP			100					100.13		

Results gained from chemical oscillating reaction processing under lab conditions with proportioned solutions are shown in figure 7. From the picture, we can demonstrate that the software analysis system is able to display a dynamic waveform diagram of each parameter at the same axis while completing an analysis of dynamic data in the meantime. Another merit of the software analysis system is that is capable of saving experimental data for subsequent analysis.

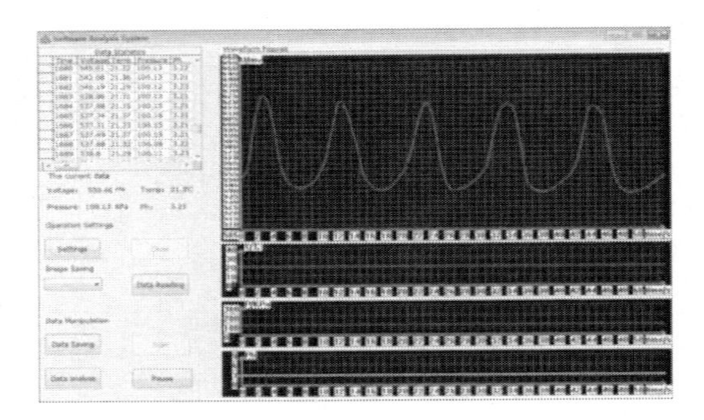

Figure 7. System experiment wave.

6 CONCLUSION

The measurement system offering high precision, complete functions, and strong flexibility has been already able to detect and analyze multiple parameters. The system can well adapt to the development trend of modern chemical experiment methods. Its great efforts for tracing and detecting amounts of materials help each field dramatically and, thus, will certainly have wide application prospects.

ACKNOWLEDGMENTS

Throughout this work, the authors would like to express their particular gratitude to their colleagues for helpful conversations and inspiration in the course of this research. The support from the Chinese National Nature Science Foundation Grant (61272025) for this research is gratefully acknowledged as well.

REFERENCES

[1] Wang, H. B.; Dong, Y. Y. (2003) *Chem. Ind. T.*, 06:8–11.

[2] LI, X. H. (2013) *Nonlinear Analysis of Different Time Scale Coupling Oscillatory Chemical Reactions*, Jiangsu: Jiangsu University.

[3] Sun, D. G.; Li, W. P.; Yang, W. J.; Hu, B. S.; Cao, Y. A. (2014) Teaching Experiment Device of Oscillatory Chemical Reaction. CN Patent 203786993u, 2014-08-20.

[4] Gao, J.Z., Wei, X. X., Yang, W., Lv, D. Y., Qu, J., Chen, H., Dai, H. X. (2007) Determination of 1-naphthylamine by using oscillating chemical reaction. *Journal of Hazardous Materials*, 144(1–2):67–72.

[5] Noyes, R. M. *J. Phys. Chem.* 1990, *94*, pp:4404–4412.

[6] Field, R. J., Boyd, P. M. *J. Phys Chem.* 1985, *73*, pp 3707–3714.

[7] Hao, W., Shen, J. X., Mei, C. *Elec. D. Eng.* 2013, *17*, pp 80–82.

[8] Hou, Q. F., Li, X. H., Li, S. (2007) *Visual C++ Database Universal Module Development and System Transplantation (1st)*, Beijing: Peking University Press, pp:43–88.

[9] Sun, X. (2012) *VC++ Deep Detailed Introduction (1st)*, Beijing: Electronic Industry Press, pp:342–365.

[10] Zhang, H.L. (2008) *Serial Port Communication Technology and Engineering Fulfilment (3rd)*, Beijing: Posts and Telecom Press, pp:120–165.

[11] Zhou, M. Y., Zhao, J. L. (2004) *Master GDI and Program (1st)*, Beijing: Peking University Press, pp:266–280.

Electric, Electronic and Control Engineering – Shao, Shu & Tian (Eds)
© 2015 Taylor & Francis Group, London, ISBN: 978-1-138-02842-5

Power electronic circuits fault diagnosis method based on neural network ensemble

Tiancheng Shi & Yigang He
School of Electrical Engineering and Automation, Hefei University of Technology, Hefei Anhui, China

ABSTRACT: A novel fault diagnosis method for a power electronic circuit based on an Artificial Neural Networks (ANN) ensemble is proposed in this paper. This method presents an ensemble structure of ANN to enhance the diagnostic accuracy. Based on the adaptive boosting (AdaBoost) algorithm, this ensemble structure can combine several neural networks together to generate a diagnotor with high diagnostic performance. A three-phase IGBT rectifier circuit with inductive load is used as the Circuit Under Test (CUT) to testify the diagnostic ability of the proposed method, and the diagnostic results proved its diagnostic precision.

KEYWORDS: fault diagnosis; power electronic circuit; ANN; AdaBoost.

1 INTRODUCTION

The power electronic systems, which provide convenient and high-efficient solutions to power conversion, play an increasingly important role in many fields such as industry, military, transportation, and aerospace technology. However, the widespread application of power electronic equipments into these fields challenges the reliability of the allover system [1]. Therefore, fault diagnosis technologies for power electronic systems have been deeply studied.

Electronic circuit fault diagnosis is divided into three categories: based on the analytical model [2], signal processing [3], and knowledge [4-7]. Due to the rapid development of artificial intelligence, the method base on knowledge has been widely researched.

As a typical algorithm based on the knowledge, the ANN methods, which possess strong non-linearity mapping ability, are effective in diagnosing and locating the fault states of power electronic devices [5]. However, the traditional ANN methods, such as the BP neural network, are usually affected due to their disadvantages of local minimum value and slow converging speed. Therefore, the research of enhancing the performance of ANN methods appears to be a new hotspot [6, 7].

Network ensemble, proposed in 1990 [8], is aiming at enhancing the learning ability by training several neural networks and combining their outputs. This technique, which has been applied in many fields such as Fault Diagnosis [9], Face Recognition [10], and Speaker Recognition [11], proves its good performance. Therefore, a novel ANN ensemble method, which is based on the AdaBoost algorithm, is proposed in this paper.

2 FAULT DIAGNOSIS METHOD BASED ON ANN ENSEMBLE

ANN is an effective machine learning method in pattern recognition, function approximation, classification, and so on. The Kolmogorov theorem has already proved that any continuous function can be implemented by a forward network with three layers. However, many researches show that a BP network with three layers can hardly get the optimal solution in a complex situation. According to the PAC learning ability model, the machine learning method can be divided into the strong method and the weak method by their learning ability; furthermore, a weak method can be improved into a strong method by optimization algorithms [12].

As a typical optimization algorithm, the AdaBoost algorithm can integrate several learning machines together to generate a new learning machine with better performance [13]. Every training sample has a weight value, which represents its probability for choosing the training set. The method can adaptively change the weight value of training samples by decreasing (or rising) the weight value of those samples that are correctly (or incorrect). Thus, the method can focus on the samples with higher weights, which seem to be harder to recognize. Main steps of the proposed method are listed later, and the corresponding flow diagram is shown in Fig 1.

1 Input: Input a training set of samples $(x_1, y_1), \ldots, (x_m, y_m), y_i \in Y = \{1, \ldots, T\}$;

2 Initialize the weight value of training samples: $w_1(i) = \frac{1}{m}$;

3 Initialize the parameters of the weak classifier: Choose the BP network with one hidden layer as the weak classifier; the structure of the network is 50-51-6;

4 Generate the training set by the sample distribution;

5 $for\ k = 1, 2, \ldots, K$, training and testing the weak classifier;

6 Calculate the error rate:

$$E(k) = \sum_i \left[w_k(i) * \left| h_k(x_i) \neq y_i \right| \right] \tag{1}$$

If $E(k) \geq 0.5$, go back to step 5); otherwise, go down to step 7)

7 Calculate the weight value of the current classifier:

$$\alpha_k = \frac{1}{2} \ln\left(\frac{1 - E(k)}{E(k)} \right) \tag{2}$$

8 Renew the weight value distribution of the training sample:

$$w_{k+1}(i) = \begin{cases} w_k(i) * e^{-\alpha k}, & h_k(x_i) = y_i \\ w_k(i) * e^{\alpha k}, & h_k(x_i) \neq y_i \end{cases} \tag{3}$$

9 Normalize the distribution of the training sample:

$$w_{k+1}(i) = \frac{w_{k+1}(i)}{\sum_i^n w_{k+1}(i)} \tag{4}$$

10 Output:

$$H(x) = \arg\max \sum \alpha_k * \left| h_k(x) = t \right| \tag{5}$$

3 SIMULATION TEST OF THE PROPOSED METHOD

A three-phase IGBT rectifier circuit with an inductive load, which is shown in figure 2, is used as an example to test the performance of the fault diagnosis method proposed in this paper. The circuit model simulated by MATLAB consists of three-phase AC voltage sources, a rectification bridge composed of six IGBT switches, and an inductive load. The necessary gating signals to the IGBT switches have been provided by the pulse generators operating by following the order of T1-T2-T3-T4-T5-T6.

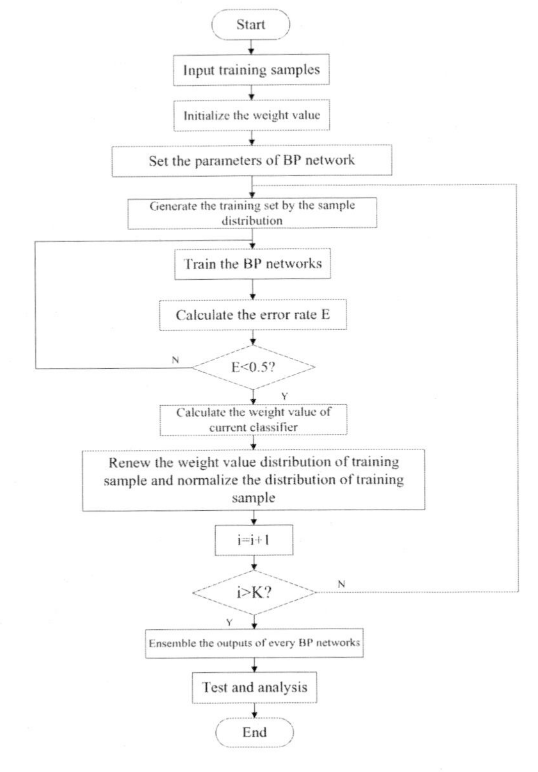

Figure 1. Flow diagram of proposed method.

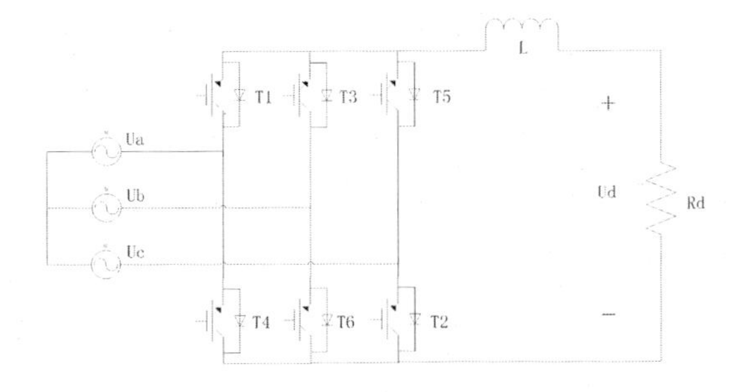

Figure 2. Three-phase rectification circuit.

The open faults, including open fault on IGBT, fusing on series fuse, and the missing of trigger pulse, which occur on the arms of rectifier bridges, are the most common faults in the power electronic circuits. Therefore, the faults of bridge arms open will be analyzed in this paper. The types of faults in the circuit under test are shown in Table 1.

Table 1. Types and causes of faults in the three-phase rectification circuit.

Type No.	Fault Cause	Type No.	Fault Cause
1	No fault	12	T1, T5 open
2	T1 open	13	T3, T5 open
3	T2 open	14	T4, T6 open
4	T3 open	15	T4, T2 open
5	T4 open	16	T6, T2 open
6	T5 open	17	T1, T2 open
7	T6 open	18	T1, T6 open
8	T1, T4 open	19	T3, T4 open
9	T3, T6 open	20	T3, T2 open
10	T5, T2 open	T1, T3 open	T5, T4 open
11	T1, T3 open	22	T5, T6 open

The output pulsating voltage u_d contains the information that can be used to determine whether there is any fault in the circuit, so u_d is used as the key sample for fault diagnosis. In this paper, with the three-phase bridge type, the controlled rectification circuit of trigger pulses $\alpha=0°, 30°, 60°, 90°$; 50 points of the output voltage u_d for all of the 22 different fault types are obtained from simulation, which lasts during the 5-cycle period. The waveforms of output voltage are shown in figure 3.

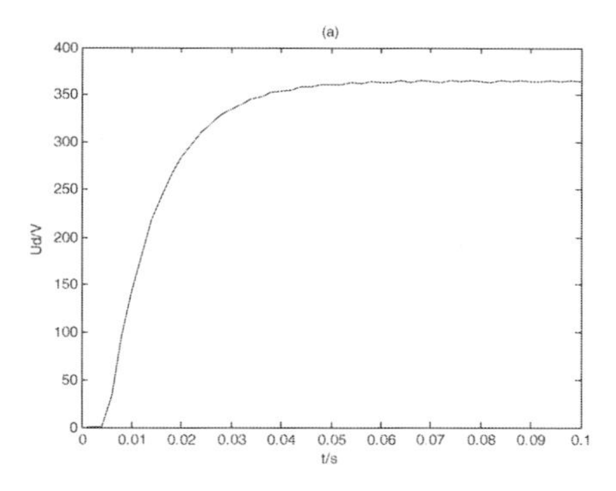

(a)$\alpha=0°$, type 1

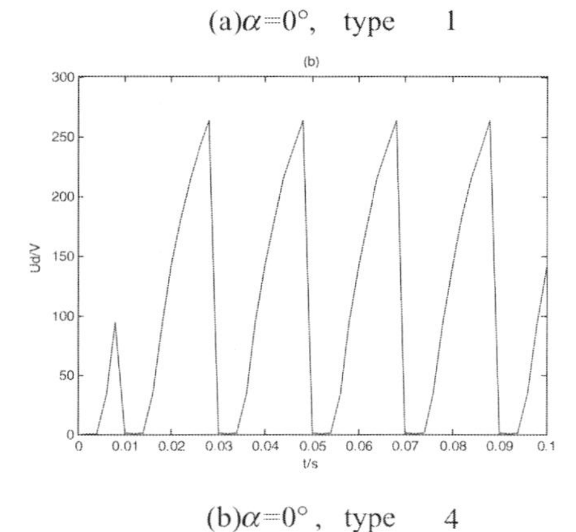

(b)$\alpha=0°$, type 4

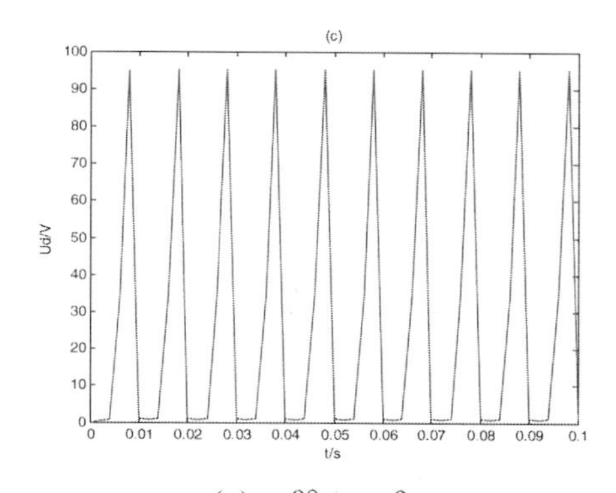

(c)$\alpha=0°$, type 9

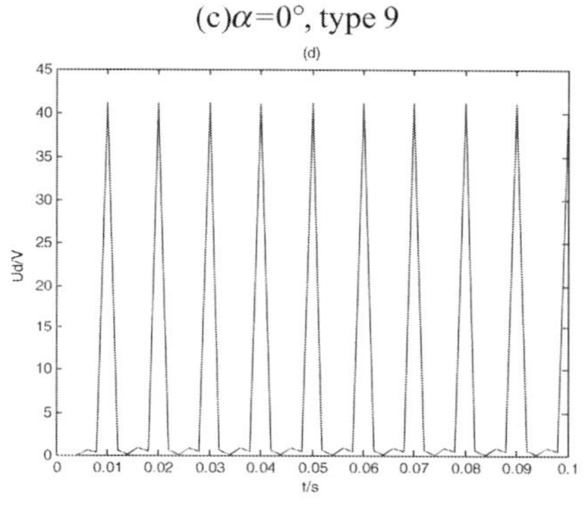

(d)$\alpha=60°$, type 9

Figure 3. Waveform of output pulsating voltage u_d with different fault type

Following the previously proposed steps, the sample data of all types of fault are used for training and testing the method. Results of the test and its analysis show the following:

Choosing the faults of T1 and T3, open with trigger pulse $\alpha=60°$, whose type number is 11; as an example, the results of the diagnosis are shown in Table 2.

Table 2. Results of diagnosis by vote mechanism with rights.

Type No.	Vote rate	Type No.	Vote rate
1	0	12	0
2	11.03%	13	3.34%
3	1.86%	14	0
4	0	15	3.36%
5	0	16	0
6	0	17	1.89%
7	0	18	0
8	0	19	1.83%
9	0	20	0
10	0	21	0
11	76.68%	22	0

According to the voting result, the method proposed in this paper believes that the most probable fault number is 11, which proved that this method has already had ability for correct diagnostic. Table 3 shows the performance comparison of the proposed method and other methods.

Table 3. Performance comparison of proposed method and other methods.

Methods	Diagnostic Accuracy Rate/%
BP network	60.8%
Reference [6]	93.3%
Reference [7]	92.9%
This Paper	97.9%

The simulation results show that, compared with other references and the traditional BP network, the diagnosis method proposed in this paper has a high diagnostic accuracy in power electronic circuits.

To test the diagnostic performance of the proposed method on the condition with noise, this paper adds different voltage fluctuations to the AC voltage sources. Table 4 shows the performance comparison of BP networks and ensemble networks on the condition with noise.

Table 4. Performance comparison on the condition with noise.

Voltage fluctuation range	BP networks	Ensemble networks
±5V	60.8%	97.9%
±10V	55.9%	94.8%
±15V	50.6%	90.6%
±20V	43.2%	87.7%

The comparison results show that, with the increase of voltage fluctuation range, the diagnostic capacities of both algorithms have varying degrees of decline. With the voltage fluctuation of ±20V, the BP networks algorithm has a diagnostic accuracy at 43.2%; whereas the proposed method based on network ensemble has a higher diagnostic accuracy at 87.7%. The results proved that the proposed method still can get a high diagnostic accuracy rate on the condition with voltage fluctuation, which indicated its robustness.

4 CONCLUSION

This paper proposed a fault diagnosis method of the power electronic circuit based on an improved neural network algorithm. Using the AdaBoost principle, we generated a diagnotor with high diagnostic accuracy combined together with several BP neural networks. The simulation results for a three-phase IGBT rectifier circuit with inductive load indicated that the proposed method has higher fault diagnostic precision than BP networks and other diagnosis methods. The neural network ensemble can highly improve its diagnostic performance, but the basic diagnotor is still the traditional BP network. Hence, the method proposed is still affected due to the shortcomings of the local minimum value, so choosing a weak classifier with better diagnostic performance will be the research issue in future.

ACKNOWLEDGMENTS

This work was supported by the National Natural Science Funds of China for Distinguished Young Scholar under Grant (50925727), The National Defense Advanced Research Project Grant (C1120110004), (9140A27020211DZ5102), the Key Grant Project of Chinese Ministry of Education under Grant (313018), Anhui Provincial Science and Technology Foundation of China under Grant (1301022036), and Hunan Provincial Science and Technology Foundation of China under Grant (2010J4), (2011JK2023).

REFERENCES

[1] Baker N, Liserre M, Dupont L, *et al.* (2014) Improved reliability of power modules: A review of online junction temperature measurement methods. *Industrial Electronics Magazine, IEEE*, 8(3):17–27.

[2] An Q T, Sun L, Sun L Z. (2013) Hardware-circuit-based diagnosis method for open-switch faults in inverters. *Electronics Letters*, 49(17):1089–1090.

[3] Seshadrinath J, Singh B, Panigrahi B K. (2014) Investigation of vibration signatures for multiple fault diagnosis in variable frequency drives using complex wavelets. *Power Electronics, IEEE Transactions on*, 29(2):936–945.

[4] Lifeng W, Xueyan Z, Yong G, *et al.* (2013) A survey of fault diagnosis technology for electronic circuit based on knowledge technology. *Control and Decision Conference (CCDC), 2013 25th Chinese. IEEE, 2013*, pp:3555–3559.

[5] Hu Q, Wang R, Zhan Y. (2008) Fault diagnose technology of power electronic based on neural network. *Innovative Computing Information and Control, 2008. ICICIC'08. 3rd International Conference on. IEEE*, 2008, pp:430–430.

[6] Xiang X, Ruiping Z, Zhixiong L. (2010) Virtual simulation analysis and experimental study on gear fault diagnosis based on wavelet neural network, *Machine Vision and Human-Machine Interface (MVHI), 2010 International Conference on. IEEE, 2010*, pp:55–58.

[7] Wang X, Wang T, Wang B. (2008) Hybrid PSO-BP based probabilistic neural network for power transformer fault diagnosis, *Intelligent Information Technology Application, 2008. IITA'08. Second International Symposium on. IEEE, 2008*, 1: 545–549.

[8] Hansen L K, Salamon P. (1990) Neural network ensembles. *IEEE Transactions on Pattern Analysis and Machine Intelligence*, 12(10):993–1001.

[9] Lei Z, Jinhui W, Ligang H, *et al.* (2009) Analog IC Fault Diagnosis based on Wavelet Neural Network Ensemble, *Intelligent Systems, 2009. GCIS'09. WRI Global Congress on. IEEE, 2009*, 4: 45–48.

[10] Wang J, Yin J. (2009) Face recognition by global optimal discriminant features and ensemble artificial neural networks, *Computer Network and Multimedia Technology, 2009. CNMT 2009. International Symposium on. IEEE, 2009*, pp:1–4.

[11] Qian B, Tang Z, Li Y, *et al.* (2007) Neural network ensemble based on vowel classification for Chinese speaker recognition, *Natural Computation, 2007. ICNC 2007. Third International Conference on. IEEE, 2007*, 3: 141–145.

[12] aliant L G. (1984) A theory of the learnable. *Communications of the ACM*, 27(11):1134–1142.

[13] Freund Y, Schapire R E. (1995) *Computational Learning Theory.* Springer Berlin Heidelberg, pp:23–37.

Electric, Electronic and Control Engineering – Shao, Shu & Tian (Eds)
© 2015 Taylor & Francis Group, London, ISBN: 978-1-138-02842-5

The model, simulation and verification of wireless power transfer via coupled magnetic resonances

Hongguang Zhang & Minxuan Zhang
Computer Academy, National University of Defense Technology, Changsha Hunan, China

ABSTRACT: The subject firstly conducted the theoretical analysis and mathematical modeling of the wireless power transfer via coupled magnetic resonances, and found the general expressions of the transfer efficiency and the output power. Then we built the two-coil model by COMSOL and analyzed the influences of transmission distance, resonant parameters, frequency, load, etc. Next step we calculated the parameters of the class-E power amplifier, and simulated its running states by MATLAB. Finally we developed the wireless power supply locomotor system, whose efficiency reached the expected value and running state was good, which might be of high value in the applications of wireless power transfer technology.

KEYWORDS: coupled magnetic resonances; wireless power transfer; two-coil model; locomotor system.

1 INTRODUCTION

Wireless power transfer (WPT) is a kind of energy transfer technology, and has better security and convenience compared with the traditional mode. Power is transferred from the input to the output without direct electrical contact by WPT[1].

During the past decade, society has witnessed a dramatic surge of use of autonomous electronic devices, electro-mobiles, artificial organs, etc. The traditional power transfer mode can never meet the needs of people. Take the electro-mobile for example, if it's charged by wire, the charging must be finished in the charging stations, which brings about great inconvenience to the users. As a consequence, interest in WPT has re-emerged and we confirm that this kind of problems will be solved by WPT.

There are three primary modes of WPT: 1) Inductive Coupled Power Transfer; 2) Micro- wave Power Transfer; 3) Coupled Magnetic Resonant Wireless Power Transfer (CMR- WPT). The preceding two modes have limited range of applications because of the unsolved contradictions between the distance and the efficiency of the transmission. At present, the third one is the main research topic.

In the current work, we focus on the CMR-WPT, which is particularly suitable for everyday applications because most of the common materials do not interact with magnetic fields. We analyze the basic theory of CMR-WPT, identify the mathematical modeling of magnetic resonances, and adopt the class-E power amplifier (PA) in the high-frequency inverter circuit, which improves the efficiency of the transfer. Besides, based on the circuits we have designed, we test every parameters and characters, and develop the wireless power supply locomotor system.

2 THEORY AND ANALYSIS

2.1 *Basic theory*

In ideal conditions, the equivalent diagram of CMR-WPT is shown as Fig.1.

In Fig.1, U is the AC source, Z_S is the source internal resistance, L_1 is the transmitting coil inductance, C_1 is the transmitting resonant capacity, I_1 is the transmitting current, L_2 is receiving coil inductance, C_2 is receiving resonant capacity, I_2 is the receiving current, R_L is the load, M_{12} is the coefficient of mutual induction.

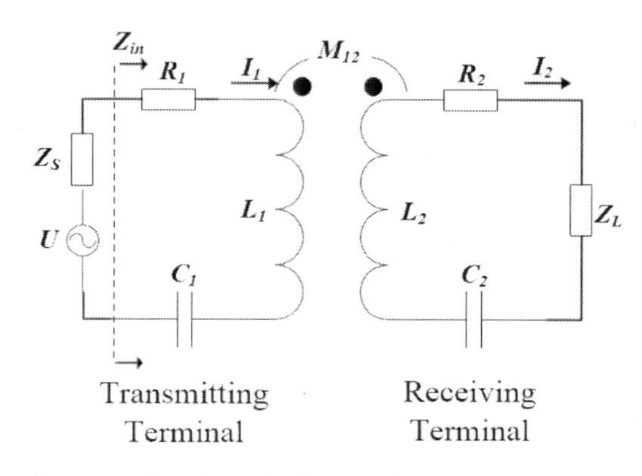

Figure 1. The schematic diagram of CMR-WPT.

If the frequency of source is f and the resonant frequency of the transmitter and the receiver, which is determined by the coil inductance and the resonant capacity, is also f, the whole system of two coupled magnetic resonances will be in resonant state. Then the energy stored in C_1 exchanges with the magnetic energy stored in L_1, meanwhile, a part of energy stored in L_1 is transmitted to L_2 so that the alternating magnetic field produces alternating current in the receiver. Finally the energy is transmitted to the receiver and the load [2].

In systems of coupled resonances, there is a "strong coupled" regime and if we can operate in that regime in a given system, the efficiency of energy transfer is expected to be very high. To better understand the distributions of the magnetic field, we can divide the space around the transmitter to near-field and far-field. Near-field is the space whose center is the center of the transmitting coil and radius is $\lambda/2\pi$, far-field is the space from $\lambda/2\pi$ to infinity where λ is the wavelength of electro-magnetic wave.

Generally, the strength of electro-magnetic field in near-field is much stronger than that in far-field. Efficient mid-range power transfer occurs in near-field so that the main working region of CMR-WPT is near-field. Besides, the waves going into the far-field region can never come back, so the magnetic resonances can only be engendered in near-field.

According to the analysis above, it's of great concern that we must select suitable resonant frequency. The paper [3] presented that the best working frequency for the CMR-WPT of mid-range is 1~50MHz. So we choose 1MHz as the resonant frequency.

2.2 *Mathematical model*

According to Fig.1, we obtain the state equation of CMR-WPT:

$$\begin{bmatrix} \dot{U}_S \\ 0 \end{bmatrix} = \begin{bmatrix} R_S + jX_S & -j\omega M \\ -j\omega M & R_L + R_D + jX_D \end{bmatrix} \begin{bmatrix} \dot{I}_S \\ \dot{I}_D \end{bmatrix} \quad (1)$$

Input voltage U_S, Transmitting Current I_S, Receiving Current I_D, Coefficient of Mutual Induction M, Resonant Radian Frequency ω, Reactance of Transmitter and Receiver X_D, X_S, Internal Resistance of Transmitting Coil and Receiving Coil R_D, R_S, Load R_L.

X_S is the reactance of transmitter:

$$X_S = \omega L_S - \frac{1}{\omega C_S} \quad (2)$$

The resonant condition of transmitter is X_S=0, so $\omega^2 L_S C_S$=1.

X_D is the reactance of receiver:

$$X_D = \omega L_D - \frac{1}{\omega C_D} \quad (3)$$

The resonant condition of transmitter is X_D=0, so $\omega^2 L_D C_D$=1

From Eq.1, we find the relationship between the current of transmitter and receiver:

$$\dot{I}_D = \frac{j\omega M}{Z_D} \dot{I}_S \quad (4)$$

According to Eq.1 and Eq.3, we find the relationship between the input voltage and input current:

$$\dot{U}_S = \left[Z_S + \frac{(\omega M)^2}{Z_D} \right] \dot{I}_S \quad (5)$$

From Eq.5, we find the equation of impedance Z_{eq}:

$$\begin{aligned} Z_{eq} &= \frac{\dot{U}_S}{\dot{I}_S} = Z_S + \frac{(\omega M)^2}{Z_D} \\ &= R_S + jX_S + \frac{(\omega M)^2}{R_L + R_D + jX_D} \end{aligned} \quad (6)$$

According to the basic theory of circuit, resistance consumes energy while inductance and capacity just transfer the reactive power. If the input power is constant, the efficiency and the output power decrease along with the increasing of the reactive power. In resonant state, X_S and X_D are 0, and the reactive power is minimized.

For CMR-WPT system, the efficiency and the output power is maximized when the transmitter and the receiver are both in resonant state.

The output power is:

$$P_{out} = I_S^2 R_L$$

According to Eq.5, we obtain the equation of P_{out}:

$$P_{out} = \frac{(\omega M)^2 U_S^2 R_L}{\left[R_S (R_L + R_D)^2 - X_D X_S + (\omega M)^2\right]^2 + \left[R_S X_D + (R_L + R_D) X_S\right]^2} \quad (7)$$

There are many factors which have influences on P_{out}, including M,ω, X_D,X_S,R_D,R_S,R_L.

We conduct partial differential on the denominator of Eq.7, and find that if M is small enough to match the condition $R_S(R_W + R_D)/(\omega M)^2 \geq 1$, the denominator has one minimum so that P_{out} has only one maximum while the system is in resonant state; if M is large enough to match the condition $R_S(R_W + R_D)/(\omega M)^2 < 1$, the denominator has two minimums so P_{out} has two maximums.

From the analysis, we have summarized the following rules: M has a threshold x. If $M < x$, the output power is maximized when the system is in resonant state; if $M \geq x$, the output power has two maximums while system isn't in the resonant state. In fact, M represents the distance between the transmitter and the receiver, M increases while the distance decreases. So x also represents a distance threshold d, which is thought to be very small. Generally, we assume the distance between two objects is further than d, and the system has only one maximum.

The equation of efficiency η is obtained from Eq.5:

$$\eta = \frac{(\omega M)^2 R_L}{R_S\left[(R_L + R_D)^2 + X_D^2\right]^2 + (\omega M)^2 (R_L + R_D)} \times 100\% \quad (8)$$

The efficiency is determined by the resonant parameters of the receiver and has no business with transmitter. The efficiency is maximized while the receiver is in resonant state,.

2.3 *Two-coil model*

We use COMSOL software to build and analyze the two-coil model of coupled magnetic resonances, and study the influences of every parameters. The two-coil model in COMSOL is shown as Fig.2.

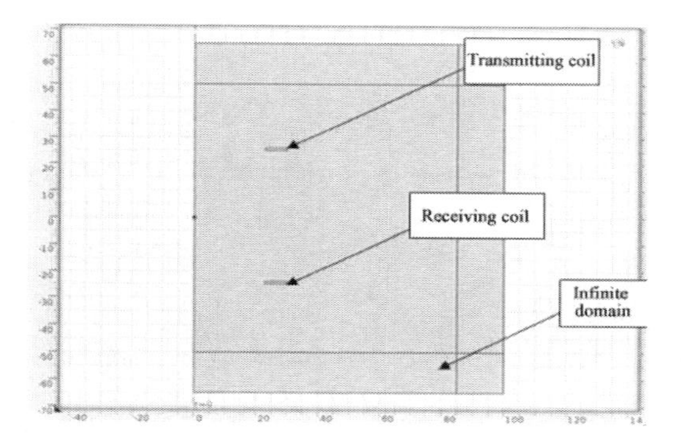

Figure 2. Two-coil model.

The settings of the working environment are as follows: the input voltage is $10V$, the power internal resistance is 0.01Ω, the load is 10Ω.

The basic parameters of two-coil model are shown in Table 1.

Table 1. The parameters of two-coil model.

Item	Value
Configuration Width	7mm
Configuration Height	1mm
Sectional Area of Wire	6mm^2
Conductivity of Wire	6e^7
Number of Turns	5
Outmost Radius	30mm
Inductance	2.28uH
Absorptive Resistance	0.315
Resonant Capacity	13nF
Resonant Frequency	1MHz

By COMSOL, we find the influence of frequency, transmission distance, load, etc. The waveforms obtained from the simulation are shown as the following figures.

From Fig.3 and Fig.4, we know that the efficiency of the transfer decreases while the transmission distance increases, and has no maximum. The efficiency approximately becomes 0 when the transmission distance reaches $70mm$. The load power firstly increases and then decreases along with the increasing of the transmission distance and is maximized when the distance is $20mm$.

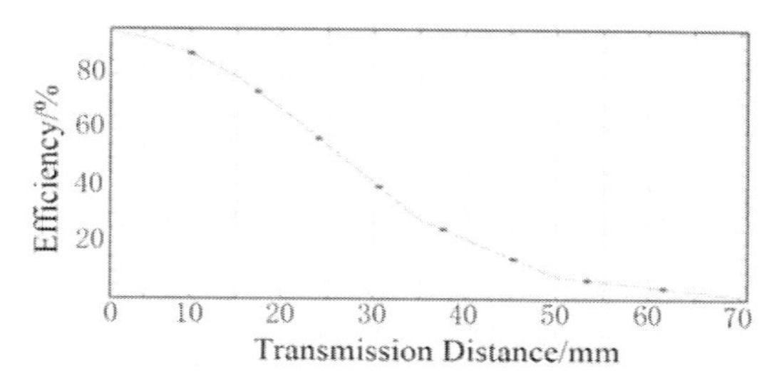

Figure 3. The variation of the efficiency along with the increasing of the transmission distance.

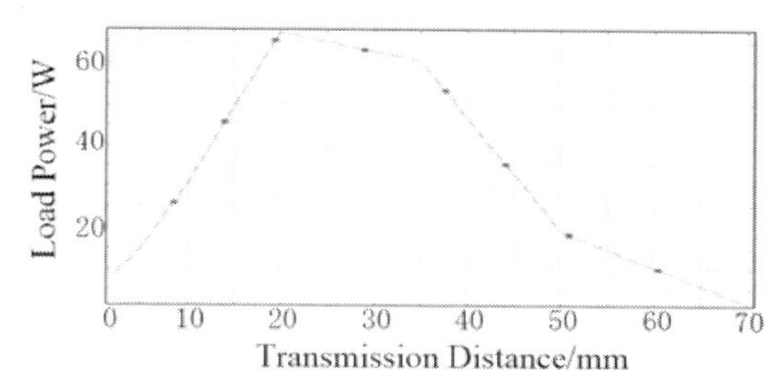

Figure 4. The variation of the transfer power along with the increasing of the transmission distance.

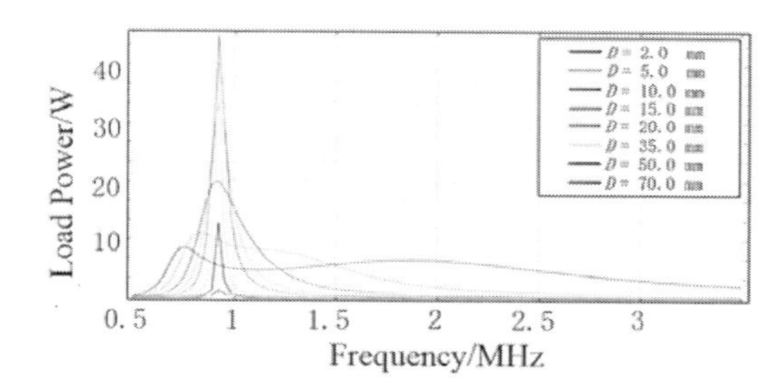

Figure 5. The variation of the load power in different frequencies.

According to Fig.5, we find the load power has two maximums when $D \le 0.005mm$; if $D \ge 0.01mm$, the load power has only one maximum, which is correspond to the conclusions of the previous chapter.

According to Fig.6, when the load increases from 0.01Ω to 10Ω, the efficiency of transfer firstly increases and then decreases along with the increasing of frequency.

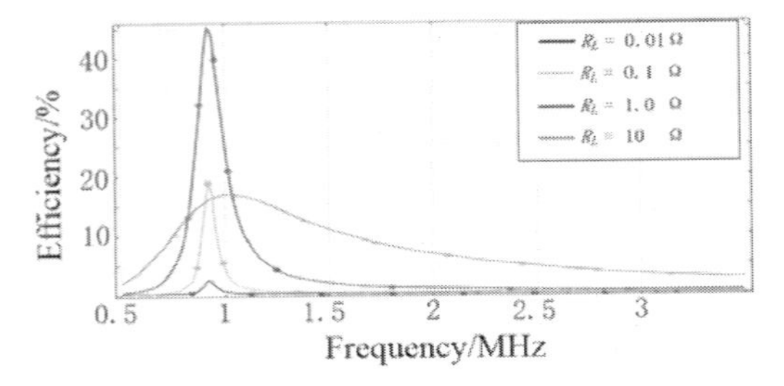

Figure 6. The variation of the efficiency in different frequencies.

Based on the results obtained from the simulation, we find that the efficiency of the transfer decreases sharply along with the increasing of the transmission distance, so it is suitable for the transfer system of mid-power and shorter distance.

3 DESIGN

The overall frame of our experiment is shown as Fig.7.

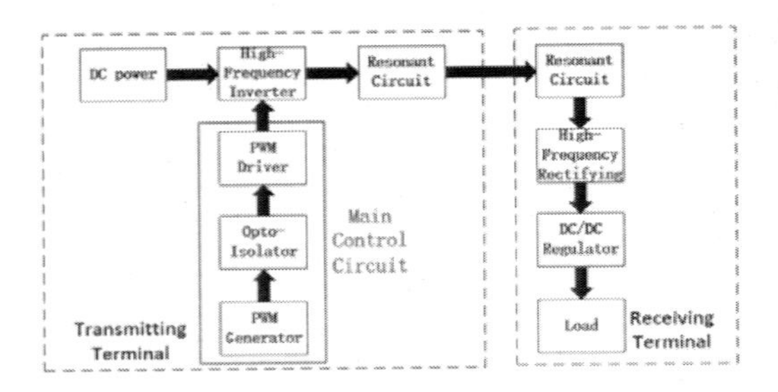

Figure 7. The overall framework of system.

The transmitter includes the DC source, the high-frequency inverter circuit, the transmitting resonant circuit, the optical coupler isolation, the PWM driver, the PWM generator and the switching power source modules.

The receiver includes the receiving resonant circuit, the high-frequency rectifying circuit and the switching power source modules.

The resonant circuits whose resonant frequency is 1*MHz* are designed as the series connection of inductance and capacity. The resonant inductance of transmitter is a 4 turns' coil made by litz wires, whose value is $48.0\mu H$. And the matched resonant capacity is $500pF$. The receiving inductance is a 7-turns' coil whose value is $24.0\mu H$ and the matched resonant capacity is $1000pF$.

The high-frequency inverter circuit is the key point among all parts of the system. We adopt class-E PA of soft switching mode as the inverter circuit. The PA has been tested by MATLAB.

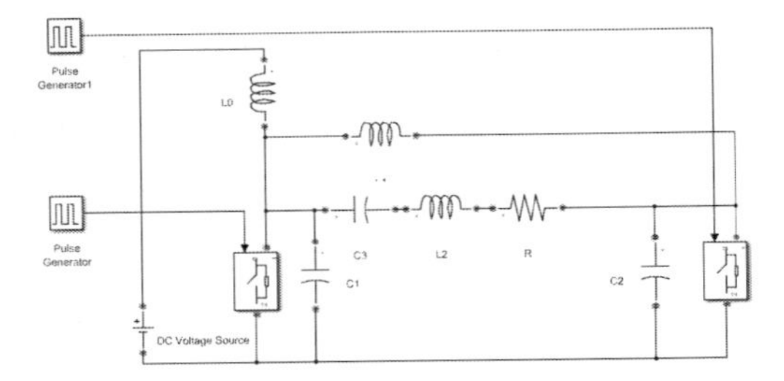

Figure 8. The simulation of the class-E PA.

In this simulation, the input source is $30V$, the frequency of PWM is $1MHz$ whose duty cycle is 50% and value of peak-to-peak voltage is $8V$; the resonant inductance L_2 is $48\mu H$, the matched resonant capacity is $527.7pF$. The waveforms obtained from the simulation are shown in Fig.9.

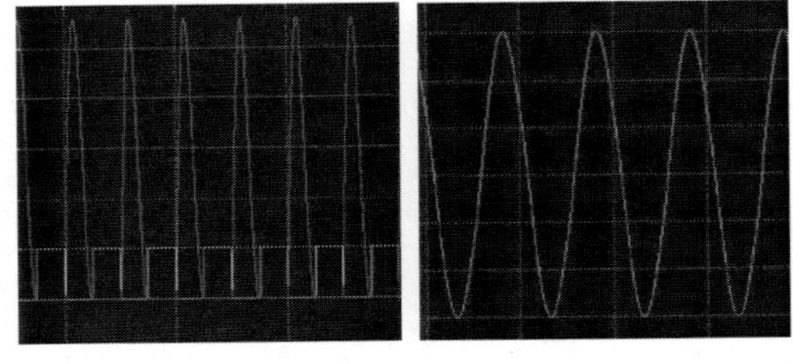

(a)MOSFET terminal voltage (b)Resonant current

Figure 9. The waveforms of the MOSFET terminal voltage and the resonant current in Matlab.

According to the results of the simulation, we find the parameters of the class-E PA work well, so we choose the same parameters in our design, and finally we get the same waveforms shown in Fig.10 from the actual circuits.

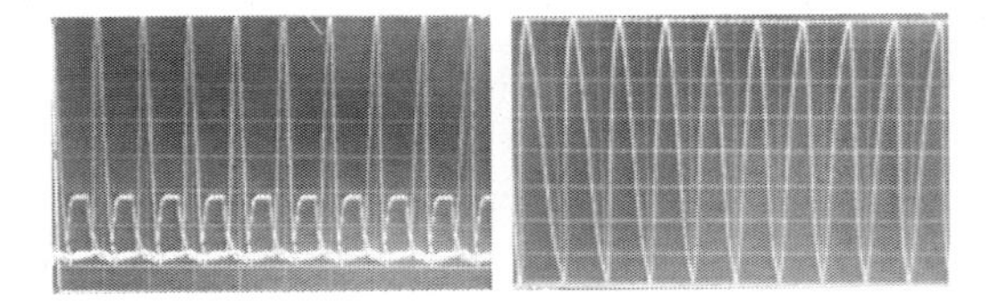

(a)MOSFET terminal voltage (b)Resonant current

Figure 10. The actual waveforms of the MOSFET terminal voltage and the actual resonant current.

Based on the high-frequency inverter, we have developed the wireless power supply locomotor system.

What we have done are as follows: Firstly, we get rid of the original power supply system, and put a rectangular coil on the ground, whose size is $15cm \times 180cm$, and then place a smaller rectangular coil on the bottom of the car, whose size is $5cm \times 15cm$. The distance between two coils is $6cm$. The output of the receiver is filtered and adjusted to supply power to the remote control car so that the car is driven without batteries while it's just over the transmitting coil.

If the input voltage reaches $30V$, the car works well. The maximum of input current is $1A$, the maximum of power is $30W$, and the efficiency of the system is about 60%.

In these experimental devices, we change the speed of the car by adjusting the output voltage of the receiver. However, the car may be not driven normally because of the low driving current if the output voltage is adjusted too high.

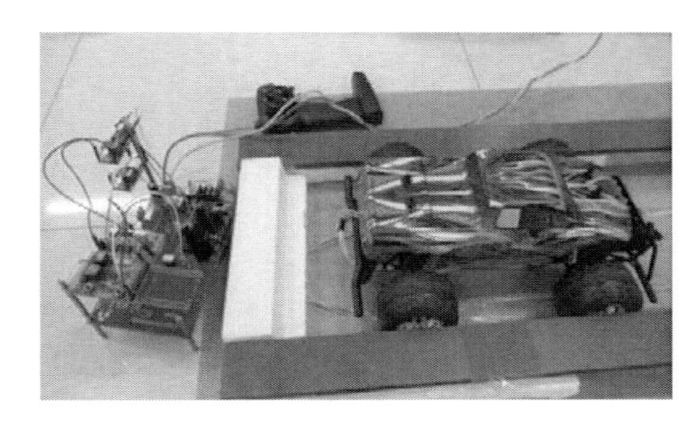

(a) Transmitter

(b)The whole system (c)Receiver

Figure 11. Wireless power supply locomotor system.

4 CONCLUSION

The main purpose of our subject is to design an application of wireless power transfer via coupled magnetic resonances.

This paper analyses the basic theory and the mathematic model of CMR-WPT in detail. We simulate the two-coil model by COMSOL, and find the correlated characteristics of the model. The locomotor system developed by ourselves works well, the remote control car responses sensitively and runs stably. In one word, the system is of higher value in the applications of wireless power transfer.

However, there are still some defects. For example, the efficiency isn't maximized at the mercy of the shape and the area of the coils. So we should devote more effort to analyze the influences of the two factors on COMSOL next step to optimize the efficiency. Besides, close transmission distance is required in our design to guarantee the smooth running of the car while the distance is likely to be further in actual applications, so we should consider that how to increase the transmission distance with high efficiency. These defects show that our design still has room for improvements, and is worthy of further studies.

REFERENCES

[1] Yang Qingxin, Chen Haiyan, Xu Jiazhi. (2010) Research progress in contactless power transmission technology. *Transactions of China Electrotechnical Society*, (7):6–13.

[2] Shinohara N, Kawasaki S. (2009) Recent wireless power transfer technologies in Japan for space solar power station/satellite. *Radio and Wireless Symposium. San Diego, USA: IEEE, 2009*, pp:13–15.

[3] P. E. Glaser. (1968) Power from the sun: its future. *Science*, pp:857–861.

[4] Zhao Zhengming, Zhang Yiming, Chen Kainan. (2013) New progress of magnetically-coupled resonant wireless power transfer technology. *Proceedings of the CSEE 2013*, 33(3):1–13.

[5] Chen Junbin, Zhu Xia. (2010) The approximate analytic formula of the mutual inductance between two discretionary coaxial loops. *Journal of Logistical Engineering University*, (5):86–91.

[6] Terman L M. (1995) Proceedings of the Fourth International Conference on Solid State and Integrated Circuit Technology, pp:7–12.

[7] Liu Feng. (2004) *Research on Main Circuit Topological Structure of High Frequency High Power Inverter Based on MOSFET*. Xi'an: Xi'an University of Technology.

[8] Imura T, Okabe H, Hori Y. (2009) Basic experimental study on helical antennas of wireless power transfer for electric vehicles by using magnetic resonant couplings. *Vehicle Power and Propulsion Conference, 2009. VPPC'09. IEEE, 2009*, pp:936–940.

[9] Zhu Yunzhong, Pu Ling, Zeng Haiqiang. (2009) Efficient wireless power transfer system based on switch circuit. *Acta Scientiarum Naturalium University Sunyatsei* , (48):228–230.
[10] Zhang Xiaozhuang. (2009) *Research on the Distance Characteristics and Experimental Device of Magnetic Resonances Based on Wireless Energy Transfer Technology*. Harbin: Harbin Institute of Technology, pp:7–12.
[11] Zou Lin, Zheng Wei, Li Li. (2013) Research on the Frequency Characteristics of Wireless Energy Transmission System Based on Magnetic Resonance Coupling[A]. *Journal of Hubei University of Technology*, 28(1):82–85.

Electric, Electronic and Control Engineering – Shao, Shu & Tian (Eds)
© 2015 Taylor & Francis Group, London, ISBN: 978-1-138-02842-5

Spatial distribution characterization of A-grade tourist attractions in Guizhou Province by GIS

Hongli Fu
School of Geography & Resource Science, Neijiang Normal University, Neijiang Sichuan, China

ABSTRACT: Guizhou contains nine prefecture-level administrative regions, each of which has a number of A-grade and higher tourist attractions. The spatial distribution of these attractions could, to some extent, decide the tourist routes of tourist groups, thereby leading to unbalanced tourist distribution among different attractions and limiting the coordinated development of regional tourist economy within the province. Against this backdrop, studying the spatial structure of tourist attractions in Guizhou has great implications. In this paper, quantitative analysis is performed on the spatial point distribution of A-grade and higher tourist attractions in the nine cities/prefectures of Guizhou with the help of GIS, using indicators including the nearest neighbor index, geography concentration index, Gin coefficient, and attraction distribution density. The purpose is to look into the spatial distribution characteristics of these tourist attractions. Our results indicate that the A-grade tourist attractions in Indonesia, Tongren, Indiana, and Upanishads are spatially uniform whereas those in the other regions are spatially causal; A-grade and higher tourist attractions are spatially more centralized and mostly found in Guiyang and Zuni; A-grade attractions in Central Guizhou are highly centralized, whereas attractions in East and West Guizhou are not so densely distributed.

KEYWORDS: A-grade tourist attractions; spatial distribution; GIS technology.

1 INTRODUCTION

Tourist attractions are an important part of the tourist industry. A-grade tourist attraction identification is the national criterion that measures the overall quality and grade of a tourist attraction[1]. In Guizhou Province, where different ethnic minorities co-exist and the natural environment is beautiful, according to incomplete statistics, as of December 2013, there were 101 A-grade tourist attractions. Chang J., ET AL (2013) noted that Guizhou receives 39 million tourists every year and ranks the 22nd after China. Of the 2,926 million tourists received by the A-grade and higher attractions across China, Guizhou contributes only 1.33% [2]. This suggests that there is still great room for the local tourist industry. As the spatial structure of tourist attractions is a constraint on the tourist flow to some extent, it is necessary to characterize the spatial structure of the A-grade and higher tourist attractions in Guangzhou.

The spatial structure of tourist attractions has long been a topic for authors both in and out of China. Wu B.H., ET AL (2003) investigated the spatial structure characteristics of the first 4A tourist attractions in China and the distance differentiation between different sizes of source markets [3]. Ma X.L., ET AL (2003) measured the spatial characteristics and correlation of 4A tourist attractions[4]. Thu H., ET AL (2008) examined the distribution and variation of the 2286 A-grade tourist attractions in China, particularly using the entropy evaluation method [5]. Au H.Q.(2013) analyzed the situation of tourist attractions in Guamanian in 1978, 1990, 2000, and 2012 using GIS technology [6]. Wang W.X., ET AL (2012) studied the spatial structure of the 125 A-grade tourist attractions in Huber and their relations with resources, administrative regions, transportation, and water systems using the graphic method and spatial quantitative analysis [7]. Wang H., ET AL (2010) examined the spatial structure of tourist attractions in Dalian using the network analysis method of geographical mathematics combined with GIS technology [8]. Besides, Chang D., ET AL (2014) made a fractal study on the spatial structure of the tourist attraction system of Xi amen [9]. GAO Y.S. (2011) analyzed and evaluated the spatial distribution of the tourist resources in Zhengzhou[10]. In this paper, after a review on previous findings, the spatial point distribution of the 101 A-grade and higher tourist attractions in the nine cities/prefectures of Guizhou Province is investigated using indicators, including nearest neighbor index, geographic concentration index, attraction distribution density, and Gin coefficient. The purpose is to provide theoretical reference for optimizing the spatial structure of the local tourist attractions and developing the local tourist industry.

2 STUDY AREA OVERVIEW

Guangzhou, called Qian or Gui for short, lies on the eastern slope of Southwest China's Yungui Plateau, adjacent to Hunan to the east, Guamanian to the south, Yunnan to the west, and Sichuan and Chongqing to the north. In Southwest China, it is an important hub connecting the east to the west and the north to the south. It is also an important corridor for sea transport and the land traffic terminal in Southwest China. Within the province, the land relief is higher in the west than in the east. The landform is grand and steep and characterized by many mountains and hills. Karst landform covers 61.9% of the province, making the province one of the regions with the most typical karst landform. Guizhou ranks the third after China in terms of the proportion of ethnic minorities and the third of the nation in terms of the ethnic minority population. The unique geography and diverse minorities offer the province colorful natural landscapes and national cultures, laying the resources basis for the rapid development of the local tourism. As of December 2013, the province had 101 A-grade and higher tourist attractions or 11.27% of the total number of A-grade and higher tourist attractions in Southwest China (Yunnan, Guangzhou, Sichuan and Chongqing). These include three class 5A, 33 class 4A, 52 class 3A, and 13 class 2A tourist attractions.

3 METHODOLOGY

Figure 1 shows the spatial distribution of A-grade and higher tourist attractions in Guizhou from map digitalization. On the basis of this map, quantitative analysis was carried out of the 101 A-grade and higher tourist attractions in the nine cities/prefectures of Guizhou on ArcGIS10.2 and Excel as the data analysis platforms, using ArcGIS spatial analysis tool.

For concrete analysis of the spatial structure of tourist attractions, the nearest neighbor index (NNI) is used to measure the spatial structure type of the attractions; the geographical concentration index is used to measure the geographical concentration of the attractions; and Gin coefficient is used to measure the distribution uniformity of the attractions. Besides, the distribution density ρ, that is, the number of attractions per ten thousand kilometers within a region, is also analyzed. The purpose is to characterize the spatial distribution of A-grade and higher tourist attractions in Guizhou and provide a theoretical reference for making a structural optimization plan and restructuring the space of the tourist attractions.

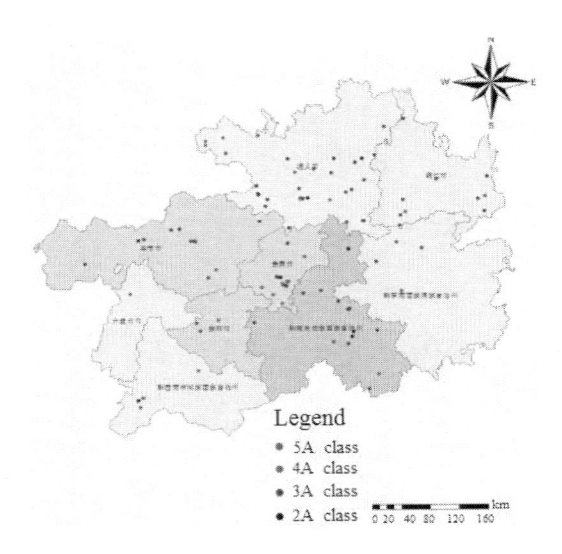

Figure 1. Spatial distribution of A-grade and higher tourist attractions in Guizhou.

The nearest neighbor index ρ is a measure that defines the point layout pattern and identifies the accurate spatial distribution type of the attractions, calculated according to formula (1):

$$\begin{cases} R = \dfrac{\bar{r}}{r_E} = 2\sqrt{Dr_1} \\ r_E = \dfrac{1}{2\sqrt{n/A}} = \dfrac{1}{2\sqrt{D}} \end{cases} \tag{1}$$

Here, $\bar{r}$- average of the actual nearest neighbor distance r; r_E - theoretical nearest neighbor distance; A- area within the study area; and n- number of point features in the study area [11]. When $R<1$, $\bar{r}<r_E$, the point distribution pattern is causal; when $R=1$时, $\bar{r}=r_E$, the point distribution pattern is random; and when $R>1$时, $\bar{r}>r_E$, the point distribution pattern is uniform.

The geography concentration index G is an important index that measures the concentration level of the study subject and can reflect the accurate concentration level of A-grade and higher tourist attractions of Guizhou in different regions, calculated according to formula [12] (2):

$$G = 100 \times \sqrt{\sum_{i=1}^{n}\left(\frac{x_i}{N}\right)^2} \tag{2}$$

Here, x_i - number of attractions in prefecture-level city i; N- total number of attractions; and G is between 0 and 100. A larger G means the attractions are more centralized, whereas a smaller G means they are more decentralized.

The Gin coefficient J is an important index that describes the distribution of spatial features. $1-J$ can indicate the distribution uniformity of the attractions, calculated according to formula[12] (3):

$$\left.\begin{aligned} &J=\frac{H}{Hm} \\ &Hm=\ln(N) \\ &H=-\sum_{i=1}^{n} p_i \ln(p_i) \\ &C=1-J \end{aligned}\right\} \Rightarrow C=\frac{\ln(N)+\sum_{i=1}^{N} p_i \ln(p_i)}{\ln(N)} \quad (3)$$

Here, P_i - proportion of the number of attractions in region i in the total of the province; N - number of regions; C - uniformity of distribution; and J is between [0, 1]. The larger this value, the higher the centralization. In other words, the larger the C, the more uniform the distribution of the attractions.

4 SPATIAL DISTRIBUTION OF A-GRADE AND HIGHER TOURIST ATTRACTIONS IN GUIZHOU PROVINCE

4.1 *Spatial distribution type of A-grade tourist attractions*

Using formula (1), the NNIs of A-grade and higher tourist attractions in the nine cities/prefectures of Guizhou are calculated. From Table 1, of the nine cities/prefectures of Guangzhou, the distribution types of the attractions in Qiandongnan, Tongren, Qianxinan, and Liupanshui are uniform; whereas all those of the other regions are causal. The NNI of Qiannan, however, is quite close to 1. Hence, we can assume that the A-grade and higher tourist attractions in this city are gradually moving toward a uniform distribution.

Table 1. List of NNI estimates.

Variable Area	Qiannan	Qiadongnan	Tongren	Guiyang	Zuni	Anshun	Qianxinan	Liupanshui	Bijie
A(km^2)	26197	30223	18023	8034	30762	9267	16804	9965	26853
N	14	5	11	16	35	3	4	1	12
r_E(km)	21.739	38.874	20.231	11.204	15.503	8.789	32.41	49.91	23.65
$\bar{r}$(km)	28.38	35.91	16.63	16.73	22.74	13.68	26.73	21.7	41.18
R	0.823	1.083	1.217	0.670	0.682	0.642	1.212	2.300	0.574
Type	causal	Uniform	Uniform	causal	causal	causal	Uniform	Uniform	causal

4.2 *Spatial distribution equilibrium of A-grade tourist attractions*

Using formula (2), the geographical concentration index of the A-grade and higher tourist attractions in Guizhou is calculated: $G = 41.90$. Let us assume that the 101 A-grade and higher tourist attractions in the province are equally distributed in the nine cities/prefectures of the province, that is, the number of tourist attractions in each city/prefecture is 101/9≈11.22; the geographical concentration index G=33.66, 41.90>33.60, suggesting that the spatial distribution of A-grade and higher tourist attractions in the province are quite centralized. From Table 2, the A-grade and higher tourist attractions in Guizhou are mostly distributed in Guiyang and Zuni; the 3A tourist attractions are mostly found in Tongren, Zuni, Bijie, and Qiannan; and the 4A tourist attractions are typically located in Zuni and Guiyang, each of which contributes 28.57% and 31.43% of the total number of 4A tourist attractions across the province.

Table 2. Summary A-grade and higher tourist attractions in Guizhou.

	Number of attractions						
Region	2A	3A	4A	5A	Total	Proportion/%	Aggregate/%
Tongren	0	8	3	0	11	10.89	10.89
Zuni	6	19	10	0	35	34.65	45.54
Bijie	1	9	1	1	12	11.88	57.42
Qiandongnan	0	2	3	0	5	4.95	62.38
Guiyang	1	4	11	0	16	15.84	78.22
Liupanshui	0	0	1	0	1	0.99	79.21
Qiannan	4	8	2	0	14	13.86	93.07
Anshun	0	0	1	2	3	2.97	96.04
Qianxinan	1	0	3	0	4	3.96	100.00

4.3 *Spatial distribution density of A-grade tourist attractions*

Table 3 estimates the distribution densities of tourist attractions in the nine cities/prefectures of Guizhou Province.

Table 3. Distribution density estimates of A-grade and higher tourist attractions.

Variable Area	Qian nan	Qian dongnan	Tong ren	Gui yang	Zuni	Ans hun	Qianxi nan	Liupan shui	Bijie
A (10^4km²)	2.6197	3.0223	1.8023	0.8034	3.0762	0.9267	1.6804	0.9965	2.6853
n	14	5	11	16	35	3	4	1	12
ρ (个/10^4km²)	5.344	1.654	6.103	19.920	11.377	3.237	2.380	1.004	4.469

From Table 3, the distribution densities of the attractions appear to be highly different. The region with the greatest attraction density is Guiyang at $19.92/10^4$km². The region with the smallest attraction density is Upanishads, the density of which is merely 0.05 times that of Guiyang. Besides, from Table 3, we can also see that the average distribution density of A-grade and higher tourist attractions in Guizhou is $6.057/10^4$km². Within the province, therefore, only Tongren, Guiyang, and Zuni are higher than the province's average in terms of the distribution density of A-grade and higher tourist attractions, whereas all other cities/prefectures are lower than the province's average.

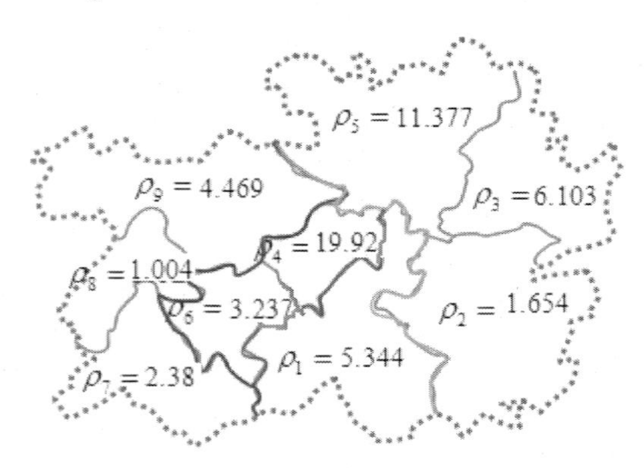

Figure 2. Nine divisions of Guizhou Province.

From this map, spatially, the attraction distribution density is larger at the center and smaller in the east and west. In the light of the density distribution of the attractions in the cities/prefectures of Guangzhou, we divided the nine cities/prefectures into three major regions (East Guizhou (Tongren, Indonesian), Central Guizhou (Zuni, Guiyang, and Qiannan), and West Guizhou (Bijie, Liupanshui, Anshun, and Qianxinan). On this basis, we conducted Gin coefficient analysis on the number of A-grade and higher tourist attractions in order to identify the uniformity level of attractions in the three major regions of the province.

Using formula (3), the Gin coefficient of the tourist attractions in East, Central, and West Guizhou is calculated to be $J = 0.816$ and the distribution uniformity is calculated to be $C = 0.184$. These values show that the A-grade and higher tourist attractions in Guizhou are centralized, that is, their distribution uniformity is low. Among the re-divided four regions, Central Guizhou has the most centralized tourist attractions and gathers more than half of the A-grade tourist attractions, followed by West and South Guizhou (Tab.4).

Table 4. Distribution uniformity estimates of tourist attractions in the three major regions.

Region	Number of Attractions	Ratio (%)	Uniformity (%)	Actual Aggregate (%)	Uniformity Aggregate (%)
East Guizhou	16	15.842	33.333	15.842	33.333
Central Guizhou	65	64.356	33.333	80.198	66.666
West Guizhou	20	19.802	33.333	100	167
Total	101	100	100		

5 CONCLUSIONS

Guizhou is endowed with privileged natural tourist resources and cultural tourist resources. Its tourist industry will surely create new highs along with its socioeconomic development. The spatial distribution of the 101 A-grade and higher tourist attractions in the nine regions of the province is characterized. The spatial point distribution of the tourist attractions in Qiannan, Qiandongnan, Tongren, Guiyang, Zuni, Anshun, Qianxinan, Liupanshui, and Bijie is examined. The spatial distribution of the local A-grade and higher attractions are re-examined after redividing the nine cities/prefectures into three regions: East, Central, and West Guizhou according to the spatial distribution of the attractions in these regions that is centralized at the central and sparse in the surrounding area. The results indicate the following:

1 Among the nine regions of Guizhou, the spatial distribution of the A-grade tourist attractions

in Qiandongnan, Tongren, Qianxinan, and Liupanshui is uniform. The distribution of the A-grade and higher tourist attractions in Qiannan is gradually moving toward a uniform distribution. This is causal in all the other cities/prefectures.

2 The geographical concentration index of the A-grade and higher tourist attractions in Guizhouis $G = 41.90$, suggesting that the spatial distribution of the local A-grade and higher tourist attractions is quite centralized and mostly located in Guiyang and Zuni.
3 The attraction distribution density appears to be highly different. The region with the greatest attraction density is Guiyang at $19.92/10^4 km^2$. The region with the smallest attraction density is Liupanshui, the density of which is merely 0.05 times that of Guiyang.
4 The spatial distribution density of A-grade tourist attractions, on the whole, appears to be greater in Central Guizhou and smaller in East and West Guizhou.

ACKNOWLEDGMENT

This paper is supported by Sichuan Provincial Department of Education Fund Project (15ZB0269).

REFERENCES

[1] Chen, H.F., Zhengzhou, J.G., Tang, F.P., Wu, G.X. (2013) Spatial distribution of A-grade tourist attractions in Alhena Province. *Economic Geography*, 33(2):179–183.
[2] Chang, J., Chang, X.S. (2013) A discussion on the construction of tourist complexes in Guizhou- In the case of Shijiazhuang Scenic Area. *Journal of Alhena Business College*, 26(4):89–91.
[3] Wu, B.H, Tang, Z.Y. (2003) A study on spatial structure of national 4A grade tourism attractions in China. *Human Geography*, 18(1):1–5.
[4] Ma, X.L. (2004) A correlating analysis on high grades tourism area. *Journal of Northwest University (Natural Science Edition)*, 34(2):233–236.
[5] Thu, H., Chen, X.L. (2008) Study on the spatial distribution structure of A-grade tourist attractions in China. *Geographical Science*, 28(5):607–615.
[6] Au, H.Q. (2013) The analysis of tourism spatial structure in Guamanian. Guamanian: Guamanian Normal University.
[7] Wang, W.X., Die, S.Y. (2012) The spatial pattern of the A-grade tourist sites in Huber Province and its optimization. *Areal Research and Development*, 31(2):124–128.
[8] Wang, H., Li, Y.Z. (2010) A study on spatial structure of scenic areas in Dalian City and its optimization. *Areal Research and Development*, 29(1):84–89.
[9] Chang, D., An, X.H. (2014) Fractal research on tourist attractions system space structure of Xi amen City. *Tourism Research*, 06(4):46–52.
[10] GAO, Y.S. (2011) Spatial structure and optimization of the tourist system of Zhengzhou City. *Journal of Zhengzhou University*, 27(6):82–85.
[11] Chi, Z.Y., Chou, B.H., Chang, Y.N. (2013) Spatial structure differences and optimization of national tourism areas above 4A level in Inhuman Province. *Journal of Changchun University*, 24(11):1509–1514.
[12] Au, J.H. (2006) Quantitative Geography. *Higher Education Press*, pp:29–34.

Electric, Electronic and Control Engineering – Shao, Shu & Tian (Eds)
© 2015 Taylor & Francis Group, London, ISBN: 978-1-138-02842-5

A novel method for fundamental component detection based on wavelet transform

Huan Chen & Yigang He
School of Electrical Engineering and Automation, Hefei University of Technology, Hefei Anhui, China

ABSTRACT: Active Power Filter (APF), as a kind of effective method used for harmonic suppression, has been applied widely in recent years. The accurate, real-time detection of harmonic current is a vital part of APF, which directly determines the performance of APF. Harmonic current can be obtained by subtracting the fundamental component from the total distorted current. Therefore, the precision of harmonic detection can be ensured once the fundamental component is detected accurately. A novel method for fundamental component detection based on wavelet transform is proposed in this paper. And two models, including only the harmonics contained and harmonics plus transient and noise disturbance, are simulated, respectively. The simulation results show that this method can detect fundamental component accurately, and the process is simple, which has certain practical value in the harmonic detection link of APF.

KEYWORDS: Active power filter; harmonic detection; fundamental component detection; wavelet transform.

1 INTRODUCTION

With the development of the power system, especially the wide application of power electronic equipment in the power system, the harmonic pollution problem has become increasingly serious. Harmonics will not only bring a bad influence to the user terminal equipment but also increase the loss of the transmission line and interfere with communication. Therefore, harmonic suppression and elimination in the power system has significant socioeconomic benefits [1].

In recent years, active power filter (APF), as a harmonic suppression device, has extensive application. The work principle of APF is shown in figure 1.

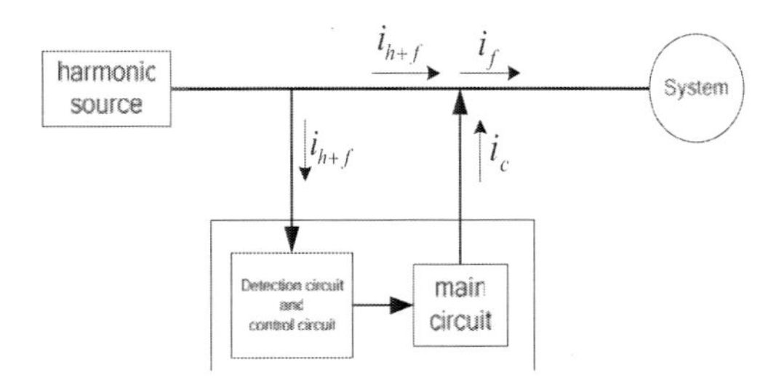

Figure 1. Block diagram of work principle for APF.

The harmonic source is generally a nonlinear load, such as a rectifier, that generates the harmonic current *i*h and the power supply system is generally a protected object. The function of APF is to generate the compensation current i_c, which has the same amplitude and opposite phase of the harmonic source current to eliminate the harmonic current and retain the fundamental component i_f[2]. And the accurate and reliable detection of harmonics contained in the power signal is an essential part in APF, which is an important premise and basis for harmonic suppression, compensation, and improvement of the quality of electric energy. Harmonics can be reasonably compensated as long as it is detected accurately first. At present, the research on the harmonic detection algorithm has become one of the hot-spot issues in the field of APF. Currently, the analysis methods based on FFT and the instantaneous reactive power theory and its derivatives are the most widely used common methods. The harmonic detection method based on FFT performs well in the case of static harmonic and full-period sampling, but it cannot achieve accurate detection once the infiltration of transient or noise disturbance takes place[3]. Adopting p-q operation mode based on the instantaneous reactive power theory will cause a big detection error once the voltage waveform is distorted[2]. It is not necessary for APF to detect each harmonic content, which is different from the general harmonic detecting instrument, and only the total harmonics needs to be detected except the fundamental wave. Therefore, harmonics can be obtained by subtracting the fundamental component from the total distorted current as long as the fundamental component is extracted. Wavelet analysis has good time-frequency localization characteristics. So, it can calculate the frequency distribution of a particular time and decompose the spectrum signal with different-frequency components into different-frequency signal blocks[6]. Therefore, the fundamental frequency can be detected by the wavelet transform to realize the detection of the harmonic current. A novel method for fundamental component detection based on the wavelet transform is proposed in this paper. And two models, including only the harmonics contained and harmonics plus transient and noise disturbance, are simulated, respectively. The simulation results show that this method can detect the fundamental component accurately, and the process is simple, which has certain practical value in the harmonic detection link of APF.

2 WAVELET TRANSFORM

2.1 *Mallat algorithm based on multi-resolution analysis*

Set up $f(t) \in L^2(R)$ and assuming its rough image $A_j f \in V_j, \{V_j\}_{j\in Z}$ in the resolution of 2^{-j} has been got, which constitutes the multi-resolution of $L^2(R)$, thus $V_j = V_{j+1} \oplus W_{j+1}$, that is:

$$A_j f = A_{j+1} f + D_{j+1} f \tag{1}$$

where $A_j f = \sum_{k=-\infty}^{\infty} C_{j,k}\phi_{j,k}(t)$, $D_j f = \sum_{k=-\infty}^{\infty} D_{j,k}\psi_{j,k}(t)$

Then, formula (1) can be expressed as follows:

$$\sum_{k=-\infty}^{\infty} C_{j,k}\phi_{j,k}(t) = \sum_{k=-\infty}^{\infty} C_{j+1,k}\phi_{j+1,k}(t) + \sum_{k=-\infty}^{\infty} D_{j+1,k}\psi_{j+1,k}(t) \tag{2}$$

According to the two-scale equation of scaling function, there is

$$\phi_{j+1,m}(t) = \sum_{k=-\infty}^{\infty} h(k-2m)\phi_{j,k}(t) \tag{3}$$

Utilizing the orthogonality of scaling function, there is

$$\langle \phi_{j+1,m}, \phi_{j,k} \rangle = h(k-2m) \tag{4}$$

According to the two-scale equation of wavelet function, there is

$$\langle \psi_{j+1,m}, \phi_{j,k} \rangle = g(k-2m) \tag{5}$$

By formulas (2), (3), and (4), the following formulas can be obtained:

$$C_{j+1,m} = \sum_{k=-\infty}^{\infty} C_{j,k} h^*(k-2m) \tag{6}$$

$$D_{j+1,m} = \sum_{k=-\infty}^{\infty} C_{j,k} g^*(k-2m) \tag{7}$$

$$C_{j,k} = \sum_{m=-\infty}^{\infty} h(k-2m) C_{j+1,m} + \sum_{m=-\infty}^{\infty} g(k-2m) D_{j+1,m} \tag{8}$$

By introducing the infinite matrix:

$$H = \left[H_{m,k}\right]_{m;k=-\infty}^{\infty} \qquad G = \left[G_{m,k}\right]_{m;k=-\infty}^{\infty}$$

where $H_{m,k} = h^*(k-2m), G_{m,k} = g^*(k-2m)$; then, formulas (6), (7), and (8) can be, respectively, expressed as follows:

$$\begin{cases} C_{j+1} = HC_j \\ D_{j+1} = GC_j \end{cases} \qquad j = 0,1,\cdots,J \tag{9}$$

$$C_j = H^* C_{j+1} + G^* D_{j+1} \ , \quad j = J, J-1, \cdots, 1, 0 \tag{10}$$

where H^*, G^*, respectively, are the conjugate transpose matrix of H and G.

Formula (9) represents the decomposition algorithm, and formula (10) represents the reconstruction algorithm, as shown in figure 2.

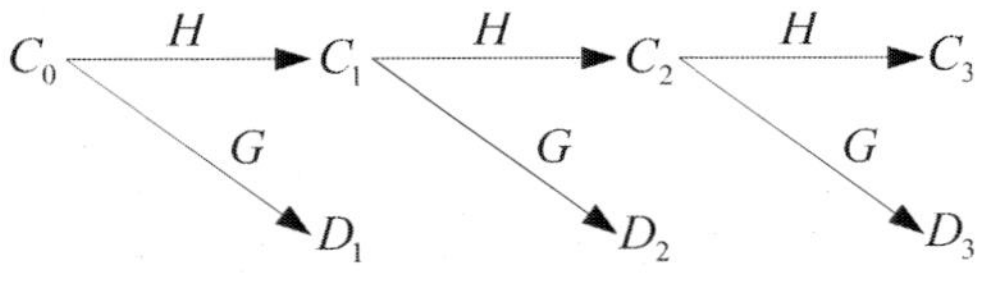

(a) Decomposition algorithm

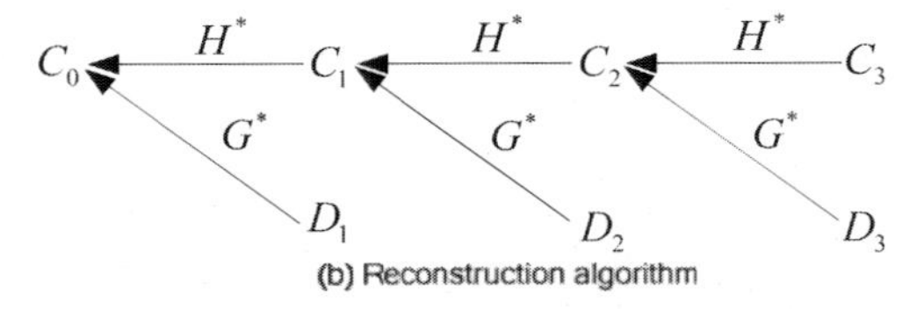

(b) Reconstruction algorithm

Figure 2. Schematic diagram of Mallat wavelet decomposition and reconstruction algorithm.

Figure 2 (a) shows that the highest frequency f contained in signal s(t) will be decomposed into the low-frequency part (approximate part) containing frequency bands $a1$: 0~$f/2$ and the high-frequency part (detail part) containing frequency bands $d1$: $f/2$~f by filter bank G and H. Second, $a1$ is decomposed into approximate part $a2$: 0~$f/4$ and detailed part $d2$: $f/4$~$f/2$ by filter bank G and H. After j time decomposition, low-frequency part containing frequency bands aj: 0~$f/(2j)$ and high-frequency part dj: $f/(2j)$~$f/[2(j-1)]$ can be obtained. The reconstruction process is contrary.

2.2 *Fundamental detection method based on wavelet transform*

The filtered high-frequency components are preserved for reconstruction. Then, the total harmonic current can be obtained by subtracting the reconstructed fundamental component from the original distorted signal to achieve the purpose of harmonic detection. A three-layered wavelet decomposition tree is shown as figure 3. It can be seen from figure 3 that only lowrequency signal decomposition is analyzed, without considering the high-frequency signal. It can be obtained from figure 3 that $s= a3+d3+d2+d1$, where $a3,a2,a1$ represent approximate parts and $d3,d2,d1$ represent detailed parts. Total harmonics can be obtained by subtracting the reconstructed fundamental component $a3$ from signal s.

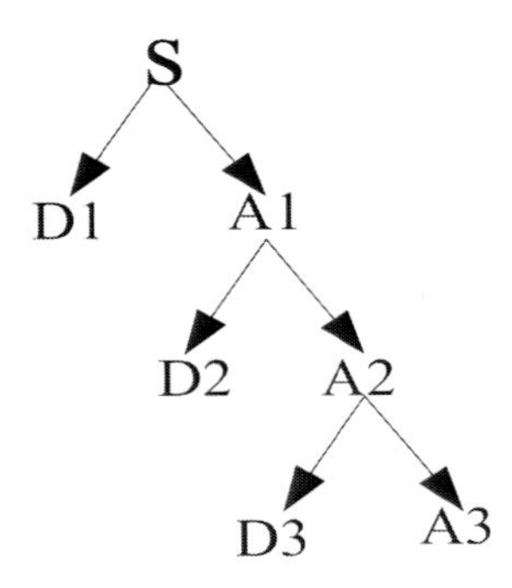

Figure 3. Three-layered wavelet decomposition structure.

3 SIMULATION

The fundamental frequency 50Hz is set up, and simulation verification is performed in two cases: only harmonics contained and harmonics plus transient and noise disturbance. The simulation results are as follows:

1 only harmonics contained :
simulation signal is

$s1=\sin(2\pi\times50t)+ 1/3\times\sin(2\pi\times150t+\pi/3)+1/5\times\sin(2\pi\times250t+\pi/6)+1/7\times\sin(2\pi\times350t+\pi/8)+1/9\times\sin(2\pi\times450t+\pi/5)+1/11\times\sin(2\pi\times550t+\pi/6)+1/13\times\sin(2\pi\times650t+\pi/8)+1/15\times\sin(2\pi\times750t+\pi/9)$;

Signal $s1$ contains the fundamental component and 3th, 5th, 7th, 9th, 11th, 13th, and 15th harmonics and its time domain waveform is shown in figure 4.

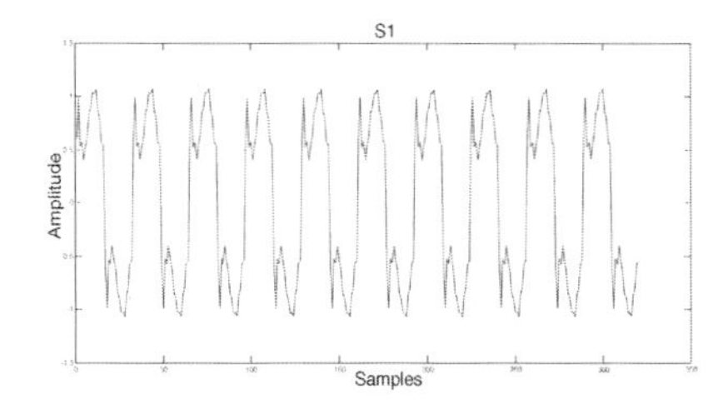

Figure 4. Time domain waveform of $s1$.

First, the decomposition level must be selected. And the level of decomposition is proportional to the sampling frequency. If the sampling frequency is equal to 800 Hz, 1600 Hz, and 3200 Hz, respectively, decomposition level should be equal to 2, 3, and 4, respectively. In order to compare the accuracy of the detection of fundamental wave against different decomposition level, the results in the three conditions mentioned earlier are compared with the standard sine signal $s0=\sin(2\pi\times50t)$ to get their own relative error curve, as shown in figure 5. The deviation of simulation signals and the standard sinusoidal signal are $\triangle1$, $\triangle2$, and $\triangle3$, respectively. It can be seen that the precision of fundamental component detection is high when the decomposition level is selected to be 3. Considering the accuracy of fundamental wave detection and the amount of calculation, decomposition level 3 and the sampling frequency 1600Hz are set up. According to Nyquist sampling theorem, the frequency band of original signal is 0~800Hz. The band information of layers of the wavelet transform is shown as Table 1.

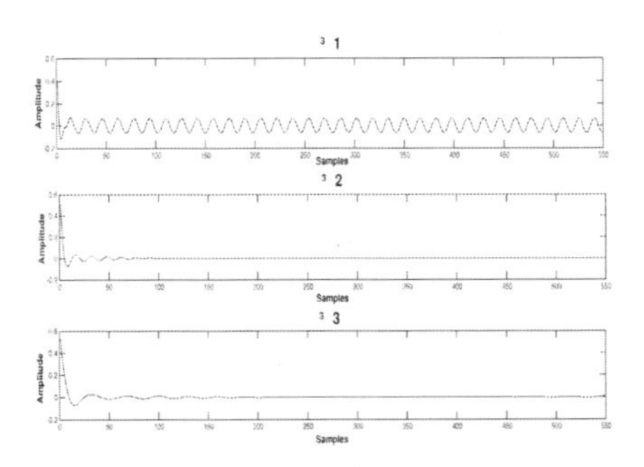

Figure 5. Relative error $\triangle1,\triangle2,\triangle3$.

Table 1. Band information of layers of wavelet transform.

Decomposition level	$a3$	$d3$	$d2$	$d1$
Frequency band	0~100Hz	100~200Hz	200~400Hz	400~800Hz

The next step is the selection of the wavelet base. The detection results of the fundamental wave against different wavelet bases are compared with the standard sine signal $s0=\sin(2\pi\times50t)$ to get their own relative error curve, as shown in figure 6. This paper selects Daubechies wavelet with the vanishing moment 43 as the wavelet base.

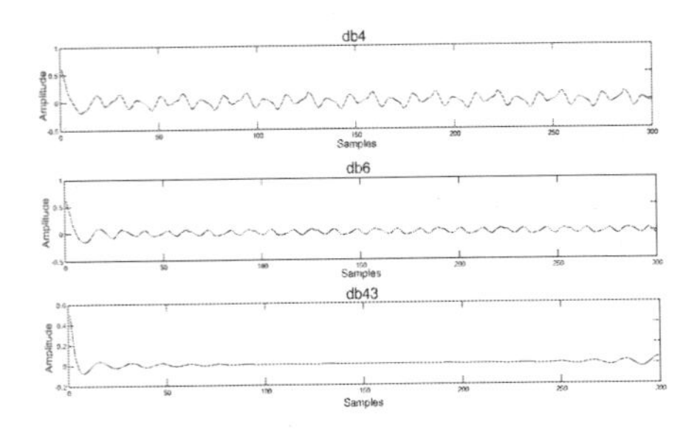

Figure 6. Relative error of fundamental component detection against different wavelet bases db4, db6, and db43.

The decomposition results of the discrete wavelet transform are shown as figure 7. It can be seen from figure 7 that although each harmonic component fails to be completely separated in detailed parts ($d3$, $d2$, and $d1$), the approximate part $a3$ basically is the fundamental component required, which has centralized energy and small-frequency leakage. Apparently, it does not lose the fundamental component, which is the main concern. The reconstructed waveform of band $a3$ approaches the fundamental component perfectly. Then, the harmonics waveform is available by subtracting the fundamental wave from $s1$.

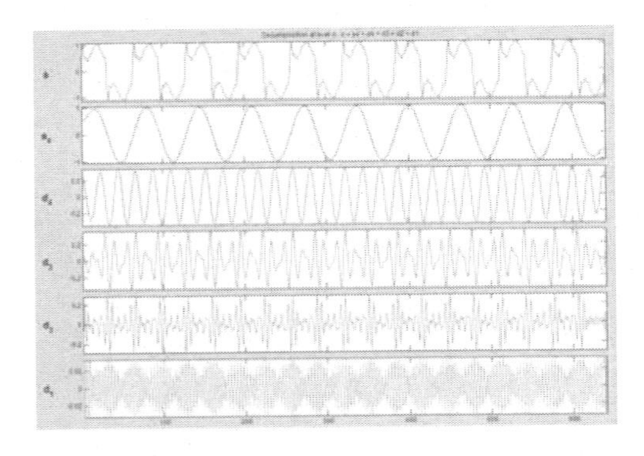

Figure 7. Result of wavelet transform decomposition.

2 harmonics + transient disturbance + noise disturbance:simulation signal is

$$s2=s1+0.7\times\sin(2\pi\times950(t-0.06))\times e^{-20t}+n(t),$$

where n(t) represents noise disturbance and the transient disturbance signal is mixed after the sampling time 0.06s.

The time domain waveform of signal $s2$ is shown as figure 8.

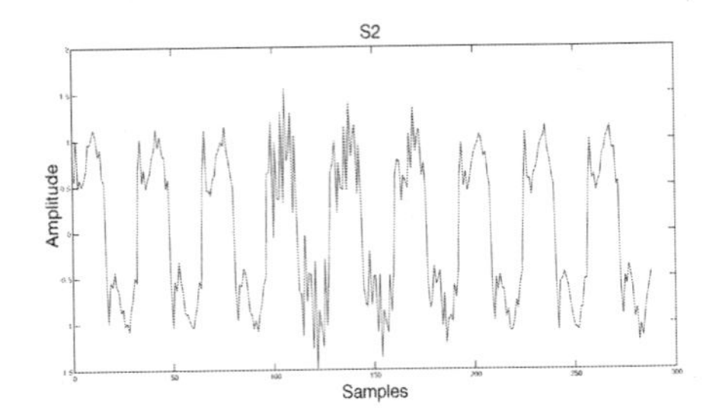

Figure 8. Time domain waveform of $s2$.

The decomposition results of the discrete wavelet transform is shown as figure 9. Then, harmonics plus transient and noise disturbance waveform is available by subtracting the fundamental wave from $s2$.

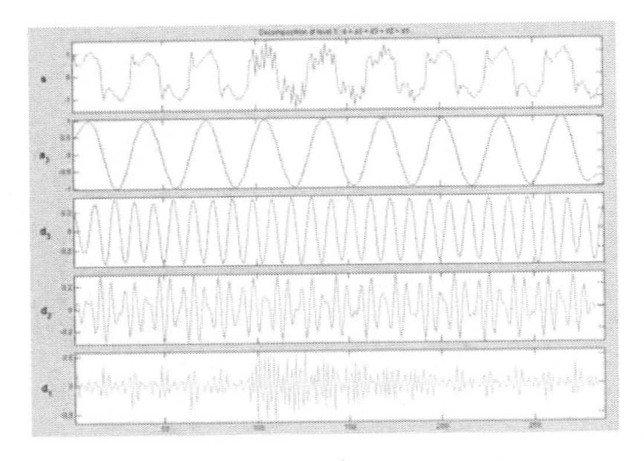

Figure 9. Result of wavelet transform decomposition.

In order to verify the accuracy of the detection of fundamental component, the result is compared with the standard sine signal $s0=\sin(2\pi\times50t)$ to get its relative error curve, as shown in figure 10. It can be seen that the algorithm of fundamental component detection based on the wavelet transform has high precision.

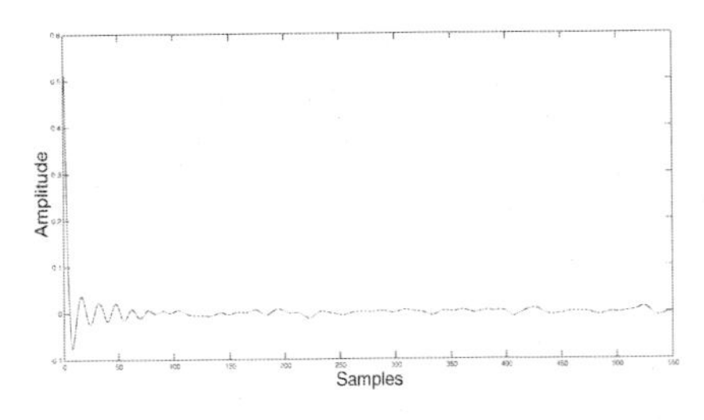

Figure 10. Relative error of fundamental component detection of signal $s2$.

4 CONCLUSION

What needs to be detected in the detection circuit of the active power filter is only the total harmonic content, not the frequency and corresponding amplitude of each harmonic. Therefore, as long as the fundamental component can be detected accurately, the accuracy of harmonic detection can be ensured. A novel method for fundamental component detection based on the wavelet transform according to its good time-frequency localization characteristic is proposed in this paper. And two cases are simulated. The simulation results show that this method can accurately detect the fundamental component not only in steady harmonics signal but also in the transient interference and noise disturbance environment, which has certain practical value.

ACKNOWLEDGMENTS

This work was financially supported by National Outstanding Youth Science Foundation (50925727), Major Project of the Education Ministry of science and Technology (313018), Anhui Province Science and Technology Key Project (1301022036), and Hunan Province Science and Technology Plan Projects (2010J4, 2011JK2023).

REFERENCES

[1] Lin Haixue, Xiao Xiangning et al. (2012) The Power Quality of Power System. Beijing: Chinese Power Press.

[2] Wang Zhaoan, Yang Jun, Liu Jinjun, et al. (2009) *Harmonic Suppression and Reactive Power Compensation*, Beijing: Machinery Industry Press.

[3] Pang Hao, Li Dongxia, Zu Yunxiao, et al. (2003) An improved algorithm for harmonic analysis of power system using FFT technique. *Proceedings of the CSEE, 2003*, 23(6):50–54.

[4] Zhang Fusheng, Geng Zhongxing, Ge Yaozhong. (1999) FFT algorithm with high accuracy for harmonic analysis in power system. *Proceedings of the CSEE, 1999*, 19(3):63–66.

[5] Hong-seok Song, Hyun-gyu Park, Kwanghee Nam. (1999) An instantaneous phase angle detection algorithm under unbalanced line voltage condition. *Power electronics specialists Conference*.

[6] Hu Guangshu. (2004) *Modern Signal Processing Tutorial*. Beijing: Tsinghua University Press.

[7] Zhou Lin. (2006) Review of harmonic measurement method based on wavelet transform. *Transactions of China Electrotechnical Society*, 21 (9):67–74.

[8] Zhou Wenhui, Lin Lili, Zhou Zhaojing. (2001) Harmonic detection method based on wavelet transform. *Chinese Journal of Scientific Instrument*, 22 (3):5–10.

[9] Xiao Yanhong, Mao You, Zhou Jinglin et al. (2002) Review on measuring method for harmonics in power system. *Power System Technology*, 26 (6):61–64.

[10] Zhang Peng, Li Hongbin. (2012) A harmonic analysis method based on discrete wavelet transform. *Transactions of China Electrotechnical Society*, 27(3):252–259.

[11] Wang Xiaofen, Xu Kejun. (2005) Fundamental Wave extraction and frequency measurement based on wavelet transform. *Chinese Journal of Scientific Instrument*, 26 (2):146–151.

[12] YOON Weon-Ki, DEVANEY M J. (2000) Reactive power measurement using the wavelet transform. *IEEE Transactions on Instrumentation and Measurement*, 49(2):246–252.

[13] He Yingjie, Zou Yunping, Li Hui, et al. (2008) A new algorithm for harmonic detection of active power filter. *Power Electronic Technology*, 40(2):56–59.

[14] Liu Hongchun, Sun Shuguang, Wang Jingqin, Hou Shichang. (2012) Comparison of the instantaneous reactive power theory with the wavelet transformation method for harmonic current detection. *Power Electronics*, 46 (9):46–49.

Electric, Electronic and Control Engineering – Shao, Shu & Tian (Eds)
 ISBN: 978-1-138-02842-5

Low-noise, low-power geomagnetic field recorder

Haifeng Wang, Ming Deng, Kai Chen & Hong Chen
China University of Geosciences, Beijing, China
Key Laboratory of Geo-detection, Ministry of Education, Beijing, China

ABSTRACT: How to realize a long-term geomagnetic signal observation with high precision and low noise is one of the problems that are in urgency of solutions. The existing technical scheme has deficiencies such as high system noise, large power dissipation, and large clock error. Technical problems such as low noise, low power dissipation, and low time drift are solved by the construction of recorder hardware composed by low-noise fluxgate sensor and mainframe of collection. Reliability and field adaptability are also further enhanced. Measured data of indoor units indicate the advancement and availability of the scheme.

KEYWORDS: Geomagnetic field; fluxgate sensor; low noise; low power.

1 INTRODUCTION

It has always been a research hotspot for geologists and geophysicists to know the interior material structure of the earth by observing the geomagnetic field. How to realize a long-term observation of the geomagnetic field with high precision is also one of the challenges currently faced by developers of geophysical instruments. In order to solve this problem, it is obliged to overcome difficulties such as low noise, low-power dissipation, and high reliability and stability of front-end sensor and back-end acquisition circuit. Common magnetic sensors used for the observation of the geomagnetic field mainly include proton magnetometer, optical pump, superconductivity, fluxgate, inductive coil, and so on. A proton magnetometer, similar to the optical pumping magnetometer, can only obtain the geomagnetic total field instead of the unidirectional magnetic-field component. The noise level of the superconducting magnetometer is extremely low. But it requires the support of the superconducting cryogenic box due to its large volume, so the field construction is limited. The inductive coil has excellent noise features, whereas it is insufficient in geomagnetic field observation. It can only observe the alternating geomagnetic field and the magnetic-field components of a certain direction. The fluxgate sensor features small volume, multiple components, and low noise, so it is an ideal choice for the observation of the geomagnetic field.

Existing geomagnetic recorders have deficiencies such as high noise and large-power dissipation. In order to meet the requirements of low-noise observation and long-term field observation, it is necessary to upgrade current technical schemes so as to improve indicators such as noise, power dissipation, and time synchronization error.

2 HARDWARE DESIGN

2.1 *Overall design*

The geomagnetic recorder is used for high precision observation of the three-component geomagnetic field. It is required that this equipment should have features such as low noise, low-power dissipation, small time error, usability, high reliability, and so on. Hardware of the geomagnetic recorder includes front-end fluxgate sensor, acquisition box, external GPS antenna, sensor connecting cable, and so on. The acquisition box is internally installed with lithium battery pack, power-switching circuit, chopper amplifier, analog-to-digital conversion circuit, micro controller, SD memorizer, temperature-compensation crystal oscillator, GPS module, temperature sensor, and so on.

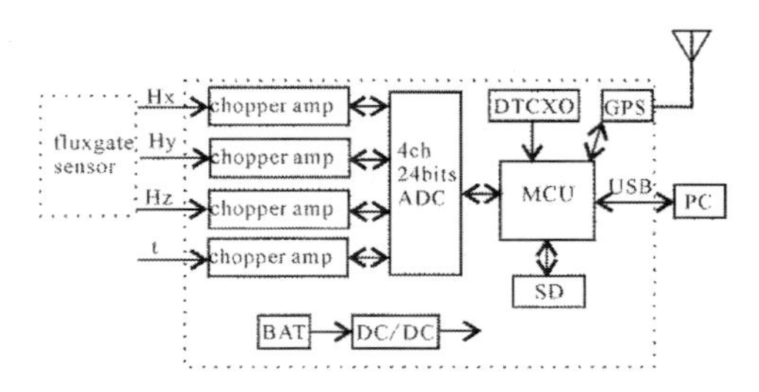

Figure 1. Block diagram of geomagnetic recorder hardware.

The front-end fluxgate sensor is designed with low noise, low-power dissipation, and three-axis orthogonal transformation, of which the measuring range is ±100000nT, the transformation coefficient is 100uV/nT, linearity precedes 15 ppm, and the power dissipation is 400mW. Thermal resistance is adopted by the temperature sensor, of which the typical temperature accuracy is ±0.5. The chopper amplifier is used for the low-noise amplification of voltage signals of a magnetic sensor, of which the gain is -12dB. Compared with the ordinary operational amplifier, the chopper amplifier features low drift. It guarantees DC accuracy perfectly. The ADC of 4 channels and 24 bytes completes the analog-digital conversion of voltage signals. Σ-ΔADC of 24 bytes has advantages of large dynamic range and synchronous conversion, of which the typical value of the dynamic range is 120dB@ (fs=250Hz). Featuring small volume, low-power dissipation, and high-frequency stability, DTCXO provides the system with a stable clock source, of which the typical power dissipation is 3mA and the frequency stability is ±50ppb. GPS provides the system with accurate time service and precise PPS. The typical error of PPS is 20ns. SD provides the system with memory space. Compared with other memorizers, SD has advantages such as small volume, low-power dissipation, plugging support, and wide operating temperature range, the standard configuration of which is 16GB. MCU uses SiLabsUSB F320 single-chip microcomputer with low dissipation. This controller completes operations such as sampling rate control, gain configuration, ADC data reading, GPS clock comparison, time drift measurement, SD card reading and writing, and USB communication of the upper computer. It also supports the FAT32 file system. DC/DC power-switching circuit completes the switching of high frequency and low noise from 9–12.6V lithium battery to ±2.5V analog power and 3.3V digital power. The standard configuration of the lithium battery pack is 20Ahr battery so that the complete machine can work continuously for more than 200hr at full speed.

2.2 *Main technical parameters*

Number of channels: 4 (Hx, Hy, Hz, t);
Measuring range: ±100000nT;
Noise: 6pT/rt(Hz)@1Hz;
Power dissipation: 1000mW;
Degree of linearity: superior to 15ppm;
Continuous operating hours: more than 200hr;
Memory space: 16GB;
Sampling rate: 1/15/150Hz;
Volume: 30*20*5cm;
Weight: 2kg (with built-in lithium battery pack)

2.3 *Low noise*

The low design is mainly reflected in low-noise processing of front-end magnetic sensor, sensor temperature correction, low-noise amplifier design, large dynamic range ADC, low noise of power supply, reasonable electrical noise shielding, and so on.

The typical noise of the fluxgate sensor is 6pT/rt(Hz)@1Hz, and the degree of linearity is superior to 15ppm. In order to reduce the non-linear error caused by temperature changes in operation, the front-end fluxgate sensor buries the magnetic sensor 1m underground, on the one hand, and records temperature changes of the underground magnetic sensor with the temperature sensor for temperature correction, on the other hand. The background noise of the chopper amplifier is 10nV/rt(Hz)@1Hz. Low-frequency noise is flat with no 1/f noise. The current ripple of ±2.5V analog power is 100uVrms. An aluminum alloy seal box with the waterproofing grade of IP67 is adopted for acquisition circuit package in order to further reduce interference on the acquisition circuit brought about by the external environment. It effectively enhances the reliability, stability, and applicability of the complete machine in field operations.

2.4 *Low-power dissipation*

The design of low-power dissipation mainly involves low-power dissipation sensor, model selection of components and parts, high-efficiency power supply design, and reasonable power management. The power dissipation of the magnetic sensor is only 400mW. Components and parts include an amplifier with low-power dissipation, analog-digital converter, MCU, and so on. Power conversion uses switching power supply, the efficiency value of which is as high as 96%. Power management is reflected in power-off of the magnetic sensor before acquisition, power-down processing of parts such as GPS during acquisition, and reduction of MCU for less current consumption. The power dissipation of the complete machine is approximately 1000mW.

2.5 *Low time drifting*

Continuous operation for a few days in field operation requires multiple recorders to maintain the consistency of time. Therefore, it is necessary to place emphasis on time drifting. GPS is needed for time service at the beginning of acquisition, and PPS is needed for triggering before the start of acquisition. PPS is used to calibrate TCXO during acquisition so as to maintain relatively high frequency stability and time consistency. The difficulty lies in measuring the time difference with standard PPS and calculating TCXO frequency error in accordance with time drifting features of TCXO so as to adjust the control voltage of TCXO and make it approach the standard value.

3 SOFTWARE DESIGN

Software design mainly includes the user's interaction program of the PC-end upper computer and the MCU control program of the acquisition box.

3.1 *PC-end program*

The PC-end program has functions such as USB communication, parameter setting, waveform display, data transmission, data storage, data playback, status display, and so on. Parameter setting includes sampling rate setting, on-off control of acquisition, and GPS clock comparison. The function of waveform display realizes real-time three-component magnetic field display and one-channel temperature waveform display. Data transmission realizes the download of parameters and the upload of real-time data. Data storage realizes local disk writing of upload data. Data playback helps browsing and analysis of previous data. Status display includes basic status information, such as ID of the current instrument, battery voltage, temperature, memory space, and so on. The user's interaction program is developed by the VS platform of Microsoft.

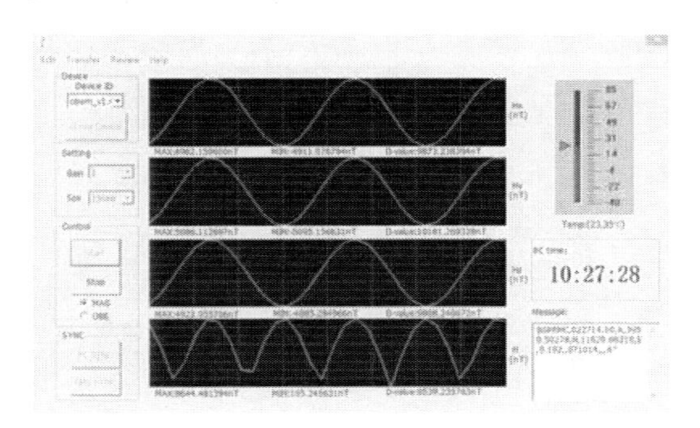

Figure 2. Screenshot of the control software.

3.2 *MCU control program*

The MCU control program achieves functions such as USB communication with the upper computer, gain control, sampling rate setting, ADC data reading, GPS clock comparison, battery voltage measurement, time drift measurement, TCXO calibration, SD card reading and writing, LED status indicator, and so on. The development platform is Keil uV4. The development of the single chip microcomputer program follows the principle of being stable, simple, and efficient so that it is able to operate continuously for a few days in the field condition stably and reliably with its instantaneity and fault tolerance.

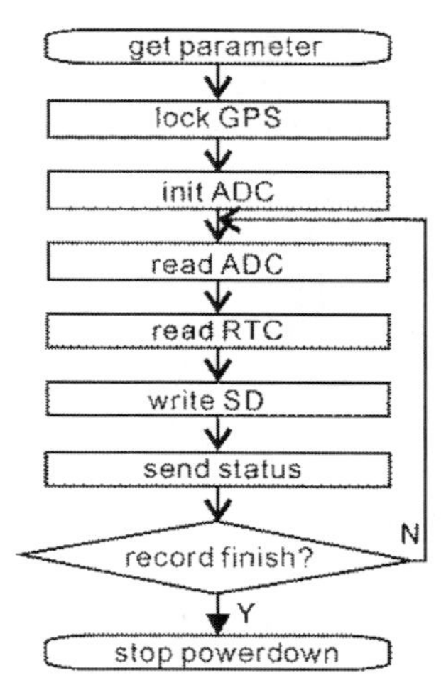

Figure 3. MCU software flow chart.

4 TEST RESULTS

4.1 *Dynamic range test of circuit*

The channel range is set as 20Vpp. The sampling rate is set as 150Hz. Input the sinusoidal signal with an amplitude of -60dBFS (20mVpp) and a frequency of 0.75Hz to the input end and conduct Fourier transformation to time series after continuous observation for a period of time. Results are shown in figure 4. SNR, SINAD, and SFDR are, respectively, calculated as 60.1dB, 60.1dB, and 81.3dB. It can be concluded that SNR reaches 120.1dB under the condition of 150Hz sampling rate and full range input.

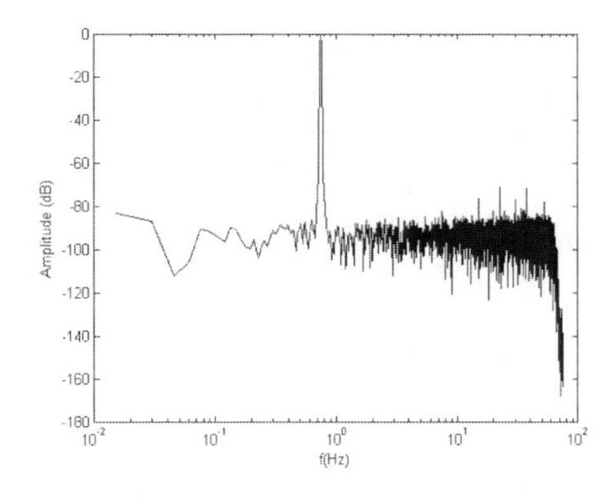

Figure 4. Results of dynamic range test.

4.2 *Background noise test under the condition of magnetic shielding*

Install the acquisition box and the fluxgate sensor in a magnetically shielded room (zero magnet space) in order to collect the system background noise of geomagnetic recorder. Observe the time series of noise continuously so as to calculate PSD (noise power spectral density) of the system. The range of the collector is 20Vpp, and the sampling rate is 15Hz. Conduct PSD calculation on noise time series after continuous observation for about 10hr. Results are shown in figure 5. The PSD of 1Hz frequency point is 6pT/rt (Hz). The noise increases with the decrease of frequency. If the frequency decreases by 100 times, the noise increases by 10 times, which is superior to 1/f.

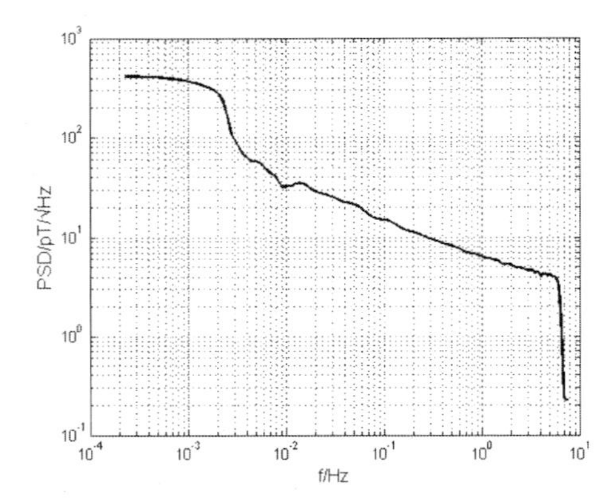

Figure 5. Power spectral density of channel E's background noise.

4.3 *Results of field measurement*

Site tests are carried out in field condition in order to further evaluate the performance of the recorder. Measured data of a certain point are provided in figure 6. Information of diurnal variation for 10 days of continuous operation is recorded in temperature. There is a certain correlation between the three-component magnetic field signal and temperature. It shows that high-quality data are obtained.

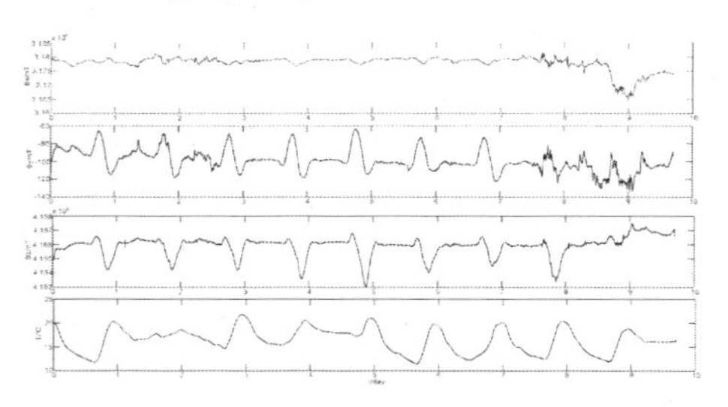

Figure 6. Field data.

5 CONCLUSION

Low-noise fluxgate sensor and recorder hardware system of acquisition machine are constructed, and relevant MCU control program and PC-end user software are developed in order to realize long-term geomagnetic field observation with high precision and low-power dissipation. Technical difficulties such as low noise, low-power dissipation, small synchronous error, and long-term stability are also overcome. Indoor test and field measurement verify the validity and correctness of this scheme. It is hoped that this scheme becomes practical and provides high precision geomagnetic observation with new methods.

REFERENCES

[1] Xu, W.Y. (1997) International geomagnetic network plan INTERMAGNET. *Progress in Geophysics*, 12(3):15–21.
[2] Hao, Y.L., Zhao, Y.F. & Hu, J.F. (2008) A preliminary analysis on geomagnetic matching in terms of underwater navigation, *Progress in Geophysics*, 23(2):594–598.
[3] Zhang, J.H. (2005) Dynamic membership function of geomagnetic precursory anomaly, *Progress in Geophysics*, 20(1):142–146.
[4] Guo, F.X., Zhang, Y.J. & Yan, M.H. (2007) Analysis on secular change features and mechanism of geomagnetic field, *Journal of Geophysics*, 50(6):1649–1657.
[5] Laurentiu ASIMOPOLOS, Agata Monica PESTINA, Natalia Silvia ASIMOPOLOS. (2010) Thoughts on geomagnetic data analysis, *Journal of Geophysics*, 53 (3):765–772.
[6] Wang, X.L. & Tian, Y.F. (2010) A research on autonomous navigation based on geomagnetic field, *Journal of Geophysics*, 53 (11):2724–2731.
[7] Alken, P., Maus, S. & Luehr, H. (2014) Geomagnetic main field modeling with DMSP, *Journal of Geophysical Research-space Physics*, 119(5):4010–4025.
[8] Wang, Y.B., Liu, S.B. & Feng, W.G. (2014) Fluxgate magnetometer with low-cost and high-resolution based on ARM. *Chinese Journal of Sensors and Actuators*, 27(3):308–311.
[9] Lei, J., Lei, C. & Wang, T. (2013) A MEMS-fluxgate-based sensing system for the detection of Dynabeads, *Journal of Micro-mechanics and Micro-engineering*, 23(9).
[10] Constable, S. (2013) Review paper: Instrumentation for marine magnetotelluric and controlled source electromagnetic sounding. *Geophysical Prospecting*, 61(1):505–532.
[11] Ripka, P. (1992) Review of fluxgate sensors. *Sensors and Actuators A-Physical*, 33(3):129–141.
[12] Tumanski, S. (2007) Induction coil sensors - a review, *Measurement Science & Technology*, 18(3):R31–R46.

Electric, Electronic and Control Engineering – Shao, Shu & Tian (Eds)
© 2015 Taylor & Francis Group, London, ISBN: 978-1-138-02842-5

Development of calibration module for OBEM

Hong Chen
China University of Geosciences, Beijing, China
Key Laboratory of Geo-detection, Ministry of Education, Beijing, China

Hongmei Duan
China University of Geosciences, Beijing, China

Ming Deng & Kai Chen
China University of Geosciences, Beijing, China
Key Laboratory of Geo-detection, Ministry of Education, Beijing, China

ABSTRACT: This article introduces an instrument calibration method of an ocean controllable-source electromagnetic acquisition station. In order to achieve accurate calibration of the instrument's acquisition channel and magnetic sensor, this article provides a design of the calibration signal generator based on the acquisition circuit, and it also represents Matlab processing program. This calibration method, based on bipolar square signal, adopts piecewise calibration for full brand scope of the acquisition station, so as to correspond to various sampling rates, with fundamental frequency being 15/64Hz, 2Hz, and 20Hz separately (adopting effective harmonic). Through an analysis on the frequency domain of acquisition of signals by Matlab, we can arrive at two calibration curves. Field testing shows that this method aims at effectively calibrating the same sensors, which coincide with the curve of Phoenix Company MTU-5 and can also work steadily in a different field environment.

KEYWORDS: OBEM; calibration; channel; sensor.

1 INTRODUCTION

One of the key technologies of Seabed MT and sea CSEM is researching the seabed electromagnetic acquisition station with high-reliable and high-precision observation. So-called high-reliable and high-precision has the following three understanding aspects: 1. The instrument has high-precision acquisition accuracy itself, as well as ideal background noise and SNR (signal-to-noise ratio); 2. The instrument is highly reliable, and it can work for continuous days or even months; 3. Different instruments are of the same height. Seabed controllable source magneto-electrotelluric exploration utilizes a single emission source and multiple acquisition stations to work simultaneously, so as to conduct acquisition for artificial source and natural source. Compared with the artificial source, the natural source's signals are relatively weak with more complex changes, and they have higher requirement of acquisition circuit noise. There are some more mature instruments in the international field, namely American SIO and Norway EMGS seabed acquisition stations. The SIO acquisition station has a typical index, that is, low noise (electric field 0.1nV/m/sqrt(Hz) @1Hz,magnetic field 0.1pT/sqrt (Hz) @1Hz, low power consumption (500mW,60days), high-reliable (95% recovery rate), and low time drift (1~5ms/day). However, the EMGS acquisition station has a similar technical index as that of SIO, is run as a commercial company that has a professional marine operation group and a data processing group, and plays an important role in the application area of seabed oil, gas, and hydrate exploration. Compared with its international counterparts, China University of Geosciences (Beijing) had a late start in marine electromagnetic method research, which conducted different levels of research during the Ninth-, Tenth- and Eleventh Five-Year[1-3]. On the zero basis, they successfully developed the sea controllable-source electromagnetic surveying system's hardware equipment and data processing interpretative system. Due to foreign blockade on new techniques and at the domestic technical level, the instrument index is relatively low and interpretation errors are relatively bigger; with the help of the Ministry of Science and Technology (MOST) Twelfth Five-Year 863 plan, China Geological Survey Bureau, and basic scientific research business funding of central universities, China University of Geosciences (Beijing) started a new round of development of seabed electromagnetic acquisition station [4], and through laboratory test and marine experimentation, they achieved an index increase in lower noise (electric field 0.1nV/m/sqrt(Hz)@1Hz, magnetic field 0.1pT/sqrt(Hz))@1Hz, lower power consumption (2000mW,21+days), higher recovery rate (electro erosion and acoustic releaser double release mechanism), lower time drift 1ms/day, and lower volume (four 17-inch glass balls), which are comparable to the level of international counterparts, that is, SIO, EMGS.

In respect of general measuring instrument research, calibration is necessary, which includes the following 3 aspects:

1 Determine input–output relation of instrument or measuring system;
2 Determine static characteristic index of instrument or measuring system;
3 Eliminate system error; improve accuracy of instrument or system.

Seabed MT and ocean CSEM method calculating medium impedance needs to know real information of the electricity magnetic field amplitude and phase position; electric field signals are received directly by the electrode, which has no impact on amplitude and phase position. However, for magnetic field signals, the actual record is an electrical signal converted by the sensor; for this reason, we need to calibrate the sensor to get a curve of sensitivity and phase-frequency characteristic. Impacted by components, circuit parameter, installation technology, and so on, differences exist in various receiving systems and different acquisition channels of the same receiving system. In order to guarantee the inversion operation's result accuracy, we need to calibrate the acquisition channel itself as well. Therefore, calibration is an important means of eliminating instrument-itself errors, getting a great real electromagnetic response; and calibration accuracy has a significant impact on final result.

In conclusion, calibration of the acquisition station needs two curves, namely, the curve of various channels' responses to specific frequency signals; the other is the curve of magnetic sensor responses to specific frequency signals, which we can call channel calibration and sensor calibration. In general, calibration signals have 3 choices, namely: 1). Single-frequency sinusoidal wave. This method has a single frequency and needs DAC, can only calibrate 1 frequency point each time, and has low efficiency; 2). Square signal, which, based on easiest theory, has rich harmonic waves. Frequency points are distributed linearly at regular intervals; but with serious attenuation of harmonic-order increase, we can take limited harmonics; and 3). 2n sequence pseudo random signal, including more frequency. Amplitude is basically the same; frequency point logarithms are distributed at regular intervals, but their generation needs an accurate code[6]. On a comparison of these 3 methods and in consideration of the hardware and software structure of the available acquisition circuit, the author selects the second method finally, namely the square signal. Meanwhile, as restricted by attenuation (as per rule of 1/n) of sampling frequency, sampling number, harmonic amplitude, and so on, a single frequency point's ultraharmonics has more serious attenuation, which cannot be processed. Thus, we need to design calibration signals for 3 kinds of frequencies, namely 15/64Hz, 2Hz, and 20Hz corresponding to 3 sampling rates of the acquisition station, namely 15Hz, 150Hz, and 2400Hz separately, so as to cover all frequency bands of the acquisition station. During signal analysis, we make calculations according to sampling rate, and then process 3 results to arrive at the final calibration result.

2 HARDWARE DESIGN

The seabed controllable-source electromagnetic acquisition station observes seabed electromagnetism 5-component signals; 2 magnetic-field components (Hx, Hy) can be acquired by 2 magnetic-field sensors (using ferromagnetism flux amplifier and flux feedback's low power consumption and small-sized complicated equipment); 3 electric field components (Ex, Ey, Ez) can be acquired by non-polarizing electrode Ag-AgCl developed by us[7]. All the signals inputted to AD acquisition circuit are voltage signals. According to requirements of the seabed MT and ocean CSEM method, the acquisition station needs to overcome a series of technical problems, such as noise, power consumption, clock drift, and so on. Therefore, noise mainly depends on the noise level of electric field sensor, low-noise chopping amplifier, and magnetic sensor; power consumption depends on the control circuit of low power consumption, analog processing circuit, and power management; and clock drift has requirements for consistency of clock system and high-frequency stability. Apart from the acquisition circuit and sensor, the entire instrument also includes the main control unit, GPS synchronization clock module, instrument recovery control module, acoustic communication module, gesture recording module, and so on. In the case of limited volume of the pressure cabin, we need to realize seabed high-precision electromagnetic data acquisition and storage; meanwhile, we can have a safe and effective recycling from the seabed.

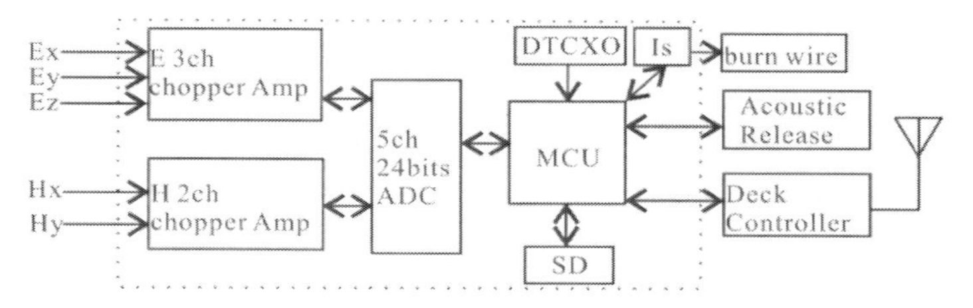

Figure 1. Diagram of seabed electromagnetic acquisition station's hardware design.

2.1 *Hardware design of acquisition station*

As shown in the earlier diagram, the acquisition circuit is in the scope of a dotted line, inputting 5 voltage signals, namely Ex, Ey, Ez, Hx, and Hy. The front end of the electric field is a fixed-gain 3-channel chopper amplifier that has a very low noise level and was developed by us, and the noise is less than 1nV/rt(Hz)@(1——100Hz); the magnetic field front-end amplifier achieves 1–64 times programmable gain (increasing by 2^n times), and the noise is less than 6nV/rt (Hz)@(1——100Hz); the ADC circuit consists of 5 single-channel ADCs and has 32-bit output data; the actual AD performance is no less than 22 bits, in order to adapt the output signal of the amplifier in an electric field; the input range of the ADC electric field channel is adjusted from 3Vrms to 3uVrms; the dynamic range is superior to 120dB(fs=150Hz); and the AD sampling frequency is configurable. In view of the research frequency range of Seabed MT and sea CSEM being less than 1KHz, the actual sampling frequency usually picks up 15Hz, 150Hz, and 2400Hz. The Main Processing Unit uses the 32-bit ARM9 processor that has a 400MHz clock frequency; the operating system uses Embedded Linux with 2.6.13 version of the Linux kernel and integrates the 16GB SD card. It also supports the FAT32 file system, and data download speed exceeds 10MB/s. OCXO provides a high stable clock for the system, and typical frequency stability is 10ppb. The MCU control circuit integrates CPLD, RTC, achieves GPS synchronous clock, and provides time marker for the data. Deck control instrument achieves functions such as parameter setting of acquisition circuit, GPS timing, data downloading, charging, self-inspection, and so on.

2.2 *Calibration signal generator*

According to the design principle, the calibration signal generator outputs three kinds of frequencies: 15/64Hz, 2.5Hz, and 20Hz, respectively, corresponding to three sampling frequencies (15Hz, 150Hz, and 2400Hz), which are beneficial for the post-calibration data process. The output waveform is of bipolar ±2.5V, and the square waveform is of duty cycle 50%. Therefore, the designed signal generator is based on TI high-precision reference power supply REF5025. The chip's output is 2.5V; it has 3ppm/°C temperature drift coefficient, 3uVpp/V noise level, and 0.05% maximum precision; meanwhile, it can output ±10mA electric current. The analogue switch adopts single-pole ADI and single-throw ADG419, which have the on-resistance less than 35Ω, switching time less than 160ns, and very low power consumption less than 35uW.

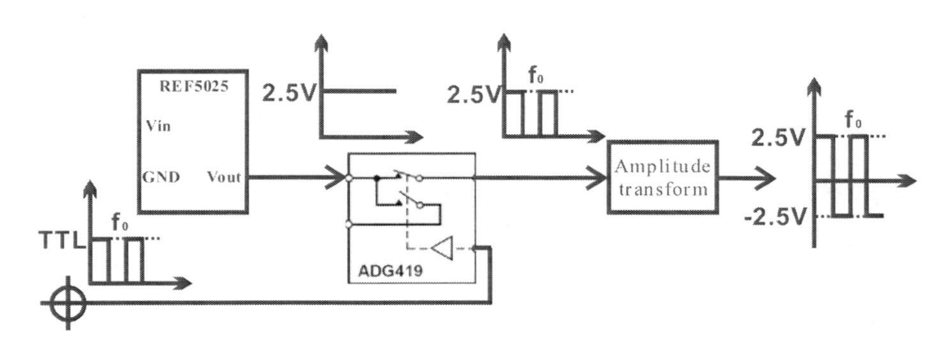

Figure 2. Calibration signal generator.

The principle of the calibration signal generator is as shown in figure 2; the control unit through CPLD outputs the square signal with frequency f0 to the analogue switch. Considering the phase position, this square signal has a special requirement, which needs to trigger the PPS synchronizing signal of the GPS module. The analogue switch input terminal connects high precision reference power supply output 2.5V; output frequency of the output terminal is f0, and amplitude of the square wave is 2.5V. Then through amplitude conversion circuit, it changes to bipolar ±2.5V, frequency f0 signal. Figure 3 shows the output of the control signal and signal generator, and a frequency of 20Hz.

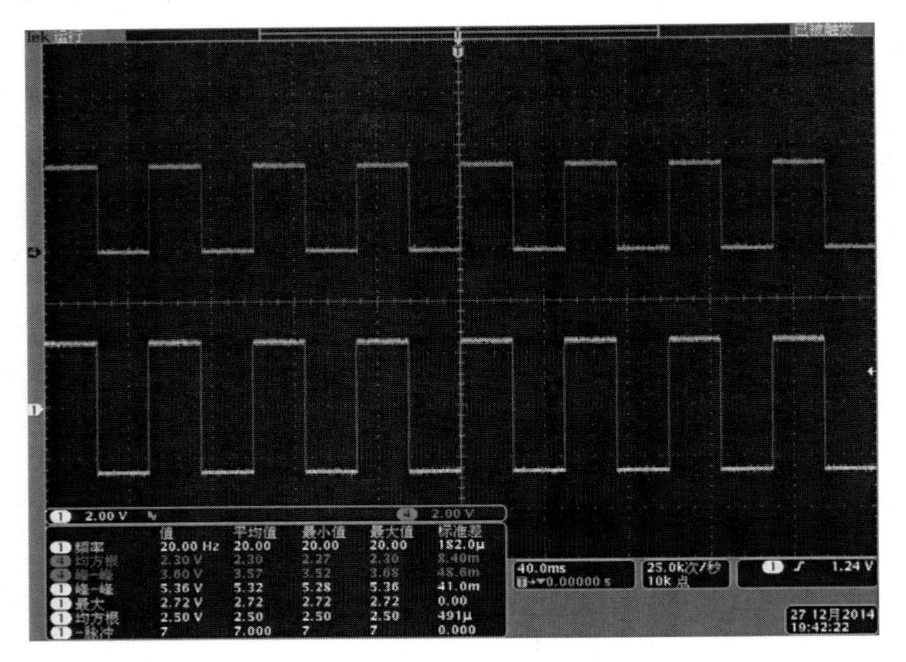

Figure 3. Output of the calibration signal generator.

3 FREQUENCY DOMAIN CALIBRATION OF ACQUISITION CHANNEL AND MAGNETIC SENSOR

3.1 *Calibration process*

As shown in figure 4, when in channel calibration, all the calibrated channel input signals switch to the same calibrated signal sources Vin; the frequencies output 15/64Hz, 2.5Hz, and 20Hz in sequence according to the variation of sampling rates. The instrument uses 15Hz, 150Hz, and 2400Hz sampling rates for sample durations of 100s, 8s, and 10s, respectively. Then, we make an analysis and process the sample data to acquire a channel calibration curve.

When using a probe for calibration, the calibrating signal Vin, as drive signal, outputs to the exciting coil in the magnetic sensor through sample resistance Ro. The generated current I will motivate the coil to generate a certain intensity of magnetic field, which generates an induced magnetic field close to uniform. The magnetic induction intensity B can be decided by formula 1: Permeability of vacuum μ is given; for the specific type of magnetic sensor, turns per coil N, length l, and winding radius R are given. As a result, the magnetic induction intensity is totally decided by I [8-9]. The magnetic induction intensity B can lead the sensor to generate output Vout. The instrument needs to acquire drive signal Vin and output signal Vout; sampling rate and sampling time should be consistent with channel calibration. Conducting some processing till the sample date will obtain the magnetic sensor frequency response curve and sensitivity.

$$B = \mu_0 I \frac{N}{l} \frac{1}{\sqrt{1 + 4R^2 / l^2}} \tag{1}$$

3.2 *Obtaining calibration curve*

Obtaining the calibration curve is the key and difficult point of the calibration process, which involves the relevant content of digital signal processing. The

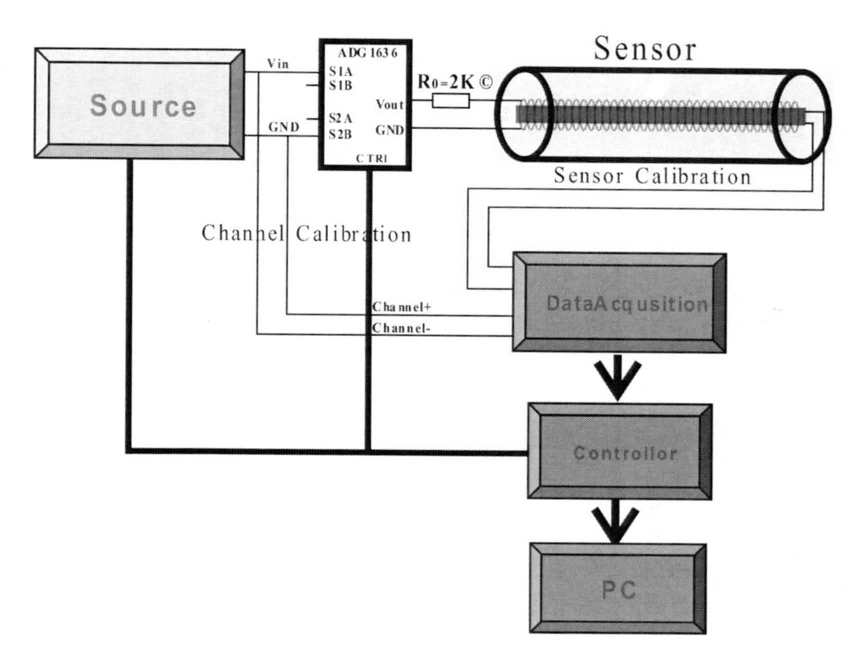

Figure 4. Diagrammatic sketch of calibration.

author achieves the calibration curve by using matlab program and data processing software SSMT2000 of Phoenix Geophysics Company. The detailed process is as follows: The matlab program acquires measured frequency response information (range and phase position etc) through processing the original data; meanwhile, it generates SSMT2000 compatible file format (.CLB and .CLC); SSMT2000 draws the corresponding calibration curve after reading the file and conducts internal processes [10] such as smoothing, and so on.

For the channel calibration, it is assumed that the input calibrating signal is an ideal square wave (there are errors; the output accuracy of voltage reference source and analogue switch time decide it is not the ideal square wave, but the author believes that can be neglected). Windowing the x0(n), the ideal digitized square wave digitization, and making FFT calculation get X0 (k); meanwhile, windowing the acquired digital signal x(n) in the same way and making the FFT calculation get X(k)[11] should be done. When reaching FFT, the information of frequency and phase position will be obtained, making the conversion according to the CLB format requirement, writing all the information to the new file, and then submitting it to SSMT2000 for processing. In the ideal conditions, the channel calibration curve should be coincided with 1. No details about probe calibration to channel calibration type are provided in this paper.

4 CALIBRATION TEST

In September 2014, Research Group conducted screening calibration test in one base in the Institute of Electronic, Chinese Academy in Huailai County, Zhangjiakou City, Hebei province, P. R. China. This base owns a giant magnetic shielding cylinder that can isolate the external magnetic field interference. Meanwhile, the research group also conducted an actual field magnetotelluric test. Both tests calibrated the acquisition channel and magnetic sensor of the acquisition station.

4.1 *Channel calibration*

Floating each channel input terminal in the acquisition station, the acquisition station enters into the channel calibration mode, and calibrating by 3 groups of sample frequencies 2400Hz, 150Hz, and 15Hz. Figure 5 shows the frequency response characteristic in two acquisition channels of the magnetic field in a field test. From this, we can conclude that two channels have good consistency.

4.2 *Calibration of magnetic sensor*

Calibration of the magnetic sensor should be conducted based on the channel calibration, motivating different-frequency signals (drive signal) to the calibrating coil of the magnetic sensor, meanwhile,

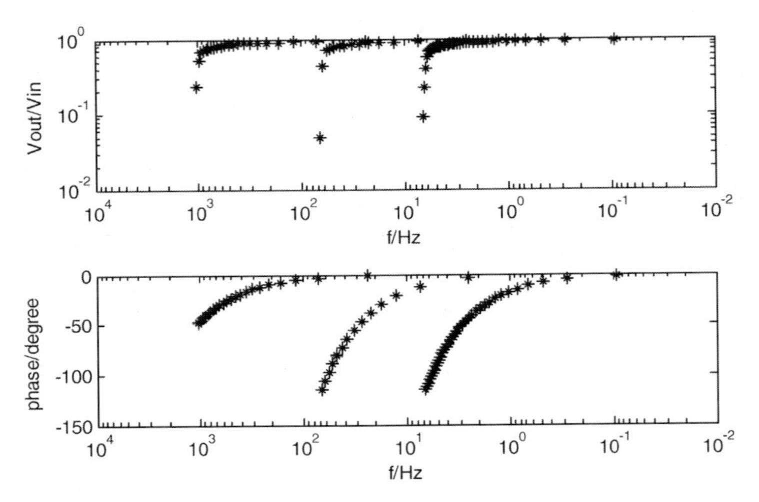

Figure 5. Calibration result of magnetic field channel.

simultaneously observing the induction signal (response signal), and carrying out calculation and analysis for the responses signal and drive signal to obtain frequency response characteristic of the magnetic sensor. The CS-10M magnetic sensor used by the research group is developed by the Institute of Electronic, Chinese Academy of Sciences, which has a sensitivity of 300mV/nT. Figure 6 shows the calibration result of a magnetic sensor. From the drawing, we can see that the parameter of sensitivity of the magnetic sensor is 300mV/nT. Meanwhile, in order to prove calibration accuracy, the result from the same magnetic sensor is also given, with the help of MTU5. After a comparison, it shows that the consistency exists at a low frequency, but for the frequency and phase position above 100Hz, there are some discrepancies. Further improvement for the calibration scheme of related software and hardware should be conducted.

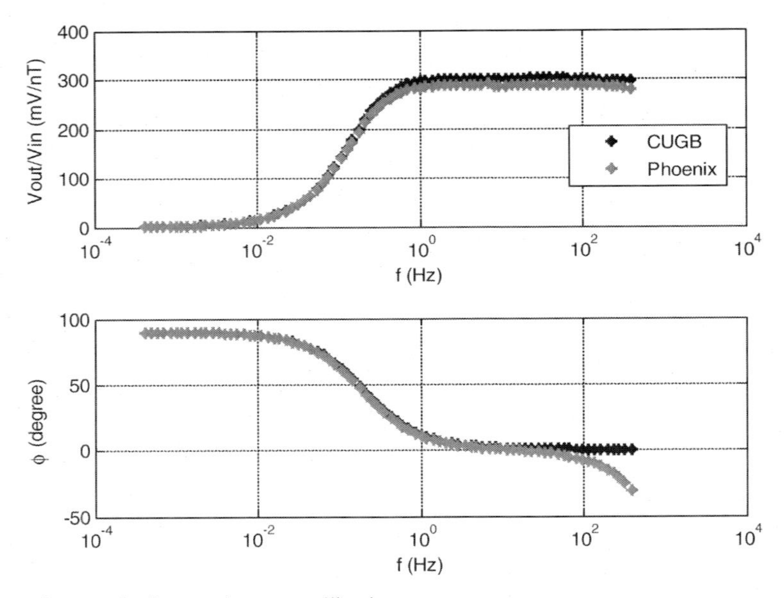

Figure 6. Comparison result of magnetic sensor calibration.

5 CONCLUSION

Calibration is critical to know about channel characteristics of the instrument itself, and to gain frequency response characteristics; a large amount of laboratory tests and screening experiments show that the calibration method introduced in this article is effective, the acquired magnetic sensor sensitivity parameter 300mV/nT is in line with the sensor's factory parameter and coincides well with Phoenix Company curve MTU 5. Meanwhile, this method is simple and useful, easy to operate and add modules in hardware, without increasing seabed electromagnetic acquisition station's pressure cabinet volume, which conforms to the design purpose - low power consumption and small volume of the seabed electromagnetic acquisition station. So, it can be used for calibration of the seabed electromagnetic acquisition station.

REFERENCES

[1] Chen, K. (2013) New progress in the design and development of AMT instruments, *Geophysical and Geochemical Exploration*, 37(1).

[2] Wei, W. B. (2002) New advance and prospect of magnetotelluric sounding (MT) in China. *Progress in Geophysics*, 17(2):245–254.

[3] Deng M, Wei W B, Sheng Y, et al. (2013) Several theoretical points and instrument technology of magnetotelluric data acquisition in deep water. *Chinese J. Geophys.*, 56(11):3610–3618.

[4] Deng, M, Zhang, Q. S, Qiu, K. L, et al. (2004) Technical issues of telluric electricity field exploration in marine environment. *Instrument Technology and Sensor*, 259(9):48–50.

[5] Wang, Y. Z, Cheng, D. F, Wan, Y. X, Lin, J. (2008) Design for Hybrid-source magneto telluric receiving and calibration system. *Academic Journal of Instrument and Apparatus*, 29(2).

[6] Zhang, W. X, Lin, J, Zhou, F. D, Liu, L. C. (2012) Generation and Inspection for distributed electromagnetic receiving system's multiple frequency calibration signal, *Optics and precision Engineering*, 20(8).

[7] Deng, M, Wei, W. B, Jin, S, et al. (2003) Seabed MT signal acquisition circuit design. *Earth Science Frontiers*, 10 (1).

[8] Zhao, J. (2013) *Design and Implementation of Induction-type Magnetic Sensor 0.003Hz10kHz*, Jilin: Jilin University.

[9] Wan, Y. X, Wang, Y. Z, Chen, D. F, (2013) Research on magnetic field calibration method by induction-type magnetic sensor. 26(4).

[10] Phoenix Geophysics. System 2000. Net User Manual. Canada: Phoenix Geophysics, 2006[2015-1-4].

[11] L, W. H, (2006) Frequency-domain analysis of signal and system based on Matlab. *Journal of Wuhan Institute of Technology*, 19(5).

Electric, Electronic and Control Engineering – Shao, Shu & Tian (Eds)
© 2015 Taylor & Francis Group, London, ISBN: 978-1-138-02842-5

Novel sampling frequency synchronization approach for PLC system in low-impedance channel

Yi Wang
Chongqing Electric Power Research Institute, Chongqing, China

Zhenli Deng & Yiyuan Chen
Chongqing Electric Power College, Chongqing, China

ABSTRACT: We address the sampling frequency synchronization method of high-speed Power-Line Communication (PLC) in a low-impedance channel using OFDM (Orthogonal Frequency Division Multiplexing) modulation. The synchronization approach is based on feed-forward frequency recovery loop and feedback control loop for VCO (Voltage Controlled Oscillator). It recovers the sampling frequency offset directly from the received samples. Results of simulations prove that this synchronization approach proposed satisfies the PLC system requirements.

KEYWORDS: Sampling frequency synchronization; PLC; feed-forward frequency recovery loop; feedback control loop.

1 INTRODUCTION

The power-line communication (PLC) system uses the existing power supply cable as a communication channel. High-speed PLC that uses the high-frequency band of the power line commonly uses the OFDM modulation scheme, and the data rate is beyond 1Gbps [1]. Since the PLC channel is a complex system, it always degrades the PLC performance, especially bringing harm to system synchronization. In the PLC system, the PLC channel is 2 wires of live (L) and neutral (N) line cables. If electrical equipments of low impedance are connected in the PLC channel, transmitted signal power is mostly absorbed and the received power is attenuated [2]. The channel distance also makes PLC signal-to-noise ratio (SNR) bad in the receiver. Due to attenuation, channel distortion, and additive impulsive noise effects from the instruments according to the operation such as switching, PLC system sampling synchronization needs to be more sensitive and stable. However, even a small noise component in the synchronization will lead to a large FFT window drift and the wrong FFT window position will significantly degrade the channel estimation if some of the channel impulse response components are outside the window for the channel estimation [3]. To avoid Inter-Carrier Interferences (ICI), the correction of occurring sampling frequency deviations is also a mandatory requirement. Most sampling frequency correction algorithms are implemented in the PLC system by using a feed-forward or feedback control loop structure. In previous works [4], pilot symbols injected in a transmitted signal can improve the loop convergence time; meanwhile, the spectrum efficiency should be degraded. This is a drawback in the high-speed PLC system such as home networks. Therefore, a non-data-aided algorithm is more desirable to avoid lowering data rate.

In this paper, we propose a non-data-aided synchronization algorithm for the estimation of sampling frequency offset based on feed-forward and feedback control loop. Compared with known feedback techniques, it has many advantageous features. Its major advantage is that it provides rapid synchronization and guaranteed stability, which are important especially in a high-speed PLC system.

This paper is organized as follows: Section 2 gives a brief description of the PCL system using OFDM modulation. In Section 3, we describe the structure of sampling frequency control and analyze the detection method. In Section 4, some results of simulation such as estimation performance versus sampling clock offset and bit error rate (BER) characteristics against CNR are presented. Section 5 concludes this paper.

2 PLC SYSTEM USING OFDM MODULATION

The conventional PLC system is shown in figure 1; since capacitors are used in order to block the main frequency, it is also called the "capacitive coupler system." If a low-impedance equipment is connected, as shown in figure 1, it will make the SNR at the receiver severe and lead to the symbol or sampling synchronization failure. We propose a novel synchronization structure in the receiver, as shown in figure 2. The feedback loop and feed-forward loop compose the synchronization system. Timing error detector characteristics (TEDC) followed by a loop filter can control the output frequency of VCO that provides the clock for ADC (Analog-to-Digital Converter). Meanwhile, the NCO (numeric controlled oscillator) also can be controlled by the filtered TEDC, and the NCO controls the interpolator for retiming the data.

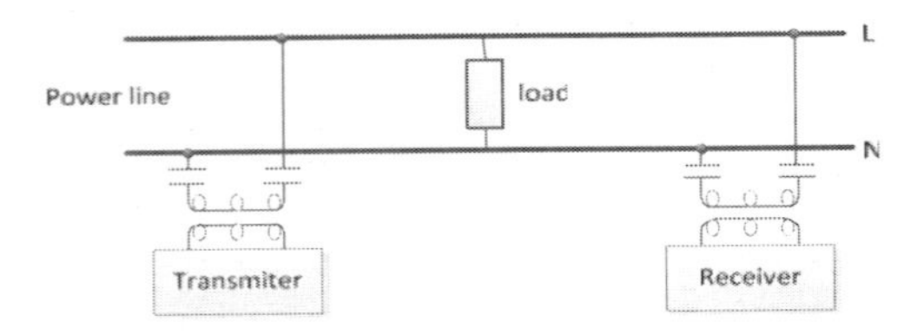

Figure 1. Conventional cap-coupling PLC system.

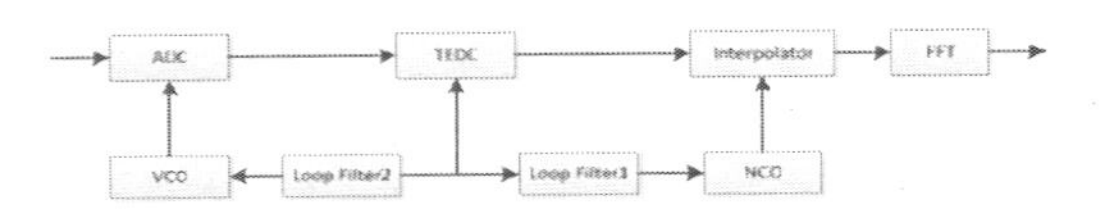

Figure 2. Sampling frequency recovery structure.

3 SAMPLING FREQUENCY RECOVERY

The sampling frequency synchronization system comprises two main tasks. (TEDC), which are usually performed by the interpolator. For the TEDC to compensate the sampling frequency deviation, it must be detected and estimated accurately. In the next section, we propose a method for the detection of the sampling frequency deviation, allowing feedback and feed-forward synchronization without any pilot data in the transmitter. The OFDM symbol has a cyclic prefix, for example, Ng samples in figure 3. So, for an ideal sampling frequency at the receiver, two same sequences per OFDM symbol (period is T) can be detected. For simplification, these sample sequences will be named Seq-1 (guard interval) and Seq-2 in figure 3[5].

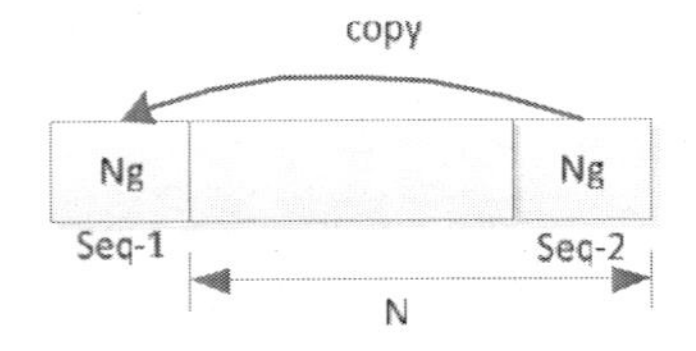

Figure 3. OFDM symbol.

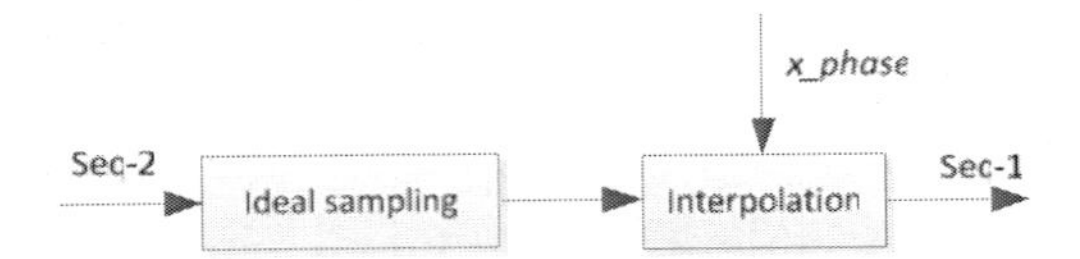

Figure 4. Model of imperfect sampling.

The output frequency of VCO is drifted slowly in a short period, and we assume that it is a constant sampling frequency deviation for one OFDM symbol period. After one OFDM symbol period, the constant phase (*x_phase* in Fig.4) is added to Seq-2 and causes the sample sequences Seq-1 and Seq-2 to differ fractionally. Because corresponding sample pairs of Seq-1 and Seq-2 are located N samples apart, the amplitudes differ by the effect of *x_phase*. These differences can be used for the calculation of the sample phase draft (*x_phase*/N). Figure 4 shows a simplified model of imperfect sampling for an OFDM symbol comprising N+Ng samples. It illustrates the sample sequences Seq-1and the repetition in Seq-2. The primary objective of the new sampling frequency detection is to calculate the interpolation parameter *x_phase*. For this purpose, a mathematical description of Seq-1 and Seq-2 according to figure 4 must be found. Considering efficiency and numerical effort, a solution can be achieved by using the Lagrange cubic interpolator. The interpolation parameter μ in the paper [6] is equal to *x_phase*/N. When we obtain the samples from ADC, we first find the start of Seq-1 and Seq2. Using the cubic interpolation equation, we can get μ.

Here, we need to find the start of the sequence using time delay correlation of the received time domain signal, $R(t) = \int_0^{T_G} x(t-\tau)x(t-\tau-T)d\tau$, x (t) means the received signal, Tg means the guard interval, and T means the total duration of one OFDM symbol. Figure 5 shows the scheme of time delay correlation for the received signal.

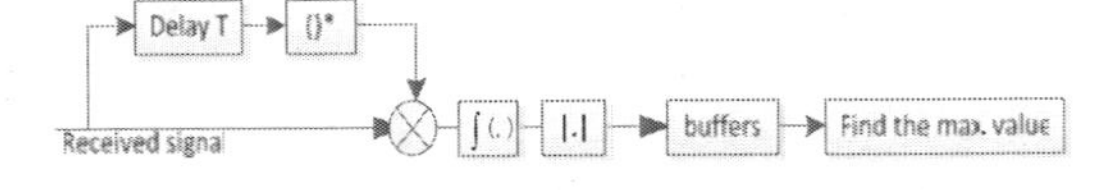

Figure 5. Structure of find the start of OFDM symbol.

It is the start of the OFDM symbol after the maximal value periodically. Of course, for real transmission scenarios, different channel influences must be considered. An essential property is the multipath propagation. The resulting channel delay leads to larger symbol duration. Depending on this reason, each sample at the receiver is influenced by adjacent samples. That disturbs the results of the TEDC detection. The loop filter is important for capture performance, and a proportional plus integral structure filter is used. The two-loop filter filters the phase noise of the TEDC output and extracts the stable component to control the NCO and VCO. The loop filter determines the stability of the loop and the speed of achieving synchronization. The NCO is responsible for determining the parameter μ and the interpolation base index of the interpolator. In the next section, the feasibility and performance of the Sampling Frequency Synchronization scheme will be proved by simulation.

4 SIMULATIONS AND DISCUSSION

We verify the algorithm proposed in the system next. It has 872 out of 1024 sub-carriers with QPSK,16QAM or 64QAM modulation; carrier interval is 100kHz; and the lowest sub-carrier frequency is 40MHz. The channel model assumed a multipath channel that has one delay wave against the direct wave, the desired-to-undesired ratio is 10dB, and the delay time of the delayed wave is 3 us. The sampling frequency offset and jitter will be characterized via several parameters such as [6] TEDC, maximum of TEDC, timing error variance, and so on.

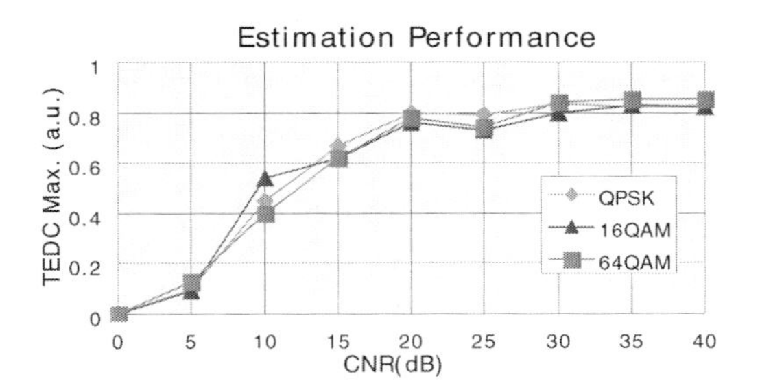

Figure 6. Estimation performance of sampling frequency recovery.

Here, we will use the maximum of TEDC to value the sampling frequency recovery loop performance. We set the 200ppm frequency offset in the transmitter side by the DAC (Digital to analog converter) sampling clock. Figure 6 shows the normalized maximum TEDC values against the CNR. Usually, if the value is beyond 0.3, we can recover the sampling frequency offset and low-frequency jitter.

The PLC system BER performance based on the proposed sampling frequency synchronization approach is shown in figure 8.

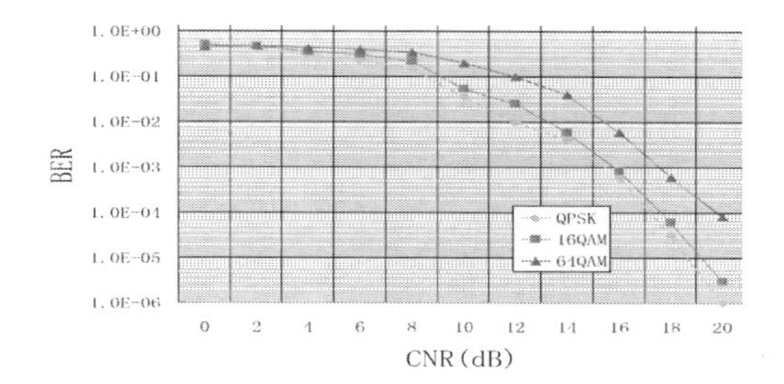

Figure 8. BER characteristics against CNR.

The BER performance gets worse when the modulation format order is high. There is 2dB CNR penalty when comparing 64QAM and 16QAM due to phase noise of the constellation in demodulation processing.

5 CONCLUSION

The PLC systems modulated by OFDM are highly sensitive to incorrect synchronization. To prevent a significant CNR degradation, efficient sampling frequency synchronization must be processed in the digital domain and should fulfil the requirements of receiver implementations. Therefore, we have proposed a novel feedback and feed-forward sampling frequency recovery loop so that we could use the feed-forward loop to compensate the high-frequency jitter and frequency offset and use the feedback loop to handle the low-frequency jitter and frequency draft. The results of simulations prove that the sampling frequency synchronization approach in this paper can be successfully used in the high-speed PLC system.

REFERENCES

[1] Tonello, A.M. (2008) Challenges for 1Gbps power line communications in home networks. *Personal, Indoor and Mobile Radio Communications, PIMRC. IEEE 19th International Symposium on*, pp:1–6.

[2] Shiji TSUZKI, et al., (2014) One Wire PLC System for Inter-floor Connectivity. *ISPLC 2014*. pp:121–126.

[3] T. Pollet, et al. (1994) The BER performance of OFDM systems using non-synchronized sampling, *GLOBECOM* 94, 1:253–57.

[4] Hirotomo Yasui, et al. A study on channel estimation using pilot symbols under impulsive PLC channel. *IEEE 19th International Symposium on PLC and Its Applications*, pp:322–327.

[5] Oswald, E. et al. (2004) NDA based feed-forward sampling frequency synchronization for OFDM systems. *Vehicular Technology Conference*, 2:1068+1072.

[6] F.M. Gardner (1993) Interpolation in digital modems-part I: Fundamentals, *IEEE Trans. On Communications*, 41(6):501–507.

[7] H. Meyr, et al. (1998) *Digital Communication Receiver*, John Wiley &Sons.

Electric, Electronic and Control Engineering – Shao, Shu & Tian (Eds)
© 2015 Taylor & Francis Group, London, ISBN: 978-1-138-02842-5

Research and application of distribution network data topological model

Jianjun Song
State Grid Kaifeng Power Supply Company, Kaifeng Henan, China

Jianli Guo & Yongheng Ku
State Grid Henan Electric Power Company, Zhengzhou Henan, China

ABSTRACT: With the ever expanding distribution network, to accommodate the management of the power distribution system, a distribution network data topological model is constructed. Topological structure, graph storage and graph traversal are examined from the topological model structure and topological graph theory. The rationale of distribution network data is investigated by constructing a topological analysis model. Distribution network equipment should be modeled by global topological analysis and local topological analysis according to the area of analysis. This increases the response speed to topological analysis programs to the distribution network structure variation and allows realization and successful application of the distribution network data topological model.

KEYWORDS: Distribution network; global topology; local topology; graph traversal.

1 BACKGROUND

Smart network construction has made remarkable contribution to information-based and smart network, but is also accompanied by many problems like steep increase of information volume that results in increased application data patterns, which prevents the sharing of data models and development platforms as they are mostly developed by different manufacturers. This increases repeated tasks for the workers on the one hand, and causes waste of information resources and low working efficiency on the other hand.

Distribution network topological analysis proves to be the most effective way of solving these disadvantages, since it sets up nodes and branches by analyzing the relation of each element in the distribution network and finally makes up a tree structure that reflects the connection and interconnection between different equipments as well as the paths of these nodes. It also allows coloring and inquiry of the equipment operation state. Besides, a change in the equipment under distribution network topological analysis will lead to a change in the tree structure too so that it is immediately updated to allow monitoring and positioning of the network elements. As such, data topological analysis provides effective basis for monitoring distribution network equipments and correctly identifying fault sources.

2 CONSTRUCTION OF A DISTRIBUTION NETWORK DATA TOPOLOGICAL MODEL AND FEATURE REALIZATION

2.1 *Distribution network equipment capacity*

Before modeling distribution network equipment, the equipment and its operation state are identified with an identifier and assigned by giving a unique ID. Next, the equipment operation state is assigned and identified with "1" for live equipment, "0" for dead equipment, on for equipment switch-on, indicated by "1", off for equipment switch-off, indicated by "0". Next, the current direction is identified with "1" for the direction from the source point end to the upstream supply equipment end of the switch element, "0" for the direction from the upstream supply equipment end of the switch element to the source point end; "2" for uncharged. Besides, the switch type including the outgoing switch, feeder switch and interconnection switch are also identified. The attributes of other types of equipments are assigned in a similar way before the distribution network equipments are modeled.

2.2 *Global and local topological analyses*

In the distribution network equipment model, the database management system PostgreSQL stores the operation state of the distribution equipments and the interconnection between them. Before starting topological analysis, the distribution information of the distribution equipment is read out of the database and the interconnection between the equipment subjects is refined. Then, mapping relation is established between the equipment subjects and the corresponding features using ID. Next, to facilitate finding and management of electrical equipments, linked list data structure is used in the distribution network system to manage the electrical equipments. Each type of equipment subject is managed by one linked list.

1 Global topological analysis
After modeling the distribution equipments, global topological analysis is performed in the event of material changes in the distribution network structure. For the purpose of our study, global topological analysis means to traverse the distribution network starting from the bus, and set each of the equipments traversed to live state. If the equipment is not traversed, that means it is not connected to the bus and is therefore power off. During the graph traversal of the equipment, we can tell if the graph traversal is completed according to the link state of the switch, i.e. whether the two elements are correlative to each other. If the switch is disconnected, we must stop traversal and set the switch to OFF. If the switch is closed, we must continue to traverse and set the switch to ON. During the traversal, we must also set the current direction mark bit and the upstream power supply equipment of the electrical equipment. The analysis speed will be increased when performing bus tracking and other features. The traversal process is shown in the chart below.

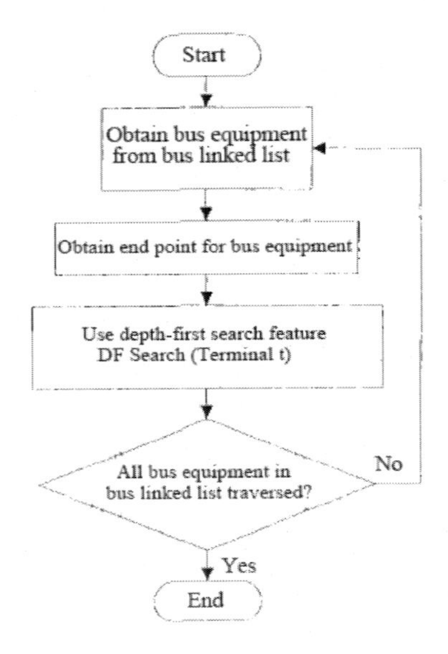

2 Local topological analysis.

When performing local topological analysis, first, the electrical equipment subject is obtained from the distribution network system and processed using different methods suited to the type of the equipment. So far, processing methods used include depth-first search (DFS) feature and breadth-first search (BFS) feature. By reference to the actual operation state and attribute of the grid equipment, DFS feature is used to obtain the traversal of the downstream end points according to the current direction identification. If the equipment fails, all the equipment elements connected to it is power off and therefore not correlative. The graph traversal by DFS is drawn to an end. Now we obtain the connection node where the bus is located and start BFS feature, performing graph traversal to all connection points of the bus. Besides, no search is performed to equipment with single end point as it has no downstream connection points and therefore will not make difference to the operation state of other equipment. The local topological analysis flowchart is shown in Figure 1.

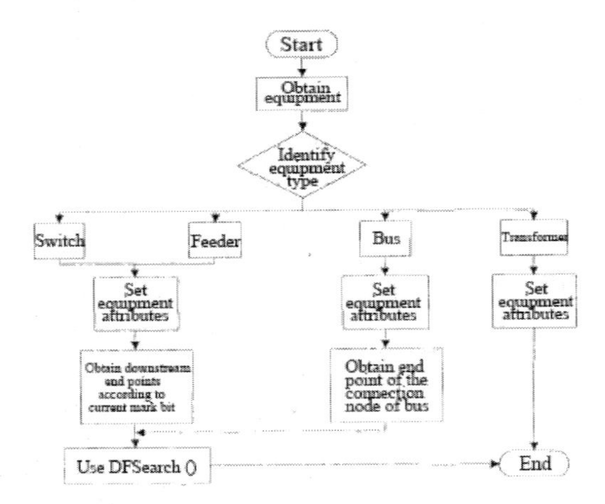

Figure 1. The local topological analysis flowchart.

3 APPLICATION OF THE DISTRIBUTION NETWORK DATA TOPOLOGICAL MODEL

The application of topological analysis include topology coloring, source point tracking, service scope analysis, dispatch task simulation, network reconstruction and load transfer. Our model provides topology coloring, source point tracking, service scope analysis and dispatch simulation features. It also allows real-time display and accommodates the correctness and timeliness requirements of the power system.

3.1 *Topology coloring*

Distribution network topology coloring is use different rendering color to the equipment primitives so that the dispatcher is able to update the grid operation timely and directly and distinguish identify the voltage rating, electrical equipment state, electrical equipment tracking and supply station area according to the response criteria.

The powerful rendering ability of GIS system itself allows easy realization of the topology coloring features discussed above, including direct display of the distribution network topological analysis results and real-time presentation of the operation state of the electrical equipment. Depending on the scope of coloring, global topological analysis and local topological analysis may be used. By switchover between these two methods, we can greatly save analysis time and improve the timeliness of the analysis.

3.2 *Source point tracking*

Source point tracking is a base application of the distribution management system that performs topological tracking of the distribution network equipment through to the supply source of the equipment and highlights the tracking path on the GIS images. When the distribution equipment is power off, it is not connected to the source point and therefore it is impossible to perform source point tracking of this equipment. As such, this operation only applies to the live operation equipment. The topological method for tracking source points can also be global topological analysis and local topological analysis.

To track the source point of the selected equipment, first judge the operation state of the equipment. If it is power off, an error prompt will be displayed. If it is power on, traverse the upstream equipment subject stored in the equipment subject. If the bus is found, the source point has been tracked. Then store the equipments traversed into a temporary linked list and then highlight these equipments. If the bus is not found, there's error in the program analysis. In such case, the traverse process is ended and an error prompt is displayed. When tracking source points, we can do it quickly according to the upstream equipment information stored in the equipment to increase the tracking speed.

3.3 *Service scope analysis*

Service scope analysis is to track the downstream of the electrical equipment and obtain all electrical equipments electrically connected to it. The main operation subjects of service scope analysis are the substation and distribution switch. The service scope of the substation covers all the electrical equipments contained in this substation. The service scope of the switchgear covers all the electrical equipments connected to this switch. The service scope analysis of the substation is to operate on the substation bus and track all the electrical equipments that prove to be connected to the bus. The service scope analysis of the switchgear is to determine the downstream nodes of the equipment according to the current direction mark bit of the equipment subject, and then perform topological tracking using DFS algorithm.

Figure 2. Service scope analysis result.

3.4 *Dispatch simulation feature*

The main task of distribution network dispatch simulation is to simulate the switching of the switchgear to find the electrical equipments affected by this switching operation, and use the topological coloring feature to update the operation state of the distribution network equipments. According to the color display of the equipment primitives, it is possible to tell whether the operations are correct. If the dispatch simulation result is the same as expected, actual operation can be performed on the distribution network to improve the accuracy and safety of grid operation. The result is shown in the figures below.

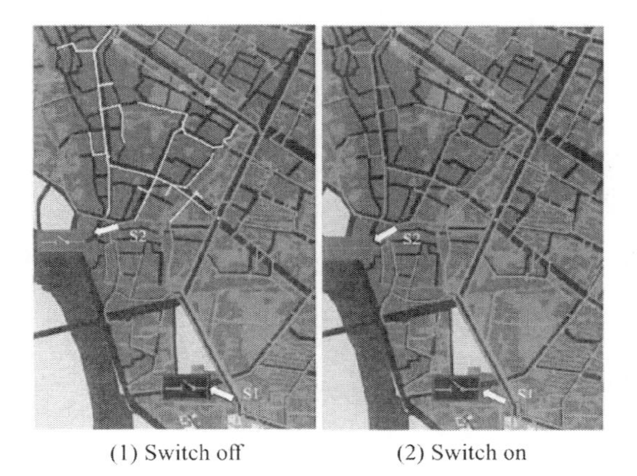

(1) Switch off (2) Switch on

Figure 3. Dispatch simulation result.

4 CONCLUSIONS

A topological analysis method combining global and local topologies is proposed and applied to the distribution network of Kaifeng Power Company. By distribution network topological analysis, the connection state of the distribution network and the operation information of the electrical equipment are obtained and provide basis for higher-level application of distribution network like load flow computation and state estimation.

REFERENCES

[1] Fan, M.T., Zhang, Z.P. (2008) *Research on Issues Relating to the Power Distribution Network Development Strategy in China*. Beijing: *China Electric Power Press*, pp:1–30.

[2] Yu, M., Ai, T.H. (2009) *Introduction to Geographic Information Systems*. Beijing: *Tsinghua University Press*, pp:1–30.

[3] Li, F.W., Li, X.X., Yuan, L., et al. (2010) City traffic information platform based on open source GIS. *Traffic Information and Security*, 20(3):25–29.

[4] Li, C., Liu, M., Xu, W.W. (2009) Research on service recommendation for network based data mining platform. *Anhui Geology*, 19(2):151–154.

[5] Wang, L., Xie, G.S., Xiao, H.Q. (2009) Real estate GTS graphic management system based on open-source technique. *Bulletin of Surveying and Mapping*, (12):57–59+64.

Electric, Electronic and Control Engineering – Shao, Shu & Tian (Eds)
© 2015 Taylor & Francis Group, London, ISBN: 978-1-138-02842-5

Micro-grid storage configuration based on wind PV hydro-storage comprehensive optimization

Bei Cao & Yue Yang
State Grid Jiangxi Electric Power Research Institute, Jiangxi Nanchang, China

ABSTRACT: Considering the features of energy storage systems, the optimal configuration mathematical model of the micro-network is established as containing Wind PV Hydro and storage to meet the needs of micro-grid load. Storage capacity is redefined, the objective function is optimized, and the genetic algorithms are used to solve it. The following conclusions were analyzed by a Suichuan Daijiapu area; by optimizing a new energy capacity ratio relative to the maximum load and the ratio of new energy, the configured storage capacity can be greatly reduced, which provides the guidance to the new energy configuration planning of the micro-grid.

KEYWORDS: micro-grid; SOC constraints; minimum storage configuration; new energy ratio optimization.

1 INTRODUCTION

With the fossil energy increasingly declining and the environment climate deteriorating, exploration of a clean and renewable energy resource will be the main task in the world. Wind, solar, hydroelectric plants, and so on became the hot researching areas with the merit of being inexhaustible and pollution free [1] [2]. Different kinds of renewable energy sources have nature complementarities in various periods and regions. The wind and PV hybrid energy complemental generation systems can avoid the dependence of seasons and can also reduce the influence of seasonal load. They can significantly reduce the energy storage by rationally allocating the proportion of new energy resources in the off-grid state, when compared with the single power generation mode.

In the essay [3], W. Kelloggt researched the optimal allocation issue of wind, PV, and woods storage energy system by a simple iterative algorithm, regarding the constraint of minimum rate of load loss and the objective function of system cost. Researchers in Bangladesh studied the optimal allocation of the wind and PV hybrid micro-grid, based on the Quasi–Newton algorithm [4] [5]. The domestic researchers also studied plenty of the research conducted on the new energy hybrid power systems, and the essay[6]; did deeper research of the polyergic hybrid system by proposing the new project integration of the wind and PV hybrid; built the system analysis; and optimized the mathematical model. The Hong Kong Polytechnic University united with the Guangzhou energy institute of Chinese Academy of Sciences and the semiconductor institute of the Chinese Academy of Science and found out a system configuration that meets the users' requirements with the minimum cost of investment of devices by adopting a comparison of optimization methods [6]. The essay [9] built the optimal micro-grid distributed power configuration with the minimum cost of investment of devices, fuel cost, operating maintenance cost, and environmental conversion cost.

In conclusion, the domestic and foreign research of polyergic hybrid technology mainly focuses on the wind and PV hybrid, wind, PV, and wood hybrid (pumped storage or others) and this kind of co-generation systems. We should mainly consider the investment cost as the optimization objective, instead of lacking the research in the small hydropower station, new energy complementary system, and capacity configuration. Especially, in the research of the full use of the complementarity of wind, solar, and water resources, and the comprehensive optimization, let us consider the storage character itself and the reduction of storage configuration. Under the requirement of load, this essay considers the restriction of characters of the storage system itself (charge discharge efficiency, SOC restriction) to analyze the optimal configuration of wind, solar, and water with the minimum objective function of the storage capacity configuration.

2 OPTIMAL CONFIGURATION MODEL OF WIND, SOLAR, AND WATER STORAGE

2.1 *Optimal objective function*

The maximal per unit per hour of relative load of wind, solar, water, and load is $W(t)$, $S(t)$, $H(t)$, and $L(t)$, respectively. So, the power mismatch per hour can be expressed as follows:

$$\Delta(t) = \gamma[aS(t) + bH(t) + cW(t)] - L(t) \tag{1}$$

This expression also decides the storage capacity directly needed, where γ means the relative maximum load ratio of installed capacity in the micro-grid (the general value is 1.15~1.3); $\gamma - 1$ is the percentage reserve of the system; a, b, and c are the portion of PV, hydroelectric, and wind power, respectively; and $c = 1 - a - b$, where the $L(t)$ means the value per hour of load; $S(t)$ means the contribution per hour only when the new energy is 100% PV in the micro-grid.

When the power mismatch is a positive value, it means the superfluous electricity can be stored by an energy storage device, and the storing efficiency is η_{in}. When the power mismatch is a negative value, the scanty electricity can be released by the energy storage device, and the releasing electricity efficiency is η_{out}. The expression of the energy storing device storing and releasing energy is as follows:

$$H\_store(t) = H\_store(t-1) + \begin{cases} \eta_{in}\Delta(t) & \Delta(t) \geq 0 \\ \eta_{out}^{-1}\Delta(t) & \Delta(t) < 0 \end{cases} \tag{2}$$

where $H\_store(t)$ is the change of energy stored by the storage device overtime without restriction. $H\_store(t)$ will be created in the case that the wind, solar, water, load, and the relative maximum load rate of new energy installed capacity, charge discharge efficiency factor of the energy storing device are decided. As cited earlier, the expression of the minimum energy storing configuration capacity chosen is as follows:

$$E_H = \max_t H\_store(t) - \min_t H\_store(t) \tag{3}$$

Because of the result of the average power minutes, the consumption of storing energy is greater than the load. Thus, the value of configuration storing required will increase as time passes. In this case, by adopting the method with the maximum storing capacity of all the time minutes, the minimum value is meaningless. Therefore, the expression of storing capacity can be refined as follows:

$$E_H = \max_T \left(H\_store(t) - \min_{t' \geq t} H\_store(t') \right) \tag{4}$$

where E_H is energy storing capacity; X is the optimal variable set X=[γ,*a*,*b*,*c*]; T is the number of simulation hours; and *H_store*(*t*) is the energy stored of the energy storage device without restriction at time t. Taking PV configuration rate a, small hydro-configuration rate b, and wind power configuration rate c as optimization variable, as $a+b+c=1$, the optimization variable can be reduced to 3, X=[γ,*a*,*b*]; with the restriction of load requirement and the objective function of minimum storage capacity, we can build a capacity optimal configuration model of independent power supply micro-grid, including wind, PV, water and electricity, and energy storage.

When the configuration capacity is expressed by the expression cited earlier, focusing on the situation that reduces the energy storage when more power is only generated by new energy, this means enabling the energy storage devices to not have a negative value when discharging energy. The storing capacity level is $H\_store(t)$ without restriction at the time t. During the period $t' \geq t$, the storage capacity level is not lower than $\min_{t' \geq t} H\_store(t')$, without restriction. The difference value of the two-energy storage capacity level reflects the storage energy needed to be released at any time t. The maximum value of all the time is the energy capacity E_H needed to be configured.

2.2 *Constraint condition*

$$\begin{aligned} &1 \leq \gamma \leq 1.5 \\ &0 \leq a,b,c \leq 1 \\ &a+b+c=1 \end{aligned} \tag{5}$$

$$C_{low} \leq SOC \leq C_{up} \tag{6}$$

$$W_i \geq 0 \tag{7}$$

where *SOC* is the hopower status of the energy storage system, C_{up} is the upper limit of the energy storage system operation; C_{low} is the lower limit of the energy storage system operation; the value range of C_{up} and C_{low} is [0,1]; and W_i is the amount of unbalance of energy every day, which is the reliability index of the power supply of the micro-grid.

1 *SOC* constraint

$$SOC = \frac{H_c(t)}{E_H}. \tag{8}$$

Here, $H_c(t)$, E_H, respectively, is the real and rated storing capacity of the energy storage system.

According to the definition of the energy storage capacity, the configuration required in the expression (4), the expression of energy storage capacity level without constraint can be translated to energy storage capacity level with constraint, and its changing status with time can be expressed as follows:

$$H_c(t)=\begin{cases} C_{up}E_H & C_{up}E_H - H_c(t-1) < \eta_{in}\Delta t \\ H_c(t-1)+\eta_{in}\Delta t & C_{up}E_H - H_c(t-1) \geq \eta_{in}\Delta t > 0 \\ H_c(t-1)+\Delta t/\eta_{out} & \Delta t \leq 0 \\ C_{low}E_H & C_{low}E_H - H_c(t-1) > \Delta t/\eta_{out} \end{cases} \quad (9)$$

The depth of discharge (DOD) means the ratio of energy that the energy storage system discharges to the rated capacity, and it is often expressed in percentage.

$$\text{DOD}=1\text{-SOC} \quad (10)$$

The relationship between DOD and charging lifetime is an inverse one. The charging lifetime of the energy storage system gets shorter as the DOD gets deeper The lower the DOD the longer the charging lifetime is. But it will lead to the storage capacity of the energy storage system discharging less, which will waste the electricity and increase the cost of the system. Thus, the SOC of this energy storage system should be kept between 0.6 and 1, that is, $C_{low} = 0.6$, $C_{up} = 1$.

2 Independent power supply reliability index of the micro-grid

Define the amount of unbalanced energy every day:

$$W_i = E_{B(i-1)}\eta_{out} + E_{W(i)} + E_{P(i)} + E_{H(i)} - Q_{L(i)} \quad (11),$$

where $E_{B(i-1)}$ is the available electricity left in the battery the previous day; the η_{out} is the charge efficiency of the battery; $E_{W(i)}$ is all the electricity generated by the all the fans that day; $EP_{(i)}$ is the electricity of the solar cell that day; $EH_{(i)}$ is the electricity of the hydropower station that day; and $QL_{(i)}$ is the load quantity demanded that day.

When $W_i>0$, it means the generating capacity of the present day and the energy stored in the battery can meet the requirement of the load, there is electricity left in the battery, and there is no lost electricity caused by load shedding; when W_i 错误！未找到引用源。 0, it means the generating capacity of the present day and the energy stored in the battery cannot meet the requirement of the load, so the only thing that can be done is load shedding: $|W_i|$ is the electricity shed that day, and simultaneously, the electricity left in the battery that day is $EB_{(i)}=0$.

The optimization goal is that the micro-grid power supply can meet the requirement of the load, the energy storage configuration is the minimum, and the genetic algorithm is used to solve the equation in this essay. We need to comprehensively consider the charge discharge efficiency of the energy storage system, the SOC constraint, and the amount of unbalanced energy every day; the minimum storage energy is the aim; and we should analyze the configuration of new energy: wind, solar, and water.

3 NEW ENERGY OPTIMAL CONFIGURATION ALGORITHM BASED ON THE GENETIC ALGORITHM

The three main steps that are used for solving all the problems using the GA algorithm are as follows: encoding and decoding, individual fitness evaluation and selection, crossover, mutation, and these kinds of genetic manipulation.

3.1 *Encoding*

The method of real number encoding is adopted in this essay, which express maximum load ratio installed capacity of the new energy in the micro-grid and the ratio of all kinds of new energy, as well as chromosomes are expressed: $[X_1 \cdots X_n]$. For example, chromosomes [1.2 0.2 0.5] means the maximum load is 1.2 times, the PV ratio is 0.2, and the water electricity ratio is 0.5, which means the wind electricity ratio is 0.3 in the new energy installed capacity.

3.2 *Individual fitness evaluation*

The precondition of this essay is the requirement of load, and the objective function is the minimum of the storage capacity, which can be expressed as follows:

$$\text{Min} f(x) = \text{MinE}_H(x) \quad (12)$$

$$x = [X_1 \cdots X_n] \quad (13)$$

Constraint condition

$$\begin{aligned} & 0 \leq X_1 \cdots X_n \leq 1 \\ & X_1 + \ldots + X_n \leq 1 \\ & 0.6 \leq SOC \leq 1 \\ & W_i \geq 0 \end{aligned} \quad (14),$$

where the $E_H(x)$ is the energy storage capacity, corresponding to the chromosome x.

The individual fitness of the genetic manipulation is the storage capacity, corresponding to the chromosome x:

$$F(x) = E_H(x) \quad (15)$$

3.3 *Adopting the ordinary genetic operation*

The method adopted in this essay is the championships strategy operator, which is relatively simpler and easy to operate. Picking up two individuals from the group (parent group), we compare their individual fitness. Then, we choose the bigger one, building the new-generation group (subgroup) until the individual number elected from the parent group and subgroup is equal.

4 EXAMPLE RESULTS AND ANALYSIS

We choose an area of Daijiapu Suichuan in Jiangxin province as an example. All the loads of this area are 2600kW, including common load 1455kW, and the average daily load curve has some fluctuations in one year. The peak value of the average daily load is around 1455.86kW, but the valley value of the average daily load is 938.54kW. This paper derived the meteorological data, including wind speed, radiation intensity, temperature, and so on, from the software—Trnsys. Through analyzing the full-year data setting 1min as the sampling period, and the output power of PV and wind being simulated by HOMER, the curve graph of wind, solar, water, and load is shown in figure 1.

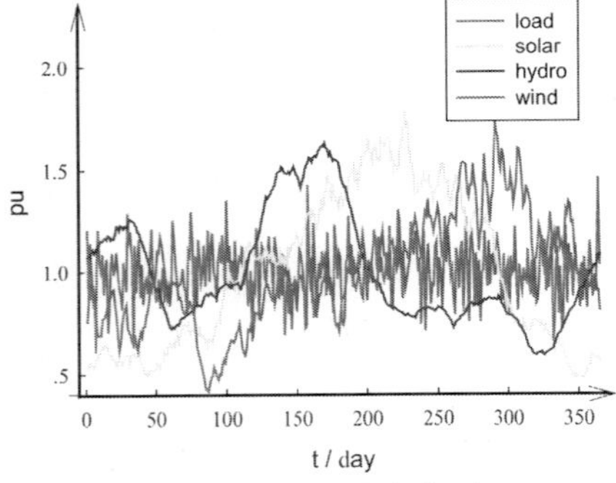

Figure 1. Variation graph of wind, solar, water, and load in the typical year.

We analyze the new energy and energy storage optimal configuration in this area by the optimal configuration model of wind, solar, and water storage described earlier. The micro-grid's power supply can meet the requirement of load. We consider the charge discharge efficiency of the energy storage system, the SOC constraint, and the amount of daily unbalanced energy and analyze the minimum energy storage configuration. The rate of the energy storage configuration required has a direct relation with the decision variable PV configuration, small hydro configuration, and wind power configuration, respectively, a, b, and c. We assume that the comprehensive charge discharge efficiency of real storage energy is 0.86 (the energy storage devices lose some power when storing and discharging), and the charge and discharge efficiency is equal; then, $\eta_{in} = \eta_{out} = \sqrt{\eta_{ES}}$=0.927. We discussed the influence of energy storage configuration caused by all kinds of new energy configuration rates.

Combining the natural conditions of wind, solar, water, and so on in this area, and analyzing the new energy storage optimal configuration by wind, solar, and water storage optimal configuration model, we get the relationship between the relative maximum load ratio γ of the new energy installed capacity and the optimal configuration of wind, solar, and water storage.

4.1 *Relationship between relative maximum load ratio γ of different new energy installed capacity and most optimal configuration of new energy*

The relationship between the relative maximum load ratio γ of the different new energy installed capacity and the optimal configuration of new energy is shown in figure 2, where the yellow sign means the ratio of photovoltaic power generation, the green sign means the ratio of wind power generation, and blue sign means the ratio of small hydropower generation. From the figure, we can find that in order to minimize the energy storage configuration to the minimum when the relative load ratio is the maximum of a different new energy installed capacity, the proportion of photovoltaic power generation to the new energy should be about 7%–10%, which has the assistant effect.

The configuration proportion of small hydropower generation to new energy is relatively high, which is a little higher than that of wind power generation. When the relative maximum load ratio γ of the different new energy installed capacity is smaller, the proportion of configuration of the small hydro to wind and electricity is about 50%. Then, with the increase of γ, the configuration ratio of the small hydro increases a little, and the stable value is about 70%. However, the configuration ratio of wind and electricity decreases a little, and the stable value is about 25%.

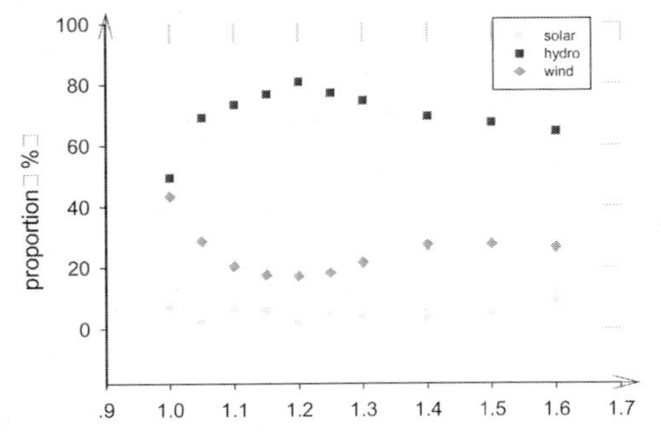

Figure 2. Optimal new energy configuration ratio of relative maximum load ratio γ of different new energy installed capacity.

4.2 *Relationship between relative maximum load ratio γ of different new energy installed capacity and energy storage capacity configuration*

The optimal energy storage capacity configuration relation of the relative maximum load ratio γ of the different new energy installed capacity and energy storage capacity configuration is shown in figure 3. From this figure, we can find that the storage energy configured will decrease with the increase of γ. The reason is that increasing the relative optimal load ratio of the new energy installed capacity means installing more power generation devices of new energy. With significantly increasing γ, the capacity configuration of energy storage devices decreased, but the investment of new energy power generation facilitates a great increase. Meanwhile, the electricity generated by the new energy is abandoned in most conditions of off-grid running, because the capacity configuration of the energy storage device is small. In conclusion, an accurate increase of the γ will ensure the decrement of the energy storage configuration capacity without simultaneously increasing the generation facilities. From the figure, it is appropriate to choose from the 1.1–1.2 that is more.

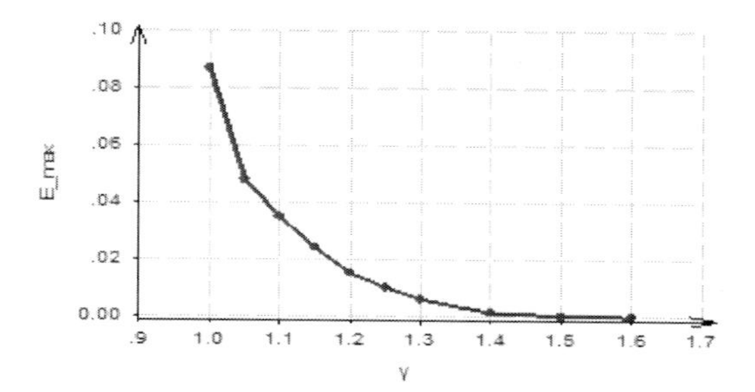

Figure 3. Optimal energy storage capacity configuration corresponds to the maximum load ratio of different new energy installed capacity.

The relation of all kinds of optimized ratio and energy storage configuration of energy corresponding to the relative maximum load ratio γ of the different new energy installed capacity is summarized in the following Table 1.

Table 1. All kinds of optimized ratio and energy storage configuration of energy corresponding to the relative maximum load ratio γ of different new energy installed capacity.

The relative maximum load ratio γ of different new energy installed capacity	Photovoltaic power generation portion (%)	Wind power generation portion (%)	Small hydropower stations portion (%)	Stored energy configuration capacity
1	7.05	43.41	49.54	0.0873
1.1	6.2	20.41	73.39	0.0352
1.2	1.92	17.19	80.89	0.0153
1.3	3.82	21.55	74.63	0.0061
1.4	3.31	27.23	69.46	0.0015
1.5	4.88	27.7	67.42	0.000392

4.3 *Comprehensive analysis of the wind, solar, and water optimal configuration*

The energy storage configuration ratio that corresponds to the relative maximum load ratio γ of the different new energy installed capacity in the different new energy power generation portion is shown in figure 4. According to the figure, we observe that the energy storage capacity decreases significantly by optimizing the configuration ratio of different new energy, with the same γ. Meeting the power distribution reliability only using single new energy such as wind, solar, small hydro, and so on needs a larger energy storage capacity. Besides, it not only increases the configuration cost of the energy storage capacity but also increases the investment of new energy power generation facilities and the waste of the redundant new energy. Therefore, the optimized project of the new energy configuration proposed in this essay can realize the optimization of new energy power generation configuration ratio and further minimize the capacity configuration ratio of the energy storage devices.

To analyze in-depth the causes of new energy power generation configuration ratio optimization, we should minimize the capacity configuration ratio of the energy storage by the method described earlier. Two kinds of new energy ratios are chosen for analysis, on the condition that the new energy installed capacity relative maximum load ratio γ=1.2, where the first ratio that

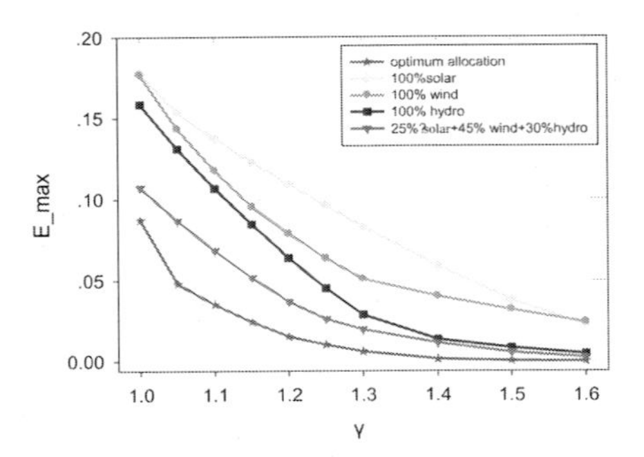

Figure 4. Energy storage configuration ratio that corresponds to the different γ of different new energy power generation configuration ratio.

is random appointed for that photovoltaic power generation is 25%, the wind power generation is 45%, and the small hydropower generation is 30% of the new energy, respectively. With the variation of the γ, the energy storage devices that needed to be configured are the pink curve and the red curve shown in figure 4. From this, we know that the new energy configuration ratio can realize the minimum of configuration energy storage capacity by using the optimization method.

The daily change of the energy storage is shown in figure 5, under the different ratio of two kinds of new energy: the upper one in which the proportion of the photovoltaic power generation, wind power generation, and small hydropower generation, respectively, is 25%, 45%, and 30% of the new energy; and the lower one is the results of the energy storage configuration on the condition that the new energy ratio is decided by the optimal project. According to the upper one, we can know that the storage devices not only fail to store enough energy but also need to discharge enough energy to meet the normal supply of the load. Therefore, before the installation of the energy storage devices, we need to charge them to meet the normal supply of load when the system running is in off-grid status. Moreover, there are some requirements about the initial capacity of the energy storage devices. Simultaneously, with the status of new energy getting better, the large redundant electricity and new energy generated that cannot be absorbed by energy storage devices will be deserted, which results in a waste of resources. But the wind, solar, water, and so on can be combined well by the project of optimizing configuration ratio of the new energy. Therefore, before the load is larger and the new energy is not so abundant (see Fig 5 6th ~130th days), we charge the energy storage devices adequately by new energy. It will also reduce the configuration of the energy storage devices. So, it can optimize the new energy ratio of the new energy installed capacity relative maximum load ratio c, and the energy storage capacity is significantly reduced. This will benefit the economic efficiency of the investment project and the reality of power supply to a great extent.

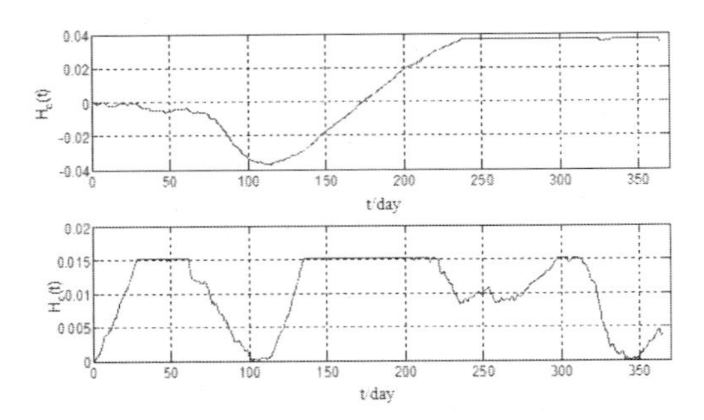

Figure 5. Storage energy changes every day with regard to two different ratios of the new energy (upper: 25%PV+45%wind power+30%small hydro, below: optimize proportion).

5 CONCLUSIONS

The optimal configuration of wind power, solar power, small hydro, and energy storage device is an important project in the pre-planning of the micro-grid.

1 The essay proposed an optimal configuration method of wind, solar, and water storage capacity. It can first meet the load requirement of the micro-grid load and comprehensively consider the influence of the charge discharge efficiency of the energy storage system and the SOC constraint. It also builds the mathematical model based on the minimum energy storage configuration corresponding to the optimal ratio of wind, solar, and water, thus solving the problem by a genetic algorithm.
2 The effect of the configuration proportion optimization of new energy installed capacity relative maximum load ratio is obvious. It also significantly reduces the energy storage configuration, and the example results verify the correctness and effectiveness of this method.
3 Based on the real new energy installed capacity relative maximum load ratio γ of the off-grid running and the new energy installed status of the example area, the new energy configuration calculated can minimize the energy storage capacity. It has guiding significance for new energy configuration planning of the micro-grid.

ACKNOWLEDGMENT

This paper is funded by the State Grid Corporation of China headquarters management technology project (52182013000V).

REFERENCES

[1] Heide D, Greiner M, et al. (2011) Reduced storage and balancing needs in a fully renewable European power system with excess wind and solar power generation. *Renewable Energy*, 36(9): 2515–23.

[2] Liu Hunhua, Chau K T, Zhang Xiaodong. (2010) An efficient wind PV hybrid generation system using doubly excited permanent-magnet brushless machine. *IEEE Transactions on industrial Electronics*, 57(3):831–839.

[3] Usa Boonbumroong, Naris Pratinthong, et al. (2009) Model-based optimization of St and a lone hybrid power system. *World Renewable Energy Congress*.

[4] Liang Youwei, Hu Zhijian, et al. (2003) Distributed generation and its application analysis in the electrical power system summary. *Power Grid Technology*, 27(12):71–75.

[5] Zhu Fang, Wang Peihong. (2009) Wind power and solar photovoltaic complementary power generation application and optimizing. *Shanghai Electric Power*, (1):23–26.

[6] Wan Jiuchun. (2003) *Ngari Prefecture Energy Application Project and Many Energy Complementary System Research*. Sichuan: Sichuan University.

[7] Yu Bo, Wang Chengshan, XIAO Jun, et al. (2012) Smooth renewable energy power generation power generation system output fluctuant energy storing system capacity optimizing method. *Proceedings of the CSEE2012*, 32(16):1–8.

[8] Zhang J, Chung H S H, Lo W L. (2007) Clustering-based adaptive crossover and mutation probabilities for genetic algorithms. *IEEE Transactions on Evolutionary Computation*, 11(3):326–335.

[9] Ma Xiyuan, Wu Yaowen, Fang Hualiang, et al. (2011) Wind solar storing mixed micro-grid power optimal configuration adopting the bacteria foraging method. *Proceedings of the CSEE 2011*. 31(25):15–25.

[10] Wang Haihua, Han Xuedong, Li Jianfeng. (2013) Application system design wind solar wood storing integration. *Electric Power Construction*, 34(8):81–6.

Electric, Electronic and Control Engineering – Shao, Shu & Tian (Eds)
© 2015 Taylor & Francis Group, London, ISBN: 978-1-138-02842-5

Analysis model and data-processing method on vertical spatial characteristics of ship-radiated noise

Yansen Liu & Xuemeng Yang
Dalian Scientific Test & Control Technology Institute, Dalian, China

ABSTRACT: This paper mainly focuses on measuring and analyzing planar-spatial characteristics of underwater radiated noise originating from ship targets in the vertical section at a specified part of the ship target under test. A corresponding mathematic model is presented based on linear array measurements. Combined with the cubic spline interpolation algorithm, a data-processing and analysis method is brought up regarding sound pressure spatial distribution and directivity in the single plane. The effectiveness of the mathematic model and data-processing method is further verified based on processing and analyzing sea experiment data. The proposed method is simple, practical, and reliable. And the analysis results can be further applied in the studies on noise source analysis, identification, and far-field feature extrapolation for underwater radiated noise.

KEYWORDS: ship radiated noise; spatial characteristics; vertical section distribution.

1 INTRODUCTION

The radiated noise characteristics of various underwater targets play a very important role in supporting the development of underwater equipment. Therefore, it is essential to carry out studies regarding underwater noise feature analysis and modeling and to acquire numerous effective noise characteristics [4-11]. Based on the previous studies[12-15], combined with cubic spline interpolation algorithm and linear array measurement technology, this paper presents a modeling method for analyzing sound pressure distribution and directivity in the vertical section at a specified part of the ship target under test. The validity of the proposed mathematic model and data-processing method is verified by a sea experiment.

2 MATHEMATIC MODEL AND ANALYSIS METHOD

2.1 *Spatial distribution model*

In this section, we will focus on the spatial directional characteristics of underwater radiated noise of ship targets in the vertical section at a specified part. Based on the data acquired by the vertical linear array, the corresponding mathematic model and the sound pressure directional distribution analysis method will be elaborated in detail. As shown in figure 1, an xyz-coordinate system is set up with the target's geometric center O as the origin. The x-axis is set in the fore-and-aft direction of the target, with the positive side pointing to the bow. The y-axis is normal to the x-axis, with the positive side pointing in the direction of the target's port. The z-axis is normal to the xOy-plane, with the positive direction pointing upward. The linear array is located at a certain distance from the target's port. At the time of data recording, the target is passing by the linear array at a constant speed v, which is equivalent to the condition of treating the target as the reference while the linear array is moving in the opposite direction at a constant speed $-v$. The resulting scenario is coincident with the concept of spatial scanning and synthetic aperture, such that the rectangular plane passed by the measurement array is defined as "the effective measurement plane." The measurement array comprises N equally spaced hydrophone units. The total length is L , and the unit's spacing distance is d. The array body is vertical with the horizontal lengthways section plane of the target under test, and the distance between the first array unit and the plane is H. The target track is parallel with the effective measurement plane, and the distance between them is D. The target track is coplanar with the effective measurement plane. Under normal measurement conditions, $x = x_b$ and $x = x_e$ represent the target's initial and terminal positions, respectively. The precise values are given by the ranging and positioning system. $|x_e - x_b|$ is considered the effective maneuver distance. $Q_{n,x}$ is the nth spatial position $(r_{n,x}, \theta_n)$ on the linear track passed by the nth hydrophone unit, n=1, 2,......, N; $x_b \le x \le x_e$, and Q_n is the spatial position (r_n, θ_n) at the vertical section of $x = x_W$. Within a certain effective measurement time interval ΔT, $p_n(t)$ is the transient sound pressure obtained in position Q_n at time t.

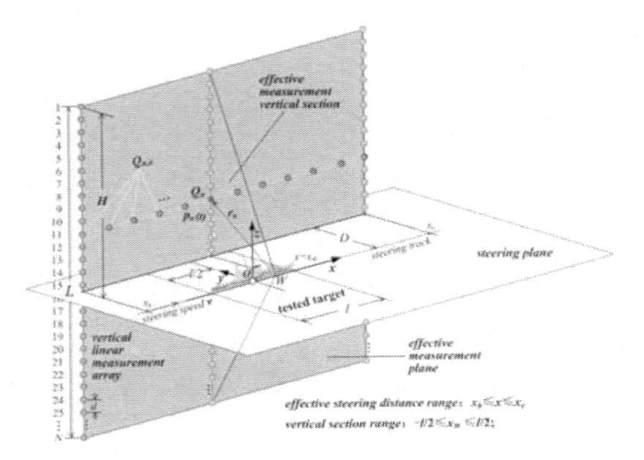

Figure 1. Mathematic model of left broadside spatial distribution of target's underwater radiated noise.

Based on the mathematical model and conditions mentioned earlier, the spatial distribution $L_{p,n}(x_W, Q_n)$ of radiated noise originating from the target under test in the vertical section at the specified part $x = x_W$ can be obtained by data processing and analysis. In addition, the corresponding spatial directional distribution $L_{p,n}(x_W, r_v, \theta_n)$ in the same plane can be further extrapolated by modifying the effect of transmission loss generated by the environment.

2.2 *Spatial distribution analysis method*

In the vertical section at the specified part $x = x_W$ of the ship target under test, the sound pressure $p_{Q_n,e}(t_o, \Delta T)$ and the corresponding sound pressure level $L_{p,n}(x_W, Q_n)$ recorded within a continuous effective measurement time interval ΔT by the nth unit at position Q_n can be calculated according to the mathematical model mentioned in section 1.1. The equations are as follows:

$$p_{Q_n,e}(t_o, \Delta T) = \sqrt{\frac{1}{\Delta T}\int_{t_o}^{t_o+\Delta T} p_n^2(t)\,dt} \quad (1)$$

$$L_{p,n}(x_W, Q_n) = 20\lg\left[p_{Q_n,e}(t_o, \Delta T) / p_{ref,w}\right] \quad (2)$$

where t_o is the initial time of measurement within ΔT. $p_{ref,w} = 1\mu Pa$ is the reference sound pressure value in underwater acoustics.

According to Equations (1) and (2), several sound distribution parameters can be yielded regarding different frequency analysis requirements, such as sound pressure level (SPL) in a selected frequency band (eg. 1/1oct & 1/3oct), sound pressure spectrum level, and broadband sound pressure level (total SPL).

In the vertical section at the specified part $x = x_W$ of the ship target under test, as shown in figure 2, the spatial coordinate parameters of the point Q_n can be calculated as follows:

$$x = x_W,\ y = D,\ z = H - d(n-1)\ (n=1, 2, \ldots\ldots, N) \quad (3)$$

$$r_n = \sqrt{D^2 + [H - d(n-1)]^2},\ \theta_n = 90^\circ - \tan^{-1}\{[H - d(n-1)]/D\} \quad (n=1, 2, \ldots N) \quad (4)$$

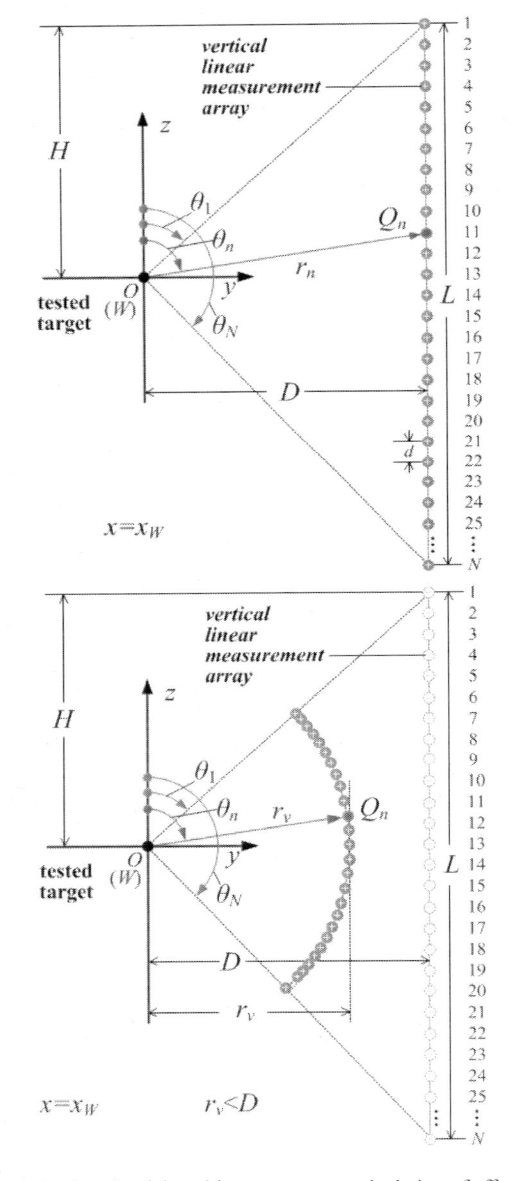

Figure 2. Spatial position parameters calculation of effective measurement vertical section of ship's hull.

As shown in figure 3, the radiated sound pressure $L_{p,n}(x_W, Q_n)$ in the vertical section at the specified part $x = x_W$ of the ship target under test is calculated definitely according to the earlier-mentioned mathematical model, leading conditions, spatial position parameters, and measurement value of radiated noise.

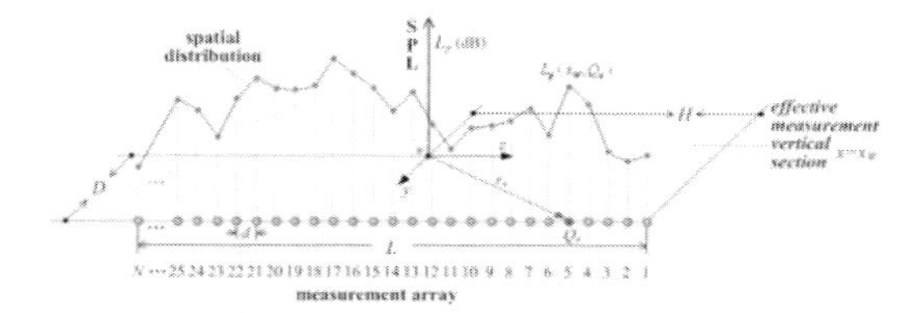

Figure 3. Spatial distribution analysis of underwater radiated noise of single vertical section of ship's hull.

At the time of data recording, the range of the sound pressure distribution plane is determined by the effective maneuver distance. The spatial resolution is co-decided by the number of array units and data analysis constraints, which may result in the problem of spatial sampling deficiency. Therefore, in this section, we propose to apply the cubic spline interpolation algorithm and the 2nd class boundary condition to improve the spatial resolution and geometric smoothness of the distribution curve. Based on calculations, we can obtain the relationship between SPL $L_{p,n}(x_W, Q_n)$ and coordinates z at the corresponding position Q_n, which is given in the following function:

$$L_{p,n}(x_W, Q_n) = L_p(z_n),\ z_n = H - d(n-1)\ (n=1, 2, \ldots\ldots, N) \quad (5)$$

As shown in figure 4, under the 2nd class boundary condition, according to the cubic spline interpolation theory and algorithm, the interpolation functions $S(\gamma)$ ($[a,b]$, $a = z_N = H - d(N-1)$, $b = z_1 = H$) of spatial distribution $L_p(z_n)$ for radiated noise in the vertical section at the specified part $x = x_W$ of the target under test can be established as follows:

$$S(\gamma) = M_k \frac{(\gamma_{k+1} - \gamma)^3}{6h_k} + M_{k+1} \frac{(\gamma - \gamma_k)^3}{6h_k} + \left[L_p(\gamma_k) - \frac{M_k h_k^2}{6}\right] \frac{\gamma_{k+1} - \gamma}{h_k} + \left[L_p(\gamma_{k+1}) - \frac{M_{k+1} h_k^2}{6}\right] \frac{\gamma - \gamma_k}{h_k}$$

$$\gamma \in [\gamma_k, \gamma_{k+1}]\ (k=0,1,\ldots,N\text{-}2) \quad (6)$$

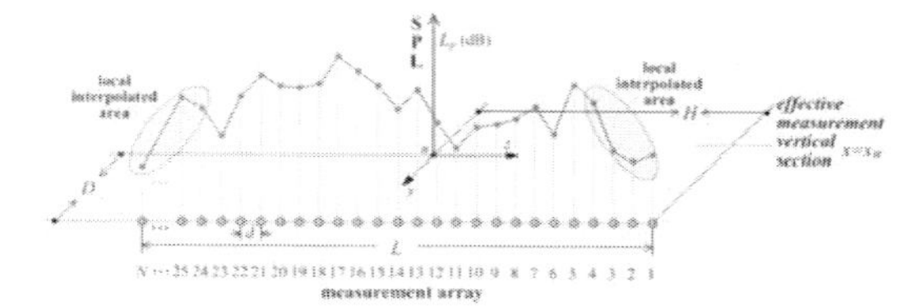

Figure 4. Calculation of second-class boundary condition of cubic spline interpolation algorithm for underwater radiated noise spatial distribution of single vertical section of ship's hull.

2.3 *Directional analysis method*

According to the attenuation rule of radiated noise, the spatial directional distribution $L_{p,n}(x_W, r_v, \theta_n)$ and directional index $D(x_W, r_v, \theta_n)$ in the vertical section at the specified part $x = x_W$ of the target under test are extrapolated by modifying the effect of environment transmission loss, $r_v \leq D$, as shown in figure 5. Therefore, the corresponding radiated sound pressure level and directional index are given, respectively, as follows:

$$L_{p,n}(x_W, r_v, \theta_n) = L_{p,n}(x_W, Q_n) + A \cdot \lg\left(\frac{r_n}{r_v}\right)\ (n=1, 2, \ldots N) \quad (7)$$

$$D(x_W, r_v, \theta_n) = L_{p,n}(x_W, r_v, \theta_n) - \max\{L_{p,n}(x_W, r_v, \theta_n), n = 1, 2, \cdots, N\} \quad (8),$$

where constant A is the correctional coefficient with transmission loss of the environment, and the $\max\{L_{p,n}(x_W, r_v, \theta_n), n = 1, 2, \cdots, N\}$ is the local maximum value of the sound pressure level extrapolated by the measurement value of radiated noise in the range of effective measuring angle.

If the numerical value of parameter r_v is specified, Equations (7) and (8) can be also simplified as follows:

$$L_p(\theta_n) = L_{p,n}(x_W, Q_n) + A \cdot \lg\left(\frac{r_n}{r_v}\right)\ (n=1, 2, \ldots\ldots, N) \quad (9)$$

$$D(\theta_n) = L_p(\theta_n) - \max\{L_p(\theta_n), n = 1, 2, \cdots, N\} \quad (n=1, 2, \ldots, N) \quad (10)$$

The interpolation function $\Gamma(\theta)$($[\theta_1 = 90^\circ - \tan^{-1}(H/D)$, $\theta_N = 90^\circ - \tan^{-1}\{[H - d(N-1)]/D\}]$) for spatial directional distribution $L_p(\theta_n)$ of underwater radiated noise of the target under test is also established based on the second-class boundary condition, and that is

$$\Gamma(\theta)=M_{\theta,k}\frac{(\theta_{k+1}-\theta)^3}{6h_{\theta,k}}+M_{\theta,k+1}\frac{(\theta-\theta_k)^3}{6h_{\theta,k}}$$
$$+\left[L_p(\theta_k)-\frac{M_{\theta,k}h_{\theta,k}^2}{6}\right]\frac{\theta_{k+1}-\theta}{h_{\theta,k}}$$
$$+\left[L_p(\theta_{k+1})-\frac{M_{\theta,k+1}h_{\theta,k}^2}{6}\right]\frac{\theta-\theta_k}{h_{\theta,k}}$$

$$\theta\in[\theta_k,\theta_{k+1}]\,(k=1,2,\ldots,N\text{-}1) \tag{11}$$

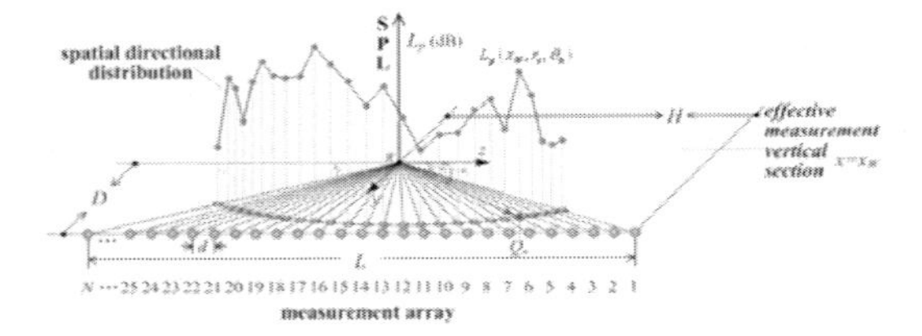

Figure 5. Spatial distribution directional analysis of underwater radiated noise of single vertical section of ship's hull.

3 EXPERIMENTAL RESEARCH AND ANALYSIS

In this section, the correctness and feasibility of the earlier-mentioned model and method is verified by processing and analyzing sea experiment data. The measuring target is a surface ship with a total length l=56m. The underwater wet end device for data acquisition is a linear acoustic array with 9 equally spaced hydrophone units. The distance between adjacent units is d=6m. The target is located above the linear array with a vertical distance of D=12m. At the time of data recording, the target is moving horizontally at a constant speed of v=7kn. The geometric configuration of the target and the linear array is indicated in figure 6. The underwater radiated noise recorded by different array units is shown in figure 7.

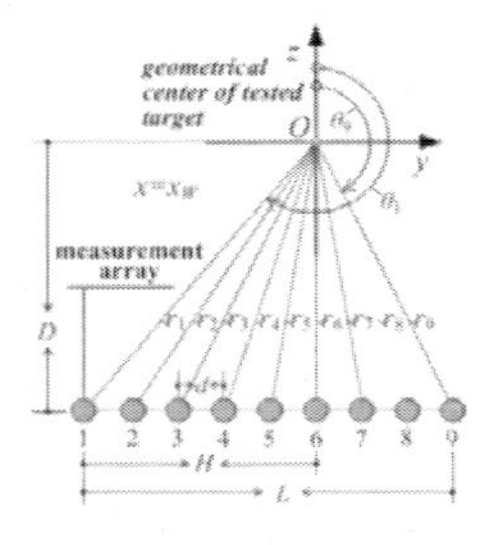

Figure 6. Geometrical configuration of tested target and measurement array.

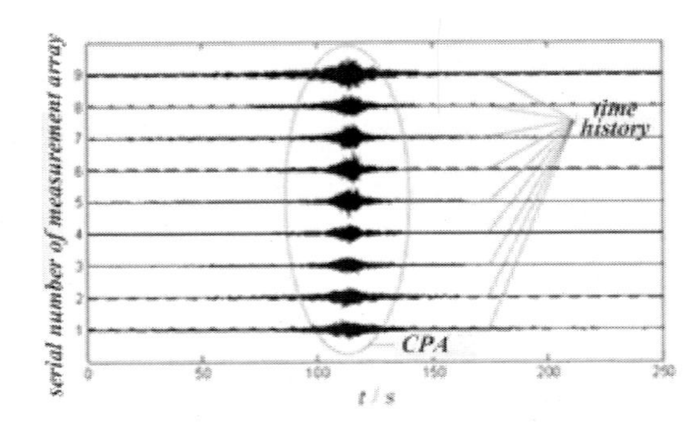

Figure 7. Time history of underwater radiated noise received by different unit of linear array.

According to power spectrum analysis results, the radiated noise level of the measured target under the measurement condition is relatively high, and its energy is dominant below 1kHz. The line spectrum components are fairly rich with the magnitude standing out at 123Hz, 222Hz, and 618Hz. Based on the frequency distribution characteristics yielded earlier, the sound pressure distribution characteristics within the effective measurement plane will be further explored according to the mathematic model and analysis method proposed. Figure 8(a) shows the measured result and the corresponding cubic spline interpolation analysis result of the spatial sound pressure distributed in the specified vertical section below 5kHz broadband frequency interval, and the analysis result of line spectrum components at 123Hz is shown in figure 8(b).

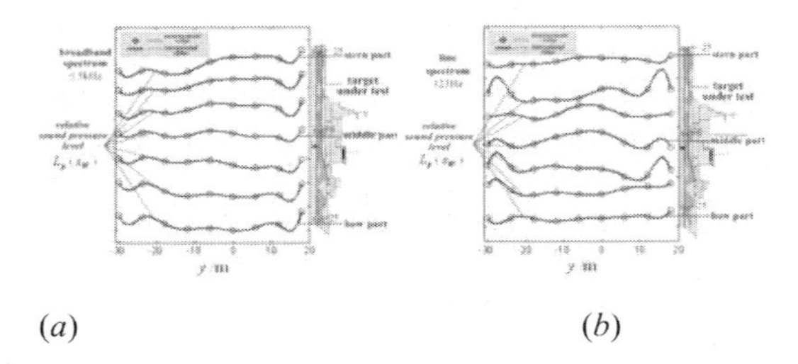

(a) (b)

Figure 8. Spatial distribution of wide frequency band and line spectrum for underwater radiated noise of different vertical section of tested target.

Figure 9 shows the measured result and the corresponding cubic spline interpolation spatial directional analysis result of the sound pressure distributed in the vertical section at a specified part (including stern, middle, and bow) below the 5kHz broadband frequency interval.

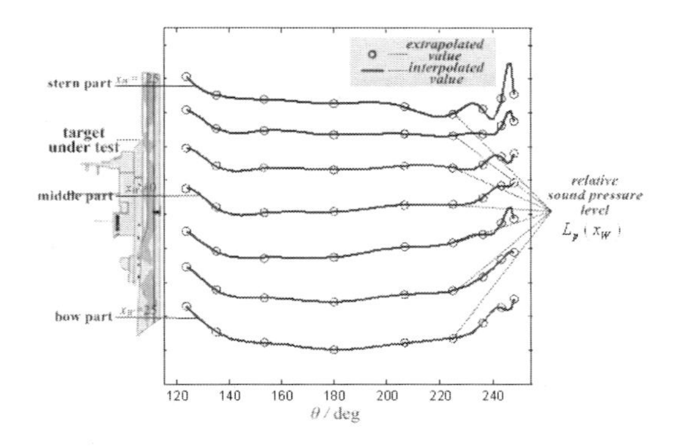

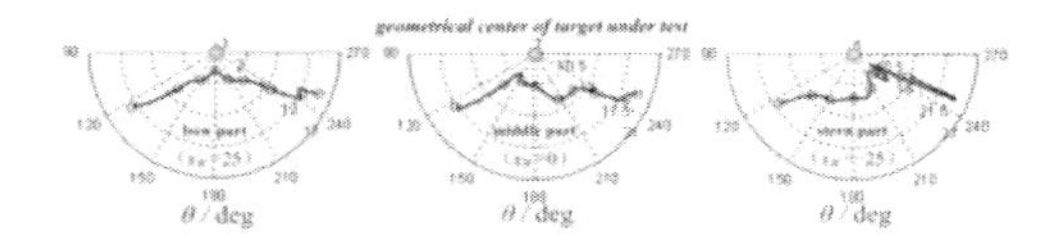

Figure 9. Spatial directional distribution of wide frequency band for underwater radiated noise of typical vertical section of tested target.

4 CONCLUSIONS

To solve the problem of directional characteristics measuring and analyzing for underwater radiated noise of ship targets in the vertical section, this paper provided a mathematic model based on linear array measurements. According to the cubic spline interpolation theory and algorithm, a mathematical analysis model and data-processing method was established to describe the spatial distribution and directional characteristics in the vertical section at the specified part. The effectiveness and robustness of the proposed method was testified by data acquired on the sea. Overall speaking, the mathematical model and the data analysis method used in this paper have the following two important features: (1) Simple, practical, and robust. The mathematical model for underwater radiated noise spatial distribution of ship targets is established easily based on the target's azimuth mode, the type of wet end to acquire data, and the number of its unit array. The spatial distribution and direction results with different frequency and wideband range in the effective measurement vertical section can be given in detail, and the rules and features of relevant spatial distribution can be analyzed. (2) The analysis results can be further applied in the studies on noise source analysis, identification, and far-field extrapolation under different sound transmission conditions. (3) The aforementioned model and analysis method can be also applied to measurement, analysis, and evaluation of underwater radiated noise for other types of underwater and surface targets.

REFERENCES

Robert J. Urick, (1985) *Principles of Underwater Sound*. Harbin: Harbin Engineering University Press, pp:263–264.

Paul Scrimger, Richard M. Heitmeyer, (1991) Acoustic source-level measurements for a variety of merchant ships, *JASA*. (2):691–699.

Liu Bosheng, Lei Jiayu, (1993) *Underwater Acoustical Principle*, Harbin: Harbin Engineering University Press, pp:224–232.

Liu Xun, Xiang Jinglin, Zhou Yue, et al., (2000) Research on longitudinal of the radiated noise of distribution characteristics a ship as a volume object, *Journal of Northwestern Polytechnical University*, 18(3):409–412.

Xiang Jinglin, Liu Xun, (2002) The longitudinal distribution of source intensity spectrum of ship-radiated noise, *Journal of Detection & Control*, 24(2):5–17.

Wang Zhicheng, Chen Zongqi, Yu Feng, et al., *Measurement and Analysis for Radiated Noise Vessels*, Beijing: National Defense Industry Press, pp:20–24.

Zou Chunping, Chen Duanshi, Hua Hongxing, (2004) Study on characteristics of ship underwater radiation noise, *Journal of Ship Mechanics*, 8(1):112–124.

Mark A. Hallett, (2004) Characteristics of merchant ship acoustic signatures during port entry/exit, *Proceedings of acoustics, Gold Coast, Australia*, pp:577–580.

Ignacy Gloza, (2004) Transmission of acoustic energy from ships into water environment, *ICA2004*, pp:1543–1546.

Eugeniusz Kozaczka, (2007) Identification of hydro-acoustic waves emitted from floating units during mooring tests, *Polish Maritime Research*, 14(4): 40–46.

Mark V. Trevorrow, Boris Vasiliev, (2008) Directionality and maneuvering effects on a surface ship underwater acoustic signature, *JASA*, 124(2):767–778.

Zhou Fei, Zhang Mingzhi, (2009) An analysis method for horizontal spatial distribution of vessel radiated noise, *Vessel Science and Technology*, 31(1).

Luo Xuefeng, Zhang Mingzhi, (2009) An analysis method for horizontal spatial distribution of vessel radiated noise based on linear array, *Vessel Science and Technology*, 31(7).

Yang Xuemeng, Liu Yansen, Du Peng, (2012) Analysis and discussion on vertical directivity of ship radiated underwater noise, *Noise and Vibration Control*, 32(5):72–75.

Yang Xuemeng, Liu Yansen, Du Peng, (2012) Measurement and analysis of the spatial distribution of underwater noise radiated from tugboat, *Technical Acoustics*, 31(6): 611–614.

Electric, Electronic and Control Engineering – Shao, Shu & Tian (Eds)
 ISBN: 978-1-138-02842-5

Research of measuring technology of pulse current measurement based on embedded systems

Yongbo Yang, Xin Chen, Haiou Yan, Jia Sun, Maozhou Wang, Xiaoliang Wang, Xueyong Ai, Xiaofei Huang & Yang Gao
State Grid Zhengzhou Electric Power Supply Company, Zhengzhou, China

Xiangyong Yang
Shenzhen Power Supply Bureau Co Ltd, Shenzhen, China

ABSTRACT: Using virtual instrument technology, this paper designed a pulse current measurement based on embedded computer systems, which uses the embedded computer as the processing core, Rogowski coil as the signal sensor, high-speed data acquisition module to achieve signals, and Visual Basic to develop human-machine interface software. The measurement has functions of system settings, waveform acquisition and display, data processing, database management, and report generation. With the two-pulse current signal representative of the measurement, the measuring instrument with the Tektronix TDS2012 makes a comparison to meet the high voltage test technical GB/T in the measurement system of the pulse current requirements. Research has practical and academic values.

KEYWORDS: Embedded systems; Pulse current; Virtual instrument.

1 INTRODUCTION

Impulse has the characteristics of a high voltage, high current. In modern science, high technology, and scientific research in the field of military defense, it has important scientific significance and application value, and its application is gradually expanded to industrial, civilian areas. Impulse measurement technology is one of the key issues of the impact of current experiments[1,2].

In order to ensure the efficient and stable operation of the pulse experiments, we need to accurately and reliably measure the discharge signals, including the impact of the current waveform amplitude, rise time, and pulse width waveform parameters. There are some methods of measurement of the pulse signal, such as high-voltage electronic oscilloscope, digital storage oscilloscope, the peak table method, and so on. High-voltage electronic oscilloscope waveform can only be recorded by imaging, which is too cumbersome, time consuming, and has low accuracy. The DSO needs to copy data to a dedicated computer for secondary treatment. Heafely company's digital recorder Heafely DMI 551 with the maximum sampling rate of 120MHz[3,4] cannot measure the pulse signal waveform display, and it must be stored and processed with IPC.

It is necessary to develop a kind of dedicated instrument with functions of waveform acquisition and display, data processing, data storage, database management, and report generation.

2 SYSTEM DESIGN PRINCIPLE

The system consists of signal acquisition module, high-speed acquisition modules, embedded host processing module, and touch-screen peripheral circuit. It is commonly used in the circuit in the series shunt method to measure impulse current. However, when the string into the shunts has difficulties, only non-contact measurement methods can be used. The system uses Rogowski coil signal acquisition module, which is a measure of the impact of the high current transformer, characterized by a simple structure, without an electricity connection between the measured signal and the circuit under test, and it has a wide measurement range. High-speed data acquisition module using FPGA and high-speed A/D as the core communicates high-speed sampling of signals with an embedded motherboard. The embedded host system, as the control and processing center, has a powerful computing capability, a graphical environment, and storage capabilities. Therefore, we need to establish a virtual instrument panel with a good performance of human-computer interaction, to complete the transfer of data, processing, storage, analysis, display, and other functions. System components are shown next:

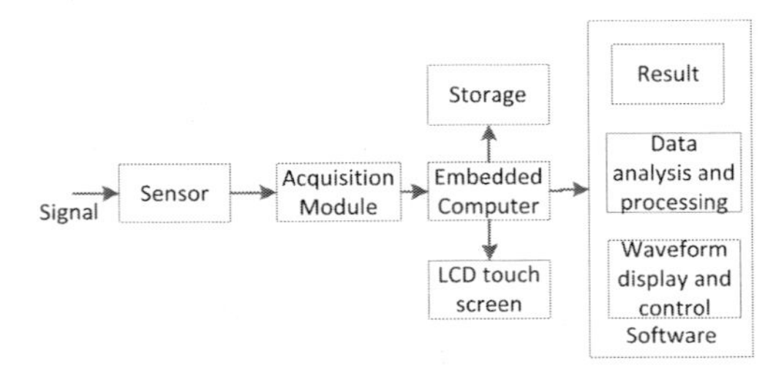

Figure 1. Overall block diagram of system.

2.1 *Embedded computer*

Reliability and the ability to intelligently control the embedded control system is the main goal of technology development of embedded computer systems. In addition to the intelligent design of the embedded computer control, often with strict limits and requirements, we also have the following: small size, light weight, and low power consumption; safe and reliable; adaptability in harsh environments; and sensitivity to costs. Therefore, the system must be designed according to the needs and preferred choice of the embedded computer.

The embedded computer dedicated ultra-low-power embedded CPU - AMD Geode LX 800, with the operating frequency till 500MHz. Schematic embedded computers are as follows. By AMD GEODE CS5536 auxiliary controllers, the embedded computer interface design has an interaction with the outside world, including hard disk interface, USB interface, PCI BUS, Ethernet port, COM port, and Flash BIOS interface.

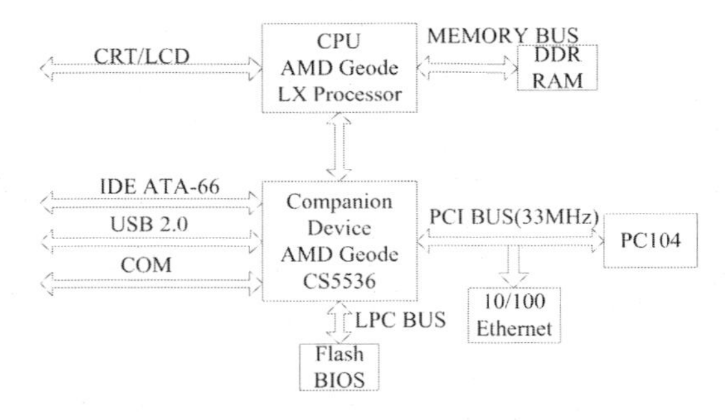

Figure 2. Schematic of embedded computer.

Embedded systems that use Windows XP Embedded are easy-to-use data acquisition card drivers and develop control programs and software handlers.

2.2 *Acquisition module*

Basic features of the acquisition module include data synchronization acquisition, range control, time base control, trigger control, and LAN mode of communication module. Data collected in the presence of SDRAM, USB communication port, or via LAN communication port SDRAM write control registers to read data.

The basic principle of the acquisition module is shown next:

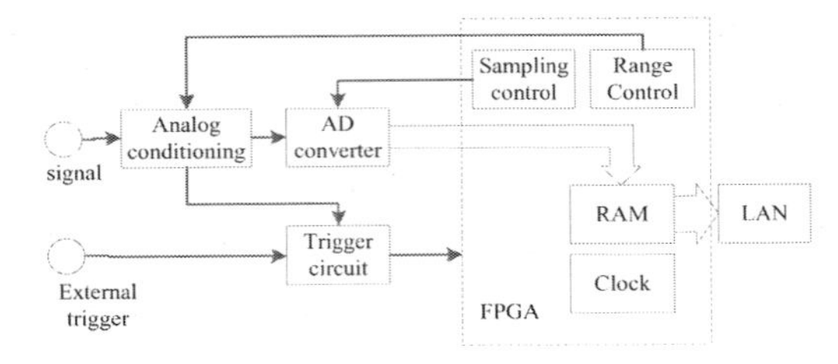

Figure 3. Functional block diagram of acquisition module.

An acquisition module with a sampling rate of high-speed 200Msps has a sampling accuracy of 10 and an Ethernet data bus. Application of optical isolation module isolation measuring instruments and pulse signal source to avoid high voltage signal by measuring the channel conduction to the embedded computer ensures the safety of the measuring system and the operator.

2.3 *Software design*

Since the data acquisition module is unified, system software design is very important. System software is a modular program design that is easy to update while helping extend the program functionality. The entire system software is divided into three modules: acquisition module, processing module, and storage module.

The collection function module aims at controlling the acquisition card and returns to the data to the embedded computer. The processing module includes drawing with the resultant data, thereby displaying the waveform; for data processing, calculating waveform parameters, including amplitude parameters, time parameters; and analysis of the collected signal to the greatest degree of the reduction wave. The data processing module includes a display waveform, a waveform scaling, measurement, and display parameters (including peak lightning wave, the rise time, the half-peak time, the peak value of the square wave, duration, and other parameters).

We need to design a data storage module, using ADO database technology, SQL database management language, to control Access database; include the original data stored directly in real time, waveform picture storage, and measurement results storage; and use Visual Basic to operate two types of databases, Excel and Access. Data acquisition cards will be collected separately and will output the result to Excel. The result of the calculation output is sent to Access. Simultaneously, reporting features are designed to facilitate the transfer of historical data and printing.

2.4 *EMC design*

GB / T 17626.5-1999, electromagnetic compatibility test, and measurement technology provide an interference test (impact measurement system): The test is carried out with the maximum operating voltage or current; the magnitude of the measured interference should be less than 1% of the measuring system output voltage or current.

In the impact of the current measurement, a high-frequency pulsed interference signal, by the measuring circuit, is coupled to the acquisition module, thus affecting the accuracy of the measurement results. In order to improve the electromagnetic compatibility characteristics measurement system, we need to study the impulse current electromagnetic environment and electromagnetic shielding measures [6].

The system uses several electromagnetic compatibility measures: (1) power supply module using the isolation transformer and surge protection devices; (2) measurement module with a surge protection device; (3) transfer process using an optical isolator; (4) the sensor using a Rogowski coil; and (5) the further use of the software anti-jamming measures. With the use of these measures, it has greatly improved the electromagnetic compatibility characteristics of the system, improving the reliability of the whole system.

3 SYSTEM TEST AND ANALYSIS

To make accurate and reliable measurement results, we first need to describe the standard device used for assessment by the TEK TDS2012 oscilloscope digital oscilloscope. The uncertainty of the scope is ± 0.5%.

Measured 8/20μs standard lightning current impacts the index wave generator. The generator produces a 4KA impulse current, converting the means for collecting ratio of 200: 1 Rogowski coil. It converts the signal to the measuring system and the oscilloscope with a voltage that can be collected directly. By software processing, results are displayed on the LCD screen. The measurement results were as follows:

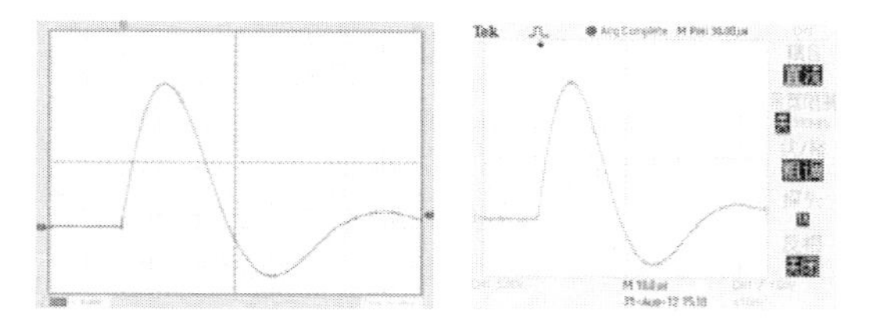

Figure 3. Impulse current measurement results (a---system, b---TDS2012).

Based on the analysis of these figures, we can see that this system can measure and display the impact of the 8/20 current generator. Analyzing the data, the peak of the system tested is 21.472V, whereas the Tektronix TDS2012 peak measured is 21.8V. The available peak error is below -1.5%. Repeating measurements of the system, recording ten groups pulse peak measurements, record and processing are as follows:

Table 1. Comparison of current peak measuring impact

Time	1	2	3	4	5	6	7	8	9	10
SYS	22.118	22.118	21.804	20.863	21.02	21.178	21.648	22.025	21.49	21.177
TDS2012	22.4	21.6	22	20.6	21.2	21.2	21.4	22.2	21.2	21

The system measured the peak amplitude of the measured parameters compared with the Tektronix TDS2012, available peak error of ±1.5%, and the standard oscilloscope has a better consistency.

4 CONCLUSION

In this paper, using virtual instrument technology, it designed a pulse current measurement based on embedded computer systems, with functions of waveform acquisition and display, data processing, database management, report generation, and good electromagnetic compatibility characteristics to meet the pulse measurement environment anti-electromagnetic interference, high reliability requirements. On comparing the measuring instrument with the calibrated Tektronix oscilloscopes, the measurement results were as follows: peak error of ± 1.5%, the error time parameters of ± 1%. Meeting the requirements of the impulse measuring system, the research results have both practical value and academic value.

This paper conducted a pulse signal tester development work. In the future, to further improve the pulse signal tester control function, complete communication should be conducted with the PLC, in order to achieve automatic measurement.

REFERENCES

[1] Zheng Jianyi, He Wen. (2008) Pulsed power technology research status and development trend overview. *Mechanical and Electrical Engineering*, 04: 1–4.

[2] Yaacob M. M., Sin L. Y., Aman A. (2010) A new polyvinyl chloride cable insulation using micro and nano filler materials. *Power Engineering and Optimization Conference, 4th International. IEEE, 2010*, pp:221–225.

[3] Hassan Rahman. (2008) Generation of High Voltage Impulse and Test on Insulator (Plywood). pp:41–45.

[4] Ramboz John D. (1996) Machinable Rogowski coil, design, and calibration. *Instrumentation and Measuremen, IEEE Transactions on*. 45(2):511–515.

Electric, Electronic and Control Engineering – Shao, Shu & Tian (Eds)

ISBN: 978-1-138-02842-5

Application research of improved PSO algorithm in BLDCM control system

Panfeng Yan, Yelin Hu, Xiaoliang Zheng, Chun Yin, Haigang Zang, Yiwei Tao & Gaolei Li
Electronic and Information Engineering College, Anhui University of Science & Technology, Huainan, China

ABSTRACT: With the BLDCM's speed control system with time-varying, nonlinear, and strong coupling characteristics, using the traditional PID control, it is hard to obtain satisfactory results. Based on the improved PSO algorithm, this paper presents a PID control strategy that is implemented at the control of BLDCM. The Matlab/Simulink simulation results show that using the control system-introduced shrinkage factor PSO algorithm PID controller for BLDCM exhibits better dynamic and static performance than the traditional PID control method.

KEYWORDS: PSO algorithm; shrinkage factor; Matlab/Simulink simulation.

1 INTRODUCTION

The BLDCM makes a quick response; has a wide speed range, a larger starting torque, and many other advantages. It has wide application in various fields. The BLDCM speed control system, with time-varying, nonlinear, and strong coupling characteristics of the traditional PID control strategy, cannot achieve the system's precision in terms of static and dynamic performance requirements. [1-2]

Built on the problems cited earlier, this paper presents an improved PSO algorithm PID control strategy. By introducing the shrinkage factor in the standard PSO algorithm for optimization of the parameters, the parameter optimization can improve the process of searching for speed. Meanwhile, in the stable operation of the process, when subjected to external disturbance, the system can quickly and smoothly return to a stable state. Through this control strategy simulation, the results show that in the case of load change, the kind of control strategy has fast response, tracking ability, high control accuracy, and excellent adaptability characteristics.

2 PARTICLE SWARM OPTIMIZATION

2.1 *Particle Swarm Optimization(PSO). [3]*

In 1995, the particle swarm optimization (PSO) algorithm was proposed by Dr. Kennedy and Dr. Eberhart for studying the foraging behavior of birds. The researchers found that the birds in flight will often change direction, spread, or gather. So their behavior is unpredictable. However, throughout the process, to maintain good overall consistency, a suitable distance is kept between individuals.

The basic idea of the PSO algorithm is as follows: The solution of the optimization problem is called a "particle"; each particle moves in the multi-dimensional search space flight at a constant speed. Through the fitness function, we measure the merits of each particle; the particles move according to their own experiences and with other particles flying through the sharing of flying experience, they dynamically adjust their flight speed according to the group of the best flight particle position and, ultimately, the search, to find the optimal solution of the optimization problem.

The standard PSO algorithm is as follows: Assuming a D-dimensional search space, there is a community of M particles; groups of particles in i positions in the t generation are represented as D-dimensional vectors $x_i^t = (x_{i1}^t, x_{i2}^t, \cdots, x_{id}^t)$. The i particle velocity is given in $v_i^t = (v_{i1}^t, v_{i2}^t, \cdots, v_{id}^t)$. The i particle best position is presented as $p_i^t = (p_{i1}^t, p_{i2}^t, \cdots, p_{id}^t)$. So far, the best place searches for $p_g^t = (p_{g1}^t, p_{g2}^t, \cdots, p_{gd}^t)$. Particle velocity and location update follow the following formula:

$$v_{id}^{t+1} = w \bullet v_{id}^t + c_1 \bullet rand() \bullet (p_{id}^t - x_{id}^t) + c_c \bullet rand() \bullet (p_{gd}^t - x_{id}^i) \quad (1)$$

$$x_{id}^{t+1} = x_{id}^{t} + v_{id}^{t+1} \quad (2)$$

Among them, t is the number of iterations for the particle update. The acceleration factor is the expression of c_1 and c_2. The *rand*() is the two uniformly distributed number within the range of [0,1]. With $v_{id} \in [-v_{max}, v_{max}]$, v_{max} as a user to set a constant, the purpose is to reduce the likelihood of particles leaving the search space during evolution.

2.2 *Improved particle swarm optimization*

In accordance with the updated formula PSO algorithm, pertaining to its global and local properties, there are two main parameters: One is the inertia weight, and the other is called "learning factor." In the renewal of each particle's speed and position, not only the information of its own individual extremum and global extremum is considered but also the information of other particles is considered. By means of changing inertia weight, the diversity of the particle swarm can be retained. In order to resolve the problems, we can more effectively control the particle speed. The strategy adopted is to introduce a shrinkage factor of PSO to enable the algorithm to achieve an effective balance between global exploration and local search as soon as possible.

$$v_i = \lambda(v_{id} + c_1 \bullet rand() \bullet (p_{id} - x_{id}) + c_2 \bullet rand() \bullet (p_{gd} - x_{gd}) \quad (3)$$

where $\lambda = \frac{2}{|2 - C - \sqrt{C \bullet C - 4 \bullet C}|}$, $C = c_1 + c_2$ and C>4 (4)

On comparing inertia weight, the shrinkage factor can be considered a more effective restraint of particle speed. Although the control target is disturbed, the dynamic response of the system will be greatly improved, thus enabling the system to more quickly reach a steady operating state.

3 PID CONTROL STRATEGY BASED ON IMPROVED PSO ALGORITHM

3.1 *Traditional PID control strategy*

The conventional analog PID control system consists of a traditional PID controller and controlled objects. The deviation of the PID control system is made up of the given value and the actual value of the difference between the outputs.

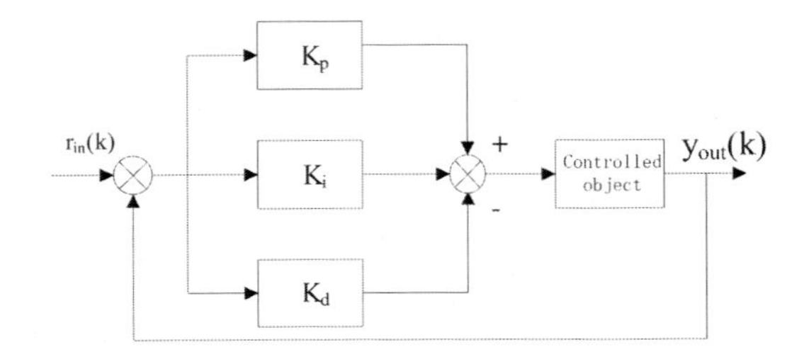

Figure 1. Conventional PID controller structure diagram.

The relationship between the controller input and output is expressed as follows:

$$y_{out}(k) = K_p \left[e(t) + \frac{1}{T_i} \int_0^t e(t) + T_d \frac{de(t)}{dt} \right] \quad (5),$$

where K_p is the proportional coefficient, T_i is the integral coefficient, and T_d is the differential coefficient.

3.2 *Improved PSO algorithm's PID control strategy*

3.2.1 *Control principle*

The PID control strategy based on the improved PSO algorithm involves the use of the self-learning ability of PSO to improve the search capabilities of particles by introducing a shrinkage factor. Through the self-learning, adaptive parameters, we should find the optimum value, so as to achieve the best adjustment results.

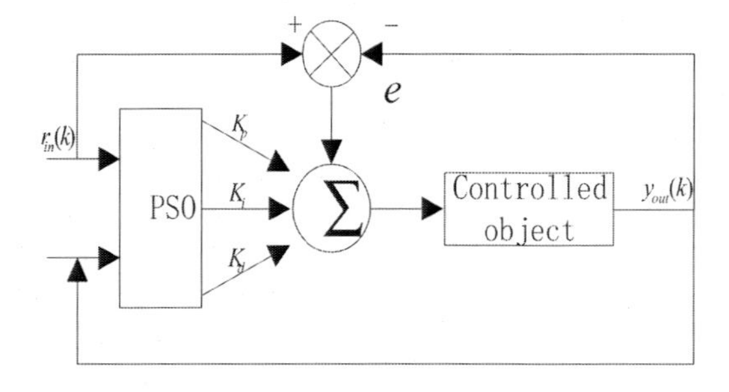

Figure 2. Introduces shrinkage factor PSO algorithm's PID controller structure.

3.2.2 *Algorithm flow*

The improved PSO algorithm's PID control strategy implementation steps are as follows:

1 **Step1:** PSO is initialized to determine the scale population.
2 **Step2:** According to the evaluation of the fitness $f(x)$ function of each particle.

3 **Step3:** Update particle's position.
4 **Step4:** Use of (1), (2) and (3) formula updating the particle.
5 **Step5:** If the end conditions are met, we have the end of operation, the output value; if the conditions are not met, we need to return till step 2 continues until it meets the conditions for the termination of the operation.

4 SIMULATION AND ANALYSIS[4-5]

This design of the PID controller is applied to the BLDCM control system. The actual parameters of the simulation are as follows: The number of pole pairs is $p = 4$, the rated voltage is UN=36V, the phase resistance is R=0. 85Ω, the moment of inertia $J = 5.0 \times 10^{-3} kg \bullet m^{-2}$, the inductance is $L - M = 5.7mH$, and the back electromotive force is $K_e = 0.072 V/rad \bullet s^{-1}$.

The relevant parameters of the PSO in the kinds of control strategy are selected. $D = 3$. $M = 30$, the inertia weight is $w = 0.6$. The most iteration number is 600 times. $c_1 = 2.8$, $c_2 = 1.3$, According to formula (4), shrinkage factor can be calculated: $\lambda = 0.73$. $K_p \in [0.01, 1]$, $K_i \in [0.01, 1]$, $K_d \in [1, 10]$.

Based on Matlab/Simulink, we build the BLDCM control system model. We are using the traditional PID control and improved PSO algorithm PID control, through the merits of the two comparison tests between the simulation results of the control method in the case of mutation load to observe which one method enables system stability.

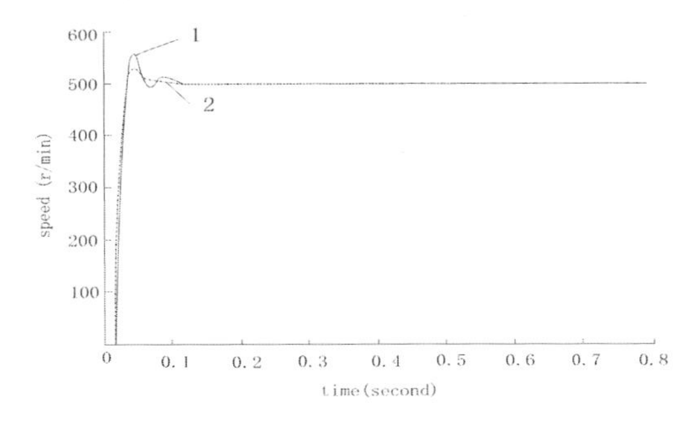

Figure 3. Two kinds of no-load control system simulation diagram.

1–Traditional PID control system simulation. 2–PSO algorithm is introduced shrinkage factor PID control system simulation.

We can observe the initial state of the system-introduced shrinkage factor PSO algorithm PID control system overshoot miniature from the earlier figure. And it can reach a steady state in a very short time. Compared with the traditional PID control system, this method has a good control effect.

When the system is disturbed: The no-load system moves into a stable state after startup, in $t = 0.4s$, suddenly applying a load of $2.5N \bullet m$ perturbation, and when the load is instantly removed, the system re-enters the no-load condition. Figure 5 shows a simulation diagram when the two control strategies result in system disturbances.

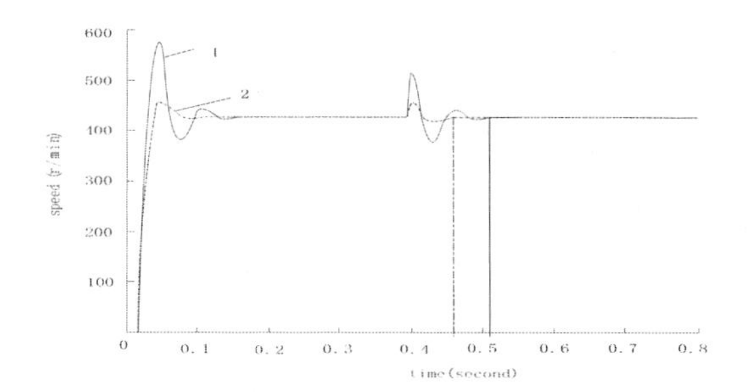

Figure 4. Two kinds of control modes add the disturbance simulation diagram. 1–traditional PID control system simulation. 2–PSO algorithm is introduced shrinkage factor PID control system simulation.

From the diagram of the system simulation seen in figure 5, the traditional PID control disturbance has a relatively poor dynamic response. About 0.51 seconds after the system reached re-stabilization, there was a large amount of overshoot. The shrinkage factor is introduced using the PSO algorithm PID controller tuning parameters, since the dynamic response has been significantly improved. The system takes around 0. 46 seconds to reach the steady state again and has a less overshoot.

5 CONCLUSIONS

This article will present the shrink factor into the PSO algorithm and the traditional PID control strategy by combining better achieved BLDCM speed control. Design experiments in the motor are characterized by no load and stable operation of the sudden perturbations load. The system can more smoothly and quickly reach a steady state and obtain satisfactory results. The Matlab/Simulink simulation is shown, in which the introduction of the contraction factor PSO Algorithm PID controller BLDCM control system has a smaller overshoot, quick response, no steady-state error, and so on. The BLDCM control system has good dynamic, static performance, robustness, and application value.

REFERENCES

[1] Song Hailong, Yu Yong, Yang Ming *et al.* (2002) A hybrid adaptive fuzzy variable structure speed controller for brushless DC motor. *IECON 02: Industrial Electronics Society, Sevilla, Strain.*

[2] Awadallah M A, Morcos M M. (2002) Adaptive-fuzzy-based stator-winding fault diagnosis of PM brushless DC motor drive by monitoring supply current. *Power Engineering Review*, 22(12): 46–49.

[3] Tang, Jun, Zhao Xiaojuan. (2009) An enhanced opposition based particle swarm optimizations. *2009 Global Congress on Intelligent Systems (GCIS 2009)*. Xiamen, China: [s. n.], pp:149–153.

[4] Ji Zhicheng, Shen Yanxia, Jiang Jianguo. (2003) A novel method for modeling and simulation of BLDC system based on Matlab. *Journal of System Simulation*, 15(12):1745–1749.

[5] Ushakumari S, Sankaran R, Nair P S C. (2001) Adaptive neuro-fuzzy controller for improved performance of a permanent magnet branchless DC motor. *Fuzzy Systems, Melbourne, Australia.*

Electric, Electronic and Control Engineering – Shao, Shu & Tian (Eds)
ISBN: 978-1-138-02842-5

Passivity-based control for doubly-fed induction generator with variable speed and constant frequency in wind power system

Junrui Wang & Piaopiao Peng
School of Electrical and Information Engineering, The North University For Ethnics, Yinchuan Ningxia, China

ABSTRACT: A Passivity-Based Control (PBC) strategy with energy consideration for Ddoubly-Fed Induction Generator (DFIG) with variable speed and constant frequency in wind power system is proposed. The model of DFIG by Euler equations is presented in this paper. The system is decomposed into two feedback interconnected passive subsystems, an electrical subsystem and a mechanical subsystem. The electrical subsystem is the unique one to be considered in designing the controllers for the torque and speed of DFIG, thus, simplifying the control algorithm. In addition, we also consider in the design the machine resistance and inductance variation during operation, thus developing an adaptive controller which makes the system robust in performance. Simulation results show that, under variable wind-speed the proposed controller guarantees the wind power system to achieve the maximal absorption of wind power, and to output the electrical power with constant frequency and constant voltage for the grids. The system output tracks the reference speed quickly, with desirable static and dynamic performance, global stability as well as high robustness.

KEYWORDS: doubly-fed wind power generation; passivity-based control; adaptive control.

1 INTRODUCTION

For a doubly-fed wind power generation system with variable speed and constant frequency, stator flux oriented vector control is generally used and forms the basis for the rotor current PI controller discussed in references [1-3] which allows the decoupling control of the active and reactive power of the system. A doubly-fed induction generator (DFIG), however, is a high-dimension, time-varying parametric nonlinear system, the high precision control of which is very complex due to the presence of strong coupling among variables. So far, decoupling control methods like vector control all rely on the accurate offset of system nonlinearity and are therefore non-robust in nature. The control law design is generally not globally stable and the control algorithms are also quite complex. On the basis of vector control, many new nonlinear control methods have been developed. These include feedback linearization, robust control, backstepping, sliding mode control and passivity-based control (PBC).

PBC is a nonlinear feedback control strategy that forces the system total energy to track the desired energy function by configuring the reactive power component, workless force, in the system energy dissipation characteristic equation so as to guarantee system stability and make the system state variable to converge asymptotically to the set point, i.e. the output of the controlled subject to converge to the desired value. The superiority of the passivity method over linear decoupling control lies in that the control design is independent of the design process of accurately linearizing the subject model, which effectively increases the robustness of system control and the control law design is also globally stable without divergence singularities of control [4-6].

In this paper, a passivity-based control (PBC) strategy with for doubly-fed induction generator (DFIG) with variable speed and constant frequency in wind power system is proposed. The system is decomposed into two feedback interconnected passive subsystems, an electrical subsystem and a mechanical subsystem. The electrical subsystem is the unique one to be considered in designing the controllers for the torque and speed of the DFIG. An adaptive identification process for rotor resistance is designed to allow asymptotical tracking and control of the DFIG flux and speed. Simulation results show that the designed controller is effective.

2 PASSIVITY OF DFIG

According to the control characteristics of Euler-Lagrange (EL) system, we can regard DFIG as an EL system consisting of an electrical subsystem and a mechanical subsystem with feedback interconnection. Under two-phase synchronous rotational coordinates, the electrical and mechanical systems of the

DFIG can be expressed by a fourth-order electrical differential equation and a first-order mechanical differential equation respectively [7-9] , i.e.

$$\begin{bmatrix} u_{sd} \\ u_{sq} \\ u_{rd} \\ u_{rq} \end{bmatrix} = \begin{bmatrix} R_s + L_s p & -\omega_1 L_s & L_m p & -\omega_1 L_m \\ \omega_1 L_s & R_s + L_s p & \omega_1 L_m & L_m p \\ L_m p & -\omega_s L_m & R_r + L_r p & -\omega_s L_r \\ \omega_s L_m & L_m p & \omega_s L_r & R_r + L_r p \end{bmatrix} \bullet \begin{bmatrix} i_{sd} \\ i_{sq} \\ i_{rd} \\ i_{rq} \end{bmatrix} \quad (1)$$

$$Jp\omega_m + k\omega_m = T_e - T_L \quad (2)$$

$$T_e = \frac{3}{2} n_p L_m (i_{sq} i_{rd} - i_{sd} i_{rq}) \quad (3)$$

Here: u and I are the voltage and current; R and L *are* the resistance and inductance; subscripts s and r are the stator and rotor components; subscripts d and q are the d and q axis components; subscripts m is the interaction between rotors; p is the differential operator; J is the rotational inertia; k is the damping factor; n_p is the number of pole pairs of the motor; T_e and T_L are the electromagnetic torque and load torque; ω_1 is the synchronous angular speed of stator; ω_s is the slip electrical angular speed; and ω_m is the mechanical angular speed of rotor.

Modify (1) into the EL form:

$$D\dot{q} + Bq + Rq = Mu \quad (4)$$

Here:

$u=[u_{sd}\ u_{sq}\ u_{rd}\ u_{rq}]^T$, $q=[i_{sd}\ i_{sq}\ i_{rd}\ i_{rq}]^T$,

$$D = \begin{bmatrix} L_s I & L_m I \\ L_m I & L_r I \end{bmatrix}, R = \begin{bmatrix} R_s I & 0 \\ 0 & R_r I \end{bmatrix},$$

$$B = \begin{bmatrix} 0 & -\omega_1 L_s & 0 & -\omega_1 L_m \\ \omega_1 L_s & 0 & \omega_1 L_m & 0 \\ 0 & -\omega_s L_m & 0 & -\omega_s L_r \\ \omega_s L_m & 0 & \omega_s L_r & 0 \end{bmatrix},$$

$$I = \begin{bmatrix} 1 & 0 \\ 0 & 1 \end{bmatrix}, M = \begin{bmatrix} I & 0 \\ 0 & I \end{bmatrix}.$$

In the EL equation (4) for DFIG system model, the right side is the applied force; the third term on the left is the dissipative force, the second is the configurable "workless force", where matrix B is antisymmetric, i.e. it satisfies $q^T Bq \equiv 0$. To the extent that the winding capacitance effect is ignored, we can define the energy function of the motor's electrical part as:

$$H = \frac{1}{2} q^T Dq \quad (5)$$

Derive this equation and substitute (4) into it, we get:

$$\dot{H} = -q^T Bq + q^T(-Rq + Mu) \quad (6)$$

As B is antisymmetric, "Bq" makes no difference to the energy variation of the system and thereby to the system stability, there's no need to offset this part of the nonlinearity in the motor control. This progress can therefore be regarded as configuring the reactive power components of the system.

Integrate the sides of (6), we have:

$$\begin{aligned} H(t) - H(t_0) &= \int_{t_0}^{t} (q^T Mu)\mathrm{d}t \\ &- \int_{t_0}^{t} (q^T Rq)\mathrm{d}t < \int_{t_0}^{t} (q^T Mu)\mathrm{d}t \end{aligned} \quad (7)$$

Here, the left side is the energy increment of the electrical subsystem and the right side is the energy supplied to the motor from the power source. Given that $[u_{sd}\ u_{sq}\ u_{rd}\ u_{rq}]$ is the input of the electrical subsystem and $[i_{sd}\ i_{sq}\ i_{rd}\ i_{rq}]$ is the output of the electrical subsystem, then the mapping $u \mapsto i$ is output strictly passive. That is, the electrical subsystem of the DFIG is strictly passive.

Assume that the motor shaft is rigid, i.e., the mechanical part of the motor only stores dynamic energy, and then its energy function is

$$H_m = \frac{1}{2} \omega_m^T J \omega_m \quad (8)$$

Derive it and substitute it into (2), and integrate the sides, we have

$$\begin{aligned} H_m(t) - H_m(t_0) &= \int_{t_0}^{t} \left[\omega_m^T (T_e - T_L)\right]\mathrm{d}t \\ &- \int_{t_0}^{t} (\omega_m^T k\omega_m)\mathrm{d}t < \int_{t_0}^{t} \left[\omega_m^T (T_e - T_L)\right]\mathrm{d}t \end{aligned} \quad (9)$$

Here, the left side is the energy increment of the mechanical subsystem and the right side is the input energy of the mechanical subsystem. Given that (T_e-T_L) is the input of the mechanical subsystem and ω_m is the output of the mechanical subsystem, then the mapping (T_e-T_L) ω_m is output strictly passive. That is, the mechanical subsystem of the DFIG is strictly passive.

The DFIG can be expressed as the feedback interconnection of the electrical and the mechanical passive subsystems. According to the passivity principle, the entire DFIG system can be considered strictly passive as shown in Fig.1.

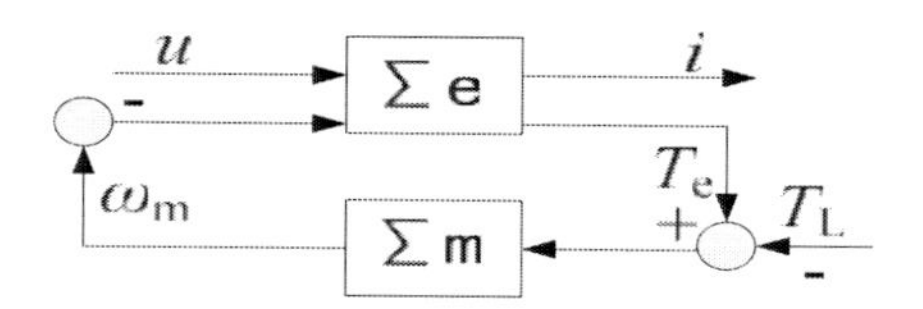

Figure 1. Decomposition of DFIG.

By decomposition, we can take the mechanical subsystem is a passive interference term of the electrical subsystem and handle the electrical subsystem alone as the controlled system, thus simplifying the controller design.

3 CONTROLLER DESIGN

3.1 *Torque controller design*

Assume the desired output torque of the system is T_e^* and the stator flux is ψ_s, in a two-phase rotational coordinate system, the output torque of the system can be expressed as:

$$T_e^* = \frac{3}{2} n_p L_m (i_{sq}^* i_{rd}^* - i_{sd}^* i_{rq}^*) \tag{10}$$

To allow asymptotic vector control of stator flux and asymptotic tracking of electromagnetic torque, the set the control target as:

1 Asymptotic tracking of electromagnetic torque:

$$\lim_{t\to\infty}(T_e - T_e^*) = 0$$

2 Asymptotic orientation of stator flux:

$$\lim_{t\to\infty}\Psi_{sq} = \lim_{t\to\infty}(L_s i_{sq} + L_m i_{rq}) = 0$$

3 Asymptotic tracking of stator flux amplitude:

$$\lim_{t\to\infty}\Psi_{sd} = \lim_{t\to\infty}(L_s i_{sd} + L_m i_{rd}) = \psi_s .$$

Hence, define the tracking error between the actual state and the reference state is $e = q - q^*$, from equation (4) we derive the system error equation:

$$D\dot{e} + (B + R)e = \xi \tag{11}$$

Here, ξ is the disturbance quantity in the form of:

$$\xi = Mu - \left\{ D\dot{q}^* + (B + R)q^* \right\} \tag{12}$$

If we select Lyapunov function $H_d = \frac{1}{2} e^{\mathrm{T}} De$, its derivative is

$$\dot{H}_d = e^{\mathrm{T}} D\dot{e} = e^{\mathrm{T}}\xi - e^{\mathrm{T}} Be - e^{\mathrm{T}} \mathrm{Re} .$$

From the above we know $e^{\mathrm{T}}Be = 0$, then

$$\dot{H}_d = e^{\mathrm{T}} D\dot{e} = e^{\mathrm{T}}\xi - e^{\mathrm{T}} \mathrm{Re} \tag{13}$$

According to Lyapunov law, if $\xi \equiv 0$, as R is positively definite, then we have

$\lim_{t\to\infty} e \to 0$, i.e. $T_e \to T_e^*$. This process is called energy shaping [7].

From the control target of flux asymptotic vector control, we have

$$\begin{cases} L_s i_{sd}^* + L_m i_{rd}^* = \psi_s \\ L_s i_{sq}^* + L_m i_{rq}^* = 0 \end{cases} \tag{14}$$

Assume that the desired ψ_s is a constant value, if we want to control the reactive power at the DFIG stator side is zero, take $i_{sd}^* = 0$.From (10) and (14) we have

$$i_{sd}^* = 0,\ i_{sq}^* = 2T_e^*/(3n_p L_m i_{rd}^*)$$

$$i_{rd}^* = \psi_s/L_m,\ i_{rq}^* = -L_s i_{sq}^*/L_m \tag{15}$$

So, from the assumed stator-side reactive power, we can define i_{sd}^* and then get the other three currents from the given instruction torque. Finally, from $\xi = 0$, we can derive the equations for controllers u_{rd} and u_{rq}:

$$\begin{cases} u_{rd} = L_m p i_{sd}^* + L_r p i_{rd}^* - \omega_s L_m i_{sq}^* - \omega_s L_r i_{rq}^* \\ \quad + R_r i_{rd}^* - k_1 (i_{rd} - i_{rd}^*) \\ u_{rq} = L_m p i_{sq}^* + L_r p i_{rq}^* + \omega_s L_m i_{sd}^* + \omega_s L_r i_{rd}^* \\ \quad + R_r i_{rq}^* - k_2 (i_{rq} - i_{rq}^*) \end{cases} \tag{16}$$

To ensure that the entire control system is strictly passive, improve the system dynamic response and reduce the control system sensitivity to parameter variation, we add a damping term into equation (16), where k_1 and k_2 are the damping factors. This process is called damping injection [7]. By properly adjusting k_1 and k_2, we can allow the actual stator flux and electromagnetic torque to track the reference values quickly and control the desired system dynamic and static behaviors when the time variability of the load torque is unknown.

3.2 *Speed controller design*

As torque control is able to track time-varying torques asymptotically, a speed controller can be designed by simply establishing a speed error feedback in the torque control structure. Using a PI regulator, we get a reference torque:

$$T_e^* = k_p(\omega_r^* - \omega_r) + k_i \int (\omega_r^* - \omega_r)\mathrm{d}t \tag{17}$$

Here: ω_r is the rotor angular speed; k_p and k_i are the proportional gain and integral gain, which are highly correlative to the system stability.

The structure of the entire control system is given in Fig.2.

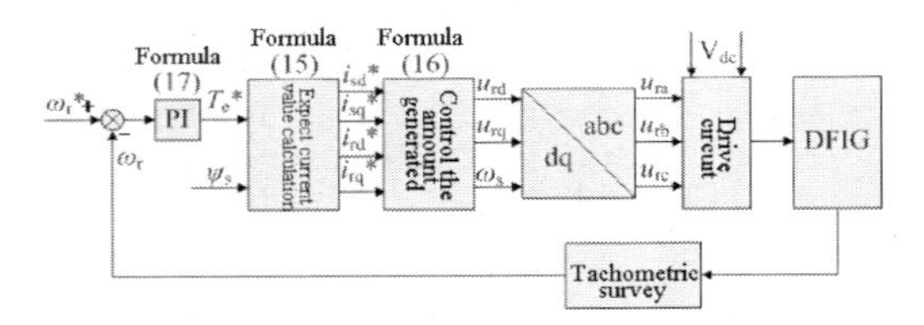

Figure 2. Block diagram for passivity-based DFIG control system.

3.3 *Adaptive PBC controller design*

As a DFIG is subject to parameter variation during operation, it is necessary to use an adaptive control strategy to improve the system robustness. If the uncertain parameters of the DFIG system are the stator and rotor resistances R_s and R_r, then we define the uncertain parameters as:

$$\mathrm{Re} = [\mathrm{Re}1\theta, \mathrm{Re}2\theta, \ldots, \mathrm{ReN}\theta] \tag{18}$$

Here: θ is the unknown parameter vector; R_{ei} ($i = 1,\ldots, N$) is the know state variable parameter; N is the number of phases. Substitute θ with the observed dynamic parameter $\hat{\theta}$, the DFIG state error equation including for the rotor resistance variation is:

$$D_e\dot{\tilde{q}}_e + B_e\tilde{q}_e + R_e\tilde{q}_e + \tilde{R}_e q_e^* = 0 \tag{19}$$

Here: $\tilde{q}_e = q_e - q_e^*$; $\tilde{R}_e = R_e - \hat{R}_e$, $\hat{R}_e$ is the estimated stator/rotor resistance.

Select the Lyapunov function

$$V = \frac{1}{2}\tilde{q}_e^{\mathrm{T}} D_e \tilde{q}_e + \frac{1}{2}(\hat{\theta} - \theta)^{\mathrm{T}}(\hat{\theta} - \theta) \tag{20}$$

Following the trajectory differential equation of (19), (20), we get

$$\dot{V} = -\tilde{q}_e^{\mathrm{T}} R_e \tilde{q}_e - \tilde{q}_e^{\mathrm{T}} \tilde{R}_e q_e^* + (\hat{\theta} - \theta)^{\mathrm{T}} \dot{\hat{\theta}} \tag{21}$$

Using (18), we have

$$\tilde{q}_e^{\mathrm{T}} \tilde{R}_e q_e^* = \left[\sum_{i=1}^{N} q_{ei}^* \tilde{q}_e^{\mathrm{T}} R_{ei}\right](\theta - \hat{\theta}) \tag{22}$$

The update rate of the design parameter is

$$\dot{\hat{\theta}} = -\left[\sum_{i=1}^{N} q_{ei}^* \tilde{q}_e^{\mathrm{T}} R_{ei}\right]^{\mathrm{T}} \tag{23}$$

Then (21) can be simplified as

$$\dot{V} = -\tilde{q}_e^{\mathrm{T}} R_e \tilde{q}_e \tag{24}$$

$V > 0$, $\dot{V} < 0$, from Lyapunov stability law, $\tilde{R}_e \to 0$, $\tilde{q}_e \to 0$. This allows self adjustment of the motor rotor parameters and effectively overcome the adverse effect of rotor resistance variation on the PBC performance.

4 SIMULATION AND RESEARCH

To verify how effective the designed controller is, we simulated with MATLAB/Simulink at system parameters: DFIG n_p=2, R_s=0.435Ω, R_r=0.816Ω, L_s=L_r=0.07131H, L_m=0.06931H, J=0.089kg.m^2. The blade radius of the wind turbine used is 2.4m; the best tip speed ratio is 9; the maximum wind utilization coefficient is 0.4; the gear ratio is 7.8.

The initial wind speed was set at 5.6m/s. After the machine stabilized, at t = 2s, this is steeply increased to 6.6m/s. At t = 4s, it is returned to 5.6m/s. According to the best wind utilization principle, when the wind speed is 5.6m/s, the DFIG rotor electrical angular speed should track to 328rad.s^{-1}; and when the wind speed is 6.6m/s, this should track to 386 rad.s^{-1}. Fig. 3 shows the waveform of the DFIG speed response: during wind speed variation, the turbine followed the maximum wind capture mechanism with quick response, no overshoot, little static error and good control effect. Fig. 4 shows the simulation waveform of the DFIG stator-side active and reactive power during sudden wind speed variation: during wind speed variation, the reactive power stayed unchanged at zero while the active power varied with the wind speed, suggesting that the active power and reactive power were independent from each other during the adjustment and achieved good decoupling regulation. Fig.5 gives the waveform of the stator voltage and stator current which are frequency constant with opposite phases, suggesting that electric power is transmitted into the power grid at the stator side. Fig.6 shows the waveform of the rotor two-phase current during the wind speed variation,

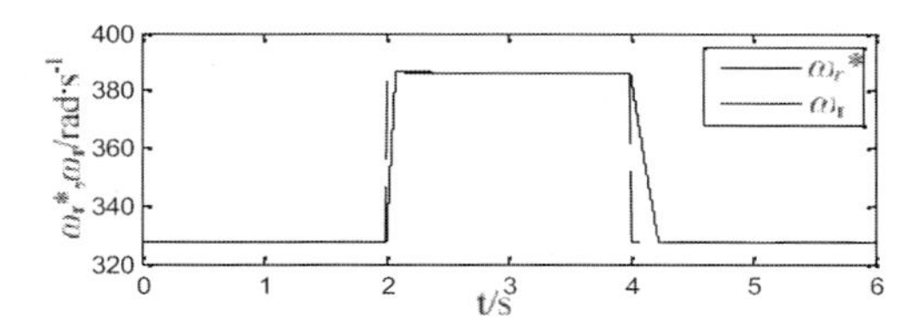

Figure 3. Speed of the DFIG and its desired value.

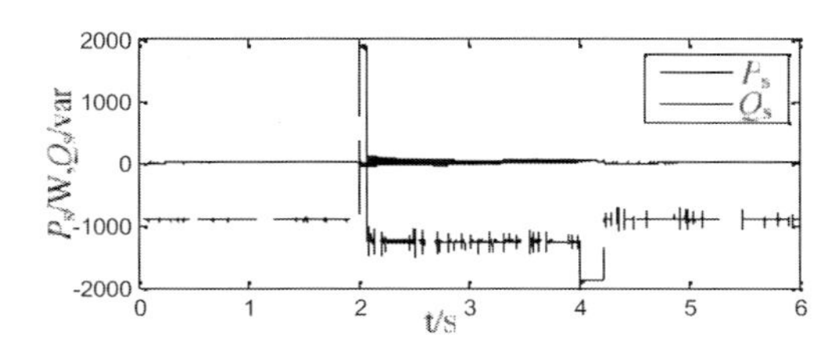

Figure 4. The simulation curves of active power and reactive power of stator.

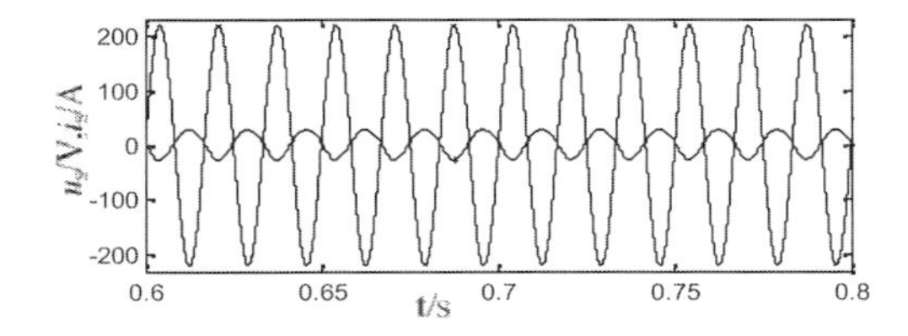

Figure 5. Line-to-neutral voltage and current of stator.

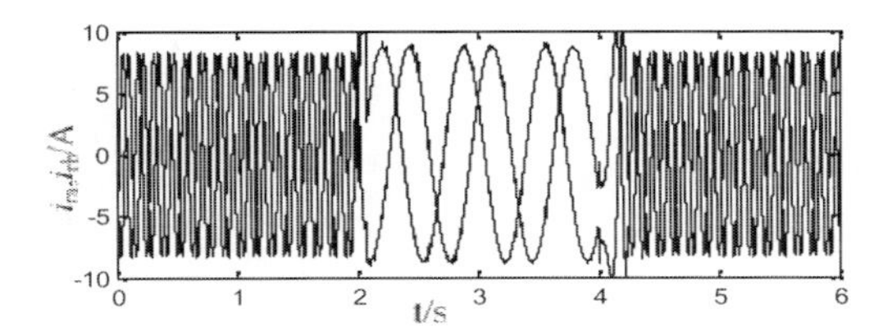

Figure 6. Two-phase current waveform of rotor.

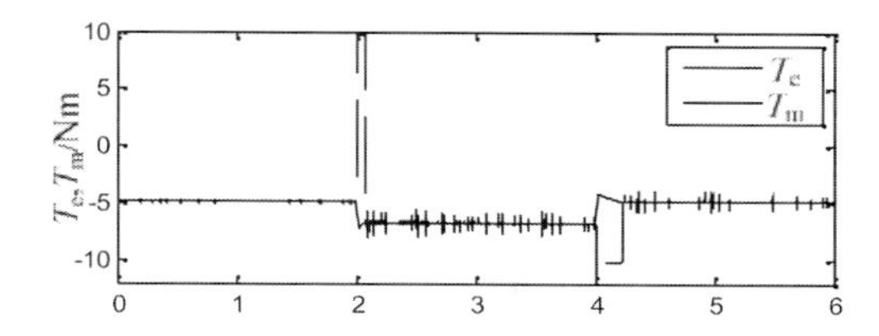

Figure 7. Torque waveform.

where the rotor current frequency varied with the generator speed, thereby guaranteeing the stator voltage to be frequency constant. Fig. 7 is the waveform of the DFIG drag torque T_m and electromagnetic torque T_e, where a negative T_e indicates that the DFIG is under power generation. If the rotor resistance varies by 50% due to heating, i.e. $\Delta R_r = 0.5R_r$, the simulation result of the rotor angular speed and stator flux is shown in Fig.8. From this chart, the motor speed and stator flux can still be highly stable even if the motor parameters vary.

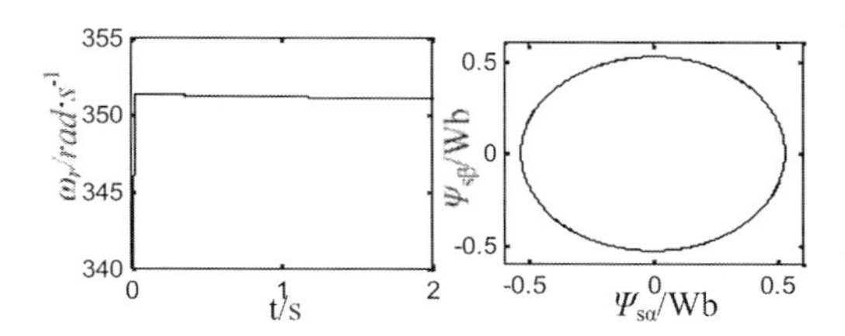

Figure 8. Simulation curves of rotor angular speed and stator flux.

5 CONCLUSIONS

After examining the EL model of a DFIG wind power generator, PBC method is used to make the system strictly passive by properly configuring the reactive power components of the system without offsetting the nonlinearity of the controlled subject, and allow quick tracking control of torque and speed while guaranteeing system stability. On this basis, a rotor resistance identification process is added and achieves adaptive control of a passive system. Simulation results show that the waveforms conform to the theoretical analysis; the system is able to track the desired stator flux and reference speed asymptotically without the singularity issues common along general input/output decoupling controls and provides good dynamic and static properties and high robustness.

ACKNOWLEDGEMENTS

This study is supported by State Ethnic Affairs Commission Science Research Project (12BFZ003); Ningxia Natural Science Fund Project (NZ14106); and The North University for Ethnics Internal Key Project (2014XZZ03).

REFERENCES

[1] R Ortega, A J Van Der Shaft, I Mareels, et al. (2001) Putting energy back in control. *IEEE Control Systems Magazine*, 21(2):18–33.

[2] Pena R, Clare J C, Asher G M. (1996) Doubly fed induction generator using back-to-back PWM converters and its application to variable-speed wind energy generation. *IEE Proceedings Electric power Applications*, 143(3):587–591.

[3] Yuan Guofeng, Chai Jianyun, Lin Yongdong. (2005) Study on excitation converter of variable speed constant frequency wind generation system. *Proceedings of the CSEE, 2005*, 25(8):90–94.

[4] C Cecati, N Rotondale. (1999) Torque and speed regulation of induction motors using the passivity theory approach. *IEEE Trans. on Industrial Electronics*, 46(1):119–127.

[5] Wang Xiaohong, Wu Jie, Yang Jinming, et al. (2008) Passivity control for current-loop of matrix converter. *Control Theory & Applications*, 25(2):341–347.

[6] Zhong Qing, Wu Jie, Yang Jinming. (2003) Application of passivity-based control in active power filters. *Control Theory & Applications*, 20(5):713–718.

[7] Chen Feng, Xu Wenli. (1999) Passivity based speed control of induction motors. *Journal of Tsinghua University (Science and Technology)*, 39(7):29–32.

[8] Ji Zhicheng, Xue Hua. (2004) Adaptive passivity-based control strategies of induction motors using dSPACE. *Journal of Xi'an Jiaotong University*, 38(12):1220–1223.

[9] Hu Jiabing, He Yikang, Liu Qihui. (2005) Optimized active power reference based maximum wind energy tracking control strategy. *Automatic of Electric Power System*, 29(24):32–38.

Electric, Electronic and Control Engineering – Shao, Shu & Tian (Eds)
© 2015 Taylor & Francis Group, London, ISBN: 978-1-138-02842-5

System structure of IEC61850-based standard digital hydropower station

Renwen Lu, Yonghua Gui & Qin Tan
HNAC Technology Co., Ltd. Changsha, Hunan, China

ABSTRACT: The text mainly explains IEC61850 standard and its application in the hydropower station integrated automatic system and focuses on the introduction about the structure, key technologies and technical advantages of the integrated automatic system in the digital hydropower station

KEYWORDS: IEC61850, digitalization, intelligentization, the integrative automation for the hydropower station.

1 INTRODUCTION

As we all know, hydropower, as a sustainable and low-carbon resource, plays an important role in coping with climate change and energy shortage. However, the water resource is limited, so how to promote the water utilization efficiency becomes particularly important. Now, before we gain a great breakthrough in terms of the energy conversion technology, promoting the automatic technology becomes the only way to improve the utilization efficiency, and to be digitalized is an effective means that can grandly enhance the automation level of the hydropower station. Now when the smart power network and the digital substation technology are becoming more mature and completed, the automatic technology of the hydropower station is developing forward in a direction of digitalization. The continuous development of smart equipment, the mature use of high-speed communication technology, together with the wide implementation of IEC61850 standard in the digital substation provide the solid foundation for boosting the construction of the integrative automatic system in the digital hydropower station.

2 INTRODUCTION TO IEC61850 STANDARD

International Electrotechnical Commission's Technical Committee 57 (IEC TC 57) worked on IEC61850 from 1995. By now, IEC61850 standard has consisted of 14 parts, all of which have been adopted as the international standards. The Chinese Standardization Committee kept following them and translated them into Chinese simultaneously. The equivalent version in China is the D/L860 standard. Having studied latest international technology and quoted the international standard in several fields, the IEC61850 standard has become the base to make the digital hydropower station come true, and by the object-oriented modeling technology and the future-communication-oriented scalable architecture, it always aims at the realization of *One World, One Technology, One Standard*.

IEC61850-ED1 (1st edition): Communication Networks and Systems in Substations

IEC61850-ED2 (2nd edition): Communication Networks and Systems for Power Utility Automation, including hydropower, wind power and other distributed power;

IEC61850-7-410 (Hydroelectric power plants - Communication for monitoring and control) supervision and communication system for hydropower stations

IEC61850-7-510 Logic Node Model of Hydropower Stations

IEC61850 divides the communication system into 3 layers: station layer, bay layer, and process layer. It features smart primary equipment, networking secondary equipment, and automatic operation and management systems.

2.1 *Smart primary equipment*

The primary equipment (including transformer, generator, motor, capacitor, circuit breaker, isolator, grounding switch, voltage transformer, current transformer, arrestor, etc.) are designed with the microprocessor and the photoelectric technology in their signaling circuits and control driving circuits, so they can automatically monitor the status. In addition, they are provided with the optical communication interface.

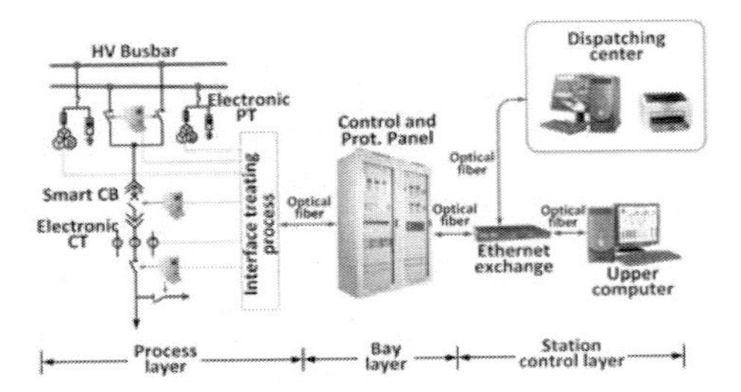

Figure 1. Smart primary equipments

The primary equipment communicate with the secondary equipment through the optical Ethernet and by IEC61850 protocol.

The conventional conductors such as signal cable and control cable are replaced by the common signal network, simplifying the structure of the conventional electromechanical relay and control loops.

In other words, in the secondary circuits, the software program logic works instead of the conventional relay and its logic loops, and the photoelectric digit and the optical fiber replace the conventional strong current analog signal and control cable. All the primary equipment are plug-and-play equipment.

2.2 *Networking secondary equipment*

The standardized and modularized microprocessor is applied in the design and manufacture of secondary equipment of the hydropower station, such as relay protective devices, anti mis-operation blocking device, measurement control device, telecontrol devices, fault wave recorder, voltage reactive power control device, sync operation device, the under-developing on-line status detector, and so on. These equipment use the high-speed Ethernet to communicate with each other.

IEC61850 standard is applied to all equipments.

Without the I/O site interfaces, the secondary equipment can truly share data and resources via the network. The conventional functional device equipped here serves as the logic functional module. All the secondary equipment are plug-and-play ones.

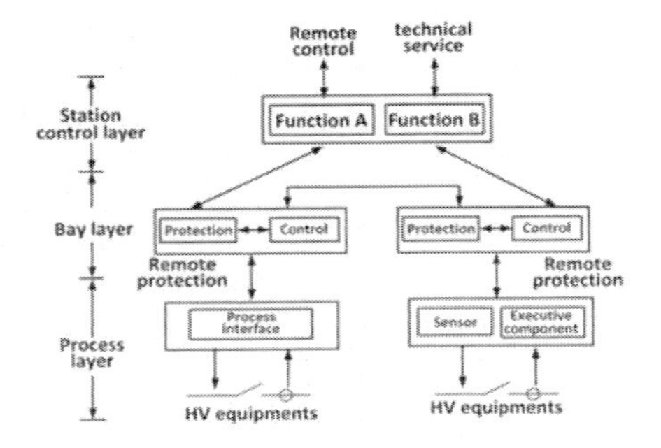

Figure 2. Structure of secondary equipment network.

2.3 *Automatic operation and management system*

The automatic operation and management system of the hydropower station mainly can furnish the paperless record and statistics of the power production and operation data and status, and the automatic data layering and distribution exchange.

If any faults are detected in the operation, the fault analyzing report will be generated instantly to indicate the fault cause and meanwhile propose solutions;

Next, the system will automatically send the equipment overhaul report and change the equipment status from *Periodic Maintenance to Condition Based Maintenance.*

2.4 *Data modeling objectification*

The modeling process of IEC61850 standard is, in essence, the process of modular decomposition. The IEC61850 model is constituted by logic device (LD), logic node (LN), data objection (DO), and data attribute (DA), with the latter being, in turn, contained in the former. In reality, the physical device is broken up into several logic devices (LD), such as Relay 1 in figure 3. The logic node (LN) is the key part of IEC61850 objection-oriented modeling, and one logic node (LN) stands for a functional module. Several logic nodes work cooperatively to finish the function that a single logic device should finish. For example, the Overcurrent Protection Logic Node PTOC1 and Measurement Logic Node MMXU1 in P3 provide the overcurrent protection and the measurement, respectively. Each logic node is formed by several data objects (DO); each data object (DO) has different data attributes (DA) to describe the attribute of data objects in every aspect.

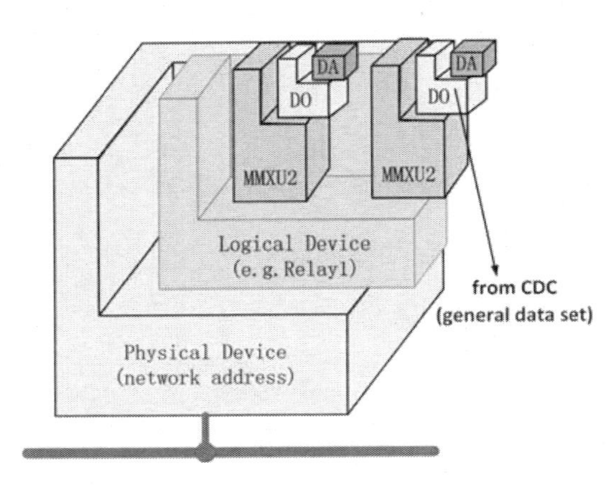

Figure 3. Inner logic module inside the physical unit.

Each function module of the Turbine Control System can be broken up into logic nodes, as shown in figure 4:

FPID: PID control
FSPT: Setting value control function
FCSD: Curve shape description
TPOS: Position indication
HCOM: Optimize coordination

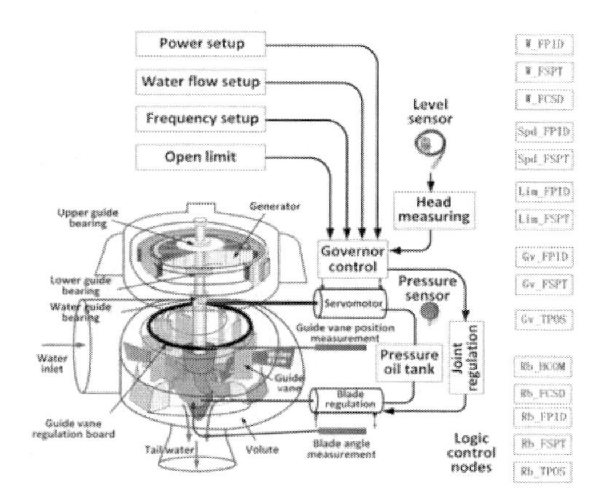

Figure 4. Logic module of turbine control system.

Each function module of the Excitation Control System can be broken up into logic nodes (LN), as shown in figure 5:

FPI: Power regulation
FLIM: Limiter
FPID: PID control
FSPT: Setting value control
FRMP: Slope control
FFIL: Filter
ZSCR: SCR
PTHF: SCR protection
MMDC: DC current and voltage measurement
CALH: Alarm processing

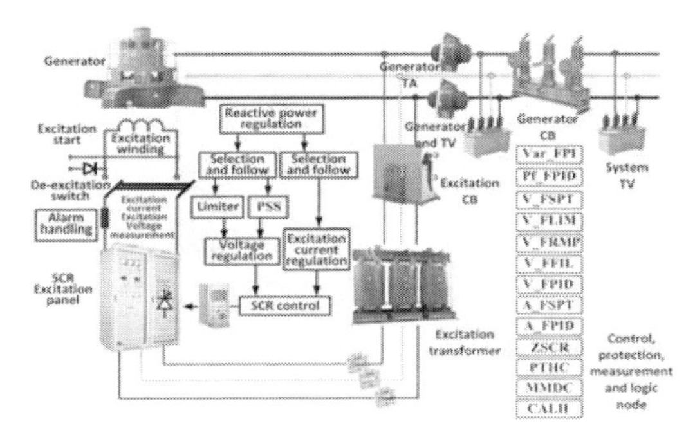

Figure 5. Logic module of excitation system.

Each function module of the Generator-Transformer Protection System can be broken up into logic nodes (LN), as shown in figure 6:

PDIF: Differential protection
PTOF/PTUF: HF/LF protection
PVPH: Over excitation protection
PTOV: Over-voltage protection
PDUP: Field loss protection
PDOP: Inverse function protection
PDIS: Distance protection
PTOC: Overcurrent protection

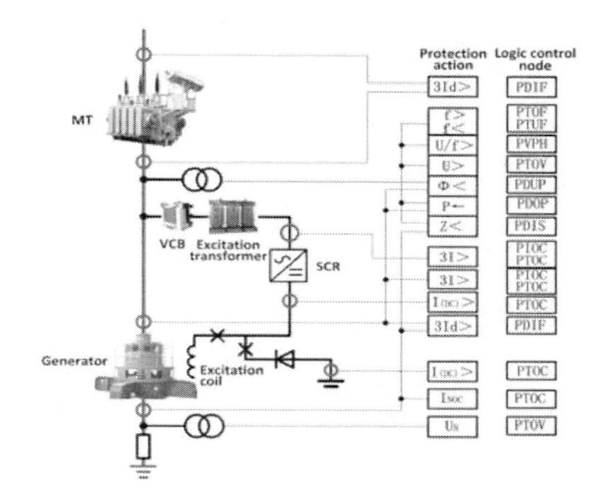

Figure 6. Logic module of g-t unit protection system.

3 COMPOSITION AND STRUCTURE OF DIGITAL HYDROPOWER STATION

Logically, the digital hydropower station is classified into station layer, bay layer, and process layer. They communicate through 2 types of networks: station-level network (IEC61850/NMS) and process-level network (GOOSE, SMV/IEC61850-9-2). The network throughout the station is linked by high-speed optical Ethernet, and the communication standard conforms to DL/T860 (IEC61850).

SCADA system master, operator workstation, server, and other equipment constitutes the station layer to fulfil functions such as status supervision, operation optimization, and so on.

The bay level consists of different subsystems, including relay protection subsystem, generator control subsystem, and common subsystem, and contains equipment, such as PLC, smart unit, protection unit, and sync devices.

The process layer is composed of equipment such as electronic transformer ETV/ETC, merging unit (MU), smart terminal, transmitter, and so on. They work together to intelligentize the primary equipments, that is, convert the traditional analog quantity into the digital one before sending it to the bay layer by the optical Ethernet.

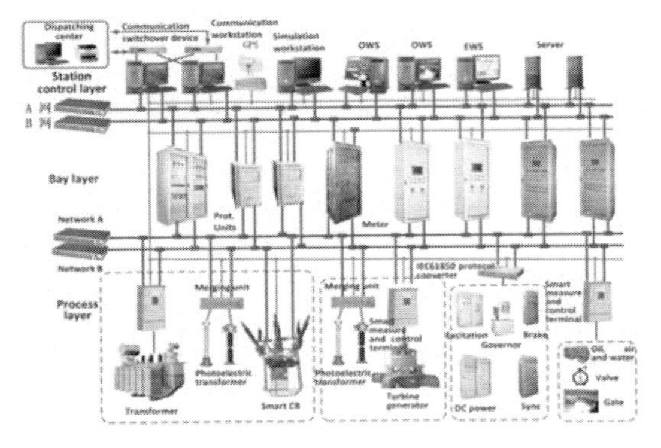

Figure 7. Structure of digital hpp integrative automatic system.

4 TECHNOLOGICAL REALIZATION OF THE DIGITAL HYDROPOWER STATION

So far, the IEC61850-based digital substation has been widely promoted and recognized in the electric power field. Compared with the digital substation, the electrical part of the digital hydropower station is the same as that of the digital substation and thus some relevant technology, such as IEC61850 optical networking system, smart transformer, smart circuit breaker, smart voltage transformer, and current transformer, can also be applied to the digital hydropower station.

The most important step to digitalize the hydropower station is to intelligentize the primary equipment and its accessories, including the turbine-generator unit, governor, the excitation system, and the oil-gas-water system.

5 ADVANTAGES COMPARED WITH THE CONVENTIONAL HYDROPOWER STATION

5.1 *Better reliability and security*

The smart primary equipment have the completed self-diagnostic function, which can set off an alarm promptly once the fault is detected and thus facilitate the fault location. Besides, having no problems such as high voltage insulation and TA open circuits, the smart electronic transformer will not endanger the personnel.

5.2 *Simpler control cable*

Instead of the complicated control cable, the digital hydropower station takes the smart primary and the secondary equipment, and the communication among the secondary equipment is fulfilled by the communication network. In such a case, logic information flow can be transmitted in a single physical communication network, which greatly reduces the amount of the secondary cables adopted and makes all the smart equipment plug-and-play ones.

5.3 *Higher measurement accuracy*

In the digital hydropower station, the smart transformer can output the digital signal directly. Therefore, no transmission error will be caused. Therefore, the accuracy of protection and measurement is promoted.

5.4 *Improved anti-electromagnetic interference capability*

The digital hydropower station adopts the optical fiber to link the primary equipment and the secondary equipments together, effectively avoiding the electromagnetic interference.

5.5 *Enhanced interoperability*

All the equipment in the digital hydropower station network are based on the IEC61850 standard to carry on modeling and communication, and they can be integrated seamlessly.

5.6 *Promoted automation level*

The smart primary equipments adopted in the digital hydropower station make the transmitted data more complete and the reliability and the real-time performance in the communication are greatly enhanced, which lays a solid foundation to further automatize and intelligentize the hydropower station, and this, thus, substantially raises the automation level for the whole station.

6 CONCLUSION

The extensive application of IEC61850 standard and the creative effort of smart primary equipment will certainly push the hydropower station to progress in the digital and smart direction. Predictably, in the near future, there will be more and more digital hydropower stations of high quality, reliability, and security built in our country.

REFERENCES

[1] Ren Yanming, Qing Lijun, Yang Qixun , (2000) Introduction and analysis of IEC61850. *Automation of Electric Power System*, (04).

[2] Xuning, Zhu Yongli, Di Jian, (2006) Substation automation object modeling based ion IEC61850. *Electric Power Automatic Device*, (03).

[3] Zhao Anguo, Yang Xiaoming, Chou Xinhong, (2009) Inner modeling method of relay protective smart device. *Power System Protection and Control*, (22).

[4] Zhang Jie, (2004) Structure model analysis for IEC61850. *Automation of Electric Power System*, 28(18):90–93.

Electric, Electronic and Control Engineering – Shao, Shu & Tian (Eds)
© 2015 Taylor & Francis Group, London, ISBN: 978-1-138-02842-5

Information hiding method based on line spectrum frequency of AMR-WB

Da Teng
Signal & Communication Research Institute, China Academy of Railway Science, Beijing, China
Railway Technology Research College, China Academy of Railway Science, Beijing, China

Haonan Feng & Jianjie Yu
Signal & Communication Research Institute, China Academy of Railway Science, Beijing, China

ABSTRACT: The information hiding algorithm based on AMR-WB (Adaptive Multi-Rate-Wideband) LSF (Line Spectrum Frequency) is presented, which chooses one or several LSF indexes first, and then hides information by making modifications in these indexes. Simulation experiments show that modifications of certain LSF indexes result in information hiding with little degradation of speech quality.

KEYWORDS: line spectrum frequency; AMR-WB; information hiding.

1 INTRODUCTION

Adaptive Multi-Rate Wideband (AMR-WB) is a wideband speech audio coding standard specified by 3GPP and ITU-T, using similar methodology as Algebraic Code Excited Linear Prediction (ACELP) [1]. The 9 different bit rates that AMR-WB operates with are 6.60 kbps, 8.85 kbps, 12.65 kbps, 14.25 kbps, 15.85 kbps, 18.25 kbps, 19.85 kbps, 23.05 kbps, and 23.85 kbps[2]. The lowest bit rate that provides excellent speech quality in a clean environment is 12.65 kbps. Higher bit rates are applied in background noise environments, music, combined speech, and multi-party conferencing[3]. Lower bit rates such as 6.60 and 8.85 kbps provide reasonable quality compared with narrow band codecs. All modes are sampled at 16 kHz and processed at 12.8 kHz[4].

Linear Prediction (LP) is a crucial technology in Speech Signal Processing. LP coefficients (LPC) play an important role in information tracing of voice recognition[5]. In Speech Signal Processing implementation, the LSP (Line Spectral Pair) coefficient, which equals the LP coefficient from a mathematical view, is used to represent LPCs. LSPs have several properties (e.g. better quantitative characteristics, better robustness, and better interpolation features) that make them superior to direct quantization of LPCs[6]. Moreover, LSPs ensure stability of filters owing to their smaller sensitivity to quantization noise, which is a key reason of the wide application of LSPs in Speech Signal Processing[7].

In conversion from LPC to LSP, Line Spectrum Frequencies (LSFs) are used to represent frequency domain values of LSPs[8]. The LSFs can also be applied in information hiding. First, one or several LSF indexes is chosen to be embedded in. Then, indexes that have little influence on the quality of communication is selected in information hiding.

2 DEFINITION OF LSF PARAMETERS

In the AMR-WB coding phase, first, the LPC is calculated; then, the LSF coefficients are calculated by the LPCs for translation into the LSPs. Finally, the LSP coefficients are quantized and encoded. The LSP expressions are used in the frequency domain to quantify the low-pass filter coefficients, as listed next:

$$f_i = \frac{f_s}{2\pi}\arccos(q_i), i = 0,\ldots 14 \tag{1}$$

$$f_i = \frac{f_s}{4\pi}\arccos(q_i), i = 15 \tag{2}$$

f_i represents the LSF parameter, of which the frequency is from 0 to 6400Hz, and f_s represents the sampling rate of 12.8 KHz. $[f_1 \ldots f_{15}]$ form an LSF vector f_t. f_t is the transpose of the vector.

Then, the residual LSF vector needs to be quantified. A first-order moving average predictor is applied in the combination of both split vector quantization method and multi-stage vector quantization method. The specific process of prediction and quantization is as follows: ($r(n)$ represents the prediction residual vector; $z(n)$ represents the LSF vector that has removed the mean of the current frame; *average*(n) is statistical data; and $r(n-1)$ is a quantized residual vector of the speech frame before the current one:

$$r(n) = z(n) - p(n) \tag{3}$$

$$z(n) = f_n - average(n) \quad (4)$$

$$p(n) = \frac{1}{3}\hat{r}(n-1) \quad (5)$$

Next, residual vector r(n) is quantified by using S-SMVQ quantizer, so that r(n) is split into two sub-vectors r_1(n) and r_2(n). The length of r_1(n) is 9 bit, and the length of r_2(n) is 7 bits. Quantization of the two sub-vectors is as follows:

r_1(n) and r_2(n) are quantized as a length of 8 bits, respectively;

r_1(n) is split into three 3-dimensional sub-vectors of length of 6 bits, 7 bits, and 7 bits, respectively, for quantization coding. r_2(n) is split into one 3-dimensional and one 4-dimensional sub-vectors of length that are 5 bits, respectively.

Thus, LSF index has a total length of 46 bits.

3 METHODOLOGY

After quantization encoding, LSF parameters are quantized and coded for a total of 5 or 7 pieces. In the mode of non-6.60 kbps, LSF parameters are quantized for 7 segments, the length of which are 8 bits, 8 bits, 6 bits, 7 bits, 7 bits, 5 bits, and 5 bits, respectively. Hence, the total length is 46 bits. In the mode of 6.60 kbps, LSF parameters are quantized for 5 segments, the length of which are 8 bits, 8 bits, 7 bits, 7 bits, and 6 bits. Hence, the total length is 36 bits.

Pitch index parameters, filter index parameters, fixed codebook index parameters, gain vector index parameters, LSF index parameters, and high frequency gain index parameters will be retrieved after decoding the received bit stream of speech signal. Thus, LSF index parameters can be retrieved by order. The LSF parameters in different rate mode are divided into several segments, and each section has a certain length, so our idea is to change the LSF parameters to hide information. On the decoder end, the receiver acquires secret information by analyzing the received LSF index parameters. Because of changes taking place in LSF after information hiding, there will be loss of sound quality after voice reconstruction. So how to effectively use these seven segments of LSF parameters for information hiding is a key issue.

In the non-6.60 kbps mode, there are 7 segments numbered as 1 to 7 in the LSF index. No.1 segment is the quantized r_1(n) vector of length 8 bit; No.2 is the quantized r_2(n) vector of length 8 bit; No.3-5 are the split 3 sub-vectors of r_1(n), of which lengths are 6 bits, 7 bits, and 7 bits, respectively; and No. 6 and 7 are the split 2 sub-vectors of r_2(n), whose lengths are 5 bits and 5 bits, respectively. In the 6.60 kbps mode, there are 5 segments numbered as 1 to 7 in the LSF index. No.1 segment is the quantized r_1(n) vector of length 8 bit; No.2 is the quantized r_2(n) vector of length 8 bit; and No.3-5 are the split 3 sub-vectors of r_1(n), whose lengths are 7 bits, 7 bits, and 6 bits, respectively. Therefore, information hiding is implemented from the following two aspects:

Change the 7 segments of the LSF index parameter, respectively, and choose the segment that has the best performance by the PESQ Algorithm test; change a combination of two or more segments of the LSF index parameter and finally examine the effect of information hiding by PESQ Algorithm tests.

4 EMBEDDING ALGORITHM

4.1 *Pre-processing*

First, an offset value table that contains 50 offset items is generated. The size of each in $Table_i$ em is 2 bits. The value of each item is randomized between 0 and 3. Then, the offset table is hidden into the pitch period of the i-th second speech signal as an agreement in advance by the sender and the receiver.

4.2 *Single LSF index at 50 bps*

First, the secret information is encoded as a binary bit stream marked as *Strm*. Then, *Strm* is divided into segments of 50-bit length. Each segment is marked as $Strm_i$. By binary addition of $Strm_i$ and the corresponding $Table_i$, a bit stream *BitL* of 50-bit length is acquired. The value of each bit (marked as $BitL_i$) of *BitL* is 0 or 1. Next, one of the seven LSF index segments is chosen for information hiding, marked as I_i $(i = 1, 2, ..., 7)$. B_j^i represents the i-th LSF segment of the j-th speech signal frame. In terms of the embedding algorithm, a certain segment of the LSF index of every frame is able to hide 1-bit information. The length of each frame is 20 ms so that it is able to hide 50-bit data per second, that is, the embedding rate is 50 bps. Figure 1 is the pseudo code of the algorithm.

For any bit $BitL_j$ in *BitL* stream, if $BitL_j = 0$, let $m = B_j^i \bmod 2$. If m equals 0, B_j^i remains unchanged; if m equals 1, B_j^i minors 1. If $BitL_j = 1$, B_j^i mod 2. If m equals 0, B_j^i minors 1; if m equals 1, B_j^i remains unchanged. Finally, the revised LSF index segments with other index parameters are encoded and sent to the receiving terminal.

4.3 *Single LSF index at 100 bps.*

First, the secret information is encoded as a binary bit stream marked as *Strm*. Then, *Strm* is divided into segments of 100-bit length marked as $Strm_i$. By the binary addition of $Strm_i$ and the corresponding $Table_i$, a bit stream $FbitL$ of 50-bit length is acquired. The value of each bit (marked as $FbitL_i$) of $FbitL$ is 0 or 1. F_j^i represents the i-th LSF segment of the j-th speech signal frame. In terms of the embedding algorithm, a certain segment of the LSF index of every frame is able to hide 2-bit information. So, it is able to hide 100-bit data per second, that is, the embedding rate is 100 bps.

For any bit $FbitL_i$ and $FbitL_{i+1}$ ($i \bmod 2 = 0$) in $FbitL$ stream, choose the LSF index F_j^i of the j-th speech frame. A quaternary value m is generated using $FbitL_i$ as a higher bit and $FbitL_{i+1}$ as a lower bit. Let $c = F_j^i \bmod 4$, if c equals m, F_j^i remains unchanged; if c does not equal m, corresponding changes are carried out to F_j^i, that is,

$$F_j^i = F_j^i - (c - m), (c > m) \tag{6}$$

$$F_j^i = F_j^i - (4 - m + c), (c \leq m) \tag{7}$$

Finally, the changed LSF index is encoded and transmitted.

4.4 *Multiple LSF indexes at 100 bps.*

First, the secret information is encoded as a binary bit stream marked as *Strm*. Then, *Strm* is divided into segments of 100-bit length marked as $Strm_i$. By binary addition of $Strm_i$ and the corresponding $Table_i$, a bit stream $MbitL$ of 100-bit length is acquired. The value of each bit (marked as $MbitL_i$) of $MbitL$ is 0 or 1. Two random indexes i and j of the 7 LSF indexes are chosen ($i = 1,2, \ldots, 7, j = 1,2, \ldots, 7$, and $i \neq j$). M_j^i represents the i-th LSF segment of the j-th speech signal frame. In terms of the embedding algorithm, a certain segment of LSF index of every frame is able to hide 2-bit information. So, it is able to hide 100-bit data per second, that is, the embedding rate is 100 bps. Figure 3 is the pseudo code of the algorithm.

In terms of the j-th speech signal frame, continuous two-bit information $MbitL_i$ and $MbitL_{i+1}$ is read from $MbitL$. If $MbitL_i = 0$ and $M_j^i \bmod 2 = 0$, M_j^i remains unchanged; if $MbitL_i = 0$ and $M_j^i \bmod 2 \neq 0$, M_j^i minors 1. If $MbitL_i = 1$ and $M_j^i \bmod 2 = 0$, M_j^i minors 1; if $MbitL_i = 1$ and $M_j^i \bmod 2 \neq 0$, M_j^i remains unchanged. In terms of $MbitL_{i+1}$, the operations are similar to those of $MbitL_i$. Thus, information is hidden using two segments of LSF indexes in each speech signal frame.

5 EXTRACTION ALGORITHM

5.1 *Extraction algorithm of single LSF index.*

When the bit stream is received, the LSF indexes that are used for information hiding can be retrieved. Then, the extraction algorithm is applied for speech signal reconstruction. The extraction algorithm for the single LSF index is as follows: In terms of a certain speech signal frame, the LSF index that is used for information hiding does modular arithmetic of modulus $m(m = 2$ or $m = 4)$ accordingly. The result (n(n=0,1 or n = 0,1,2,3) is the hidden data. Offset value table can be obtained by decoding a specific speech signal frame. The full speech signal can be acquired by binary addition with the offset value table, permutation, and encoding.

5.2 *Extraction algorithm of multiple LSF indexes.*

When the bit stream is received, the two LSF indexes of each single frame are retrieved—the i-th and j-th indexes. The extraction algorithm for single LSF index is as follows: In terms of a certain speech signal frame, the LSF indexes that are used for information hiding do modular arithmetic of modulus 2. The result b1 and b2 are the hidden data. The offset value table can be obtained by decoding the specific speech signal frame. The full speech signal can be acquired by binary addition with the offset value table, permutation, and encoding.

6 EXPERIMENTAL RESULTS

6.1 *Experimental results of single LSF index at 50 bps*

6.60 kbps Mode. In the 6.60kbps rate mode, LSF parameters are quantized into five segments. Use these five segments, respectively, for information hiding, and calculate the MOS value using the PESQ algorithm. The results are as follows:

Table 1. Experimental results of single index in 6.60 kbps.

LSF Index No.	MOS Difference
1	0.0187
2	0.0276
3	0.3154
4	0.0917
5	0.7652

The experimental results show that in the 6.60kbps rate mode, the first, second, and fourth LSF indexes for information hiding have little impact on voice quality, whereas the third and fifth segments have a greater impact on voice quality.

Non-6.60 kbps Mode: In the non-6.60kbps rate mode, the LSF parameters are quantized into five segments. Use these five segments, respectively, for information hiding, and calculate the MOS value using the PESQ algorithm. The results are as follows:

Table 2. Experimental results of single index in non-6.60 kbps.

Rate Mode	LSF Index No.						
	1	2	3	4	5	6	7
8.85	0.0158	0.0145	0.0235	0.0457	0.6482	0.0285	0.9662
12.65	0.0246	0.0293	0.0127	0.0752	0.7841	0.0242	0.9832
14.25	0.0385	0.0151	0.0432	0.0756	0.8387	0.0135	1.2512
15.85	0.0214	0.0268	0.0372	0.0846	0.8114	0.0146	1.2756
18.25	0.0129	0.0158	0.0359	0.0672	0.8443	0.0224	1.2563
19.85	0.0157	0.0162	0.0194	0.0831	0.8564	0.0285	1.2721
23.05	0.0211	0.0161	0.0362	0.0824	0.8552	0.0364	1.2675

The experimental results show that in the non-6.60kbps rate mode, the first, second, third, fourth, and sixth LSF indexes for information hiding have little impact on voice quality, whereas the fifth and seventh segments severely degrade voice quality.

6.2 *Experimental results of single LSF index at 100 bps*

6.60 kbps Mode. The experimental results show that in the 6.60kbps rate mode, the first, second, and fourth LSF indexes for information hiding have little impact on voice quality.

Non-6.60 kbps Mode. The experimental results show that in the non-6.60kbps rate mode, the first, second, third, fourth, and sixth LSF indexes for information hiding have little impact on voice quality.

Table 3. Experimental results of single index in 6.60 kbp.

LSF Index No.	MOS Difference
1	0.0756
2	0.0176
3	0.3763
4	0.1027
5	0.5245

Table 4. Experimental results of single index in non-6.60 kbps.

Rate Mode	LSF Index No.						
	1	2	3	4	5	6	7
8.85	0.0204	0.0185	0.0156	0.0314	0.0376	8.85	0.0204
12.65	0.0266	0.0164	0.0236	0.0676	0.0194	12.65	0.0266
14.25	0.0265	0.0215	0.0442	0.0581	0.0386	14.25	0.0265
15.85	0.0275	0.0184	0.0521	0.0646	0.0186	15.85	0.0275
18.25	0.0167	0.0123	0.0275	0.0514	0.0291	18.25	0.0167
19.85	0.0269	0.0136	0.0564	0.0672	0.0164	19.85	0.0269
23.05	0.0273	0.0154	0.0476	0.0666	0.0223	23.05	0.0273

6.3 *Experimental results of multiple LSF indexes at 100 bps.*

6.60 kbps Mode. In the 6.60 kbps rate mode, using the first section, using the combinations of any two of the first, the second, and the fourth LSF index segments for information hiding will not significantly decline the sound quality.

Table 5. Experimental results of multiple indexes in 6.60 kbp

Rate Mode	LSF Index No.		
	1 and 2	1 and 4	2 and 4
8.85	0.0164	0.0932	0.086

Table 6. Experimental results of multiple indexes in non-6.60 kbps

Rate Mode	LSF Index No.		
	1 and 2	1 and 6	2 and 6
8.85	0.0156	0.0392	0.0386
12.65	0.0306	0.0525	0.0518
14.25	0.0272	0.0848	0.0792
15.85	0.0318	0.0588	0.0548
18.25	0.0091	0.0284	0.0434
19.85	0.0276	0.0256	0.0189
23.05	0.0182	0.0494	0.0326

Non-6.60 kbps Mode. In the non-6.60kbps rate mode, using the first section, using the combinations of any two of the first, the second, and the sixth LSF index segments for information hiding will not significantly decline the sound quality. The embedding rate is 100bps in the multiple LSF index algorithm.

7 CONCLUSION

In the AMR-WB codec process, the LSF is quantized into five segments in the 6.60kbps rate mode; in the non-6.60kbps rate mode, the LSF is quantized into seven segments. This paper proposes an information hiding algorithm based on LSF index parameters in different rate mode. The algorithm first selects one or several LSF index segments and then modifies the parameters for information hiding. Experimental results show that using a specific LSF index parameter has little impact on voice quality, which is difficult to be distinguished during normal conversation.

REFERENCES

[1] Zhang Chunling, Zhao Shenghui, Xiao Hongyuan, et al., (2013) An Improved method for AMR-WB speech codec. *Advanced Materials Research*, 756–759:1259.

[2] Yang Zhihua, Qi Dongxu, Yang Lihua. (2006) Detecting pitch period based on Hilbert-Huang Transform. *Chinese Journal of Computers*, 29(1).

[3] 3GPP. TS 26.194 V8.0.0 Adaptive Muti-Rate-Wideband (AMR-WB) speech codec [EB/OL]. http://tZhou Jijun, Yang Zhu, Niu Xinxin, et al., (2004) Research on the detecting algorithm of text document information hiding. *Journal of China Institute of Communications*, 25(12).

[4] Yu Zhengshan, Huang Liusheng, Chen Zhili, et al., (2009) High embedding ratio text steganography by Ci-Poetry of Song Dynasty. *Journal of Chinese Information Processing*, 23(4).

[5] Feng Dengguo. (2002) Status quo and trend of cryptography. *Journal of China Institute of Communications*, 23(5):18–26.

[6] Faisal Alturki, (2001) *Theory and Applications of Information Hiding in Still Images*, 2001, UMI Number: 3003824.

[7] Kadambe S, Boudreaux-Barrels G.F. (1992) Application of the wavelet transform for pitch detection of speech signals. *IEEE Transactions on Information Theory*, 38(2):91.

Electric, Electronic and Control Engineering – Shao, Shu & Tian (Eds)
© 2015 Taylor & Francis Group, London, ISBN: 978-1-138-02842-5

CMOS-Integrated accelerometer sensor for passive RFID applications

Maoxu Liu, Yigang He, Fangming Deng, Shan Li & Yuzhu Zhang
School of Electrical Engineering and Automation, Hefei University of Technology, Hefei, China
School of Electrical and Electronic Engineering, East China Jiaotong University, Nanchang, China

ABSTRACT: This paper designs an integrated accelerometer using 0.35-μm complementary metal oxide semiconductor technology for passive RFID applications. The accelerometer unit uses positive processing technology of bulk silicon based on Deep Reactive Ion Etching (DRIE) and then carries out two steps of dry etching after the standard IC process. The sensor interface is based on phase-locked architecture, which allows the use of fully digital blocks. Measurement results show that the proposed integrated accelerometer has benefits of good linearity and stability, covering 0.24 mm2 chip area, and consuming only 1.8μW power. The proposed accelerometer covers 0.23 mm2 chip area and consumes only 1.8 μW power under 1.2 V supply voltage, which is very suitable for passive RFID applications.

KEYWORDS: accelerometer; passive RFID tag; interface circuit.

1 INTRODUCTION

Micro-electro-mechanical-system (MEMS) accelerometers can detect tilt and vibration in real time, and they are playing an increasingly important role in the area of intelligent robot, auto control, modern consumer electronics, and so on. According to the detection principle, several types of MEMS accelerometer have been divided, for instance, capacitive, piezoelectric, piezoresistive, tunnel current type, and heat convection type, of which the capacitive accelerometer is one of the most widely studied devices due to its advantages of low power consumption, high sensitivity, low temperature sensitivity, and easy integration[1]. As the main fabrication technology of the integrated circuit, the complementary-metal-oxide–semiconductor (CMOS) technology can easily integrate the sensor with the memorizer and interface circuit, which provides higher accuracy, smaller chip area, and lower power consumption. So, it is very attractive to fabricate the capacitive MEMS accelerometer using CMOS technology [2].

Radio-frequency-identification (RFID), as the key technology of the Internet of things, is widely applied in traffic management, logistics transportation, medicine management, food production, and so on[3]. Generally, RFID tags can be divided into two groups: active and passive. Passive RFID tags are widely used due to their advantages such as battery-less operation, wireless communication, high flexibility, low cost, and fast deployment, so it becomes a trend to add sensing functionality to passive RFID tags, which have benefits of low fabricating cost, small circuit area, and high stability[4]. Hence, there are broad market prospects to integrate the accelerometer with passive RFID tags; however, the power consumption must be low as it determines the maximum working distance of the tags. For the main power consumption of the accelerometer coming from the interface circuit, the design of the interface circuit is the key in the design of the integration of the accelerometer with passive RFID tags.

The traditional interface is based on the operational amplifier. In [5], the conversion starts with a capacitance-to-voltage converter followed with a voltage-to-digital converter. This technique can achieve high speed and high resolution performance. However, due to the use of the operational amplifier in switched-capacitor amplifiers (SCA), this technique costs too much power dissipation, which is unsuitable for passive design. Accordingly, an inverter is proposed to replace the operational amplifier in SCA [6,7], which reduces the entire power dissipation significantly. However, they still employ a relatively high supply voltage. In [8,9], the pulse-width modulation technique commences with a capacitance-to-time converter followed with a time-to-digital converter. This technique is suitably applied in the field of large-scale capacitance variation. However, it always employs complicated architecture and still consumes high power.

This paper aims at designing a CMOS accelerometer with an integrated interface circuit for passive RFID application. The rest of this paper is organized as follows. Section 2 presents the principle and fabrication process of a CMOS MEMS accelerometer.

Section 3 introduces an ultra-low-power fully digital capacitive sensor interface based on an oscillator. The measurement results are presented and discussed in Section 4. Finally, some conclusions are drawn in Section 5.

2 PRINCIPLE AND FABRICATION PROCESS

2.1 *The principle of accelerometer*

The MEMS accelerometer has been divided into three types, according to its structure, including a finger-shaped, cantilever beam, and a pendulous accelerometer, on which the finger-shaped accelerometer can connect several capacitor plates in parallel to form a relatively large capacitance and get high sensitivity, low power consumption, and easy integration[10]. So it quite meets the requirements of the passive RFID. The following is the principle analysis:

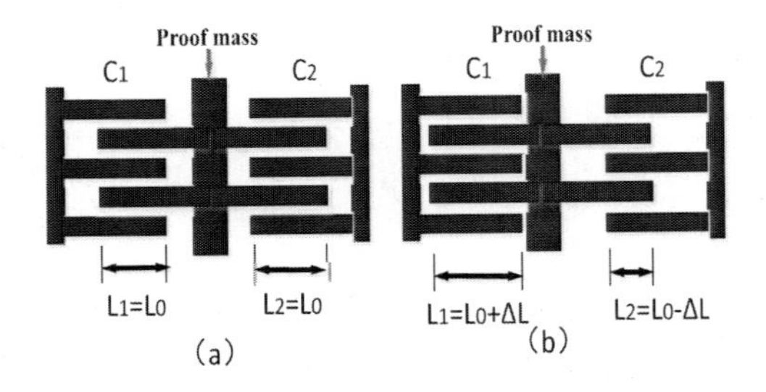

Figure 1. Detection principle of comb capacitive accelero meter.

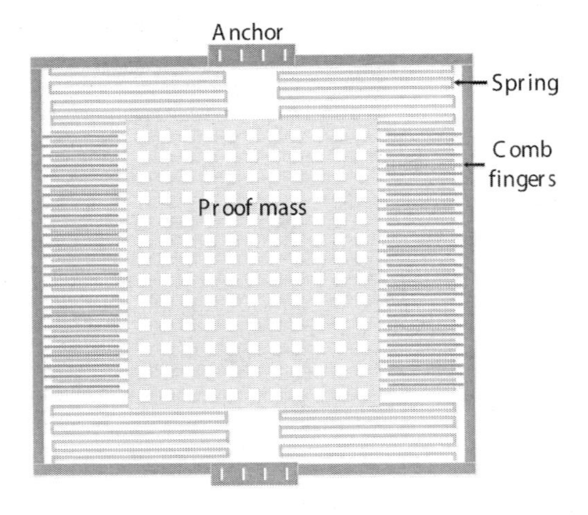

Figure 2. Top view of the CMOS-MEMS accelerometer.

As shown in figure 1 (a), the proof mass is in the balance position when there is no acceleration. And the differential capacitance is given by

$$C_1 = C_2 = C_0 = N \cdot \frac{\xi_0 \xi_r \cdot h \cdot L_0}{d} \quad (1)$$

$$L_1 = L_2 = L_0 \quad (2),$$

where ξ_0 is the permittivity of vacuum, ξ_r is the relative dielectric constant, N is the number of the capacitor plates, h represents the thickness of the capacitor plates, L_0 is the overlap length of the capacitor plates, and C_0 is the capacitance when the proof mass is in the balance position. As shown in figure 1 (b), the proof mass is offset from the balance position when the acceleration is not zero and the overlap length changes as follows:

$$L_1 = L_0 + \Delta L \quad (3)$$

$$L_2 = L_0 - \Delta L \quad (4)$$

So, the differential capacitance becomes

$$C_1 = N \cdot \frac{\xi_0 \xi_r \cdot h \cdot (L_0 + \Delta L)}{d} \quad (5)$$

$$C_2 = N \cdot \frac{\xi_0 \xi_r \cdot h \cdot (L_0 - \Delta L)}{d} \quad (6)$$

And the total capacitance variation is

$$\Delta C = C_1 - C_2 = N \cdot \frac{\xi_0 \xi_r \cdot h}{d} \cdot 2 \cdot \Delta L \quad (7)$$

According to Equation (7), we know that the capacitance variation ΔC has a completely linear relationship with the overlap length variation ΔL.

At frequencies much lower than the resonant frequency, the proof mass displacement to acceleration ratio can be expressed as follows:

$$\frac{\Delta L}{a} = \frac{m}{k} = \frac{1}{\omega_r^2} \quad (8),$$

where ΔL is the displacement, a is the acceleration, m is the proof mass, k is the spring constant, and ω_r is the resonant frequency of the microstructure. As m, k, and ω_r are constant, the displacement also has a completely linear relationship with the acceleration.

So, the spring will deform under inertia force and the overlap area will change subsequently, causing the change of the capacitance. The acceleration can be ascertained through measuring the capacitance, and theoretically, the nonlinearity will be zero.

2.2 *Fabrication process*

The microstructures, as shown in figure 2, contain springs, a proof mass, and an anchor, on which four serpentine springs are attached to the corners of the proof mass; the comb electrodes are distributed on both sides of the proof mass, and each side has 26 pairs.

Figure 3 shows the process steps for sensor fabrication using the Taiwan Semiconductor Manufactory Company (TSMC) 2P4M 0.35 μm CMOS process. (1) As shown in figure 3 (a), a 0.8-μm-thick silicon dioxide film and a 0.9-μm-thick aluminum thin film are deposited and patterned on the silicon substrate. A 1.5-μm-thick photoresist is then coved on the aluminum film. (2) Then, a masking layer whose opaque region includes springs, proof mass, anchor, and comb fingers is coved on the photoresist. The photoresist and aluminum without the protection of a masking layer is then removed and etched as illustrated in figure 3 (b). (3) Repeating the earlier described steps, four aluminum films can be obtained, as shown in figure 3 (c) and anisotropic reactive ion etching of silicon dioxide using CHF_3: O_2 for defining structural sidewalls is shown in figure 3 (d). (4)Then, isotropic deep silicon etching using XeF_2 and structure release is illustrated in figure 3 (e).

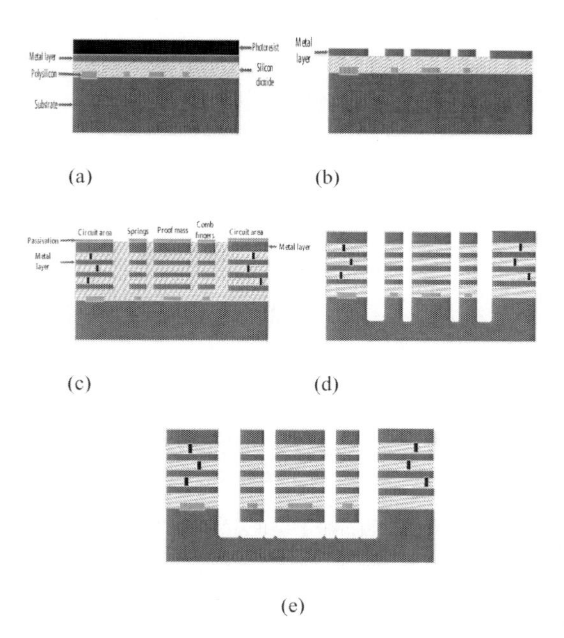

Figure 3. CMOS-MEMS process steps: (a) the photoresist is patterned (b) the first layer of aluminum is patterned (c) four layers of aluminum and passivation layer forming (d) anisotropic etching of SiO_2 layer (e) isotropic deep silicon etching and structure release.

Table 1. Main parameters of the accelerometer.

Parameter (unit)	Value
Sensor mass (kg)	2.4×10^{-9}
Quality factor	2.5
Spring constant ($N \cdot m^{-1}$)	2.0
Sensing capacity (fF/g)	2.1
Resonant frequency (kHz)	4.7
Comb finger length (μm)	71
Comb finger width (μm)	3.5
Comb finger gap (μm)	2.2
Sensor size (μm×μm)	400×560

Table 1 summarizes the primary design parameters of the accelerometer.

3 INTERFACE CIRCUIT DESIGN

CMOS technology has been mainly developed for digital Integrated Circuit (IC). Benefiting from technology scaling, digital IC achieves great improvements in speed, power dissipation, chip area, and so on. However, analog IC is less scalable and suffers more when dealing with nanometer CMOS technology. Apart from the issues of matching and noise, the threshold voltage is reduced less remarkably compared with the power supply, resulting in reduced voltage headroom for traditional analog amplitude-based sensor interfaces [11]. To cope with this challenge, a novel method transferring sensor signal processing from the traditional voltage domain to the frequency domain has been put forward in recent years [12,13]. This method allows the use of fully digital circuits rather than analog circuits, overcoming the limitation of the decreased voltage headroom. Hence, the interface circuit with this approach can work with ultra-low supply voltages and is especially suitable for low-power interface design.

3.1 *PLL-based sensor interface theory*

The architecture of the PLL-based sensor interface, developed from Danneels [13], is shown in figure 3. It consists of two main blocks: a frequency-modulating block, which converts the sensor information to the frequency domain, and a frequency-demodulating block, which converts the frequency to the digital domain, resulting in a complete sensor-to-digital flow. The frequency-modulating block consists of a Sensor-Controlled Oscillator (SCO) and directly converts the capacitive value of the sensor to corresponding frequency fs. The frequency-demodulating block is a digital first-order Bang-Bang-Phased-Locked-Loop (BBPLL), consisting of a single-bit phase detector and a Digitally Controlled Oscillator (DCO).

This BBPLL measures whether fs leads or lags fd (the frequency of the DCO), and, hence, the DCO is only steered by a single-bit signal bo. When the entire feedback loop is locked, fs shifts between a maximum and minimum value that corresponds to the maximal and minimal value of the sensor (seen in figure 4). The average digital frequency fd will correspond to the sensor frequency fs. Therefore, the over-sampled output bo represents the digital value of the sensor value.

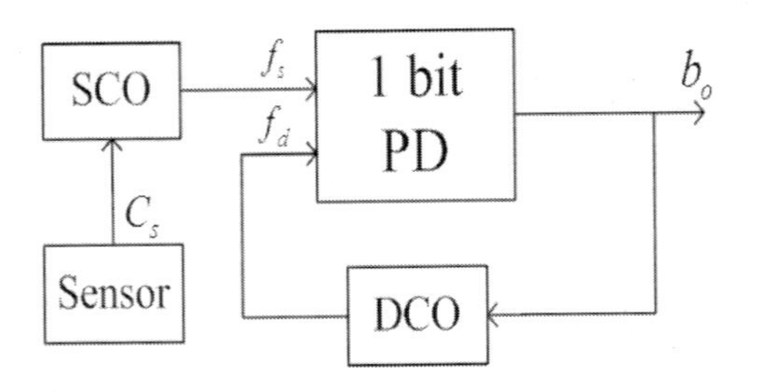

Figure 3. Architecture of PLL-based sensor interface

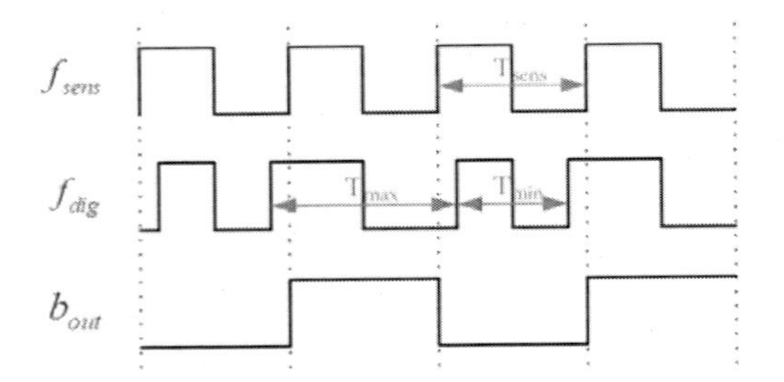

Figure 4. Corresponding waveforms in locked-state.

This architecture employs the BBPLL block, which has advantages ranging from low voltage capabilities, low power consumption, small chip area, and scalability to smaller technologies and robustness to process variation [14]. Another significant improvement of this architecture is the direct conversion from the capacitive sensor signal to the frequency domain, avoiding any intermediate transformation of the capacitive information to the voltage domain.

3.2 *Implementation of fully digital capacitive sensor interface*

The implementation of the proposed fully digital capacitive sensor interface is shown in figure 5a. Both the SCO and the DCO are implemented as 3-stage inverter-based ring oscillators (seen in figure 5b). The PD is implemented as a d flip-flop (seen in figure 5c). The sensor capacitor C_s acts as the variable load on a single stage of the SCO, thereby generating sensor-controlled frequency f_s. The variable capacitive load on a single stage of the DCO consists of two capacitors, C_o and C_m. The capacitor C_o, designed equal to the quiescent value of C_s, is always connected to the DCO. But the capacitor C_m, designed slightly larger than the maximum variation of C_s, is swapped in or out of the DCO depending on the feedback from the single-bit phase detector. Considering the issues of system linearity and process variation, C_m is usually designed as a programmable capacitor.

The oscillating frequency of the multistage relaxation oscillator depends on the time it takes a transition to propagate around the loop. For a standard ring oscillator and assuming an equal rise and fall time for the different stages, respectively, the oscillating frequency f_o can be expressed as follows:

$$f_0 \approx \frac{I_l}{V_m C_l} \quad (9),$$

where I_l is the current flowing through the inverter, V_m is the swing range of the output voltage, and C_l is the load capacitor, which equals C_s or C_o+C_m for the SCO or the DCO, respectively.

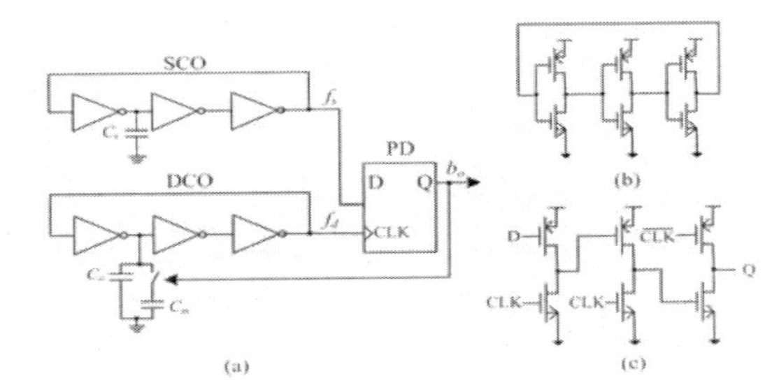

Figure 5. Proposed fully digital capacitive sensor interface. (a) entire architecture. (b) schematic of ring oscillator. (c) schematic of phase detector

The matching between the SCO and DCO plays an important role in this design. Mismatch between the SCO and DCO can be expressed as an offset between the coefficients of the polynomials and the characteristics of the oscillators, resulting in a deterioration of the effective number of bits (ENOB). Considering the issues of system linearity and process variation, C_m is usually designed as a programmable capacitor. When the loop is stable, the average values over time of f_s and f_d are equal, thus the two frequency controls of the SCO and DCO are correlated. Since both oscillators are implemented identically, the control of the SCO, the sensor value, is correlated to the control of the DCO, the single-bit output of the phase detector. Hence, the value of the sensor is digitized in this single-bit DCO control signal b_o. In order to increase resolution, b_o is oversampled by taking its average duty cycle overtime.

4 MEASUREMENT RESULTS AND DISCUSSION

The 0.35 μm CMOS process of TSMC is employed to fabricate the proposed accelerometer and sensor interface whose microphotograph is shown in figure 9, in which (a) represents the accelerometer and (b) is the interface circuit.

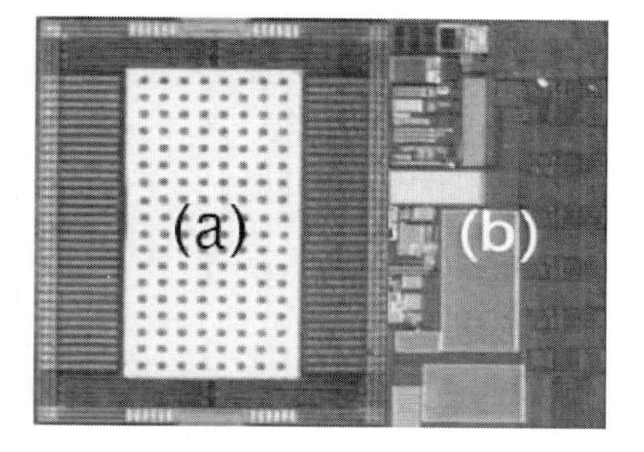

Figure 9. Microphotograph of the fabricated chip.

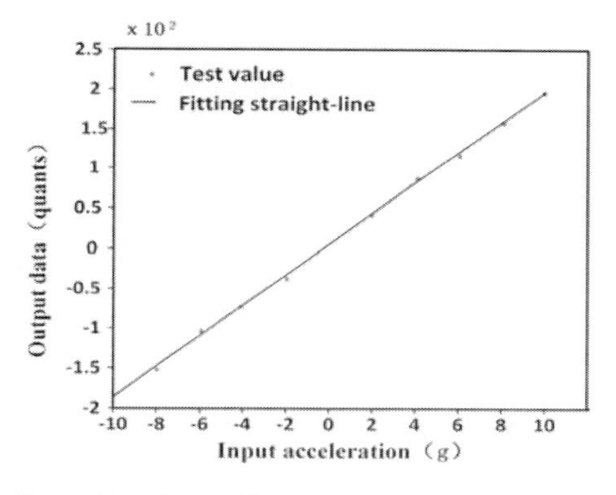

Figure 10. The nonlinear test curve of accelerometer.

This paper tests the performance of the accelerometer, for instance, nonlinearity, sensitivity, frequency response, and stability using the sinusoidal excitation method. The accelerometer chip is fixed in the central position of the vibration platform, keeping the measured direction perpendicular to the vibration table, and is then motivated by a single sinusoidal vibration excitation. And we can get the parameters of the accelerometer through measuring the output of the counter $D_{out.}$

As illustrated in figure 10, after the quadratic fit to the data using the least-square method, the sensitivity of the accelerometer is 1.86 × 104 quants/g within -10g and +10g, and the nonlinearity is about 0.76%, so the proposed accelerometer has a good linearity.

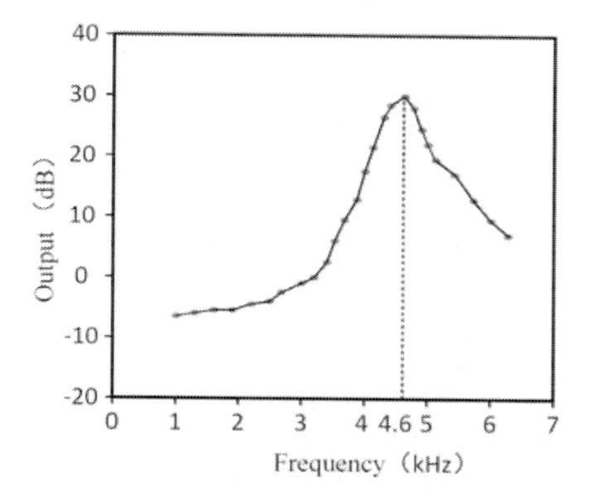

Fig.11 Test charts of resonance frequency.

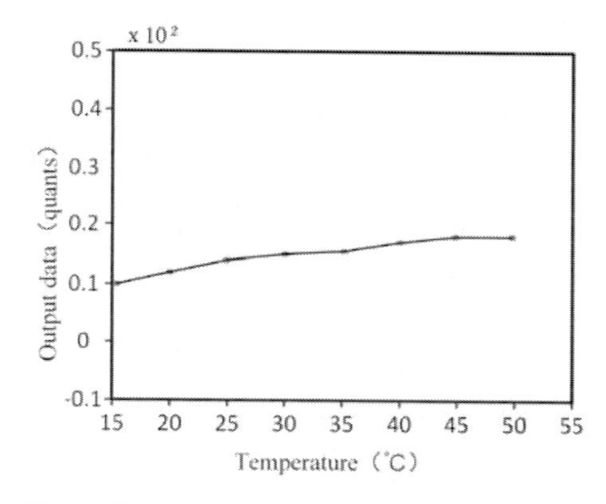

Figure 13. Test of temperature characteristic.

The measured resonant frequency of the accelerometer, as illustrated in figure 11, is 4.6kHz, which is slightly lower than the design value, as there is a line loss with photoetching, resulting in the narrowing down of the springs and the decrease of the spring constant *k*. From Equation (8), it is observed that *k* is proportional to the resonance frequency squared ω_r^2, so the resonance frequency also decreases.

As the change of the temperature has a great influence on zero offset for changing offset voltage and offset current, we test the temperature stability of the zero offset. The accelerometer system is smoothly settled in a high-precision water oil tank with a constant temperature, and it is then placed horizontally on the horizontal vibration isolation platform.

We should ensure that the input of acceleration is zero at a temperature of 15°C, and we should collect data every second for a total of 4000 seconds. After data processing, the variance is about 2.15quants and the zero offset is about 0.42mg. Then, we should change the temperature of the water oil tank and collect data every 5°C, as illustrated in figure 13. The stability of zero offset is not affected by the change of the system temperature, as the design adopts a full digital circuit that is less influenced by temperature.

Though the design in [7] and [12] achieves the comparable power dissipation, they employ a more expensive fabrication process than our design and [12] only achieve 8.1 bits of Effective Number Of Bits(ENOB).

Table 2. Performance comparison of integrated sensor interfaces.

	Process /μm	Supply voltage/V	ENOB / bits	Area / mm^2	Power /μW
[6]	0.16	1.2	12.5	0.15	10.3
[7]	0.09	1.0	10.4	0.45	3.0
[8]	0.35	3.0	9.3	0.09	54.0
[9]	0.32	3.0	9.8	0.52	84.0
[12]	0.18	0.5	8.1	0.13	1.1
This paper	0.35	1.2	10.5	0.23	1.8

Table 2 shows a comparison of the performance parameters of this sensor interface with previous designs. This fully digital sensor interface consumes only 1.8 μW power at 1.2V supply.

5 CONCLUSION

According to the rapid development of RFID, this paper presents an integrated accelerometer using the Taiwan Semiconductor Manufactory Company (TSMC 2P4M 0.35 μm CMOS process. The accelerometer unit uses positive processing technology of bulk silicon based on DRIE and then carries out two steps of dry etching after the standard IC process. The sensor interface is based on phase-locked architecture, which allows the use of the fully-digital blocks. Measurement results show that the proposed integrated accelerometer achieves excellent linearity and stability. The proposed accelerometer covers 0.24 mm2 chip area and consumes only 1.8 μW power under 1.2 V supply voltage, which proves especially suitable for passive RFID tag design. However, the proposed sensor interface shows only moderate ENOB, which encourage us to design a higher ENOB interface for low-power RFID application in future.

ACKNOWLEDGMENTS

This work was supported by the National Natural Science Funds of China for Distinguished Young Scholar under Grant No. 50925727, The National Defense Advanced Research Project Grant No.C1120110004, 9140A27020211DZ5102, the Key Grant Project of Chinese Ministry of Education under Grant No.313018, Anhui Provincial Science and Technology Foundation of China under Grant No. 1301022036, and Hunan Provincial Science and Technology Foundation of China under Grant No.2010J4 and 2011JK202.

REFERENCES

[1] Liu Yu, Ju Weibin, Liu Yuxi, (2010) Research progress of gauge check technology of MEMS accelerometers. *Metrology & Measurement Technology*, 30(4):5–8.

[2] Chihming Sun, Minghan Tsai, Yuchia Liu, Weileun Fang, et al. (2010) Implementation of a Monolithic Single Proof-Mass Tri-Axis Accelerometer Using CMOS-MEMS Technique. *IEEE Transactions on Electron Devices*, 57(7):1670–1678.

[3] Li B, He Y G, Zuo L, et al. (2014) Metric of the application environment impact to the passive UHF RFID system. *IEEE Transactions on Instrumentation and Measurement*, 63(10):2387–2395.

[4] Wang B, Law M K, Bermak A. et al. (2014) A passive RFID tag embedded temperature sensor with improved process spreads immunity for a -30°C to 60°C sensing range. *IEEE Circuits and System Magazine*, 61(2):337–346.

[5] Xia S, Makinwa K, Nihtianov S. (2012) A capacitance-to-digital converter for displacement sensing with 17b resolution and 20μs conversion time. *International Solid-State Circuits Conference Digest of Technical Papers, IEEE, 2012*, pp:198–200.

[6] Tan Z, Daamen R, Humbert A, et al. (2013) A 1.2-V 8.3-nJ CMOS Humidity Sensor for RFID Applications. *IEEE Journal of Solid-State Circuits*, 48(10):2469–2477.

[7] Nguyen T T, Hafliger P. (2013) An energy efficient inverter based readout circuit for capacitive sensor. *Biomedical Circuits and Systems Conference, IEEE, 2013*, pp:326–329.

[8] Sheu M L, Hsu W H, Tsao L J. (2012) A capacitance-ratio-modulated current front-end circuit with pulsewidth modulation output for a capacitive sensor interface. *IEEE Transactions on Instrumentation and Measurement*, 61(2):447–455.

[9] Nizza N, Dei M, Butti F, et al. (2013) A low-power interface for capacitive sensors with PWM output and intrinsic low pass characteristic. *IEEE Transactions on Circuits and Systems I: Regular Papers*, 60(6):1419–1431.

[10] Weigold, J.W., K. Najafi. & S.W. Pang. (2001) Design and fabrication of submicrometer, single crystal si accelerometer, *Journal of Microelectro-Mechanical Systems*, 10(4):518–524.

[11] Chatterjee S, Tsividis Y, and Kinget P. (2005) 0.5 V analog circuit techniques and their application in OTA and filter design. *IEEE J. Solid-St. Circ.*, 40:2373–87.

[12] Danneels H, Piette F, de Smedt V, Dehaene W, and Gielen G. (2011) A novel PLL-based frequency-to-digital conversion mechanism for sensor interfaces. *Sens. Actuators A Phys.*, 172:220–27.

[13] Danneels H, Coddens K, and Gielen G. (2011) A fully-digital 0.3V 270nW capacitive sensor interface without external references. *Proc. 37th European Solide-State Circuits Conference (Helsinki, Finland)*, pp:287–90.

[14] Shulaker M M, Rethy J V, Hills G, Wei H, Chen H Y, Gielen G, Wong H S P and Mitra S. (2013) Sensor-to-digital interface built entirely with carbon nanotube FETs. *IEEE J. Solid-St. Circ.*, 49:1–12.

Electric, Electronic and Control Engineering – Shao, Shu & Tian (Eds)
© 2015 Taylor & Francis Group, London, ISBN: 978-1-138-02842-5

Ship course control based on humanoid intelligent control

Yue Zhao
College of Navigation, Jiangsu Maritime Institute, Nanjing, China

Yuelin Zhao
College of Navigation, Dalian Maritime University, Dalian, China

Renqiang Wang
College of Navigation, Jiangsu Maritime Institute, Nanjing, China

ABSTRACT: The ship course intelligent control algorithm is designed in this paper. First, the objective of ship course control is proposed by the input of control rudder angle and the output of ship heading; Then, according to the deviation trend of the ship course output and its ideal trajectory, information about course deviation and its rate is extracted. Finally, in order to achieve ship course control, rudder angle control strategy is enacted under a specific state by using human thinking, reasoning, and control strategy. The simulation is executed with MATLAB software, whose results show that the controller, which is stable, has strong robustness.

KEYWORDS: humanoid intelligent control; control strategy; ship; course control; simulation.

1 INTRODUCTION

The ship is a typical pure lag, non-linear, and time-varying uncertain system, and therefore the conventional numerical PID control could no longer precisely conduct the ship course control. The sliding mode variable structure control theory[1] provides the non-linear and uncertain control of ship motion with an effective solution, but such a method possesses problem of serious control input vibration, which is against the engineering realization; the adaptive control[2] based on the adaptive control law makes a fundamental achievement in the non-linearity and uncertainty aspects, but its defect is the reliance on the mathematical model of the controlled object. Therefore, the exploration of the effective and practical project realization control algorithm is the main part of future research.

Humanoid control[3-4] works on the basis of the human thought mode and reasoning-control strategy; without the reliance on the model of controlled objects, it processes the decision control of the controlled object's featured state (system deviation and deviation rate) and then reaches the desired objective. At present, humanoid control has be widely applied in the non-linear time-varying uncertain system, such as the vehicle parking control[5], electro-hydraulic positioning servo system automatic control[6], temperature automatic control of hot working production line with time delay characteristics[7], and robot route planning control[8].

With such a purpose, this paper adopts the humanoid intelligence control algorithm and directly conducts decision control to featured states in order to realize the rapid, stable, and precise control of the ship course.

2 RESPONSIVE MODEL OF SHIP MOTION

In the 1950s, the responsive mathematical model of ship motion based on input of ship rudder angle and output of heading was proposed by Nomoto Kensaku[9] from the perspective of control theory. Bech[9], who developed the ship maneuvering mathematical model that is applicable to ship course, has drastically changed, whereas it is able to explain unstable and nonlinear phenomena of ship heading based on the second-order Nomoto model. Bech's model also developed the consideration of external disturbance and ship system parameter perturbation, with the mathematical expression of

$$T_1T_2\dddot{\varphi}+(T_1+T_2)\ddot{\varphi}+KH(\varphi)=K(T_3\dot{\delta}+\delta)+\Delta \tag{1}$$

In the expression, $H(\dot{\varphi})=\alpha_0+\alpha_1\dot{\varphi}+\alpha_2\dot{\varphi}^2+\alpha_3\dot{\varphi}^3$, Δ is the uncertain item.

Therefore, the aim of this paper is to find a control law (δ), which makes the ship heading output (φ) asymptotic tracking expected course (φ_r), that is, when $t\rightarrow\infty$, the tracking error $e=y-\varphi_r\rightarrow 0$.

3 SHIP COURSE HUMANOID INTELLIGENT CONTROL STRATEGY

3.1 *Humanoid intelligent control concept*

The humanoid control algorithm was first introduced by Professor Bo Jian-guo and Professor Zhou Qi-jian[4] from Chongqing University in 1979. The research was based on the measurement, memorizing, differentiation, decision, and intelligent behaviors of the controlled target. Such an algorithm adopted the hierarchical structure, and it simulated human experience techniques and control, reasoning logic, instinct identification, and other forms of knowledge experience.

The design process of intelligent control algorithm[10-11] is based on the acquirement of system input and deviation information, establishing multi-mode control law and featured state relation. Its control algorithm could be described in the form of production rule "IF <condition>, THEN <result>".

3.2 *Ship course humanoid intelligent control strategy*

1 Ship course humanoid control concept:

Based on the trend of deviation between ship course output and desired trajectory, as illustrated in figure 1, the course deviation information and deviation changing information are extracted. And the ship course control is realized by human thought mode, reasoning, and the control strategy, which are designed in the specialized state. The control system diagram is described in figure 2.

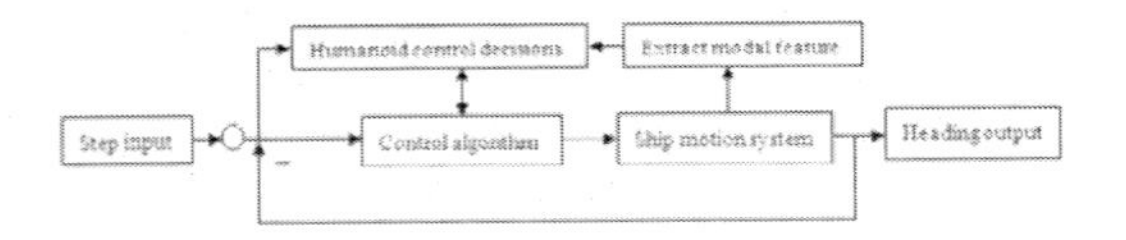

Figure 1. Ships course humanoid intelligent control system diagram.

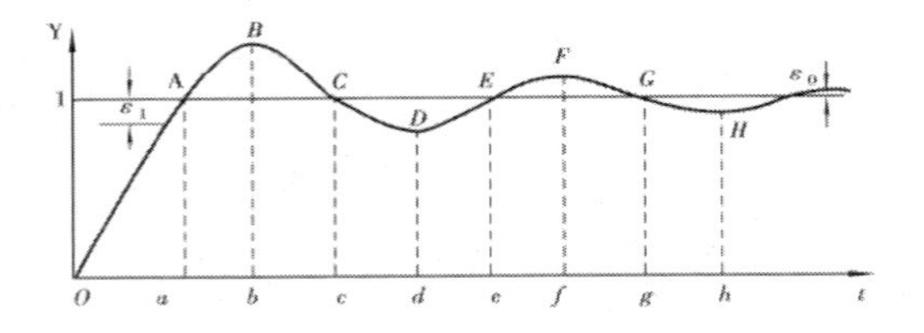

Figure 2. Ship course step response curve.

2 Ship course humanoid control strategy:

Based on the position of object trajectory on the error phase plane ($e-\dot{e}$), as illustrated in figure 3, on account of the deviation (between the desired trajectory and system motion) generated when the system motion is in a certain featured state of the featured model, and the motion trend of the desired trajectory, the humanoid control decision behavior is simulated; then, the control mode or rectify mode is designed; and, finally, the specific parameters[8] of the modal are designed.

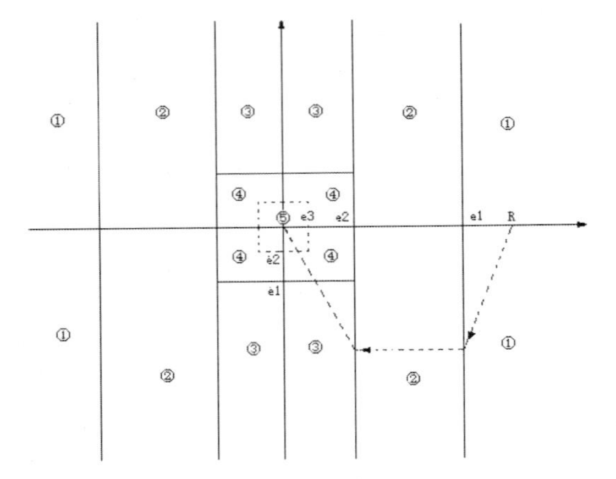

Figure 3. Ideal target trajectory of error.

Aiming at the realization of uniformity between the error trajectory and the desired error object trajectory, the humanoid control strategy is introduced as given next:

A, in region ①, when the deviation is obvious, adopting the possible strongest control, such as the bang-bang control, namely,

$$\text{IF: } |e_n| > e_1, \text{ Then: } u_n = \text{sgn}(e_n)\cdot U_{\max} \tag{2}$$

B, in region ⑤, when the deviation and the deviation changing rate is considerably small, in order to eliminate the error, the PID control could be adopted, namely,

$$\text{IF: } |e_n| < e_3 \cap |\dot{e}_n| < |\dot{e}_2|, \text{ Then:}$$

$$u_n = k_{p5}\cdot e + k_{d5}\cdot \dot{e} + k_{i5}\int edt \tag{3}$$

C, in region ②; if the deviation is relatively obvious, then adopt the proportional modal control; in the meantime, to guarantee the error changing rate is not over scale, the weak differential control is introduced, namely,

$$\text{IF: } |e_n| < e_1 \cap |e_n| > e_2, \text{ Then: } u_n = k_{p2}\cdot e + k_{d2}\cdot \dot{e} \tag{4}$$

D, in region ④, in the process of deviation diminishing, if the deviation changing rate is lower or equal to the scheduled one, the combined proportional modal and differential modal control could be adopted, namely,

$$\text{IF: } A \cap B \cap C \text{, Then: } u_n = k_{p4} \cdot e + k_{d4} \cdot \dot{e} \quad (5),$$

$$A: |e_n| < e_2$$
where $B: |\dot{e}_n| < |\dot{e}_1|$
$$C: |e_n| < e_3 \cap |\dot{e}_n| < |\dot{e}_2|.$$

E, in region ③, in the process of deviation diminishing, if the deviation changing rate is higher than the scheduled one, then the strong differential control should be introduced on the basis of the proportional modal, to make the deviation changing rate diminish as fast as possible, namely,

$$\text{IF: } |e_n| < e_2 \cap |\dot{e}_n| \geq |\dot{e}_1| \text{, Then: } u_n = k_{p3} \cdot e + k_{d3} \cdot \dot{e} \quad (6)$$

3 Ship course humanoid control input limits

According to the principle of ship course control, the control input is made from the servo rudder machine onboard the ship. Therefore, the feature of the rudder machine, namely the limitation of the rudder turning rate of the rudder machine, has to be considered. The rudder machine feature formula is shown as follows:

$$T_E \dot{\delta} = \delta_E - \delta \quad (7),$$

where T_E is the rudder machine feature index, δ_E is the command rudder angle, and with the limitation of $|\delta_E| \leq 35^0$, $\dot{\delta}$ is the rudder speed, with the limitation of $|\dot{\delta}| = 3^0/s$.

4 SIMULATION

With the models of Dalian Maritime University training ship "Yulong" (Ships parameters refer to literature 9) and the COSCO container ship "COSCO ROTTERDAM" (Ships parameters refer to literature 12), the Matlab/Simulink ship course humanoid simulation platform is established, as illustrated in figure 4.

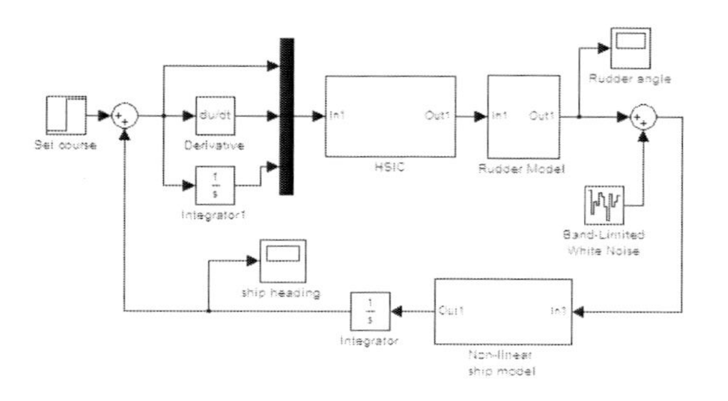

Figure 4. MATLAB simulation platform.

4.1 *Alter course experiment based on humanoid control*

We should implement the alter course experiment on M/V "Yulong" and "COSCO ROTTERDAM" and set the course at 20°. After adding the external random disturbance, the experiment generates the ship course output and control rudder angle input, as illustrated in figures 5 and 6. From these two figures, it could be witnessed that the ship course output is rapid, stable, has no overshoot, and is not sensitive to the external disturbance, and the control rudder angle output is smooth and reasonable.

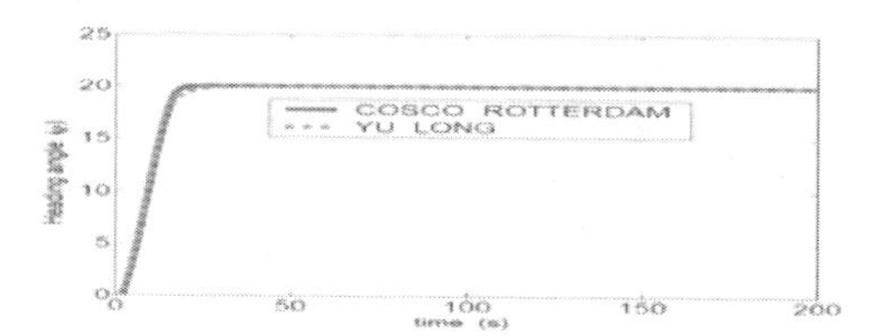

Figure 5. Output of ship heading.

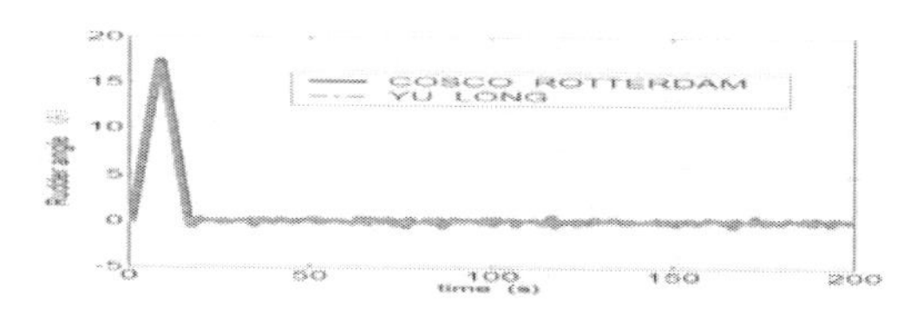

Figure 6. Input of rudder control.

4.2 *Alter course control comparison experiment*

The comparison simulation experiment is implemented with the ordinary PID and adaptive control algorithm. After setting the course at 20° and adding the external random disturbance, the ship course outputs, which are illustrated in figure 7 (the comparison simulation experiment on M/V "YU LONG") and figure 8 (the comparison simulation experiment on M/V "COSCO ROTTERDAM"), are achieved. From these two figures, it could be witnessed that the ship ourse output is rapid, stable, and with no overshoot and vibration.

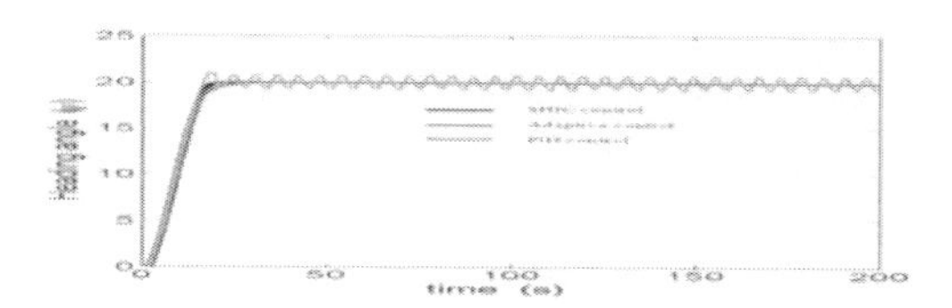

Figure 7. Ship heading output of YU LONG.

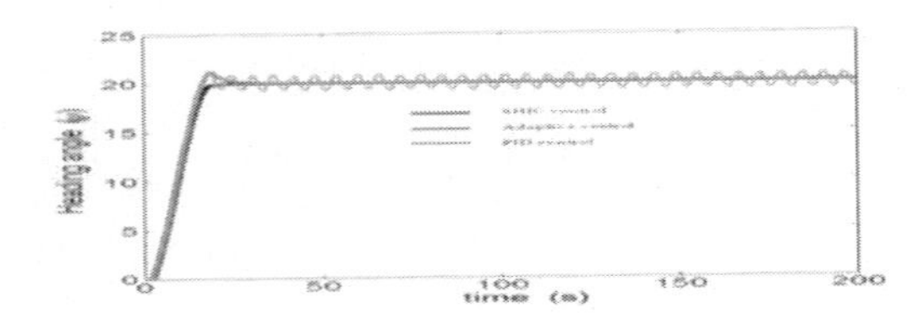

Figure 8. Ship heading output of COSCO ROTTERDAM

In summary, the controller designed in this paper has superior performance and strong robustness

5 CONCLUSIONS

The rudder angle control mode is established on the basis of ship rudder angle input and course output, and with the reference of the humanoid control concept, the course output error and its changing state, and fundamentally implementing the ship course multi-mode control. Such a control mode is free of the reliance on the ship model, and it possesses high flexibility and desired effects.

ACKNOWLEDGMENT

This research work was supported by the Jiangsu Maritime Vocational and Technical College project funds under Grant No. 2013QNZX-01.

REFERENCES

[1] Chen Zhimei, Wang Zhenyan, Zhang Jinggang. (2012) Sliding mode variable structure control theory and applications. Beijing: Electronic Industry Press.

[2] Ge S S, Li G Y, Lee T H. (2003) Adaptive NN control for a class of strick-feedback discrete-time nonlinear systems. *Automatic*, 39(5):807–819.

[3] Cai Zixing, Zhou Xiang, Li Meiyi. (2000) A novel intelligent control method evolutionary control[C]//*Pro 3th World Contgress Intelligent Control & Automation*. Heifei, China, 1:387–390.

[4] Liu Shujun, Gai Xiaohua, Zhang Nanlun. (2008) Research and simulation of anti-disturbance problem on improved characteristic model algorithm of HSIC. *Journal of System Simulation*, 6(20):2905–2908.

[5] Tu Yaqing, Chen Hao, Yang Huiyue. (2014) Autonomous parking method based on human-simulated intelligent control. *Control Engineering of China*, 3(21):161–167.

[6] Guoqing Zhou. (2014) Simulation of muti-modal control in electro-hydraulic. *Hydro-Mechanics Engineering*, 6(42):115–119.

[7] Zhang Zuying. (2014) Temperature control technology of hot working production line with time delay characteristics. *Foundry Technology*, 2(35):386–388.

[8] Ke Wende, Peng Zhiping, Cai Zesu, Chen Ke. (2013) Constraint and optimization control on similar stepping upstairs for humanoid robot. *Robot*, 3(36):233–240.

[9] Jia Xinle, Yang Yansheng. (1999) *Ship Motion Mathematical Model*. Dalian: Dalian Maritime University Press.

[10] Liang Dongwu. (2008) *A Thesis Submitted to Chongqing University in Partial Fulfillment of the Requirement for the Degree of Master of Engineering*. Chongqing: Chongqing University.

[11] Chen Guiqiang. (2007) *Parameter Optimization and Structure Automation Design of Human-Simulated Intelligent Controller*. Chongqing: Chongqing University.

[12] Shen Zhipeng, Guo Chen, Zhang Ning. (2007) Mathematical modeling and simulation of 5446TEU ultra large container Ship. *J. Cent. South Univ.* (Science and Technology), 8(38):808–812.

Electric, Electronic and Control Engineering – Shao, Shu & Tian (Eds)
© 2015 Taylor & Francis Group, London, ISBN: 978-1-138-02842-5

Integrated evaluation method for transmission grid safety and economy and its application

Hongjun Fu, Hongyin Wang, Jun Chen, Gaichao Xue & Zhiheng Li
State Grid Henan Electric Power Company, Zhengzhou Henan, China

ABSTRACT: With the rapid growth of China's national economy and the upgrading of smart grid technology, transmission grids have upgraded in terms of both economy and safety. How to maintain a balance between its economy and its safety has become the most important topic for our study.

KEYWORDS: transmission grid safety; evaluation system.

1 INTRODUCTION

With the rapid growth of China's national economy and the upgrading of smart grid technology, China's electric power industry has been substantially upgraded. The rapid development of the electric power industry, however, is also accompanied by a series of concerns including grid safety, grid pollution and energy quality. The most important issue is, of course, grid safety, which is subdivided into grid safety, transmission grid safety and distribution grid safety. Grid safety is decided by many factors including internal and external factors. The former involve management and equipment while the latter involve safety in connection with natural disasters. Our study is focused on transmission grid safety and ways to improve the economy of the power grid by solving its safety problems and to further reduce the operating cost of the power grid by solving the safety problems of the transmission grid.

As State Grid Henan Electric Power Company, we are responsible for the planning, design of the province's power grid and the maintenance of the transmission grid system and distribution grid system. When planning the construction of a transmission grid, the first thing we must consider is the safety aspect of the transmission grid, since the safety operation of the transmission grid is directly decisive to the people's liveliness as well as a number of issues relating to the national economy and infrastructure construction. As a provincial electric power company, when designing and planning a transmission grid, the first thing we must consider is the safety defense of the transmission grid or how to avoid safety accidents instead of how to remove them afterwards. Hence, the transmission grid should be designed and planned from the macro perspective and include for a number of related aspects including the surroundings of the transmission line, material for building the transmission line, the monitoring network and communication network.

2 TRANSMISSION GRID SAFETY EVALUATION METHOD

Power system safety means the system's ability to continue stable operation when encountering a failure and avoid extensive power failure. Power system safety includes steady operation and transient operation. So far, evaluation methods for power system safety typically include DC load flow method, compensation method and sensitivity analysis method that evaluate the transmission line safety. As DC load flow method ignores reactive power and ground capacitance and does not involve repeated iteration of values during the computation process, the computation process is simpler and occupies less memory. In the paragraphs below, we are going to describe how to evaluate power system safety using DC load flow method.

When calculating the load flow of an anticipated accident, a DC load flow model is superior over an AC model. For the safety evaluation of a high-voltage or an ultrahigh-voltage grid system, DC load flow modeling has lower error rate, which is 3% ~5 % maximum.

When a DC load flow model is used, we generally make the following assumptions:

1 Assume X >> R, where R is ignorable, the conductor susceptance is:

$$b_{ij} = -\frac{1}{x_{ij}} \tag{1}$$

Here: X_{ij} represents the unearthed branch reactance.

2 Assume the voltage phase angular differences between the branch ends is approximately zero, we have:

$$\sin\theta_{ij} = \sin(\theta_i - \theta_j) \approx \theta_i - \theta_j \qquad \cos\theta \approx 1 \tag{2}$$

3 Ignore all ground susceptances of the branch, then: $b_{i0}=b_{j0}=0$

4 Assume the voltage mode values at all nodes are the same, order it equals to the per unit value 1.0, then: $U_i=U_j=1$

Substitute the simplified condition into formula (1), we get:

$$\begin{cases} p_{ij} = -b_{ij}(\theta_i - \theta_j) = \dfrac{\theta_i - \theta_j}{x_{ij}} \\ Q_{ij} = 0 \end{cases} \tag{3}$$

From Kirchhoff's current law we know that the sum of the current inflow into a node is constantly equal to the current outflow. This is expressed with power as:

$$p_i = \sum_{j \in i} p_{ij} \tag{4}$$

Assume the equilibrium node $\boldsymbol{\theta}^s=0$ phase angle, then, for n- 1 nodes beyond equilibrium node s, we can obtain a DC load flow model composed of these nodes $\mathbf{P}=\mathbf{B}\boldsymbol{\theta}$.

3 TRANSMISSION GRID SAFETY EVALUATION INDICES

In power system operation, it is necessary to set safety evaluation indices to rate the grid safety. Grid operation safety indices are very important as each of them must be able to rate the safety of the grid operation system both in a qualitative and a quantitatively manner. It must be able to guide the grid operation as well as remove potential safety hazards. The evaluation indices for transmission grid safety include danger index, power supply loss rate index and risk index.

1 Danger index can be expressed by a formula, $DI = {}^{I}\!/\!{}_{L}$, where I is the sum of system disconnected circuits and L is the sum of the total branch circuit values. The larger the number of the system disconnected circuits, the higher the danger coefficient of the grid operation.

2 Power supply loss rate index is related to the failure in one of the generator units, when the power output is reduced. Also, reduced current output and decreased voltage can also lead to extensive power failure. The calculation method is shown in the chart below:

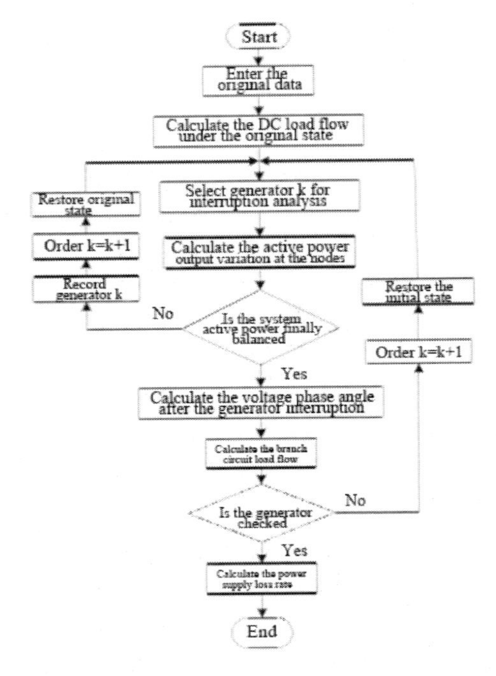

Figure 1. The calculation method.

3 Risk index

By anticipated accident calculation, we select the elements that may give rise to system overload after breaking. To reflect the cases of system overload, we introduce a performance index k_{PI} to quantify the grid safety.

$$PI_k = \sum_{l=1}^{\beta} w_l \left(\frac{p_l'}{p_l^{lim}} \right)^2 \tag{5}$$

Here: p_l' is the over-limit branch circuit load flow of element k after breaking; p_l^{lim} is the ultimate capacity of this circuit; L is the number of over-limit branch circuits; w_l is the weight coefficient of branch circuit 1, which represents the effect of the branch circuit failure on the system and can be selected according to operating experience.

As index PI_k merely represents the adverse effect of a certain occasional accident on the system and the actual system safety is subject to all possible accidents, to describe this overall effect, we define the risk index *RI* to reflect the overall risk level of the system.

$$RI = \sum_{k=1}^{N_B} PI_k \tag{6}$$

4 GRID ECONOMY ANALYSIS UNDER DIFFERENT SAFETY RATINGS

As electric power resources are becoming more and more marketized, the conflict between the economy of electric power construction and the safety of electric power operation is also becoming more and more outstanding. In the grid construction and operational management across Henan province, this has been, and will always be, the core of concern for the management level. In the early grid system construction, people used to pay attention only to the safety other than the economy, driven by power marketization, stakeholders in electric power construction are more concerned about the economy of the electric power system in pursuit of greater economic benefits.

As a result, electric power experts in and outside China are all seeking to find a balance between the economy and safety of grid construction. So far, China has opened the generation-side market, and State Grid's role is an energy purchaser. To achieve the maximum benefit and to the extent that the user-side power demand is satisfied, what we have to consider is how we can purchase the power demand of the user side from the generation side with the minimum money. To achieve this, we have to try to minimize the loss of the power resources during transmission, i.e. the network loss. As such, network loss has become an important economic and technical index that measures the safety factor of the transmission grid, which not only measures the rationality level of grid planning and design, but also reflects the electric power production technique and marketing management level.

Research on how to minimize network loss has great implications on the effective and economical operation of the power grid. As a result, an integrated coordination and optimization model and algorithm for electric power systems that both reduces the energy purchase cost and network loss is established to maximize the overall performance of energy purchase and transmission. We generally establish an energy purchase cost – network loss coordination and optimization model for the economical operation of the power system, then applies the modified multi-objective bacterial colony chemotaxis (MOBCC) algorithm to the coordination and optimization of energy purchase cost and network loss. After the optimal solution is derived, the decision maker can obtain the optimal satisfactory solution or compromise solution from the Pareto optimal solution set, and then determine the overall optimization of grid safety and economy according to the satisfactory solution and compromise solution.

After Pareto optimal solution set is obtained, it is necessary to rate the safety of the operation state of each of the optimal solutions and group the optimal solutions of the same safety rating. This will allow the decision maker to compare the economy levels under the same safety level. Finally, the optimal compromise solution is obtained from these solutions according to the multi-attribute decision method or the network loss index, and includes for both the safety and economy requirements of the system. The entire optimization flowchart and steps are shown below.

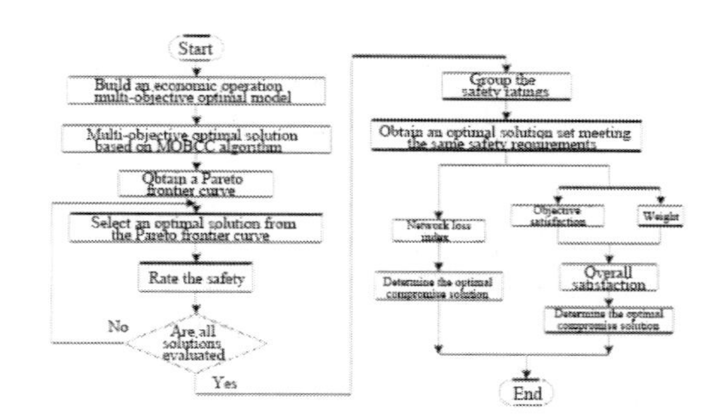

Figure 2. The entire optimization flowchart and steps.

5 CONCLUSIONS

As different regions and different industries have different requirements for grid safety: some grid companies are more concerned about the economy and some others are more interested in the safety, different grid safety ratings have to be defined and different measures used for the safety rating management and economy of each grid. When our company is concerned, we perform integrated evaluation on a number of factors of the transmission grid including the grid structure, the operation mode, the element failure rate and the compensation rate, which successfully improved both the safety and economy of the power grid and reduced the management cost at the same time.

REFERENCES

[1] Li, W.W., Wang, H., Zou, Z.J., et al. (2005) Function optimization method based on bacterial colony chemotaxis. *Journal of Circuits and Systems*, 10(1):58–63.

[2] Zhang, C., Huang, W. (2008) Daily active power economic dispatch for power generating units in the electric power market based on BCC algorithm. *Modern Electric Power*, 25(3):77–80.

[3] Zhuang, L.R., Cai, J.D., Li, T.Y. (2006) Model and algorithm of daily power purchase optimization for regional grid. *Electric Power Automation Equipment*, 26(12):13–17.

[4] Liang R H, Liao J H. (2007) A Fuzzy-optimization Approach for Generation Scheduling with Wind and Solar Energy Systems. *IEEE Trans on Power Systems*, 22(4):1665–1673.

[5] R. Raineri, S. Rios, D. Schiele. (2006) Technical and economic aspects of ancillary services markets in the electric power industry: an international comparison. *Energy Policy*, 34:1540–1555.

Electric, Electronic and Control Engineering – Shao, Shu & Tian (Eds)
© 2015 Taylor & Francis Group, London, ISBN: 978-1-138-02842-5

Improved digital image steganography algorithm

Xudong Wan & Rener Yang
Faculty of Information Science and Technology, Ningbo University, Ningbo Zhejiang, China

ABSTRACT: In order to reduce modifiers of the cover image to improve the security of steganography system during steganographic process, this paper proposes a digital image steganography algorithm that is based on the block. First, the cover image is grouped by 8 pixels, and takes its last to the steganographic operation, using the eighth as the flag bit, and the rest is the embedded secret information. We need to flag bit controls regardless of whether they are flapping secret information before it is embedded. The final result depends on the fact that block image pixel modifiers are always less than half of the block after embedding information. Experimental results show that the improved steganographic algorithm keeps the LSB algorithm's characteristics of large capacity, easy to implement with higher signal-to-noise ratio, and large embedding efficiency, which can resist the chi-square test simultaneously. In addition, compared with the matrix embedding algorithm proposed by Qian Mao, the new algorithm has relatively low algorithm complexity and high computing speed.

KEYWORDS: LSB; steganography; block; embedding efficiency.

1 INTRODUCTION

With the rapid development of the Internet, information technology is changing people's lives and working mode in recent years. Simultaneously, the problem of information security is highlighted day after day. And it has been one of the conclusive factors of influencing the country's security and so on. Information security includes many branches, and steganography has been the focus of the information security study [1]. Compared with encryption, the most important advantage of steganography is that it not only hides content but also conceals the fact. It transfers attackers' sights and ensures the security transformation of information. In order to get higher safety, steganography is often combined with encryption in actual application [2]. Encryption technology has a long history, and it has formed many mature algorithms, such as DES, AES, and so on. However, steganography is still in the initial period. Therefore, the algorithms still have some defects. But as time passes, steganography will form a safe system and become an important part of communication.

LSB is a common algorithm of steganography, and it suits gray images. It hides secret information by embedding encrypted data into every carrier's least significant bit[3,4,5]. The change of image pixels' least significant bit will not make a big change to the image because of the human eyes' redundancy in recognizing images[6]. However, the shortcoming of the LSB algorithm is that it can be easily detected by the steganography analytical methods, such as R/S Analysis and the chi-square[7]. There are many steganography algorithms such as the LSB matching steganographic method[8,9,10], which has been promoted a lot in safety and still maintains very high steganographyrate. The matrix embedding steganography algorithm, put forward by Crandall, is famous for its high embedding efficiency[11]. Meanwhile, the variation of the pixels of carrier is small and the safety of the steganography method is promoted. But the algorithm's shortcomings are too complex, and the embedding speed is too low. Through the research of the steganography algorithm mentioned earlier and analysis of their advantages and disadvantages, this paper comes up with an improved LSB algorithm. It changes the traditional LSB embedding method of the secret information, which is the continuous and sequential embedding method. It divides the image pixels into 1 block and 8 pixels for one block. The last one of every block is the flag bit that controls the reversal of the secret information, but it has no direct relationship with the secret information. The introduction of the flag bit can keep the balance between embedding efficiency and carrier modifier. The test data indicate that the method keeps a high embedding speed. Meanwhile, it improves the SNR of the images and has high embedding efficiency.

2 DIGITAL IMAGE STEGANOGRAPHY ALGORITHM BASED ON BLOCK

The traditional LSB algorithm directly embeds secret data into carriers without preprocessing carriers and ciphertext. According to this algorithm, all SNRs of modified images are around 51. The less the modified pixels of carrier images, the higher SNRs of the image. The algorithm presented in this paper is based on the idea described earlier.

The carrier images are blocked according to the length of 8 pixels. The last bit of each block is a flag bit, and the remaining bits are regarded as informative bits. Preprocessing steps are as follows:

Step 1: Bit-wisely compare the informative bit of the carrier with the secret data whose length is the same as the former, and mark the number of difference as d_t.

The pixel value that is in i row and j column can be shown as $x[i \times col + j]$. Assuming the length of the block is 8, the largest number of the block is $(M \times N / 8)$. Assuming the current number of the block is t (t∈{0,1,2....$row \times col / 8 - 1$}), the corresponding carrier pixel value of the current block is $x[t \times 8 + j]$ $(j = 0,1,....7)$.When $j = 7$, it is the flag of the current block image among them. The corresponding secret informative bit is as follows:

$$m[t \times 7 + j] \quad (j = 0,1,....6). \tag{1}$$

The block is shown in Table 1.

Table 1. Expression of carrier pixels and secret data's memory of block NO. t.

Pixels of block NO. t	$x[t \times 8 + 0]$	$x[t \times 8 + 1]$	...	$x[t \times 8 + 6]$	$x[t \times 8 + 7]$
least significant bit	$x[t \times 8 + 0]^{\%2}$	$x[t \times 8 + 1]^{\%2}$	...	$x[t \times 8 + 6]^{\%2}$	$x[t \times 8 + 7]^{\%2}$
secret informative bit	$m[t \times 7 + 0]$	$m[t \times 7 + 1]$	...	$m[t \times 7 + 6]$	$m[t \times 8 + 7]$

Step 2: There are two cases as follows:

1 First case, if the minimum meaningful bit of the last bit of the block pixel is "0," and the number of difference d_t is less than the half of the block, the LSB method will directly embed secret data into corresponding position and the flag bit will continue to be "0." Otherwise, the flag bit will be "1" and secret information will be inverted.
2 Second case, if the minimum meaningful bit of the last bit of the block pixel is "1," we will first consider inverting informative bits and then embed them. At this time, the number of difference is 7-d_t . If 7- d_t is less than the half of the length of the block, secret data will be inverted before embedding, and the flag will continue to be "1." Otherwise, the secret data will be embedded directly, and the flag bit will be "0."
3 Step 3: According to the LSB method, embed eventual secret data and flag bits into the block.

All contents of preprocessing have been mentioned earlier. Preprocessing can ensure that in one block, embedded information that renders the variation of carrier pixels is less than the half length of the block. Apparently, citing this flag bit can decrease the variation of carrier pixels, reduce the distortion of images, improve the SNR of images, and make the performance of ratio higher than the original LSB.

The $M \times N$ experimental image's pixel value composes a two-dimensional array, and its internal memory is stored according to row order.

Before embedding data, we compare the least significant bit with the secret informative bit to find whether one is equal to the other, and record the number of difference, which is expressed by d_t. For convenience, the flag bit value of the block that is $x[t \times 8 + 7]$ is recorded as the flag.

If flag=0 and $d_t <= 4$, secret informative bits are embedded into pixels of block NO. t. according to the traditional LSB algorithm, and the changed pixel value is as follows:

$$x[t \times 8 + j] - x[t \times 8 + j]\%2 + m[t \times 7 + j] \quad (j = 0,1,....6). \tag{2}$$

The pixel value of the flag bit is constant. Otherwise, invert all secret informative bits and embed them into the block No. t , then mark the flag bit as "1."

If flag=1, when $8 - 1 - d_t <= 4$, invert all secret informative bits and then embed new secret informative bits $\tilde{m}[t \times 7 + j]$ $(j = 0,1,....6)$ into the pixels of block NO. t directly according to the LSB algorithm. The changed pixel value is as follows:

$$x[t \times 8 + j] - x[t \times 8 + j]\%2 + m[t \times 7 + j] \quad (j = 0,1,....6). \tag{3}$$

Meanwhile, the flag bit is still "1." Otherwise, directly embed secret informative bits into corresponding pixels, and mark the flag bit as "0." The flow chart of the improved algorithm is as shown in figure 1.

3 EXPERIMENTAL RESULTS AND ANALYSIS

Simulation Experiment Environment is under the Windows 7 operating system of Microsoft Visual c + + 6.0, and the secrete information is the pseudo-random sequence of 0 and 1 bit stream that is produced by the random function in C language. The experimental images use a digital camera to collect the different contents, different colors, and different texture natural images, and they unify them into 512* 512 gray images.

The experiment selects the eight 512*512 standard grayscale images. Then, these images embed secret data by the traditional LSB algorithm and the improved algorithm, respectively. Lastly, compared with original images, it is easy to find the difference between modified SNR and the number of the modified carriers' pixels. The experimental data are shown in Table 2.

According to the table, it is clear that the SNR of modified images by using the LSB algorithm is about 51.14. However, the SNR of using the improved algorithm is about 52.5. For the same image, using the

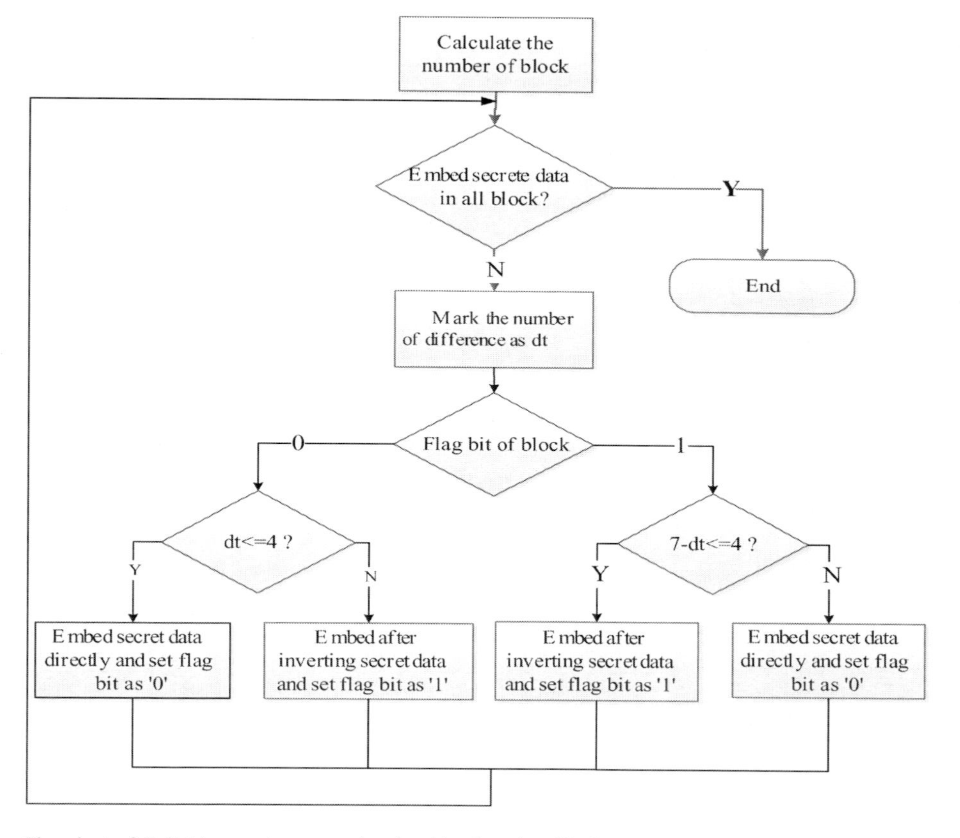

Figure 1. Flowchart of digital image steganography algorithm based on block.

Table 2. Comparison of LSB algorithm and improved algorithm with SNR and modifiers.

Test image		1	2	3	4	5	6	7	8
LSB algorithm	SNR	51.14	51.15	51.14	51.14	51.14	51.14	51.14	51.15
	modifiers	130816	131424	131378	130899	131050	131206	130917	131423
improved algorithm	SNR	52.53	52.54	52.53	52.53	52.52	52.54	52.54	52.54
	modifiers	95252	95406	95311	95108	95304	95413	95025	95349

Table 3. Comparisons of embedding rate, average embedding time, and average embedding efficiency of two algorithms.

The type of steganography algorithm	Embedding rate	Avg. embedding time[s] (10^4 bits per embedding)	Avg. embedding efficiency
Digital image steganography algorithm based on blocking	0.875	0.000203	2.41
Fast matrix embedding algorithm	0.75	0.0985	2.29

improved algorithm's SNR is higher than the common LSB algorithm.

In addition, compared with the common LSB algorithm, the improved algorithm achieves a better performance at modifying the pixels of images. Thus, it has a higher visual imperceptibility. The improvement of the improved algorithm's performance is at the expense of embedding rate. The block length is 8, of which seven bits are embedding secret data, including a flag bit that controls the reversal of secret data. Therefore, the embedding rate is 7/8.

Qian Mao uses a type of matrix embedding algorithm that is of higher embedding efficiency but slower embedding rate in the paper that named a fast algorithm for matrix embedding steganography[12], where she quickly finds the leader of cosets with LUT algorithms that makes the computational complexity simple. However, it still keeps the same embedding efficiency by contrasting with the traditional matrix embedding algorithm. We get the experimental results from data that are embedding rate, embedding time used, and embedding efficiency by using the fast algorithm. Comparing the three data from the paper with the three from the experiment, we can find the new algorithm whose embedding efficiency and embedding rate exceed the fast algorithm that Qian Mao comes up with, whereas the embedding time apparently is less than Qian Mao's. Thus, it shows that the digital image steganography algorithm based on blocking from this paper has obvious superiority. The comparison of results is shown in Table 3.

4 CONCLUSION

In order to improve the security of communication, designers of the steganography algorithm consider ensuring the information transportation rate besides reducing the revision of carriers as much as possible. And it also aims at increasing the embedding efficiency under the premise of ensuring embedding rate. On the basis of various application requirements, it is necessary to find an optimal balance between embedding efficiency and embedding rate. This paper puts forward an improved digital image steganography algorithm. The method involves dividing the block as 8 and taking out a space from every block as a flag bit. The results indicate that the algorithm remarkably reduces the revision of carrier images and improves the SNR. Most importantly, the method shortens the embedding time, improves the embedding efficiency, and lowers the level of operation difficulty, which is efficient to indicate that most mobile devices, including cellphone and so on, can be suitable for safe communication. In the light of the importance of application on steganography, resisting types of steganography analysis of an improved LSB algorithm will be a light in the future.

ACKNOWLEDGMENTS

This work was supported in part by the National Science Foundation of China (No.61271399), Zhejiang Science and Technology Innovation Team Fund Project (No.2012R10009-04), Ningbo University Project (XKXL1302), and Ningbo Innovation Team Fund Project (No.2011B81002).

REFERENCES

[1] Niels Provos, (2001) Defending against statistical steganalysis. In: *10th USENIX Security Symposium*, 10:323–336.

[2] Li You, Zhang Dinghui. (2010) Information hiding techniques based on steganography. *Information Technology*, (7):119–122.

[3] Petitcolas F A P, Anderson R J, Kuhn M G. (1999) Information hiding: A survey. *Proceeding of the IEEE (USA)*, 87(7):1062–1078.

[4] Fridrich J. (2003) Quantitative steganalysis of digital images: estimating the secret message length. *ACM Multimedia Systems Journal*, 9(3): 288–302.

[5] Fridrich J, Goljan M. (2001) Detecting LSB steganography in color and gray-scale images. *Proc. of Magazine of IEEE Multimedia Special Issue on Security*. (S1):22–28.

[6] Huang Haibo, Yang Sen. (2011) Research on image information hiding technology based on LSB. *Office Automation*, 3:32–34.

[7] Zhang Hongjuan, Zhu Chenming. (2008) Novel LSB steganography algorithm of against statistical analysis. *Computer Engineering*, 34(23):144–146.

[8] Chang CC, Lin M H, Hu Y C. (2002) A fast and secure image hiding scheme based on LSB substitution. *International Journal of Pattern Recognition and Artificial Intelligence*, 16(4):399–416.

[9] Chen Dan, Wang Yumin. (2007) Universal steganalysis algorithm for additive noise modelable steganography in spacial domain. *Journal of Southeast University (Natural Science Edition)*, (S1).
[10] Zhang Liang, Liu Hong, Wu Renbiao, Yang Guoqing. (2007) Wavelet domain steganography for JPEG2000, *Communications, Circuits and Systems Proceedings, 2006 International Conference on*. pp:27–30.
[11] Gupta A, Mumick I. (1995) Maintenance of materialized views: problems, techniques, and applications. *IEEE Data Eng. Bull.*, 18(2): 3–19.
[12] Qian Mao. (2014) A fast algorithm for matrix embedding steganography. Digital Signal Processing, (25):248–254.

Electric, Electronic and Control Engineering – Shao, Shu & Tian (Eds)
© 2015 Taylor & Francis Group, London, ISBN: 978-1-138-02842-5

Characteristic of HV insulators' leakage current on surface discharge

Yun Tian
State Grid Shanxi Electric Power Research Institute of SEPC, Taiyuan Shanxi, China

Lijun Feng
State Grid Shanxi Electric Power Dispatch and Control Center, Taiyuan Shanxi, China

Tianzheng Wang & Ming Jiang
State Grid Shanxi Electric Power Research Institute of SEPC, Taiyuan Shanxi, China

ABSTRACT: On-line monitoring technology is significant to establish a scientific efficient anti-flashover mechanism, to improve the reliability of the grid. Choosing a suitable detected signal and analysis method is the key of on-line monitoring technology. In this paper, according to the high false-alarm rate and inexact result in evaluating the risk of flashover, the mean, variance, skewness, kurtosis of the envelope of leakage current, and the relationship between the four indexes with surface discharge were studied. The method of quadrature was employed for synthesizing multi-information of the indexes and obtained the comprehensive indexes. It was proved that comprehensive indexes can decrease the false-alarm rate in evaluating flashover risk more than single indexes, meaning that they were more suitable for flashover risk evaluation.

KEYWORDS: pollution flashover leakage current characteristic signal surface discharge.

1 INTRODUCTION

The high-voltage insulator pollution flashover with the sudden characteristic of complex mechanism and severe harm significantly influences the reliability of a high-voltage grid, and it becomes the most threatening factor of grid operation far beyond lightning overvoltage and switching overvoltage. With the development of grid capacity and the increase of voltage level, the harm of pollution flashover will be aggravated[1].

Current anti-pollution flashover measures that are extensively used, such as choosing a new insulator, adjusting creepage distance, employing RTV paint, and regular cleaning, are passive[2,3,4]. They are blindly chosen during the implementation process due to the lack of understanding on insulator surface states, which results in the flashover accidents that are not totally avoided in special circumstances[5,6,7,8].

In this paper, on the basis of the artificial pollution test, the leakage current was chosen as the analysis and processing object, and its variation characteristics during the process from surface discharge to flashover breakdown were researched for characteristic data closely related to arc discharge states. The data can be employed for dielectric strength assessment and flashover prediction, and they also decrease the false-alarm rate for providing the basis for on-line monitoring.

2 TEST APPARATUS AND METHOD

2.1 *Test apparatus*

The tests employed the solid chromatography method for preliminary product pollution, including constant pressure and clean flog test. The size of salt spray cabinet was 4m×4m×5m, and the transformer was 125/250kV no-free transformer with rated capacity of 125kVA, rated voltage of 0.4/250kV, and rated current of 312.5/0.5 A. The test insulator type was XP-70. According to the first insulator of insulator chain composed of 7 pieces of insulator employed in 110kV power transmission line withstanding maximum voltage as about 12.662kV in the case of a no-grading ring, the test voltage of 26kV was chosen in the XP-70 insulator test.

The tests employed self-made leakage current sensor and regulation circuit, and they collected leakage current signal, temperature signal, and humidity signal with an acquisition card of Advantech 9111DG, in the meantime through IPC.

2.2 *Test arrangement*

The tests employed the solid chromatography method for preliminary product pollution, including constant pressure and clean flog test. The preliminary product pollution adopted the smearing method. The ash density value of pollution degree was selected for 2.0 mg / cm^2, and the salt density value, respectively, for 0.10 mg / cm^2, 0.15 mg / cm^2, and 0.20 mg / cm^2. The tests were given to three groups with different salt density values, and the group with a better flashover developing process was chosen for analysis.

3 TEST RESULTS AND ANALYSIS

3.1 *Data preprocessing*

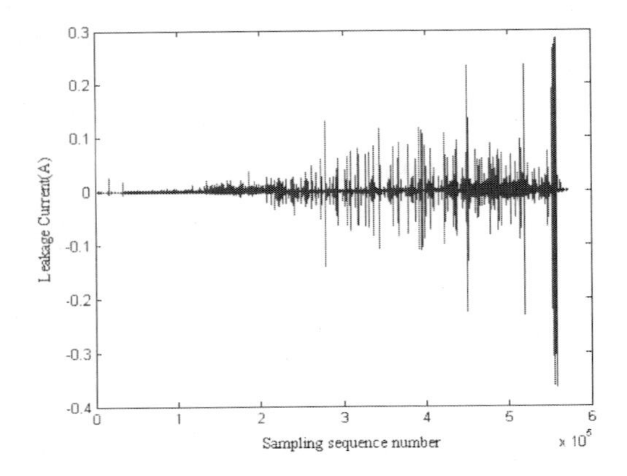

Figure 1. Real leakage current.

Figure 1 depicts the leakage current wave of the test, and it illustrates that the electric insulating strength shows a downward trend as the leakage current increases in the test process. Because the test of this paper is the AC test and the blowing out of electric arc may exist in a power frequency cycle, the current of figure 1 included two kinds of physical phenomenon: One kind was the conductive current equal to the arc discharge current in the residual sludge layer, ignoring the influence of the parasitic capacitance; the other kind was the residual sludge layer current without arc to react with the arc state. However, the arc discharge details should be supplied for judging the pollution flashover risk. The leakage current envelope covering the arc discharge current information was studied in this paper, and lots of zero passage data of arc blowout were removed.

This paper employed the digital filtering method to process leakage current data, and the filter was designed as a low-pass filter with a direct frequency of 165Hz, gain of 0dB, cut-off frequency of 180Hz, and gain of -200dB to filter out the high-frequency interference signal. The frequency response characteristics of filter are shown in figure 2.

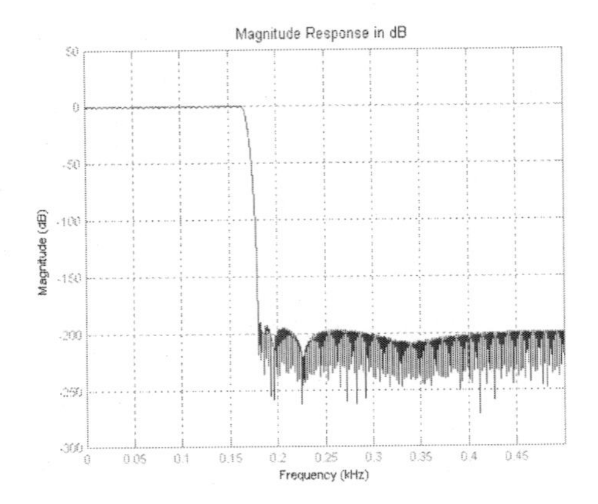

Figure 2. Amplitude-frequency of digital filter.

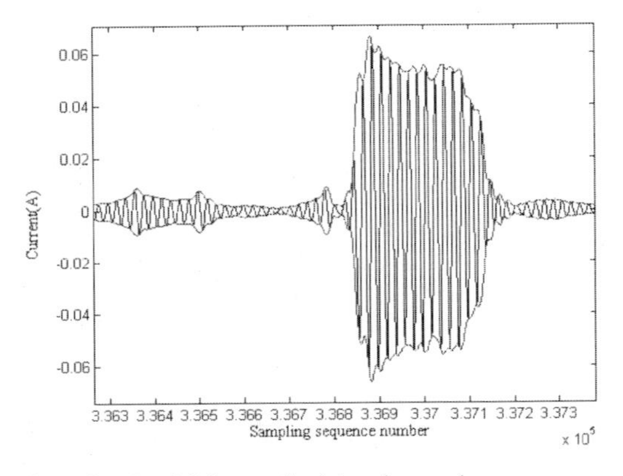

Figure 3. Partial figure of gaining the envelope.

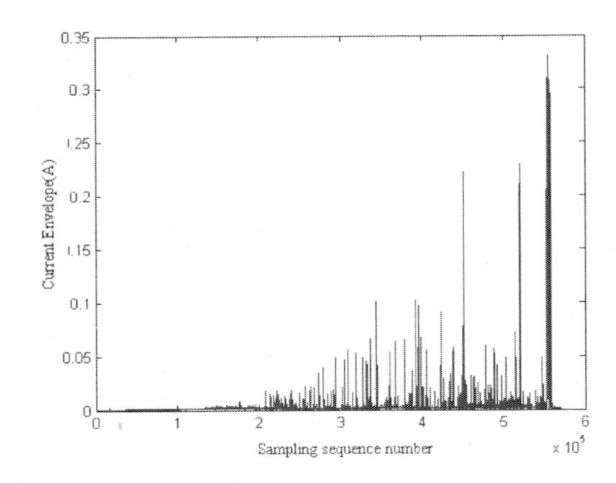

Figure 4. The envelope of leakage current.

The current data after filtering was processed with Hilbert transformation to extract the current envelope[6,7,8,9]. Because the leakage current deviated from sine wave, the envelope extracted with Hilbert transformation would be influenced by phase shift, which could be eliminated through the compensation method. Figure 3 was a partial enlarged drawing of envelope extraction, and it indicated the method was correct. Figure 4 showed the leakage current envelope.

3.2 *Data analysis*

The study result showed that the leakage current increased as the surface electric arc developed and the flashover drew near, so it was deduced that the mean leakage current contained state information of arc discharge. However, leakage current could not be simply used for distinguishing discharge state, as a result that higher continuous discharge was worthy of observation and a short-time large current could easily cause an unnecessary false alarm.

The variance of leakage current could be employed to measure the level of current surge, but it was infeasible to only employ variance to evaluate flashover risk[10]. At the beginning of the test, the variance was very small because of the small variation of current, and it could be small as well as a result that the arc maintains as large a current in the critical state. However, the variance could be larger during the intermediate stage with large current surge on account of the surface arc becoming strong and weak alternately under the influence of surrounding humidification and high temperature action of the arc, as the dry area was at transition between the formation and disappearing period.

Skewness and kurtosis are two common indexes in statistics for mainly judging distribution characteristics of data samples. When the skewness is greater than 0, it shows that the data sample is left skewed and the trailing is on the right, whereas it shows that the data sample is left skewed and the trailing is on the left. Skewness is mainly employed for revealing that the sample distribution is either left skewed or right skewed. Kurtosis is mainly employed for revealing sample steepness. When the kurtosis is greater than 0, it shows that the peak of the sample is higher and steeper, whereas it shows that the peak is lower and gentler[12,13,14]. This paper holds that the strength of large current discharge and discharge duration time were changed with the discharge strength, resulting in the change in skewness and kurtosis of the leakage current envelope sample.

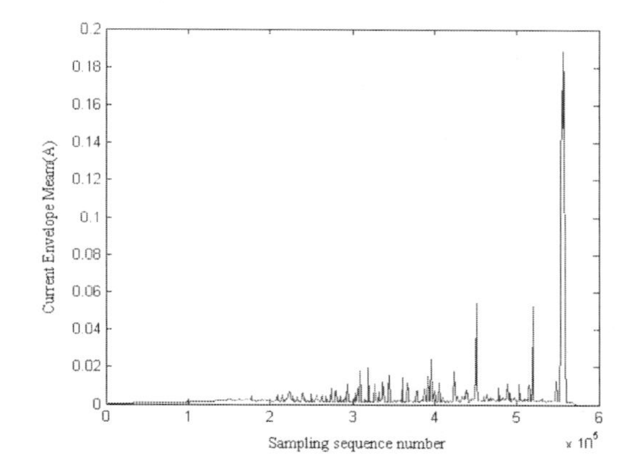

Figure 5. Mean of current envelope.

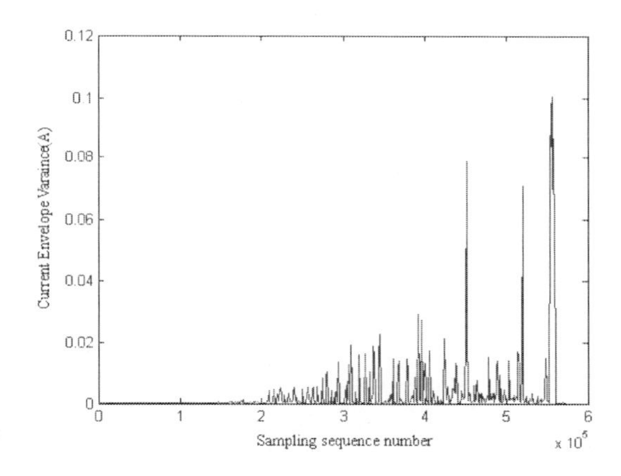

Figure 6. Variance of current envelope.

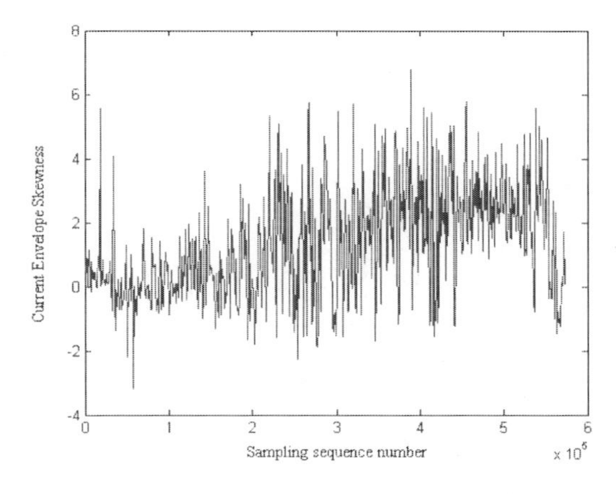

Figure 7. Skewness of current envelope.

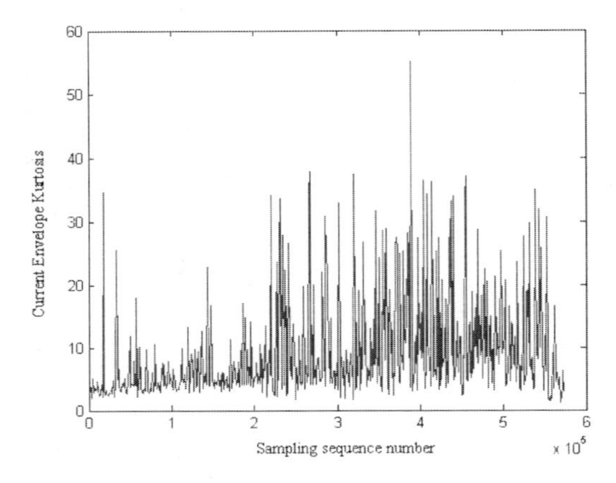

Figure 8. Kurtosis of current envelope.

This paper chose time-varying characteristics of mean, variance, skewness, and kurtosis of the leakage current for the analysis object[15]. The envelope data were divided into second section for the unit, and 1000 data samples in the range were employed for calculating mean, variance, skewness, and kurtosis and for obtaining each index changing features with time. Figures 5~8 were time-varying plots of mean, variance, skewness, and kurtosis of the envelope.

According to figures 5~8, the mean, variance, skewness, and kurtosis of the current changed with the development of surface discharge. However, there was a very wide gap between the index values, especially the kurtosis vary by 4 orders of magnitude with the mean and variance. This paper was based on the idea of comprehensively utilizing four indexes, so that every index needed to be preprocessed; otherwise, the index of large magnitude contained a large weight that was obviously contrary to comprehensive judgment. This paper employed index data to divide the range for preprocessing.

For the mean, variance, skewness, and kurtosis, multiplication was employed for extracting summarized information, such as in Equations 1 and 2.

$$\text{Index 1} = \text{mean} \times \text{variance} \times \text{skewness} \tag{1}$$

$$\text{Index 2} = \text{mean} \times \text{variance} \times \text{kurtosis} \tag{2}$$

Figures 9 and 10 were plots of indexes 1 and 2. On comparing figures 9~10 with figures 1, figures 5~8, it could be observed that figures 9 and 10 had an obvious shock for greater discharge strength, and they had an inconspicuous shock for less discharge strength. Indexes 1 and 2 that synthesized the features of mean, variance, skewness, and kurtosis reinforced the strong discharge signal and restrained the weak or short-time signal. Therefore, for the comprehensive evaluation on flashover risk, it largely reduced the false-alarm risk. Figure 9 showed that skewness was obviously less than 0 before flashover; it indicated that the current envelope sample was right skewed, and the values of most samples were large inosculated with a physical phenomenon of arc persistence burning. Moreover, it showed that once the skewness is obviously less than zero, the flashover risk is great.

Figures 9 and 10 showed that the sample indexes could have a potentiation effect on strong impulse discharge signal and an inhibition effect on weak discharge signal. For the evaluation on flashover risk, choosing index 2 as an analysis object would possess a lower false-alarm risk. However, index 1 was more conductive to catching critical flashover state, because its skewness was less than 0 before flashover[16].

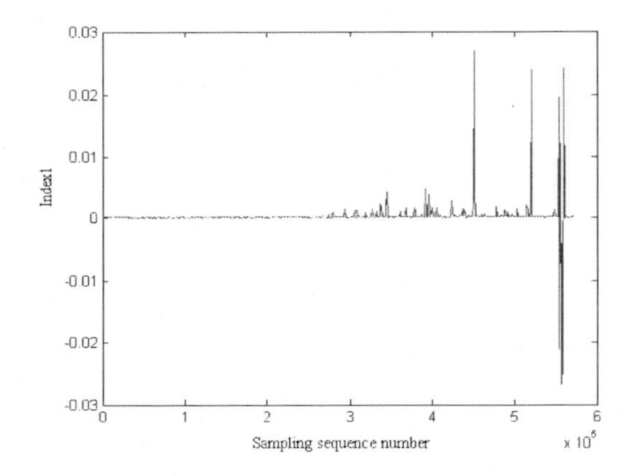

Figure 9. Plot of the index1.

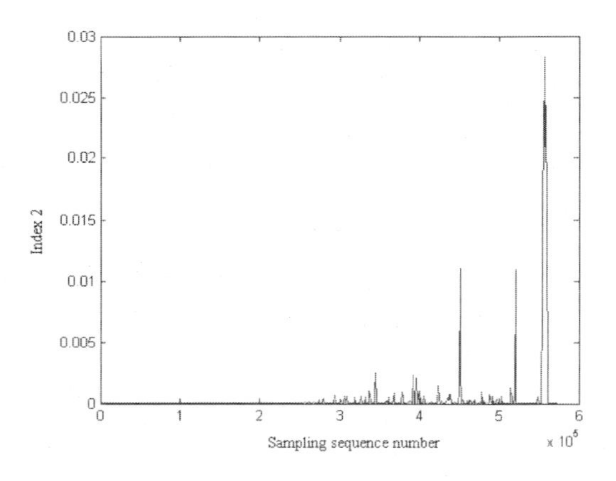

Figure 10. Plot of the index2.

4 CONCLUSION

This paper was based on an artificial pollution test, chose leakage current data as the analysis object to extract the envelope of the leakage current, and studied the relationship between the time-varying plot of mean, variance, skewness, kurtosis, and surface discharge states. The method of quadrature was employed for a comprehensive analysis to get two types of composite index. The result shows that indexes 1 and 2 can be used for evaluating flashover risk and largely reduce false-alarm rate because of features of strengthening impulse discharge and inhibiting weaker discharge. Meanwhile, index 2 is more accurate for predicting flashover risk than index 1, and index 1 is more conductive to catch critical states of flashover. Moreover, the study result has positive significance for prediction and forewarning of pollution flashover.

REFERENCES

[1] Qiu Shuchang, Shi Wei, Zhang Wenyuan. (1995) *High Voltage Engineering*. Xi'an: Xi'an Jiaotong University Press.

[2] Hu Yi. (2001) The analysis of "the pollution of network on feb. 22th" and counter measures from pollution flashover. *Insulators and Surge Arresters*, 182(4):3–6.

[3] He Bo, Lin Hui. (2006) Present status and future prospects for flashover research of contaminated insulators. *Insulators and Surge Arresters*, 44(2):7–11.

[4] Guan Zhicheng, Wang Shaowu, Liang Xidong. (2000) Application and prospect of polymeric outdoor insulation in China. *High Voltage Engineering*, 26(6):37–39.

[5] P. Claverie. (1971) Predetermination of behaviour of polluted insulator. *IEEE Transactions*, PAS-90(4):1902–1908.

[6] George G. Karady, Felix Amarh, Raji Sundararajan. (1999) Dynamic Modeling of AC insulator Flashover Characteristics. *High Voltage Engineering Symposium, August 22–27, 1999*, pp:1999–2003.

[7] Felix Amarh. (2007) *Electric Transmission Line Flashover Prediction System*. Arizona: Arizona State University, pp:30–100.

[8] George Karady, Felix Amarh. (1999) Signature analysis for leakage current wave-forms of polluted insulators. *IEEE Transmission and Distribution Conference, New Orleans, April 1999*, pp:806–811.

[9] Zheng Nanning. (1995) *Digital Signal Processing*. Xi'an: Xi'an Jiaotong University Press.

[10] Lou Shuntian, Li Bohan. (1998) *System Analysis and Design Based on Matlab*. Xi'an: Xi'an Electronic Science and Technology Press.

[11] J. P. Holtzhausen. (1997) AC pollution flashover models: Accuracy in predicting insulator flashover. *10th International Symposium on High Voltage Engineering, Montreal, Canada, August 1997*, pp:68–72.

[12] R. W. S. Carcia. N. H. C. Santiago. C. M. M. Portela. (1991) A mathematical model to study the influence of source parameters in polluted insulators tests. *Proceedings of the 3th International Conference on Properties and Applications of Dielectric Materials. Tokyo, Japan, July 8–11, 1991*, pp:953–956.

[13] Qin Chaoying, Zhao Xuanming, Shi Yimin. (1999) *Mathematical Statistics*. Xi'an: Northwestern Polytechnical University Press.

[14] Zhang Wentong. (2002) *SPSS11 Senior Statistical Analysis Tutorial*. Beijing: Beijing Hope Electronic Press.

[15] He Bo, Lin Hui. (2006) Statistical characteristic of insulators' leakage current in artificial pollution tests. *High Voltage Engineering*, 32(7).

Electric, Electronic and Control Engineering – Shao, Shu & Tian (Eds)
© 2015 Taylor & Francis Group, London, ISBN: 978-1-138-02842-5

New assessment method of internal model control performance and its application

Liye Liu & Qibing Jin
Institute of Automation, Beijing University of Chemical and Technology, Beijing, P.R. China

ABSTRACT: Currently, there are many assessment methods of IMC (Internal Model Control) performance. In most of those methods, the sigular performance indicator is used to assess the controller performance. In this paper, a new assessment method of IMC performance is presnted, which can assess the performance of IMC, both objectively and roundly. In the meantime, a comprehensive assessment indicator is presented to avoid the partial optimization in the sigular performance indicator. In this paper, the analytic hierarchy process is used in the comprehensive assessment of IMC. The three different internal model controllers are treated as the object of this study. After programming in MATLAB, the comprehensive assessment indicator is obtained. The results show that this method can assess the IMC performance, both objectively and effectively.

KEYWORDS: internal model control, performance indicator, analytic hierarchy process, comprehensive assessment.

1 INTRODUCTION

The transient quality and the static performance are meaured by the relevant performance indicator in the control system. The performance of system typically includes the stability of system, the performance of static state, and the performance of transient state. The specific performance mainly includes overshoot (δ), setting time (t_s), rising time (t_r), peak time (t_p), peak, and so on.

In the actual industry production, only one or several performances are used to assess the performance of the control system. So, when the researchers design the controller, they often consider and optimize one or several performance indicators. The PID control algorithm is improved in [1]. In [1], Yan et al. considered and optimized the overshoot and setting time. Anakwa and Pearson (1974) proposed the design methods of the multivariable controller in [2]. They took the smoothness of simulation curve as a performance indicator to assess the performance of the controller. Garcla (1987) and Swanda (1999) proposed the design method of IMC-PID controller in [3-4]. They adopted the method of the connection of overshoot and ISE (Integrated Square Error) to assess the performance of the controller. Gong (2000) improved the method of the setting parameter of the IMC-PID controller and took the robust stability indicator and ISE as the preformance indicator of the controller assessment [5]. For the process of the time lag of the first and second order, Huang (2002) developed the IAE optimization with the generalized controller and the structure of the PI/PID controller [6]. Then, he took the IAE optimization to assess the step tracking of the controller. In reference [7], Lee (2008) proposed a new method to track the performance of the IMC controller and to assess the robust performance. He adopted IAE and sensibility (M_s) as the performance indicator to analyze the controller performance. On the whole, most scholars adopted the ISE as the performance indicator since Garcla proposed the method of the designing IMC controller.

For the reasons given earlier, if we consider only one or several indicators in the designing process of the IMC controllers, this will probably result in practical optimization. In this paper, a new comprehensive assessment of the control system is proposed to assess the IMC performance, both objectively and effectively.

2 PRELIMINARIES

Define 1. Assume that A is a $n \times n$ matrix, and if $A=(a_{ij})_{n\times n}$ satisfies the following equations

$$\begin{cases} a_{ij} > 0 \\ a_{ij} = \dfrac{1}{a_{ij}} \quad (i, j = 1, 2 \cdots, n) \end{cases} \tag{1}$$

Then, A is the positive reciprocal matrix.

Define 2. Assume that A is a $n \times n$ and the positive reciprocity matrix, and if A satisfies the following equations

$$a_{ii}a_{jk} = a_{ik} \ \forall i, j, k = 1,2\cdots n \tag{2}$$

Then, A is the consistent matrix

Theorem 1 If A is the positive reciprocity matrix, the maximum eigenvalue (λ_{max}) of A must be the positive real. The other modulus of the eigenvalues of A is less than λ_{max}.

Theorem 2 If A is the positive reciprocity consistent matrix, then

a. A is the positive reciprocity matrix;
b. A^T is the consistent matrix;
c. The maximum eigenvalue is equal to n, which is the order of the matrix eigenvalue. Other eigenvalues are zero;
d. $W = (\omega_1, \omega_2 \cdots \omega_n)^T$ is the eigenvector of the maximum eigenvalue .

$$a_{ij} = \frac{\omega_i}{\omega_j}, \ \forall i, j = 1,2\ldots n \tag{3}$$

That is,

$$A = \begin{bmatrix} \frac{\omega_1}{\omega_1} & \frac{\omega_1}{\omega_2} & \cdots & \frac{\omega_1}{\omega_n} \\ \frac{\omega_2}{\omega_1} & \frac{\omega_2}{\omega_2} & \cdots & \frac{\omega_2}{\omega_n} \\ \vdots & \vdots & \ddots & \vdots \\ \frac{\omega_n}{\omega_1} & \frac{\omega_n}{\omega_2} & \cdots & \frac{\omega_n}{\omega_n} \end{bmatrix} \tag{4},$$

where

$$A\omega = \lambda_{max}\omega \tag{5}$$

Theorem 3 Assume that A is a $n \times n$ and the positive reciprocity matrix, then λ_{max} satisfies the following formula:

$$\lambda_{max} \geq n \tag{6}$$

When λ_{max} is equal to n, A is the consistent matrix.

3 COMPREHENSIVE ASSESSMENT

3.1 *Building the hierarchical structure*

The hierarchical structure is composed of the object level, the performance index level, and the controller plan level. The hierarchical structure of the comprehensive assessment is shown in figure 1.

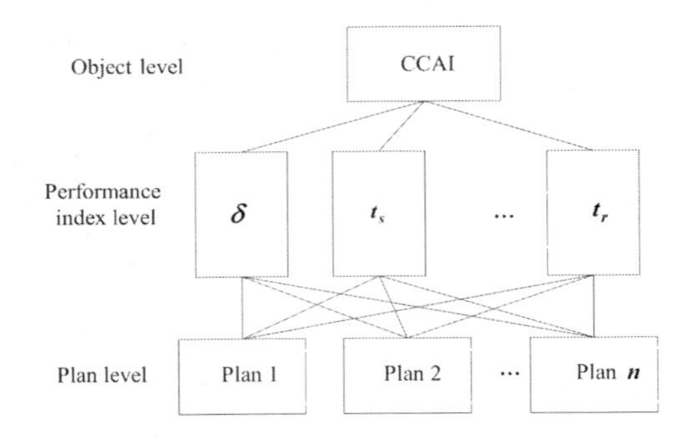

Figure 1. Hierarchical structure of comprehensive assessment.

The first level of the model is the object level, which is the primary goal of the analysis. The goal is to estimate the controller comprehensive assessment index (CCAI). The second level is the performance index level, which includes the criteria used to analyze the goal (e. g. t_s, t_r, t_p, M_p). The controller plan level is the different design schemes of the IMC controller. This information is considered as the decision alternative of the hierarchical structure in level 3. The decision made in level 3 will be instrumental in setting up the hierarchical structure of the comprehensive assessment. Then, it is necessary to construct the judgment matrix.

3.2 *Constructing the judgment matrix*

Assume that $X = \{x_1, x_2, \cdots, x_n\}$ is the factor of the performance index level and Z is the CCAI. And a_{ij} is the ratio of (x_i, Z) to (x_j, Z). ((x_i, Z) is the impact factor of x_i to Z.) All of the comparison results compose the matrix $A=(a_{ij})_{n\times n}$. That is,

$$A = \begin{bmatrix} a_{11} & a_{12} & \cdots & a_{1n} \\ a_{21} & a_{22} & \cdots & a_{2n} \\ \vdots & \vdots & \ddots & \vdots \\ a_{n1} & a_{n2} & \cdots & a_{nn} \end{bmatrix} \tag{7}$$

where

$$a_{ji} = \frac{1}{a_{ij}}, \ i, j = 1,2\cdots n \tag{8}$$

According to Definition 1, it can be observed that the matrix $A=(a_{ij})_{n\times n}$ is the positive reciprocal matrix about determining the value of a_{ij}.

3.3 *Sublevel sequence and consistency check*

After calculating the maximum eigenvalue and the eigenvector, the consistency check is taken by the consistency index, the random index, and the consistency ratio. If the result of the consistency check is less than 0.10, the eigenvalue is the weight vector; if the result of the consistency check is more than 0.10, the matrix A must be rebuilt. The maximum eigenvalue is given.

Step 1. Normalize the column vectors of the matrix A.

$$\overline{w}_{ij} = \frac{a_{ij}}{\sum_{i=1}^{n} a_{ij}} \tag{9}$$

Step 2. Sum $\overline{w}_{ij}$ in the row.

$$\overline{w}_i = \sum_{j=1}^{n} \overline{w}_{ij} \tag{10}$$

Step 3. Normalize $\overline{w}_i$

$$\overline{w}_i = \frac{w_i}{\sum_{i=1}^{n} w_i} \tag{11}$$

$\overline{w}_i$ is the approximate eigenvector.

Step 4. Calculate the approximate maximum eigenvalue. The equation of maximum eigenvalue is given by

$$\lambda = \frac{1}{n}\sum_{i=1}^{n}\frac{(Aw)_i}{w_i} \tag{12}$$

According to theorems 1, 2, and 3, it can be observed that the matrix A is the consistent matrix when the maximum eigenvalue is equal to n. Due to the eigenvalue dependent on a_{ij}, the eigenvector of the maximum eigenvalue does not exactly reflect the impact factors of (x_i, Z), when the maximum eigenvalue is far greater than n. So the consistency check for the matrix A has to be taken into account.

The process of consistency check is shown as follows:

Step 1. Calculate the consistency index CI. According to theorem 3, it is observed that the maximum eigenvalue is dependent on a_{ij}. Then, the value of $\lambda_{max} - n$ is adopted to adjust the consistency of the matrix A.

$$CI = \frac{\lambda_{max} - n}{n-1} \tag{13}$$

Step 2. Search for the random consistency index RI. The randomization is used to construct 500 sample matrixes, and the scales are randomly selected to compose the positive reciprocal matrix. Then, the average value of the maximum eigenvalue (λ'_{max}) is calculated as follows:

$$RI = \frac{\lambda'_{max} - n}{n-1} \tag{14}$$

where $n = 1, \cdots 9$, the value of the random consistency index is shown in Table 1.

Table 1. Value of the random consistency index.

N	1	2	3	4	5	6	7	8	9
RI	0	0	0.58	0.90	1.12	1.24	1.32	1.41	1.45

Step 3. Calculate consistency ratio CR.

$$CR = \frac{CI}{RI} \tag{15}$$

The consistency check of the judgment matrix is accepted when $CR < 0.10$. Equations (13-15) are used to make the consistency check.

3.4 *Total level sequence and consistency check*

According to the model studied in this paper, it is known that a group of factors correspond to the weight of a factor of the last level. In order to obtain the sequence weight of the plan level to the object level, the sublevel weight is combined by the top-down approach. The total level sequence is shown in Table 2.

Table 2. Combining total level sequence.

	A_1	A_2	…	A_n	CCAI
	a_1	a_2	…	a_n	
B_1	b_{11}	b_{12}	…	b_{1n}	$\sum_{j=1}^{m} b_{1j}a_j$
B_2	b_{21}	b_{22}	…	b_{2n}	$\sum_{j=1}^{m} b_{2j}a_j$
⋮	…	…	⋱	…	
B_n	b_{n1}	b_{n2}	…	b_{nm}	$\sum_{j=1}^{m} b_{nj}a_j$

There are m factors in the performance index level. The factor weights are $\{a_1, a_2, \cdots, a_n\}$. There are n factors in the plan level. The factor weights are $\{b_{1j}, b_{2j}, \cdots, b_{nj}\}$. Then, the total level sequence value $\{b_1, b_2, \cdots, b_n\}$ is calculated. The total level sequence value is shown in the following equation:

$$b_i = \sum_{j=1}^{m} b_{ij}a_j \text{ , } i = 1, 2 \cdots n \tag{16}$$

It is possible for all inconsistencies of the level to be cumulative. So, it is possible for the final analysis result to be inconsistent. The consistency check of the total level sequence value needs to be taken into account.

We should assume that the consistency check of the judgment matrix of the performance index level is accepted and calculate the consistency index **CR** (*j*) and the average random consistency index **RI** (*j*) ($j = 1,2\cdots m$). Then, the random consistency ratio of the total sequence is shown in the following equation:

$$CR = \frac{\sum_{j=1}^{m} CI(j)a_j}{\sum_{j=1}^{m} RI(j)a_j} \tag{17}$$

When $CR < 0.10$, the consistency check of the total level sequence value is accepted.

4 SIMULATION EXAMPLE

The system performance is dependent on the structure of the system and the parameters of the system. In order to study the controller performance, the typical controller of IMC-PID is adopted. In this paper, three different controllers of IMC-PID are exampled at the identical control structure and the identical control model. The control object is the second order plus dead time.

$$G(s) = \frac{Ke^{-\theta s}}{(T_1 s + 1)(T_2 s + 1)} \tag{18}$$

Three methods are adopted to be equivalent to the time delay, including the first-order Pade approximation, the second-order asymmetry Pade approximation, and the full pole approximation.

Assuming $T_1=6$, $T_2=1$, $\theta = 2$, $K = 1$, and $\lambda = 2$, the unit step input is imposed in this control system. When $t=50$, the step disturbance (0.2) is added. According to (18-20), the parameters of the three different IMC-PID controllers can be obtained, which are shown in Table 3.

Table 3. Parameters of the IMC-PID controller.

	T_f	T_i	T_d	K_c
First order Pade	0.5000	7.5000	1.2330	1.8750
Second order asymmetry	0.6782	7.0950	0.9303	4.5934
Full pole	7.0000	0.5762	0.1646	1.7500

After programming and simulating in MATLAB, the simulation curve from time can be obtained, which is shown in figure 2. In the figure, y_1 is the simulation curve of the first-order Pade approximation; y_2 is the simulation curve of the second-order asymmetry Pade approximation; and y_3 is the simulation curve of the full pole approximation.

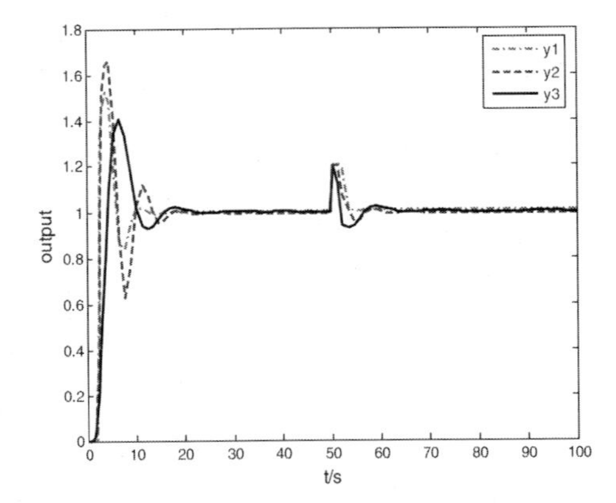

Figure 2. Output simulation curves of the IMC-PID.

In this paper, the time domain is adopted to analyze the system. The nine performance indexes are selected as the factors of the performance index level. According to figure 2, the details of the performance index parameters are obtained. They are shown in Table 4. In Table 4, ε is the steady-state error, and μ is the shake time.

Table 4. Performance indexes of the control system.

	δ /%	t_s/s	t_r/s	t_p/s	M_p	ε	μ	ISE
y_1	51.87	21.222	2.535	3.734	1.5237	0.0133	4	0.0176
y_2	65.45	27.838	2.711	3.802	1.6459	0.0152	5	0.0231
y_3	40.74	22.469	4.263	6.554	1.4066	0.0113	3	0.0127

According to the details of the performance index parameter and the description in Chapter 3, the matrix *A* is obtained through comparing the weights of the performance index level with the object level.

Table 5. Judgment matrix table of the criterion level (*A*).

A	B_1	B_2	B_3	B_4	B_5	B_6	B_7	B_8
B_1	1	5	3	7	3	1	5	1
B_2	1/5	1	1/3	3	1/3	1/5	1	1/5
B_3	1/3	3	1	5	1	1/3	3	1/3
B_4	1/7	1/3	1/5	1	1/5	1/7	1/3	1/7
B_5	1/3	3	1	5	1	1/3	3	1/3
B_6	1	5	3	7	3	1	5	1
B_7	1/5	1	1/3	3	1/3	1/5	1	1/5
B_8	1	5	3	7	3	1	5	1

Table 6. Judgement matrix table of the plan level (B).

B_1	C_1	C_2	C_3	B_2	C_1	C_2	C_3
C_1	1	3	1/3	C_1	1	5	3
C_2	1/3	1	1/5	C_2	1/5	1	1/3
C_3	3	5	1	C_3	1/3	3	1
B_5	C_1	C_2	C_3	B_6	C_1	C_2	C_3
C_1	1	3	1/3	C_1	1	3	1/3
C_2	1/3	1	1/5	C_2	1/3	1	1/5
C_3	3	5	1	C_3	3	5	1
B_3	C_1	C_2	C_3	B_4	C_1	C_2	C_3
C_1	1	3	7	C_1	1	3	7
C_2	1/3	1	5	C_2	1/3	1	5
C_3	1/7	1/5	1	C_3	1/7	1/5	1
B_7	C_1	C_2	C_3	B_8	C_1	C_2	C_3
C_1	1	3	1/3	C_1	1	7	1/3
C_2	1/3	1	1/5	C_2	1/7	1	1/9
C_3	3	5	1	C_3	3	9	1

Then, the matrix $\boldsymbol{B}$ (the judgment matrix of the plan level to the performance index level) is obtained in the same manner. The tables of the matrix $\boldsymbol{A}$ and the matrix $\boldsymbol{B}$ are shown in Tables 5 and 6, respectively.

From Table 7, it is known that the CCAI of the full pole approximation is equal to 0.5555, which is the maximum. So the conclusion is arrived at that the performance of the IMC-PID controller of the full pole approximation is the best in these three IMC-PID controllers with the identical control structure and the identical control model. From the simulations of the time domain and the frequency domain, it is known that the performance of the IMC-PID controller of the full pole approximation is better and the robustness is stronger; that is, corresponding to the results of the total level sequence. It is verified that CCAI reflects the performance of IMC controllers, both objectively and effectively.

It is known that CR is equal to 0.0443, which is less than 0.10. So, the consistency check is accepted. Then, the total level sequence is obtained, which is shown in Table 7.

Table 7. Total level sequence of the IMC-PID controller.

	δ 0.0332	t_s/s 0.0332	t_r/s 0.0559	t_p/s 0.0559	M_p 0.0332	ε 0.0332	μ 0.0332	ISE 0.0692	CCAI
y_1	0.25828	0.63699	0.64912	0.64912	0.25828	0.25828	0.25828	0.28974	0.3299
y_2	0.10473	0.10473	0.27895	0.27895	0.10473	0.10473	0.10473	0.05490	0.1145
y_3	0.63699	0.25828	0.07193	0.071927	0.63699	0.63670	0.63670	0.65536	0.5555

5 CONCLUSION

In this paper, a new assessment method of internal model control performance is proposed. It is possible that this new assessment method of internal model control performance is used to assess the performance through the theory analysis and example. According to the simulation and the total level sequence, it is known that the new assessment method reflects the controller performance. But there are some limitations in the application of this assessment method, which is that the application of the assessment method must be in the identical control structure and the identical control model.

ACKNOWLEDGMENTS

The authors would like to acknowledge the financial support of the National Natural Foundation of China (61273132) and higher school specialized research fund for the doctoral program (20110010010) and partly acknowledge the Automation Institute Beijing University of Chemical Technology. The authors are also grateful to the anonymous reviewers for their valuable recommendations.

REFERENCES

[1] K. X. YAN, G. Q. Tan. (2012) An improved algorithm of the digital PID. *China Computer and Communication*, 7: 113–114.

[2] W. N. Anakwa, A. E. Pearson. (1974) A design procedure for multivariable control systems. *Journal of Dynamic Systems, Measurement and Control*, 96(3):287–300.

[3] Garcia CE, Morari M. (1982) Internal model control-1. A unifying review and some new results. *Industrial Engineering Chemical Process Design Development*, 21:308–323.

[4] Swanda A. P, Seborg D. E. (1999) Controller performance assessment based on setpoint response data.

Proceedings of the American Control Conference. NJ, USA: IEEE, pp:3863–3867.

[5] J. P. Gong, F. C. Zou. (2000) Improvement of extended turning method for IMC controller parameter. *Journal of Beijing University of Chemical Technology*, 27(3):9–10.

[6] H. P. Huang, J. C. Jeng. (2002) Monitoring and assessment of control performance for single loop systems. *Industrial Engineering and Chemistry Research*, 4(5):1297–1309.

[7] G. Lee, Q. L. Wang. (2008) Internal model controller performance assessment, diagnosis and tuning based on step tracking response. *Information and Control*, 37(4):459–464

[8] Ordys, A., Uduehi, D. and Johnson, M. A. (2007) Process Control Performance Assessment: From Theory to Implementation. Berlin: Springer.

[9] Sendjaja, A. Y. and Kariwala, V. (2009) Achievable PID performance using sums of squares programming. *Journal of Process Control*, 19:1061–1065.

[10] Damouras, S. Chang, M. D. Sejdic, E., and Chau, T. (2010) An empirical examination of detruded fluctuation analysis for gait data. *Gait & Posture*, 31: 336–340.

[11] B. Srinivasan, T. Spinner, R. Rengaswamy. (2012) Control loop performance assessment using detruded fluctuation analysis (DFA). *Automatic*, 48:1359–1363.

[12] Yucai Zhu, Paul P. J. (2000) Optimal-loop identification test design for internal model control. *Automatic*, 36(8):1237–1241.

[13] Jin Qibing Zhao Liang, Hao Feng, Liu Siwen. (2012) Design of a multivariable internal model controller based on singular value decomposition. *The Canadian Journal of Chemical Engineering*, 16(8):67–76.

[14] Zhao Zhicheng, Liu Zhiyuan, Zhang Jinggang. (2011) IMC-PID tuning method based on sensitivity specification for process with time-delay. *J. Cent South Univ. Technol.*, 18:1153–1160.

Electric, Electronic and Control Engineering – Shao, Shu & Tian (Eds)
 ISBN: 978-1-138-02842-5

The research on how to control a three-phase four-leg inverter

Tong Jia, Wenguang Luo, Niu Liu & Yang Yang
Guangxi University of Science and Technology, Liuzhou Guangxi, China

ABSTRACT: To solve the asymmetric output voltage from a three-phase four-leg inverter under unbalanced or nonlinear load conditions, a novel decomposing sequence-based feed-forward decoupling control algorithm is proposed. First, a mathematical model for such inverter is established. Next, asymmetric components are decomposed into positive, negative and zero sequence components for separate feed-forward decoupling control. Then, a dual closed-circuit voltage and current control system is designed. Finally, simulation experiment is carried out with MATLAB/SIMULINK and proves the correctness and superiority of the control algorithm proposed.

KEYWORDS: Three-phase four-leg inverter; asymmetric component; feed-forward decoupling; dual closed-loop control.

1 INTRODUCTION

Three-phase three-leg inverters are widely used in areas like reactive power compensation, active power filtering and new energy power generation. The output power from a three-phase three-leg inverter, if symmetric, provides very good power supply to balanced three-phase loads. Three-phase inverters used in some specific locations, however, have to be able to carry unbalanced loads, whereas a traditional three-phase three-leg inverter sometimes outputs unbalanced voltage when carrying unbalanced loads [1-2]. To solve this problem, a three-phase four-leg inverter is used. Three-phase four-leg inverters are a hotspot among experts across the globe and widely used in engineering practices as a good solution to carrying unbalanced loads and nonlinear loads. As the addition of a fourth leg complicates the decoupling of the inverter as well as the modulation of spatial vectors, a simple but effective control algorithm is an indispensable prerequisite for keeping the three-phase four-leg inverter normally functional. So far, two of the most commonly used control algorithms are symmetric component method that decomposes the unbalanced voltage or current into symmetric positive, negative and zero sequence components before multi-loop control is implemented, and spatial vector control of current for the inner loop and PI control of voltage for the outer loop [3-4]. While both methods are able to improve the symmetric level of the output voltage from a three-phase four-leg inverter in a way, the control design is too complicated to be widely used in engineering practices [5]. For this reason, we propose a decomposing sequence-based control strategy that decomposes asymmetric load voltages and currents into three groups of symmetric positive, negative and zero sequences, then transform the positive/negative abc/dq coordinates into DC volumes under a rotating coordinate system, and use a PI dual closed-loop controller to achieve floating control. This control algorithm is able to control the output voltage waveforms from the four legs of the inverter both easily and effectively with low voltage distortion.

2 CHAIN TRUNK FEEDER AND ITS MATHEMATICAL MODELING

Fig.1 shows the circuit topology of a three-phase four-leg inverter, in which the midpoint n of the fourth leg is connected to the neutral point of the load via inductor L_f, which is designed to inhibit the switching ripple that flows past the zero sequence current of the neutral wire, and L and C are the inductance and capacitance of the three-phase filter. This circuit can carry unbalanced loads or nonlinear loads or any of their combinations [6-7].

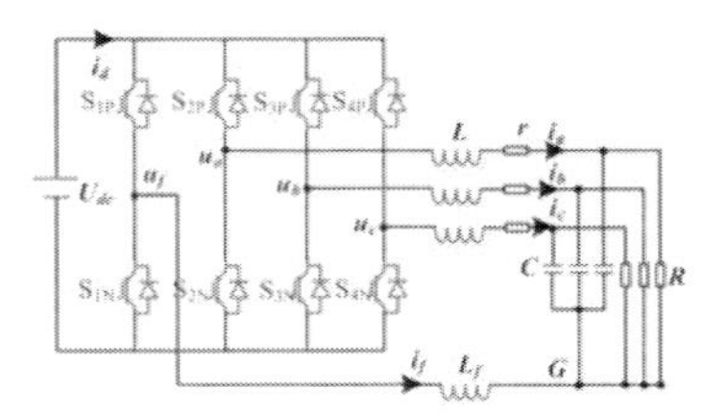

Figure 1. The main circuit of a three-phase four-leg inverter.

From circuit knowledge we can have a mathematic model of a three-phase four-leg inverter under a three-phase static coordinate system, and then a mathematic model under a synchronous rotating coordinate system by transforming the abc/dq coordinates:

$$\begin{cases} L\dfrac{di_{Ld}}{dt}=U_{dc}D_d-u_d+\omega Li_{Lq}-ri_{Ld} \\ L\dfrac{di_{Lq}}{dt}=U_{dc}D_q-u_q-\omega Li_{Lq}-ri_{Lq} \\ C\dfrac{du_d}{dt}=i_{Ld}-i_d+\omega Cu_q \\ C\dfrac{du_q}{dt}=i_{Lq}-i_q+\omega Cu_d \end{cases} \quad (1)$$

$$\begin{cases} (L+3L_f)\dfrac{di_{Lo}}{dt}=U_{dc}D_0-u_0 \\ C\dfrac{du_0}{dt}=i_{Lo}-i_{Lo} \end{cases} \quad (2)$$

Also, we can have the DC-side current as:

$$i_d=[\frac{3}{2}D_d \quad \frac{3}{2}D_q]\bullet[i_a \quad i_b \quad i_c] \quad (3)$$

3 CONTROL STRATEGY ANALYSIS

3.1 *Asymmetric component decomposition*

By using symmetric component method, we can decompose a group of unbalanced phasors (unbalanced voltage or unbalanced current) into positive, negative and zero components. The algorithm is expressed as [8]:

$$[\overline{x_{ip}}]=[C_p]\bullet[\overline{x_i}]$$

$$[\overline{x_{in}}]=[C_n]\bullet[\overline{x_i}] \quad (4)$$

$$[\overline{x_{i0}}]=[C_0]\bullet[\overline{x_i}]$$

Here: $\overline{x_{ip}}$, $\overline{x_{in}}$ and $\overline{x_{i0}}$ are the positive , negative and zero sequence components; $\overline{x_{i0}}$ (i=a, b, c) is the three-phase voltage or current vector transformation matrix; C_p, C_n and C_0 are:

$$[C_p]=\frac{1}{3}\begin{bmatrix} 1 & a & a^2 \\ a^2 & 1 & a \\ a & a^2 & 1 \end{bmatrix} \quad (5)$$

$$[C_n]=\frac{1}{3}\begin{bmatrix} 1 & a^2 & a \\ a & 1 & a^2 \\ a^2 & a & 1 \end{bmatrix} \quad (6)$$

$$[C_0]=\frac{1}{3}\begin{bmatrix} 1 & 1 & 1 \\ 1 & 1 & 1 \\ 1 & 1 & 1 \end{bmatrix} \quad (7)$$

Here: $a=e^{\frac{2\pi}{3}}$,

From the transformation matrix, we know the zero sequence components $\begin{bmatrix} \overline{x_{a0}} & \overline{x_{b0}} & \overline{x_{c0}} \end{bmatrix}$ are a group of phasors having the same amplitude with the same phase and frequency which cannot be brought under a synchronous rotating coordinate system using abc/dq transformation. To transform AC zero sequence vectors into DC volumes, a PI controller is used. In our study, zero sequence vectors are reformed into three-phase symmetric AC volumes by rotating the zero sequence component of phase b 120° clockwise and that of phase c 120° anticlockwise so that the newly constructed zero sequence components are symmetric three-phase zero sequence components that are 120° between each other and can thereby be controlled under a synchronous rotating coordinate system using abc/dq transformation. The transformation matrix for this group of zero sequence components so constructed is:

$$[C_0^*]=\frac{1}{3}\begin{bmatrix} 1 & 1 & 1 \\ a^2 & a^2 & a^2 \\ a & a & a \end{bmatrix} \quad (8)$$

Then, by positive sequence abc/dq coordinate transformation of the resulting positive and negative components, we get two groups of DC components x_{dqp} and x_{dq0} under the rotating coordinate system. The transformation matrix for the positive sequence abc/dq coordinates is:

$$T_{3/p}=\begin{bmatrix} \cos(\omega t) & \cos(\omega t-2\pi/3) & \cos(\omega t+2\pi/3) \\ -\sin(\omega t) & -\sin(\omega t-2\pi/3) & -\sin(\omega t+2\pi/3) \end{bmatrix} \quad (9)$$

Here: ωt is the angular velocity of the positive sequence abc/dq coordinates, which is the same as the fundamental wave angular frequency of the output voltage.

By transforming the resulting negative component to the negative abc/dq coordinates, we get a group of DC components x_{dqn} and x_{dq0}. The transformation matrix for the negative sequence abc/dq coordinates is:

$$T_{3/n}=\begin{bmatrix} \cos(\omega t) & \cos(\omega t+2\pi/3) & \cos(\omega t-2\pi/3) \\ -\sin(\omega t) & -\sin(\omega t+2\pi/3) & -\sin(\omega t-2\pi/3) \end{bmatrix} \quad (10)$$

This way, floating control is possible for the positive, negative and zero sequence components with the help of a PI controller.

The mathematic model of the inverter under a rotating coordinate system has a coupling relationship with the dq axis, which has to be eliminated in order to facilitate control. For this purpose, we used a feed-forward decoupling control strategy. Fig.2 shows the block diagram of feed-forward decoupling for positive sequence components. For negative and zero sequence components, the same decoupling control algorithm can be used and is not discussed here.

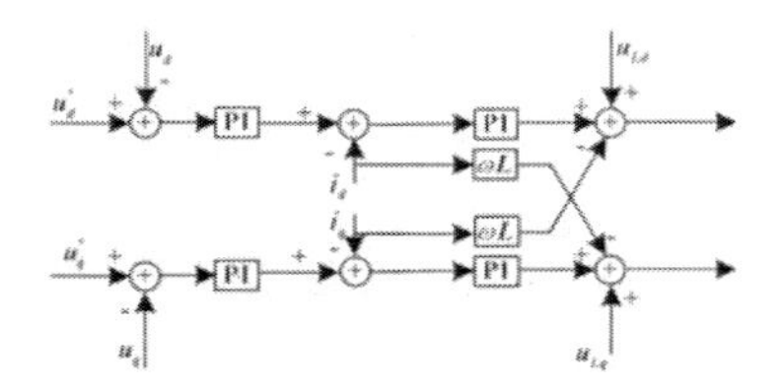

Figure 2. The feed-forward decoupling control of positive-sequence component.

3.2 *Inner current loop control*

When this feed-forward decoupling control algorithm is used, all the current loops in the control system become a typical single-input, single-output system [9-10]. In this section, the designed control algorithm is further described using positive sequence d-axis current control as an example. Fig.3 shows the block diagram of the inner current loop.

I-type systems are characterized by quick tracing compared with other types systems. To ensure quick tracing of current, we consider designing the inner current loop into an I-type system. Among the many considerations involved in designing an inner current loop regulator, two aspects are worth particular attention: (i) to overcome the potential effect of load current harmonics, the inner current loop frequency f must not be too high and should generally be far lower than the switching frequency fa; (ii) to the contrary, to increase the real current so as to enable quick tracing of its reference value, the inner current loop frequency f must be high enough and should generally be higher than the output frequency $f0$(f_0=50Hz). Based on this analysis, we can set the inner current loop system frequency f to close to 1kHz.

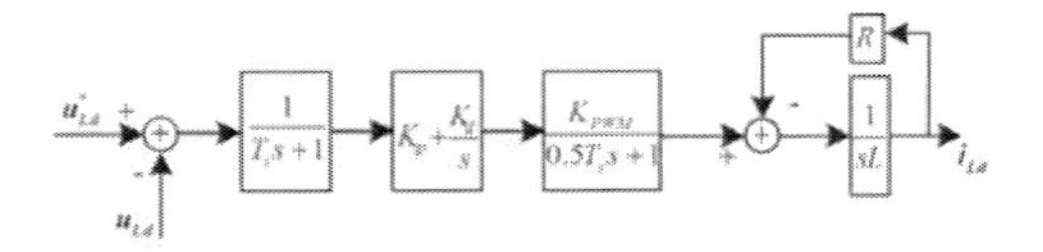

Figure 3. The block diagram of an inner current.

3.3 *Outer voltage loop control*

Fig.4 shows the block diagram of the outer voltage loop controlled by a typical PI controller. The open-loop transfer function is:

$$W_{ol} = \frac{K_{uI}(\frac{K_{uP}}{K_{uI}}s+1)}{Cs^2(T_u s+1)(\frac{R\tau_i}{K_{iP}K_{PWM}}s+1)} \tag{11}$$

As the purpose of outer voltage loop control is to ensure steady load voltage, the outer voltage loop control system should be set to both guarantee the quick tracing behavior and improve the interference immunity of the outer voltage loop. With these considerations, we use a typical II-type system to design the transfer function of the outer voltage loop at the parameters $K_{uI} = 9.6$ and $K_{uP} = 1.2 \times 10^{-2}$.

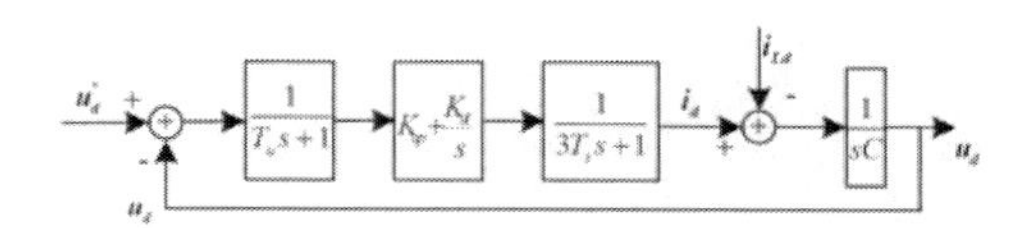

Figure 4. The control diagram of an outer voltage.

4 VERIFICATION BY SIMULATION EXPERIMENT

To verify how good our control algorithm is, comparative simulation is made using the traditional positive sequence rotating coordinate system control strategy and our decomposing sequence feed-forward decoupling control strategy, some of the result of which is given below. Fig.5 shows the simulation waveform when the traditional positive sequence synchronous rotating coordinate system alone is used. As the negative sequence component became 2-fold frequency component after positive sequence abc/dq coordinate transformation, the PI controller was only able to perform floating control on the DC volumes and not on the AC volumes. As a result, the feedback values do not provide ideal performance in tracing the reference values, as can be seen from the simulation diagram. Hence, when the inverter carries unbalanced loads and the output current is asymmetric, a dual closed-loop control strategy under the positive sequence synchronous rotating coordinate system is not able to control the inverter effectively or accurately.

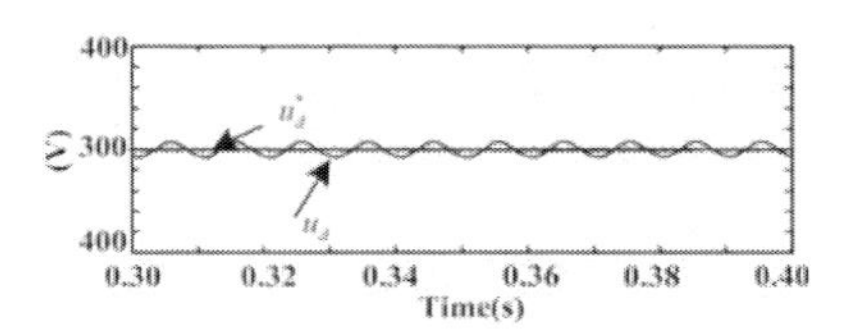

Figure 5. The simulation waveform by the traditional control strategy.

Fig.6 shows the simulation waveform by our decomposing sequence decoupling control strategy. From this chart, when the system comes to the steady state, the real feedback values are to trace given values very satisfactorily, suggesting that our control strategy is significantly superior over its traditional counterpart.

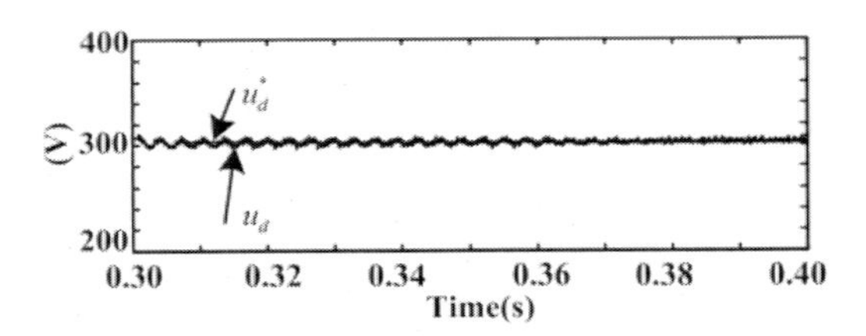

Figure 6. The simulation waveform by the proposed control strategy.

Fig 7 shows the simulation waveform of at the load end of a three-phase four-leg inverter when carrying unbalanced loads. From this chart, when the system comes to the steady state, the waveform of the output voltage is quite symmetric. Fourier analysis on MATLAB indicates the total distortion harmonic rate THD=0.67%. As the output voltage is three-phase symmetric and the three-phase loads are unbalanced loads, the output voltage waveform of the inverter is three-phase asymmetric current as shown in Fig.8.

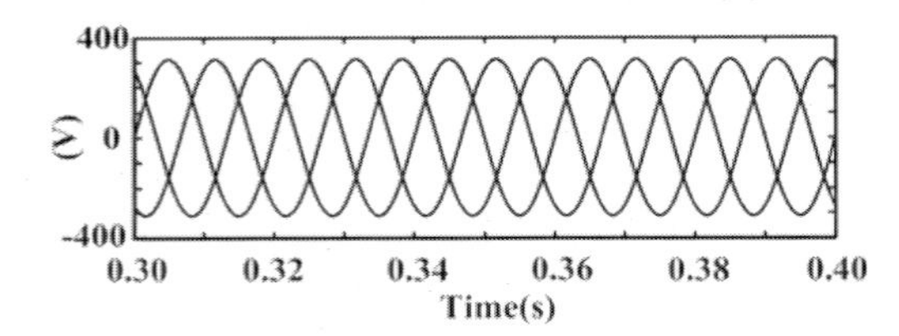

Figure 7. The three phase voltage.

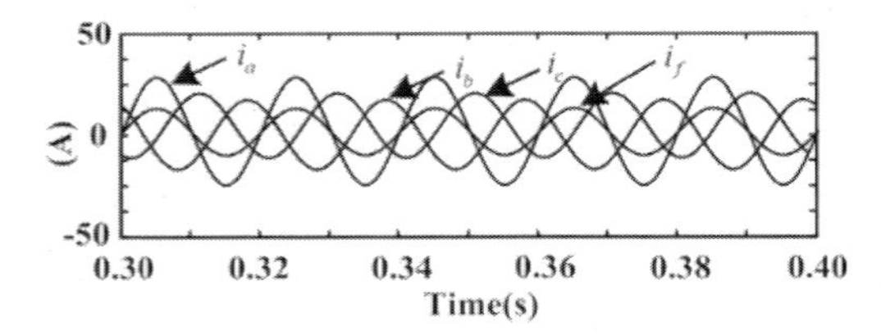

Figure 8. The three phase current.

5 CONCLUSIONS

For three-phase four-leg inverters, a novel decomposing sequence feed-forward decoupling control algorithm is proposed and provides for floating control of current. On this basis, a control system for inverters under a synchronous rotating coordinate system is designed. This control algorithm does not only overcomes the impossibility for the PI controller to achieve floating control of the negative sequence component as it appears to be 2-fold frequency component under a negative sequence synchronous rotating coordinate system, but also decomposes asymmetric components into three symmetric positive, negative and zero sequence components using the symmetric decomposition theory before turning them into DC volumes under the rotating coordinate system using abc/dq coordinate transformation and achieving floating control by using a PI controller. Simulation experiment confirms that under the proposed control strategy, the three-phase four-leg inverter provides quite symmetric load voltage when carrying unbalanced loads.

REFERENCES

[1] Sun Chi, Bi Zengjun, Wei Guanghui. (2004) Modeling and simulation of a three-phase four-leg inverter based on a novel decoupled control technique. *Proceedings of the CSEE* 2004, 24(1):124–130.

[2] Liu Xiuchong, Zhang Huaguang, Chen Hongzhi. (2007) Control strategy of fourth leg in four-leg inverter. *Proceedings of the CSEE*, 2007, 27(33):87–92.

[3] Chen Yao, Jin Xinmin, Tong Yibin. (2007) Study of the unification of SPWM and SVPWM in three-phase four-wire systems. *Transactions of China Electrotechnical Society*, 22(12):122–127.

[4] Sun Chi, Ma Weiming, Lu Junyong. (2006) Analysis of the Unsymmetrical output voltages distortion mechanism of three-phase inverter and its corrections. *Proceedings of the CSEE*, 2006, 26(21):57–64.

[5] Kaifei W, Fang Z, Yandong L, et al. (2004) Three-phase four-wire dynamic voltage restorer based on a new SVPWM algorithm. *IEEE 35th Power Electronics Specialists Conference*, 2004, 5:3877–3882.

[6] Kim S, Enjeti P N. (2000) Control strategies for active power filter in three-phase four-wire systems. *IEEE 15th Applied Power Electronics Conference and Exposition*, 2000, 1:420–426.

[7] Wang Huizhen, Ding Yong, Zhang Fanshua, et al. (2008) Four-leg three-phase inverter based on switching-node preset. *Proeeedings of the CSEE*, 2008, 28(3):73–76.

[8] Ojo O, Kshirsagar P M. (2004) Concise modulation strategies for four-leg voltage source inverters. *Power Electronics, IEEE Transactions on*, 2004, 19(1):46–53.

[9] Yang Hong, Ruan Xinbo, Yan Yangguang. (2002) PWM control of a four-leg three-ohase inverter. *Journal of Nanjing Univemity of Aeronautics & Astronautics*, 34(6):575–579.

[10] Wang Xiaogang, Xie Yunxiang, Huang Shaohui, et al. (2010) Unification of SPWM and SVPWM in four-leg inverter[J]. *Electric Machines and Control*, 14(1):23–34.

Electric, Electronic and Control Engineering – Shao, Shu & Tian (Eds)
© 2015 Taylor & Francis Group, London, ISBN: 978-1-138-02842-5

Hybrid anti-collision algorithm based on RFID system

Jinan Zhang, Yigang He, Huan Chen & Maoxu Liu
School of Electrical Engineering and Automation, Hefei University of Technology, Hefei Anhui, China

ABSTRACT: A novel hybrid anti-collision algorithm is proposed based on the analysis of the binary tree algorithm and ALOHA algorithm that is the grouping dynamic frame slot AHOLA collision tacking algorithm (GDCA). Firstly, the proposed algorithm combines Dynamic Frame Slot ALOHA (DFSA) algorithm, Collision Tracking Tree (CTT) algorithm and grouping idea together. Secondly, the dynamic tag estimation method is used to estimate the number of existing tags. Then tags will be grouped and frame size within each group is determined according to the estimation results. At last, the CTT algorithm is used to identify tags in the collision slots within each group. The comparison of simulation shows that GDCA algorithm has higher throughput rate and less total number of slots than CTT algorithm and DFSA algorithm. The Radio Frequency Identification (RFID) system gets better performance with the proposed algorithm.

KEYWORDS: RFID system; anti-collision algorithm; DFSA; CTT; GDCA.

1 INTRODUCTION

As a kind of non-contact automatic identification system, RFID technology has been widely used in electronic payment, supply chain management, traffic management and road toll systems and so on[1].

A typical RFID system consists of four parts: reader, tags, antennas and computer processing system[2]. Tags are generally attached to the surface or inside of objects which need to be identified and store some data of those objects. When tags appear in the identification scope of a reader, the reader will send a query command through antennas to read out specific data information stored in the tags. If two or more tags send data to a reader at the same time, tag collision occurs. In this situation, the tags cannot be successful identified, thus seriously affecting the identification efficiency of RFID system[3]. Therefore, in order to have a faster identification speed and higher accuracy in RFID system, it is very vital significance to study the tag anti-collision algorithm.

Time division multiple access (TDMA) is the most widely used method in tag anti-collision algorithm of RFID system[4], including two most commonly used algorithms: tree-based deterministic algorithms and ALOHA-based probabilistic algorithms[5].Tree-based algorithms include: binary search (BS), dynamic binary search (DBS), query tree (QT), collision tracking tree(CTT). ALOHA-based algorithms include: basic ALOHA (BA), slot ALOHA (SA), frame slot ALOHA (FSA) and dynamic frame slot ALOHA (DFSA).

In this paper, an advanced grouping dynamic frame slot AHOLA collision tracking algorithm (GDCA) is proposed. This algorithm combines advantages of DFSA and CTT[6,7]. So, GDCA has less total slots, higher throughput rate and faster identification speed than that of above two algorithms.

2 IDEAS AND STEPS OF THE PROPOSED ALGORITHM

2.1 *Ideas of the proposed algorithm*

Though the identification rate of tree-based tag anti-collision algorithms is 100%, but the total time consumed to identify all tags becomes very long with the increase of tags. ALOHA-based algorithms are easy to implement and have a small time delay. However, there is a problem that a tag may not be identified over a longer period of time.

An improved algorithm—grouping dynamic frame slot ALOH collision tracking (GDCA) algorithm is proposed in this paper by combining tree-based algorithm, ALOHA-based algorithm and grouping idea. Firstly, the GDCA algorithm uses Vogt estimation method to estimate the number of tags according to the principle of the DFSA algorithm [8]. Tags are grouped according to the estimated number of tags, then the most suitable frame size L is assigned to tags to be identified in each group. Each tag counter selects any one slot from slot 0 to slot (L-1) randomly. If only one tag responds in any one slot, the tag is identified directly. If two or more tags response, the CTT algorithm is used to identify all tags belonging to this slot. Don't need to jump to the next frame for identification. The frame structure allocation principle is shown in Fig.1.

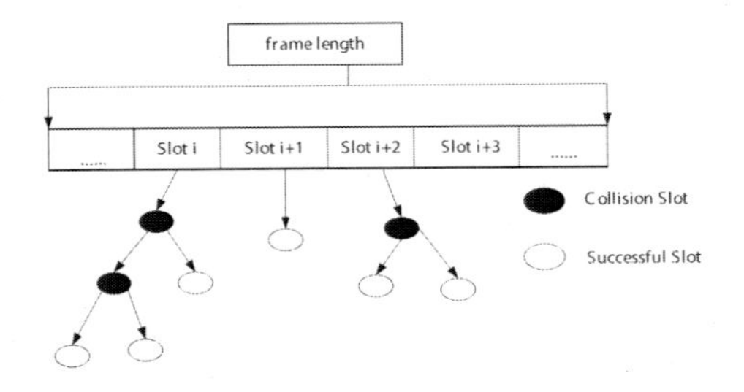

Figure 1. Principle of frame structure distribution.

In practice, when the number of tags to be identified unlimited increases, the reader can't allocate the number of slots equal to the number of tags objectively. The reader usually has its own frame size range (1, 8, 16, 32, 64, 128, 256). Therefore, when the number of tags within a certain range, the maximum threshold P_{th} is defined. When the number of unidentified tags is far greater than 256, the unidentified tags are grouped according to the rules of Table 1.

Table 1. Tag grouping rules.

Number of unidentified tags	Frame size (P_{th})	Number of group
1417–2831	256	8
708–1416	256	4
355–707	256	2
177–354	256	1
82–176	128	1
41–81	64	1
20–40	32	1
12–19	16	1
1–11	8	1

2.2 *Steps of the proposed algorithm*

The algorithm discussed in this paper doesn't consider the influence of capture effect and all kinds of noise in real environment or other factors. Tags to be identified are assumed to be static.

Now, the state of tags within the identified scope of reader could be divided into three types: standby, silent and dormant.

Standby state: Tags is waiting for communication with the reader, belonging to the state to be identified.

Silent state: Tags have communicated with the reader. They are successfully identified and don't involve in tag competition of the latter part.

Dormant state: Tags belong to a non-activated state and don't participate in tag competition temporarily.

The process of GDCA algorithm is shown in Fig.2.

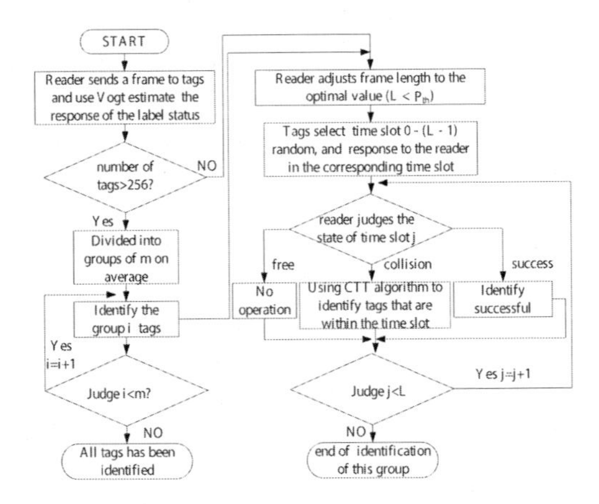

Figure 2. The process of GDCA algorithm.

3 PERFORMANCE ANALYSIS AND SIMULATION OF THE PROPOSED ALGORITHM

3.1 *Performance analysis*

Analysis of GDCA algorithm mainly includes the total number of slots and throughput rate in the process of identification. Firstly, the total number of slots is researched. According to the principle of GDCA algorithm, the identification process of this algorithm can be divided into two parts, that is tag identification stage and the tag estimation stage. Therefore, the total number of slots T_{Total} used in GDCA algorithm is consisted of the number of slots T_{est} in tag estimation stage and the number of slots T_{iden} in tag identification stage.

$$T_{Total} = T_{est} + T_{iden}. \tag{1}$$

The Vogt estimation method is used to estimate tags. L' is represented the frame size $T_{est} = L'$ sent by readers in the first round. Reader dynamically estimates the number of tags to be identified according to the state of each one slot in this frame. Thus the most appropriate frame size L can be accurately estimated. If the tags collision occurs in a slot while the frame size is L, the CTT algorithm is used to identify the collision tags. Assuming that the number of tags can be identified within the identification scope is n.

c_0, c_1, $c_{\geq 2}$ represent actual idle slot number, successful identification slot number and collision slot number in a frame respectively. According to the principles of binomial, the probability of i tags randomly distributes in the same slot, as follows

$$P(L,n,c_i) = C_n^i (\frac{1}{L})^i (1-\frac{1}{L})^{n-i} \tag{2}$$

When i tags select the same slot in a frame, the expected value is

$$E(L,n,c_i) = LC_n^i (\frac{1}{L})^i (1-\frac{1}{L})^{n-i} \tag{3}$$

When there are k tags occurring collision in collision slot, CTT algorithm is used to process the collision. The number of slots consumed by the CTT algorithm is $T_{CCT}(n) = \sum_{j=0}^{C}(2n_k - 1)$. The number of idle slots is I. The number of successful slots is S and the number of collision slots is C.

The total number of slots in tag identification stage is

$$T_{iden} = \sum_{k=0}^{n} E(L,n,c_i) T_{CTT}(k) = \sum_{k=0}^{n} LC_n^k \frac{1}{L}\left(1-\frac{1}{L}\right)^{n-k} T_{CTT}(k) \tag{4}$$

Therefore, the total number of slots used in GDCA algorithm is

$$T_{total} = T_{est} + T_{iden} = L' + \sum_{k=0}^{n} LC_n^k \frac{1}{L}\left(1-\frac{1}{L}\right)^{n-k} T_{CTT}(k) \tag{5}$$

When $L \to \infty$, the number of slots used to estimate the number to tags is small and can be ignored. So, $T_{total}' \approx \sum_{k=0}^{n} \frac{L}{k!e} T_{CTT}(k)$.

Next, the throughput rate namely slot utilization of GDCA algorithm is analyzed. Slot utilization is defined as $\eta_{slot} = \frac{\text{number of successful slot + number of collision slot}}{\text{total number of slots}}$.

Namely, $\eta_{slot} = \frac{S + \sum_{j=0}^{C} n_k}{T_{total}}$. Let $S' = S + \sum_{j=0}^{C} n_k$ (the number of slots completing tag identification). So,

$$\eta_{Slot} = \frac{S + \sum_{j=0}^{C} n_k}{T_{total}} = \frac{S'}{L' + (L-C) + \sum_{j=0}^{C}(2n_k - 1)} = \frac{S'}{L' + 2S' - (L-2I)} \tag{6}$$

When the number of tags is infinite, L' can be ignored. So,

$$\eta_{max} = \frac{S'}{L' + 2S' - (L-2I)} \geq \frac{S'}{2S' - 1} \tag{7}$$

Due to $S' = n$, $\eta_{max} \geq \frac{n}{2n-1}$. The slot utilization of DFSA algorithm is

$$\eta_{DFSA} = \frac{S}{L} = \frac{n}{L}\left(1-\frac{1}{L}\right)^{n-1} = \left(1-\frac{1}{n}\right)^{n-1} \tag{8}$$

The slot utilization of CTT algorithm is

$$\eta_{CTT} = \frac{n}{2n-1} \tag{9}$$

It can be seen that the total slot utilization of the proposed GDCA algorithm is higher than that of CTT algorithm.

3.2 *Simulation*

This paper focuses on the simulation comparison among GDCA, DFSA and CTT algorithms in system performance of throughput rate and total number of slots. Firstly, the simulation of the system about throughput rate among GDCA, DFSA and CTT algorithms is on the condition of a relatively small number of tags, as shown in Fig.3.The throughput rate of the system with CTT and DFSA algorithms is high when the number of tags is small. But with the number of tags gradually increases, systematic throughput rate decreases and flattens. The systematic throughput rate of CTT algorithm reduces close to 50%, while the systematic throughput rate of DFSA algorithm gradually achieves the best condition 36.8% with the gradual increase of tags. Systematic throughput rate simulation graph of GDCA algorithm shows volatility change, because the principle of GDCA algorithm adopts the tag grouping method. As shown in Fig.4, at the outset, when the number of tags is small, the systematic throughput rate of GDCA algorithm is less than CTT algorithm. But with the increase of the number of tags, the throughput rate significantly higher than CTT algorithm and DFSA algorithm.

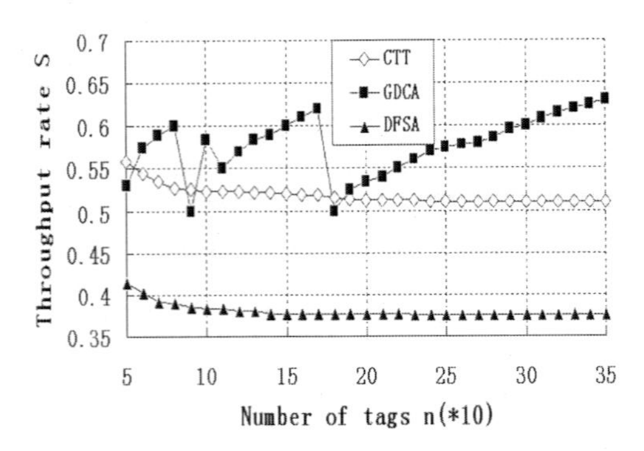

Figure 3. Throughput rate contrast when the number of tags is small.

In the case of a large number of tags, the systematic throughput rate simulation results of three algorithms above are shown in Fig.4.The systematic throughput rate of CTT and DFSA algorithms maintains constant at 50% and 36.8% while the systematic throughput rate of GDCA algorithm increases fast. The throughput rate is maintained at over 54%, maximum 60% after the number of tags reaching 650.

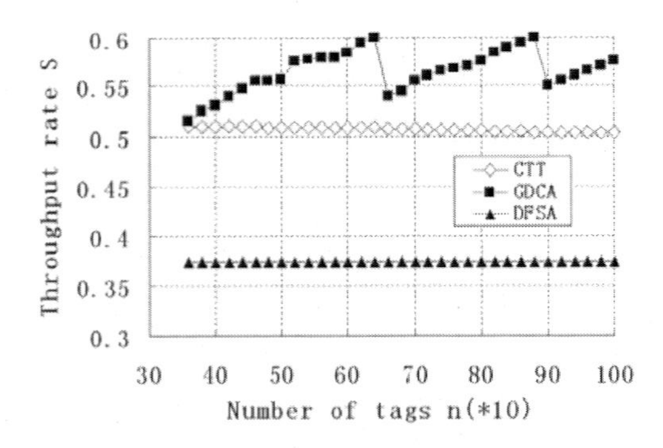

Figure 4. Throughput rate contrast when the number of tags is large.

The comparison of the total number of query slots among GDCA, DFSA and CTT algorithms is shown in Fig.5.The initial frame size of DFSA algorithm is set to 16 while the initial frame size of GDCA algorithm is set to 128.It can be seen from Fig.5 that the total number of slots of CTT algorithm and DFSA algorithm is less when the number of tags is small. However, with the gradual increase of the number of tags, the total number of slots of DFSA algorithm dramatically increases while the total number of slots of CTT algorithm increases linearly. At the same time, the number of slots of GDCA algorithm increase slowly. When the number of tags are over 100, the total number of slots of GDCA algorithm are less than or equal to that of CTT and DFSA algorithms.

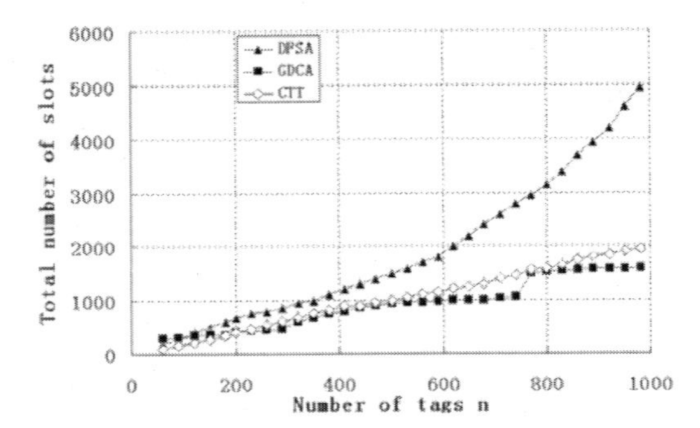

Figure 5. Comparison of the total number of slots.

4 CONCLUSIONS

A novel RFID anti-collision algorithm is proposed in this paper, which is GDCA. Firstly, a dynamic tag estimation method is introduced, and then DFSA algorithm and CTT algorithm are briefly introduced. The proposed algorithm is analyzed theoretically and compared with CTT algorithm and DFSA algorithm after describing the basic principles and process steps in detail. Simulation results show that the systematic throughput rate of GDCA algorithm is significantly higher than the previous two algorithms and the total number of slots is relatively small in the case of a large number of tags.Therefore, the performance of the new algorithm is superior to the above two algorithms.

ACKNOWLEDGEMENTS

This work was financially supported by National Outstanding Youth Science Foundation (50925727); Major Project of the Education Ministry of science and Technology (313018); Anhui Province Science and Technology Key Project (1301022036); Hunan Province Science and Technology Plan Projects (2010J4, 2011JK2023) and National Defense Science and Technology Plan Projects (C1120110004, 9140A27020211DZ5102).

REFERENCES

[1] Yan Hui. (2013) *Research of Anti-collision Algorithm Based on RFID System*. Nanjing: Nanjing University of Science and Technology.

[2] Li Meng. (2011) *Research on Anti-collision Algorithm Based on ALOHA for RFID System*. Jilin: Jilin University.
[3] Xia Zhiguo, He Yigang, Hou Zhouguo. (2010) Binary-tree bit-detecting RFID tag anti-collision algorithm. *Computer Engineering and Applications*, 46(20):245–248.
[4] Li Weiwei. (2012) *The Research of Tag Anti-Collision Based on the Dynamic Frame Slotted*. Hebei: Hebei University of Technology.
[5] Kalache Mohamed Aissa, Fergani Lamya. (2014) Performances comparison of RFID anti-collision algorithms. *2014 International Conference on Multimedia Computing and Systems, ICMCS 2014*, pp:808–813.
[6] Guo Zhitao, Chen Linlin, Zhou Yancong and Gu Junhua. (2012) Improve of dynamic framed slotted ALOHA algorithm. *Application Research of Computers*, 29(3):907–909.
[7] Wang Shoufeng, Zhang Dongchen, Xu Xiaoyan, Shi Shumeng and Wang Tianlan. (2014) A novel anti-collision scheme for RFID systems. *2014 IEEE World Forum on Internet of Things (WF-IoT)*, pp:458–461.
[8] Mohammed J. Hakeem, Kaamran Raahemifar and Gul N. Khan. (2014) Novel modulo based aloha anti-collision algorithm for RFID systems. *2014 IEEE International Conference on RFID (IEEE RFID)*, pp:97–102.

Electric, Electronic and Control Engineering – Shao, Shu & Tian (Eds)
 ISBN: 978-1-138-02842-5

Research on disaster recovery policy of dual-active data center based on cloud computing

Xiao Chen & Longjun Zhang
Department of Information Engineering, Engineering University of CAPF, Xi'an Shaanxi, China

ABSTRACT: Cloud computing is now applied in various fields more and more widely due to its advantages. The data size will become larger and larger with the increasing of users. The exponential growth of mass data makes cloud disaster recovery more important. Focusing on the disaster recovery environment of cloud computing, this paper designs and implements the "double-active" scheme and the storage strategy based on "double active" data center with the combination of virtualization technology, realizing zero loss of users' data and zero interruption of operation on the basis of "double-active" network and operation.

KEYWORDS: Data center; cloud disaster recovery; double-active; storage strategy.

1 INTRODUCTION

In recent years, cloud computing is emerging worldwide with the rapid development of virtualization and network technology, triggering off the wave of the third information technology revolution. Cloud computing provides a whole new information service mode with huge market space for people. Economic benefits brought to users are becoming increasingly remarkable with the gradual development of cloud computing technology. Meanwhile, the emergence of cloud storage technology provides users with a third party IT resource and service supporting usage on demand and flexible architecture. It is able to solve the problem of growing demand for IT resource brought by the rapid development of information technology.

However, there are still lots of users hesitating between cloud computing and the traditional software architecture in spite of the huge value created by cloud computing and cloud storage. The primary reason is that the problem of integrity and availability of users' data hasn't been solved properly. The problem of data integrity and availability is the largest obstacle for users to transfer data and applications to the cloud computing architecture, so to speak. The integrity and availability of users' data depends on the disaster recovery system to a large extent. So data security of IT system is now put to a new height in the existing environment. Traditional disaster recovery is unable to meet the requirements of disaster recovery in the new environment. The problem of effective data storage of cloud disaster recovery has become a major bottleneck of the development of disaster recovery. The construction of a new "double-active" data center is an urgent need for cloud disaster recovery. This paper mainly introduces the construction of a "double-active" data center based on traditional network architecture and designs the "double-active" scheme of network and operation, realizing zero loss of users' data and zero interruption of operation by constructing a virtual experimental environment.

2 CLOUDING COMPUTING AND CLOUD DISASTER RECOVERY

2.1 *Cloud computing system*

Peng Liu, a Chinese expert in grid computing and cloud computing, defines cloud computing as follows: cloud computing distributes computational task in the resource pool constituted by a large number of computers so that various application systems are able to obtain computing power, storage space and all kinds of software services as needed. Simply speaking, cloud computing is to establish a huge computational system by a large-scale data center and a super computer cluster so that users can acquire various kinds of software and services through the internet instead of buying an independent operating system and various software for each computer[1].

The structure of cloud computing system consists of five parts, namely the application layer, the platform layer, the resource layer, the users' access layer and the management layer. The essence of cloud computing is to provide services through the network. So, the core of the system structure is service. The application layer, the platform layer and the resource layer are services of different levels provided by cloud computing, shown in Figure 1:

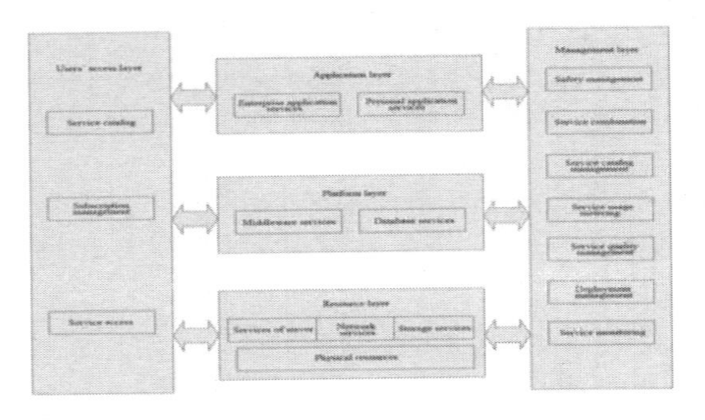

Figure 1. Cloud computing system.

The resource layer refers to cloud computing service of the basic structure level. This service provides virtualized resources so as to conceal the complexity of physical resources. The platform layer provides users with the encapsulation of the resource layer so that users are able to construct their own applications by more advanced services. The application layer provides users with software services. The users' access layer is convenient of users to use a variety of support services of cloud computing. Cloud computing services in each layer should provide relevant interfaces. The management layer provides management functions for cloud computing of all layers[2].

2.2 *Cloud disaster recovery*

As a new concept in the disaster recovery field, cloud disaster recovery provides an effective solution of disaster recovery for enterprises. Cloud disaster recovery is a mode, in which disaster recovery is regarded as a service, basic installations are provided by the third party manufacturers of constructing cloud, and users pay for the disaster recovery service[3]. With this mode, users are able to realize disaster recovery goals, reduce operational cost and working intensity, and reduce the total cost of ownership of disaster recovery by taking advantage of technical resources, plentiful disaster recovery experience and mature operational management process offered by service providers.

Traditional cloud disaster recovery provides manageable and operational disaster recovery services through currently advanced, safe and reliable data backup and data replication technologies. It provides users with city-wide or remote disaster recovery services of different levels so as to ensure that business data and major application systems of users can be recovered rapidly and accurately after disasters and guarantee the continuous operation of users' operations. At present, the cloud disaster recovery service mainly includes disaster recovery of data and disaster recovery of application.

Cloud disaster recovery has such distinct advantages as high degree of specialization, low invest cost, resource sharing and high quality of service. These advantages endow "cloud disaster recovery", a "mainstream trend" of socialized services, with strong vitality. Take America with relatively mature disaster recovery industry for instance. "Cloud disaster recovery" is now accounting for 56% of the disaster recovery industry. It can be seen that cloud disaster recovery is becoming a mainstream trend.

3 "DOUBLE-ACTIVE" DATA CENTER

Data center provides users with multiple web services like network resources, service trusteeship, broadband access, etc. It is the place with the most intensive information system resources and the most frequent data exchange. With the deepening of cloud computing application, the operational environment of data center is now transforming from traditional clients/servers to larger server clusters of network connection.

"Double-active" data center is created with "cloud" for the purpose of disaster recovery. Generally, the two data centers are in operating state and able to undertake services at the same time so as to improve the overall service capacity and system resource utilization of data centers. Besides, the two data centers provide backup mutually. If there is a malfunction in the data center that provides operational services, the operation can be switched to another data center automatically so as to realize zero loss of data and zero interruption of service [4].

The construction of "double-active" data center is a complex system project, including resources of infrastructure like data, system, service, network, server, etc. In addition, requirements of users are mixed together, so the construction is extremely complicated. The demand of "double-active" data center is mainly reflected in three aspects, namely network link of wide area, server/storage and infrastructure of machine room. According to different requirements, the deployment pattern of data center includes double-active network, double-active service, double-active resource, etc. There is no necessary link among the three. They can be constructed independently or assembly so as to meet the needs of different users on performance, investment protection and the flexibility of operation deployment. The scheme of "double-active" network and "double-active" service is studied and designed in this paper.

4 DESIGN AND IMPLEMENTATION OF "DOUBLE-ACTIVE" SERVICE BASED ON CLOUD COMPUTING

4.1 *Double-active of network and operation*

In order to realize "double-active" network on the premise of "double-active" service, it is required that address fields of "double-active" service in different data centers should be different as well. Network technologies of DNS and global load balance are adopted for the schematic design, which is shown in Figure 2.

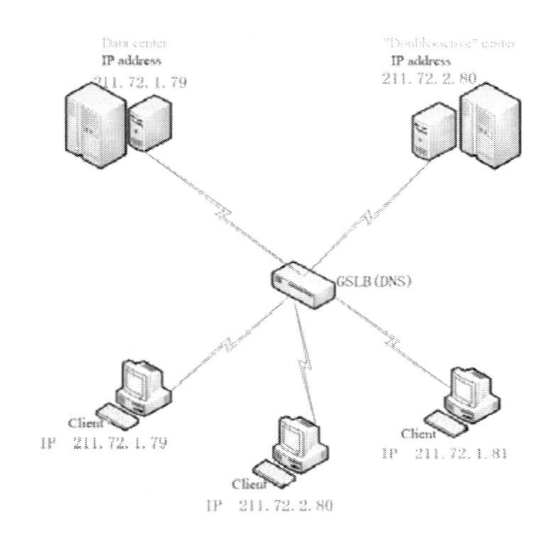

Figure 2. Deployment scheme of "double-active" network and operation.

Both the data center and the double-active center are able to respond to operations independently. But the two centers are distributed in different network segment. Because different gateway addresses can be deployed by a server, it is required by the model that operations of the same kind in the operational system should support IP addresses and gateway addresses of different network segments. Clients' automatic access to websites with the optimum performance and the consistency of operations can be guaranteed by network technologies of DNS and global load balance.

The front end of data center adopts GSLB (global server load balance) based on intelligent DNS analysis, which is in charge of load balancing requested by DNS analysis, server status monitoring and client access path optimization, so as to make sure that requests from clients can be distributed to available nodes[5,6]. GSLB is responsible for domain name analysis request, providing clients with the nearest node addresses through a set of predefined strategies so that clients can be served immediately. GSLB adopting DNS is able to find service interruption timely and switch to accessible websites automatically. In order to guarantee the capability of "double-active" data center of providing continuous services, it is necessary to equip the back end with technologies of server load balance and HA so as to coordinate with the "double-active" network of the front end and realize deployment and high availability of resources.

4.2 *Storage service design based on "double-active" center*

Design and realize a solution of storage service based on "double-active" center on the basis of cloud computing and "double-active" service. The design is shown in Figure 3.

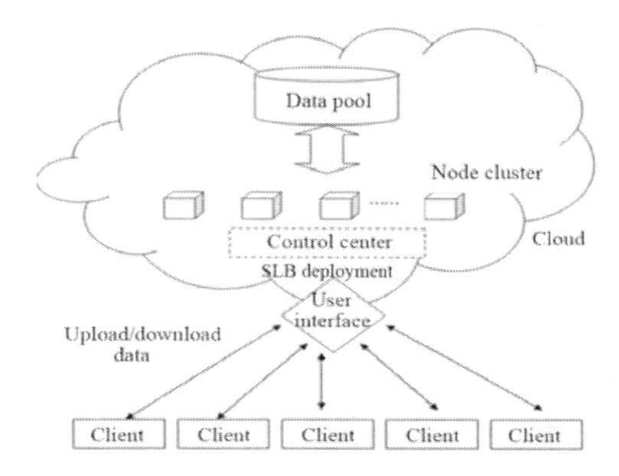

Figure 3. Design of storage service of data center.

Server virtualization plus storage virtualization and virtualized resource pool plus dynamic applications and data movement are adopted in the scheme. Storage virtualization is realized by VPLEX and server virtualization is realized by VMware. The client realizes upload and download of data by calling cloud interfaces [9]. The data pool is just a place for data storage, saving all the data for clients. There is no need for the data pool to have computational capability.

In order to satisfy the service needs of high performance and high reliability, the part of cloud computing centers on nodes. A node cluster is a server cluster virtualized through the virtualization technology. Each virtual server provides same or similar network storage services and is managed by the control center located at the front end of the node cluster. SLB deployed by the control center distributes client requests in the server cluster according to the configured equilibrium strategy, provides clients with services, and maintains the server. With the coordination of GLSB at the front end of the data center, SLB can not only perfectly realize load balancing of the full path from the front end to the inside of the data center but also realize health detection of the server. In the process of computing, nodes are adjusted to elastic scalable clouds dynamically to be in charge of the handling of client data, which are ultimately saved in the data pool.

The main data center and the "double-active" data center require L2 interconnection to satisfy the requirements of L2 communication among cluster members. Meanwhile, SAN interconnection is also needed to realize data synchronization.

5 EXPERIMENT SIMULATION

Carry out the simulation and implementation of the data center system solution of cloud computing architecture with 3 servers, 16 computers and 4 routers in the laboratory environment. Data are distributed in two servers with different network segments and subsystem modules to simulate the cloud data center. There is also a management server serving as the control center and 16 computers serving as clients. Results of experiment simulation are shown in Figure 4.

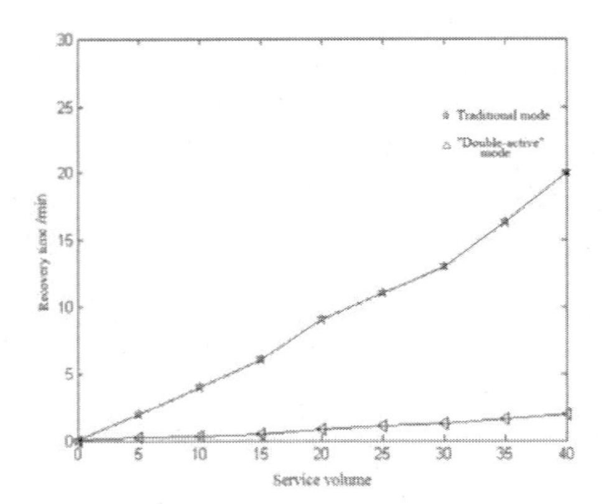

Figure 4. Results of experiment simulation.

It can be seen from the figure that, if the traditional disaster recovery scheme is adopted, the service recovery time is obviously prolonged with the increase of service volume when the main data center is out of order. With the increase of service volume, the service interruption time of the disaster recovery system deployed with "double-active" scheme becomes extremely short after the main data center stops operating. Experiment results show that the "double-active" center can realize the effect of disaster recovery more perfectly.

6 CONCLUSION

As a new form of disaster recovery, cloud disaster recovery has drawn extensive attentions. However, the traditional disaster recovery is unable to meet the needs of the development of information age due to the rapid growth of data. "Double-active" data center is created with "cloud", compensating for shortcomings of the traditional ways of disaster recovery.

This paper designs and implements the "double-active" network and service and the storage scheme based on "double-active" center. However, a real "double-active" data center should also contain "double-active" data. There is normally one data center of the "double-active" scheme existing as Active due to the current technical limitations and the consistency of data. Other data centers realize data backup function through synchronous replication or asynchronous replication according to geographical distances. Therefore, the "double-active" data center scheme discussed in this paper revolves only basic installations and "double-active" applications. Meanwhile, the construction direction and technological selection of data center in future should be in line with the trend of virtualization and cloud computing. Factors like standard, openness, flexibility and maintainability should be fully taken into account in technological selection.

REFERENCES

[1] Peng Liu, (2011) *Cloud Computing*, Peking: Electronic Industry Press.

[2] A. Bajpai, P. Rana, and S. Maitrey, (2013) Remote mirroring: A disaster recovery technique in cloud computing, *International Journal of Advance Research in Science and Engineering*, 2(8).

[3] Yu Gu, Dongsheng Wang, and Chuanyi Liu, (2014) DR-cloud: Multi-cloud based disaster recovery service, *Tsinghua Science and Technology*, 19(1):13–23.

[4] Ferdaus, M. H., & Murshed, M. (2014) *Energy-aware Virtual Machine Consolidation in IaaS Cloud Computing*, pp:179–208.

[5] Kamal, J. M. M., & Murshed, M. (2014) *Distributed Database Management Systems: Architectural Design Choices for the Cloud Computing*, pp:23–50.

[6] Kiran, M. (2014) *A Methodology for Cloud Security Risks Management Cloud Computing*, pp:75–104.

[7] Kronabeter, A., & Fenz, S. (2013) Cloud Security and Privacy in the Light of the 2012 EU Data Protection Regulation Cloud Computing, pp:114–123.

[8] Lombardi, F., & Di Pietro, R. (2014) Virtualization and Cloud Security: Benefits, Caveats, and Future Developments, *Cloud Computing*, pp:237–255.

[9] Peng, G. C. A., Dutta, A., & Choudhary, A. (2014) Exploring critical risks associated with enterprise cloud computing, *Cloud Computing*, pp:132–141.

[10] Wei Den, Fangming Liu, Hai Jin, et al. (2013) Leveraging Renewable Energy in Cloud Computing Datacenters: State of the Art and Future Research, *Chinese Journal of Computer*, 03:582–598.

[11] Information on http://aws.amazon.com/simpledb/.

Electric, Electronic and Control Engineering – Shao, Shu & Tian (Eds)
© 2015 Taylor & Francis Group, London, ISBN: 978-1-138-02842-5

Defect identification in pipes by chirp signals

Fei Deng
School of Electronic and Information Engineering, Ningbo University of Technology, Ningbo, China

Honglei Chen
College of Electrical and Electronic Engineering, Shanghai Institute of Technology, Shanghai, China

ABSTRACT: To realize defect identification in pipes, chirp signals were chosen to excite guided waves for pipe inspection. The response signals of different single-frequency tone burst, which were calculated based on the original wide-frequency response signals, are processed with Hilbert Huang transform, and then the value of damage index is calculated for forming the feature curves. The results show that the method can distinguish the elliptical hole and rectangular crack defects with different size and tilted angles.

KEYWORDS: Defect identification; pipe; chirp signal.

1 INTRODUCTION

Much of the research in pipe inspection with guided waves in the last decade has focused on how to identify the shape and size of defects. Since the 90s, many scholars have examined some typical defects, especially circumferential cracks in their research[1-4]. J. L. Rose[5] studied the reflection echoes from the corrosion defects. J. Davies[6] investigated the corresponding relationship between the size of a crack or a hole and the amplitude of the reflection echo. Recently, A. Velichko [7] examined the scattering of guided waves for complex defects in pipes. These works had emphasized the importance of analyzing the scattering waves from defects. In addition, there are lots of works interested in finding characteristic parameters standing for the features of a defect. P. Rizzo[8] presented a defect classification method in pipes by neural networks using multiple guided wave features extracted after wavelet processing. D. Fei[9] demonstrated that the variation curves of time reversal window width as a function of the reflection coefficients of the defect signals could be used for defect identification. However, either these methods are based on the premise of known defect types, or are very time hungry.

After J. E. Michaels [10] introduced the chirp excitation for guided wave inspection in plates, it is easier to analyze the scattering waves from defects at any designated exciting frequency[11]. The multi-modes and dispersive characteristics are differ from different defect conditions, which results in scattering waveforms are diverse corresponding to an arbitrary frequency. So, this paper investigated the relationship between the amplitude of reflection echoes from defects and the exciting frequency based on the chirp excitation in guided wave inspection of pipes. According to the relationship curves, a defect identification method has further come into being.

2 CHIRP SIGNAL PROCESS METHOD

The chirp signal is depicted in *Eq.*1:

$$S_c(t) = w(t)\sin(2\pi f_0 + \pi Bt^2 / T) \tag{1}$$

Where $w(t)$ is a rectangular window function during T. f_0 and B are the initial phase frequency and bandwidth in *kHz*.

Fig.1 shows a chirp signal with frequency sweep from 50 *kHz* to 300 *kHz* over a 120μs rectangular window:

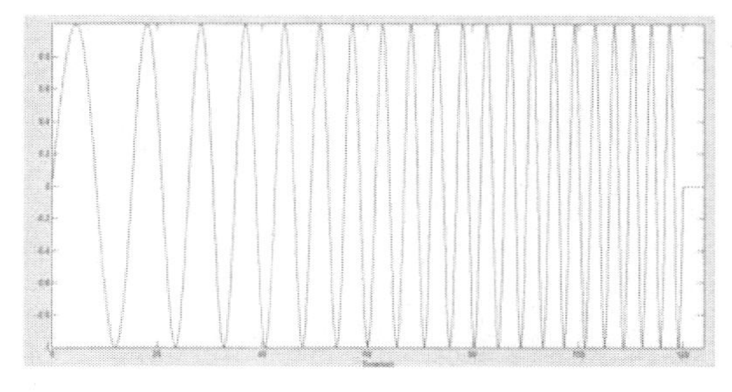

Figure 1. Chirp signal with sweeping curve from 50 to 300 *kHz*.

According to paper [10], responses of tone burst can be calculated from *Eq*.2:

$$r_d(t) = ifft\left[R_c(\omega)\frac{S_d(\omega)}{S_c(\omega)}\right] \quad (2)$$

Where $R_c(\omega)$ is the chirp response; $S_d(\omega)$ and $S_c(\omega)$ are the excitation of tone burst and broadband signal in the frequency domain. $r_d(\omega)$ is a tone burst response that we desired.

3 NUMERICAL SIMULATIONS

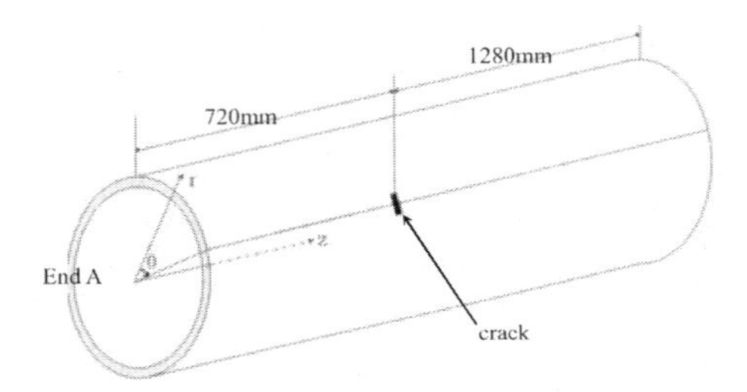

Figure 2. The pipe model with one crack.

The commercial finite element analysis software, Ansys, is used in this study. A pipe model with one defect is illustrated schematically in Fig.2. All pipes considered in this paper are designed with 70mm in external diameter, 3.5mm in wall thickness, 2m in length. The axial distance between the center of the defect and the pipe end A is 0.72m. Young's modulus is 2.16906e^{11} N/m^2, the density of material is 7932kg/m^3, and Poisson's ratio is 0.286543.

Two types of defects, including rectangular and elliptical standing for crack and corrosion, are investigated by this pipe model by removing the full wall thickness. Except for auto-meshing in the vicinity of the defect, meshes of linear quadrilateral membrane elements with identical sizes are used. There are 48 elements around the 360-deg circumference of the model. The element axial length is 2mm. With the reference cylindrical coordinate system in Fig.2, the tilted angle of defects is defined as the angle between defect's main axis and the circumferential generatrix.

In order to eliminate the affection from environmental factors, a baseline signal is obtained in an intact pipe model with the same parameters as before except for no defect. The chirp signals are excited with the intact pipe end A simultaneously, and then the baseline response signal is stored in the current environment. This process is an essential first step and the detailed procedure had mentioned in paper [11].

4 FEATURE SELECTION AND PROCESS

The aim of this section is to define a damage index that is robust against noise and is related to the characteristics of the defect. That robustness against noise is an important aspect in real guided wave detecting, especially for small defect.

Fig.3 presents two typical wide-frequency waveforms simulated on the pipe models with the chirp excitation on end A, one model with 20*2mm 45degree crack, another with 20*2mm 0degree crack. The response signals received on the nodes of the pipe end A had been superimposed. Obviously, defects result in the scattering echoes and the reflection echoes are clear and distinct. The discrimination stemming from different defects implies the characteristics of these defects, which is propitious to defect identification in the next stage.

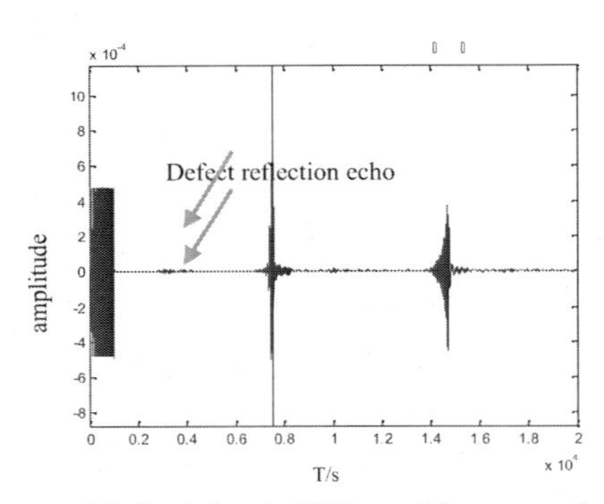

(a) the defect is 20*2mm 0degree crack

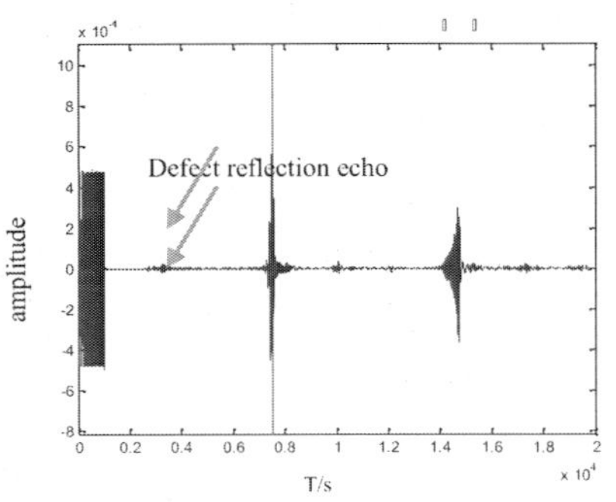

(b) the defect is 20*2mm 45degree crack

Figure 3. Typical wide-frequency response signal simulated on the pipe model with defect.

For the defect identification research in this paper, all simulated signals are based on the same chirp excitation signal (with frequency sweeping from 50kHz to 500kHz over a 200μs rectangular window and the amplitude normalized to 1). We choose the first reflection echo's peak amplitude, which is from the Hilbert transform of the single-frequency time waveform calculated based on the wide-frequency time waveform results, standing for the characteristics of defects. So, the proposed damage index (D.I.) is the ratio between the peak amplitude $A_{reflection}$ and the amplitude of the chirp signal A_{chirp}.

$$D.I. = \frac{A_{reflection}}{A_{chirp}} \tag{3}$$

5 REPRESENTATIVE D.I. RESULTS

112 samples are considered in this research, including oblique cracks changed in length and tilted angle, elliptic holes changed in length-width ratio, and circular holes changed in radius. Detailed information is listed in Table 1.

Due to the chirp signal with frequency sweeping from 50khz to 500khz, the single-frequency time waveforms are calculated from 100kHz to 400kHz, with the interval of 20kHz. So, 16 feature parameters of D.I. to frequency are plotted together in a curve and used to recognize defect's features. We chose to illustrate this problem with different diagrams.

Table 1. The sample list.

Shape			Rectangular (fixed the width to 2mm)								
No.	Length (mm)	Tilted angle (degree)	No.	Length (mm)	Tilted angle (degree)	No.	Length (mm)	Tilted angle (degree)	No.	Length (mm)	Tilted angle (degree)
1	11	0	15	17	0	29	23	0	43	29	0
2	11	15	16	17	15	30	23	15	44	29	15
3	11	30	17	17	30	31	23	30	45	29	30
4	11	45	18	17	45	32	23	45	46	29	45
5	11	60	19	17	60	33	23	60	47	29	60
6	11	75	20	17	75	34	23	75	48	29	75
7	11	90	21	17	90	35	23	90	49	29	90
8	14	0	22	20	0	36	26	0	50	32	0
9	14	15	23	20	15	37	26	15	51	32	15
10	14	30	24	20	30	38	26	30	52	32	30
11	14	45	25	20	45	39	26	45	53	32	45
12	14	60	26	20	60	40	26	60	54	32	60
13	14	75	27	20	75	41	26	75	55	32	75
14	14	90	28	20	90	42	26	90	56	32	90

Shape	Elliptic hole (fixed the minor axis to 2mm, ratio: major axis-minor axis ratio)								Shape	Circular hole
No.	ratio	Tilted angle (degree)	No.	ratio	Tilted angle (degree)	No.	ratio	Tilted angle (degree)	No.	Radius (mm)
57	2	0	71	6	0	85	10	0	99	3
58	2	15	72	6	15	86	10	15	100	4
59	2	30	73	6	30	87	10	30	101	5
60	2	45	74	6	45	88	10	45	102	6
61	2	60	75	6	60	89	10	60	103	7
62	2	75	76	6	75	90	10	75	104	8
63	2	90	77	6	90	91	10	90	105	9
64	4	0	78	8	0	92	12	0	106	10
65	4	15	79	8	15	93	12	15	107	11
66	4	30	80	8	30	94	12	30	108	12
67	4	45	81	8	45	95	12	45	109	13
68	4	60	82	8	60	96	12	60	110	14
69	4	75	83	8	75	97	12	75	111	15
70	4	90	84	8	90	98	12	90	112	18

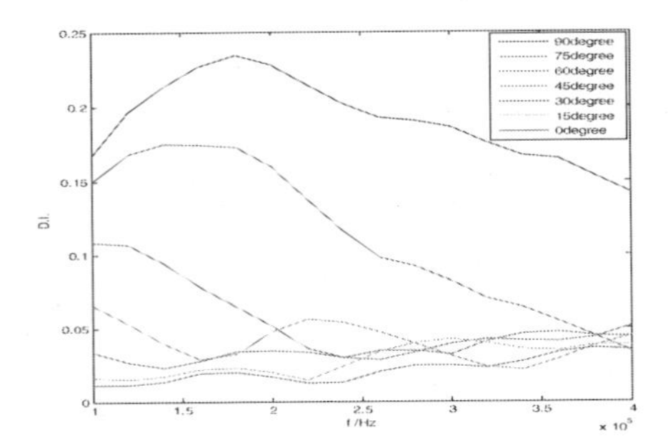

Figure 4. 7 feature curves calculated based on the same pipe models with 23*2mm crack, except for crack tilted angles, differed from 0 degree to tilted angle90 degree with 15 degree interval.

A total of 7 feature curves calculated based on different pipe models with the same crack size 23*2mm, except for crack tilted angles differed from 0 degree to 90 degree with 15 degree interval, are plotted as Fig.4. We concluded that the crack tilted angle should affect the trend of the feature curve. When the tilted angle is close to 90 degree, the values of D.I. in the low frequency band are significant higher than that in the high frequency band. When the tilted angle is close to 45 degree, the curve of D.I. is oscillation in the whole frequency band. When the tilted angle is close to 0 degree, the values of D.I. in the high frequency band are significant higher than that in the low frequency band. However, with the tilted angle decreases, the varying extent depresses too.

Furthermore, we compared some feature curves derived from pipe models with circular hole defects in different radius. As Fig.5 shows that the trend for these curves are similar. The bigger the defect's radius is, the value of D.I. to a certain frequency is bigger. This conclusion is also suited to the cracks. When the tilted angle is constant, the length of the crack is longer, the D.I. value is bigger, and the trend for feature curves is identical.

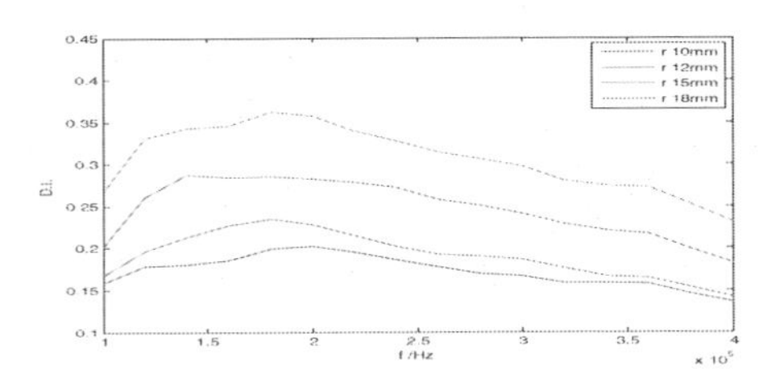

Figure 5. 4 feature curves calculated on the same pipe model with circular hole, just the radius is different.

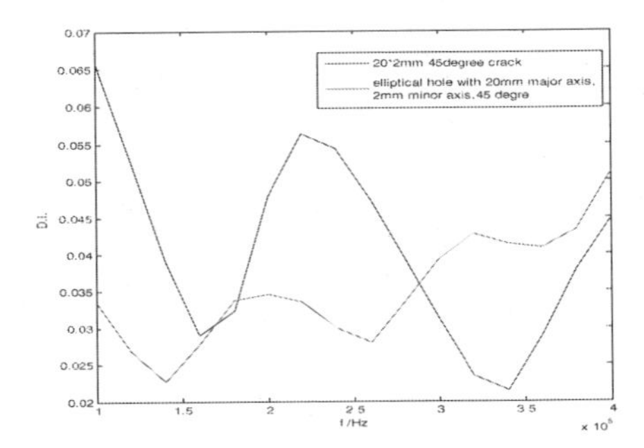

Figure 6. 4 feature curves calculated on the same pipe model with circular hole, just the radius is different.

In this section, we further compared feature curves for crack defects and elliptical holes. Interestingly, although the length-width ratio for cracks and elliptical holes is very close, such as 20*2mm 45degree crack and elliptical hole with 20mm major axis, 2mm minor axis, 45 degree, the trend of feature curves are quite different (as shown in Fig.6). It shows that the difference in the changing trend of feature curves indicate the discrimination between the sharp edges of the crack and smooth edges of elliptical hole.

112 feature curves for all samples are concerned. How many kinds of defects, just how many unique feature curves accordingly. So, if construct sample feature curves library for different characteristic defects, the parameters of the unknown sample, such as shape, size and direction, could be given by using the method of feature matching, neural networks, etc.

6 CONCLUSION

The aim of this paper is to provide a method for identifying the characteristics of different defect in pipes with guided waves. Chirp signals are excited on numerical pipe models with different defects, and then the wide-frequency response signals are collected for calculating single-frequency response result in some designated frequency points. After that, all values of D.I. to these frequency points are reckoned on the envelopes of these single-frequency time waveforms obtained by Hilbert Huang transform. The last step is to give the unique feature curves standing for each defect with different characteristics. The feature curve is essential and critical to defect identified, and more researches are still required before the final goal of defect feature recognition can be completed.

ACKNOWLEDGMENTS

Gratefully acknowledge and appreciate the support of the project of the National Natural Science Fund (No.11202137), and Shanghai City Board of Education Science and Technology Innovation Fund Project (No.13YZ119), as well as scientific research start-up fund of Ningbo University of Technology (No. 0080011540161).

REFERENCES

[1] M. Lowe, D. Alleyne, P. Cawley. (1998) The mode conversion of a guided wave by a part-circumferential notch in a pipe. *Journal of Applied Mechanics*, 65:649–656.

[2] D. Alleyne, M. Lowe, P. Cawley. (1998) The reflection of guided waves from circumferential notches in pipes. *Journal of Applied Mechanics*, 65:635–664.

[3] A. Demma, P. Cawley, M. Lowe. (2003) The Reflection of the Fundamental Torsional Mode from Cracks and Notches in Pipes. *Journal of the Acoustical Society of America*, 114(2):611–625.

[4] W. Zhu. (2002) An FEM simulation for guided elastic wave generation and reflection in hollow cylinders with corrosion defects. *Journal of Pressure Vessel Technology*, Transactions of the ASME., 124(1):108–117.

[5] J. L. Rose, S. Pelts, W. Zhu. (2000) Flaw sizing potential with guided waves. *Review of Progress in Quantitative Nondestructive Evaluation*, 19:927–934.

[6] J. Davies, P. Cawley. (2007) The application of synthetically focused imaging techniques for high resolution guided wave pipe inspection. *Review of Quantitative Nondestructive Evaluation*, 26:681–688.

[7] A. Velichko, P. D. Wilcox. (2013) Scattering of guided waves from complex defects in plates in pipes. *Review of Progress in Quantitative Nondestructive Evaluation*, 39:129–136.

[8] P. Rizzo, I. Bartoli, A. Marzani, F. L. D. Scalea. (2005) Defect classification in pipes by neural networks using multiple guided ultrasonic wave features extracted after wavelet processing. *Journal of Pressure Vessel Technology*, Transactions of the ASME, 127:294–303.

[9] D. Fei, W. Bin, H. Cunfu. (2010) A time reversal guided wave defect identification method. Journal of Mechanical Engineering, 46(8):18–24.

[10] Michaels J E, Lee S J, Coxford A J, et al. (2013) Chirp excitation of ultrasonic guided waves. *Ultrasonic*, 53(1):265–270.

[11] C. Honglei, D. Fei, Z. Xi. (2013) Defect imaging via chirp signal excitation in plate. *Advanced Materials Research*, 846:826–830.

[12] D. Fei. (2012) Defect identification using time reversal guided waves based on baseline subtraction, The 2nd International Conference on Electronics, Communications and Control, pp:559–562.

Electric, Electronic and Control Engineering – Shao, Shu & Tian (Eds)
© 2015 Taylor & Francis Group, London, ISBN: 978-1-138-02842-5

A comprehensive evaluation of investment ability of power grid enterprises–Taking power grid enterprise of Zhejiang province as an example

Jiao Fan, Dongxiao Niu, Xiaomin Xu & Honghao Qin
School of Business &Management, North China Electric Power University, Beijing

Hui Xu
Zhejiang Electric Power Company Economic Research Institute, Hangzhou, China

ABSTRACT: The investment of power grid enterprises has the characteristics of large scale, long payback period and easily affected by external factors, therefore it is necessary to carry out a comprehensive evaluation of the investment ability of power grid enterprises in order to arrange an investment strategy reasonably. By building the comprehensive evaluation index system of investment ability, this paper applied Sequence Relations with Entropy Method to determine the compound index weight with the financial data of the Zhejiang Electricity Power Company. Then it used TOPSIS and Gray Correlation method respectively to evaluate the investment ability of Zhejiang in 2005-2013. Finally, the paper drew the conclusion that apart from the year of 2008 and 2009, which lead to a dramatic loss of investment ability affected by extreme factors such as the financial crisis, the investment ability realized a stable development.

KEYWORDS: Power grid enterprises, Investment ability, Index System, Comprehensive evaluation.

1 INTRODUCTION

With the surging demand of electricity in the whole society, the investment scale of the power grid has realized large increase. In recent years, the investment environment of power grid enterprises faces increasing complexity and macroeconomic regulations, such as the control of investment in fixed assets, the rise of interest rates and mandatory energy saving target. So it is necessary for enterprise companies to conduct a comprehensive evaluation of investment ability to provide the basis for reasonable grid investment strategy and improve the level of investment management.

Investment ability analysis generally includes income level, financing ability, investment efficiency analysis, etc.. Power grid enterprise investment ability refers to the electricity company finance of the supported investment scale of the ceiling, in the conditions of target profits and asset-liability ratio limit[1]. Research abroad, mainly concentrated on the electricity market financing ability and risk control. Europe and the United States of the power market mainly study from the cost-benefit analysis, portfolio and risk management[2-3]. Domestic analysis in investment ability mainly gathered in a single power engineering projects, including the use of Analytic Hierarchy Process (AHP) and fuzzy mathematics theory, incremental method[4-5]. Throughout the domestic and foreign research on investment ability, current studies on systematic and comprehensive evaluation of investment ability is not much, thus by evaluating the current power grid investment ability, it can guide the enterprise to determine a reasonable investment scale and investment projects.

2 THE CONSTRUCTION OF EVALUATION INDEX SYSTEM

Investment ability assessment needs to build multi-level indicators from multidimensional aspects, and the construction of index system should follow the main principles of scientific, comparability, systematic and dynamic. Investment ability, mainly consisting of earnings and financing, is affected by many external factors. Therefore, the index system can be built from three aspects of the profitability, financing ability and external factors. The price and sales of electricity construct the main income of power grid enterprises, and the cost mainly includes electricity purchasing cost, transmission and distribution costs, depreciation, etc.. In addition, we select total assets return rate to reflect the enterprise investment profitability and investment efficiency. Enterprise financing ability includes financing, loan interest rate and financial expenses. And external factors such as GDP growth rate and the power supply of the population also affect the investment ability. The final index system is shown in Table 1, including 3 first level indicators and 12 secondary indicators.

Table 1. Power grid enterprise investment ability evaluation index.

The target layer	First class indicators	Second class indicators
Investment ability A	Profitability B1	Sales of electricity C1
		Sale price C2
		Power purchase cost C3
		Transmission and distribution costs C4
		Depreciation cost C5
		Administration expense C6
		Return on total assets C7
	Financing ability B2	Financed amount C8
		Loan rate C9
		Financial expense C10
	External factors B3	GDP growth rate C11
		Power supply of population C12

3 THE DETERMINATION OF INDEX WEIGHT

Subjective empower method mainly depends on the logical relationship between the index and evaluation expert's subjective judgment or intuition, while objective method completely depends on the sample data and the data accuracy. This paper combines Sequence Relations and Entropy Method to determine the compound weight.

3.1 *First class indicators weight based on sequence relations*

Sequence Relations firstly sort the evaluation indexes qualitatively, then carries on the rational judgment to adjacent index, and finally uses a quantitative calculation. Compared with AHP judgment matrix, it reduces the amount of calculation. Due to the Sequence Relations act completely expresses the experts' will, the result is trustworthy. Sequence Relations steps are as follows:

1 To determine the sequence relation. When evaluators consider evaluation index x_i relative to a certainthe assessment criteriace of assessment criteria is greater than (or not less than) x_j, that is $x_i \succ x_j$

2 Determine the relative importance between x_{k-1} and x_k, set the ratio w_{k-1}/w_k of the importance of the evaluation index x_{k-1} and x_k is

$$w_{k-1}/w_k = r_k \ (k = m, m-1, \cdots, 2) \tag{1}$$

3 Calculation of weight coefficient w_k

If r_k given by experts satisfies the relationship $r_{k-1} > 1/r_k \ (k = m, m-1, \cdots, 2)$, so

$$w_m = \left(1 + \sum_{k=2}^{m} \prod_{i=k}^{m} r_i\right)^{-1} \tag{2}$$

$$w_{k-1} = r_k w_k \ (k = m-1, \cdots, 2) \tag{3}$$

Owe to the primary indext (B class) are independent of each other and there is no specific index value, we use sequence relations act to determine the weight of indicators to the target layer.

First of all, make sure the sequence of profitability B1, financing ability B2 and external factors B3. Experts thought it existed sequence relation $B2 \succ B1 \succ B3$ marking $B1^* \succ B2^* \succ B3^*$ and calculating the importance judgment, $r2 = w1^*/w2^* = 1.4, r3 = w2^*/w3^* = 1.6$, the resolve is $r2 \times r3 = 2.24, r3 = 1.6, r2 \times r3 + r3 = 3.84$.

So $w3^* = 1/(1+3.84) = 0.2066, w2^* = w3^* \times r3 = 0.3306$

In the same way, $w1^* = w2 * \times r2 = 0.4628$

The weight coefficient of evaluation index B1, B2, B3 is $w1 = w2^* = 0.3306, w2 = w1^* = 0.4628, w3 = w3^* = 0.2066$

3.2 *Second class indicators weight based on entropy method*

The entropy value method is based on the size of information loads to determine the indicators weight.

Set the decision matrix is $D = \begin{bmatrix} x_{11} & x_{12} & \cdots & x_{1m} \\ x_{21} & x_{22} & \cdots & x_{2m} \\ \cdots & \cdots & \cdots & \cdots \\ x_{n1} & x_{n2} & \cdots & x_{nm} \end{bmatrix}$, in this x_{ij} is the value of the j scheme of the i indicator attribute. The steps as follows:

1 Calculate in the j indicator, the i system weight of characteristics or contribution.

$$p_{ij} = \frac{x_{ij}}{\sum_{i=1}^{n} x_{ij}} \tag{4}$$

Therein, p_{ij} represents the contribution of the i scheme of the j indicator attribute.

2 Calculate the entropy value e_j and differential coefficient g_j in the j indicator.

The entropy value e_j represents the total contribution of the j indicator of all solutions.

$$e_j = -k\sum_{i=1}^{n} p_{ij} \ln p_{ij} \tag{5}$$

Therein, constant k is $k = 1/\ln n$, that can make sure $0 \le e \le 1$.

Coefficient variance g_j represents in the j indicator inconsistency degree of the contribution of each program, it's calculated by 3-10.

$$g_j = 1 - e_j \tag{6}$$

Apparently, the larger the g_j is; the more important the role of indicators is.

3 Normalized weight coefficient w_j.

Comprehensive weight calculation results are shown in Table 2.

Table 2. Weight calculation results.

First class	First class weight	Second class	Second class weight	Final weight
Profitability	0.3306	Sales of electricity	0.0455	0.0151
		Sale price	0.0016	0.0005
		Power purchase cost	0.0681	0.0225
		Transmission and distribution costs	0.6542	0.2163
		Depreciation cost	0.0819	0.0271
		Administration expense	0.0878	0.0290
		Return on total assets	0.0610	0.0202
Financing ability	0.4628	Financed amount	0.7654	0.3542
		Loan rate	0.0298	0.0138
		Financial expense	0.2048	0.0948
External influence	0.2066	The second industry GDP	0.9571	0.1977
		Power supply of population	0.0429	0.0089

Empowerment results can be seen that the largest weight of the first class indicators is the financial ability, following by the profitability and the external influence. In the second class indicators, factors with greater influence on the investment ability are financed amount, power transmission and distribution costs, the second industry GDP, financial expenses, etc.

4 A COMPREHENSIVE EVALUATION OF INVESTMENT ABILITY

4.1 *Comprehensive evaluation based on TOPSIS*

TOPSIS method measures the vector distance between each object and the positive ideal solution or the negative ideal solution to determine sorting. It can make full use of the origin data and lose less information. It namely sets an ideal system $(x_{i1}^*, x_{i2}^*, \cdots x_{im}^*)$ and evaluation objects $(x_{i1}, x_{i2}, \cdots x_{im})$. The weighted distance y_i between the two is:

$$y_i = \sum_{j=1}^{m} w_j f(x_{ij}, x_j^*),\ i=1,2,...n \tag{7}$$

Therein, w_j is weight coefficient, $f(x_{ij}, x_j^*)$ is a certain distance of x_{ij} and x_j^*, commonly using Euclidean distance.

With queuing indicated value to evaluate the system, its calculation formula is as follows:

$$C_i = \frac{y_i^-}{y_i^+ + y_i^-} \tag{8}$$

Among C_i, the larger indicated value is the better.

The evaluation results based on TOPSIS are shown in Table 3.

4.2 *Comprehensive evaluation based on gray correlation*

In this paper, all the indicators are on the basis of unification, and each index is the maximum value of the reference sequence. The main parameters of gray correlation are as follows:

$$\min_i \min_j |x_0(j) - x_i(j)| = 0$$

$$\max_i \max_j |x_0(j) - x_i(j)| = 0.9426$$

$$\rho = 0.5$$

Having no relation to the weight, the traditional gray correlation method is not affected by way of determining weight. But in this paper, the calculation of correlation is fixed by optimization, considering the weight coefficient by the entropy value method. Generating into the correlation fixed expressions:

$$\gamma_i = \frac{1}{n}\sum_{j=1}^{m} w_j \xi_i(j) \tag{9}$$

The evaluation results based on Gray Correlation are shown in Table 3.

Table 3. A variety of comprehensive evaluation method results summary.

Year	TOPSIS method	Rank	Grey Correlation	Rank
2005	4.7513	4	6.24862	4
2006	3.4565	7	5.03559	7
2007	3.6381	6	5.15636	6
2008	2.4537	8	4.54203	8
2009	2.3822	9	4.50985	9
2010	4.4157	5	5.68501	5
2011	5.4380	1	7.36537	1
2012	5.1174	2	6.86797	2
2013	5.0334	3	6.73889	3

From Table 3, we can see the evaluation result of the investment ability of using TOPSIS method and the Grey Correlation method are almost consistent. The first three years with the strongest investment ability are 2011, 2012 and 2013, and three years with the worst investment ability are 2006, 2008 and 2009. For the comprehensive evaluation results more intuitive to reflect the differences, drawing a line chart is shown in Figure 1.

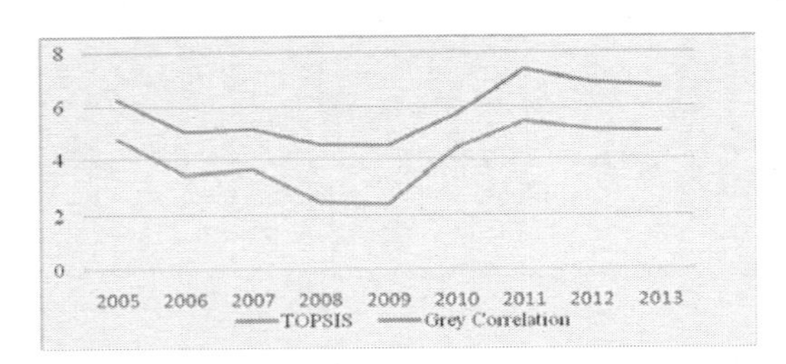

Figure 1. A variety of comprehensive evaluation results.

From Figure 1, we can see that the TOPSIS method's parameter values of discrete degree are bigger. It can be seen clearly that the Zhejiang province in 2008 and 2009 investment ability was at low levels, after 2010 the investment ability was stronger. Through comprehensive evaluation result, nearly 9 years in Zhejiang power grid investment ability showed a trend of increasing roughly. In addition to the 2008 and 2009, showing a significant decline, may be due to the influence of the global financial crisis and domestic natural disasters resulted quantity sold and financing amount reducing; But since 2011 because of strengthen management, cost control, enterprise profitability improved, at the same time, stable economic environment, security financing amount, the enterprise achieves investment ability increasing greatly.

5 CONCLUSION

In this paper, investment ability of power grid enterprises from 2005 to 2013 in Zhejiang province are evaluated comprehensively and analyzed thoroughly. First of all, from three aspects of profitability, financing ability and the external factors, it selected 12 indicators to construct relatively perfect evaluation index system. Then it used Sequence Relations and Entropy Method to execute combination empowerment and gets the conclusion that the key influence factor are financing amount and power purchasing cost. Finally, by the methods of TOPSIS and Grey Correlation, it evaluated the investment ability and analyzes the results comparatively. For investment ability reducing sharply in 2008 and 2009, power grid enterprises should strengthen financial management, power purchase cost control consciously. They also need to strengthen the prevention ability of financial crisis and extreme events of natural disaster, for achieving rational and stable development of power grid investment.

ACKNOWLEDGMENT

This paper was supported by the National Natural Science Fund (71471059) and the project "Research on the reasonable investment strategy of power grid based on the development demand and investment capacity" from State Grid

REFERENCES

[1] Zhang Yan. (2009) Analysis of media listing corporation investment capacity–Taking diananchuanmei as an example, *News Media*, 1:24–27.

[2] Zhao Huiru, Fu Liwen. (2012) Quantitative research of investment capacity of power grid enterprises. *Hydroelectric Energy*, 30(4):191–194.

[3] He Ying. (2011) Analysis and evaluation of economic benefit of power enterprise investment project. *Journal of North China Electric Power University (Social Science Edition)*, (12):32–36.

[4] Yang Yuqun, Ju Zeli, Wang Xin. (2011) Investment benefit evaluation of power grid enterprise. *Power System Technology*, 35(11):42–48.

[5] Jing Ping. (2008) The forecasting and assessing method for urban sustainable development tendency. *IEEE*. Tianjin Normal University Serving Tianjin & Binhai New District Special Research Funds.

Electric, Electronic and Control Engineering – Shao, Shu & Tian (Eds)
 ISBN: 978-1-138-02842-5

Research and application of online grid load monitoring and smart analysis system

Henggui Wang
Guang'an Power Supply Company, State Grid Sichuan Electric Power Company, Guang'an Sichuan, China

ABSTRACT: Electric load forecasting allows proper arrangement of generator operation within the power grid to guarantee safe and stable grid operation. The online grid load monitoring and smart analysis system is examined using integrated forecasting modeling to perform load analysis and load forecasting for a power supply company and achieve real-time load forecasting and effective data sharing.

KEYWORDS: Load forecasting; integrated forecast model; load analysis.

1 INTRODUCTION

As the electric power system develops and grid management moves towards modernization and complexity, research on power system load forecasting has aroused wide concern. Correct power load forecasting allows proper scheduling of generator operation within the power grid to guarantee safe and stable grid operation, proper scheduling of generator maintenance to ensure normal production of the society and normal life of the public, determination of the capacity of new installed capacities and dispatching of the power grid to effectively reduce the power production cost and improve both the economic and social benefits[1]. As such, power system load forecasting has become an important topic of study in modern power system operation and management as well as an important area in modern electric power system sciences.

2 DEVELOPMENT OF LOAD FORECAST SYSTEM SOFTWARE

As power system load is highly variable and complicated, manual empirical forecasting as a conventional practice is no longer in pace with the management modernization of power operators as it is far from meeting the required accuracy and speed. More importantly, the extensive application of computer technology in the power system has resulted in the replacement of manual means of power load forecasting by software means. In this context, developing an easy, man-machine friendly, reliable load forecasting software package convenient for user operation and accommodating power company requirements will not only improve the forecasting accuracy, but also allow the power producers to perform automatic control and operational analysis according to the forecasting result and provide the management division with basis for making management plans and development strategies[2].

2.1 *System framework*

On an integrated forecasting model, this software is a modular structure consisting of long-term load forecasting and short-term load forecasting. The software package is divided into different functional modules and the functions of, and interfaces between, these modules are determined before they are specifically programmed. Arranged as a modular structure, the software package is composed of a historic data processing module, a load forecasting module, a forecasting result processing module and an inquiry module. These modules are independent from each other and share data via the inter-module interfaces so that the package is easily debugged and extended in future. The system framework is shown in the diagram below.

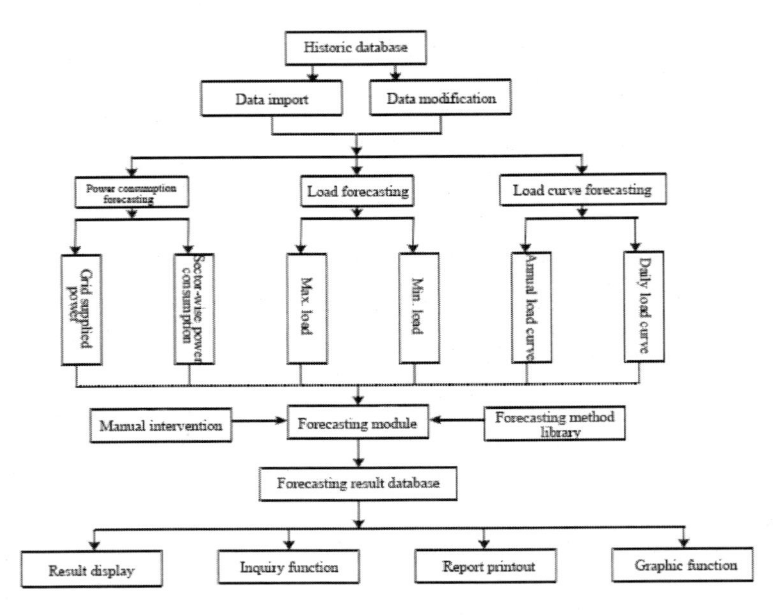

Figure 1. System framework diagram.

This system uses integrated load forecasting that forecasts loads by a number of new protocols, compares and determines the optimal forecasting result.

2.2 *System functions*

2.2.1 *System function flowchart*

The system function flowchart is shown in the diagram below (see Fig. 2)

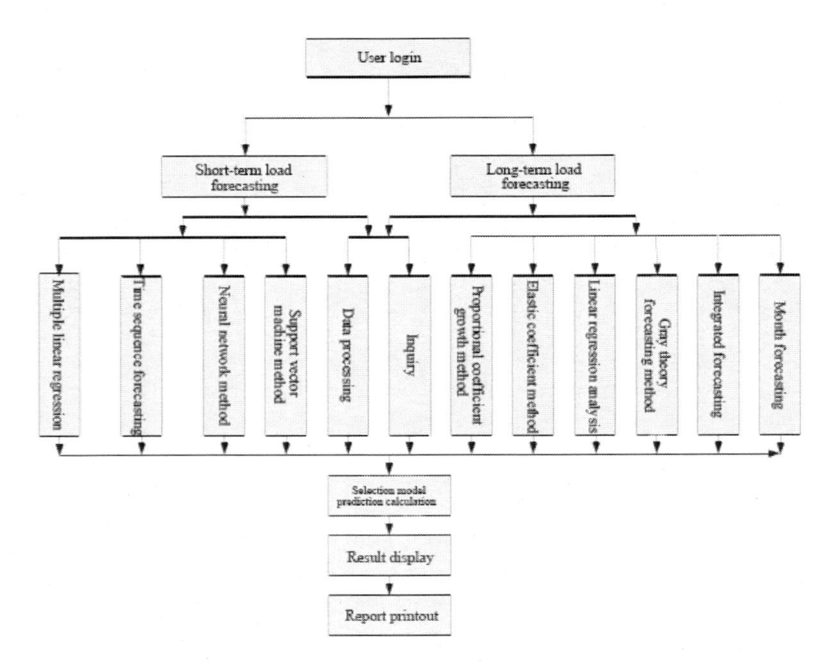

Figure 2. The system function flowchart.

2.2.2 *Software function realization*

1 Historic data entry: in the database of the software, a uniform-field datasheet has been created in which the user may directly add or modify the existing data. The user may also create a new datasheet and enter data either manually or using the automatic entry function of Access.
2 Load and power supply inquiry: the software allows inquiry on the annual grid power supply, annual maximum load, monthly power supply, monthly load and daily load.
3 Short-term load forecasting realization: in the short-term load model library, select a forecasting model, perform load calculation and finally obtain numerically and graphically displayed result.
4 Annual load forecasting realization: in the long-term load forecasting library, select a forecasting model, perform load calculation and finally obtain numerically and graphically displayed result.
5 Monthly load forecasting realization: in the monthly load model library, select a forecasting model, perform load calculation and finally obtain numerically and graphically displayed result.

3 SYSTEM KEY TECHNICAL RESEARCH

This system is a load forecasting system supported by integrated forecasting modeling[3]. In power system load forecasting, as the forecasting is performed under certain assumptions and the power load variation is subject to a number of factors, one method alone hardly gives a satisfactory result. It is therefore necessary to perform integrated analysis by combining results from different models in a proper and scientific manner to obtain an integrated forecasting result having the best forecasting performance. This is what we call integrated forecasting.

The essential of integrated forecasting modeling is to show the proportion of each single method in the final forecasting result with the weight of a targeted orientation effect. For the same issue forecasted, linear combination of a number of different forecasting models can in some cases effectively improve the model fitting ability and increase the forecasting accuracy. By forecasting loads with an integrated model, it is possible to combine different models in an organic manner so as to integrate the merits of these models and provide more accurate forecasting results[4]. So far, integrated forecasting is realized by means of integrating forecasting modeling, integrated optimal fitting modeling, integrated sub-optimal forecasting modeling and integrated optimal forecasting modeling. Generally, integrated forecasting is obtaining the weighted average of the forecasting results from a number of forecasting methods with properly selected weights or selecting a forecasting method with the optimal goodness-of-fit or with the minimum standard deviation.

4 SYSTEM FUNCTION REALIZATION OF SOFTWARE

4.1 *Composition and acquisition of base data*

Base data are automatically collected. Data is acquired and the data format is automatically transferred to adapt to the system requirements and automatically stored into the system database to allow subsequent analysis and forecasting.

Data interface with the dispatch automation system and statistics system is provided to allow automatic acquisition of load, power consumption and other base information.

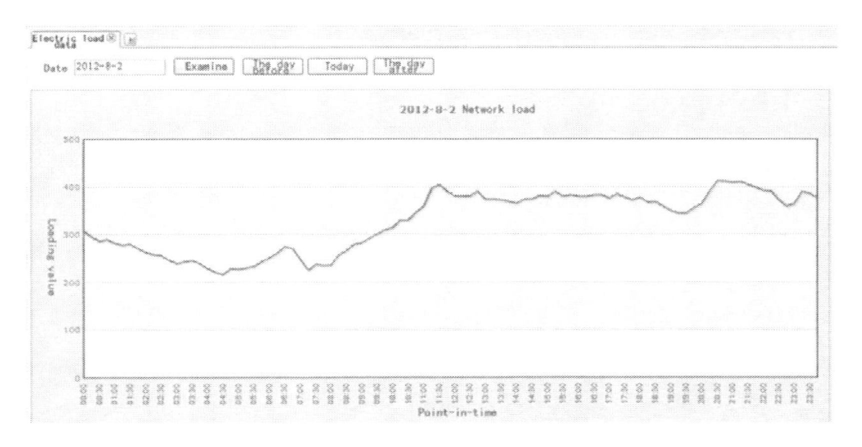

Figure 3. Electric load data.

4.2 *Load analysis*

Load analysis is the fundamental means of updating load variation. By displaying load curves and analyzing load characteristics, it is possible to reflect the load variation pattern and influencing factors so as to better support the work of load forecasting specialists.

Build an outgoing low-frequency load shedding hourly database for the loads, build the annual average maximum load database and provide recommended adjustment plans according to the low-frequency load shedding scheme.

Build a standard load forecasting analysis and decision database, design a proper data structure, allow full sharing of electric power data, minimize data redundancy and provide high-reliability, high-consistency information sources for operators at all tiers and all levels.

1. Comprehensive load analysis

Calculate the load characteristics at one or more load points within one date or at one or more date points and display the load time-sequence curves of one or more days. Compare the load variation within this period. Select any date and load point as shown in the chart below:

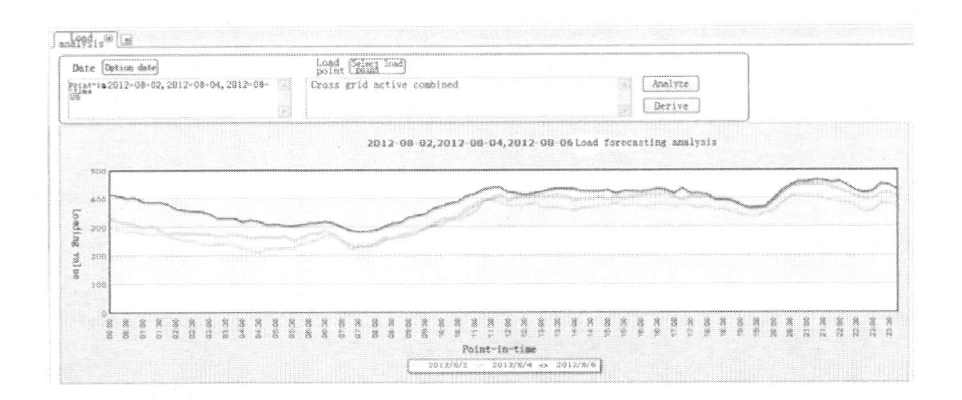

2. Daily load analysis

- Daily load curve analysis

Calculate and display the load data and curve within a continuous period of time and compare the time-sequence load curves.

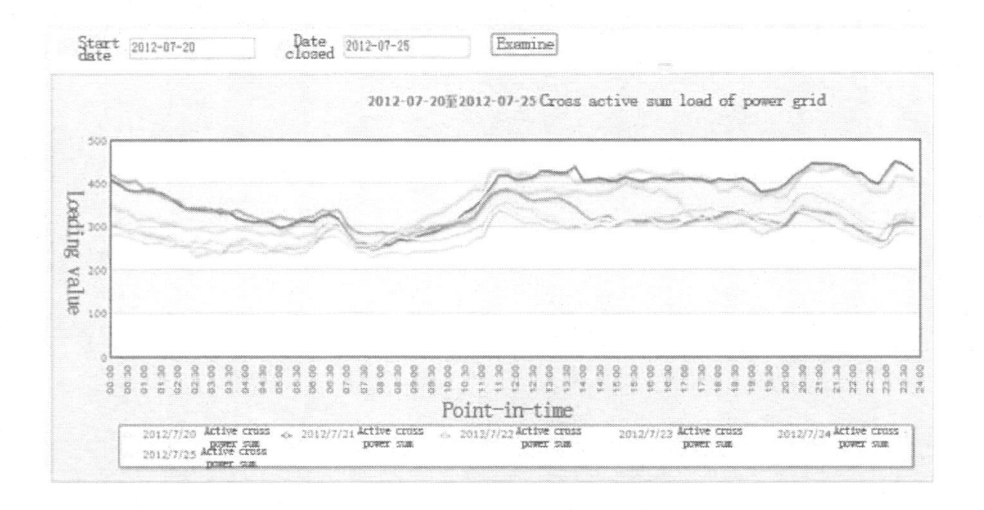

- Daily load curve comparison

Select the daily load curves of any two days and compare them.

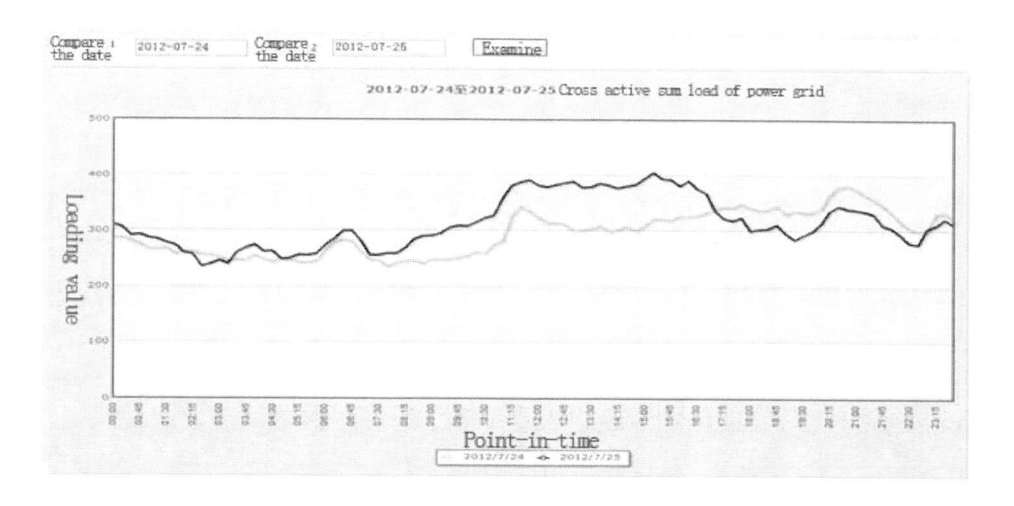

- Multi-day load curve comparison

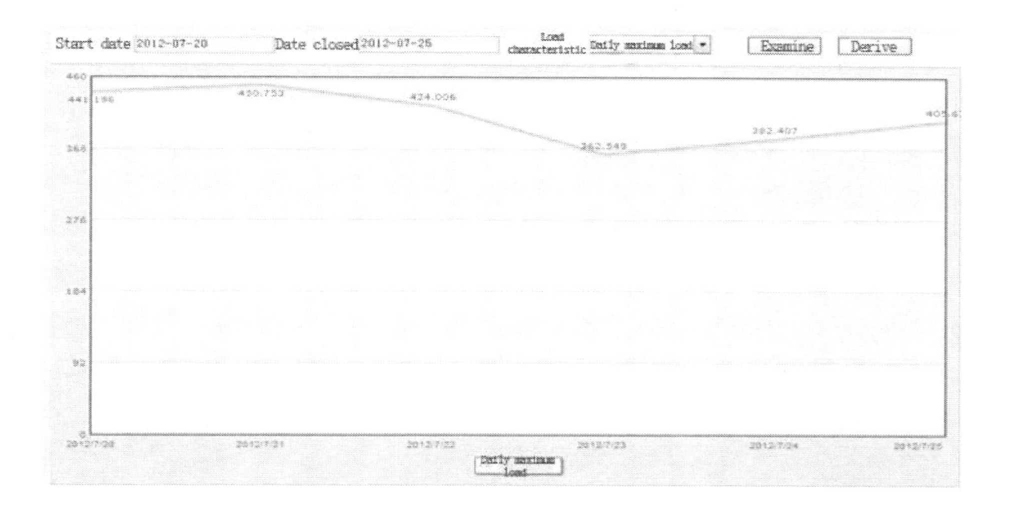

3. Weekly load characteristic analysis

Calculate and display the maximum load, minimum load, load rate, maximum valley-to-peak difference, maximum valley-to-peak different rate, average valley-to-peak difference and average valley-to-peak difference rate of each week within a period of time.

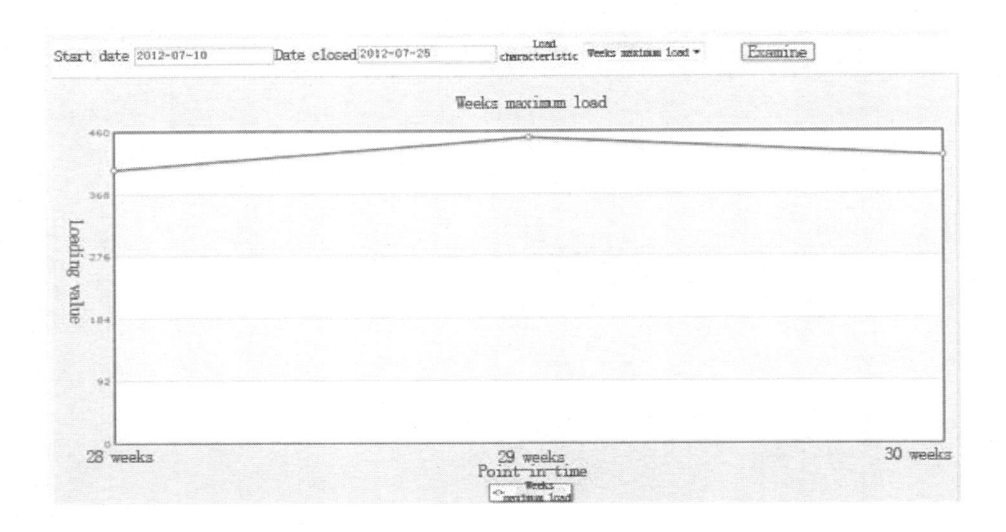

4. Monthly load characteristic analysis

Calculate and display the maximum monthly load, minimum monthly load, monthly load rate, maximum monthly valley-to-peak difference and minimum monthly valley-to-peak difference of each month within a period of time.

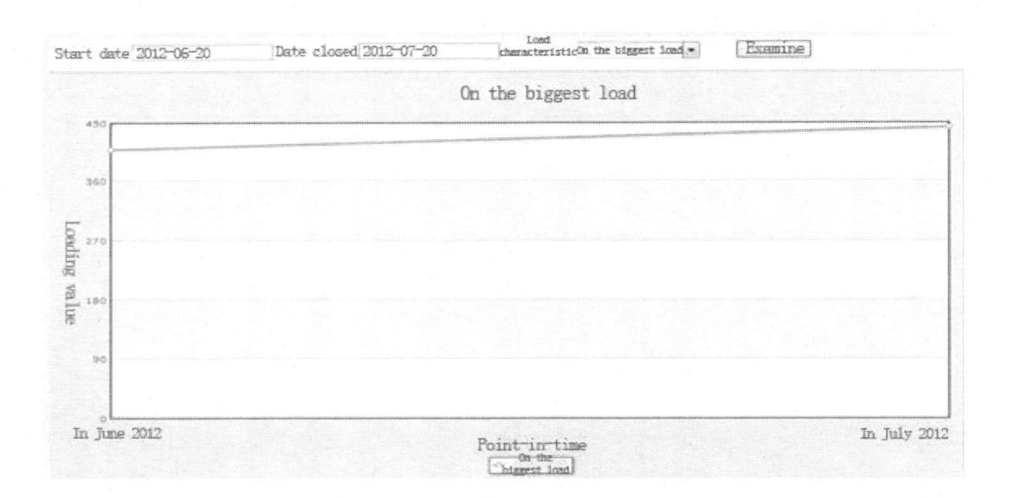

5. Annual load characteristic analysis

Display the annual maximum load, annual minimum load, annual load rate and annual maximum valley-to-peak difference of each year within a period of time.

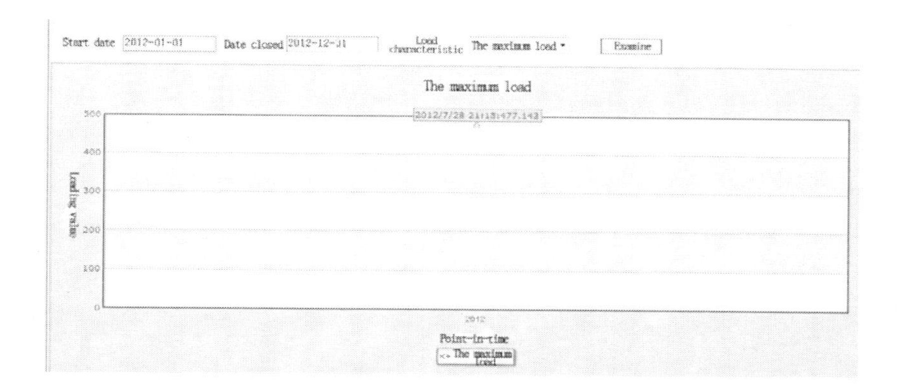

4.3 *Load forecast*

Build a load forecasting method library suited to the local grid particularities, provide advanced forecasting strategies, provide accurate forecasting data and automatically generate integrated models to effectively improve the forecasting accuracy and provide technical support for the grid load forecasting work.

Analyze historic load data, related weather information and load control, find out sensitive factors with methods suitable for the forecasting purposes, quantify the extent of influence and automatically calculate errors.

1.Manual load forecast

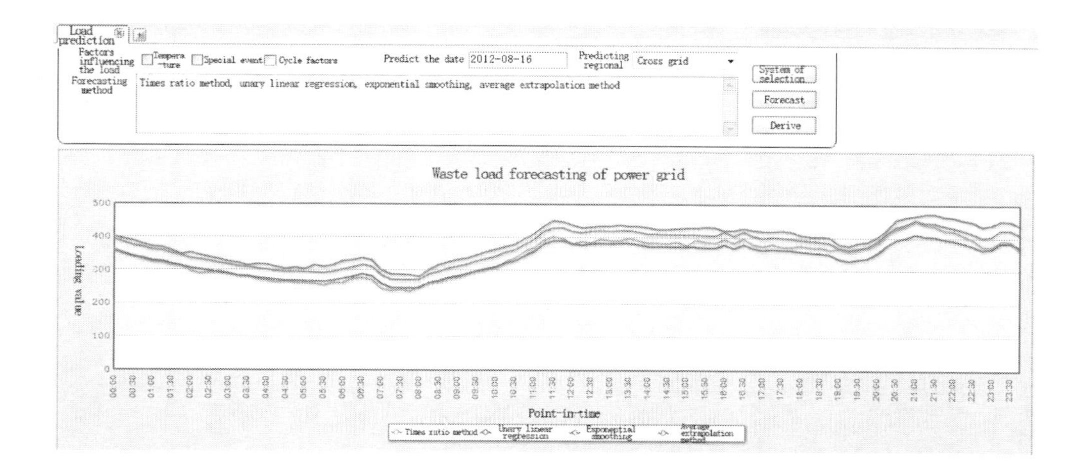

4.4 *Automatic load forecast*

Automatic load forecasting is performed according to the user defined forecasting area, forecasting model and forecasting time.

Related factors like temperature, duration and/or special social events may be added according to the type of the forecasted date.

4.5 *Sensitivity analysis*

Display the correlation between the effective temperature and load within a period of time. Analyze how the load is affected by temperature variation.

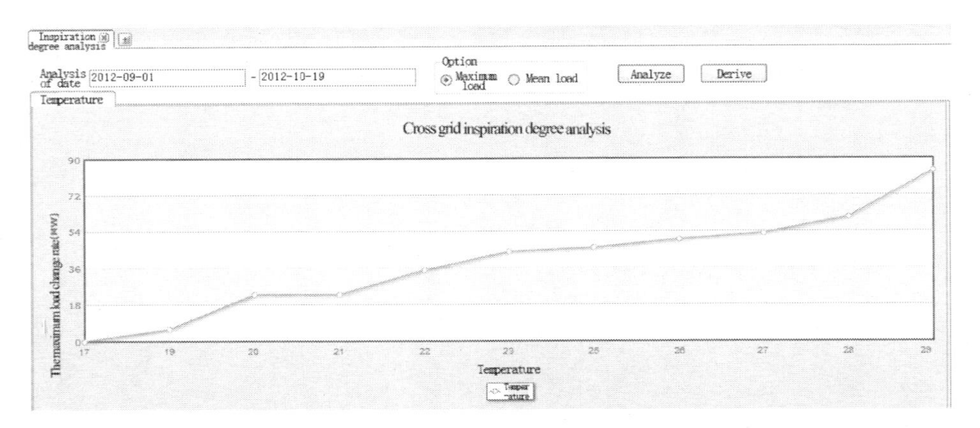

Figure 4. The correlation between the effective temperature and load within a period of time.

4.6 *Network loss computation*

By calculating and analyzing the power grid by grade and by region, properly arrange the operation mode to minimize network loss.

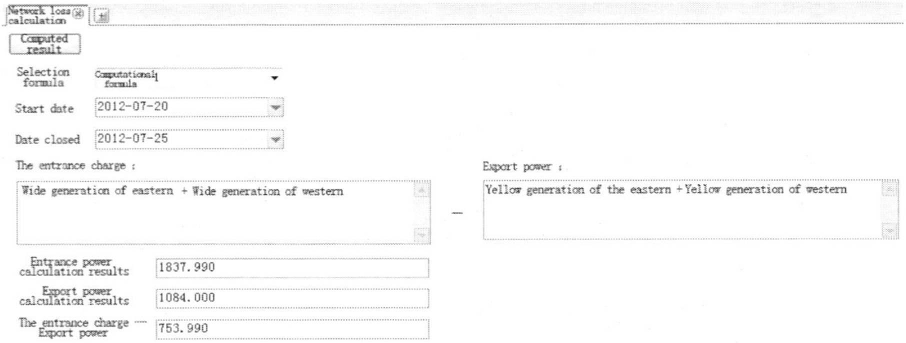

Network loss calculation

Computed result

Selection formula: Computational formula

Start date: 2012-07-20

Date closed: 2012-07-25

The entrance charge:

Wide generation of eastern + Wide generation of western

–

Export power:

Yellow generation of the eastern + Yellow generation of western

Entrance power calculation results: 1837.990

Export power calculation results: 1084.000

The entrance charge — Export power: 753.990

Figure 5. Network loss.

5 CONCLUSION

This system allows real-time load forecasting and effective data sharing through real-time data acquisition. It also allows local business integration so that business communication between the prefecture-level dispatch and the county-level dispatch. By allowing accurate, timely, reliable and smart short-time load forecasting, this system improves the timeliness of data collection, the working efficiency and the accuracy of forecasting, provides reliable guarantee for the safe and economical operation of the power system, and achieves very good economic and social benefits.

REFERENCES

[1] Wei, W., Niu, D.X., Chang, Z. (2002) New Development of Load Forecast Technology. *Journal of North China Electric Power University*, 29(1):10–15.

[2] Wang, L.M., Wang, Y.S. (2007) Short term load forecasting based on RBF neural network, *Electrical Technology*, 4: 53–55.

[3] Du, X.H., Zhang, Y. (2009) Application of neural network and support vector machine in short-term load forecasting, *Electrical Technology*, 9:17–21.

[4] Wang, S.X. (2007) *Mastering MATLAB Interface and Programming*. Beijing: *Publishing House of Electronic Industry*.

[5] Liu, X., Luo, D.S., Yao, J.G, et al. (2009) Short-term load forecasting based on load decomposition and hourly weather factors. *Power System Technology*, 12(33):94–100.

[6] Kang, C.Q., Xia, Q., Liu, M. (2007) *Electric System Load Forecasting*. Beijing: *China Electric Power Press*.

[7] Mao, L.F., Yao, J.G., Jin, Y.S. (2010) Theoretical study of combination model for medium and long term load forecasting. *Proceedings of the CSEE*, 2010, 30(16):53–59.

Electric, Electronic and Control Engineering – Shao, Shu & Tian (Eds)

A Study on sports video analysis based on motion mining technique

Xiaopeng Chi
Department of P.E. Research, Beijing Foreign Studies University, Beijing, China

ABSTRACT: A large number of sports videos can be acquired through different channels in daily life of the modern society. So, sports video is widely accepted by the public. Information with values can be extracted in the process of analyzing videos of this kind through applications of the motion mining technique method. The application prospect is extremely wide. With the analysis on sports videos as the research object and the method of motion mining technique as the starting point, this paper carries out a brief analysis on the theoretical framework of sports video mining and discusses the implementation of the sport mining method in sports video analysis, aiming at drawing attentions of the public.

KEYWORDS: Sports video; motion mining technique; analysis.

Sports competition has gained an unprecedented development with the gradually accelerated development of the society and economy. Requirements of various competitive sports events on athletes are also enhanced constantly. The auxiliary analysis technology of sports video data mining has drawn extensive attentions so that athletes can gain better development in terms of the competitive level. In the meantime, video analysis technology is also needed in some large sports events so as to provide necessary image support for judgments of referees. There are certain limitations for the existing DVCoach software in popularization and application due to the high equipment cost. It is an extremely important problem in this circumstance that how to develop a more popular video mining technology being able to provide the function of assistant decision. Thus, it can be seen that the method of motion mining technique in sports video analysis is very important. This paper carries out a detailed analysis and study on this problem.

1 INTRODUCTION

In the society of modern information, the public has different levels of requirements on information acquisition, storage, processing and expression. According to the current development situation, multimedia has become the most important platform for information. It has become one of the current hot research topics that how to realize data mining on the platform of multimedia. Sports video and news video are the most important two research directions from the aspect of video mining. Major bases are as follow: firstly, both news video and sports video are extensively accepted by the public. They exist in daily life of the public and we need to serve the public with these news videos through constant mining. Secondly, structural features of news video and sports video are very distinct with strong regularity, being able to reflect implicit knowledge effectively and directly and obtain deeper mining theories. In particular, as one of the most important constituent parts of the whole video, sports information reflects the changing rule of video images from the aspect of time axis. It is of great significance for the description of video content. In order to obtain meanings of corresponding characteristics, moving object in video can be segmented and tracked so as to extract relevant sports features and provide support for information analysis. In this research field, existing research results at home and abroad can be concluded as follows.

2 THEORETICAL FRAMEWORK OF MOTION MINING TECHNIQUE IN SPORTS VIDEO

The basic concept of motion mining technique in sports video analysis is: extract relevant data from the database, acquire relevant sports characteristics through sports analysis and then obtain implicit knowledge modes in video images through various methods of data mining. These knowledge modes can not only be inquired by users but also provide the support of assistant decision for users. In sports video mining, sports characteristics can be expressed not only in basic sports characteristics without semantic meanings but also in sports characteristics with semantic features. Compared to the traditional multimedia mining technology and the video mining technology, the technology of motion mining has a certain particularity. Firstly, sports feature is the main feature type of motion mining technique. Secondly, the range of motion mining technique is restricted to sports video files. Thirdly, the purpose of motion mining technique is to satisfy the requirements of knowledge acquisition and query. Finally, methods of motion mining technique mainly include statistics, correlation and cluster analysis.

The overall basic framework of sports video mining is shown in Figure 1. Constituent elements of the whole system include a basic feature layer, a model-event layer and a knowledge layer. The basic feature layer provides services for the model-event layer as well as the knowledge layer. Relevant modes can be derived and the extraction of events with features can be achieved after the acquisition of basic characteristics of moving objects. The acquired knowledge has the feedback effect so as to provide users with the function of query and assistant decision.

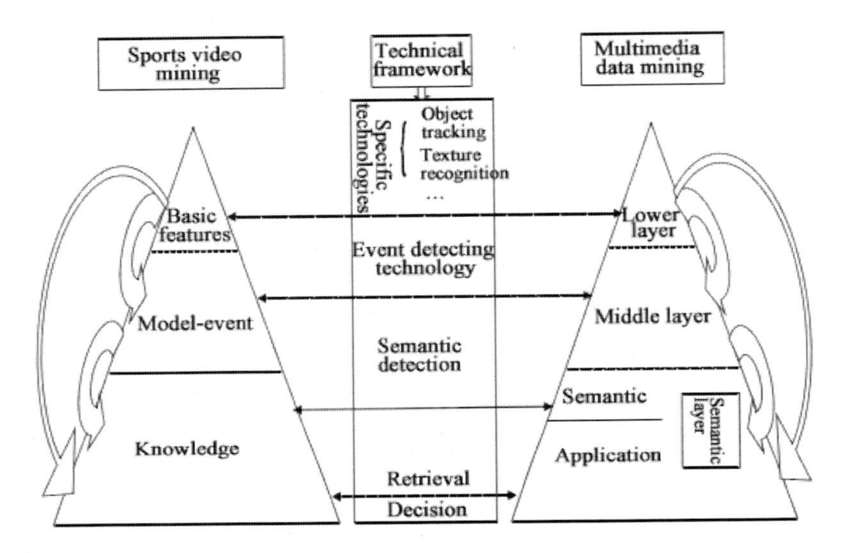

Figure 1. Basic framework of sports video mining.

3 PRACTICAL APPLICATIONS OF MOTION MINING TECHNIQUE

3.1 *Detection of moving objects*

In sports video analysis, the purpose of detecting objects in moving state is to separate independent moving objects from image sequences. The target of visual processing of most computers is nothing but the moving region in a scene as well as the relevant pixel. Time and calculation amount of video processing can be reduced if moving regions can be detected accurately, which is convenient for video processing of higher level. However, with the support of current technologies, there are still quite a lot problems and insufficiencies existing in the detection of moving objects, including interference of fixed objects on detection, extraction of light and shadow in a scene, extraction of other moving objects, sheltering of scenes, shaking and stretching of camera lens, etc. According to the above-mentioned problems, thoughts on traditional moving objects detection are adjusted in the sequential study of this paper. The detection of moving objects is divided into two parts for respective extraction and analysis, namely objects in the target region and objects of moving detection. The specific detection method can be divided into:

3.1.1 *Determination of moving region*

At present, the method of decision binary tree can be used to solve the multi-classification problem of moving detection. This method is able to simplify complex problems of multi-classification and solve problems gradually in different levels. The basic structure is shown in Figure 2. This method has relatively higher classification accuracy with a simple structural mode. It is deeply applied in the inductive algorithm.

Sports video can be detected by temporal difference under the condition that the method of decision binary tree is introduced. As shown in Figure 3, divide an image in line with the dynamic/static standard and mark moving regions in the video with rectangular boxes, which are labeled as interested regions. On this

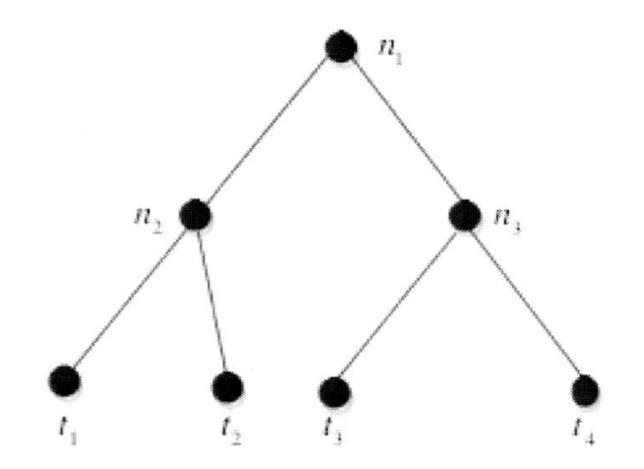

Figure 2. Structure chart of decision binary tree.

basis, sizes and colors of marked regions can be taken as prior knowledge to construct a relevant decision binary tree and process interested regions further. Video images containing motion information can be ultimately obtained by eliminating regions that have interference on analysis results.

Figure 3. Marked moving parts and interested regions.

3.1.2 *Extraction of moving objects*

After determining moving regions on the video image, corresponding edge characteristics of moving objects can be further mined with the edge detection algorithm. Moving objects on this image can be ultimately obtained by separating characteristics further (the introduced contour extraction algorithm is Open CV algorithm in this process).

According to the above-mentioned way, interference on extraction results brought by other objects on the video image can be solved so as to avoid inaccurate detections that might be caused by the above-mentioned factors and increase the accuracy of extracting moving images.

3.2 *Tracking of moving objects*

The tracking of moving objects is of great importance in the field of computer vision. The main content is: detect independent moving objects on sports videos constituted by images of various frames, designate interested regions of users, and locate targets and regions on images of sequential frames accurately. As for sports videos studied in this paper, moving objects mainly concentrate upon referees or athletes, postures of whom vary tremendously in moving process. Both region size and moving direction are random. It is suggested to introduce CamShift algorithm in the mining and tracking process of moving information in sports videos if the above-mentioned features are taken into account. Corresponding analysis is shown below:

Under the condition of using traditional technologies, there are a large number of problems in the tracking of moving objects, including the complexity of movement of moving objects, the complexity of scenes, the interference of objects and the mobility of image recording equipment. It can be presumed that, in a sports video image, the velocity of a moving object maintains constant or the accelerated velocity is constant during the whole moving process. Under this circumstance, the task of tracking moving objects can be simplified and the effectiveness of tracking can be improved through the method of prior knowledge like size and color of moving objects.

Under the background of introducing CamShift algorithm, the whole tracking process of moving objects is shown in Figure 4. With the combination of Figure 4 and CamShift algorithm, the tracking window will choose the region with the largest probability density like the region of the coat of an athlete on the video. However, the detected moving region is the moving region of an athlete, including the athlete. So, targeted regions can be acquired by moving detection and it can be determined whether the original tracking window of CamShift is in this region. If it locates in this region, it can be regarded as correct tracking and can be continued. It might be unable to continue the tracking if the tracking fails or extracted objects become background regions due to the instability of images. In order to solve this problem, it is suggested to add an accumulator in video analysis. The initial value is 0. Set the initial value as 1 if the tracking window of a moving object cannot be correctly located in the athlete region. Tracking loss caused by camera shaking can be solved in 1-2 frames. If the problem still occurs after three frames, the value of the accumulator should be larger than 3. Problems are very likely to occur in tracking deviation. The solution is to adjust the tracking window. It can be seen that a good tracking effect can be acquired by combining CamShift with moving detection in scene switching.

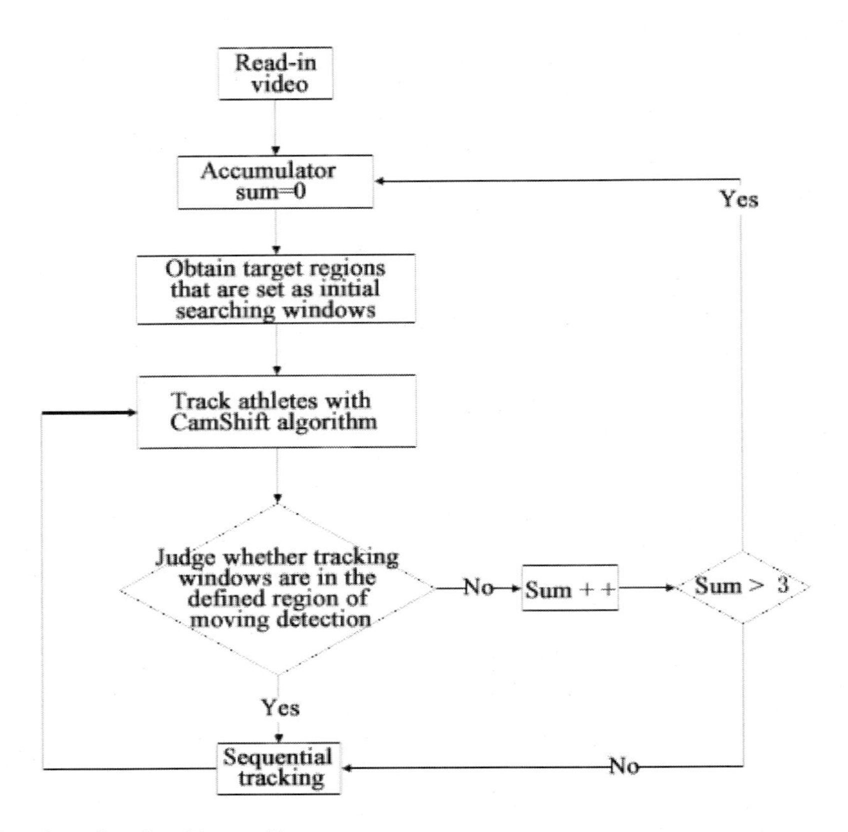

Figure 4. Flow chart of moving object tracking.

3.3 *Experimental verification*

Both moving object detection and moving object tracking applied in this paper have achieved good results. Accurate tracking and extraction of moving objects are completed based on the two aspects. Moving object detection can be divided into moving detection and moving object extraction. Moving detection is the initial part that involves no semantic motion mining technique, which not only provides moving object tracking with an initial tracking window but also recalibrate the moving region if the tracking object is lost. Accurate moving object tracking is the basis of moving object extraction. Although the whole scene of a sports video is complicated, the competition field is relatively simple. Accurate tracking windows provide accurate windows for moving object detection, ensuring the implementation of the two detection methods proposes in this paper. The analysis on part of sports videos is carried out in order to further verify the validity of the algorithm through experiments. The calculation on precision ratio of the algorithm is conducted in accordance with the following formula:

Precision ratio of the algorithm = Detection of the correct moving object region / Detection of the entire moving object region * 100%;

Collect 40 videos on the Internet, including tennis, badminton, basketball, football and table tennis, for moving objects analysis after downloading. The frame number should be within 500-4500. Take CamShift algorithm as the research object and MeanShift algorithm as the control object, precision ratios of which are calculated at the same time. Five typical videos are taken as the analysis objects. Specific experimental data are provided in the following table (Table 1).

It can be known from data in Table 1 that the algorithm based on CamShift and athlete tracking system proposed in this paper not only can be applied in the processing of static videos to make sure that the precision ratio is improved compared to traditional algorithms but also have higher accuracy compared to general methods as for videos involving scene switching and camera shaking. The athlete detection and tracking algorithm proposed in this paper is able to achieve a perfect effect in practical applications.

Table 1. Experiment results analysis (%).

Experimental videos	Precision ratios of moving regions	Precision ratios of MeanShift	Precision ratios of CamShift	Precision ratios of CamShift and moving detection
1	89.00	61.00	77.00	87.00
2	91.00	75.00	80.00	88.00
3	90.00	67.00	79.00	92.00
4	93.00	81.00	89.00	95.00
5	94.00	71.00	80.00	94.00

4 CONCLUSION

This paper carries out an analysis on motion mining technique from the perspective of sports video analysis. It is believed that analytical results of the traditional mining technology are influenced by internal and external factors in different levels. The above-mentioned problem is solved by CamShift algorithm. Detect regions of athletes (or other major moving objects) with the moving detection technology after obtaining regions of moving objects, which can be used as tracking windows in CamShift algorithm. Detection results are verified and calibrated timely when problems of tacking loss are compensated so as to increase analytical precision.

REFERENCES

[1] Liu, Y., Huang, Q.M., Gao, W. *et al.* (2006) Adaptive Gaussian mixture model of pitch detection algorithm and its application in sports video analysis, *Journal of Computer Research and Development*, 43(7): 1207–1215.

[2] Mei, T., Zhou, H.Q., Feng, H.Q. *et al.* (2005) Unsupervised mining of sports video structure with Mosaic, *Journal of University of Science and Technology of China*, 35(2): 250–257.

[3] Han, B., Hu, Y.C., Wu, W.G. *et al.* (2007) Camera switching detection of sports video based on unified characteristic model, *Television Technology*, 31(8): 76–79.

[4] Yi, D. & Guo, L.F. (2009) Sports video mining via multichannel segmental hidden Markov models, *IEEE Transactions on Multimedia*, 11(7): 1301–1309.

[5] Automatic detection and analysis of player action in moving background sports video sequences, *IEEE Transactions on Circuits and Systems for Video Technology*, 2010, 20(3): 351.

[6] Jungong Han, Farin D.,de With P. *et al.* (2008) Broadcast court-net sports video analysis using fast 3-D camera modeling, *IEEE Transactions on Circuits and Systems for Video Technology*, 18(11): 1628–1638.

[7] Jui-Hsin Lai, Chieh-Li Chen, Chieh-Chi Kao, *et al.* (2011) Tennis Video 2.0: A new presentation of sports videos with content separation and rendering, *Journal of Visual Communication & Image Representation*, 22(3): 271–283.

[8] Lu, Wei-Lwun, Ting, Jo-Anne, Little, James J. *et al.* (2013) Learning to track and identify players from broadcast sports videos, *IEEE Transactions on Pattern Analysis and Machine Intelligence*, 35(7): 1704–1716.

[9] Raffay Hamid, Ramkrishan Kumar, Jessica Hodgins, *et al.* (2014) A visualization framework for team sports captured using multiple static cameras, *Computer Vision and Image Understanding: CVIU*, 118: 171–183.

[10] Zhang, Y.F., Xu, C.S., Zhang, X.Y. *et al.* (2009) Personalized retrieval of sports video based on multi-modal analysis and user preference acquisition, *Multimedia Tools and Applications*, 44(2): 161–186.

Electric, Electronic and Control Engineering – Shao, Shu & Tian (Eds)
© 2015 Taylor & Francis Group, London, ISBN: 978-1-138-02842-5

Study on the influence of two-navigation holes cross-sea bridge construction on tide environment of the existing channel

Chenyang Wang & Zhe Liu
Tianjin Research Institute for Water Transport Engineering, Tianjin, China

ABSTRACT: With the cross-sea bridge with two navigation holes - Pu Qian bridge as the object of study, and using the irregular triangular net finite difference method, this paper establishes the two-dimensional tide mathematical model in the bridge sea area, carries out the simulation study to the tide power in this sea area through the means of the tide numerical simulation, analyzes the influences of the bridge construction on the tide environment of the existing channel waters, and the changes f the navigation condition in the two bridge navigation holes. Research results show that the construction of the Pu Qian Bridge has little influence on the tide in the Pu Qian bay. If the mean velocity change 0.02 m/s is served as the criteria with significant influence, the project influence on tide is limited within 1km scope in north and south of the bridge, the flow regime of the sea area far away from the bridge is rarely influenced, the mean velocity of the rising/falling tide of the water area near the bridge increases to some extent, the flow velocity of the lane water area at both sides of the bridge slightly increases. The research results can provide certain reference for the design optimization of the similar two-navigation holes bridge.

KEYWORDS: Tide field, triangular net, Pu Qianpp Bridge, bilateral navigation.

1 INTRODUCTION

The problems of the cross-sea bridge influence on the dynamic environment of the sea area water is mainly researched via physical modelling experiment, numerical simulation calculation and other methods. In the numerical simulation of the two-dimensional tide, the tidal movement is controlled through the continuity equation and momentum equation, which simulates the tidal movement via calculating approximate solution with the numerical dispersing method to the governing equation. The horizontal size of the estuary and coast aequilate shallow waters is far more than the vertical depth, which is suitable for the two-dimensional model for simulation, such kind of model has approached mature, and been widely applied in the strait tide and calculation, and the calculation of the offshore engineering tide field[1]. The former studies on the bridge construction influence on the water dynamic environment are mainly researched via the physical modelling experiment, numerical simulation calculation and other methods, and it has obtained abundant achievements. Tang Shifang and Li Pei[2] have obtained the resistance coefficient formula of the single pile and pipe groups through testing and stated how to consider the influences of the pipe and pipe group in mathematical model. Wang Xiaoshu[3] has simulated under the condition of regarding the pile pier as waterproof area on the basis of two-dimensional hydrodynamic model, as partial mesh generation of the pipe pier is not sufficiently delicate, resulting that the obtained partial flow-regime result is not too accurate. And the research result is only applied in small scope of waters. Wen Xian [4] also treated the pile pier as the impervious boundary, and triangular mesh is used to partially refine the pile pier, and discussed the sensibility of the roughness factor and turbulent fluctuation viscosity coefficient to backwater value, but only considered the flow rate of the water and the perpendicular and orthogonality of the pile pier. With the two-navigation holes cross-sea bridge - Pu Qian Bridge as the object of study, and based on the measured data, according to the character tics of the bridge and its sea area, this paper carries out the tide numerical simulation study to the bridge project via establishing the two-dimension tide mathematical model based on the irregular triangular net finite difference method, analyzes the influence of the bridge construction crossing the bay on tide field of the Pu Qian bay, emphasizes on the analysis of the influence of the bridge construction on the tide environment of the existing channel sea area, and the changes of the navigation condition within two navigation holes of the bridge.

2 BRIDGE OVERVIEW

This study serves the proposed Pu Qian cross-sea bridge in Hainan province as the object of this study. The Pu Qian bridge is located in the marine outfall of the Dong Zhai harbor, Pu Qian bay east-south of the Qiongzhou Strait, from the Pu Qian two of Wenchang city, the bridge passes through Bei Gang island up to the Ta city, Yan Feng Town, Haikou City, and border on the planning and construction Jiang Dong Bridge in Haikou City. The length of the cross-sea bridge is about 3.85km, the length of the wiring roadbed of both shores is about 1.9km. The layout of the bridge site is as shown in Figure 1

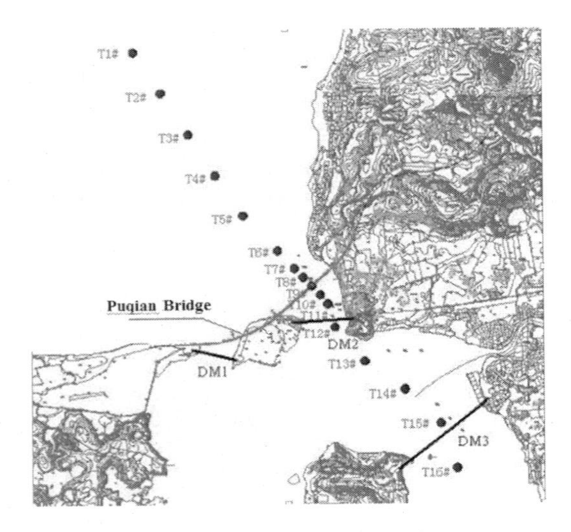

Figure 1. Layout of the bridge.

3 TWO-DIMENSIONAL TIDE MATHEMATICAL MODEL

3.1 *Basic equation*

Continuity equation

$$\frac{\partial \zeta}{\partial t}+\frac{\partial[(h+\zeta)u]}{\partial x}+\frac{\partial[(h+\zeta)v]}{\partial y}=0 \tag{1}$$

Kinematic equation

$$\frac{\partial u}{\partial t}+u\frac{\partial u}{\partial x}+v\frac{\partial v}{\partial y}-fv=-g\frac{\partial \zeta}{\partial x}-\frac{gu\sqrt{u^2+v^2}}{C^2H}+\varepsilon(\frac{\partial^2 u}{\partial x^2}+\frac{\partial^2 u}{\partial y^2}) \tag{2}$$

$$\frac{\partial v}{\partial t}+u\frac{\partial v}{\partial x}+v\frac{\partial v}{\partial y}+fu=-g\frac{\partial \zeta}{\partial y}-\frac{gv\sqrt{u^2+v^2}}{C^2H}+\varepsilon(\frac{\partial^2 v}{\partial x^2}+\frac{\partial^2 v}{\partial y^2}) \tag{3}$$

In the formula: x、y are the rectangular coordinate system coordinates overlapped with the static sea level (or base level), u、v are respectively the velocity components in x、y directions, h is the depth of water (the distance between base level to bed surface), ζ is tide level (the distance between base level to free water surface), H is total depth of water, $H=h+\zeta$, f is Ke's coefficient, g is gravitational acceleration, C is Chezy coefficient, $C=H^{1/6}/n$, is Manning roughness coefficient, t is time, and ε is viscosity coefficient of the horizontal vortex motion.

3.2 *Definite condition*

1 Boundary conditions

There is the following formula in the opening boundary Γ_1 where the computational domain is connected with water areas

$$\zeta(x,y,t)|_{\Gamma_1}=\zeta^*(x.y.,t) \tag{4}$$

or

$$\begin{cases} u(x,y,t)|_{\Gamma_1}=u^*(x,y,t) \\ v(x,y,t)|_{\Gamma_1}=v^*(x,y,t) \end{cases} \tag{5}$$

There is the following formula in the fixed boundary Γ_2 where the computational domain is border on land

$$\vec{U}\cdot\vec{n}\Big|_{\Gamma_2}=0 \tag{6}$$

In the formula: $\vec{n}$ is the formula of the fixed boundary, $u^*(x,y,t)$ and $v^*(x,y,t)$ are given values (the measured, or accurately measured or assay value). The physical significance of the formula is that the normal component of the flow velocity vector is zero along with the fixed boundary.

2 Initial conditions

$$\begin{cases} \zeta(x,y,t)\big|_{t=t_0}=\zeta_0(x,y,t_0) \\ u(x,y,t)\big|_{t=t_0}=u_0(x,y,t_0) \\ v(x,y,t)\big|_{t=t_0}=v_0(x,y,t_0) \end{cases} \tag{7}$$

In the formula: $\zeta_0(x,y,t_0)$, $u_0(x,y,t_0)$ and $v_0(x,y,t_0)$ are the given values of the initial moment t_0.

3.3 *Compute field and mesh generation*

The calculation range of this mathematical model: west boundary is from Deng Lou angle to Yu Bao angle, about 25.6km in length, the east boundary is from the near Yan Jin angle to the north, and to the Hannan angle to the south, about 32km in length. The east-west distance of the whole compute field is about 72km. The calculation foundation level is the local theoretical low water level.

In order to summarize the complex coastline, island and terrain characteristics of the sea area better, this study uses the scalene triangle mesh generation compute field. In view of large scope of compute field and relatively small pier size, the method of partial mesh encryption is adopted to encrypt to each pier boundary, carry out mesh encryption to the partial sea areas of the project, during the calculation process, treat the pier as land [11,12]. The maximum spatial lag of the model is 800m, minimum spatial lag is 6m, triangle grid node is 10657, grid cell is 20282, and partial grid of the model is as shown in Figure 2.

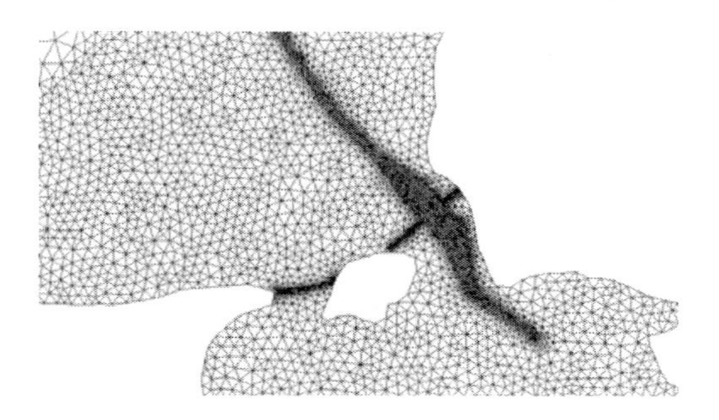

Figure 2. Partial mesh of the bridge.

4 RESULTS

4.1 *Change analysis for tide field of bridge zone*

Qiongzhou Strait is located between the Leizhou Peninsula and Hainan Island, nearly east-west trend, and connected with the west Guandong sea area and Beibu Gulf area. The tide of the Qiongzhou Strait is manly controlled by the South China Sea in east port and the two independent tide bulge systems in Beibu Gulf, the east port is mainly influenced by the semi-diurnal tide wave, and the west port is influenced by the diurnal tide wave, and considering the topographic factor, the local area tide is very complex. The strait tide is mixed tide, mean range is less than 2m, the maximum surface flow rate of the tide is up to above 4m/s. The entire tide in the strait is diurnal tide, giving re-flowing as priority, and the flow direction changes from NE (SW) to the E (W) from east to west.

From outer bay, top bay to the inner bay (Dong Zhai bay), the tide process of the Pu Qian bay transits irregular semi-diurnal tide from the irregular diurnal tide, the flow direction of the tide gradually changes into north-south from east-west from the outer bay to inner bay. Figure 3 is the fluctuation torrent diagram near the back and forth bridge of this project.

It can be seen from the diagram above that after the completion of the Pu Qian Bridge, the flow regime near the bridge changes, and the flow regime of sea area far away from the bridge is not influenced. Further compare the changes of the fluctuation torrent velocity in the front and back of the bridge to obtain that (1) compared with flood tide, the range of the velocity in the sea areas of the project bridge under falling tide state is more extensive and its amplitude is larger. (2) Influenced by the water resistance of the pier, the torrent velocity change of the rising and falling tide near the main pier is slightly large. (3) According to the rising/falling torrent velocity contour map of the sea areas in front and back of the project, if 0.05m/s is treated as the definition standard with significant change of the rising/falling torrent velocity, the bridge influence on the tide flow speed of the project water areas is basically within the scope of 2km both on south and north, in which, within each 1km scope of upstream/downstream of the two major navigation holes, the flow rate increases to some extent. (4) With the flow rate correct to 0.02m/s as the definition standard with obvious significance, and according to the mean velocity of the rising/falling tide in front and back of the project, the bridge influence on the tide flow rate of the channel water areas is within each 1km scope of the north and south.

4.2 *Bridge influence on tide level*

In order to analyze the changes of the tide level within this sea area after building the Pu Qian bridge project, there are 16 feature points successively arranged from north to south along the existing central line of the channel, as shown in Figure 1, in which, T9# point is over the bridge location, the distances between the T1# to T16# feature points are respectively 1km, 1km, 1km, 1km, 1km, 0.5km, 0.25km, 0.25km, 0.25km, 0.25km, 0.5km, 1km, 1km, 1km, 1km. The changes statistics of the high tide level and low tide level for each feature point before and after building the bridge construction are as shown in Table 1, according to the calculation results, it can be seen:

1 After the implementation of the bridge project, the variation of the high tide level and low tide level in the engineering area is generally within 1.7cm;

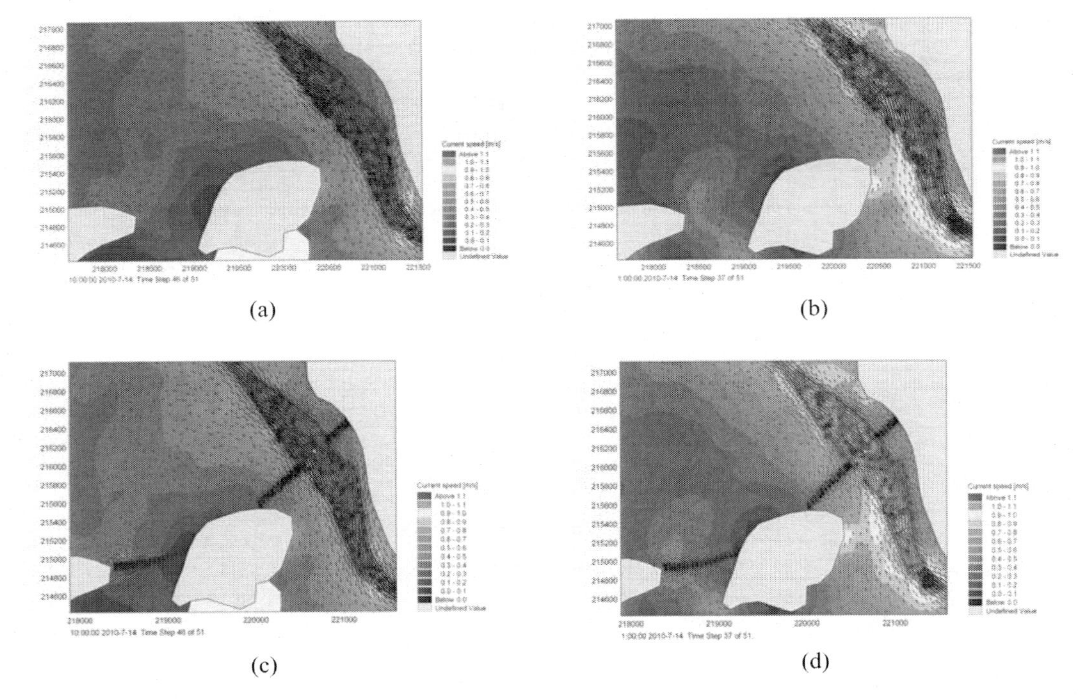

Figure 3. Fluctuation torrent diagram near the back and forth.

2 After the implementation of the bridge project, the tidal range variation in the engineering area is generally within 2cm;
3 In general, after the implementation of the bridge project, it has smaller influence on the tide level of the engineering area, and the bridge construction does not cause the phenomenon of obstructing tide level due to obvious water blocking.

4.3 *Bridge influence on the tidal prism of Pu Qian bay*

For the analysis of the bridge project influence on tidal volume, take 3 strips of analysis section (as shown in figure 1) in the project zone, DM1 section (Xin Xi angle to Bei Gang island), DM2 section (Bei Gang island to Pu Qian town) and DM3 section (Xia Chang village to Gao Wei village), the section width is respectively 698m, 911m and 1752m. The tidal volume of rising/falling tide before and after the project is as shown in table 2. It can be seen from the table 2 that for the existing channel water area (DM2, DM3) from Pu Qian bay to Dong Zhai harbor, the tidal volume of rising/falling tide before and after the project changes less, the amplitude is basically about 1.5%, in which, the amplitude of the rising tide volume is more than one of the falling tide, which is consistent with the slight increase of the low water in the water area south of the bridge line after construction, for the water area (DM1) from Xin Xi angle to Bei Gang island, the depth of the water is relatively shallow, the tidal prism

Table 1. Feature points tidal fluctuation value calculation (cm).

Tide level station	T1#	T2#	T3#	T4#	T5#	T6#	T7#	T8#
High tide level	-0.4	-0.5	-0.5	-0.6	-0.6	-0.7	-0.8	-0.8
Low tide level	-0.1	-0.2	-0.3	-0.2	-0.2	-0.3	-0.4	-0.3
Tide level station	T9#	T10#	T11#	T12#	T13#	T14#	T15#	T16#
High tide level	-1.3	-1.2	-1.2	-1.4	-1.5	-1.6	-1.7	-1.1
Low tide level	-0.2	-0.3	-0.3	0.1	0.2	0.1	0.1	0.1

is smaller, the sense of water resistance to the bridge should be more obvious. The amplitude of the falling tide volume can be up to 5.6%, and the amplitude of the rising tide volume changes slightly (about 1.5%).

Table 2. The tidal volume of rising/falling tide before and after the project ($10^4 m^3$),

Falling tide				Rising tide			
Status	After	Difference	Amplitude (%)	Status	After	Difference	Amplitude (%)
330	312	−18	−5.57	276	272	−4	−1.52
4328	4329	1	0.02	4010	4071	61	1.53
2729	2713	−16	−0.57	2484	2516	32	1.3

5 CONCLUSIONS

According to the measured data, this paper simulates the two-dimensional tide value to the sea area of the Fu Qiao Bridge project of the structureless triangular mesh, analyzes the changes of the flow regime, flow and channel tide field environment of the project sea area before and after bridge construction, especially analyzes the sea area near the two navigation holes of the bridge zone. And now, it is summarized as follows:

1 After the implementation of the bridge project, the variation of the high tide level and low tide level in the engineering area is generally within 1.7cm, the tidal range change is within 2cm, the bridge construction does not cause the phenomenon of obstructing tide level due to obvious water blocking.
2 From the point of mean velocity of the rising/falling tide before and after project, with the flow velocity change correct to 0.02m/s as the definition standard with obvious significance, the bridge influence on the tide rate of the channel water area is basically within each 1km scope of north and south, and from the point of the torrent rate of the rising/falling tide before and after project, with 0.05m/s as the definition standard of the rising/falling torrent with obvious significance, the bridge influence on the tide flow rate of the project water area is basically within each 2km scope of the .
3 According to the tidal volume of the rising/falling tide from the bridge-site waters section, It can be seen that after the construction of the bridge navigation, the tidal volume amplitude of the rising/falling tide before and after the exiting channel waters project is about 1.5%, in which, the amplitude of the flood tidal volume is more than the amplitude of the ebb tidal volume, which is consistent with the slight increase of the low water in the water area south of the bridge line after construction, for the water areas from Xin Xi angle to Bei Gang island, the depth of water is relatively shallow, the tidal prism is smaller, it has more obvious sense of water resistance to the bridge, being influenced by the bridge project, the tidal volume amplitude of the rising/falling tide is about 5%.

REFERENCES

[1] S F Tang, B Li. (2001) Study on numerical simulation of tidal flow influenced by pile group resistance. *China Harbour Engineering*, (5):25–29.

[2] X S Wang, W Zhang, P Liao. (2006) Study on Tidal Current around Piers of Shanghai Offshore Wind Power Plant [EB/OL]. [2006-06-20]. Available from: http://www.paper.edu.cn.

[3] M X Xie, W Zhang. (2008) Numerical simulation of open channel flow influenced by pile piers. *Advances in Science and Technology of Water Resources*, (3):20–24.

[4] Xie M X, Zhang W, Xie H J. (2008) Simplification method in numerical modeling of bridge pier group. *Chinese Journal of Hydrodynamics*, A23(4):464–471.

Electric, Electronic and Control Engineering – Shao, Shu & Tian (Eds)
© 2015 Taylor & Francis Group, London, ISBN: 978-1-138-02842-5

Simulation analysis for heat balance of groundwater heat pump in multi-field coupling condition

Xiaoyu An, Wendong Ji & Yue Zhao
Tianjin Research Institute for Water Transport Engineering, Tianjin, China

ABSTRACT: By taking up the demonstration project of Ground-Water Heat Pump (GWHP) in Zhengding base of the Institute of Hebei Hydrological Engineering Geological Investigation, this paper built the three-dimensional coupling numerical model of groundwater seepage and heat migration to simulate the coupling process of groundwater seepage and heat migration during the system operation, predicted the design scheme on the development trend of heat balance of GWHP, and analyzed the heat transfixion phenomenon of groundwater with the various changed factors during the heat migration and operation. Based on the three-dimensional coupling model of unsteady seepage and heat migration of groundwater, the scope of influence (namely, 16.3 isotherm range) for heat of groundwater when the refrigeration period was finished in summer was taken as the basis, to evaluate the influence of change of cooling and heating load design, pumping, recharge well spacing, and other factors on the heat migration process of groundwater. The simulated result showed that the temperature of groundwater in the whole area was gradually reduced as the heat pump system was run continuously in the design condition, and the weak cold accumulation phenomenon (namely heat transfixion phenomenon) arose. When the total load capacity was changed, the change of circulating water amount had a big impact on the intensity of heat transfixion phenomenon. Therefore, it will adopt the mode of "Big temperature difference and small discharge" when the GWHP is run.

KEYWORDS: Ground-Water Heat Pump, thermal transport, heat transfixion, seepage field, numerical simulation.

1 INTRODUCTION

As our urbanization process is promoted and people's living standard is improved, the proportion of building energy consumption in the total energy consumption of society is increased year by year. According to this, the building heating and air-conditioning energy consumption account for 50%–70% of the building energy consumption, and these continue to rise; thus, it can be seen that the building energy efficiency has an important role in reducing the energy consumption of society, in which the heating and air-conditioning energy saving located in an important position will be affected[1][2].

Groundwater Heat Pump (GWHP) belongs to the branch of heat pump air-conditioning system engineering[3][4], namely, transferring the heat in the low-grade heat source to the place needing the heat or warming with the relatively stable characteristics of underground soil, surface water, and underground water temperature through consuming the electric energy, and transferring the indoor waste heat to the low-grade heat source in summer, so as to play the role of heating in winter and refrigerating in summer[5-7].

A large number of engineering applications play a significant role in promoting the study of GWHP, but there are few studies on the transfer of groundwater of GWHP. Xin Changzheng[8] and others carried out the annual operation simulation on the velocity field and temperature of typical twin-well laminated aquifer with the HST3D program prepared by the United States Geological Survey. The annual fixed discharge and fixed temperature were used due to the limitation of the program, and the result showed that two fully penetrating wells having a distance of 100m had the "heat transfixion" phenomenon in the winter and summer conditions. He Manchao[9] and others built the mathematical model of seepage field of geothermal recharge according to the geothermal recharge of seepage field in the recharge process, and they deduced the recharge formula of the single well under the different conditions of constant and variable permeability coefficient. Zhang Kunfeng[10] and others simulated the winter operation condition of GWHP of a large-diameter well (of which the diameter is 1.2m and 0.8m, and the depth is 10m), and the results showed that the well water flow in the larger-diameter well was dropped uniformly. Specific to the GWHP project, there are also many studies on the problem of heat migration of groundwater abroad, and Paksoy[11] and others studied the minimum distance between the groups of pumping and recharge wells by taking the range of water level of the pumping well and recharge well as the constraint condition and ensuring not to have the heat transfixion. Tenma[12] and others carried out the quantitative contrast simulation on the mining and recharge amount, and length, position, and run cycle of filter tube in the ideal well-to-well model with the FEHM program.

2 THREE-DIMENSIONAL COUPLING NUMERICAL CALCULATION FOR UNSTEADY SEEPAGE OF AND HEAT MIGRATION OF GROUND WATER

The author built the mathematical model that is capable of reflecting the actual state of groundwater on the basis of carrying out a detailed study on the seepage theory and heat migration theory of groundwater. The aquifer in the model was heterogeneous anisotropy, and the flow state of groundwater was three-dimensional unsteady flow.

2.1 *Mathematical model for unsteady seepage of three-dimensional groundwater*[13-14]

$$\begin{cases} \dfrac{\partial}{\partial x}\left(K_{xx}\dfrac{\partial h}{\partial x}\right)+\dfrac{\partial}{\partial y}\left(K_{yy}\dfrac{\partial h}{\partial y}\right)+\dfrac{\partial}{\partial z}\left(K_{zz}\dfrac{\partial h}{\partial z}\right)+W=\mu_s\dfrac{\partial h}{\partial t} & (x,y,z)\in\Omega \\ h(x,y,z,t)\big|_{t=t_0}=h_0(x,y,z,t_0) & (x,y,z)\in\Omega \\ h(x,y,z,t)\big|_{\Gamma}=h_1(x,y,z,t) & (x,y,z)\in\Gamma_1 \\ K_{xx}\dfrac{\partial h}{\partial x}\cos(n,x)+K_{yy}\dfrac{\partial h}{\partial y}\cos(n,y)+K_{yy}\dfrac{\partial h}{\partial z}\cos(n,y)\bigg|_{\Gamma_2}=q(x,y,z,t) & (x,y,z)\in\Gamma_2 \\ h(x,y,z,t)=z(x,y,t) & (x,y,z)\in\Gamma_3 \\ K\dfrac{\partial h}{\partial n}\bigg|_{\Gamma_3}=-\mu\dfrac{\partial h}{\partial t}\cos\theta & (x,y,z)\in\Gamma_3 \end{cases} \quad (1)$$

In the formula: K_{xx}, K_{yy}, and K_{zz} are the permeability coefficient (m/d) of anisotropic main direction, respectively; h is the water head value (m) of points (x, y, z) at t time; W is the source sink term (1/d); μ_s is the water storage rate (1/m); t is the time (d); Ω is the calculation zone; $h_0(x, y, z, t_0)$ is the initial water head value (m) at point (x, y, z); Γ_1, Γ_2, and Γ_3 are the first boundary, the second boundary, and the free surface boundary, respectively; $h_1(x, y, z, t)$ is the water heat value (m) of the first boundary; $\cos(n, x)$, $\cos(n, y)$, and $\cos(n, z)$ are cosine of angles in the coordinate axis direction and flow boundary outer normal direction, respectively; $q(x, y, z, t)$ is the replenishment quantity of unit area on the second boundary (m/d); K is the permeability coefficient (m/d) of the free surface boundary; μ is the saturation deficit while the free surface rises or specific yield while the free surface falls (dimensionless); and θ is the angle of outer normal of the infiltration curve of free surface boundary with a vertical direction.

2.2 *Mathematical model for heat migration of groundwater*[15-17]

$$\begin{cases} \frac{\partial}{\partial x}\left(\lambda_x \frac{\partial T}{\partial x}\right)+\frac{\partial}{\partial y}\left(\lambda_y \frac{\partial T}{\partial y}\right)+\frac{\partial}{\partial z}\left(\lambda_z \frac{\partial T}{\partial z}\right)-c_w\left[\frac{\partial(v_x T)}{\partial x}+\frac{\partial(v_y T)}{\partial y}+\frac{\partial(v_z T)}{\partial z}\right]+Q_c=c\frac{\partial T}{\partial t} & (x,y,z)\in\Omega \\ T(x,y,z,t)\big|_{t=0}=T_0(x,y,z) & (x,y,z)\in\Omega \\ T(x,y,z,t)\big|_{\Gamma_1}=T_1(x,y,z,t) & (x,y,z)\in\Gamma_1 \\ \lambda_x\frac{\partial T}{\partial x}\cos(n,x)+\lambda_y\frac{\partial T}{\partial y}\cos(n,y)+\lambda_z\frac{\partial T}{\partial z}\cos(n,z)\Big|_{\Gamma_2}=Q(x,y,z,t) & (x,y,z)\in\Gamma_2 \end{cases} \tag{2}$$

In the formula: λ_x, λ_y, and λ_z indicate the dispersion coefficient of heat power of water in all directions, respectively; ($W/(m\cdot K)$) is obtained by the heat conduction coefficient of groundwater and water-bearing medium skeleton, horizontal and longitudinal heat dispersity of groundwater, and seepage velocity of groundwater; c_w indicates the heat capacity of water, and it is equal to the product ($J/(m^3\cdot K)$) between the specific heat capacity c_d and density; c indicates the heat capacity ($J/(m^3\cdot K)$) of the water-bearing medium; v_x, v_y, and v_z indicate the seepage velocity component (m/d) of groundwater; $T_0(x, y, z)$ indicates the initial temperature value (K) of point (x, y, z); $T_1(x, y, z, t)$ indicates the temperature function (K) of the first-class boundary; Q_c indicates the heat source term $Q_c = c_w W(T_Q - T)$; T_Q indicates the temperature of the source sink term; and Γ_1 indicates the Class I boundary.

The equation of motion for groundwater flow is as follows:

$$\vec{v}=-K_x\frac{\partial h}{\partial x}-K_y\frac{\partial h}{\partial y}-K_z\frac{\partial h}{\partial z} \tag{3}$$

Definite solution problems couple Formulas (1) and (2) through the equation of motion of groundwater flow (3), and they constitute the three-dimensional coupling mathematical model of unsteady seepage and heat migration of groundwater in the field.

3 ENGINEERING APPLICATION

3.1 *Project profile*

The demonstration project is located at the east of Church Village, Yanzhao North Street, Zhengding, Shijiazhuang, Hebei. It is located in the east of Taihang Mountain and the upper part of the alluvial-proluvial fan of the mountain front, which is the inclined plain of the mountain front. Its general trend is high in the northwest and low in the southeast, inclined from northwest to southeast. The attitude of the project is about 70.0m. The field covers the thin tillage soil, silty soil, and sandy soil. The lithology of aquifer mainly consists of sand gravel-cobble, sand gravel, and coarse sand, and its thickness is 1560m. The stable aquiclude is not provided between the Holocene series and upper Pleistocene series; therefore, they have a close hydraulic connection and a uniform groundwater level. This aquifer has good connectivity, easily supplied by the atmospheric precipitation, strong in water abundance, and rich in water volume; and the specific field is generally 30~60m^3/h•m, and it can reach 100m^3/h•m at the shaft part of the alluvial-proluvial fan.

One pumping well and one recharge well are arranged in this project, and the distance between them is about 60m; the well depth is 120m, and three ground temperature monitoring holes are arranged in the field. The total load of winter heating is 469.3kw; and the total load of summer refrigerating is 603.4kw. See figure 1.

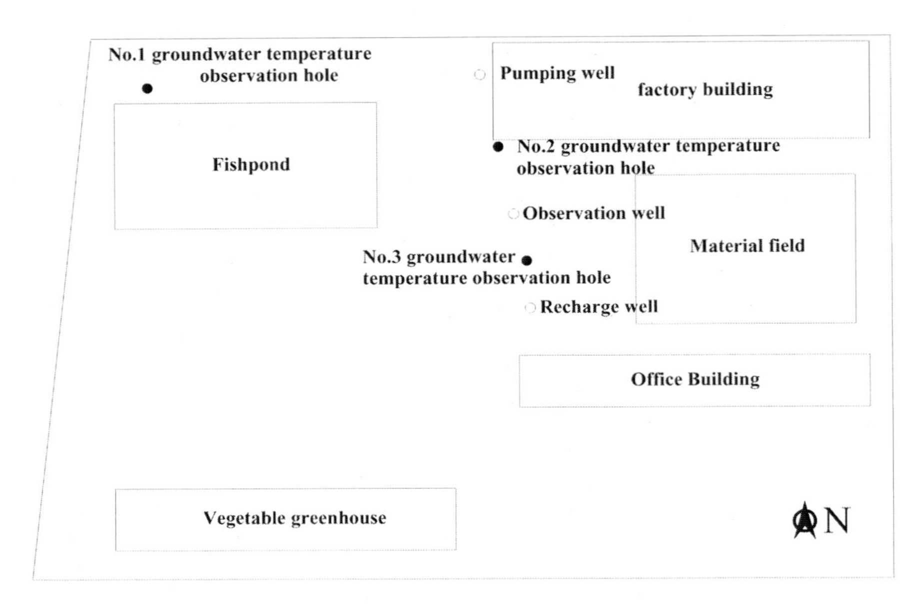

Figure 1. Project layout chart.

3.2 *Conceptual model for groundwater seepage and heat migration of field*

The conceptual model for groundwater seepage and heat migration of field is built on the basis of analyzing the hydrogeological condition, dynamic characteristic of groundwater and conversion for heat migration of the groundwater system in the field according to the data collection, field survey, and research results of the test.

By taking the center of pumping-recharge well on the simulated calculation planes as the starting point, it should extend 1000m toward four directions, respectively, that is, east, south, west, and north, and the total calculating area is $4*10^6$ m^2. It is divided into the superficial silty soil and silty aquifer (I aquifer), middle fine sand, coarse sand aquifer (II aquifer), and bottom sand gravel-cobble aquifer (III aquifer) vertically from top to bottom. On the one hand, the top is supplied by the rainfall, which is the contour of recharge; on the other hand, the groundwater is evaporated through it, which is the drainage boundary; and the bottom of the system is the confining boundary, its periphery is the boundary of fixed water level, and the flow state of groundwater is three-dimensional unsteady flow. It is set that the thermodynamic equilibrium of groundwater and water-bearing medium skeleton in the calculation range is completed instantaneously, namely, the water-bearing medium skeleton and groundwater around have the same temperature, the influence of natural convection up and down caused by the different densities of water due to the temperature difference is ignored, and the periphery of the aquifer is conceptualized as the Class I heat boundary.

3.3 *Model identification and verification*

The computational domain is divided into 924 triangular units on the plane, 493 for the node of each layer; considering the influence of the pumping well, observation well, and thickness of the layer in vertical direction, it is totally divided into three aquifers, four calculation levels, 2772 units, and 1972 nodes in total. The three-dimensional space subdivision of the computational domain is as shown in figure 2.

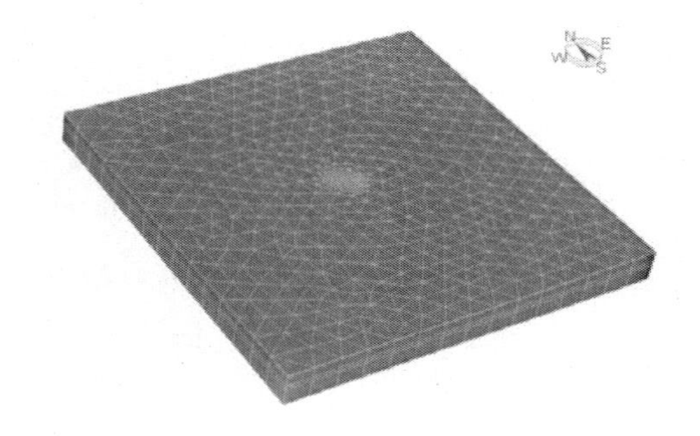

Figure 2. 3-D imaging of study area.

The author uses the pumping well, observation well, and recharge well to carry out water level fitting, and uses 3 groundwater temperature observation hole data to carry out water temperature fitting. According to the actual observation data, the time intervals from August 28, 2012, 12:00 to September 6, 2012, 12:00 are selected as the identification verification period of the model; there are 166 stress periods. The initial water level of the model in each layer is the actual-measured value of each aquifer at 12:00 on August 28, 2012, 12. Based on the actual-measured surface atmospheric temperature at 12:00 on August 28, 2012, 12, 25°C is taken as the initial temperature of Ith aquifer; the many-year atmospheric mean temperature is taken as the initial temperature of IIth aquifer; and according to the geothermal gradient formula, 16.3°C is taken as the initial temperature of IIIth aquifer.

The fitting situations of the pumping well, observation well, and recharge well during identification validation phase are as shown in figure 3, and the temperature fitting situation for each geothermal monitoring hole is as shown in figure 4.

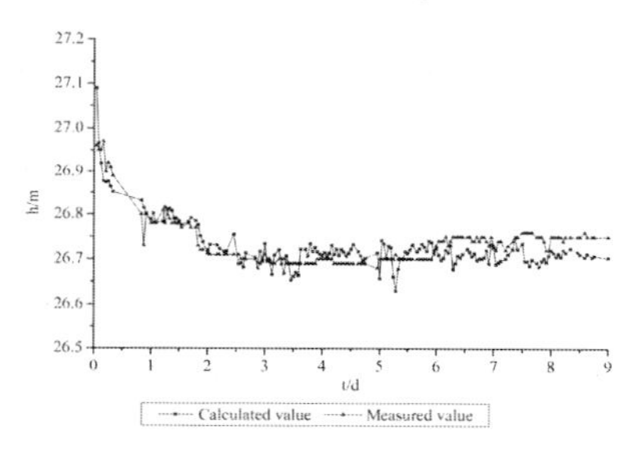

(a) Pumping well

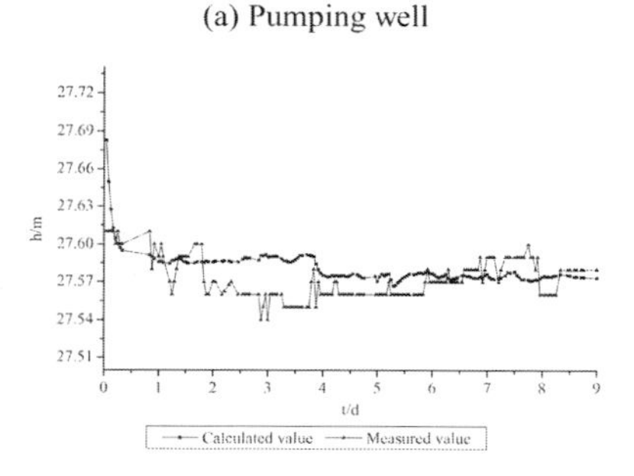

(b) Observation well

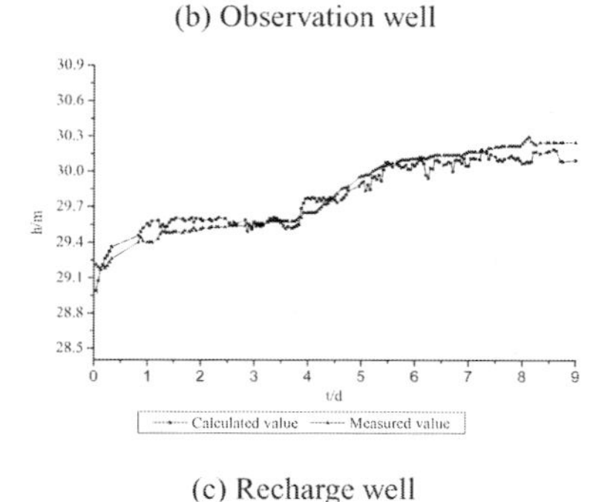

(c) Recharge well

Figure 3. Fittings of groundwater level.

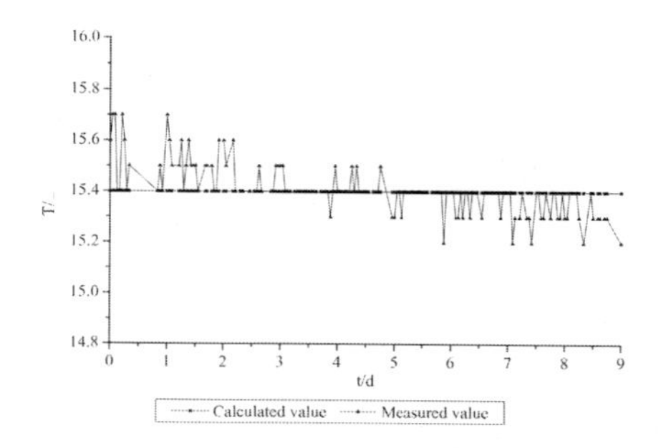

(a) No.1 observation hole

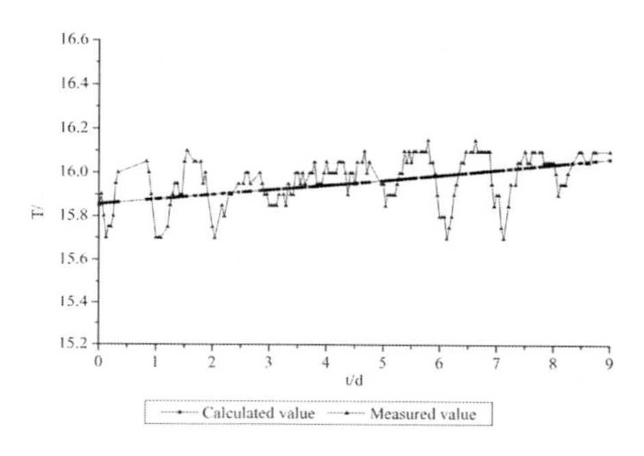

(b) No.2 observation hole

(c) No.3 observation hole

Figure 4. Fittings of groundwater temperature.

From the fitting results, the fitting accuracy of the calculated curve and measured curve is better, the overall trend is consistent, the error of the underground water level is below 0.2m, and the temperature calculation error is within the tolerance range, which proves that the built model is correct and reliable, has reasonable parameters, can accurately reflect the essential characteristics of the groundwater system in the study area, and can be used in the simulation prediction of groundwater seepage field and temperature field.

4 HEAT BALANCE ANALOG COMPUTATION OF GROUNDWATER HEAT PUMPS

Based on the validated model, the running effect and cold/heat transfixion phenomenon of the GWHP demonstration project are evaluated. In order to evaluate the running state of the demonstration project GWHP under different working conditions better, and to provide a decision basis for the long-term and effective operation of the groundwater heat pumps, the author simulates the four schemes of gross load design constant, gross load design increase, gross load design decrease, and the pumping/recharge well distance change.

The starting time for each scheme is the beginning of the summer refrigeration in late July, and the early July of the next year is selected as the end time; the groundwater seepage and heat migration for the GWHP are predicted. There are 24 stress periods in total, and each one has a time step. The initial water level for each aquifer is given by pumping test data, the atmospheric precipitation infiltration recharge in each stress period takes the mean value for many years, and the initial temperature for the I, II, and III aquifers takes 25°C, 15.5°C, and 16.3°C, respectively.

4.1 *Constant load design gross*

When the load design gross is constant, there is an inverse relation between the temperature difference of the pumping/recharge water and the circulation water volume, namely, the temperature difference increases, but the circulation water volume decreases and vice versa. The variation trend of the groundwater seepage and heat migration of the demonstration project is stimulated and analyzed under the three working conditions when the water temperature difference of the pumping well and the recharge well is, respectively, the working condition I (10°C), working condition II, namely the temperature difference drops by 20% (8°C), and the working condition III, increases by 20% (12°C). Under this circumstance that the pumping well and recharge well work simultaneously, the water volume is 100% recharge, and at the end of the summer refrigeration in each working condition, the heat influence range (namely the isotherm of 16.3°C) is as shown in figure 5(a).

After running a cycle under such three working conditions with temperature differences at 8°C, 10°C, and 12°C, respectively, the water temperature of the pumping well is, respectively, 16.38°C, 16.37°C, and 16.37°C, and there is a weak cold heat transfixion phenomenon; compared with the initial temperature, the temperature rise is less than 0.1°C, and the cold/heat transfixion phenomenon is not obvious.

Figure 5(a) shows that in the case of constant load, with the increase of the temperature difference of the pumping/recharge water, the 16.3°C curve scope shortens gradually, and the cold/heat transfixion phenomenon gradually weakens. Therefore, when the load is constant, the method for decreasing recycled water volume and increasing the temperature difference of the pumping/recharge water is beneficial to ease heat transfixion phenomenon, and it is beneficial to increase sustainable running of the underground water source heat pump system, which is in line with the operation suggestion of large temperature difference, small flow advocated in the industry.

4.2 *Load design increasing*

When there is uncertainty in running factors, the actual cold and heat load design changes should be considered while the underground water source heat pump system is running. Overall, 20% of load gross increase can be achieved via two methods of 20% increase of constant temperature-difference water volume (working condition IV) and 20% increase of water constant temperature difference (working condition V). This section simulates and analyzes the variation trends of the groundwater seepage and heat migration of the demonstration project under two such working conditions. Under this circumstance that the pumping well and recharge well work simultaneously, the water volume is 100% recharge, and at the end of the summer refrigeration in each working condition, the heat influence range (namely the isotherm of 16.3°C) is as shown in figure 5(b).

At the end of the summer refrigeration, the temperature of the pumping well is 16.39°C in working condition IV, the temperature of the pumping well is 16.37°C in working condition V; it is relatively higher, and the latter one has a more obvious cold/heat transfixion phenomenon. Figure 8 further shows that the working condition IV has greater thermodynamic influence scope. It objectively reflects that within the allowed range of the underground water source heat pump system, the load design increase of the heat pump can ease the heat transfixion phenomenon via increasing the temperature difference of pumping/recharg water, thus increasing the sustainable running of the underground water source heat pump system.

4.3 *Different well spacing*

When the cooling and heating load design, circulating water capacity, and temperature difference of pumping/recharge water (the temperature difference of design condition is 10°C) are determined, aiming at the year-round cold/heat transfixion phenomenon and the well spacing optimization problem of the underground water source heat pump system, the vibration trends of the groundwater seepage and heat migration of the underground water source heat pump system are stimulated and discussed under different pumping and recharge well spacing (20m, 40m, 60m, 80m, and

100m). Under this circumstance that the 1 pumping well and 1 recharge well work simultaneously, and the water volume is 100% recharge, the circulating water capacity statistics are as shown in figure 5(c).

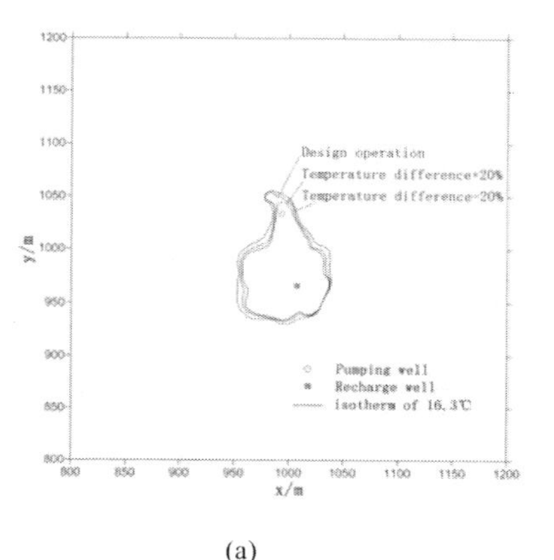

(a)

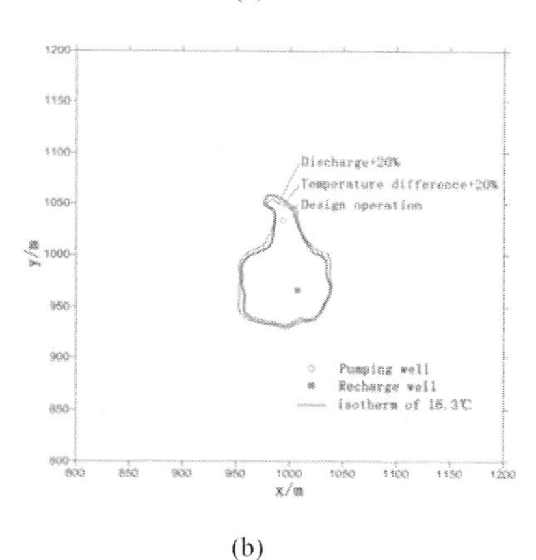

(b)

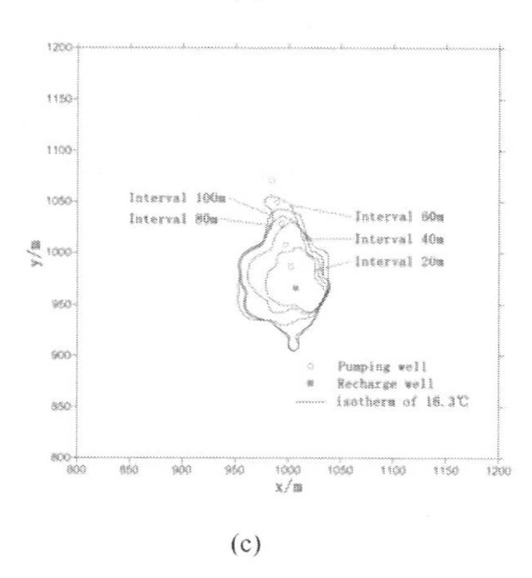

(c)

Figure 5. Water temperature isotherm of 16.3°C for Layer 3 after summer refrigeration.

According to the calculation results, when the well spacing is 20m, the water temperature of the pumping well is 17.47°C at the end of the summer refrigeration; when the well spacing is 40m, the water temperature of the pumping well is 16.47°C; when the well spacing is 60m, the water temperature of the pumping well is 16.32°C; when the well spacing is 80m, the water temperature of the pumping well is 16.30°C; and when the well spacing is 100m, the water temperature of the pumping well is 16.30°C, which shows that with the increasing of the well spacing, the effect of pumping/recharge flow on the groundwater heat migration weakens.

The isotherm of 16.3°C in figure 10 also shows that when the well spacing is equal to or less than 60m, the heat influence scope reaches the pumping well, and the cold and heat transfixion phenomenon exists. When the well spacing is 80m, the recharge hot water has no influence on the outlet water temperature of the pumping well, and there is no cold and heat transfixion phenomenon. To achieve the sustainable running of the underground water source heat pump system well, the pumping-recharge well spacing should be more than or equal to 80m. Therefore, the pumping-recharge well spacing has a significant effect on the operating performance of the underground water source heat pump system; the essence is that the increasing of the pumping-recharge well spacing decreases the current influence on the heat migration process of the groundwater.

According to the results of simulation evaluation, the change of the running scheme has certain influence on the running effect and sustainable operation of the underground water source heat pump system. The circulating water capacity and pumping-recharge well spacing jointly determine the degree of the heat transfixion phenomenon between pumping and recharge well spacing; the reason is that the effect of groundwater flow has a decisive influence on the groundwater heat migration process in water-bearing media.

5 CONCLUSIONS

1 As a new type of development and utilization of shallow ground temperature energy, under the condition of specific water flow and the heat source, the migration law study of the groundwater heat during the operation of the underground water source heat pump system is the theoretical basis and premise foundation of the sustainable running for its project. Simulation and evaluation results show that the change of the cooling and heating load design scheme has significant influence on

the sustainable operation of the underground water source heat pump system.

2 Under the working condition of the design operation (working condition I), with the continuous running of the GWHP in Zhengding base, the temperature of the underground water in the entire region gradually decreases, and the cold accumulation phenomenon appears.
3 The simulated results of the heat pump system show that in each working condition while cooling and heating load changes, the pumping-recharge circulating water volume is the major factor influencing the degree of the sustainable running of the heat pump system; the reason is that the effect of the pumping-recharge groundwater flow has a decisive influence on the groundwater heat migration process in water-bearing media.
4 When the load design scheme is determined, the drops of the circulating water and the increase of the temperature difference are beneficial to ease the phenomenon of thermal penetration; moreover, when the pumping-recharge well spacing is greater than or equal to 80m, the utilization efficiency of the GWHP can be effectively improved.

REFERENCES

[1] Stefano Lo Russo, Glenda Taddia, Giorgia Baccino, Vittorio Verda. (2011) Different design scenarios related to an open loop groundwater heat pump in a large building: Impact on subsurface and primary energy consumption. *Energy and Buildings*, 43(2/3).

[2] Stojanovic B, Akander J. (2010) Build-up and long-term performance test of a full-scale solar-assisted heat pump system for residential heating in Nordic climatic conditions. *Applied Thermal Engineering*, 30(2/3):188–195.

[3] Long Ni, Jiaping Feng, Zuiliang Ma. (2004) The state of research and development of groundwater heat pump systems. *Building Energy & Environment*, 23(2):26–31.

[4] Dejan Milenic, Petar Vasiljevic, Ana Vranjes. (2010) Criteria for use of groundwater as renewable energy source in geothermal heat pump systems for building heating/cooling purposes. *Energy and Buildings*, 42(5).

[5] Aifeng Zhang, Weiping Zhao, Xianghua Liu, *et al.* (2008) Ground-source heat pump technology and its application. *Journal of Hefei University of Technology*, 31(12):2028–2030.

[6] Rui Fan, Yiqiang Jiang, Yang Yao, Zuiliang Ma. (2008) Theoretical study on the performance of an integrated ground-source heat pump system in a whole year. *Energy*, 33(11).

[7] Qunli Zhang, Jin Wang. (2003) Research and Development and Practical Problem Analysis of Ground Source Heat Pump and Groundwater Source Heat Pump, 31(5):50–54.

[8] Changzheng Xin, Yingxin Zhu. (2002) Study on transfer-storage performance of aquifer when deep well recharge heat pump well in motion. *National HVAC Refrigeration 2002 Academic Essays*, Beijing: China Building Industry Press, pp:263–270.

[9] Manchao He, Bin Liu, Leihua Yao, *et al.* (2003) Study on the theory of seepage field for geothermal single well reinjection. *Acta Engergiae Solaris Sinica*, 24(2):197–200.

[10] Kunfeng Zhang, Fang mei Ma, Liuyi Jin. (1998) Numerical simulation and analysis of the water heat pump system operating in winter condition. *Journal of Huazhong University of Science and Technology*, 26(5): 1–4.

[11] Paksoy H Q, Andersson Q, Abaci S, *et al.* (2000) Heating and cooling of a hospital using solar energy coupled with seasonal thermal energy storage in an aquifer. *Renewable Energy*, 19:117–122.

[12] Tenma N, Yasukawa K, Zyvoloski G, (2003) Model study of the thermal storage system by FEHM code. *Geothermics*, 32: 603–607.

[13] Zujiang Luo, Lin Yang, Zhanjun Li. (2012) Three-dimensional numerical simulation for immersion prediction of left bank of Songyuan reservoir project. *Transactions of the Chinese Society of Agricultural Engineering (Transactions of the CSAE)*, 28(3):129–134.

[14] Yuequn Xue, Chunhong Xie. (2007) *Numerical simulation of Groundwater Flow*. Beijing: Science Press. pp:402–413.

[15] Long Ni, Haorong Li, Yiqiang Jiang, Yang Yao, Zuiliang Ma. (2011) A model of groundwater seepage and heat transfer for single-well ground source heat pump systems. *Applied Thermal Engineering*, 31(14/15)

[16] Aly A H, Hwangbo C K. (2009) Electro-thermal transport in quantum point contact nanodevice. *International Journal of Thermophysics*, 30(2):661–668.

[17] Demirel Y. (2009) Thermodynamically coupled heat and mass flows in a reaction-transport system with external resistances. *International Journal of Heat and Mass Transfer*, 52(7–8):2018–2025.

Electric, Electronic and Control Engineering – Shao, Shu & Tian (Eds)
© 2015 Taylor & Francis Group, London, ISBN: 978-1-138-02842-5

Research on lighting withstand performance of HVDC power transmission line based on the new electrical geometry model

Xiaogang Gao*, Jinqiao Du, Kexue Liu, Xiaoying Xie & Yang Yuan
Qinhuangdao Power Supply Company, State Grid Jibei Electric Power Company, Qinhuangdao Hebei, China

ABSTRACT: The long-term operation results show that: lightning stroke is the important causes of the accidents that occur on the transmission line. With the increase of the voltage level of transmission line, the lightning accident is also more and more serious to EHVDC and UHVDC transmission lines. Based on several methods to lightning shielding failure rate of the classical electrical geometry model, a new electrical geometry model is developed. The model takes into account the terrain and the angle of incidence of lightning leader. In this paper, we do a variety of calculations, draw new electrical geometry model calculation results are more accurate. Through new electrical geometry model lightning shielding failure rate calculated in different conditions. The results show that the tower height, the line protection angle and the tower ground tilt angle changes in the rate of lightning shielding failure.

KEYWORDS: HVDC; lightning protection performance; the new electrical geometry model; shielding failure.

1 INTRODUCTION

HVDC transmission shows its previous advantages in the distant and the great electric power system, which not only improves the economic standard and technical capabilities of the electric system, but also makes the electric system more reliable. Especially in the last century, as the rapid development of high-power electric technique and computer control technique, the HVDC transmission can be developed more rapidly. According to the statistical data of grid faults, the total number of trips caused by lightning strike in AC high voltage and ultra high voltage transmission line was accounted for 40%-70%. The operating experience of ±500kV HVDC transmission line also indicates Lightning strike of DC transmission lines are very serious in our country.

There are two ways of lightning transmission tower: strike back and shielding failure. Although the probability of shielding failure is small and the lightning current amplitude is low as the shielding failure happened, the impulsive voltage still can lead the shielding insulator arc-over, and endanger the electric equipment. With the voltage degree of the power transmission increases continuously, the lightning accidents of shielding failure account for an increasing proportion. The earliest formula for computation of ratios of Shielding Failure for Transmission Lines simply considered the protection angle. In assessing the effects of ground wire shielding conductor, we do the same thing. Now our study is centrally concerned with the new electrical geometry model, which can explain the failure of shielding by physical model, which is used widely in the computation of Shielding Failure ratio on Transmission Line.

Electrical geometry models are verified in the computation of Shielding Failure Flashover ratio from the classical model to the improved electrical geometry model. A kind of new electrical geometry model is presented in this paper, and we will find the reasonableness of the new electrical geometry model by comparison with the present electrical geometry model, and use this way to continue our study about the factors affecting the shielding failure flashover ratio.

2 PARAMETER OF SHIELDING FAILURE RATIO COMPUTATION

2.1 *Shielding failure lightning withstand level*

The equivalent current method is used to calculate the lightning withstand level. In DC transmission line,

*Corresponding author: *gaofei7072483@163.com*

detour lightning equivalent lines after calculating the working voltage can be shown in Fig.1:

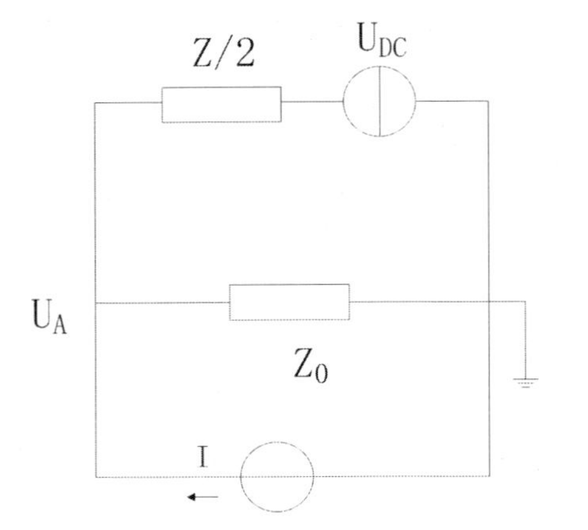

Figure 1. Detour lightning equivalent transmission line determination.

Z_0 is channel wave resistance, U_{DC} is equivalent DC power, Z is polar line resistance. With computation, we can get the electric potential A:

$$U_A = \frac{Z_0}{Z+2Z_0}(IZ + 2U_{DC}) \geq U_{50\%} \tag{1}$$

As the U_A potential > $U_{50\%}$ flashover voltage of insulators, Shielding accident occurs. The lightning current amplitude is simply represented by the channel wave resistance amplitude, which is $Z_0 > Z$. The simple formula is shown as:

$$I_{\min} \geq \frac{2(U_{50\%}\text{-}U_{DC})}{Z} \tag{2}$$

2.2 *Determination of the height from conductor and grounding wire to the ground*

As the electrical geometry model is used to calculate the shielding failure, the average height of the conductor can be get from calculating the longitudinal profile. E.R Whitehead, the author of Lightning, which mentioned, terrain falls into 3 sections: plains, hilly lands and mountains.[11] For the 3 sections, we will adopt different ways to determine their height of conductors. The formulas are shown as Formula (3), (4), (5).

Plains: $h_d = h_{dt} - {}^{2S_d}\!/_3$ (3)

Hilly lands: $h_d = h_{dt}$ (4)

Mountains: $h_d = kh_{dt}$ (5)

The average height h_b of the grounding wire from different terrains to ground can be calculated by the formula 2.6:

$$h_b = h_d + (h_{bt} - h_{dt}) + \frac{2}{3}(S_d - S_b) \tag{6}$$

h_{dt}——the height of conductor on poles and towers
h_{bt}——the height of grounding wire on poles and towers
S_d——conductor arc.
S_b——grounding wire arc

This method is easy and proper to be used in the plains and the similar hilly lands.

2.3 *Determination of striking distance factor*

We adopt the formula 2.7[12] from IEEE Std1243-1997:

$$\begin{aligned} &k=0.36+0.17\ln(43-h) && h \leq 40\text{m} \\ &k=0.55 && h > 40\text{m} \end{aligned} \tag{7}$$

2.4 *Computation of striking distance*

Before striking distance determined, the point of lightning current is not stable. When it reached the striking distance, the discharge will occur.

We adopt the formula from IEEE, Formula (8), (9).

$$r_s = 10I^{0.65} \tag{8}$$

$$r_g = \begin{cases} [3.6+1.7\ln(43-y_c)]\,I^{0.65} & y_c \leq 40m \\ 5.5I^{0.65} \end{cases} \tag{9}$$

I——lightning current;
r_s——striking distance of grounding wire;
r_g——striking distance of ground;
y_c——average height of conductor (m).

2.5 *The effect of working voltage*

Because 90% of the lightning in our country are almost negative, the DC working voltage of the positive wires intensifies the attraction to the negative. For the high DC working voltage, the effect will be more serious than before. Therefore, we must consider the working voltage from another conductor striking distance, which is shown in Formula (10):

$$r_c = 1.63 \times (5.015I^{0.578} + U_{dc})^{1.125} \tag{10}$$

In the formula, r_c——the striking distance from lightning to every conductor with working voltage, m;

U_{dc}——the working voltage manipulates from lightning to the conductor, MV.

2.6 *The computation of the longest striking distance*

$$r_{sm}=\left[k(h_b+h_d)+\sin(\alpha+\beta)\sqrt{(h_b+h_d)^2-G}\right]\cos\theta/2F \quad (11)$$

$$F=k^2-\sin^2(\alpha+\theta) \quad (12)$$

$$G=F\left[(h_b-h_d)/(\cos\alpha\cos\theta)\right]^2 \quad (13)$$

h_b, h_d——the height from the grounding wire and conductor to the ground;
α——the average angle for the grounding wire;
θ——the slope angle on the poles and towers;
k——the striking distance factor.

2.7 *The scope of the lightning current*

As the shielding failure trips are calculated, we need to consider 3 lightning manipulates, I_{min} (the minimum lightning current); I_c (the critical lightning current); I_m (the maximum lightning current). The maximum current is determined by the maximum distance, the shielding failure will occur in $\left[I_{min}, I_{max}\right]$. I_c is related to insulativity, so the shielding failure level will be expressed by the critical lightning current. If $I_{min} < I_{max} < I_c$, shielding failure flashover will be zero. If $I_{min} < I_c < I_{max}$, the lowest point will be I_c, the highest point will be I_{max}; If $I_c < I_{min} < I_{max}$, the lowest and highest point will be I_{min}, I_{max}.

2.8 *The probability density of the lightning current*

$$f(I)=\frac{1}{I\sigma\sqrt{2\pi}}\exp\left\{-0.5\times\frac{\left[\ln(I)-\ln(\bar{I})\right]^2}{\sigma^2}\right\} \quad (14)$$

$\bar{I}$——average lightning current;
σ——standard deviation 0.72 [13].

3 THE NEW ELECTRIC GEOMETRY MODEL

3.1 *Construct the new electric geometry model*

The paper considered the disadvantage of the classical geometry model and the improved model, based on the computation of exposure arc ratio, included two different scenes resulted by the slope angle on the ground of the poles and towers and the lightning current not vertical lunching by conducting first, and use a new computation, lessening the calculation error.

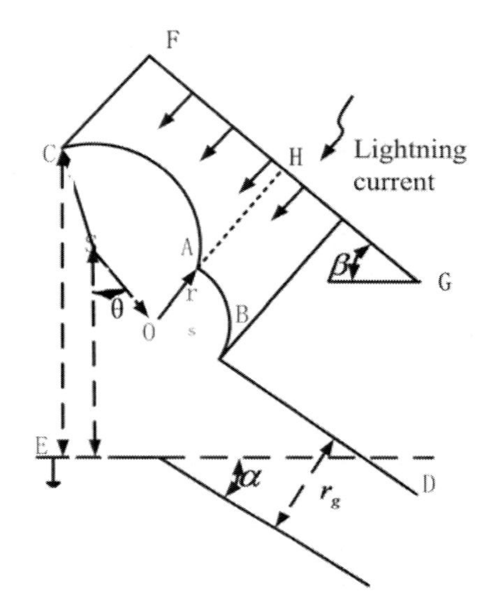

Figure 2. The new electric geometry model.

By Fig. 2, we can find that, with the effect of the ground slope angle and incident angle, the expanding exposure district, the different results of lightning merely caused by the probability of the incident angle β current striking the exposure arc and the shielding arc. Then the exposure arc and the shielding arc will not be looked as they will only project to the horizontal plane. The new electric geometry model presented we should conduct the sheaf projection to the lightning current, as the plane FG. According to the computation, we think the shielding failure ratios P is the ratio of line GH and GF.

$$P_r=\int_{I_c}^{I_{max}}\frac{L_{GH}}{L_{FG}}f(I)dI \quad (14)$$

L_{GH}——exposure arc projection to the sheaf;
L_{FG}——exposure and sheaf arc projection to the sheaf
P_r——shielding failure ratios
$f(I)$——lightning current manipulate function
I_{max}——the max manipulation of shielding failure
I_C——the critical manipulation of shielding failure

3.2 *Example calculation and analysis*

There are 4 computation methods to calculate the shielding failure trip ratios, including exposure distance, exposure arc ratio, equivalent withstand lightning width, exposure projection. The 4 methods can calculate the arc ratio, but they are failed at the lightning current conduct the incident angle, or ignored the slope angle of the ground. So we need to use the fifth method to calculate it. Computation institutes adopt CG lightning density, lightning days, the most representative Z27-31suspension poles and towers; the protection angle of grounding wire to polar line is -10°;practical height is 40m,lightning days is 80 per year. The ground flashing density is $\gamma = 0.023T_d^{0.3}$; the model of lightning wave is2.6/50μs; the striking distance factor is k=0.36+0.17ln(43-h); the striking formula is $r_s = 10I^{0.65}$; the present probability of lightning current is

$$f(I) = \frac{1}{I\sigma\sqrt{2\pi}} \exp\left\{-0.5 \times \frac{\left[\ln(I) - \ln(\bar{I})\right]^2}{\sigma^2}\right\};$$

The standard deviation σ is 0.72. The results are shown in Table 1.

Table 1. Shielding failure ratio calculation.

Methods	Exposure distance	Exposure arc ratio	Equivalent withstand lightning width	Exposure projection	New electric geometry model
Shielding failure trips ratio	0.0177	0.0131	0.1131	0.0237	0.0161

Through analyzing the results of calculation, you can find the ratio calculated by the exposure arc ratio method is the minimum, but calculated by the equivalent withstand lightning width method is the maximum, and the mistake is maximum comparability with the other 4 methods. The reason is it included the grounding wire district, but the scope doesn't include the grounding area, and then the ratio is upper. The other 4 ratios are similar, but the exposure arc projection calculation is upper, which is because they mistake the arc to the straight line, but the practical arc is longer than the line, so the ratio is upper. However the exposure arc ratio calculation is minimum, since they didn't consider the probability of lightning striking the ground directly, so the calculated ratio is minimized. Exposure distance method considers the lightning current as the vertical incidence, which doesn't fit to the slope angle electric model. Although the exposure considers the geometric size of it, it still makes the result not accurate by its former ideal calculation and its narrowing. The ratio calculated by the new electric geometry model between the maximum and the minimum, making the calculation more accurate and can be used widely.

4 SHIELDING FAILURE CALCULATION PERFORMANCE ANALYSIS

4.1 *The effect of the height of poles and towers to the shielding failure performance*

The expanding of the poles and towers will increase the width of withstanding lightning for the transmission line, and meanwhile the ground will lessen its shielding ability from the conductor. The effects of conductor to attract lightning become stronger, which affect the shielding failure withstand-lightning capability. We set the slope angle 20°, protection angle -10°, the height of the poles and towers are 35m, 40m, 45m, 50m, 55m, 60m to be calculated, the result is shown as below:

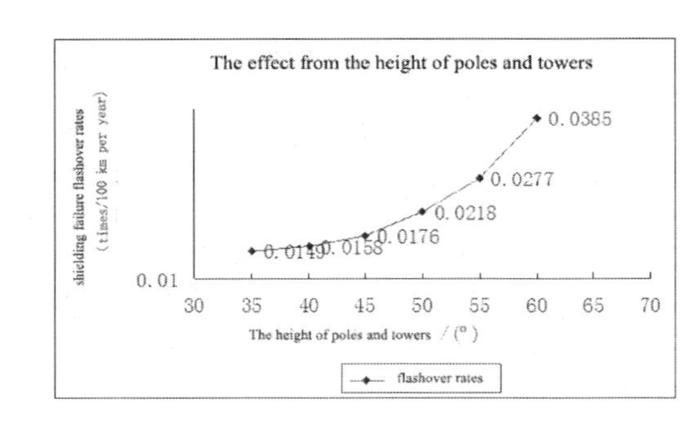

Figure 3. The affects from the height of poles and towers.

In Fig. 3, with the increasing height of poles and towers, the shielding failure flashover rates are increasing and the speed of rate increases more quickly. Meanwhile the probability of transmission lines attracting lightning becomes stronger, but the ground not, which lead the transmission line district gets smaller, and the flashover rates get stronger. To lessen the rates, we should reduce the height of poles and towers. With the increasing level of electric voltage, high poles and long distance has been used popularly. Therefore, we need to adopt other shielding failure methods to improve the lightning protection performance.

4.2 *The effects of protection to the shielding flashover rates*

By Fig. 4, we can find the shielding failure flashover rates are increasing with the increasing transmission line protection angles. Because of the direction of the lightning current down to the ground, the narrowed protection angle will control the conductor up, and the grounding wire will take the advantage by their competitive situations, the protection function will become stronger, the rates will be reduced. Therefore, we will narrow the protection angle as much as possible when we design the poles and towers, so as to increase the performance of DC power transmission line against shielding failure.

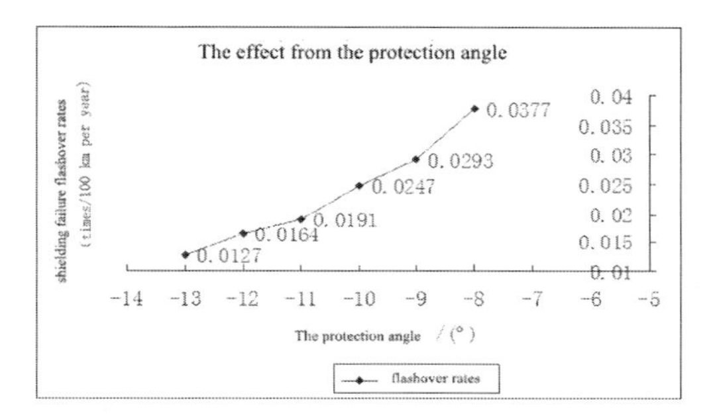

Figure 4. The effects of protection angle to the shielding flashover rates.

By Fig. 4, we can find the shielding failure flashover rates are increasing with the increasing transmission line protection angles. Because of the direction of the lightning current down to the ground, the narrowed protection angle will control the conductor up, and the grounding wire will take the advantage of their competitive situation, the protection function will become stronger, the rates will be reduced. Therefore, we will narrow the protection angle as much as possible when we design the poles and towers, so as to increase the performance of DC power transmission line against shielding failure.

4.3 *The effects of slope angle of the poles and towers*

The places of poles and towers are not always planes. When the poles and towers are set in the mountains, the conductor of the outside slope will have a frequent flashover rate by its increasing shielding failure arc and the lower distance to the ground. But the inside slope conductor will have a small probability of flashover rates. According to the above, we need to choose the height is 40m,protection angle -10° of the poles and towers, and the slope angle need to be 0°, 5°, 10°, 15°, 20°, 25 °. The calculation is shown as below:

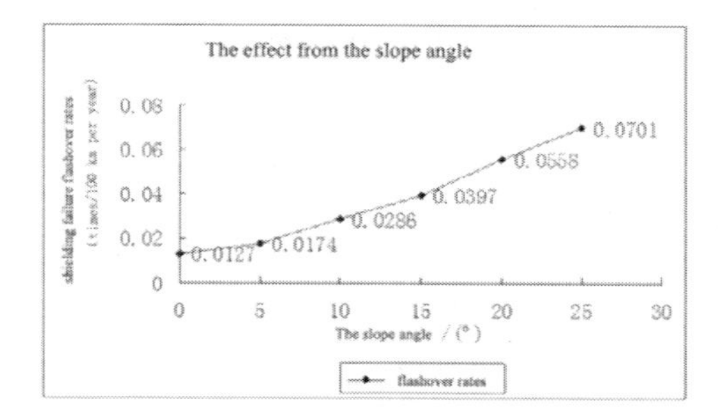

Figure 5 The effect from the slope angle to the flashover rates.

By Fig.5, we can find the flashover rates have become lower as the slope angle is zero or smaller. However, with the increasing of ground angle, the flashover will happen frequently. As the slope angle is 30°, and the flashover is almost 10 times of the zero angle. So we need to strengthen the protection of poles and towers when the transmission lines pass the mountains and hilly lands and the other steep gradient, so that we can lower its shielding failure rates to maintain the transmission line in a good condition.

5 CONCLUSIONS

The main point of this paper is to study the performance of the DC power transmission line. We present a kind of new electric geometry model, compared with the trips difference between the present classical electric geometry calculation and the new calculation to determine the advantage of the new geometry model. The new calculation not only include the slope angle of conductor and ground, but also make the result of trips more accurate, and fit for the mountains.

To analyze the study of the shielding failure trip rates affecting DC power transmission line, we find some regulations from the height, the protection angle and the ground slope angle of the poles and towers affecting the shielding failure trip rates.

REFERENCES

[1] Zhao Wanjun. (2004) *UHVDC Technology*. (1st version). Beijing: China Electric Power Press, pp:20–25.

[2] Liu Zhenya. (2009) *UHVDC Theory*. Beijing: China Electric Power Press, pp:1–3.

[3] Shu Yinbiao. (2004) Present status and prospect of HVDC transmission in China. *High-voltage Engineering*, 30(11):1–3.
[4] Kim Shawwillie, Wu Weihan. (1998) The analysis of lightning protection for EHV and UHV Transmission in Russia. *High-voltage Technology*, 24(2):76–79.
[5] Zhang Weibo, Gao Yuming. (1987) *Overvoltage Protection and Insulation Coordination*. Beijing: Tsinghua University Press, pp:193–195.
[6] L. Dellera, E. Garbagnati. (1990) Lightning strokes simulation by means of the lead progression model (Part 1: description of the model and evaluation of exposure offree-standing structures). *IEEE Trans. on Power Delivery*. PWRD-5(4):2009–2022.
[7] F. A. M. Rizk. (1989) Switching impulse strength of air insulation: Leader inception criterion. *IEEE Trans. on Power Delivery*. PWRD-4(4):2187–2195.
[8] Brow n G W, Whit ehead E R. (1969) Field and analytical studies of transmissi on line shielding: part 2. *IEEE Transaction on Power Apparatus and System*, 88(5):617–626.
[9] R. H. Golde. (1983) *Lightning (Vol. 1)*. Beijing: China Water Power Press.
[10] IEEE Working Groupon Estimating Lightning Performanee of Transmission Lines. (1993) Estimating lightning performance of transmission lines 11 UP dates to analytical models. *IEEE Trans.*, PWRD-8(3):1254–1267.
[11] R. B. Anderson, A. J. Eriksson. (1980) Lightning parameters for engineering application. *Electra*, (69):65–102.
[12] A. J. Eriksson (1987) An improved electrogeometrie model for transmission line shielding analysis. *IEEETPWRD*, 2(3):859–970.
[13] IEEE Working Group on Estimating Lightning Performance of Transmission Line. (1985) A simplified methods on estimating lightning performance of transmission line. *IEEE Transactions on Power Apparatus and System*, 104(4):919–931.
[14] Sima Wenxia, Chen Ning, Xu Gaofeng, *et al.* (2002) Lightning protection performance of transmission line with high tower and great span. *Journal of Chongqing University*, 25(9):25–28.
[15] Rizk FAM. (1990) Modeling of transmission line exposure to direet lightning strokes. *IEEE TPWRD*, 5(4):1983–1997.

Electric, Electronic and Control Engineering – Shao, Shu & Tian (Eds)
© 2015 Taylor & Francis Group, London, ISBN: 978-1-138-02842-5

Design and development of online photoelectric detection turbidimeter for water environment

Hui Zhang*, Yuewen Huang, Yi Yu & Biao Xu
Instruments and automation Institute, The Changjiang River Scientific Research Institute, Wuhan Hubei, China

ABSTRACT: This paper describes the design principles of online photoelectric turbidimeter, including schematic design, hardware design, software design and functional design. It also describes Advanced Grapher, a curve simulation software. The turbidimeter performs with high-speed data acquisition, MODBUS bus transmission, high-performance CPU data processing and high-capacity data storage. A new add-in function Bluetooth allows real-time data transmission from device to computer. It implements the purpose of continuous online control of water quality in the production process, highlighting the real-time and online detection, easy to operate, apply to turbidity measurement requirements in online water environment monitoring system.

KEYWORDS: Online photoelectric turbidimeter; Advanced Grapher; water environment monitoring system

1 INTRODUCTION

Turbidity is one of the significant basis for water quality evaluation, commonly used to characterize the cleanliness of the water turbidity[1]. With the improvement of life level, the requirements on water quality become more and more stringent. Therefore put forward higher requirements of turbidity measurement. According to the national standard GB5749-2006 "Standards for drinking water quality" & CJ/T206 "Water quality standards for urban water supply", drinking water turbidity is not more than 1 NTU, treated water turbidity of water plant is not more than 0.5 NTU[2].

This paper describes the design principles of online photoelectric detecting turbidimeter, hardware design, software design, functional design and Advanced. Grapher. Curve simulation software, high-speed data acquisition and MODBUS bus transmission, high-performance CPU data processing, high-capacity data storage, newly added Bluetooth, test data can be transferred to the computer in real time, it implements the purpose of the production process online continuous control of water quality, highlighting the real-time and online detection, easy to operate, suitable for water environment monitoring system for online detection of water turbidity measurement requirements. Different from traditional turbidimeter, the measurement process of online photoelectric detecting turbidimeter does not require contact with the measured object, thus greatly reducing the maintenance of the turbidimeter. A photodetector line turbidimeter which utilises high-speed CPU processing chips can measure the turbidity of a relatively wider range. Online and real-time sampling data collections are the highlights of this turbidimeter. It is easy to operate and will be an important direction for future development of turbidity measurements.

2 MEASURING PRINCIPLE

2.1 *Domestic and international background*

The earliest turbidity measurement is "Jackson candle turbidimeter" in 1900 which used candles as the light source[3]; there were transmitted light measurement method and scatter light measurement method in 1966[4]. The rapid development of LED and laser technology promotes the development of turbidimeter. The use of silicon photodiode, condenser and filter and integrated control chip, greatly improves the measurement range and accuracy.

2100A laboratory turbidimeter by HACH (USA), CUE23/CUE24 turbidimeter by E+H (Germany), Inpro8000 series turbidimeter by METTLER TOLEDO (Germany) and 7997 series turbidimeter by ABB-KENT (UK) are the most popular brands on the market[5].

China's 70 to 80 years, industrial on-line turbidity measurement instrument is almost a blank spot, after nearly 10 years of technical research and the introduction of advanced technology and experience of foreign countries, turbidity meter has been great changes made in China, there were many manufacturers and some and some technical indicators had reached a high level of domestic, such as TSZ series laboratory turbidimeter by Xiamen Feihua instrument

*Corresponding author: *diudiuzh@163.com*

company, WGZ series by Shanghai Shanke, WSZ series by Shanghai Leici and WT_OT3 series by Wuhan Wote.

2.2 *Measuring principle*

Turbidity measurement is mainly the use of Rayleigh scattering theory (strong forward scattering) and Mie scattering theory (fluctuate intensity of scattered light from different angles on the side)[6]. The relationship between scatter light and incident light can be expressed as:

$$I_s = \alpha N I_o \exp\text{-}(\tau l) \quad (1)$$

I_s is the intensity of scatter light; Io is the intensity of incident light; N is the number of particles contained in water samples which proportional to the turbidity; α is the coefficient associated with the scattering function; τ is the attenuation coefficient which is independent to luminous intensity, l is the scatter light distance.

2.3 *Schematic design*

Figure 1 is the schematic design. The sample water enters the detector by inlet K1, flow out by outlet K2 according to the direction that indicated by arrow T; D1, D2, D3 are photoemissive tubes, D3 receives the transmission light beams emitted by D1, D2 receives the light beams from D1 which scattered by solids in the water; the received signals of D3 and D2 changes with the turbidity of the measured water and it completes the water turbidity parameter detection.

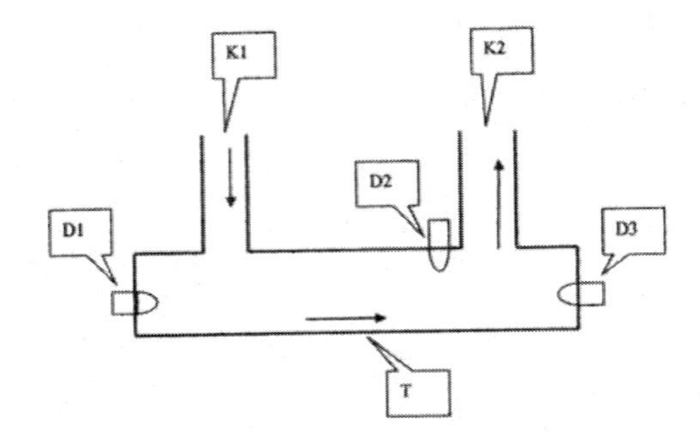

Figure 1. Design schematics.

3 TURBIDITY DETECTION SYSTEM

3.1 *Hardware design*

Figure 3 shows the block diagram of hardware design. Bubble separator will be applied when water samples go through the valve to photoelectric sensor, and optical signal is converted to an electrical signal by photoelectric sensor, after preamplifier, the electrical signal convert to digital signal by AD converter and transfer to CPU for data processing using bus transmission. CPU connected power module, switching circuit, Bluetooth module, display control and auxiliary control, display control is connected to the monitor.

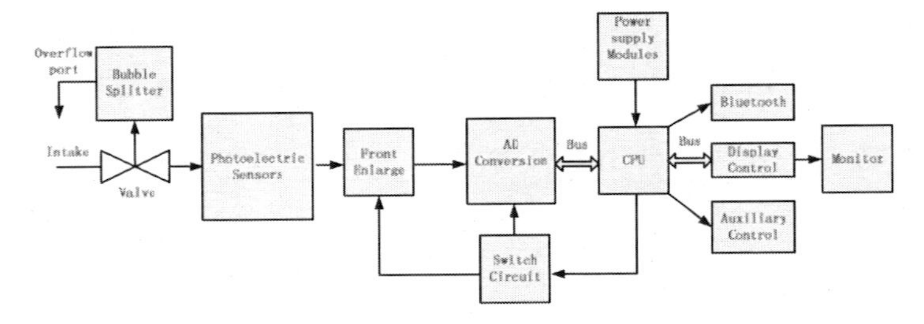

Figure 2. Hardware design diagram.

3.2 *Software design*

Initialize all the modules when the host is on, including 485 buses, a Bluetooth module, timer, etc., and then build the display UI, send a command to acquisition board requesting for real-time data, the data received by acquisition board, including the light source voltage, the reference voltage, photosensitive voltage at 90°, photosensitive voltage at 180°, and then analyze and process these data to get current turbidity value by finding turbidity voltage in the preset curve equation. In the end, analyze the current turbidity value for alarm processing, and convert the turbidity value into current signal and digital signal for the output. The software flowchart is shown below in Figure 3.

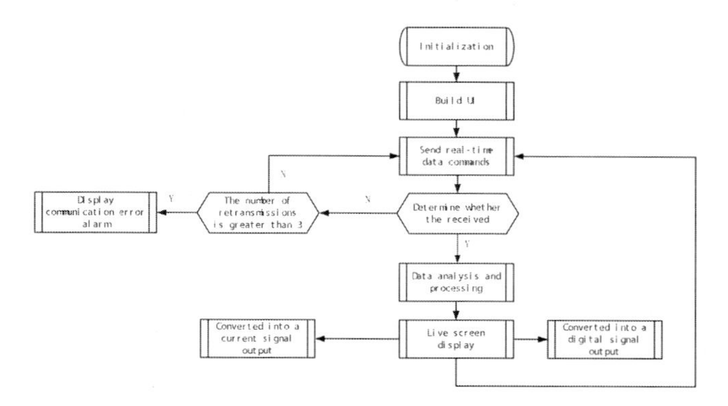

Figure 3. Flow chart of software.

3.3 *Function design*

Bubbles in water are the largest interference source of turbidity measurements, scatter light increases when the light beam pass through the bubbles in the water, which causing interference on the measurement results, it should be removed. The sample water will go through a series of inner diaphragm before sensor that effectively eliminate the bubbles, ensuring the stable turbidity readings and prevent the phenomenon of false high turbidity readings.

Unique design of flow path avoids the impact of bubbles on the determination, meanwhile the design uses ultrasonic cleaning technology to ensure measuring chamber is always in a clean state, it will clean the inner wall on setting time automatically to prevent steam condensation affecting scattering. Newly added Bluetooth function helps transmit data to the computer for easy storage and management of data.

4 DEVELOPMENT AND TESTING

4.1 *Data comparison*

Table 1 for the preparation of standard water samples, the turbidity were 0NTU, 1NTU, 2NTU, 4NTU, 6NTU, 8NTU, 10NTU.The design model for CW, the test data compared with the United States HACH2100N desktop turbidity, the data in Table 1 below.

Table 1. Test data sheet.

Standard water (NTU)	Display value of 2100N (NTU)	Display value of CW (NTU)
0	0.007	0.009
1	0.986	1.013
2	1.963	2.028
4	3.948	4.067
6	5.890	6.083
8	7.831	8.118
10	9.792	10.168

The standard deviation of the data was displayed as: 0.002NTU0.027NTU, 0.065NTU, 0.119NTU, 0.193NTU, 0.287NTU, 0.376NTU. The relative standard deviation was displayed as: 2%, 2.7%, 3.2%, 2.9%, 3.2%, 3.5%, 3.7%. The data show that CW has good accuracy.

4.2 *Data fitting*

The design uses Advanced. Grapher for simulation of Curve fitting, a specific algorithm is added to fit out a closer to the actual turbidity curve, and the obtained turbidity values are accurate and stable. The specific algorithm is the corresponding relationship between turbidity and voltage that obtained by calibrating the standard solution repeatedly. Use the specific algorithm to analyze these data, fitting a N-th equation curve, this equation can be a good representation of this turbidity trend curve, and divided the curve into 1000 (or more) straight lines, for each acquisition of turbidity corresponding voltage, determine which interval this voltage is in on the straight line, an accurate actual value can be directly obtained by using the linear relationship. See for Figure 4.

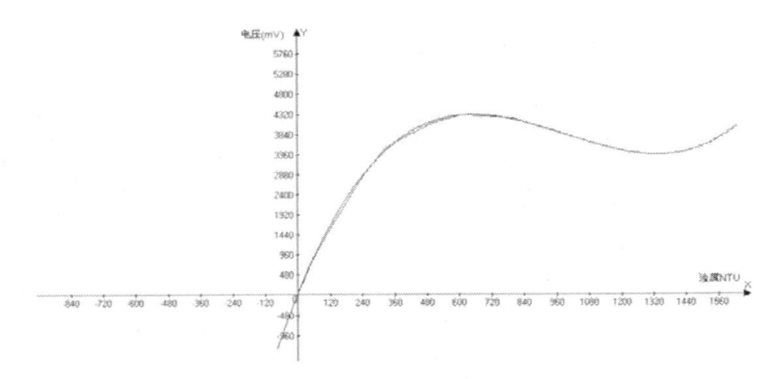

Figure 4. Simulation graph.

Simulation tests indicate: the linear correlation is good in the range of 0 ~ 400NTU, linear of scattered light intensity vs. turbidity decreased in the range of 400 ~ 1000NTU where the compensation of the transmitted light is needed. 0 ~ 1000NTU simulation fitting curve comply with the design, the equipment is stable for continuous operation, the measured value is in the tolerance range.

4.3 *Performance testing*

The design of the test prototype is shown in Figure 5.

Figure 5. Test prototype.

Prototype performance:
Measurement range: 0~1000NTU
Tolerance: ≤5%(0~10NTU)≤1%
Resolution: 0.01NTU
Operate mode: Continuous long-time online measurement
Output voltage: 4~20mV
Data storage: 64G memory
Display: Color TFT-LCD
Output data: MODBUS protocal, bluetooth

5 CONCLUSIONS

Different from traditional turbidimeter, the measurement process of online photoelectric detecting turbidimeter does not require contact with the measured object, thus greatly reducing the maintenance of the turbidimeter. Real-time transmission of online test data to computer lead to easy data storage and management can realize the continuous on-line control of water quality in the production process, suitable for a water environment monitoring system for online detection of water turbidity measurement requirements.

ACKNOWLEDGMENT

This paper is funded by Technology Development and Transformation Projects to Promote Achievement of the Yangtze River Academy (CKZS2014004 / YQ).

REFERENCES

[1] Liu Zhenqi. (2003) Utility model turbiditor. *Modern Scientific Instruments*, (06).
[2] Xu Hanchen. (1995) Intelligent design scattering turbidity. *Instrumentation Technology*, (06).
[3] Gao Ping. (2001) Underwater scattering line turbidity meter. *Practical Testing Technology*, 5:14–1.
[4] Elliott T. (1986) Turbidity: the final element in assessing water quality. *Power*, 12:65–66.
[5] Hong Z. (1998) On-line turbidity measurement. *Using Surface Scattering. SPIE*, 3558:20–30.
[6] Zhang Li. (2006) Research and function test of a new-type column integrating on-line intelligent turbidimeter. *Chinese Journal of Scientific Instrument*, (08).

Electric, Electronic and Control Engineering – Shao, Shu & Tian (Eds)
 ISBN: 978-1-138-02842-5

Implementation of accurate attention on students in classroom teaching based on big data

Yiwen Zheng
Department of Film and Television Technology, Zhejiang Vocational Academy of Art, Hangzhou Zhejiang, China

Wenhong Zhao & Hongxing Chen
School of Mechanical Engineering, Zhejiang University of Technology, Hangzhou Zhejiang, China

Yunhui Bai
Department of Basic Teaching, Zhejiang Vocational Academy of Art, Hangzhou Zhejiang, China

ABSTRACT: Teaching activities in current classroom teaching are normally carried out by a teacher and dozens of students. There are certain difficulties for teachers to pay accurate attention to all students. The advent of the big data era has made effective attention on students possible with digital media technology by extracting facial expressions and posture characteristics for data decoding and analysis. A recognizing and tracing system of students' learning activities is designed in this paper, which decodes and analyzes learning situations of students accurately by collecting and processing status data of students' sitting postures, eyes and voices. The experiment shows that the scheme is feasible and effective.

KEYWORDS: Sitting posture; eyes; voice; accurate attention.

1 INTRODUCTION

It is of great importance for teachers to pay attention to students in teaching activities. This is an effective way of improving teaching effect. Teaching activities in current classroom teaching are normally carried out by a teacher and twenty or thirty students, even forty or fifty students. How to effectively and accurately pay attention to students' learning state, such as concentration, attendance, presentation, reflection, class exercise, emotional and physical condition, is of great significance for the improvement of teaching efficiency. All these states can be presented by facial expressions and posture characteristics. There are three methods of acquiring data of students' learning state. The first method is the detection based on physiological parameters, which acquires electrical activity of brain cells, eye movement and heart activity through electroencephalogram, electro-oculogram and electrocardiogram. The cost of this method is relatively high and the difficulty is large as well. The second method is the detection based on behaviors, conducting monitoring on learning behaviors of students, such as attendance, classroom learning time and classroom discussion and presentation. The third method is the detection based on images, capturing images of real-time learning situation with external equipment like camera and infrared facility so as to know learning attention of students.

A recognizing and tracing system of students' learning activities is designed in this paper, including a sitting posture measuring system, an eyes recognizing system, and a noise recognizing system. By acquiring big data of students' states in classroom, it is able to accurately decode and analyze real-time learning situations of students, such as attendance, concentration, classroom activeness and physical tiredness, so that accurate and effective attention on each student becomes possible.

2 ACQUISITION OF BIG DATA IN CLASSROOM

2.1 *Eyes recognizing system*

The eyes recognizing system is composed of image acquisition module, illumination pretreatment module, eyes locating module and eyes state module. In the image acquisition module, digital image signals are transmitted to FPGA after CCD sensor conducting double sampling, dark level truncation, variable gain amplification and A/D conversion to analog image signals. FPGA converts digital image signals to low voltage differential signals that are then transmitted to the upper computer. Thus, the operation of image acquisition and transmission is completed [1] [2]. Noises exist in facial images obtained in natural conditions. For example, noises brought by sensors are presented as isolated and discrete spots on images. The technology of median filtering can be adopted in order to eliminate these noises. The basic idea of median filtering technology is to sequence pixels in a local area in line with the gray scale. The median of the gray scale in this area is taken as the gray value of the current pixel. It performs well in inhibiting impulse and spot noise and maintains image edge perfectly. In addition, the gray scale of image might be confined to a small range under the condition of insufficient or excessive exposure, producing blurring images. Nonlinear gray scale transformation technology is adopted in this experiment to eliminate the influence. Gray scale transformation is carried out with an exponential function as a mapping function.

$$g(x, y) = a + \frac{\ln[f(x, y)+1]/b}{\ln c} \quad (1)$$

Here, f(x,y) is the pixel gray value at the position of (x, y) of the original image, g(x, y) is the gray value after transformation, and a, b, c are adjustable parameters. The low gray scale area is expanded while the high gray scale area is compressed through transformation, so that the distribution of the gray scale is uniform [3] [4] [5].

With the dipolar location algorithm from thick to thin, eyes detection module divides the eyes searching region into several pieces in fixed sFize. Calculate the complexity of each piece and find those pieces with the highest complexity. And then delete and combine these candidate pieces in line with certain rules so as to acquire the position of human eyes. The formula of the algorithm is given below:

$$C(A) = \frac{1}{S(A)} \Sigma_i \Sigma_j Edge(A) \quad (2)$$

Edge is the edge detection operator and S (A) stands for the area of region A [6].

Extract the characteristic value from facial image. The key is the binary processing, which sets the gray scale value of each spot on image as 0 or 255 so that the entire image presents an obvious effect of black and white. Black pieces of human eyes are shown in Figure 1.

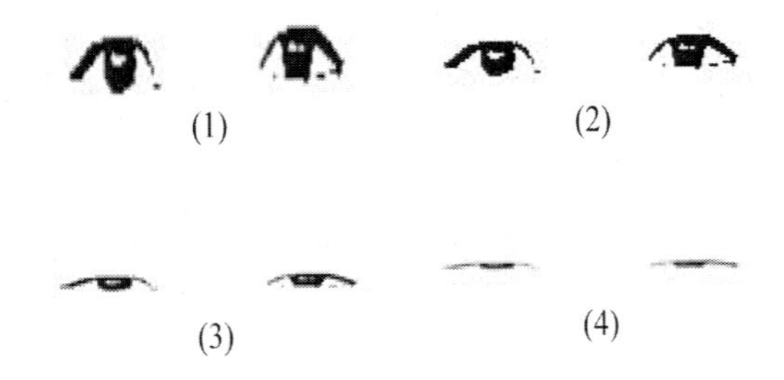

Figure 1. Black pieces of human eyes.

Sizes of black pieces of human eyes are not fixed. They change with the closure state of human eyes. After a great deal of experiments combined with learning situation of students, black pieces of human eyes can be mainly divided into: (1)staring eyes, (2)normal eyes, (3)squinting eyes, (4)closed eyes.

2.2 *Sound pick-up system*

Elite KO-280 high-performance sound pick-up is adopted in this system for audio acquisition. The vibration membrane of this sound pick-up is a capacitance microphone, featuring large scope of sound pick-up, high sensitivity and wide frequency range (20-20KHz). High-speed signal processing circuits like DSP digital noise reduction and AGC automatic gain control are internally installed so as to prevent distortion and attenuation of voice signals effectively. Self-adaptive dynamic noise reduction processing and high-speed processor are internally installed in the system. The U.S. BOURNS potentiometer for exclusive use features high reliability of volume control and high accuracy with functions of protection against lightning strike, reverse polarity protection and electrostatic protection. It is widely used in classroom synchronous recording system, meeting recording system and interrogation room due to its excellent performance [7].

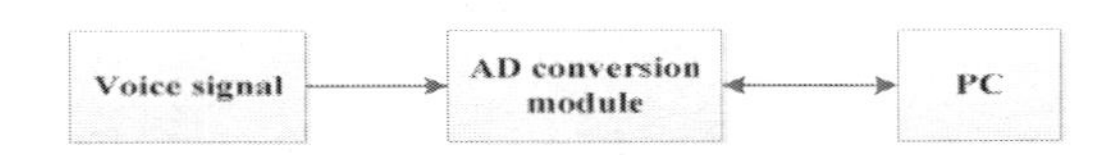

Figure 2. Diagram of audio acquisition.

Voice transmission mainly includes the following two parts:

1 Voice signals of students in class are transformed into analog audio signals by sound pick-up, amplified by audio power amplifier, and then restored by sound equipment.

2 Extract audio signals of the computer from the Line Out port of sound card, lead to the amplification route through the voice frequency of center control, and input to Aux In compositely. Voices are restored by sound equipment.

Here, the high-quality and adjustable front feedback device is equipped with an equalizer with full frequency, being able to adjust the sound pick-up volume and line volume independently. The RS232 communication interface adjusts volume on the central dashboard so as to meet the requirements of different tone qualities [8].

2.3 *Measurement system of sitting posture*

Use the body pressure measurement system of Tekscan Company to measure sitting postures of students. The system includes an A/D switching circuit based on PC and a pressure-distribution and cushion-type sensor that can be used repeatedly. Pressure display and analysis software based on MS Windows are also adopted. The A/D switching circuit based on PC scans each sensing point and measures the resistance value of each force bearing point. Pressure display and analysis software based on MS Windows are able to realize such functions as real-time acquisition and control of pressure distribution data, visual real-time display, recording and playback of pressure distribution, analysis on pressure distribution data, etc. The pressure distribution data reading and analysis module conducts analysis on pressure distribution data, calculates the sensitive region of pressure distribution, and outputs parameters like pressure of specific location [9] [10] [11] [12]. The module is shown in Figure 3.

Figure 3. Process structure of pressure distribution measurement data.

Data collected by the body pressure measurement platform are analyzed by Labview software. It is easy to know sitting postures of students in class at some point. Five common sitting postures are recorded as: upright, bending forward for 15°, bending forward for 45°, left-sided body, and right-sided body.

3 DATA ANALYSIS OF STUDENTS IN CLASS

Dataflow computation originates from a concept: the value of data decreases as time goes by. Therefore, events must be processes immediately instead of being stored for batch processing. The core value of stream computation lies in real-time integration of data from a variety of heterogeneous data sources and continuous real-time processing on huge amounts of "dynamic" data [13]. The diagram of dataflow computation is shown in Figure 4.

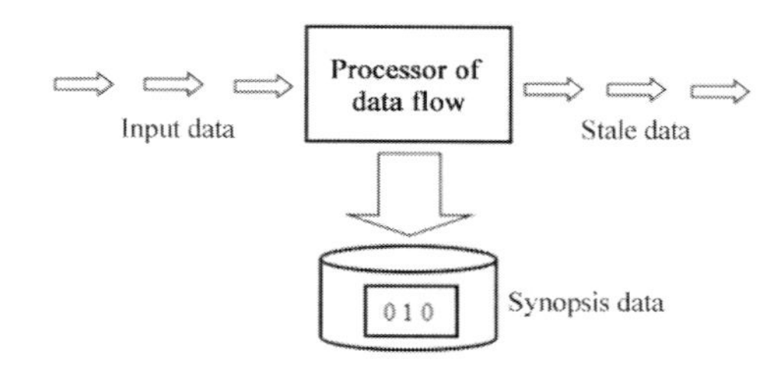

Figure 4. Stream computation system.

As for streaming big data featuring volatility, instantaneity, randomness, abruptness and infinity, an ideal stream computation system for big data should have such characteristics as high throughput, low latency, continuous and stable operation, elasticity, etc. All these cannot do without reasonable planning and good design of key technologies like system structure, data transmission and programmatic interface. System structure, serving as the compound mode of subsystems, is a mutual key technology of big data computation. Stream computation of big data requires a specific system structure for task deployment. Data transmission is the data transmitting mode among computation nodes after the deployment from directed task graphs to physical computation nodes. In the environment of big data streaming computation, task graphs and the mapping mode from task graphs to physical computation nodes need to be optimized more systematically in order to achieve high throughput and low latency. Programmatic interface provides users with the convenience of describing internal logic and dependency of task through task graph according to features of stream computation. Processing functions of each node in task graph are realized through programming. Big data batch computation stores data in a persistent device in advance so that data replay can be easily realized after node failure.

Sample data are processed by the Storm system of Twitter Company on account of the fact that data collected in this experiment feature large quantity and diversity. Storm system is a real-time dataflow computation system of recent Open Source. It provides distributed real-time computation with a set of general primitive and processes messages and updates database in the way of "stream processing". Major characteristics of Storm are presented below:

1 Simple programming model: reduce the complexity of parallel batch processing and real-time processing.
2 Support a variety of programming languages: various programming languages can be used in Storm. The support for multiple languages can be realized through a simple Storm communication protocol.
3 Fault-tolerant: Storm manages work progress and node failures.
4 Support horizontal scaling: computations are conducted among multiple threads, processes and servers.
5 Reliable message processing: Storm guarantees that each message can be processed completely at least for one time. It reads the message source again when the task fails [14].

4 ANALYSIS OF EXPERIMENTAL RESULTS

It is effective to conduct real-time processing on data of students' state in class through dataflow computation. Eyes, voices and sitting postures of students in class are drawn in Table 1. Carry out a comprehensive analysis through the three major indicators so as to represent students' state in class.

Table 1. Common state of students in class.

Students	Eyes	Voices	Sitting postures
A	Normal size and normal frequency of wink	No noise around	Upright, few changes
B	Often beyond detection range	Sporadic noises, fast and unstable speed	Normal changes
C	Many times of closing eyes	No noise around	Incorrect posture, few changes

Student A sits in an upright posture with eyes staring at the blackboard. The state is natural and stable without noise. Students of this kind are typically concentrated in class with serious attention.

The posture of students B is constantly changing and the eyes are often beyond the detection range with continuous noises around. Students of this kind cannot focus attentions and always look around. They are not interested in the teaching content or lost interest in learning. It is suggested that effective communications between teachers and students of this kind should be enhanced. Appropriate attentions should be paid on these students so that they can be lead to normal learning designedly.

It is quiet around student C who is in a stable posture but not upright. The frequency of wink is lower than that of normal people (normal people blink 10 to 15 times per minute). Students of this kind are in a fatigue state and unable to focus on class. As for this kind of students, it is suggested that parents should be contacted immediately so as to make clear reasons of the fatigue state. Parents should supervise the normal schedule of students.

5 CONCLUSION

Effective attentions on students will influence performance of students directly or indirectly. Each student is eager to get the attention of teachers. This is an effective means of helping, motivating and guiding students' self-development and self-improvement. Acquiring more accurate information of students from big data generated in the teaching process will be the direction of quality education research in future. This paper makes the first attempt at judging the learning state in class by collecting three important indicators, namely eyes condition, sitting posture and noise condition, solves the embarrassing state that teachers are unable to pay attention to each student, helps students to overcome difficulties and obstacles pointedly, and guide students to transform from potential possibility to practical development.

REFERENCES

[1] Xu, W.H. & Wu, D.H. (2012) Design of CCD imaging system of ultrahigh resolution, *Optics and Precision Engineering*, (7):1603–1608.
[2] B.Skarman, J.Becher, K.Wozniakk. (2004) Simultaneous 3D-PIV temperature measurements using a new CCD-based holographic interferometer, *Flow Measurement and Instrumentation*, 33(10):1711–1716.
[3] Karin Sobottka, Ioannis Pitas. (1997) *A Fully Automatic Approach to Facial Feature Detection and Tracking*, In: AVBPA 97 Department of Informatics, University of Thessaloniki, Greece, pp:77–84.
[4] Feng, J.Q., Liu, W.B., Yu, S.L. (2005) Human eyes location based on gray scale integral projection, *Computer Simulation*, 22 (4).
[5] Zhang, J., Yang, X.F., Zhao, R.L. (2005). A study on precise location of human eyes based on integral projection and HOUGH transform circle detection method, *Electronic Devices*, 28(4):706–709.
[6] Tang, X.S., Ou, Z.Y., Su, T.M., et al. (2006) Rapid location of human eyes in a complex background, *Journal of Computer Aided Design and Graphics*, 18(10):1535–1540.

[7] Shen, X.Q. & Yang, J.P. Installation and connection of sound pick-up for monitoring, *Practical Science and Technology*.

[8] Shen, X.Q. et al. (2014) Design and implementation of voice transmission system in intelligent multimedia classrooms, *Electronic Design Engineering*, 22(9).

[9] Sun, J.G. (2008) New progress in clustering algorithms research, *Journal of Software*, 19(1):48–61.

[10] K. Kasten, A. Stratmann, M. Munz, K. Dirscherl. (2006) IBolt technology-a weight sensing system for advanced passenger safety, *Advanced Microsystems for Automotive Applications*, pp:171–186.

[11] Xiao, Z.H. & Gao, Z.H. *Physical Signs Recognition of Passengers Based on Body Pressure Distribution Detection and Classification of Support Vector Machine*, Changchun: Jilin University.

[12] Miao, S.S. & Gao, Z.H. *Multiple Sensor Fusion Algorithm of Physical Signs Recognition of Passengers*, Changchun: Jilin University.

[13] Sun, D.W., Zhang, G.Y., Zheng, W.M. (2014) Stream computation of big data: key technologies and living examples, *Journal of Software*, 25(4):839–862.

[14] Meng, X.F. & Ci, X. (2013) Management of big data: concept, technology and challenge, *Computer Research and Development*.

Electric, Electronic and Control Engineering – Shao, Shu & Tian (Eds)
© 2015 Taylor & Francis Group, London, ISBN: 978-1-138-02842-5

A neural network method for measuring plate based on machine vision

Yan Liu & Ruijian Yang
School of Automation & Electrical Engineering, University of Science and Technology Beijing, Beijing, China

Yinhu Wang
Ansteel Engineering Technology Corporation Limited in Anshan Iron and Steel Group, Anshan, China

Jiangyun Li*
School of Automation & Electrical Engineering, University of Science and Technology Beijing, Beijing, China

ABSTRACT: The length, width and diagonal lines of the wide plate strip are key indexes on quality, but these data are hard to obtain precisely due to complex production conditions. This paper proposed a method of a single CCD-array for a single edge to measure length, width and diagonal lines of the plate in a row. Meanwhile, a neural network algorithm was put forward to correct the measurement errors. These errors are affected by several elements such as the height difference of camera installation, out-of-level platform, and other system errors as well as the lens distortion when imaging with a single lens with a huge field of view. The practice in a real world shows that the length precision measured by the proposed method can reach ±2 mm and diagonal line precision is within ±5 mm. The results satisfied the requirement of wide plate's quality.

KEYWORDS: Wide plate; dimension measurement; machine vision; neural network.

1 INTRODUCTION TO SYSTEM

The size measurement system is the important link which guarantees the geometric precision of plate in the production of wide plate. With the increasing fierce competition in the market, end users of product have higher requirements of the dimension of the plate [1]. The precise sizing of length and width of end product, especially including the diagonal line, can significantly reduce waste materials and waste plates in the process of material processing. However, due to the big extending range of length and width during the rolling of wide plate, complex measurement environment and other factors, the completion of high-accuracy measurement has always been a difficult technical problem in the field.

At present, there are mainly two types of systems which can complete length and width measurement of wide plate. The first type of system is a laser ranging method. This method uses the distance of the laser beam from the plate edge to measure the material size. It actually uses the laser ranging principle [2,3]. The highest precision can reach ±1mm. The disadvantages include poor stability, demand of plate stability and operation parallel to processing table. Otherwise, the laser beam is easy to deviate from the edge, causing measurement errors. For production line of wide plate, it is hard to meet this condition, so it is not widely spread.

The other type of solution is a CCD measurement method using machine vision [4-6]. The basic principle is as follows. Photos are taken for the plate with the camera. Then the corresponding relationship between the image pixels and the size of the actual subject is used to determine the edge and calculate the size[7]. At present, on one hand, for the practical system using area array CCD solution, in case of the restricted large field of view, the lens distortion causes error. On the other hand, owing to the cost of high-resolution camera, low-pixel and multi-camera splicing method is commonly used to measure the size. The highest stabilization precision is only within ±4mm. The camera should be often calibrated. The maintenance and use cost are very high.

This paper applied the high-definition CCD-array to form a measure system featured with a single camera for a single edge. Through the neural network learning, the errors were limited within ±2 mm for length and ±5 mm for diagonal line.

2 SYSTEM DESIGN

2.1 *System introduction*

The measurement system uses 4 CCD cameras with 5 million pixels. The parameter is 2448×2048.

*Corresponding author: *jiangyunlee@gmail.com*

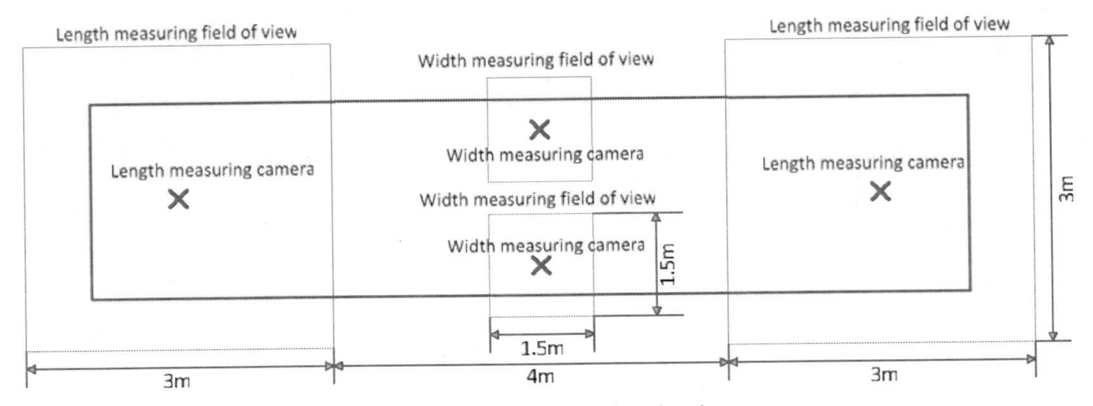

Figure 1. Top view of the position of the camera measuring the size.

Among them, the width measuring camera uses the focal length of 12 mm. The length measuring camera uses the focal length of 5mm. The layout is as shown in Figure 1.

Because of the principles of length measurement, width measurement and diagonal line measurement are similar. Each camera imaging has the matrix of 2448×2048 pixels. The actual specification corresponding to each pixel is 1.275 mm × 1.274 mm. After the pixel segment detection, the minimum distinguishable distance of the system can reach 0.12 mm.

2.2 *Calibration processing*

After setting up the system, calibration plate shall be used for the camera calibration. There are many methods to calibrate camera lenses [8-12]. For this paper, the calibration plate is composed of a lattice with known spacing. The corresponding relationship between the image coordinates and the coordinates of the calibration plate of measuring plane shall be established. Because the distortion affecting the measurement is the mirroring distortion of the camera lens, and the mirroring distortion of pixels is proportional to the square of the center distance of the field of view, the binary quadratic function is used to simplify the model.

$$x' = a_1x^2 + b_1y^2 + a_2x + b_2y + c_1 \tag{1}$$

$$x'' = a_3x^2 + b_3y^2 + a_4x + b_4y + c_2 \tag{2}$$

$$y' = a_5x^2 + b_5y^2 + a_6x + b_6y + c_3 \tag{3}$$

$$y'' = a_7x^2 + b_7y^2 + a_8x + b_8y + c_4 \tag{4}$$

In which, x and y are lattice coordinates of the image. x' and x'' are horizontal ordinates of the lattice plane of the calibration plate in the front and the rear field of view. y' and y'' are vertical ordinates of the lattice plane of the calibration plate in the front and the rear field of view. The least square method is adopted to define the factors in the formula.

3 ERROR CORRECTION

3.1 *Causes of system error*

1 The distortion of camera lens will affect the measurement results. The thickness of the steel plate can also cause the change of edge imaging coordinates. As shown in Figure 2(a), we can know that, due to the different thickness, steel plates with the same length will change the coordinates of image edges. A lot of literatures have studied the solution.
2 In the practical application, during the online measurement of the steel plate, the steel plate will have deformation because of the gravity, as shown in Figure 2 (b).
3 The production of the calibration plate in the wide field of view is quite difficult. The calibration board horizontally placed on the roller way is just for the approximate calibration of work plane.

3.2 *Neural network training corrects errors and improves detection precision*

On condition that the measurement position of steel plates with the same specification remains the same, the preliminary measured length of the diagonal line, the pixel of the edge of the front of the steel plate in the field of view and the thickness of the steel plate shall be considered as the input of neural network. After error correction, with length detection error of diagonal line of training steel plate, high-precision test results can be obtained.

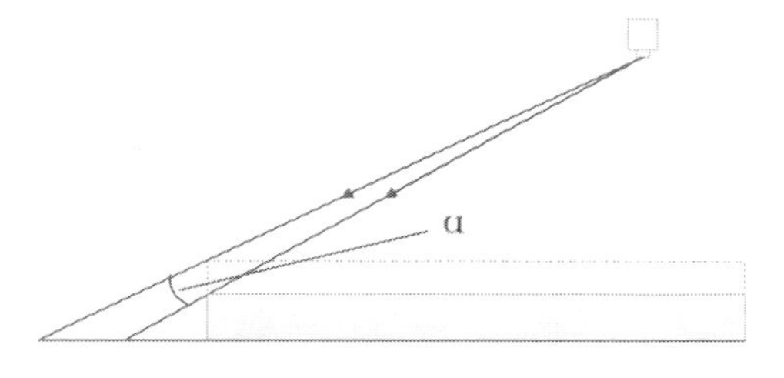

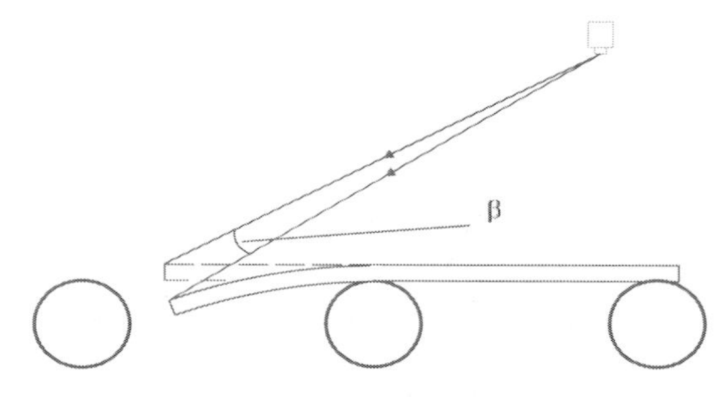

Figure 2. The schematic diagram of measurement error.

Ga Daqi obtained and determined the experience formula of the number of LBF hidden layers after summarizing a large number of application examples [13].

$$s = \sqrt{0.43mn + 0.12n^2 + 2.54m + 0.77n + 0.35} + 0.51 \quad (5)$$

In which, s is the number of nodes of the hidden layers. m is the number of input nodes. n is the number of output nodes. This paper adopts the above empirical formula to determine the number of nodes in the hidden layers of BP neural network. The results show that error correction to the network training system has a good effect. After rounding of calculated values of m=3 and n=1 in this system, s=4 is obtained. It is determined that the number of nodes of hidden layers of the network is 4, as shown in Figure 3.

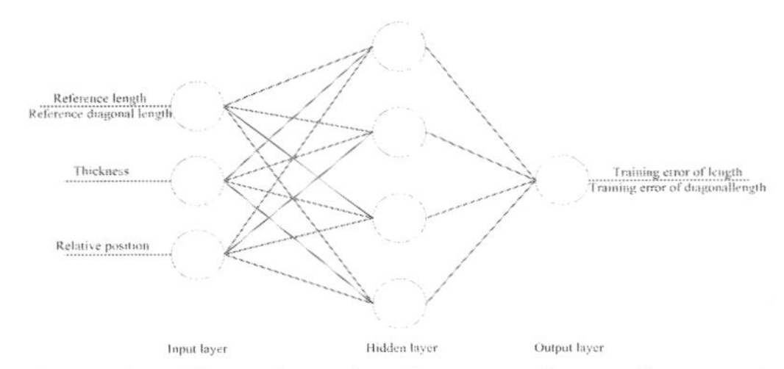

Figure 3. The schematic diagram of neural network structure.

The network uses the Levenberg-Marquardt optimization method. Its weight and update formula is:

$$X_{k+1} = X_k - (J^T J + \mu I)^{-1} J^T e \quad (6)$$

The calculation formula of gradient is:

$$g = J^T e \quad (7)$$

In which, J is a Jacobin matrix of error and weight differential. e is the error vector. μ is a scalar. Depending on the amplitude of μ, this method changes between the Newton method (when $\mu \to 0$) and steepest descent method (when). Levenberg-Marquardt optimization method has the shortest learning time and best actual application effect.

3.3 *Preliminary termination improves the generalization ability*

In the process of training, 10 groups of statistic results of length measuring data before and after error compensation of steel plates with different specifications were randomly chosen. The maximum deviation of measured value and reference value of the length is as shown in Figure 4. The maximum deviation of measured value and reference value of diagonal length is as shown in Figure 5.

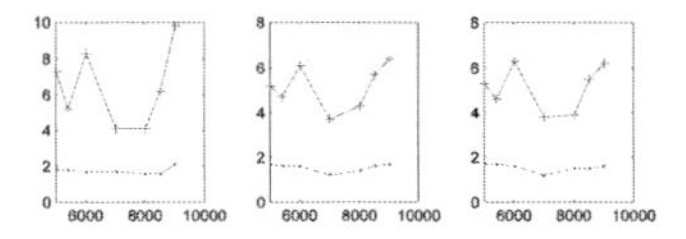

a. 8mm b.12mm c.20mm

x-axis is the length of plate (mm);y-axis is the maximum deviation of different length (mm)

"+" is original value before training; "." is correct value after training

Figure 4. Comparison of maximum deviations of different thickness before and after detecting length training.

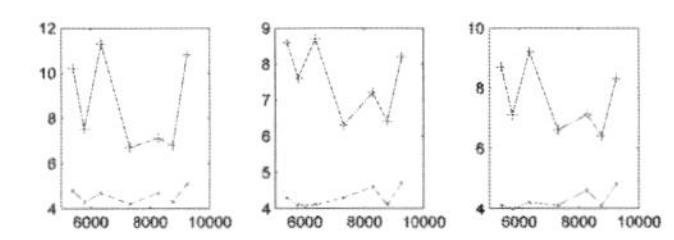

a. 8mm b.12mm c.20mm

X-axis is the length of the plate (mm) ;y-axis is the maximum deviation of different length (mm)

"+" is the original value before training; "." is the correct value after training

Figure 5. Comparison of maximum deviations of different thickness before and after training on detecting the length of the diagonal line.

The measurement results have the following features:

1 Without error correction of neural network system, the directly measured value is not stable enough. The error is quite beyond the limit. It does not meet the accuracy requirement;
2 After correction with the method, the error meets the accuracy requirements and the effect is good;
3 The smaller the distance between the measured edge and center of the field of view is, the more accurate the measurement is.
4 The thinner the steel plate is and the poorer the rigidity is, the larger the vibration during movement on the roller way is, and the greater the measurement standard deviation is.

4 CONCLUSIONS

This paper uses high-pixel digital camera and image processing technology to realize an automatic online measurement of length, width and diagonal lines of the wide plate production line. This paper discussed the measurement errors caused by wide-field lens distortion, equipment installation error, out-of-level platform and other relative system errors are reduced based on the proposed neural network learning. It has been used in the production line of a large steel group and excellent effect has been obtained. With the development of digital cameras, this method can achieve higher precision and will have more applications.

ACKNOWLEDGEMENT

The authors gratefully acknowledge the support of the Fundamental Research Funds for the Central Universities (No. FRF-SD-12-008B). The authors are grateful to the anonymous referee for a careful checking of the details and for helpful comments that improved this paper.

REFERENCES

[1] Ge L, Du P. Gong Y.M. (2002) Non-contact continuous casting slab measurement and cutting control system. *Chinese Journal of Scientific Instrument*, 23(4):427–430.

[2] Qin G, He D, Shang H. (2009) Method and implementation of laser-aided steel width measure. *Journal of Hubei Automotive Industries Institute*, 23(2):52–55.

[3] Qin G, He D, Shang H. (2009) Implementation of a new steel width measurement system with the auxiliary of lazer. *Electronic Design Engineering*, 17(10):31–33.

[4] He D. (2006) Development and application of online width, length and thickness measurement system for steel plate. *Wide and Heavy Plate*, 12(4):26–30.

[5] He S, Zhong S, Wang X, et al. (1996) Hardware system for measuring length and width of steel-plates. Journal of Wuhan Technical University of Surveying and Mapping, 21(2):186–189.

[6] Li Y, Liu H, Wang S, et al. (2003) The application of slab width measurement with CCD. *Measurement and Control Technology*, 5:21–23.

[7] Jurkovic J, Korosec M, Kopac J. (2005) New approach in tool wear measuring technique using CCD vision system. *International Journal of Machine Tools and Manufacture*, 45(9): 1023–1030.

[8] J, Cohen P, Herniou M. (1992) Camera calibration with distortion models and ac-curacy evaluation. *IEEE Transactions on Pattern Analysis and Machine Intelligence*, 14(10):965–980.

[9] Salvi J, Armangué X, Batlle J. (2002) A comparative review of camera calibrating methods with accuracy evaluation. *Pattern Recognition*, 35(7):1617–1635.

[10] J, Silvén O. (1997) A four-step camera calibration procedure with implicit image correction. *Proceedings of IEEE Computer Society Conference on Computer Vision and Pattern Recognition*, pp:1106–1112.

[11] Z. (2000) A flexible new technique for camera calibration. *Pattern Analysis and Machine Intelligence, IEEE Transactions on*, 22(11):1330–1334.

[12] Remondino F, Fraser C. (2006) Digital camera calibration methods: considerations and comparisons. *International Archives of Photogrammetry, Remote Sensing and Spatial Information Sciences*, 36(5):266–272.

[13] Gao D. (1998) On structures of supervised linear basis function feed forward three-layered neural networks. *Chinese Journal of Computers*, 1:011.

Electric, Electronic and Control Engineering – Shao, Shu & Tian (Eds)
 ISBN: 978-1-138-02842-5

Framework study of accounting query and computing system under heterogeneous distributed environment

Yan Jia
Shandong Women's University, Jinan Shandong, China

ABSTRACT: The core of accounting computerization is electronic computer. To solve the timely sharing and updating of isolate information challenging the accounting query and computing system, the collaborative framework, system general framework and XML-based heterogeneous database synchronization system for accounting query and computing system are investigated taking electronic tax service system as the carrier and by referring to the system design base model under heterogeneous distributed environment; the Agent communication at the server and customer sides of the database synchronization system is also characterized with a view to providing theoretical guidance for further pushing forward China's accounting computerization.

KEYWORDS: Accounting computerization; Mobile Agent; XML technology; heterogeneous; query and computing system.

1 INTRODUCTION

Accounting computerization is a process that keeps cots and report accounts on computer instead of by hand, replaces some of the processing, analysis and judgment of accounting information that used to be completed by human brains [1]. As is known to all, the effective application of computer technology to information management of different areas greatly improves their working efficiency. However, Zhang Y., et al (2007) noted that, as it is impossible to share and timely update data among individual separate "information islands", regulation of this technology in areas where it is applied can be quite difficult. They also suggested that it is possible to effectively improve the data synchronization among heterogeneous databases under distributed network application environment [2]. Wu Q.S. (2008) made a framework study on the distributed accounting query and computing system based on mobile Agent and constructed an Agent-based, open, loose coupling, re-extendable accounting query and computing system that contributes to the construction of a systematic, intelligent accounting query and computing process [3]. Jin S.H., et al (2009) developed a new data synchronization method for hospital management system and validated in practice the effectiveness and extendability of the data synchronization system based on distributed heterogeneous database [4]. Yang H.G., et al (2013) made standardized abstraction of data query according to the demand of distributed heterogeneous data source standard query system and realized the sharing of distributed heterogeneous data sources by shielding differences [5]. Shen Y., et al (2014) designed an XML-based heterogeneous database synchronization system and realized data synchronization among distributed heterogeneous databases by means of Web interfaces [6]. Qiu L.J. (2014) analyzed the functional modules needed for designing corporate computerization systems and made intensive studies on the construction and realization of China's corporate computerization system by reviewing related literature [7]. All these findings provide good guidance for our study on the accounting query and computing framework under heterogeneous distributed environment here.

After reviewing previous findings, this paper looks into the accounting query and computing system under heterogeneous distributed environment in the light of the characteristics of the base model from the perspective of the accounting query and computing system demand.

2 DEMAND FOR ACCOUNTING QUERY AND COMPUTING SYSTEM

Social, economic and technological development brings along new demand for the accounting, auditing, inventory and taxation management of the accounting system. The current accounting management system, however, is limited by the monotonous functions, low integration and weak security of the computerized system, which will inevitably become even more prominent as global economic integration takes shape and the Internet-cored network technology further develops. As such, only by obtaining support from network accounting software that incorporates network technology and the state-of-the-art management philosophy can we remedy our limitations and meet the demand of the accounting system. It is important that the accounting query and computing system include features like remote operation, online payment, collaboration with the business departments and remote information exchange. The integrated application of accounting software with electronic tax service system has already become an inevitable development trend for the accounting query and computing system.

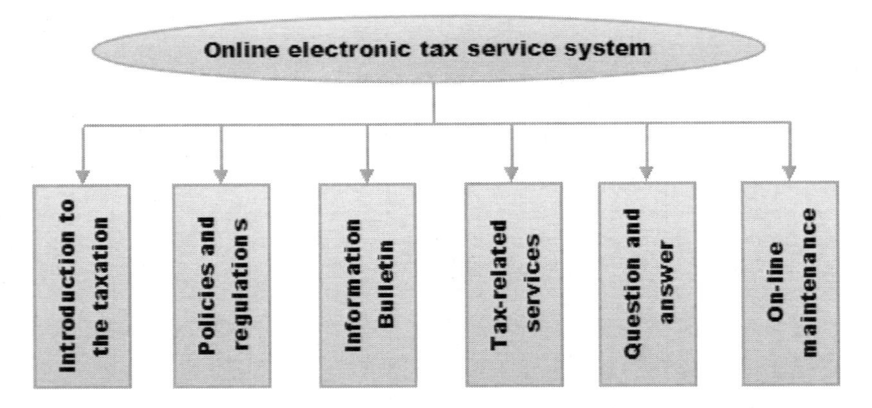

Figure 1. Functional structure of the service system software based on online electronic tax service.

According to the principal objective of our study, the demand of the accounting query and computing system, the accounting software model for electronic tax services to be reconstructed under heterogeneous distributed environment has to provide five features: openness (provides external interfaces for data exchange), support for online tax payment (provides online payment interfaces for smooth interconnection between traditional bookkeeping vouchers and electronic bills of the accounting software), support for electronic bills (supports standard data exchange message structure), realization of remote processing (able to perform digital signature for electronic bills on the Internet and realizes layer-to-layer auditing and signature of accounting vouchers) and security (allows safe and reliable information exchange between external sources and the organization and safeguards the safe operation of this exchange).

3 BASE MODEL FOR ACCOUNTING QUERY AND COMPUTING SOFTWARE

By operating characteristic tendency, the functions of accounting query and computing system software include inward and outward functions. Here, inward refers to those targeted at minor systems, including the software itself, and outward refers to those targeted at banks or tax authorities. Both are realized by processing electronic vouchers. To enable the system software to realize these functions, it is necessary to access to Internet to construct a Web-based application which is a three-layer network structure [7]. A processing-type firewall is incorporated to isolate Intranet from Extranet to perform safety control of the network. The resulting isolation allows the application server of the Intranet software system to respond to the requests of the external function realizations. Fig.2 shows the network structure of accounting query and computing system software.

As shown by the B/W/D three-layer structure in Fig. 3, our accounting query and computing system software uses the Web-based three-layer system structure whose merits lie in that it effectively follows the simplification principle of the system, reflects the number of client browsers and also serves as a data client. Not only can it be used to receive a service request, the position transmitter of which is the Web-based client browser's UR, it can also transform the service request information into SQL statements, transmit them to the database server and pass the result from the database service to the client in HTML texts [9].

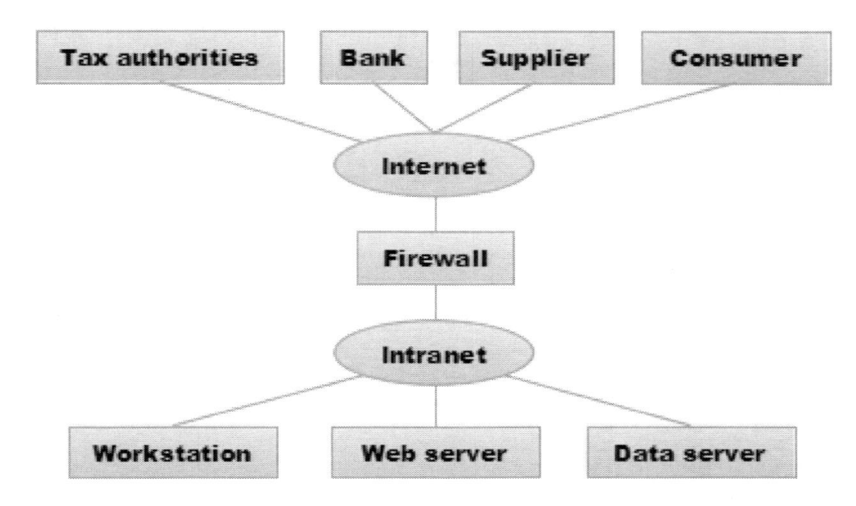

Figure 2. Network structure of accounting query and computing system software.

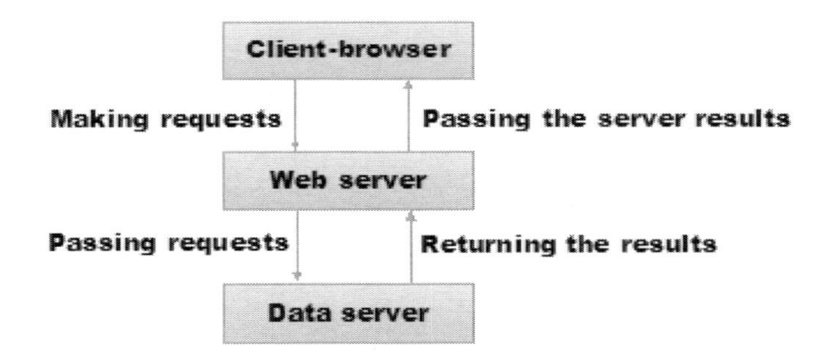

Figure 3. B/W/D structure of accounting query and computing system software.

4 FRAMEWORK STUDY OF ACCOUNTING QUERY AND COMPUTING SYSTEM

The distributed calculation of heterogeneous databases is normally realized by one of the three patterns below:

1. As there are different nodes in a heterogeneous database, each of which corresponds to a different sub-database, to establish interaction among different sub-databases, it is necessary to establish user interaction interfaces integrated from different patterns [10] to allow separate management of global data in different regions.
2. Manage physically distributed, logically related databases in the form of distributed database system structure [3].
3. Provide a guarantee for data consistency by modifying the original mapping relation among original databases [3].

However, none of these patterns is able to ultimately solve the "information island" inherent to distributed heterogeneous data [11].

Fig.4 shows the structural framework we built after analyzing the demand and examining the base model of an accounting query and computing system. Fig. 5 shows the general framework of our accounting query and computing system.

Fig. 6 shows the hierarchical structure of the Mobile Agent platform composition. Here, "region" is the Agent platform server that manages the entire distributed Agent environment.

In this paper, the application program layer refers to the XML-based heterogeneous database synchronization system that consists of a client and a server. The former is responsible for capturing varying data, transforming data types, writing in data documents, compressing and encoding data documents and transmitting data. The latter is responsible for decoding,

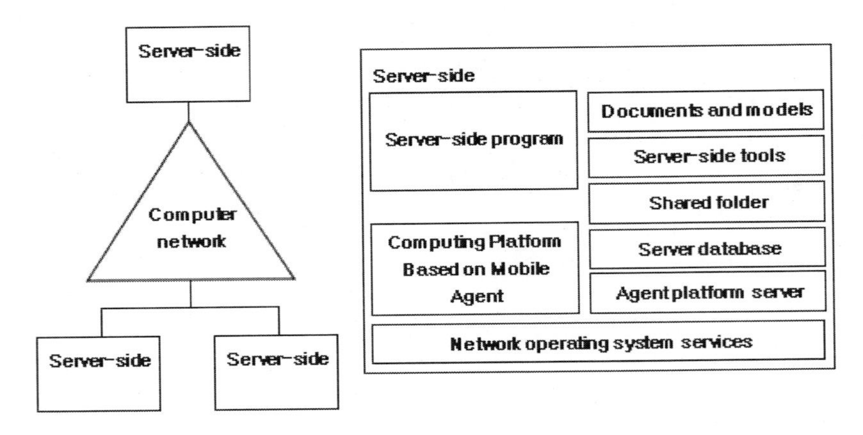

Figure 4. Structural framework of the collaborative design system.

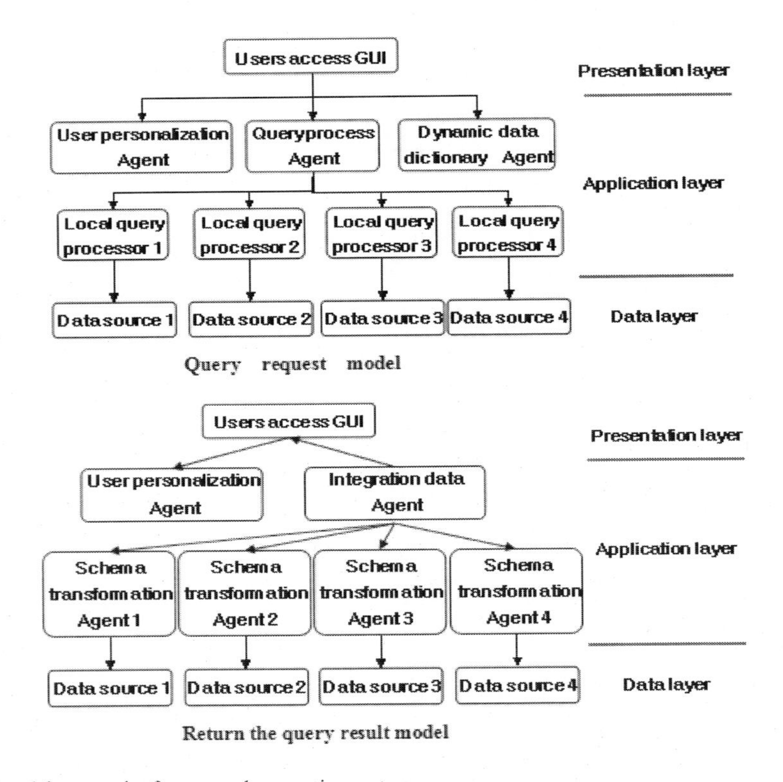

Figure 5. General framework of query and computing system.

decompressing, analyzing and storing data documents. To supply the attribute gap among heterogeneous databases generally resulted from field names, data types and data structures [13], it is necessary to use store synchronous data having the same data type and data structure with the same semantics using different data names. To minimize data transmission, after capturing varying data and generating XML heterogeneity-shaded documents, the client will compress the data, encode the data document with an encoding algorithm agreed with the server, and then transmit the encoded data document to the server in the form of FTP according to the server FTP information defined by the synchronization rule.

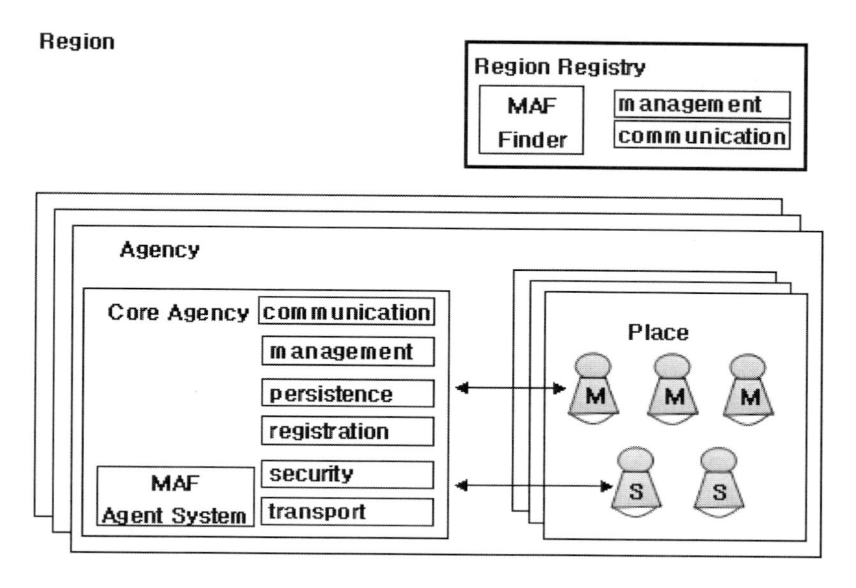

Figure 6. Hierarchical structure of Mobile Agent platform composition.

Essentially, the communication knowledge between Agents, KQML (Knowledge Query and Manipulation Language) and the transformation and execution processes of XML are as shown by the communication process in Agent in Fig. 7. Fig. 6 describes how an Agent receives a KQML communication message from the communication interface, encapsulates it into an XML document and transmits this instruction to the XML transformer. The KQML executing module analyzes and executes the KQML request and returns KQML encapsulation result realized. After that, the message is encapsulated in a standard

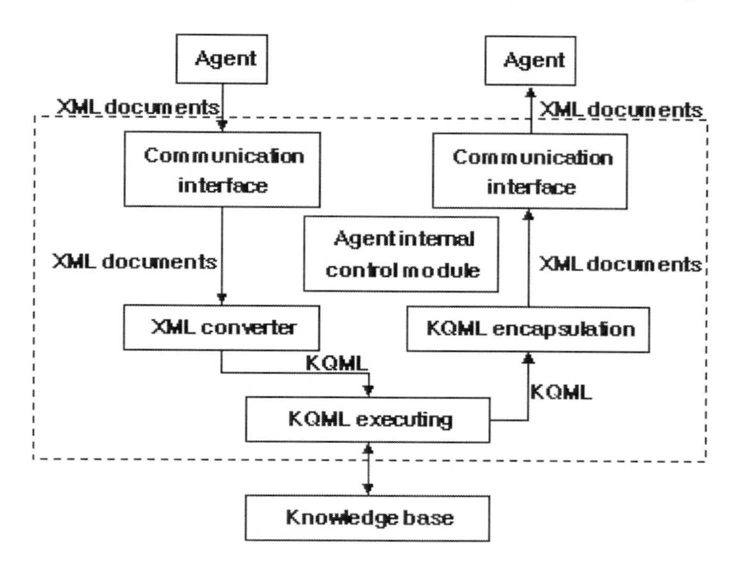

Figure 7. Communication process of Agent.

XML document according to the specific DTD and transmitted to the communication interface before it is eventually transmitted in the form of XML to the Agent responsible for receiving messages. The Agent communication service control center is responsible for directing and coordinating all parts throughout the communication process.

5 CONCLUSIONS

After analyzing the demand of an accounting query and computing system, the framework of XML and Mobile Agent-based accounting query and computing system under heterogeneous distributed environment is investigated in the light of the system software network structural diagram and Web-based B/W/D three-layer structure diagram. The collaborative design system, the general framework structure and Mobile Agent communication process are emphatically analyzed with a view to providing a better solution to the "information island" problem.

ACKNOWLEDGEMENT

The paper is supported by the Key Project of Enterprise Training and Employee Education in Shandong Province (Project No: 2014-065).

REFERENCES

[1] Chen, X.L. (2007) Strengthening accounting computerization of salt system. *CO-Operative Economy & Science*. 32(11):121–123.

[2] Zhang, Y., Xia, K.J., Zhang, F.M., Jiang, C.X. (2007) Research and realization of distributed heterogeneous database data synchronization system. *Mini-Micro Systems*, 28(10):1803–1806.

[3] Wu, Q.S., Zhao, J.C. (2008) *Framework Study on Distributed Accounting Query and Computing System Based on Mobile Agent*. Zhejiang: Journal of Zhejiang University.

[4] Jin, S.H., et al. (2009) Research of data synchronization system based on distributed heterogeneous database. *Journal of Yunnan Nationalities University*, 18(2):162–164.

[5] Yang, H.G. (2013) Design and realization of distributed heterogeneous data source standard query. *Journal of Information Technology in Civil Engineering and Architecture*, 5(4):135–138.

[6] Shen, Y. (2014) Implementation and optimization of distributed heterogeneous database data synchronization system. *Journal of Guilin University of Electronics Technology*, 34(4):253–255.

[7] Qiu, L.J. (2014) Construction and realization of China's accounting computerization system. *Financial Service Sector (Academic Edition)*. (7):95–97.

[8] Ao, H.W., Liu, R. (2010) Architecture based on J2EE design of online tax service office. *Ningxia Engineering Technology*, 9(03):226–228.

[9] Zeng, J.X., Tan, J.J. Zeng, J.X., Tan, J.J. Zeng, J.X., Tan, J.J. (2002) Development of Three-layer C/S Structure and Books Search System. *Information Research*. (2):41–44.

[10] Qiu, H.Y. (2002) Research and realization of B/W/D network computing model for hospital information system [J]. *Chinese Journal of Medical Device*, 15(2):16–17.

[11] Ying, M. (2010) Application of distributed database in hospital management system. *Computer CD Software and Application*. (6):141.

[12] Li, P. (2008) Research on data synchronization system based on distributed heterogeneous database. *Study of Science and Engineering at RTVU*. (1):63–65.

[13] Chen, H. (2009) A study of mobile agent technology in CSCD based on P2P platform. *CD Technology*. (4):41.

[14] Wang, G.W., Chen, S. (2012) Research on heterogeneous database extraction and conversion based on XML. *Modern Computer*. (18):37–39.

Electric, Electronic and Control Engineering – Shao, Shu & Tian (Eds)

LABVIEW-based simulation training system of Chinese medicine bone-setting manipulation

Haiyan Mo, Jing Liu, Hui Cao & Chen Ni
School of Science and Engineering, Shandong University of Traditional Chinese Medicine, Jinan Shandong, China

Junzhong Zhang
Affiliated Hospital of Shandong University of Traditional Chinese Medicine, Jinan Shandong, China

Xiaorui Song
Taishan Medical University, Taian Shandong, China

ABSTRACT: To quantify Chinese medicine bone-setting manipulation accurately and inherit effectively, we built simulation training system based on LABVIEW platform. Using a graphical programming language, sensors, data acquisition card and Kinect to show angle changes and target trajectory visually in the process of manual reduction, which can increase the manipulation visibility and promote the application of virtual reality in the way of traditional Chinese medicine

KEYWORDS: LABVIEW, sensor, Kinect, manipulation, virtual simulation.

1 INTRODUCTION

Virtual reality (VR) technology was to create a simulation of real-world effect of the special circumstances in the computer [1]. We could use the technology to provide users with visual, auditory, tactile and other real-time sensing means, which could produce an immerse experience by varieties of sensor devices. Nowadays, VR was applied in surgical simulation and rehabilitation [2-4], while rarely in the field of Chinese Medicine Bone-setting Manipulation (hereinafter called "CMBM").

CMBM is a featured subject in Traditional Chinese Medicine (TCM) academic system and one of the chief means in treating orthopedic diseases [5]. However, due to the lack of scientific selection criteria and standard operating basis, which greatly influenced the heritage and CMBM development. We were committed to apply VR, Multi-sensor Information Fusion (MSIF), signal processing and other modern technology to CMBM research. The approach aimed to build a virtual simulation training system to obtain data in the process of bone-setting and then come to standard operation basis.

The system includes acquisition target trajectory and virtual skeleton simulation two modules. Fig.1 shows a flow chart of the entire CMBM system. In order to obtain the movement trajectory of fracture point, we designed one upper limb mechanical model, and collected information by installing multiple sensors on it. Then we inputted the collected information to PC and analyzed the data in the LABVIEW platform. In the virtual skeleton simulation module, using the Kinect sensor to obtain grayscale images of human upper limb bones, and then designing algorithms in LABVIEW to get related data during manual action and display in the form of waveform graph.

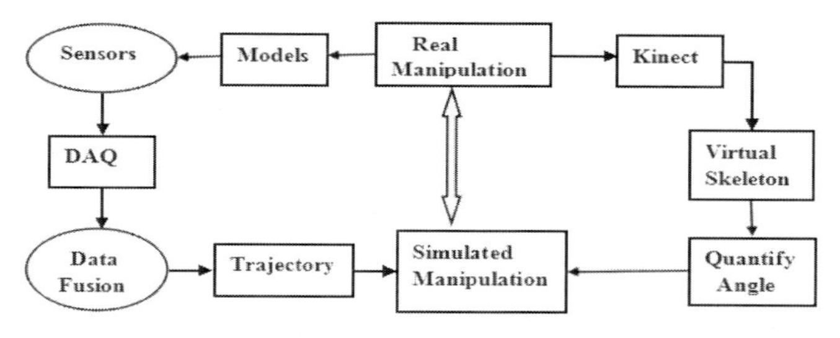

Figure 1. Flow chart of the entire CMBM system.

2 SYSTEM HARDWARE

As illustrated in Fig. 1, the major components of our LABVIEW-based simulation training system are as follow: Windows-based PC, mechanical model, sensors, Data Acquisition card (DAQ) and Kinect. A brief description of the hardware is given as follows:

A. PC

As a user interface of our work, we used Windows based PC to accomplish all data processing.

B. Mechanical model

Taking the upper limb as an example, we made the mechanical model in series connecting rod form, whose structure was shown in Fig. 2, including 6 joints (NO.1-6) , broken bone model operating end (NO.7), broken bone model fixed end (NO.8), model base (NO.9) and sensor fixing device. To make sure that the model completes various manipulation types, the degree of model freedom was set to six.

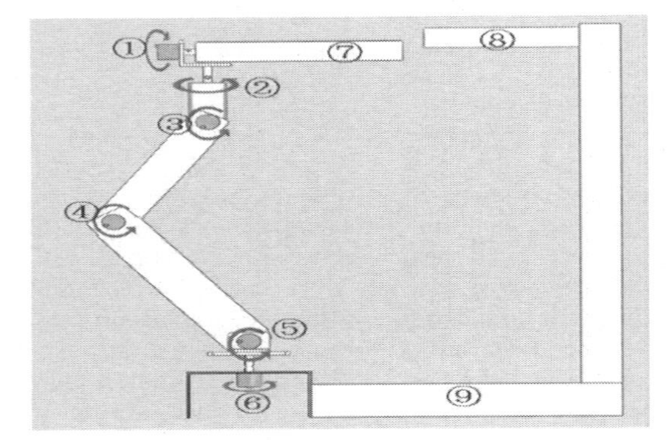

Figure 2. Mechanical model schematic diagram.

C. Sensors

Sensors are data collection source of the system. Only by getting magnitude and direction of manipulation force, mobile displacement and upper limb model trajectory and other physical quantities, can we simulate manipulation better. We use a multi-sensor fusion technique to ensure the data accuracy. The system uses WDG-AM23-360 as angle sensors, whose parameters were shown in Table 1. The relative value of each joint angle changes combined model length could calculate real-time location model to simulate the trajectory.

Table1. Sensors parameters.

Model	Working Voltage	Range	Output Voltage	Resolution	Starting Torque
WDG-AM23-360	5V	0~360°	0.5~4.5V	12bit	≤0.09N

D. DAQ

DAQ is used for data acquisition, A/D conversion and data storage. We used USB-6008 DAQ from NI Company in this system. which is a multipurpose DAQ based on USB interface.Its parameters are as follows:12bit, Sampling rate 10ks/s,8 analog input channels ,2 analog output channels,12digital I/O channels and 1counter.

E. Kinect

Kinect is a gesture sensing input device produced by Microsoft Corp, which is XBOX external 3D somatosensory camera [6-7]. Fig. 3 illustrates that it contains an RGB color camera, one pair of 3D depth sensors (infrared camera left and CMOS camera right), a multi-array mic and a rotating motor. Kinect could trace 20 skeletal points as shown in Fig. 4, to get the relative position and angle between them by locating each point, and to identify human body posture [8-11]. We adopted hand let, hand right and spine bone points to simulate proximal and distal fracture model, taking spine to fix fracture point. As we all know, doctors' hands played a key role in CMBM, for which we only selected 10 skeletal points to simulate their upper body. Points were marked in Fig.4.

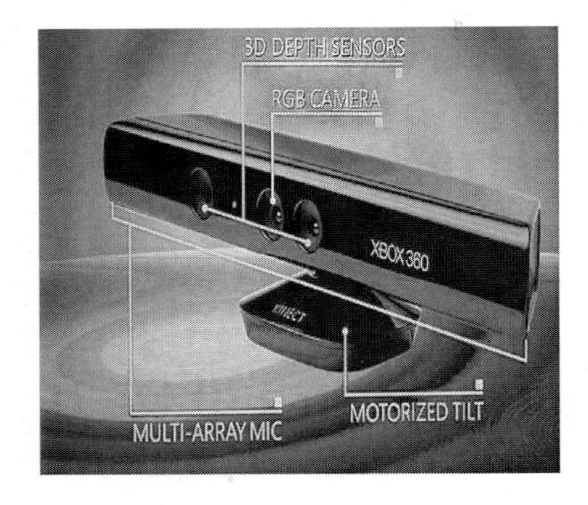

Figure 3. Kinect.

3 MAIN FUNCTION

3.1 *Data processing*

The design idea of this part was using DAQ to read data from specified channels to send to PC after preparation. Then we have the real-time waveform displayed, data analyzed and data stored by calling signal processing sub-template, writing procedure and designing front panel based on LABVIEW. This part included the following modules.

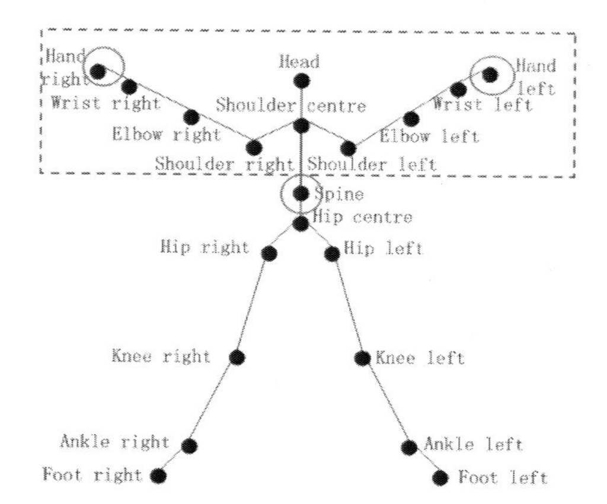

Figure 4. Skeletal image.

1 Collection.
Several parameters needed to be set by NI-DAQ mx before data collection, such as sampling frequency, points, channels, voltage limits and so on. The system used a 6-channel, continuous acquisition mode, the frequency of 500Hz, the input voltage range of 0-5V.

2 Analysis.
The analysis object was the collected data by multiple sensors from the mechanical model. As we used several sensors, we needed to combine sensors data by certain criteria to generate consistent environmental description, called multi-sensor fusion technology. The main task of data fusion was to estimate target state. Kalman filter, α-β filter and their related extension were widely used for fusion. Since the mechanical model and motion model were non-linear and sensors' data were unstable, we chose the Extended Kalman Filter (EKF) algorithm to estimate the target location [12-13]. It could obtain the current state of the system and forecast the future state, in this way, could we get target positioning accuracy and trajectory during manual reduction.

3 Display.
When CMBM was applied on the mechanical model, data collected by 6 sensors would display on the front panel in waveform. Fig.5 showed the waveform of bone-setting reduction operated by a doctor.

4 Storage and Playback.
Data storage was to save acquired waveform for analysis and playback calls, and data playback was to reread and display the stored data for analysis.

3.2 *Trajectory simulation*

Fig.6 illustrates that A was the distal fracture point, whose real trajectory in correction process was shown

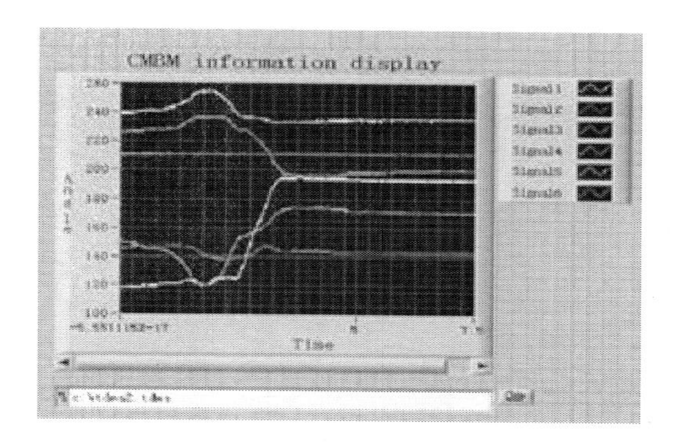

Figure 5. Bone-setting reduction waveform.

in Fig.7. Ideal trajectory was free from any interference, and actual trajectory was true motion track by outside interference. We aimed to get the trajectory as close to the actual movement as possible. Point A was established in three dimensional space coordinate system, and its equations of motion spatial coordinates were as follows:

$X_k = R(k)\sin\beta\cos\theta$
$Y_k = R(k)\cos\beta$
$Z_k = R(k)\ \sin\beta\sin\theta$

where R(k) represented the distance from observers to target position, θ/β represented horizontal/ vertical angle.

Assuming the spatial position of A corresponding to the measured values from 6 angle sensors, we used multiple linear regression to seek a linear relationship between the spatial position and measured voltage to obtain regression equation($p<0.05$). Values of test statistics R, F and P verified validity of the linear regression model. Meanwhile, we used EKF mentioned above to fuse sensors information, to estimate moving target, and ultimately realized trajectory simulation.

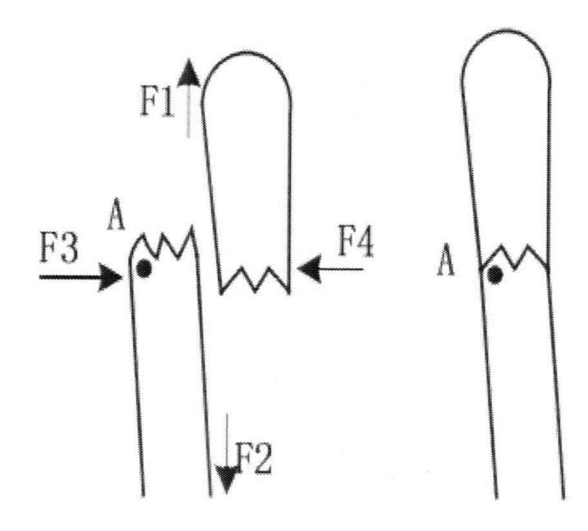

Figure 6. Manual therapy.

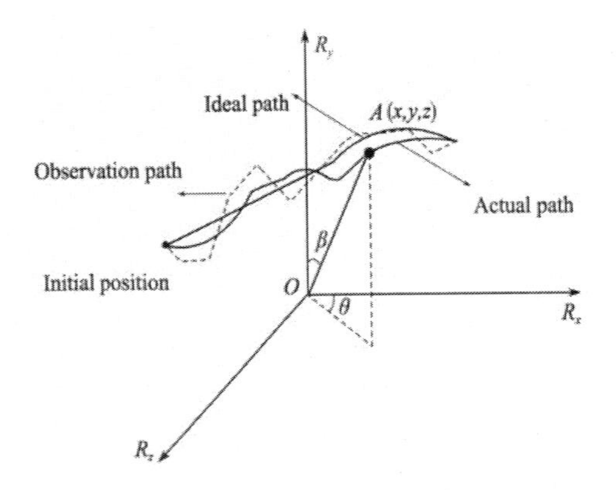

Figure 7. Space coordinates of moving target.

In order to verify the feasibility of EKF, we conducted a simulation trajectory state estimation and error analysis. Results in Fig. 8, 9, 10 respectively showed target trajectory observation data before and after EKF filtering in X-axis, Y-axis, Z-axis was convergent, and fused data is closer to real target movement.

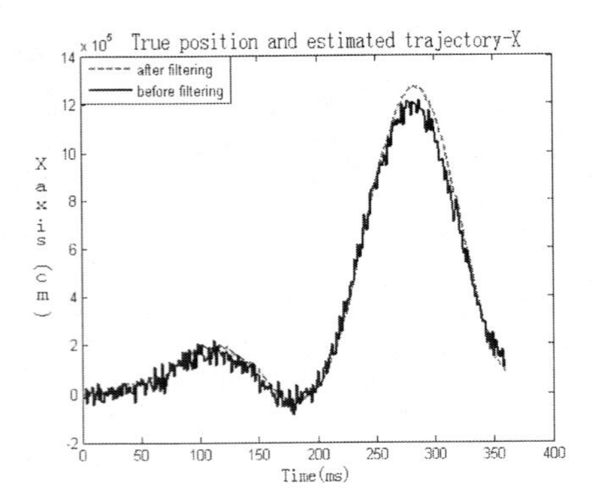

Figure 8. Target trajectory in X-axis.

3.3 *Bone simulation*

As Fig.4 showed, human skeleton image traced by Kinect was composed of 20 points and 19 bones. We used Kinect Configure VI, Kinect Initialise VI, Kinect Close VI and Kinect Read VI to achieve human grayscale images and skeleton images. And we chose sphere to simulate skeletal points, and cylinder to bones. The fracture model was simulated by hand left, hand right and spine 3 bone points, and the doctor's upper body model was simulated by 10 bone points, which were head, shoulder center, shoulder

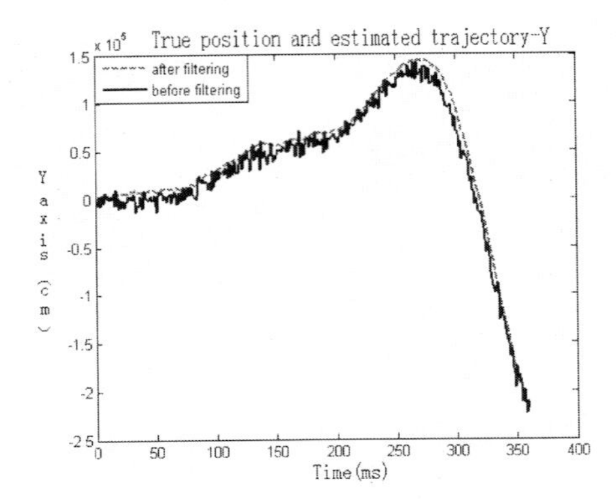

Figure 9. Target trajectory in Y-axis.

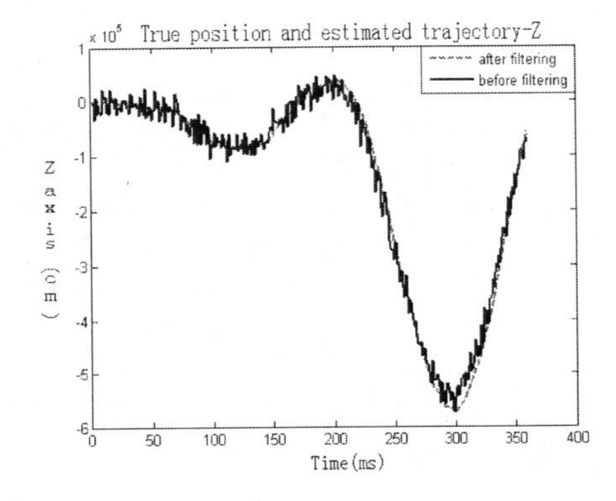

Figure 10. Target trajectory in Z-axis.

left, shoulder right, elbow left, elbow right, wrist left, wrist right, hand left, hand right. Programming in LABVIEW we obtained proximal and distal fracture model (Fig.11) and a doctor's upper body model (see Fig.12). While the subjects of CMBM were doctors and patients, the two models should be combined together (see Fig.13).

To quantify operation parameters, we researched rotatory manipulation with wrist rotation angle as parameters [14]. Joint Coordinates VI in Kinesthesia kit helped us to get three-dimensional coordinate of bone points. Then angle θ1 between the hands was calculated by hand left and hand right two coordinate, and right wrist rotation angle θ2 was calculated by hand right, elbow right and wrist right three coordinate, and left wrist rotation angle θ3 was got in the similar way with θ2. Through real-time curve change, we could observe the fluctuations range before and after a certain time intuitively, so that learners could grasp manipulation more accurately.

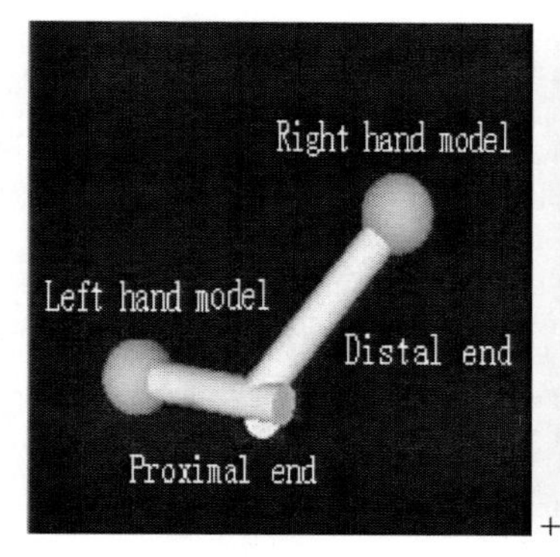

Figure 11. Fracture model.

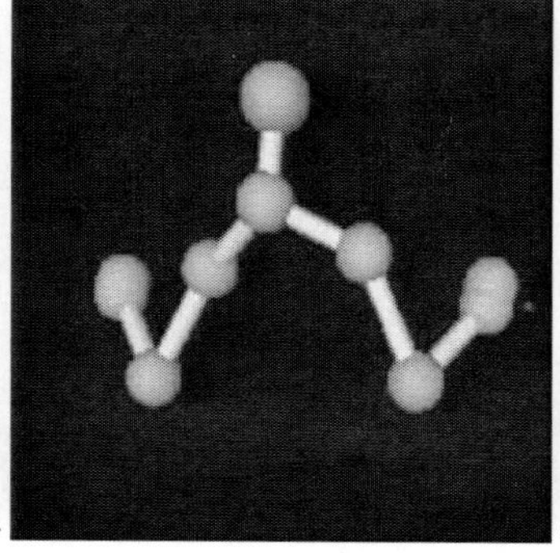

Figure 12. Upper body model.

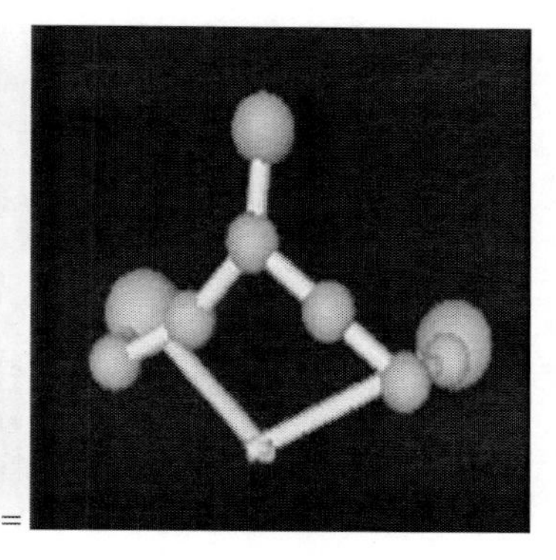

Figure 13. Combined model.

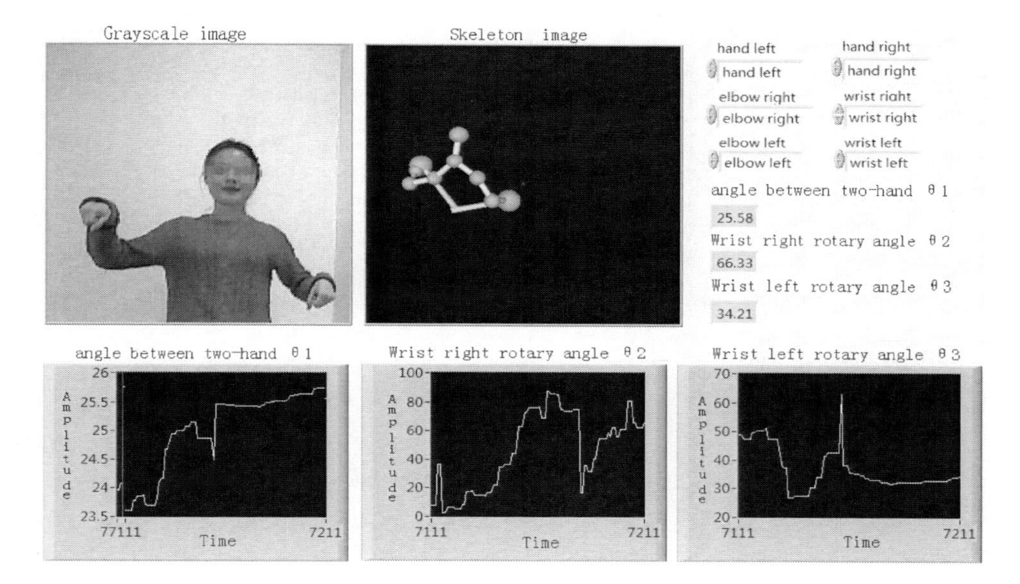

Figure 14. Angle θ1, θ2, θ3.

4 CONCLUSION

We build the simulation training system of Chinese medicine bone-setting manipulation based on LABVIEW by combining sensors, DAQ, EKF and Kinect together, which provides a convenient platform for learners. It contributes to the manual reduction development by means of changing the traditional teaching model. However, there are still many issues waiting for settlement. For instance, a more realistic virtual model and force and audio feedback applications are all problems coming in the next to be solved.

ACKNOWLEDGEMENTS

This work was financially supported by the National Natural Science Foundation of China (NO. 81373661), National Natural Science Foundation of China (NO. 81473708).

REFERENCES

[1] Yang C. (2006) Olfactory display: development and application in virtual reality therapy. *Proceedings of the 16th International Conference on Artificial Reality and Telexistence Workshops, IEEE*, pp:580–584.

[2] Sangwoo Cho, Jeonghun Ku, Yun Kyung Cho, *et al.* (2014) Development of virtual reality proprioceptive rehabilitation system for stroke patients. *Computer Methods and Programs in Biomedicine*, 2(3):258–265.
[3] Caglio M, Latini-Corazzini L, D'Agata F, *et al.* (2012) Virtual navigation for memory rehabilitation in a traumatic brain injured patient. *Neurocase*, 18: 123–131.
[4] Asit Arora, Chloe Swords, Sam Khemani, et al. (2014) Virtual reality case-specific rehearsal in temporal bone surgery: A preliminary evaluation. *International Journal Of Surgery*, 12:141–145.
[5] Liheng Wang, Chaohui Chen. (2009) The application of the principles and practices orthopedic applications syndrome. *Chinese Journal of Basic Medicine in Traditional Chinese Medicine*, 15(4):310–311.
[6] Enrique Ramos Meglar. (2012) *Arduino and Kinect Projects*. American: Apress, pp:23–34.
[7] Chang Y J, Chen S F, Huang J D. (2011) A Kinect-based system for physical rehabilitation: A pilot study for young adults with motor disabilities. *Res. Dev. Disabil*, 32(6):2566–2570.
[8] Biswas K K, Basu Saurav Kumar. (2011) Gesture recognition using Microsoft Kinect. *Proceedings of the 5th International Conference on Automation, Robotics and Applications: IEEE, 2011*, pp:100–103.
[9] Bo Antonio Padilha Lanari, Hayashibe Mitsuhiro, Poignet Philippe. (2011) Joint angle estimation in rehabilitation with inertial sensors and its integration with Kinect. *33rd International Conference Of The Ieee Engineering In Medicine And Biology Society: IEEE, 2011*, pp:3479–3483.
[10] Kondori Farid Abedan, Yousefi Shahrouz, Li Haibo, *et al.* (2011) 3D head pose estimation using the Kinect. *International Conference on Wireless Communications and Signal Processing: IEEE, 2011*, pp:1–4.
[11] Alexiadis Dimitrios, Daras Petros, Kelly Philip, *et al.* (2011) Evaluating a dancer's performance using Kinect-based skeleton tracking. *Proceedings of the 2011 ACM Multimedia Conference and Co-located Workshops, 2011*, pp:659–662.
[12] Shademan A, Janabi-Sharifi F. (2005) Sensitivity analysis of EKF and iterated EKF pose estimation for position-based visual servoing. *Canada: Proceeding Of 2005 IEEE CCA on Control Applications*, pp:755–760.
[13] Wang Xiaoyu, Yan Jihong, Qin Yong, Zhao Jie. (2007) Attitude estimation based on extended Kalman filter for a two-wheeled robot. *Journal of Harbin Institute of Technology*, 8(5):1920–1924.
[14] Lin Haibo, Mei Weilin, Zhang Yi, *et al.* (2013) Design and implentation of robot arm somatosensory interaction system based on kinect skeletal information. *Computer Application and Software*, 30(2):157–160.

Electric, Electronic and Control Engineering – Shao, Shu & Tian (Eds)
 ISBN: 978-1-138-02842-5

Object detection based on Gaussian mixture background modeling

Jing Liu
College of Technology, Shandong University of Traditional Chinese Medicine, Jinan Shandong, China

Dawei Qiu
College of Technology, Shandong University of Traditional Chinese Medicine, Jinan Shandong, China
College of Computer Science and Technology, Nanjing University of Aeronautics and Astronautics, Nanjing Jiangsu, China

Hui Cao
College of Technology, Shandong University of Traditional Chinese Medicine, Jinan Shandong, China

Junzhong Zhang
Department of Orthopedics, Shandong Hospital of Traditional Chinese Medicine, Jinan Shandong, China

Haiyan Mo
College of Technology, Shandong University of Traditional Chinese Medicine, Jinan Shandong, China

ABSTRACT: For background modeling, the conventional Gaussian Mixture Model (GMM) is a popular approach. However, because of the inappropriate parameters updating method, GMM often suffers from a problem that it cannot classify a pixel into background or foreground correctly for longtime. In the paper, we proposed a new parameters updating method for GMM, and built background model for every pixel and global foreground model for the entire image. We presented an improved object detecting and tracking scheme based on the proposed approach. The experimental results show the proposed GMM parameters updating method, together with the object detection and tracking framework, give better performance than the conventional Gaussian Mixture Modeling algorithm.

KEYWORDS: object detection, Gaussian Mixture Model, maximum likelihood.

1 INTRODUCTION

The detection of regions of interest is the important step in many computer vision applications, such as visual surveillance, event detection, image understanding etc. A general object detection technique is desired, but it is very difficult to detect objects even unknown objects under complex environments. So many computer vision applications assume stable environments or few changes in scenes, in which the background model is trained and foreground regions are obtained via the difference between current frame and the background model, the task of object detection implementation is straightforward. The process is called background subtraction.

In recent years, many researchers have proposed various background modeling and subtraction algorithms. Background modeling methods consists of pixels-based and blocks-based (Yongzhong Wang *et al.*, 2009). Background modeling based on pixels builds models according to every pixel's distribution information in time fields. In (C. R. Wren *et al.*, 1997) each pixel is modeled by a Gaussian model. From that, Gaussian models are more and more used in background subtraction. Friedman et al. N. Friedman and S. Russell (N. Friedman & S. Russell. 1997) proposed Mixture of Gaussians in traffic surveillance, and each pixel consists of three Gaussian components corresponding to car, road and shadow respectively. Stauffer et al. C. Stauffer (C. Stauffer *et al.*, 2000) proposed more general Gaussian Mixture Model (GMM) method. The adaptive background subtraction approach models each pixel as a mixture of Gaussians and uses on-line approximation to update the model. Elgammal et al. (A. Elgammal et al., 2002) proposed a nonparametric background modeling method, which utilizes general nonparametric kernel density estimation techniques for building statistical representations of the foreground and the background

*Corresponding author: *liujing@mail.sdu.edu.cn; dwqiu@foxmail.com*

without any assumptions about the underlying distributions. Y. Ren et al. (Y. Ren *et al.*, 2003) proposed a new background subtraction approach, which can deal with non-stationary scenes. The method classified a pixel in the current frame into foreground or background based on the temporally and spatially modeled distribution of each background pixel. After that, object detection with nonstationary background is referred to by more and more researchers. In (Xiangfeng Bai et al., 2011), the authors proposed a new Gaussian matching criteria and reduced the calculation burden.

Background modeling based on blocks segments the frame into blocks, background model is then built using features of each block. Compared with pixel-based method, block-based method can utilize more spatial distribution information and is insensitive to local changes in the scene. Matsuyama et al. T. Matsuyama et al. (T. Matsuyama *et al.*, 2000) used normalized vector distance measure correlation between different image blocks. In (M. Heikkila & M. Pietikainen. 2006), LBP (Local Binary Pattern) histogram in each pixel's neighborhood is used to represent the background model. The drawback of these algorithms is that the obtained moving objects regions are coarse and are not suitable for precise object segmentation. In (Y. Chen *et al.*, 2007), the authors proposed an efficient hierarchical method for background subtraction, the method combines pixel-based and block-based approaches into a single framework, and these two approaches are complementary to each other.

In this paper, we propose a new object detection algorithm based on an improved Gaussian Mixture Model. The algorithm uses Maximum Likelihood (ML) approach to update the GMM parameters and builds background and foreground models with GMM respectively. The new pixel is classified as background or foreground according to these two models. Experimental results show that the proposed algorithm can improve the precision and performance than the conventional GMM approach.

The rest of the paper is organized as follows. Section 2 describes the conventional GMM algorithm, the proposed algorithm is detailed in Section 3, and the experimental results are presented in Section 4. Finally, brief conclusions are given in Section 5.

2 THE CONVENTIONAL GAUSSIAN MIXTURE MODEL

For object detection in an image sequence, every pixel is viewed as K independent Gaussian mixture model, for a new observation $X_{i,t}$ at position i at time t, $X_{i,t}$ can be (R, G, B) vectors or intensity values, the probability can be modeled as K GMM:

$$P(X_{i,t}) = \sum_{k=1}^{K} \gamma_{k,t} N(X_{i,t} \mid \mu_{k,t}, \Sigma_{k,t}). \tag{1}$$

K is the number of Gaussian components (In general, K is 3 or 5). $\gamma_{k,t}$ is the weight of the kth Gaussian component at time t, and $N(X_{i,t} \mid \mu_{k,t}, \Sigma_{k,t})$ is Gaussian function:

$$N(X_{i,t} \mid \mu_{k,t}, \Sigma_{k,t}) = \frac{1}{(2\pi)^{d/2} \mid \Sigma_{k,t} \mid^{1/2}} \cdot \exp\left(-\frac{1}{2}(X_{i,t} - \mu_{k,t})^T \Sigma_{k,t}^{-1}(X_{i,t} - \mu_{k,t}) \right) \tag{2}$$

And for simplicity,

$$\Sigma_{k,t} = \sigma^2 \cdot I \tag{3}$$

The algorithm selects the biggest B Gaussian components as the background model according to γ_k / σ_k.

$$B = \arg\min_b \left(\sum_{k=1}^{b} \gamma_{k,t} > T \right) \tag{4}$$

T is the weight threshold.

For a new image, the algorithm compares every observation $X_{i,t+1}$ with the K Gaussian component, if $| X_{i,t+1} - \mu_{k,t} | \le 2.5\sigma_{k,t}$, then the pixel is classified as background and set $M_{k,t+1} = 1$, else it belongs to foreground and set $M_{k,t+1} = 0$. If the pixel matches with a Gaussian component, the parameters of the model are updated, the weights of the other $K-1$ components are altered, and means and variances are unchanged.

$$\gamma_{k,t+1} = (1-\alpha) \cdot \gamma_{k,t} + \alpha \cdot M_{k,t+1} \tag{5}$$

$$\mu_{k,t+1} = (1-\rho) \cdot \mu_{k,t} + \rho \cdot X_{t+1} \tag{6}$$

$$\sigma_{k,t+1}^2 = (1-\rho) \cdot \sigma_{k,t}^2 + \rho \cdot (X_{t+1} - \mu_{k,t+1})^T (X_{t+1} - \mu_{k,t+1}) \tag{7}$$

$$\rho = \alpha \cdot N(X_{i,t+1} \mid \mu_{k,t}, \Sigma_{k,t}) \tag{8}$$

Gaussian Mixture Model can adapt slow change of the scene, and depict periodic motion, so it is regarded as one of the best background modeling method. But most of the GMM methods adopt fixed or changed learning rate, these learning rates can not reflect the real background variation.

3 PROPOSED OBJECT DETECTION ALGORITHM

In this paper, we propose a new object detection algorithm based on an improved Gaussian Mixture Model. The algorithm uses Maximum Likelihood (ML) approach to update the parameters of the Gaussian Mixture Models. Then every pixel is classified as background or foreground, the algorithm maintain background and foreground model, respectively. The flow chart of the entire algorithm is shown in Figure 1:

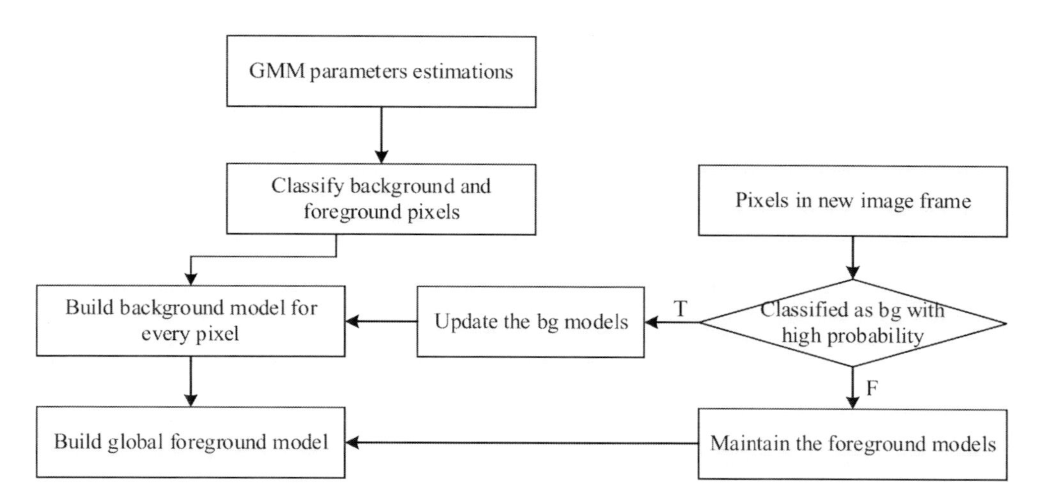

Figure 1. The flow chart of the proposed algorithm.

For every pixel, the observation sequence is $X_1, X_2, ..., X_t$, the maximum likelihood function of GMM is:

$$\sum_{j=1}^{t} \log \left\{ \sum_{k=1}^{K} \gamma_k N(X_{j,t} \mid \mu_k, \Sigma_k) \right\} \tag{9}$$

The above equation is hard to optimize, so we adopt the following two steps:

Step 1. Estimate the probability of the pixel belongs to each Gaussian component, it can be computed as:

$$\gamma_k = \frac{\pi_k N(X_t \mid \mu_k, \Sigma_k)}{\sum_{j=1}^{K} \pi_j N(X_t \mid \mu_j, \Sigma_j)} \tag{10}$$

Step 2. Estimate the parameters of each Gaussian component:

$$\mu_k = \frac{1}{N_k} \sum_{j=1}^{t} \gamma_k X_j \tag{11}$$

$$\Sigma_k = \frac{1}{N_k} \sum_{j=1}^{t} \gamma_k (X_j - \mu_k)(X_j - \mu_k)^T \tag{12}$$

In the equations, $N_k = \sum_{j=1}^{t} \gamma_k$, $\pi_k = \frac{N_k}{N}$. The algorithm iterates the two steps until convergence.

The proposed algorithm is as follows.

Step 1. Build background and global foreground models with the first t image sequences according to Eq. 4 and Eq. 10-12.
Step 2. For new observations, classify it as background pixel if the pixel matches its background models with predefined high probability, and update the corresponding Gaussian component parameters and weights of other components, go to step 5, else go to step 3.
Step 3. If the pixel matches foreground model with high probability, update the foreground model, go to step 5, else go to step 4.
Step 4. If the pixel does not match the foreground model with high probability, classify it as background model, and build new Gaussian component to add it to its background model or replace the Gaussian component with lowest weight.
Step 5. For new image frames, repeat Steps 2-4 until end.

4 EXPERIMENTAL RESULTS

In order to evaluate the performance of the proposed algorithm, we run our algorithm and the conventional GMM method on several widely used datasets. The

Figure 2. Tracking results of the conventional GMM and the proposed algorithm.

algorithms are implemented using MATLAB 2013a on a Pentium IV, 2.40GHz, 4G RAM computer. The datasets come from PETS2001 (Pets 2001), the resolution of the images is 768×576. The experimental results are shown in Figure 2, the left columns are the results of the conventional GMM, and the right columns below are the results of the proposed algorithm.

From the experiment, we can see that for the inappropriate parameters updating method of the conventional GMM approach, the algorithm cannot track objects longtime and precisely. Because of accurate parameters updating method, our proposed algorithm can track object precisely.

Due to limited space, we show an experiment here, and several other experiments also demonstrate the efficiency and precision of our algorithm.

5 CONCLUSION

In this work, we adopt a new parameters updating method for Gaussian Mixture Model, and build global foreground model to detect objects in image sequences. The experimental results show the effectiveness of the proposed approach. But the method needs more computation resources to fulfil the parameters updating process than the conventional GMM. In the future, we will make improvements to speed up the algorithm without loss of precision.

ACKNOWLEDGEMENTS

The authors would like to thank the anonymous reviewers for their valuable comments. This work was financially supported by the National Natural Science Foundation (NO. 81373661; NO. 81473708), and Youth Researchers fostering Foundation, Shangdong University of Traditional Chinese Medicine (NO. ZYDXY1349; NO. ZYDXY1358).

REFERENCES

[1] Yongzhong Wang, Yan Liang & Quan Pan, Yongmei Cheng & Chunhui Zhao. (2009) Spatiotemporal background modelling based on adaptive mixture of Gaussians. *Acta Automatica Sinica*, 35(4), 371–378.

[2] C. R. Wren, A. Azarbayejani, T. Darrel & A. P. Pentland. (1997) Pfinder: real-time tracking of the human body. *IEEE Transactions on Pattern Analysis and Machine Intelligence*, 19(7):780–785.

[3] N. Friedman & S. Russell. (1997) Image segmentation in video sequences: a probabilistic approach. *USA: Morgan Kaufmann 1997: Proceedings of the 13th Conference on Uncertainty in Artificial Intelligence. Providence*. pp:175–181.

[4] C. Stauffer & W. Eric L. Grimson. (2000) Learning patterns of activity using real-time tracking. *IEEE Transactions on Pattern Analysis and Machine Intelligence*, 22(8):747–757.

[5] A. Elgammal, R. Duraiswami, D. Harwood & L. S. Davis. (2002) Background and foreground modeling using nonparametric kernel density estimation for visual surveillance. *Proceedings of the IEEE*, 90(7):1151–1163.

[6] Y. Ren, C. Chua & Y. Ho. (2003) Motion detection with non-stationary background. *Machine Vision and Applications*, 13:332–343.

[7] Xiangfeng Bai, Aihua Li, Xilai Li & Renbing Li. (2011) A novel background Gaussian mixture model. *Journal of Image and Graphics*, 16(6):983–988.

[8] T. Matsuyama, T. Ohya & H. Habe. (2000) Background subtraction for non-stationary scenes. *Taipei, Taiwan: Proceedings of the 4th Asian Conference on Computer Vision. University Trier Press*. pp:114–116.

[9] M. Heikkila & M. Pietikainen. (2006) A Texture-based method for modelling the background and detecting moving objects. *IEEE Transactions on Pattern Analysis and Machine Intelligence*, 28(4):657–662.

[10] Y. Chen, C. Chen, C. Huang & Y. Hung. (2007) Efficient hierarchical method for background subtraction. *Pattern Recognition*, 40(10):2706–2715.

[11] Pets2001.[Online]. Available from: http://www.anc.ed.ac.uk/demos/tracker/pets2001.html.

Electric, Electronic and Control Engineering – Shao, Shu & Tian (Eds)
 ISBN: 978-1-138-02842-5

Design and implement of monitoring system for marine environment based on Zigbee

Liangcheng Wang & Zhiyong Liang
Sanya University, Sanya Hainan, China

ABSTRACT: The paper compares the advantages and disadvantages of Zigbee communication technology with other wireless communication technology, the use of IEEE802.15.4 technology to the standard Zigbee2007 protocol consists of a wireless sensor network. The system uses TI Company's Zigbee2007 protocol stack based peripheral circuit design, using E-201-C rechargeable PH composite electrode and temperature sensor DS18B20 to simulate the ocean water environment monitoring parameter acquisition sensor, data is sent to the CC2530 as the communication module to the core. After testing in open spaces between the module with power amplifier data transmission distance is 300 meters, achieved multiple routing data jump between multiple router module, and data acquisition and transmission coordination controller to control the whole network.

KEYWORDS: Zigbee, Wireless sensor network, marine environment.

1 INTRODUCTION

China has a vast maritime territory, as the size of the sea is about more than 3 million square kilometers, and the total length of the eastern coastal coastline is more than 32000 kilometers. Economic development has brought the issue of the coastal marine environment what has been destroyed, such as the coastal of tourism and real estate. The importance of marine environmental monitoring has been emerging gradually. Wireless sensor network, which based on Zigbee is widely applied in military reconnaissance, environmental monitoring, and target orientation and so on. It can sense, collect and process the information about the object in the range covering timely. It also has the advantage of large coverage, monitoring remotely, high monitoring accuracy, distributing network quickly, and low power consumption, low cost and so on. A wireless sensor network technology, which used in monitoring system for marine environment has become a hot research at home and abroad in recent years.

Nowadays, there are two main ways to monitor the marine environment at home and abroad. One approach is to use portable water quality monitoring instrument. Researchers collected samples at the scene and then analyzed the data in the laboratory. Unable to monitor the parameters of water quality remotely at the same time, It must have a lot of disadvantages, including low-level real-time, time-consuming and very uneconomical. Another approach is chemical testing method. The detection process of the method was very complex and non-real-time. It brought secondary pollution simultaneously.

2 COMMON TECHNOLOGY FOR SHORT-RANGE COMMUNICATION

The technology for short-range wireless communication is developing rapidly at present. They include Bluetooth, IEEE 802.11b (Wi-Fi) and Zigbee which are used commonly and rarely.

Bluetooth is an application for short-range wireless communication. The main feature of Bluetooth is the low rate of data transmission. The transmission rate is 1Mbps, and the distance is 10-100m. Bluetooth can be used in a variety of communication situations which has a strong portability. It also can be flexible roaming, transport the voice and data after the introduction of identity. So in practice, Bluetooth is used to access detection in parking system and PAD, mobile phone.

The maximum transfer rate of IEEE 802.11b is up to 11Mbps, and the distance is 30-100m. This technology provides wireless access to the Internet which is used in laptop computer.

ZigBee is a specification for a suite of high-level communication protocols used to create personal area networks built from small, low-power digital radios. Though its low power consumption limits transmission distances to 10–100 meters line-of-sight, depending on power output and environmental characteristics, ZigBee devices can transmit data over long distances by passing data through a mesh network of intermediate devices to reach more distant ones. ZigBee is typically used in low data rate applications that require long battery life and secure networking. ZigBee has a defined rate of 250 kbps, best suited for intermittent data transmissions from a sensor or input device. Applications include wireless light switches, electrical meters, traffic management systems, and other consumer and industrial equipment that require short-range low-rate wireless data transfer.

Compared with the several short-range communication technologies, the cost of the system based on Zigbee will be lower. So using ZigBee in the monitoring system for the marine environment is more appropriate. There are several characteristics of the 3 short-range wireless communications in the Table 1.

Table 1. The comparisons of the short-range wireless communication technology.

Communications Technology	Transmission Rate	Transmission Distance	Applications
ZigBee	20-250kbps	10-100m	Wireless Sensor Networks
IEEE 802.11b(Wi-Fi)	1-3Mbps	2-10m	Wireless Handheld Devices
Bluetooth	1-11Mbps	30-100m	Wireless Internet Access

3 HARDWARE DESIGN

3.1 *The composing of the hardware system*

This hardware system consists of RFD (Reduced Function Device), FFD (Full Function Device) and Router (FFD). FFD (Full Function Device) can provide all of the IEEE 802.15.4 MAC service to this system; it is responsible for organizing networks and the network management. At the same time, FFD is responsible for data transmission to the central management computer. So FFD is the only in the system. RFD as the data acquisition node for Zigbee can be configured as required, including the size of the system, the number of data acquisition nodes, the distance of communication and so on. However, there is only one FFD in one system.

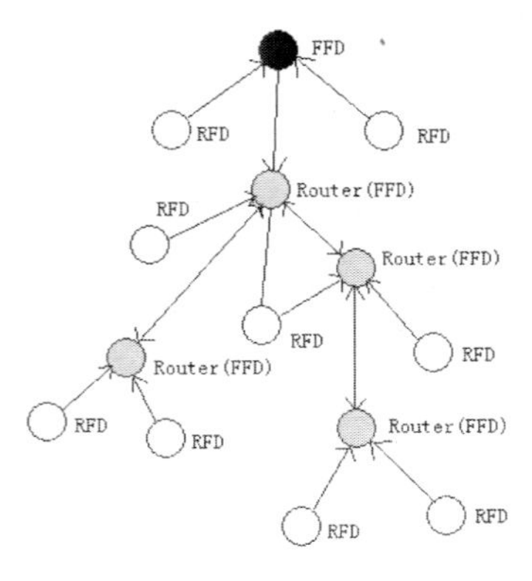

Figure 1. Network structure of the Zigbee.

3.2 *Design of data acquisition node*

The data acquisition node is the most basic and simple data node in the system. This node needs to collect data from sensor and process it. Then the node sends the processed data to its previous level node (router node or coordinator node). Due to the lack of the functions for data forwarding and routing, RFD uses less resources. It is in sleep mode when the data has been collected and sent, so energy consumption is very low. It mainly consists of the power circuit, sensor circuit, antenna, serial circuit, keyboard, LED and other components. Design diagram is shown as Figure 2

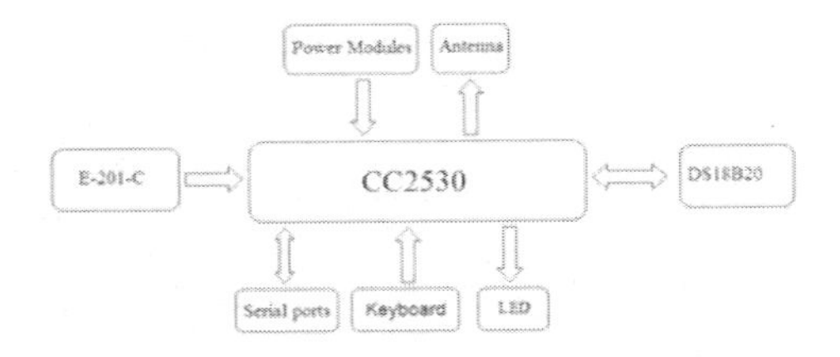

Figure 2. Design diagram of data acquisition node.

3.2.1 *Design of power circuit*

The biggest characteristics of Zigbee are low cost and stable performance. Because this system uses solar panels as power supply, so the power consumption of nodes needs to maintain very low. Solar panel output voltage of 5V, so they can provide power for LED. A Texas Instruments Inc's chip which named CC2530F256 has power management features; its standard supply voltage is 3.3V. In order to obtain a suitable chip supply voltage, 5V voltage of Solar panels is regulated by a voltage stabilizing module which named LM1117-3.3V, 3.3V DC output voltage obtained in the end. Voltage regulator circuit is shown as Figure 3. In this voltage regulator circuit, electrolytic capacitors (C4, C5) are mainly used for high frequency filtering. In order to improve the reliability of power supply, we use tantalum capacitors. Ceramic capacitor (C1, C6) is mainly used for low frequency filtering, its role is to increase the supply stability and reduce interference.

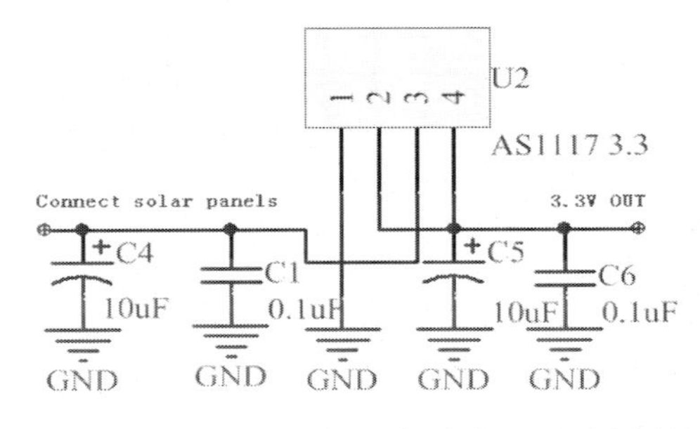

Figure 3. Voltage regulator circuit for LM1117-3.3V.

3.2.2 *Design of the sensor circuit*

The sensor circuit provides data for acquisition parameters of seawater from the system. Sensor converts water information to an electric signal (voltage or current). By enlargement processing and AD conversion, signal is sent to the computer, and then analyzed the data. In order to test system performance, we can use the Temperature Sensor (DS18B20) and PH sensor (E-201-C). The sensor is collected the temperature of water and PH environment the information on PH environment, tested the performance indicators of system, and provided evidence for the parameters of the marine environment. Temperature data acquisition uses the temperature sensor to complete. The sensor through a single bus protocol to transmit data and passes the directive to the processor. It needs an IO port at work. So it has the advantages of reliable working temperature, less resource consumption, packaging variously and so on. Measurement temperature is -55°C to +125°C, and its accuracy is ±0.5°C when it works at -10~+85°C.So it is suitable for temperature measurements in harsh environment. The circuit is shown in Figure 4.

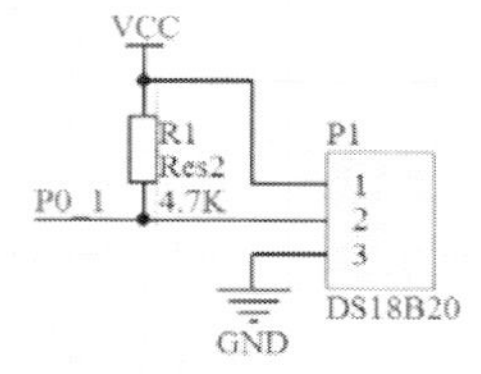

Figure 4. Circuit of temperature sensor.

The PH data collection is chosen E-201-C rechargeable composite PH electrode what's produced by the Shanghai Leici Company. The PH value of the electrode may vary depend on the measured pH of the water quality, and output of different potentials. It is through the two-stage amplifier circuit, the amplified signal PH electrodes, and then sent to the data processing CC2530. The PH value acquisition circuit is shown as Figure 5.

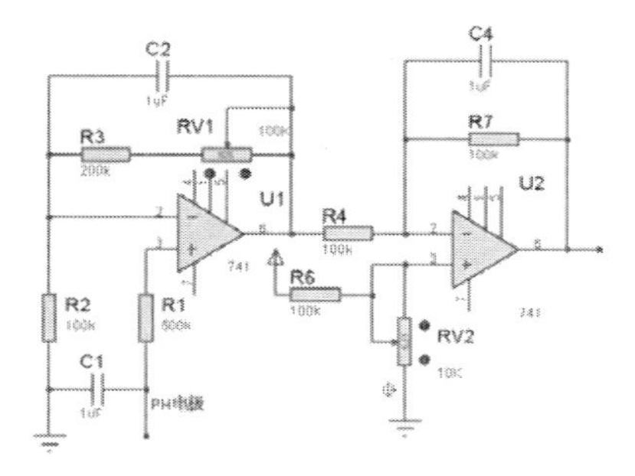

Figure 5. PH value acquisition circuit.

3.2.3 *Design of serial debugging circuit*

In order to facilitate PC communication software and computer systems design and system integration of the USB-to-serial circuit. Conversion devices use PL2302. The circuit is shown as Figure 6.

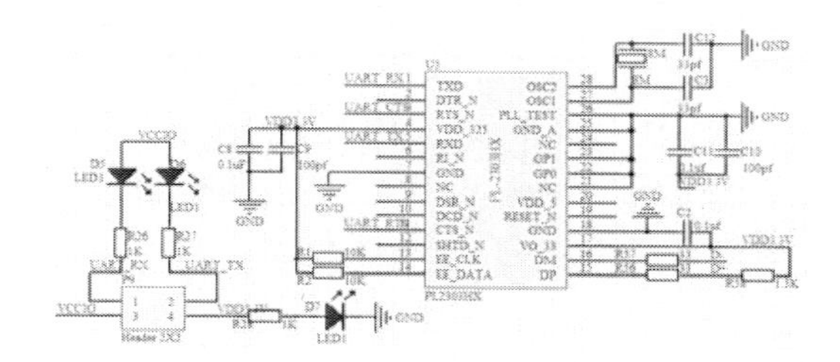

Figure 6. USB-to-serial circuits.

4 SOFTWARE DESIGN

4.1 *Design of coordinator node*

The coordinator is the only node within network management functions in the network. Its main role is responsible for creating the entire ZigBee network, network management, and forwarding sensor data. Program design ideas are as follows: first of all, the coordinator is energized and initialized; three acquisition nodes (with three nodes, for example) are turned off when they gather information which is shown as Figure 7. The system does not send or receive any data dormancy, thereby reducing system power consumption. Then coordinator key1 is pressed, the coordinator will perform a broadcast command (#start1000), and store the broadcast commands. At this time, the system is beginning to work. Data acquisition module (at the same time as the router function) is acquiring data once every second, and sent the coordinator, which is shown as Figure 8. After the network coordinator is receiving data from the router, it is to analyze the data type: water parameter data or coordination command. If data type is the command, command (#start1000) is sent by the router when conducted to a unicast broadcast, and then the router is begun collecting data. If data type is the data of sensor, data are displayed in the specified area of the LED. Router lost timing is cleared to 0 at the same time. If the router lost timing is 3 seconds, the word" disconnect" is displayed in the designated area of the LED. It represents the collection of information is failed, and node had dropped which is shown as Figure 9.

Figure 7. Turned off.

Figure 8. Began collecting.

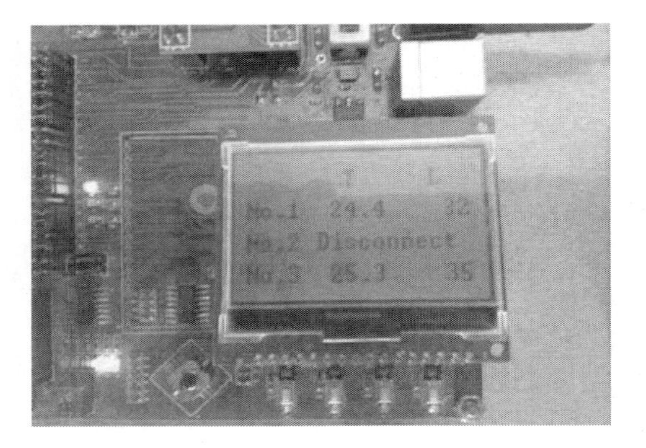

Figure 9. Lost node display.

This program is designed to make the idea of either collecting data initialization added router, and can make the switch control personnel through coordinated control of the entire network data collection capabilities. In a subsequent development, we can also collect data by PC software at any node or any time. We can also choose un-real-time acquisition, acquisition time and location. This will not only save the energy consumption of the entire system, but also easy to set according to the different needs of different acquisition modes, the system can display the working status of each routing node in the PC in real time to display the working status of the entire system network.

4.2 *Programming of router nodes*

Router node design ideas are: the router is powered at first, coordinator within the scope of radiation, the router is joined in a network. The router is determined whether to apply to join the network successfully when network status has been changed. After the addition is successful, the router has sent a request command to the coordinator. Coordinator issue commands to the router based on data from various system settings after command is received. When the router is getting the beginning-collecting-command from the coordinator, if the command is # start1000, the router begins collecting data once per second and sends to the coordinator, or goes sleep in order to reduce energy consumption and wait coordinator sends a valid command.

5 SYSTEM TESTING

The effective transmission distance and the packet loss rate of systems are very important indicators in Zigbee. The system was tested at two natural lakes which separated by 230-380 meters in Sanya University. There were no hills, no buildings, but only two rows of trees around the lakes. The overall environment was fit for testing system functions. In order to verify the relationship between communication node distance packet loss rates, we used two ZigBee nodes at different intervals for data transmission, and also calculated the loss rate. The transmission power of CC2530 was set to 4dBm. During the test, it was sent 10,000 byte packets at a time. In order to ensure the reliability of the system work, we set the actual packet transmission rate is 20KB/S. We tested it and put the data collection node near the lakes, and we used the mobile routing node to test. Test point was set 12, followed by 1 m, 20 m, 50 m, 80 m, 150 m, 200 m, 220 m, 240 m, 260 m, 280 m, 300 m, 320 m. Because, as the distance increased, the loss rate increased, so the final test points were set denser. Finally, we get the test data as shown in Table 2.

Table 2. Relationship between packet loss rate and communication distance.

No.	Communication Distance	Send	Receive	Lost
1	1m	10000	10000	0.00%
2	20m	10000	10000	0.00%
3	50m	10000	10000	0.00%
4	80m	10000	10000	0.00%
5	150m	10000	9980	0.20%
6	200m	10000	9981	0.19%
7	220m	10000	9975	0.25%
8	240m	10000	9856	1.44%
9	260m	10000	9805	1.95%
10	280m	10000	9602	3.98%
11	300m	10000	9404	5.96%
12	320m	10000	0	100%

Through the test, the packet loss rate of the system was less than 0.5% within 220m, the system showed good performance. The system could work within 300 meters. However, the system began unstable work over 300m.So within 300 meters basically reached the system design requirements.

6 CONCLUSIONS

We completed system based on Zigbee water environment detection around the lakes. The system could be easily ported to the marine environment monitoring. The system also could collect real-time information based on pH and water lighting information, and it is stable and better performance. In the peripheral circuit design, we should try to save the system interface resources to greatest extent increase parameter monitoring species and improve system efficiency.

ACKNOWLEDGEMENT

The sea water environment monitoring system solutions based on a Zigbee wireless sensor network (Item Number: 2012YD49).

REFERENCES

[1] Zhang Bing. (2007) *Research of Data Management and Index Technology in Wireless Sensor Networks*. Nanjing: Nanjing University of Aeronautics and Astronautics.

[2] Cheng Chunrong. (2010) *A Design of Water Quality Monitoring System Based on ZigBee*. Hangzhou: Hangzhou Dianzi University.

[3] Xu xiaoxiao. (2008) *Research and Design of Embedded Bluetooth Gateway*. Tsingtao: Tsingtao University of Science & Technology.

[4] Wang Xiaoqiang. (2012) *Design of Wireless Sensor Network Based on Arm and Zigbee Technology*. Changsha: University of Electronic Science and Technology.

[5] Wang Yanhui. (2006) *Basie Research on Embedded Wireless Sensor Networks*. Dalian: Dalina University of Technology.

[6] Zhang Haisheng, Tu Jinglu. (2012) A Design of Home Security System Based on ZigBee. *Computer Knowledge and Technology*, 12.

[7] Gao Yuqin. (2010) The System for field information detect and manage–Based on ZigBee and GPRS, *Journal of Agricultural Mechanization Research*.

[8] Li Ye, Dong Shoutian, Huang Dandan. (2015) Rice automatic irrigation control system based on ZigBee technology design. *Journal of Agricultural Mechanization Research*, 2.

[9] Li Aimin. (2015) Implementation of intelligent monitoring system for temperature and humidity based on Zigbee/GPRS. *Journal of Agricultural Mechanization Research*, 3.

[10] Cai Yihua, Liu Gang, Li Li. (2009) Design and test of nodes for farmland data acquisition based on wireless sensor network. *Transactions of the CSAE*, 4.

Electric, Electronic and Control Engineering – Shao, Shu & Tian (Eds)
© 2015 Taylor & Francis Group, London, ISBN: 978-1-138-02842-5

RFID sensor network for rail transit intelligent supervisory control

Shanshan Yu & Yigang He
School of Electrical Engineering and Automation, Hefei University of Technology, Hefei Anhui, China

ABSTRACT: Smart Rail Transit represents the future trend of rail transportation. RFID and wireless sensor network are regarded as the core technology for the Internet of Things. Based on the introduction about Smart Rail Transit and Integrated Supervisory and Control System (ISCS), this paper puts forward the application idea of RFID sensor network in Rail Transit intelligent Supervisory Control. Combined with the mature ISCS, it can effectively monitor the train running information and the surrounding environment condition. This improves the processing capacity of the train to respond to emergencies, service level and reliability. The scheme is feasible and promising according to the existing RFID and WSN application case. The construction of RFID sensor network will play an important significance for Smart Rail Transit, the independent innovation of railway technology and the acceleration of economic development.

KEYWORDS: smart rail transit; the internet of things; RFID; wireless sensor network.

1 INTRODUCTION

At the international conference on Intelligent Rail Transportation 2013, experts from all over the world around the "green railway, security and intelligence" discussed the development trend of intelligent rail transportation. This shows that rail transport as an environmentally friendly, safe and efficient means of transportation, has attracted the attention of the world. At present, rail transportation puts forward higher requirements on security and service level, the construction of Smart Rail Transit will be the development trend. And its key depends that ISCS has a more thorough perception, a wider range of connectivity, a more intelligent information processing capability to the surrounding environment [1].It is also an important issue facing rail technology about the research on a new generation of ISCS for Smart Rail Transit.

2 SMART RAIL TRANSIT

Smart Rail Transit is the abstract concept of rail traffic development trend and the mode of operation. And it is the macro concept and target for the construction of rail transit. As a real-time accurate, safe and efficient integrated management and control system, it is composed of advanced sensor technology, communication technology, data processing method and network technology etc. Three layer architecture includes intelligent infrastructure, the whole railway traffic network, management decision-making and command system[2]. The entire system completes the intelligent information processing process, through four steps which include data collection, data fusion, data mining and command decision[3].

Integrated supervisory and control system (ISCS) consists of a central monitoring system, station monitoring system and backbone network. Through the integration of various monitoring subsystem and the interconnection of other independent system, it monitors related equipment such as :the rail communication equipment, electric power equipment, station equipment, vehicle state, the line state etc.. By the data sharing platform, it realizes the unified management of information. It also improves the effective linkage between systems and enhances the capacity of handling all kinds of emergency events. At present, China's Guangzhou Metro Line 3 in Figure 1, has the largest domestic ISCS which includes almost all the subsystem of train operation[4].

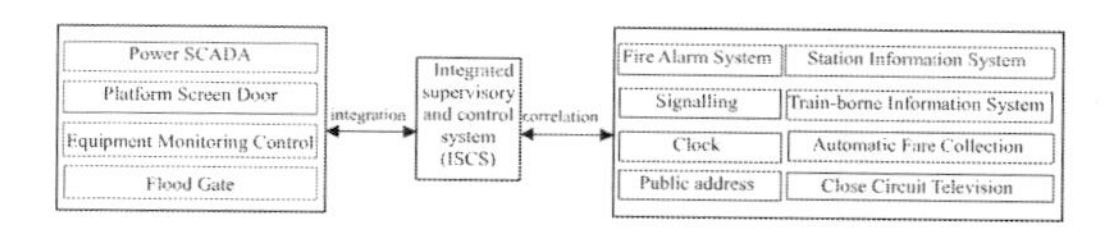

Figure 1. The ISCS framework of guangzhou metro line 3.

3 THE PERCEPTUAL LAYER TECHNOLOGY OF THE INTERNET OF THINGS

3.1 *Radio Frequency Identification (RFID)*

RFID is an automatic identification technology. The RFID system consists of host computer, reader and a series of tags. The tag includes an antenna, a radio frequency module, control module and storage module. The reader is composed of an antenna, a radio frequency module, reader module, clock and a power supply. Each tag has a unique electronic code and stores object information. The RFID tag antenna transmits data to a nearby reader through radio waves and the reader sends the information to users. Thus, this achieves the purpose of object recognition and data acquisition. David describes the application of RFID in ornithological research[5]. RFID has been widely applied in the field of supply chain management.

Application of RFID in rail transit can solve the massive task of information transmission between the train and the ground. It is too difficult for track circuit. In order to ensure the safe operation, it is very necessary to have safe and reliable information transmission channels between the train and the ground. The RFID system can provide for the train or control center some informations such as: interval signal point, the track curve, the flow of personnel, cargo tracking, the train position, and it efficiently and accurately identifies the targets and network information sharing, improve the monitoring capacity of rail transit.

3.2 *Wireless Sensor Network (WSN)*

WSN is a Multi-hop network through wireless communication and deployed in the monitoring area. It consists of large numbers of sensor nodes. The node contains sensor module, processor module, wireless communication module and energy module. Every node monitors environmental information and send it to the monitoring center. Monitoring center processes the received data, then make the decision and judge whether there are abnormal conditions in the deployment area. WSN has potential application value in agricultural, industrial and food monitoring field [6].

Rail transit may have some natural disasters including earthquakes, geological disasters, severe weather, fires, floods, and lightning. In addition, there are some traffic accidents due to human factor and equipment failure. WSN can be adapted and deployed in harsh environment. In order to improve the control accuracy, some areas often need to deploy a large number of sensor nodes. WSN can use various types of sensors to perceive the environment parameters including temperature, humidity, roadbed, environmental climate, seismic, electromagnetic, pressure, speed, etc., and it can play a effective warning role to the above disasters.

4 RFID SENSOR NETWORK SUPERVISORY CONTROL MODEL

According to the features about application of RFID and WSN in rail transit, there is a clear complementary relationship between them. RFID sensor network is a new network integrating these two technologies. It will inherit the characteristics of RFID about automatic target recognition, to achieve identification tracking ability while it can also monitor real-time environmental information, to realize the active perception and communication function. In the RFID system, the communication mode is single hop between tag and reader, without communication ability between tags. But the RFID sensor network is able to provide a multi hop communication for RFID in the broader region[7]. The followings are two suitable RFID sensor network Supervisory Control model for rail traffic.

4.1 *RFID tag sensor network*

This monitoring framework integrates RFID tag with wireless sensor node and calls it tag sensor node, structure of the node is shown in Figure 2.It is needed to set up a special interface and data access mechanism between the control module and the node microprocessor. The tag can communicate with RFID reader through the tag antenna, and communicate with other nodes by node antenna. The data transmitted between the tag sensor nodes include a unique electronic code information and dynamic environmental information.

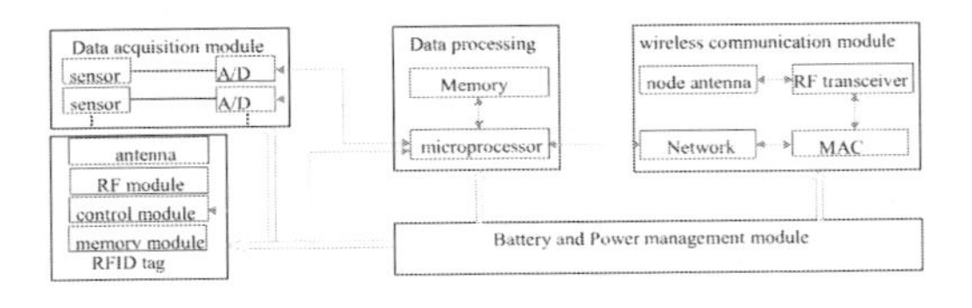

Figure 2. The framework of RFID tag sensor node.

IRFID tags is an example about tag and sensor fusion designed by Machine talker, it is a kind of active tag integrated with sensor used to measure the environmental parameters[8]. At present, a series of targets detection are completed by track circuit, such as trains occupying line, control signal devices, speed level information, etc.. Deployment of RFID tags sensor network can identify the rail line curvature, deformation, stress, speed limit information etc.

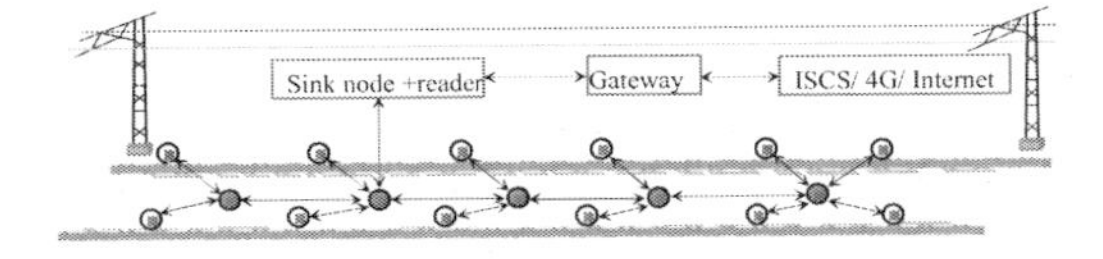

Figure 3. The framework of RFID tag sensor network.

Figure 3 describes the monitoring structure based on the multi-hop wireless communication mode.

According to the function, nodes can be divided into the sensing nodes, the cluster head nodes and the sink nodes in the network. In the figure, the hollow circles represent the sensing nodes, and the solid circles represent the cluster head nodes. Monitoring information is the main function of sensor nodes. The data transmitted inter-nodes include their identity and dynamic information. The cluster head nodes display the data aggregation and routing function, and forward the data from the sensing node to the sink node. The cluster head node can be specified in advance, and also be selected using the cluster algorithm based on energy effective principles [9]. The sink nodes process and classify the received data, and finally transmit them to the monitoring center or the Internet through the gateway.

4.2 *RFID reader sensor network*

This monitoring framework integrates RFID reader with wireless sensor node and calls it reader sensor node, and the structure of the node is shown in Figure 4. In this way, the reader sensor node, with the function of network communication, can read the tags beyond the coverage of one reader. The microprocessor can control the reader and RF transceiver into sleep mode to save energy during the free time.

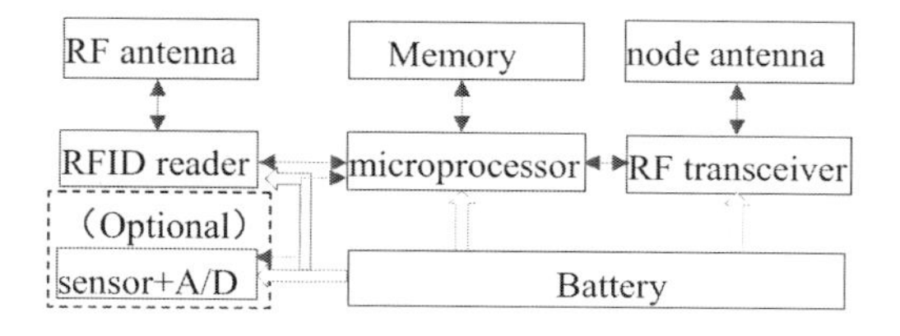

Figure 4. The framework of RFID reader sensor node.

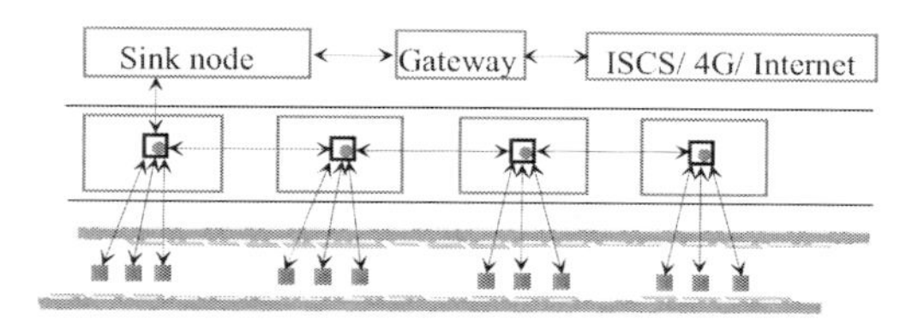

Figure 5. The framework of RFID reader sensor network.

Zhang and Wang studied several compositive methods about reader and sensor [10]. Pereira et al. used this idea in the study of animal tracking and monitoring [11]. They sticked tags on animals, and then deployed the sensor with RFID reader in the environment. When the animal enters the coverage of reader, reader can obtain the tag information of animal and sensor perceives dynamic data, such as temperature, food intake.

Figure 5 presents three kinds of equipment: RFID tag, reader sensor node and sink node in RFID reader sensor network of rail transportation. The reader sensor node collects tag data in its coverage area and sends it to the sink node through multi hop communication. For example, in order to monitor the information of train station, the position tracking and the flow of the crowd, one method is to deploy RFID tags stored with train and personnel information, then install reader sensor nodes in the vicinity of railway station. The reader sensor node can be integrated with a variety of sensors to monitor the train running state, such as acceleration, speed, the crowd. These informations can be shared on the Internet, and it is convenient for the public to query the current congestion in a certain section. In addition, the motion sensor module can trigger the reader module into the active state to read tag data from the train according to the real-time speed, and the reader may turn into dormancy or shut down to conserve energy until the train leave.

4.3 *Wireless communication standard*

Now ZigBee and IPv6/6Lowpan are the mainly communication standards for wireless sensor network. ZigBee is a wireless communication protocol with the Characteristics of low cost, low complexity, low power consumption, low data rate and high security. It is established according to the standard of IEEE802.15.4. The protocol stack supports the network topology like point-to-point, star, mesh etc. It can accommodate more than 60,000 nodes and achieve the ability of network and communication with each other in a short time. Importantly, its energy consumption is very low. As the technical standard of wireless sensor network in network layer and application layer, ZigBee conforms to the development of the Internet of things. In Beijing Metro Line 4, ZigBee wireless positioning system has been applied in the early warning system. In the construction of Metro Line 9, ZigBee positioning system improved the management level.

IPv6/6Lowpan is proposed by IETF in 2007.It is a standard for the transmission of IPv6 data packet on IEEE802.15.4 network. The purpose is to promote the use of the IP protocol stack for wireless sensor network and to accelerate the integration of wireless sensor networks and Internet. IPv6/6Lowpan has the obvious disadvantages compared to the ZigBee in terms of the maturity of technology and product market. But the advantage of 6Lowpan is that it can provide a lot of IP addresses for the wireless sensor node. Wireless sensor network and the Internet can be directly interconnected or communication without the gateway. Some researchers consider dealing with emergencies by using wireless sensor network based on 6Lowpan standard. In the study of WSN, Lakshmi et al confirmed the validity of this view by experiment [12]. IPv6/6Lowpan has the potential to replace ZigBee in the future. With the development of RFID and wireless sensor network, IPv6/6Lowpan will be more perfect.

5 CONCLUSION

RFID sensor network has a good application prospect in rail transit. This control architecture can be used as an independent monitoring subsystem, interconnected with integrated supervisory control system, and be applied to the design of it as a new technology standard. RFID sensor networks face three challenges in technology. It requires the reliable communication, energy efficiency and network lifetime. These problems need to be further studied. With the development of Internet of things and the wireless communication standard, RFID sensor network will be widely used. This monitoring system will enrich the monitoring mode of rail transit, and it provides the certain reference value to the construction about Smart Rail Transit.

ACKNOWLEDGMENTS

This work was supported by the National Natural Science Found of China for Distinguished Young Scholars (50925727), the National Defense Science and Technology Project of China (C1120110004, 9140A27020211DZ5102), the Key Grant Project of Chinese Ministry of Education (313018), and the Key Science and Technology Project of Anhui Province of China (1301022036).

REFERENCES

[1] Chen X D,Yang B. (2012) Smart Rail Transit, achieve a more thorough perception. *Journal of Computer Applications*, 05:1196–1198.

[2] Zeng H S,Zhu H F. (2012) Smart Rail Transit and its system architecture. *Journal of Computer Applications*, 05:1191–1195.

[3] Yang Y,Zhu Y,Dai Q,Li T R. (2012) Smart Rail Transit, Achieve deeper intelligence. *Journal of Computer Applications*, 05:1205–1207+1216.

[4] Wei X D. (2011) *Urban Mass Transit Automation System and Technology (2nd edition)*. Beijing: Publishing House of Electronics Industry.

[5] Bonter D N, Bridge E S. (2011) Applications of radio frequency identification (RFID) in ornithological research: a review. *Journal of Field Ornithology*, 82(1):1–10.

[6] Ruiz-Garcia L, Lunadei L, Barreiro P, et al. (2009) A review of wireless sensor technologies and applications in agriculture and food industry: state of the art and current trends. *Sensors*, 9(6):4728–4750.

[7] Liu H, Bolic M, Nayak A, et al. (2007) Integration of RFID and wireless sensor networks. *Proceedings of Sense ID 2007 Worksop at ACN SenSys, 2007*, pp:6–9.

[8] Zhang Y, Yang L T, Chen J, eds. (2010) *RFID and sensor Networks: Architectures, Protocols, Security, and Integrations*. CRC Press.

[9] Shang F J. (2011) *Communication Protocol for Wireless Sensor Network*. Beijing: Publishing House Of Electronics Industry.

[10] Zhang L, Wang Z. (2006) Integration of RFID into wireless sensor networks: Architectures, opportunities and challenging problems. *Grid and Cooperative Computing Workshops, 2006. GCCW'06. Fifth International Conference on. IEEE, 2006*, pp:463–469.

[11] Pereira D P, Dias W R A, Braga M, et al. (2008) Model to integration of RFID into wireless sensor network for tracking and monitoring animals. *Computational Science and Engineering, 2008. CSE'08. 11th IEEE International Conference on. IEEE, 2008*, pp:125–131.

[12] Ele S L, Kothari M R, Kota D N N. (2014) 6LoWPAN Based wireless sensor network to monitor temperature. *International Journal of Advanced Electronics and Communication Engineering*, 1(1): 1–6.

Electric, Electronic and Control Engineering – Shao, Shu & Tian (Eds)
 ISBN: 978-1-138-02842-5

An algorithm of digital image denoising based on threshold

Yiwen Shi & Rener Yang
Faculty of Information Science and Technology, Ningbo University, Ningbo Zhejiang, China

ABSTRACT: This paper proposes a digital image denoising algorithm based on threshold. This algorithm is based on salt and pepper noise, by setting the threshold, leaving pixel point untreated when the gap between its gray value and the average of its surrounding pixels is less than the threshold. Otherwise, it will be replaced by the average. Meanwhile, the paper analyzes the relationship between the best threshold and the average, variance, entropy and complexity of the image, providing a theoretical foundation for finding the best threshold quicker. By simulation experiment, compared with other denoising algorithm, this method is better at removing noise without making the image unclear.

KEYWORDS: digital image de-noising; salt and pepper noise; selective mask smoothing de-noising algorithm.

1 INTRODUCTION

In the process of a digital image generation or transmission, it will be inevitably mixed impulse noise[1]. Salt and pepper noise are a main type of impulse noise. As the image is contaminated, a part of the pixel values will be changed, namely the emergence of pollution point of greater or smaller gray value. It's just like white powder salt and black pepper sprinkled on the image[2]. In order to acquire accurate image information, it's necessary to take appropriate methods for denoising.

At present, domestic and foreign scholars have made a number of denoising methods, such as median filtering[3], mean filtering, selective masking smoothing denoising algorithm and so on. The median filter algorithm can effectively restrain salt and pepper noise as well as having high efficiency of computation. But it cannot completely filter out the noise and restore image edge information from highly corrupted images [4]. The mean filtering algorithm is simple and easy to implement, but images' pixels are taken as the average pixel leaving the field to blur the image [5] which has great limitations. Although other algorithms, such as selective masking smoothing denoising algorithm, which are improved on the basis of the mean denoising, denoising effect is still not ideal.

This paper proposes a nonlinear algorithm for denoising--based on the threshold mean denoising method. As long as selecting the appropriate threshold, it will get better denoising results. This algorithm can overcome the shortcomings of traditional methods and deal with the pixels that are corrupted by noise as far as possible. Meanwhile, the gray value of other pixels should keep unchanged, so that images aren't blurred after denoising and edge information isn't lost, while points, lines and the details of steeple processing are strengthened.

2 THE CHARACTERS OF DIGITAL IMAGE NOISE

The noise of digital image mainly comes from the images' generation and transmission, and it becomes an unwanted part of an image or a certain part of the case there is no research value[6]. The general noise is as followings:

1 The additive noise. The noise does not change as the intensity of the image signal changes and its existence has no relation to the image signal.
2 Multiplicative noise. The noise is associated with the image signal. It usually changes as the image signal changes, and their relation is shown in formula (1):

$$g(x,y)=f(x,y)\times \eta(x,y) \tag{1}$$

3 Gaussian noise. The noise of n-dimensional distribution obeys Gaussian distribution, which is easy to be treated in mathematics. Formula is shown as follows:

$$p(z)=\frac{1}{\sqrt{2\pi}\sigma}e^{-(z-a)^2/2\sigma^2} \tag{2}$$

4 Salt and pepper noise. Salt and pepper noise refer to two kinds of noise. One is salt noise, and the other is pepper noise. The former is the high gray-scale noise, while the latter belongs to the low gray

noise. Usually two noises occur simultaneously, which presents black white noise on the image. The noise is caused due to image segmentation and noise points evenly distribute in the whole image.

3 THE METHODS OF DENOISING

3.1 *Selective mask smoothing denoising algorithm*

Selective mask smoothing denoising algorithm is an improved mean filter algorithm. When using the mean filtering algorithm eliminates noise, it's inevitable to bring shortcomings on average, making big changing lines and edges blurred. Selective mask smoothing filter can maintain a certain degree of edge information in order to keep the details of the edge contour of the images.

Using the selective mask method selects the 5×5 model window. Then select a reference pixel from the model window. Based on the pixel, make the drawing shown in Figure 1 in the surroundings, which has a total of nine windows, including a square, four pentagons and four hexagons. Calculated the average and variance of pixel value of each screen window, and then the window with minimum variance is adopted to mean processing. The variance of each image's pixel δ can estimate the average of each image's pixels in the window value M, by the following equation

$$\delta = \sum_{k=1}^{N} (f_k - M)^2 \tag{3}$$

where f_k is pixel value and N is the number of images' pixels.

0 0 0 0 0 0 1 1 1 0 0 1 1 1 0 0 1 1 1 0 0 0 0 0 0	0 0 0 0 0 1 1 0 0 0 1 1 1 0 0 1 1 0 0 0 0 0 0 0 0	0 1 1 1 0 0 1 1 1 0 0 0 1 0 0 0 0 0 0 0 0 0 0 0 0
square	pentagon 1	pentagon 2
0 0 0 0 0 0 0 0 1 1 0 0 1 1 1 0 0 0 1 1 0 0 0 0 0	0 0 0 0 0 0 0 0 0 0 0 0 1 0 0 0 1 1 1 0 0 1 1 1 0	1 1 0 0 0 1 1 1 0 0 0 1 1 0 0 0 0 0 0 0 0 0 0 0 0
pentagon 3	pentagon 4	hexagon 1
0 0 0 1 1 0 0 1 1 1 0 0 1 1 0 0 0 0 0 0 0 0 0 0 0	0 0 0 0 0 0 0 0 0 0 0 0 1 1 0 0 0 1 1 1 0 0 0 1 1	0 0 0 0 0 0 0 0 0 0 0 1 1 0 0 1 1 1 0 0 1 1 0 0 0
hexagon 2	hexagon 3	hexagon 4

Figure 1. Nine kinds of mask model windows.

3.2 *Mean denoising algorithm based on threshold*

3.2.1 *Implementation of the algorithm*

Though compared with the traditional denoising method, the selective mask smoothing denoising algorithm has been improved. It still cannot solve the problem of image blur. Therefore, on the basis of mask denoising, this paper proposes a mean filtering algorithm based on threshold. In general, an image, corrupted by salt and pepper noise, should be filtered out noise points whose value is greatly different from the image's pixels mean, while pixel points which aren't corrupted can be preserved if we select an appropriate threshold. For example, we assume a threshold as B, and select 3*3 sizes of $H = \frac{1}{9}\begin{pmatrix} a_1 & a_4 & a_7 \\ a_2 & a_5 & a_8 \\ a_3 & a_6 & a_9 \end{pmatrix}$ as the reference template. Next step is to select a_5 as the pixel point processing. Then the output of the mean filter is $M = \frac{1}{9}[a_1 + a_2 + ... + a_8 + a_9]$ after the pixel processed. Naturally, we can obtain a mean value M, and treat the pixel point a_5 according to the threshold value B. Therefore, double cases are considered in the following.

case 1. $|a_5 - M| \geq B$

In this case, the filter outputs the pixel value $A=M$. In other word, the pixel value is equal to the average value M instead of original gray value at this point.

case 2. $|a_5 - M| \leq B$

In this case, the filter outputs the pixel value $A=a_5$. So the pixel point value remains unchanged.

According to the character of the algorithm, it's clear that the denoising algorithm doesn't treat all each pixels in an image, but chooses to retain the original gray value selectively. So it effectively avoids the destruction of those pixels which are not polluted by noise. Also, the algorithm will preserve the details and edge information of images as far as possible, so that the images are not vague. Obviously, the result of PSNR obtained by the denoising algorithm based on the threshold is excellent, which further proves the reliability of the method.

3.2.2 *Selection of threshold*

Selecting appropriate threshold plays an important role in denoising method proposed in this paper. If the threshold is too small to degenerate into mean denoising method, the images will become very blurred. Also, large threshold influences the denoising effect. By the priori knowledge, different statistical parameters maybe have relations with a threshold value[8]. So in this paper, we calculate the relevant statistical parameters first before denoising. Then, according to the relation between each parameter and an optimal threshold, this step can determine the approximate range of an optimal threshold value of an image. After that, the final threshold can be received.

To study the relationship of four characteristic parameters, including mean, entropy, variance and complexity, and the optimal threshold, the article adopts controlled variable method. For example, we select some images whose parameters are the same except the mean. We first add the same number of salt and pepper noise points (N) to these images, and choosing the 200 integers between 1 and 200 regards as a prospective value of threshold B. Then compared with PSNR, the optimal threshold and the relation of the mean and threshold can be received by the way of different denoising methods proposed in this paper. Fig. 2 shows how to have an image's optimal threshold.

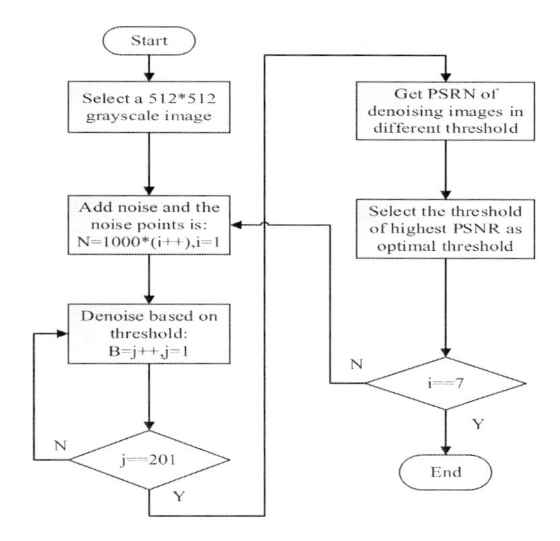

Figure 2. The flowchart of an image's optimal threshold.

4 EXPERIMENTAL RESULTS AND ANALYSIS

4.1 *Comparison of three methods for removing salt and pepper noise*

The experiment selects the 512*512 images in 256 gray levels. Fig.3 (a) is just an unprocessed image. Fig.3 (b) based on the Fig.3 (a) is the image added salt and pepper noise (N=12000). Fig.3 (c), 3(d) and 3 (e) all based on the Fig.3 (b) is the effect diagram applied for mean method, selective masking algorithm and the denoising method proposed in this paper.

Figure 3. Effect diagram of plus-noise and denoising image (N=12000). (a) Unprocessed image. (b) Plus noise image (N=12000). (c) Mean denoising algorithm. (d) Selective masking denoising algorithm. (e) Proposed denoising algorithm.

The effect of mean denoising method is terrible as shown in Fig.3. In other word, this algorithm not only cannot remove salt and pepper noise, but also make the image blurred. However, by the selective mask denoising algorithm, the effect of image denoising has been greatly improved. The shortcomings of the method are the image distortion, which mainly are lack of some points, lines and the details of the steeple. As can be seen from Fig.3 (e), the mean flittering algorithm based on a threshold that is proposed in this paper achieves great performance at denoising. It's obviously better than the former mean algorithm, which not only removes salt and pepper noise, but also effectively maintains the image of the important features and edge information.

4.2 *The relationship between images' related statistical parameters and the optimal threshold*

In the experiment, in order to study the relationship between characteristic parameters of images and optimal denoising threshold, we use 12 pieces of 512*512 gray images divided into 4 groups to ensure that one parameter for each set of 3 images in different value. Then, by making use of the controlling variable method above, we add the same salt and pepper noise to the 3 images respectively, getting the best threshold. Meanwhile, we change the number of salt and pepper noise. Then comparing the variation of threshold in different number of noise points, the experiment can receive data as shown in Table1.

As can be seen from Table 1, when the images' mean A small to large increases, the optimal threshold decreases first, then increases, which mean is close to 0 or 255 and the optimal threshold achieves the maximum. Obviously, when the entropy S or the variance D becomes larger, the optimal threshold value will grow smaller. But when the complexity C value becomes larger, the optimal threshold is also correspondingly large. At the same time, for the same image, the optimal threshold has a decreasing trend with the salt and pepper noise points increasing. After knowing the relationship between these four parameters and the optimal threshold, we can quickly select the optimal threshold value's range according to the parameters of the images, and make the denoising effect better through adjusting the optimal threshold.

Table 1. The optimal threshold of statistical parameters of images with different denosing points (N).

Image parameter value	N=1000	N=2000	N=3000	N=4000	N=5000	N=6000
A=26	184	170	168	156	146	140
A=84	88	81	77	75	74	75
A=200	190	190	190	190	189	188
S=2.69	118	109	111	97	103	96
S=5.51	81	79	79	72	71	72
S=6.20	77	75	71	68	69	69
D=13.36	129	129	129	128	128	128
D=17.11	113	108	106	103	105	106
D=34.76	101	111	96	116	110	98
C=11.03	75	70	68	70	74	74
C=19.29	173	173	172	171	171	171
C=25.63	188	184	180	178	170	154

5 CONCLUSION

The results of VC ++ simulation platform show that the mean filtering algorithm based on a threshold which is proposed by this paper makes the images not blurred as far as possible, and it also strengthens the processing of the details of the images' points, lines and steeple, which ensures that the image details and edge information are not lost. Compared with the mean algorithm and selective mask denoising smoothing method, this algorithm achieves better performance at denoising. Of course, the algorithm is at the cost of time in order to obtain better denoising effect. So the authors will focus on improving the effectiveness of the algorithm research in the future.

REFERENCES

[1] Bovik A. (2000) *Handbook of Image and Video Processing*. America: Academic Press.

[2] Yiqiu Dong, Shufang Xu. (2006) An efficient salt-and-pepper noise removal. *Acta Scientiarum Naturalium Universitatis Pekinensis*, 42(5).

[3] Haixia Guo, Kai Xie. (2007) An improved method of adaptive median filter. *Journal of Image and Graphics*,12(7):1185- 1188.

[4] Ron Zahavi. (2000) *Enterprise Application Integration with CORBA*. America: John Wiley & Sons.

[5] Pinquan Huang, Xuben Wang. (2005) An efficient threshold based image filtering algorithm and its realization. *Computer Simulation*, 22(5):111–114.

[6] Ying Wang, Guangyu Zeng. (2011) Research on image de-noising algorithm. *Computer and Information Technology*, 19(4):8–12.

[7] Caihong Tang, Lidong Cai. (2006) A weighted mean filtering algorithm based on histogram. *Control & Automation*, (13):209–211.

[8] Do M N, Vetterli M. (2005) The contourlet transform: An efficient directional multi-resolution image representation. *IEEE Trans on Image Processing*, 14(12):2091–2106.

Electric, Electronic and Control Engineering – Shao, Shu & Tian (Eds)

A Study on the construction status of smart power distribution network planning and the method of improvement

Meng Qi
State Grid Kaifeng Power Supply Company, Kaifeng Henan, China

Na Zheng
Henan Enpai High-tech Electric Power Group Co., Ltd, Zhengzhou Henan, China

Lei Peng
State Grid Henan Electric Power Company, Zhengzhou Henan, China

ABSTRACT: This paper expounds the connotation and features of smart power distribution network as well as the development goal of smart power distribution network in China and proposes the technical framework of smart power distribution network on this basis: the improvement of grid infrastructure, the construction of technological support of information and communication, and the development of grid smart applications. The establishment of smart power grid meeting the economic development needs of the city is an indispensable link of economic construction and development. The city mainly contains one city, four counties and five districts, including many remote rural areas. The landform is quite complicated. For this reason, reasonable and improved planning on a technical framework should be considered as the precondition for the construction of advanced smart power grid in the city.

KEYWORDS: power distribution network planning; smart power distribution network; GIS.

1 BACKGROUND

In the 21th century, traditional power distribution network technologies fail to meet the needs of industry and life with the improvement of people's living standard, the enhancement of industrial technologies and the rapid development of economy. However, the technology of smart power distribution network experiences a qualitative leap with the development of computer technology and communication technology. Electric power sectors have conducted upgrading and reconstruction of smart power distribution network based on the original power distribution network at the end of the 20th century. Power grid planning in China is not perfect and the overall and superior consciousness is not strong enough. Only geology and material cost are taken into account in power distribution network planning of industry and life. Monitoring on power grid operation and operating data of the system is not taken into account. Therefore, it is of vital importance for us to carry out a study on the planning and construction of smart power distribution network.

2 RESEARCH STATUS AT HOME AND ABROAD

Electrical Power Research Institute defined future power grid as smart power grid formally at the beginning of the 21st century. Besides, Electrical Power Research Institute cooperated with several institutes of power grid in the construction of experimental projects of smart power grid. This is the earliest research project on smart power grid of the world. Developed countries like Europe and Japan also started the research and implemented the power grid planning with the core of smart power grid in the corresponding period. China started researches on smart power grid in 2007. It is a difficulty needed to be overcome urgently for us that how to combine the strong power grid system in our country. The power grid system in China is a power supply system coordinated by power distribution networks of all levels with the main part of extra-high voltage power transmission network. Consequently, studies on smart power distribution network should be carried out with the main body of strong power grid, the basis of smart power distribution technology and automation technologies of communication, sensor, computer network and measurement, so as to construct a smart power grid of two way communication and visual monitoring.

The construction and research of smart power grid in China can be divided into three stages. The first stage is the preliminary research on smart power grid technologies and the formulation of industrial standard; the second stage is the overall planning layout of the construction and application of smart power grid; the third stage is the technical promotion and overall construction of smart power grid in China.

3 THE CONSTRUCTION AND PLANNING STATUS OF SUBSTATIONS BASED ON GIS PLATFORM IN KAIFENG CITY

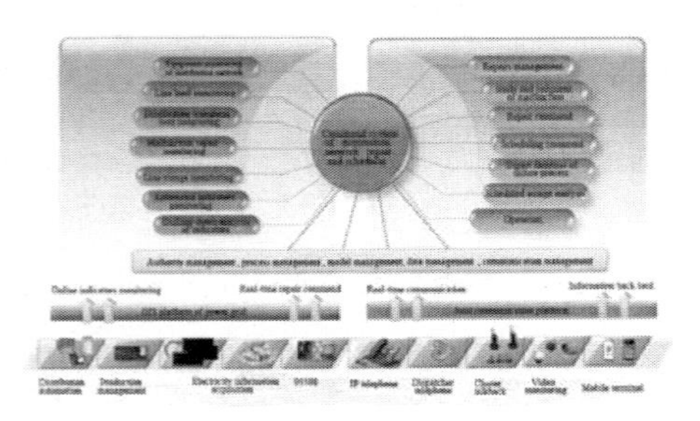

3.1 *The concept of GIS*

GIS platform is an integrated software platform, where geographic space information is showed to users authentically and effectively in forms of topological data, attribute data and temporal data through a series of computer technologies like modern surveying and mapping technology and graphic processing technology. Users are able to know local geological conditions and geographic information directly from the platform.

So, GIS platform is widely used in power system and it is of great significance for urban power distribution network management and smart power distribution network management and planning.

1 GIS platform can conduct an integrated and unified management on various graphics, images, geological data and other data of regions of smart power grid planning. Besides, it is able to write all kinds of programs with the aid of computer instructions. This is to the benefit of comprehensive analysis and processing on complicated data.
2 GIS platform provides relevant miscellaneous functions in terms of site selection of substation. Through various complicated data, it provides geological and geographical conditions of sites for selection and relevant auxiliary analyses like reasonable distributing line layout in neighboring regions, largely reducing rules for site selection personnel in the optimization of computational procedure.
3 The phenomenon of information isolated island appears frequently in network distribution. GIS platform can avoid information isolated island completely and maintain the consistency of data.

3.2 *The minimum cost model of substation planning based on GIS platform construction*

1 Both geographical locations of substations and economic benefits should be taken into account in the construction of a substation. We should try to reduce network loss rate while the pace of modernization is speeding up and the land cost is increasing. So, it is an optimal choice for us to build substations on load points. The increment of the land cost for substations should be added in the minimum cost model: $C_4 = \mu(x_i, y_i, s_i)$.
Here, Si stands for the floor space of the ith new substation.
2 There are another two constraint conditions should be considered in the construction of substations. One condition is the capacity load rate, which means that a substation cannot bear negative load. The other is that the requirements of power supply radius should be satisfied. There are also many problems in applications. A major problem is the change of geographical location of power load, which means that the load of the former substation is removed to the supply radius of another substation for various reasons. Besides, load rates of some substations are high while load rates of some other substations are relatively low. As for problems in practical applications, a penalty function is added in the objective function to substitute constraint conditions.
The penalty function is expressed as:

$$\lambda\Phi\left(\sum_{j\leq J_i} P_j - \frac{S_i}{r_i}\right) + \gamma(\sum_{i=1}^{N}\sum_{j\leq J_i} \varphi\,(d_{ij} - d_{max}))$$

The function with the largest positive integer is defined as:

$$\varphi(x) = \begin{cases} 1, x>0 \\ 0, x\leq 0 \end{cases}$$

3 Site selection and capacity determination of substation based on GIS have become the planned

development direction with the economic development of the city and the improvement of communication technology, network technology, surveying and mapping technology and geological exploration technology of the nation. How to combine the optimization theory with GIS information perfectly with strong operability is a problem worth discussing. In this paper, geographic information is taken as a constraint condition of the objective function of optimization calculation. The calculation quantity of this condition is quite small. The processing of geographic information at an earlier stage is realized by the method of offline and visualization. The improved objective function is shown in the formula:

$$\min C = C_1 + C_2 + C_3 + C_4 + \lambda_\varphi(\sum_{j \le ji} p_j - S_i / r_i) + r\sum_{i=1} \varphi(d_{ij} - d_{max})$$

The constraint condition is:

$$V_i = (x, y) \in A\{A_1, A_2, A_3 \cdots\}$$

Here: $v_i(x, y)$ stands for the real geographic location of the ith new substation. Locating at a geographic layer, A stands for the collection of available lands in the geographic layer. This collection is created by data processing of layers contained in site selection factors of multiple substations.

4 TECHNICAL SCHEME DESIGN OF THE CONSTRUCTION OF SMART POWER DISTRIBUTION NETWORK IN KAIFENG CITY

The establishment of smart power grid meeting the economic development needs of the city is an indispensable link of economic construction and development. The city mainly contains one city, four counties and five districts, including many remote rural areas. The landform is quite complicated. For this reason, reasonable and improved planning on a technical framework should be considered as the precondition for the construction of advanced smart power grid in the city.

A reasonable and strong grid structure is the safe, economic and reliable basis for the economic operation of power grid. Rational smart power distribution equipment and an improved application system constitute the smart power distribution network architecture of the city. The original power distribution framework is relatively weak if the fundamental

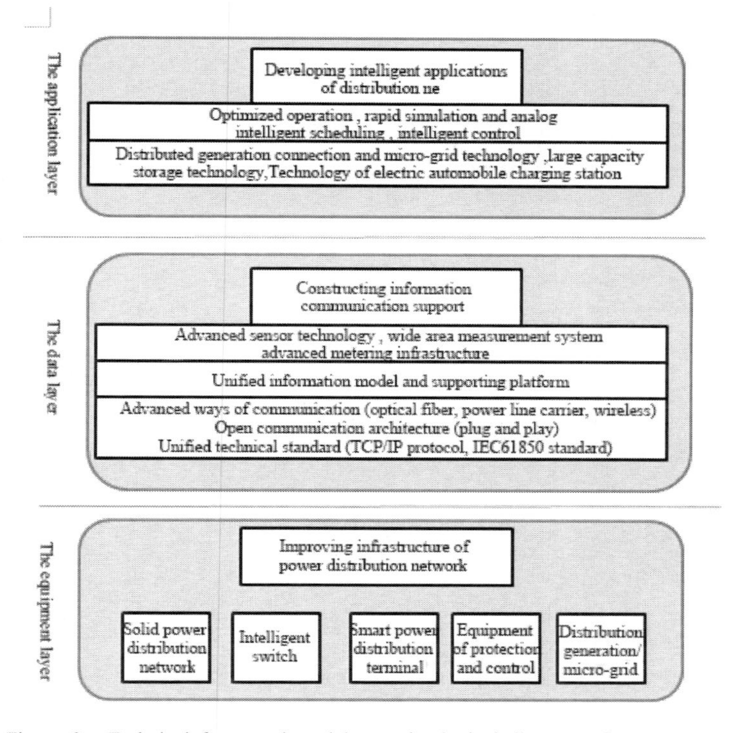

Figure 2. Technical framework and key technological diagram of smart power distribution network.

realities of the city are taken into account. For this reason, in the construction of smart distribution network, the original weak power distribution network structure is upgraded to a new type of power distribution network supported by a strong network structure. In this way, the future power distribution network structure can respond to emergency incidents like natural disasters more flexibly so as to lay a solid foundation for smart power distribution network. Equipment like intelligent switch, smart power distribution terminal, protection and control, and distributed generation/micro-grid connection is installed with the support of s strong network structure. Thus, we can further improve the infrastructure of power distribution network and construct a solid hardware structure.

We should strengthen studies on supporting point in order to make the development of power distribution network in the city more intelligent, which is able to not only conduct real-time monitoring on operating states of power distribution equipment and surroundings of power distribution network but also transmit data to the monitoring center. With the development of smart power grid technology, more and more smart micro-grids will be incorporated into smart power grid under the encouragement of national policies. Hence, in the planning of smart power grid, we should take into account such problems as the calculation and storage of mass data and the construction of data platform. It is necessary to build up a sensing measurement system, an open communication platform and a unified information platform at the data layer.

In the three-layer technical framework, the application layer is the nearest layer from end-users. In order to control and manage interconnected micro-grids and distributed generations more perfectly and guarantee the safe, economic, reliable and stable operation of power distribution network, we should carry out analyses and calculations on data of the data layer so as to ensure the stable operation of the application layer of power system.

5 CONCLUSION

The study on smart power distribution network is of historical significance, which will bring immeasurable economic benefits and strategic importance for the development of national economy in China. Therefore, analyses and studies on the status of smart power grid will provide a substantial basis for the construction of smart power grid. The smart power grid technology will be another qualitative leap of power distribution network technology. Complete coverage of smart power grid will make the power distribution network system in our country more and more economic and reliable.

REFERENCES

[1] Q/GDW 156–2006. *Guidelines on the Planning and Design of City Power Grid*, [S].

[2] Fan, M.T. (2008). *A study on relevant issues of the development strategy of power distribution network in China*, Beijing: China Electric Power Press.

[3] H. L. Willis., H. Tram., M. V. Engel., *et al.* (1995). Optimization applications to power distribution, *IEEE Computer Application Sinpower*, 18(4): 25–31.

[4] Wang, P.Y. (1998). A challenge of the new century——new trend of the power system, *Power System Technology*, 22(11): 1–3.

[5] Zhou, L. *A Study on Intelligent Decision Support System and Calculation of Power Distribution Network Planning Based on GIS*: [Doctoral Dissertation].

[6] Chow, M.Y., Tram, H. (2007). Application of fuzzy multi-objective decision making in spatial load forecasting, *IEEE Transactions on Power System*, 12(3): 1360–1366.

Electric, Electronic and Control Engineering – Shao, Shu & Tian (Eds)
© 2015 Taylor & Francis Group, London, ISBN: 978-1-138-02842-5

An approach for improving the resolution of MUSIC

Wei Feng, Yuwen Wang, Zhenzhen Li & Zhili Wang
School of Aeronautics and Astronautics, University of Electronic Science and Technology of China, Chengdu Sichuan, China

ABSTRACT: At present, Resolution of the Multiple Signal Classification (MUSIC) algorithm decreases under severe environments such as low SNR, a small number of snapshots, and incidence from a close angle. This paper puts forward the concept of subspace spectrum, the difference between the MUSIC spatial spectrum and sub space spectrum forms a new spatial spectrum. In the new spatial spectrum, a peak which is complete aliasing will be divided into two peaks. This is verified experimentally. It is proved to be a good method to improve resolution.

KEYWORDS: resolution, MUSIC, Spatial spectrum, Peak separation.

1 INTRODUCTION

Multiple signal classification (MUSIC) algorithm was put forward by Schmidt in 1979[1]. It has caused a tremendous upsurge of scientific interest due to its high resolution, high estimation accuracy and good stability. Based on MUSIC, a weighted MUSIC algorithm (WMUSIC) and Root MUSIC algorithm (Root-MUSIC) were proposed in [2] and [3]. In order to realize Direction of Arrival (DOA) estimation of Correlated or Coherent Signals, the decorrelation algorithms have been proposed in [4]-[6].

There is a big deviation when MUSIC is applied in the scene with low SNR and a small sample. The performance of the algorithm rapidly declines. MUSIC has been improved and promotion in [7]-[10]. An algorithm named SSMUSIC was presented in [7]. In this paper, weighted signal subspace projection and noise subspace projection were used to obtain synthetic spectrum to improve resolution. The authors have improved the resolution of algorithms by the theory of random matrix and correlation matrix in [8]. On SSMUSIC basis, an improved algorithm was proposed that can make full use of noise subspace (ISSMUSIC) in [9] and [10].

2 SYSTEM MODEL

Consider a uniform linear array (ULA) composed of M sensors and K narrowband signals of different DOAs $\theta_1,\theta_1,\cdots,\theta_K$. Received data model

$$\boldsymbol{X}(t) = \boldsymbol{AS}(t) + \upsilon(t) \tag{1}$$

Where $\boldsymbol{X}(t) \in C^{M\times 1}$, $\boldsymbol{A} \in C^{M\times K}$ is a direction vector matrix of array, $\boldsymbol{S}(t) \in C^{K\times 1}$ space signal vector, $\upsilon(t) \in C^{M\times 1}$ noise vector.

Covariance matrix of the received data

$$R = E\left[X(t)X^H(t)\right] = AR_S A^H + \sigma^2 I \tag{2}$$

Where R_s is source covariance matrix, σ^2 is the power of noise. Noise signal and source signal are uncorrelated, so the eigen-decomposition of covariance matrix $\boldsymbol{R}$ showed as (3)

$$R = U_S \Sigma_S U_S^H + U_N \Sigma_N U_N^H \tag{3}$$

Where U_S is signal subspaces consist of eigenvectors which correspond large eigenvalues, U_N is the noise subspaces consist of eigenvectors corresponding the small eigenvalues. In the ideal condition, signal subspaces and noise subspaces are orthogonal. In practical application, the actual receives the data matrix is based on a finite number of observed data, and so U_N are incomplete orthogonalised. Spatial spectrum function is as (4).

$$\boldsymbol{P}_{MUSIC}(\theta) = \frac{1}{a^H(\theta) U_N U_N^H a(\theta)} \tag{4}$$

3 IMPROVED MUSIC ALGORITHM

3.1 *Aliasing of spectrum peak*

The cause of high spatial resolution of MUSIC is that it obtained receiving data from a number of sensors and through accumulation of data to approximate the array covariance matrix. The more data are received, the more accurate of the covariance matrix and, the better performance of the algorithm. However, at low SNR or smaller snapshots, the covariance matrix estimation error is a large value, even aliasing in spectrum peak is caused, especially when the two signals incident from a close angle. The appearance of aliasing influenced the performance of the MUSIC algorithm seriously.

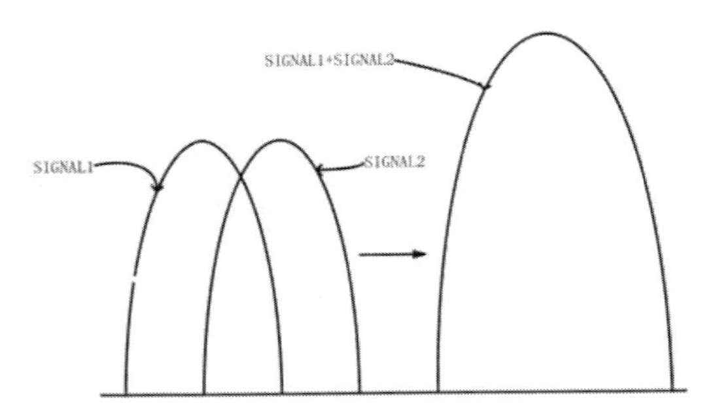

Figure 1. A complete aliasing peak.

If two peaks mix into one peak, we call it complete aliasing. Fig.1 is a complete aliasing peak. Otherwise, we call it incomplete aliasing. We can get two peaks for incomplete aliasing. In this paper, we just study complete aliasing peaks.

3.2 *Synthesize signal*

We know that the two vectors can be synthesized one vector by parallelogram law. If the same theory applies to the signals, two signals with close incident angle can be synthesized one signal, this is a virtual signal.

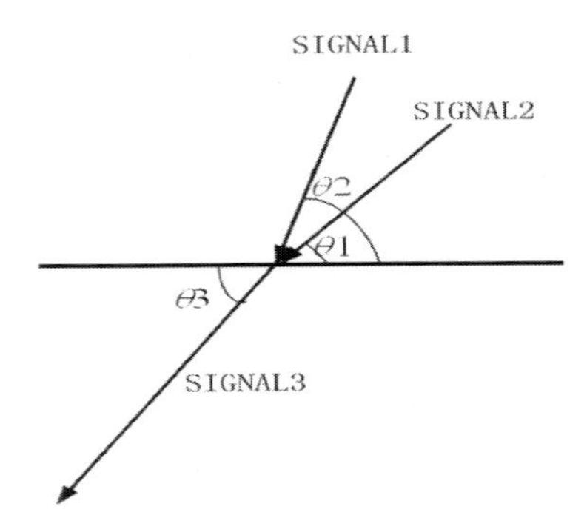

Figure 2. Synthesize virtual signal.

As shown in Fig. 2, virtual signal SIGNAL3 is synthesized the real signal SIANAL1 and SIGNAL2. From the parallelogram law, $\theta 3 \in (\theta 1, \theta 2)$. The power of SIGNAL3 is approximately equal to the sum of SIGNAL1 and SIGNAL2 when the two signals have close angle.

3.3 *Signal peak separation*

Supposing that there are two signals coming from afar, *signal1* and *signal2*, the difference α in the incidence angle of two signals was a small value, $\theta 1$ and $\theta 2$ are the incident angle of the two signals. With low SNR and small sample data, $\boldsymbol{R}$ is the covariance matrix of the received data, direction vector $a(\theta)$, $u_1, u_2, \ldots, u_M$ are eigenvectors of $\boldsymbol{R}$, noise subspaces $\boldsymbol{U}_N = [u_3, u_4, \cdots, u_M]$, Spatial spectrum

$$\boldsymbol{PMUS}(\theta) = \frac{1}{a^H(\theta) U_N U_N^H a(\theta)} \tag{5}$$

We assume that *signal3* is a virtual signal synthesized by *signal1* and *signal2*. $\boldsymbol{R}$ is the covariance matrix of the received data of *signal3*. *Its* noise subspace $U_N{}' = [u_2, u_3, \cdots, u_M]$, and we get the spatial spectrum of the virtual signal *signal3*.

$$\boldsymbol{PMUS}1(\theta) = \frac{1}{a^H(\theta) U_N' U_N'^H a(\theta)} \tag{6}$$

In this case, $\boldsymbol{PMUS}(\theta)$ and $\boldsymbol{PMUS}1(\theta)$ had only one peak. We make the following hypothesis. $\theta 1$ and $\theta 2$ are the incident angles of *signal1* and *signal2*.

Hypothesis 1: The angle θ that corresponded to the peak for *signal3* located at $(\theta 1, \theta 2)$.

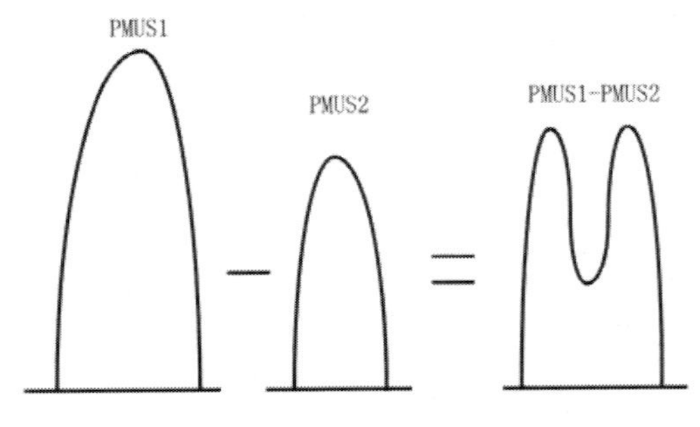

Figure 3. Peak separation.

$\boldsymbol{PMUS}(\theta)$-$\boldsymbol{PMUS}1(\theta)$, we get two peaks in the new spatial spectrum, as shown in Fig.3. In the latter experiment, we will verify the correctness of the two hypotheses.

$\boldsymbol{PMUS}1(\theta)$ is called subspace spectrum, now we give the definition of subspace spectrum.

If we assign some signal eigenvalues to noise subspace, then we will get a new spatial spectrum. We call it subspace spectrum.

The angle $\hat{\theta}$ corresponded a peak in spatial spectrum ***PMUS*(θ)**, ***PMUS*1(θ)** is a subspace spectrum of ***PMUS*(θ)**, now we give the definition of Peak separation.

If there are two peaks in the new spatial spectrum ***PMUS*(θ)– *PMUS*1(θ)**, one belongs to $[\hat{\theta}-\alpha, \hat{\theta}]$, the other lies in $[\hat{\theta}, \hat{\theta}+\alpha]$, but only one peak in ***PMUS*(θ)** when $\theta \in [\hat{\theta}-\alpha, \hat{\theta}+\alpha]$, we call it Peak separation of ***PMUS*(θ)**.

Where α is the resolution threshold of MUSIC algorithm. If the angle difference of two incident signal is greater than α, there are two peaks of the two signals in the MUSIC spatial spectrum. Otherwise there is just only one peak in the MUSIC spatial spectrum.

3.4 *Improved MUSIC algorithm based on peak separation*

The previous section, we discussed that aliasing peak minus non- aliasing peak. In practical applications, we will encounter the case of non-aliasing peaks minus non-aliasing of spectral peaks.

Supposing that there are two signals coming from afar–*signal1* and *signal2*. The difference α in the incidence angle of two signals is a large value. The spatial spectrum ***PMUS*(θ)** has two peaks, ***PMUS*1(θ)** has just one peak in the ideal condition. So, ***PMUS*(θ)– *PMUS*1(θ)** is two non-aliasing of spectral peaks in the subtraction. Under ideal conditions, the peak of ***PMUS*1(θ)** is less than the corresponding peak in ***PMUS*(θ)**, because one signal is divided into noise subspace, the power of noise increase, signal power decrease. The peak's angles of ***PMUS*1(θ)** and ***PMUS*(θ)** are equal. So there is no Peak separation in the new spatial spectrum of ***PMUS*(θ)– *PMUS*1(θ)**.

There are two possibilities of ***PMUS*(θ)– *PMUS*1(θ)**, if the peak of ***PMUS*1(θ)** is a complete aliasing peak, Peak separation occurs, otherwise, no Peak separation in the new spatial spectrum.

Description of MUSIC algorithm based on peak separation is as follows:

1. Get the MUSIC spectrum (*PMUS*) by the Covariance matrix of the received data.
2. If the number of peaks in *PMUS* is equal to the number of signals, end the algorithm. Otherwise let $n=0$.
3. $n=n+1$, dividing the last M-n eigenvectors into noise subspace, computing spectrum subspace *PMUS*(n).
4. Check the number i of occurrence of Peak separation of *PMUS*- *PMUS*(n), marking the peak, which occurring Peak separation and a peak no longer participates in the examination once it is marked.
5. If $m>K$, the algorithm fails, end the algorithm; if $m=K$, end the algorithm.
6. If n = K-1, the algorithm fails, otherwise return to Step 3.

Where M is the number of sensors, K is the number of signals. Finally, an unlabeled peak of *PMUS* represents an incident angle, if a peak is marked, the two peaks after Peak separation of the marked peak represent two incident angles.

4 COMPUTATIONAL EXAMPLES AND ANALYSIS

In order to illustrate the feasibility and effectiveness of the algorithm, we construct two examples.

Example 1 We consider a simple example of K=2 equi-power sources impinging on the array from $\theta 1=0°$ and $\theta 2=-5°$, SNR=0dB in Fig.4, the ULA is constituted of M=10 sensors, the sample size is set to L=100. We have applied MUSIC, WMUSIC, ISSMUSIC, the method in this paper (BPSMU) for comparison.

As shown in Fig.4, Peak separation occurred with the peak for *PMUS*, we can distinguish the two signals directly. In Fig.5, the method in this paper has better performance, it has a high success rate when SNR=-5dB.

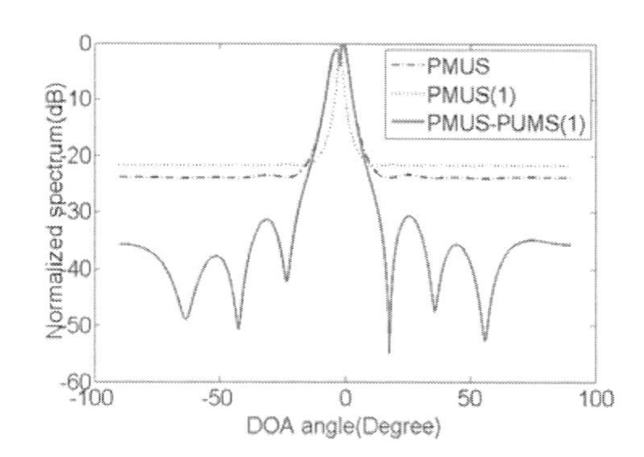

Figure 4. K=2, Peak separation.

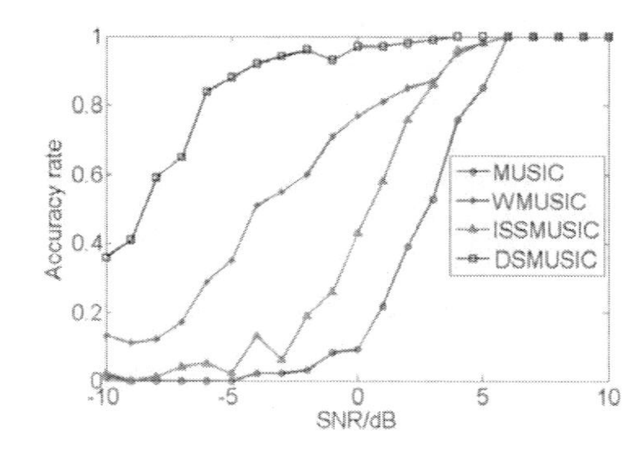

Figure 5. K=2, success probability of DOA estimation.

Example 2 We consider a simple example of K=3 equipower sources impinging on the array from $\theta 1=0$, $\theta 2=-5°$ $\theta 3=-25°$, SNR=0dB in Fig.5, the ULA is constituted of M=10 sensors, the sample size is set to L=100.We have applied MUSIC, WMUSIC, ISSMUSIC, method in this paper (BPSMU) for comparison.

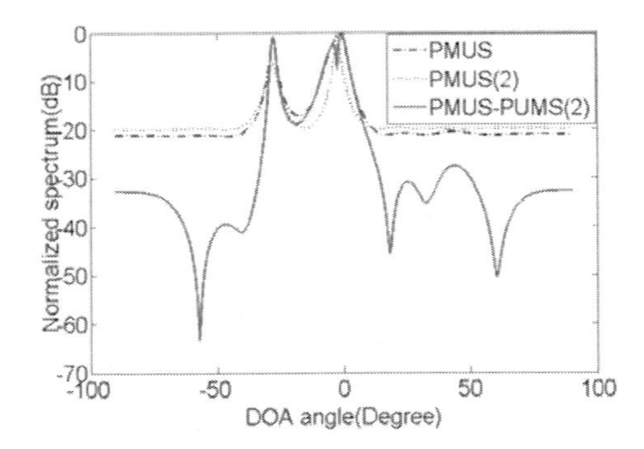

Figure 6. $K=3$, Peak separation.

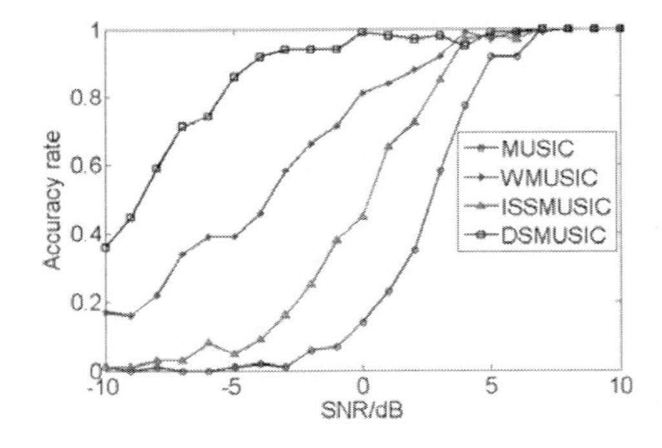

Figure 7. K=3, success probability of DOA estimation.

In Fig.6, Peak separation occurred with the peak for *PMUS*, we can distinguish the three signals directly, but it needs the second iteration. In Fig.7, the method in this paper has the best performance, and it has a high success rate as in example 1.

5 CONCLUSIONS

In this paper, an improved MUSIC algorithm based on peak separation is proposed. It is mainly applied in the places with low SNR and a small number of snapshots. Compared with MUSIC, WMUSIC and ISSMUSIC, algorithm in this paper has better performance in low SNR. However, it requires a large amount of calculation and need to be improved.

REFERENCES

[1] Ralph O. Schmidt. (1986) *Multiple Emitter Location and Signal Parameter Estimation*. IEEE Trans. On AP, pp:276–280.

[2] Stoica P, Nehorai A. (1989) MUSIC, maximum likelihood, and Cramer-Rao bound. IEEE Trans. On ASSP, 37(5):720–741.

[3] Barabell A J. (1983) Improving the resolution performance of eigenstructure-based direction finding algorithms. ICASSP, pp:336–339.

[4] Shan T J, Kailath T. (1985) Adaptive beamforming for coherent signals and interference[C]. IEEE Trans. Acoust. *Speech Signal Process*, pp:527–536.

[5] Di A. (1985) Multiple sources location a matrix decomposition approach[C]. IEEE Trans. on ASSP, pp:1086–1091.

[6] Gao Shiwei, Bao Zheng. (1988) A family of data-based matrix decomposition for high-resolution array processing of coherent signals. *Journal of China Institute of Communications*, 9(1):4–13.

[7] MeCloud M L, Scharf L L. (2002) A new subspace identification algorithm for high resolution DOA estimation. *IEEE Transaction on Antennas and Propagation*, 50(10):1382–1390.

[8] Mestre X. (2006) Estimating the eigenvalues and associated subspaces of correlation matrices from a small number of observations. *Processing of 2nd International Symposium on Communications, Control and Signal Processing*. Marrakech: IEEE Press, pp:121–125.

[9] Lan Xiaoyu. (2012) *Research on Super-resolution Direction Finding Algorithm for Improving the Resolution of Spatial Spectrum Estimation*. Harbin: Harbin Engineering University, pp:42–46.

[10] Shi Weijian, Lan Xiaoyu, Liu Xue. (2013) *Direction of arrival estimation for improving resolution performance of spatial spectrum algorithm. Journal of Applied Sciences*, 31(4):381–386.

Electric, Electronic and Control Engineering – Shao, Shu & Tian (Eds)
© 2015 Taylor & Francis Group, London, ISBN: 978-1-138-02842-5

Crashworthiness optimization design of triangular honeycombs under axial compression

Qiang He* & Dawei Ma
School of Mechanical Engineering, Nanjing University of Science and Technology, Nanjing Jiangsu, China

ABSTRACT: The full-scale elaborator finite element (FE) model of triangular honeycomb is created through LS-DYNA. Focus on the energy absorption of triangular honeycomb under axial crushing, the cell length and foil thickness of the triangular honeycomb are chosen as the design parameters. Optimal Latin Hypercube Design (OLHD) method is used for the selection of sampling design points from the design space. For all the cases, the stable and progressive folding deformation patterns are developed. An accurate surrogate modelling method named as response surface method (RSM), is adopted to achieve maximum specific energy absorption (SEA) capacity and minimum peak crushing stress (σ_{peak}) are optimized. The results demonstrate that the method is effective in solving crashworthiness design optimization problems.

1 INTRODUCTION

Metal honeycomb structures have been applied in a wide range of fields, including aerospace, marine and railway engineering. As the energy absorber, the honeycomb structure is expected to absorb as much impact energy as possible per unit mass or volume, especially when used in aerospace system. However, designing optimization using non-linear tools for evaluating the objectives and constraints is expensive and time consuming. In the design process, statistical methods are incorporated to enhance the effects of various parameters in order to optimize the crashworthiness characteristics. Therefore, meta-models are developed as surrogates of the expensive simulation processes of model approximation, optimization and space exploration design[1]. Kurtaran et al.[2] used response surface methodology (RSM) to construct approximations and applied the approximate optimization method to solve the crashworthiness design problems. The design of experiments (DOE) of the factorial design and D-optimal criterion techniques was employed to construct the response surface (RS) by Hou et al. [3] in the crashworthiness of the regular hexagonal thin-walled columns for different sectional profiles. S. S. Esfahlani et al. [4] identified the performance of various meta-models to optimize the energy absorption characteristic of hexagonal honeycomb. Li et al. [5] found the most optimized alternative square honeycomb structure via employing the RSM in crashworthiness design.

However, most of these above-mentioned crashworthiness optimizations were focused on the hexagonal honeycomb. To facilitate the application of triangular honeycomb in energy-absorbing systems, it is essential to understand its energy absorption characteristics. To date, there are limited researches on mechanical characteristics of triangular honeycomb under axial impact. In this study, full-scale elaborator triangular aluminum honeycomb is constructed based on explicit nonlinear finite element simulation method. Then, optimal Latin hypercube design (OLHD) of experiments method is used to determine the sampling design points. Based on the design of experiments (DOE) results, the RSM is applied to build surrogate models that relate SEA_m and σ_{peak} to the geometric variables. Optimization is then carried out by using multi-objective particle swarm optimization (MOPSO) algorithm.

2 CRASHWORTHINESS INDICATORS

There are a number of indicators available to evaluate the energy absorption capabilities of different structures. SEA_m is defined as the amount of absorbed energy absorbed by per mass of the structure and can be formulated as follows [3]:

$$SEA_m = \frac{E_{total}}{M} = \frac{\int_0^{\delta} F(\delta)d\delta}{M} t \tag{1}$$

*Corresponding author: *18260098162@163.com*

Where δ is the axial crushing distance. M is the total mass of the structure and E_{total} is total absorbed energy by the honeycomb during crushing, $F(\delta)$ is the axial crushing force.

When axially loaded, the peak compression stress (σ_{peak}) in the honeycomb structure is equally important for the buffer structural design as the SEA_m. For the safety design of energy absorbers, σ_{peak} should be constrained under a certain level. Consequently, σ_{peak} is taken as an essential crashworthiness indicator. And it is defined as,

$$\sigma_{peak} = PCF / A_0 \tag{2}$$

Where PCF is the peak compression force, A_0 is the total area of the structure. For the safety design of energy absorbers, σ_{peak} should be constrained under a certain level [5].

3 FINITE ELEMENT ANALYSIS

The full-scale elaborator models of three novel honeycomb structures are simulated by ANSYS/LS-DYNA. The honeycomb structure is placed between two rigid plates. Clamped boundary conditions are applied at the bottom of the honeycomb. The lower plate is fully fixed, forming the supporting platform for the impacting, while the upper one falls at a speed of 10m/s. H denotes the honeycomb structure height. All walls are modeled using the Belytschko-Tsay four-node shell elements with three integration points through the thickness and one integration point in the element plane. An automatic single-surface contact algorithm is defined to account for the contact of the column itself during deformation, while automatic node-to-surface contact is used between honeycomb structure and rigid wall. For both static and dynamic friction, the friction coefficient of 0.2 is adopted for all contact condition. The honeycomb structure is made of aluminum alloy AA6060 T4 with mechanical properties: Young's modulus $E = 68.2GPa$, Poisson's ration $\upsilon = 0.3$, initial yield stress $\sigma_y = 80MPa$, ultimate stress $\sigma_u = 173MPa$, and power law exponent $n = 0.23$, so $\sigma_0 = 106MPa$. Since the aluminum is insensitive to the strain rate effect, this effect is neglected in the finite element analysis.

Figure 1. Full-scale elaborator model of honeycomb.

For out-of-plane impact analysis, the size sensibility of FE model is mainly embodied in the cell number of reinforced honeycomb. With H being kept 50 mm, which is enough to cause the steady state in collapse mode for observing the deformation pattern and determining the crush strength, simulations of the triangular honeycomb are carried out with 5 × 5, 7 × 7, 9 × 9, 11 × 11, 13 × 13, 15 × 15 in width and length. The calculated mean out-of-plane dynamic plateau stresses are listed in Table.1. The results show that the values of mean out-of-plane dynamic plateau stresses become stable when the cell number is more than or equal to 9 × 9.

Table 1. Mean out-of-plane dynamic plateau stress of triangular honeycomb.

Cell number	5 × 5	7 × 7	9 × 9	11 × 11	13 × 13	15 × 15
$\sigma_d(MPa)$	2.71843	2.76492	2.831928	2.848235	2.840231	2.837219

The corresponding full-scale elaborator model of triangular honeycomb structures is presented in Fig.1.

Examples of deformed patterns ($t = 0.06mm$, $l = 4mm$) in numerical simulation are shown in Fig.2. As can be seen in the figure, the progressive plastic buckling deformation waves initiate from the top and move downwards; the top plastic fold is followed by the next cycle of plastic fold; some of the outer cells have collapse patterns with shear, however the inner cells deform completely in the axial collapse mode because of the constraint by the neighbouring cells. The stable and progressive folding deformation pattern provides confidence in the extension research for further study about triangular honeycomb.

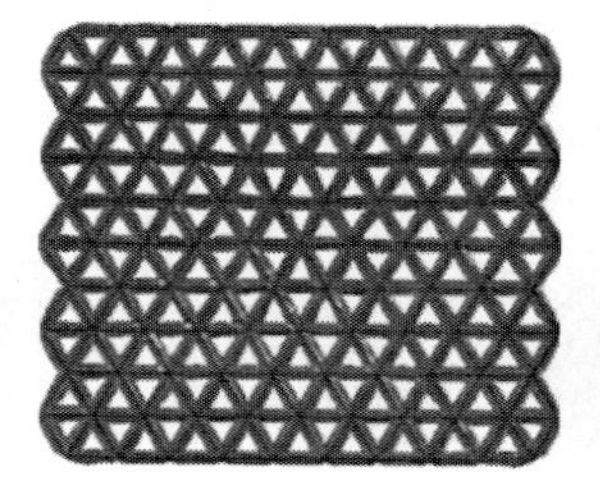

Figure 2. Deformation patterns of honeycombs.

4 OPTIMIZATION PROBLEM SET-UP

There is no doubt that the higher the SEA_m, the better the capacity of energy absorption for a structure. However, the increase of SEA_m often leads the increase of σ_{peak}. As a result, it is difficult to have two concurrently optimized objectives.

Meanwhile, it is also essential to identify the most important parameters for the crashworthiness design. Based on the above parametric study about the reinforced honeycomb, the cell wall thickness t and cell length l are taken as the design variables, with a range of 0.04 to 0.1 mm and 4 to 10 mm respectively.

Mathematically, with two objectives of SEA_m and σ_{peak}, the multi-objective optimization problem for maximizing SEA_m and minimizing σ_{peak} can be formulated as follows:

$$\begin{cases} \max \quad SEA_m \\ \min \quad \sigma_{peak} \\ s.t. \quad 0.04mm \le t \le 0.1mm \\ \quad\quad 4mm \le l \le 10mm \end{cases} \tag{3}$$

5 RESPONSE SURFACE METHOD

RSM fits a polynomial function to a set of experimental and numerical design points by making it a suitable method in design optimization. RSM has the objective of approximating the best functional relationship between dependent variable y and independent design variables ($x_1, x_2, ..., x_n$), where this relationship can be defined in a general form and matrix form as follows:

$$y(x) \approx \tilde{y}(x) = \sum_{i=1}^{N} \beta_i \varphi_i(x) + \varepsilon \Leftrightarrow \mathbf{Y} = \beta \mathbf{X} + \mathbf{E} \tag{4}$$

Here N is the number of basis functions $\varphi_i(x)$that depends on the design variables x_i and is used to approximate the model. The β_i coefficients are constant values and ε is a random error, which includes effects of random variables, measurement errors and noise factors. In addition, ε is also known as statistical error with a normal distribution [6]. The coefficients β_i is estimated using the least-squares method in order to minimize the sum of square error $E(\beta)$ between the calculated response Y and approximation function $\tilde{Y}$ at the selected experimental design points p,

$$\begin{aligned} E(\beta) &= \sum_{i=1}^{P} \varepsilon_i^2 = \varepsilon^T \varepsilon = (y - X\beta)^T (y - X\beta) \\ &= \sum_{i=1}^{P} \left[y(x_i) - \tilde{y}(x_i) \right]^2 \end{aligned} \tag{5}$$

The least square estimator of β is defined as

$$\beta = \left[\mathbf{X}^T \mathbf{X} \right]^{-1} \mathbf{X}^T \mathbf{Y} \tag{6}$$

To check the accuracy of the fitted model, the root mean square error ($RMSE$), R^2 and the relative error (RE) are applied to evaluate the meta-model's accuracy respectively, which can be written as [7]:

$$RMSE = \sqrt{\frac{1}{N_{val}} \sum_{i=1}^{N_{val}} D(y_i - \tilde{y}_i)^2} \tag{7}$$

$$R^2 = 1 - \frac{\sum_{i=1}^{m} (y_i - y_i^r)^2}{\sum_{i=1}^{m} (y_i - \overline{y}_i^r)^2} \tag{8}$$

Where N_{val} is the number of validation points. y_i and y_i are the FEA result and the approximate result by meta-model respectively.

According to Eq.(3), the crashworthiness optimization of honeycomb structure is a typical multi-objective problem. We have to impose the optimum on a Pareto set. The multi-objective optimal problem is defined by the multi-objective particle swarm optimization (MOPSO) methods.

Compared with other multi-objective optimization algorithms such as NSGA and PEAS approaches, the MOPSO that incorporates the mechanism of crowding distance computation has been widely used recently due to its relatively fast convergence and well-distributed Pareto front. The details of MOPSO can be consulted from Ref. [8]. In this paper, the MOPSO algorithm is employed to obtain the Pareto front of the two conflicting objectives of SEA_m and σ_{peak}. The Pareto front is actually a set of the non-dominated optimal solutions. The engineers can make their own decision according to the actual demand from the obtained Pareto front.

6 OPTIMIZATION OF TRIANGULAR HONEYCOMB

In order to establish the meta-models of SEA_m and σ_{peak} for the triangular honeycomb, 20 initial design points are sampled respectively using the optimal Latin hypercube design (OLHD) method [9]. The OLHD method adds a distribution criterion to the LHD. Thus, the distribution of OLHD sample points is really evener than that of LHD in the design space. These design points and the corresponding response values are listed in Table. 2.

Table 2. Sampling design points of triangular honeycomb for crashworthiness.

			RRH-H	
N	t (mm)	l (mm)	σ_{peak} (MPa)	SEA_m (kJ/kg)
1	0.0842	7.16	1.174	5.530
2	0.04	7.47	0.361	3.732
3	0.0621	9.05	0.523	4.224
4	0.0684	7.79	0.758	4.779
5	0.0526	8.11	0.481	4.107
6	0.0747	10	0.59t5	4.408
7	0.0779	5.26	1.660	6.206
8	0.1	4.95	2.644	7.248
9	0.0811	8.74	0.823	4.912
10	0.0905	9.68	0.833	4.931
11	0.0716	6.53	1.057	5.340
12	0.0937	6.21	1.707	6.264
13	0.0558	6.84	0.679	4.606
14	0.0653	4.32	1.711	6.270
15	0.0968	8.42	1.135	5.468
16	0.0432	5.89	0.578	4.367
17	0.0874	4	2.974	7.538
18	0.0463	9.37	0.320	3.585
19	0.0589	5.58	0.999	5.239
20	0.0495	4.63	1.018	5.273

Table 3. 10 additional sampling design points of triangular honeycomb for crashworthiness.

			RRH-H	
N	t (mm)	l (mm)	σ_{peak} (MPa)	SEA_m (kJ/kg)
1	0.0933	8.67	1.028	5.290
2	0.0667	4.67	1.572	6.095
3	0.0867	4	2.939	7.508
4	0.08	6.67	1.210	5.585
5	0.04	8	0.326	3.606
6	0.0733	9.33	0.641	4.520
7	0.1	6	1.982	6.584
8	0.0533	10	0.358	3.723
9	0.0467	5.33	0.755	4.773
10	0.06	7.33	0.682	4.614

In order to evaluate the accuracies of the meta-models, additional 10 points shown in Table.3 are used.In the process of generating these RS models, the quartic polynomial functions are selected as basis functions. The *RMSE* and R^2 for SEA_m are $RMSE$ = 0.0261, R^2 = 0.9726 and σ_{peak} are $RMSE$ = 0.0386, R^2 = 0.9813. Hence the surrogate models used here have been proven to be accurate for this study.

Fig.3 shows the crushing property maps of the triangular honeycomb. It is clearly seen that SEA_m is negatively correlated to σ_{peak}, which means the increase of SEA_m will always lead to increase of σ_{peak}. It is easy to optimize the design of triangular honeycomb as buffer structures based on the meta-models of SEA_m and σ_{peak}.

The Pareto frontiers of SEA_m vs. σ_{peak} is plotted in Fig.4. At the same time, the Pareto frontiers show evenly distributed solution points in the Pareto space. In fact, any point on Pareto frontiers can be an optimum, and a range of optimal solutions are supplied for the decision-maker. However, in the design of buffer structure, the σ_{peak} is required not to exceed 1.21MPa [5].

Then the optimal structural parameter for the triangular honeycomb is marked as the solid symbols in Fig.3 and the detailed design parameters are $t = 0.064mm$ and $l = 5.371mm$. We then establish the FE model of these optimal designs. The RE of RS approximate value and simulation value are also summarized in Table.4. The RE indicates that the approximation models for the dynamic responses under axial compression are accurate.

7 CONCLUSIONS

To find the optimal design of triangular honeycomb, the design optimization using response surface method (RSM) together with multi-objective particle swarm optimization(MOPSO) algorithm have been

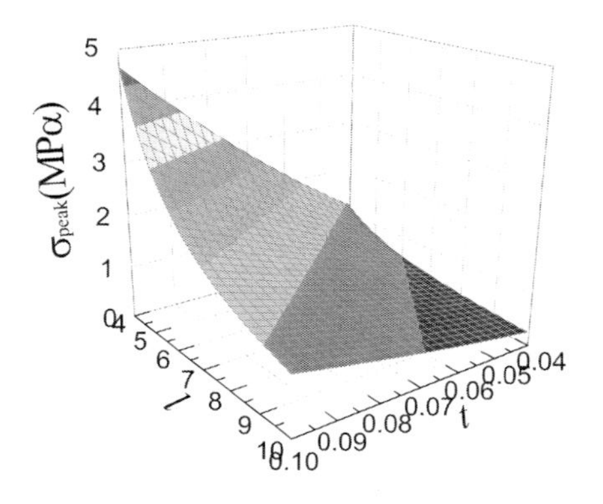

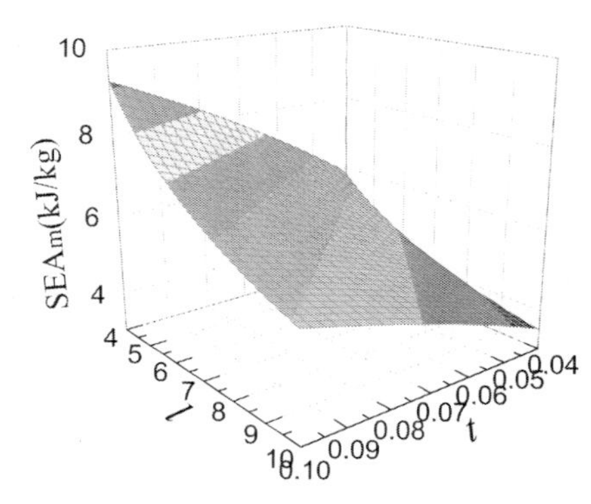

(a) Response maps of σ_{peak} (b) Response maps of SEA_m

Figure 3. Response surface maps on crashworthiness indicators.

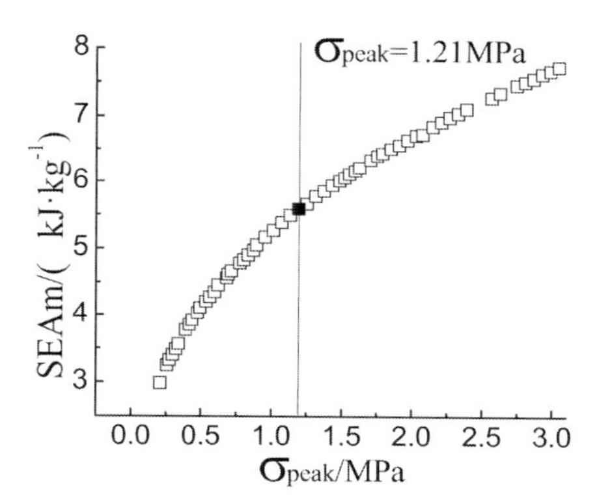

Figure 4. Pareto fronts of triangular honeycomb.

Table 4. Optimal results of triangular honeycomb structures.

Type	Terms	SEA_m (kJ/kg)	σ_{peak} (MPa)
Triangular honeycomb	Approximate value	5.582	1.204
	FEsimulation value	5.469	1.192
	RE (%)	−2.06	1.42

presented. The cell wall thickness t and cell length l are taken as the design variables while the specific energy absorption per mass (SEA_m) and the peak crushing stress (σ_{peak}) are defined to be analytic objectives for crashworthiness optimization design. The optimal design parameters are got under the same limitation of σ_{peak}. Finally, the FE model of these optimal designs is established. The optimization results correlate with the simulation results very well. This further indicates that the crashworthiness optimization of honeycombs based on surrogate models and the finite element analysis technique are feasible and effective.

REFERENCES

[1] Sun Guangyong, Li Guangyao, Zhou Shiwei, Li Hongzhou, Hou Shujuan, Li Qing. (2011) Crashworthiness design of vehicle by using multi-objective robust optimization, *Struct. Multidiscip. O.*, 4:99–110.

[2] Kurtaran H, Eskandarian A, Marzougui D, Bedewi NE. (2002) Crashworthiness design optimization using successive response surface approximations. *Comput. Mech.*, 29(4–5):409–421.

[3] Shujuan Hou, Lili Ren, Duo Dong, Xu Han. (2012) Crashworthiness optimization design of honeycomb sandwich panel based on factor screening. *J. Sandwich Struct. Mater.*, 14(6):655–678.

[4] S.S. Esfahlani, H. Shirvani, A. Shirvani, S. Nwaubani, H. Mebrahtu, C. Chirwa. (2013) Hexagonal honeycomb cell optimization by way of meta-model techniques. *Int. J. Crashworthiness*, 18(3):264–275.

[5] LI Meng, Deng Zongquan, Guo Hongwei, Liu Rongqiang, Ding Beichen. (2014) Optimizing crashworthiness design of square honeycomb structure. *J. Cent. South Univ.*, 21:912–919.

[6] Jin R, Chen W, Simpson TW. (2001) Comparative studies of meta-modeling techniques under multiple modeling criteria. *Struct Multidiscipl Optim*, 23:1–13.

[7] Fang H, Rais R, Liu Z. (2005) A comparative study of meta-modeling methods for multi objective crashworthiness optimization. *Comput Struct*, 83(25–26):2121–2136.

[8] Raquel C, Naval P. (2005) An effective use of crowding distance in multi-objective particle swarm optimization. *In: Proceedings of the 2005 conference on genetic and evolutionary computation. Washington (DC), USA; 2005*. pp:257–64.

[9] Park JS. (1994) Optimal Latin-hypercube designs for computer experiments. *J. Stat. Plan Infer.*, 39(1):95–111.

Electric, Electronic and Control Engineering – Shao, Shu & Tian (Eds)
© 2015 Taylor & Francis Group, London, ISBN: 978-1-138-02842-5

Research on service matching of single resource in cloud manufacturing

Ming Zhang, Yuling Shang*, Chunquan Li, Yuhe Chang & Neng Zhang
Guilin University of Electronic Technology, Guilin Guangxi, China

ABSTRACT: Service matching for manufacturing resource is an important issue in cloud manufacturing, which is one of the latest modes of network manufacturing. There are two ways of service matching: service matching for single resource and service matching for aggregated resources. In this paper, service matching for single resource is studied and an algorithm of the service matching is proposed. After the purposing of the algorithm a matching example is demonstrated. It is indicated in the results of the example that the algorithm works efficiently and rapidly.

KEYWORDS: cloud manufacturing; service matching; single resource; matching algorithm.

1 INTRODUCTION

Cloud computing (CComputing) is a model for enabling ubiquitous, convenient and on-demand network access to a shared pool of configurable computing resources (e.g. Networks, servers, storage, applications, and services) that can be rapidly provisioned and released with minimal management effort or service provider interactions [1,2]. Many successful CComputing business cases are found worldwidely[2]. As an excellent business service model, there are various characteristic, such as pay-as-you-go, on-demand service and multi-renter, which is very suitable for the business environment of network manufacturing. Thus, on the basis of CComputing model, Cloud Manufacturing (CM) is a new concept extending and adopting the concept of cloud computing for manufacturing [4,5]. A CM system, which serves multiple companies to deploy and manage manufacturing information and sustainable management services for accessing and exploiting over the Internet, can provide a cost-effective, flexible and scalable solution to global manufacturing enterprises by sharing complex database, software and various manufacturing resource with lower support costs [6].

In CM system, resource services are widely distributed and exist in heterogeneous networks. And the customer's requests are various and unpredictable. Under this condition, the resource service matching (RSM) is highly complicated and one of the key issues of CM. There are two ways to achieve service matching: service matching for single resource and service matching for aggregated resources. In this paper, the algorithm of service matching for single resource is studied to locate the best resource service base on the request of user efficiently and rapidly, under the assumption that the service of single resource will match the customer's request.

The rest of the paper is organized as follows: In Section II, the service matching algorithm of single resource is presented. In Section III, the simulation experiments on the example of service matching are given, and the usability of the algorithm is proved. Some conclusions are made in Section V.

*Corresponding author: *syl@guet.edu.cn*

2 SERVICE MATCHING ALGORITHM OF SINGLE RESOURCE

In many occasions single resource will meet the customer's request in CM system. Under this assumption, we focus on the study of service matching of single resource in the paper. At first, the parametric mathematical model of service resource is established. Then matching algorithm base on resource service level for single resource is presented.

2.1 *The mathematical model of service resource*

Definition 1: description of service resource. The service resource is provided by the cloud service provider (CSP). CSP registration information is unified descripted when it access CM. The description is described as:

$$S = \{N_s, I_s, P_s, F_s, R_s\}$$

Where, N_s represents the manufacturing class of CMS which is provided by CSP.

I_s represents the CSP description information which includes id, name, location, contact and remark. P_s represents the parametric description of S and is defined as:

$$P_S = \{x_i, a_i, b_i | i = 1, 2, 3, \ldots, n\}$$

Where x_i is the name of the parameter; a_i is the lower limit of the parameter; b_i is the upper limit of the parameter.

F_s represents the feedback of customers and is defined as:

$$F_s = \{x_i, f_i \mid i = 1, 2, 3, \ldots, n\}$$

Where x_i is the name of the parameter, f_i is the value of feedback level.

Definition 2: description of service request. The description is described as:

$$R = \{N_R, I_R, G_R, P_R\}$$

Where, N_R represents the manufacturing class of service request.

S_R represents the information description of service request;

G_R represents the required accuracy level for matching of the customers and is defined in (1);

P_R represents the parametric description of CMR and is defined as

$$P_R = \{x_i, c_i, d_i, e_i, p_i | i = 1, 2, 3, \ldots, n\}$$

Where x_i is the name of parameter and corresponds to the x_i of P_S; c_i is the acceptable lower limit of the parameter for the customer; d_i is the acceptable upper limit of the parameter for the customer; e_i is the expectations of the parameter; p_i is the priority level of the parameter.

$$G_R = \begin{cases} 1 \ \ level\ A \\ 2 \ \ level\ B \\ 3 \ \ level\ C \end{cases} \quad (1)$$

Define 3: manufacturing class matching.

$$M(N_S, N_R) = \begin{cases} 1, \ N_S = N_R \\ 0, \ N_S \neq N_R \end{cases} \quad (2)$$

If the manufacturing class of the service resource equals the class of the service request, they are matching in the manufacturing class.

2.2 *Matching algorithm*

At first, some definitions are made. The account of the elements of P_S is n. Define cP_k $(0 < cP_k \leq 1)$ is the matching degree of one of the service $S_k (1 \leq k \leq m)$, cP_{xi} $(0 < cP_{xi} \leq 1)$ is the weight value of the parameter x_i, which is the corresponding element in P_R and P_S.

Step 1, Calculate $M(N_S, N_R)$ with equation (2) for every service in S. The set of S_{step1} only includes the services meet the condition of $M(N_S, N_R) = 1$.

$$S_{step1} = \{S_1, S_2 \ldots\ldots S_m\} \ , m \geq 1$$

Step 2, calculate cP_{xi} on traverse of all the parameters of one of the service S_k's attribute P_S as:

If $G_R = 1$, it means service customer expect more precision in the service matching. The expectations e_i of the parameter i determines the value of cP_{xi}. cP_{xi} is the maximum value when $e_i = (a_i + b_i)/2$, and is the minimum value when $e_i = a_i$ or $e_i = b_i$. According to the characteristics of the exponential function, the expression of cP_{xi} can be defined in equation Eq. 3.

$$cP_{xi} = \frac{1}{e^{|e_i - (a_i + b_i)/2|}} \quad (3)$$

If $G_R = 2$, it means service customer expect normal precision in the service matching. If $e_i \in [a_i, b_i]$, it means that service resource meets the request of service request in the parameter i and $cP_{xi} = 1$. Otherwise, cP_{xi} still can be defined in equation Eq. 3; cP_{xi} can be defined as Eq. 4 in both cases.

$$cP_{xi} = \begin{cases} 1 & e_i \in [a_i, b_i] \\ \dfrac{1}{e^{|e_i - (a_i + b_i)/2|}} & e_i \notin [a_i, b_i] \end{cases} \quad (4)$$

If $G_R = 3$, it means service customer expects lower precision in service matching. In this case, cP_{xi} is associated with the coincidence degree of $[a_i, b_i]$ and $[c_i, d_i]$. cP_{xi} can be defined in equation Eq. 5.

$$cP_{xi} = \sqrt{\left|\frac{\min(b_i, d_i) - \max(a_i, c_i)}{\max(b_i, d_i) - \min(a_i, c_i)}\right|} \quad (5)$$

Step 3, After all the cP_{xi} have been calculated, calculate cP_k by centroid method according to Eq. 6.

$$cP_k = \frac{\sum_{i=1}^{n} cP_{xi} * P_R.p_i * G_R.f_i}{\sum_{i=1}^{n} P_R.p_i * G_R.f_i} \quad (6)$$

Step 4, After all the cP_k of every service have been calculated, sort all the services with their cP_k. The service with the maximum cP_k is the best service for matching results.

3 SERVICE MATCHING EXAMPLE

In this paper, we demonstrate the service matching for NC machine resources. The restrictive relation between parameters is ignored. The elements of P_R is listed in Table 1. Assume customer sets $G_R = 2$ and there are three services ($S = \{S_1, S_2, S_3\}$) and all three services match the request in manufacturing class. The elements of P_S is listed in Tables 2 through 4.

Table 1. Elements list of P_R.

[Parameter]	[Name]	($[c_i, d_i]$)	[e_i]	[p_i]
[x_1]	[Accuracy class]	([7,8])	[7]	[3]
[x_2]	[Cost]	([11,12])	[11]	[2]
[x_3]	[Length of the machine bed]	([8,10])	[10]	[1]
[x_4]	[Width of the machine bed]	([1.5,2])	[2]	[1]

Table 2. Elements list of P_{S1}.

[Parameter]	($[a_i, b_i]$)	[f_i]
[x1]	([8,10])	[3]
[x2]	([13,14])	[4]
[x3]	([2,8])	[3]
[x4]	([1,2])	[3]

Table 3. Elements list of P_{S2}.

[Parameter]	($[a_i, b_i]$)	[f_i]
[x1]	([6,10])	[5]
[x2]	([10,12])	[4]
[x3]	([2,10])	[3]
[x4]	([1,3])	[4]

Table 4. Elements list of P_{S3}.

[Parameter]	($[a_i, b_i]$)	[f_i]
[x1]	([7,10])	[5]
[x2]	([12,13])	[3]
[x3]	([2,9])	[4]
[x4]	([1,3])	[3]

cP_{xi} and cP_1 of S_1 can be calculated from (4) and (6):

$$cP_{x1} = e^{-2}, cP_{x2} = e^{-2.5}, cP_{x3} = e^{-5}, cP_{x4} = 1$$

$$cP_1 = \frac{e^{-2}*3*3 + e^{-2.5}*2*4 + e^{-5}*1*4 + 1*1*3}{3*3 + 2*4 + 1*3 + 1*3} = 0.21 \quad (7)$$

As the same, cP_{xi} and cP_2 of S_2 can be calculated as:

$$cP_{x1} = 1, cP_{x2} = 1, cP_{x3} = 1, cP_{x4} = 1$$

$$cP_2 = \frac{1*3*5 + 1*2*4 + 1*1*3 + 1*1*4}{3*5 + 2*4 + 1*3 + 1*4} = 1 \quad (8)$$

cP_{xi} and cP_3 of S_3 can be calculated as:

$$cP_{x1} = 1, cP_{x2} = e^{-1.5}, cP_{x3} = e^{-4.5}, cP_{x4} = 1$$

$$cP_3 = \frac{1*3*5 + e^{-1.5}*2*3 + e^{-4.5}*1*4 + 1*1*3}{3*5 + 2*3 + 1*4 + 1*3} = 0.69 \quad (9)$$

After all the calculation, cP_2 is determined to be the maximum among all the cP_i. It means S_2 is the best service for the customer.

4 CONCLUSIONS

In this paper, an algorithm of single service matching is purposed and a matching example is demonstrated. The service matching algorithm works effectively in matching example for searching the best service for customer. For further research, resource matching for the aggregated resources will be studied when single service resource cannot meet the requirement of service customer.

ACKNOWLEDGMENT

Authors would like to acknowledge the support of the following projects: National Natural Science Foundation of China (51165004, 51465013); Natural Science Foundation of Guangxi Provi nce (2012GXNSFBA053176, 2012GXNSFD A053029); Guangxi Key Laboratory of Manu facturing System & Advanced manufacturing technology (10-046-07_004, 12-071-11-61_003).

REFERENCES

[1] P Mell, T Grance. (2009)Perspectives on cloud computing and standards. National Institute of Standards and Technology (NIST). *Information Technology Laboratory*.

[2] Xi Vincent Wang, Xun W. Xu (2013) An interoperable solution for cloud manufacturing. *Robotics and Computer Integrated Manufacturing*, 29(4):232–247.

[3] Xu Xun (2012) From cloud computing to cloud manufacturing. *Robotics and Computer Integrated Manufacturing*, 28(1):75–86.
[4] Tao F., Zhang L., Venkatesh V. C., Luo Y., Cheng Y. (2011) Cloud manufacturing: a computing and service-oriented manufacturing model. *Proceedings of the Institution of Mechanical Engineers, Part B: Journal of Engineering Manufacture*, pp:1969–1976.
[5] W. D. Li, J. Mehnen (2013) *Cloud Manufacturing Distributed Computing Technologies for Global and Sustainable Manufacturing*, Springer-Verlag.
[6] Li B. H., Zhang L., Wang S. L., Tao F. (2010) Cloud manufacturing: a new service-oriented networked manufacturing model. *Computer Integrated Manufacturing Systems*, 16(1):1–7.
[7] Li Weiping, Gao Fuliang, Zhu Xuwei, Chu Weijie, Lin Huiping. (2011) Semantic based service discovery and matching Method, *Journal of Chinese Computer Systems*, 32(9):1728–1733.
[8] Wu L, Yang C. A. (2010) Solution of manufacturing resources sharing in cloud computing environment. *7th International Conference on Cooperative Design, Visualization, and Engineering, CDVE 2010*, Sept. 19–22, 2010, Calvia, Mallorca, Spain, pp:247–252.
[9] Guo H, Zhang L, Tao F, et al. (2010) Research on the measurement method of flexibility of resource service composition in cloud manufacturing. *2010 International Conference on Manufacturing Engineering and Automation, ICMEA2010, December 7–9, 2010, Guangzhou, China*, pp:1451~1454.

Electric, Electronic and Control Engineering – Shao, Shu & Tian (Eds)
 ISBN: 978-1-138-02842-5

Research on cloud storage technology of a grouting monitoring system based on the Internet of Things

Shan Gao
Guodian Dadu River Dagangshan Hydropower Development Co., Ltd., Ya'an, China

Hui Zhang, Yuewen Huang & Xiangwei Yu
Instruments and automation Institute, The Changjiang River Scientific Research Institute, Wuhan, China

ABSTRACT: The rapid development of the Internet of Things provides an international leading brand new platform for water conservancy and hydropower engineering industry development in our country, bas778ed on the advantages and characteristics of the Internet of things, cloud storage technology of the grouting monitoring system was studied. The grouting monitoring system adopts a variety of wireless network transmission technology, grouting data collecting to form many sets of grouting recorder were sent to the storage server. In the Internet environment, cloud storage technology of a grouting monitoring system of large-scale data effectively monitoring abnormal state, which provide a scientific basis for grouting construction monitoring and the technical support for the development of a grouting monitoring system based on the Internet of things.

KEYWORDS: The Internet of things; cloud storage; grouting recorder; grouting monitoring system.

1 TECHNOLOGY OVERVIEW

The development of water conservancy project construction in China has now been in the transient process from traditional to modern. The rapid development of the Internet of things provides a brand new leading international platform for synchronizing China Water Conservancy and hydropower engineering industry development and the world[1]. The Internet of things (IoT) is an important part of the new generation of information technology, it is the interconnection of uniquely identifiable embedded computing devices within the existing Internet infrastructure. Typically, IoT is expected to offer advanced connectivity of device, system, and service that goes beyond machine-to-machine communications (M2M) and covers a variety of protocols, domains, and applications [2]. The interconnection of these embedded devices (including smart objects), is expected to usher in automation in nearly all fields, while also enabling advanced applications like a Smart Grid[3].

There will be contradictions between high precision and large-scale state when grouting monitoring data in large-scale state, especially the monitoring of grouting abnormal state, has important significance for cloud storage. This paper focuses on the research of the grouting monitoring system of large-scale data and grouting cloud storage in an abnormal state, for effective monitoring of the abnormal state of large-scale data, for the small-scale cloud storage case, mainly sends a heartbeat message from each storage server to the master server periodically [4], or use the master server to register with all storage servers are polled cloud storage platform.

Master server collects the way by polling each storage server load, this research using a common low-performance PC cluster to build high-performance, highly reliable cloud storage system consists of three modules: the master server module, storage server module and client proxy module], the system architecture consists of two main server, using Active-Standby work mode, that one server is active for some services, while the other the server is in standby status, cloud storage system architecture shown in Figure 1.

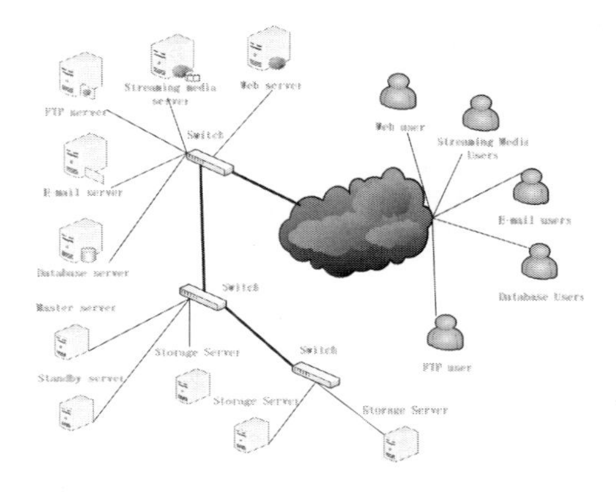

Figure 1. Structure diagram of cloud storage system.

In cloud storage system, database servers, email servers, FTP servers, streaming media servers and Web servers connect to the database user, e-mail users, FTP users, streaming media users, Web users with IoT cloud network system after the switch. Build data platform based on cloud storage technology has the following advantages:

1. High reliability. Cloud storage data stored on different servers, in the case of damaged hardware, the system will automatically read the instruction and store the file on another server. This is the main advantage of cloud storage technology, namely hardware redundancy and automatic fault switching
2. Massively parallel expansion, and low cost. Expansion of cloud storage technology is very simple, take the architecture is a parallel expansion, capacity expansion is almost unlimited, even in the case of insufficient capacity, as long as the purchase of new storage server on it.
3. Easy-management, easy-maintenance. Storage management was always very complex previously; different storage vendor management interface is different, data center personnel often need to face different storage products, to understand the various memory capacities, loads, etc.. But cloud storage is virtual, multiple storage servers can be considered as one. All the software and data of cloud storage users are stored in information center for centralized maintenance, and the maintenance is easy for the user machine as a terminal.

2 GROUTING MONITORING SYSTEM

2.1 *Structure of grouting monitoring system*

The grouting monitoring system is designed in accordance with the hierarchical distributed control system, the hierarchical distributed system is a distributed control system (DCS), the entire system can be divided into field control, process management and operational management, in the grouting monitoring system, various sensors of the grouting recorder as the field control level which is directly facing the grouting, it is the basis for all data information. The level of various sensors form a wireless sensor network collected by the respective sensor signals wirelessly to the host recorder, then collects and reprocess the recorded grouting data, and wireless transmission of collecting data to grouting Logger host, the system structure shown in Figure 2.

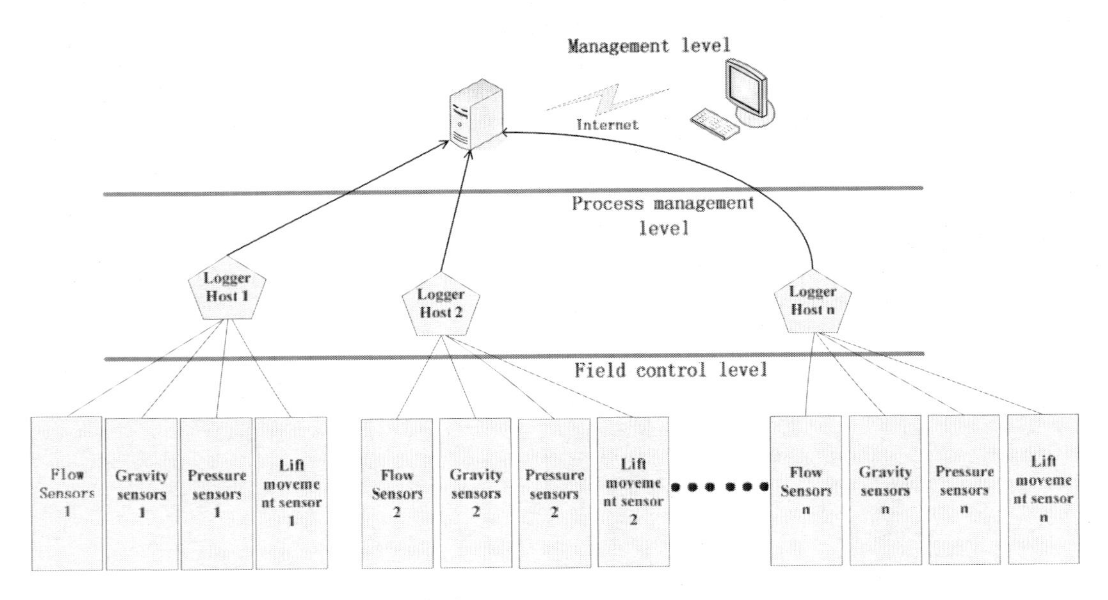

Figure 2. Structure diagram of grouting monitoring system.

Managers can browse real-time grouting data and query the historical data via browsing the grouting data management system client, sum all grouting recorder data to obtain a series of effective information, such as construction progress and quality, in order to assist management personnel to manage grouting.

2.2 *Function of grouting monitoring system*

1 Real-time and historical grouting data
According to the field grouting recorder collected data recording, real-time data can be browsed Grouting Monitoring System Based on the Internet of things in different equipment number grouting recorder, can also through the selection of equipment, the number and time information to query the historical grouting recording.

2 Abnormal data alarm notification
Data anomalies appear in the grouting data, such as high pressure, excessive flow, etc., the grouting data monitoring system will send text messages to set user to inform the specific abnormal data, so that the user can make quick decisions according to the actual situation.

3 Report of grouting data result
In the grouting data results report, the grouting data monitoring system can collect all kinds of grouting data through channels to the network server for data fusion, to form grouting result list, grouting hole sequence statistics and the sequence of holes grouting results table, and authorize different user level with different permissions, different levels of users to view and generate reports that are different.

3 DESIGN AND DEVELOPMENT

3.1 *Systematic design*

The grouting monitoring system is based on water conservancy construction-related technical specifications, based on the use of distributed database technology, wireless network technology and three-dimensional visualization technology to establish grouting management system, for grasping and controlling the progress and quality of the site grouting to achieve grouting management standardization, institutionalization and procedures. The system is divided into three subsystems which are data acquisition and process subsystem, grouting query and analysis subsystem, grouting image visualization subsystem; the system design of the grouting monitoring system is shown in Figure 3.

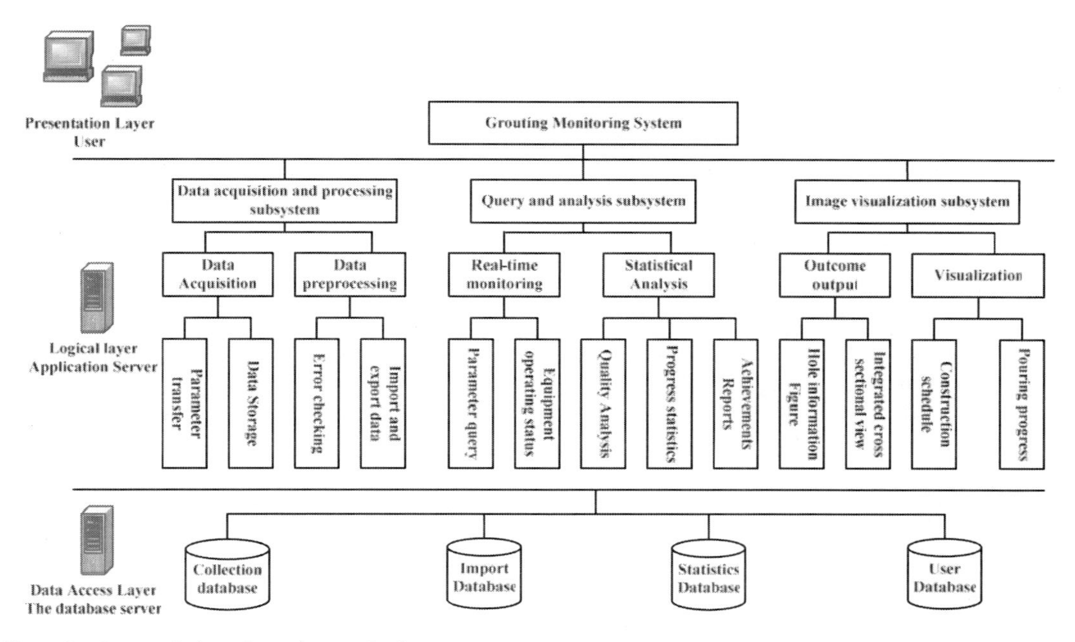

Figure 3. System design of grouting monitoring system.

3.2 *Software platform*

Grouting data monitoring system using Windows 2003 Server as storage server clusters, support for multiple instances running and multi-session, multi-desktop support multiple user terminals simultaneously linked to a storage server, it supports multiple sharing software installed on a storage server. Structure of the cloud storage software platform is shown in Figure 4.

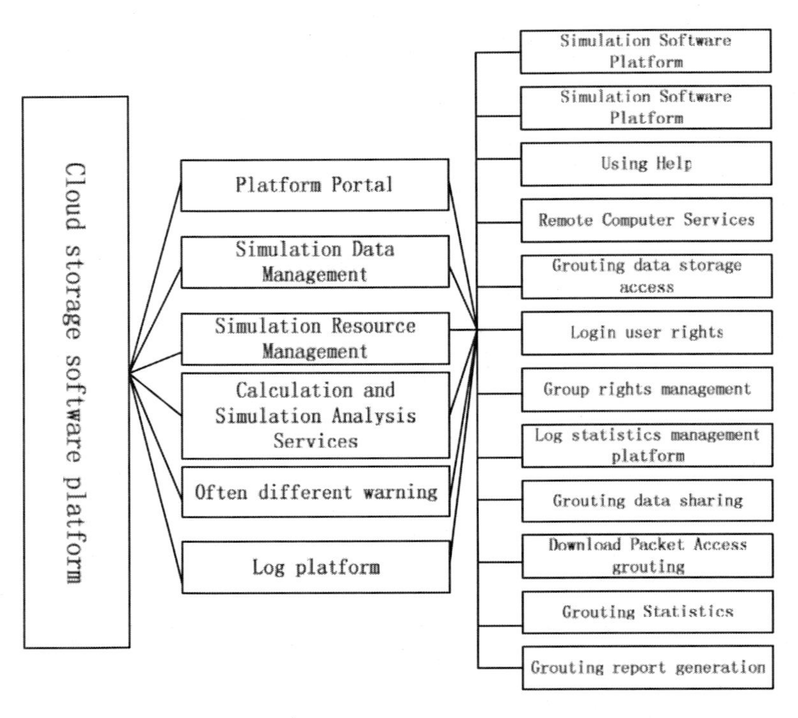

Figure 4. Structure of cloud storage software platform.

The following function platforms have been achieved by cloud storage software:

1 User and system privileges: privileges and priorities of system administrators, owners, the construction side, cloud administrator and tourist.
2 Sharing grouting data: grouting data include grouting time, flow, pressure, density, moving-up value, water-cement ratio, the total amount of grouting, the total consumption of cement, unit grouting volume, the unit consumption of cement, etc., the user via a terminal program unified login system, the system simultaneously records the user's login events.
3 Remote access: support multi-user remote access.
4 Abnormal warning: alarm condition end-user policy settings and SMS notification alarm. Including exception pages highlighted for input and overfall, pressure, density abnormalities in the form on the page to highlight tips; live SMS alert users by setting lugeon, moving-up, consumption and other threshold, and set reporting The phone number that is beyond the scope of the exception message is sent to the user's mobile phone.
5 System log statistics and report forms management: report of the query, download and print original reports and statistical reports, engineering reports, curves and three-dimensional histogram, including the result list, grouting sequence report, grouting completion report, comprehensive statistics, frequency graph, comprehensive profile maps. Figure 5 is the unit cement injection amount curve in statistical report.

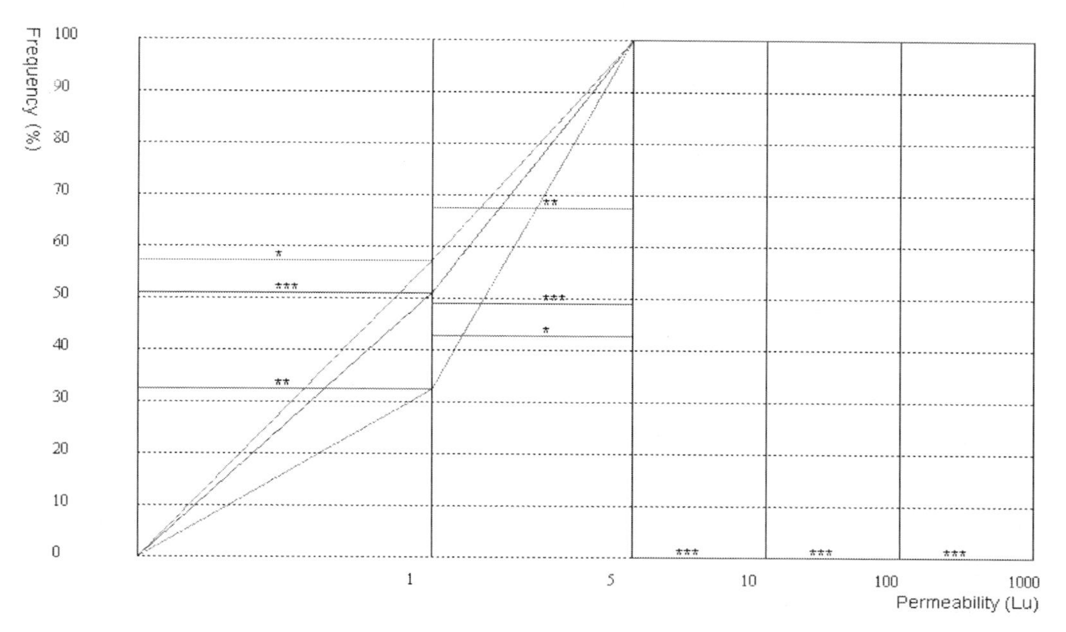

Figure 5. Graph of unit ash injection rate.

3.3 *Database file code*

Grouting monitoring database file consists of four parts: real-time data template file, grouting process database template file, grouting setting information database template files and grouting result database template file. System for grouting database template file, follow a low layer in wireless network transmission protocol, formulation of the template file, all kinds of template files associated with each other, the organic combination formed grouting monitoring basic database. Grouting real-time data template file as shown in the table:

[dataid] [int] IDENTITY (1, 1) NOT NULL, Number
[sendid] [varchar] (50) COLLATE Chinese_PRC_CI_AS NULL, Sender ID
[datatime] [datetime] NULL ,Time
[type] [char] (10) COLLATE Chinese_PRC_CI_AS NULL, Practices
[hole] [varchar] (50) COLLATE Chinese_PRC_CI_AS NULL, Hole No.
[hole_seq] [char] (10) COLLATE Chinese_PRC_CI_AS NULL, Hole No.
[block] [int] NULL ,sections No.
[total_q] [decimal](18, 2) NULL ,Cumulative flow
[q] [decimal](18, 2) NULL, Flow
[p] [decimal](18, 2) NULL, Pressure
[w] [decimal](18, 2) NULL, Density
[delta] [decimal](18, 0) NULL, Micro displacement
[qin] [decimal](18, 2) NULL, Inflow of slurry
[qout] [decimal](18, 2) NULL, Effluent of slurry
[status] [char] (10) COLLATE Chinese_PRC_CI_AS NULL, Status
[section] [varchar] (200) COLLATE Chinese_PRC_CI_AS NULL, Position

4 CONCLUSION

The rapid development of the Internet of Things (IoT) provides a brand new leading international platform for synchronizing China Water Conservancy and hydropower engineering industry development and the world, based on the advantages and characteristics of IoT, study cloud storage technology in the grouting monitoring system. Grouting monitoring system using a variety of wireless network transmission technology, the project construction site multiple grouting recorder acquisition of the grouting data sent to the storage server, network setup grouting material. In IoT environment, cloud storage technology for large-scale data grouting abnormal condition monitoring system for effective monitoring, to provide a scientific basis for grouting monitoring and management, to provide technical support for the development of the Internet of Things monitoring system based grouting.

REFERENCES

[1] Kung H T, Lin C K, Vlah D. (2011) Cloud Sense: continuous fine-grain cloud monitoring with compressive sensing. *Proceedings of the 3rd USENIX Workshop on Hot Topics in Cloud Computing*, Portland or USA, pp:15–19.

[2] J. Höller, V. Tsiatsis, C. Mulligan, S. Karnouskos, S. Avesand, D. Boyle. (2014) *From Machine-to-Machine to the Internet of Things: Introduction to a New Age of Intelligence*. Amsterdam: Elsevier.

[3] O. Monnier (2013) *A smarter grid with the Internet of Things*. Texas: Texas Instruments.

[4] Wu Haijia, Chen Weiwei, Hu Guyu. (2011) FFs: A PB-level cloud-storage system based on network. *Journal on Communications*, 32(9):24–33.

Electric, Electronic and Control Engineering – Shao, Shu & Tian (Eds)
© 2015 Taylor & Francis Group, London, ISBN: 978-1-138-02842-5

Object tracking algorithm based on feature matching under complex scenes

Jing Liu
College of Technology, Shandong University of Traditional Chinese Medicine, Jinan, Shandong Province, China

Dawei Qiu
College of Technology, Shandong University of Traditional Chinese Medicine, Jinan, Shandong Province, China
College of Computer Science and Technology, Nanjing University of Aeronautics and Astronautics, Nanjing, Jiangsu Province, China

Hui Cao
College of Technology, Shandong University of Traditional Chinese Medicine, Jinan, Shandong Province, China

Junzhong Zhang
Department of Orthopedics, Shandong Hospital of Traditional Chinese Medicine, Jinan, Shandong Province, China

Haiyan Mo
College of Technology, Shandong University of Traditional Chinese Medicine, Jinan, Shandong Province, China

ABSTRACT: Object tracking under complex scenes, especially when targets occluded, the appearances and scales changed, attracts interests of more and more researchers. In the paper, we proposed a new object tracking algorithm based on Speeded-Up Robust Features (SURF). Because of invariant to image scale, illumination changes and rotation, the algorithm firstly detected the SURF points of the tracking objects, the targets were represented by RGB histogram of these points and their neighborhood. Inspired by the fact that the objects have high similarity in successive frames, the algorithm searched the optimal SURF points in the current image, and updated the objects model according to the new states of the targets. The experimental results show that the proposed method can detect the objects' location under occlusions and the appearances changes.

KEYWORDS: Object tracking, SURF, complex scenes, occlusion.

1 INTRODUCTION

Object tracking is an important problem in computer vision, and has wide range applications in visual meeting, object surveillance, and human-computer interaction. The main goal of object tracking is to continuously and reliably acquire the targets' precise positions under changing circumstances (Weiming Hu *et al.*, 2004). For recent years, researchers make efforts to reach the goal, and proposed many object detection and tracking algorithms. These algorithms can be classified into four types (R. V. Babu *et al.*, 2010). Gradient based methods locate target via minimizing cost functions, Knowledge based methods track objects using prior information, e.g. appearance and contour. Learning based approaches build objects model, and then search the target in the coming frames. Feature based techniques extract characteristic such as luminance, color, and edges to track the moving targets.

Active Contour Model (M. Kass *et al.*, 1988) (Snake) has been hot field in image segmentation since it was been proposed in 1988. But its initialization and running speed is hard to solve when applying it to object tracking.

Learning based methods (R. V. Babu *et al.*, 2010, M. Musavi 1992, Shai Avidan 2004, Shai Avidan 2007) can model comprehensive and nonlinear objects, but these approaches need large amounts of offline samples.

In recent years, feature based methods attract people's attention. Mean shift tracking (D. Comaniciu *et al.*, 2003) and its improved algorithms represent objects with histogram, because of its fast running and effectiveness, the algorithms are one of the widely used methods. But the algorithm needs to search the tracking objects in the entire image, and the object model update need to be improved too.

Invariant features are introduced in object tracking and have obtained better performance. In (Feng Tang & Hai Tao 2005), the authors used attributed relational graph (ARG) to represent object, features detected in several successive frames are regarded as reliable features. But in real scenarios, these features are rare to detect. In (G. Carrera 2007), the authors detected interest points with Harris method, and represented it with SURF. Experimental results showed that it has good robustness under rotation and scale variation, but when the appearances changed, the method cannot update the model timely.

In the paper, we propose an object tracking method based on SURF, in which the objects are represented by histogram around SURF points, according to the variation, the objects models are updated dynamically.

The rest of the paper is organized as follows. The proposed algorithm is detailed in Section 2, the experimental results are presented in Section 3, and finally, brief conclusions are given in Section 4.

2 OBJECT TRACKING ALGORITHM BASED SURF

In object tracking field, object representation, target occlusion, scale variation, update of object model are several important problems that need to be solved.

The algorithm takes the objects in rectangles in the first frame as its input, and detects the objects in the successive frames as its output. The flow chart of the entire algorithm is showed in Figure 1.

After acquire the input, for every new frame, in order to decrease the amount of computation, the algorithm firstly determine the searching area according to the object's current state, not the entire image. The algorithm detects SURF points

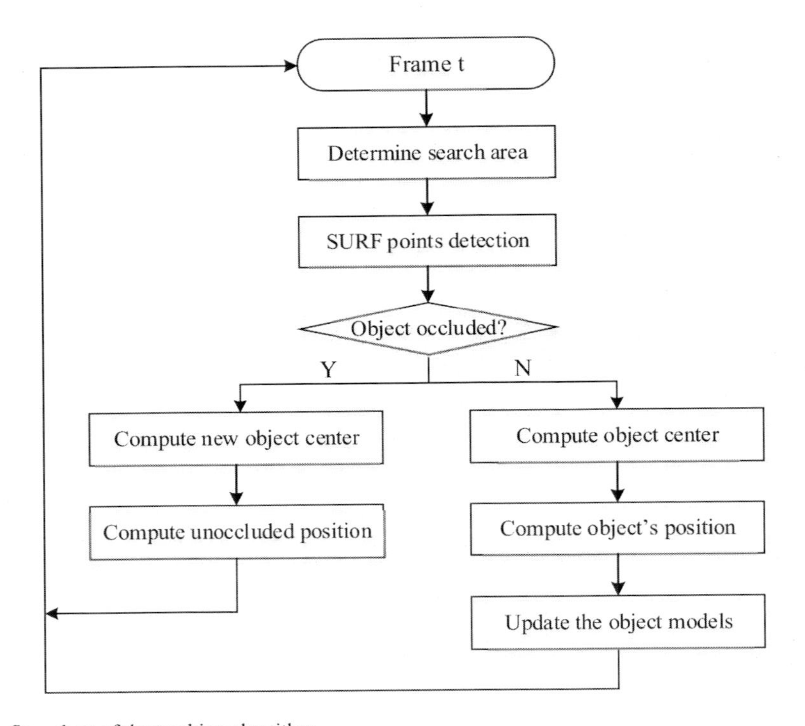

Figure 1. The flow chart of the tracking algorithm.

and compares them with those in the object model, and selects the optimal matching points. Then the algorithm determine if the occlusion is took place, if it is so, the algorithm compute the actual position of the object, else the approach computes the optimal scale and position, and updates the object model.

Object Representation. According to the input, the algorithm detects SURF points in the target area, then calculates histogram around each feature point. The object is represented by these SURF points and

Figure 2. Target representation.

their corresponding RGB histogram. It is depicted as Figure 2. The weights of histograms in different rectangles are calculated according to the distance to the object center. This can decrease the influence of the peripheral background pixels and improve the tracking precision.

Object Detection. The main task of the object detection is to calculate the searching area, then determine the position of the object according to the detected SURF points.

The determination of the searching area is based on the object's velocity and position in the previous frame. The searching area is calculated as:

$$testRect(1) = targetRect(1) - step\_w \tag{1}$$

$$testRect(2) = targetRect(2) - step\_h \tag{2}$$

$$testRect(3) = targetRect(3) + 2 \times step\_w + dx \times 2 \tag{3}$$

$$testRect(4) = targetRect(4) + 2 \times step\_h + dy \times 2 \tag{4}$$

Where *testRect* and *targetRect* represent object searching area in the current frame and target area in the previous frame, respectively. The first and second elements represent the coordinate of the top left corner of the area, and the third and fourth element represent width and height of the region. dx and dy represent the velocity of the target on the x and y axis. $step\_w$ and $step\_h$ stand for the incremental amount on x and y direction, respectively. In general, we set the incremental amount 10%-30%. The computation of the searching area is described as Figure 3.

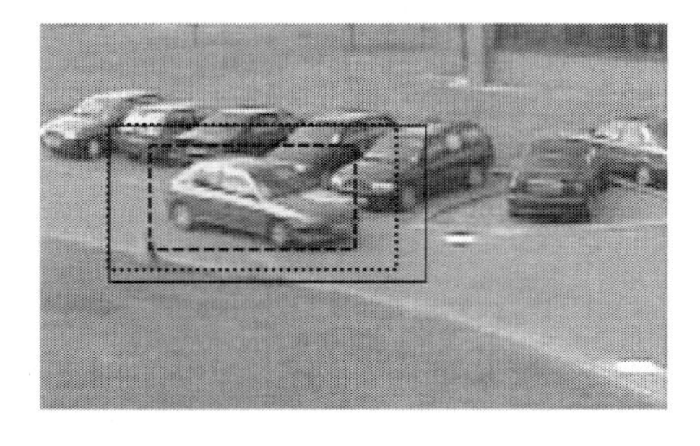

Figure 3. The computation of the target searching area.

SURF Detection. After determination of searching area, we adopt the SURF algorithm to detect all SURF points in the searching area. In order to acquire the optimal matching points, the algorithm computes the similarity between feature points in searching area and target model. The points with higher similarity above predefined threshold are selected as the optimal matching points. The similarity if calculated as Eq. 5.

$$dist(i) = \min_k\left(\sqrt{\sum_{j=1}^{n}(\boldsymbol{TI}(i).desc(j) - \boldsymbol{TS}(k).desc(j))^2}\right) \tag{5}$$

Where $\boldsymbol{TI}(i)$ represents the i th point in the searching area, $\boldsymbol{TS}(k)$ stands for the k th point in the target model.

Object Occlusion. For simplicity, the algorithm determines if the occlusion took place according to the following criteria, and takes corresponding actions to deal with:

If no matching points are detected, this means that the object is occluded totally, if the SURF points are detected in the successive frames, it means the object emerged again.

If the matching points are less than 75 percent number of optimal points in the previous frame, this means that the occlusion took place. In this case, the target center is computed according to the distance between matching points and the object center. For ease of description, we assume that the target center is $\vec{O}$, the optimal matching points are $\vec{P_i}$, the distance from target center to that matching point is:

$$\overrightarrow{OP_i} = \vec{P_i} - \vec{O} \Rightarrow \vec{O} = \vec{P_i} - \overrightarrow{OP_i} \tag{6}$$

As described in Figure 4.

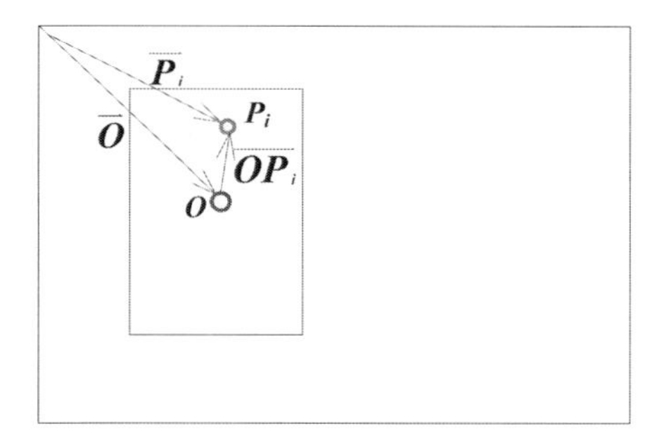

Figure 4. The computation of the target center with occlusions.

The target center is estimated as Eq. 7:

$$\hat{O} = \frac{1}{n}\sum_{i=1}^{n}(\vec{P_i} - \overrightarrow{OP_i}) \tag{7}$$

Model Update. Because of appearances, illumination and circumstances changes, the features changed continuously, so the object model should reflects these variations. In the paper, we propose an online model update scheme.

Before occlusion took place, the system stores previous M (In the paper, we set $M = 10$) frames object features (SURF points and adjacent histogram). If occlusion didn't take place, we compute the object features and discard the last $M + 1$ frame ahead of the current frame, else if occlusion took place, the update process stopped.

3 EXPERIMENTAL RESULTS

In order to evaluate the performance of the proposed algorithm, we compare the algorithm with the classic mean shift tracking (MST) algorithm. The algorithms are implemented using MATLAB 2013a on a Pentium IV, 2.40GHz, 4G RAM computer. The datasets come from PETS2007 (H. Bay 2006), the resolution of the images is 720×576. The experimental results are shown in Figure 5, the left columns are the results of the classic mean shift tracking algorithm, and the right columns are the results of the proposed algorithm.

From the experimental results, we can see that from frame 47, the classic MST algorithm shifted from the object center, and in frame 75, it lost the target. But the proposed algorithm can track the person effectively.

Due to limited space, we only show one experiment here, other experiments can also demonstrate the precision and effectiveness of our proposed algorithm.

Figure 5. The comparison of tracking results using MST and our algorithms under occlusion.

4 CONCLUSIONS

In this work, we propose a new object tracking algorithm based on SURF, which can decrease the computation afford without loss of precision. The experimental results show the effectiveness of the proposed approach. In the future, we will make efforts to enhance the robustness of the algorithm.

ACKNOWLEDGEMENTS

The authors would like to thank the anonymous reviewers for their valuable comments. This work was financially supported by the National Natural Science Foundation (NO. 81373661 and NO. 81473708), and Youth Researchers fostering Foundation, Shangdong University of Traditional Chinese Medicine (NO. ZYDXY1349 and NO. ZYDXY1358).

REFERENCES

[1] Weiming Hu, Tieniu Tan, Liang Wang & Steve Maybank. (2004) A survey on visual surveillance of object motion and behaviors. *IEEE Transactions on Systems, Man and Cybernetics*, 34(3):334–352.

[2] R. V. Babu, S. Suresh, and A. Makur. (2010) Online adaptive radial basis function networks for robust object tracking. *Computer Vision and Image Understanding*, 114(3):297–310.

[3] M. Kass, A. Witkin & D. Terzopoulos. (1988) Snakes: Active contour models. *International Journal of Computer Vision*. 1(4):321–331.

[4] M. Musavi, W. Ahmed, K. Chan, K. Faris & D. Hummels. (1992) On the training of radial basis function classifiers. *Neural Networks*, 5(4):595–603.

[5] Shai Avidan. (2004) Support Vector Tracking. *IEEE Transactions on Pattern Analysis and Machine Intelligence*, 26(8):1064–1072.

[6] Shai Avidan. (2007) Ensemble Tracking. *IEEE Transactions on Pattern Analysis and Machine Intelligence*, 29(2):261–271.

[7] D. Comaniciu, V. Ramesh & P. Meer. (2003) Kernel based object tracking. *IEEE Transactions on Pattern Analysis and Machine Intelligence*. 25(5):564–577.

[8] Feng Tang & Hai Tao. (2005) Object tracking with dynamic feature graph. *Visual Surveillance and Performance Evaluation of Tracking and Surveillance, 2005. 2nd Joint IEEE International Workshop on*, pp:25–32.

[9] G. Carrera, J. Savage & W. Mayol-Cuevas. (2007) Robust Feature Descriptors for Efficient Vision-Based Tracking. *Progress in Pattern Recognition, Image Analysis and Applications (CIARP)*, pp:251–260.

[10] H. Bay, T. Tuytelaars, and L. Van Gool. (2006) Surf: Speeded up robust features. *Computer Vision – European Conference on Computer Vision 2006*, pp:404–417.

Electric, Electronic and Control Engineering – Shao, Shu & Tian (Eds)
© 2015 Taylor & Francis Group, London, ISBN: 978-1-138-02842-5

Wide area backup protection algorithm based on fault area detection

Shan Li, Yigang He & Maoxu Liu
School of Electrical Engineering and Automation, Hefei University of Technology, Hefei Anhui, China

ABSTRACT: In order to reduce the communication amount of wide area system, and improve the fault-tolerance of wide area backup protection, a wide area backup protection algorithm based on fault area detection is proposed. Protection information of local line and adjacent lines are weighed to get the comprehensive value, then comparing with threshold to identify the faulty component. Meanwhile, the bus sequence voltage of promoter stations is sorted to obtain the association faulted bus. Then suspected faulted line and fault area are determined according to correlation matrix to reduce the communication amount of system and accelerate the speed of fault element identification. The simulation tests based on IEEE 4 generator 8-bus system illustrate the effectiveness and high fault-tolerance capability of the algorithm with the mis-operation and refuse-operation of multiple protections.

KEYWORDS: wide area backup protection; fault area detection; weighted comprehensive; fault-tolerance performance.

1 INTRODUCTION

The grid structure is complex, the traditional backup protection based on local measurements of information exist difficult setting, action duration, and it is difficult to adapt to non presupposition of change or the operating condition of the power grid structure. Furthermore, it is difficult to distinguish between line internal fault and overload when the flow transfer happens .Then it is easy to cause the cascading trip accident and increase the instability risk of the power grid under perturbation[1-2]. In recent years, with the development of wide area measurement technology (wide area measurement system, WAMS), network protection based on wide area information has received extensive attention [3]. The existing high voltage power grid main protection mainly uses different kinds of pilot protection, because of easy setting, the correct action rate high, fast speed and it is not affected by the load transfer, and the wide area information existing limitations of synchronization, propagation delay, reliability. Therefore, improving the performance of relay protection based on wide area information is mainly concentrated on the back-up protection [4].

Using wide area information recognizing faulty components reliably is the core of the current wide area backup protection research. The wide area backup protection system mainly uses two kinds of wide area information to realize the protection criterion: a class is the protection relay and circuit breaker status information; another kind is the voltage, current and other electrical information[5]. Literature [6] presented a wide area protection algorithm based on longitudinal direction principle, which is simple in principle, without strict synchronous sampling. However, high resistance grounding, non whole phase operation, power backward and conversion fault have great impact on the direction element, and is the lack of a clear solution to this problem. Literature [7] presented a wide area differential protection algorithm based on current information, the algorithm is simple in principle, but extremely high demands on the synchronization. And the higher, unbalanced current caused by the accumulated error of measuring multi -point current value in wide area and capacitance, current compensation cause problems in the wide area current differential protection. Literature [8] determined fault location through fault-tolerant discrimination. The algorithm required less information, can correctly identify faulty components when any protection information about protection element is missing. However, multiple protection malfunction or refuse to move, and the algorithm is difficult to get the correct result of the judgment.

Wide area information acquisition range is larger, and it is very difficult to have precise anti-interference performance as a traditional protective device. Due to the device or channel fault there will be part of the protection information is missing or wrong. Therefore, the principle of wide area protection must have strong fault tolerance. Wide-area backup protection system requires the integration of multiple sub-station information to identify faulty components. For complex, large grid systems, if information of all intelligent electronic devices within the region is uploaded to the central station, system decision-making center traffic must be larger. Therefore, communication traffic should be reduced on the premise of reliable to identify faults. Literature [9] presented the limited wide area protection division based on graph theory through the correlation matrix of power grid topology, reduced communication traffic. However, the relationship of fault points and protected area can not be reflected dynamically. Literature [10] presented the fault area detection strategy, but the sensitivity of start criterion and the reliability of bus voltage sorting need to be improved further.

In order to reduce communication and realize a simple and higher fault tolerance protection, this paper presents a wide area backup protection algorithm, using traditional protection information to identify faulty components. Based on the degree of influence of each protection fault identification, protection action values are weighted comprehensive to obtain comprehensive protection value and compared with threshold to identify faulty components. The algorithm reduces the amount of wide area information communication through fault area detection, and accelerates the speed of fault element identification by suspected faulted line detection. The simulation tests based on IEEE 4 generator 8-bus system illustrate the effectiveness and high fault-tolerance capability of the algorithm with the mis-operation and refuse-operation of multiple protections.

2 FAULT AREA DETECTION OF WIDE AREA PROTECTION SYSTEM

The core idea of wide area protection is to use wide-area measurement information in the grid and determine the specific location of the fault element through information fusion calculation [10]. In order to realize the wide area backup protection function, it is necessary to obtain information from the relevant region associated with a protected object, and combine multiple sub-station information to identify faulty components. In fact, the scale of the power grid is large. If each sub-station information within protection area uploads to the central station, it will cause obstruction of the wide area communications network, and reduce the speed of the decision-making center for information processing. Therefore, to reduce the amount of wide area information transmission and improve the speed of fault recognition, the first step has determined fault area. Then only the required information about fault area is uploaded to decision-making center.

2.1 *Association faulted bus detection*

When a fault occurs in the power grid, only the voltage and current of the bus and line close to the point of failure change obviously, and the current and voltage variation of buses away from the point of failure is small. Then sub-stations start criterion can be created according to this principle. Start criterion is as follows:

$$(|\dot{U}_{m2}| \geq K_N U_N) \cup (|\dot{U}_{m0}| \geq K_Z U_N) \cup (|\dot{U}_{m1}| \leq K_P U_N) \quad (1)$$

Where $\dot{U}_{m2}$, $\dot{U}_{m0}$ and $\dot{U}_{m1}$ are the negative-sequence, zero-sequence and positive-sequence bus voltages of bus "m", and U_N is the rated voltage magnitude of the bus. K_N and K_Z are the proportional coefficients of negative-sequence and zero-sequence components to decide the thresholds. The thresholds for K_N and K_Z are set at 0.1 for an unbalanced fault. The reason of choosing low threshold settings for K_N and K_Z is that the combined information of negative-sequence and zero-sequence voltage magnitudes could improve the sensitivity of start criterion during a high-resistance earth fault. K_P is the positive-sequence proportional coefficient. The threshold of K_P is set at 0.5 for a balanced fault. Higher threshold is chosen in order to avoid frequent start of sub-stations under normal switching of the power system.

2.2 *Suspected faulted line detection based on correlation matrix*

After bus sequence voltages uploaded to the decision-making center, they will be sorted by the decision-making center. Under normal circumstances, if the negative-sequence or zero-sequence amplitude of a bus is maximum or its positive-sequence amplitude is minimum, it is the nearest bus from the point of failure. However, because of the impact of short line, complex fault, and measurement errors, the selection of associated bus may be wrong. In order to ensure redundancy ordering, select top 3 as fault correlation bus. Then suspected fault line can be detected through the following steps.

Step 1: The formation of correlation matrix based on bus and line

A is defined as correlation matrix of bus and line, describes the connection relationship of bus and line.

$$A_{ij} = \begin{cases} 1 & \text{bus } i \text{ is connected to line } j \\ 0 & \text{bus } i \text{ is not connected to line } j \end{cases} \quad (2)$$

According to the principle, A can be calculated as follows for the systems of Fig.1.

$$A = \overbrace{\begin{bmatrix} 1 & 1 & 1 & 0 & 0 & 0 & 0 & 0 & 0 & 0 \\ 0 & 0 & 1 & 1 & 0 & 0 & 0 & 0 & 0 & 0 \\ 0 & 0 & 0 & 1 & 1 & 0 & 0 & 0 & 0 & 0 \\ 0 & 0 & 0 & 0 & 1 & 1 & 1 & 0 & 0 & 0 \\ 0 & 1 & 0 & 0 & 0 & 0 & 1 & 1 & 0 & 0 \\ 0 & 0 & 0 & 0 & 0 & 1 & 0 & 1 & 1 & 0 \\ 0 & 0 & 0 & 0 & 0 & 0 & 0 & 0 & 1 & 1 \\ 1 & 0 & 0 & 0 & 0 & 0 & 0 & 0 & 0 & 1 \end{bmatrix}}^{Line:\ 1\ 2\ 3\ 4\ 5\ 6\ 7\ 8\ 9\ 10} \begin{matrix} B1 \\ B2 \\ B3 \\ B4 \\ B5 \\ B6 \\ B7 \\ B8 \end{matrix} \quad (3)$$

Step 2: The formation of fault correction matrix

After selecting fault correlation buses, all elements of the rows which fault correlation buses located in the A will be retained, all the remaining elements are defined as zero. Setting B_1, B_2, B_8 are detected as fault correlation buses, do the above treatment for A. Then fault correction matrix A_1 can be obtained as follows:

$$A_1 = \overbrace{\begin{bmatrix} 1 & 1 & 1 & 0 & 0 & 0 & 0 & 0 & 0 & 0 \\ 0 & 0 & 1 & 1 & 0 & 0 & 0 & 0 & 0 & 0 \\ 0 & 0 & 0 & 0 & 0 & 0 & 0 & 0 & 0 & 0 \\ 0 & 0 & 0 & 0 & 0 & 0 & 0 & 0 & 0 & 0 \\ 0 & 0 & 0 & 0 & 0 & 0 & 0 & 0 & 0 & 0 \\ 0 & 0 & 0 & 0 & 0 & 0 & 0 & 0 & 0 & 0 \\ 0 & 0 & 0 & 0 & 0 & 0 & 0 & 0 & 0 & 0 \\ 1 & 0 & 0 & 0 & 0 & 0 & 0 & 0 & 0 & 1 \end{bmatrix}}^{Line:\ 1\ 2\ 3\ 4\ 5\ 6\ 7\ 8\ 9\ 10} \begin{matrix} B1 \\ B2 \\ B3 \\ B4 \\ B5 \\ B6 \\ B7 \\ B8 \end{matrix} \quad (4)$$

Step 3: The formation of confirmed matrix for suspected faulted line

Searched each column of fault correction matrix, and made all elements of each column do OR operation, confirmed matrix of suspected faulted line A_2 is formed.

$$A_2 = \overbrace{\begin{bmatrix} 1 & 1 & 1 & 1 & 0 & 0 & 0 & 0 & 0 & 1 \end{bmatrix}}^{Line:\ 1\ 2\ 3\ 4\ 5\ 6\ 7\ 8\ 9\ 10} \quad (5)$$

1 in A_2 represents suspected faulted line, 0 in A_2 represents a normal line. Setting up the total number of independent lines connected to the association faulted bus is N, and a set of suspected faulted line is constructed.

$$L = \{Li, i = 1, \cdots, N\} \quad (6)$$

For the system shown in Figure 1, if B_1, B_2, B_8 are detected as fault correlation buses, then the set of suspected faulted line is $L = \{L1, L2, L3, L4, L10\}$.

3 FAULT IDENTIFICATION

In order to increase the redundancy of information, and improve the fault-tolerance of system, this paper using the line protection information such as main protection, distance protection, and directional element, using the adjacent line protection information such as directional element. The protection information is fused to obtain comprehensive protection value, then comparing with threshold to identify the faulty component.

3.1 *Protection element action coefficient*

Action coefficient of direction element is defined as follows:

$$f = \begin{cases} +1 & \text{Fault in the positive direction} \\ -1 & \text{Fault in the opposite direction} \\ 0 & \text{No fault} \end{cases} \quad (7)$$

Action coefficient of distance element is defined as follows:

$$d = \begin{cases} 1 & \text{Distance protection element is action} \\ 0 & \text{Distance protection element is not action} \end{cases} \quad (8)$$

3.2 *Fault identification algorithm based on a variety of protected information*

This paper defines two correlation coefficients: The line correlation coefficient M_1 and the adjacent line correlation coefficient M_2.

M_1 represents the line protection information on their comprehensive reaction of line fault, and the calculation formula is as follows:

$$M_1 = 6(A + A') + 6(B_I + B_I') + 3(B_{II} + B_{II}') + 2(B_{III} + B_{III}') + 3(C + C') \quad (9)$$

Where A and A' are the main protection action values of the side and contralateral, B_I and B_I' are the first distance protection action values of the side and contralateral, B_{II} and B_{II}' are the second distance protection values of the side and contralateral, B_{III} and B_{III}' are the third distance protection values of the side and contralateral, and C and C' are the directional protection action values of the side and contralateral. The five weights of main protection, first zone protection, second zone protection, third zone protection, and directional element represent the degree of each protection information to fault confirmation. They were taken as 6, 6, 3, 2, and 3 according to literature [11].

M_2 represents the adjacent line protection information on their comprehensive reaction of line fault, and the calculation formula is as follow:

$$M_2 = \sum_{i}^{N_n}(-3C_{iN} + 3C_{iF}) \tag{10}$$

Where C_{iN} and C_{iF} are the directional element action values of adjacent line proximal and distal, i is serial number of other adjacent line, N_n is the number of adjacent lines.

1 Calculation of comprehensive protection values and threshold values for lines
Comprehensive protection value of each line:

$$F_{out.Li} = M_{1Li} + M_{2Li} \tag{11}$$

Threshold value of each line

$$F_{set.Li} = M_{1set.Li} + M_{2set.Li} \tag{12}$$

If the following criterion is satisfied, then the line is identified as faulted line:

$$F_{out.Li} \geq F_{set.Li} \tag{13}$$

Threshold value can be obtained through the following calculation according to thinking of literature [12-13]: Assuming the IED located at the head of line is failure, when fault is located in the head, $M_1 = 6\times1+6\times0+3\times1+2\times1+3\times1=14$, when the fault is located in the end $M_1 = 6\times1+6\times1+3\times1+2\times1+3\times1=20$. Then $M_{1set}=\frac{1}{2}(14+20)=17$. When half of adjacent line directional, elements refuse to move, the value of M_2 is M_{2set}, that is $M_{2set}=\frac{1}{2}N_n$. Threshold value is shown as :$F_{set}=M_{1set}+M_{2set}=17+\frac{1}{2}N_n$.

4 SIMULATION ANALYSIS OF THE ALGORITHM

In order to validate the algorithm, IEEE 4 generator 8-bus system is used for analysis. Using PSCAD / EMTDC to build the system shown in Fig.1. Setting A phase fault occurs in the middle of line L3. At this time, B1, B2, and B8 are detected as fault correlation buses. L1, L2, L3, L4, and L10 are detected as suspected faulted line.

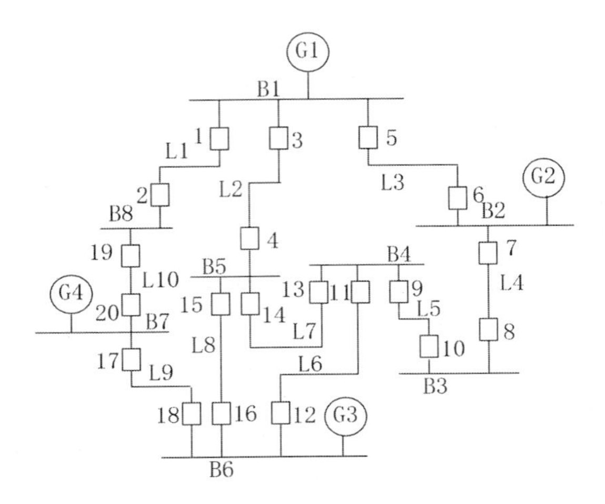

Figure 1. IEEE4-generator 8-bus system.

In Fig.1, Number 1-20 are serial number of circuit breaker IED, L1-L10 stand for lines, and B1-B8 stand for buses.

4.1 *All information of one IED in faulted line misses*

Setting A phase fault occurs in the middle of line L3, and all protection information of IED5 is missing. In this case, judgment results of suspected faulted lines are shown as Tab.1.

Table 1. Judgment results of suspected faulted lines when IED5 fails to operate.

Line	M_1	M_2	F_{out}	F_{set}	Judgment result
L1	2	15	17	26	Normal
L2	2	9	11	29	Normal
L3	20	18	38	26	Fault
L4	2	3	5	23	Normal
L10	0	0	0	26	Normal

From the results of fault diagnosis, it can be learned that even if one IED in faulted line is fail, faulty line can still be accurately identified.

4.2 *All information of one IED in faulted line and one IED in normal line miss*

Setting A phase fault occurs in the middle of line L3, and all protection information of IED4 and IED5 are missing. In this case, judgment results of suspected faulted lines are shown as Tab.2.

Table 2. Judgment results of suspected faulted lines when IED4 and IED5 fail to operate.

Line	M_1	M_2	F_{out}	F_{set}	Judgment result
L1	2	12	14	26	Normal
L2	-3	9	6	29	Normal
L3	20	15	35	26	Fault
L4	2	3	5	23	Normal
L10	0	0	0	26	Normal

From the results of fault diagnosis, it can be learned that even if one IED in faulted line and one IED in normal line are fail, faulty line can still be accurately identified.

4.3 *All information of one IED in faulted line is wrong*

Setting A phase fault occurs in the middle of line L3, and all protection information of IED5 is wrong. At this time, 5 kinds of protective action for IED5 are as follows: 0(refuse-operation), 0(refuse-operation), 0(refuse-operation), 0(refuse-operation), -1(mis-operation). In this case, judgment results of suspected faulted lines are shown as Tab.3.

Table 3. Judgment results of suspected faulted lines when all information of IED5 is wrong.

Line	M_1	M_2	F_{out}	F_{set}	Judgment result
L1	2	18	20	26	Normal
L2	2	12	14	29	Normal
L3	17	18	35	26	Fault
L4	2	0	2	23	Normal
L10	0	0	0	26	Normal

From the results of fault diagnosis, it can be learned that even if all information of one IED in faulted line is wrong, faulty line can still be accurately identified.

4.4 *All information of one IED in normal line is wrong*

Setting A phase fault occurs in the middle of line L3, and all protection information of IED4 in line L2 is wrong. At this time, 5 kinds of protective action for IED4 is as follows: 1(mis-operation), 1(mis-operation),1 (mis-operation),0(refuse-operation), -1 (mis-operation). In this case, judgment results of suspected faulted lines are shown as Tab.4

Table 4. Judgment results of suspected faulted lines when all information of IED4 is wrong.

Line	M_1	M_2	F_{out}	F_{set}	Judgment result
L1	2	6	8	26	Normal
L2	9	6	15	29	Normal
L3	40	12	52	26	Fault
L4	2	6	8	23	Normal
L10	0	0	0	26	Normal

From the results of fault diagnosis, it can be learned that even if all information of one IED in normal line is wrong, faulty line can still be accurately identified.

5 CONCLUSION

This paper proposes a wide area backup protection algorithm based on fault area detection, using protection information of local line and adjacent lines for fault identification. Protection information of local line and adjacent lines are weighed to get the comprehensive value, then comparing with threshold to identify the faulty component. In this paper, the bus sequence voltage of promoter stations is sorted to obtain the association faulted bus. Then suspected faulted line and fault area are determined according to correlation matrix to reduce the communication amount of system and accelerate the speed of fault element identification. The simulation tests based on IEEE 4 generator 8-bus system illustrate the effectiveness and high fault-tolerance capability of the algorithm with the mis-operation and refuse-operation of multiple protections. Even if one IED in faulted line and one IED in normal line are failed, faulty line can still be accurately identified. Meanwhile, when all information of one IED in faulted line or in normal line is wrong, faulty line can still be accurately identified.

ACKNOWLEDGMENTS

This work was supported by the National Natural Science Funds of China for Distinguished Young Scholar under Grant No. 50925727, The National Defense Advanced Research Project Grant No.C1120110004, 9140A27020211DZ5102, the Key Grant Project of Chinese Ministry of Education under Grant No.313018, Anhui Provincial Science and Technology Foundation of China under Grant No. 1301022036 and Hunan Provincial Science and Technology Foundation of China under Grant No.2010J4 and2011JK202.

REFERENCES

[1] Xu Huiming, Bi Tianshu, Huang Shaofeng, et al. (2007) Study on wide area measurement system based control strategy to prevent cascading trips. *Proceedings of the CSEE, 2007*, 27(9):32–38.

[2] Yi Jun, Zhou Xiaoxin. (2006) A survey on power system wide-area protection and control. *Power System Technology*, 30(8):7–14.

[3] Adamiak M G, Apostolov A P, Begovic M M, et al. (2006) Wide area protection-technology and infrastructures. *IEEE Transactions on Power Delivery*, 21(2):601–609.

[4] Li Zhenxing, Yin Xianggen, Zhang Zhe, et al. (2011) Study on system architecture and fault identification of zone-division wide area protection. *Proceedings of the CSEE, 2011*, 31(28):95–103.

[5] Wang Rui, Wang Xiaoru, Huang Fei, et al. (2013) Wide area backup protection algorithm for power grid based on correlation matrix. *Automation of Electric Power Systems*, 37(4):69–74.

[6] Yang Zengli, Shi Dongyuan, Duan Xiaozhong. (2008) Wide-area protection system based on direction comparison principle. *Proceedings of the CSEE, 2008*, 28(22):77–81.

[7] Renan Giovanini, Kenneth Hopkinson, et al. (2006) A primary and backup cooperation protection system based on wide area agents. *IEEE Transactions on Power Delivery*, 21(3):1222–1230.

[8] Zhang Baohui, Zhou Liangcai, Wang Chenggen, et al. (2010) Wide area backup protection algorithm with fault-tolerance performance. *Automation of Electric Power Systems*, 34(5):66–71.

[9] Li Zhengxing, Yin Xianggen, Zhang Zhe, et al. (2010) A Study of zone division on limited wide area protection system. *Automation of Electric Power Systems*, 34(19):48–51.

[10] Eissa M M, Masoud M E, Elanwar M M. (2010) A novel back up wide area protection technique for power transmission grids using phasor measurement unit. *IEEE: Transactions on Power Delivery*, 25(1):270–278.

[11] Tan J C, Crossley P A, Mclaren P G, et al. (2002) Application of a wide area backup protection expert system to prevent cascading outages. *IEEE Trans on Power Delivery*, M(2):375–380.

[12] Cong Wei, Pan Zhencun, Zhao Jianguo. (2006) A wide area relaying protection algorithm based on longitudinal comparison principle. *Proceedings of the CSEE, 2006*, 26(21):8–14.

[13] Tian Congcong, Wen Minghao, Liu Hang, et al. (2012) Wide-area backup protection system based on information fusion with adjacent substation. *Automation of Electric Power Systems*, 36(15): 83–90.

Electric, Electronic and Control Engineering – Shao, Shu & Tian (Eds)
 ISBN: 978-1-138-02842-5

Fuzzy comprehensive evaluation of university network security risks

Tao Li
Weifang University of Science and Technology, Shouguang Shandong, China

ABSTRACT: As the development of computers and smart phones promotes the rapid development of network, university network security has become a social issue of common concern. This paper first determines the evaluation index of university network security risk, then fixes the weight of each index using the AHP, and finally establishes a fuzzy mathematics evaluation model of university network security risk. The risk is divided into five levels, which can be determined according to the expert group evaluation. And the university network security management could be promoted.

KEYWORDS: Colleges and universities; network security; fuzzy mathematics; evaluation.

With the rapid development of computers and smart phones, network technology continues to grow. The wireless network basically covers the whole area, especially in the high- knowledge, high-tech colleges and universities. Student dormitories, office buildings, and the campuses have realized wireless high-speed Internet access, which is increasing the university reliance on network. Once the campus network emerges safety problems, the impact on the colleges will be great.

The influence factors of university network include physical factors, user factors, external factors and management factors. In this paper, the university network security risk is divided into five levels according to the four main factors. By its various indexes, the university network security risk is evaluated.

1 DETERMINATION OF UNIVERSITY NETWORK SECURITY RISK EVALUATION INDEX

By looking up literature, combined with the actual situation in some colleges and universities in Shandong area, university network security risk evaluation index system is determined as follows (see Table 1).

Table 1. University network security risk evaluation index of P.

First-level index	The second-level index
Physical factors P1	Computer hardware equipment P11;Computer network equipment P12; Computer server environment P13;
The user factor P2	Users security sense P21;The number of users P22; The distribution of the users P23;
External factors P3	The hacker attacks P31;Virus invasion P32;Unauthorized access P33;
Management factors P4	The administrator security sense P41;The number of the administrators P42;The administrator management skills P43; Administrators' management permissions P44;

2 DETERMINATION OF THE WEIGHT OF EACH INDEX BY USING AHP

Firstly, establish a hierarchical structure as shown in Figure 1:

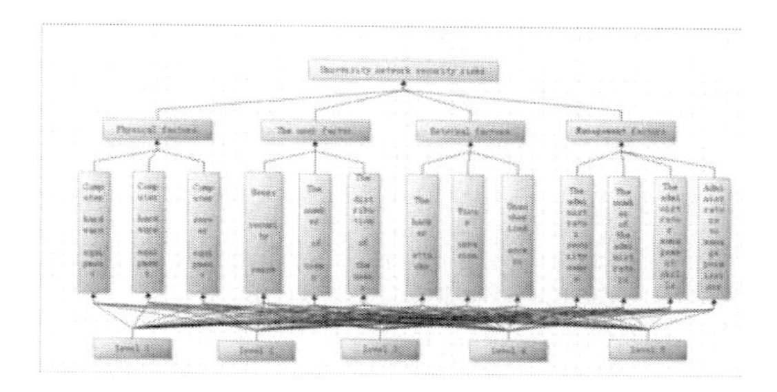

Figure 1. The hierarchical structure

Secondly, structure comparative matrix.

Comparing the factor effect on the same level to the corresponding factors above it by using 1-9 scale value (see Table 2), determine the comparative matrix.

Table 2. 1-9 scale value.

Scale value a_{ij}	compared results between i and j
1	The influence of index i and j is the same
3	The influence of index i and j is a bit strong
5	The influence of index i and j is strong
7	The influence of index i and j significantly stronger
9	The influence of index i and j is absolutely strong
2, 4, 6, 8	The ratio of the influence of index i and j is between the two above adjacent level
$\frac{1}{2},\ldots,\frac{1}{9}$	The ratio of the influence of index j and i is the reciprocal number of the above a_{ij}

Comparative matrix effected by index P1, P2, P3, P4 to index P:

$$P = \begin{bmatrix} 1 & 1/3 & 1/7 & 1/7 \\ 3 & 1 & 1/2 & 1/3 \\ 7 & 2 & 1 & 2 \\ 7 & 3 & 1/2 & 1 \end{bmatrix};$$

Comparative matrix effected by index P11, P12, P13 to index P1:

$$P_1 = \begin{bmatrix} 1 & 1/5 & 1/3 \\ 5 & 1 & 3 \\ 3 & 1/3 & 1 \end{bmatrix};$$

Comparative matrix effected by index P21, P22, P23 to index P2:

$$P_2 = \begin{bmatrix} 1 & 5 & 3 \\ 1/5 & 1 & 1/2 \\ 1/3 & 2 & 1 \end{bmatrix};$$

Comparative matrix effected by index P31, P32, P33 to index P3:

$$P_3 = \begin{bmatrix} 1 & 1 & 7 \\ 1 & 1 & 7 \\ 1/7 & 1/7 & 1 \end{bmatrix};$$

Comparative matrix effected by index P41, P42, P43 , P44 to index P4

$$P_4 = \begin{bmatrix} 1 & 5 & 1/2 & 1 \\ 1/5 & 1 & 1/9 & 1/5 \\ 2 & 9 & 1 & 2 \\ 1 & 5 & 1/2 & 1 \end{bmatrix};$$

Thirdly, calculate the size of each index weight by using the geometric mean value method.

Specific process is as follows:

1 For the product of elements in each row in Comparative matrix, get vector α;
2 To root each component of vector α, get vector β;
3 Normalize vector β, get vector γ of its corresponding index weight.

The calculation process of index P1, P2, P3, P4 weight is as follows:

1 For the product of elements in each row in matrix P, get vector $\alpha = \begin{pmatrix} \frac{1}{147} & \frac{1}{2} & 28 & \frac{21}{2} \end{pmatrix}^T$;
2 To root each component of vector α, get vector $\beta = \begin{pmatrix} 0.287 & 0.841 & 2.3 & 1.8 \end{pmatrix}^T$;
3 Normalize vector β, get vector $\gamma = \begin{pmatrix} 0.055 & 0.161 & 0.44 & 0.344 \end{pmatrix}^T$ of its corresponding index weight

The same procedure may be easily adapted to obtain the weight vector of each index.

The weight vector of P11, P12, P13 is $\gamma_1 = \begin{pmatrix} 0.105 & 0.637 & 0.258 \end{pmatrix}^T$;

The weight vector of P21, P22, P23 is $\gamma_2 = \begin{pmatrix} 0.648 & 0.122 & 0.23 \end{pmatrix}^T$;

The weight vector of P31, P32, P33 is $\gamma_3 = \begin{pmatrix} 0.467 & 0.467 & 0.066 \end{pmatrix}^T$;

The weight vector of P41, P42, P43, P44 is $\gamma_4 = \begin{pmatrix} 0.241 & 0.049 & 0.469 & 0.241 \end{pmatrix}^T$.

Finally, carry out a consistency check of comparative matrix.

Owing to the certain human factors effecting the size of comparative matrix, it is necessary to carry out a consistency check. The specific methods are as follows:

1 Calculate the consistency index $CI = \frac{\lambda_{max} - n}{n-1}$, and $\lambda_{max} = \frac{1}{n}\sum_{i=1}^{n} \frac{\sum_{j=1}^{n} a_{ij} r_j}{r_i}$;
2 Calculate the consistency ratio index $CR = \frac{CI}{RI}$, and RI is random consistency index, specific values are shown in table 3.

Table 3. Random consistency index.

n	1	2	3	4	5	6
RI	0	0	0.58	0.90	1.12	1.24

Generally, as $CR < 0.10$, comparative matrix passes consistency check.

The results of P, P1, P2, P3, P4 consistency check are shown in table 4.

Table 4. Results of each comparative matrix consistency check .

comparative matrix	P	P1	P2	P3	P4
λ_{max}	4.114	3.038	3.004	3.000	4.002
CI	0.038	0.019	0.002	0.000	0.0007
RI	0.90	0.58	0.58	0.58	0.90
CR	0.042	0.033	0.003	0.000	0.0008

Thanks to $CR < 0.10$, P,P1, P2, P3, P4 all pass the consistency check

The summary of university network security risk index weight is shown in table 5.

Table 5. The university network security risk index weight.

First- level index	The weight	Second - level index	The weight
Physical factors P1	0.055	Computer hardware equipment P11	0.105
		Computer network equipment P12	0.637
		Computer server environment P13	0.258
The user factor P2	0.161	User security sense P21	0.648
		The number of users P22	0.122
		The distribution of the user P23	0.230
External factors P3	0.440	The hacker attacks P31	0.467
		Virus invasion P32	0.467
		Unauthorized access P33	0.066
Management factors P4	0.344	The administrator security sense P41	0.241
		The number of the administrators P42	0.049
		The administrators' management skills P43	0.469
		Administrators' management permissions P44	0.241

3 ESTABLISH THE FUZZY EVALUATION MODEL OF UNIVERSITY NETWORK SECURITY RISK

Combined with the situation of university network security risk in Shandong area, the second-level index of university network security risk evaluation is divided into five evaluation sets. Approved by experts, we establish the fuzzy judgment matrix, calculate the grade membership function of first-level index according to front calculated weight, and then get the grade evaluation vector of university Network security risk. Based on the principle of maximization, the risk levels of university network security is determined. Specific process is as follows:

1 To establish the five evaluation sets of second-level index

Specific details are shown in table 6.

Table 6. Second-level index evaluation sets of university network security risk.

second-level index	Level					Hierarchical classification standard
Computer hardware equipment P11	Very good	good	general	poor	very poor	Equipment quality and update
Computer network equipment P12	Very good	good	general	poor	very poor	Equipment quality and update
Computer server environment P13	Very good	good	general	poor	very poor	Waterproof and fireproof lightning protection magnetically moisture-proof anti-static, etc.
User security sense P21	Very strong	strong	general	weak	very weak	Whether to use anti-virus, whether it is safe surfing for Web, etc.
The number of users P22	Very few	few	general	more	very more	How many users
The distribution of the user P23	Very focused	concen trated	general	The dispersion	scattered	Distance, whether dispersion, etc.
The hacker attacks P31	Very few	few	general	more	very more	Attack times
Virus invasion P32	Very few	few	general	more	very more	invasion times
Unauthorized access P33	Very few	few	general	more	very more	visits times
The administrator security sense P41	Very strong	strong	general	weak	very weak	Whether proper operation, whether security attention , etc.
The number of the administrators P42	Very few	few	general	more	very more	How many people
The administrators' management skills P43	Very strong	strong	general	weak	very weak	Crisis management skills, etc.
Administrators' management permissions P44	Very reasonable	reasonable	general	unrea sonable	Very unreasonable	Whether distribution is reasonable, permissions size, etc.

2 Determine the fuzzy evaluation matrix

If the judgement result of the i th second-level index in first-level index P1 is

$\begin{pmatrix} r_{i1}^{(1)} & r_{i2}^{(1)} & r_{i3}^{(1)} & r_{i4}^{(1)} & r_{i5}^{(1)} \end{pmatrix}$, and $i = 1, 2, 3$, the fuzzy evaluation matrix of corresponding second-level index is as follows:

$$R_1 = \begin{bmatrix} r_{11}^{(1)} & r_{12}^{(1)} & r_{13}^{(1)} & r_{14}^{(1)} & r_{15}^{(1)} \\ r_{21}^{(1)} & r_{22}^{(1)} & r_{23}^{(1)} & r_{24}^{(1)} & r_{25}^{(1)} \\ r_{31}^{(1)} & r_{32}^{(1)} & r_{33}^{(1)} & r_{34}^{(1)} & r_{35}^{(1)} \end{bmatrix},$$

r_{ij} for the P1 i level two indexes of the j grades of membership.

and $r_{ij} = \frac{\text{the number of experts graded j}}{\text{the total number of experts}}$ is the j th grade membership of second-level index in P1

The same procedure may be easily adapted to obtain the results, just as follows:

$$R_2 = \begin{bmatrix} r_{11}^{(2)} & r_{12}^{(2)} & r_{13}^{(2)} & r_{14}^{(2)} & r_{15}^{(2)} \\ r_{21}^{(2)} & r_{22}^{(2)} & r_{23}^{(2)} & r_{24}^{(2)} & r_{25}^{(2)} \\ r_{31}^{(2)} & r_{32}^{(2)} & r_{33}^{(2)} & r_{34}^{(2)} & r_{35}^{(2)} \end{bmatrix},$$

$$R_3 = \begin{bmatrix} r_{11}^{(3)} & r_{12}^{(3)} & r_{13}^{(3)} & r_{14}^{(3)} & r_{15}^{(3)} \\ r_{21}^{(3)} & r_{22}^{(3)} & r_{23}^{(3)} & r_{24}^{(3)} & r_{25}^{(3)} \\ r_{31}^{(3)} & r_{32}^{(3)} & r_{33}^{(3)} & r_{34}^{(3)} & r_{35}^{(3)} \end{bmatrix},$$

$$R_4 = \begin{bmatrix} r_{11}^{(4)} & r_{12}^{(4)} & r_{13}^{(4)} & r_{14}^{(4)} & r_{15}^{(4)} \\ r_{21}^{(4)} & r_{22}^{(4)} & r_{23}^{(4)} & r_{24}^{(4)} & r_{25}^{(4)} \\ r_{31}^{(4)} & r_{32}^{(4)} & r_{33}^{(4)} & r_{34}^{(4)} & r_{35}^{(4)} \\ r_{41}^{(4)} & r_{42}^{(4)} & r_{43}^{(4)} & r_{44}^{(4)} & r_{45}^{(4)} \end{bmatrix}.$$

Based on the weight of second-level index, the fuzzy evaluation matrix of first-level index is:

$$R = \begin{bmatrix} r_1^T R_1 & r_2^T R_2 & r_3^T R_3 & r_4^T R_4 \end{bmatrix},$$

Again, on the basis of the first-level index weight, the evaluation vector of campus network security risk is:

$$w = r^T R.$$

Based on the principle of maximization, we could determine the risk levels of university network security. And the grade of campus network security risk is also divided into five levels, that is, level one(very safe), level two(safe), level three (general risk), level four (dangerous), level five (very dangerous).

4 ANALYSIS AND APPLICATION OF THE MODELS

It is reasonable, comprehensive, fast, and simple of the models to make an accurate evaluation on university network security risk, which makes them feasible. But the grade judgement of second-level index is given by the group of experts. The results given by different experts are different. And this depends on the experts' perception of the university network security risk of.

Based on the above models, a group of 10 experts rate the network security risk index of Weifang University of Science and Technology, the results in Table 7:

Table 7. the second-level index results of network security risk of Weifang University of Science and Technology.

The second- level index	Evaluation results				
Computer hardware equipment P11	0	2	5	2	1
Computer network equipment P12	1	1	5	2	1
Computer server environment P13	1	3	4	2	0
User security sense P21	2	2	2	3	1
The number of users P22	0	1	3	4	2
The distribution of the user P23	1	2	3	3	1
The hacker attacks P31	0	2	2	5	1
Virus invasion P32	0	1	5	3	1
Unauthorized access P33	0	3	4	2	1
The administrator security sense P41	1	3	4	2	0
The number of the administrators P42	1	4	5	0	0
The administrators' management skills P43	1	3	3	2	1
Administrators' management permissions P44	0	2	4	4	0

The results showed that

$$R_1 = \begin{bmatrix} 0 & 0.2 & 0.5 & 0.2 & 0.1 \\ 0.1 & 0.1 & 0.5 & 0.2 & 0.1 \\ 0.1 & 0.3 & 0.4 & 0.2 & 0 \end{bmatrix}, \quad R_2 = \begin{bmatrix} 0.2 & 0.2 & 0.2 & 0.3 & 0.1 \\ 0 & 0.1 & 0.3 & 0.4 & 0.2 \\ 0.1 & 0.2 & 0.3 & 0.3 & 0.1 \end{bmatrix},$$

$$R_3 = \begin{bmatrix} 0 & 0.2 & 0.2 & 0.5 & 0.1 \\ 0 & 0.1 & 0.5 & 0.3 & 0.1 \\ 0 & 0.3 & 0.4 & 0.2 & 0.1 \end{bmatrix}, \quad R_4 = \begin{bmatrix} 0.1 & 0.3 & 0.4 & 0.2 & 0 \\ 0.1 & 0.4 & 0.5 & 0 & 0 \\ 0.1 & 0.3 & 0.3 & 0.2 & 0.1 \\ 0 & 0.2 & 0.4 & 0.4 & 0 \end{bmatrix}.$$

Obtained

$$R = \begin{bmatrix} 0.0895 & 0.1621 & 0.4742 & 0.2 & 0.0742 \\ 0.1526 & 0.1878 & 0.2352 & 0.3122 & 0.1122 \\ 0 & 0.1599 & 0.3533 & 0.3868 & 0.1 \\ 0.0759 & 0.2808 & 0.358 & 0.2384 & 0.0469 \end{bmatrix},$$

So the evaluation vector of campus network security risk of Weifang University of Science and Technology is as follows:

$$w = \begin{bmatrix} 0.056 & 0.206 & 0.343 & 0.313 & 0.071 \end{bmatrix}.$$

Based on the principle of maximization, the network security risk level of Weifang University of Science and Technology is level three that is general risk.

5 CONCLUSION

At present many colleges and universities do not pay enough attention to the campus network security. I suggest that colleges and universities should take positive protective measures, make corresponding strategies according to the different risk levels to ensure the normal operation of the campus network.

REFERENCES

[1] Han Zhonggeng. (2005) Mathematical Modeling Method and its Application. Beijing: Higher Education Press.

[2] Wei Jie. (2011) Analysis of computer network security. *Guangxi Light Industry*, (2):64–65.

[3] Liu Bingji. (2014) Factors affecting the safety of computer network and the countermeasures. *Software*, (3):152–154.

[4] Shen Jingwei, Wang Xinyi. (2002) Fuzzy mathematics in the application of the physical and chemical laboratory quality comprehensive evaluation. Shanghai Journal of Preventive Medicine, 6(14):265–266.

[5] Lijuan Zhang, Qingxian Wang. (2011) Fuzzy analytic hierarchy process (AHP) in the application of the network security situation assessment. *The Computer Simulation*, (12):138–140.

Electric, Electronic and Control Engineering – Shao, Shu & Tian (Eds)
© 2015 Taylor & Francis Group, London, ISBN: 978-1-138-02842-5

Analysis on incremental transmission loss and voltage level of wind power system with Doubly-Fed Induction Generators (DFIGs)

Tao Wang & Yigang He
Hefei University of Technology, Hefei, China

Mingyi Li
Ningguo Electric Power Supply Company, Ningguo, China

ABSTRACT: Grid integration of wind power changes power flow distribution and incremental transmission loss. Based on the Maximum Power Point Tracking (MPPT) strategy, the power flow model of Doubly-Fed Induction Generators (DFIGs) are applied to build the relationship between incremental transmission loss, voltage level and input wind speed. According to the incremental transmission loss and voltage level, the best location for wind power integration is determined to reduce the incremental transmission loss, and improve the transmission efficiency and the power quality of the power grid by decreasing the fluctuation of the voltage level. This has important significance for researching to reduce the transmission loss and improve economic efficiency.

KEYWORDS: DFIG; power flow; incremental transmission loss; voltage level.

1 INTRODUCTION

Currently in the field of wind power, the kinds of wind turbines being used more are asynchronous turbines, doubly- fed induction wind turbines, etc. [1]. As the DFIG rotor has an independent power supply, has a certain ability to regulate reactive power, to some extent, can control power flexibly and has a lower cost, so it gets a wide range of applications.

Since wind energy has great randomness, wind power is also led to great instability and randomness. After the grid integration of DFIG, lots of complex issues being aroused. So, find a proper location for wind power integration has critical significance. In [2], the access point fluctuations of voltage and system frequency over time have been discussed, located the relatively reasonable access point. In [3], when asynchronous turbine is used, the effects of different access point to the node voltage have been discussed. When it comes to DFIG, in addition to fluctuations in voltage and frequency over time, the impact of wind speed on the node voltage of the access point and incremental transmission loss also needs to be considered. This requires the relationship between wind speed, the incremental transmission loss and the voltage level of the access point.

Under the model of MPPT, active power output increases with the rise of wind speed after DFIG integration, as for reactive power output, it can be approximately considered to remain unchanged, but the fluctuation of node voltage still exists with the change of wind speed. In addition, the incremental transmission loss and the node voltage of the access point are influenced by the access location. By establishing relationships between wind speed and incremental transmission loss, combined with the impact of different access points for access point node voltage levels at different wind speeds, determining the best access points, in order to reduce transmission losses.

Based on the power flow model of DFIG, we established a model of incremental transmission loss and node voltage level to the wind speed, we probe the best access point location depended on the changes in incremental transmission loss and voltage level with the growth of wind speed under different access point, to get a smaller incremental transmission loss and voltage fluctuations as much as possible, thereby improving the efficiency of the power transmission after the integration of DFIG.

2 POWER FLOW MODEL OF DFIG

2.1 *The mechanical power model of wind turbine*

The mechanical power obtained by capturing wind energy with wind turbine [4, 5]

$$P_W = 0.5\rho\pi R^2 V_W^3 C_P \quad (1)$$

The parameters used in this equation are as follows:
ρ----- Air density
R----- Area wind turbine swept
V_W---- Wind speed
C_P----- Wind energy conversion factor

When using the exponential form, C_P expressed as (2), c_1 to c_9 are coefficients, λ_i is Intermediate variables, λ is blade top speed ratio in (3).

$$\begin{cases} C_P = c_1\left(\frac{c_2}{\lambda_i} - c_3\beta - c_4\beta^{c_5} - c_6\right)e^{-\frac{c_7}{\lambda_i}} \\ \frac{1}{\lambda_i} = \frac{1}{\lambda + c_8\beta} - \frac{c_9}{\beta^3+1} \end{cases} \quad (2)$$

$$\lambda = \frac{\omega_w R}{V_w} \quad (3)$$

ω_w in (3) is the speed of the wind turbine.

Figure 1 shows the typical power characteristics of DFIG [6], when the wind speed is higher than the cut-in speed and lower than the rated wind speed, the active power output changes in a larger range showed in figure 1, wind turbine can take to the optimal speed $\omega_{w.opt}$, That is the way to run in accordance with MPPT, getting the maximum wind energy conversion factor. When running under the model of MPPT, in this case, we get the optimal slip ratio s_{opt}:

$$s_{opt} = 1 - \frac{p\eta}{60f}\omega_{w.opt} \quad (4)$$

2.2 *Power balance equation of DFIG*

The Configuration and equivalent circuit of DFIG are shown in figure 2 and figure 3 [7, 8].

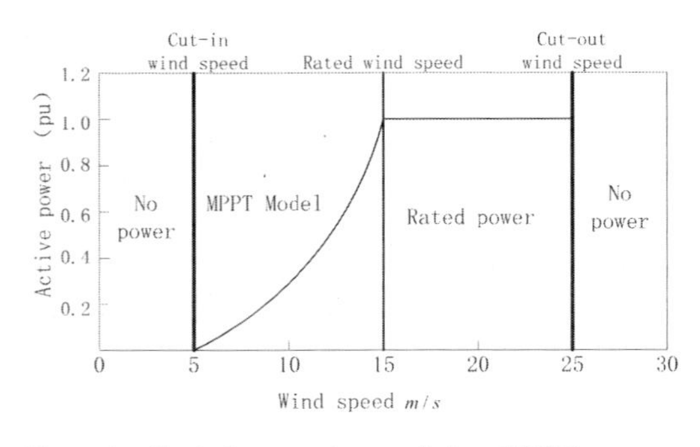

Figure 1. Typical power characteristics of DFIG.

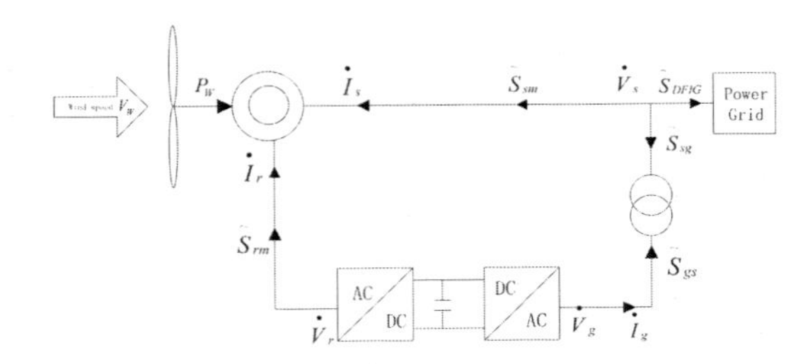

Figure 2. Configuration of DFIG-based wind power unit.

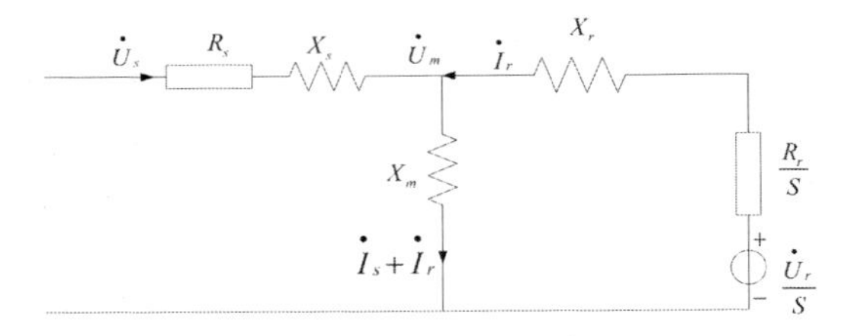

Figure 3. Equivalent circuit of DFIG-based wind power unit.

In the two figures above, U, I, R, X, Z, S respectively represent voltage, current, resistance, reactance, impedance and apparent power. Subscript s, m, r, g, T respectively represent stator node, excitation circuit, rotor, grid-side converter and transformer. For stator node, the power balance equations are as follows:

$$\Delta P_s = -P_{DFIG} - P_{sm} - P_{sg} = 0 \quad (5)$$

$$\Delta Q_s = -Q_{DFIG} - Q_{sm} - Q_{sg} = 0 \quad (6)$$

Excitation circuit:

$$\Delta P_m = -P_{ms} - P_{mr} = 0 \quad (7)$$

$$\Delta Q_m = -Q_{ms} - Q_{mr} = 0 \quad (8)$$

Converter:

$$\Delta P_g = -P_{rm} - P_{gs} = 0 \quad (9)$$

$$\Delta Q_g = Q_{g.set} - Q_{gs} = 0 \quad (10)$$

Torque equilibrium condition necessary:

$$\Delta T = -\frac{P_w}{1-s} - P_{em} = 0 \quad (11)$$

2.3 *The revise for the calculation equation of power flow with DFIG*

The revise for the traditional calculation equation of power flow can be represented as [9]:

$$\begin{bmatrix} \Delta P \\ \Delta Q \end{bmatrix} = \begin{bmatrix} H & N \\ L & J \end{bmatrix} \begin{bmatrix} \Delta \theta \\ \Delta U \end{bmatrix} \tag{12}$$

Add (5)-(11) as constraints to (12), forming the revised equation:

$$\begin{bmatrix} \Delta P_{sys} \\ \Delta Q_{sys} \\ \Delta P_m \\ \Delta Q_m \\ \Delta P_g \\ \Delta Q_g \\ \Delta P_s \\ \Delta Q_s \\ \Delta T \end{bmatrix} = \begin{bmatrix} H' & N' \\ L' & J' \end{bmatrix} \begin{bmatrix} \Delta \theta_{sys} \\ \Delta U_{sys} \\ \Delta \theta_m \\ \Delta U_m \\ \Delta \theta_g \\ \Delta U_g \\ \Delta \theta_s \\ \Delta U_s \\ \Delta U_r \end{bmatrix} \tag{13}$$

3 THE INCREMENTAL TRANSMISSION LOSS AND VOLTAGE LEVEL WITH DFIG UNDER THE MODEL OF MPPT.

3.1 *Incremental transmission loss*

According to relevant material of incremental transmission loss [10, 11], the following equation can be derived:

$$\Delta P_\Sigma = \sum_{i=1}^{n} \sum_{j=1}^{n} U_i U_j G_{ij} \cos \theta_{ij} \tag{14}$$

For the derivation of θ_k and U_k (k represents any node subscript) respectively, we can get:

$$\frac{\partial \Delta P_\Sigma}{\partial \theta_k} = 2 \sum_{i=1}^{n} U_i U_k G_{ik} \sin \theta_{ik} \tag{15}$$

$$\frac{\partial \Delta P_\Sigma}{\partial U_k} = 2 \sum_{i=1}^{n} U_i G_{ik} \cos \theta_{ik} \tag{16}$$

From (15), (16) we can get the sensitivity of transmission loss to node voltage:

$$\Delta \Delta P_\Sigma = \begin{bmatrix} \frac{\partial \Delta P_\Sigma}{\partial \theta} & \frac{\partial \Delta P_\Sigma}{\partial U} \end{bmatrix} \begin{bmatrix} \Delta \theta \\ \Delta U \end{bmatrix} \tag{17}$$

The incremental transmission loss can be calculated by the method transposing Jacobi matrix [9]:

$$\frac{\partial \Delta P}{\partial P} = \begin{bmatrix} 1 & 0 \end{bmatrix} \left[\begin{bmatrix} H & N \\ J & L \end{bmatrix}^t \right]^{-1} \begin{bmatrix} \frac{\partial \Delta P}{\partial \theta} \\ \frac{\partial \Delta P}{\partial U} \end{bmatrix} \tag{18}$$

3.2 *Node voltage*

The node voltage can be got in the calculation of power flow.

4 ANALYSIS OF EXAMPLE

4.1 *Original data*

This example use a doubly-fed wind turbine consisted of 10 DFIG with the same parameters. Following are the original data: the parameter of wind turbine c_1 =0.73, c_2=151, c_3=0.58, c_4=18.4, c_5=2.14, c_6=13.2, c_7=18.4, c_8=-0.02, c_9=-0.003; ρ =1.225 kg/m^3 ; R =35.5 m; the wind turbine's speed ω_w range from 6~21.5r/min, the wind speed range from 6~10 m/s, speed increasing ratio η =94. The parameters of DFIG: V_N =690V, P_N=2MW, the number of pole pairs P=2, the per-unit value of impedance: R_s=0.0078, X_s =0.0794, R_r=0.025, X_r=0.4, X_m =0.0078, R_T =0.03, X_T=0.05. Select the 5 node electric power system as the test system. Node 5 is the balanced node, voltage amplitude is 1.00, and the rest are PQ node. The test system showed in figure 4.

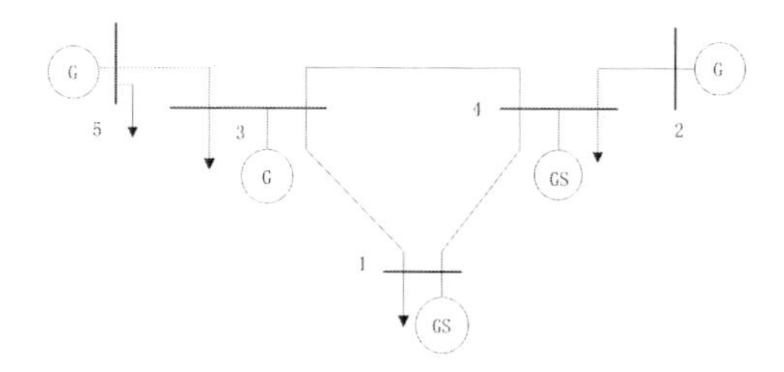

Figure 4. Configuration of test system.

4.2 *The incremental transmission loss and node voltage level when DFIG accesses different location*

Accessing DFIG to 1, 2, 3, 4 respectively. The data of incremental transmission loss and node voltage to different node are showed below.

Table 1. Incremental transmission loss and voltage level with DFIGs at bus 1.

Wind speed m/s	6	6.5	7	7.5	8
Incremental transmission loss (pu)	−0.0201	−0.0202	−0.0202	−0.0203	−0.0203
Node voltage (pu)	1.0011	1.0022	1.0023	1.0025	1.0028
Wind speed m/s	8.5	9	9.5	10	10.5
Incremental transmission loss (pu)	−0.0204	−0.0205	−0.0207	−0.0209	−0.0302
Node voltage (pu)	1.00301	1.00312	1.00318	1.00322	1.0033

Table 2. Incremental transmission loss and voltage level with DFIGs at bus 2.

Wind speed (m/s)	6	6.5	7	7.5	8
Incremental transmission loss (pu)	0.0405	0.0407	0.0408	0.0408	0.0409
Node voltage (pu)	0.9925	0.9919	0.9913	0.9907	0.9901
Wind speed (m/s)	8.5	9	9.5	10	10.5
Incremental transmission loss (pu)	0.0412	0.0414	0.0417	0.0419	0.0501
Node voltage (pu)	0.9892	0.9887	0.9881	0.9776	0.9769

Table 3. Incremental transmission loss and voltage level with DFIGs at bus 3.

Wind speed (m/s)	6	6.5	7	7.5	8
Incremental transmission loss (pu)	−0.104	−0.106	−0.107	−0.109	−0.111
Node voltage (pu)	1.0339	1.0341	1.0344	1.0346	1.0347
Wind speed (m/s)	8.5	9	9.5	10	10.5
Incremental transmission loss (pu)	−0.113	−0.116	−0.119	−0.121	−0.124
Node voltage (pu)	1.0349	1.0350	1.0353	1.0357	1.0361

Table 4. Incremental transmission loss and voltage level with DFIGs at bus 4.

Wind speed (m/s)	6	6.5	7	7.5	8
Incremental transmission loss (pu)	0.0562	0.0563	0.0563	0.0564	0.0565
Node voltage (pu)	0.9745	0.9757	0.9766	0.9771	0.9778
Wind speed (m/s)	8.5	9	9.5	10	10.5
Incremental transmission loss (pu)	0.0566	0.0568	0.0570	0.0573	0.0577
Node voltage (pu)	0.9807	0.9845	0.9866	0.9901	0.9976

In this example, when access to node 1, we get the minimum fluctuation of node voltage, and at the same time, the incremental transmission loss is also the lowest, so the best location for DFIG to integration is node 1 in this example.

5 CONCLUSION

According to the power balance equation of DFIG, this paper probe the best location to integration by the influence of wind speed on the parameters of incremental transmission loss an node voltage, based on the calculation of power flow after DFIG integration, getting the lower fluctuation of node power and incremental transmission loss.

ACKNOWLEDGEMENT

This work was supported by the National Natural Science Funds of China for Distinguished Young Scholar (GN: 50925727); The National Defense Advanced Research Project (GN: C1120110004, 9140A27020211DZ5102); the Key Grant Project of Chinese Ministry of Education (GN: 313018); Anhui Provincial Science and Technology Foundation of China (GN: 1301022036) and Hunan Provincial Science and Technology Foundation of China (GN: 2010J4, 2011JK2023).

REFERENCES

[1] Wu Zhenjing, Yu Chunshun, Zhang Xiangzhi, et al. (2011) Simulation of inverter in wind-power generation with back-to-back topology. *Transactions of Beijing Institute of Technology*, 31(9): 1075–1079.

[2] Gu Feng, Cao Youletu. (2010) Choice of best access point for connecting wind farms to power systems. *Power System and Clean Energy*, 26(3):73–76.

[3] Liu Yang, Li Zhi, Zhang Bo, Han Xueshan. (2008) The research on power flow calculation after grid integration of wind power. *Chinese Universities Power System and Automation Twenty-fourth Annual Conference Proceedings*. China Agricultural University.

[4] Li Shenghu, Sun Shasha, Wang Zhengfeng. (2013) Initialization algorithm of doubly-fed induction (DFIGs) under spinning reserve. *Automation of Electric Power Systems*, 27(11):7–12.

[5] Ding Ming, Wu Yichun. (2004) Summary on studies of operation and planning of wind power generation systems. *International Electric Power*, 7(3): 36–40.

[6] Huang Xueliang, Liu Zhiren, Zhu Ruijin, et al. (2010) Impact of power system integrated with large capacity of variable speed constant frequency wind turbines. Journal of Electrician Technique Analysis, 25(4):142–149.

[7] Su Ping, Zhang Kaoshe. (2010) Processing method for the node of the doubly-fed induction generator wind farm in flow calculation of the power system. *Renewable Energy Resources*, 28(2):105–109.

[8] Li S. (2013) Power flow modeling to doubly-fed induction generators (DFIGs) under power regulation. *Power Systems, IEEE Transactions on*, 28(3):3292–3301.

[9] Chen Heng. (2007) *Steady-state Analysis of Electric Power System*. Beijing: China Electric Power Press.

[10] Shi Lei, Cheng Ken, Zhang Boming, et al. (2001) Deduction of Jacobian matrix method for calculating incremental transmission losses. *Journal of Nanchang University (Engineering & Technology)*, 23(2): 55–58.

[11] Wang Erzhi, Zhao Yuhuan. (1992) *Sensitivity Analysis and Power Flow Calculation of Electric Power System*. Beijing: Machinery Industry Press.

Electric, Electronic and Control Engineering – Shao, Shu & Tian (Eds)
© 2015 Taylor & Francis Group, London, ISBN: 978-1-138-02842-5

Research and prevention of illegal intrusion of digital television network

Xiangfeng Hu
College of Physics, Tonghua Normal University, Tonghua, Jilin, China

ABSTRACT: Digital television is now developing rapidly in China. The momentum is thrilling, which also brings numerous network security issues. This paper carries out a study on illegal intrusion of digital television network. Starting from detection model, it proposes the intrusion detection method based on ESTQ and the intrusion detection method based on expert system and BP neural network algorithm on the basis of the analysis on potential safety hazards of digital television network. It also designs a realization approach of linkage alarm in method research and analysis so as to lay a theoretical foundation for the sound development of radio and television undertakings in China.

KEYWORDS: illegal intrusion detection; alarming device; ESTQ four components; expert system; BP neural network algorithm.

1 INTRODUCTION

Requirements on radio and television undertakings are very strict in China. With the increase of demand quantity of digital television and the development of information technology, there are numerous factors that have influences on reliability and security of television network, especially the illegal intrusion of network. The key of preventing illegal intrusion lies in detection means. This paper carries out a study on detection means and precautionary measures of illegal intrusion of digital television network so as to provide theoretical guidance for the healthy development of radio and television undertakings in China.

Many people have made efforts in the detection and prevention of network illegal intrusion. Youwei Sun (2005) proposed a security strategy that is suitable for centralized safety management and graded safety control of digital cable TV network of the next generation and expounded briefly the automatic network threat defense technology that can be used as the BOSS security means [1]. Renshang Zhang (2012) designed an intelligent intrusion detection system, which is able to improve the network intrusion detection accuracy and reduce false alarm rate and missing alarm rate of network intrusion, and conducted a simulated realization with the database of network intrusion detection KDD CUP99 [2]. Xiaolan Xie (2013) carried out a research on the network security deign based on the technology of system vulnerability scanning and antivirus, indicating that the network antivirus technology is realized mainly through the deployment of network antivirus system and the management is conducted mainly through the method of cross-WAN Web which distributes antivirus system strategy intensively [3]. Zhao Jin (2014) carried out a study on the application of the first security line of network, namely network firewall [4]. Zhi Tian (2014) elaborated the architecture of computer network firewall, analyzed types of computer network firewall, and proposed a setting method of computer network firewall in order to enhance the security level of computer network system [5]. Lu Zhuang (2014) introduced how to prevent illegal signals of the front end of television station by monitoring TS flow and abnormal condition of video image [6].

Based on previous studies, this paper analyzes potential safety hazards of digital television network, studies the basic model of illegal intrusion detection of digital television network, and emphasizes the illegal intrusion detection of digital television network based on ESTQ as well as detection and prevention methods of illegal intrusion based on two detection technologies.

2 SAFETY ANALYSIS OF DIGITAL TELEVISION NETWORK

Digital TV is a television in which the whole program is displayed in digital signals or transmitted in digital stream formed by binary digit strings [7]. Cable digital TV technology processes analog signals of cable TV transmission based on computer technology so as to convert them into scrambling of TV signals, and Cable digital TV technology is a TV transmission and exchange technology that processes analog signals of cable TV transmission based on computer technology so as to convert them into scrambling of TV signals and scrambles TV signals to digital signals with the help of limited network transmission technology and terminal receiving technology [8]. Digital cable TV network system is composed by data broadcaster, satellite receiver, live video, external interaction and user. Potential safety hazards of digital television network mainly come from [9] such eight aspects as physics, network structure, network equipment, network system, network resource sharing, data information, network management and link transmission. Threats of the above-mentioned eight aspects are presented in Table 1:

Table 1. Potential safety hazards of digital television network.

Types	Sources of potential safety hazards
Physics	Destruction of environmental accidents, data loss caused by the stealing of equipment, data leakage caused by electromagnetic radiation, etc.
Structure	Abuse of internal LAN, staff workers' checking or reception of offensive materials, insiders' deliberate divulgence, transmission of destructive programs on intranet, stealing of classified information of others.
Equipment	Inappropriate application of network equipment, communication faults of remote line provided by telecom, network equipment malfunction, counterfeit and shortage of network identity, detection and alarm mechanism of network intrusion.
System	Unnecessary functional modules and unnecessary ports in the system. There might be potential safety hazards in these unnecessary ports.
Resource sharing	Staff workers share important information catalog in hard disk intentionally or unintentionally. It might be stolen by external staff or other internal staff due to the long-term exposure to network neighborhood.
Data information	Data is easily stolen during the transmission on public lines. Hackers or some industrial espionage try to acquire data information on lines with many advanced technologies.
Management	Internal management staff or staff workers use the same login name and command. Random access to machine rooms and destructions in the form of network jokes.
Link transmission	If a LAN is wired in the form UTP, it might be able to install wiretaps on transmission lines so as to steal important data.

It can be known from Table 1 that network security issues of digital television can be divided into intranet issues and extranet issues. According to the situation that the extranet is intruded, Xiaosheng Cheng (2013) pointed out that main reasons of external intrusion of network are unauthorized access, destruction of data integrity, server attacking and virus transmitting through network, webpage and email [10]. Security issues of the intranet are mainly reflected in the security of the physical layer, access control policy of resource sharing, intranet virus aggression, unauthorized access of legal users, illegal access of counterfeit users, data destruction, etc. This paper mainly studies the illegal intrusion of digital television network in order to explore relevant precautionary measures.

3 ILLEGAL INTRUSION DETECTION MODEL OF DIGITAL TV NETWORK

Intrusion detection is used for illegal attacking of computer system and network system or information system in a broad sense. It is the second defense line of network information system security and the supplement to security infrastructure as well. It is a safety technology that detects behaviors violating security policy and signs of being attacked in network or system by collecting and analyzing information of computer network and system based on key points [11]. It can be known from the analysis on digital TV network security in the previous chapter that objects of illegal intrusion include host and network. So, the design of the detection system should also be divided into the two aspects. In this paper, the illegal intrusion detection system of digital TV network aiming at host and network is designed in accordance with the standard format of security incidents alarm shown in Figure 1:

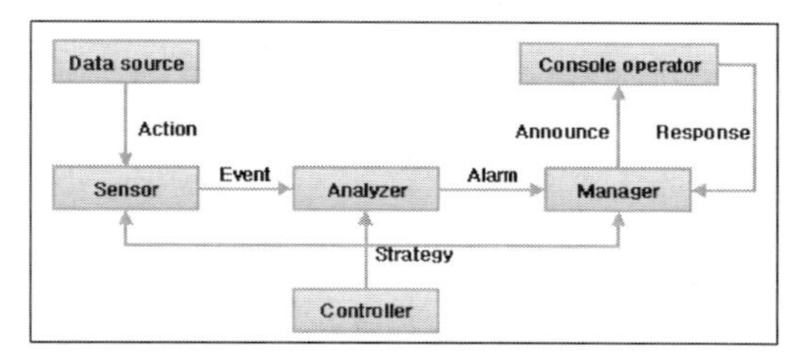

Figure 1. Standard frame model of security incidents alarm and detection.

Functional requirements and logical structure of the intrusion detection system are reflected in Figure 1. The process of intrusion includes information collection, data analysis and incident response. The intrusion detection system of host takes audit and log files like system logs and application programs as data and compares audit and log files with attack signatures so as to figure out whether they are matched. If they match, the detection system will send out an intrusion alarm to the system administrator and take corresponding actions [13]. The intrusion detection system structure based on host is shown in Figure 2:

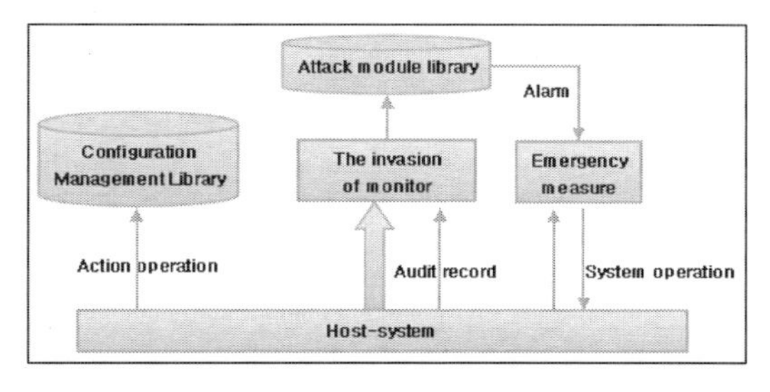

Figure 2. Structural diagram of the intrusion detection system based on host.

The intrusion detection system based on host shown in Figure 2 is able to judge intrusion incidents of the application layer according to characteristics of different operating systems. It is obliged to guarantee that audit data of the system is not modified because data auditing is a major way of collecting user behavior information. The network intrusion detection system is installed on the protected network. Original network messages are taken as data sources for attack analysis. A network adapter is frequently adopted to carry out monitoring and analysis on all data transmitted through the network. Once an attack is detected, relevant modules of the intrusion detection system will respond to the attack by notification, alarm or disconnection. The intrusion network system structure based on a network is shown in Figure 3:

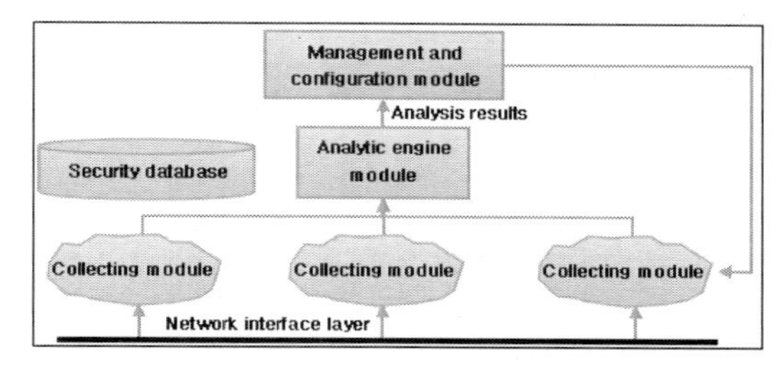

Figure 3. Structural diagram of the intrusion detection system based on network.

The major function of the management and configuration module in Figure 3 is to manage configurations of other modules and notify the network administrator with output results from the analytical engine module in effective ways. The data collecting module consists of three parts, namely filter, network interface engine and detector. The following functions can be realized:

1 Obtain data package related to security incidents from the network according to certain rules and transmit the data package to the analytical engine module for security analysis of this module.
2 Analytical engine module of intrusion will conduct an analysis according to the data package transmitted from the collecting module and network security database. Analytical results will be transmitted to the management and configuration module.

4 RESEARCH ON THE ILLEGAL INTRUSION DETECTION METHOD OF DIGITAL TV NETWORK

4.1 *Research on the illegal intrusion detection method of digital TV network based on ESTQ*

Xiaocheng Wang (2001) indicated that all network attacks taking advantage of network protocol security holes will abnormal network actions to generate unusual states or behaviors of network protocol so as to deceive and utilize protocols. The essence is the abuse of network protocol [14]. In this paper, network protocol attack is described in four components <event, state of protocol, time relation, quantity relation>. Event refers to the specific manifestation of behaviors of attackers in network; state of protocol refers to network protocol state or state transition when attacks occur; time relation refers to the logical relation between specific times of event and state of protocol; quantity relation refers to all quantities related to event and state of protocol. The above-mentioned description method is called ESTQ four components [14]. According to the description method of ESTQ four components, network protocol attacks can be divided into the type of event, the type of event-protocol and the type of event-protocol state-time-quantity relation. The classification of network protocol attacks should follow the principle of consistency and simplification.

If E stands for the set of events, S stands for the set of protocol states, T stands for the set of time relation, Q stands for the set of quantity relation, and $E \times S \times T \times Q$ stands for the ordered pair set of the above-mentioned four sets, the function *EventDetect* can map network protocol attacks that satisfy conditions to the type of *Event* of ESTQ. The description of the method ESTQ is shown in Figure 4.

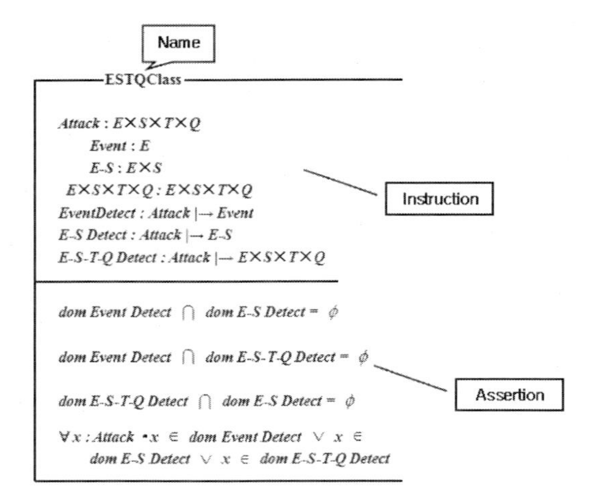

Figure 4. Description of the method ESTQ.

Assertion in Figure 4 refers to the definition of variable values. Instruction refers to the explanation on variables. Three functions that map network attacks to each attack type of ESTQ are described in Figure 4. There is no intersection among function domains. All network attacks that can be described with <*E, S, T, Q*> four components are within the range of three function domains. Any network attack that can be described with <*E, S, T, Q*> four components only belongs to one type of ESTQ. The description of adding a new attack to ESTQ is shown in Figure 5:

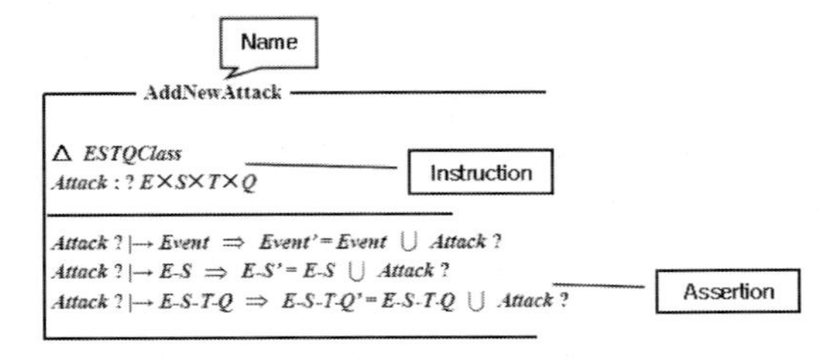

Figure 5. Description of adding a new attack to ESTQ.

It can be known from Figure 5 that the new attack changes the current ESTQ attack classification. The number of the type to which the new attack belongs increases correspondingly. The process of attack detection is shown in Figure 6.

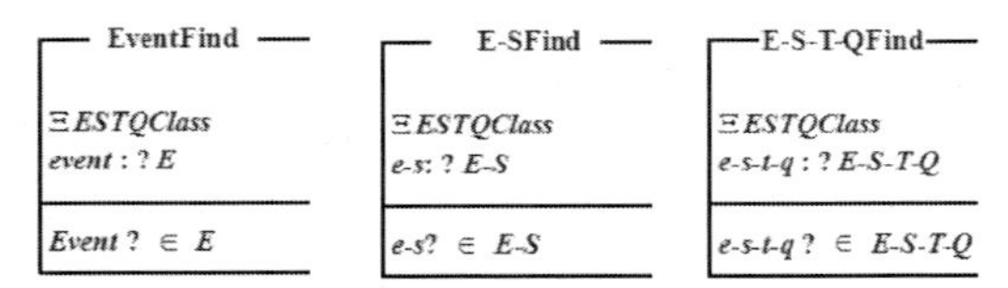

Figure 6. Description of attack detection process.

Figure 6 suggests that attack detection is to find out attack events in corresponding types, namely event-protocol state or event-protocol state-time relation-quantity relation.

The intrusion detection process based on network protocol attack includes the following two steps:

STEP1. Data collection: the collection of network package, protocol state, specific events and quantity that stand for specific events.

STEP2. Data analysis and processing: the relation between event and protocol state, logical time relation and the analysis on quantity relation match with the final attack description. It also includes the control on data collection.

4.2 *Research on the illegal intrusion detection method of digital TV network based on expert system and BP neural network*

Two detection technologies based on expert system and BP neural network are analyzed in this section. Efficiency and speed of network detection can be improved by serial combination shown in Figure 7.

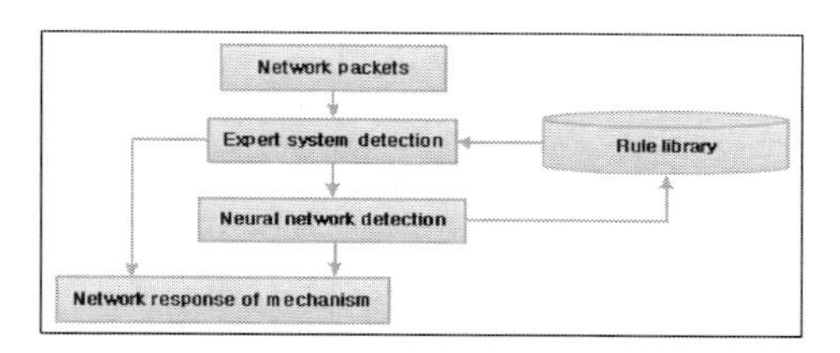

Figure 7. Serial combination of two detection technologies.

Expert system is the first defense line of intrusion detection, which matches network packages with intrusive behaviors of rule base. The detection technology of neural network is the second defense line of intrusion detection system, which is able to detect abnormal behaviors and improve the rule base of expert system.

Expert system of network intrusion detection is a program system with the detection capability of expert level. The structure is shown Figure 8. The knowledge base of intrusion detection is the key of expert system, which can be divided into fact base, rule base and execution rule base. Each network intrusion detection rule includes two parts, namely rule head and rule option. Reasoning mechanism of data driving is adopted in expert system of network intrusion detection. The reasoning process is shown below:

STEP1. Extract characteristics of captured network data messages.

STEP2. Match message characteristics with rules of the rule head. If the two match, network intrusion occurs. Otherwise, continue the searching.

STEP3. Search the rule tree of network intrusion. If it matches with a certain node, network intrusion occurs. Otherwise, it is regarded as a normal behavior.

If the characteristics of captured data messages match with a certain rule of the intrusion knowledge base, describe the data according to rules and output detection results to the response module of network intrusion detection system.

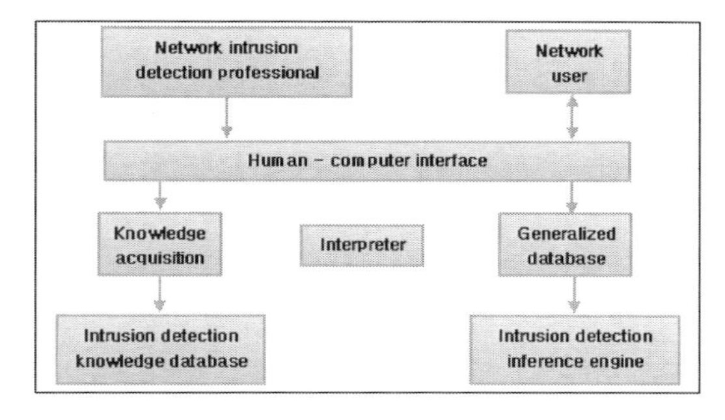

Figure 8. Expert system structure of network intrusion detection.

The structure of BP neural network is shown in Figure 9:

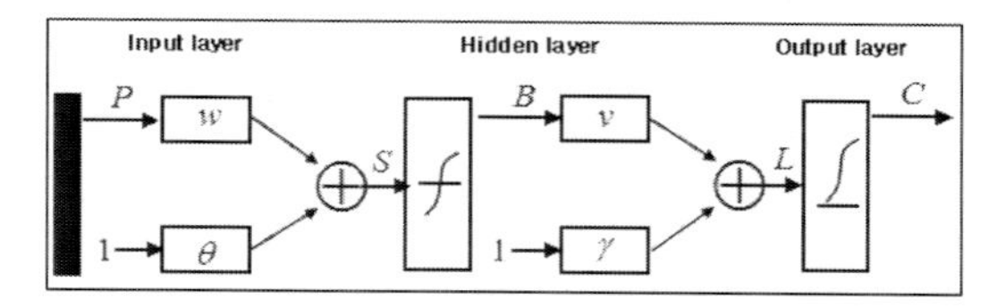

Figure 9. Diagram of BP neural network structure.

In order to reduce the complexity of network structure, this paper carries out a principal component analysis on data of network intrusion detection. Analytical steps are shown below:

STEP1. Standardization of network data is completed by Formula (1). In Formula (1), x_{ij} is the primitive character of network, $\overline{x}_j$ is the mean of characteristic components, and s_j is the standard deviation of each characteristic component.

$$y_{ij} = \frac{x_{ij} - \overline{x}_j}{s_j} \in Y \tag{1}$$

STEP2. Covariance matrix υ of computer network intrusion can be acquired by Formula (2)

$$\upsilon = \frac{1}{N}\left[Y - \overline{Y}_i\right]\ \left[Y - \overline{Y}\right]^T \tag{2}$$

STEP3. Calculate characteristic values and eigenvectors with the characteristic equation $(\lambda I - S)U = 0$ and then choose principle components of network intrusion.

After the feature extraction of network data with principal component analysis, extracted characteristics can be input to BP neural network for learning and training. Problems of neural network structure and parameter optimization should be taken into account in the training process. If there are altogether m characteristics extracted by principal component analysis and k types of network intrusion, the number of nodes of the input layer of neural network should be m, the number of nodes of the output layer is k, and the number of nodes of the hidden layer is $m+1$.

The realization process of the illegal intrusion detection method of digital TV network based on expert system and BP neural network is shown below:

STEP1. Collect network message data and carry out relevant preprocessing.

STEP2. Input network message data for detection. If the datum is an intrusive behavior, output the detection results to relevant modules of network intrusion detection system and send out an alarm.

STEP3. Conduct principal component analysis on normal data according to expert system and extract main characteristics of data.

STEP4. Input extracted characteristics to BP neural network for learning and establish an intrusion detection model.

STEP5. Detect network data with the network intrusion detection model based on BP neural network. Intrusion behaviors are transmitted to the response module of the network intrusion detection system, and the knowledge base of expert system is upgraded. Otherwise, behaviors are regarded as normal network behaviors.

5 CONCLUSION

The key of illegal intrusion prevention of digital TV network lies in detection technologies. And the core content of detection technologies is the analytical algorithm on network data of detection equipment. This paper not only introduces the intrusion detection system model based on host and the intrusion detection system model based on network with the standard frame model of security incident alarm detection, but also proposes the illegal intrusion detection method of digital TV network based on ESTQ and the illegal intrusion detection method of digital TV network based on expert system and BP neural network algorithm. The former mainly focuses on intrusion detection of network attacks that take advantage of network protocol security holes while the latter focuses mainly on intelligent analysis on illegal signals. Besides, this paper proposes the realization approach of linkage alarm when analyzing the method of illegal intrusion detection.

REFERENCES

[1] Sun, Y.W. (2005) Security policy of digital cable TV network BOSS of the next generation, Television Technology, (9):04–06.

[2] Zhang, R.S. (2012) Network intrusion detection system based on expert system and neural network, Computer Simulation, 29(9):162–165.

[3] Xie, X.L. (2013) Network security design based on system vulnerability scanning and antivirus technology, Technological Development of Enterprise, 32(5):71–72.

[4] Jin, Z. (2014) Firewall technology application based on enterprise network information security, Network Security Technology & Application, (2):14–15.

[5] Tian, Z. (2014) A study on firewall settings of computer network, Software Guide, 13(3): 140–141.

[6] Zhuang, L. (2014) A study on monitoring and prevention technology of digital TV illegal signals, Broadcasting Realm (Broadcasting & Television Technology), (5):51–54.

[7] Bie, Z.D. (2013) Safety construction of bilateral network of digital television, Technological Development of Enterprise, 32(6):69–70.

[8] Zhu, B.H. (2012) Design and maintenance of cable digital TV network, Silicon Valley, (4):19–20.

[9] Lu, Y.S. (2007) Basic network security design of digital TV center, Modern Television Technology, (10):115–118.

[10] Chen, X.S. (2013) Network management and implementation of cable digital television, Technological Development of Enterprise, 32(1):74–76.

[11] Niu, S.Y. (2004) Introduction of Information Security. Beijing: Beijing University of Posts and Telecommunications Press.

[12] Ding, G.Q. et al. (2011) Network intrusion detection system design of improved BM algorithm strategy, Computer Measurement & Control, 19(11):2661–2664.

[13] Zhang, Z.F. (2010) Quantum evolutionary algorithm of optimizing the feature library of network intrusion, Computer Application, 30(8):2142–2145.

Electric, Electronic and Control Engineering – Shao, Shu & Tian (Eds)
 ISBN: 978-1-138-02842-5

A study on numerical control transformation of milling machine based on interpolation algorithm

Chun Sha & Ji Luo
School of Electrical and Mechanical Engineering, Nantong Polytechnic College, Nantong Jiangsu, China

ABSTRACT: The superiority of interpolation algorithm and its integrating degree with controller determine the numerical control level of milling machine to some extent. This paper studies the self-adaptive control of numerical control combining the parametric cubic spline interpolation algorithm of five-axis linkage and the self-adaptive control theory so as to achieve high precision and high efficiency of numerical control processing of milling machine. Results indicate that the self-adaptive controller based on parametric cubic spline interpolation algorithm can effective avoid the disadvantage of error amplification. The meshing degree of the actual interpolation curve path and the ideal curve path is relatively high. The matching relation between machining precision and machining efficiency can be effectively solved by automatic compromise of the control algorithm.

KEYWORDS: five-axis linkage; parametric cubic spline interpolation; self-adaptive control; simulated analysis; error analysis.

1 INTRODUCTION

The development of computer microprocessor technology has greatly improved the data handling capability. The computer numerical control system has also developed rapidly as an application of the machining industry. The control of NC machine is supposed to achieve a higher precision under the background of computer technology development [1]. Yanjie Yin et al. (2011) indicated that old machine tools not only have a low machining efficiency but also consume a great deal of energy. Improving the numerical control level of processing and manufacturing equipment is a major way of changing the situation [2]. Aiming at improving the efficiency and precision of numerical control machining, this paper carries out a study on the self-adaptive controller based on the cubic spline interpolation algorithm so as to provide theoretical guidance for the high-precision and high-efficiency development of manufacturing industry on the basis of realizing the transformation of CNC machine.

Interpolation technology is the basis for the control system of NC milling machine to realize path control. In the control system of NC milling machine, linear interpolation and circular interpolation are the most basic functions of interpolation. However, for complex curves, the single use of linear interpolation or circular interpolation would lead to the consequence that efficiency, precision and speed are severely influenced [3]. Qun Shi et al. (2004) completely solved the uniform distribution of randomly specified pulses in an interpolation cycle with an open-loop NC system interpolated with the time-splitting method [4]. According to the idea of time-splitting method, Guodong Cui et al. (2008) proposed a spiral-line interpolation algorithm and introduced the basic principle and derivation process of the algorithm [5]. Ruiyang Tong et al. (2009) carried out a study on the application method of motion controller in numerical control transformation of milling machine. Zhenhua Zheng et al. (2010) designed a new type of numerical control platform with AVR single chip microcomputer as the controller core. It feeds back with grating scale in a closed-loop control system. The interpolation algorithm of integral comparison is adopted as the control algorithm, which is significantly improved in terms of stability and precision compared to traditional interpolation algorithms [7]. Yan Peng (2011) realized the spiral shape of work pieces by controlling the motion curve of cutter with spiral interpolation algorithm and simulated the algorithm with LabVIEW software [8]. Although research results of the above scholars have improved the machining efficiency and precision of NC machine to some extent, there are also certain limitations, which can be roughly divided into the cumulative increase of precision error, the complexity of control algorithm realization and the control idea of low level.

In order to avoid the above-mentioned errors, this paper studies the self-adaptive control of numerical control combining the parametric cubic spline interpolation algorithm of five-axis linkage and the self-adaptive control theory so as to achieve the goal of no amplification of errors, uniform precision and high-precision automation.

2 PRINCIPLE DESCRIPTION OF CNC MACHINING

The machining process of NC machine is shown in Figure 1 [9]. Firstly, the part program obtained from off-line programming is read in the numerical control system through an intermediate medium. After the coding of computer numerical control (CNC), the part program is transformed into the language that can be identified by the system and processed with cutter radius compensation and computer assistant functions. Secondly, conduct the compacting of data point according to the interpolation cycle. Position increment calculated in this interpolation cycle is transmitted to the servo control program for calculation. Finally, drive the servo device to complete the cutter spacing feeding in this interpolation cycle.

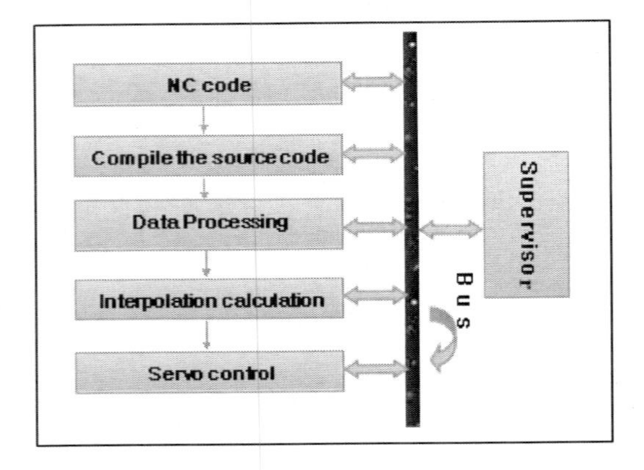

Figure 1. The machining process of NC machine.

It can be known from Figure 1 that the whole process of NC machining is conducted under the integrated deployment of Supervisor. It is composed of feedings of data point in several interpolation cycles. Therefore, the minimum system of the machining process of NC machine is the subsystem of interpolation calculation. In order to directly accept spline information of curve and curved surface acquired from CAD design or described in control point sequence and data point sequence so as to generate orders of cutting path and motion of work pieces and realize the control on machining process, this paper mainly develops and studies the interpolation calculation of CNC system.

In the machining of contour surface, machining path on the machined surface is left by the motion of milling cutter driven by the servo system. Specifically speaking, data information of machined curve and surface is calculated in the interpolation cycle of the system. A series of interpolation points are obtained in accordance with values of system programming error. NC system connects these interpolation points successively as the composition of numerous broken lines. The cutter is continuously moving along these broken lines. The milling machine studied in this paper is a kind of horizontal milling machine. The motion of the milling cutter follows the model of five-axis linkage. The feeding mode of the milling cutter of five-axis linkage is shown in Figure2. It can be known from the figure that the motion of the milling cutter also includes the rotation of X axis and Y axis except the translational motion on the three-dimensional coordinate axis.

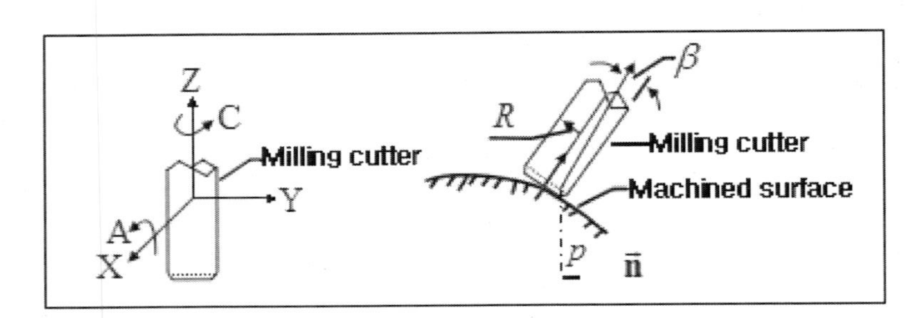

Figure 2. Feeding mode of milling cutter of five-axis linkage.

The main milling function of milling cutter of five-axis linkage is to produce milling force moment along the Z axis[10]. This moment of force provides milling with rotational speed. The function of A axis is to guarantee that the direction of cutter axis is a normal vector with a curved surface. The angle of back part is β, which aims at avoiding interference and collision in machining.

3 THE PRINCIPLE OF INTERPOLATION ALGORITHM AND THE ESTABLISHMENT OF CONTROL MODEL

In order to realize the numerical control transformation of five-axis linkage, this section studies a self-adaptive control model based on parametric cubic spline interpolation and self-adaptive control from the perspective of definition and properties of spline interpolation function, and designs a self-adaptive controller of five-axis machining of free curve.

3.1 *Parametric cubic spline interpolation algorithm*

Suppose coordinates of n points of p are $p_1(x_1, y_1)$, $p_2(x_2, y_2), \cdots, p_n(x_n, y_n)$, $(x_1 < x_2 < \cdots < x_n)$, the function $S(x)$ satisfies all data points in the defined section, each point of $S(x)$ within the definition domain has continuous first derivative and second derivative, and each subinterval $[x_i, x_{i+1}]$ of $S(x)$ is a cubic polynomial, $S(x)$ can be regarded as a cubic spline function with $x_i\ (i = 1, 2, \cdots, n)$ as nodes in the interval $[x_1, x_n]$[11]. The disadvantage of this cubic polynomial can be reflected in Figure 3.

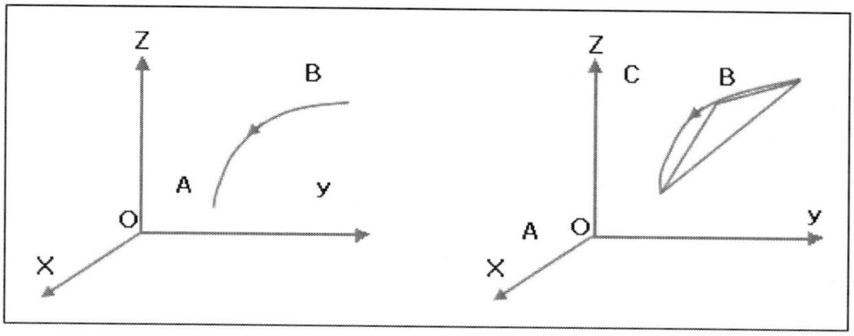

Figure 3. Negation of spline interpolation direction (left), arc substitution and subdivision (right).

Introduce the cubic spline function into the space coordinates. The chord length calculation formula is shown in Formula (1) if n data points are $(x_j, y_j, z_j)(j = 1, 2, \cdots, n)$.

$$s_j = \begin{cases} 0 & j = 1 \\ \sum_{l=2}^{j} \sqrt{(x_l - x_{l-1})^2 + (y_l - y_{l-1})^2 + (z_l - z_{l-1})^2} & j = 2, 3, \cdots, n \end{cases} \tag{1}$$

It can be known from Formula (1) that $s_1 < s_2 < \cdots < s_n$. Suppose $\Delta s_{j+1} = s_{j+1} - s_j$ and $\Delta x_{j+1} = x_{j+1} - x_j$, Formula (2) can be derived from continuous first derivative and second derivative of each point that satisfies the cubic spline function.

$$\Delta s_{j+1} x_j'' + 2(\Delta s_{j+2} + \Delta s_{j+1}) x_j'' + \Delta s_{j+2} x_{j+2}'' = 6\left(\frac{\Delta x_{j+2}}{\Delta s_{j+2}} - \frac{\Delta x_{j+1}}{\Delta s_{j+1}}\right) \tag{2}$$

Add end slopes s_x and ω_x on the basis of Formula (2). Undetermined coefficient of the cubic spline function of data points coordinates and parameter s can be obtained through the chasing method, which is shown in Formula (3):

$$\begin{cases} x_j(s) = A_j + B_j(s - s_j) + C_j(s - s_j)^2 + D_j(s - s_j)^3 & j = 1, 2, \cdots, n-1 \\ y_j(s) = A_j' + B_j'(s - s_j) + C_j'(s - s_j)^2 + D_j'(s - s_j)^3 & j = 1, 2, \cdots, n-1 \\ z_j(s) = A_j'' + B_j''(s - s_j) + C_j''(s - s_j)^2 + D_j''(s - s_j)^3 & j = 1, 2, \cdots, n-1 \end{cases} \tag{3}$$

In conclusion: any data point (x_j, y_j, z_j) of the parametric cubic spline function is the function of parameter s, which is also calculated from original data points. Changes of interpolation points can be realized by altering s.

3.2 *Model establishment and controller design*

It can be known from the analysis in section 3.1 that changes of interpolation points can be realized by altering s. Establish the control model with self-adaptive control as the theory that the machining contour length is constant in free calibration of curve from the perspective of self-correcting and self-following feature of self-adaptive control. The self-adaptive control model of contour length is shown in Figure 4:

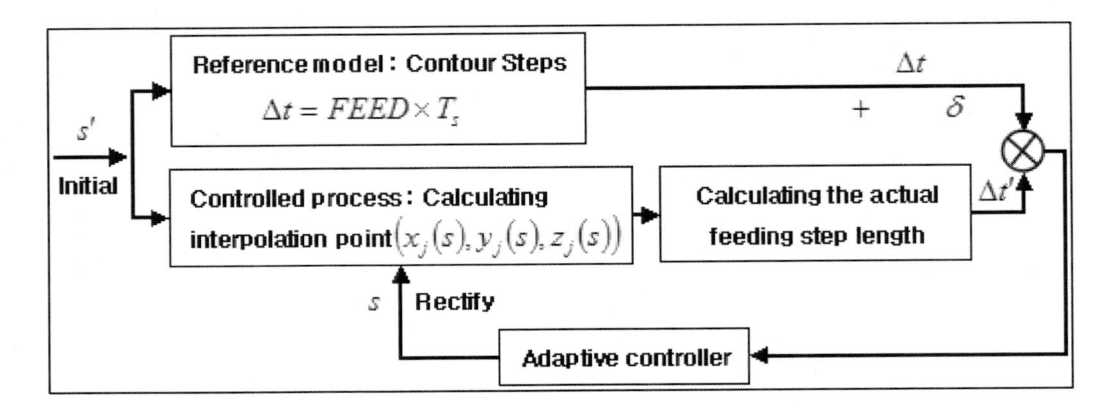

Figure 4. The frame model of adaptive control.

Calculation formulas of $\Delta t'$ and δ in Figure 4 are respectively shown in Formula (4) and (5):

$$\Delta t' = \sqrt{(x_j - x_{j-1})^2 + (y_j - y_{j-1})^2 + (z_j - z_{j-1})^2} \quad (4)$$

$$\delta = \Delta t' - FEED \times T_s \quad (5)$$

After the subdivision in line with programming accuracy, s' can be obtained from initial parameters derived from original data points. In the controlled process model, s' calculates relevant interpolation points $(x_j(s), y_j(s), z_j(s))$ through rough interpolation. T_s stands for the time of the milling cutter passing every two points, namely the interpolation cycle. The actual feeding length $\Delta t'$ can be calculated with Formula (4). After the comparison of two comparators, the error value δ can be calculated through Formula (5). The error value is transmitted to the adaptive controller as the standard of adjusting parameter s'. After the adjustment of the adaptive controller, the new parameter s' is taken as an independent variable to recalculate the feeding length. Transmit the error value obtained from the comparison of comparators and specified contour lengths to the adaptive controller so as to obtain the parameter s after calibration. Repeat the process until the value of δ is in the permitted error range. $(x_j(s), y_j(s), z_j(s))$ is transmitted to the driving device as the control point of cutter position and then to the next point for the calibration of feeding length.

The step length of contour in cubic spline interpolation refers to the length of line segment cut by the milling cutter in each interpolation cycle. However, each engage point is calculated through the parametric cubic spline interpolation algorithm in line with programming error, so that the distance between two points and the required actual step length of contour are not constant. Errors are not inevitable. Therefore, in this paper, the distance between every two points is changed by altering the initial value of the interpolation point. According to the feature that the self-adaptive control can be automatically calibrated with changes of the controlled object, the design process of the adaptive controller, with the step length as the controlled object, the ideal step length of contour determined by cutting speed and interpolation cycle as the reference model, and parameter s as the controlling object, is shown below:

STEP1. Suppose interpolation points of the interpolation j are $(x_j(s), y_j(s), z_j(s))$ and the step length between point j and point $j-1$ is $\Delta t'_j$, the deviation between $\Delta t'_j$ and the ideal step length is $\delta = \Delta t' - FEED \times T_s$. If the deviation is far larger than 0, interpolate h point evenly in (s_{j-1}, s_j). Take the intermediate point $i = 0.5 \times h$ for interpolation calculation so as to acquire the feeding step length Δt_i. Suppose $\delta = \Delta t_i - \Delta t$, the analysis can be carried out as follows:

Analysis1. If $\delta > 0.001$, take the point $i = 0.25 \times h$ for interpolation calculation so as to acquire the relevant feeding step length Δt_{i1}, which is then compared to Δt. Suppose $\delta = \Delta t_{i1} - \Delta t$, repeat Analysis1 if $\delta > 0.001$ until $i = 1$. If $\delta > 0.001$, make $s_j = s_{j1}$ and interpolate h points evenly in (s_{j-1}, s_j). Repeat the process of Analysis1 until $\delta < 0.001$. Make the judgment whether $\delta < -0.001$ exists. If $\delta < -0.001$ exists, repeat Analysis2. Cycle these separate points until the value of δ satisfies the condition $-0.001 < \delta < 0.001$, and then take the interpolation point s_{j-1}, s_j.

Analysis2. If $\delta < -0.001$, take the point $i = 0.75 \times h$ for interpolation calculation so as to acquire the feeding step length Δt_{i2}, which is then compared to Δt. Suppose $\delta = \Delta t_{i2} - \Delta t$, repeat Analysis1 if $\delta > 0.001$. Repeat Analysis2 if $\delta < -0.001$. Cycle these separate points until $j = h-1$. If $\delta < -0.001$ still exists, make $s_j = s_{jh-1}$ and execute STEP2.

STEP2. If $\delta < -0.001$, interpolate h points evenly in (s_{j-1}, s_j). Repeat STEP1 until the value of δ meets the condition $-0.001 < \delta < 0.001$, and then take the interpolation point s_{j-1}, s_j.

The self-adaptive control process of Module 1 that is suitable for the calibration of contour step length in spline interpolation is shown in Figure 5. Module 2 is similar to Module 1.

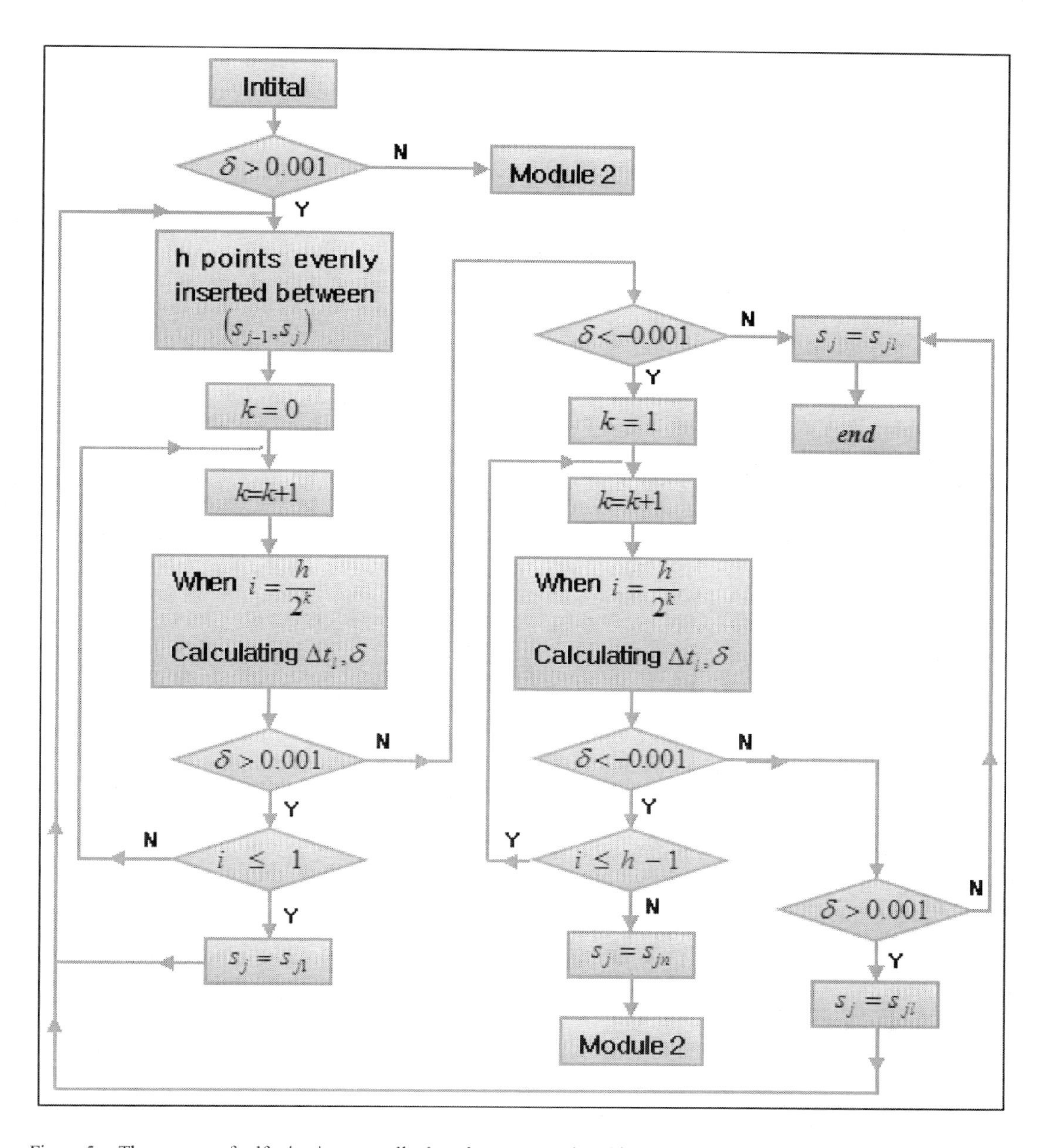

Figure 5. The process of self-adaptive controller based on parametric cubic spline interpolation.

4 ANALYSIS ON FREE CURVE INTERPOLATION SIMULATION BASED ON PARAMETRIC CUBIC SPLINE

The process of free-form surface NC machining carried out by milling cutter on work pieces is actually completed in the form of free curves. The distance between two adjacent curves is called machining width. So, the smooth degree of curved surface depends on the width. Qualified machining of free-form surface can be only achieved by the reasonable control of width in NC machining. In order to verify the transformation effect of milling machine based on the combination of parametric cubic spline interpolation and self-adaptive controller, this paper takes the spatial curvilinear equation of higher degree, shown in Formula (6), as an example and carries out 90 interpolation simulations on specified data points. The starting point of interpolation is set as (9.9,49.0,38.06) and the terminal point is set as (99.9,4989.96,3865.10). The sampling period of interpolation is 10ms, the slope of endpoint is 0, and the feeding velocity of cutting is 4940mm/min. The value of interpolation j ranges between 1 and 90. Coordinates of each interpolation point is (x_i, y_i, z_i). Coordinate values range between the starting point and the terminal point. Here, $|z_i - z|$ stands for the deviation between the interpolation i and the ideal point. $\Delta t = \sqrt{\Delta x^2 + \Delta y^2 + \Delta z^2}$. Contour error is $e_i = \sqrt{\Delta e_x^2 + \Delta e_y^2 + \Delta e_z^2}$. Calculation results of spatial free curve interpolation are shown in Table 1.

$$Z = \sqrt{|8 - 0.1X^4 + Y^2|} \tag{6}$$

Table 1. Calculation results of spatial free curve interpolation.

Times of interpolation	$\sqrt{\Delta x^2 + \Delta y^2 + \Delta z^2}$	$\sqrt{\Delta e_x^2 + \Delta e_y^2 + \Delta e_z^2}$	x_i	y_i	z_i
Last time	0.814240	0.000854	99.899580	4989.9018	3865.1308
⋮	0.816314	0.000910	99.893417	4989.2603	3864.6310
⋮	0.823028	0.001000	99.889784	4988.6874	3864.0581
⋮	⋮	⋮	⋮	⋮	⋮
7	0.823349	0.000364	10.286061	52.902183	41.075327
6	0.823304	0.000380	10.222712	52.252612	40.573383
5	0.823285	0.000378	10.158975	51.603085	40.071505
4	0.823294	0.000348	10.094844	50.953582	39.569676
3	0.823329	0.000285	10.030309	50.304081	39.067883
2	0.823392	0.000172	9.965364	49.654561	38.566109
1	…	0.000000	9.900000	49.005000	38.064340

The deviation between the ideal curve trajectory and the actual interpolation trajectory acquired in machining spatial curve of higher degree with parametric cubic spline interpolation is shown in Figure 6. Solid lines are ideal trajectories and (*) stands for part of points of the actual interpolation trajectory.

Four simulation curves in Figure 6 are respectively: three-dimensional interpolation of free curve (upper left); the projection of free curve interpolation on XOZ plane (upper right); the projection of free curve interpolation on XOY plane (lower left); the projection of free curve interpolation on YOZ plane (lower right). The figure shows that the fitting error of the algorithm is also the main source of contour error in spatial curve fitting of higher degree of cubic spline interpolation function. The fitting error of the algorithm increases sharply in the last interval of interpolation due to limitations of spline function. In this paper, the parametric cubic spline interpolation algorithm blends the self-adaptive theory into spline interpolation and achieves the desired results of error control by the combination of data point encryption and self-adaptive calibration.

There are certain contradictions between machining precision and machining efficiency in actual NC machining process. The machining precision increases accordingly when interpolation points are encrypted. But the machining efficiency decreases. Therefore, the relation between interpolation points and machining efficiency (system time-consumption) is explored in this paper. System time-consumption of the experimental simulation that carries out machining on different interpolation points with the same curve is shown in Figure 7. It can be concluded from Figure 7 that the interpolation calculating time increases gradually with the increase of interpolation points. But the interpolation cycle is fixed. This will inevitably bring impact to the implementation of other functions of system management, reducing the working efficiency of the whole system.

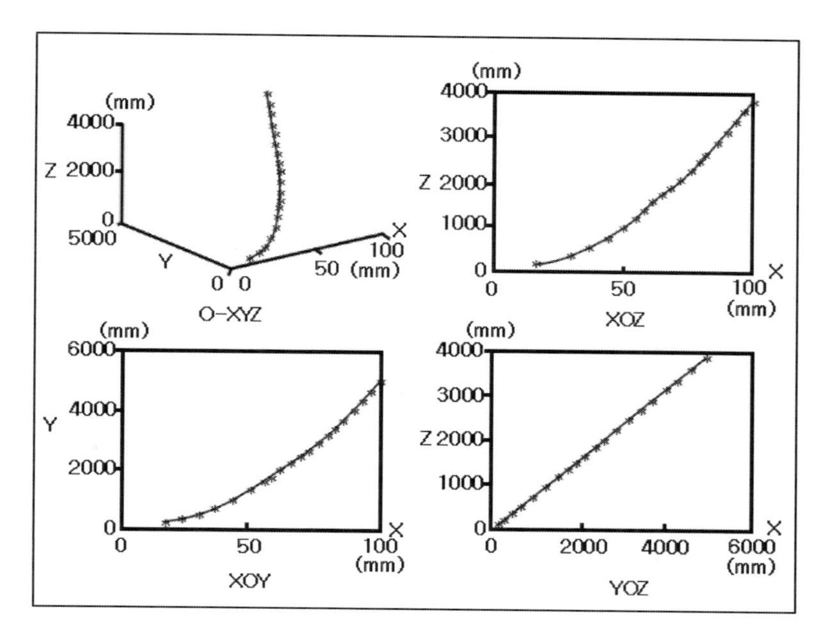

Figure 6. Simulation curve of interpolation based on cubic spline algorithm.

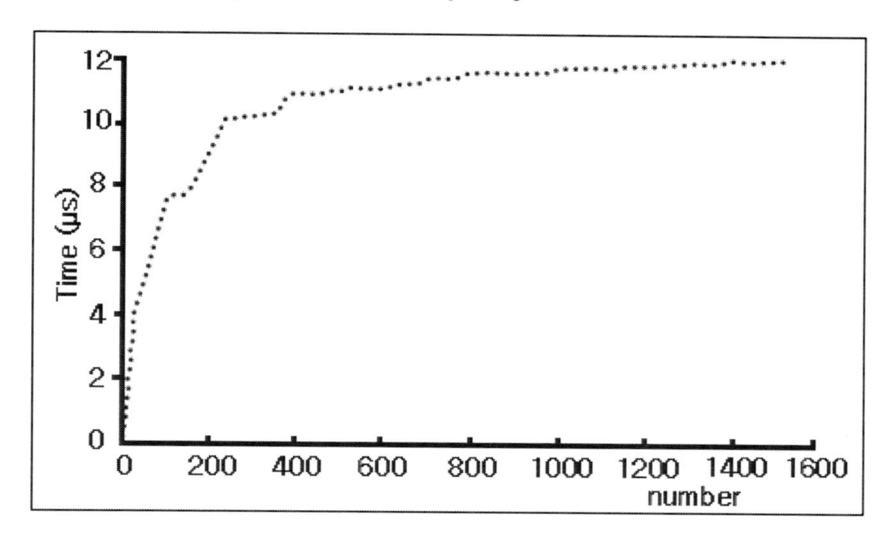

Figure 7. Relation between interpolation points and system time-consumption.

5 CONCLUSIONS

Reading and decoding of core programs of the traditional NC machining have wasted a great deal of manpower and machine hours. And the rough algorithm accuracy also has certain effect on NC machining of milling machine. Disadvantages of the traditional NC machining can be effectively avoided by the self-adaptive controller designed through the combination of parametric cubic spline interpolation algorithm and the self-adaptive control theory. Based on the study on parametric cubic spline interpolation algorithm, this paper designs a self-adaptive controller of parametric cubic spline interpolation algorithm through the combination of the algorithm and the self-adaptive control theory and carries out a simulated analysis on design results. Conclusions are drawn as follows:

1. The parametric cubic spline interpolation algorithm of five-axis linkage can effective make up for the deficiency that the spline interpolation algorithm can be only applied in a two-dimensional surface.
2. Results of the interpolation algorithm studied in this paper indicate that the matching degree of the actual interpolation curvilinear path and the ideal curvilinear path is relatively high during the whole interpolation period.
3. Replace the arc feeding step length with the chord feeding step length and replace the trivial deviation caused by the ideal contour step length with the actual feeding step length. This is very practical in actual cutting process, which is able to improve the interpolation precision effectively.
4. The encryption of interpolation points helps to improve the machining precision. However, the increase of interpolation points leads to the extension of time consumed by the system while the interpolation period of the system is constant, resulting the fact that the improvement of machining precision reduces the machining efficiency of NC machine. The matching relation between machining precision and machining efficiency can be effectively solved by the control of the parametric cubic spline interpolation algorithm, realizing the compromise control of machining precision and machining efficiency in NC machining.

REFERENCES

[1] Luo, X.K. (2002) *Numerical Control Technology and Manufacturing Automation*, Beijing: China Machine Press.

[2] Yin, Y.J., Li, Z.L., Liao, X.B., Zhang, R. (2011) NC transformation design of universal tool miller, *Machine Tool & Hydraulics*, 39(10): 134–136.

[3] Yang, Y.J. (2005) Numerical Control Technology, Beijing: China Machine Press.

[4] Shi, Q., Wang, X.C. (2004) A study on fine interpolation of NC system based on high-speed and uniform pulse distribution algorithm, *Industrial Instrumentation & Automation*, 32(2): 69–72.

[5] Cui, G.D., Zhao, D.B. (2008) A study on a practical spiral line interpolation algorithm, *Machinery & Electronics*, 05(12): 14–16.

[6] Tong, R.Y., Shao, G.J. (2009) A study on NC milling machine based on embedded motion controller, *Microcomputer Information*, 25(1): 190–192.

[7] Zheng, Z.H., Guo, Q., Wu, G.C. (2010) The study and design on a two-dimensional NC platform based on interpolation algorithm, *Technology of Measure and Control*, 29(6): 58–61.

[8] Peng, Y. (2011) Spiral interpolation algorithm of NC milling machine based on LabVIEW, Journal of Wuyi University, 25(2): 61–64.

[9] Liu, L.B. (2006) *Technical Transformation of NC Electrical System of SC55 Milling Machine*, Liaoning Project Technology University.

[10] Zhan, Y., Zhou, Y.F., Zhou, J. (2000) Spatial circular interpolation of five-axis NC milling machine, *Journal of Huazhong University of Science and Technology*, 28(5): 4–6.

[11] Han, Z.Y., Wang, Y.Z., Fu, H.Y., Fu, Y.Z. (2003) A study on the interpolator of cubic B spline curve data sampling, *Machinery Design and Manufacture*, 15(1): 81–82.

Electric, Electronic and Control Engineering – Shao, Shu & Tian (Eds)
© 2015 Taylor & Francis Group, London, ISBN: 978-1-138-02842-5

Study on HCI of mobile internet—take graphic design of IOS7 windows as an example

Fei Chang
Pingdingshan Institute of Education, Pingdingshan Henan, China.

ABSTRACT: The emergence of Intelligent User Interface (UI) with its strong adaptability has greatly driven the development of HCI. The research on HCI, including its tasks and challenges, as well as the methods to maximize its convenient service for users, needs to be conducted and studied further and further. The paper at first introduces the research background and purposes, the basic concepts and design principles of HCI, and then implies the HCI design process and related studies on IOS7 provisional windows. The last part of the paper presents in detail the graphic design of IOS7 windows so as to study HCI of mobile Internet.

KEYWORDS: Mobile Internet; Human-Computer Interface (HCI); graphic design of IOS7 windows.

1 INTRODUCTION

In the age of information explosion, people always expect that the advanced technology can meet their needs and serve their lives more conveniently and economically. Under such circumstance, it is evidently important to study HCI of mobile internet. With social development, advancement of the ages and improvement of living standards, especially the popularization of computer, traditional Internet mode cannot satisfy people any longer. That's why Internet HCI has produced and become a hot pot being studied. However, most researches pay more attention to catch peoples' eyes, but ignore user prompts, warnings and useful guidance. The study on provisional window design targets at guiding users to avoid possible risks in appropriate ways, and weakening users' discomfort with system interruption in the most elegant form. The author will put forward possible problems and solutions of graphic design of IOS7 windows in accordance to the analysis and conclusion.

2 STUDY PURPOSES & BASIC CONCEPTS AND DESIGN PRINCIPLES OF HCI

2.1 *Study purposes*

Nowadays, users are no longer choosing mobile phones for its being of new technology, like a touch screen or voice communication, nor can they buy phones because they are pursuing one kind or another kind of hardware or software facilities. More developed technology becomes, higher expectations people will have. The real driven force for users to purchase is the smooth and novel user experiences. User experiences contain products' design, various types and satisfied service, like buy service and after-sale service. In addition, it also includes operation introductions. An excellently designed window can warn the users the possible risks that may cause the disclosure of personal private information and property loss. The bilateral communication between human and computer and information transferring can be achieved through effective interaction between users and system by studying HCI while the interface is a specific manner to achieve such interaction. In the process of interaction design, interactions are the content and spirit while interface is the body and form. Interface achieves the establishment of human-computer relationships in most researches, which, however, is not the final target of studies. The final aim is to solve users' problems and to meet their requirements and expectations so as to serve and benefit people better.

2.2 *Basic concepts of HCI*

HCI refers to the field where human and machine can exchange a variety of information and affect each other functionally, also called human-machine integration or information exchange. This combination surface includes not only the direct contact of points, lines and planes, but also distant information transmission and related functional space that can be controlled. The combined surface of human and machine is the core part of the system, the main branch of which is for Safety, Ergonomics to research related problems that are put forward by Engineering, and to find specific solutions and measures with the help of Equipment Engineering, Security Management Engineering and safety system.

Thus, it is possible that internal information can exchange with the forms accepted by humans so as to achieve more accurate and fast operations for users.

HCI, in other words, the user interface (UI), is a subject to study the interaction between systems and users. The system can be various machines, or some computerized systems and software, and interface are the parts that human is willing to see. Users can exchange with and operate the system through HCI. This is the science that studies interface design, related evaluation and interactive computing systems.

2.3 *Design principles*

There are four principles: User-centered design principle, design orders principle, design functions principle and the principle of consistency. User-centered design principle requires researchers to understand users' psychology, aims and their needs when designing. Design orders principle requires researchers to design from largest to smallest and from top to bottom according to time order of processing and visiting. Functions Principle refers to connecting the relevant control types of different kinds of electronic systems and the unified interface of various management objects to design functional areas and a multi - level menu and to prompt messages and various interfaces hierarchically in accordance with the application environments of the objects and their specific requirements, the purpose of which is to make users grasp its rules and relevant characteristics. The Principle of Consistency means that color, operating area and words need to be consistent. That is to say, on the one hand, the shape and color of the interface and its fonts should be consistent with those of the country and the world; on the other hand, it should have its own character so as not to distract users.

3 DESIGN PROCESS OF HCI

The design flow of HCI contains four steps in general: to build an external model of system functions; to make sure the relevant tasks for the completion of system functions and computer; to think about common problems met in the process; to establish an interface prototype. To build an external model requires researchers to understand HCI and to be familiar with software structures, design principles and overall design structure. Meanwhile, designers must know users' information, like their educational levels, mental age and personal characteristics. Based on this information, researchers can design a UI that satisfy users, and in return, users' satisfaction drives research conduction and thus to motivate the development of technology. To make sure that man and computer are clear about their tasks and take their responsibilities for them can help develop the research. The tasks can be analyzed by two means: one is to study system's needs; another is to reflect the manual application system on HCI according to the reality and practices. Through accurate analysis, the tasks are segmented clearly and besides, the relevant objects and actions are clarified and recognized. There are no doubts some problems in design. The system response time of HCI must be taken into consideration. The mostly complaint, problem by users is too much time taken for the system to respond and to solve the information errors. The researcher must take users' problems seriously and solve these problems in integrated forms in case they deal with them repetitively, waste time and decrease the work efficiency. When error messages appear, designers can give warnings through ringing sounds and special colors. To build the software prototype not only in the UI, but also in the users' verification until the consistency of the designed UI and the expected UI is achieved.

4 GRAPHIC DESIGN OF IOS7 WINDOWS

4.1 *Basic constituents of IOS windows*

The emerging order of IOS window can be described as follows: a message from institution to UI, the event that happens next; the running program that downloads the data; the events that have changed. The concrete process is: first to check the object that receives messages and judge it in energy-saving IOS system; and then to choose to call the function, to bind the corresponding function until the message generates and starts to run, and to be handled by the compiler; followed by the structure of message object; finally to decompose and analyze the related objects and to release to the APP.

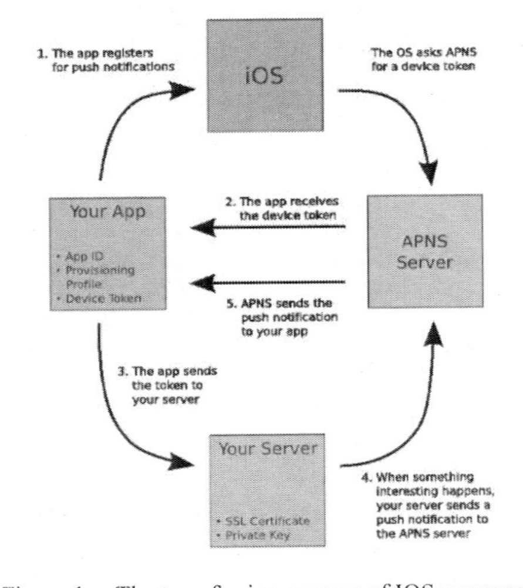

Figure 1. The transferring process of IOS message.

4.2 *Types of IOS7 windows*

4.2.1 *Commonly used banner advertisements*

The differences between IOS6 icons and IOS7 icons are: IOS7 icons are designed to be flat and being of many functions, like viewing 3D tab page in full screen, previewing and putting man voice into music so as to control Bluetooth with Siri if possible; the control box comes out when sliding up in IOS7, and picture processing like colors and images in IOS7 is more vivid and lively.

Figure 2. The comparison between IOS6 icons and IOS7 icons.

The banner advertisements are black translucent strips at first and then disappear themselves in few seconds, which operation is comprised by notification message and the related system parts. As Figure 2 shows above, it aims to make users timely and efficiently know what time APP sends those messages and notifications. Besides, once these banner advertisements emerge, can users click and operate directly and easily.

4.2.2 *Warning window*

Warning window, as the name suggests, is what appear above the screen automatically when the system is threatened. This window can be put into one title and one or more buttons. However, in actual operations, IOS cannot achieve such appearance because the designers have no right to change orders, like background, fonts, appearance and any other order buttons under IOS7 system. Some titles and more warning windows can be put into the space between two warning windows so as to display more clearly and evidently when users are clicking. Highlights in IOS7 are shown in bolded fonts in the same background. If a warning button provided, the situation will be easier for users to achieve the desirable effect by clicking on the OK button. When the phone screen being locked and APP not open, the titles of custom button will be displayed in the form of sliding.

4.2.3 *Action bar*

The action bar can carry out the selection of related tasks according to the action list memory.

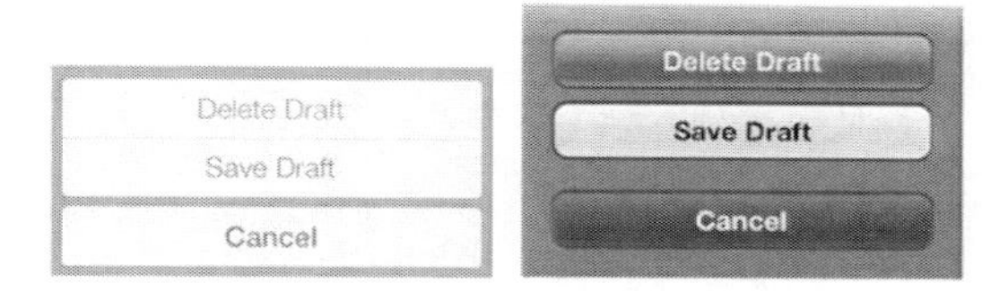

Figure 3. The action bar of IOS7 and IOS6.

The function of action bar has the following characteristics: Firstly, it can provide options to execute related task commands. In order to avoid the appearance of some interfaces that disturb the users' normal operation when clicking the space, action bar mainly works for communication with users positively, and for obtaining users' approval before completing related tasks where various risks may exist. If users know what they do next, they will realize what the outcome may be brought by his task, which requires their consent first. Secondly, action bar has a Cancel button and Back button which aims to protect the feasibility of all act retrograde. Therefore, you also take the placement of Cancel button into account when designing. Generally, the placing order is from top to bottom, left to right, so that users can easily know where they are and thus ensure that they have read all the options described above. Thirdly, one significant feature of all systems designed by Apple is that the red button represents the action of execution but its operation destroys common habits. Some operation options are shown as the red buttons and appear in the users' vision field rather than in some uneasy going areas, which has double effects. Meanwhile, it can avoid pulling the scroll bar because if there are too many options during the operation process, to some extent, it will cause the window out on display. Under this circumstance, users have to pull the scroll bar manually to read the unseen parts. As a result, the designed UI is not ideal and perfect and users are hardly satisfied with it.

4.3 *Advantages of IOS7 windows design*

The advantages include users first, information focus, tasks simplification and interaction reversibility, etc. Products are developed for users. Even if IOS7 provides various windows for users, these temporary windows can be closed by them if not needed. In that way, the occupied space by windows can be spared and the prompt frequency can also decrease. In fact, most provisional windows of IOS7 options can be closed. The core idea of IOS7 is information focus and information being brief, clear and understandable for users' quick and efficient reading. Moreover, there is one centralized Exit option for windows. In order to reduce operational errors as much as possible, Apple allows users to regret at any time, which is humane and safe. IOS7 provides users simple and clear tasks with keeping provisional windows always displaying ahead and advocates not building provisional windows in application system with the purpose of keeping users remembering their major tasks. ISO7 is being of huge compatibility of users' expectations and interface design, of strong adaptability that users take the controlling position and whole system adapts itself to users, of a clear hierarchic structure with its distinct graphic design that lowers its responsibility.

4.4 *Design period and user verification period of HCI of IOS7 windows*

Any system development has a complicated process, including research conduction, overall design, and specific plan implementation. The design flow researchers devise HCI is as follows: collection and analysis of the needs — description of users' characteristics— defining objectives— analysis of tasks— interface analysis of IOS7 windows— user verification— experts' evaluation— implementation and assessment.

Users' verification is a direct standard of products' validity. Availability is to examine the usability, efficiency and satisfaction comprehensively from the users' perspective.

5 CONCLUSION

The paper is divided into two parts generally, one introducing the basic concepts and features of HCI as well as its usual design flow. Based on the basic knowledge, the author focuses on the graphic design, including the types and constituents of IOS7 windows, the advantages of IOS7 windows design as well as design period and user verification period of HCI of IOS7 windows. Another presents that the design of HCI, in convenient and fast ways, achieves the perfect combination of human and computer. During the design process, though the HCI technology has developed rapidly, the researchers should pay most of their attention to clients' psychological characteristics and expectations so as to make sure their research targets in designing HCI.

REFERENCES

[1] Ronald W. Langacker. (2009) Reference-point constructions. *Cognitive Linguistics (includes Cognitive Linguistic Bibliography)*, (1):32–34.

[2] Louis Goossens. (2009) Metaphtonymy: the interaction of metaphor and metonymy in expressions for linguistic action. *Cognitive Linguistics (includes Cognitive Linguistic Bibliography)*, (3):56–57.

[3] Anna Abraham, Sabine Windmann. (2006) Creative cognition: The diverse operations and the prospect of applying a cognitive neuroscience perspective. *Methods*, (1):43–46.

[4] Jean-François Petiot, Bernard Yannou. (2004) Measuring consumer perceptions for a better comprehension, specification and assessment of product semantics. *International Journal of Industrial Ergonomics*. (6):25–27.

[5] Nathan Crilly, James Moultrie, P. John Clarkson. (2004) Seeing things: consumer response to the visual domain in product design. *Design Studies*, (6):87–88.

[6] Shih-Wen Hsiao, Hsi-Ping Wang. (1998) Applying the semantic transformation method to product form design. *Design Studies*, (3):54.

[7] Wu Zenghong. (2011) *Study on the Theory and Methods of Personalized Map Service*. The PLA Information Engineering University, (08):24–26.

[8] Li Rui. (2009) *Assessment Study on Availability of ECDIS HCI*. Harbin: Harbin Engineering University. (03):14.

[9] Shen Ping. (2010) *Elevator System Design Based on the Concept of Human Design*. Tianjin: Tianjin University. (04):27.

[10] Su Huiming. (2011) Research on Virtual Reality Display Design Based on Web 3D Technology. *Central South University of Forestry and Technology*. (06).

Electric, Electronic and Control Engineering – Shao, Shu & Tian (Eds)
© 2015 Taylor & Francis Group, London, ISBN: 978-1-138-02842-5

Research on data acquisition for medical air quality detection

Zhigang Liu & Baoqiang Wang
School of Electronic Engineering, Chengdu University of Information Technology, Chengdu, China

ABSTRACT: The medical air quality plays a crucial role in human health. A new detection method and monitor for medical air quality detection are researched on the subject. The air quality monitor is designed with multi-channel data acquisition and is used to simultaneously detect a variety of gases and display indoor air circumstances. Meanwhile, it can be used to control, disinfection devices for keeping a healthy environment. The air quality monitor has solved the problem of air monitoring and purification simultaneously. The device has the feature of high stability, real-time, easy extension and portable carrying.

KEYWORDS: Medical air quality detection, Multi-channel data acquisition, Gas sensors.

1 INTRODUCTION

Some small spaces with the intensive populations, such as hospital, conference rooms and movie theaters usually gather a lot of microorganisms and bacteria. In these environments, bacteria easily breed and spread by droplets or mosquito bites. If a man works frequently in these kinds of situations, the health of human will be affected seriously. The human life and production process will inevitably produce a variety of harmful gases. Indoor decorative paint, man-made wood furniture and decorative textiles can long-term release harmful gases, such as formaldehyde, benzene series, gas leakage, the combustion of various fuels, cooking oil fume and SO_2.

Taking indoor bacteria and harmful gases into account, indoor air quality detection is raised in the paper. The air quality monitor may be applied to some small spaces with the intensive populations in real-time, detect indoor air quality, if the indoor harmful gas concentration exceeds the standard concentration, the monitor will issue control order to start airing disinfection machine. If so, indoor space may form a healthy environment.

2 ARCHITECTURE OF INDOOR AIR DETECTION SYSTEM

The air quality detection system based on FPGA and ARM mainly includes an AD (ADS6424) with four channels, sampling rate 100 MBPS and serial differential input and output, a piece of ARM (AM3359) controller and a DDR2 (MT47H128M16) memory with 128 M address space. Air detection system architecture is shown in Figure 1.

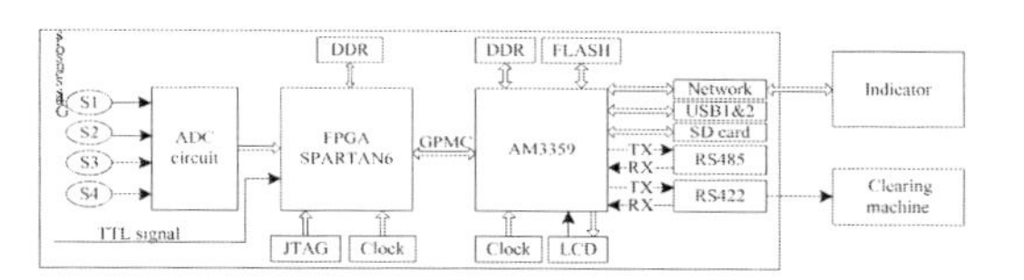

Figure 1. Architecture of system.

S1~S4 is 4 gas sensors, which may sample gas signal to issue them to AD converter. Signal conditioning mainly realizes the input analog signal amplification, attenuation, filter and impedance matching to meet the requirements of AD converter. The network may connect with computer to show relevant data, RS422 mainly control air disinfection machine. The monitor may use following sensors to acknowledge requirement. Used sensors are shown in Table 1.

Table 1. Technical indicators of sensor

Model	Monitoring object	Loop voltage	Detection range
MQ-138	Formaldehyde	5V±0. 1V	1ppm~10ppm
MQ-137	Ammonia	5V±0. 1V	10ppm~300ppm
MQ-7	Carbon monoxide	5V±0. 1V	10ppm~1000ppm
MQ-216	Natural gas	6V±0. 1V	500PPm~10000PPm

3 IMPLEMENTATION OF PROJECT

3.1 *Indoor harmful gases and characteristics*

There are many kinds of indoor virus, bacteria and harmful gas, the research involves carbon monoxide, carbon dioxide, formaldehyde, benzene, ammonia and radon.

3.1.1 *Formaldehyde*

1 Chemical formula: CH_2O. Molecular weight: 30.
2 Main source: Indoor formaldehyde comes mainly from indoor decorative, block board, medium density fiberboard, particleboard, artificial board manufacturing furniture and other decorative materials containing formaldehyde content of interior decoration textiles.
3 Characteristics: Under normal circumstances, formaldehyde is a flammable, colorless and irritant gas, relative density (air =1 g/ml) is 1.081-1.085 g/ml.
4 Main harm: Effects of formaldehyde on human health are mainly in the olfactory stimulation, allergy, pulmonary function abnormalities, liver function abnormalities and abnormal immune function. High doses of formaldehyde can cause chronic respiratory diseases and cause nasopharyngeal cancer, colon cancer, brain tumors, menstrual disorders and nucleus of the gene mutation.

3.1.2 *Ammonia*

1 Chemical formula: NH_3. Molecular weight: 17.
2 Main source: Ammonia is mainly from concrete admixtures, indoor decorative materials, additives and brighteners in the construction.
3 Characteristics: Physical properties: the ammonia typically is a colorless gas pungent, small density, easily soluble in water and easily liquefied; chemical properties: it can generate a hydrous ammonia reaction in water.
4 Main harm: Ammonia on the skin has corrosion and stimulation, organization of the skin can absorb moisture, so ammonia destroys the structure of cell membranes. The high solubility of ammonia is harmful to the body of the animal or the upper respiratory tract irritation and corrosion. The gas is often adsorbed on the skin, mucous membrane and conjunctiva, resulting in irritation and inflammation. Ammonia gas is usually absorbed by the human body and easily into the blood, so it destructs oxygen transportation function.

3.1.3 *Carbon monoxide*

1 Chemical formula: CO. Molecular weight: 28.
2 Main source: The major source of carbon monoxide is a gas leak.
3 Characteristics: Under ordinary conditions, carbon monoxide is a kind of colorless, tasteless, no gas odor, poisonous and neutral gas, melting point is -199°C, boiling point is 191.5°C. The density of the gas is 1.25g/L at standard conditions. Carbon monoxide molecule valence is + 2 and can be further oxidized into +4, so that carbon monoxide is flammable and reductive and burns in the air or oxygen, producing carbon dioxide.
4 Main harm: Carbon monoxide (CO) is a kind of pollutants of blood and nervous system. In the air, carbon monoxide (CO) entering the blood by the respiratory system not only reduces the ability of blood to carry oxygen, analysis and release, but also inhibits and delays the oxygen hemoglobin. Severe cases may endanger human health.

3.1.4 *Benzene*

1 Chemical formula: C_6H_6. Molecular weight: 78.
2 Main source: The main source of indoor benzene is indoor decorative plywood, furniture paint, paint mucosa, medium density fiberboard, particleboard, other man-made board and artificial board manufacturing etc.
3 Characteristics: Benzene is a colorless liquid mixture, inflammable and volatile under normal temperature.
4 Main harm: High concentration can damage the hematopoietic system and the nervous system to release chronic benzene poisoning, which is mainly manifested as headache, dizziness, fatigue, poor sleep, loss of appetite and decreased white blood cells. If the condition further aggravates, skin can be bleeding and produce anemia and leukemia in serious case.

3.2 *Design of indoor air detection process*

The air quality detection process is shown as follows. After the system powers on, firstly, FPGA initializes ADS6424 chip, PC issued configuration instruction to FPGA by ARM, configuring work mode, sampling rate of FPGA, secondly, FPGA will cache series data processed to DDR2, and it will send interrupted command to ARM, ARM extract data in DDR2 through GPMC bus, lastly ARM will transmit data to a PC through a network or USB. According to the data collection process, and the feature FPGA may implement, FPGA logic design can be divided into 5 modules, namely initializing the ADC module, serial to parallel conversion module, GPMC bus interface module, processing data module, data storage interface module, the relationship between the modules is shown in Figure 2.

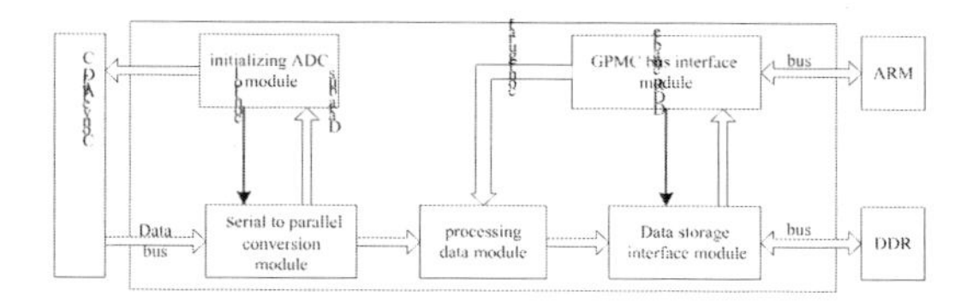

Figure 2. Logic design diagram of FPGA.

3.2.1 *Initializing ADC module*

Initializing ADC module is used to reset and initialize the ADS6424 chip to meet the AD to work properly. The module is called only once, when the system is powered or reset, and it coordinates work with serial converter module for 600M serial low-voltage differential signaling (LVDS) data.

3.2.2 *Serial to parallel conversion module*

Serial to parallel conversion module receives 600M bit clock and 4-channel differential serial data from ADS6424, and it converts serial data for 12-bit parallel data via ISERDES (series-parallel converter). Using ISERDES2 may simplify the complexity of the design, ensure temporal logic accuracy and shorten the development cycle.

In addition, it has FIFO between serial to parallel conversion module and processing data module to synchronize 100MSPS sampling clock.

3.2.3 *GPMC bus interface module*

The GPMC bus interface module is used to resolve the input command ARM, to configure the FPGA's sampling rate, mode and trigger samples, or it reads the DDR2 memory under the ARM indicates the read signal, and address signals. This module can bridge FPGA and ARM.

3.2.4 *Data storage interface module*

The module mainly realizes reading and writing for DDR2 memory. It may convert 4 groups of 16 bit data into 64 bits to write into DDR2 memory by the NPI interface of MPMC, when it transmits data from DDR2 by valid read-enable signal, 64 bit data is split into 4 groups of 16 bit data and the output in accordance. The data storage module has not addressed, input interface, its interior is simple address-inducted, when reset, address makes zero.

3.2.5 *Processing data module*

Processing data module includes 3 modules for the three work states respectively, each work state has independent data and control signals, interface may select the switch state, according to work status, three kinds of work status is shown in the following text.

1 Normal work mode:

FPGA anti-interference algorithm is shown as follows.

1 According to the set threshold, filtering out background noise.
2 Compared with noise channel, judging it is a useful signal or noise signal.
3 Judging whether it is a square wave signal, if it is a square wave signal, then it will be removed.

Design method: according to the set valve, it compares the signal with noise, if the signal channel valve is greater than the noise channel value, then signal channel is signal information, or, it is noisy and is filtered. If the signal channel is square wave and also filtered. For the useful pulse, when it is transmitted to the ARM, we will extract 55 points before the first threshold crossing, and we will extract 200 points after the first threshold-crossing, we get 256 points together, which guarantee that extracted pulse is a complete pulse paragraph.

4 Long time recording mode:

After the system has started, it will transmit data into DDR2 in terms of the input clock and enable signal, until it samples 30M points, and sends interrupt signals to the ARM. Long time recording mode sends the collected points to the ARM without any treatment depending on the set sampling rate and sampling time. ARM transmits signal information to the PC through a network or USB. Long time recording mode aims to get the background noise value of each channel to set thresholds conveniently.

5 Triggering recording mode:

After the system has started, it will transmit data into DDR2 in terms of the input clock and enable signal, until it samples set points. Judging input data one by one, when it gets to triggering condition, the mode will calculate the first reading DDR2 address, and count the remaining sampling points, when the counting value gets to sampling points, it will stop storing data, and send interrupt signals to the ARM.

3.3 *Experiment and application*

The air detection system is applied to air quality detection through different sensors in small medical spaces. We may configure graphical user interface in Figure 4 to intelligently control gas sensors' working model, channel and some information. This air quality monitor has a lot of peripheral interfaces and internal available resources. Users can develop more applications that meet their own needs.

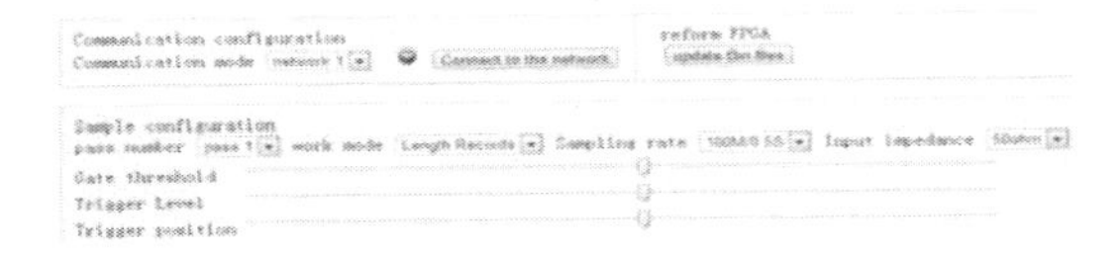

Figure 4. Graphical user interface.

4 CONCLUSION

A new system for multi-channel data acquisition for air quality detection is elaborated. The process of design, work patterns and software testing by using different sensors are shown in this air detection program. After a long period of testing and inspection, the air quality monitor can achieve multi-channel gas signal and show proper result. This device can be applied to a variety of situations about data collection and analysis. Nowadays, the monitor has been used to indoor air quality detection.

ACKNOWLEDGEMENT

The paper is supported by SC object "Air Quality Detection for Medical Environment".

REFERENCES

[1] Wei Zhang, Yiming Han, Xinling Wu, (2002) Design of high speed data acquisition system based on FPGA, *Information on Electric Power*, 4(3):46–49.
[2] Xiaolin Liu, Yubing Fan, Chunhui Luo, (2009) Design of multi-channel data acquisition system based on FPGA, *Electronic Technology and Application*, 7:42–44.
[3] Yanfeng Teng. (2006) Design and implementation of 100MHz high speed data acquisition system with USB interface, Micro Computer Information, No. 20, PP. 227–229,.
[4] Wenbo Xu, Yun Tian, Development of Xilinx FPGA tutorial (Second Edition), BeiJing, Tsinghua University press, 2012.
[5] Zonghai Li, Shuyu Chen, Haiwei Li, (2010) Building of embedded Linux system on ARM platform, *Computer Systems & Applications*, 19(10):153–157.
[6] Rubini A, Corbet J, (2006) *Linux Device Driver Version Third*, Beijing: Beijing Chinese Power Press.
[7] Yuan Yong. (2013) *Principle of Sensor and Detection Technology*. Beijing: Science Press.
[8] Yanhua Liu. (2013) *Detection and Control of Indoor Air Quality*. Beijing: Chemical Industry Press.
[9] Yunge Fang. (2007) *Indoor Air Quality Testing and Practical Technology*. Beijing: Quality inspection China press.
[10] Bmant G Maynard C, Bekal S, *et al*. (2006) Development and validation of an oligonucleotide microarray for detection of multiple virulence and anfimicrobial resistance genes in Escherichia coli. *J. Appl. Environ. Micmbiol.*, 72(5):3780–3784.
[11] Yu X, Zhang D, Li T, *et al.* (2003) 32D microarrays biochip for DNA amplification in polydimethylsiloxane (PDMS) elastomer. *Sens. Actuators A*, 108 (1-3):103–107.
[12] Robyr D, Grunstein M. (2003) Genomewide histone acetylation microbar-rays. *Methods*, 31 (1):83-89.
[13] Arlinghaus HF, Ostrop M, Friedrichs O, *et al.* (2003) Genome diagnostics with TOF-SIMS. *Appl. Surf. Sci.*, (203–204):689–692.
[14] Wang J, Bai Y, Li T, *et al.* (2003) DNA microarrays with unimolecular hairpin double2stranded DNA probes: fabrication and exploration of sequence-specific DNA/protein interactions. *J. Biochem. Biophys. Methods*, 55 (3):215–232.
[15] Hamels S, Gala J L'Dufour S, *et al.* (2001) Consensus PCR and microarray for diagnosis of the genus Staphylococcus, species, and methicillin resistance. *Biotechniques*, 31(6):1364–1366+1368+1370–1372.

Electric, Electronic and Control Engineering – Shao, Shu & Tian (Eds)
 ISBN: 978-1-138-02842-5

A new type of photonic crystal fiber with high nonlinearity and high birefringence based on microstructure fiber core

Qingchao Meng & Yukun Bai
Engineering Research Center of Communication Devices and Technology, Ministry of Education, School of Computer and Communication Engineering, Tianjin University of Technology, Tianjin, China

ABSTRACT: According to the application requirements of the communication system with high nonlinearity and high birefringence, a new type of photonic crystal fiber with high nonlinearity and high birefringence is designed in this paper. It carries out numerical simulations on fundamental mode field distribution, birefringence characteristics, effective mode field area and nonlinear coefficient of optical fiber structure through the full-vector finite element method. The numerical analysis suggests that nonlinear coefficient and birefringence characteristics can be adjusted effectively by altering the innermost fiber core air hole, cladding air hole pitch and inner-cladding air hole size. Through the optimization of structure parameters, the minimum effective mode field area of the optical fiber is $2.1816\mu m^2$, the nonlinear coefficient reaches as high as $59.46 km^{-1} \cdot W^{-1}$ and the birefringence value is 1.679×10^{-2} when the wave length is $1.55\mu m$. This design provides a new structure for obtaining high nonlinearity and high birefringence at the same time. It has good application prospects in fields of nonlinear optics, polarization control and super-continuum spectrum.

KEYWORDS: photonic crystal fiber; microstructure fiber core; high birefringence; nonlinear coefficient.

1 INTRODUCTION

Photonic crystal fiber (PCF), also known as microstructure fiber, has drawn extensive attentions of scientific researchers at home and abroad since it was first created in 1996 by Knight et al. [1] due to its features of infinite single-mode transmission, high birefringence, high nonlinearity and extremely low loss[2–3].

At the present stage, PCF obtains relatively high birefringence mainly by altering structural symmetry, such as size and shape of cladding air holes, distribution mode of holes and the introduction of elliptical holes. Optical fields are highly concentrated locally in PCF, greatly improving the interaction efficiency of nonlinear optics. The nonlinear effect of PCF can be effectively enhanced by reducing fiber core area or increasing filling rate of cladding air holes. In 2013, Ye Cao et al. [4] designed a PCF with high nonlinearity and high birefringence by introducing two elliptical holes in a fiber core and large air holes in a cladding. Acquired birefringence value and nonlinear coefficient were respectively 1.53×10^{-2} and $42 km^{-1} \cdot W^{-1}$. In 2013, Yani Zhang et al. [5] designed a PCF with high nonlinearity by introducing a rectangular crystal structure with elliptical air holes as the cladding, of which the nonlinear coefficient was as high as $46 km^{-1} \cdot W^{-1}$ while the birefringence was only 3.2×10^{-3}. In 2014, Jianquan Yao et al. [6] designed a PCF with high nonlinearity and high birefringence by installing two elliptical air holes and four tangent large air holes at the inner-cladding symmetrically, of which the birefringence reached 1.0929×10^{-2} and the nonlinear coefficient is $53.5 km^{-1} \cdot W^{-1}$. Although the nonlinear coefficient of this design is relatively high, the combination of nonlinear coefficient and birefringence remains to be improved.

The new type of photonic crystal fiber proposed in this paper is introduced with 10 same-sized circular air holes arranged in a semicircle in the fiber core and circular air holes of different sizes in the inner-cladding so as to form an elliptical core with the symmetry of dual rotation. The influence on characteristics of optical fiber brought by small circular air holes in the fiber core, air hole pitch of the cladding and the size of air holes of the inner-cladding can be analyzed by the numerical analysis software of finite element COM-SOL. The PCF structure proposed in this paper is able to not only maintain high birefringence and high nonlinearity at the same time but also effectively avoid manufacturing difficulties brought by elliptical air holes.

2 THEORETICAL BASIS AND OPTICAL FIBER STRUCTURE

2.1 *Theoretical basis*

This paper analyzes properties of the fundamental mode of optical fiber through the finite element method with the combination of the numerical analysis software COM-SOL and boundary conditions of perfectly matched layer. It also studies birefringence features, effective mode field area and nonlinear coefficient of photonic crystal fiber respectively.

Model birefringence [7] is an important parameter for measuring fiber polarization characteristics. The birefringence of the fundamental mode is obtained through the subtraction of real parts of fundamental mode effective refractive indexes in different polarization directions. Model birefringence is expressed as:

$$B(\lambda) = |\mathrm{Re}(n_{\mathrm{eff}}^{y}) - \mathrm{Re}(n_{\mathrm{eff}}^{x})| \qquad (1)$$

In the formula, n_{eff}^{x} and n_{eff}^{y} represent corresponding effective refractive indexes of fundamental modes in polarization direction X and Y. Re stands for the real part.

The parameter for measuring the optical fiber nonlinearity is the nonlinear coefficient γ. The nonlinear coefficient [8] γ is defined as:

$$\gamma = \frac{n_2 \omega_0}{c A_{eff}} = \frac{2\pi}{\lambda} \frac{n_2}{A_{eff}} \qquad (2)$$

In the above formula, c is the velocity of light in a vacuum, λ is the wave length, and n_2 is the coefficient of nonlinear refractivity. In this paper, n_2=3.2×10^{-20} m^2 / W, A_{eff} stands for the effective mode field area. It can be known from Formula (2) that the effective mode field area has a direct impact on the strength of nonlinear effect. A_{eff} [9] of the fundamental mode of optical fiber is defined as:

$$A_{eff} = \frac{\left[\iint_S |E(x,y)|^2 \, dx\, dy\right]^2}{\iint_S |E(x,y)|^4 \, dx\, dy} \qquad (3)$$

Here, E is the distribution of transverse electric field and S is the cross section of optical fiber.

2.2 *Optical fiber structure*

The structure of PCF with high nonlinearity and birefringence designed in this paper is shown in Figure 1. The effective refractivity in the background of silica glass is 1.45. The cross section includes a fiber core region and a cladding region. The fiber core consists of 10 same-sized circular air holes arranged in a semicircle instead of air holes in the original center. The diameter of the small circular air holes is d. The circle center distance of 10 circular air holes on X axis is fixed as Λ_1=0.22μm while the circle center distance on Y axis is Λ_2=0.1μm. The diameter of air holes of the inner cladding on Y axis is d_1 while the diameter of air holes on X axis is d_2=0.62μm. The diameter of air holes of other claddings is d_3=0.76μm. The air hole pitch of the cladding is Λ. The refractive index of air holes of the cladding is set as 1.

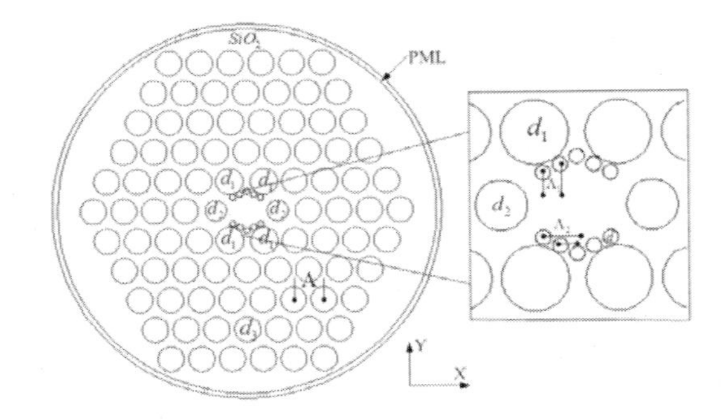

Figure 1. Cross section of the PCF.

3 NUMERICAL SIMULATION AND RESULT ANALYSIS

3.1 *Distribution characteristics of fundamental mode field*

The fundamental mode is the lowest order mode of optical wave transmission in PCF. Energy transfer of optical wave in PCF is conducted in the form of fundamental mode transmission. So, distribution characteristics of fundamental mode field serve as an important standard for the measurement of fiber optical features. Structure parameters are as follows: the diameter of the small circular air holes is d=0.22 μm; the radius of air holes of the inner cladding on Y axis is r_1=0.4μm; the radius of air holes on X axis is r_2=0.31μm; the diameter of air holes of other claddings is d_3=0.76μm; the air hole pitch of the cladding is Λ=0.9μm. Distribution characteristics of fundamental mode field can be acquired through the numerical solution of the PCF structure. Fundamental mode field distributions of two orthogonal polarizations HE_{11}^{x} and HE_{11}^{y} at 1.55μm are respectively shown in Figure 2(a) and 2(b). It can be concluded from figures that the energy of optical fiber mainly concentrates in the transmission of fiber core. Mode fields are elliptical, extending along X axis. The analysis suggests that the effective refractivity of X axis is larger than that of Y axis due to the introduction of 10 same-sized circular air holes arranged in a semicircle, the fact that the diameter of air holes of the inner cladding on X axis is small while that on Y axis is large, and the dissymmetry of fiber core and cladding. The mode field is distributed elliptically and extends along X axis.

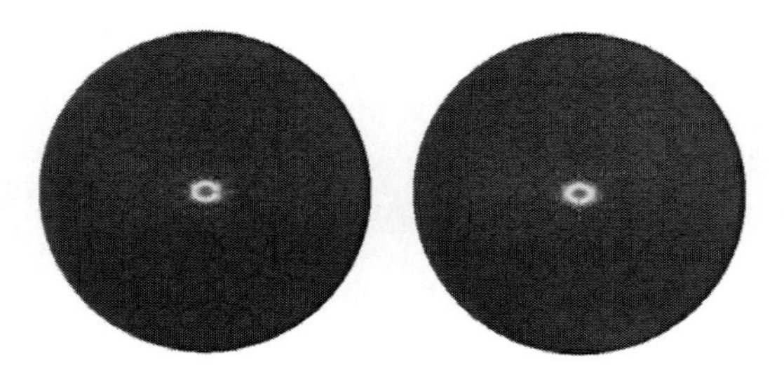

(a) Mode field in x-direction (b) Mode field in y-direction

Figure 2. Distribution of fundamental mode field at 1.55μm.

3.2 *Properties of birefringence*

Modal birefringence B is an important indicator for measuring characteristics of optical fiber polarization. Wave lengths within 1.0~1.8μm are respectively provided in Figure 3(a), 3(b) and 3(c) while other parameters are constant. The diameter of the small circular air holes d increases from 0.18μm to 0.22μm, the air hole pitch of the cladding Λ increases from 0.9μm to 0.11μm, and the radius of air holes of the inner cladding on Y axis increases from 0.36μm to 0.40μm. Changing curves of PCF birefringence with the wavelength are shown below. It can be known from figures that the birefringence of optical fiber increases with wavelength when the three parameters vary. As for Figure 3(a), the birefringence increases if diameter of the small circular air holes d increases as well. It is believed through the analysis that there are two factors for the change of birefringence, namely the size of fiber core and the asymmetry of cladding. When parameters of fiber core and cladding are fixed, mode field permeates to cladding and the effect of air holes becomes strengthened gradually with the increase of incident wavelength. However, the difference between the permeation speed of X polarization mode and the permeation speed of Y polarization mode results in the increase of birefringence. When the wavelength is fixed, the overlapping degree increases gradually with d, so the birefringence increases as well. As for Figure 3(b), the birefringence decreases when the air hole pitch of the inner cladding Λ increases. Because the increase of Λ would lead to the increase of fiber core area, the effect of mode field and inner cladding air holes weakens and the birefringence decreases. As for Figure 3(c), birefringence increases if the radius of air holes of the inner cladding on Y axis r_1 increases as well. Increase rates of birefringence of different air holes r_1 are basically the same when $\lambda \leq 1.2\mu m$. This is because birefringence is mainly influenced by the diameter of the small circular air holes and the air hole pitch of cladding Λ in a relatively short wavelength range. When $\lambda > 1.2\mu m$, birefringence grows rapidly with the increase of air holes r_1 and the curve becomes gentle gradually after the increasing to a certain degree. This is because the permeation of mode field to cladding becomes strengthened in a relatively long wavelength range. The asymmetry between X axis and Y axis is further enhanced by the introduction of a large r_1 under the condition that the influence of the diameter of the small circular air holes d and the air hole pitch of cladding Λ on birefringence is definite. Through the optimization of structure parameters, the high birefringence of 1.679×10^{-2} can be acquired at $\lambda=1.55\mu m$ when d=0.22μm, Λ=0.9μm and r_1=0.4μm, which is higher than those in literatures, such as 1.0929×10^{-2} [6] and 1.53×10^{-2} [4]. To sum up, as for specified wavelength, properties of PCF birefringence can be improved by increasing the diameter of the small circular air holes, reducing the air hole pitch of cladding and increasing the radius of air holes of the inner cladding on Y axis.

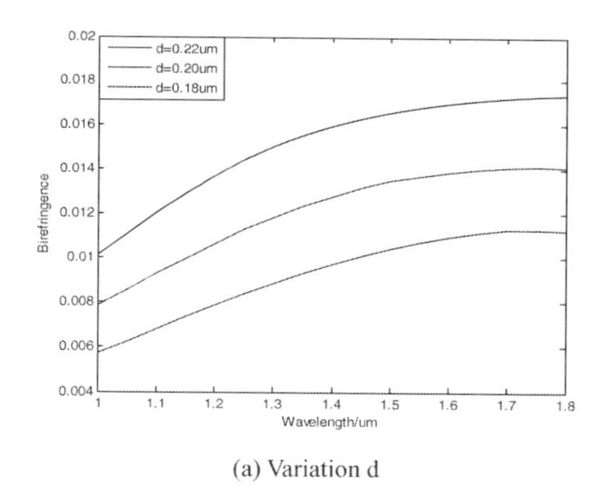

(a) Variation d

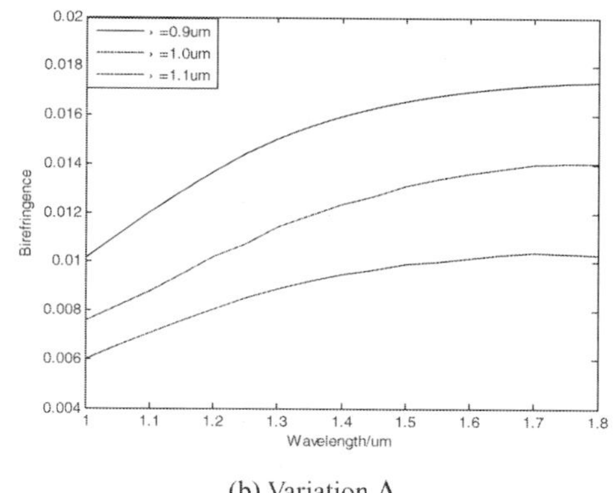

(b) Variation Λ

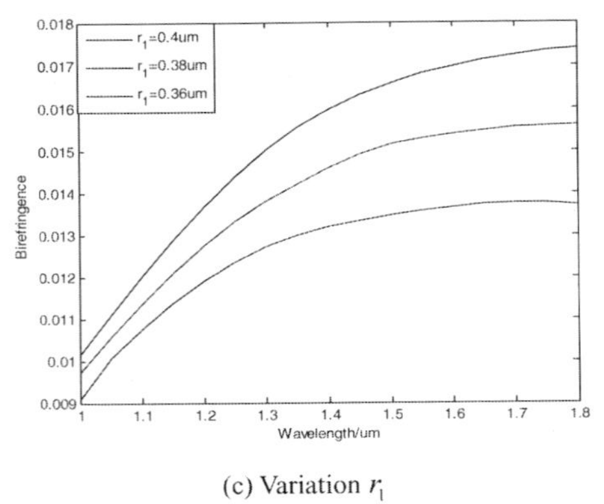

(c) Variation r_1

Figure 3. Birefringence characteristics as function of wavelength.

3.3 *Properties of effective mode field*

Changing curves of the effective mode field A_{eff}, which changes with transmission wavelength and structure parameter, are provided in Figure 4(a), 4(b) and 4(c). It can be known from these figures that the effective mode field A_{eff} of optical fiber increases with wavelength when the three parameters vary. As for Figure 4(a), A_{eff} decreases with the increase of d when $\lambda \leq 1.65\mu m$; A_{eff} increases with the increase of d when $\lambda > 1.65\mu m$. This is because PCF has a stronger binding force for light of short wavelength. Mode field energy of optical fiber can be perfectly confined to the fiber core and A_{eff} is correspondingly small. The leakage of optical field increases when the wavelength increases as well. Part of the energy spreads to the cladding gradually, leading to a sharp increase of A_{eff}. Generally speaking, the effective mode field area of PCF decreases with the increase of air holes of the cladding. So, when $\lambda > 1.65\mu m$, small air holes of the inner cladding of optical fiber should be regarded as a constituent part of the fiber core, also known as the micro-structural fiber core, instead of air holes of the cladding. As for Figure 4(b), when the wavelength is specified, the binding force of the cladding for light decreases with the increase of air hole pitch of the cladding Λ, and A_{eff} increases consequently. As for Figure 4(c), the increase of r_1 has small influence on the variation of A_{eff} when $\lambda \leq 1.2\mu m$. The permeation of mode field to the cladding is enhanced with the increase of wavelength when $\lambda > 1.2\mu m$. The increase of r_1 alleviates the permeation of energy to the cladding, so the increase of r_1 restrains the increase of A_{eff}. Through the comparative analysis, the effective mode field area of $2.1816\mu m^2$ can be acquired at $\lambda = 1.55\mu m$ when d=$0.22\mu m$, $\Lambda = 0.9\mu m$ and $r_1 = 0.4\mu m$. The area is smaller than areas in literatures, such as $3\mu m^2$[4], $2.86\mu m^2$[10] and $2.42\mu m^2$[6]. Such a small effective mode filed area lays a solid foundation for the realization of high nonlinearity.

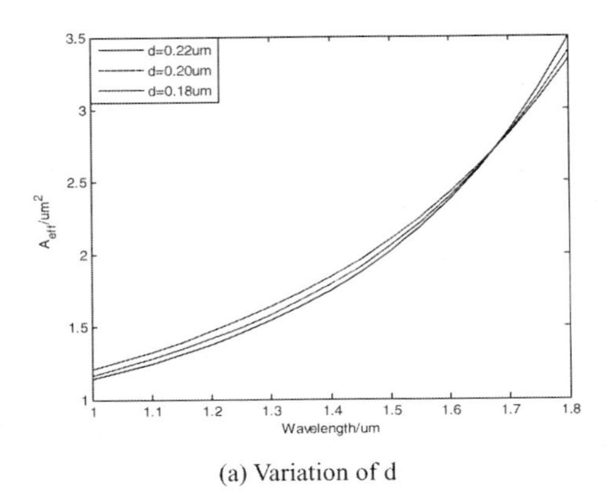

(a) Variation of d

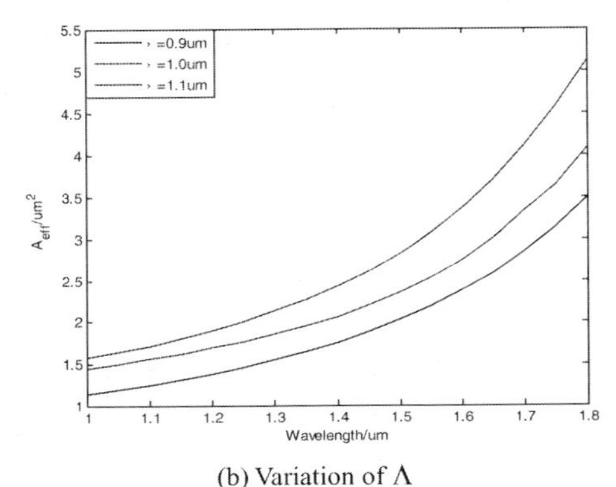

(b) Variation of Λ

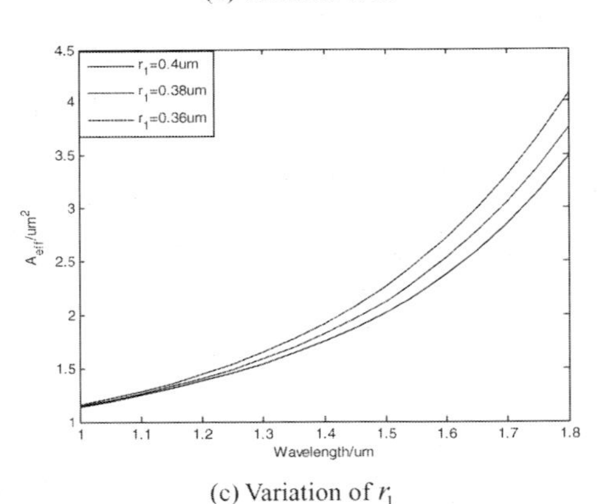

(c) Variation of r_1

Figure 4. Effective mode field area as function of wavelength.

3.4 *Properties of nonlinearity*

Properties of nonlinearity of PCF can be applied to such aspects as optical soliton communication, the generation of broad super-continuum spectrum, high power impulse compression and the production of optical fiber devices. Changing curves of the nonlinear coefficient γ, which changes with transmission wavelength and structure parameter, are provided in Figure 5(a), 5(b) and 5(c). It can be known from these figures that the nonlinear coefficient γ of optical fiber decreases with the increase of wavelength when the three parameters vary. As for Figure 4(a), γ increases with the increase of d when $\lambda \leq 1.65\mu m$; γ decreases with the increase of d when $\lambda > 1.65\mu m$. As for Figure 5(b), when the wavelength is specified, the binding force of the cladding for light decreases with the increase of air hole pitch of the cladding Λ, so the effective mode field area increases and the nonlinear coefficient decreases. As for Figure 5(c), the nonlinear coefficient γ decreases with the increase of wavelength. When the wavelength is specified, the increase of r_1 is able to alleviates the decrease of γ. Through the optimization of structure parameters, the nonlinear coefficient of $59.46 km^{-1} \cdot W^{-1}$ can be acquired at λ=1.55μm when d=0.22μm, Λ=0.9μm and r_1=0.4μm. The coefficient is higher than coefficients in literatures, such as $46 km^{-1} \cdot W^{-1}$[5], $50.22 km^{-1} \cdot W^{-1}$[10] and $53.5 km^{-1} \cdot W^{-1}$[6].

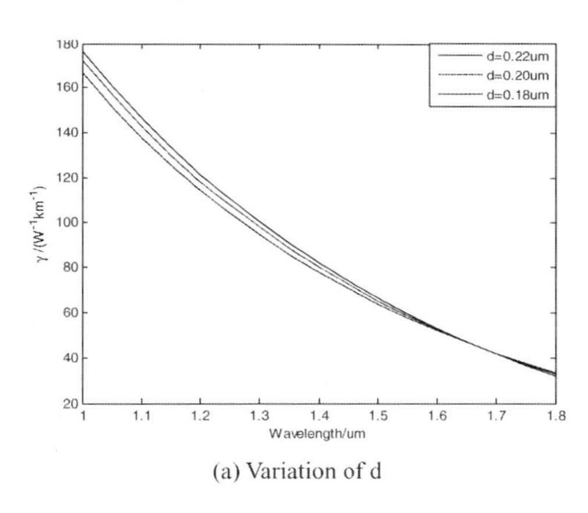

(a) Variation of d

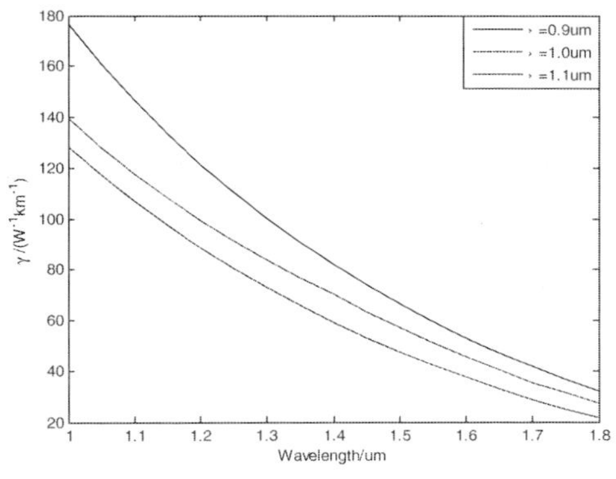

(b) Variation of Λ

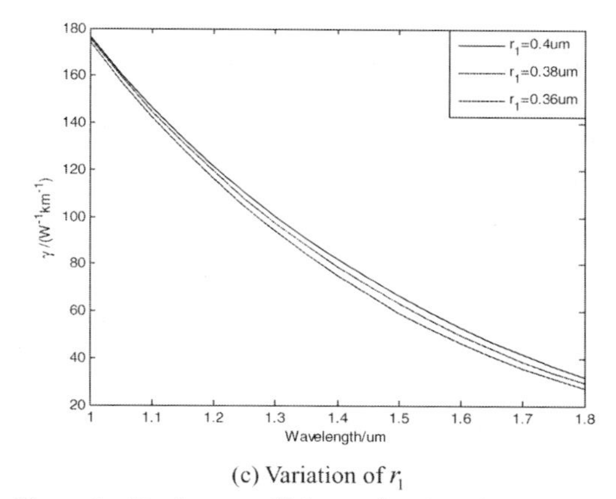

(c) Variation of r_1

Figure 5. Nonlinear coefficient as function of wavelength.

Generally speaking, the larger the diameter of PCF air hole is, the easier the PCF can be drew. The PCF with the small air hole diameter of 0.11μm is designed in Literature [11]. The diameter of fiber core air hole of the PCF designed in this paper is 0.22μm and the air holes are circular. Therefore, it is not difficult to realize the production process of PCF designed in this paper.

The influence of the air hole radius r_2 of the inner cladding on X axis brought on the nonlinear coefficient and birefringence characteristics of the whole PCF structure is further analyzed in this paper. Changing curves of the nonlinear coefficient and birefringence characteristics when the wavelength is 1.55μm and r_2 increases from 0.31μm to 0.35μm are provided respectively in Figure 6(a) and 6(b). It can be concluded from the comparison between Figure 6(a) and 6(b) that the nonlinear coefficient of PCF increases and the birefringence decreases with the increase of r_2. So, the nonlinear coefficient and birefringence characteristics of optical fiber can be altered flexibly through the adjustment of r_2. High birefringence is 1.679×10^{-2}, $1.584\times10^{-2}$and 1.511×10^{-2} and nonlinear coefficient is 59.46, 60.78 and $61.94 km^{-1} \cdot W^{-1}$ respectively when r_2 increases gradually from 0.31$\mu m$to 0.33μm. It can be known from the above analysis that nonlinear coefficient and birefringence properties are mutually restricted in terms of the adjustment of r_2. Both of the two factors should be taken into account in studies so that they can be adjusted to an ideal state simultaneously. Optimal structure parameters in this paper are as follows: the diameter of small air holes is d=0.22μm; the radius of air holes of the inner cladding on Y axis is r_1=0.4μm; the radius of air holes of the inner cladding on X axis is r_2=0.31μm; the diameter of air holes of other claddings is d_3=0.76μm; the air hole pitch of the cladding is Λ=0.9μm; the ultimately obtained high nonlinearity coefficient is $59.46 km^{-1} \cdot W^{-1}$ and the high birefringence is 1.679×10^{-2}.

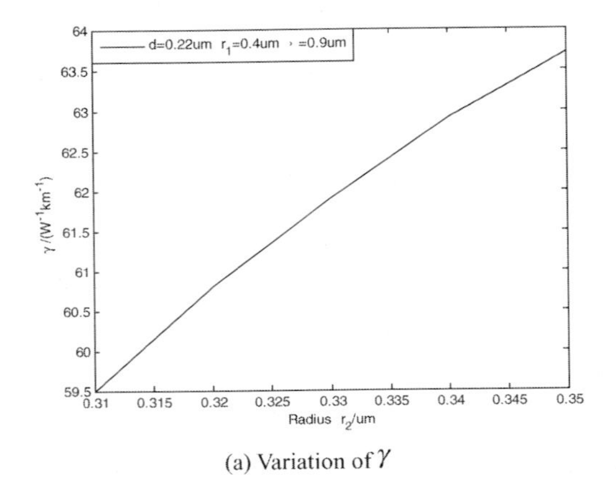

(a) Variation of γ

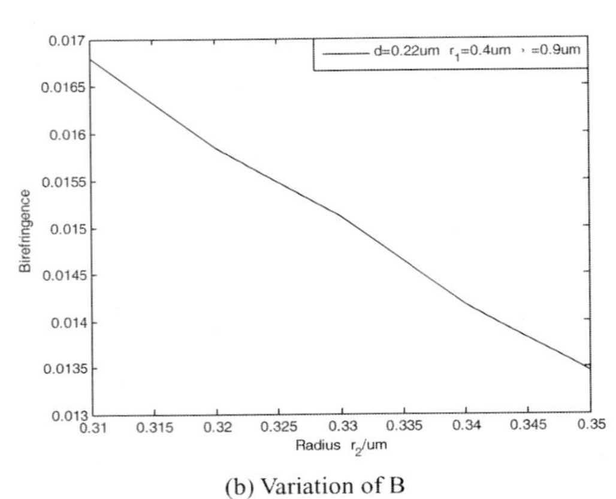

(b) Variation of B

Figure 6. Nonlinear coefficient and birefringence as function of r_2.

4 CONCLUSION

With the photonic crystal fiber of total internal reflection as the research object, this paper designs a new type of photonic crystal fiber with high nonlinearity and high birefringence. The optical fiber is introduced with 10 same-sized circular air holes arranged in a semicircle in the fiber core. The structure of the cladding is arranged in a hexagon by circular air holes of different sizes. This paper carries out numerical simulations on fundamental mode field distribution, birefringence characteristics, effective mode field area and nonlinear coefficient of optical fiber structure through the full-vector finite element method. It also analyzes deeply the influence on these properties brought by the change of structure parameters of optical fiber. The numerical analysis suggests that nonlinear coefficient and birefringence characteristics can be adjusted effectively by altering the innermost fiber core air hole, cladding air hole pitch and inner-cladding air hole size. Through the optimization of structure parameters, the minimum effective mode field area of the optical fiber is $2.1816\mu m^2$, the nonlinear coefficient reaches as high as $59.46 km^{-1} \cdot W^{-1}$ and the birefringence value is 1.679×10^{-2} when the wave length is $1.55\mu m$. Meanwhile, the nonlinear coefficient and birefringence characteristics of optical fiber can be altered flexibly through the adjustment of r_2. In practical applications, parameters can be set according to actual demands so as to be in accord with the required performance indexes. Compared with the existing structural design of PCF, the design in this paper is able to maintain high birefringence and high nonlinearity at the same time. The structure is simple, which can be easily realized in terms of production process. This fiber is of great importance for fields of communication system that has high requirements on nonlinearity and birefringence and the production of nonlinear optical devices and optical fiber devices of super-continuum spectrum.

REFERENCES

[1] Knight, J.C., Birks, T.A., Russell, P.S.J., *et al.* (1996) All-silica single-mode optical fiber with photonic crystal cladding, *Optics Letters*, 21(19):1547–1549.

[2] Ortigosa-Blanch, A., Knight, J.C., Wadsworth, W.J., *et al.* (2000) High birefringence photonic crystal fibers, *Optics Letters*, 25(18):1325–1327.

[3] Finazzi, V., Monro, T.M., Richardson, D.J. (2003) Small-core silica holey fibers: nonlinearity and confinement loss trade-offs. *JOSA B*, 20(7):1427–1436.

[4] Cao, Y., Li, R.M. & Tong, Z.R. (2013) A study on properties of a new type of photonic crystal fiber with high birefringence, *Chinese Journal of Physics*, 62(8):84215–084215.

[5] Ya-Ni, Z. (2013) Optimization of highly nonlinear dispersion-flattened photonic crystal fiber for super-continuum generation, *Chinese Physics B*, 22(1):014214.

[6] Yao, J.Q., Lu, Y. & Miao, Y.P. (2014) Photonic crystal fiber of 1.55 μm with high nonlinearity and high birefringence, *Optics and Precision Engineering*, 22(3):588–596.

[7] Ju, J., Jin, W., Demokan S. (2003) Properties of a high birefringence photonic crystal fiber, *IEEE Photonics Technology Letters*, 15(10):1375–1377.

[8] Limpert, J., Schreiber, T., Nolte, S., et al. (2003) High-power air-clad large-mode-area photonic crystal fiber laser. Optics Express, 11(7):818–823.

[9] Mortensen, N.A. (2002) Effective area of photonic crystal fibers, *Optics Express*, 10(7):341–348.

[10] Wang, E.L., Jiang, H.M., Xie, K., *et al.* (2014) Photonic crystal fiber of zero dispersion wavelength with high nonlinearity and high birefringence, *Chinese Journal of Physics*, 63(13):134210–134210.

[11] Wiederhecker, G.S., Cordeiro, C.M.B., Couny, F., *et al.* (2007) Field enhancement within an optical fiber with a sub-wavelength air core, *Nature Photonics*, 1(2):115–118.

Electric, Electronic and Control Engineering – Shao, Shu & Tian (Eds)
© 2015 Taylor & Francis Group, London, ISBN: 978-1-138-02842-5

The application of LABVIEW-based virtual instrument technology in electro-hydraulic servo test system

Gang Zhao, Junjie Shi, Lihua Sun & Xianda Wang
Troop 63981 of Peoples' Liberation Army, Wuhan Hubei, China

ABSTRACT: Using the virtual instrument technology, proportional control and many other new advanced technologies, a certain turret electro-hydraulic servo system, which makes up the limits of traditional test method for simulation of separation instrument and the limitation of traditional simulation testing method, has constructed its hardware platform. It has the advantages of high precision, sensitive reaction, and high technical requirements. With computer data collection and processing system as the development platform, and LabVIEW-based virtual instrument as the development environment, it constructs the virtual test system equipped with electro-hydraulic servo device performance which is to achieve efficient and convenient state monitoring, performance test, data analysis and report-printing function.

KEYWORDS: electro-hydraulic servo systems; LabVIEW virtual instrument; data collection and analysis; performance testing system; modular software design.

1 INTRODUCTION

A turret of electro-hydraulic servo system and its implementation mechanism is a set of high speed and high precision differential servo system to control and drive the launcher, which can recurrent the command radar position and movement rules in two directions of azimuth and height. The core part is a set of valve controlled hydraulic cylinder in height and a set of valve controlled motor in azimuth. The mechanical part is composed of servo valve, servo motor, oil cylinder and the hydraulic energy system and the electrical control signal coming from the dynamic differential electrical signal given by the command radar. The principle of valve controlled hydraulic is shown in Figure 1.

As the core part of the turret emission system, the stability and state are directly related to the real-time tracking and ammunition on the target. The traditional testing system which is huge and complicated cannot meet the needs of current performance testing which is characterized with high technology, difficult testing and its specialty. As a result, new testing system

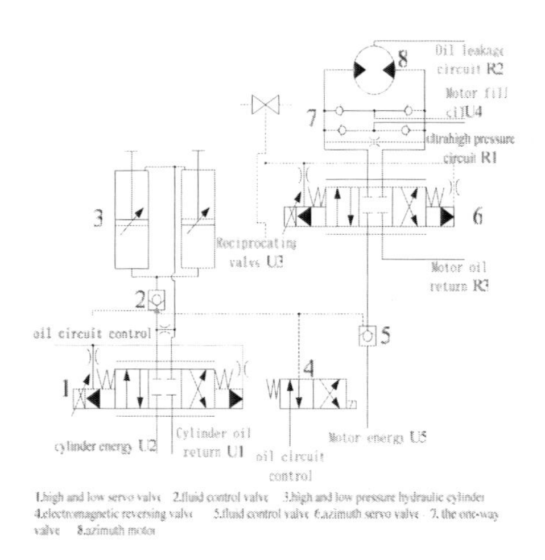

Figure 1. The principle of turret electro-hydraulic servo system.

should have the ability of detecting the circuit pressure, flow, temperature of the servo system, testing and controlling the current, leakage and drift performance of servo valve. The functions of automatic testing of system state signal acquisition, data processing, fault alarm and report output also should be acquired.

As the development and application of computer technology in the field of hydraulic pressure, the servo performance automatic testing system based on LabVIEW virtual instrument platform, characteristic of high integration, simple structure, low cost of development has acquired wide application. To complete the turret performance test and repair tasks, the research and development of the test system of electro-hydraulic servo turret based on the virtual instrument technology has obtained good results in practice. In the system, the NI6025E data acquisition card, the signal conditioning module, IPC and hydraulic load, etc have been used to construct the hardware testing platform while the graphical programming language (G language), GPI and VXI equipment have been used to write the professional testing software.

2 THE MAIN PROJECTS OF ELECTRO-HYDRAULIC SERVO VIRTUAL TEST

1 The initial state test: to test the state of machinery and electric state before startup.
2 The ready state test: with the electrical control signal, the test system starts and comes into the working state and displays the current parameters.
3 Pressure test: to test the pressure of the bypass circuit after giving electricity to the bypass combination.
4 Main pump pressure flow test: to test the performance indicators of hydraulic source, including pressure and output flow, which can be used to judge the power of the hydraulic pump.
5 The leakage test in electro-hydraulic servo valve: to test the leakage value of the servo value spool not at zero when the electro-hydraulic servo valve is zero offset on no load condition.
6 Zero-load test of electro-hydraulic servo valve: to test the output load flow at the given pressure drop and zero load pressure, making the input current between positive and negative current rating with no effect on dynamic characteristics.
7 Frequency response test of electro-hydraulic servo valve: to test the percentage of no-load flow and the input current as well as the frequency width of the servo valve while the input current is in the constant amplitude frequency sinusoidal variation at a certain range of frequencies.
8 Performance test of safety valve: to test whether the safety valve can work normally when the electro-hydraulic servo system is overpressured.

3 HARDWARE PLATFORM CONSTRUCTION OF ELECTRO-HYDRAULIC SERVO TEST SYSTEM

3.1 *Hydraulic loading platform design*

To ensure the response sensibilty and the reliability in greate power and high speed, the system not also tests its no-load performance, but also its full-load performance, the latter as the main point. Techonolgies as the proportional relief valve, the high-pressure stop valve, auxiliary pump,sensor data acqusition have been adpoted to desigh its hydraulic loading and test system which are butted with the electro-hydraulic servo system and simutlate the load and control of the elctro-hydraulic servo system.The singal acqusition devices of pressure, flow, and temperature sensor have been set up in the bypasses to provide the control computer the important parameters of the test system in the form of electrical signal, which can be used to acqurie and judge the working state of the system. Test system is shown in Figure 2.

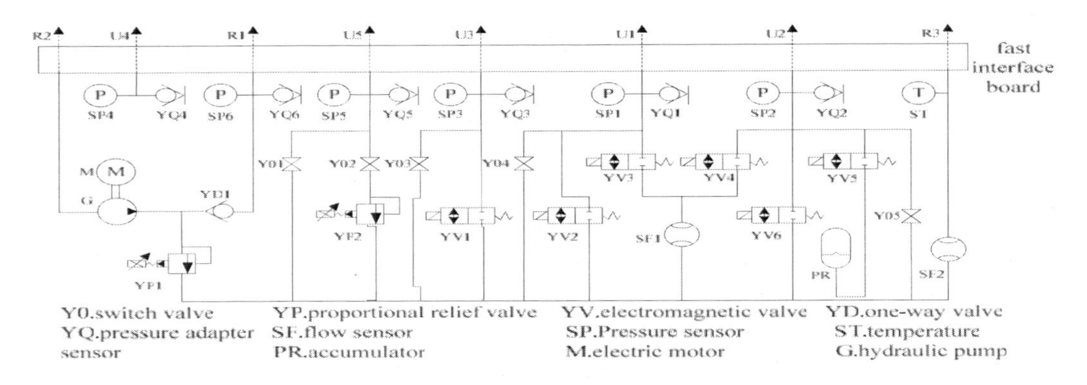

Figure 2. Sketch of hydraulic load and data collection system.

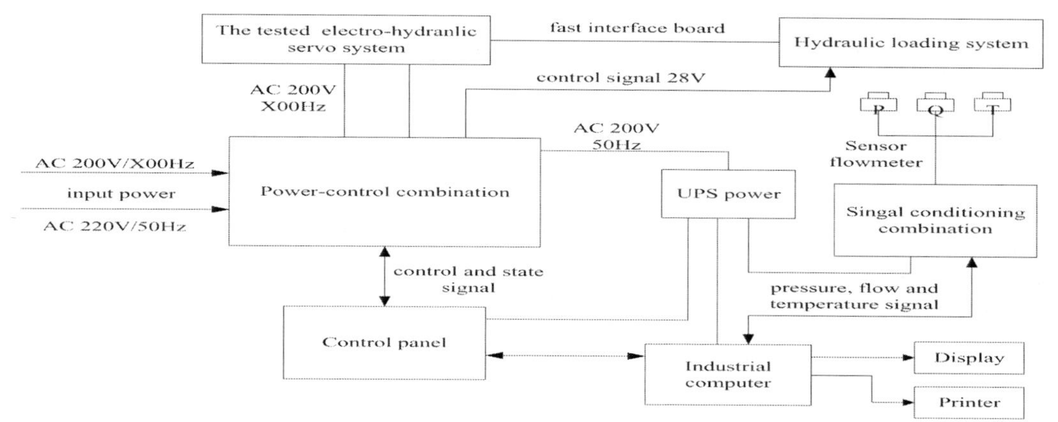

Figure 3. The platform of the electro-hydraulic servo system.

3.2 *Data collection and controll hardware design of the electro-hydraulic servo system*

The hardware of the LABVIEW-based virtual instrument test system mainly includes devices such as the control computer,data acqusiton card, display, sensor, signal conditoning module and control singal input panel, etc. As the structure and state of the system cannot be altered, the function of the data collection has been designed on the loading part, in which the collection of data is achieved by conncetion with the measured system.

2.0L/min and the latter parameter 65L/min, aiming to test the oil return ablity and energy source performance.(The technical indicators are shown below.)

3.2.2 *The seletion of pressure sensor*

Pressure data collection mainly focuses on mearsuring the pressure parameters of circuits (oil source circuit, working circuit and back pressure loop,respectively) and delivering the test data to singal conditioning module to realize first acquisiton. (Technical indicators can be seen below.)

mode	supply voltage	flow range(L/min)	maximum working pressure(MPa)	Temperature	measuring accuracy
CLL-8	10±0.01V DC	0~6	6.3	−50°C~125°C	≤±3%
CLG-4	10±0.01V DC	20~80	10	−50°C~125°C	≤±3.5%

model	range	accuracy	output	power
CY-YZ-174(1MPa)	0-3.5MPa	≤0.2%F.S	≤10mA	DC 24V
CY-YZ-174(25MPa)	0-35MPa	≤0.2%F.S	≤10mA	DC 24V

3.2.1 *Selection of flow sensor*

Flow data collcection is to test the small flow parameters of the high-voltage circuit and the large flow parameters of the low-voltage flow with the former paremter

3.2.3 *Selection of temperature sensor*

To prevent system damage caused by friction, throttle and overflow leading to too high temperature, the system collects the temperature signal and

Name	Model	Measuring Range	Temperature	Temperature Coefficient	Insulation Resistance (Electrical Characteristic)	Output Mode (Static Characteristic)
Platinum Thermistor	WZP533	−50~200°C −50~400°C	−50°C~200°C −50°C~400°C	$(3850\pm15)\times10^{-6}$/°C	≥100MΩ	Resistance Value

conducts the analysis of oil temperature changes. When the oil temperature goes beyond the warning line, it will alarm. (Technical indicators are seen above).

3.2.4 *Selection of signal conditioning module*

A type of 5B series module design is used in signal conditioning. In the same system, different test objects can be mixed and configured with any type of I/O module while test signal types and quantities can be arbitrarily combined so as to reduce surplus I/0 channels. The module mainly conducts the amplification of signal collection, filter, signal excitation, linear compensation, cold end compensation, common mode signal suppressing circuit, normal mode signal suppressing circuit, high noise suppression, anti-radio frequency interference and anti-electromagnetic interference. It also has access to various kinds of analog signal, such as voltage (mV/V), current (mA/A), frequency, strain, temperature (thermocouple/thermal resistance), revolution, etc. The signals of voltage and current in wide range can be provided for collection card. (Technical indictors are shown above).

model	function	output choice
5B30	millivolt voltage input, bandwidth 4Hz	0V~+5V or -5V~+5V
5B34	2 lines , 3 lines, 4 lines RTD input, 100Ω platinum, 10Ω copper, 120Ω nickel	0V~+5V
5B35	4 lines RTD input, 100Ω platinum,10Ω copper,120Ω nickel	0V~+5V

3.2.5 *Selection of data collection card*

In the test system, the data collection card uses PCI-6025E series produced by National Instrument. Specific indicators are as follows, (1) single simulation input with 16 output ports and 24-bit accuracy

of analog, (2) sampling rate is 200kS/s, (3) simulation output with 2 output ports and 12-bit accuracy of analog, (4) 32 lines of digital I/O, (5) 2 24-bit 20MHz timebase (the collection card is shown above).

3.2.6 *Selection of industrial control computer*

With advanced fuctions such as automatic frequcny reduction when overtemperatured, ESD protection, BIOS anti-virus,the IPC series chassis, produced by a certaiin corporation has been empolyed in the system. Specific modular design of the chassis makes it possible to maitain the system in the shortest period when it is shutdown.

3.2.7 *The principle of signal colletion and analysis*

Singal collection is the premise of data processing. To achieve computer processing of digital signal,it shoudl solve the issue of signal digitalization. It is to convert the continous time domain simulation siganl by A/D converter to digital signal which can be handled by computers. Through multiplying the time domain signal x(t) by periodic unit pulse sequence function g(t) (the interval isT_s). The results of data colection can be sampling signal, adn it can be expressed as following,

$$x(n)= x(Nt_s)= x(n/f_s) \quad n=0,1,2......N-1 \text{(Formula 1)}$$

(n stands for the sampling sequnce)

In the sampling theorem, for the sampling of continuous siganl x(f) of limited spectrum X(f), it shows that when the sampling frequecny is $f_s \geq 2fmax$ (2fmax is the hignest frequency), the sampling signal $X_s(Nt_s)$ can restore to the orignal singal X(t) without distortion. It is shown below.

$$X(t)=\sum_{-\infty}^{+\infty} X_s(nT_S)\frac{\sin[\frac{\pi}{T_S}(t-nT_s)]}{\frac{\pi}{T_s}(t-nT_s)}$$

The selection of sampling interval is in accord with the sampling theorem,that is,the sampling frequency f_s must greater or equal to two times the highest frequcney $f_{h.}$ ($f_s \geq 2 f_h$ 或 $T_s \leq 1/2 f_h$) 16-bit high precision data collection card of PCI bus is used in sampling hardware. Isolation conditioning module is adopted in the primary signal conditioning which can conduct comprehensive conditioning to directly accessed sensor signals and provide three-levels isolation protection in order to protect the safety and reliablity and greatly impove the anti-interferce ablity.

4 SOFTWARE DESIGN OF THE ELECTRO-HYDRAULIC SERVO SYSTEM

4.1 *The design of main control process*

As all the working process of the electro-hydraulic servo test system is realized by its test function under the supervision of the software, the test software is the core part of the whole test system. The software employs graphical programming language (G language) of LabVIEW virtual instrument devlopment platform. (The main process of the test software is shown in Figure 4.)

4.2 *The design of software structure*

The software is develpoed in the form of modularized strucutre and each module possesses specific funtion in which both separate implemententition and subroutine are possible. The design, easy to read and easy to debug, greatly imporves the efficiency of program development. Monitoring and testing are paralled processed in the process. (as shown in Figure 5.)

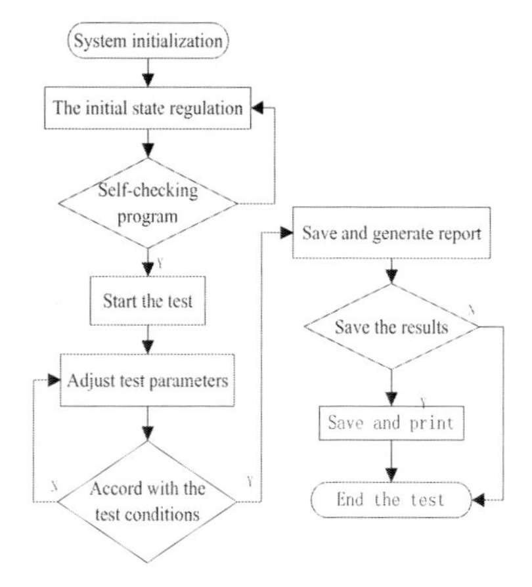

Figure 4. The flowchart of overall software design.

4.3 *The design of functional module*

Module design principle and visual programming technology have been used in function testing part, in which the visual programming produces high readability and the design of virtual instrument makes it easy to test and operate.

1 Self-checking module: after startup,self-checking program firstly dectects whether the key parameters is noraml or not adn then it comes into main test cycle.
2 Data collection adn data processing module: After choosing different test items in the main interface,it goes into data collection and processing program. The module judges the state at first, then read data from specific channel of the collection card after entering the testing state and return to the main interface after corresponding processing.
3 File operation module:conduct the operatioins as open,edit,disk adn output to the files of the test data.

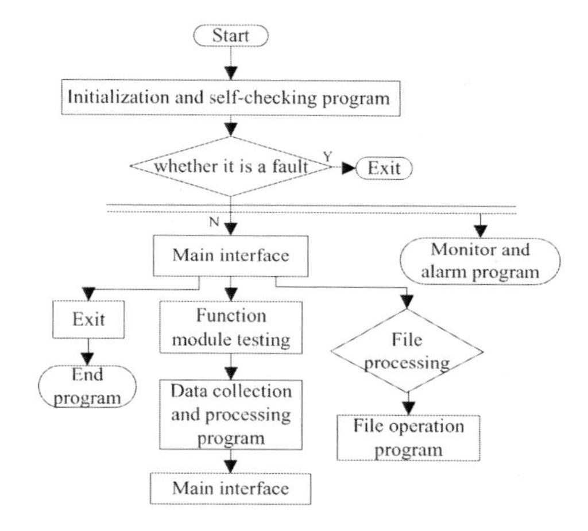

Figure 5. Flowchart of software function structure.

4 Monitoring module: while entering the main test procedure, it monitors the testing process from beginning. When abnormal state parameters occur, it will timely suspend the test procedure in order to protect the safety of test system and the measured objects.

4.4 *Graphical programming of the test system*

Using the LabVIEW professional development system as the develpment platform and G language graphic programming scheme, the main program source file of the test system is shown as below (see Figure 6).

5 CONCLUSIONS

Combination of functions as computer data collecting and processing based on LabVIEW virtual instrument, proportional relief technology, real-time

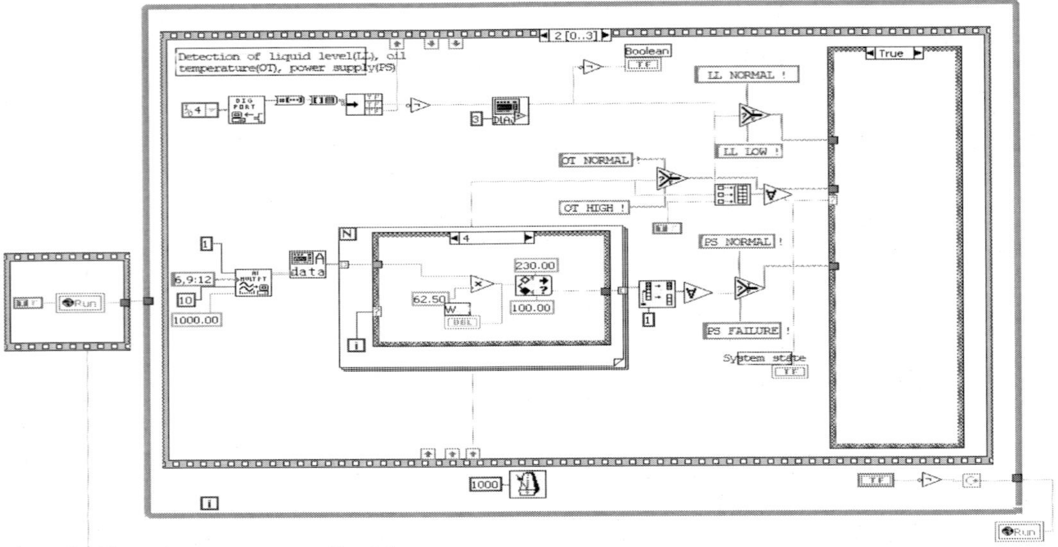

Figure 6. The main program source file of the test system.

monitoring, fault alarming etc, the turret electro-hydraulic servo system runs stably and possesses an intuitive and friendly user operation interface in real test. The performance test on the servo valve meets the needs of actual test requirement while great value in actual use is produced in fields as the performance test, technical maintenance and repair of the electro-draulic servo system. It satisfies the technical needs of performance test, fault interpretation and repair debugging.

REFERENCES

[1] Li Kejie (2002) *New Handbook of Senor Technology*. Beijing: National Defense Industry Press.

[2] Yang Leping, Li Haitao, Xiao Kai (2003) *Introduction to Virtual Instrument Technology*. Beijing: Publishing House of Electronic Industry.

[3] Wang Haibao et al. (2004) *Design and Application of LabVIEW Virtual Instrument*, Xi'an: Xi'an Jiaotong University Press.

[4] National Instruments. (2006) *LabVIEW control Design Toolkit User Manual*.

[5] Li Yang, Zhu Zhengtao (1999) Dynamic characteristic test and Simulation of hydraulic device based on virtual instrument, *Machine Tool & Hydraulics*.

[6] Dai Pengfei, Wang Shengkai et al. (2006) *Test Engineering and the Application of LabVIEW*, Publishing House of electronics industy,

[7] Lei Yong, (2005) *Design and Practice of Virtual Instrument*, Beijing: Publishing House of electronics industy,

[8] National Instruments Corporation. (2002) *LabWindows/CVI SQL Tookkit Reference Manual*. Austion Texas US: National Instruments Corporation.

[9] Li Xianming, Wu Hao, (2009) Automatic detection technology. Beijing: Machinery Industry Press.

Electric, Electronic and Control Engineering – Shao, Shu & Tian (Eds)
© 2015 Taylor & Francis Group, London, ISBN: 978-1-138-02842-5

The design and implementation of virtual oscilloscope based on LabVIEW

Lihua Sun, Hang Bai, Gang Zhao & Jiangfeng Ma
Troop of Peoples' Liberation Army, Wuhan Hubei, China

ABSTRACT: This paper introduces the characteristics of Virtual Instrument (VI), describing the features of advanced virtual instrument technology platform-Lab VIEW in the field of current test. With the technical background of Lab VIEW, the design of a virtual oscilloscope system is constructed in view of the test principle of oscilloscope and its implementation will be discussed in detail. The system can basically achieve the oscilloscope measurement and digital storage function and owns friendly man-machine interface.

1 INTRODUCTION

Along with social progress and the development of science and technology, the demand for product testing has been constantly improved in modern society which makes test system become more and more complicated as the inherent disadvantages of traditional test in itself is very difficult to adapt to the requirements of modern test and people also gradually pay much attention to cost control and investment protection of the test system. Virtual Instrument is a new technology along with the development of computer software technology, fully uses strong software and hardware functions of computers and combines computer technology and test technology closely which prompts traditional test instruments to develop in digital, intelligent, modular and networked direction, widely being applied to multiply fields such as military, electrical measurement, aerospace, engineering measurement, medical treatment, vibration analysis, acoustic analysis, entertainment, etc. It is an important technology in the field of current computer aided testing (CAT).

As a kind of specific application of Virtual Instrument technology, virtual oscilloscope will utilize the combination of computer geographic simulation technology with oscilloscope real functions to achieve various functions of the oscilloscope in computer virtual simulation and to extend the system. The performance design of virtual environment can make integrated test equipment greater degree and reduce system cost more effectively. The paper makes an analysis of the implementation of virtual oscilloscope software and hardware based on the background of virtual instrument- LabVIEW.

2 FEATURES OF VIRTUAL INSTRUMENT

As a kind of integrated measurement and control system formed by the combination of applied computing technology with instrumentation, virtual instrument consists of three key elements: computer, modularization instrument hardware and software for data acquisition, process analysis and graphical user interface. The essence of virtual instrument technology is making full use of the latest computer technology to implement and extend functions of traditional instruments, of which its outstanding advantage is to strengthen the new concept -"software is instrument", free from the limitation of traditional instruments regular maintenance on the volume, precision and hardware and it makes use of strong graphical development environment to establish intuitive, flexible and efficient virtual instrument panel which can be defined by users themselves to set up test functions needed by them, fully reflecting virtual feature . Virtual instrument changes the way of using traditional instruments, breaks through the limitations on data collection, analysis, process, transmission, storage, etc, improves the efficiency of use and functions of instruments and substantially reduces the cost of use which can not be compared by traditional instrument.

Lab VIEW is a programming development environment (PDE), consisting of three main parts: front panel, block diagram program and icon/connection ports similar to C and Pascal language, however, Lab VIEW language uses graphics language to write programs in the form of block diagram such as various switches, knobs, thermometers and so on and Lab VIEW language interface is an audio-visual image compared with traditional programming language. Therefore, its program is also called Virtual instruments (VIs). Virtual instrument structure is as shown in figure 1.

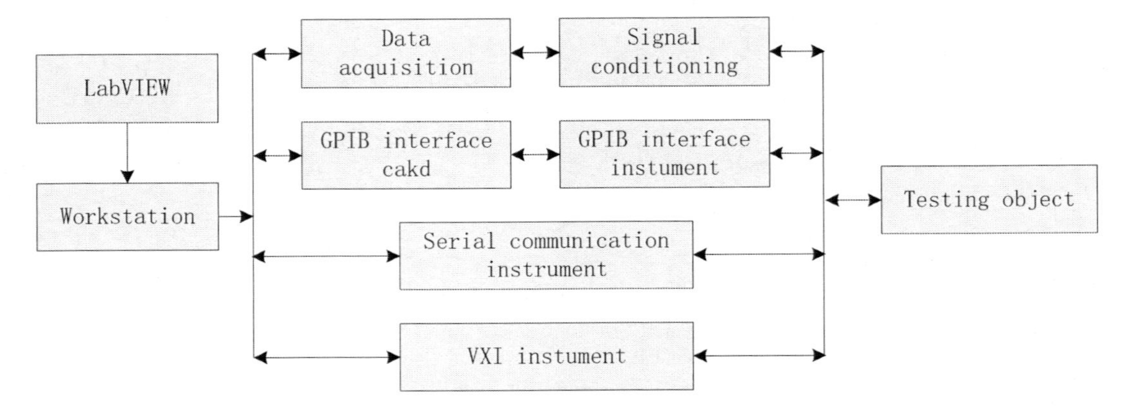

Figure 1. Virtual instrument system block diagram.

3 THE PRINCIPLE OF VIRTUAL OSCILLOSCOPE

In this paper, virtual oscilloscope is composed of a virtual instrument panel, hardware device and interface and device driver software based on Lab VIEW virtual instrument for technical background and PC as its controlling platform and sustained by modular hardware. Virtual instrument links with existing instrument by means of device driver software and shows various switch controls corresponding to operating elements of a traditional realistic instrument panel on the PC screen in the form of a virtual instrument panel. Hardware device and interface consist of computer and instrument interface equipments and device driver software is driver directly controlling various front hardware interfaces. The signal detection circuit is detected from some relevant part to multiple measured signals through the sensor, converts the output signal to standard signal after the conversion of amplifying filter and electric level via data conditioning module. Data acquisition card (DAC) collects the signals and sends them to the PC so that the setup and performance test of measured signals can be completed under the background of virtual instruments. The operating panel of an oscilloscope can be simulated through a graphical user interface so that three main functions-signal acquisition, signal process and display output can be completed.

In this paper, the specific design of virtual oscilloscope generally consists of six modules: data acquisition, signal processing, parameter measurement, data storage, waveform display and print, its structure diagram shown in figure 2.

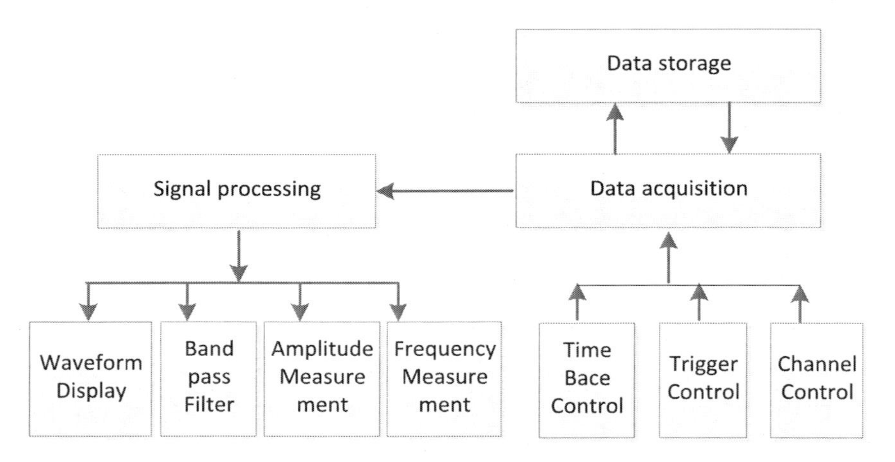

Figure 2. Virtual oscilloscope structure diagrams.

4 THE OVERALL STRUCTURE DESIGN OF VIRTUAL OSCILLOSCOPE

4.1 *The main functions completed by virtual oscilloscope are as follows:*

1 Lab VIEW software can be applied to the trigger, can continually adjust the trigger level and can trigger rising edge along with the falling edge to complete the positive level trigger or negative level trigger function.
2 The oscilloscope time base (T. B.) can be realized to be adjustable so that input signals of different frequencies can be accurately displayed.
3 Field waveform can be stored as date files, can be long-term preserved or can call at any time if necessary and can deal with backup from the printer.
4 Data processing of arbitrary combination and analysis function can be completed according to need through a programming template.
5 The practical testing of two channels shows waveform as a main performance index, which can simultaneously be reflected in different windows of display screen with signal average function and signal-noise rate effectively improving.
6 Its adjustable gain of vertical sensitivity is for six grades and corresponding gain is for six classes.
7 It has multiply display modes and can observe single process or minute changes of date signal.

4.2 *Hardware structure of virtual oscilloscope*

In this paper, virtual oscilloscope system hardware is made up of sensor system, data acquisition card, signal conditioning module, PC, etc. Hardware structure is shown in figure 3.

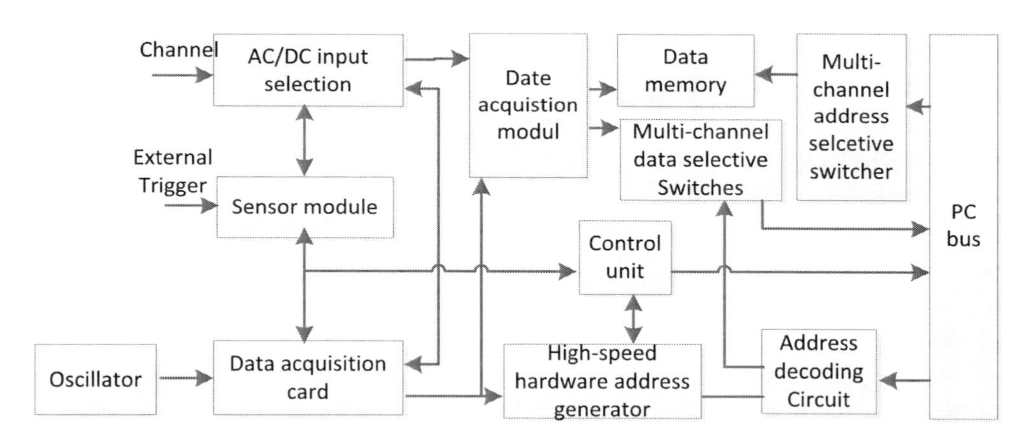

Figure 3. Virtual oscilloscope hardware structure block diagram.

4.2.1 *Data acquisition card*

The information and advancement of signal processing technology to a large extent determine developing trend of information society; however, playing an essential role in digital signal processing system is a data acquisition system. Automatically collect angle, flow, temperature, force and speed existing in external reality and other non-electric signals or electric signals from analogue and digital unit under test like sensor and other equipment and send the signals into computer for analysis. The process of transmission, display and saving is called as “data acquisition”.

The data acquisition system is generally composed of three parts.

1 Data collector: it builds in embedded operating systems and communication module, including samplings to hold amplifier, measuring amplifier, analog-to-digital converter, multi-channel switch and so on, owns real-time acquisition, automatic storage, real-time display, automatic processing, etc. and will be multiplied analogue signals after quantification for digital signals sent to the PC.
2 Microcomputer interface circuit: I/O interface and memory interface circuit are two types of interface equipments used by computers linking with and changing data with external equipment and storage to transmit data, status information and controlling signals to meet the need for a data acquisition system.
3 Digital- to- analogue converter: It converts digital signal to analogue digital and consists of four parts: weighted resistance network, operational amplifier, reference power supply and analogue switch to implement manipulation, display, storage and other tasks required by the system.

Data acquisition card (DAC) is computer extension card to realize the data acquisition function and accesses a computer through PCI, ISA, Ethernet and other bus, generally including four parts: multi-channel switch, A/D converter, amplifier and use/holder. Data acquisition card is located in the fore channel of PC while D/A converter located in subsequent channel. Users cannot use a data acquisition card until they need to configure DAC hardware index.

Each parameter index integrated, the virtual oscilloscope in the paper adopts PCI-6025E data acquisition card of NI corporate. The specific index is as follows: two-way twelve-bit analogue output, eight channel analogue inputs, twelve-bit resolution, 200Ks/s sampling rate, digital timing trigger/simultaneous trigger, 100ppm time base stability, 20MHZ largest signal source frequency, 24mA single-channel current driving ability and 252mA total current driving ability.

4.2.2 *Computer*

This set of virtual oscilloscope uses an industry person computer (IPC) with strong anti-jamming performance and good stabilizing performance as a virtual instrument carrying platform and IPC chassis selects Advantech IPC-610 type case with rack installed case with 14U high and 14 slots of ATX motherboard selection which owns good module design and better ability of dustproof, anti-vibration and preventing electromagnetic interference, suitable for a variety of field use.

4.3 *Software design of virtual oscilloscope system*

4.3.1 *Basic frame structure*

The set of virtual oscilloscope is constructed in the form of modular structure and Lab VIEW has an efficient compiling effect. The created program not only operates alone, but also can be called as subprogram by other higher programs with powerful function, strong visibility, program development cycle reducing and work efficiency improving. According to the tasks that virtual oscilloscope needs to implement, we design system structure software process as shown in figure 4.

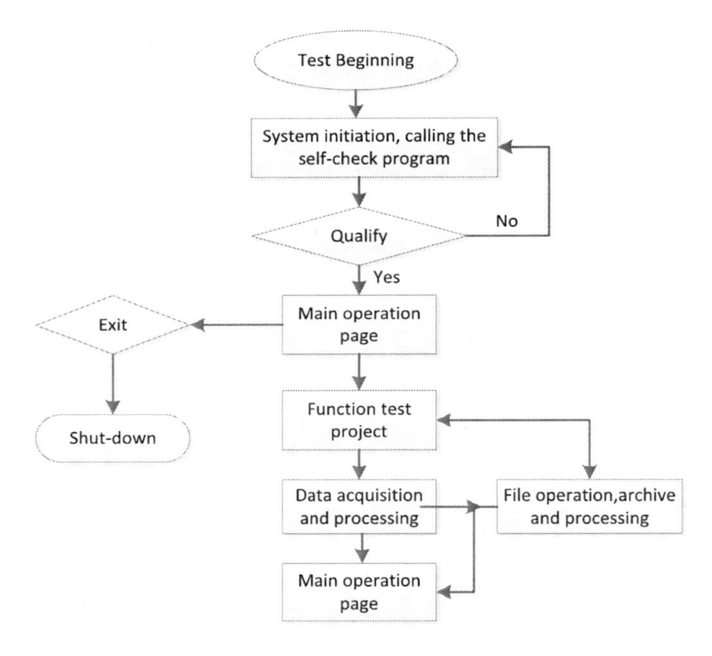

Figure 4. Test system application software structure flowcharts.

4.3.2 *Each module test principle and software design*

In this set of analogue oscilloscope, we set up channel option control, trigger mode selection control, horizontal and vertical gain control, switch control, display and other button controls on panel to put in front panel of analogue oscilloscope which can be directly faced with users to operate and use and control the work of analogue oscilloscope. Modular structure for software development, its front panel is as shown in figure 5 and virtual oscilloscope flowchart program design as shown in figure 6.

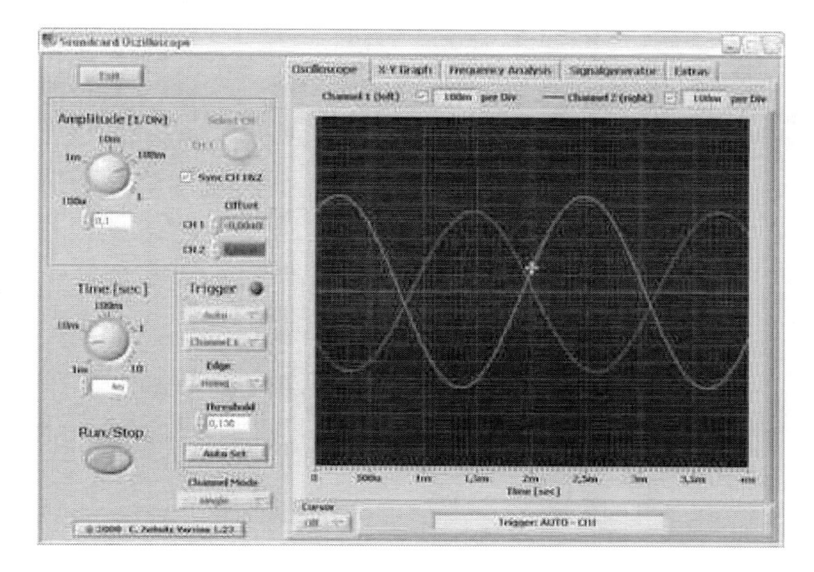

Figure 5. Virtual oscilloscope front panel program design.

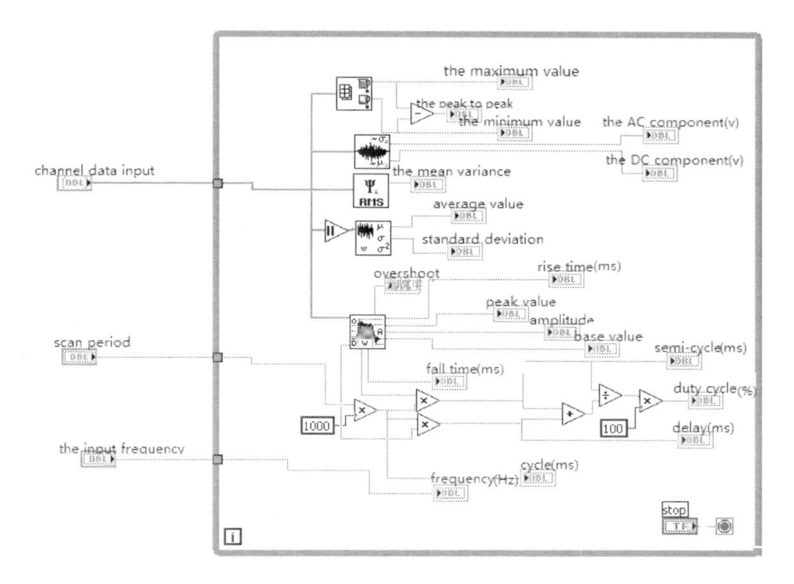

Figure 6. Virtual oscilloscope flowcharts program design.

1 Data acquisition module

Select different testing projects entry into the main interface in terms of test requirements and subsequently goes to data acquisition program and processing program. After program enters into testing state, first of all, predict program from current equipment state, then read data based on the channel specified by data acquisition card and display data on the interface in sequence. The main body of testing software is mainly constructed by data acquisition and processing module.

2 Waveform display module

The main role of the waveform display module is to display the waveform of virtual oscilloscope, usually consisting of five display modes: A mode, B mode, A+B mode, A-B mode and A&B mode. This type of mode mainly implements zoom-in and zoom- up of waveform in proportion, controls the adjusted frequency and other aspects and can simultaneously display the input signal waveform of single or two-channel.

3 Parameter measuring module

Parameter measuring module mainly implements the measurement and display of a virtual oscilloscope for as many as eleven kinds of voltage parameters, seven kinds of frequencies and periodical parameter. The main measuring parameters are AC voltage, DC voltage, average voltage, root-mean-square voltage, maximum voltage, minimum voltage, maximum differential pressure, differential pressure, peak voltage, base voltage, overshoot voltage, sampling frequency, sampling period, rise time, fall time, overshoot, undershoot and duty cycle.

4 Spectrum analysis module

To complete frequency domain analysis spectrum analysis module uses fast Fourier basic Transform algorithm and completes A/D sampling data input of the signal data acquisition card. Then do subsequent processing within a computer and make proper displaying through nine kinds of spectrum analysis windows to achieve a required instrument function to a complete oscilloscope function which provides two kinds of coordinate display mode. According to a different selection of users, sub module can be converted between amplitude and phase. For data input, at certain times of interpolation or extract can be made to achieve the variety of back stage to display scanning rate change of waveform. In terms of analysis functions chosen by users, which display with different statistical units (Vrms, Vpk, Vrms 2, Vpk2) and automatically different and appropriate abscissas. If displayed amplitude need changing, it can be achieved by only input data multiplies corresponding coefficient.

The spectrum analyzer functions can be completed by means of two ways. The first way is to call formula node from the virtual instrument Lab VIEW through discrete Fourier transformation formula. The second way is to convert input data to digital transformation from time domain to frequency domain so as to call corresponding digital signal processing module to program, completing analysis of signal spectra.

5 Data storage and playback module

The main function of the module is to save and mount real-time result of the system as well as to write disks and to read disks. It saves relevant information of measured data to meet the need of subsequent inquiry and statistics analysis. The documentary operation and management play an important part in testing system software development. Data saving, parameter input and system management cannot do without documentary setup, operation and maintenance. Virtual oscilloscope can directly set up to save documents, automatically measure, display wave and save data which can be consistent with real-time data during the test.

Press "save" button to control when data storage is needed, press "reading disk" button when date needs reading from a data file and display wave in other windows. Two signal wave input by A and B channels can be displayed at the same time in normal display. "Save" button can be pressed to save date of A channel as well as to save input signal wave of A channel, in the state of B, it is the same as A.

6 Filter module

The main function of Filter module is to turn an input sequence into the output sequence by means of a certain operation and also plays a filter's role at the same time, improving measurement accuracy through filtering superfluous frequency components of the signal. The basic structure of the filter consists of an analogue low pass filter and digital filter. Analogue filter plays a role in removing high-frequency noise and interference of the signal path among A/D converter, usually putting in front of A/D converter. Digital output converted is able to remove aliasing wave information through an analogue filter. However, digital filter plays a role in reducing noise and interference of frequency within the pass band by means of average technology, usually putting on the back of A/D converter. Digital filter can also remove noise and interference in the process of conversion because of its position.

5 CONCLUSION

The paper develops oscilloscope system based on virtual environment aiming to performance features of the oscilloscope. Virtual oscilloscope not only has a desktop digital oscilloscope function, but also gives full play to computer powerful function and flexibility of software design. The test shows the system is simple, convenient, intuitive, high reliability and good use effect. The lab VIEW virtual oscilloscope is programmed with its graphical and modular features so users can call modules with different functions and conveniently and create their own application interfaces with a stronger function based on their own testing requirements. The design philosophy will play more active role in measurement and control system along with the development of computer technology.

REFERENCES

[1] National Instruments Corporation (2006) *National Instruments, LabVIEW Control Design Toolkit User Manual.*

[2] National Instruments Corporation. (2002) *LabWindows/CVI SQL Toolkit Reference Manual.*
[3] Ma Mingjian, (2005) *The Data Acquisition and Processing Technology*, Xi'an: Xi'an Jiaotong University Press.
[4] Zhou Xingpeng, Chou Guofu, Wang Shourong, (2004) *Modern Testing Technology,* Beijing: Higher Education Press.
[5] LeiYong, (2005) *Design and Practice of Virtual Instrument*, Beijing: Publishing House of Electronics Industry.

Electric, Electronic and Control Engineering – Shao, Shu & Tian (Eds)
© 2015 Taylor & Francis Group, London, ISBN: 978-1-138-02842-5

High level semantic image retrieval based on Ontology

Shouqian Wang, Gongwen Xu* & Chen Zhang
School of Computer Science and Technology, Shandong Jianzhu University, Jinan, China

Binbin Chen
School of Mathematical Sciences, Heilongjiang University, Harbin, China

Chunxiu Xu
School of Computer Science and Technology, Shandong Xiehe College, Jinan, China

ABSTRACT: The images on the Internet are emerging in a rapid speed, so how to retrieve the interesting images is a focus on the present, and the accurate and effective retrieval method is in urgent need. In this paper, higher level semantic image retrieval based on Ontology is proposed. The Ontology technique is introduced firstly. Then the method of calculating the similarity between image and the center point is introduced. At last the retrieval method based on Ontology is carried out. The semantic features about the image are described via Ontology method, and the user's quest is normalized before the image retrieval. The target image can be retrieved accurately and quickly with the help of the retrieval system, which was designed and implemented based on Ontology. The retrieval method based on keywords was taken as the comparison with that based on Ontology. After the experiment, the proposed method is shown to outperform the method based on keywords.

KEYWORDS: ontology; semantic features; image retrieval; retrieval model.

1 INTRODUCTION

The digital image is an important form of multimedia business and there includes abundant information in an image. In our daily life or work time, we will get in touch with a lot of pictures. If we cannot organize these pictures efficiently, we cannot browse and retrieve them high efficiently and quickly[1]. The image retrieval technology involves multiple domains such as database, image processing, pattern recognition, artificial intelligence, so it is a challenging task to build an efficient and effective retrieval system for image retrieval[2].

The traditional retrieval system is based on a text which is named text-based retrieval. Comparing with the text-based retrieval, the multimedia retrieval is more complicated, such as content-based retrieval. This retrieval type has formed a set of mature theory and typical systems. The content and context semantics of the media are all considered in the retrieval [3]. This method is based on image processing, pattern recognition, computer vision, and image understanding and so on.

Taking image features analysis as an example, the features include color, texture, shape, space relationship, and so on. The color histogram can describe the color features and each color's share in the whole image. But this method can not tell the color's space and allocation in the image.

The gray level co-occurrence matrix can be used to analyze the texture of the image, and then we can obtain four key features: energy, inertia, entropy, and relevance. The texture and orientation can be extracted by calculating the image energy spectrum function. The attributes of the texture also include coarseness, contrast, regularity and direction[4].

In recent years, plenty of studying achievements about image retrieval spring up, but the majority of them is limited to a single domain and small database system [5].

2 KEY TECHNOLOGIES ABOUT THE PAPER

2.1 *Brief introduction about Ontology*

In the artificial intelligence domain, the Ontology was firstly defined by Neches. An ontology is an explicit specification of a conceptualization [6]. The properties, types, and interrelationships of the entities in an image domain are named and defined. The Ontology has great potential in information organization and management; it is also helpful in image processing and image retrieval at the same time.

*Corresponding author: *xugongwen@163.com*

The Ontology is important in the links of the retrieval. An abstract concept is given to each node in the Ontology system. High level node contains the low level node in the conceptual category[7]. The nodes and the node relationship construct the Ontology. The Ontology node structure is shown in figure 1. One concept has many parts. The center points, under the Ontology node, represent the conceptual aspects, and different feature of the concept can be described by each aspect, so the image concept can be obtained by Ontology. Under the center point there are images which meet the according rules. At last the images are arranged on the basis of similarity between them.

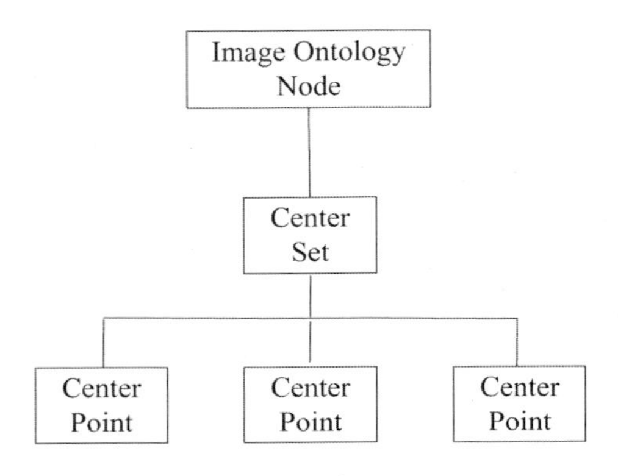

Figure 1. The Ontology node structure.

In addition, the MPEG 7 standard was carried out by "Moving picture experts group", which has provided a structured and broad description for the image content.

Some scholars realized image annotation by means of Ontology in the keyword image retrieval. The Ontology is also used in information retrieval based on content to solve the content matching problem in network documents. The image retrieval under the help of Ontology and the knowledge containing in the Ontology need to be further researched[8].

2.2 *The similarity algorithms*

The similarity between the image and center point can be calculated using the follow method.

1 According to the selected m features, calling the corresponding features extracting procedure. The features vector group v1 is obtained, which has m vectors.
2 Comparing the feature kinds of the center point, the vectors which have not the features in the center point will be gotten rid of. So features vector group v2 is obtained which has n vectors.
3 The Euclid distance is calculated between n vectors and the vectors at the center point.

The above algorithm is used to calculate the similarity between the retrieved image and the center point.

Users can feedback when he gets the retrieval results. The inner structure of the Ontology node can be adjusted according the user advice. When a picture P is retrieved, the satisfactory images are expected. The above algorithms can be used to calculate the distance between P and the center point, then the image will return to the user. The P will be added into Ontology[9].

As for the returned files, the user selects the correct results and records the according results in P. When P is retrieved in the next time, the former results can be returned at the same time.

2.3 *The design of system flow*

The system flow is described below.

1 The construct of the image Ontology.

The Ontology plays an important role in the intelligent information system. It is the core of the system and it defines the conceptual level of the field knowledge and the relationship between the concepts and the semantic reasoning rules. The main functions of the Ontology are shown below. Firstly, when labeling and extracting the information from the web, the Ontology will be taken as the reference to the acquisition of semantic metadata; Secondly, the Ontology will be taken as the reference to semantic expansion and retrieval construction in the preprocessing period. Thirdly, the Ontology provides the rules for semantic model reasoning. So in order to build intelligent information retrieval system, the field knowledge of the Ontology must be built under the help of the field experts. The suitable storage method and the evolutionary strategy should be selected. The functions of this part are completed by the Ontology management module.

2 The acquisition of information.

The aim of the intelligent information retrieval system is the image information. So there should be enough information to be used. In this paper, there are enough images in the database.

3 The semantic annotation and metadata extraction.

In this step, the image content features will be revealed and extracted. The semantic annotation means that the images are analyzed and given a certain amount of annotation according to certain rules and procedures. Because the semantic metadata is organized via the Ontology structure, the image information and the relationship between the images are reflected. The traditional retrieval technology can not reflect the situation and relationship of the image database. The function of this part is completed by semantic annotation and abstraction.

4 The semantic encoding of metadata.

The metadata is the data about data, which describes the data content, coverage, quality, management mode, data owner. It is the bridge between the data and data users. In the intelligent information retrieval system, semantic metadata is the carrier of the images and the object of semantic reasoning and querying.

5 The preprocessing of retrieval.

After the above four steps, the information retrieval system has completed the acquisition, annotation, metadata abstraction, semantic coding, which are the base of querying and reasoning. The querying model contains three sub modules, querying preprocessing processor, querying processor, and the querying result processor. When there is not the data which matches the user's querying information, the semantic extension will be automatically operated.

6 Semantic querying and reasoning.

The querying preprocessor will submit the completed querying method to the querying operation. In the beginning the querying operation will carry out the semantic reasoning based on certain rules by the Ontology inference machine, in order to produce the hidden information in the images.

7 The processing of the query results.

This step includes the order and display mode of the retrieval results. The order of the results is important to the information retrieval engine. An excellent order algorithm means the success of the research engine. It directly determines the importance and usefulness of the research results.

Through the cooperation of the above-mentioned modules, the intelligent information retrieval based on Ontology is finished.

The whole process can be shown in figure 2.

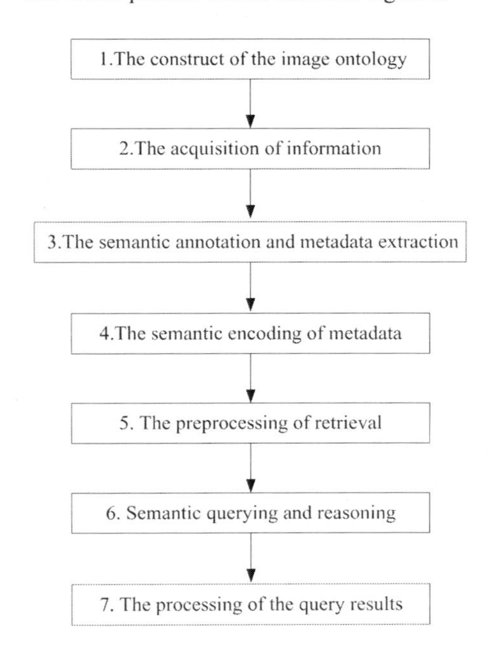

Figure 2. The Ontology retrieval mode.

3 THE RETRIEVAL METHOD BASED ON ONTOLOGY

3.1 *The Ontology retrieval mode*

To meet with the need of image retrieval, the image retrieval model is built up, as shown in figure 3. In this mode, there are five intelligent subjects: interface, pre-process, information process, information research, and management. The interface subject is used to deal with the user's request. The management subject is used to coordinate the whole information obtaining procedure.

In this retrieval method, there are two steps: retrieving image and obtaining image. The image retrieval server manages the information server directly and accepts the user's request, providing the interface between user and server. In the image retrieval procedure, there are also two steps: submitting retrieval request and retrieving image.

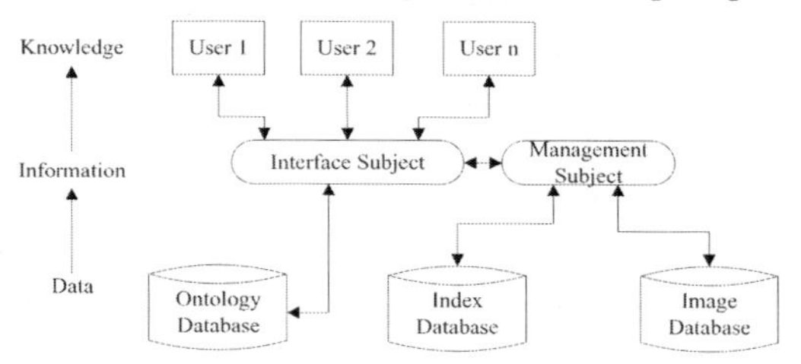

Figure 3. The Ontology retrieval mode.

When submitting a retrieval request, the according interface subject is established on the basis of the user's retrieval request, and the suitable retrieval request is constructed by user under the help of direction. The whole procedure is described below.

1 The user submits the retrieval request to the image retrieval server.
2 The image retrieval server carries out an interface subject to make an interaction with the user. When receiving the user's request keywords, the interface sends out each domain containing the keywords after searching the Ontology database.
3 The user can determine the searching domain and the meanings in the interface according to his own intention.
4 The interface subject normalizes request using Ontology technique, then sends it to management subject.

Then comes the second steps of image retrieval that management subject leads the searching work.

1 The management subject access index database directly when receiving the retrieval request from the interface subject. The associated index information is searched out after matching the domain and keywords. Then the index information is submitted to the interface subject.

2 The interface subject presents the retrieved information to the user in the form of abstraction. User can select and examine it.

3.2 *The experimental analysis*

In the experiment, the images in the Corel image database are selected out to test the performance of the image retrieval based on Ontology.

In the first experiment, a request 'a train passes over the bridge' was submitted in the retrieval system. The results show that this method can select out the pictures meeting with our will.

Figure 4. Some pictures of the retrieval result.

In order to verify the effectiveness of this method, the retrieval method based on keywords was taken as the comparison.

In the keywords retrieval method, the testing images and retrieving images are described by the keywords accurately. While in the Ontology retrieval method, the image semantic features are described by the Ontology method and classified into different domains and the query information is normalized before retrieval.

From fig. 5, the method based on Ontology is superior to the method based on keywords. As the recall rate keeps the same degree of 0.7, the precision of Ontology method is 0.762 while the precision of keywords method is 0.719.

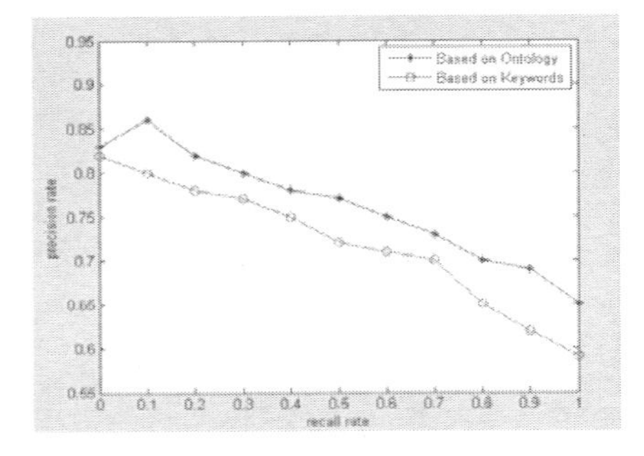

Figure 5. The performance of the two methods.

4 CONCLUSIONS

In this essay, an image retrieval method based on Ontology is proposed, and an image retrieval system based on Ontology is described and structured. This system can classify the images and normalize the user request. In this system, the target image can be retrieved accurately and quickly. The more the Ontology is described accurately, the better the retrieval system performance will be. But the process to build up the accurate domain Ontology is difficult and complex. Determining how to construct the domain Ontology is the further research.

The retrieval system based on Ontology is effective and accurate, so we expect even greater successes to follow.

ACKNOWLEDGMENT

The work is supported by the School Fund project of the Shandong Xiehe College (XHXY201431) and Jinan Higher Educational Innovation Plan (201401214, 201303001), the Project of Shandong Province Higher Educational Science and Technology Program (J14LN59, J13LN11, and J12LN31), the natural science foundation of Shandong Province (ZR2012GQ010).

REFERENCES

[1] Neches R, Fikes RE, Gruber T R, et al. (1991) Enabling technology for knowledge sharing. *AI Magazine*, 12(3):36–56.

[2] Wu Fei, Zhang Hong, Zhuang Yueting. (2006) Learning Semantic Correlations for Cross-media Retfieval. *Proc. of International Conference on Image Processing. Atlanta*, USA: IEEE Press. pp:1465–1468.

[3] Zijun Yang, CC Jay Kuo. (1999) A semantic classification and composite indexing approach to robust image retrieval. *IEEE Int. Conf. on Image Processing*. pp:134–138.

[4] Hong Wu, Mingjing Li, et al. (2002) Improving image retrieval with semantic classification using relevance feedback. *Sixth Working Conference on Visual Database Systems: Visual and Multimedia Information Management*. Netherlands: KLuwer, B. V. pp:327–339.

[5] C. Zhang, T. Chen. (2002) An active learning framework for content based information retrieval. *IEEE Transactions on Multimedia*, 4(2):260–268.

[6] Bo Hu, S. Dasmahapatra, et al. (2003) Ontology-based medical image Annotation with Description Logics. *Proc. 15th IEEE Int. Conf. on Tools with artificial Intelligence, IEEE Computer Society*, pp:77–82.

[7] C. Breen, L. Khan, A. Prnnusamy. (2002) Ontology-based image classification using neural networks. *SPIE Internet Multimedia Management Systems III, Boston*. pp:198–208.

[8] Shuqiang Jiang, Jun Du, Qingming Huang, et al. (2005) Visual ontology construction for digitized art image retrieval. *Journal of Computer Science and Technology*, 20 (6):855–860.

[9] Xu Gongwen, Zhang Zhijun, Yuan Weihua, Xu Li'na. (2014) On medical image segmentation based on wavelet transform. *ISDEA 2014. 15-16*, pp:671–674.

Electric, Electronic and Control Engineering – Shao, Shu & Tian (Eds)
© 2015 Taylor & Francis Group, London, ISBN: 978-1-138-02842-5

A new method of pulse comparison detect in IR-UWB ranging

Lin Zhu, He Zheng & Hao Zhang
Chongqing Communication Institute, Chongqing, China

ABSTRACT: Here we describe a ranging method based on the 2PPM (2 Pulse Position Modulation) modulated IR-UWB (Impulse Radio-Ultra Wideband). This method samples the signals on the positions which the pulse takes the '1' and '0' information, then synchronizes the pulse by comparing the maximum values of the samples, and finally complete the ranging by TOA (Time of Arrival) estimation. The simulation result shows the validity of this method. It is not necessary to set a decision-threshold to detect the position of the pulse using this method, and it is more accurate, improves the precision of the ranging.

KEYWORDS: IR-UWB; pulse detection; TOA ranging.

1 INTRODUCTION

There are limitations in the traditional location technique such as GPS or positioning Based on Cell ID [1]. And the IR-UWB technique provides noise immunity capability, strong anti-interference ability, high multipath resolution and good penetration ability. So the ranging and positioning method based on IR-UWB has the incomparable advantage over traditional methods.

This article introduces a ranging method based on IR-UWB. This method completes the ranging process by the estimated value of TOA based on the position detection of the received pulses. A remarkable feature of this method is avoiding the difficulty of how to choose the threshold when the channel parameters change. So the method reduces the changing channel's influence to the ranging precision. The simulation verified the validity of this method.

2 RANGING METHODS

In our method, a TOA measurement is made for ranging. All nodes create a coordinate system. Fig.1 is the Block diagram of the system.

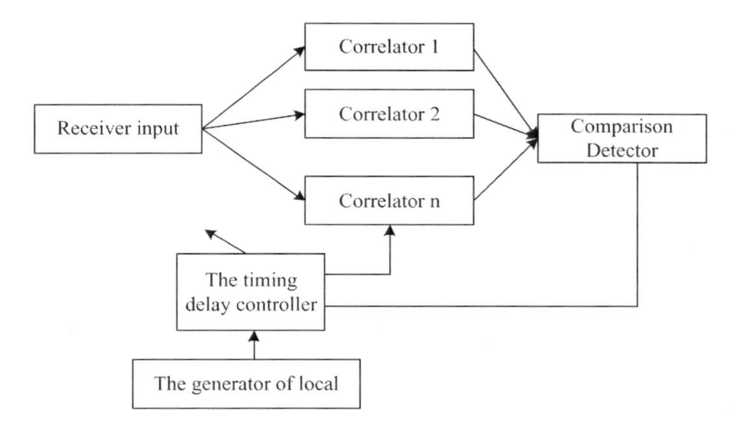

Figure 1. Block diagram of the system.

Receiver input: The signal at the transmitter is the pseudo-random codes used in ranging after 2PPM-DS-UWB modulating, which is expressed by Eq. (1)[2]:

$$S^{(k)}(t)=\sum_{j=-\infty}^{\infty}\sum_{n=0}^{N_s-1} a_{j,1}^{(k)} g_n^{(k)} p(t-jT_f-(n+\frac{a_{j,0}^{(k)}+1}{2})T_c) \tag{1}$$

Among them: $p(t)$ is a pulse wave function, $g_n^{(k)} \in \{-1,+1\}$ is the pseudo random sequence of the k-th user. T_c is the pulse repetition period, T_f is information period, $a_{j,0}^{(k)}, a_{j,1}^{(k)} \in \{-1,+1\}$ are two independent binary sequences of the k-th user. $N_s = T_f / T_c$, N_s pulses represent an information symbol.

After propagation over the UWB multi-path wireless path, the input signal at the receiver can be expressed by Eq. (2) [3]:

$$r(t) = A(D)s(t - \tau(D)) + n(t) \quad (2)$$

Where $A(D)$ is the amplitude of the signal which is the function of transmission distance D, $\tau(D)$ is the timing delay of the signal after propagation over the distance D, $n(t)$ is the channel noise.

The generator of local pulse: Offer series of pulses with the same waveforms as those at the transmitter.

The timing delay controller: Delay local pulses some time, then send them to each correlator.

Correlator: Compute the correlation between local pulses after time delay and the input signal at the receiver. The operation in correlator can be expressed as follows:

$$S_C(t) = \int_0^{T_S} r(t) \cdot S_t(t - ((n-1)\cdot\tau + (T-1)\cdot C\cdot\tau))d\tau \quad (3)$$

Where upper limit of the integral is the length of $s_L(t)$, n is the serial number of correlators, C is the total number of correlators, T is the times of decisions for the decider to make, whose function will be described in the following contents. From Eq. (3), we can see that there is a fixed time delay τ between the delays of two neighboring correlators.

Decider: It has two functions: first, when the output of certain correlator is higher than the threshold, TOA is estimated by $(n-1)\cdot\tau + (T-1)\cdot C\cdot\tau$. Second, when there is certain output of a correlator higher than the threshold, the decider will control the 'Timing delay controller' to finish working, thus correlators are stopped too, and then the estimation of the TOA is terminated. The decision times T decider has made is used to compute TOA.

The function that 'Decider' can control the ending time of 'Timing delay controller' and 'Correlators' leads to a decrease of time used to estimate TOA. When signals to direct path or signals on the path with shortest transmission distance in the case of there is no direct path arrive the receiver, the output of certain correlator is higher than the threshold, therefore, the estimation of the TOA is terminated, from which we can see that this function can also decrease the influence of multi-path on estimating TOA, and the ranging precision is greatly improved correspondingly.

Comparison decision device: decision device has two functions: one is to sample the signals on the positions which the pulse takes the '1' and '0' information, then compare the values of the two samples to detect the position of the pulse without the decision-threshold. The basic principle is shown in fig.2. The latency of correlator $(n-1)\cdot\tau + (T-1)\cdot C\cdot\tau$ is the estimated value of TOA; the other one is to stop the 'delay controller' and the correlator when finishing the estimation of the value of TOA.

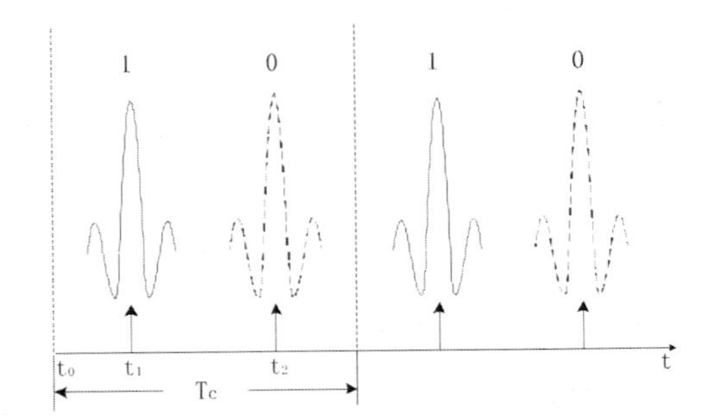

Figure 2. The basic principle of comparison decision.

3 THE BASIC PRINCIPLE OF COMPARISON DECISION

This article describes the basic principles of comparison decision detection based on a 2PPM modulation. The pulse may appear in one of the two positions in each pulse cycle of the 2PPM modulation. As the figure 2 shows, the position of the solid line pulse means the information is '1', and the position of the dashed line pulse means '0'.

Whenever the pulse is synchronized with the receiver, the start position of the pulse cycle is obtained, which is the position of t_0 in figure 2. Also, we can calculate the positions of the pulse carried the information '1' should appear, which is t_1, and the position of the pulse carried the information '0' should appear, which is t_2. After that, we compare the two sampling values of the two positions, if the value of the t_1 position is greater than the value of the t_2 position, we determine the received information is '1', otherwise we determine the received information is '0'. We determine the information of every cycle, according to this principle, finally we can obtain the TOA of the ranging pulse sequence.

4 SIMULATION EXPERIMENT

Through the simulation experiment, the validity of the proposed method is proved and the performances of it are evaluated.

4.1 *Experiment model*

The simulation parameters are given as follows:

1 Sampling frequency: 400MHz;
2 Pulse duration: 8ns;
3 Pulse repetition period: 40ns;
4 The length of pseudo-random codes: Gold codes with a length of 127;
5 Pulse is the 2th derivative of the Gaussian pulse;
6 Suppose there are four multi-path signals, in which the energy of the signal on the direct path is 10% of the total energy;
7 Suppose there are 10 nodes, distributing on a square which is 100m×100m in dimensions randomly.
8 SNR: -10dB.

We simulated the algorithm with System Generator and Simulink. Figure 3 shows the time domain waveform after the synchronization and filtering of the received pulse sequence which is sent from another node. As the figure shows, we can determine the position of pulse by comparing the sampling values of the t_1 and t_2 time, then complete the TOA estimation.

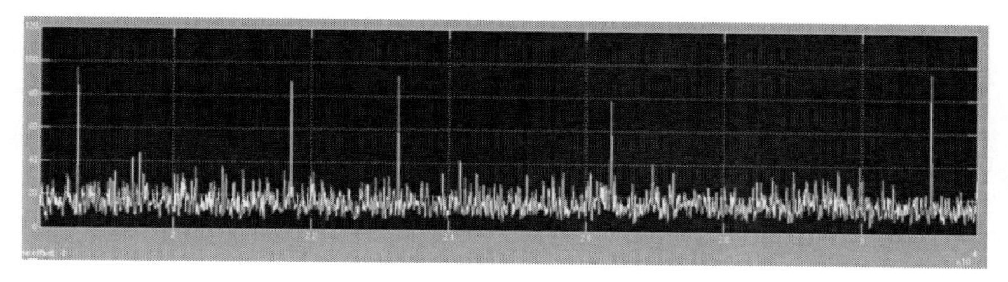

Figure 3. The time domain waveform of receiving pulse sequence.

4.2 *The influence of side lobes of correlation function on ranging*

In the following simulation, Average power of transmitting signals: -30dBm, Power of white Gaussian noise: 0dBm, other simulation conditions are the same with the figure 3.

When analyzing the influence of side lobes of correlation function on ranging, we suppose there are only a direct path and another path.

The simulation conditions described by the solid line in fig.4 are: the energy of the signal on the direct path is 10% of the total energy, and the energy of the signal on the path 2 is 0.5% of the total energy. The latter arrives 1478ns later. The peak of correlation function at 246ns is the main lobes of the signal on the direct path, where 246ns is the transmission time.

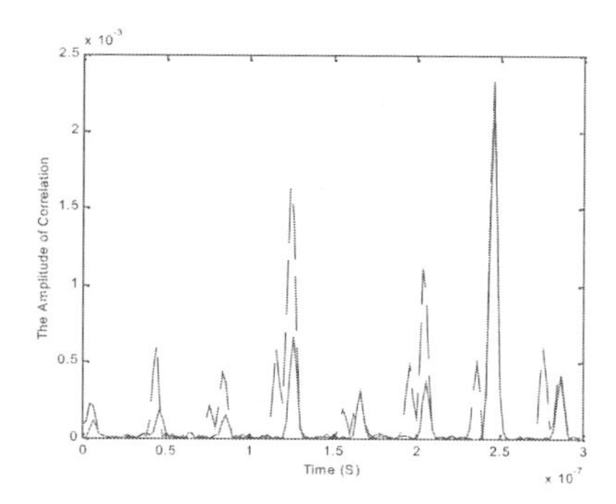

Figure 4. Side lobes versus TOA.

The total energy on the two paths is less than the energy of transmitting signal is because there are attenuation and spoilage in propagation. Since different paths have different levels of spoilage in the actual communication, it is likely that the strength of the signal on the direct path is smaller than that of signal on another path.

From fig.4 we can see clearly: At 124ns the side lobes of the correlation function of signal on path 2 is approximately equal to the main lobes of the correlation function of signal on the direct path, which can increase the difficulties to decide the location of the main lobes of the correlation function of signal on direct path. Therefore, in order to acquire a correct TOA, a type of pseudo-random code with low correlation function side lobes must be selected.

4.3 *The influence of τ on ranging error*

From fig.1 and Eq. (2) we can see that τ is timing delay between the two neighboring correlators. In fact, τ is the largest theory error in estimating TOA in the proposed system.

The correlation functions of signal on direct path when $\tau = 2ns$ and $\tau = 1ns$ are described in fig.5 and fig.6. Distinctly, the correlation functions when $\tau = 1ns$ is more accurate than that at $\tau = 2ns$.The data acquired from repeating the simulations for 100 times shows that the average ranging error is 0.52m when $\tau = 2ns$, while when $\tau = 1ns$,it is 0.23m.

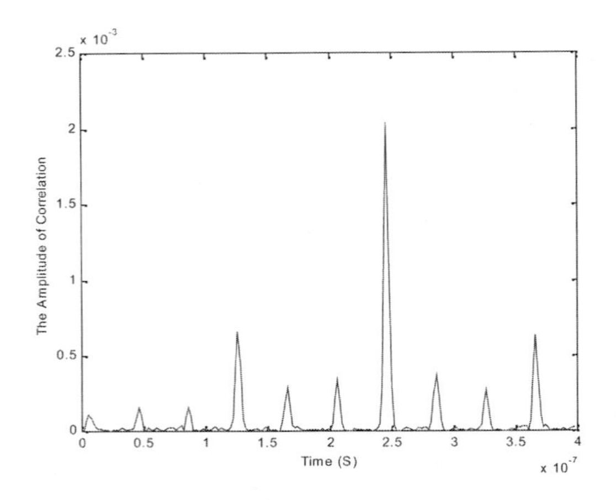

Figure 5. The correlation function when $\tau = 2ns$.

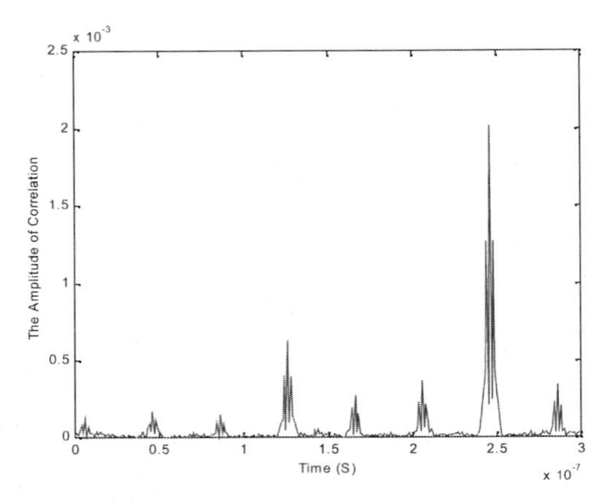

Figure 6. The correlation function when $\tau = 1ns$.

5 SUMMARY

A new ranging method based on IR-UWB is investigated in which the comparison decision device is used to estimate the TOA. The simulation verified that the average ranging error can be lowered to 0.23m, we can conclude that:

1 The autocorrelation function of pseudo-random codes used in ranging must be with lowest side lobes;
2 It is not necessary to set a decision-threshold to detect the position of the pulse using the method provided in this article. This method reduces the changing channel's influence to the ranging precision;
3 Decreasing the timing delay between two neighboring correlators can lead to a decrease in the ranging error.

REFERENCES

[1] Xie Z. P, Xiong S. M. and Xu Z. Q. (2004) Wireless positioning technologies and its development, *Modern Communications* (3):7–9+(4):7–8.

[2] Li jia Ge, Fan xin Zheng, Yu Li Liu & Guang Rong Yue (2005) In: *Ultra Wide Band Wireless Communications*, Chapter 4, Beijing: National Defense Industry Press.

[3] Maria-Gabriella Di Benedetto, Guerino Giancola (2004) In: *Understanding Ultra Wide Band Radio Fundamentals*, Chapter 10, published by Pearson Education, Inc, publishing as Prentice Hall PTR.

Electric, Electronic and Control Engineering – Shao, Shu & Tian (Eds)
© 2015 Taylor & Francis Group, London, ISBN: 978-1-138-02842-5

Research on the influence of wind power integration on transmission service price

Hongjun Kan & Lunnong Tan
School of Electrical and Information Engineering, Jiangsu University, Zhenjiang, Jiangsu, China

Yijun Zhang
Shandong Gold Power Company, Yantai, Shandong, China

ABSTRACT: Wind power has been developed rapidly during recent years. However, costs of power transmission services have been rising due to their instability of working, which has become a barrier restricting their developments. To optimize the wind power capacity is an effective means to improve the value of the wind power resource. According to the characteristics of wind power system, the topic of influences on the price of power network transmission service caused by the wind power integration has been discussed. Based on the calculation model of transmission price, comparing different prices of power transmission service under the different proportion of wind power capacity, point out that it is more economical after the optimization of wind power system, which has been tested and verified through the 6-bus system.

KEYWORDS: Power market, wind power, power transmission service price.

1 INTRODUCTION

With the rapid development of wind power generation, assessment of wind power resources value has played a vital role. Wind farm generally established in Mongolia, Gansu and other underdeveloped regions and local couldn't consume, must be delivered to the load center area through long distance transmission line [1]. However, the volatility of wind is easier to bring various stability problems for long distance transmission network and even the whole system [2].

From the perspective of transmission service charges, study the optimal value of wind power capacity is an effective method. Establish the transmission service fee pricing model and assessed the contribution factor of transmission service cost in consideration of wind power transmission cost volatility risk factors [3]. To investigate the influence of wind power capacity accounts for the proportion of the transmission service cost.

In this paper the economic influences on power transmission service prices caused by the wind power system interconnection have been analyzed in terms of power output, which has laid a foundation for promoting renewable resources like wind energy to achieve the performance of conventional energies, as well as making a further evaluation on the wind energy resources.

2 TYPICAL OUTPUT CURVE OF WIND POWER AND PHOTOVOLTAIC POWER

In this paper, select a typical summer day for analysis, get short or medium-term wind power curve. A large number of experiments show that the relationship between the output power of wind turbine and wind speed is:

$$P_W = C_e(T_t)\pi\rho R_t^2 v^3 / 2 \quad (1)$$

R_t is the radius wind turbine, ρ is the density of air, $C_e(T_t)$ is the energy utilization coefficient of wind turbine, T_t is the tip speed ratio, v is the wind speed.

In the practical application of engineering, wind farm output is a power curve based on the fan and wind speed. Select a wind farm output curve of a region typical day (as shown in Figure 1).

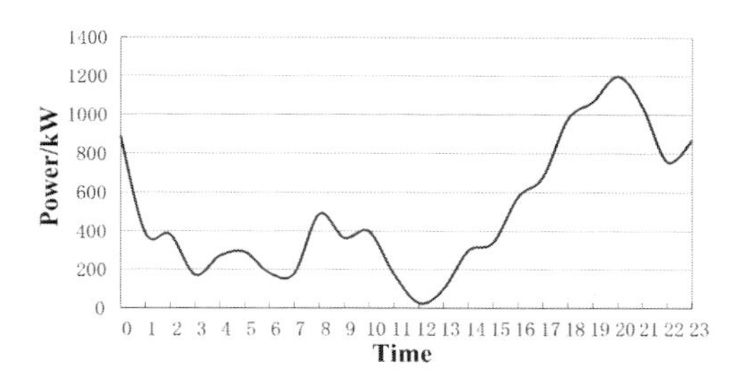

Figure 1. Typical output curve of wind power.

Judging from the wind power output process, the random variation of its daily output is larger, with a strong random and intermittent. The instability of energy brought difficulty in the use of wind energy. The process of wind power output is uncontrollable, wind power sent separately in the same area is wasteful, and the usage ratio of transformation equipment and lines is low.

3 TRANSMISSION SERVICE COST IN ELECTRICITY MARKET ENVIRONMENT

Transmission service means the overall process of conveying the electric energy from power stations to users with safety and high quality. Transmission cost, from the board sense, contains not only production running costs, but also capital costs, which includes operation and maintenance cost, depreciation cost, salary and additional charge, transmission losses, taxes, relative financial cost and return on investment. The calculation of transmission cost is the upper-level problem of transmission charge and the base of pricing.

In the power market environment, the transmission cost can be expressed as:

$$C_{tol} = C_{oc} + E_m + E_l + E_{iv} + E_{mg} + E_{as} \tag{2}$$

C_{oc} is network equipment costs, E_m is network maintenance costs, E_l is transmission loss, E_{inv} is the power grid investment returns, E_{mg} is the management service fee, E_{as} is ancillary service costs.

4 INFLUENCE OF TRANSMISSION SERVICE COST OF WIND POWER RESOURCE'S VALUE

In the large capacity of wind power integration, its intermittent and volatility pose a severe test for the power system security, stability and power quality, which will make the operation cost of power system greatly increased.

4.1 *The reserve capacity cost*

For wind power generation uncertainty, we should consider the economic effects of increased spinning reserve capacity cost brought to the whole system required in the calculation of wind power transmission price. The reserve capacity cost is:

$$C_{WRC} = C_W R_{Wind,h} \tag{3}$$

C_W is the reserve capacity cost price, $R_{Wind,h}$ is the reserve capacity cost for h time.

4.2 *Reactive power equipment cost*

When large-scale wind power connects to the grid, it will consume a large amount of reactive power. Reactive power cost is composed of two parts, reactive power generation costs and compensator operating cost.

Reactive power generation cost function:

$$Q_{gqi}(Q_{Gi}) = k[C_{gi}(S_{Gi}) - C_{gi}P_{Gi}] \tag{4}$$

Reactive power compensation device running cost function:

$$C_{Bj}(Q_{Bj}) = \frac{11600 Q_{Bj} h}{15 \times 365 \times 24} = 0.059 Q_{Bj} \tag{5}$$

$C_{Bj}(Q_{Bj})$ is the cost function of the reactive power compensation device on J nodes. h for the use of frequency, approximate to 2/3.

Reactive power cost:

$$C_Q = \sum_{i \in N_G} C_{gi}(Q_{Gi}) + \sum_{j \in N_C} C_{Bj}(Q_{Bj}) \tag{6}$$

N_G is a system Generator set of nodes, N_C is a system with a reactive power compensator set of nodes.

By the above cost, the total cost of wind or photoelectric power transmission cost is the sum of the above fee:

$$C_{tol} = C_{oc} + E_m + E_{loss} + E_{inv} + E_{mag} + E_{as} + C_{WRC} + C_Q \tag{7}$$

4.3 *Transmission cost allocation based on the power flow tracing method*

The power flow tracing method, the tracking algorithm is used to obtain the occupy share of the unit or the user on transmission lines. Then, according to the occupy share and certain rules determine the transmission cost and network loss cost allocation.

The contribution factor λ is defined as the contribution ratio for generator power running through each line or node. When combined wind power transmission, different wind speeds results in a corresponding proportion of the various parts are different.

The smaller wind speed change, the wind effects on power system stability is smaller. The risk of fluctuations in wind speed function is:

$$\lambda_{rk} = \left| \frac{v_{av}}{v} - 1 \right| \times \lambda_{wd} \times \lambda_{wd\%} \tag{8}$$

λ_{wd} is the wind power contribution factor, $\lambda_{wd\%}$ is the proportion of Wind power accounted for power, λ_{rk} is the risk value for wind speed fluctuation , v_{av} is the annual average wind speed, v is the basic wind speed.

Considering the risk cost of wind power long distance transmission and transmission network sharing costs, obtain the network service share price for wind transmission based on the analysis of risk and power flow tracing:

$$C_{wind} = (\lambda_{wd} + \lambda_{rk})C_t \quad (9)$$

5 ANALYSIS OF OPTIMAL WIND POWER CAPACITY

By the wind power output on a typical day for a month, according to local load curve (as shown in Figure 2) characteristic selection proportion of wind power output,

After organizing the historical data, we get the total output power function for Wind power system interconnection:

$$P_Z = kP_W + (1-k)P_T \quad (10)$$

Through the analysis of the transmission service pricing model, calculate the value of the different K, and then get the value of K when the transmission service price is the most economic.

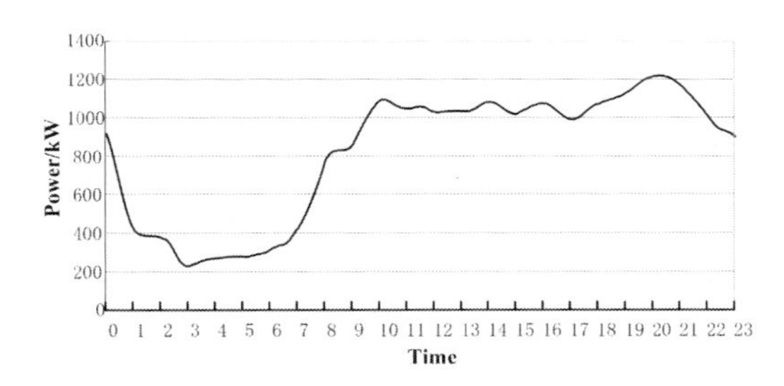

Figure 2. Local load curve.

Optimal wind power capacity of grid connected power system can contribute to improve the power quality, reduce the supporting transmission project investment, optimize the selection of electrical equipment and simplify the operation of the project management etc. Reducing the transmission service costs and improve the resource value of wind power resources.

6 EXAMPLE

Using Power World software to improve the 6-bus system, to verify the research done on this paper. Set the pending units to variable-capacity wind turbines. Gain the conclusion by comparing the transmission service price under different proportions of wind power capacity. 6-bus system with wind unit is shown in Fig. 3.

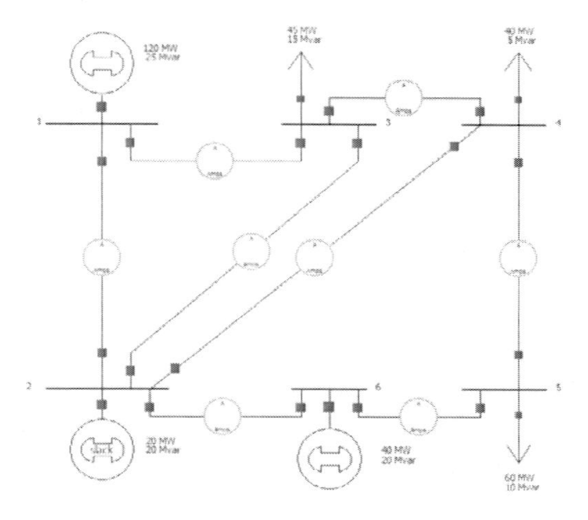

Figure 3. 6-bus system.

To process 6-bus system, defines node 1 unit as wind units. Node 2 units are for conventional units and balance node. The results of transmission service price are shown in Fig. 4.

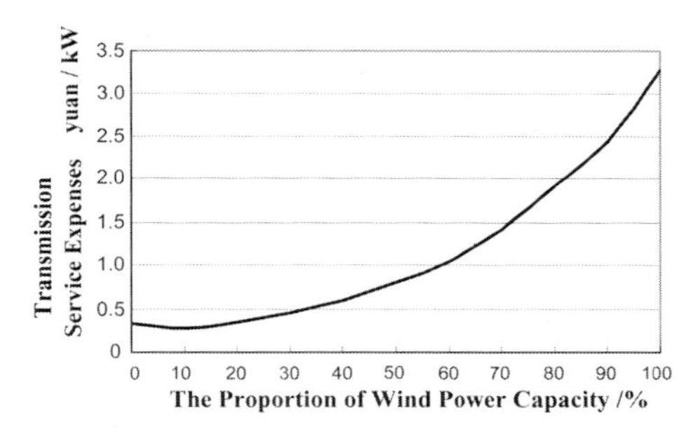

Figure 4. Transmission service price trends.

The transmission service cost borne by the wind power system is lowest when the K value is 9%. By contrast verify the conclusions: compared with ordinary wind power systems, after the optimization of wind power system transmission service price allocation is less. Power curve and power load curve change tendency basic consistent, can alleviate network load pressure.

7 SUMMARY

Electric power transmission is the bridge between the power generation and power usage. We analyze the influences on power transmission price caused by the combination of wind energy, through 6-bus system model to calculate transmission service price shared by wind unit under different wind power system grid capacity ratio. And the conclusion that reduction of the electric power transmission service cost after optimizing the wind power capacity has been made in this paper.

The economic and reasonable transmission price will provide correct economic signals for market participants (including the generators and users), which will optimize the usage of electric power transmission resources, improve the values of wind power resources, strengthen the ability of new energy power generation's entering the market and its competition partly, and also favor the popularization of new energy resources in the development of our modern society.

REFERENCES

[1] Songhuai Du. (2008) *Electricity Market*. 285:43.

[2] Cui Yang, Mu Gang, Liu Yu, et al. (2011) Spatiotemporal distribution characteristic of power fluctuation. *Power System Technology*, 35(2):110–114.

[3] Lun Nongtan, Zhang Baohui. (2002) Problems of using proportion of transmission line and loss allocation. *Transactions of China Electrotechnical Society*, 17(6):97–101.

[4] Wang Qingran, Zhang Lizi, Xie Guohui. (2010) A transmission pricing mechanism for cross-regional electricity trading. *Electric Power Automation Equipment*, 34(13):11–15.

Electric, Electronic and Control Engineering – Shao, Shu & Tian (Eds)
© 2015 Taylor & Francis Group, London, ISBN: 978-1-138-02842-5

A recoverable color image blind watermark scheme and its application system based on internet

Xinlong Chen & Guoqing Hu
School of Communication Engineering, Chongqing University, Chongqing, China

ABSTRACT: This paper releases an invisible watermark scheme, including embedding, extracting and recovering for color image based on internet. By converting the color image from RGB to YUV, this algorithm, based on DCT domain, is carried out by modifying some coefficients according to the characteristics of Human Visual System. The watermarked color image can be gained by visiting the website and uploading the original image and the original watermark. By detecting or modifying some coefficients, the detection or removal of watermark, which doesn't need an original watermark image, also can be done on the web. Comparing with other algorithm, this paper first release the algorithm for removal of watermark; the algorithm is very simple and quite suitable for running on the internet. The web site of the system as follows: http://dgdz.ccee.cqu.edu.cn/watermark/invisiblewatermarkb.aspx

KEYWORDS: recoverable watermark; color image; Internet.

1 INTRODUCTION

In the knowledge times, the intellectual property rights, which are the concentrated expression of the core competitiveness of enterprises, is the important wealth and resource of factors of production. More and more problems of intellectual property rights infringement are generated by digital image transmission. The copyright protection and content authentication method and its application system for digital image, based on the Internet, have important significance. Digital image watermarking technology, a powerful tool for digital image copyright protection and content authentication, has been widely concerned by scholars.

There have been many research results in digital image watermark technology, including visible and invisible watermarks. The invisible watermark requires that the embedded watermark should be not only transparent to observers [1], but also robust enough so that it cannot be easily destroyed or removed after some digital image processing or attacks [2]. According to the watermark embedding technology, the digital watermark is divided into spatial domain and transform domain[3].The existing frequency transformation methods for watermark embedding include discrete Fourier transform (DFT) [4], discrete cosine transform (DCT)[5], and discrete wavelet transform (DWT)[6]. In recent years, watermarking techniques have been improved using optimization algorithms such as genetic algorithm (GA) which is a popular evolutionary optimization technique invented by Holland [7]. In the field of watermarking, GA is mainly used in the embedding procedure to search for locations to embed the watermark [8-11].

As humans entered twenty-first century, the internet has changed people's life. Designing a recoverable color image watermark scheme and its application system, based on the internet, are of great significance to the further promotion of digital watermarking technology. In this paper, the digital watermark scheme, which has been authorized Chinese invention patent (Patent No.: ZL200910104746.6), is a transform domain algorithm, and the texture complexity [12] and edge feature [13] of the carried image is not considered.

2 THE ORIGINAL IMAGE AND THE ORIGINAL WATERMARK

The original image and the original watermark are obtained by File Upload component of ASP.Net. Considering the need to use by Internet platform, the original image, supporting JPG, PNG, GIF, BMP format, is a color image. When its width has more than 800 pixels, or the height has more than 600 pixels, the application system adjusts it to 800 pixels or 600 pixels automatically.

The original watermark, supporting JPG, PNG, GIF, BMP format, is a two value image. Its width and height are 32 pixels. When the image has more than 2 colors, or the width and height is not equal to 32 pixels, the application system adjusts it to two value images with 32 pixels width and height automatically.

3 WATERMARK EMBEDDING

DCT (discrete cosine transform), proposed by Ahmed, etc. in 1974, is an orthogonal transformation method. The real part of the Fourier transform has many advantages, such as a high compression ratio, a small bit error rate, concentrate information and low computational complexity [14].

In this paper, the Scheme of watermark embedding, considering the need to use by Internet platform, is based on DCT transform and described in following figure.

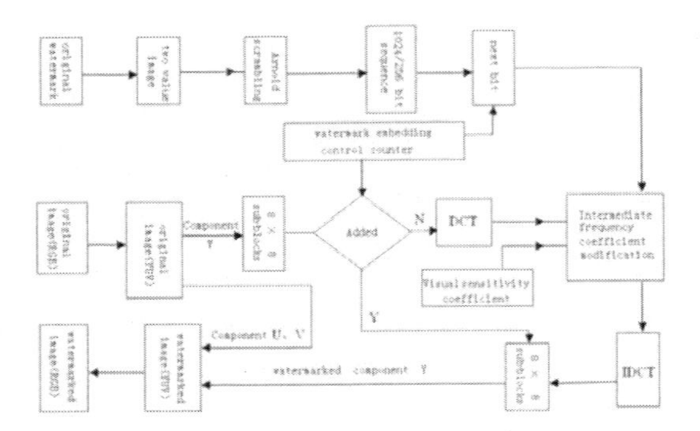

Figure 1. The scheme of watermark embedding.

The Scheme uses .NET framework GDI+ technique to obtain the original image and finish the processing of the image. When the width of the original image has less than 256 pixels, or the height has less than 256 pixels, the application system adjusts the image of watermark to two value images with 16 pixels width and height automatically.

The application system also uses the.NET framework of GDI+ technology to obtain the original watermark. The original watermark is a two value image and scrambled by Arnold transformation.

Follows the steps of using the .NET framework GDI+ technique to obtain the image and finish the processing of the image.

Using the bitmap class of the.NET GDI+ to object a new instance of the Bitmap object box, the Width, Height and other attribute information of the original image and original watermark will be obtained. Creating a new instance of the Color object, then you can gain the original image pixels and the original watermark image pixels of RGB component.

After that, the RGB image of the original image is converted to YUV space.

Y: brightness

U, V: color difference

The relationship from RGB color space to YUV color space, as follows:

$$\begin{cases} Y = 0.299R + 0.587G + 0.114B \\ U = 0.147R + 0.289G + 0.436B \\ V = 0.615R - 0.515G + 0.100B \end{cases} \quad (1)$$

The Y component data are extracted from the original image and divided into many 8*8 subblocks. Each subblock does not overlap each other and performs DCT transform respectively.

Follows the DCT transform and inverse DCT transform:

$$X_{uv} = \alpha_u \alpha_v \sum_{i=0}^{N-1} \sum_{k=0}^{N-1} x_{ik} \cos\left[\frac{(2i+1)u\pi}{2N}\right] \cos\left[\frac{(2k+1)v\pi}{2N}\right] \quad (2)$$

$$X_{ik} = \sum_{u=0}^{N-1} \sum_{v=0}^{N-1} \alpha_u \alpha_v X_{uv} \cos\left[\frac{(2i+1)u\pi}{2N}\right] \cos\left[\frac{(2k+1)v\pi}{2N}\right] \quad (3)$$

The watermark is embedded by modifying two intermediate frequency coefficients, which is named D1 and D2.

Following the reference of the two intermediate frequency coefficient:

$DCT\_block(4,3), DCT\_block(5,2)$.

The method for modifying the middle frequency coefficients is amended as follows.

$$\begin{aligned} &Water(m,n) = 0,\ \text{D1=N, D2=-N;} \\ &Water(m,n) = 1,\ \text{D1=-N, D2=N;} \end{aligned} \quad (4)$$

The value of N is greater, the robustness of watermark image is better, but more distortion will come. According to the actual situation to set the value of N, maybe 10 is the right value of N.

Y component will be gained by performing inverse DCT transform for each sub watermarked block. Using Y component, U components and V components, converted them to RGB space, the invisible watermarked image has gained.

The relationship of YUV color space to RGB color space as follows:

$$\begin{cases} R = Y + 1.140V \\ G = Y - 0.395U - 0.581V \\ B = Y + 2.032U \end{cases} \quad (5)$$

4 WATERMARK EXTRACTING AND RECOVERING

The algorithm of watermark extracting is shown in Figure 2, explained briefly as follows:

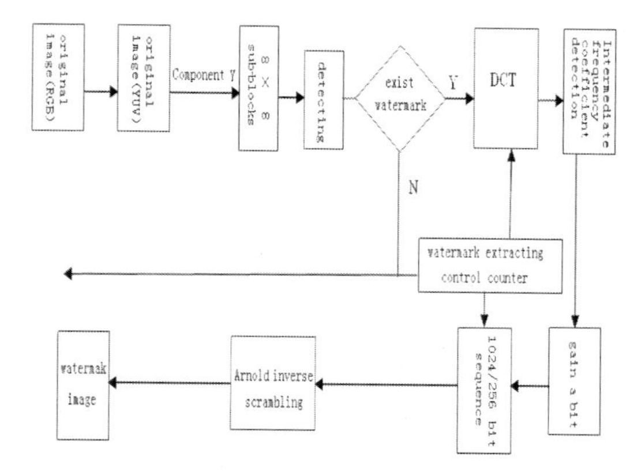

Figure 2. The scheme of watermark extracting.

The application system uses the .NET framework of GDI+ technology to obtain the Y component of the original image with watermark, divides the Y component into many 8*8 subblocks, and performs DCT transform for each subblock.

Got the values of the D1 and D2 and compared them to the threshold of T, if the image has embedded watermark, got the bit of watermark image.

The method for getting the bit of watermark image is amended as follows.

D1>T, D2<-T: $Water(m,n)=0$;

D1<-T, D2>T: $Water(m,n)=1$;

According to the actual situation to set the value of T, maybe 4 is the right value of T.

The bit sequences, extracted from the D1 and D2, are scrambled inversely by Arnold transformation, the watermark, which has embedded in the original image, can be obtained.

The algorithm of watermark recovering is as shown in Figure 3.

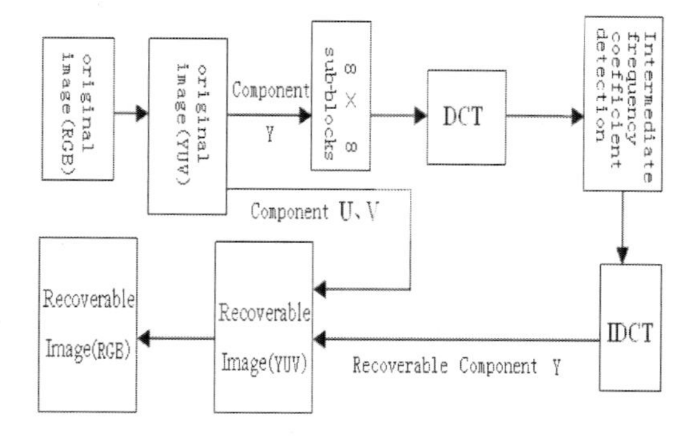

Figure 3. The algorithm of watermark recovering.

Got the values of the D1 and D2 and compared them to the threshold of T, if the image has embedded watermark, modify the values of the D1 and D2.

The method for modifying the values of the D1 and D2 is amended as follows.

D1>T, D2<-T:D1=D2 =0;
D1<-T, D2>T:D1=D2 =0;

5 RESULTS AND CONCLUSIONS

The first interface from the website of the application system is as shown in Figure 4.

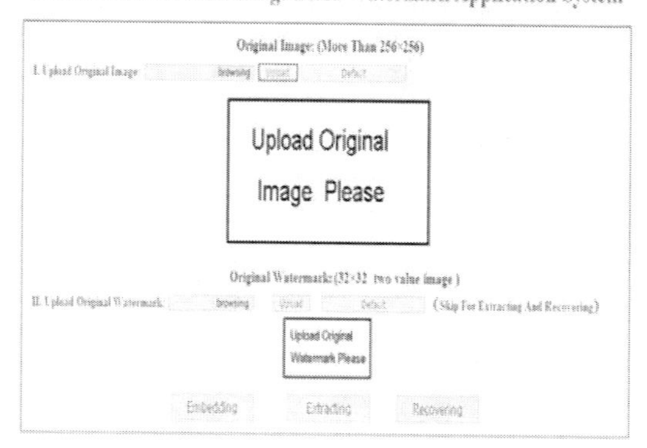

Figure 4. The first interface from the website.

When you have uploaded the original image and the watermark image to the server, you can click the Embedding button to add a watermark in the original image. You can also click the extracting button to extract the watermark and click the recovering button to recover the watermark from the watermarked image.

You can click the default button to get the default original image (Fig. 5 a) and the default watermark image (Fig. 5 c).

Some test data is as shown in Figure 5. The test data shows that the method of watermark embedding has little influence on the original image and the image of watermark extracting is visible clearly.

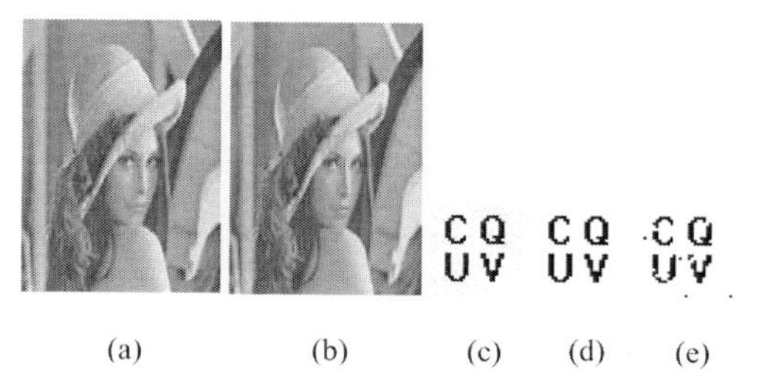

(a) (b) (c) (d) (e)

Figure 5. The test images.
A: original image b: watermarked image
C: original watermark d: watermark extracting

This method can also be applied to gray image. The original gray image as shown in Figure 1, the watermarked image is shown in figure 6, the watermark extracting is shown in Fig. 5e.

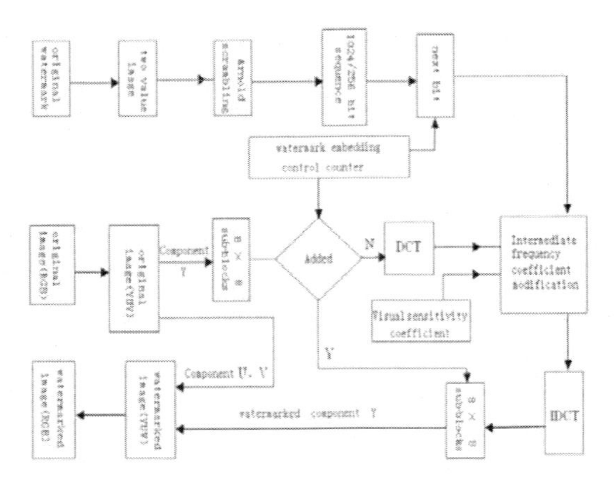

Figure 6. The gray watermarked image.

The experimental images from the attack of shearing are as shown in Figure 7.

(a) watermarked images

CQ UV CQ UV CQ UV CQ UV CQ UV

(b) The images of watermark extracting

Figure 7. The experimental of the attack of shearing.

Through the experimental site and test data, it is not difficult to get the conclusions as follows:

The recoverable color image blind watermark scheme, based on the internet, has good imperceptibility. The effect on the color image is small and the watermark extracting is clearly visible. It is very suitable for implicit identification of the copyright for digital image through the Internet and the practicability of the scheme is very strong.

REFERENCES

[1] W. Wang, W.H. Li, Y.K. Liu, Z. Borut, (2014) A SVD feature based watermarking algorithm for gray-level image watermark, *Journal of Computers*, 9:1497-1502.

[2] Wei Wang, Chengxi Wang, (2008) A watermarking algorighm for gray-level watermark based on local feature region and SVD, *International Congress on Image and Signal Processing*, pp:650–654.

[3] R B Wolfgang, C I Podilchuk, E J Delp. (2007) Perceptual watermark for digital image and video. *Pro. IEEE, 2007*, 87(1):1108–1126.

[4] M. David, S. R. Jordi, and F. Mehdi, (2010) Efficient self-synchronised blind audio watermarking system based on time domain and FFT amplitude modification. *Signal Processing*, 90:3078–3092.

[5] W. Liu and C. H. Zhao, (2009) Digital watermarking for volume data based on 3D-DWT and 3D-DCT, *The 2nd International Conference on Interaction Sciences: Information Technology, Culture and Human*, pp:352–357.

[6] B. Deepayan and A. Charith, (2010) Video watermarking using motion compensated 2D+t+2D filtering, *The 12th ACM Workshop on Multimedia and Security*, pp:127–136.

[7] J. Holland, (1975) *Adaptation in Natural and Artificial Systems*. Ann Arbor, MI: University of Michigan Press.

[8] P. Kumsawat, K. Attakitmongcol, and A. Srikaew, (2005) A new approach for optimization in image watermarking by using genetic algorithms, *IEEE Transactions on Signal Processing*, 53:4707–4719.

[9] Y. T. Wu and F. Y. Shih, (2006) Genetic algorithm based methodology for breaking the steganalytic systems, *IEEE Transactions on Systems, Man, and Cybernetics*, 36:24–31.

[10] H. C. Huang, J. S. Pan, Y. H. Huang, F. H. Wang, and K.C. Huang, (2007) Progressive watermarking techniques using genetic algorithms, *Circuits Systems Signal Processing*, 26:671–687.

[11] S. C. Chu, H. C. Huang, Y. Shi, S. Y. Wu, and C. S. Shieh, (2008) Genetic watermarking for zero tree-based applications, *Circuits Systems Signal Process*, 27:171–182.

[12] H Wang, B Wang. (2011) Digital watermarking algorithm based on the complexity of the image texture. *Computer Engineering*, 37(17):102–104.

[13] B Hou, Y Hu, L Jiao. (2010) Improvement of SAR image by Shearlet water edge detection. *Journal of China Image*, 15(10):1549–1554.

[14] T.Y. Wang, H.W. Li, (2013) A novel robust color image digital watermarking algorithm based on discrete cosine transform, *Journal of Computers*, 8:2507–2511.

Electric, Electronic and Control Engineering – Shao, Shu & Tian (Eds)
© 2015 Taylor & Francis Group, London, ISBN: 978-1-138-02842-5

A new high-frequency transmission line ice-melting technique

Xiaogang Gao*, Yantao Peng, Kexue Liu, Xiaoying Xie & Changyuan Li
Qinhuangdao Power Supply Company, State Grid Jibei Electric Power Company, Qinhuangdao Hebei, China

ABSTRACT: This paper explores a new continuous type of high frequency excitation ice-melting method and composition principle of its device, and it has great application value to improve the safety and reliability of power transmission system. Through investigation and study, it gets the parameters of iced conductor, builds the simulation model of high frequency excitation ice-melting system, uses ANSYS software to simulate by computer and demonstrates that 18kV, 40kHz is the best incentive ice-melting power source at 0.1 icing loss angle tangent. Finally, Mobile high frequency excitation ice-melting device applying to 500 V transmission line is designed in this paper. The relative parameters of phase-shifting transformer and power unit are calculated. Result of the simulation in cascaded high frequency and high voltage excitation ice-melting system verifies that using cascaded power unit can improve the harmonic and output voltage and adopting multiple-phase shifting principle can achieve the desired voltage and frequency in the pressurization range of IGBT.

KEYWORDS: transmission line; high frequency excitation; skin effect; ice-melting device; cascade.

1 INTRODUCTION

Accidents arising out of transformer line icing greatly limits the safe operation of the power system. These accidents have continued to take place since the first transmission line icing in the United States in 1932 [1]. So far, commonly accepted DC or AC ice-melting techniques for transmission lines require power interruption to the transmission line. Mechanical DE-icing methods, while simple, are operationally difficult.

As such, exploring new ice-melting techniques has become an important topic for electric system research. In 2001, Charles ET AL proposed a new high-frequency excitation ice-melting technique. As high-frequency ice is a lossy dielectric, using the media heat from ice itself together with the resistance loss-induced heat resulted from the skin effect of the conductor surface current can help melt the ice on transmission lines [2-5]. On the basis of this theory, we propose a new continuous-type high-frequency excitation ice-melting technique with the composition principle of its device, which supplies the gap of conventional ice-melting methods, utilizes the line high-frequency skin effect and electric loss principle to achieves high heating power with small ice-melting current and allows online ice melting. This offers an effective solution to transmission line icing and will have significant application value for improving the safety and reliability levels of the transmission system [6,7].

2 DESIGN OF A HIGH-FREQUENCY ICE-MELTING DEVICE

2.1 *Design scheme*

When developing a movable high-frequency ice-melting device for 500 V transmission lines, we determined the following design principles:(1) perform high-frequency excitation ice-melting to the 500 V line with AC/DC inverter circuit;(2) to ensure that the output voltage of the device is adjustable and the device itself is movable, we cascaded the sectionalized transformer with the diode rectifier to allow cascade regulation and ensures that each transformer corresponds to a tap;(3) to minimize maintenance and improve reliability, a diode uncontrolled rectifier is used for the device;(4) to provide sufficient capacity to meet the ice-melting demand of the 500 V line as well as reduce the insulation level of the device, we selected a 35 V main transformer as the power supply for excitation ice-melting;(5) to ensure fit the capacity needed by the ice melting of the 50 km long line to the device, we used a step-down transformer to reduce voltage and increase current; and(6) to minimize reactive power loss and harmonic content of the device during operation, we decided to use 12-pulse full-wave rectification.

On grounds of these design principles, we eventually decided to use the 35 V main transformer secondary in the 500 V substation as the power supply

*Corresponding author: *gaofei7072483@163.com*

for high-frequency ice melting [14]. Fig.5 shows the main circuit structure of the ice-melting device. Main merits of this device include: (1) each individual power unit is able to carry the power averagely, which greatly reduces the rated power; individual appliances are used to omit problems in connection with parallel connection; (2) high pressure is equally shared by the IGBT in each power unit so that high-voltage output is achieved simply by using a low voltage-withstand switch tube. This is not only technically mature, the drive circuit is also more reliable; (3) by using carrier phase shifting, a cascaded device can make its equivalent switch frequency equal to N times of the switching frequency of the IGBT in each power unit(N is the number of the single-phase cascaded power units). In other words, this type of device does not enlarge the outputted current and voltage bandwidths by direct means and does not need to increase the switching frequency of the IGBT in each power unit. The appliances cost is therefore low and the switching loss is modest, making it easier to achieve high-frequency voltage and current from the device; (4) error tolerance technique is used so that any fault in a power unit will cause the control unit to bypass the faulted unit automatically and keep the device unaffected by the fault; and (5) each unit provides for modular design, allowing uniform design, production and maintenance.

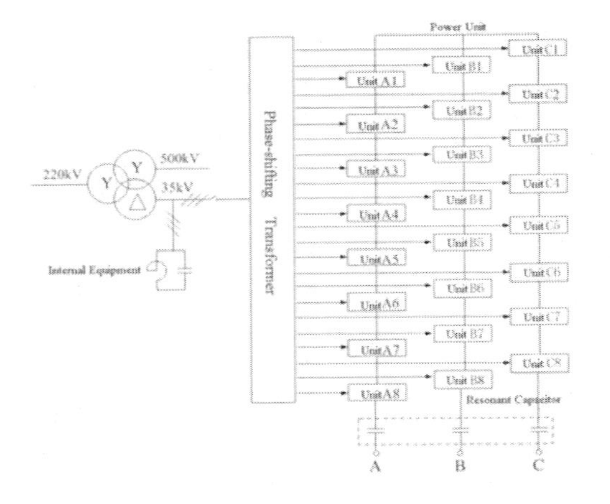

Figure 1. Main circuit structure of the high-frequency ice-melting device.

As discussed above, this type of structure can serve as the main circuit structure of the high-frequency excitation ice-melting device. It is also easily controllable and convenient for fabrication. Its operation is also reliable. The device is composed of a phase-shifting transformer and a power unit part.

2.2 *Phase-shifting transformer*

An extended delta cascaded phase-shifting transformer is used. Extended delta method is used to shift the phase between the secondary-side windings to achieve multiple rectification. When the phase-shifting transformer is in an Ydll connection, by adjusting the secondary-side extended delta, it is possible to locate the phase-shifting range to any angle at which the primary-side voltage lags behind the secondary-side voltage by 0°~60°.

In our study, the line input voltage is 35kV+10%, the frequency is 50 Hz, and the ice-melting excitation source output power is 2.52 MW.

1 Transformer capacity calculation

$$S_{TN} = P_{CN}(\cos\phi_{CN} \times \eta)(kVA) \tag{1}$$

Here: S_{TN} is the capacity of the transformer; P_{CN} is the output power of the excitation source; $\cos\phi_{CN}$ is the input power factor of the excitation source; η is the efficiency of the excitation source. The power factor of the excitation source is 0.8~0.85 in the presence of an input AC reactor and 0.6~0.8 in the absence of an input AC reactor. So taking $\cos\phi_{CN} = 0.85$, $\eta = 0.95$, $P_{CN} = 2.52$MW, we have:

$$S_{TN} = P_{CN}(\cos\phi_{CN} \times \eta)$$

$$= 2.52/(0.85\times0.95)$$

$$= 3.12\text{MVA} \tag{2}$$

As for a general manufacturers' device, the capacity is normally empirically taken as around 130%, i.e.

$$S_{TN}' = 1.3\times S_{TN} = 4.05(MVA) \tag{3}$$

Thus the capacity of the phase-shifting transformer is

$$S_{TN}' = 4.05(MVA)\ .$$

2 Phase-shifting angle calculation

The calculation formula for the phase-shifting angle between adjacent windings is:

$$\Delta\theta = 60° / N \tag{4}$$

Here: $\Delta\theta$ is the phase-shifting angle between adjacent windings; N is the number of power units per phase series.

For a 35 V high-frequency excitation source, each phase has 8-unit cascade. The phase-shifting angle between adjacent windings of the transformer is:

$$\Delta\theta = 60° / N = 60° / 8 = 7.5° \tag{5}$$

Hence the phase-shifting angles of the input transformer are ±26.25°, ±18.75°, ±11.25° and ±3.75°.

2.3 *Power unit*

The power units are arranged in cascaded structure. The transformation part is composed of 24 power units. Fig.6 shows the structure of each of these power units. In the input part of the circuit, three-phase diode uncontrolled rectification is used. In the DC circuit at the center, large capacity is used to smooth surges. The entire output part of the power unit presents the characteristics of the voltage source. In the output circuit part of the power unit, conventional H-bridge inverter circuit is used. Each bridge is composed of an inverse series power diode and an upper and a lower IGBT [15,16].

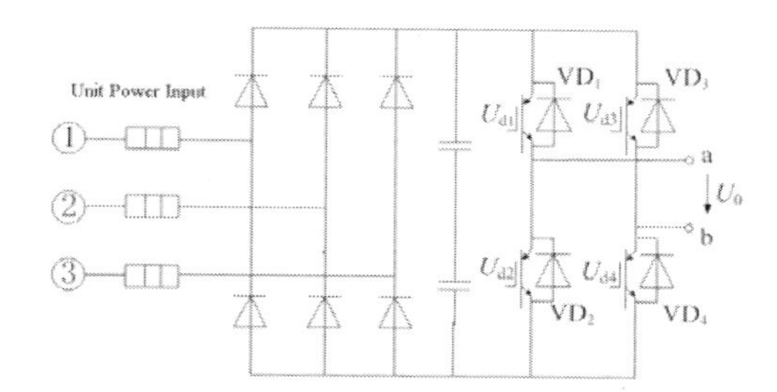

Figure 2. Power unit structure.

Besides, as the device has to increase the amplitude of the output phase voltage, it can connect more than one power units in series, thereby increasing the single-phase voltage output. Fig.7 shows the structure of the series circuit.

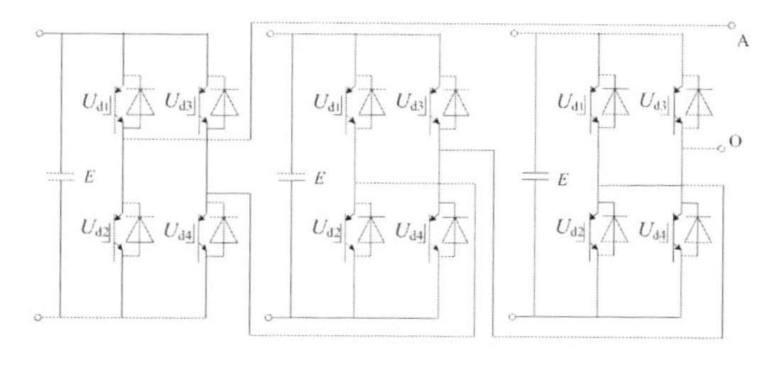

Figure 3. Single-phase series structure.

If a number of power units are outputted in series, the voltage rating of each phase output will be greatly increased. For example, if N power units are connected in series, there are 2N+1 voltage ratings outputted from each phase, which are respectively +NE, +(N-1)E, ..., -(N-1)E, -NE. And there will be 4N+1 voltage ratings outputted from the line voltage of the device. This will increase the outputted voltage ratings and the outputted current and voltage waveforms will be closer to sine wave, which reduces the harmonic content. When N power units are outputted in series, the current flowing through each power unit is phase current whereas the actual voltage accounts for only 1/N outputted phase voltage. Hence the output power from each power unit contributes only 1/(3N) to the power of the entire high-frequency power source.

In our study, the power units used are two-level structure inputted by 35 V power supply. Each phase consists of eight power units. The secondary output line voltage U of the phase-shifting transformer is 18 V.

$$U_{DC} = \frac{3\sqrt{2}}{\pi}U = 1.35 \times 18000 = 24.3kV \tag{6}$$

When a filter capacitor is added, the U_{DC} value is the maximum AC line peak voltage, i.e.

$$U_{DCP} = \sqrt{2}U = 1.414 \times 18000 = 25.5kV \tag{7}$$

The current flowing through each power unit is the output line current of the device, i.e.

$$I = \frac{P}{\sqrt{3}U\cos\phi} \tag{8}$$

The design output of the high-frequency excitation source is P=2.52MW. The design line voltage is $\sqrt{3}U$=35kV. The design power factor cosϕ is 0.95, i.e.

$$I = \frac{P}{\sqrt{3}U\cos\phi} = \frac{2520000}{35000 \times 0.95} = 75.8A \tag{9}$$

When the inputted three-phase current is rectified, plus the work of the filter capacitor, DC bus voltage is formed. By inputting the 18 kV rated voltage into the power unit, we get 1.35~1.44-fold, approximately 24.2 V DC bus voltage. The power unit single inverter forms single-phase inverter circuit consisting of eight IGBTs in series connection. Controlled by SPWM, the inverted AC current outputs 0~40kHz, 0~18kV single-phase alternating current from the output end of the power unit.

3 SYSTEM SIMULATION OF AN EXCITATION ICE-MELTING DEVICE

System simulation was performed on the Simulink platform of MATLAB7.9 on the high-frequency excitation ice-melting system cascaded from H-bridge power units.

3.1 *System simulation model*

In this section, system simulation is performed after detailed design of the excitation ice-melting device. If each phase consists of eight power units in series connection, for example, the three phases consist of 24 power units. Encapsulate an H-bridge power unit into a subsystem and reproduce it into eight subsystems: Subsystem, Subsystem 1, ... Subsystem 7, which represent the eight units of A phase of the three phases. Connect the outputs of the eight units in connection as shown in Fig.8.

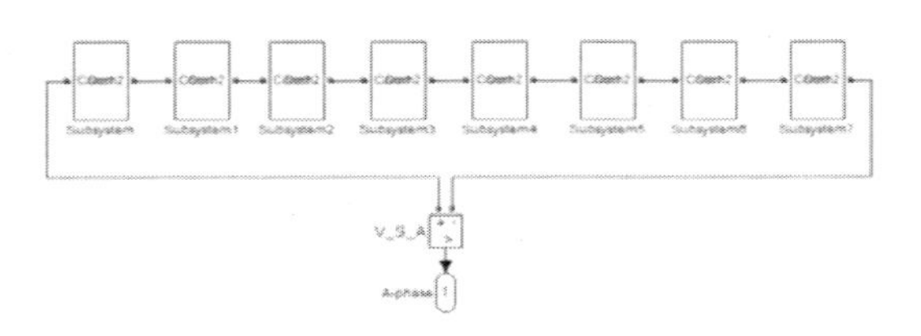

Figure 4. Simulated A phase structure.

In our simulation system, the main transformer secondary-side 35 V of the 500 V substation is used as the ice-melting source. The inputted 35 V AC current is rectified into pulsating DC. The intermediate current filtered by the capacitor is inputted into the inverter unit. By controlling the connection of the IGBT, PWM transformation and output is achieved. Fig.9 shows the structure an individual power unit in the device [17]. The power unit cascaded high-frequency excitation ice-melting system incorporates multi-level phase-shifting carrier technology. Its carrier signals are each phase-shifted by 120/N=120/8=15°. The eight power units of each phase in series connection use the same sine wave.

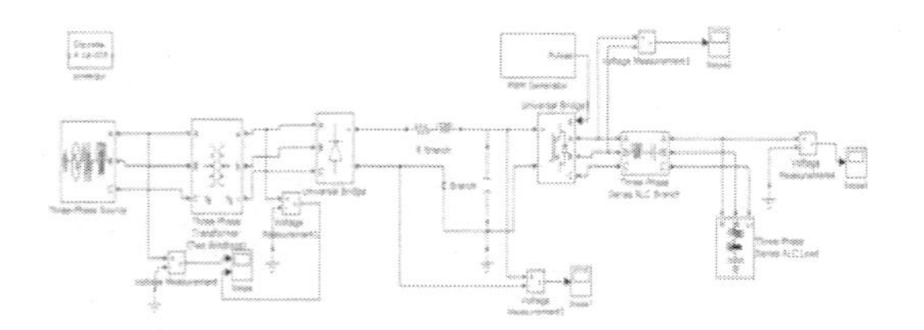

Figure 5. Simulated power unit structure.

3.2 *Analysis of the simulation result*

In the power unit, the measured voltage of the rectified three-phase power through Scope1 is approximately 3050 V. The voltage waveform is shown in Fig.10. After IGBT inversion, the measured voltage through Scope2 is approximately 2800 V. The frequency is 40 kHz. The waveform is shown in Fig.11. As the frequency after inversion is as high as 40 kHz, the voltage wave period is only 2.5×10-5s. The sampling is quite dense.

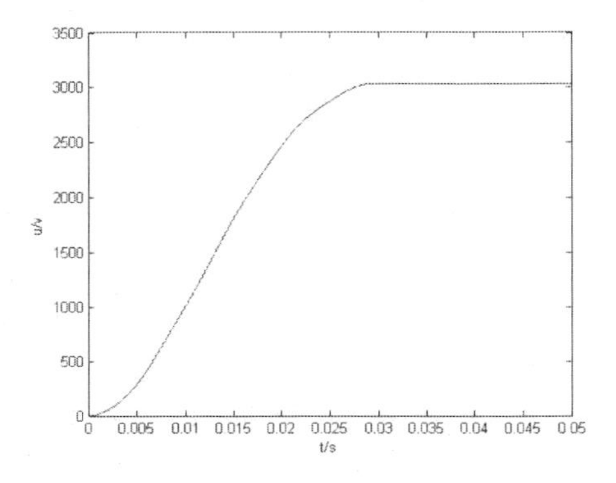

Figure 6. Rectifier voltage waveform of the power unit.

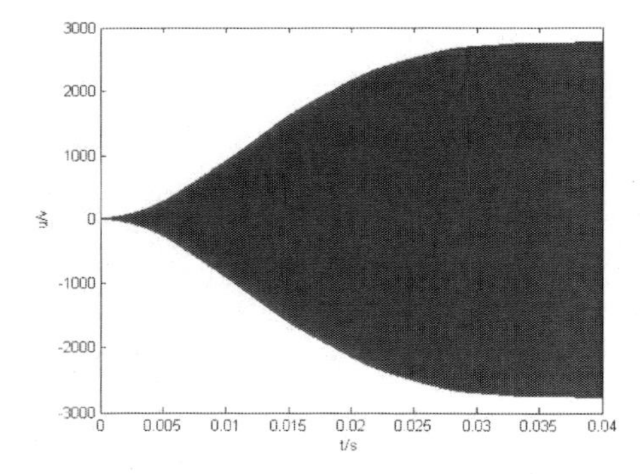

Figure 7. Inverter voltage waveform of the power unit.

The A-phase voltage output cascaded from eight H-bridge power units is eight times that of each of the power unit. Encapsulate the A-phase circuit system consisting of H-bridge power units in series connection shown in Fig.9 into a subsystem, i.e. A-phase module. Generate B-phase and C-phase modules in the same way, and set the PWM output waveform phase difference of each phase at 120°. Fig.12 shows the structure as a cascaded high-requency excitation ice-melting system. By operating this system, we can finally obtain the 18 V, 40 kHz high-frequency excitation needed to melt ice. Fig.13 shows the three-phase voltage output waveform. Fig.14 shows the three-phase voltage output waveform at 0.028~0.03s.

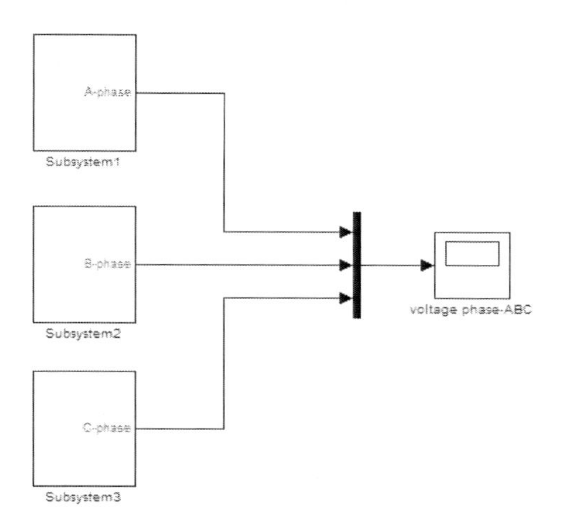

Figure 8. Cascaded high-frequency excitation ice-melting system.

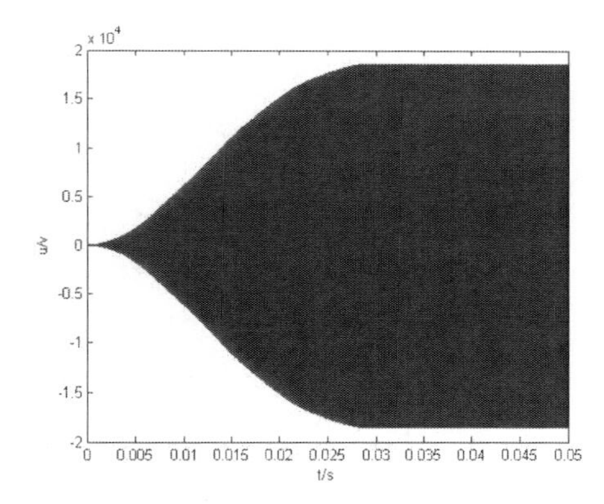

Figure 9. Three-phase power waveform of the power unit series.

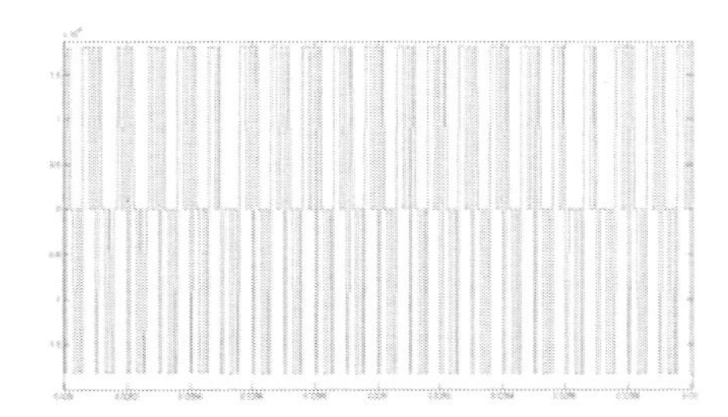

Figure 10. Three-phase power stability waveform of the power unit series.

As shown in Fig.15, by performing FFT analysis of this system, we can see the resulting excitation frequency is 40 kHz.

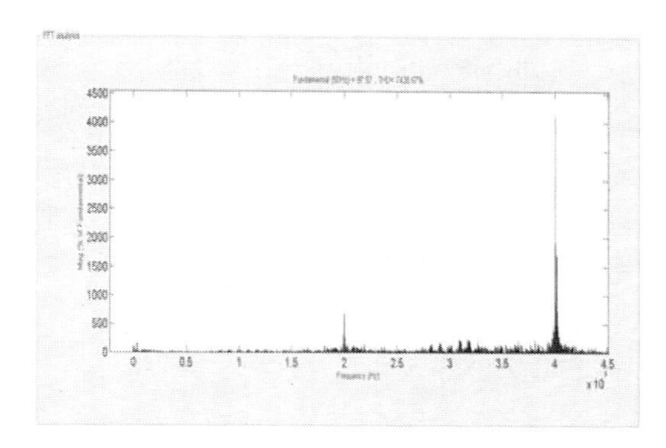

Figure 11. FFT analysis of the power unit series.

4 CONCLUSIONS

1. Parameters needed for ANSYS simulation, including transmission line and environment and weather parameters, material properties, structural parameters and convective heat transfer coefficients, are collected. It is determined that, when the ice loss angle tangent is $\tan\delta = 0.1$, the optimum ice-melting excitation is 18 V, 40 kHz. ANSYS finite analysis and simulation is performed. The result indicates that, after excitation is applied for 80min, the ice is completely melted, thus validating the feasibility of our method.
2. The design scheme for a movable high-frequency excitation ice-melting device for 500 V lines is worked out. The phase-shifting transformer and power unit part of the device are examined.
3. By performing Simulink simulation to the cascaded high-frequency, high-voltage ice-melting system, it is concluded that using H-bridge power unit cascading can increase the output voltage and greatly improve the overlapped harmonic. Using phase-shifting multiplication technology, it is possible to design an ideal frequency modulator within the voltage withstand limit of IGBT and obtain ideal voltage waveform and frequency. When simulating the H-bridge power unit, it is very important to select the right values for the DC-side filter capacitor L1 and capacity C1. By adjusting the L1 and C1 values, it is possible to adjust the stability time and overshoot of DC voltage as well as the DC voltage fluctuation. Adding capacitance to the inverter load side can make the output waveform stabilize within a short time. This demonstrates that capacitance plays a significant role in rectifying circuit waveform.

The 18 V, 40 kH excitation ice-melting technique proposed here takes 20min more time to melt ice than the 33 V, 100 kHz high-frequency excitation ice-melting method proposed by authors outside China, but needs substantially lower frequency for the excitation source, especially that for ice melting. This reduces the design difficulty of a high-frequency ice-melting device, the cost of the device and environmental impact of high frequencies. As such, our 18 V, 40 kH high-frequency excitation ice-melting technique is very advantageous. Its simulation result is satisfactory. The design and research of this device also has great potential.

REFERENCES

[1] Huang Xinbo, Liu Jiabing, Cai Wei, et al. (2008) Present research situation of icing and snowing of overhead transmission lines in China and foreign countries. *Power System Technology*, 32(4): 23–28.

[2] Shan Xia, Shu Naiqiu. (2006) Discussion on methods of deicing for overhead transmission lines. *High Voltage Engineering*, 32(4): 25–27.

[3] Paul C R. (1992) *Introduction to Electromagnetic Compatibility*. John Wiley & Sons Inc.

[4] Kaiser K L. (2004) *Electromagnetic Compatibility Handbook*. Boca Raton: CRC Press.

[5] McCurdy J D, Sullivan C R, Petrenko V F. Sullivan and Victor F. (2001) Using dielectric losses to DE-ice power transmission lines with 100 kHz high-voltage excitation. *In Conf. Rec. IEEE Industry Applications Society Annu. Meeting*, 2001, pp:2515–2519.

[6] Li Chengrong, Lü Yuzhen, Cui Xiang, et al. (2008) Research issues for safe operation of power grid in China under ice-snow disasters. *Power System Technology*, 32(4): 14–22.

[7] Li Zaihua, Bai Xiaomin, Zhou Ziguan, et al. (2008) Prevention and treatment methods of ice coating in power networks and its recent study. *Power System Technology*, 32(4):8–13.

[8] Jiao Chongqing,Qi Lei, et al. (2010) Powerful line DE-Icing using medium frequency power source. *Transactions of China Electrotechnical Society*, 25(7):159–162.

[9] Liu Zhenya. (2005) *Ultra-high Voltage Grid*. Beijing: China Economic Publishing House.

[10] Matthias C Perz. (1968) Analytic determination of high-frequency propagation on ice-covered power lines. *IEEE Transactions on Power Apparatus and Systems*, 87(4):695–703.

[11] Petrenko V F, Whitworth R W. (1999) *Physics of Ice*. London: Oxford Univ. Press.

[12] Zhou Yusheng, Chen Peiyao, Gai Xiaogang, et al. (2011) Research on transmission line DE-icing method based on high-frequency and high-voltage excitation. *Insulators and Surge Arresters*, (6):1–4.

[13] Li Ning, Zhou Yusheng. (2010) Analysis on transmission line icing elimination method by high frequency source. *High Voltage Apparatus*, 46(12):18–21.

[14] Wu Jiang, Li Weiguo, Ma Jixian, et al. (2009) Simulation research on transient over-voltage at 500 kV substation secondary equipment during transient over-voltage lightning strike. *Insulators and Surge Arresters*, (6):22–25

[15] Mao Chengxiong, Li Weibo, Lu Jiming, et al. (2003) Study of the common-mode voltage in a high-voltage Asd's System. *Proceedings of The CSEE*, 2003, 23(9): 57–62.

[16] Lu Jiming, Li Weibo,Mao Chengxiong, et al. (2002) Study of the filter at the motor terminals in a high voltage inverter drive system. *High Voltage Engineering*, 28(11): 13–16

[17] Li Weibo. (2007) *Application of MATLAB in Electric Engineering*. Beijing: China Electric Power Press.

Electric, Electronic and Control Engineering – Shao, Shu & Tian (Eds)
 ISBN: 978-1-138-02842-5

A consistency evaluation and maintenance method of electric vehicle Lithium-ion+ battery based on resistance

You Xu & Yong Yang
School of Automotive Engineering, Guangdong Polytechnic Normal University, Guangzhou, China

ABSTRACT: The efficiency of an electric vehicle may be affected by the consistency of electric vehicle Lithium-ion+ Battery. In order to prolong the life of the battery and ensure the mileage, the consistency of battery must be detected. A consistency evaluation method based on the standard deviation of battery resistance was proposed. According to the change of battery resistance, the consistency of battery can be divided into four levels, which correspond to four kinds of maintenance strategies. The electric vehicle Lithium-ion+ battery resistance charging/discharging tests were used in the experiments, and the results showed that the four consistency curves provided a reference for the application and maintenance of batteries.

KEYWORDS: resistance of electric vehicle Lithium-ion+ battery; consistency; standard deviation; normal distribution.

1 INTRODUCTION

In order to reduce the dependence of oil and the vehicle exhaust, many countries have paid more and more attention to the development of electric vehicles. However, the capacity and the power of battery are lower than the fuel. It caused that the driving distance of electric vehicle is limited by the battery [1]-[6]. The inconsistency of battery is one of the most important problems about the short-driving distance of electric vehicle [7]-[11]. The problem will become worse together with the growth of using time and the increase of aging degree. It affects the battery characteristic and the operating efficiency of electric vehicle. The battery consistency detectors and the battery maintenance are important in the City-Business-Scale of electric vehicle. Without battery consistency detection and battery maintenance, the battery may damage and the economic performance of electric vehicle operation may be affected. In order to define the battery consistency, Reference [12] proposed a method about the battery discharging capacity consistency and the battery discharging voltage consistency base on standard deviation. However, it didn't solve the problem about the evaluation of battery consistency. Reference [9] built a mathematical model about the relationship between battery consistency and the service life and defined the damage coefficient of battery capacity. The method can be used in the estimation of battery life, but it can't be used in real-time operation. Reference [13] evaluated the circulation characteristics of battery and determined the battery cycle life formula. However, it was also difficult to solve the problem of real-time detection and maintenance. Reference [14] evaluated the SOC of the battery and the battery life state by the Kalman filter method. The method can reflect the intrinsic characteristics of the battery. But it was difficult to accomplish in vehicle control unit (VCU). Reference [15] proposed a consistency evaluation method based on the standard deviation of voltage. Because the working voltage of battery may change with the current and the temperature, error appears in real-time operation. According to the different characteristics of battery in different aging levels, four maintenance strategies in different aging levels were introduced in this paper. And the Lithium-ion+ battery consistency based on the resistance was studied. A consistency evaluation method based on resistance standard deviation was proposed. The electric vehicle test was carried out to verify the effectiveness of the new method.

2 BATTERY MAINTENANCE STRATEGY IN DIFFERENT STAGES

In order to ensure that the electric vehicle works normally, it is necessary to maintain the battery regularly. According the degree of battery consistency, the methods of battery maintenance can be divided into 4 levels: (1) Charge and Discharge normally; (2) Battery Equalizing; (3) Battery Consistence Matching; (4) Battery-swapping and Consistence Matching. As shown as Fig.1, the consistency index σ can be used corresponding to the 4 levels of battery maintenance: If $\sigma \leq d_1$, it belongs to the first level of battery maintenance; If $d_1<\sigma\leq d_2$, it belongs to the second level of battery maintenance; The third level of battery maintenance may be carried out when $d_2<\sigma\leq d_3$; And the last level of battery maintenance may be used if $d_3<\sigma$.

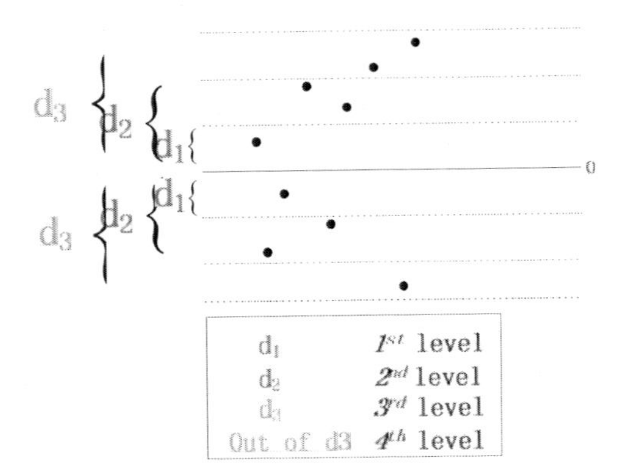

Figure 1. Levels of battery consistency.

3 BATTERY CONSISTENCY EVALUATION METHOD

The battery consistency includes the capacity consistency, the voltage consistency and the resistance consistency. Theoretically, the inconsistency of battery can be described accurately by the capacity consistency methods. However, the capacity consistency methods are complex and difficult to carry out. The voltage consistency methods can describe the battery consistency intuitively. But it may be affected by the temperature and charge/discharge current. The resistance consistency methods can express the battery characteristics in essence. Compared with the other two methods, the resistance consistency methods are more realistic and available.

3.1 *Battery consistency evaluation method based on resistance*

The battery resistance can be calculated as formula (1).

$$R_i(SOC,I,T)=\frac{U_{io}(SOC,I,T)-U_i(SOC,I,T)}{I} \quad (1)$$

Where $R_i(SOC,I,T)$ is the resistance of the ith battery, it is related to the percentage of battery charge state (SOC), the current I, and the temperature T. $U_i(SOC,I,T)$ is the working voltage in the percentage of battery charge state (SOC), the current I, and the temperature T. $U_{io}(SOC,I,T)$ is the EMF, which is related to the percentage of battery charge state (SOC), the current I, and the temperature T.

The battery consistency standard deviation σ based on resistance can be expressed by formula (2) and (3).

$$R_{av}=\frac{\sum_{i=1}^{n}R_i}{n} \quad (2)$$

$$\sigma=\sqrt{\frac{\sum_{i=1}^{n}(R_i-R_{av})^2}{n-1}} \quad (3)$$

Where n is the total number of batteries, R_i is the resistance of the ith battery. R_{av} means the average resistance of batteries.

In order to build the relationship between the d_{max}, which is the maximum value of the differences between single resistance and the average resistance, and the σ_e, which is the resistance standard deviation in different levels. The distribution of battery resistance can be considered to conform to normal distribution. Assuming that there are $\boldsymbol{n}$ batteries in a pack, the resistance standard deviation $\boldsymbol{\sigma_e}$ in different levels can be calculated as the following steps:

1 The normal distribution function can be expressed by formula (4):

$$f(x)=\frac{1}{\sqrt{2\pi}}e^{(-\frac{x^2}{2})} \quad (4)$$

2 d_{max} can be considered to conform to "3σ-*paradigm*". It means that d_{max} appears when $x=\pm3$ in formula (4). d_i, which is the difference between the $\boldsymbol{i}^{th}$ battery resistance and the average resistance, is distributed evenly in [-3,3] as shown as formula (5).

$$x_i=-3+(i-1)\times\frac{6}{n-1}$$
$$d_i=\frac{(\frac{1}{\sqrt{2\pi}}e^{(-\frac{x_i^2}{2})}-\frac{1}{\sqrt{2\pi}})}{(\frac{1}{\sqrt{2\pi}}e^{(-\frac{9}{2})}-\frac{1}{\sqrt{2\pi}})}\times d_{\max} \quad (5)$$

3 According to formula (5), the resistance standard deviation $\boldsymbol{\sigma_e}$ in different levels can be expressed by formula (6).

$$\sigma_e=\sqrt{\frac{\sum_{i=1}^{n}d_j^2}{n-1}}(j=1....n) \quad (6)$$

4 In order to evaluate the battery consistency, the battery consistency standard deviation $\boldsymbol{\sigma}$ can be

calculated according to formula (2) and formula (3). σ can be compared with the resistance standard deviation σ_e in different levels so that the level of battery consistency can be determined and the corresponding maintenance method can be used.

Considering that the battery resistance may change with the percentage of battery charge state (SOC), the current *I*, and the temperature *T*. Especially, the battery resistance changes greatly at the end of the charging and discharging process. Therefore, the battery consistency evaluation of electric vehicle must be combined with the whole process of charging and discharging.

3.2 *Battery resistance consistency detection in different current*

Due to the effect of the temperature and current, the battery resistance consistency rule must be tested in different temperature and different current as the following steps:

1. Place the battery pack in the position that the ambient temperature is *T*. Keep no-charge/discharge state until the temperature of battery pole is close to the ambient temperature.
2. Charge the battery in the current *I* and stop when the maximum voltage of the single battery is close to the maximum charging voltage. Then Keep no-charge/discharge state until the temperature of battery pole is close to the ambient temperature.
3. Discharge the battery in the current *I* and continue for a short time of t_d. Then Keep no-charge/discharge state until the temperature of battery pole is close to the ambient temperature. Calculate the resistance R_i of each battery in the pack by formula (1).
4. If the minimum voltage is close to the minimum discharging voltage, go to step (5), or else return to step (3).
5. Keep no-charge/discharge state until the temperature of battery pole is close to the ambient temperature.
6. Charge the battery in the current *I* and continue for a short time of tc. Then Keep no-charge/discharge state until the temperature of battery pole is close to the ambient temperature. Calculate the resistance R_i of each battery in the pack by formula (1).
7. If the maximum voltage is close to the maximum charging voltage, go to step (8), or else return to step (6).
8. Keep no-charge/discharge state until the temperature of battery pole is close to the ambient temperature.
9. Draw the battery consistency curve in charging and discharging, which is related to the SOC and current.
10. series of 110Ah Lithium-ion+ battery were used in the test. And the ambient temperature was 20°C. The current was set to 0.3C, 0.4C, 0.7C and 1C. And the battery resistance standard deviation curve in charging and discharging was as shown as Fig.2.

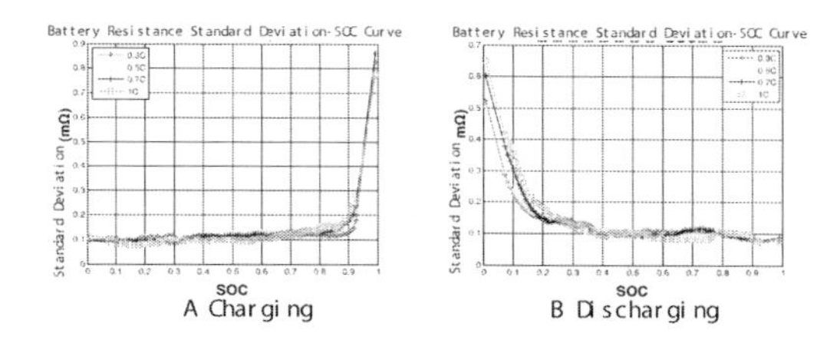

Figure 2. Battery resistance standard deviation curve in charging and discharging.

Fig. 2 showed that the resistance standard deviation may be considered as unaffected by the current when the value of SOC was between 20% and 90%. Therefore, a certain current can be used in detecting the resistance standard deviation. Generally, the current may be set to 0.5C to ensure that the battery is not damaged and the detection efficiency is reasonable.

3.3 *Evaluation method for different level of battery consistency*

Generally, the consistency of the new battery is better than the used one, so that the consistency evaluation curve of new battery can be used as the benchmark. Therefore, the evaluation method for different level of battery consistency can be shown as the formula (7) and formula (8).

$$g_{dn}(\lambda(1-x)) = r_{dn}(1-x) + f_{d1}(1-x) \tag{7}$$

$$g_{cn}(\lambda x) = r_{cn}(x) + f_{c1}(x) \tag{8}$$

Where x is the percentage of battery charge state (SOC); $g_{dn}(x)$ is the function of d_{max} in the nth ($n>1;n\leq4$) level of discharge cycle; $f_{dl}(x)$ is the function of d_{max} in the first level of discharge cycle; $g_{cn}(x)$ is the function of d_{max} in the nth ($n>1;n\leq4$) level of charge cycle; $f_{cl}(x)$ is the function of d_{max} in the first level of charge cycle; λ is the battery attenuation coefficient which describes the degradation of battery performance; $r_{dn}(x)$ is the function that expresses the difference between d_{max} in the nth ($n>1;n\leq4$) level of discharge cycle and the one in the first level; $r_{cn}(x)$ is the function that expresses the difference between d_{max} in the nth ($n>1;n\leq4$) level of charge cycle and the one in the first level.

In order to solve λ, $r_{dn}(x)$ and $r_{cn}(x)$, battery charge-discharge cycle test is necessary. The 110AH battery was used in the test, and the change of battery capacity was shown as Fig.3. From the figure, the percentage of battery capacity reduced to 95% when the number of cycle times was about 440, and it reduced to 90% when the number of cycle times was about 840.

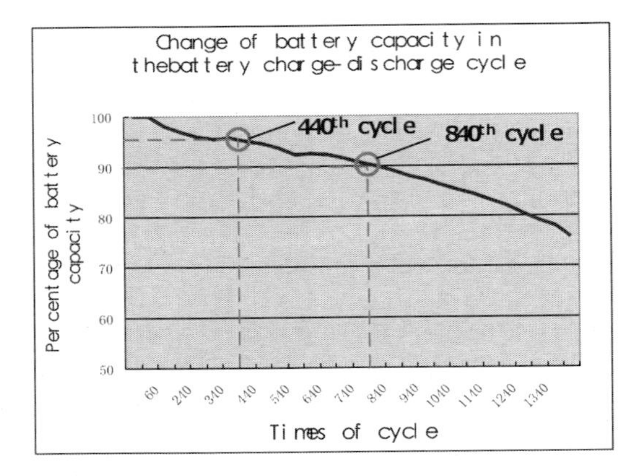

Figure 3. Capacity change curve in charge-discharge cycle.

Due to the processing techniques, the environment and the current, the capacity of the battery in electric vehicle may reduce faster than that in the experiment. Based on experience, battery equalizing is necessary when the percentage of battery capacity is close to 95%, and battery consistence matching is necessary when the percentage of battery capacity is close to 90%. When the percentage of battery capacity is less than 90%, it is necessary to consider the battery-swapping and consistence matching.

4 EXPERIMENT AND RESULT

In order to verify the effectiveness of the method, 10 series of 110Ah new Lithium-ion+ battery were used in the experiment. And the steps were as follows:

1. Carry out the charge-discharge cycle test. In the test, the current was 50A; the minimum voltage of the battery was 2.9V; the maximum voltage of the battery was 3.8V.
2. According to Chapter 2, draw the battery consistency evaluation curve in different level in charging and discharging.

The battery consistency evaluation curve in different level can be shown as Fig.4 and Fig.5 after the test.

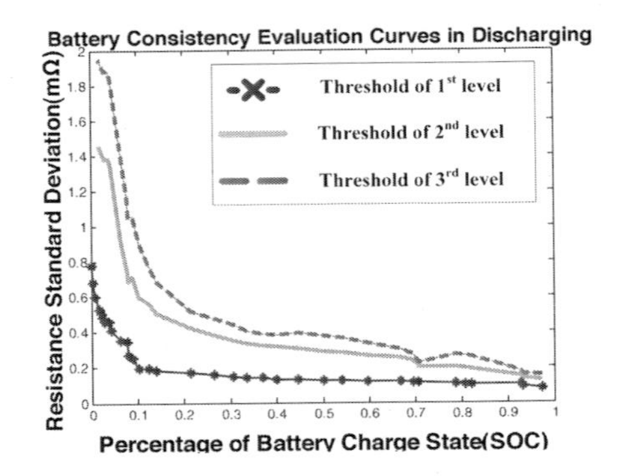

Figure 4. Battery consistency evaluation curves in discharging.

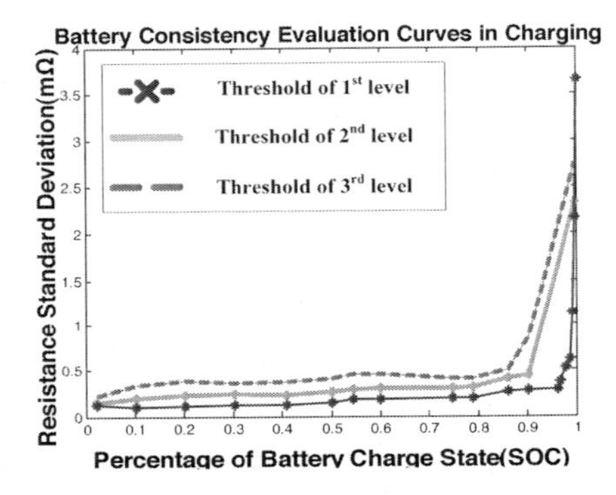

Figure 5. Battery consistency evaluation curves in charging.

According to Fig.4 and Fig.5, the battery consistency evaluation curves in charging/discharging can be used in regular tests and the evaluation method can be chosen according to the threshold of different level.

The lightweight electric vehicle FDG6601 (Fig.6), which was developed by the Institute of Dongguan-Sun Yat-Sen University, was used in the experiment.

Figure 6. FDG6601.

Comparative testing, which was about the consistency of the new battery in the lightweight electric vehicle and that used for over a month, was carried out. Fig.7 and Fig.8 showed the results of the battery consistency.

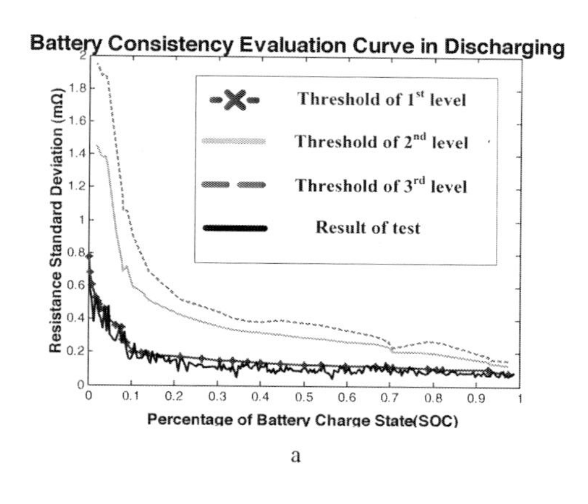

a

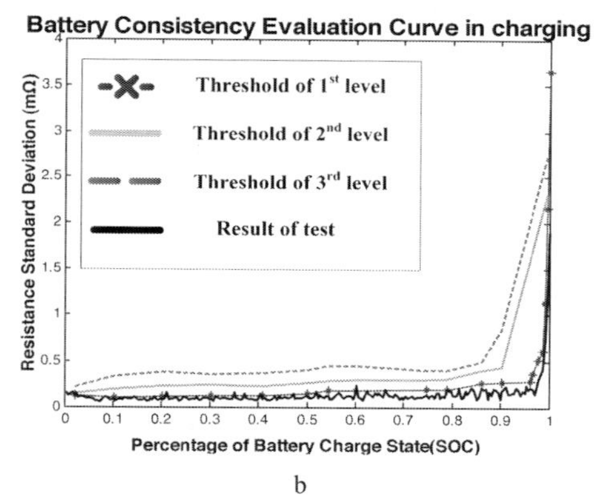

b

Figure 7. Consistency evaluation curves of new battery.

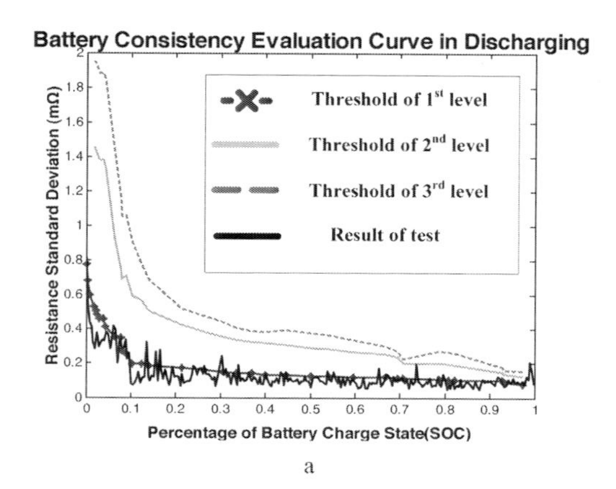

a

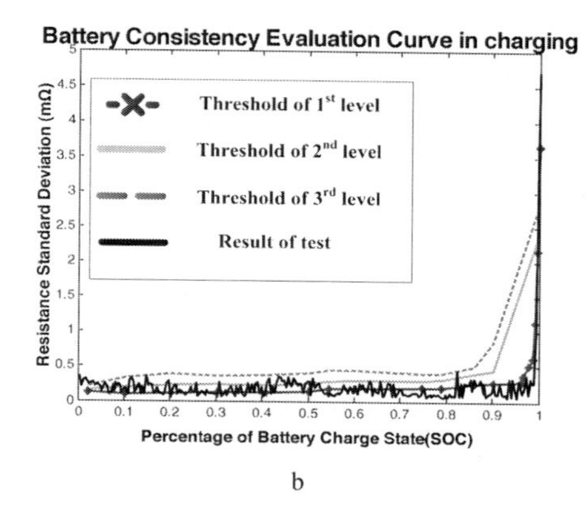

b

Figure 8. Consistency evaluation curves of battery which was used after one month.

From Figs.7 and 8, the consistency evaluation curves of battery used for over a month were above the threshold of 1st level and it meant that the battery equalizing maintenance method was necessary.

5 SUMMARY AND CONCLUSION

This paper discussed the relationship between the Lithium-ion^{+} battery consistency and the battery maintenance method. A consistency evaluation method based on resistance standard deviation was proposed. According to the methods of battery maintenance, the consistency evaluation curves were divided into 4 levels. Then the battery consistency evaluation curve in different level was obtained after the charging/discharging cycle test. In order to verify the effectiveness of the method, Lithium-ion^{+} battery in the electric vehicle were used in the experiment. Results showed that the method is simple and effective. It can be the reference to the battery maintenance of electric vehicle.

REFERENCES

[1] Xu Guanghua, Chen Qingtai. (2011) *Small Electric Vehicle and Automobile Industry Transformation.* [R]. Green Car of The Year China.

[2] Zha Hongshan. (2011) *Research on BEV's Powertrain Energy Management and Optimization.* Sun Yet-sen University.

[3] Tan Xiaojun. (2011) *Diandong Qiche Dongli Dianchi Guanli Xitong Sheji.* Sun Yet-sen University Press.

[4] Xiong Qi, Tang Donghan. (2003) Research Progress on Super capacitor in Hybrid Electric Vehicle. *Acta Scientiarum Naturalium Universitatis Sunyatseni*, 42:130–133.

[5] Zha Hongshan, Zong Zhijian, Liu Zhongtu, et al. (2010) Matching design and simulation of power train

parameters for electrical vehicle. *Acta Scientiarum Naturalium Universitatis Sunyatseni*, 49(5):47–51.
[6] Liu Zhongtu, Wu Qinglong, Zong Zhijian. (2011) Study on the energy consumption economy of electric vehicle based on test bench simulation. *Acta Scientiarum Naturalium Universitatis Sunyatseni*, 50(1):46–52.
[7] Wang Zhenpo, Sun Fengchun, Zhang Chengning. (2003) Study on Inconsistency of electric vehicle battery pack. *Chinese Journal of Power Sources*, 27(5):438–441.
[8] Ma Youliang, Chen Quanshi. (2001) Inconsistent influence analysis of battery for Hybrid Electric Vehicle. *Auto Electric Parts*, (2):5–7.
[9] Wang Zhenpo, Sun Fengchun, Lin Chen. (2006) An analysis on the influence of inconsistencies upon the service life of power battery packs. *Transactions of Beijing Institute of Technology*, 26(7): 577–580.
[10] CHAN Chung Chow, CHAN K.T. (2001) *Modern Electric Vehicle Technology*. Oxford: Oxford University, pp:35–47.
[11] Chen Quanshi, Lin Chengtao. (2005) Summarization of studies on performance models of batteries for electric vehicle. *Automobile Technology*. (3):1–5.
[12] QC/T-743-2006, Lithium-ion Batteries for Electric Vehicles [S].
[13] Wang Fang, Fan Bin, Liu Shiqiang, et al. (2012) Attenuation and duplication of power battery's cycle life. *Automotive Safety and Energy*, 3(1):71–76.
[14] Dai Haifeng, Sun Zechang, Wei Xuezhe. (2009) Used on electric vehicles by dual extended Kalman filter. *Journal of Mechanical Engineering*, 45(6):95–100.
[15] Xu You, Zong Zhijian, Gao Qun, et al. (2014) A consistency evaluation and maintenance method of electric vehicle Lithium-ion+ battery. *Acta Scientiarum Naturalium Universitatis Sunyatseni*, 53(5):25–28.

Electric, Electronic and Control Engineering – Shao, Shu & Tian (Eds)
 ISBN: 978-1-138-02842-5

A machine learning practice to improve the profit for a Chinese restaurant

Xiaobin Li
College of Software Engineering, Lanzhou Institute of Technology, Lanzhou, China
Jozef Stefan International Postgraduate School, Ljubljana, Slovenia

Nada Lavrac
Department of Knowledge Technologies, Jozef Stefan Institute, Ljubljana, Slovenia

ABSTRACT: This paper presents a basic analysis of the data from a Chinese restaurant using data mining techniques. The main purpose was, to investigate and analyze any business pattern of the restaurant and where possible represent them using existing data models in Weka software. The algorithms J48, Naïve Bayes classifier, and Jrip decision rule were used to perform the analysis. The 10-fold cross validation model was used to evaluate the performance of the algorithms which were applied to the data. Some interesting results have been found by the analysis. Although These findings are not the direct guidance for the manager of the restaurant to adjust the marketing strategy, it is still helpful to some extent.

KEYWORDS: data mining; business pattern; data acquiring; data preprocessing; data analyzing.

1 INTRODUCTION

The analysis was performed on the sales records from January 01, 2007 to December 31, 2007 of the restaurant, provided in November, 2008 by Peking d.o.o which is located in Ljubljana, Slovenia.

Three tables have been created, derived from sales records, which are beijing_winterSeason, beijing_summerSeason and beijing_weekends, then the analysis was performed by using models in Weka. The results could be useful for the management of the restaurant to analyze the business, and use the information to improve its business.

2 DATA

As for the three datasets beijing_winterSeason, beijing_summerSeason, beijing_weekends, each of them consists of 359 instances which represent 359 days' sales record. Each instance has 214 attributes, the first 213 attributes are all sales items which include 213 categories of food and drink whilst the attributes are all numerical, the name is the item, their values are quantities of each kind of item that had been sold. Three datasets are all sparse which means for each day only a few items out of all possible 213 items were sold. The last attribute has been added by myself, which is the binary target variable "class". The target variable of three tables are all nominal. The value distribution of the target variable are shown in Table 1 and Figure 1.

Table 1. The value distribution of the target variable.

	yes	no
beijing_winterSeason	31	328
beijing_summerSeason	32	327
beijing_weekends	103	256

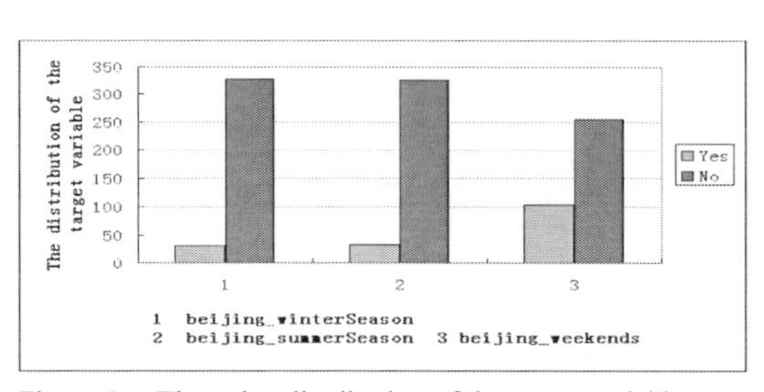

Figure 1. The value distribution of the target variable.

2.1 *Data preprocessing*

The data was extracted from the charging system which runs in a DOS environment into a text file to correspond to the sales records of one day. The data were a collection of very strictly formatted text files which were quite easy to transform because of the regular structure. A short program was written in the Python to transform all the data into a Weka format.

There are two attributes totalSum(the gross income in one day) and mostExpensive (the most expensive item was sold in the day) were added, so as to do more analysis of the data.

2.2 *Data understanding*

Generally, the restaurant had better business (means the higher gross income which is measured by attributes totalSum in this work.) on the weekends and in the "winterSeason" (meant december), while they had worse business in the "summerSeason" (meant from July 15 to August 15) compared to the other dates. These are shown in the Figure 2, 3 and 4 which are visualized by Weka.

But the factors which influenced the business are complicated, the income had a significant impact in the different periods. In this study, several experiments were processed to testify the general business pattern in the restaurant.

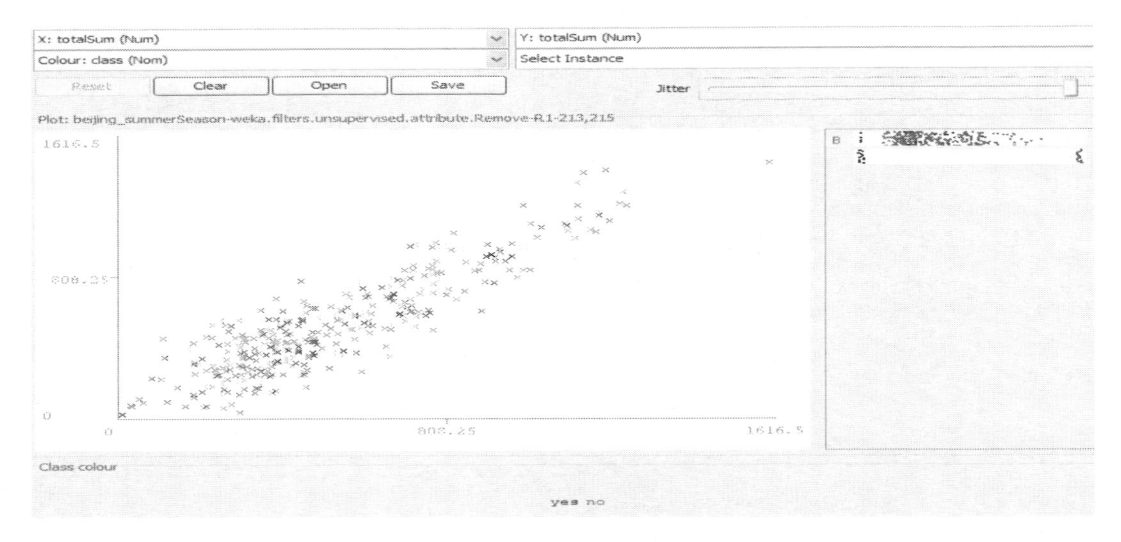

Figure 2. A scatterplot of the totalSum attribute and the class attribute in beijing_winterseason.

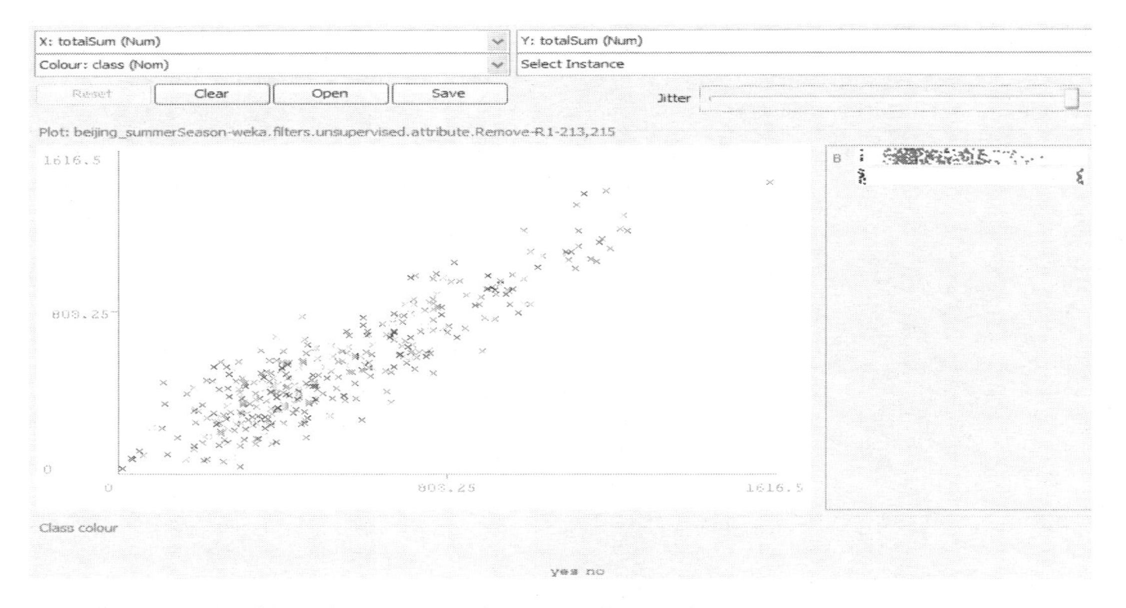

Figure 3. A scatterplot of the totalSum attribute and the class attribute in beijing_summerseason.

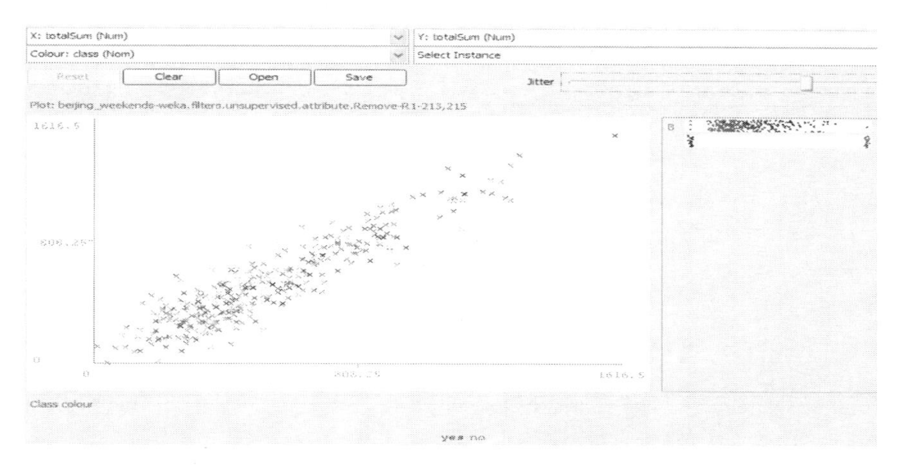

Figure 4. A scatterplot of the totalSum attribute and the class attribute in beijing_weekends.

3 MACHINE LEARNING METHODS USED

After the data was organized, the next step was to select the suitable methods. Numerous machine learning methods have been considered. Taking into consideration the differing effects of each method on the data, the following three predictive methods were selected: decision tree, naïve bayes classifier[13,18], Jrip decision rules.

3.1 *Decision tree: J48*

A decision tree is a predictive model which maps observations about an item to conclusions about the item's target value[3, 6, 16, 17].

In this experiment, the algorithm C4.5 (J48) was used to build a decision tree.

The learning algorithm decides which attribute to put in the node of the tree based on its information gain. In information theory, the amount of information is measured using entropy. The concept of entropy describes how much information there is in an event. Entropy can be defined in terms of discrete random variable X, with possible states $x_1, x_2, ..., x_n$ As:

$$H(X) = \sum_{i=1}^{n} p(x_i)\log_2(\frac{1}{p(x_i)}) = -\sum_{i=1}^{n} p(x_i)\log_2(p(x_i)$$

Using the entropy as a measure for information gain, the approach can be described as:

- Chose an attribute that has the highest entropy value.
- Create a separate tree branch for each value of the chosen attribute.
- Divide the instances into subgroups so as to reflect the attribute values of the chosen node.
- For each group, terminate the attribute selection process if all members of a subgroup belong to the same class or a subgroup contains a single node.
- For each subgroup that has not been labeled as terminal repeat the above process.

3.2 *Naïve bayes classifier*

A naïve bayes classifier is a simple probabilistic classifier based on applying Bayes' theorem with strong (naive) independence assumptions[4,5,9,12,14,15]. It assumes the conditional independence of attribute values given the class:

$$p(v_1, v_2, ...v_n \mid c) = \prod_i p(v_{il} \mid c)$$

Naive Bayes formula:

$$p(c \mid v_1, v_2, ...v_n) = p(c) \bullet \prod_i \frac{p(c \mid v_i)}{p(c)}$$

When classifying a new instance$(v_1, v_2, ..., v_n)$, suppose the dataset has m classes $(c_1, c_2, ..., c_m)$ (target variable with m values). The naïve bayes classifier calculates for each class c_i the conditional probability:

$p(c_i \mid v_1, v_2, ..., v_n)$ of class c_i given evidence $(v_1, v_2, ..., v_n)$ according to the naïve bayes formula. It classifies the example into the class with the highest probability.

3.3 *Decision rules: Jrip*

Jrip implements a propositional rule learner, Repeated Incremental Pruning to Produce Error Reduction (RIPPER), which was proposed by William W. Cohen as an optimized version of IREP[7, 8, 10, 11].

4 EVALUATION

All algorithms were tested by using Weka software, version 3.6.

4.1 *Evaluation of the J48 decision tree*

The first algorithm that had been tested is the J48 decision tree.

4.1.1 *Experiment on the dataset beijing_winterSeason*

The J48 algorithm was tested using the default parameter values. The visualization of the tree is shown in Figure 5.

Figure 5. The visualization of the tree.

```
Correctly Classified Instances         321               89.415  %
Incorrectly Classified Instances        38               10.585  %
Kappa statistic                          0.2144
Mean absolute error                      0.122
Root mean squared error                  0.3224
Relative absolute error                 76.3073 %
Root relative squared error            114.7593 %
Total Number of Instances              359

=== Detailed Accuracy By Class ===

               TP Rate   FP Rate   Precision   Recall  F-Measure   ROC Area  Class
                 0.226     0.043      0.333     0.226     0.269      0.564    yes
                 0.957     0.774      0.929     0.957     0.943      0.564    no
Weighted Avg.    0.894     0.711      0.878     0.894     0.885      0.564

=== Confusion Matrix ===

   a   b   <-- classified as
   7  24 |   a = yes
  14 314 |   b = no
```

Figure 6. Results of 10-fold cross validation using J48 algorithm.

The results of 10-fold cross validations are shown in Figure 6. The algorithm achieved 89.415 % accuracy.

This is not a good result because if it is always predicted majority which says "no" that has the biggest number of examples, then without using any model, can still achieve an accuracy of 91.3649 %.

The ReliefF Ranking Filter[1, 2] was then used to rank the attributes. Attributes which have a ranking lower than 21 were removed and a new, simpler classifier was compared to the old one which works on all attributes. It was found that the new classifier which was performed on the 21 highest ranking attributes (shown in Figure 7) achieved an accuracy of 91.6435 %, which was more accurate than the original and it

```
Ranked attributes:
 0.10935933147632218    142 RADENSKA_025
 0.10877437325905313     25 GOVEDINA_S_PAPR
 0.10538532961931353      3 A_GOVEDINA_V_CES
 0.09275766016713158    189 T13_OCVRTI_KROMPIR
 0.0926183844011148     185 SVINJINA_V_CESN
 0.0922934076137428     138 RACA_Z_MANDELJN
 0.08554064320081017      8 CAJ_KITAJSKI_JASMI
 0.08226555246053921      6 BELA_KAVA
 0.0760445682451253     123 PISCANEC_NA_VR
 0.07548746518105903     29 GOVEJI_CHOPSUI
 0.07158774373259155     99 OCVRTO_JAJCEVCI
 0.07116991643454103    154 RIBA_V_RJAVI_OM
 0.06935933147632335    179 SUHO_PECENA_GOV
 0.06824512534819037    134 PRAZEN_BAMBUS_Z
 0.062228412256266956   107 PECENI_RIZ_S_P
 0.0612116991643451 2     4 A_PISCANEC_V_CES
 0.061002785515320825   117 PIKANTNA_RIBA
 0.05974930362117002    209 ZELENI_CAJ
 0.056824512534818995    24 GOVEDINA_S_CURR
 0.0545961002785515 6    70 LIGNJI_S_PAPRIK
 0.054503249767873754   153 RIBA_NA_SECUANS
```

Figure 7. Ranked attributes of the dataset beijing_winterSeason.

proved that the removed attributes carried no important information about the classification problem. The results are presented in Figure 8 and 9.

4.1.3 *Experiment on the dataset beijing_weekends*

The same experiment was also applied on the dataset beijing_weekends by using J48 algorithm with the

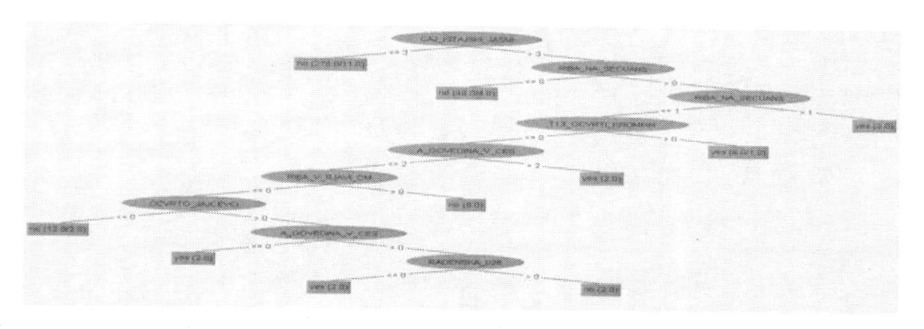

Figure 8. Visualization of the tree performed on the 21 highest ranking attributes.

```
Correctly Classified Instances         329               91.6435 %
Incorrectly Classified Instances        30                8.3565 %
Kappa statistic                          0.2519
Mean absolute error                      0.1274
Root mean squared error                  0.2722
Relative absolute error                 79.6728 %
Root relative squared error             96.8785 %
Total Number of Instances              359

=== Detailed Accuracy By Class ===

               TP Rate   FP Rate   Precision   Recall  F-Measure   ROC Area  Class
                 0.194     0.015      0.545     0.194     0.286      0.708    yes
                 0.985     0.806      0.928     0.985     0.956      0.708    no
Weighted Avg.    0.916     0.738      0.895     0.916     0.898      0.708

=== Confusion Matrix ===

   a   b   <-- classified as
   6  25 |   a = yes
   5 323 |   b = no
```

Figure 9. Results of 10-fold cross validation using J48 algorithm performed on the 21 highest ranking attributes.

4.1.2 *Experiment on the dataset beijing_summerSeason*

The same experiment was applied on the dataset beijing_summerSeason by using the J48 algorithm with the default parameter values. An accuracy of 88.8579 % was achieved.

A higher accuracy of 91.3649 % was obtained, when the J48 algorithm was performed on the 13 highest ranking attributes which was achieved by using the ReliefF Ranking Filter as shown in Figure 10.

```
Ranked attributes:
 0.05860724233983243     142 RADENSKA_025
 0.05205913649025068     191 T15_TOFU_BAO
 0.04933147632311937     143 RADENSKA_STEKLENIC
 0.04046282265852449      99 OCVRTO_JAJCEVCI
 0.03899721448467958     127 PISCANEC_V_PIKA
 0.0377437325905294       25 GOVEDINA_S_PAPR
 0.03613839612959969      55 KISLO_PEKOCA_JUH
 0.03546889507892284     108 PECENI_RIZ_S_S
 0.03384401114206157     163 SEZAMOVE__KROGLICE
 0.03231197771587758      52 KAVA
 0.031476323111977175     93 OCVRTE_RAKOVE_R
 0.03036211699164318     151 REFOSK
 0.030252686032630136     13 COCA_COLA_025
```

Figure 10. Ranked attributes of the dataset beijing_summerSeason.

default parameter values. An accuracy of 80.5014 % was achieved.

A higher accuracy of 81.6156 % was obtained, when the J48 algorithm was performed on the 24 highest ranking attributes which was achieved by using the ReliefF Ranking Filter as shown in Figure 11.

4.2 *Evaluation of naïve bayes classifier*

The next algorithm to be tested was the naïve bayes classifier.

4.2.1 *Experiment on the dataset beijing_winterSeason*

Parts of the Weka outputs of 10-fold cross validation are shown in Figure 12. The algorithm achieved 82.7298 % accuracy.

A higher accuracy of 88.8579 % was obtained, when the naïve bayes classifier was performed on the 21 highest ranking attributes which was achieved by using the ReliefF Ranking Filter as shown in Figure 7.

```
Ranked attributes:
 0.0795264623955438     29 GOVEJI_CHOPSUI
 0.0723769730733531 8  158 RIZ
 0.0560584958217267 56 135 RACA_NA_SECUANS
 0.0558321727019498 8  206 ZA_HARMONY
 0.0543175487465181 25  24 GOVEDINA_S_CURR
 0.0488857938718666 3  166 SLADKO_KISLA_RA
 0.0473537604456823 75 179 SUHO_PECENA_GOV
 0.046995622276163941   91 OCVRTA_ZELENJAV
 0.0434540389972143 74 181 SUHO_PRAZENI_RA
 0.0433147632311977 3  104 PECENI_KITAJSKI
 0.0431754874651813 16  78 MENU_ZA_STIRI
 0.0423651557356292 6   93 OCVRTE_RAKOVE_R
 0.0422469823584028 5  167 SLADKO_KISLA_RI
 0.0403899721448471 5  187 SVINJINA_Z_BAMB
 0.0401671309192200 8  142 RADENSKA_025
 0.0399721448467961 9  143 RADENSKA_STEKLENIC
 0.0387186629526461 3   26 GOVEDINA_S_SAMP
 0.038866295264623 94  160 SAMOSTANSKA_POJ
 0.0385515320334261 9   96 OCVRTI_SLADOLE
 0.0352566653402307 3  199 UNION_0.33
 0.0330680461599680 9  144 RADLER_0.50
 0.0321262766945217 5  102 OMAKA
 0.0311977715877439 78  95 OCVRTI_LIGNJI
 0.0302228412256267 85  25 GOVEDINA_S_PAPR
```

Figure 11. Ranked attributes of the dataset beijing_weekends.

4.2.2 *Experiment on the dataset beijing_summerSeason*

The same test was processed on the dataset beijing_summerSeason by using naive Bayes classifier with the default parameter values. An accuracy of 64.9025% was achieved.

A higher accuracy of 87.4652 % was obtained, when the naïve bayes classifier was performed on the 21 highest ranking attributes which was achieved by using the ReliefF Ranking Filter as shown in Figure 12.

4.2.3 *Experiment on the dataset beijing_weekends*

The same test was also processed on the dataset beijing_weekends by using naïve bayes classifier with the default parameter values. An accuracy of 82.1727 % was achieved.

A higher accuracy of 84.6797 % was obtained, when the naïve bayes classifier was performed on the 21 highest ranking attributes which was achieved by using the ReliefF Ranking Filter as shown in Figure 11

4.3 *Evaluation of the Jrip decision rule*

When the datasets were tested using the Jrip decision rule, the short rules were achieved which are explained as follows.

4.3.1 *Experiment on the dataset beijing_winterSeason*

Parts of the Weka outputs of 10-fold cross validation are shown as follows:

(CAJ_KITAJSKI_JASMI >= 4) and (KISLO_PEKOCA_JUH >= 24) and (RIBA_NA_SECUANS >= 1) => class=yes (13.0/4.0)

(RDECI_LEDENI_CAJ >= 1) => class=yes (5.0/0.0)

=> class=no (341.0/17.0)

An accuracy of 89.9721 % was achieved.

A lower accuracy of 89.1365 % was obtained, when the Jrip decision rule was performed on the 21 highest ranking attributes which was achieved by using the ReliefF Ranking Filter as shown in Figure 7.

4.3.2 *Experiment on the dataset beijing_summerSeason*

Parts of the Weka outputs of 10-fold cross validation are shown below:

(OCVRTE_BANANE <= 0) and (KISLO_PEKOCA_JUH <= 4) => class=yes (5.0/0.0)

=> class=no (354.0/27.0)

An accuracy of 89.415 % was achieved.

If two attributes attributes totalSum and mostExpensive were added, an accuracy of 88.8579 % and the following rules were achieved:

(RADENSKA_STEKLENIC >= 1) and (CAJ_KITAJSKI_JASMI <= 0) and (mostExpensive <= 18) and (A._POMFRI <= 1) => class=yes (14.0/4.0)

(KAVA_Z_MLEKOM <= 0) and (OCVRTI_SLADOLE <= 0) and (KORUZNA_JUHA <= 0) => class=yes (8.0/1.0)

=> class=no (337.0/15.0)

```
Correctly Classified Instances         297               82.7298 %
Incorrectly Classified Instances        62               17.2702 %
Kappa statistic                          0.2232
Mean absolute error                      0.1749
Root mean squared error                  0.4104
Relative absolute error                109.3449 %
Root relative squared error            146.0756 %
Total Number of Instances              359

=== Detailed Accuracy By Class ===

               TP Rate   FP Rate   Precision   Recall  F-Measure   ROC Area  Class
                 0.452     0.137      0.237     0.452     0.311      0.685    yes
                 0.863     0.548      0.943     0.863     0.901      0.684    no
Weighted Avg.    0.827     0.513      0.882     0.827     0.85       0.684

=== Confusion Matrix ===

   a   b   <-- classified as
  14  17 |   a = yes
  45 283 |   b = no
```

Figure 12. Results of 10-fold cross validation using naïve bayes classifier.

The attribute "CAJ_KITAJSKI_JASMIN" (which means "Chinese jasmine tea") is presented in the rules for the winter and the summer seasons. In the winter season the value is larger than 4, and in the summer season the value is 0. This could indicate that guests in the restaurant did not drink as much Chinese jasmine tea in the summer.

A lower accuracy of 89.1365 % was obtained, when the Jrip decision rule was performed on the 13 highest ranking attributes which was achieved by using the ReliefF Ranking Filter as shown in Figure 10.

4.3.3 *Experiment on the dataset beijing_weekends*

Parts of the Weka outputs of 10-fold cross validation are shown as follows:

(SOJINI_KALCKI_V >= 6) and (KISLO_PEKOCA_JUH >= 17) and (UNION_0.33 >= 1) => class=yes (58.0/8.0)

(SAMOSTANSKA_POJ >= 2) and (OCVRTE_RAKOVE_R >= 6) => class=yes (17.0/3.0)

(OCVRTI_SLADOLE >= 6) and (UNION_0.50 >= 6) => class=yes (13.0/2.0)

(HRUSTLJAVA_RACA >= 5) and (JUHA_S_PISCANCEM >= 2) => class=yes (13.0/3.0)

=> class=no (258.0/18.0)

An accuracy of 81.337 % was achieved.

A lower accuracy of 79.9443 % was obtained, when the Jrip decision rule was performed on the 21 highest ranking attributes which was achieved by using the ReliefF

Ranking Filter as shown in Figure 11.

In three cases, the majority rule e.g. " => class=no (341.0/17.0)) " has far the greatest coverage (the first number in the brackets), simply because the "no" class have so many instances (about 80% in all cases).

4.4 *Analysis with additional attributes*

Two attributes "totalSum" and "mostExpensive" were added, to further analyze the data. The accuracy are shown in Table 2.

Table 2. Results of 10-fold cross validation on the dataset only with the attributes totalSum and mostExpensive.

	beijing_ winterSeason	beijing_ summerSeason	beijing_ weekends
J48	91.3649 %	91.0864 %	81.0585 %
Naïve bayes	88.8579 %	91.0864 %	79.3872 %
Jrip	90.8078 %	91.922 %	79.6657 %

4.5 *Contrast of the accuracy*

The contrast between the accuracy that was achieved when using different algorithms and the accuracy always predicted as majority class without using any algorithm is shown in Table 3, and Figure 13, 14, 15 on the dataset beijing_winterSeason, beijing_summerSeason and beijing_weekends respectively.

Column A corresponds to the accuracy achieved from the original data,column B is the accuracy achieved from the data only with the attributes totalSum and mostExpensive, Column C is the accuracy received from the data after the lower rank attributes had been removed (as previously mentioned), whilst column D is the accuracy always predicted as majority class without using any algorithm. The majority class in the datasets beijing_winterSeason and beijing_summerSeason are all "NO", whilst in the dataset beijing_weekends is "YES".

From Table 3 and Figure 13, 14, 15, it was found that only the test using Weka algorithms on the dataset beijing_weekends got the higher accuracy,

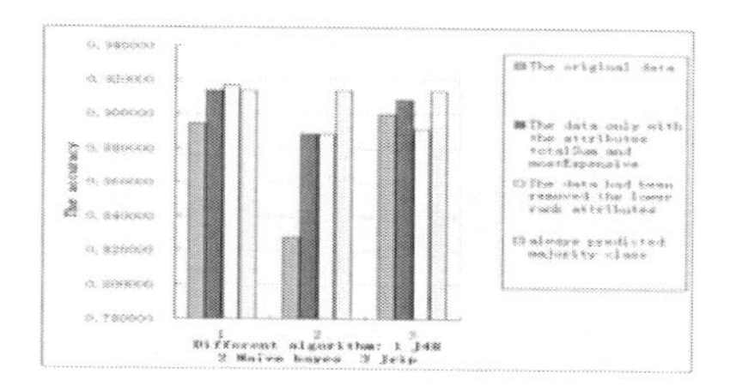

Figure 13. Contrast of the accuracy of the dataset beijing_winterSeason.

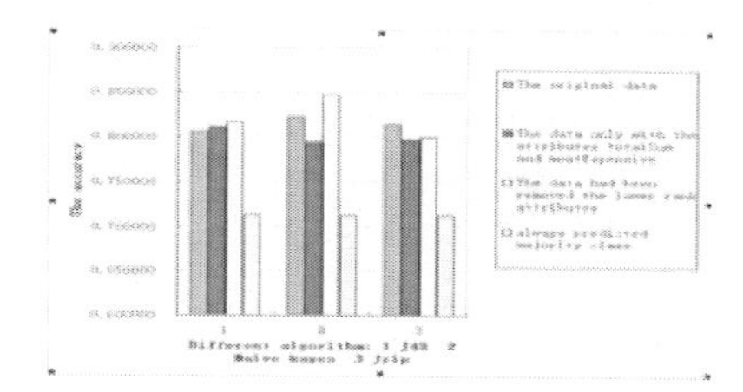

Figure 14. Contrast of the accuracy of the dataset beijing_SummerSeason.

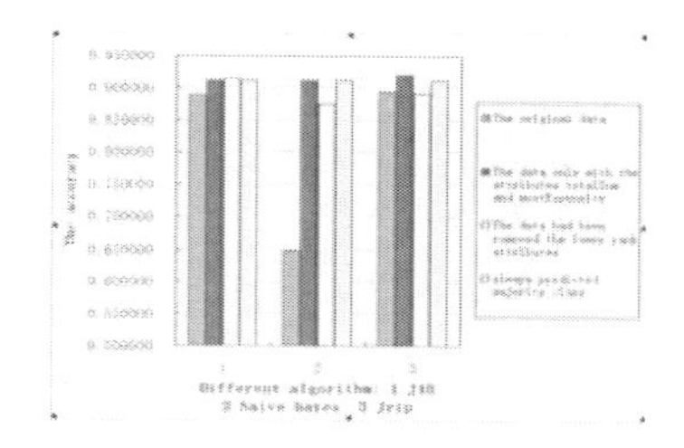

Figure 15. Contrast of the accuracy of the dataset beijing_weekends.

Table 3. Contrast of the accuracy.

	beijing_winterSeason				beijing_summerSeason				beijing_weekends			
	A	B	C	D	A	B	C	D	A	B	C	D
J48	89.415%	91.3649%	91.6435%	91.3649%	88.8579%	91.0864%	91.3649%	91.0864%	80.5014%	81.0585%	81.6156%	71.3092%
Naïve bayes	82.7298%	88.8579%	88.8579%		64.9025%	91.0864%	87.4652%		82.1727%	79.3872%	84.6797%	
Jrip	89.9721%	90.8078%	89.1365%		89.4150%	91.9220%	89.1365%		81.3370%	79.6657%	79.9443%	

however on the dataset beijing_winterSeason and beijing_summerSeason a lower accuracy was achieved compared to always predicted as majority class without using any algorithm.

It makes no sense to use the classifier which gives lower accuracy as better results be achieved by always predicting as the majority class. This is because of the data, not because of the algorithm.

In general, machine learning algorithms often have problems when there are large numbers of attributes. Such results imply that the attributes do not related directly to the class, possibly a lot of irrelevancy is present or that the attributes are related by a complex function which can not be approximated by the model, ...

5 CONCLUSION

In this work three different classifiers have been used which were implemented in the Weka environment and compared in performance to classification of the data from the restaurant. Initially it was anticipated to find some interesting business pattern by analyzing the restaurant data and then provide some useful suggestions for the management of the restaurant to improve its profitability. However, during the analysis, although the information was accurate, the results were more interesting than useful. The machine learning rules functioned correctly and were accurate. Comparatively higher accuracy was achieved on the dataset beijing_weekends after the lower rank attributes had been removed (as previously mentioned), this is because only a few attributes (the top on the ranking list) carry information that is actually related to the class attribute. Although it was difficult to understand how these results could be used profitably by the restaurant management, it is still useful to some extent.

REFERENCES

[1] A. Ilin and T. Raiko. (2010) Practical approaches to principal component analysis in the presence of missing values. *Journal of Machine Learning Research*, 11:1957–2000.

[2] D. Heckerman, D. M. Chickering, C. Meek, R. Rounthwaite, and C. Kadie. (2000) Dependency networks for inference, collaborative filtering, and data visualization. *Journal of MachineLearning Research*, 1:49–75,.

[3] D. Lowd and J. Davis. (2014) Improving narkov network structure learning using decision trees. *Journal of Machine Learning Research*, 15:501–532.

[4] D. Rusakov and D. Geiger. (2005) Asymptotic model selection for naive Bayesian network. *Journal of Machine Learning Research*, 6:1–35.

[5] M. Aoyagi and K. Nagata. (2012) Learning coefficient of generalization error in Bayesian estimation and Vandermonde matrix-type singularity. *Neural Computation*, 24(6):1569–1610.

[6] M. Bishop. (2006) *Pattern Recognition and Machine Learning*. New York: Springer.

[7] M. Drton. (2010) Reduced rank regression. In *Workshop on Singular Learning Theory*. American Institute of Mathematics.

[8] M. E. Tipping and C. M. Bishop. (1999) Probabilistic principal component analysis. *Journal of the Royal Statistical Society*, 61:611–622.

[9] M. Seeger and G. Bouchard. (2012) Fast variational Bayesian inference for non-conjugate matrix factorization models. In *Proc. of AISTATS*, La Palma, Spain.

[10] O. Luaces, J. Diez, J. Barranquero, J. del Coz, and A. Bahamonde. (2012) Binary relevance efficacy for multilabel classification. *Progress in Artificial Intelligence*, 1(4):303–313.

[11] R. Hardoon, S. R. Szedmak, and J. R. Shawe-Taylor. (2004) Canonical correlation analysis: An overview with application to learning methods. *Neural Computation*, 16(12):2639–2664.

[12] S. D. Babacan, M. Luessi, R.Molina, and A. K. Katsaggelos. (2012) Sparse Bayesian methods for low-rank matrix estimation. *IEEE Trans. on Signal Processing*, 60(8):3964–3977.

[13] S. Ji and J. Ye. (2009) An accelerated gradient method for trace norm minimization. In *Proceedings of International Conference on Machine Learning*.

[14] S. Lin. (2011) *Algebraic Methods for Evaluating Integrals in Bayesian Statistics*. PhD thesis, Ph.D. dissertation, University of California, Berkeley.

[15] S. Nakajima, M. Sugiyama, S. D. Babacan and R. Tomioka. (2013) Global Analytic Solution of Fully-observed Variational Bayesian Matrix Factorization. *Journal of Machine Learning Research*, 14:1–37.

[16] S. Watanabe. (2009) *Algebraic Geometry and Statistical Learning Theory*. Cambridge: Cambridge University Press.

[17] T. M. Mitchell. (1997) *Machine learning*. McGraw-Hill.

[18] W. Cheng and E. Hullermeier. (2009) Combining instance-based learning and logistic regression for multilabel classification. *Machine Learning*, 76(2–3):211–225.

[19] Y. Wu, H. Zhang, and Y. Liu. (2010) Robust model-free multiclass probability estimation. *Journal of the American Statistical Association*, 105:424–436.

Electric, Electronic and Control Engineering – Shao, Shu & Tian (Eds)
© 2015 Taylor & Francis Group, London, ISBN: 978-1-138-02842-5

The control system research of X-ray generator in medical diagnostic

Yu Liu*, Beibei Dong, Jingjing Yang & Benzhen Guo
Hebei North University, Zhangjiakou, Hebei, China

ABSTRACT: As an important equipment of a hospital diagnosis system, to some extent, the control system of X-ray affected the operational effectiveness of the X ray generator. In order to improve the emission efficiency and capability of non-war time routine maintenance, this paper designed a closed-loop control system, and the design scheme aimed at controlling the high frequency and high voltage generator system. The control model included the high frequency power control, voltage control, a clock control, error control and feedback control, and all of the subsystems integrated with ATmegal28 MCU. So that we can get such verdicts: this control system can realize different links of X ray generator to use in public one same platform, and reduce the occupied space of the system in the equipment; The system can effectively improve the efficiency of X ray generator maintenance and launching performance.

KEYWORDS: control system; operational performance; design scheme; feedback control.

1 INTRODUCTION

Since1895, when X-ray was first discovered by a German scientist, Wilhelm Konrad Rontgen in laboratory of institute of physics at the Wurzburg University, X rays is extremely important for the development of human medical history meaning, and has been applied to different parts of the human body detection, such as bone and joints, head and neck, heart and great vessels, respiratory system, digestive tract, oral cavity, breast, etc., the application of X-ray on CT especially success. With the progress of technology, the requirements of doctors and patients to X-ray control precision and quality was constantly improved, so that promote the development and progress of X-ray generator .

In recent years, the technology of X-ray generator was developing rapidly, and the key technology mainly involved in the following categories: intelligent electronic technology, high frequency inverter technology, automatic control technology and new materials and modular construction technology, etc.. At present, the X-ray generator was mainly composed of the power system module, control system module, inverter circuit module, high frequency, high voltage transformer, multiple doubler rectifier, filter, filament circuit, sampling feedback circuit modules, and it constantly develops in the direction of the information and the intellectualized. Aiming at the high frequency, high voltage transmitter, one kind of closed-loop microcomputer control system, which included power control, the clock control, operation control and feedback control was designed to effectively realize automation, intelligence and informatization of the X-ray generator control system.

2 THE STRUCTURE & OPERATION THEORY OF X-RAY MACHINE

It is well-known that X-ray has some energy, and can make a form of energy conversion for X rays to release energy by the conservation of energy, so the extranuclear electron of atoms on the target surface can come into being step through putting high-speed electron beam to bombard target surface made in some kind of specific materials, for the moment, the atoms keep in the excited state ,when the atoms be in excited states return to ground state ,it will release energy in the form of emission X-ray photons. Therefore, the production of X-ray only need to meet three conditions on a macro level, the first is the cathode electron source, the second is anode target, and the third is a high pressure electric field between the anode and the cathode.

The structure of X-ray machine mainly has composed of the host and peripheral parts, the host also called an X-ray generator, and included X-ray tube equipment, high-voltage generator, control equipment, etc., the periphery of X-ray machine was assembled with machinery and auxiliary equipment according to the actual requirements of CT equipment, its structure is shown in figure 1:

*Corresponding author: *liuyu0729@yeah.net*

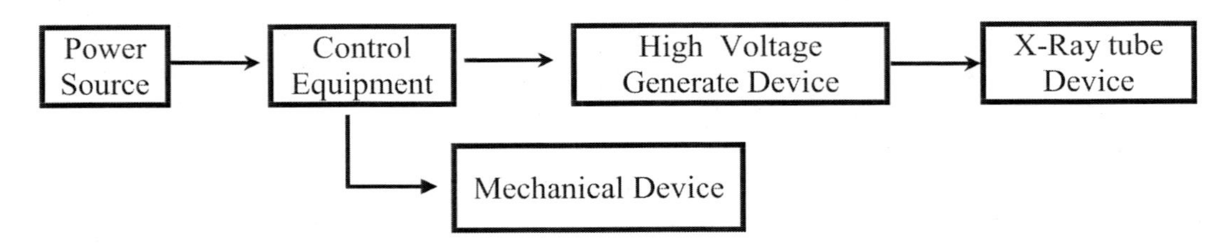

Figure 1. Composition diagram of X-ray machine.

3 THE STRUCTURE ANALYSIS OF CONTROL SYSTEM

There is no doubt that the function of the X-ray machine is producing X-ray, only the control of tube voltage, tube current, exposure time length, and the size of focal point were reliable, that X-ray machine would work normally and stably. When X-ray machine is working, the higher voltage will obtain greater X-ray energy and stronger ray penetration, as well as the greater current can get a greater density of X ray, also, the size of the focus directly affects heat dissipation and clarity, therefore, to make X-ray machine work stably, it is necessary to precisely control the various working parameters.

In order to meet the demand for CT in the medical and health industry to X-ray machine, this paper mainly analyzes the following functions of the control system:

1 The intelligent and sensitive control of power can be reliable and effective to avoid unnecessary halting problem caused by overload and open circuit.
2 The continuous adjustable control of high frequency and voltage source, electric current and exposure time can meet the requirements that different working conditions need different X-ray intensity.
3 The stability of the clock controller can effectively adjust the exposure time of CT image.
4 The accuracy of error control can make the influence which was taken on quality by impossible, absolutely avoid system error to the lowest feasible levels.
5 The real-time feedback loop can improve control circuit, and ensure the stability of the system.

In conclusion, the function in the form of a block diagram of X-ray machine's controller was shown in figure 2.

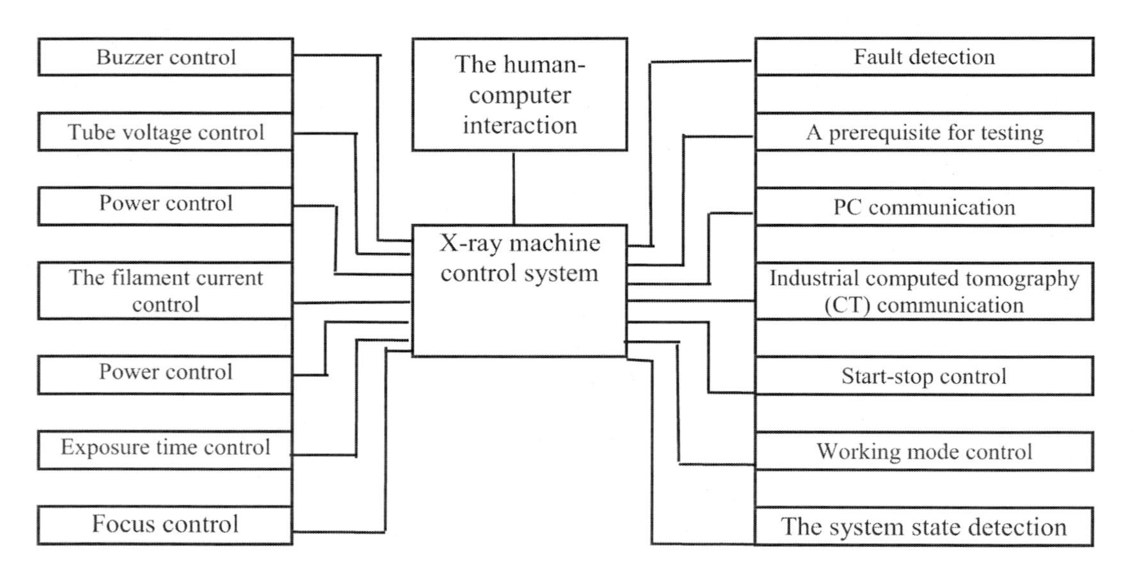

Figure 2. The function of the X-ray machine's controller.

3.1 *The power switch control*

Control power supply is also known as auxiliary power supply system, it has two functions: one is controlling start-up and shutdown of the high frequency and voltage generator, and another is controlling the other powers which were used to adjust high frequency and voltage generated by the relay, and the control circuit schematic diagram was shown in figure 3:

As the power button of control system,when "S2" was closed, "R3" would be energized, and triode "Q2" was trigger conduction, then current flows through"R4"and "R5", also "Q3" was trigger conduction, eventually, all circuit components formed a circulation loop, the components included relay "K1", "K2", "K3", diode "D2", and transistor "Q3", so relay "K1", "K2", "K3" were closed, the circuits controlled by relay were working normally, so the system boot was completed successfully.

While, as the shutdown button, when "S1"was closed, "R2"would be energized, and triode "Q1" was trigger conduction, then base working voltage was destroyed, so that make "Q2" and "Q3" stop working at the same time, and destroyed the normal circuit loop, eventually, the system shutdown was completed successfully.

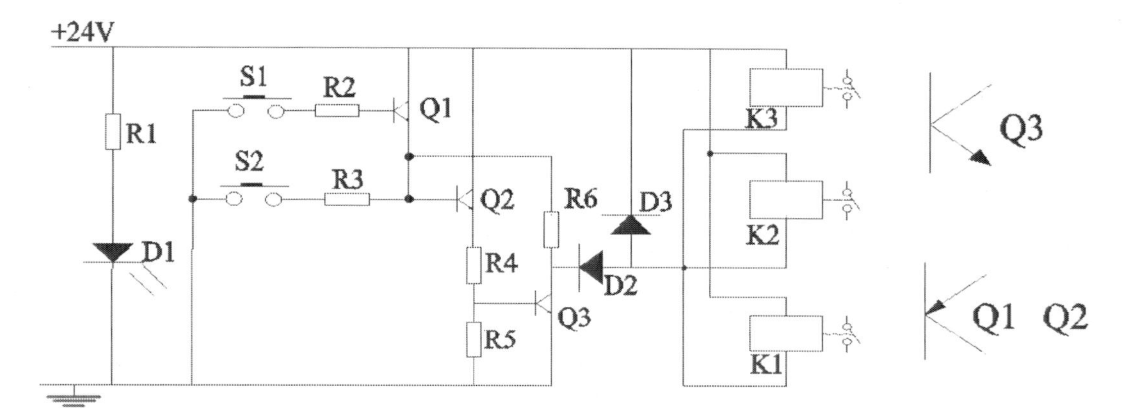

Figure 3. The control circuit of function of the power switch.

3.2 *The clock control*

The clock system adopts serial real-time clock chip DS1302S, the circuit has two independent power supplies to provide electricity, and the active crystal oscillation was set within the circuit, it allowed input clock frequency range of 15.625 ~ 387 MHZ, and provided real-time timing, also had a millisecond error compensation function, the data exchange was completed successfully by I/O interface, so that it can guarantee that clock chip never power outages and work accurately, the control circuit was shown in figure 4:

3.3 *The error operation control*

it is inevitable that error is existent for a control system, and the error operation circuit can realize the

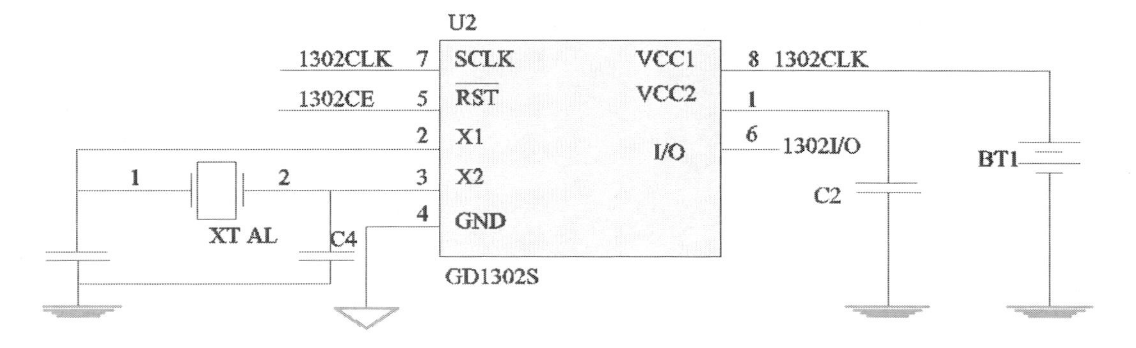

Figure 4. The clock control circuit diagram.

function that monitoring and controlling the system error real-time, so that can effectively guarantee the stability and reliability of system steady-state error, and its processing circuit diagram was shown in figure 5:

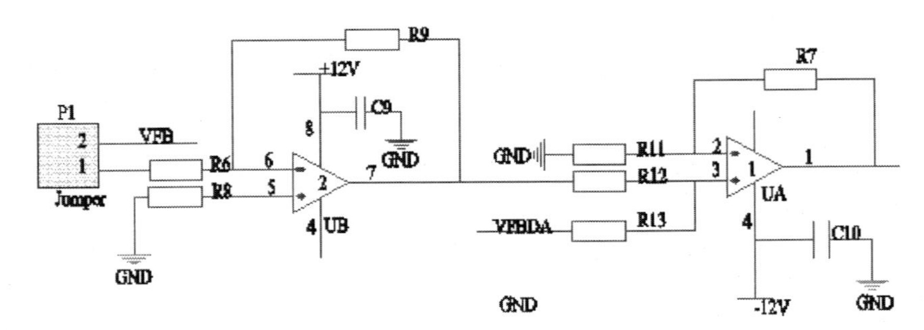

Figure 5. The error operation circuit diagram.

The error operation circuit in figure 5 is composed of operational amplifier, its relational expression between input and output is shown in formula (1):

$$V_{Err} = V_{FBDA} - V_{FB} \tag{1}$$

In the whole error operation circuit, VFBDA was a ADC results, which was obtained by quantitative output voltage through setting parameters, the value corresponding to the set parameters are real output, VFB was the quantitative sample back feedback signals.

3.4 *The feedback control*

In order to obtain a stable and precision circuit system which is high frequency, high voltage, high current system, it is necessary that hold a reliable and effective feedback control modulator for automatic control. Ultimately, the control system realizes automatic adjustment.

It is the closed loop feedback PID control mode that used to maintain a stable system, and through adjusting proportion, integral and differential parameters to ensure the stability of circuit, also this circuit can efficaciously realize adjusting and controlling complex system which has the characteristic of nonlinear, time-varying, coupling, uncertain parameters and structure. Introducing integral item in the feedback system, as the error depends on the time integral, with the increase of time, the integral will be escalated. In this way, integral item will also be increased with the time, even if the error was very small, and the steady-state error will be diminished further, until is equal to zero with increasing of drive controller output. In the same way, introduction of differential item, it can predict the error's change trend, in this way, it can advance to make control function of suppressing error is equal to zero by having the proportion and differential controller, so that it can avoid the serious overshoot of a controlled object, the control circuit diagram is shown in figure 6:

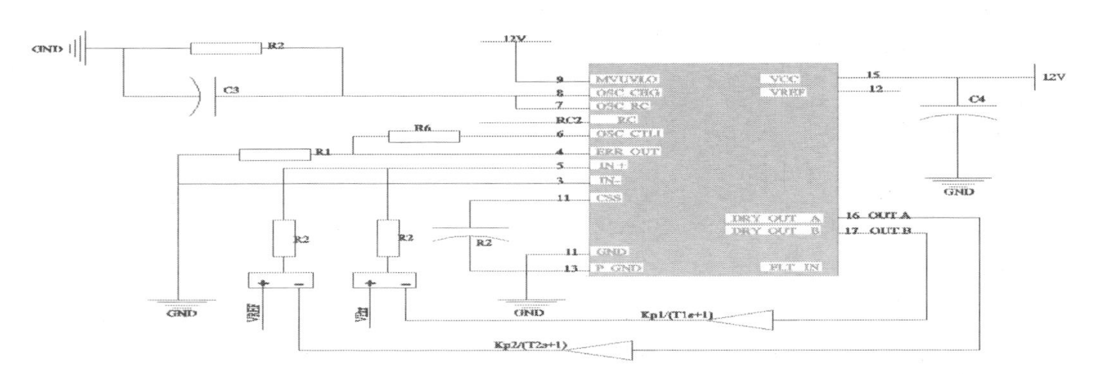

Figure 6. The feedback control circuit diagram.

3.5 *The encapsulation of control system*

we have clearly understood the structure of control system and analyzed the function of the control system, on this basis, the overall structure of the X-ray generator is introduced, and the general program of control system is put forward.

The control system's components were integrated in the tiny chips by using the core-board of ATmegal28 micro-computer, and the key of the control system was achieving the integrated control system according to the circuit principle of each link. This system used JTAG as download way, and able to complete online simulation, also the data transfer and instruction dispatch were completed through the asynchronous serial communication.

The main function of X-ray generator control circuit was completed the accommodation of high frequency and high voltage what were needed to produce X-ray, and this system didn't need to conduct a man-machine interactive operation, but a complete calculation of various data, feedback and control logic. When each module circuit component was welded on core board, and linked them together, it realized that the whole control system module was enveloped on MCU chip. The encapsulation structure was shown in Figure 7:

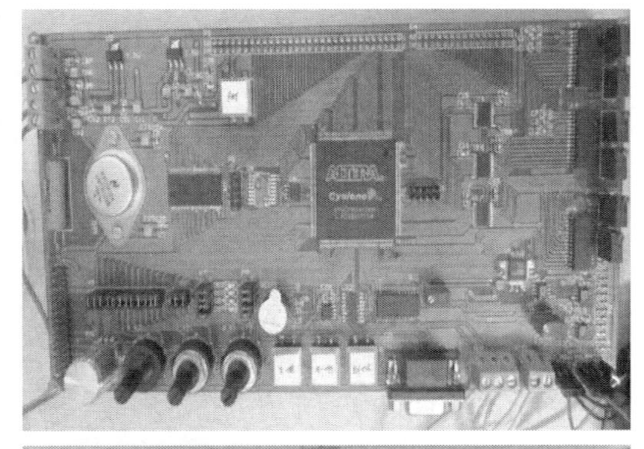

Figure 7. The encapsulation structure of the control system.

The control system of X ray generator regards the microprocessor as the core, this system not only reduce the system equipment space occupied, but also can effectively improve the automation, intellectualization and informatization of X-ray generator control system.

4 CONCLUSIONS

This paper puts forward an effective control system' design Project of X-ray machine based on the principle of microcomputer control and integrated chip technology by analyzing the structure of X-ray generators and the demand of control system, and this system has the following advantages:

1. This controller system not only has the characteristics of high integration density, shorter development cycle, low design cost, but also it is reconfigurable, so that their features make the system more active and better scalability.
2. This system not only effectively controls the source system of X-ray, but also realizes the automation, intellectualization and informatization of the control system.
3. It is the characteristic that the control system is small in size that make this system can provide favorable conditions for the system structure design of the X ray generator.

ACKNOWLEDGEMENTS

The authors thank Prof. Zhang Xiao and Prof. Hao Shangfu for the valuable discussion and recommendation. This project was supported partially by the Hebei province department of education youth fund projects (QN2014182) and Hebei North university field project (No. Q2014002).

REFERENCES

[1] Tojo Fumio, Hirakawa SyunZou. (2007) Evaluation of Plastic Film Thickness Measuring System Using an X-ray Slit Beam. *IEEE Instrumentation and Measurement Technology Conference, Synergy of Science and Technology.*

[2] Shyh-Shin Liang, Ying-Yu Tzou. (2001) DSP control of resonant switching high voltage power supply for X-ray. *IEEE Power Electron and Drive Systems Conf.* Oct. 2001, 2:522–526.

[3] Sunj Ding, Nakaoka M. (2000) Series resonant high-voltage PFM DC-DC circuit with voltage multiplier based a two-step frequency switching control for medical-use X-ray power generator. *IEEE Power Electronics and Motion Control Conf.*, 2000, 2:596–601.

[4] G.T. Herman. (1980) *Image Reconstruction from Projections: the Fundamental of Computerized Tomography*. New York: Academic Press.
[5] Henry Neugass. (1991) An interactive development environment and real-time kernel for GMICRO TM. *IEEE.1991*, pp:196–207.
[6] Sakabe N, Ohsawa S, Sugimura T, et al. (2008) Highly bright X-ray generator using heat of fusion with a specially designed rotating anticathode. *J. Synchrotron Radiat.*, 15(Pt3):231–234.

Electric, Electronic and Control Engineering – Shao, Shu & Tian (Eds)
© 2015 Taylor & Francis Group, London, ISBN: 978-1-138-02842-5

Study on the magnitude-frequency response for RLC series circuit

Huanyin Zhou, Lirong Li & Yanhui Xie
College of Chemical Defense, Beijing, China

ABSTRACT: Measuring magnitude-frequency response of an RLC series circuit accurately is generally difficult, because AC voltage reading often fluctuates in a range of about 1%. The measurement precision of magnitude-frequency response of an RLC series circuit is studied based on magnitude-frequency response difference analysis. Through this study, a conclusion is drawn that measurement precision of magnitude-frequency response is different at different frequencies. At some special frequencies, the magnitude change is very sensitive to frequent change. In this case, the magnitude-frequency response can be measured accurately. However, at some other frequencies where the magnitude change is insensitive to the frequency change, the measuring will lead to a larger error. If error sources are analyzed thoroughly, the measurement precision can be improved greatly with an appropriate method.

KEYWORDS: RLC series circuit; magnitude-frequency response; difference; measurement precision.

1 INTRODUCTION

An RLC series circuit consisting of a resistor (R), an inductor (L), and a capacitor (C) is a very typical circuit in college electric course. The circuit is widely used in radio, television, jamming and anti-jamming for it has a frequency-selective property. Alternating voltage is not often measured accurately because of sampling error. The reading error often reaches to 1%. Therefore, measuring magnitude-frequency response accurately is difficult. For an RLC series circuit with low quality factor, the resonance frequency f_0 cannot be measured accurately through measuring the maximum voltage at the resistor, because the magnitude change of resistor voltage is not sensitive to the frequency change at f_0. But the measurement of cutoff frequencies f_L and f_H is more accurate. The magnitude-frequency difference response for the RLC series circuit is studied as follows. Through this difference response study, the measurement precision of magnitude-frequency response is analyzed.

2 THE CHARACTERS OF MAGNITUDE-FREQUENCY RESPONSE[1]

An RLC series circuit is shown in Fig. 1. This circuit is an actual experimental circuit in an experimental box branded "Tianhuang".

When the frequency of the AC voltage source U_s is zero, that is to say, at $f=0$, the capacitor C behaves as an open circuit, and the inductor L behaves as a short circuit. In this case, the output voltage at the resistor R is zero, namely $U_R=0$. Conversely, at $f=\infty$, the capacitor C behaves as a short circuit, and the inductor L behaves as an open circuit. In this case, the output voltage is also zero.

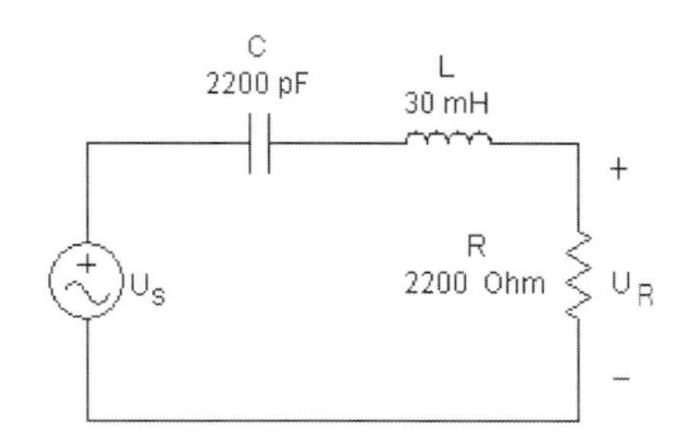

Figure 1. RLC series circuit.

Between $f=0$ and $f=\infty$, both capacitor and inductor have finite impedances Z. In this frequency region, some voltage will reach R. Impedances Z for RLC series circuit is expressed as Equation 1.

$$Z = Z_R + Z_L + Z_C \tag{1}$$

Z_R stands for the impedance of the resistor R, and it equals to R. Z_L stands for the impedance of the inductor L, and it equals to jωL. Z_C stands for the impedance of the capacitor C, and it equals to 1/jωC. The relation between ω and f is expressed as Equation 2. At a certain frequency, Z_L and Z_C have equal magnitudes and opposite signs, they will cancel out. That causes the output voltage maximum. This special frequency f_0 is called resonance frequency. This resonance frequency f_0 is expressed as Equation 3.

$$\omega = 2\pi f \tag{2}$$

$$f_0 = \frac{1}{2\pi\sqrt{LC}} \tag{3}$$

The quality factor Q of RLC series circuit is expressed as equation 4.

$$Q = \frac{1}{R}\sqrt{\frac{L}{C}} \tag{4}$$

The magnitude-frequency response curve is shown in Fig.2. The highest point of the curve is at resonance frequency f_0.

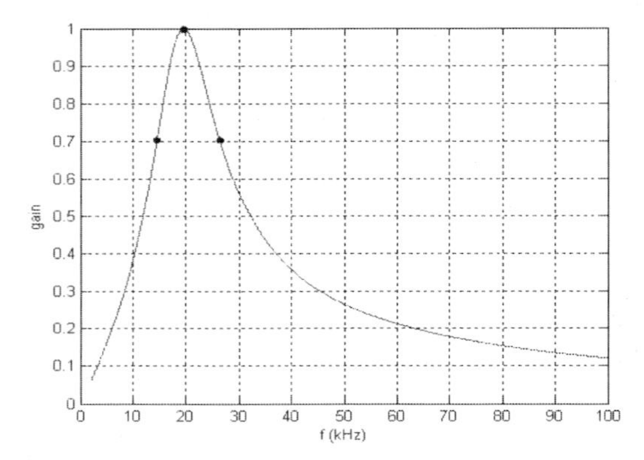

Figure 2. The magnitude-frequency response of the RLC series circuit.

Though Fig.2, for the curve of magnitude-frequency response, at some frequency points such as $f=0$, $f=f_0$, and $f=\infty$,the slopes of the curve are all low. At these frequency points, the output voltage change is not sensitive to frequency change. The magnitude-frequency response measurement will be inaccurate because a large frequency change can only lead to a little output voltage change. Between $f=0$ and $f=f_0$, there will be one frequency point at which the slope has a positive maximum. At the maximum slope frequency point, the output voltage change is more sensitive to the frequency change. So the magnitude-frequency response measurement is most accurate at this frequency point. Likewise, between $f=f_0$ and $f=\infty$, there will be another frequency point, at which the magnitude-frequency response measurement is most accurate too.

For circuit of Fig.1, the output voltage U_R at R can be expressed as Equation 5:

$$U_R = Us \times \frac{R}{\sqrt{(|Z_L + Z_C|)^2 + R^2}} \tag{5}$$

We can get the *gain* of the circuit according to Equation 6.

$$gain = U_R / U_S = \frac{R}{\sqrt{(2\pi fL - \frac{1}{2\pi fC})^2 + R^2}} \tag{6}$$

From Equation 6, the maximum slope points can be calculated by derivation. However, this method is somewhat difficult. If we analyze the magnitude-frequency response curve by means of simulation based on Matlab. The analysis will be simpler, and the result will be more intuitive[2].

3 ANALYSIS OF THE RLC SERIES CIRCUIT'S MAGNITUDE-FREQUENCY DIFFERENCE RESPONSE

3.1 *Magnitude-frequency response curve*

According to Equation 3, the resonance frequency f_0 can be calculated according to Equation 3, and f_0 is 19.591 kHz. So we set the analyzed frequency scope as 2~200 kHz, and the frequency difference step as 0.01 kHz. The frequency array can be expressed as f=2:0.01:200 in Matlab. In this case, gain can be calculated according to Equation 6. When *gain* is the maximum 1.0000, f is 19.59 kHz, so the resonance frequency is about 19.59 kHz. When gain is equal to 0.7071, f is 14.60 kHz and 26.28 kHz respectively. These two frequencies are called low cutoff frequency and high cutoff frequency respectively. The magnitude-frequency response curve is shown in Fig. 2. This magnitude-frequency response curve is also called Bode magnitude plot.

3.2 *Magnitude-frequency first difference response*

According to Fig.2, at f_0, the gain is max and the curve slope is zero. Zero slope means frequency change can hardly lead to gain change. So this frequency cannot be measured accurately.

If the gain is handled with first difference, we can get the slope curve of magnitude-frequency response shown in Fig.3. Gain's first difference (*gfd*) can be expressed as Equation 7:

$$gfd = {}^{\Delta gain}\!/_{\Delta f} = (gain(i+1) - gain(i)) / 0.01 \tag{7}$$

The gain's first difference curve is the same as the one of the gain's slopes. From this curve, there are two extreme value points at two frequency points. At these two frequency points, the slopes get extreme values, so the output voltage change is most sensitive to frequency change, thus the magnitude-frequency response can be measured more accurately. The larger gfd is, the larger slope is, the more sensitive gain change, the more accurately magnitude-frequency response can be measured.

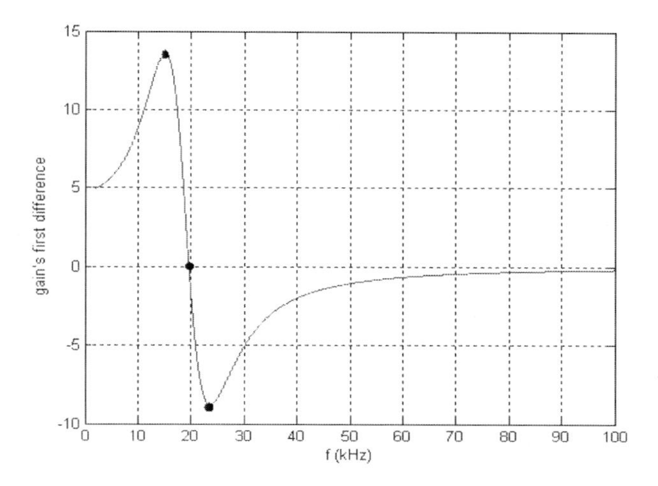

Figure 3. The magnitude-frequency first difference response of the RLC series circuit.

3.3 *Magnitude-frequency response's second difference*

If the gain is handled with second difference, we can get the gain's second difference curve shown in Fig 4. Gain's second difference (*gsd*) can be expressed as Equation 8:

$$gsd = \Delta gfd / \Delta f = (gfd(i+1) - gfd(i)) / 0.01 \tag{8}$$

When *gsd* is equal to zero, at these frequency points, the curve slopes of magnitude-frequency response will be extreme value. Based on the above analysis, the frequencies can be found easily, at which the magnitude-frequency response can be measured more accurately.

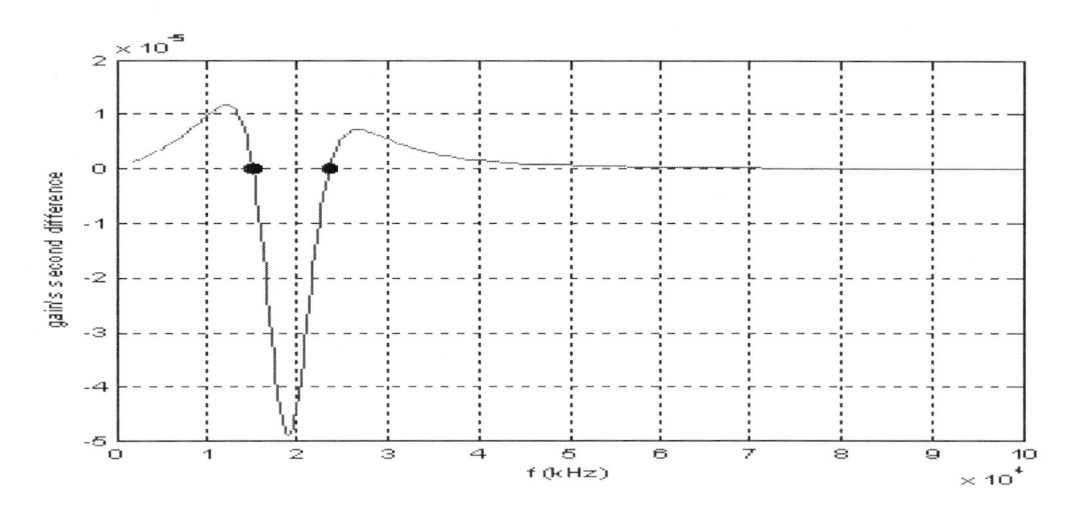

Figure 4. The magnitude-frequency second differential response of RLC series circuit.

From this second difference analysis, at two frequency 15.10 kHz and 23.61 kHz, the second differences response are both zero, and the gain's slopes are extremum, so the output voltage change is most sensitive to frequency change, and the measurement of magnitude-frequency response is most accurate. These two frequency points are called the magnitude-frequency response curve's inflexion points.

3.4 *Magnitude-frequency response's contrast at some typical points*

As analyzed above, the measurement precision of output voltage at R is different at different frequencies. Some typical points are analyzed based on Matlab simulation. The result is shown in Tab. 1.

As is known, the relation among f_0, f_L and f_H is expressed as Equation 9.

$$f_0^2 = f_L \times f_H \tag{9}$$

If we measure f_0 by means of measuring f_L and f_H, the measurement precision of f_0 will be improved about ten times, because f_L and f_H can be measured much more accurately.

4 CONCLUSION

According to the difference analysis above, measurement precision of magnitude-frequency response at one frequency may be different from another one. To measure a parameter in experiments, if a more rational method is selected, the precision can be improved greatly. So choosing a smart experimental method is very important.

Table 1. Magnitude-frequency response's contrast of some typical points.

Typical Points	Gain	Frequency Change Range when Output Voltage Change 1% (kHz)	Center Frequency(kHz)	Scale of Frequency Change Range (kHz)	$\frac{\Delta gain}{\Delta f}$(kHz^{-1})	Output Voltage Change Sensitivity	Measurement Precision
Resonance frequency	1	18.78–20.44	19.59	1.66	0	low	low
Low cutoff frequency	0.7071	14.56–14.65	14.60	0.09	0.0008479	high	high
High cutoff frequency	0.7071	26.20–26.36	26.28	0.16	-0.0004714	high	high
Plus maximum slope	0.7492	15.06–15.14	15.10	0.08	0.0008547	maximum	maximum
Minus maximum slope	0.8460	23.54–23.69	23.61	0.15	-0.0005521	high	high

From Tab. 1, when the frequency changes between 18.78 and 20.44 kHz near the resonance frequency f_0, the output voltage at R changes 1%. The scale of the frequency change range is 1.66 kHz. When frequency changes between 14.56 and 14.65 kHz near f_L, the voltage at R changes 1% too. In this case the scale of the frequency change range is 0.09 kHz. Though the voltage error range is same, the frequency error range is different. The measurement error of f_L is much less than that of f_0. That is to say, the cutoff frequency f_L can be measured much more accurately. The traditional method through measuring the maximum voltage will lead to a large error. If we measure f_0 through the following method, the measurement precision can be improved.

REFERENCES

[1] James W. Nilsson, Susan A. Riedel. (2005) *Electronic Circuits Seventh Edition*. Publishing House of Electronics Industry.

[2] Steven T. Karris. (2003) *Circuit Analysis II with Matlab Applications*. California: Orchard Publications, Fremont.

Electric, Electronic and Control Engineering – Shao, Shu & Tian (Eds)
© 2015 Taylor & Francis Group, London, ISBN: 978-1-138-02842-5

Research on grey neural network based on genetic algorithm used in the air pollution index model

Binbin Chen* & Haoran Wu
School of Mathematical Sciences, Heilongjiang University, Harbin, China

ABSTRACT: This article lay emphasis on API with complex non-linear characteristics and established gray neural network model. We also use a genetic algorithm to optimize it and tested it using the API in the spring of Harbin. The results show that: Compared with traditional neural network model, the genetic algorithm optimized gray neural network model has improved much to predict more accurately. So the method is effective.

KEYWORDS: gray system; neural network; air pollution index; genetic algorithm.

1 INTRODUCTION

Air quality is heavily dependent on local weather condition[1]. Under different weather conditions, the concentration of pollutants on the ground caused by the same pollution emissions can vary several times or even hundreds of times and artificial neural network (ANN) is the effective tool to describe and characterize the nonlinear phenomenon, but on the condition that the amounts of data are small, ANN lacks the precision of prediction. In addition, its training is slow and is easy to fall into local optimum. Taking advantages of the gray system under a smaller amount of data, we establish a gray neural network and applied genetic algorithm optimization to the forecast of the air pollution index.

2 MODEL ASSUMPTIONS

1 We do not take into consideration the influence of difference in sources of pollution and uneven distribution of pollution
2 We do not consider the impact of sudden natural disasters on air quality.

3 SOURCES OF DATA

This article selects the conventional meteorological data from the Harbin station from March 1, 2010 to May 31, 2010. The meteorological data include the average pressure, average temperature, average relative humidity and API on the day. When constructing the BP neural network, we need to determine the input and output data, the input data is the predicted conventional meteorological data that day and the output data is API on the day. Also taking into account the transport of contaminants to migrate will take some time, the general air condition in the day before has the greatest impact on API in the next day, so the prediction model input data should contain API in the day before and we have collected most information that is needed.

Input data for forecasting day routine meteorological data, output data was the day API. Having considered that the API index and meteorological conditions in Harbin have a significant seasonal variation[2], we take spring in Harbin for example to establish grey neural network forecast model

4 GREY THEORY

A Grey system theory is a new method applied in research with small amount of data, poor information and an uncertain problem minority, according to a research, poor information and uncertain problems. It is put forward by Professor Julong Deng in 1986.

In the mid 1980s, the world sees the re-emergence of research about the artificial neural network theory [5] and its application. Because the neural network is composed of a large number of neurons in a non-linear model of the system of organic combination and it can express the complex nonlinear relationship without the requiring analysis objects meet certain rules, the neural network method used for the analysis of air meteorological data and processing will be very effective.

*Corresponding author; This research was supported by the Dawn commune social welfare practice.

Gray behavior problem refers to the prediction of changes in characteristic uncertainty of the grey system. After accumulating The original series of the uncertain system characteristic $x_t^{(0)}$(t=0,1,2,...,N-1), the sequence $x_t^{(1)}$ received observes the exponential growth law, which can be fitted and forecasted using a continuous function or differential equation. In order to facilitate the expression, we can redefine the symbol. Original series $x_t^{(0)}$ can be represented as x (t) and obtained columns after an accumulation $x_t^{(1)}$ is expressed as y (t), the prediction result $x_t^{*(1)}$ is expressed as z (t). Differential equations of gray neural network model parameters for n is

$$\frac{dy_1}{dt}+ay_1=b_1y_2+b_2y_3+\quad +b_{n-1}y_n \tag{1}$$

In the formula, $y_1,y_2,\ldots,y_n$ are system input parameters, and $a,b_1,b_2,\ldots,b_{n-1}$ are coefficients of differential equation.

The time response formula of formula (1) is

$$z(t)=\left(y_1(0)-\frac{b_1}{a}y_2(t)-\frac{b_2}{a}y_3(t)-\quad -\frac{b_{n-1}}{a}y_n(t)\right)e^{-at}+ \frac{b_1}{a}y_2(t)+\frac{b_2}{a}y_3(t)+\quad +\frac{b_{n-1}}{a}y_n(t) \tag{2}$$

We can make

$$d=\frac{b_1}{a}y_2(t)+\frac{b_2}{a}y_3(t)+\quad +\frac{b_{n-1}}{a}y_n(t) \tag{3}$$

formula(2)can be transformed as:

$$z(t)=\left(\left(y_1(0)-d\right)\bullet\frac{e^{-at}}{1+e^{-at}}+d\bullet\frac{1}{1+e^{-at}}\right)\bullet\left(1+e^{-at}\right)= \left(\left(y_1(0)-d\right)\bullet\left(1-\frac{1}{1+e^{-at}}\right)+d\bullet\frac{1}{1+e^{-at}}\right)\bullet\left(1+e^{-at}\right)= \left(\left(y_1(0)-d\right)-y_1(0)\bullet\frac{1}{1+e^{-at}}+2d\bullet\frac{1}{1+e^{-at}}\right)\bullet\left(1+e^{-at}\right) \tag{4}$$

Reflect transformed Equation (4) into an expanded BP neural network and we can get n input parameters as well as an input parameter gray neural network. The network topology structure is shown in figure 1.

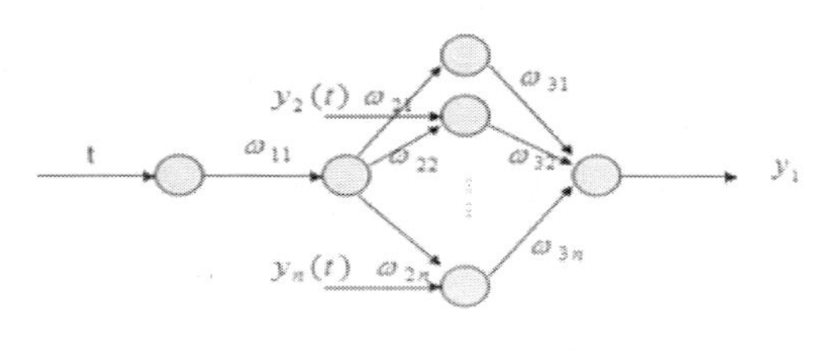

Figure 1. The topology structure of the gray neural network.

In figure 1, t stands for the serial number of input parameters: $y_2(t),\ldots,y_n(t)$are the network input parameters; $\omega_{11},\omega_{21},\omega_{22},\ldots,\omega_{2n},\omega_{31},\omega_{32},\ldots,\omega_{3n}$ represents the network weights; $y_1$is the predicted value of the network; LA, LB, LC, LD, respectively, represents the four layers of gray neural network.

We make $\frac{2b_1}{a}=u_1,\frac{2b_2}{a}=u_2,\ldots,\frac{2b_{n-1}}{a}=u_{n-1}$, and then the optimal network weights is:

$$\omega_{11}=a,\omega_{21}=-y_1(0),\omega_{22}=u_1,\omega_{23}=u_2,\ldots,\omega_{2n}=u_{n-1}$$

$$\omega_{31}=\omega_{32}=\quad =\omega_{3n}=1+e^{-at} \tag{5}$$

The threshold of output node in LD layer is

$$\theta=(1+e^{-\omega t})(d-y_1(0)) \tag{6}$$

The grey neural network learning process is as follows:

Step 1: According to the characteristics of training data to initialize the network structure, decide initialization parameters a, b, and calculate u according to a and b.

Step 2: Calculate with the network weights definition:

$$\omega_{11},\omega_{21},\omega_{22},\ldots,\omega_{2n},\omega_{31},\omega_{32},\ldots,\omega_{3n} \tag{7}$$

Step 3: For each input sequence (t, y (t)), t = 1, 2, 3, ... , N, calculate the output of each layer.

Layer LA:

$$a=\omega_{11}t \tag{8}$$

Layer LB:

$$b=f(\omega_{11}t)=\frac{1}{1+e^{-\omega_{11}t}} \tag{9}$$

Layer LC:

$$c_1=b\omega_{21},c_2=y_2(t)b\omega_{22},c_3=y_3(t)b\omega_{23},\ldots,c_n =y_n(t)b\omega_{2n} \tag{10}$$

Layer LD:

$$d=\omega_{31}c_1+\omega_{32}c_2+\ldots+\omega_{3n}c_n-\theta_{y1} \tag{11}$$

Step 4: Calculate network prediction error between the output and the expected output, and make adjustment of weights and thresholds. Error in layer LD:$\delta=d-y_1(t)$

Error in layer LC:

$$\delta_1 = \delta(1+e^{-\omega_{11}t}), \delta_2 = \delta(1+e^{-\omega_{11}t}), \ldots,$$
$$\delta_n = \delta(1+e^{-\omega_{11}t}) \quad (12)$$

Error in layer:

$$\delta_{n+1} = \frac{1}{1+e^{-\omega_{11}t}}\left(1-\frac{1}{1+e^{-\omega_{11}t}}\right)(\omega_{21}\delta_1 + \omega_{22}\delta_2 + \quad + \omega_{2n}\delta_n) \quad (13)$$

Adjust weights according to prediction error.

Adjustment of the linkage weights from layer LB to layer LC:

$$\omega_{21} = -y_1(0), \omega_{22} = \omega_{22} - \mu_1\delta_2 b, \cdots, \omega_{2n} = \omega_{2n} - \mu_{n-1}\delta_n b \quad (14)$$

Adjustment of the linkage weights from layer LA to layer LB: $\omega_{11} = \omega_{11} + at\delta_{n+1}$

Adjustment of the threshold:

$$\theta = (1+e^{-\omega_{11}t})\left(\frac{\omega_{22}}{2}y_2(t) + \frac{\omega_{23}}{2}y_3(t) + \quad + \frac{\omega_{2n}}{2}y_n(t) - y_1(0)\right) \quad (15)$$

Step 5: Determine whether it is the end of the training, if not, return to step 3.

5 GENETIC ALGORITHM THEORY

Genetic algorithm was founded by professor Holland at the University of Michigan in 1962, which is a parallel random search optimization method using simulation of natural genetic mechanisms and biological evolution. It introduced the nature of "survival of the fittest" principle of biological evolution to optimize the encoding parameters of a series in the group and select the individual through genetic selection, crossover and mutation screening according to the selected fitness function to make fitted individuals retained and unfitted individuals eliminated. The new group inherits the information of the previous generation and is also better than the previous generation. This cycle is repeated until conditions are met. Basic operations of Genetic algorithms include Selection operation, crossover operation and mutation operation.

The basic elements of Genetic algorithm include chromosome coding method, fitness function, genetic operation and operation parameters.

6 ESTABLISHMENT AND SOLUTION OF MODELS

6.1 *The solution of gray neural network model*

The air pollution index prediction algorithm based on the grey neural network process is shown in figure 2. Among them, the building process of the grey neural network is decided according to the dimension of the input and output data. Because of the five-dimensional input data and the one-dimensional output data, the gray neural network architecture is 1-1-6-1, that is to say, there is a node in LA layer, one node in LB layer and 6 nodes in LC layer. In addition, the input time series is t and the second to the sixth inputs are API the day before, average pressure, average temperature, average relative humidity and the average temperature factor of five normalized data. Then, the output is API on the day.

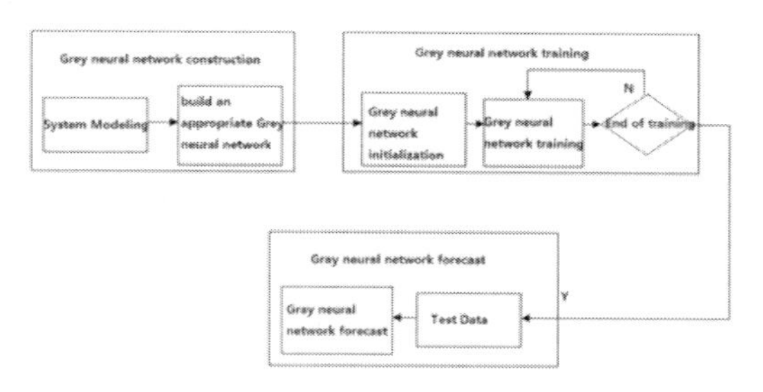

Figure 2. Neural network process.

Grey neural network training uses the training data to train the network so that the network has the ability to predict API. Grey neural network forecast API through the network, and determine the network performance based on prediction error. In our research, there are data in the past 92 days, and we choose data in the first 80 days as training data to train the network. The network has evolved through learning for100 times. At last, we use the remaining 12 sets of data to predict the performance evaluation of the network.

Neural network training process as shown in Figure 3:

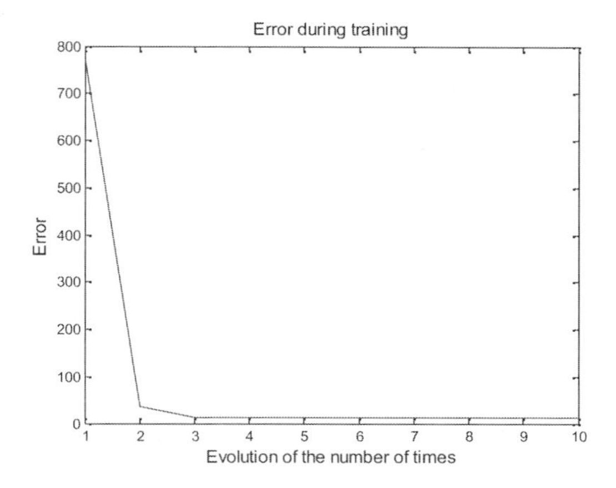

Figure 3. Grey neural network training process.

As can be seen from Figure 3, the speed of gray neural network convergence is very fast, but the network quickly falls in local optimum, unable to further correct parameters. No further correction parameters. Using trained gray neural network to predict air pollution index API, we have predicted the results as shown in figure 4.

The average error of the Grey neural network prediction is 10.21%, and that of the BP neural network is 14.85%, indicating that the grey neural network is better than BP neural network in the study the problem. The average error mentioned here is to take the absolute value of the error and then calculate its average value.

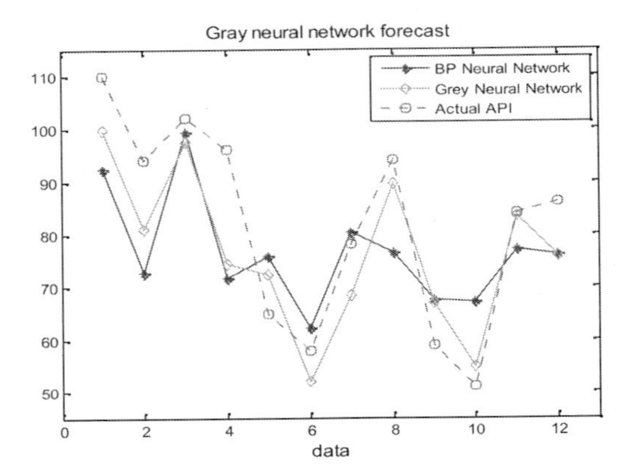

Figure 4 Grey neural network forecasting.

6.2 *Grey neural network based on genetic algorithm optimization*

The Grey neural network is better than the weight threshold random initialization as its network evolution is easy to fall into local optimum and its results are not the same each time. The Grey neural network was optimized by applying genetic algorithm to weights and thresholds in each layer and regarding individual prediction error as the fitness values of the individual. Population size is 30, the number of iterations is 100 times, the relationship between genetic algorithm the optimal individual fitness value and the change of the number of iterations is shown in figure 5.

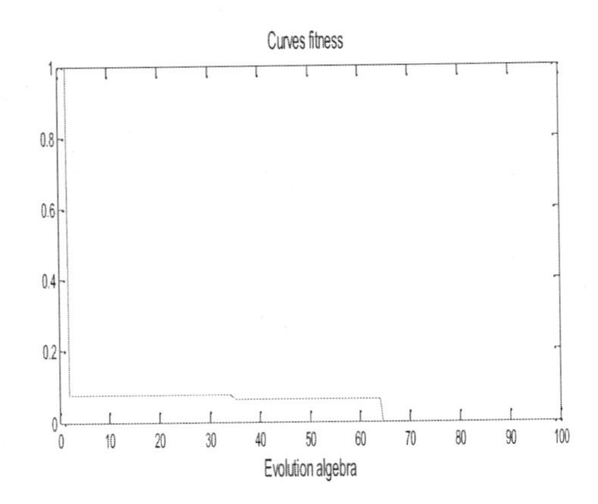

Figure 5. Genetic algorithm optimization process.

Make the optimized parameters and the threshold value the initial value of the grey neural network, and the predicted results of network after training is shown as below.

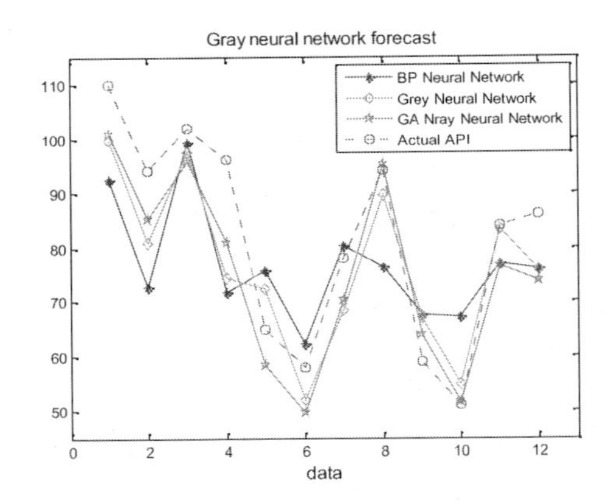

Figure 6. Grey neural network forecasting genetic algorithm optimization.

The average error of the optimized gray neural network through genetic algorithm is 8.97%, having achieved good results compared with that of an unoptimized gray neural network with error of 10.21%. Comparison of the average relative error in three neural networks is shown in Table 1.

Table 1. Error comparison of the three models.

	BP Neural Network	Grey Neural Network	GA Gray Neural Network
The average relative error	14.85%	10.21%	8.97%

7 CONCLUSIONS

1. This article takes spring in Harbin for example to build the grey neural network model. Taking into account the transport of contaminants, we add the API the day before as a predictor, improving the prediction accuracy of neural network. Having considered the randomness of the initial value during training and the disadvantage of easy to fall into local optimum, we use a genetic algorithm to optimize. After comparative analysis of the predicted results of the model, we found that model established can well describe the complex nonlinear relationship between air pollution index and influencing factors, and we have overcome the traditional neural network model dependency on large amounts of data.
2. Gray neural network optimized by the genetic algorithm has a good forecasting effect on air pollution index in Harbin. Moreover, it is simple and can be extended in studies of specific atmospheric pollutant concentrations. And we firmly believe that it will gain more popularity in researches on air pollution in the near future.

ACKNOWLEDGEMENTS

The research results benefit from Heilongjiang University of dawn commune public welfare project and received support from the Yum Heilongjiang market, Youth Development Foundation of Heilongjiang Province and Heilongjiang University. Now we are making use of this chance to express our sincere gratitude to them and thanks for the guidance of Dr. Shimo Wang.

REFERENCES

[1] Jacob D J, Winner D A. (2009) Effect of climate change on air quality. *Atmospheric Environment*, 43:51–56.
[2] Weimei Jiang, Jianning Sun, Wenjun Cao, et al. (2004) *Air Pollution Meteorology Tutorial*. Beijing: China Meteorological Press.
[3] Wang Fang, Cheng Shuiyuan, Li Mingjun, Fan Qing. (2009) Optimizing BP networks by means of genetic algorithms in air pollution prediction. *Journal of Belting University of Technology*, 35(9):1230–1234.
[4] Wang Wei, Ma Qinzhong, Lin Ming zhou, Wu Gengfeng, Wu Shaochun. (2005) Primary component analysis method and reduction of seismicity parameters. *Acta Seismologica Sinica*, 27(5):524 – 531.
[5] Bai Heming, Shen Runping, Shi Huading. (2013) Forecasting model of air pollution index based on BP neural network. *Environmental Science & Technology*, 36(3):186–189.
[6] Liu Shunkui, Li Rongfeng. (1999) The Application of Neural network in the Analysis and Prediction of Earthquakes. *Journal of Xiamen University (Natural Science)*, 04 .
[7] Hecht N R. (1987) Counterpropagation networks. *Applied Optics*, 26(12):4979–4984.
[8] Xue Zhigang, Liu Yan, Chai Fahe, Liang Guixiong, Xu Feng,Z hang Kai, Tao Jun. (2011) Schemes and Demonstrations for Air Pollution Index Improvement. *Research of Environmental Sciences*, 24(2):125–132.
[9] Zhu Guocheng, Fang Mingjian, Zheng Xuxu, Yin Zhongyi. (2010) Study on NOx concentration prediction in a street canyon based on artificial neural networks. *Chinese Journal of Environmental Engineering*, 4:875–880.

Multimedia, Communication and Computing Application – Li (Ed.)
© 2015 Taylor & Francis Group, London, ISBN 978-1-138-02775-6

Study on the influence of the cooperation network based on the PageRank algorithm

Binbin Chen* & Haoran Wu
School of Mathematical Sciences, Heilongjiang University, Harbin, China

Gongwen Xu
School of Computer Science and Technology, Shandong Jianzhu University, Jinan, China

ABSTRACT: This article lay emphasis on complex co-author network problems and improved the traditional PageRank algorithm to build an author influence model and paper influence model, and Erdos's co-author network were analyzed and discussed as an example. The results show that this model is effective, and can be extended to more social networks, breaking the limitations of the PageRank algorithm.

KEYWORDS: PageRank algorithm; Co-author cooperation network; The influence of paper.

1 INTRODUCTION

By means of vast hyperlinks in the network, PageRank can confirm the rank of a page. Google interprets the link from page A to page B as a vote from page A to page B.Google can determine a new rank according to the source of the vote (even the source of the source, that is, the page that linked to page A) and the rank of the voting objective.To put it simply, a high rank of the page can level up the other low rank of pages.

This paper improves the algorithm of PageRank and applies it to cooperation network to study its influence.

2 CO-AUTHOR NETWORK OF ERDOS

Social Network Analysis is a set of method of mathematical analysis, which was derived by Harison White, Boorman, Brieger and Linton Freeman et al. with the mathematical graph theory since the 1960s.

Essentially, aiming to achieve some specific purpose, a social network is a kind of connection network for people's communication and resource utilization. The social network is a dynamic system that includes the relationship among individuals or organizations.

In this paper, the co-author tightness and the overall structure of the network are measured respectively by using network agglomeration degree analysis, clique analysis and centrality analysis.

2.1 *Data extraction*

There are 3 preconditions when we build the co-author network. The second one can determine the co-author at the correct node of Erdos. The third one can confirm the links between them (the relation between co-authors).

1 Leave out the direction of cooperative relations, which means the cooperation between two co-authors is mutual and nondirectional.
2 When a paper has more than three (includes three) authors, consider only the cooperative relation between the other authors and the first author, relations between other non-first authors are not taken into account.
3 The correct node in the co-author network is every author, and the side is the cooperative relation between authors.

Make the adjacency matrix of the network used to construct the network by software after analysis.

*Corresponding author: *595130080@qq.com*

2.2 *Network construction*

We use Ucinet software and 1.1 get the data construction of the co-author network. Shown as figure 1:

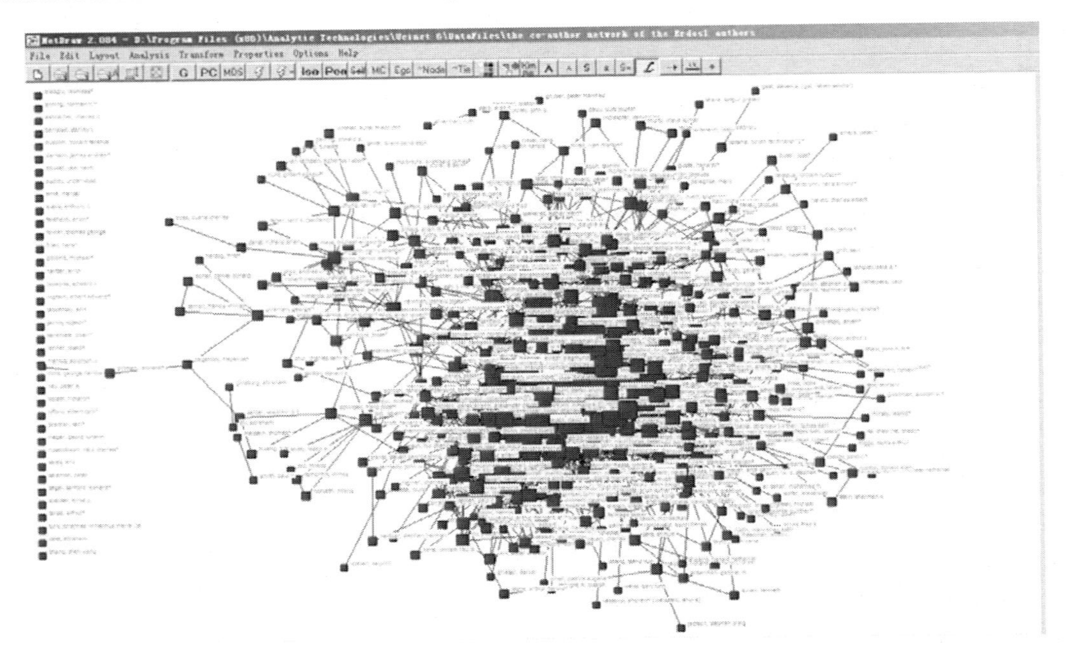

Figure 1. The co-author network of the Erdos1 authors.

2.3 *Network analysis*

Next we analyze this author's co-author network from the clique, network agglomeration degree and centrality.

2.3.1 *Identification and feature analysis of clique*

Clique (or called faction, etc.) is the subgroups combined with a small group of people with close relationship between them. It is the subgraph that belongs to the whole network graph [9, 13]. Clique analysis always adopts some methods to discuss the problem of subgraph, aiming to inspect the structural features of the complete graph itself, so that we can find out that this graph can be divided into several subgraphs that exist naturally.

There are two kinds of methods to calculate clique: one is calculated by the degree of nodes, A group of connected nodes as a clique, including k-plex, k-cor and LambdaSets methods, etc; another kind is calculated by distance, Within a certain distance can reach the point as a clique, including n-clique, n-clan and n-club methods, etc. based on the analysis of the clique of co-author network, using the k-core method of the first kind. Because this method has the advantages of recognition accuracy and easy operation, also can catch a lot of the overall nature of the diagram structure.

Table 1. The number of groups of different scale in the co-author network of the Erdos1 authors.

Scale of groups	1	2	466
Number of groups	36	4	1

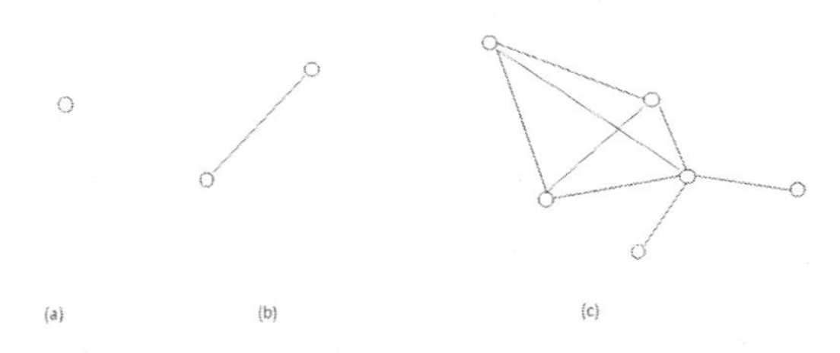

Figure 2. The category of subgroups in the author cooperation network.

After deeply analyzing on the co-author network by using Ucinet software, we find that the co-author network in figure 1 is an unconnected graph, where exists separate cooperative groups. Table 1 shows the number of different scale of groups in co-authors network, so we can see that:

1. There are 3 cliques, 510 authors in the whole co-author network, and the average number of the authors in each clique is 170.
2. Cliques are in the majority, 66.6% are under the scale of 3, but the number of people is only 7.8%. In fact, cliques don't have great effect on the network.
3. The biggest scale of the group is made up of 466 people, which means by the cooperative network of Erdos, authors are being connected to form a big network. According to the structural features of cliques, we use the k-core Method to divide cliques into 3 categories: zero degree group network, double degree group network and multi-degree group network. The first kind of group consists of only 1 person which means publishing papers independently, and it is an isolated point in the whole network as shown in figure 2(a). The second kind of group is made up of 2 people, which is a kind of connected network as shown in figure 2(b). In the third kind of group, the degree of the members is more than 2. The feature of this kind is that the core members have been cooperated as shown in figure 2(c).

2.3.2 *Analysis of agglomeration degree of the co-author network*

For complete graph and each subgraph, the graphic density is an important variable when measure the structural form of group. It can measure the close relationship of a group.

In this article, we calculate the density of the cliques in co-author network and of the whole network, as there are more cooperative behaviors in a close group, leading to an easy way of information communicating, and it is not in the interest of the spreading of knowledge in the colleague and the communicating of information when in a distant group, which will affect the output of the group at last.

The aggregation level of the co-author network is the concentration of the correct node of the authors in the network, that is, how closely the authors cooperate, and it's always been measured by graphic density. The density of a graph is the proportion of the subsistent lines and the possible number of lines in this graph as shown in the following equation:

$$\Delta = \frac{2L}{g(g-1)} \quad (1)$$

L is the number of edges actually exists, g is the number of nodes in the graph $\Delta = 1$. Fully connected co-author network means that all the authors in the network have co-operations with each other. In the unconnected network, $0 \le \Delta < 1$. The bigger the density, the greater the agglomeration degree of the network, and the wider the range of the co-operations among authors, otherwise the co-operations are lesser.

According to the data, we make an analysis of the agglomeration degree of the co-author network of the Erdos1 authors.Shown as Table 2:

Table 2. Agglomeration degree of the co-author network of Erdos.

Scale of groups	1	2	466
L	0	1	3250
g	1	2	466
Δ	0	1	0.03

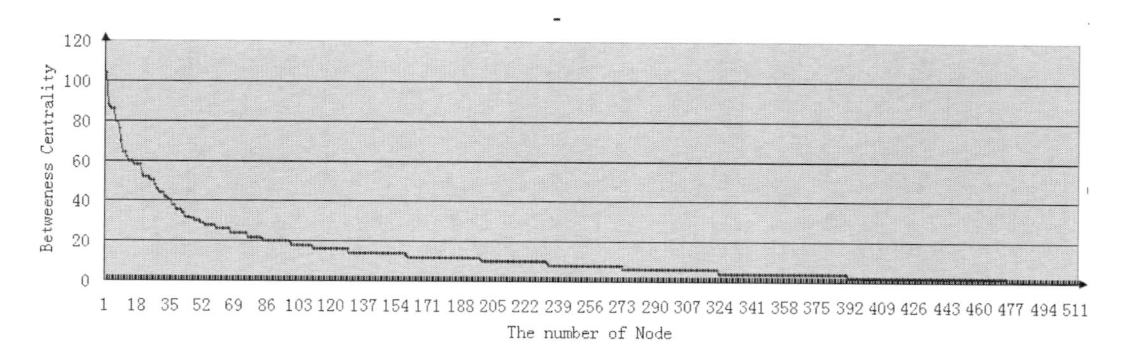

Figure 3. Trend chart of the number of connections of the nodes in co-author network.

The agglomeration degree of the co-author network reflects the level of compactness among the authors. The bigger the value is, the greater the compactness is, and otherwise the looser. Using the Ucinet software, we calculate the agglomeration degree of the whole network and each clique network. The agglomeration degree of the whole network is 0.0251, and the standard deviation is 0.2223. From the agglomeration degree of different scale of clique network in figure 3, we can see:

1 As the scale of the network is increasing, the density of the clique network has a general trend of descending, of course, there exists situations that same scales have different density or same density has different scales.
2 Two-person's co-author network has the biggest agglomeration degree, with the degree of 1. The biggest co-operation group (consists of 56 people) has the smallest agglomeration degree, with the degree of only 0.04. By analyzing the structure of the clique network, we think the reason why the agglomeration degree is small is because that even though these networks are connected, it's still far from fully connected, and although there are co-operations among authors, not all the authors in the group have co-operations with each other.

2.4 *Co-operation network analysis*

In this article, we use the social network analysis to study the co-operations of the Erdos' co-authors, and we get several conclusions.

1 There are big co-operation groups in the co-author network, and the biggest one is made up of 466 people.
2 There is a negative exponent relationship between the number of the cliques and the scale of the group, between the agglomeration degree of the group and the scale of the group, which shows a certain feature of scale-free.
3 This article has great meanings in aspect of identifying the co-operation group in the field of science, strengthening the communication and co-operation among different groups, and achieving leapfrog development in the field of science.

3 A MODEL OF THE INFLUENCE OF THE AUTHOR BASED ON THE ALGORITHM OF PAGERANK

Now we apply the concept of PageRank to the influence model.

3.1 *Downward influence model*

Through analyzing we know that the influence of A equals the sum of the average influence of other co-authors.

$$PR(A) = \frac{PR(B)}{L(B)} + \frac{PR(C)}{L(C)} + \frac{PR(D)}{L(D)} + \qquad (2)$$

At last, all these are converted to a percent .Because the PageRank that transmitted by the people who "don't have co-operations with others" will be zero, so through the mathematical system, we use Google to give every page a method of minimal value 1/ N, we attach a minimal value ∂_1 to every author.

$$PR(A) = \left(\frac{PR(B)}{L(B)} + \frac{PR(C)}{L(C)} + \frac{PR(D)}{L(D)} + \quad \right) + \partial_1 \qquad (3)$$

Explanation: in the original text of Sergey Brin and Lawrence Page, the minimal value they set for each page is 1-d, not the ∂_1 here.

So an author's PageRank is calculated by other authors' PageRank, by repeatedly calculating the PageRank of every page. If every author is given a random PageRank value (not 0), then after repeatedly calculating, the PR value of these pages will tend to be stable, that is, a state of convergence. This is the reason why we use ∂_1. However, when it comes to practical issues, authors who have made important achievement, or who have an important research relationship with Erdos will make a difference to the influence. So we give the different projects of different co-authors their own weight. The equation is as below:

$$\mathrm{PR}(p_i) = \partial_1 + d_{p_j} \sum_{p_j \in M(p_i)} \frac{\mathrm{PR}(p_j)}{L(p_j)} \qquad (4)$$

p_j: weights for p_j, p_j $p_i \in M(p_i)$, $p_1, p_2, \ldots, p_n$ the author studied, $M(p_i)$: a set links to the author p_i, $L(p_j)$: the number of authors linking from p_j, ∂_1: a random number not zero.

3.2 *Feedback influence model*

Considering that the key point of a paper lies in the main author, the main author should take up more weight. If we don't correct this, the result won't be practical enough. So we build feedback influence model to a better correction. We give main author the average PR value which comes from the secondary authors who have co-operations with the main author.

$$\text{PR}(p_i') = \partial_2 + \sum_{p_j \in M'(p_i)} \frac{\text{PR}(p_j)}{L(p_i)} \tag{5}$$

p_i' is the main author, $PR(p_j)$: the influence of secondary co-author with p_i', $L(i)$: the number of co-author with p_i', $M'(p_i)$: the set of co-author with p_i'.

3.3 *The model after correction*

$$\text{PRL}(p_i) = (1-\theta)\cdot \text{PR}(p_i) + \theta \cdot \text{PR}(p_i)$$

$$= \partial_3 + (1-\theta)\cdot d_{p_i} \sum_{p_j \in M(p_i)} \frac{\text{PR}(p_j)}{L(p_j)} + \theta \cdot \sum_{p_j \in M'(p_i)} \frac{\text{PR}(p_j)}{L(p_i)} \tag{6}$$

The θ is the weight of feedback influence and $\theta \in (0,1)$ $\partial_3 = \partial_1 \cdot (1-\theta) + \partial_2 \cdot \theta \cdot PRL(P_i)$ is the final PR value of p_j.

4 A MODEL OF THE INFLUENCE OF THE PAPER BASED ON THE ALGORITHM OF PAGERANK

Paper's influence can be considered from the influence of co-author. It can be measured by the weighted sum of influence of co-author in the network.

4.1 *Model establishment*

The influence of a paper can be expressed as the weighted sum of the influence of the co-author. We build an influence model based on the co-author's influence:

$$\text{P(i)} = d_{q_j} \sum_{q_j \in N(i)} \frac{\text{PR}(q_j)}{L(q_j)} \tag{7}$$

P(i): the ability of i-th paper, q_j: all authors of this paper, $L(q_j)$: the number of author, $N(i)$: the set links to this paper, d_{p_i} : the weight of different authors, as it is always the first author that plays an important part in the influence of a paper.

$$\text{P(i)} = \sum_{q_j \in N(i)} \frac{\text{PR}(q_j)}{L(i)} + \rho_1 \cdot PR(q_m) + \rho_2 \cdot PR(q_f) \tag{8}$$

P(i): the ability of i-th paper; q_j: all authors of this paper; $L(q_j)$ the number of author; $N(i)$:the set links to this paper; $PR(q_m)$:the ability of the most influential author; $PR(q_m)$: the influence of the first author; ρ_1:the weights of the most influential author; ρ_2: the weights of the first author, because the first author on the paper occupies an important impact on weight.

5 SOLUTION AND ANALYSIS OF MODEL

5.1 *The solution of the model after correction*

Because ∂_3 is a random value that is not 0, finally the influence will be converged to a fixed value, here we take ∂_3=1 .In addition, due to the time limits, the enormous data doesn't show the importance of the co-operation. As a result, we don't take weight into account here, but a more accurate answer can be gained if plugging the proper weight into the practical problems according to the actual conditions.

Taking advantage of the data the topic gave and our procedure (in the appendix), we obtain the specific situations about the rank of the influence (Table 6). As space is limited, we only list the top ten influences.

Let d_{p_i} =1. To simplify the calculations, we also let θ = 0.618. This is a subjective estimate, and you can also use a more scientific approach, such as regression to determine the weight. In the future, we will be more in-depth study.

By using the data of Erdos co-author network, we conclude that the specific situation of influence ranking as shown in table 3. Due to limited space, we only give the influence of the top ten rankings.

Table 3. Ranks of co-authors in the network.

Rank	author	PRL(p_i)
1	SOS, VERA TURAN	5.89461
2	TURAN, PAL*	5.04239
3	ANDRASFAI, BELA	4.68456
4	HAJNAL, ANDRAS	4.23336
5	GRUNWALD, GEZA*	4.20375
6	SZEKERES, ESTHER KLEIN*	4.05863
7	MALOUF, JANICE L.	4.00977
8	POMERANCE, CARLBERNARD	3.97602
9	SELFRIDGE, JOHN L.	3.86859
10	SARKOZY, ANDRAS	3.68321

By the analysis of Table 3, we think the most influential authors are SOS, VERA TURAN.

5.2 *Solution of model based on the author's influence*

Because of data limitations, we can't list the influence of all works in a short time. We use the model to calculate its influence and list the top ten PR value. Shown in Tab4:

Table 4. Ranking of the article.

Rank	Article	PR
1	Emergence of scaling in random networks	65.5
2	Statistical mechanics of complex networks	64.5
3	Models of core/periphery structures	4.08
4	On Random Graphs	4.08
5	Networks, influence and public opinion formation	3.327
6	Identity and search in social networks	3.20
7	Identifying sets of key players in a network	2.11
8	The structure of scientific collaboration networks	1.17
9	Scientific collaboration networks:||. Shortest paths, weighted networks and centrality	1.17
10	The structure and function of complex network	1.17

6 CONCLUSIONS

1 PageRank itself is a fairly successful algorithm based on the analysis of hyperlinks .It can evaluate the importance of the research objects effective, and hence the scientificity of our model deserves recognition.

2 Our model is based on the PageRank algorithm, and applies our idea to the model, as we introduced earlier. What's more, the maximize impact of the result is more effective and the sphere of influence is wider when using the influence model based on PageRank rather than the traditional Linear Threshold Model, Weighted Cascade Model and (Independent Cascade Mode .

3 When building the model, we consider that the different level of importance of different co-authors will make a difference for researchers, so we give them different weight, which reduce the deviation and make the result more scientific.

ACKNOWLEDGEMENT

This research was supported by the Students Innovation and Entrepreneurship Training Program Heilongjiang University (No. 2014SX10).

REFERENCES

[1] [DB/OL] http://en.wikipedia.org/wiki/Page_rank.

[2] Jaide Liu. (2005) *Social Network Analysis*. Beijing: Social Sciences Academic Press (CHINA).

[3] Jun Liu, (2005) *An Introduction to Social Network Analysis*. Beijing: Social Sciences Academic Press (CHINA).

[4] YasminH. Said, Edward J. Wegman, WalidK. Sharabat, I, John T. Rigsby. (2008) Social networks of author–co-author relationships. *Computational Statistics& Data Analysis*, (52):2177–2184.

[5] John Scott, (2007) *Social Network Analysis*. Chongqing: Chongqing University Press.

[6] Fu Yun; Niu Wenyuan; Wang Yunlin; Li Ding. (2009) Analysis on the author cooperation network in the field of science: A case of science research management from 2004 to 2008. *Science Research Management*. (5).

[7] Li Kaixuan, Lin Na, Yang Hongyong. (2007) *Modern Information*. (9):57–61.

Electric, Electronic and Control Engineering – Shao, Shu & Tian (Eds)
© 2015 Taylor & Francis Group, London, ISBN: 978-1-138-02842-5

Palmprint recognition using SURF features

Runsen Geng
National Time Service Center, Chinese Academy of Sciences, Xi'an, China
Key Laboratory for Precise Navigation, Positioning and Timing of Chinese Academy of Sciences, Xi'an, China
Graduate University of Chinese Academy of Sciences, Beijing, China

Xiaojiao Tao & Li Lei
Shaanxi Normal University, Xi'an, China

ABSTRACT: In this paper, we propose a new approach for palmprint recognition by employing the Speeded Up Robust Features (SURF). SURF feature descriptor is invariant to uniform scaling, orientation, partially invariant to affine distortion and illumination changes. The Bhattacharyya coefficient between two gray histograms of each pair matched points is calculate as features. And a fussy k-nearest neighbors' classifier is developed for matching. Experimental results demonstrate that the proposed method yields better performance in orientation, position and illumination compared with the recent methods.

KEYWORDS: SURF; SIFT; point matching; Bhattacharyya coefficient; palmprint recognition.

1 INTRODUCTION

Palmprint recognition is a relatively new biometric technology and has become an active research topic in recent years [1]. A palmprint region has stable information like principal lines, wrinkles, ridges, and minutiae points compared with other technologies, such as iris, finger-print, face, voice [2]. In addition, palmprint-based identifiers are not only user friendly, but also use less data amount and can be operated using cheap electronic imaging device.

In previous studies, palmprint recognition focus on typical structural features includes principal lines, minutia points and delta points [3]. In the palm line based methods, the slope, intercept, inclination or orientation of palm lines are used as features [3-4], and matched directly or represented in other formats for matching. These approaches yield a good identification performance for high resolution palmprint images, but are highly sensitive to the variations of orientation, position and illumination in capturing palmprint images.

For the on-line applications, in order to deal with the features extracted from palmprint images or their transform domains [5-6], appearance based approaches use principal component analysis (PCA), linear discriminant analysis (LDA) or independent component analysis (ICA). Though these approaches are suitable for both high and low resolution palmprint images, the differential ability is not strong enough for the accurate palmprint recognition applications.

Texture is also one of the most clearly and observable palmprint features in low resolution images [7]. In texture based approaches, texture features are extracted by filtering palmprint images using filters such as the Gabor filter, the derivative of Gaussian filter or the wavelet [8–12]. The sample points are encode to two bits codes in the Gabor filtered image in [10], and implement palmprint matching by selecting a nearest neighbor classifier which is based on a normalized hamming distance. Pan and Ruan [11] employ 2DPCA to deal with the Gabor filtered image in both row and column directions. The subspace coefficients are regarded as features. In our early work [12], the palmprint identification is implemented with 2D-Gabor wavelets and support vector machine (SVM). The method yield satisfactory robustness to the variations of orientation, position and illumination in capturing palmprint images and outperforms compared with the Kong's et al. [10], Pan's and Ruan [11] approaches, but its computational cost is dissatisfactory.

The SIFT descriptor [13] is one of the most successful and popular local image descriptor among all the above mentioned descriptors. It has been proven to perform better than the other local invariant feature descriptors [14] until recent time. In earlier paper, SIFT is used to extract the feature, which describes the local features of the image. Badrinath G.S. et al. [15]

applied SIFT on palmprint authentication, the number of matching points between two images is considered to be the same subject or not. Chen and Moon [16] combine SIFT with Symbolic Aggregate approximation (SAX) to 2D data for the palmprint representation and matching, the experiments show that through fusion the SIFT based method can further improve the recognition accuracy. Zhu et. al. [17] using SIFT and PROSAC based matching strategy to identify palmprints. SURF [18] is first presented by Herbert Bay et al. in 2006 inspired by the SIFT descriptor. SIFT has detected more number of features compared to SURF but it is suffered with speed. The SURF is fast and has good performance as the same as SIFT [19]. In Badrinath's later work [20], the use of SURF features in the context of palmprint verification has been investigated, and the decision is still taken based on the number of matching points between two palmprints.

In this paper, SURF is still used to extract the feature, and for each pair of matched points, the average Bhattacharyya coefficient of 10×10 neighboring region around the the point is calculated and regarded as features, and the greater the coefficient between two palmprints, the greater is the similarity between them. Experimental results show the superiority of this approach in terms of the correct classification percentages and lower computational cost compared with the recent approaches.

The rest of this paper is organized as follows. In Section 2, the procedure of extracting SURF keypoints from palmprint image is introduced; the proposed system, and palmprint extraction is described in Section 3; the experimental results are reported in Section 4; finally, the conclusions are presented in Section 5.

2 REVIEW OF SIFT/SURF

2.1 *SIFT algorithm overview*

SIFT consists of four major stages: scale-space extrema detection, key point localization, orientation assignment and keypoint descriptor.

1 Scale-space extrema detection: It is implemented efficiently by using a Difference-of- Gaussian (DoG) function to identify potential keypoints. Given a Gaussian-blurred image:

$$L(x,y,\sigma)=G(x,y,\sigma)*I(x,y) \tag{1}$$

where I(x, y) is the given image and * is the convolution operation at x and y and,

$$G(x,y,\sigma)=\frac{1}{2\pi\sigma^2}e^{-\frac{x^2+y^2}{\sigma^2}} \tag{2}$$

Stable key-point locations in scale space can be computed from the difference of Gaussians separated by a constant multiplicative scalar k:

$$D(x,y,\sigma)=L(x,y,k\sigma)-L(x,y\sigma) \tag{3}$$

2 Key point localization:
In this stage, key point candidates are localized and refined by eliminating the key points where they rejected the low contrast points. Sub-sample accurate position and scale is computed for each candidate feature point by fitting a quadratic polynomial to the scale space DoG function and finding the extremum,

$$\hat{X}=-\frac{\partial^2 D^{-1}}{\partial X^2}\frac{\partial D}{\partial X} \tag{4}$$

Where is the $\hat{X}=(x,y,\delta)^T$ is the extremum position providing accurate position and scale.

Feature points are selected based on measures of their stability and eliminated if found to be unstable. A 2×2 hessian matrix is computed to determine the curvature, giving

$$H=\begin{bmatrix} D_{xx} D_{xy} \\ D_{xy} D_{yy} \end{bmatrix} \tag{5}$$

$$stability=\frac{\left(D_{xx}+D_{yy}\right)^2}{D_{xx}D_{yy}-D_{xy}^2}<\frac{(r+1)^2}{r} \tag{6}$$

where the D denote the respective partial derivatives of $D(x,y,\sigma)$, r is the ratio of maximum eigenvalue to the minimum eigenvalue.

3 orientation assignment
This step aims to assign a dominant orientation to each feature point based on local image gradient directions. The gradient magnitude and orientation is computed as following:

$$m(x,y)=\sqrt{(L(x+1,y)-L(x-1,y))^2+(L(x,y+1)-L(x,y-1))^2} \tag{7}$$

$$\theta(x,y)=\tan^{-1}\left(\frac{L(x,y+1)-L(x,y-1)}{L(x+1,y)-L(x-1,y)}\right) \tag{8}$$

4 keypoint descriptor
This stage is to compute the local image descriptor for each key point based on image gradient magnitude and orientation at each image sample point in a region centered at key point. If the key point in the direction is selected, the feature descriptor can be

calculated by creating orientation histograms over 4×4 neighbourhood. Each descriptor contains array of 4×4=16 histograms around, and each histogram contains 8 bins. This leads to a SIFT feature vector with 4×4×8 = 128 elements. Thus the descriptor is invariant to affine changes in illumination.

2.2 *SURF algorithm overview*

SIFT and SURF algorithms employ slightly different ways of detecting features. SURF is based on multi-scale space theory and the feature detector is based on Hessian matrix. Since Hessian matrix has good performance and accuracy. In image I, x = (x, y) is the given point, the Hessian matrix H(x, σ) in x at scale σ, it can be define as

$$H(x,\sigma)=\begin{bmatrix} L_{xx}(x,\sigma) L_{xy}(x,\sigma) \\ L_{yx}(x,\sigma) L_{yy}(x,\sigma) \end{bmatrix} \tag{9}$$

Where $L_{xx}(x,\sigma)$ is the convolution result of the second order derivative of Gaussian filter $\frac{\partial^2}{\partial x^2} g(\sigma)$ with the image I in point x, and similarly for $L_{xy}(x,\sigma)$ and $L_{yy}(x,\sigma)$.

SIFT builds an image pyramids, filtering each layer with Gaussians of increasing sigma values and taking the difference. Due to the use of integral images, SURF filters the stack using a box filter approximation of second-order Gaussian partial derivatives, since integral images allow the computation of rectangular box filters in near constant time. In descriptors, SIFT uses histogram of gradient directions in the neighborhood as descriptor of the feature, while SURF again takes advantage of integral image, and uses Haar wavelet response $\sum dx$, $\sum|dx|$, $\sum dy$, $\sum|dy|$ as the feature descriptor. The SURF key points are very distinctive for different palms in both amount and location, which can be a strong evidence for discriminating different palms, as shown in Fig.1.

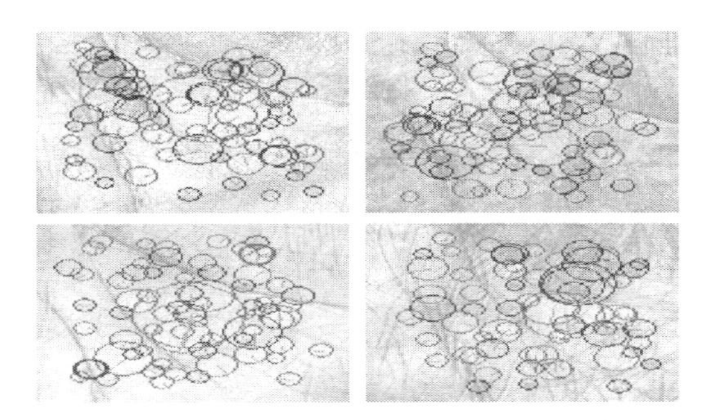

Figure 1. Different SURF key points on different palms.

3 THE PROPOSED APPROACH

The block diagram of the proposed robust approach using SURF features of palmprint is shown in Fig. 2. In the region of interest module image is firstly normalized in the orientation, position and illumination, and region of interest is extracted from the normalized images. Then features are extracted using SURF. All the SURF key points of two images are compared during matching. In keypoint matching step, the nearest neighbor is defined as the keypoint with minimum Euclidean distance for the invariant descriptor vector.

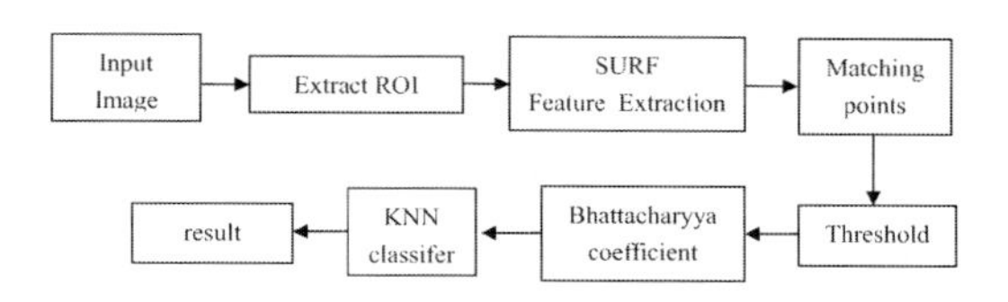

Figure 2. The block diagram of the proposed approach.

Number of matching points represents the similarity between two palmprint images. The number of associations is used as a similarity measure. As shown in Fig.3, there are very many pairs of matched key points in the different palmprint images of the same subject, and Fig.4 shows the matched key points of different subject. It is obvious that the number of matching points between different subject is less, even hardly any, and matched point pairs are always erroneous.

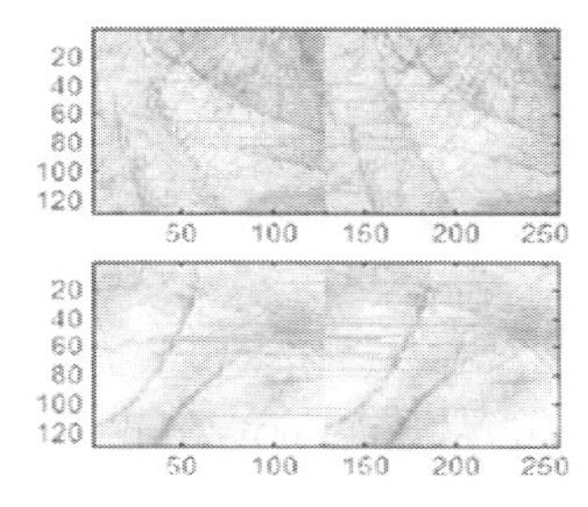

Figure 3. Matching points of same subject.

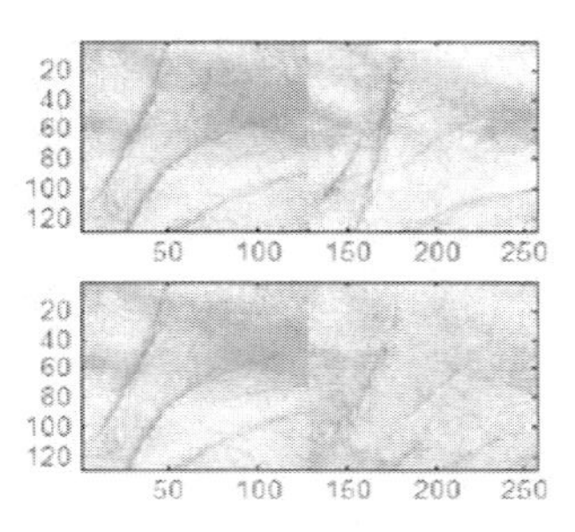

Figure 4. Matching points of different subject.

Only using SURF algorithm to calculate the similarity based on the Euclidean distance of feature points easily lead to generate some erroneous matched point pairs. In order to decrease some of these false matching, we remove the matched point pairs in which the spatial Euclidean distance between the two points are larger than a fixed threshold. Moreover, we improve the SIFT algorithm combined with the distance factor. Given two images A and B, representing the live image and the enrolled image from the database respectively. The location in the scale-space of key point P is (x_0, y_0)in imageA, the SIFT feature vector is P, the location of the best candidate key point Q1 is (x_1, y_1) and the second candidate Q2 is (x_2, y_2) in imageB, the SIFT feature vectors are Q1 and Q2. we have:

$$Ed(P,Q_1) = \sqrt{(x_0 - x_1)^2 + (y_0 - y_1)^2} < d \tag{10}$$

$$\frac{Ed(P,Q_1)}{Ed(P,Q_2)} < r \tag{11}$$

SURF algorithm only consider the global palmprint images while ignored the local features, so we put forward the Bhattacharyya coefficient [21] to match the palmprint images combined with the idea of local feature matches. The Bhattacharyya coefficient can be used to determine the relative closeness of the two samples being considered, the Bhattacharyya coefficient is defined as following formula:

$$BC(p,p') = \sum_{i=1}^{N} \sqrt{p(i)p'(i)} \tag{12}$$

where p and p' represent the image histogram data, BC is image similarity values.

In the method we proposed, the histograms of central pixel's 10×10 neighbors is computed after matching key points, then compute the average Bhattacharyya coefficient, here BC means the similarity of each pair histograms of matched key points. It is obvious to see that $BC \in [0,1]$, if p is very similar with p', we have $BC \approx 1$, whereas p is entirely different from p', i.e. there is no overlapped section between them, we have $BC = 0$.

The proposed fussy *k*-nearest neighbors' classifier consists of three steps:

Step1: Given a palmprint x and a train set $S = \{y_i, 1 \le i \le L\}$, we calculate the BC between x and y_i , and select the *k*-nearest neighbors $\{kn_i, 1 \le i \le k\}$ from the training samples.

Step 2: Sum up the BC of the *k*-nearest neighbors which belong to the same class w_i by

$$S_{w_i} = \sum_{kn_j \in w_i} BC(x, kn_j) \tag{13}$$

Step3: determine the maximum S_{w_m}, and classify x into the class w_m.

4 SIMULATION RESULTS AND PERFORMANCE ANALYSIS

The proposed system has also been tested on the database from Hong Kong Polytechnic University (PolyU) [22] which consists of 7752 grayscale images corresponding to 386 different palms. The images are collected at spatial resolution of 75 dots per inch, and 256 gray levels using CCD. For each palm, the samples were collected in two sessions, where the first 10 samples were captured in the first session and other 10 in the second session. In the second session, they changed the light source and adjusted the focus of the CCD camera so that the images collected on the first and second occasions could be regarded as being captured by two different palmprint devices.

The experiments are conducted as follows:

A. This experiment was conducted to test the classification accuracy of this approach. 60 palmprint images for each subject were randomly selected from the the first session of PolyU-II palmprint database, the first five palmprint images for each subject were regarded as training samples. The others were used for testing. We utilized our approach, principal component analysis [5], linear discriminant analysis [6], dual-tree complex wavelets [9] , Kong's et al. [10], Pan's and Ruan [11] approaches and our early work [12] was performed to implement classification. The correct classification percentages (CCPs) were determined. The resultant values are listed in Table 1.

Table 1. Comparison of the correct classification percentages conducted by our method and recent approaches.

Methods	The correct classification percentage (%)
Kong's approach	92.88
Pan's approach	96.30
Our early work	97.37
Dual-tree complex wavelets	92.35
Principal component analysis	94.25
Linear discriminant analysis	92.05
The proposed method	**97.63**

The experimental results illustrate the proposed system performs better than the earlier known approaches.

B. The objective of this experiment was to verify the robustness to the variations of orientation, position and illumination of the proposed method. The first five samples in the first session of PolyU-II palmprint database were still regarded as the training samples; the samples in the second session were used for testing. The correct classification percentages (CCPs) were determined via our approach and the recent approaches mentioned above. The resultant values are listed in Table 2. It can be concluded from these results that the proposed approach yields relatively high robustness to the variations of orientation, but compared with our early work, the robustness is dissatisfactory.

Table 2. Comparison of the correct classification percentages conducted by our method and recent approaches.

Methods	The correct classification percentage (%)
Kong's approach	88.67
Pan's approach	75.50
Our early work	95.40
Dual-tree complex wavelets	87.25
Principal component analysis	83.25
Linear discriminant analysis	84.05
The proposed method	**89.96**

C. The experiment was to test the computational cost of this approach. The simulation experiment was run on a personal computer with Intel(R) Celeron(R) CPU 3.06GHz, 1.00GB memory, and the operation system is Microsoft Windows XP professional. The first sample in the first session of PolyU-II palmprint database was used, and the computation time to extract features in the approaches mentioned in experiment A was obtained, the results were summarized in Table 4.

Table 3. The computation time to extract features from a palmprint image in different methods.

Methods	Average time (s)
Kong's approach	1.4508
Pan's approach	1.0398
Our early work	1.4436
Dual-tree complex wavelets	0.8210
Principal component analysis	0.7624
Linear discriminant analysis	1.4205
The proposed method	**0.4592**

As shown in Table 4, the computation time to extract features in our approach is less than that of Kong's approach, but it is more than that of principal component analysis, linear discriminant analysis and Pan's approach.

D. We compare the performance of the proposed method to other three SIFT based algorithm. Fig. 5 shows the score distributions of the genuine matching and the imposter matching, as well as the ROC (Receiver Operating Characteristic) curve. We observe that Badrinath G.S. et al. [15] perform not well, that the proposed fusion method over performs Chen et al. [16], and is similar to obtain by Zhu et al. [17].

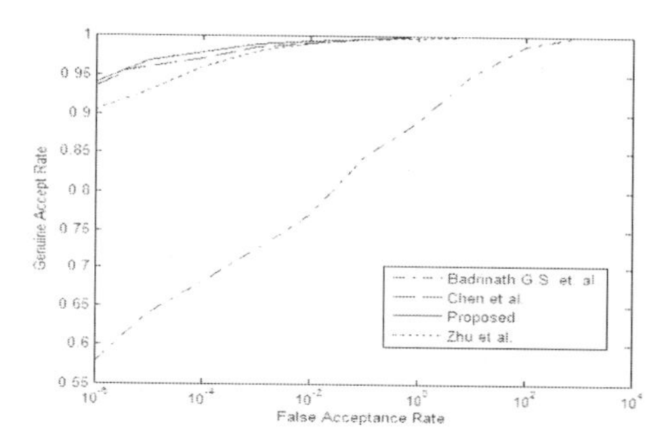

Figure 5. The ROC curve for the PolyU-II palmprint database.

5 CONCLUSION

This paper presents an approach which is robust to rotation, translation and scale of the palm image. Furthermore, it can be easily realized by FPGA. The features extracted using SURF are scale invariant and hence it makes the system to improve its robustness to either high or low resolution palmprint images. Experimental results show that the proposed approach yields a better per -formance in terms of the correct classification percentages with the recent approaches. And compared with the recent approaches, our method has lower com- putational cost. Thus the design of the approach with robustness, performance, and use of low cost scanner for acquisition of palm image suggests its use for criminal detection, personal identification, and user authentication.

REFERENCES

[1] A. Kong, D. Zhang, M. Kamel, (2009) A survey of palmprint recognition, Pattern Recognition, 42:1408–1418.

[2] N. Duta, A.K. Jain, K.V. Mardia, (2002) Matching of palmprint, Pattern Recognition Letters, 23(4):477–485.

[3] D. Zhang, W. Shu, (1999) Two novel characteristics in palmprint verification: datum point invariance and line feature matching, Pattern Recognition, 32(4):691–702.
[4] D.S. Huang, W. Jia, D. Zhang, (2008) Palmprint verification based on principal lines, Pattern Recognition, 41:1316–1328.
[5] X.Q. Wu, D. Zhang, K. Wang, (2003) Fisherpalms based palmprint recognition, Pattern Recognition Letters, 24(15):2829–2838.
[6] G. Lu, D. Zhang, K. Wang, (2003) Palmprint recognition using eigenpalms features, Pattern Recognition Letters, 24:1463–1467.
[7] A. Kong, D. Zhang, and M. Kamel, (2006) Palmprint identification using feature-level fusion, Pattern Recognition, 39:478–487.
[8] W. Li, D. Zhang, Z. Xu, (2002) Palmprint identification by Fourier transform, International Journal of Pattern Recognition and Artificial Intelligence, 16(4):417–432.
[9] G.Y. Chen, W.F. Xie, (2007) Pattern recognition with SVM and dual-tree complex wavelets, Image and Vision Computing, 25:960–966.
[10] W.K. Kong, D. Zhang, W. Li, (2003) Palmprint feature extraction using 2D Gabor filters, Pattern Recognition, 36:2339–2347.
[11] X. Pan, Q.Q. Ruan, (2009) Palmprint recognition using Gabor-based local invariant features, Neurocomputing, 72 (7–9):2040–2045.
[12] X. Wang, L. Lei, M.Z. Wang, (2012) Palmprint verification based on 2D-Gabor wavelet and pulse-coupled neural network, Knowledge-Based Systems, 27:451–455.
[13] D. G. Lowe, (2004) Distinctive image features from scale-invariant keypoints, International Journal of Computer Vision, 2:91–110.
[14] Mikolajczyk, K., Schmid, C. (2005) A performance evaluation of local descriptors, IEEE Transactions on Pattern Analysis and Machine Intelligence, 27(10):1615–1630.
[15] Badrinath G. S., Phalguni Gupta, (2008) Palmprint Verification using SIFT features, Image Processing Theory, Tools and Applications, pp:1–8.
[16] Jiansheng Chen, Yiu-Sang Moon, (2008) Using SIFT Features in Palmprint Authentication, Pattern Recognition, ICPR 2008. 19th International Conference on.
[17] Leqing Zhu, Sanyuan Zhang, Rui Xing, Yin Zhang, (2008) Using SIFT and PROSAC to identify palmprints, Proceedings of the 2008 International Conference on Image Processing, Computer Vision & Pattern Recognition, IPCV 2008, pp:136–141.
[18] Bay, H., Tuytelaars, T., & Van Gool, L. (2006) SURF: Speeded Up Robust Features, 9th European Conference on Computer Vision.
[19] P M Panchal , S R Panchal , S K Shah, (2013) A Comparison of SIFT and SURF, International Journal of Innovative Research in Computer and Communication Engineering, 1(2).
[20] Badrinath G. S., Phalguni Gupta, (2009) Palmprint based verification system using SURF features, Contemporary Computing, Communications in Computer and Information Science, 40:250–262.
[21] Aherne, F., Thacker, N. & Rockett, P. (1997) The Bhattacharyya metric as an absolute similarity measure for frequency coded data. Kybernetika, 32(4):1–7.
[22] The PolyU palmprint database [Online]. Available: http://www.comp.polyu.edu.hk/~biometrics.

Electric, Electronic and Control Engineering – Shao, Shu & Tian (Eds)
© 2015 Taylor & Francis Group, London, ISBN: 978-1-138-02842-5

A BTT missile optimal controller design based on diffeomorphism exact linearization

Changzhu Wei
Department of Aerospace Engineering, Harbin Institute of Technology, Harbin, China

Chao Cheng
Department of Aerospace Engineering, Harbin Institute of Technology, Harbin, China
Beijing Electro-mechanical Engineering Institute, Beijing, China

Yuanbei Gu
Department of Aerospace Engineering, Harbin Institute of Technology, Harbin, China

ABSTRACT: BTT missiles can swiftly steer the maximum lifting surface in the required maneuver direction, so they can better adapt to the complex and ever-changing battle environment. This paper, at first, gives a BTT missile controller based on traditional coordinate decoupling method. Then considering the fact that applying such method is limited with strong nonlinear model, the diffeomorphism transformation method is involved to exactly linearize the whole nonlinear model of BTT missile. Also analysis and demonstration of the stability of generated internal dynamics subsystem are proposed. Finally, an optimal controller combined with this exact linearization model is provided. Furthermore, nonlinear simulation results verify the feasibility and applicability of the diffeomorphism exact linearization optimal control method.

1 INTRODUCTION

In order to better adapt to the complex and ever-changing battle environment, missiles should bare the ability of fast-implemented and large-scale maneuver. Improving the aerodynamic layout and applying advanced control method are effective approaches for those implementations. Under the condition of plane-symmetric aerodynamic layout, the missiles utilizing the Bank-To-Turn (BTT) control strategy swiftly steer the maximum lifting surface in the required maneuver direction by controlling the roll channel, meanwhile obtain desired maneuver acceleration by controlling the pitch channel[1,2]. Because, during the guidance flight of BTT missile, the maximum lifting surface should turn to the optimal direction of guidance requirements with the fastest speed, high-speed rotation will bring remarkable aerodynamic coupling, kinematic coupling and inertial coupling effects. Therefore BTT missile is a typical nonlinear system with uncertain parameters[3]. In recent years, the scholars at home and abroad have done much research work in the aspects of designing BTT missile control system, in general containing classical SISO control system design method and modern control theory design method[4].

After modern control theories have been widely applied in aviation and aerospace field, the scholars worldwide, aiming at solving the problems of SISO design issues, launch studies of using modern control theory to the design of BTT missile autopilot. Lin etc. designed a general autopilot for a Bank-To-Turn (BTT) missile, which is through a generalized Linear Quadratic Gaussian and Loop Transfer Recovery (LQG/LTR) method. The detailed design procedure including two steps, Kalman filter design and an optimal control law design. The proposed controllers have a prescribed degree of stability even in the case of nonminimum phase problem[5]. Meanwhile, some scholars use feedback linearization method to linearize the nonlinear model of BTT missile and design the control system according to linear system theory. Feedback linearization is one of the important approaches of designing the nonlinear control system, dividing into nonlinear dynamic inversion method and differential geometry method[6]. In such two methods, nonlinear dynamic inversion is not limited to the specific form of system equations, which makes it available to investigate general nonlinear system directly. Also its physical concept is clear and intuitive, so this method is widely employed in nonlinear control engineering[7,8]. However, this method cannot deeply analyze the inherent characteristics of nonlinear systems. With the increase of complexity of nonlinear system, the uncertainty of nonlinear dynamic inverse system raise. Therefore in recent years, some scholars have used the differential geometry feedback linearization method to design BTT missile control system. Zhang etc. designed a BTT missile controller based on differential geometry feedback linearization method. Taking the existence of internal dynamics subsystem in the full-state dynamics model into consideration, the BTT missile dynamics model is simplified through the ignorance of lift and lateral forces generated by deflection of rudder surfaces. Also the exact linearization system whose relative order is equal to the dimension of the system is achieved. Then with the linear state feedback, the command orders which can steadily control roll angle, attack angle and side-slip angle are given[9].

The differential geometry feedback linearization method provides decoupled structure through diffeomorphism coordinate transformation and nonlinear state feedback, transforming the problem into "geometry domain". So it is more abstract to understand, but due to the completeness of such theory, this method will definitely play a bigger role in solving the nonlinear control problems. In regard to the design of BTT missile control system applied with differential geometry feedback linearization method, there are still many problems to be solved, such as the necessity of feedback linearization of BTT missile and stability analysis of internal dynamics subsystem in full-state model. This paper will carry on the related researches in the above aspects, and the specific structure of the following components is proposed as below: in the second chapter, on account of the characteristics of high-speed rotation and strong nonlinearity, the whole BTT missile model is exactly linearized based on diffeomorphism theory; in the third chapter, a BTT missile optimal controller based on diffeomorphism exact linearization model is given; in the fourth chapter, nonlinear simulations are conducted to illustrate the performances of the proposed BTT missile controller.

2 EXACT LINEARIZATION OF BTT MISSILE BASED ON DIFFEOMORPHISM TRANSFORMATION

In the process of establishing missile dynamics and kinematics models, the following coordinate systems are involved in: the ground coordinates, the ballistic coordinates, the body coordinates and the velocity coordinates. The detailed definitions of the referred coordinates can be found in Ref. [10].

The exact linearization method based on diffeomorphism transformation theory has aroused the interests of many scholars at home and abroad in recent years, and some achievements have been made during the research. The basic idea of the method is transforming the dynamic characteristics of nonlinear system into (all or part of) that of linear system through proper nonlinear transformation and feedback transformation which will implement input-to-output or input-to-state exact linearization. This procedure can change the complicated synthesis problem of nonlinear system into that of linear system; in turn the familiar linear control methods can be used in design and analysis processes[11]. Totally different from the small perturbation linearization method, the exact linearization is realized with strict state transformation and feedback transformation. In the whole linearization process, there are not any high-order nonlinear items being ignored, so this linearization method is exact. Compared to other feedback linearization method, such as nonlinear dynamic inversion, the exact linearization method based on diffeomorphism can theoretically analyze the unobservable part of system and more clearly acquire the characteristics of system[12].

2.1 *Analysis of BTT missile linearization without feedback*

The nonlinear control model of state-measurable BTT missile is[13]

$$\begin{cases} n_z = \lambda_{71}\omega_x n_y - b_4 n_z + \lambda_{72}\omega_y + \lambda_{73}\omega_x\delta_z \\ \quad + \lambda_{74}\delta_y + \lambda_{75}\delta_{yc} \\ \omega_z = \lambda_{81}n_y + \lambda_{82}\omega_x n_z + \lambda_{83}\omega_x\omega_y + \lambda_{84}\omega_z \\ \quad + \lambda_{85}\omega_x\delta_y + \lambda_{86}\delta_z \\ \delta_z = -\kappa\delta_z - \delta_{zc} \\ n_y = -a_4 n_y + \lambda_{41}\omega_x n_z + \lambda_{42}\omega_z + \lambda_{43}\omega_x\delta_y \\ \quad + \lambda_{44}\delta_z + \lambda_{45}\delta_{zc} \\ \omega_y = \lambda_{51}\omega_x n_y + \lambda_{52}n_z + \lambda_{53}\omega_y + \lambda_{54}\omega_x\omega_z \\ \quad + \lambda_{55}\delta_y + \lambda_{56}\omega_x\delta_z \\ \delta_y = -\kappa\delta_y - \delta_{yc} \end{cases} \tag{1}$$

$$\text{and} \begin{cases} \gamma = \omega_x \\ \omega_x = -c_1\omega_x - c_3\delta_x \\ \delta_x = -\kappa\delta_x - \delta_{xc} \end{cases}$$

where the specific form of each aerodynamic parameter is shown as

$$\lambda_{41} = \frac{a_4}{b_4}, \lambda_{42} = \frac{a_4 V}{g}, \lambda_{43} = \frac{V a_4 b_5}{g b_4}, \lambda_{44} = -\frac{\kappa V a_5}{g}$$

$$\lambda_{45} = -\frac{V a_5}{g}, \lambda_{51} = -\frac{b_1' g}{a_4 V}, \lambda_{52} = \frac{g}{V}\left(\frac{b_2}{b_4} - b_1'\right),$$

$$\lambda_{53} = -(b_1 + b_1')$$

$$\lambda_{54}=\frac{J_z-J_x}{J_y},\lambda_{55}=\frac{b_2b_5}{b_4}-b_3,\lambda_{56}=\frac{b_1'a_5}{a_4},$$

$$\lambda_{71}=\frac{b_4}{a_4},\lambda_{72}=-\frac{b_4V}{g},\lambda_{73}=-\frac{Vb_4a_5}{ga_4},\lambda_{74}=\frac{\kappa Vb_5}{g},$$

$$\lambda_{75}=\frac{Vb_5}{g}$$

$$\lambda_{81}=-\frac{g}{V}\left(\frac{a_2}{a_4}-a_1'\right),\lambda_{82}=-\frac{a_1'g}{b_4V},\lambda_{83}=\frac{J_x-J_y}{J_z},$$

$$\lambda_{84}=-(a_1+a_1'),\lambda_{85}=-\frac{a_1'b_5}{b_4},\lambda_{86}=\frac{a_2a_5}{a_4}-a_3$$

The model of BTT missile can be described as the standard form of affine nonlinear model as follows

$$\begin{cases} X=f(X)+g(X)u \\ Y=H(X) \end{cases} \tag{2}$$

in which, the state variable vector is $X=[x_1,x_2...,x_9]^T=[\gamma,\omega_x,\delta_x,n_y,\omega_y,\delta_y,n_z,\omega_z,\delta_z]^T$.

The system equation is defined as

$$f(X)=\begin{cases} x_2 \\ -c_1x_2-c_3x_3 \\ -\kappa x_3 \\ -a_4x_4+\lambda_{41}x_2x_7+\lambda_{42}x_8+\lambda_{43}x_2x_6+\lambda_{44}x_9 \\ \lambda_{51}x_2x_4+\lambda_{52}x_7+\lambda_{53}x_5+\lambda_{54}x_2x_8+\lambda_{55}x_6 \\ \quad +\lambda_{56}x_2x_9 \\ -\kappa x_6 \\ \lambda_{71}x_2x_4-b_4x_7+\lambda_{72}x_5+\lambda_{73}x_2x_9+\lambda_{74}x_6 \\ \lambda_{81}x_4+\lambda_{82}x_2x_7+\lambda_{83}x_2x_5+\lambda_{84}x_8+\lambda_{85}x_2x_6 \\ \quad +\lambda_{86}x_9 \\ -\kappa x_9 \end{cases} \tag{3}$$

The control variable is $u=[\delta_{xc},\delta_{yc},\delta_{zc}]^T$, and control function g_i is

$$\begin{aligned} &g_1=[0,0,-1,0,0,0,0,0,0]^T, \\ &g_2=[0,0,0,0,0,-1,1_{75},0,0]^T, \\ &g_3=[0,0,0,\lambda_{45},0,0,0,0,-1]^T \end{aligned} \tag{4}$$

The expected output variables are roll angle, overloads of pitch/yaw channel, namely

$$\begin{aligned} Y&=[y_1,y_2,y_3]^T=[h_1(X),h_2(X),h_3(X)]^T \\ &=[x_1,x_4,x_7]^T \end{aligned} \tag{5}$$

The system which can fulfill the linearization without feedback, itself may be linear relationship. However, due to the choice of coordinate system or difficulty of definition to coordinate system in which linearization is implemented, non-linearization is on the surface. In this situation, the linearization process can be handled by coordinate space transfer. If linearization is realized in this pattern, it is the most ideal way because of the advantage of avoiding extra state feedback.

2.2 *The exact linearization model of BTT missile based on diffeomorphism transformation theory*

Solve differential of each output y_i respectively. Assume that r_i is the smallest integer which will make more than one input emerge in output $y_i^{(r_i)}$, then there will be $y_i^{(r_i)}=L_f^{r_i}h_i+\sum_{j=1}^{m}L_{g_j}L_f^{r_i-1}h_iu_j$, where more than one u_j in the neighbourhood U_i of X_0 will make $L_{g_j}L_f^{r_i-1}h_i(X)\neq 0$, namely, $y_i^{(k)}=L_f^kh_i$, $k=0,1,2,...,r_i-1,y_i^{(r_i)}=L_f^{r_i}h_i+\sum_{j=1}^{m}L_{g_i}L_f^{r_i-1}h_iu_j$.

Following the steps above, each y_i is processed differentially, and there exists

$$\begin{aligned} &[y_1^{(r_1)},y_2^{(r_2)},y_3^{(r_3)}]^T=[L_f^{r_1}h_1(X), \\ &L_f^{r_2}h_2(X),L_f^{r_3}h_3(X)]^T+A(X)u \end{aligned} \tag{6}$$

where the 3×3 matrix $A(X)$ is

$$A(X)=\begin{bmatrix} L_{g_1}L_f^{r_1-1}h_1 & L_{g_2}L_f^{r_1-1}h_1 & L_{g_3}L_f^{r_1-1}h_1 \\ L_{g_1}L_f^{r_2-1}h_2 & L_{g_2}L_f^{r_2-1}h_2 & L_{g_3}L_f^{r_2-1}h_2 \\ L_{g_1}L_f^{r_3-1}h_3 & L_{g_2}L_f^{r_3-1}h_3 & L_{g_3}L_f^{r_3-1}h_3 \end{bmatrix} \tag{7}$$

Meanwhile, the relative order vector is described as: if the positive integer r_i exists, and makes that $L_{g_j}L_f^kh_i(X)\equiv 0, 0\le k\le r_i-2, i=1,2,...,m, j=1,2,...,m$, also the matrix $A(X)$ in Eqn.(6) is reversible at point X_0, then say that the system(2) has the relative order vector $\{r_1,r_2,...,r_m\}$ at point X_0.

Afterwards, the system (14) is analyzed according to the following process:

1 Analyze $h_1(X)$ and $f(X)$

When $r_1'=0$, there will be $L_{g_1}L_f^0h_1=L_{g_1}h_1=0$; $L_{g_2}L_f^0h_1=L_{g_2}h_1=0$; $L_{g_3}L_f^0h_1=L_{g_3}h_1=0$; When $r_1'=1$, there will be $L_{g_1}L_f^1 h_1 = 0; L_{g_2}L_f^1 h_1 = 0; L_{g_3}L_f^1 h_1 = 0$; When $r_1' = 2$, there will be $L_{g_1}L_f^2 h_1 = \kappa c_3; L_{g_2}L_f^2 h_1 = 0;$ $L_{g_3}L_f^2 h_1 = 0$. So in regard to $h_1(X)$ and $f(X)$, the relative order of system is $r_1 = 3$.

2 Analyze $h_2(X)$ and $f(X)$

When $r_2' = 0$, there will be $L_{g_1}L_f^0 h_2 = L_{g_1} h_2 = 0;$ $L_{g_2}L_f^0 h_2 = L_{g_2} h_2 = 0; L_{g_3}L_f^0 h_2 = L_{g_3} h_2 = \lambda_{75}$. So in regard to $h_2(X)$ and $f(X)$, the relative order of system is $r_2 = 1$.

3 Analyze $h_3(X)$ and $f(X)$

When $r_3' = 0$, there will be $L_{g_1}L_f^0 h_3 = 0;$ $L_{g_2}L_f^0 h_3 = \lambda_{45}; L_{g_3}L_f^0 h_3 = 0$. So in regard to $h_3(X)$ andv $f(X)$, the relative order of system is $r_3 = 1$.

In summary, the relative order vector of system (2) is $\{3,1,1\}$, and the total relative order is 5. Furthermore, the equations system can be achieved from Eqn. (19) as

$$y_1 = L_f^3 h_1 + \kappa c_3 u_1, y_2 = L_f h_2 + \lambda_{45} u \quad (8)$$
$$y_3 = L_f h_3 + \lambda_{75} u_2$$

where, $L_f^3 h_1 = c_1(c_1 x_2 + c_3 x_3) + \kappa c_3 x$, $L_f h_2 = \lambda_{42} x_8 + \lambda_{44} x_9 - a_4 x_4 + \lambda_{43} x_2 x_6 + \lambda_{41} x_2 x_7$, $L_f h_3 = \lambda_{72} x_5 + \lambda_{74} x_6 - b_4 x_7 + \lambda_{71} x_2 x_4 + \lambda_{73} x_2 x_9$.

The state variables is chosen as $\mu = [\mu_1, \mu_2, \mu_3, \mu_4, \mu_5]^T = [y_1, y_1, y_1, y_2, y_3]^T = [x_1, x_2, -c_1 x_2 - c_3 x_3, x_4, x_7]^T$.

In later part of this paper, demonstration to the diffeomorphism of state μ and X will be given, combined with the extended state of internal dynamics subsystem. The further equations from Eqn. (8) are shown as follows

$$\mu_1 = \mu_2, \mu_2 = \mu_3, \mu_3 = L_f^3 h_1 + \kappa c_3 u_1, \mu_4 = L_f h_2 + \lambda_{45} u_3, \mu_5 = L_f h_3 + \lambda_{75} u_2 \quad (9)$$

The control variable is changed into

$$u = \begin{bmatrix} u_1 \\ u_2 \\ u_3 \end{bmatrix} = \begin{bmatrix} \kappa c_3 & 0 & 0 \\ 0 & 0 & \lambda_{45} \\ 0 & \lambda_{75} & 0 \end{bmatrix}^{-1} \begin{bmatrix} v_1(X) - L_f^3 h_1 \\ v_2(X) - L_f h_2 \\ v_3(X) - L_f h_3 \end{bmatrix} \quad (10)$$

And diffeomorphism linearization model of system(8) is described as

$$\mu_1 = \mu_2, \mu_2 = \mu_3, \mu_3 = v_1(X), \mu_4 = v_2(X), \mu_5 = v_3(X) \quad (11)$$

The exact linearization control system of BTT missile has been established so far. After the proper control variable v which satisfies the requirements is achieved through the linearization control theory, then by Eqn. (10), the desired control variable of the original nonlinear control system (2) will be obtained.

2.3 *Analysis of the stability of internal dynamics subsystem*

In the whole analytical process of exact linearization, not only the linearized part of system should be stable, but also the internal dynamics subsystem BIBO should meet the stability requirements.

As can be seen from the former part, the relative order of exact linearization control system is smaller than the dimension of system, which indicates the existence of internal dynamics subsystem. Therefore, the stability analysis to the system is essential for ensuring the reliability of controlling the diffeomorphism model.

Because the dimension of original system is 9, extended 4-dimensional state variable of internal dynamics subsystem $\psi = [\psi_1, \psi_2, \psi_3, \psi_4]^t$ should be added up, which makes itself together with the state of exact linearization system μ constitute normal state equations. The system serves as

$$\mu_1 = \mu_2, \mu_2 = \mu_3, \mu_3 = L_f^3 h_1 + \kappa c_3 u_1, \mu_4 = L_f h_2 + \lambda_{45} \iota$$
$$\mu_5 = L_f h_3 + \lambda_{75} u_2, \psi = \omega(\mu, \psi) \quad (12)$$

where the equation of internal dynamics subsystem is $\psi = \omega(\mu, \psi)$.

The output of system is $Y = [\mu_1, \mu_3, \mu_5]^T$. The principle of selecting the equation of internal dynamics subsystem is: as long as the Jaccobi matrix achieved from the matrix $\Phi(X) = [h_1(X), L_f h_1(X), L_f^2 h_1(X), h_2(X), h_3(X), \omega_1(X), \omega_2(X), \omega_3(X), \omega_4(X)]^T$ is nonsingular at the point X_0, then the state of dynamics subsystem, which satisfies the following partial differential equation

$$L_{g_i} \psi_j = 0, 1 \le i \le 3, 1 \le j \le 4 \quad (13)$$

, can be chosen.

Substitute ψ_j into Eqn.(13), and there will be $\frac{\partial \psi_j}{\partial x_3} = 0$, $-\frac{\partial \psi_j}{\partial x_6} + \lambda_{75}\frac{\partial \psi_j}{\partial x_7} = 0$, $\lambda_{75}\frac{\partial \psi_j}{\partial x_4} - \frac{\partial \psi_j}{\partial x_9} = 0$.

There are many groups of proper ψ_j which yield to Eqn. (13). Considering the simplicity of solving Φ^{-1}, ψ_j is selected with the condition $\psi_1 = \lambda_{75} x_6 + x_7$,

$\psi_2 = x_5, \psi_3 = x_8, \psi_4 = \lambda_{45}x_9 + x_4$. Then, combining the original system(9) with internal dynamics subsystem, the corresponding diffeomorphism state is constructed as

$$
\begin{aligned}
\boldsymbol{Z} &= [\mu_1, \mu_2, \mu_3, \mu_4, \mu_5, \psi_1, \psi_2, \psi_3, \psi_4]^T \\
&= [x_1, x_2, -c_1x_2 - c_3x_3, x_4, x_7, \lambda_{75}x_6 \\
&\quad + x_7, x_5, x_8, \lambda_{45}x_9 + x_4]^T
\end{aligned}
\tag{14}
$$

whose corresponding determinant is $\det(\partial \boldsymbol{Z} / \partial \boldsymbol{X}) = c_3\lambda_{45}\lambda_{75}$.

As can be observed from $\lambda_{45} = -Va_5 / g$ and $\lambda_{75} = Vb_5 / g$, for any missile flight state that $V \neq 0$, $\partial \boldsymbol{Z} / \partial \boldsymbol{X}$ is nonsingular, that means the differentiable manifold designed in this chapter is smooth. Next, the form of $\boldsymbol{X} = \boldsymbol{\Phi}^{-1}(\boldsymbol{Z})$ is analyzed for easily getting the specific expression of internal dynamics subsystem. From Eqn. (14), the equation below is acquired

$$
\begin{aligned}
&\boldsymbol{X} = \boldsymbol{\Phi}^{-1}(Z) = [\mu_1, \mu_2, -(c_1\mu_2 + \mu_3) / c_3, \\
&\mu_4, \psi_2, (\psi_1 - \mu_5) / \lambda_{75}, \mu_5, \psi_3, (\psi_4 - \mu_4) / \lambda_{45}]^T
\end{aligned}
\tag{15}
$$

And then, the full-state equations constructed by exact linearization system and internal dynamics subsystem on new differentiable manifold are described as

$$
\begin{aligned}
&\mu_1 = \mu_2; \mu_2 = \mu_3; \mu_3 = c_1^2\mu_2 + (c_1c_3 + \kappa c_3) + c_3u_1; \\
&\mu_4 = \lambda_{42}\psi_3 + \lambda_{44}(\psi_4 - \mu_4)/\lambda_{45} - a_4\mu_4 + \lambda_{43}\mu_2(\psi_1 - \mu_5)/ \\
&\lambda_{75} + \lambda_{41}\mu_2\mu_5 + \lambda_{45}u_3; \\
&\mu_5 = \lambda_{72}\psi_2 + \lambda_{74}(\psi_1 - \mu_5) / \lambda_{75} - b_4\mu_5 + \lambda_{71}\mu_2\mu_4 + \lambda_{73}\mu_2 \\
&(\psi_4 - \mu_4) / \lambda_{45} + \lambda_{75}u_2; \\
&\mu_5 = \lambda_{72}\psi_2 + \lambda_{74}(\psi_1 - \mu_5) / \lambda_{75} - b_4\mu_5 + \lambda_{71}\mu_2\mu_4 + \lambda_{73}\mu_2 \\
&(\psi_4 - \mu_4) / \lambda_{45} + \lambda_{75}u_2; \\
&\psi_1 = -\kappa(\psi_1 - \mu_5) + \lambda_{71}\mu_2\mu_4 - b_4\mu_5 + \lambda_{72}\psi_2 + \lambda_{73}\mu_2 \\
&(\psi_4 - \mu_4) / \lambda_{45} + \lambda_{74}(\psi_1 - \mu_5) / \lambda_{75}; \\
&\psi_2 = \lambda_{51}\mu_2\mu_4 + \lambda_{52}\mu_5 + \lambda_{53}\psi_2 + \lambda_{54}\mu_2\psi_3 + \lambda_{55}(\psi_1 - \mu_5)/ \\
&\lambda_{75} + \lambda_{56}\mu_2(\psi_4 - \mu_4) / \lambda_{45}; \\
&\psi_3 = \lambda_{81}\mu_4 + \lambda_{82}\mu_2\mu_5 + \lambda_{83}\mu_2\psi_2 + \lambda_{84}\psi_3 + \lambda_{85}\mu_2(\psi_1 - \mu_5)/ \\
&\lambda_{75} + \lambda_{86}(\psi_4 - \mu_4) / \lambda_{45}; \\
&\psi_4 = -\kappa(\psi_4 - \mu_4) - a_4\mu_4 + \lambda_{41}\mu_2\mu_5 + \lambda_{42}\psi_3 \\
&\quad + \lambda_{43}\mu_2(\psi_1 - \mu_5) / \lambda_{75} + \lambda_{44}(\psi_4 - \mu_4) / \lambda_{45}.
\end{aligned}
$$

In order to analyze the characteristics of internal dynamics subsystem, the state input, except extended state, can be appointed as 0, namely the zero-dynamic subsystem

$$
\begin{cases}
\psi_1 = -\kappa\psi_1 + \lambda_{72}\psi_2 + \lambda_{74}\psi_1 / \lambda_{75} \\
\psi_2 = \lambda_{53}\psi_2 + \lambda_{55}\psi_1 / \lambda_{75} \\
\psi_3 = \lambda_{84}\psi_3 + \lambda_{86}\psi_4 / \lambda_{45} \\
\psi_4 = -\kappa\psi_4 + \lambda_{42}\psi_3 + \lambda_{44}\psi_4 / \lambda_{45}
\end{cases}
\tag{16}
$$

In the next part, analyze the characteristics of zero-dynamic subsystem alone. The equations above are transformed into the state space form as $\boldsymbol{\psi} = \boldsymbol{A\psi}$, where the matrix of zero-dynamic subsystem is

$$
\boldsymbol{A} = \begin{bmatrix}
\lambda_{74} / \lambda_{75} - \kappa & \lambda_{72} & 0 & 0 \\
\lambda_{55} / \lambda_{75} & \lambda_{53} & 0 & 0 \\
0 & 0 & \lambda_{84} & \lambda_{86} / \lambda_{45} \\
0 & 0 & \lambda_{42} & \lambda_{44} / \lambda_{45} - \kappa
\end{bmatrix}.
$$

The characteristic of stability of zero-dynamic subsystem can be analyzed with eigen values of the matrix A. After substituting relative aerodynamic data into the matrix, the solved eigen values are $s_1 = -32.359$; $s_2 = -0.551$; $s_3 = -32.561$; $s_4 = -0.135$.

It is clear that all of the eigen values of matrix A lie in the left half-plane of whole complex plane. So the zero-dynamic subsystem is asymptotically stable, further the internal dynamics subsystem of system (12) is asymptotically stable too. As long as the exact linearization system with designed control law is asymptotically stable, the original nonlinear system can be asymptotically stably controlled.

3 OPTIMAL CONTROLLER DESIGN OF BTT MISSILES BASED ON DIFFEOMORPHISM EXACT LINEARIZATION MODEL

Considering the fact that when the linear system is controlled by optimal control method, the control performance is optimal and the control energy is constrained at the same time, the method of linear quadratic regulation (LQR) is adopted in this chapter to design the optimal controller of exact linearization system. From Eqn., the following differential equations system is given

$$
y_1 = v_1(\boldsymbol{X}), y_2 = v_2(\boldsymbol{X}), y_3 = v_3(\boldsymbol{X}) \tag{17}
$$

Assume that the desired output of system is $[y_{1d}, y_{2d}, y_{3d}]^T$, and choose the error tracking state as new state $\boldsymbol{E} = [e_1, e_2, e_3, e_4, e_5]^T = [y_1 - y_{1d}, y_1 - y_{1d}, y_2 - y_{2d}, y_3 - y_{3d}]^T$.

The original system (11) is transformed into the following error tracking system

$$E = AE + BU \tag{18}$$

where the system matrix $A = \begin{bmatrix} 0 & 1 & 0 & 0 & 0 \\ 0 & 0 & 1 & 0 & 0 \\ 0 & 0 & 0 & 0 & 0 \\ 0 & 0 & 0 & 0 & 0 \\ 0 & 0 & 0 & 0 & 0 \end{bmatrix}$

and control action matrix $B = \begin{bmatrix} 0 & 0 & 0 \\ 0 & 0 & 0 \\ 1 & 0 & 0 \\ 0 & 1 & 0 \\ 0 & 0 & 1 \end{bmatrix}$.

There exists the following relation between the control variable of error tracking system and that $v = [v_1, v_2, v_3]^T$ of original system (17)

$$BU = Bv + AX_d - X_d \tag{19}$$

where $X_d = [y_{1d}, y_{1d}, y_{1d}, y_{2d}, y_{3d}]^T$. In regard to the system (18), the quadratic performance index is given as $J = \int_{t_0}^{\infty} [E^T QE + U^T RU]\mathrm{d}t$, where Q is system state regulation weight coefficient matrix; R system control energy weight coefficient matrix. This system is completely controllable and observable. The minimum optimal control variable which can implement the index is $U^* = -R^{-1}B^T \bar{P}E$, where $\bar{P}$ is the solution of Riccati equation, $-\bar{P}A - A^T\bar{P} + \bar{P}BR^{-1}B^T\bar{P} - Q = 0$. Furthermore, combined with Eqn. (19), the generalized inverse matrix transformation is made to achieve $v = B^{-1}(BU^* - AX_d + X_d)$.

Then by Eqn., the control variable imposed on original nonlinear system is acquired to realize system's exact linearization control.

4 MATHEMATICAL SIMULATIONS AND RESULTS ANALYSIS

4.1 *Simulation conditions*

1. Deflection angle of rudder of the missile is limited from -17 deg to +17 deg;
2. After control process starts, the norm of side-slip angle should be smaller than 3 deg;
3. Step response curve to meet: overshoot $\sigma < 12\%$, rise time $t_r \leq 1.5$s, number of half oscillations $N < 2$, system steady-state error ess<5%;
4. The performance of coordinate control is checked by fixed-point nonlinear simulation, and the fixed-point conditions are: the imposed longitudinal overload command $n_{yc} = 8$, lateral channel is served as coordinate channel and $n_{zc} = 0$, roll angle order $\gamma_c = 90$.

4.2 *Simulation of exact linearization optimal control*

Combined with simulation conditions, diffeomorphism exact linearization optimal control method is applied, and simulation results are shown in Fig.1 to Fig. 6.

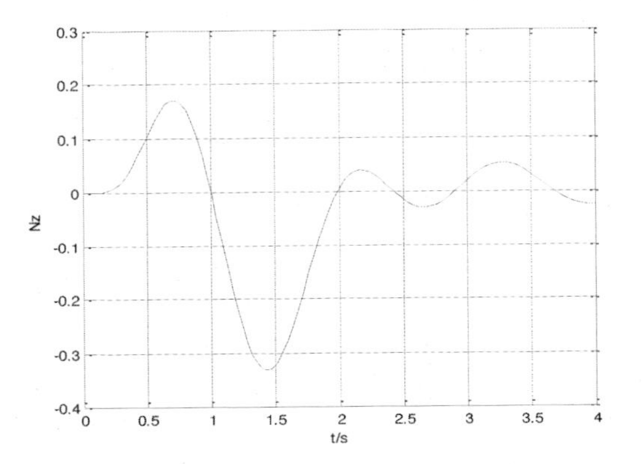

Figure 1. The side overload of missile.

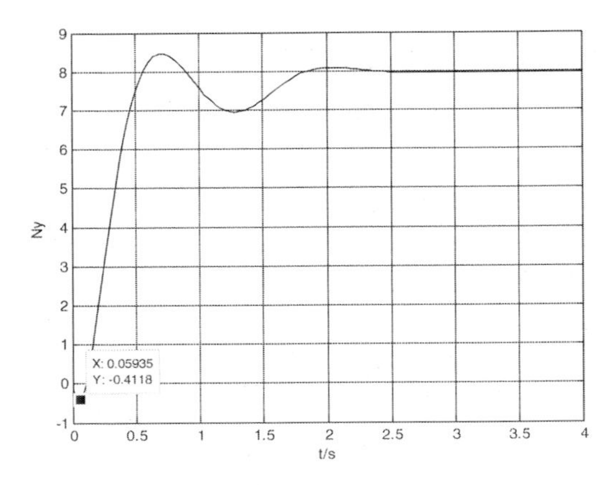

Figure 2. The normal overload of missile.

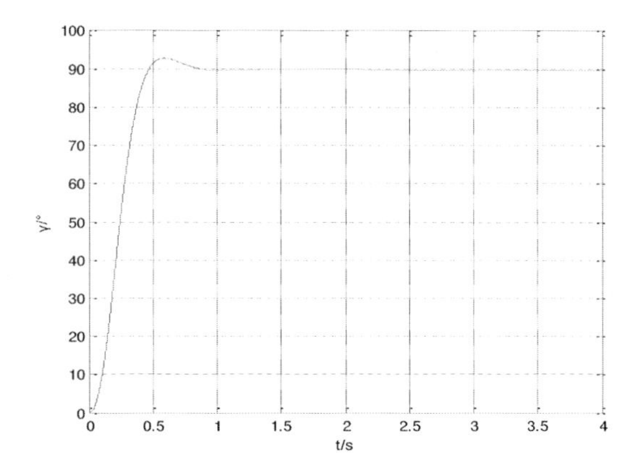

Figure 3. The curve of roll angle.

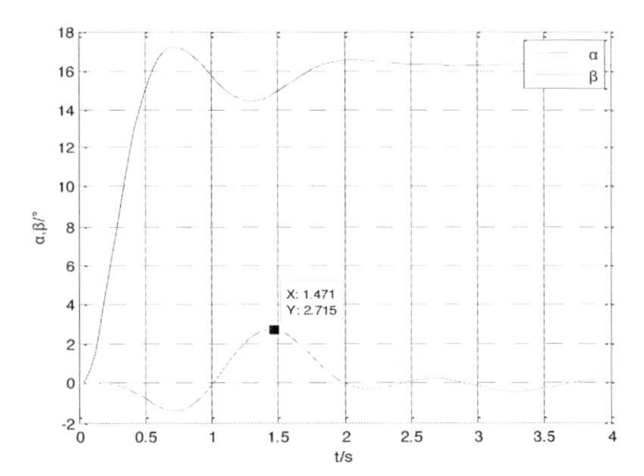

Figure 4. The curve of angles of attack and sideslip.

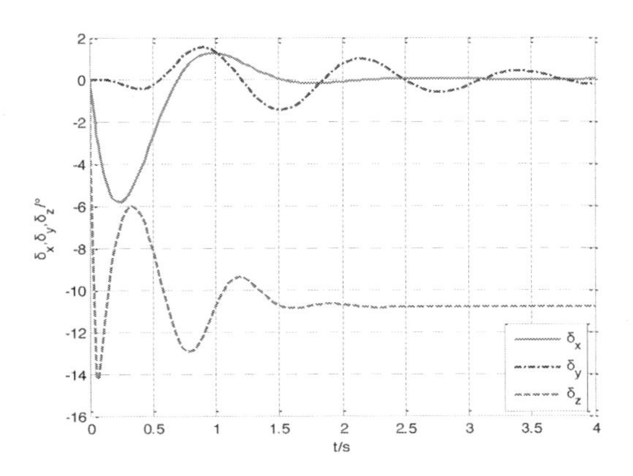

Figure 5. The curve of rudder deflection angles.

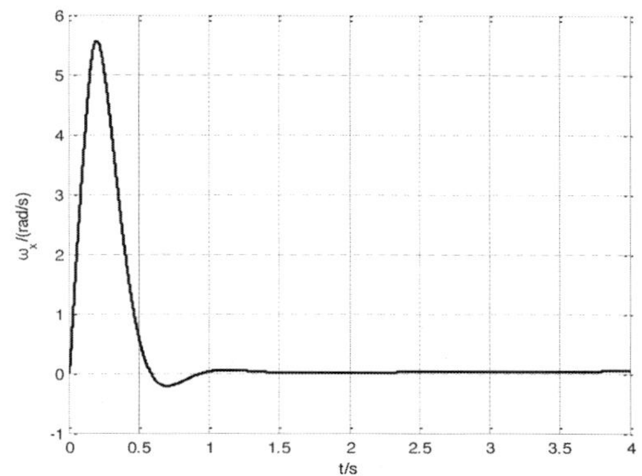

Figure 6. The angular velocity in roll of missile.

As can be seen from the tests:

(1) The optimal controller based on diffeomorphism exact linearization can control the maximum value of missile side overload within 0.2. The transient time of missile normal overload is probably around 2s. The transient time of missile roll angle is around 1.5s. Comparing with the CBTT controller, the control ability of roll channel is more powerful; (2) The missile side-slip angle is controlled as $|\beta_{max}| < 2$, which satisfies the control requirements of side-slip angle; (3) As can be observed in Fig. 2, the diffeomorphism exact linearization optimal controller of BTT missile can quickly response to pitch channel commands; (4) As can be seen from δ_x in Fig. 4, the optimal controller based on exact linearization model can implement the control process with control energy constraints; (5) As can be seen from the comparison with the missile roll angular velocity of CBTT control, because the exact linearization optimal controller can more quickly realize the control of roll channel, the bigger roll angular velocity will be generated, approximately 5.5 rad/s in Fig. 6. However, the control performance of optimal controller is better than that of CBTT controller in every channel[10], which illustrates that the exact linearization method has a better way to realize the coupling nonlinear control of BTT missile.

5 CONCLUSIONS

By means of analyzing the possibility of implementing linearization without feedback, this paper firstly gives demonstrations to the real nonlinearity of BTT missile's control system is firstly given. Then the control system of BTT missile is exactly linearized through the exact linearization theory of differential homeomorphism transform. Also the stability analysis of internal dynamics subsystem is proposed, which proves that the selected differentiable manifold is satisfied with the homeomorphism mapping condition, and stable controller containing exact linearization part can be designed to control the original nonlinear system. At last, the LQR controller is provided on the basis of established exact linearization model, fulfilling the optimal control of BTT missile. Moreover, tests are conducted to check the performance of diffeomorphism exact linearization optimal controller, and simulation results show the feasibility and applicability of the designed method.

ACKNOWLEDGMENT

This work is supported by the National Nature Science Fund of China (GN:61403100); the Fundamental Research Funds for the Central Universities (GN:HIT. NSRIF.2015037) and the open National Defense Key Disciplines Laboratory of Exploration of Deep Space Landing and Return Control Technology, Harbin Institute of Technology (GN: HIT.KLOF.2013.079).

REFERENCES

[1] Ye Zhenxin et al. (2009) An overview an development of bank-to-turn technology for tactical missiles. *Aerospace Control*, (5):106–112.
[2] Yang Jun, Yang Chen. (2005) *Design of Guidance and Control System of Modern Missile*. Beijing: Aviation Industry Press.
[3] Guo Haifen, Huang Changqiang, Ding Dali. (2014) Robust backstepping controller design for the supersonic gliding BTT missile. *Electronics Optics and Control*, (3):66–72.
[4] Arrow. (1985) Status and concerns for bank-to-turn control of tactical missiles. *Journal of Guidance, Control and Dynamics*, 8(2):267–274.
[5] Jiumming Lin and Huanliang Tsai. (2006) General Autopilot Design for BTT Missile by Generalized Linear Quadratic Gaussian / Loop Transfer Recovery Method. AIAA Guidance, Navigation, and Control Conference and Exhibit 21–24 August 2006, Keystone, Colorado.
[6] Shankar Sastry. (1999) *Nonlinear Systems Analysis, Stability, and Control*. New York: Springer-Verlag, Inc.
[7] Lberto Isidori. (1999) *Nonlinear Control Systems*. New York: Springer, pp:78–200.
[8] Zhang Yan, Duan Chaoyang, Zhang Ping. (2007) BTT missile autopilot design based on dynamic inversion. *Journal of Beijing University of Aeronautics and Astronautics*, 33(4):422–426.
[9] Zhang You'an, Cui Gutao, Cui Pingyuan. (1999) BTT missile autopilot design method based on feedback linearization theory. *Tactical Missile Technology*, (2):31–34.
[10] Naigang Cui, Changzhu Wei, Jifeng Guo and Biao Zhao, (2009) Research on missile formation control system, *Proceeding of IEEE International Conference on Mechatronics and Automation, 2009 August 9–12*, Changchun, China, pp:4197–4202.
[11] He Yuyao, Yan Maode. (2007) *Nonlinear Control Theory and Application*. Xidian University Press, pp:85–157.
[12] Duan Fuhai, Han Chongzhao. (2002) The contrast of dynamic inversion method and differential geometric feedback linearization method. *Automation & Instrumentation*, (3):4–6.
[13] Zheng Jianhua, Yang Di. (2001) *Robust Control Theory and Application for bank to Turn Missiles*. Beijing: National Defence Industry Press.

Electric, Electronic and Control Engineering – Shao, Shu & Tian (Eds)

A low power consumption indoor locating method research based on UWB technology

Huipeng Hong, Chong Shen, Yuhao Zhu, Liqiang Zheng & Fang Dong
College of Information Science & Technology, Hainan University, Haikou, China

ABSTRACT: With the rapid growth of the economy and employment pressure, more and more job fairs and exhibitions are held in the city. In order to help people find their place in the hall fast and accurately, indoor positioning technology[1] in recent years shows increasingly widespread commercial value and the prospects. With the help of the outdoor GPS (Global Position System) positioning ideas, this paper puts forward a kind of low power consumption indoor location method based on the technology of UWB. The receiver of Tag[2] (Label) will only receive information packages from the Anchor (Base Station). Thus it can reduce the number of intercommunication times between Tag and Anchor, and lower power dissipation of the Tag.

KEYWORDS: hyperbola & triangulation algorithm; UWB; TDOA; low power consumption

This paper ensures the entire Anchor keeping time synchronization by Kalman filtering algorithm[3]. The Tag will send broadcast messages to the Anchor when it enters the hall. At the same time the awakened Anchor sends data packets to all the Tag in the hall, and then each Tag records the arrival time and ID of each Anchor. Based on a TDOA localization algorithm [4] we can count the time difference, and then multiplies by the speed of light we can obtain the absolute value of the distance difference according to hyperbolic algorithm. Finally, we can calculate the coordinates of the Tag based on triangulation algorithm that is the Tag location information.

1 INTRODUCTION

With the rapid development of wireless communication technology and growing demand for location-based services, the wireless indoor positioning technology obtains more and more attention. However, GPS specifically aims at wireless outdoor positioning function through its unique structure. For the outdoor environment which signals reach easily, GPS can provide high-precision positioning information. But for indoor environments, due to the obstructions of buildings and their internal structures such as walls, doors and windows, and the real-time change of personnel walking can lead to weak signals and cannot obtain the effective location information. In this context, an indoor positioning system emerged since it can provide reliable service for the growing indoor positioning needs.

Nowadays the wireless communication technology of indoor location used includes: WIFI[5], Bluetooth[6], Infrared, GSM, Ultrasonic, RFID[7], UWB and so on. As the growing demand for the positioning, people also put forward a higher request to the accuracy of positioning. But the position accuracy of WIFI, Bluetooth technology usually is in meters, which makes little sense for the high requirements of the indoor positioning accuracy. Now the positioning chip (DW1000, made by the Irish company DecaWave) based on UWB is accurate down to 10 decimeters or less. It will greatly satisfy the indoor positioning accuracy requirements of people.

As a new spectrum sharing technology, Ultra-Wideband (UWB) can be used for any kind of communication system. Its advantages include low power consumption, low cost, anti-multipath[8] interference ability and strong penetrating power. In addition, the UWB wireless communication within 10 meters can realize hundreds of Mbit/s even several Gbit/s data transmission rate, which is faster than other communication systems. However, UWB exists in the shortcomings of large occupation bandwidth, and can cause interference to other radio system. Therefore, various countries have strict limits on UWB transmission power. The current international rule of the transmitted power is -41dbm/500Mhz. At present, there is ever a great upsurge in the study of UWB positioning technology, and some domestic manufacturers are developing UWB indoor positioning technology.

Besides above characteristics of the UWB communication, it also has the advantage of high precision, security and low system complexity which makes it the preferred wireless location technology in the future. The application of it on activities such as the mine safety, weeding robot, child safety, industrial automation[9] has a prominent significance. As the application market of wireless indoor positioning technology expanded rapidly, the development prospect of UWB positioning technology, which represents its development direction will be broader.

2 LOW THE POWER CONSUMPTION OF TAG

In indoor navigation and positioning, especially in crowded environment such as large recruitment exhibition, terminal, expo, supermarket etc., if we use the TOF (Time Of Flight) algorithms[10], it means that each Tag will intercommunicate with the Anchor in the center of conference many times for ranging and positioning. As shown in the figure 1 below, it is a general communication positioning process based on TOF. After four times communication, the Anchor and the Tag will measure the distance between each other. However, in this process of communication, there are many transmitting/receiving between Tag and Anchor, which will lead to the Tag increasing power consumption. Along with the rising demands of the location, the validity and reliability of the Anchor's communication will reduce, which will greatly lower the performance of communication systems. In order to reduce power dissipation in the Tag, we presented an idea of GPS navigation and positioning, which Tag only receives signals and Master Anchor only communicates with Slave Anchor when they do clock synchronization, Master Anchor do not receive anything in any other kinds of situations.

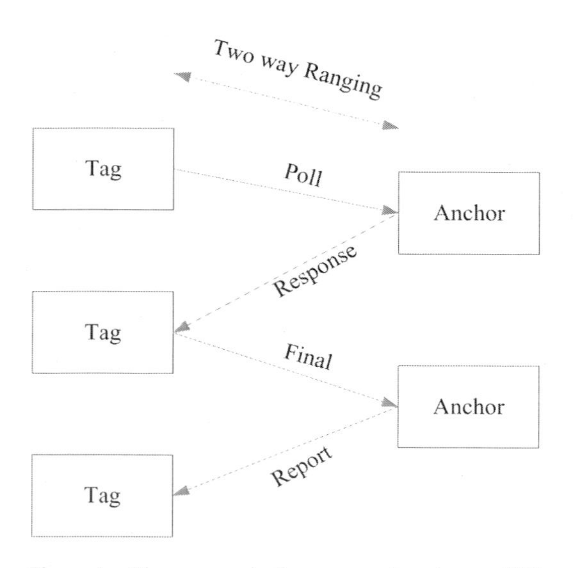

Figure 1. The communication process based on a TOF positioning algorithm.

3 THE OVERALL FRAME OF THE SYSTEM

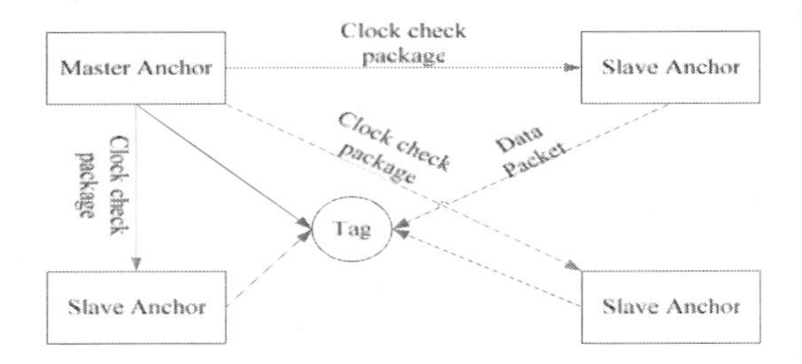

Figure 2. The communication process based on a TDOA location algorithm.

As shown in figure 2, Master Anchor sends clock check packages to each Slave Anchor on time to keep the entire Anchor realize time synchronization. In this system we will use the Kalman filter algorithm to solve the problem of time out-of-synchronization. Kalman filter estimates the desired signal through the algorithm from the observed quantity related to the extracted signal, and it is the linear, unbiased optimal state estimation filter which is as the criterion of minimum root mean square error. In the conventional GPS navigation and positioning system, it mainly uses satellites to locate and track the mobile station continuously, thus it can get the better positioning effect. But in this system, when the Master Anchor sends the clock check packages to all Slave Anchor, all the Slave Anchor will send the data packages which contain frame control information, Anchor ID, sent time etc. to Tag. At this time the Anchor plays the satellite role in the GPS, and the Tag is equal to the GPS receiver. After the Tag accepts the related data sent by the Anchor, it can get its own location information by calculation of hyperbola and triangle algorithm.

4 HYPERBOLA & TRIANGULATION ALGORITHM

The Tag is in moving state indoor which can be seen as the hyperbolic moving point. And Anchor1 and Anchor2 are static nodes in the center of the indoor

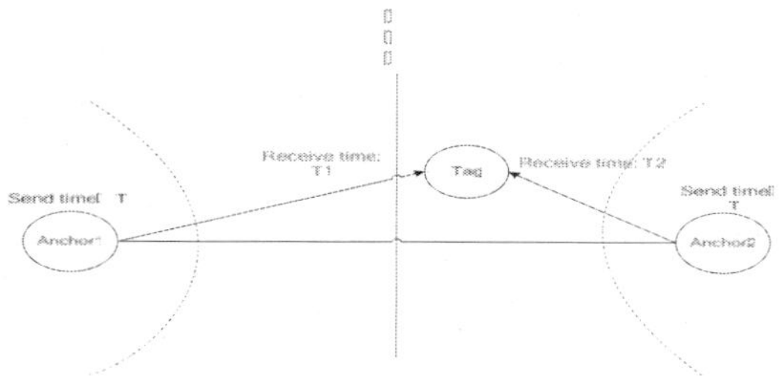

Figure 3. Get distance difference by hyperbola.

which can be seen as the hyperbolic fixed point. So, according to the hyperbolic definition, the absolute value of distance difference of Anchor1 to Tag and Anchor2 to Tag is the constant **2a**, thus we can get:

$$[(T_1 - T) - (T_2 - T)] * c = 2a \tag{1}$$

By formula (1) we can obtain the constant **a**, and get the distance difference.

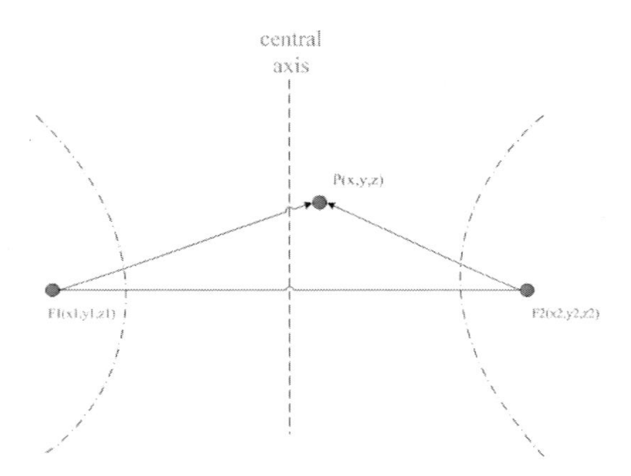

Figure 4. Get the coordinates value by the distance difference.

$$F_1P = \sqrt{(x-x1)^2 + (y-y1)^2 + (z-z1)^2} \tag{2}$$

$$F_2P = \sqrt{(x-x2)^2 + (y-y2)^2 + (z-z2)^2} \tag{3}$$

Put the formula (2) and formula (3) into formula (1) we can get:

$$\left|\sqrt{(x-x1)^2 + (y-y1)^2 + (z-z1)^2} \quad \sqrt{(x-x2)^2 + (y-y2)^2 + (z-z2)^2}\right| = 2a \tag{4}$$

According to GPS positioning ideas, each Tag can be received three or more than three Anchor information, thus available:

$$\left|\sqrt{(x-x2)^2 + (y-y2)^2 + (z-z2)^2} - \sqrt{(x-x3)^2 + (y-y3)^2 + (z-z3)^2}\right| = 2a \tag{5}$$

Similarly, we can get:

$$\left|\sqrt{(x-x3)^2 + (y-y3)^2 + (z-z3)^2} - \sqrt{(x-x4)^2 + (y-y4)^2 + (z-z4)^2}\right| = 2a$$
$$\left|\sqrt{(x-x4)^2 + (y-y4)^2 + (z-z4)^2} - \sqrt{(x-x1)^2 + (y-y1)^2 + (z-z1)^2}\right| = 2a$$
$$\left|\sqrt{(x-x4)^2 + (y-y4)^2 + (z-z4)^2} - \sqrt{(x-x2)^2 + (y-y2)^2 + (z-z2)^2}\right| = 2a \tag{6}$$

By formula (4) (5) (6) we can obtain the label node P(x, y, z), then we can determine the position of the label.

5 EVALUATION RESULTS AND DISCUSSION

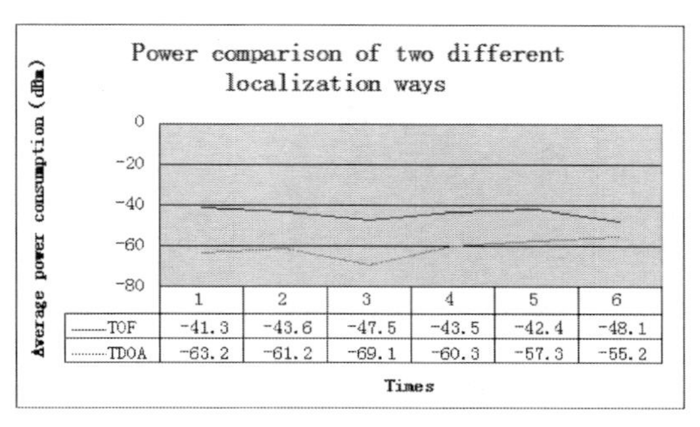

Figure 5. Power comparison of two different localization ways.

As is shown in figure 5, we get the data from the text that we respectively text the Anchor of two different patterns in the same environment (closed indoor). For the first measurement, we place four Anchor based on the TOF model in the room, and a second measurement we change the TOF model to TDOA model. By a spectrum analyzer, we measure the power consumption of two different models and get a set of data every 30 times. The above figure shows that the average power consumption of Anchor based on TDOA model is about 20 dBm higher than TOF model.

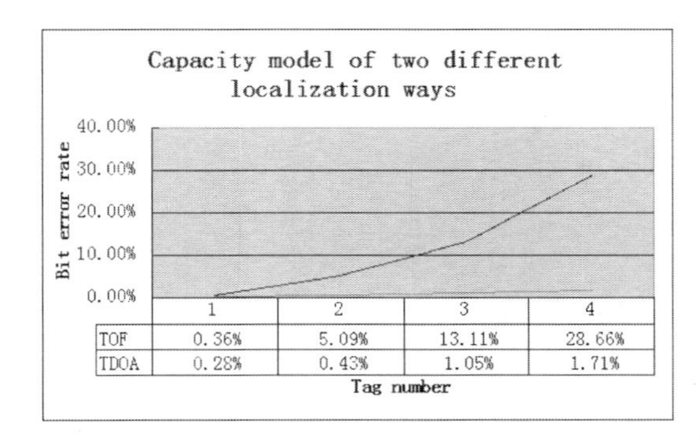

Figure 6. Capacity model of two different localization ways.

The data of figure 6 is from the experiment that we compare the Anchor's bit error rate of two different models (TOF and TODA) with the increase of the tag.

Obviously, the bit error rate of two models is much at one when placing one Tag in the room. However, the bit error rate of TOF model is higher and higher with the increase of Tag's numbers. In contrast, it has the slow growth in the TDOA model, which can prove that it suits the crowd indoor because of its low bit error rate.

6 CONCLUSIONS

Through the above analysis, we can use a TOF location algorithm for a small number of personnel and good positioning. But for a large-scale indoor positioning, TOF cannot satisfy the high-precision indoor localization requirements because of its high power consumption and bit error rate. Indoor positioning once forms the scale, it will certainly move towards the indoor navigation. In the crowded indoor, if the number of communication times between the Tag and the Anchor is too many, it will lead to the whole of UWB communication network congestion even paralysis. So the best solution in the large-scale indoor positioning is the TDOA model which can improve the performance of the indoor positioning communication system significantly.

REFERENCES

[1] Liu, H., et al. (2007) Survey of wireless indoor positioning techniques and systems. Systems, Man, and Cybernetics, *Part C: Applications and Reviews, IEEE Transactions on*, 37(6):1067–1080.

[2] Ahn, S.Y., et al. (2011) *ScaleLoc: A Scalable Real-Time Locating System for Moving Targets, in Convergence and Hybrid Information Technology*, Springer. pp: 97–103.

[3] Hamilton, B.R., et al. (2008) Aces: adaptive clock estimation and synchronization using Kalman filtering. *In Proceedings of the 14th ACM International Conference on Mobile Computing and Networking*. ACM.

[4] Gholami, M.R., S. Gezici and E.G. Strom. (2012) Improved position estimation using hybrid TW-TOA and TDOA in cooperative networks. *Signal Processing, IEEE Transactions on*, 60(7):3770–3785.

[5] Eisa, S., Peixoto, J., et al. (2013) Removing useless APs and fingerprints from WIFI indoor ositioning radio maps. *Indoor Positioning and Indoor Navigation (IPIN)*, International Conference on. 2013, pp:1–7.

[6] Yapeng W., Xu Y., et al. (2013) Bluetooth positioning using RSSI and triangulation methods. *Consumer Communications and Networking Conference (CCNC), IEEE*. pp:837–842.

[7] Kim, J., et al. (2008) Scalable RTLS: Design and Implementation of the Scalable Real Time Locating System Using Active RFID, in Information Networking. Towards Ubiquitous Networking and Services, Springer. pp:503–512.

[8] Dardari, D., et al. (2009) Ranging with ultra wide bandwidth signals in multipath environments. *Proceedings of the IEEE*, 97(2):404–426.

[9] Mahfouz, M.R., et al. (2008) Investigation of high-accuracy indoor 3-D positioning using UWB technology. *Microwave Theory and Techniques, IEEE Transactions on*, 56(6):1316–1330.

[10] Jiang, Y. and V.C.M. Leung. (2007) An asymmetric double sided two-way ranging for crystal offset. *In Signals, Systems and Electronics, 2007. ISSSE'07*. International Symposium on. IEEE.

Electric, Electronic and Control Engineering – Shao, Shu & Tian (Eds)
© 2015 Taylor & Francis Group, London, ISBN: 978-1-138-02842-5

Research on unmanned ground vehicle following system

Zhihui Wang, Nianyu Li, Yunan Zhang & Jie Zhang
Department of Control Engineering, Academy of Armored Forces Engineering, Beijing, China

ABSTRACT: This paper researches the following system of 6×6 unmanned ground vehicle designed by a certain institute. Establish virtual trailer link model and studies the reasons and methods of using parametric cubic spline to construct leader trajectory. Then gives the vehicle tracking control algorithm and proposes a tracking control law to correct the trajectory. Finally, the tracking algorithm is verified by Matlab.

KEYWORDS: unmanned ground vehicle; virtual trailer link model; tracking control law

1 INTRODUCTION

In recent years an increasing attention is attached on vehicle tracking system for industry or national defense applications. The vehicle tracking system is an important part of advanced driver assistance systems in an intelligent transport system (ITS) [1], and can improve traffic efficiency and security significantly. In Europe some scholars proposed a "road train" technology [2]. As can be seen from Fig.1, for a small fleet, the lead car is autonomous driving. Other cars join the fleet via network communication, following the front car automatically through obtaining its information.

Figure 1. Road train.

Fig.2 shows a 6×6 unmanned ground test vehicle designed by a certain institute. It is driven by six independent motors and has the Global Position System, laser radar, and others. Steering with differential velocity, it has no steering mechanism.

A perfect car tracking system needs to make sure that follower moves through the leader's path precisely as well as keeping a certain distance. The tracking system should guarantee vehicle driving safety and efficiency, as well as successful tracking in a narrow space. Dr. Ng Teck Chew of Nanyang Technological University (NTU) generalized several following strategies and proposed the concept of "Virtual Trailer Link"[3]. However, from his simulation result, we can see there exists a considerable error during vehicle turning, and due to the reason the hinged level cannot represent the travelling path of vehicles. In [4], Zhang Xiaomei estimated Chew's method and introduced a new strategy, in which she used the short arc method usually applied in the robot control system to fit the vehicle trajectory. This control method is easy to realize. Yet there are still several problems. Firstly, using short arc to fit the front car's path causes turning radius mutation and the heading angle is not continuous. Moreover, the error is still sort of big because the vehicle path is the same as clothiod curve. At last, there is no feedback control in the follower's tracking, resulting in error accumulation.

Figure 2. Unmanned ground vehicle.

This paper mainly discusses the tracking control of a 6×6 unmanned ground Vehicle. To ensure a smooth Vehicle tracking control, a strategy of route planning is applied, which uses cubic splines to fit the path. In the global coordinate system we choose a parametric cubic spline interpolation to fit the vehicle trajectory. The 6×6 unmanned Vehicle travels along the trajectory and keeps a certain length with the target. Especially, when the vehicle is turning, the Vehicle should have the ability of time delay as well as the accurate tracking path. Meanwhile, adaptive control is adopted to get a feedback control in order to cancel the accumulated error.

2 THE DESIGN OF VIRTUAL TRAILER LINK MODEL

In order to improve the tracking accuracy and ensure the accurate tracking of the car, it demands to get the precise description of the trajectory of the front vehicle first. After getting the discrete points of leader's movement, 6×6 UGV needs to finish tracking through these points. It needs a curve fitting method to the trajectory. To overcome the deficiency of piecewise linear interpolation and single higher-order polynomial interpolation, the piecewise polynomial method can be used. That is, the lower conic section is getting via piecewise polynomial Interpolation, and is connected under certain conditions. Cubic spline is a good trade off. Cubic spline has its second order derivative which ensures a continuous first derivative [5]. So the heading angle is continuous. However, the commonly used spline function is expressed by, the following problems will be occurred:

1. When considering point-group $P_i(x_i, y_i)$, the necessary condition that using a cubic spline function to fit curve is: $x_1 < x_2 <\ldots < x_n$. If the situation like which Fig.3 (a) is shown, the condition is not met.
2. If the point-ground includes two points (x_i, y_i), (x_j, y_j), which means there is a path perpendicular to the X axis, the slope of the curve is considered as infinity.

The curve cannot be expressed by $y = f(x)$ Fig.3 (b) is shown.

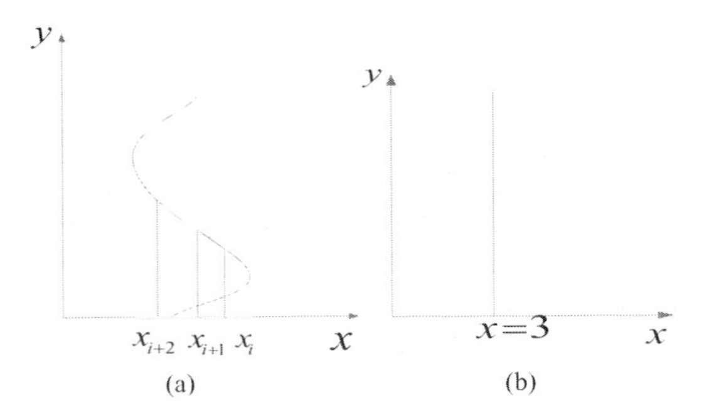

Figure 3. Commonly used cubic spline problem.

To solve this problem, the traveling path can be depicted by cubic spline curve. The parametric curve is not sensitive to the direction of these points. Curves like $x = 1$ which is perpendicular to X axis can be depicted, as well.

For the set of points $P_i(x_i, y_i)$, every two points (x_i, y_i), (x_{i+1}, y_{i+1}), $(i = 1,2,3,\ldots,n-1)$ constitute one segmental curve. Write in the form of parametric equations as follows:

$$x(t) = a_1 + a_2t + a_3t^2 + a_4t^3$$
$$y(t) = b_1 + b_2t + b_3t^2 + b_4t^3 \tag{1}$$

t is chord length: For one point (x, y) which is between (x_i, y_i) and (x_{i+1}, y_{i+1}), t could be computed:

$$t = \sqrt{(x - x_i)^2 + (y - y_i)^2} \tag{2}$$

Set t_{max} as maximum chord length of the section:

$$t_{\max} = \sqrt{(x_{i+1} - x_i)^2 + (y_{i+1} - y_i)^2} \tag{3}$$

In $[0, t_{max}]$, each t value corresponds to a point. In the curve from (x_i, y_i) to (x_{i+1}, y_{i+1}), the chord length t always monotonically increases, which guarantees the solvability of the spline function equations. So the parametric equations can describe any growth direction curve.

For a curve such as $x = 3$, it can also be described by the parameter equation:

$$x = 3$$
$$y = b_1 + b_2t \tag{4}$$

In the set of points $P_i(x_i, y_i)$, for every two points $p_i(x_i, y_i)$, $P_{i+1}(x_{i+1}, y_{i+1})$ $(i = 1,2,3,\ldots,n-1)$, define a parametric curve:

$$0 \le t \le t_i$$
$$x(t) = a_1 + a_2t + a_3t^2 + a_4t^3 \tag{5}$$
$$y(t) = b_1 + b_2t + b_3t^2 + b_4t^3$$

t_i is the length of the line segment connecting two points. Set $x(t_i)$ function as example:

$$x(0) = x_i = a_1 \tag{6}$$

$$x_i^{'} = x^{'}(0) = a_2 \tag{7}$$

$$x(t_i) = x_{i+1} = a_1 + a_2 t_i + a_3 t_i^2 + a_4 t_i^3 \quad (8)$$

$$x'_{i+1} = x'(t_i) = a_2 + 2a_3 t_i + 3a_4 t_i^2 \quad (9)$$

so:

$$x(t) = x_i + x'_i t + [\frac{3(x_{i+1} - x_i)}{t_i^2} - \frac{2x'_i}{t_i} - \frac{x'_i}{t_i} - \frac{x'_{i+1}}{t_i}]t^2 +$$

$$+[\frac{2(x_i - x_{i+1})}{t_i^3} + \frac{x'_i}{t_i^2} + \frac{x'_{i+1}}{t_i^2}]t^3 \quad (10)$$

$0 \le t \le t_m$

According constraints of cubic spline function, adjacent curve segments should maintain second continuity. That means, $P_i P_{i+1}$ and $P_{i+1} P_{i+2}$ should keep second continuous at the point P_{i+1}:

$$2a_3^i + 6a_4^i t_t = 2a_3^{i+1} \quad (11)$$

$$\text{So: } [t_{i+1} \quad 2(t_{i+1} + t_i) \quad t_i] \begin{bmatrix} x'_i \\ x'_{i+1} \\ x'_{i+2} \end{bmatrix} \quad (12)$$

$$= 3(\frac{t_i}{t_{i+1}}(x_{i+2} - x_{i+1}) + \frac{t_{i+1}}{t_i}(x_{i+1} - x_i))$$

$$\text{Set: } A^i = 3(\frac{t_i}{t_{i+1}}(x_{i+2} - x_{i+1}) + \frac{t_{i+1}}{t_i}(x_{i+1} - x_i)) \quad (13)$$

$$\text{So: } [t_{i+1} \quad 2(t_{i+1} + t_i) \quad t_i] \begin{bmatrix} x'_i \\ x'_{i+1} \\ x'_{i+2} \end{bmatrix} = A^i \quad (14)$$

x_i, x_{i+1}, t_i, t_{i+1} are all known parameters, while x_i', x_{i+1}', x_{i+2}' are all unknown. So it needs to solve n unknowns totally. i ranges for [1, n - 2]. It can get n-2 equations, and needs to add two conditions.

Like ordinary cubic spline curve, use common constraints x_1', x_n' to complement endpoint conditions ($x'_1 = \frac{x_2 - x_1}{t_1}$, $x'_n = \frac{x_n - x_{n-1}}{t_{n-1}}$). So, equations can be regarded as three -diagonal matrix, it can be solved by TSS method.

The leader's heading can be obtained by:

$$\tan\theta = \frac{y'}{x'} \quad (x_i < x < x_{i+1}, \ y_i < y < y_{i+1}) \quad (15)$$

The length of virtual trailer link can be obtained by:

$$l_i = \int_0^{t_i} \sqrt{x'(t)^2 + y'(t)^2}\, dt \quad (16)$$

3 DESIGN AND IMPLEMENTATION OF VEHICLE TRACKING CONTROL ALGORITHM

Connecting the UGV with real trailer systems, the virtual fender is equivalent to springs. The follower is pulled forward through the springs' fictitious force. Instead of straight-line distance, the present length of the virtual fender l is the distance between the path of two vehicles. The virtual fender pulls the car in an accelerated motion state. Otherwise, the follower's speed is zero. The 6×6 UGV's speed and acceleration can be adjusted simultaneously to obtain a balanced system via the force of the virtual fender.

Like the spring, assume the fender's origin length is D,

$$D = k_a a_l + k_v v_l + D_{\min} \quad (17)$$

Where a_l, v_l are the leader's acceleration and velocity separately. $D_{\min}$ is the minimum safe distance which is set according to the vehicle length beforehand. The natural length is changed when the leader's speed is changed. If the leader's speed is high, the fender's length is long enough so the follower has enough braking distance to respond to emergency situations. When the speed is low, the length is short which benefits to ease traffic congestion.

The function that represents the virtual force is:

$$F = k_b(l - D) \quad (18)$$

The acceleration of the 6×6UGV system induced by this virtual force is:

$$a_f = F / m_f \quad (19)$$

Where m_f is the follower's mass, it is not needed to be equal to its real value. The distance which 6×6 UGV will move in the next sampling time is:

$$s = v_f T + \frac{1}{2} a_f T^2 \quad (20)$$

Where T is the sampling time. The velocity of 6×6 UGV will be $v_{fnew} = v_f + a_f T$. And it will arrive at the point $A'(x_{A'}, y_{A'}, \theta_{A'})$, $\omega_{fnew} = \frac{\theta' - \theta}{t}$ (21)

We have obtained the velocity and acceleration of 6×6 UGV under virtual link. However, in the actual trailer system, there exists interference. Adopting an adaptive control law[6] to feedback control is necessary:

$$Q = \begin{bmatrix} v \\ \omega \end{bmatrix} = \begin{bmatrix} v_r \cos\theta_e + k_x X_e \\ \omega_r + v_r(k_y Y_e + k_\theta \sin\theta_e) \end{bmatrix} \quad (22)$$

Where, we adopt $v_r = \frac{v_r + v_{fnew}}{2}$, $\omega_r = \omega_{fnew}$.

The wheel speed of 6×6 both sides are:

$$\begin{bmatrix} v_L \\ v_R \end{bmatrix} = \begin{bmatrix} v + 0.5\omega d \\ v - 0.5\omega d \end{bmatrix} \quad (23)$$

4 VEHICLE TRACKING ALGORITHM SIMULATION

In Matlab simulation experiments, the trajectory of leader is set up a circle and a segment of S –curve, including straight, turning left and turning right. In the control law, set $k_x = 3$, $k_y = 2$. And set $k_b = 4$. In the trajectory of circle, the speed of leader is 0.4m/s, and radius is 4m. Initial distance between leader and follower is 1m. In the trajectory of S-curve, the speed of leader is 0.4m/s. Initial distance between leader and follower is 1m.

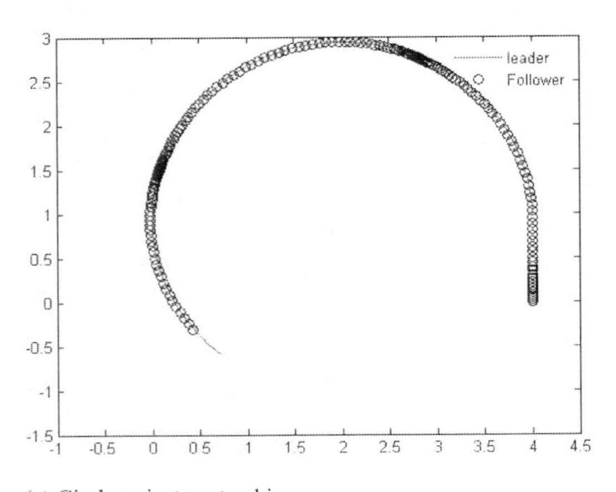

(a) Circle trajectory tracking

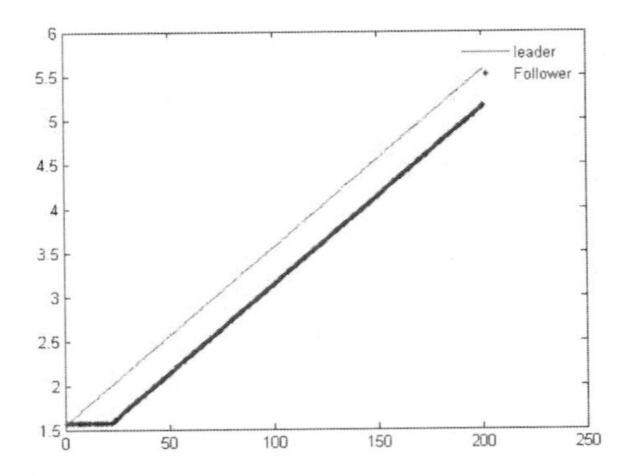

(b) Heading angle change

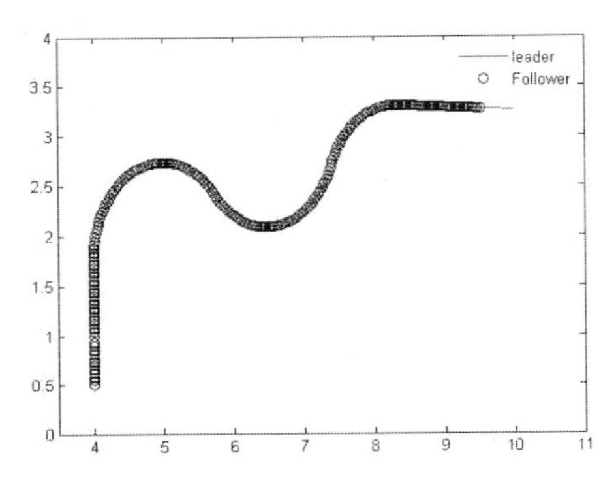

(c) S-curve trajectory tracking

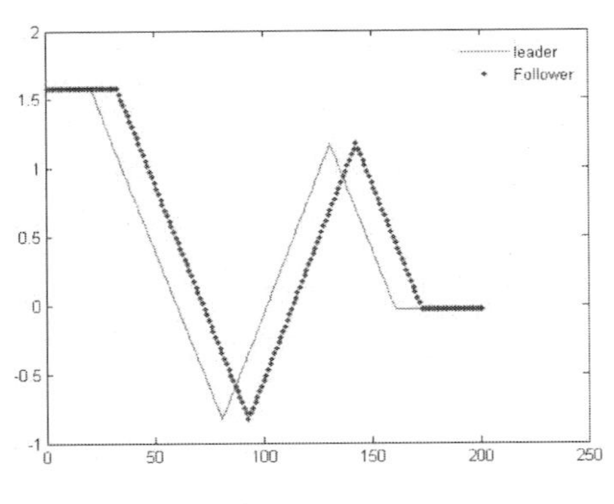

(d) Heading angle change

From this simulation result, we can see the trajectory of follower coincide with the trajectory of leader. The change of heading angle is smooth. And there is a time lag between two cars when turning around. So this algorithm is right.

REFERENCES:

[1] Mo Y K, Deng J. (2009) Fundamentals of intelligent public transportation dispatching systems planning. *2009 ISECS Inter Colloquium on Computing, Communication, Control, and Management. Sanya, China*, pp:41–44.

[2] Anonymous. (2009) Automated road train is easy on the driver. *Professional Engineering*, 22 (19):42–45.

[3] Ng T C. (2009) *Autonomous Vehicle Following-A Virtual Trailer Link Approach*. Singapore: Nayang Technological University.

[4] Zhang Xiaomei, Chen Weihai. (2010) A virtual tail model for intelligent vehicle following systems[C]. *Proceedings of the 5th IEEE Conference on Industrial Electronics and Applications, ICIEA 2010, Taichung, Taiwan, China, June, 2010.*

[5] Howard TM, Kelly A. (2006) Trajectory and spline generation for all-wheel steering mobil-e robots. *2006 IEEE/RSJ International Conference on Intelligent Robots and Systems Intelligent Robots and Systems. IEEE, 2006*, pp:4827–4832.

[6] Kanayama Y, Kimura Y, Miyazaki F. (1990) A stable tracking control method for an autonomous mobile robot. *Anon. Proceeding of the 1990 IEEE International Conference on Robotics and Automation. Los Alamitos.CA: IEEE Computer Society Press, 1990*, pp:384–389.

Electric, Electronic and Control Engineering – Shao, Shu & Tian (Eds)
 ISBN: 978-1-138-02842-5

The display and application of campus energy consumption of temperature and humidity data acquisition system

Xiaoli Wang, Yue Yu & Chuan Jiang
School of Electrical and Electronic Information, Jilin Jianzhu University, Changchun Jilin, China

ABSTRACT: Based on the virtual instrument software development platform of graphical compilation language (G language) connect STH10 temperature and humidity sensor to STM32 development board, adopt automatic data collection method, use the STM32 development board to collect and display indoor temperature and humidity data in real-time and then pass the data to host computer, use programs written by LabView to display, process, store temperature and humidity data collected by the serial port , and display the real-time temperature and humidity data and waveform figure in the STM32 LCD and panel and front panel. This system can monitor the energy situation, analyze the collected data, further build energy consumption data control system, and effectively control the school internal energy resources consumption.

KEYWORDS: energy consumption monitoring; STM32; Labview; SHT10; liquid crystal display

1 INTRODUCTION

With the development of energy using, people also meet with some inevitable problems which include shortage of resources, competition for energy, and increasing environment pollution caused by overuse of the energy. The colleges and universities as an important part of the whole social system are the main body of energy consumption. Building energy consumption monitoring system to monitor, measure the energy consumption of school, analyze energy-saving space and reduce energy waste. To provide reliable data support and effective means for further improvement of energy utilization efficiency and management level of the school.

Today's society treats the campus energy saving as one of the main research parts, and how to collect and monitor the required data effectively has become an important factor in the whole process. In order to solve this problem, a design on campus energy consumption of the display and application of temperature and humidity data acquisition system was made on the basis of the study and summarize the existing energy saving system. Using programs written by LabView to display, process, store temperature and humidity data collected from the serial port, and display the real-time temperature and humidity data and waveform figure in the STM32 LCD and panel and front panel. By using this method, can effectively collect the required data, and control it targeted to achieve the purpose of energy saving.

2 THE INSTRUCTION TO SHT10 TEMPERATURE AND HIMUIDY

STH10 temperature and humidity sensor adopt industrial CMDS processes with patented micro-machining (CMDSens® technology) ensures highest reliability and excellent long term stability. The device consists of a capacitive polymer sensing element for relative humidity and a band-gap temperature sensor. It has many advantages, including superior signal quality, a fast response time and insensitivity to external disturbances (EMC) at a very competitive price which result in widely use of STH10 temperature and humidity sensor.

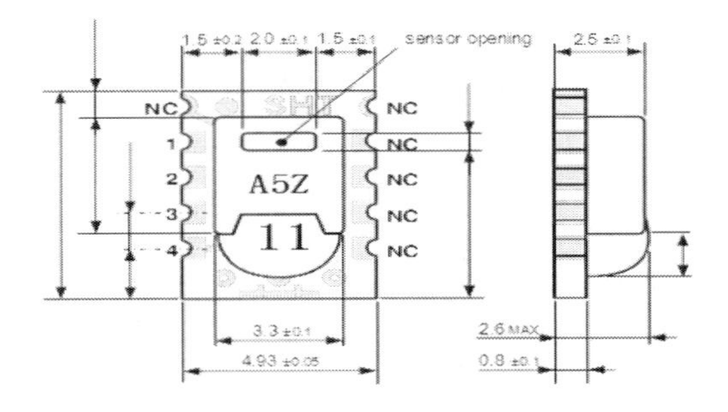

Pin1 is ground (GND), power supply voltage range is 2.4-5.5 V, add a decoupling filtering capacitor 100nF between power pin and GND in use. Pin 2 is

serial data (DATA), tristate structure used to read data. When send commands to the sensor, serial data is valid for the clock signal rising edge and keep in constant at a high level, data will be changed for the serial data falling edge. The microprocessor should drive serial data in a low level state, otherwise the signal will conflict. Pin 3 is a serial clock input port (SCK), to keep communication between the sensor and microprocessor in synchronization. Pin 4 is a power (VDD).NC must keep dangling.

3 THE SENSOSR'S COMMUNICATION

A. Start sensor
Select voltage firstly, and then power on for sensor, notice that the rate of power on should be higher than 1V/ms. The sensor works into a dormant state after power on, the duration of this process is11ms, send commands to sensor only after dormant stop.

B. Transmit command
Operating process: DATA convert to a low level when serial clock SCK is high level, and then SCK converts to low level, when clock signal SCK converts to a high level again, the DATA then will convert to high level, circulation in turn.

Subsequent command contains three address bits and five command bits. DATA convert to low level after the eighth falling edge of SCK clock, back to high level after the ninth, falling edge of SCK clock, this indicates SHT10 has received instructions correctly.

C. Temperature and humidity measurement
Send a set of measurement commands ('00000101' represents humidity RH, '00000011' represents temperature T) takes about 20/80/320 ms, corresponding to 8/12/14 bits measurement. Under the influence of an internal crystal oscillator, exact time more or less has some changes. DATA is low level marks the end of measurement, the sensor is in idle mode, tested DATA can be stored, "DATA ready" signal will read data when it's necessary, the controller can trigger the clock signal one more time. Then transmit 2 bytes measured signal data and 1 byte CRC odd-even check. Then for CRC odd-even check (optional reading), DATA pull-down is in a low level state to ensure that all bytes have been confirmed. All of the data start from MSB, right value valid. Receive confirmed bit of CRC marks the end of the communication. In condition of without using CRC-8 calibration, the controller can keep the confirmed bit at high level behind the LSB of measured value, so as to make the sensor into sleep mode, communication suspending.

D. Communication reset program
Interrupted with sensor communication, can use the following signal timing to reset signals: during the serial data is at a high level, the trigger times of clock signal are no less than nine times. There will be a "Transmission Start" timing to send before receiving new instruction, the timing is only used to realize UART reset, but the contents in status register will not be changed.

E. CRC-8 check
CRC-8 checks can ensure the reliability of data transmission and can detect and remove all mistakes in the process of checking.

4 LCD

A. Introduction
The LCD is short for liquid crystal display, which has advantages of small volume, light weight, low power consumption, large amount of carrying information and other advantages, and widely used in the information output field. There is no LCD specified control interface in STM32, so this experiment applies the ILI9341 chip controller as driver chip to control LCD, and LCD drive circuit is independent.

The control IC structure of the LCD is very complex and the most important part inside ILI9341 is in a GRAM in the middle part, known as video memory. Video memory has a lot of memory cells, also has a lot of pixels on the front panel, it is one-to-one correspondence between memory cells and pixels, as a joint result of each module on the right side, the data stored in video memory can be converted to control signal, make the pixels of LCD present a particular color.

B. The use of an LCD
8080 communication interface timing of ILI9341 uses FSMC interface, FSMC is short for flexible static memory controller. NOR FLASH, PSRAM and NAND FLASH memory chips can be used in the STM32 [2] chip control, use FSMC NOR \ PSRAM mode to control LCD in the experiment. Then it can be used to control the ILI9341 after FSMC interface initialization.

C. The reason for use of LCD
For the entire system, we use the LabVIEW as a host computer, link the data got from salve computer and judgment which can display all the collected data, and also can make a judgmental control strategy on the basis of the data. Using LCD can display the working state of the node and the collected data in real time that is convenient to system debugging, inspection, and maintenance. The data displayed in LabVIEW is mostly a period of the time data waveform, namely changing process, and the collected data of temperature and humidity is stored in the specified file,

convenient for reading and checking. So there is a big difference in the measured data displayed between LabVIEW and LCD.

5 THE SOFTWARE OF THE ENERGY CONSUMPTION DATA ACQUISITIONSYSTEM

A. The definition of Lab VIEW

The LabVIEW special graphics program and simple interface, which shortens the time of prototype development and is convenience of software maintenance in the future, therefore this has gradually grown in popularity with the system development and researchers.

LabVIEW is a kind of graphical programming language by using icons instead of text to create an application program. It uses the data flow programming way, the execution order of the program which is decided by the data flow direction between the nodes in program block, the icon represents a function and connection represents data flow direction. VI refers to virtual instrument, is the program module of LabVIEW.

B. The LabView front panel and program block diagram

LabView has two basic windows: front panel and program block diagram. The front panel is the user interface for placing the control object and display object. Place the controls on the connector according to the existing position of controls in connector. Place the input controls on the left side of the front panel and place display controls on the right side of the front panel. The error input cluster is usually in the lower left corner of the front panel, error output cluster in the lower right corner of the front panel. Program panel is used for graphics source code coding. It is a one-to-one correspondence of objects and controls on the front panel in the program block diagram, and program block diagram of, the controls will display the data in a certain way when operating the data flow controls in the program block diagram. The block diagram program is made up of four elements of node, endpoint, block and connection. The node is similar to the text language program statements and functions or subroutines. LabVIEW has two types of nodes, function nodes and sub-VI nodes. The difference between the two is: function node is the machine code compiled by LabVIEW for user to use, while the sub-VI node in the form of graphic language for user to use. The user can access and modify any codes of the sub-VI node, but can't modify the function node. VI program as is shown in the above block diagram program has two function modes: one function is to get the sum of two values, the other function is to get the difference of two values.

6 THE ENERGY CONSUMPTION MONITORING DATA ACQUISITION [3] SYSTEM BASED ON LABWIEW

A. FW design

The system is divided into three VI: login interface, welcome interface and serial port data acquisition and analysis.

B. Serial port acquisition and analysis program

The front panel settings tab

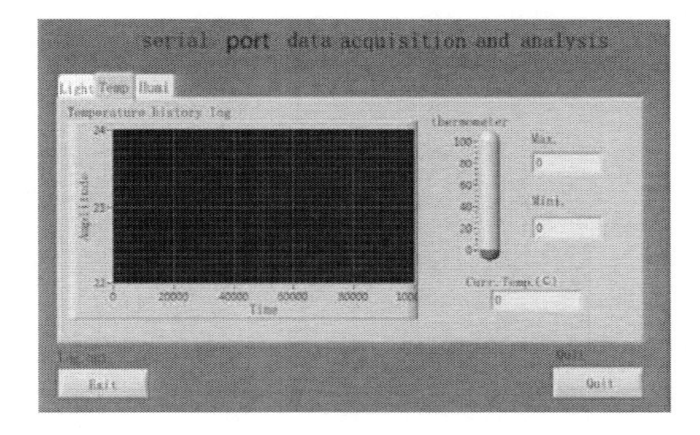

Figure 1. The temperature measured value.

Tab in figure 1 displays related temperature measured value [4], left side of the front panel displays temperature graph, when click on the stop button in lower right corner, the left side can display the current temperature value as well as minimum and maximum value of temperature during the data collection period

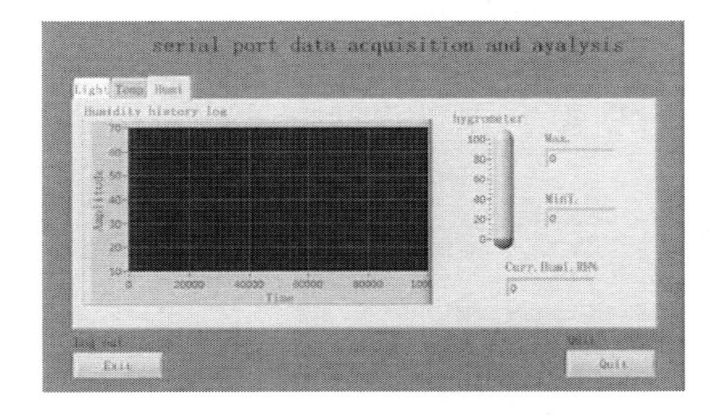

Figure 2. The humidity measured value.

Tab. in figure 2 displays related humidity measured value, similar to temperature tab that the left side is used to display humidity graph, the right side displays current humidity as well as minimum and maximum value of humidity during the data collection period.

Program block diagram of serial port acquisition and analysis system is shown in figure 3.

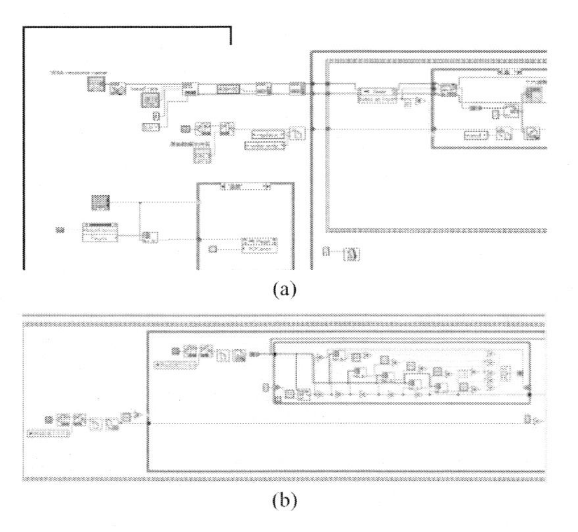

Figure 3. Program block diagram of serial port acquisition and analysis system.

First, the lower left part is the set of tabs mentioned earlier, this system uses two tabs: temperature, humidity. The temperature tab index number is 1and the humidity tab index number is 2. Place corresponding controls on corresponding tab and adjust the position of each space. Set the tab "allowed a variety of colors" attribute is to true, because the tab has three pages, there are three elements in the following settings, collect these three elements in one array, when we choose the temperature tab, the selected array elements is an element, which will execute temperature branch in the back condition block diagram, set the foreground color of the temperature tab to orange. Till this step, we have got two tabs' color and layout set up. The front panel is the window of HMI and direct window to a user, a simple and beautiful interface can help users to find a required function quickly, which save the user's time and easy to use. The users are in favor of it.

VISA resource name: that is the VISA resource name. VISA is a virtual instrument software architecture which can improve efficiency by reducing built-up time of the system. When a set of program finished, people hope that the program could be applied to any hardware interface, and the role of VISA has been just for this, when programming general program of I/O interface, it only needs to call the same VISA library functions and configure the required equipment parameters for it. In addition, VISA also defines its own data type, so as to avoid the problems caused by the inconsistent data type size when program transplantation in this way.

The upper left part is for operation of serial port read data. After VISA drive Installed, it will display effective serial port on the front panel, after serial port selected, it is needed to have serial port configuration [5], set the baud rate to 115200, data bits is 8 bits, stop bit is 1. These parameters should be accorded with slave computer, otherwise it is unable to communicate between host computer and slave computer. After serial port configured, set the buffer, the buffer size in the program is 409600, the data read from data terminal will be temporarily stored in the buffer for reading operation later. After setting the buffer empty the contents in the area, and to spare enough space for the data sent from serial port.

Labview writes data read from serial port to the binary files which needs to be aware that LabVIEW considers the contents you want to write as the characters of ASIC code, each 8 bits as an ASIC code character, and then write binary code corresponding to character to the file. We want to write binary data directly to the file, so I put each 8bits of binary data together to form an array and then convert them to decimal array, and then use the U8 functions in the LabVIEW array processing module to make it into the corresponding ASIC code, then write it to the file at last, so get the binary number we want.

Next is a serial port reading operation, into a While loop, execution once per 2ms, each cycle needs to detect how many bytes can be read in serial port, if bytes is not zero, then execute the branch of following condition block diagram, read out the bytes and data processing, if the bytes are zero, not read, wait for the next cycle until there are effective bytes to read.

When the bytes are valid, read out the data, now the data are displayed in the string, whether slave computer send single-byte or string data, host computer display in string after reading, so it should have a string to a byte array transformation, then convert the array of bytes to hexadecimal array, stored in the text file for debugging.

Text file Settings: find the path of VI firstly, analyze this path, get the VI's path at the next higher level, combined with text, will get the text file path, thus put the text files and VI files together, file path setting finished. Set the text file to only allow to write data to the file, if the text file already exists before the program running, this will replace the original file, if there is no such file before running, will generate such a file, and then write data.

The above operation completed the serial port read, and data written and save, and it will read out the data and store data in a text file for each cycle.

Read and open the file, get the file size, if the file byte is not less than 56, then performs the true branch of the condition block diagram; if the file byte is less than 56, then performs false branch; doesn't do anything if there is no code in the branch and directly to the next cycle, read data, save and judgment, until the byte is not less than 56. True branch operation is a process of data analysis. Read out the data in a text file, get temperature and humidity index according to certain protocol, and display on the front panel.

If the byte is greater than or equal to 56, convert the data in a text file in a decimal byte array. The following while loop will check the array, and justify whether it satisfy some certain condition. If yes, into the next processing.

Now look at while loop. The loop is far behind in five kinds of situation and operation, every kind of circumstance is an operation, such as the first case, there are two conditions: if some index is -1, if some data is equal to 69, any one of two conditions output is true, only five kinds of circumstance are true, the while loop will end. Assuming that the read data into the decimal array is shown in the following table

Table 1. The read data into decimal array.

Data	12	34	46	55	69	10	55	69	10	54
Index	0	1	2	3	4	5	6	7	8	9

First search from the first data, if there are 55 in array, output its index3, index number plus 1, judging data index number 4 is 69 or not, if so, the first case output is true. Continue to determine the index number of 5 data is 10 or not, if so, the second output is also true, and by such analogy, the index number of 6 data is 55, and the third output is true, the index number of 7 data is 69, the fourth output is true, the index number data of 8 data is 10, the fifth output is also true when the five outputs are all true, while

Figure 4. The screen of value of temperature and humidity.

loop stops. The While loop is used to look for a start, the "556910556910" decimal array corresponding character is "7 E line feed 7 E line feed", the data in the back is valid if this these characters found in the text file.

I breathe, as shown on the temperature and humidity sensor, the sensor detects the corresponding data, show the temperature and humidity on the front panel waveform figure a maximum and minimum values, and the screen shows the value of temperature and humidity, which shown on the interface of LabVIEW data is a period of time waveform data, and store the collected data into different files, convenient to read and check.

7 CONCLUSIONS

This study based on LabVIEW the campus energy consumption monitoring temperature and humidity data acquisition system, adopt the way of collaboration between upper-PC and the lower-PC , the lower PC collects data of temperature and humidity by the STM32 development board drive, then transmit the data via a serial port to the upper-PC, the upper-PC use development of the NI LabVIEW company. And in order to practice the observation of the current node data, the liquid crystal display was adopted, it can display the data in real-time and the upper- PC mainly used for historical data statistics, energy consumption calculation and other statistics and control functions. Schools can use this system to monitor the energy usage, and by controlling the switch of nodes to achieve the effect of energy conservation and emissions reduction, contribute to the campus energy saving. And can further build campus energy monitoring and control system, through this system, can improve accuracy, reduce artificial tedious work, can accurately and effectively monitor and collect the temperature and humidity of the room monitored, it can be applied in the different environment, and in favor of the energy saving control applications. From the point of these views, the design has a long-term significance in real life.

ACKNOWLEDGMENT

This paper was sponsored by Jilin Province Science & Technology support plan key projects.

REFERENCES

[1] Honggang Yu, Xiuzhen Zhao, Tao Wang (2011) Energy consumption monitoring system study based on campus network, *Information Technology and Information Technology*, pp:2–3.

[2] Ning. Li. (2008) *The STM32 Processor Development Based on MDK Application*, Beijing: Beijing University of Aeronautics and Astronautics Press, pp:5–8.
[3] Kefeng Xiang. (2011) The data acquisition system design and implementation based on LabVIEW, *Mechanical Management Development*, pp:1–3.
[4] Jian Xu (2013) *Temperature Control Experiment Teaching System Research and Development Based on LabVIEW*, Central South University.
[5] Yikui Liao (2013) *ARM Cortex-M4 Embedded Development Essence Solution – Based on STM32F4 China*, Beijing: Beijing University of Aeronautical and Astronautics Press, pp:40–45.

Electric, Electronic and Control Engineering – Shao, Shu & Tian (Eds)
© 2015 Taylor & Francis Group, London, ISBN: 978-1-138-02842-5

Design and implementation of vulnerability database maintenance system based on topic web crawler

Haiyan Liu, Yifei Huo, Tingmei Xue & Liuyuqin Deng
Department of Information Engineering, Academy of Armored Force Engineering, Beijing, China

ABSTRACT: Vulnerability database is an important part of information security infrastructure used for storing vulnerability information. This paper introduced topic web crawler into the vulnerability database maintenance system, in which the web pages that are related to vulnerabilities can be accessed through a topic web crawler and the vulnerability information can be abstracted so as to update the vulnerability database. This technology reduced the workload of maintenance and resolved the problems of coverage incompletion and lack of information in the existing vulnerability database. By analyzing the structure of the existing vulnerability database at home and abroad, studying the association of vulnerability attributes, and implementing group classification description method, this paper built the vulnerability database syntactic model. It also came up with a dynamic topic, maintenance system.

KEYWORDS: topic web crawler; dynamic topic; vulnerability model; vulnerability database

1 INTRODUCTION

A wide coverage, the rich content vulnerability database is conducive to provide support to the security agencies and the product manufacturers to find and fix bugs. And it's conducive to the relevant government departments to analyze the current security situation and to make the future security policies. Currently, due to different collection area, collection methods and information content, the vulnerability database has been built exists the problem of incomplete vulnerability entries and incomplete information. There is lots of vulnerability information over the internet, and some of them come from the management agencies that has built vulnerability database and some of them comes from the security organizations and vendors, as well as from the users' personal. This information can all be used as sources of information to vulnerability database, but this information generally do not have a fixed format and distribute widely, not easy to collect. Collecting this information manually is a tedious and time-consuming work. This article introduces topic web crawler technology into the internet vulnerability information gathering, through the topic web crawler gathering as much as vulnerability information over the internet, after extraction, de-emphasis and association, storing the information into the database. This method reduces staff workload and improves the coverage and the amount of information about the database.

2 ANALYSIS OF THE NOWADAYS VULNERABILITY DATABASE

Currently, some government agencies and some security companies have established a few vulnerability databases; here are some examples of the most famous vulnerability databases: MITRE CVE (Common Vulnerabilities and Exposures) [1]; NVD (National Vulnerabilities Database) [2]; CNNVD [3] and so on. In these databases, CVE is the list of the standardized definition of the currently known vulnerabilities and security deficiencies, most of the vulnerability databases provide the CVE number of the vulnerabilities it included, and it's convenient to reference and made it compatible between the vulnerability databases.

Nowadays the main problem of the domestic and overseas vulnerability databases are the followings: First, the information about the vulnerabilities is incomplete, such as CNNVD doesn't have the CVSS grade and CVE which used as standardized reference is just like the dictionary of the vulnerabilities and lack of the detailed description information of the vulnerabilities. Secondly, the coverage of the vulnerabilities is incomplete, the maintenance of the vulnerability information rely on the report from the security product manufacturers or professionals, the unreported vulnerabilities rely on the manager of vulnerabilities to gather and it's to a great extent restrict the coverage of the vulnerability database, besides CVE doesn't include the vulnerabilities of most Chinese software [4].

3 ANALYSIS OF TECHNOLOGY ON TOPIC WEB CRAWLER

3.1 *The concept of topic web crawler*

The web crawler is a program that automatically accesses the information on the webpage through the internet; a search engine is the typical application. A topic web crawler is a special web crawler program, topic web crawler can selective retention the information from the internet based on the given theme, so it needs to analyze whether the webpage is relevant to the topic. Therefore, the top two important aspects of analysis of topic web crawler are the description of the topic and the analysis of the relevant.

3.2 *Description of the topic*

Description of the topic is the method to transform the topic that the user needs to the calculation and comparison format. The common descriptions currently are ontology-based approach [5], an approach based on hierarchical division [6] and approach based on dynamic key word [7].

Ontology-based approach is the method of describing the topic through the concept of ontology; ontology is the distinct specification for describing the conceptual model [8].

An approach based on hierarchical division is the description of the topic based on dendritically structure catalogue. The common method is selecting some nodes in the setting catalogue tree as positive cases, others as negative cases [9].

An approach based on dynamic key word is the development on approach based on key words. This method dynamically enlarging the basic key word set during the crawling, it maintains the advantage of simple and strong practical [10], and make up the one-sidedness of the key word based method. This article chooses this description method in the system implementation.

3.3 *Approach of relevant analysis*

Relevant analysis is about how to judge a web page is relevant to the given topic and to figure out the degree in the relevant. Typically, there are two solutions, one is an approach based on content relevant and the other is an approach based on link relevancy.

An approach based on content relevant is to analyze the text on the webpage to see whether it is relevant to the topic. The dominating methods are matching character string through scanning and vector model. Matching character string through scanning is to compare the text to the topic dictionary to find out the overlap area, the accuracy of the analysis relies on the strong dictionary. The premise of the vector model is that the order of the key word is irrelevant to the expression of the topic, for example, vector a (windows, buffer overflow, vulnerability) is equivalent to vector b (vulnerability, windows, buffer overflow). First, express the topic dictionary into n-dimension vector k (k_1, k_2 ... k_n), k_i means the accuracy of the description of the number i key word to the topic (it means the bigger k_i is, the more accuracy of the description the number i key word to the topic), by calculating the cosine between k and a, we can gain the degree of the relevancy between the topic and the content, the range of the degree is from 0 to 1. If the result is 0, it means the least relevant. And

If the result is 1, it means the most relevant.

An approach based on link relevant rank the degree of relevant through the link between the webpage, the common algorithms are Google PageRank [11] and HITS. The drawback of this approach is that it's easy to make the theme of the webpage against the theme that the custom want, we call this phenomenon "theme migration" [12].

To improve the accuracy of the relevant analysis, some systems adopt two of the above approaches. For example, formula $k = k_1\alpha + k_2\beta$ or formula $k = k_1\alpha\beta$ can be used to rank the degree of the relevant [13, 14], α means the result of content relevant analyses and β means the result of link relevant analysis, k_1,k_2 are adjustment parameters, k is the final relevant analysis result. Our system adopts this approach, it analyzes the degree of relevance of the URL and web pages while running to improve the accuracy of the analysis.

4 DESIGN AND IMPLEMENTATION

4.1 *Design of the system*

The most common use of the Vulnerability Database Maintenance System is to update the data in the Vulnerability Database, unite with the traits of topic web crawler, the running process of the system as shown in Figure 1.

Set the topic at the beginning of the process running, it will help to rank the degree of relevance between the web page and URL to the topic. Queue URL stores the original URL and URL collected during the crawling. When system accomplishes crawling, it will provide the next URL that the system needs to let the system access the web page in the order of the URL in the queue.

The system will analyze the relevant of the accessed web pages. When the system judge the webpage is irrelevant to the topic, it will take another URL in the queue and access the webpage again. And if the system judge the webpage is relevant to the topic, it will dynamically adjust the preset topic according to the information in the webpage to improve the accuracy

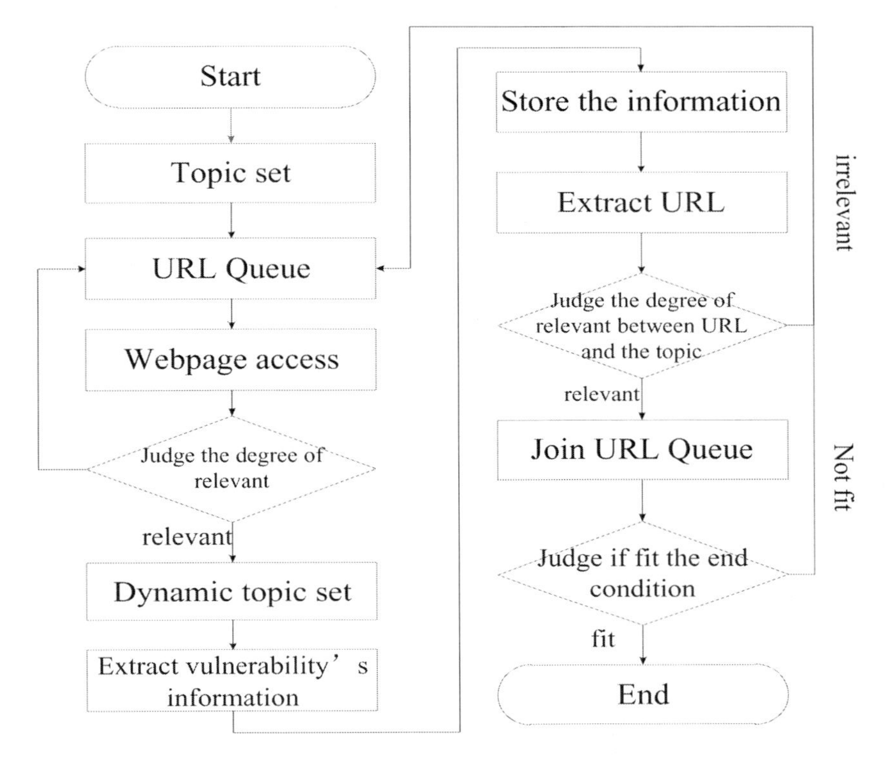

Figure 1. The running process of the system.

of the relevant calculation. The system extracts the information about the vulnerability from the webpage according to the preset vulnerability model and stores the information in the database.

The system takes all the URL from every web page, when the system judge the accessed URL is relevant to the topic, the URL will be put into the URL queue automatically and the irrelevant URL will not be put into the queue, this method guarantee the link in the URL queue all relevant to the topic.

4.2 *Design of the vulnerability database*

Currently, specifications about security vulnerabilities in our country are rare. "Specification of identification and description of security vulnerability" in effect has now ruled that the description of the vulnerability at least should include identification number, name, publish them, publish department, category, degree, the influence of the system and so on. This specification has just formulated the basic structure of the vulnerability database, if just according this specification to describe vulnerabilities, the amount of information is too little to precisely describe the vulnerabilities' information [13].

This article deep analyzes and summarizes the main vulnerability databases domestic and overseas, comprehensive all the advantages that each vulnerability database had, picking out 25 attributes, adopting group classified description to analyze the vulnerability attributes with grouping [13]. This method groups many dimensions of the vulnerability attributes into independent sets, one group including many close attributes. The structural model is shown in Figure 2.

This model covers all the needed attributes that "Specification of identification and description of security vulnerability" ruled and support CVE number, meanwhile it provide correspondent BUGTRAQ number and CNNVD number and so on. It makes the vulnerability database compatible with the main vulnerability databases domestic and overseas and can be cross-index. The rank method of the vulnerabilities adopts CVSS (Common Vulnerability Scoring System) specification. It makes the estimate of the vulnerability's risk grade more scientific and more accurate and provide a vulnerability grade to the security assess instrument. This vulnerability model even includes the attributes of the source vulnerability database as the identification of quoting the other vulnerability databases' information.

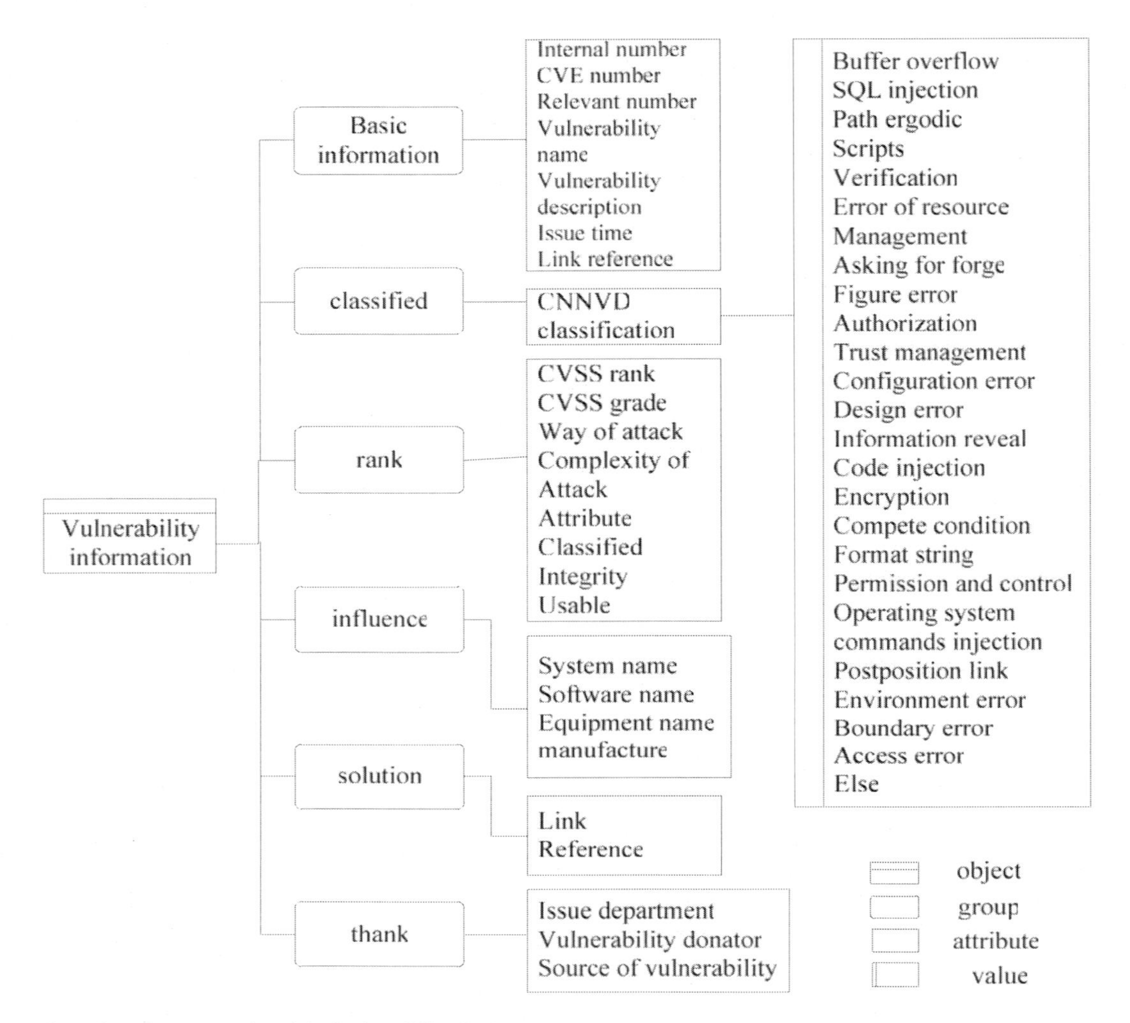

Figure 2. The structural model of vulnerability database.

Vulnerability database mainly includes vulnerability information table and vulnerability degree table.

4.3 *Implementation of the system*

The whole system can be divided into two parts, topic web crawler part and database part. Topic web crawler part mainly includes URL queue, web page access module, relevant analyze module, and dynamic topic module, information access module, and information integrate module, database access module and external interface module. The structure of the system is shown in Figure 3. The database is MySQL database.

Basic topic dictionary in dynamic topic module is collected manually and from CPE (Common Platform Enumeration) [14]. CPE is a specification that uniform named operating system, application program and so on, including widely applied products' name worldwide. Basic topic dictionary generally includes two parts, one part is a basic topic dictionary faced link and another part is a basic topic dictionary faced content. Basic topic dictionary faced link includes URL vulnerability database and URL in CPE's manufacture website. Basic topic dictionary faced content also contains two parts, one part are word set that describe the "vulnerability", such as "vulnerability", "damage", "SQL injection", "buffer overflow", "denial of service" and so on; another part is merged and extracted by CPE, the so-called merging and extracting is merging the different product name from different vision into one, such as CPE includes different visions' GNOME Display Manager

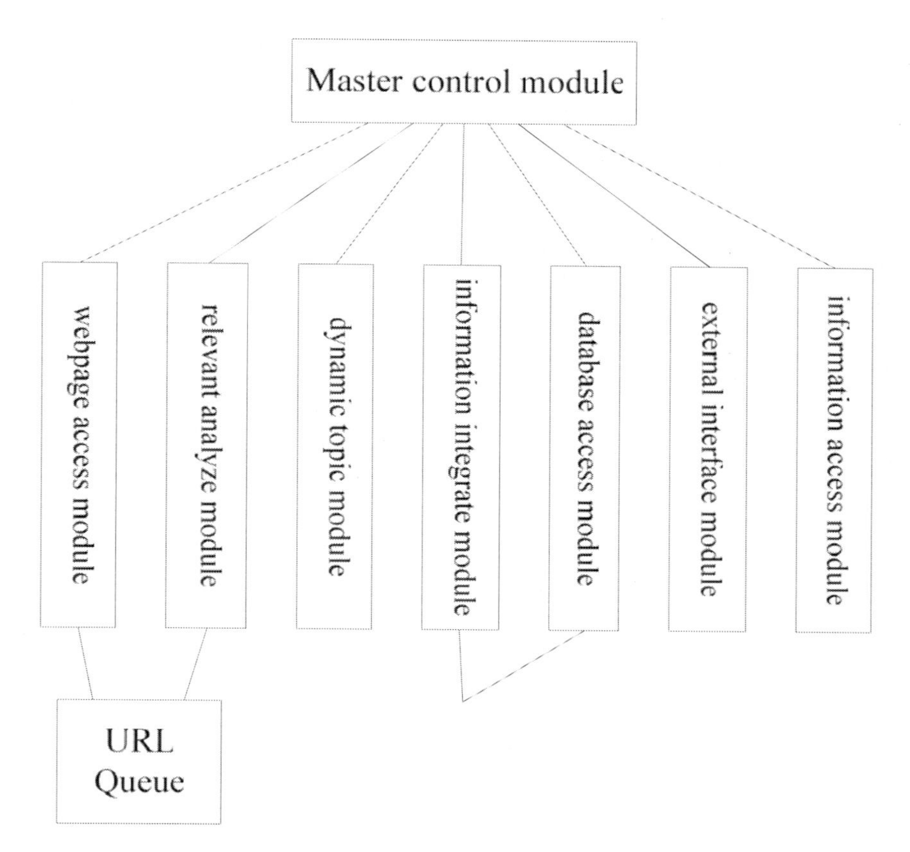

Figure 3. The structure of the system.

(GDM) X.X.X as many as 142, after the merging and extracting the product's name is GNOME Display Manager (GDM). In this way, the scale of the basic topic dictionary can be decreased, and the effective of the process can promoted. There are two approaches can be used to enlarge the dynamic topic dictionary, one way is through the unknown fields in the text corresponding to URL that connected to the topic to enlarge the dictionary, another way is analyzing the collected webpages, find the word that appears frequently and put the word into the dictionary.

Relevant analyze module analyzes the relevant between the URL link and the content on the webpage. To analyze the relevance of the URL link, we adopt the realm name match approach, it means we match the realm name of the to be analyzed URL link to the basic topic dictionary faced link to make sure the crawling website is in the scale of the topic websites. The relevant analyze of the content adopt the approach of matching string character. First, analyze the to be analyzed web page, move the sign of the webpage and extract the content, matching the content with the basic topic dictionary faced content and reach the conclusion of the relevant. Combining the result of the relevant analyze faced link, we have the final relevant degree.

An information access module based on the specification of HTML language, according to the vulnerability model to extract the vulnerability information.

The webpage access module establishes and initializes the HTTP dialogue and downloads the HTML file that allocated the URL.

Information integrates module updates and merges the vulnerability information and removes the overlapped information. Although the name and the description of the vulnerability on the internet are different, it is still may the same vulnerability. So merging and removing the overlapped information when accessing it to the database is necessary. The information about the vulnerability, mostly provides its number in the main vulnerability database. Therefore, this vulnerability database adopts a CVE number, BUGTRAQ number and CNNVD number as the sign of merging and removing overlapped information. The priority from high to low is: CVE, BUGTRAQ, and CNNVD. To the vulnerabilities that didn't use these numbers, we use the URL address to remove the overlapped information. Mostly these vulnerabilities exist in the small-scaled products and are not included in the main vulnerability databases.

Because the web sites that topic web crawlers crawled is concentrated, some particular web servers may be visited constantly at the same time. So the system must search friendly, avoiding visiting a particular web server constantly and adjusting the time gap between the two visits.

5 TESTS OF THE SYSTEM

This system adopts CVE, NVD and CNNVD as the beginning URL, ending with the security announcement issued by the flawed product's manufacturer to collect information about the vulnerabilities. The system is very stable during the test and successfully collects over 70,000 web page files, covers over 50,000 vulnerabilities.

6 CONCLUSIONS

This article accomplished Vulnerability Database Maintenance System Based on Topic Web Crawler; this system reduces the burden of the workers and can play an important role in the daily life. However, there are still some problems exist in the system, for example, it isn't detailed specification of semantic analysis in the relevant analysis and integration of vulnerability information, so it can't leave the manual work. But with the completing of the system, the automatic degree of the system will be improved and it will be applied widely.

REFERENCES

[1] Common vulnerabilities and exposures [EB/OL]. http://cve.mitre.org/.

[2] National vulnerability database [EB/OL]. http://nvd.nist.gov/.

[3] Vulnerability database of China information security [EB/OL]. http://www.cnnvd.org.cn/.

[4] Liu Xuqi, Zhang Yuqing, Gong Yafeng, *et al.* (2011) Study of Specification of identification and description of security vulnerability. *Information Security*, (7):4–6.

[5] Zheng H T, Kang B Y, Kim H G. (2008) An ontology-based approach to learnable focused crawling. *Information Sciences*, 178(23):4512–4522.

[6] Menczer F, Pant G, Srinivasan P. (2004) Topical web crawlers: Evaluating adaptive algorithms. *ACM Transactions on Internet Technology (TOIT)*, 4(4):378–419.

[7] Liu Jianming. (2013) *Study of Topic Web Crawler Technology in Vertical Search Engine*. Guangdong: Guangdong University of Technology.

[8] Liu YuSong. (2009) Study of noumenon structure and development instruments. *Contemporary Intelligence*, 8(3):417–447.

[9] P. Srinivasan, F. Menczer, G Pant. (2005) A general evaluation framework for topical crawlers. Information Retrieval, 8 (3):417–447.

[10] Lin Haixia, Yuan Fuyong, Chen Jinsen, *et al.* (2007) An improved algorithm of topic web spider. *Computer Engineering and Application*, 43(10):174–176.

[11] Zhang Qun, Li Peipei, Zhu Baoping *et al.* (2013) An importance assessment of directed weighed complicated web notes based on PageRank. *Journal of Nanjing University of Aeronautics and Astronautics*, 45(3):429–434.

[12] Yan Lei, Ma Yongnan, Ding Bin, *et al.* (2013) Topic web crawler of vertical search engine. *Fujian Computer*, 29(3):83–85.

[13] Zhang Yuqing, Wu Shuping, Liu Qixu, *et al.* (2011) Design and implementation of national vulnerability database, *Journal on Communications*, 32(6):93–100.

[14] Common Platform Enumeration Dictionary [EB/OL]. http://nvd.nist.gov/cpe.cfm.

Electric, Electronic and Control Engineering – Shao, Shu & Tian (Eds)
 ISBN: 978-1-138-02842-5

A new method of high-accuracy detection for modal parameter identification of al parameter identification of power system low frequency oscillation

Yan Zhao
School of Electrical Engineering and Automation, Harbin Institute of Technology, Harbin, China
Power Transmission and Transformation Technology College, Northeast Dianli University, Jilin, China

Zhimin Li
School of Electrical Engineering and Automation, Harbin Institute of Technology, Harbin, China

Tianyun Li
Power Transmission and Transformation Technology College, Northeast Dianli University, Jilin, China

ABSTRACT: Measure signals with Gaussian white noise of low frequency oscillation from Wide Area Monitoring Systems (WAMSs) are always processed by a low pass filter. The colored Gaussian noise produced by the low pass filter may impact the accuracy of oscillation mode identification. To solve it, this paper proposed a new method based on Tunable Q-factor Wavelet Transform (TQWT) and Matrix Pencil (MP) method for oscillation mode identification. Firstly, the wavelet function is generated by TQWT as its basis function of signal decomposition. Low Q-factor WT is used for sparse representation of the transient component. High Q-factor WT used for sparse representation of the oscillatory (rhythmic) component; secondly, Morphological Component Analysis (MCA) is used to separate the two signal components. The signal is decomposed into high resonance signal with high Q-factor and low resonance signals with low Q-factor. High-resonance is LFO signal. The residual is most colored Gaussian noise. And lastly, modal parameter is identified by the MP. After that, high-accuracy detection is achieved. Simulation has proven the effectiveness of the method.

KEYWORDS: low-frequency oscillation; tunable Q-factor wavelet transform; High Q-factor; Low Q-factor; matrix pencil

1 INTRODUCTION

Low frequency oscillation (LFO) arising from the interconnection of local power systems has caused more and more concerns these years. It may limit the power output of generators and transmission capability of tie lines, and may cause the system collapse in some serious circumstances. Therefore, a significant of work has been done in this field, and many of them focused on LFO oscillation principle, analysis methods, and suppression measures [1–3]. Among them, it is a difficult problem that accurate and effective LFO mode parameter identification.

WAMS provides a technical basis to better monitor low frequency oscillation and to identify low frequency oscillation mode. WAMS includes the Phase Measurement Unit (PMU), communication system and the control system of the EMS and SCADA. The PMU is a key unit for real-time dynamic monitoring system and WAMS. In the process of forming raw sample data, it will cause great interference that high-frequency sampling of PMU results in unwanted noise. Therefore, it is necessary that low pass filters are designed to purify high-frequency component other than the relevant frequency of the LFO. Colored Gaussian noise is produced by low pass filters of output signal added Gaussian white noise. The high approval methods of good anti-noise performance includes: Revised Prony algorithm[4], matrix pencil method[5], Stochastic Subspace Identification[6], and EMD[7] etc. Each has its own characteristics and applications. It's lacking to consider that colored Gaussian noise produced by low pass filters may impact the accuracy and stability of oscillation mode, although the general methods consider Gaussian white noise.

Selesnick recently proposed a new sparsity-enabled signal analysis named TQWT [8]. It is a new nonlinear signal analysis method based on signal resonance, rather than on frequency or scale, as provided by the Fourier and wavelet transforms. This method expresses a signal as the sum of a 'high-resonance' and a 'low-resonance' component. A high-resonance component is a signal consisting of multiple simultaneous sustained oscillations which is with high Q-factor; a low-resonance component is a signal consisting of non-oscillatory transients of unspecified shape and duration which is with low Q-factor. TQWT is a discrete-time wavelet transform, for which the Q-factor is easily specified. TQWT adaptively generates the basic functions of signal decomposition. MCA is applied to separate signal into a high resonance component and a low resonant component according to the nonlinear characteristic so that their best sparse signal representation is established.

The insufficient damping mechanism is one of many reasons of power system LFO. As a result, the LFO is an output of underdamped systems at a specific frequency. A system with high Q-factor is said to be underdamped. So, LFO signal is a high-resonance component which consists of multiple simultaneous sustained oscillations. In this paper, LFO signals with colored Gaussian noise were decomposed into high-resonant component, low-resonance component and residual by TQWT. The high resonance component is extractive LFO. At the same time, the residual is most colored Gaussian noise. Secondly, modal Parameter is identified by the MP. After that, high-accuracy detection is achieved.

2 TUNABLE Q-FACTOR WAVELET TRANSFORM

2.1 *Q factor, damping and LFO's resonance*

The quality factor or Q factor is a dimensionless parameter that describes how under-damped an oscillator or resonator is, or equivalently, characterizes a resonator's bandwidth relative to its center frequency. For high values of Q, the following definition is also mathematically accurate:

$$Q = {}^{f_c}\!/_{B_w} \tag{1}$$

where f_c is the resonant frequency, B_w is the half-power bandwidth.

Higher Q indicates a lower rate of energy loss relative to the stored energy of the resonator. The oscillations die out more slowly. Resonators with high quality factors have low damping. A system with high quality factor is said to be under damped. Underdamped systems combine oscillation at a specific frequency with decay of the amplitude of the signal.

A system with low quality factor is said to be over-damped. Such a system doesn't oscillate at all, but when displaced from its equilibrium steady-state output it returns to it by exponential decay, approaching the steady state value asymptotically. So, LFO signal is the output of underdamped systems at a specific frequency which is with high Q- factor and high-resonance property.

2.2 *Tunable Q-factor wavelet transform*

2.2.1 *Filter banks for the TQWT*

TQWT is a discrete-time wavelet trans-form for which the Q-factor is easily specified. Hence, the transform can be tuned according to the oscillatory behavior of the signal to which it is applied. The transform is based on a real-valued scaling factor (dilation-factor) and is implemented using a perfect reconstruction over-sampled filter bank with real-valued sampling factors[18–19].

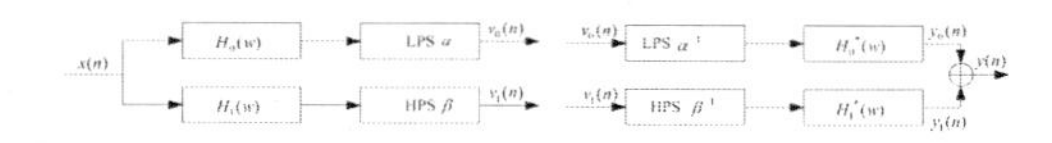

Figure 1. Analysis and synthesis filter banks for the TQWT.

In Fig.1, the subband signal v_0(n) has a sampling rate of αf_s where f_s is the sampling rate of the input signal x(n). Likewise, the subband signal $v_1(n)$ has a sampling rate of βf_s. LPS and HPS represent low-pass scaling and high-pass scaling respectively.

2.2.2 *Wavelet filter bank*

TQWT is implemented by iteratively applying the two-channel filter bank of its low-pass channel. For example, a three-stage wavelet transform is illustrated in Fig. 2. The wavelet transform inherits the perfect reconstruction property from the two channel filter bank. We denote the wavelet subband signals by w(j)(n) for j ≥ 1, with j = 1 being the high-pass subband produced by the first stage (as in Fig. 1). The sampling rate at subband j is given by $\beta\alpha^{j-1}f_s$ where f_s is the sampling rate of the input signal.

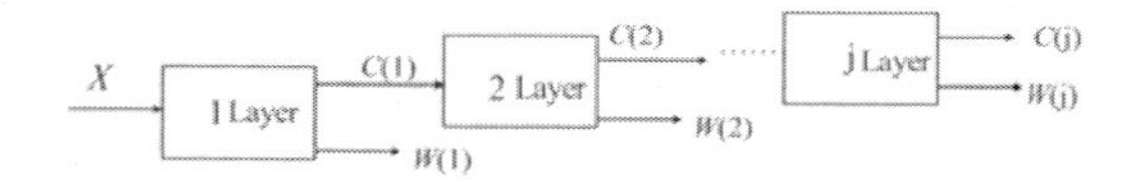

Figure 2. TQWT decomposition structure diagram.

2.2.3 *Parameters*

Oversampling rate (redundancy): The two-channel filterbank is oversampled by a factor of α+β.If the two-channel filter bank is iterated on its low-pass output ad infinitum so as to implement a wavelet transform, then the wavelet transform is oversampled by a factor of

$$\gamma = \beta / (1 - \alpha) \tag{2}$$

which we call the redundancy r of the wavelet transform.This expression is obtained by noting that the sampling rate at subband j (with $j \geq 1$) is given by $\beta\alpha^{j-1}f_s$ where f_s is the sampling rate of the input signal. The sum of the sampling rates over all subbands $j \geq 1$ gives $\beta\alpha^{j-1}f_s$ and hence the oversampling rate in (2).

Selecting α and β:

we can express α and β in terms of the Q-factor and redundancy:

$$\beta = 2/Q+1, \alpha = 1-\beta/\gamma \tag{3}$$

The specified Q-factor should be chosen subject to $Q \geq 1$. Setting $Q = 1$ leads to a wavelet transform, for which the wavelet resembles the second derivative of a Gaussian. Higher values of Q lead to more oscillatory wavelets. The specified oversampling rate r must be strictly greater than 1.

2.3 *Morphological component analysis*

Sparse signal processing exploits sparse representations of signals for applications such as denoising, deconvolution, signal separation, classification, etc. Therefore, transforms for the sparse representation of signals is a key ingredient for sparse signal processing.

In the noisy case, we should not ask for exact equality.

$$x = x_1 + x_2 + n, \quad \text{with } x, x_1, x_2 \in R^N, \tag{4}$$

Where n is noise. The goal of MCA is to estimate/determine x_1 and x_2 individually. Assuming that x_1 and x_2 can be sparsely represented in bases (or frames) Φ_1 and Φ_2 respectively, they can be estimated by minimizing the objective function,

$$J(w_1, w_2) = \lambda_1 \|w_1\|_1 + \lambda_2 \|w_2\|_1 \tag{5}$$

with respect to w_1 and w_2, subject to the constraint:

$$\Phi_1 w_1 + \Phi_2 w_2 = x \tag{6}$$

Then MCA provides the estimates

$$\hat{x}1 = \Phi_1 w_1 \tag{7}$$

and

$$\hat{x}2 = \Phi_2 w_2 \tag{8}$$

We apply variant of SALSA (split augmented Lagrangian shrinkage algorithm).to the MCA problem yields the iterative algorithm.

3 SIMULATION ANALYSES

3.1 *Test signal with white Gaussian noise*

$$x(t) = e^{-0.08t}\cos(2\pi \times 0.5t + 60°) + 0.5e^{-0.5t}(\cos 2\pi \times 1.1t + 45°) + \lambda(t) \tag{9}$$

The test signal with white Gaussian noise in Equation (9) simulates weakly damped oscillation. Where $\lambda(t)$ is white noise, signal to noise ratio (SNR) is 9.17dB.

When the TQWT is applied to test signal with white Gaussian noise, we got decomposition of a test signal with white Gaussian noise into high- and low-resonance components in Fig. 3. We used the parameters $Q_1 = 1$, $Q_2 = 4$, $\gamma_1 = \gamma_2 = 3$.

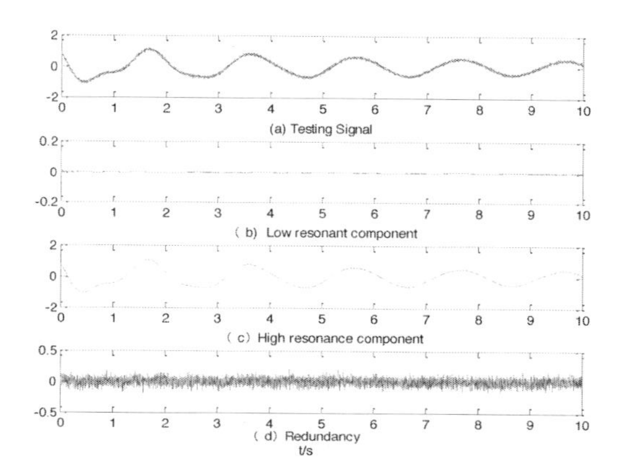

Figure 3. TQWT of the test signal with white Gaussian noise.

The high-resonance signal component is sparsely represented using a high Q-factor WT.

Similarly, the low-resonance signal component is sparsely represented using a low Q-factor WT.

The LFO is the output of underdamped systems at a specific frequency, so the high resonance component is the extracted LFO signal which consists of multiple simultaneous sustained oscillations. There is hardly any low resonance component. Note that the residual appears noise-like. The residual is the signal component that is sparsely represented in neither the high Q-factor wavelet transform nor the low Q-factor wavelet transform. The residual is most white Gaussian noise.

To verify the feasibility of the method which this article proposed, test signal with white Gaussian noise denoised by WT, EEMD and EMD respectively. The simulation number is 100. The evaluation indicators are, SNR, root mean square error, (RMSE), amplitude relative error, (ARE), and normalized correlation coefficient, (NCC). We use the NCC for feature points matching. NCC value is between 0 and 1.The higher the number, the more similar the curves. You can find the concrete evaluation indicators and methods in references [9], [10].

The table 1 proves that TQWT, EEMD and WT all remove gauss noise well. TQWT is slightly better. EMD cannot effectively remove Gaussian White Noise. A goal of EMD is to decompose a multicomponent signal into several narrow-band components (intrinsic mode functions). Gaussian white noise may cause modes mixing of EMD.

Table 1. Evaluation indicator of LFO with white Gaussian noise.

method\ indicator	SNR	RMSE	ARE	NCC
TQWT	16.944	0.4647	-1.04%	0.9807
EEMD	14.613	0.6525	-1.09%	0.9213
WT	14.7944	0.6604	-3.09%	0.8973

3.2 *Test signal with colored Gaussian noise*

For data sampled at 1000 Hz, we design a Butterworth low-pass filter with less than 3 dB of ripple in the passband, defined from 0 to 2.5 Hz, and at least 20 dB of attenuation in the stopband, defined from 150 Hz to the Nyquist frequency (500 Hz). The transfer function is:

$$H(z)=\frac{0.0487(1+Z^{-1})}{1-0.9026Z^{-1}} \tag{10}$$

Plot the filter's frequency response:

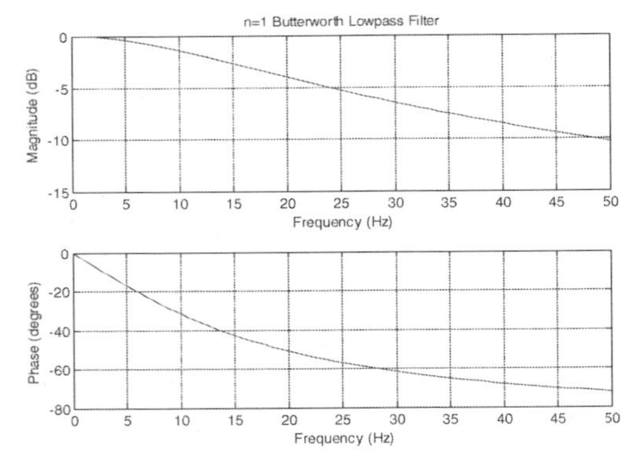

Figure 4. Frequency characteristic of Butterworth lowpass filter.

After the test signal with white Gaussian noise of equation (9) is processed through a low-pass filter, we obtained a signal with colored Gaussian noise. It can be seen in Fig7 that LFO with colored Gaussian noise were decomposed into high-resonant component, low-resonance component and residual by TQWT. The high resonance component is extractive LFO. At the same time, the residual is most colored Gaussian noise. We used the parameters Q_1 = 1, Q_2 = 4, $\gamma_1 = \gamma_2 = 3$. The table 2 showed evaluation indicators of TQWT, EEMD and WT.

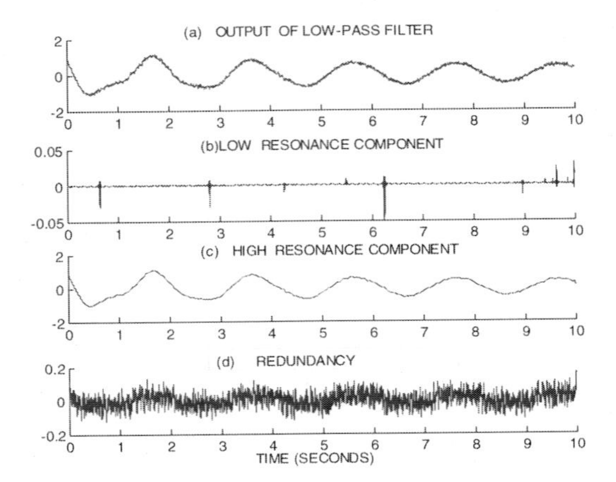

Figure 5. TQWT of the test signal with colored Gaussian noise.

Table 2. Evaluation indicator of LFO with colored Gaussian noise.

method\indicator	SNR	RMSE	ARE	NCC
TQWT	12.628	0.6198	-9.17%	0.8822
EEMD	9.1352	1.0185	-10.43%	0.5676
WT	6.8176	0.7172	-13.7%	0.5244

Data in Tab.2 show that three methods mentioned above have lower accuracy. The colored Gaussian noise produced by low pass filters impacts the accuracy of TQWT, EEMD and WT.But TQWT is best. Tab.3 shows results of modal Parameter identification by the matrix pencil method only, and TQWT plus MP. Form Tab.3, high-accuracy detection is achieved by TQWT-MP.

Table 3. Result of identification with colored Gaussian noise.

Identification Algorithm	Modality	F/Hz	Error (%)	Damping Coefficient	Error (%)
MP	1	0.4860	2.8	0.0250	1.884
	2	1.1153	1.391	0.0771	6.521
TQWT-MP	1	0.4960	0.8	0.0254	0.314
	2	1.1031	0.282	0.0733	1.271

4 CONCLUSION

TQWT is a discrete-time wavelet trans-form for which the Q-factor is easily specified. Hence, the transform can be tuned according to the oscillatory behavior of the signal to which it is applied. Adjustable Q-factor Can attain higher Q-factors than (or same low Q-factor of) the dyadic WT. Higher Q-factors Can achieve higher-frequency resolution needed for oscillatory signals. LFO signal is the output of underdamped systems at a specific frequency which is with high Q-factor and high-resonance property. Compared with other methods (WT, EMD, etc.), TQWT methods are efficient in Extracting the LFO. TQWT-MP methods achieved high-accuracy detection for modal parameter identification of power system low frequency oscillation.

REFERENCES

[1] Zhi Yong, Wang Guanhong, Xiao Yang, et al. (2011) Dynamic stability analysis of Gansu power grid after 750kV lines was put into operation. Power System Protection and Control, 39(3):114–118.

[2] Li Xun, Gong Qingwu, Jia Jingjing. (2012) Atomic sparse decomposition based identification method for low-frequency oscillation modal parameters. Transactions of China Electrotechnical Society, 27(9):124–131.

[3] Jia Yong, He Zhengyou. (2008) Review on analysis methods for low frequency oscillations based on disturbed trajectories. Power System Protection and Control, 40(11):140–148.

[4] Hauer J F. (1991) Application of Prony analysis to the determination of modal content and equivalent models for measured power system response. IEEE Transactions on Power Systems, 6(3):1062–1068.

[5] Hua Yingbo, Sarkar T k. (1991) On SVD for estimating generalized eigenvalues of singular matrix pencil in noise. IEEE Transactions on Signal Processing, 39(4):892–900.

[6] Ni J M, Shen C, Liu F. (2012) Estimation of the electromechanical characteristics of power systems based on a revised stochastic subspace method and the stabilization diagram. Sci. China Tech. Sci., 55(6):1677–1687.

[7] Li Tianyun, Xie Jiaan, Zhang Fangyan, et al. (2007) Application of HHT for extracting model parameters of low frequency oscillations in power systems. Proceedings of CSEE, 2007, 27(28):79–83.

[8] Selesnick I W. (2011) Resonance-based signal decomposition: a new sparsity-enabled Signal analysis method. Signal Processing, 91(12):2793–2809.

[9] Xie Qing, Zhang Lijun, Chen Shuyi, et al. (2012) Application of the fast ICA algorithm to PD ultrasonic array signal denoising. Proceedings of the CSEE, 2012, 32(18):160–166.

[10] X. Zhang, J. Zhou, N. Li, et al. (2012) Suppression of UHF partial discharge signals buried in white-noise interference based on block thresholding spatial correlation combinative de-noising method. IET Generation, Transmission & Distribution, 6(5):353–362.

Electric, Electronic and Control Engineering – Shao, Shu & Tian (Eds)
© 2015 Taylor & Francis Group, London, ISBN: 978-1-138-02842-5

Bifurcational and chaotic analysis of the virus-induced innate immune system

Jinying Tan, Guanglian Qin & Yanling Xu
College of Science, Huazhong Agricultural University, Wuhan Hubei, China

ABSTRACT: In this paper, we present a study of bifurcational and chaotic property of a virus-induced innate immune system. We investigate the stability of the equilibrium points, the Hopf bifurcation and its direction by the first Lyapunov value. It is also confirmed that there exists a chaotic solution for this system by calculating the maximal Lyapunov exponent.

KEYWORDS: innate immune system; dynamic model; bifurcation analysis; numerical simulation

1 INTRODUCTION

Bifurcation and Chaos, as the most important and basic feature of a nonlinear system which widely exists in various natural and social science fields, such as in meteorology, economics, engineering, and so on, has been studied since Lorenz found the first chaotic attractor in 1963 [1–5]. Life activities reveal the nonlinear complexity, which no doubt provides good examples for nonlinear dynamical studies and applications.

Interferons (IFNs) are important small molecule proteins, induced by viruses in the cells, which form the organism's own defense system with the biological functions of anti-virus, anti-tumor or immune regulation functions [6]. Perceiving the invasion of the viruses, the cells secrete a small amount of IFNs which in turn induce more IFNs through the JAK-STAT pathway (see Fig. 1). Then, these IFNs induce antiviral proteins (AVPs) which inhibit the activity of the viruses to reduce their reproductive ability and finally clear them [7, 8]. Under normal circumstances the body generally does not produce IFNs, but the body will make the appropriate response to produce IFNs after viral infection.

In the paper [9], we have established a third-order nonlinear system to display the relationships among the viruses, IFNs and AVPs based on the law of mass action. Due to the limitation of the biological experimental conditions, we just discussed the nature of the system when Hill coefficients $n_1=n_2=1$ and $n_1=n_2=2$. We find that there are more rich dynamical behaviors in the system when n_1 and n_2 are arbitrary positive integers. In the present work, we will provide the general dynamical analysis to distinguishing the chaotic, periodic and quasi-periodic orbits. Through the dimensionless method without considering the impact of time delay, the virus-induced innate immune system can be described as follows:

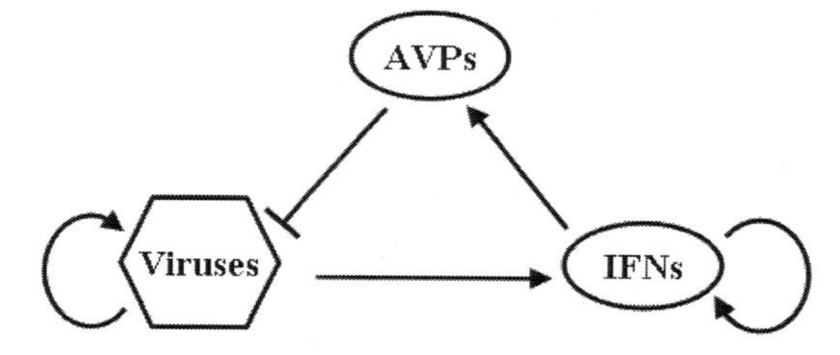

Figure 1. The regulatory network of three components: viruses, interferons (IFNs) and antiviral proteins (AVPs). (A line with arrow represents positive regulation and a line with short bar indicates negative regulation.)

$$\begin{cases} \dfrac{dx}{dt} = \dfrac{\sigma_1\alpha_2 x(t)}{1+z^{n_1}(t)} - \alpha_2 x(t) \\ \dfrac{dy}{dt} = x(t) + \dfrac{\sigma_2\alpha_4 y^{n_2}(t)}{K^{n_2}+y^{n_2}(t)} - \alpha_4 y(t) \\ \dfrac{dz}{dt} = y(t) - z(t) \end{cases} \tag{1}$$

Where $x(t)$, $y(t)$ and $z(t)$ are the dimensionless concentrations of the components, viruses, IFNs and AVPs, respectively. α_2 and α_4 represent the degradation rates of viruses and IFNs, respectively. K is the dissociation constant of IFNs. $\sigma_1\alpha_2$ is the production rate of viruses and $\sigma_2\alpha_4$ is the maximum transcription rate of IFNs. Thus, σ_1 and σ_2 can present the relative strengths of the two positive self-feedback loops, respectively. n_1 and n_2 are Hill coefficients. All parameters are dimensionless.

The organization of the paper is as follows. In Section 2 we analyze the stability conditions of the equilibrium points of the system (1). We obtain the conditions to clear all viruses. In Section 3 we use an analytical technique to analyze the Hopf bifurcation and in Section 4 we study the chaotic phenomenon of the system (1). In Section 5 we discuss and summarize our results.

2 STABILITY RESULTS [10]

Considering the biological significance, we assume that all parameters are greater than 0 and n_1, n_2 take positive integers.

2.1 *Case 1: $n_2=1$*

The system (1) can be simplified as follows

$$\begin{cases} \dfrac{dx}{dt} = \dfrac{\sigma_1\alpha_2 x(t)}{1+z^{n_1}(t)} - \alpha_2 x(t) \\ \dfrac{dy}{dt} = x(t) + \dfrac{\sigma_2\alpha_4 y(t)}{K+y(t)} - \alpha_4 y(t) \\ \dfrac{dz}{dt} = y(t) - z(t) \end{cases} \tag{2}$$

Then we easily obtain the three non-negative steady states of the system (2) (See Table 1.).

Table 1. Three non-negative steady states ($n_2=1$).

steady-states	$\overline{x}^{n_1,n_2=1}$	$\overline{y}^{n_1,n_2=1}$	$\overline{z}^{n_1,n_2=1}$
$O_1^{n_1,n_2=1}$	0	0	0
$O_2^{n_1,n_2=1}$	0	σ_2-K	σ_2-K
$O_3^{n_1,n_2=1}$	$\alpha_4(\sigma_1-1)^{1/n_1}[1-\frac{\sigma_2}{K+(\sigma_1-1)^{1/n_1}}]$	$(\sigma_1-1)^{1/n_1}$	$(\sigma_1-1)^{1/n_1}$

We have the following results about the stability conditions for the system (2):

Theorem 1: There are three non-negative steady states for system (2), and only one steady state is locally asymptotically stable under any conditions:

1. If $\sigma_1<1$ and $\sigma_2<K$, the system (2) is locally asymptotically stable at the first steady state $O_1^{n_1,n_2=1}$.
2. If $\sigma_2>\max\{K,\ K+(\sigma_1-1)^{1/n_1}\ (\sigma_1\ge 1)\}$, the system (2) is locally asymptotically stable at the second steady state $O_2^{n_1,n_2=1}$.
3. If $\sigma_1>1$, $\sigma_2<K+(\sigma_1-1)^{1/n_1}$ and $\alpha_2<C^{n_1,n_2=1}$, the system (2) is locally asymptotically stable at the third steady state $O_3^{n_1,n_2=1}$, where

$$C^{n_1,n_2=1} = \frac{\sigma_1}{n_1(\sigma_1-1)}\cdot\{1+\alpha_4-\frac{K\sigma_2\alpha_4}{[K+(\sigma_1-1)^{1/n_1}]^2}\}\cdot$$

$$\frac{K+(\sigma_1-1)^{1/n_1}-\dfrac{K\sigma_2}{K+(\sigma_1-1)^{1/n_1}}}{K+(\sigma_1-1)^{1/n_1}-\sigma_2}$$

We can easily obtain the following corollary from Theorem 1 (3):

Corollary 1: If $\sigma_1>1$, $\sigma_2<K+(\sigma_1-1)^{1/n_1}$ and $\alpha_2\le 1$, the system (2) is locally asymptotically stable at the third steady state $O_3^{n_1,n_2=1}$.

2.2 *Case 2: $n_2>1$*

Then we easily obtain the four non-negative steady states of system (1) (See Table 3.).

Table 3. Four non-negative steady states ($n_2>1$).

steady-states	$\overline{x}^{n_1,n_2}$	$\overline{y}^{n_1,n_2}$	$\overline{z}^{n_1,n_2}$
$O_1^{n_1,n_2}$	0	0	0
$O_{21}^{n_1,n_2}$	0	$\overline{y}_{21}^{n_1,n_2}$	$\overline{z}_{21}^{n_1,n_2}(=\overline{y}_{21}^{n_1,n_2})$
$O_{22}^{n_1,n_2}$	0	$\overline{y}_{22}^{n_1,n_2}$	$\overline{z}_{22}^{n_1,n_2}(=\overline{y}_{22}^{n_1,n_2})$
$O_3^{n_1,n_2}$	$\alpha_4[(\sigma_1-1)^{1/n_1}-\frac{\sigma_2(\sigma_1-1)^{n_2/n_1}}{K^{n_2}+(\sigma_1-1)^{n_2/n_1}}]$	$(\sigma_1-1)^{1/n_1}$	$(\sigma_1-1)^{1/n_1}$

We have the following results about the stability conditions for the system (1) when $n_2>1$:

Theorem 2: The system (1) exists three non-negative stable steady states under certain conditions and an unstable steady state:

1. If $\sigma_1<1$, the system (1) is locally asymptotically stable at the first steady state $O_1^{n_1,n_2}$.
2. The second steady state $O_{21}^{n_1,n_2}$ is unstable.
3. If $\sigma_2>\dfrac{n_2 K}{(n_2-1)^{\frac{n_2-1}{n_2}}}$ and $\sigma_2>\dfrac{K^{n_2}+(\sigma_1-1)^{n_2/n_1}}{(\sigma_1-1)^{\frac{n_2-1}{n_1}}}$, $(\sigma_1\ge 1+(n_2-1)^{n_1/n_2}K^{n_1})$ the system (1) is locally

asymptotically stable at the third steady state $O_{22}^{n_1,n_2}$.

4 If $\sigma_1 > 1$, $\sigma_2 < \min\{\frac{K^{n_2} + (\sigma_1 - 1)^{n_2/n_1}}{(\sigma_1 - 1)^{n_2-1/n_1}}, \frac{[K^{n_2} + (\sigma_1 - 1)^{n_2/n_1}]^2}{n_2 K^{n_2} (\sigma_1 - 1)^{n_2-1/n_1}}\}$ and $\alpha_2 < C^{n_1,n_2}$, the system (1) is locally asymptotically stable at the fourth steady state $O_3^{n_1,n_2}$, where

$$C^{n_1,n_2} = \frac{\{1 + \alpha_4 - \frac{n_2 \sigma_2 \alpha_4 K^{n_2} (\sigma_1 - 1)^{n_2-1/n_1}}{[K^{n_2} + (\sigma_1 - 1)^{n_2/n_1}]^2}\} \cdot \{1 - \frac{n_2 \sigma_2 K^{n_2} (\sigma_1 - 1)^{n_2-1/n_1}}{[K^{n_2} + (\sigma_1 - 1)^{n_2/n_1}]^2}\}}{\frac{n_1(\sigma_1 - 1)[K^{n_2} + (\sigma_1 - 1)^{n_2/n_1} - \sigma_2 (\sigma_1 - 1)^{n_2-1/n_1}]}{\sigma_1 [K^{n_2} + (\sigma_1 - 1)^{n_2/n_1}]}} \cdot$$

3 BIFURCATION ANALYSIS

As can be seen from the above analysis, when σ_1, σ_2, K and α_4 are fixed, the change of α_2 should have impact on the stability of the non-negative steady-state $O_3^{n_1,n_2}$. Taking into account that our model has a negative feedback, we conjecture that the increase of α_2 will destabilize the equilibrium point $O_3^{n_1,n_2}$ if α_2 is large enough, and usually we find that periodic solution (self-oscillation) can arise from equilibrium by Hopf bifurcation (Fig. 5).

Theorem 3: If α_2 is viewed as parameter, then a Hopf bifurcation occurs as α_2 passes through C^{n_1,n_2} for the system (1). [10]

A: When a_2=10.2347 (C = 10.2347, n_1=n_2=1), a Hopf bifurcation occurs. B: When a_2=0.6303 (C = 0.6303, n_1=n_2=2), a Hopf bifurcation occurs. The other parameters are same: s_1=4, s_2=3, a_4=4 and K=2.

We will then discuss Hopf bifurcation direction of the system (1) by numerical computation of the first Lyapunov value [11, 12].

To keep the presentation short, we will only consider the case n_1=n_2=1, we can similarly study the bifurcation of the system (1) at the other cases. We will coordinate origin shifts of O_3, i.e. let $x_1 = x - \bar{x}_3, x_2 = y - \bar{y}_3, x_3 = z - \bar{z}_3$, and for convenience, we still use x, y and z instead of x_1, x_2 and x_3. Then from the system (1), we have

$$\begin{cases} \frac{dx}{dt} = \frac{\sigma_1 \alpha_2 (x + \bar{x}_3)}{1 + \bar{z}_3 + z} - \alpha_2 (x + \bar{x}_3) \\ \frac{dy}{dt} = (x + \bar{x}_3) + \frac{\sigma_2 \alpha_4 (y + \bar{y}_3)}{K + (y + \bar{y}_3)} - \alpha_4 (y + \bar{y}_3) \\ \frac{dz}{dt} = y - z \end{cases} \quad (3)$$

We put the above system (3) by Taylor expansion to third-order terms (and ignoring higher order terms) in the form as follows: $\dot{x} = A(\alpha_2)x + F(x, \alpha_2) \qquad x \in R^n$.

$$\begin{cases} \frac{dx}{dt} = -\frac{\alpha_2 \bar{x}_3}{\sigma_1} z - \frac{\alpha_2}{\sigma_1} xz + \frac{\alpha_2 \bar{x}_3}{\sigma_1^2} z^2 + \frac{\alpha_2}{\sigma_1^2} xz^2 - \frac{\alpha_2 \bar{x}_3}{\sigma_1^3} z^3 \\ \frac{dy}{dt} = x + [\frac{K \sigma_2 \alpha_4}{(K + \bar{y}_3)^2} - \alpha_4] y - \frac{K \sigma_2 \alpha_4}{(K + \bar{y}_3)^2} y^2 + \frac{K \sigma_2 \alpha_4}{(K + \bar{y}_3)^4} y^3 \\ \frac{dz}{dt} = y - z \end{cases} \quad (4)$$

Here we use the First Lyapunov value to discuss the direction of the Hopf bifurcation about the parameter α_2. In order to facilitate the calculation, we take a set of values to be calculated (within the parameters of any one group of values calculation results the same).

We set $K = 2$, $\sigma_1 = 4$, $\sigma_2 = 3$, $\alpha_4 = 4$ then $\bar{x}_3 = 4.8$, $\bar{y}_3 = \bar{z}_3 = 3$, $C = 10.2347$ and

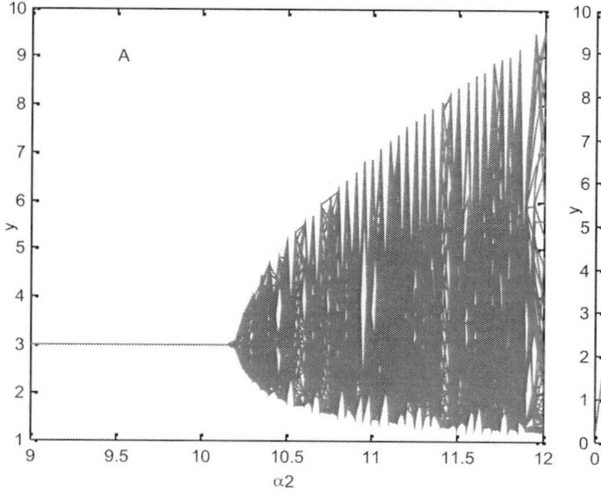

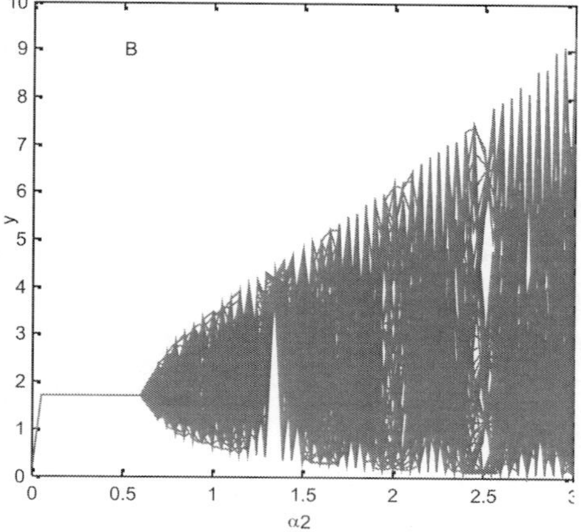

Figure 5. Bifurcation graph about a_2.

$$\begin{cases} \frac{dx}{dt} = -1.2\alpha_2 z - 0.25\alpha_2 xz + 0.3\alpha_2 z^2 \\ \quad +0.0625\alpha_2 xz^2 - 0.075\alpha_2 z^3 \\ \frac{dy}{dt} = x - 3.04y - 0.192y^2 + 0.0348y^3 \\ \frac{dz}{dt} = y - z \end{cases} \tag{5}$$

When $\alpha_2 = C = 10.2347$, then

$$A = \begin{pmatrix} 0 & 0 & -12.2816 \\ 1 & -3.04 & 0 \\ 0 & 1 & -1 \end{pmatrix} \tag{6}$$

and the eigenvalue $\lambda_1 = -4.04$ and pure imaginary roots $\lambda_{2,3} = \pm\beta i = \pm 1.7436i$. Finding eigenvectors of A for $\lambda_2 = 1.7436i$, we have

$$u = \begin{pmatrix} -0.9528 \\ -0.2358 + 0.1353i \\ 0.1353i \end{pmatrix} \tag{7}$$

and

$$v = \begin{pmatrix} -0.4424 - 0.1909i \\ -0.3329 + 0.7713i \\ 0.3329 + 2.9252i \end{pmatrix} \tag{8}$$

which satisfy $Au = i1.7436u$, $A^T v = -i1.7436v$ and $< v,u >= 1$ ($< v,u >= \sum_{i=1}^{n} \bar{v}_i u_i$ is the inner product of the vectors u, v).

For the system (1), the bilinear and trilinear functions are

$$B(X,X') = \begin{pmatrix} -5.1173xz' + 6.1408zz' \\ -0.384yy' \\ 0 \end{pmatrix} \tag{9}$$

and

$$C(X,X',X'') = \begin{pmatrix} 3.838xz'z'' - 4.6056zz'z'' \\ 0.2304yy'y'' \\ 0 \end{pmatrix} \tag{10}$$

respectively, where $X = (x,y,z)^T$, $X' = (x',y',z')^T$, $X'' = (x'',y'',z'')^T$.

By some tedious calculations, from above formulas, we have

$$B(u,u) = \begin{pmatrix} -0.1124 + 0.6595i \\ -0.0143 + 0.0245i \\ 0 \end{pmatrix} \tag{11}$$

$$B(u,\bar{u}) = \begin{pmatrix} 0.1124 - 0.6595i \\ -0.0284 \\ 0 \end{pmatrix} \tag{12}$$

and

$$A^{-1} = \begin{pmatrix} -0.2475 & 1 & 3.04 \\ -0.0814 & 0 & 1 \\ -0.0814 & 0 & 0 \end{pmatrix} \tag{13}$$

Let

$$s = A^{-1}B(u,\bar{u}) = \begin{pmatrix} -0.0562 + 0.1632i \\ -0.0091 + 0.0537i \\ -0.0091 + 0.0537i \end{pmatrix} \tag{14}$$

and then

$$B(u,s) = \begin{pmatrix} -0.0892 + 0.2542i \\ 0.0020 + 0.0053i \\ 0 \end{pmatrix} \tag{15}$$

therefore

$$< v, B(u,s) >= -0.0056 - 0.1328i \tag{16}$$

and

$$(2i\beta I - A)^{-1} = \tag{17}$$

$$\begin{pmatrix} -0.0473 - 0.3415i & 0.191 - 0.1649i & 1.1556 + 0.1649i \\ -0.0624 - 0.0408i & 0.1423 - 0.2175i & 0.191 - 0.1649i \\ -0.0156 + 0.0134i & -0.0468 - 0.0542i & 0.0468 - 0.3281i \end{pmatrix}$$

Let

$$s' = (2i\beta I - A)^{-1}B(u,u) = \begin{pmatrix} 0.2319 + 0.0142i \\ 0.0372 - 0.0299i \\ -0.0051 - 0.0121i \end{pmatrix} \tag{18}$$

we can obtain

$$B(\bar{u},s') = \begin{pmatrix} -0.0350 - 0.0549i \\ 0.0049 - 0.0008i \\ 0 \end{pmatrix} \tag{19}$$

$$< v, B(\bar{u},s') >= 0.0237 + 0.0141i \tag{20}$$

$$C(u,u,\bar{u}) = \begin{pmatrix} -0.0669 - 0.0114i \\ -0.0040 + 0.0023i \\ 0 \end{pmatrix} \tag{21}$$

$$< v, C(u,u,\bar{u}) >= 0.0349 - 0.0054i \tag{22}$$

Consequently, from above computation, we have

$$\begin{aligned} l_1(0) &= \frac{1}{2\beta}\mathrm{Re}(< v, C(u,u,\bar{u}) > -2 < v, B(u, A^{-1}B(u,\bar{u})) > \\ &+ < v, B(\bar{u}, B(\bar{q}, (2i\beta I - A)^{-1}B(u,u)) >) \\ &= \frac{1}{2\cdot 1.7436}\mathrm{Re}(0.0698 + 0.2742i) = 0.02 > 0. \end{aligned} \tag{23}$$

According to the method in [11], we know that the system (5) displays a Hopf bifurcation when $l_1(0) > 0$, which is non-degenerate and subcritical (Fig. 5(A) and Fig. 7).

4 CHAOS ANALYSIS

In Fig. 6 the case when $\alpha_2 = 5 < C$ is shown. It is seen that here the system is stable. After $\alpha_2 = 10.2347 = C$ the Hopf bifurcation occurs and induces an unstable limit cycle. These results are in accordance with results obtained in our study in previous section.

As α_2 increases, the system (1) is more and more sensitive to initial conditions, which shows the characteristics of chaos. To confirm our conclusion, we calculate the maximal Lyapunov exponent for this case. Following [13, 14], the maximal Lyapunov exponent for a given data set can be calculated by means of the sum

$$L_{\max} = \sum_{k=0}^{N} \ln(\varepsilon'_k / \varepsilon_k) \Big/ \sum_{k=0}^{N} T_k \tag{24}$$

where ε_k is the distance between the two points in two trajectories from two arbitrary initial points x_0, x'_0 at the kth iterations, i.e. $\varepsilon_k = \|x_k - x'_k\|$, ε'_k is the distance between two trajectories will exceed some value $\varepsilon_{\max}$ at the kth iterations, T_k is time between x_k and x'_k, and N is the number of algorithm's iterations. The obtained maximal Lyapunov exponent (per unit time) is: $L_{\max}$ = 0.005518 and 8.54243 when a_2 = 10.2347 and 50, respectively. In Fig. 7 and Fig. 8 we show that after a_2 = 10.2347 the chaotic behavior of system takes place.

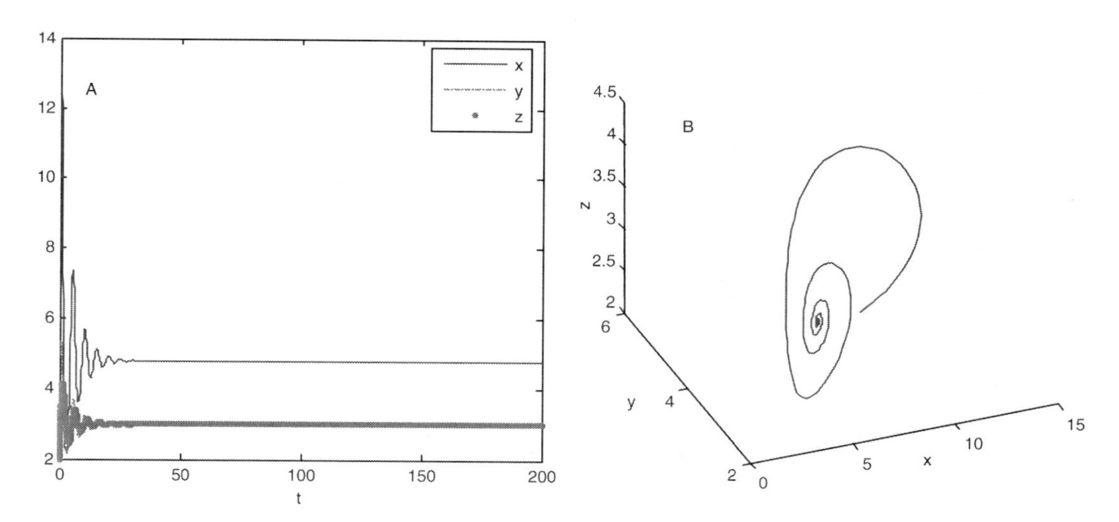

Figure 6. Stable solution (A) and its phase portrait (B) of the system (1) at a_2=5. The other parameters s_1=4, s_2=3, a_4=4, K=2 and n_1=n_2=1.

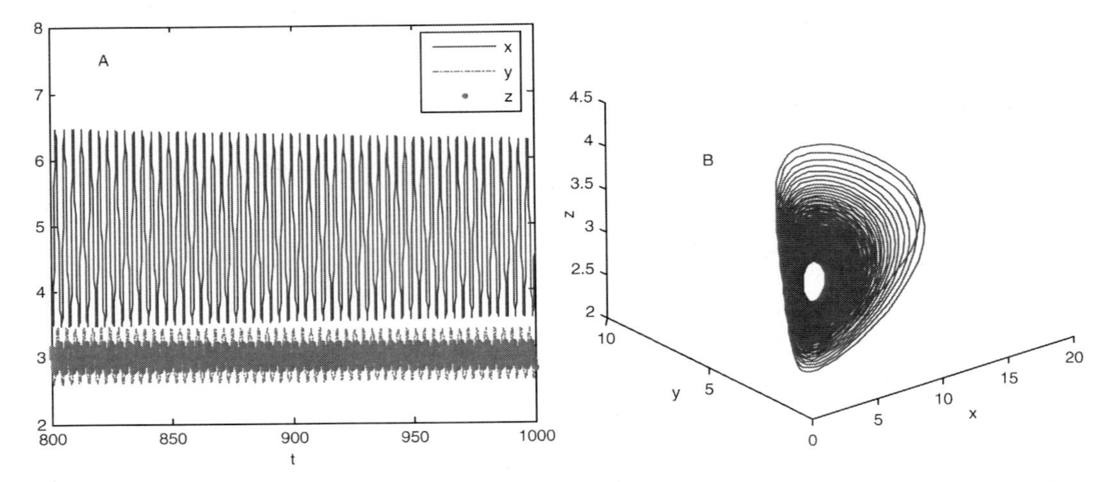

Figure 7. Unstable periodic solution (A) and its phase portrait (B) of the system (1) at a2=10.2347. The other parameters s1=4, s2=3, a4=4, K=2 and n1=n2=1.

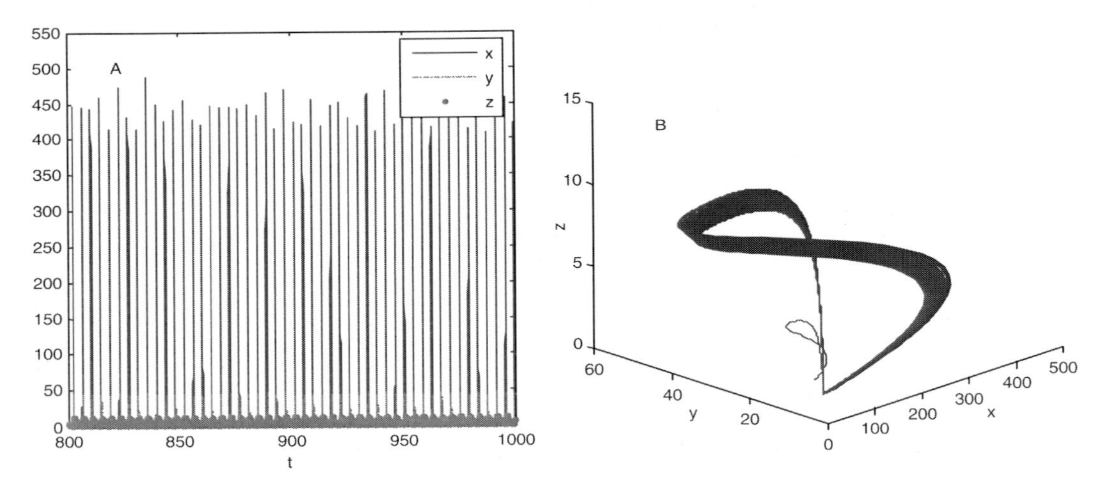

Figure 8. Chaotic solution (A) and its phase portrait (B) of the system (1) at a2=50. The other parameters s1=4, s2=3, a4=4, K=2 and n1=n2=1.

5 SUMMARY AND CONCLUSIONS

The paper presents a study of the dynamical behaviors of the virus-induced innate immune system. We investigate the stability of the equilibrium points, the Hopf bifurcation and its direction by the first Lyapunov value, and the confirmation of chaotic orbit through the maximal Lyapunov exponent.

Summarizing the results obtained in Sections 2 - 4, we can conclude that:

1 The system (1) is local stable when $\alpha_2 < C^{n_1,n_2}$.

2 A Hopf bifurcation occurs as the control parameter α_2 passes through C^{n_1,n_2} for the system (1), inducing an unstable limit cycle.

3 The system (1) may be in a chaotic state when $\alpha_2 \geq C^{n_1,n_2}$.

ACKNOWLEDGMENTS

The authors thank the anonymous reviewers for their helpful comments and suggestions. This work was supported by Huazhong Agricultural

University Scientific & Technological Self-innovation Foundation (Program No. 2662014BQ067).

REFERENCES:

[1] Lorenz E N. (1963) Deterministic nonperiodic flow. *Journal of the Atmospheric Sciences*, 20(2):130–141.
[2] Takens F. (1981) Detecting strange attractors in turbulence. *Lecture Notes in Mathematicks*, 898:366–381.
[3] Sulem J, Vardoulakis I G. (2004) *Bifurcation Analysis in Geomechanics*. CRC Press.
[4] Chen Y, Leung A Y T. (1998) Bifurcation and Chaos in Engineering. London: Springer,.
[5] Panchev S. (2001) *Theory of Chaos*, Acad. Publishing House, Sofia.
[6] Brzózka K, Finke S, Conzelmann K K. (2005) Identification of the rabies virus alpha/beta interferon antagonist: Phosphoprotein P interferes with phosphorylation of interferon regulatory factor 3. *Journal of Virology*, 79(12):7673–7681.
[7] Akira S, Takeda K. (2004) Toll-like receptor signaling. *Nature Reviews Immunology*, 4(7):499–511.
[8] Haller O, Kochs G, Weber F. (2006) The interferon response circuit: induction and suppression by pathogenic viruses. *Virology*, 344(1):119–130.
[9] Tan J, Pan R, Qiao L, et al. (2012) Modeling and dynamical analysis of virus-triggered innate immune signaling pathways. *PLoS One*, 7(10):e48114.
[10] Tan J, Zou X. (2013) Complex dynamical analysis of a coupled network from innate immune responses. *International Journal of Bifurcation and Chaos*, 23(11).
[11] Kuznetsov YA. (1998) *Elements of Applied Bifurcation Theory*. (2nd ed). New York: Springer-Verlag.
[12] Yu Y, Zhang S. (2004) Hopf bifurcation analysis of the Lü system. *Chaos, Solitons & Fractals*, 21(5):1215–1220.
[13] Bryant P, Brown R, Abarbanel H D I. (1990) Lyapunov exponents from observed time series. *Physical Review Letters*, 65(13):1523.
[14] Wolf A, Swift J B, Swinney H L, et al. (1985) Determining Lyapunov exponents from a time series. *Physica D: Nonlinear Phenomena*, , 16(3):285–317.

Electric, Electronic and Control Engineering – Shao, Shu & Tian (Eds)
© 2015 Taylor & Francis Group, London, ISBN: 978-1-138-02842-5

A multiple utility factors-based parallel packet scheduling algorithm of BWM system

Min Wang, Qiaoyun Sun, Shuguang Zhang & Yu Zhang
Beijing City University, Beijing, China

Yinlong Liu
Institute of Information Engineering, Chinese Academy of Sciences, Beijing, China

ABSTRACT: According to the mode of large and small base stations in BWM system, a multiple utility factors-based parallel packet scheduling algorithm that is suitable for real-time multimedia mixed services is proposed in this paper. This algorithm takes into account quality demands of different services and determines scheduling priority in line with channel condition and queue state of users. Besides, the method of parallel scheduling is applied to further reducing the computational complexity. Only parts of large single utility factors are selected for scheduling. Simulation results show that the scheduling algorithm performs well in terms of time delay, handling capacity and equity.

KEYWORDS: radio resource management; scheduling algorithm; quality of service

In the integration of three networks, broadcast and communication of the wireless communication system become gradually integrated through the complementation of respective technological advantages in order to meet user demand of multimedia services. Broadband wireless multimedia (BWM) network is a new type of broadband wireless mobile network integrating technical features of the mobile TV network and the broadband wireless access network. It aims at finding a highly-efficient technological approach with lower cost so as to achieve the overall objective of "three networks integration" in an all-round way from such aspects as air interface, radio access network, core network, service platform, terminal, etc. In order to make full use of current network resources with wireless coverage formed by large base stations of the traditional radio and television system and small base stations of the traditional cellular system, the mode of large and small base stations becomes the most intuitional way of realizing the integration of broadcast and communication in the wireless communication system. It is a convergence between broadcast network and cellular network. But contradictions between the growing needs of services and increasingly scanty wireless spectrum resources become gradually prominent. So, it is extremely important to realize wireless resource joint management of the mode of large and small base stations effectively, especially the joint scheduling problem between large and small base stations. However, the current modified proportional fair (MPF) algorithm based on quality of service (QoS) [1–3] and the scheduling algorithm based on utility function [4–6] fail to consider characteristics of the mode of large and small base stations. Besides, the computational complexity is so high that the two algorithms cannot be applied directly.

In order to improve the utilization of wireless resources and provide better quality of services, a multiple utility factors-based parallel packet scheduling algorithm is proposed in this paper so as to realize the joint scheduling of large and small base stations and guarantee the seamless combination between broadcast and communication in the mode of large and small base stations. Select multiple utility factors for dynamic scheduling in line with demands of QoS, queue state and channel condition and reduce the computational complexity with the method of parallel scheduling.

1 SYSTEM MODEL

BWM refers to the broadband wireless multimedia network based on OFDM, of which the mode of large and small base stations is shown in Figure 1. Multiple cellular base stations (CBS) are simultaneously covered in a broadcast base station (BBS), where radio and television services are transmitted by the large base station and broadband wireless data services are transmitted by small base stations. Thus, current resources of radio and television system and cellular system can be made full use of so as to reduce the construction cost. Meanwhile, mixed wireless network has strong flexibility in operation services and is adaptable in separate and synthetic operational mode of radio and television services and broadband wireless access services.

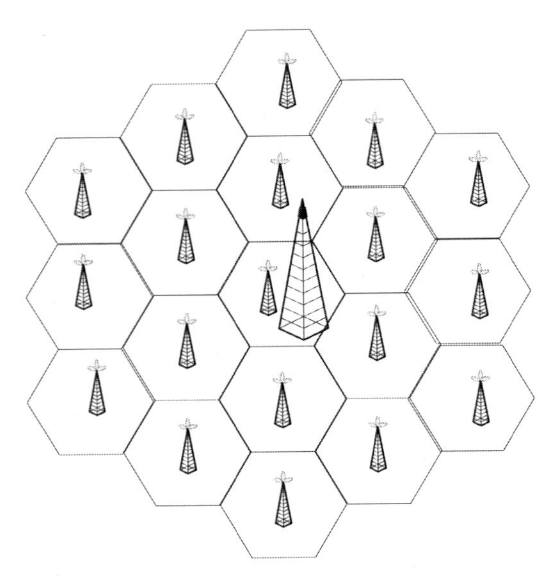

Figure 1. The mode of large and small base stations

2 RESOURCE SCHEDULING ALGORITHM

2.1 *Problem analysis*

As for the mode of large and small base stations, BBS has large transmission power and coverage in broadcasting services and the single frequency network is adopted to provide services for users so as to guarantee the equity of users. Therefore, the compromise of service equity and handling capacity should be considered more in broadcast scheduling of large base stations. In addition, influences brought by quality of services like time delay, deadline and packet loss probability and real-time factors should also be taken into account comprehensively. Being different from BBS, CBS has relatively small transmission power and coverage in single broadcasting services. It is more necessary to attach importance to users in single broadcast scheduling of small base stations. Because of the diversity of user demands in services, influences of service features and user demands should be considered comprehensively, the compromise of user equity and handling capacity should be emphasized, and influences brought by quality of services and real-time factors like time delay, delay variation and packet loss probability should be taken into account as well. Besides, channel feedback information in the cellular network can be also applied to conducting effective scheduling based on channel state of users so as to improve the utilization rate of wireless resources.

According to the characteristics of the mode of large and small base stations, MUFPS algorithm is used for guaranteeing the realization of broadcasting services of BBS and the realization of single broadcasting services of CBS respectively after service bandwidth allocation of BBS and CBS in line with features of broadcasting and single broadcasting services. As for the selection of multiple utility factors in each scheduling algorithm, such factors as handling capacity, equity, instantaneity and QoS of services and users should be considered comprehensively. The effectiveness of the algorithm can be improved by the dynamic scheduling in line with demands of QoS, queue state and channel condition. However, multiple utility factors increase the computational complexity in the meantime. In order to further reduce the operational cost, the method of parallel scheduling can be applied to simplifying the computational priority so as to improve the performance of the whole algorithm.

2.2 *Algorithm description*

In order to guarantee the service quality of broadcasting services, the MUFPS algorithm of broadcast scheduling in large base stations prioritizes the service queue based on such parameters as queue delay, occupancy of buffer, data rate, deadline, etc. The computational formula of the priority $P_j(n)$ of the jth queue in the nth TTI is:

$$P_j(n) = A_j(n) \cdot T_j(n) \cdot \Lambda_j(n) \cdot \Gamma_j(n) \cdot \Xi_j(n) \qquad j = 1,2 \cdots J = n = 1,2 \cdots N \tag{1}$$

The parameter of service priority $A_j(n)$ refers to the priority of service itself. It is a constant time-independent parameter in transmission, reflecting the corresponding priority of the service of the jth queue in transmission. It is only related to requirements of the service in terms of instantaneity and urgency. High priority of services means that if there is no such extreme case as buffer overflow or too long queue delay, corresponding services will be scheduled with priority.

The parameter of queue delay $T_j(n)$ is able to adjust the scheduling priority according to queue state so as to meet the service demand of time delay. The threshold value of queue delay τ_j^* stands for the maximum delay of services, which prevents the situation of invalid transmission of services caused by excessive delay. $T_j(n)$ can be expressed as :

$$T_j(n) = \begin{cases} 1 & if\ \overline{\tau_j}(n) \le \tau_j^* \\ \dfrac{\overline{\tau_j}(n)}{\tau_j^*} & if\ \overline{\tau_j}(n) > \tau_j^* \end{cases} \qquad j = 1,2 \cdots J = n = 1,2 \cdots N \tag{2}$$

The average delay of the service queue in the above formula is expresses as:

$$\overline{\tau}_j(n)=\frac{\sum_{k\in\Delta_j(n)}\tau^q_{j,k}(n)+\sum_{k\in\Theta_j(n)}\tau^q_{j,k}(n)}{N^l_j(n)+N^q_j(n)}\quad j=1,2\cdots J;\ n=1,2\cdots N \quad (3)$$

Here, $N^l_j(n)$ is the number of transmitted data packages in the jth queue of the nth TTI, of which the set is $\Delta_j(n)=\{1,2,\cdots N^l_j(n)\}$; $N^q_i(n)$ is the number of unsent data packages in the queue, of which the set is $\Theta_j(n)=\{N^l_j(n)+1, N^l_j(n)+2,\cdots N^l_j(n)+N^q_j(n)\}$; $\tau^q_{j,k}(n)$ stands for the time delay of data packages. The computational formula is shown below:

$$\tau^q_{j,k}(n)=\begin{cases}T^{lev}_j(k)-T^{avl}_j(k) & \text{if } k\le N^l_j(n)\\ n\cdot T_{tti}-T^{avl}_j(k) & \text{if } k>N^l_j(n)\end{cases}\quad j=1,2\cdots J;\ n=1,2\cdots N \quad (4)$$

In the above formula, $T^{avl}_j(k)$ and $T^{lev}_j(k)$ respectively refer to the arrival time and the departure time of the kth data package in the jth queue.

The parameter of buffer overflow $\Lambda_j(n$ reflects the buffer occupancy of service queue which avoids service data packet dropout caused by buffer overflow. Priority coefficients are adjusted by setting the threshold length of buffer. The expression of $\Lambda_j(n$ is shown below:

$$\Lambda_j(n)=\begin{cases}1 & \text{if } \lambda_j(n)\le\lambda^*_j\cdot\sigma_j\\ \dfrac{\lambda_j(n)}{\lambda^*_j\cdot\sigma_j} & \text{if } \lambda_j(n)>\lambda^*_j\cdot\sigma_j\end{cases}\quad j=1,2\cdots J;\ n=1,2\cdots N \quad (5)$$

Here, λ^*_j is the length of buffer and σ_j is the threshold value of buffer occupancy.

The parameter of service data rate is an indicator for the transmission rate of service data. The priority can be adjusted in line with the ratio between average speed and threshold value of service transmission so as to make sure that the transmission rate of each service can meet requirements. The expression is shown below:

$$\Gamma_j(n)=\begin{cases}1 & \text{if } \overline{\gamma}_j(n)>\gamma^*_j\\ \dfrac{\gamma^*_j}{\overline{\gamma}_j(n)} & \text{if } \overline{\gamma}_j(n)\le\gamma^*_j\end{cases}\quad j=1,2\cdots J;\ n=1,2\cdots N \quad (6)$$

Here, $\overline{\gamma}_j(n)$ stands for the average data transmission rate of the jth queue in the nth TTI.

$$\overline{\gamma}_j(n)=\frac{\sum_{k=1}^{N^l_j(n)}S_{j,k}}{n\cdot T_{tti}}\quad j=1,2\cdots J;\ n=1,2\cdots N \quad (7)$$

In the formula, $s_{j,k}$ stands for the size of the kth data package of the jth queue.

The parameter of deadline $\Xi_j(n)$ makes sure that services arrive at the user end before the deadline and avoids resource waste of invalid data transmission caused by the total delay of service transmission that exceeds the deadline. If the service delay exceeds the deadline, the service is regarded as invalid information that should be discarded timely. The expression of $\Xi_j(n)$ is:

Here, t^*_j stands for the deadline when service j becomes invalid, Υ_j stands for the ratio threshold of

$$\Xi_j(n)=\begin{cases}1 & \text{if } t^\Delta_j<t^*_j\cdot\Upsilon_j\\ \dfrac{t^\Delta_j}{t^*_j\cdot\Upsilon_j} & \text{if } t^*_j\ge t^\Delta_j\ge t^*_j\cdot\Upsilon_j\\ 0 & \text{if } t^\Delta_j>t^*_j\end{cases}\quad j=1,2\cdots J;\ n=1,2\cdots N \quad (8)$$

the deadline of service j, and t^Δ_j stands for the time delay estimation on data package in the buffer queue of service j.

$$t^\Delta=\frac{N^q_j(n)}{N^l_j(n)}\cdot\tau^q_{j,k}(n)+\tau^q_{j,k}(n)\qquad k<N^l_j(n) \quad (9)$$

Being different from the MUFPS algorithm of broadcast scheduling in large base stations, the MUFPS algorithm of single broadcast scheduling in small base stations is added with a factor parameter of channel condition $C_j(n)$, which makes full use of channel conditions of different users so as to improve the utilization of wireless resources and guarantee the service quality of single broadcasting services.

$$P_j(n)=A_j(n)\cdot T_j(n)\cdot\Lambda_j(n)\cdot\Gamma_j(n)\cdot\Xi_j(n)\cdot C_j(n)\quad j=1,2\cdots J;\ n=1,2\cdots N \quad (10)$$

The factor parameter of channel condition $C_j(n)$ is used for reflecting the indicator of signal-to-noise ratio. Users with relatively large signal-to-noise ratios on the wireless channel can be scheduled with priority. Adjust the priority by setting the threshold of signal-to-noise ratio so as to make full use of wireless resources of the system. The expression of $C_j(n)$ can be described as:

$$C_j(n)=\begin{cases}1 & \text{if } S_j(n)\le S^*_j\cdot\theta_j\\ \dfrac{S_j(n)}{S^*_j\cdot\theta_j} & \text{if } S_j(n)>S^*_j\cdot\theta_j\end{cases}\quad j=1,2\cdots J;\ n=1,2\cdots N \quad (11)$$

Here, s^*_j is the largest signal-to-noise ratio on the wireless channel and θ_j is ratio threshold of user's signal-to-noise ratio.

2.3 *Steps of the algorithm*

In MUFPS algorithm, the priority of scheduling is determined to improve the effectiveness of the algorithm by the control of such multiple utility factors as service priority, queue delay, buffer occupancy, data rate and deadline. The method of parallel scheduling can be applied to simplifying the computational priority so as to reduce the complexity of the algorithm. The utilization of wireless resources can be improved through the joint scheduling and coordinated bandwidth allocation between large and small base stations, guaranteeing the seamless combination between broadcast and communication in the mode of large and small base stations. The basic flow chart of MUFPS algorithm of broadcast scheduling in large base stations is shown below:

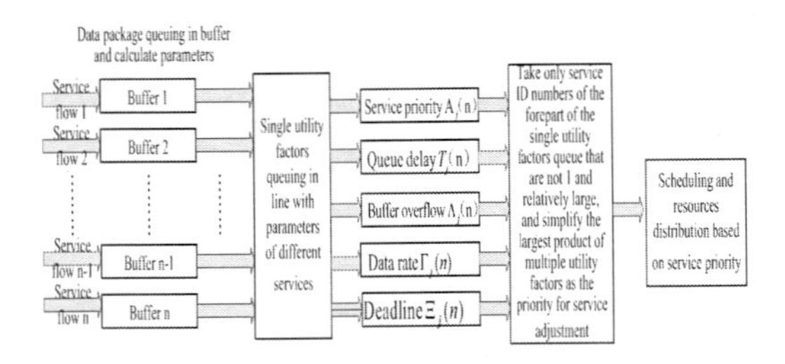

Figure 2. The basic flow chart of MUFPS algorithm.

Specific steps of MUFPS algorithm can be concluded from the figure:

1 Queue service flows in buffer and calculate parameters of each single utility factors like service priority $A_j(n)$, queue delay $T_j(n)$, buffer overflow $\Lambda_j(n)$, service data rate $\Gamma_j(n)$ and deadline $\Xi_j(n)$ according to real-time situations of services.
2 Determine the priority of single utility factors $A_j(n)$, $T_j(n)$, $\Lambda_j(n)$, $\Gamma_j(n)$ and $\Xi_j(n)$ according to various parameters and queue these single utility factors based on these parameters.
3 With the method of parallel scheduling, take only service ID numbers of the forepart of the single utility factors queue that are not 1 and relatively large, and simplify the largest product of multiple utility factors as the priority for service adjustment
4 Conduct joint scheduling and bandwidth resources distribution in line with the priority of services so as to realize effective scheduling of large base stations of broadcast and guarantee the service quality of broadcasting services.

3 SIMULATION RESULTS

The performance of MUFPS algorithm in the mixed BWM network of the mode of large and small base stations can be verified through the system-level simulation. Mixed services like VoIP, real-time video, non-real time video and full buffer are adopted in the simulation. Specific parameters of simulation are shown in Table 1.

Table 1. Parameters of system simulation

Parameters	Values
System bandwidth	10 MHz
Subcarrier frequency	15 kHz
The number of available subcarriers	600
Transmission time interval TTI	0.5 ms
Scheduling period	0.5 ms
Distance of CBS	500 m
Distance of BBS	2500 m
Total transmission power of CBS	20 W
Total transmission power of BBS	100 W
User mobile rate	30 km/h
Objective BER	10^{-6}
Path loss model	L=128.1 +37.6 log ^{10} (d)
Shadow fading index	8 dB

Through the comparison with the classic MAX C/I and the MPF algorithm based on the proportional fairness of QoS, the performance of MUFPS algorithm can be evaluated by analyzing package delay and handling capacity. Figure 3(a) and 3(b) show the comparison between the average package delay and average handling capacity in terms of CDF curve. As shown in Figure 3(a), user statistical distributions of algorithms MAX C/I, MPF and MUFPS approximates 87%, 90% and 95% respectively in the region where the average package delay is close to 0. The total package delay and the total handling capacity are respectively compared in Figure 3(c) and 3(d). Compared with algorithm MAX C/I and MPF, the total package delay of the MUFPS algorithm is reduced by 10 times and 65% and the total handling capacity is reduced by 50% and 35%. It can be seen that the MUFPS algorithm has the largest user statistical distribution in low delay region with relatively large gain in time delay performance. But the handling capacity is reduced to some extent. The reason is that the algorithm attaches more importance to equity and instantaneity of users. Queue delay and deadline are selected from multiple utility factors to enhance the influence of time delay performance on scheduling priority. Therefore, services with higher time delay requirements will be scheduled with priority so as to guarantee the QoS and obtain higher performance gain of time delay. Only one factor of channel condition is used for optimizing user's handling capacity. In order to guarantee the QoS, channels should not be distributed to users with the largest channel gain and the handling capacity should also be reduced.

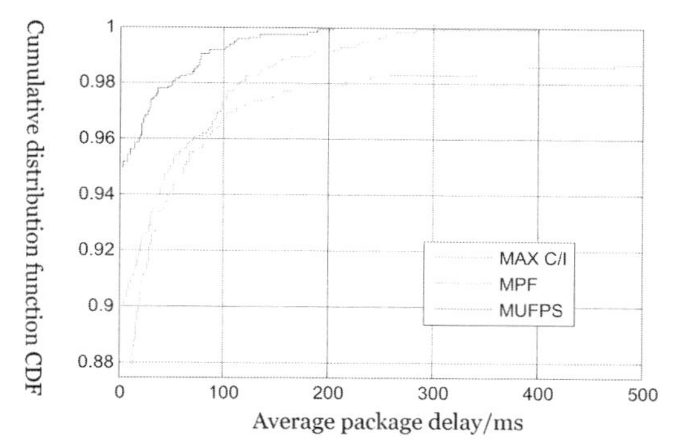

(a) Performance of average package delay

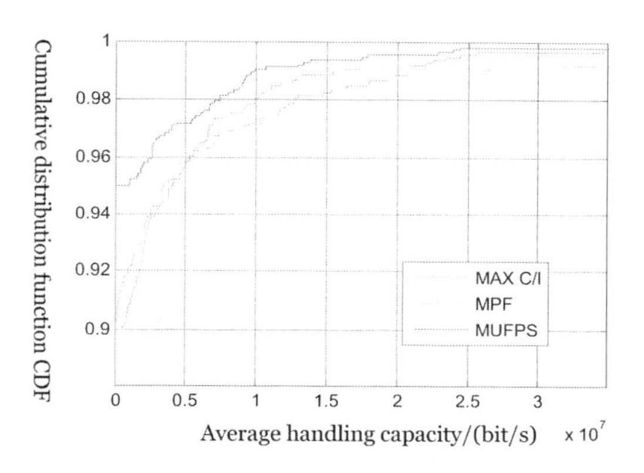

(b) Performance of average handling capacity

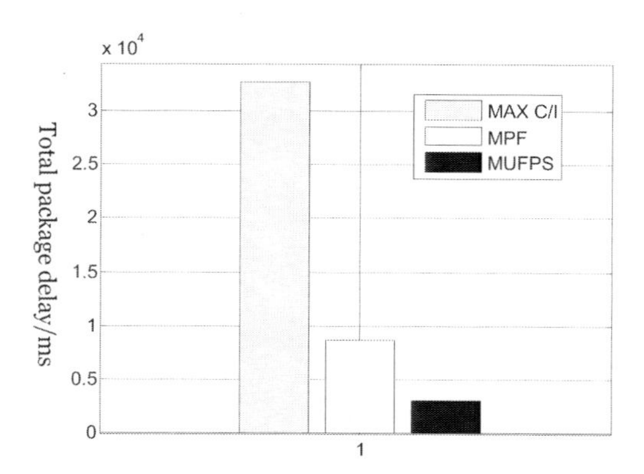

(c) Performance of the total package delay

In order to analyze the influence of the channel condition factor $c_j(n)$ on the performance of the MUFPS algorithm, threes conditions are respectively compared in the simulation, namely BBS scheduling with no regard to $c_j(n)$ (BBS), CBS scheduling with no regard to $c_j(n)$ (CBS1) and CBS scheduling regarding $c_j(n)$ (CBS1). Performance gains of the

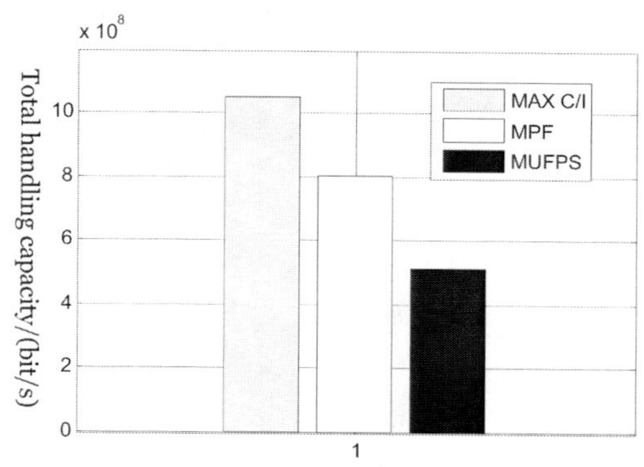

(d) Performance of the total handling capacity

Figure 3. Performance comparisons of algorithms MAX C/I, MPF and MUFPS.

total handling capacity and the total package delay are respectively shown in Figure 4(a) and 4(b). It can be seen that CBS2 increases the time delay at the same time when increasing the handling capacity. Certain performance gains of handling capacity can be acquired through the determination of the priority by adding channel condition factors. Meanwhile, the influence of the time delay factor on scheduling priority is reduced as well, leading to the fact that the total package delay performance gain of users with higher requirements of time delay and poor channel conditions is reduced due to untimely scheduling.

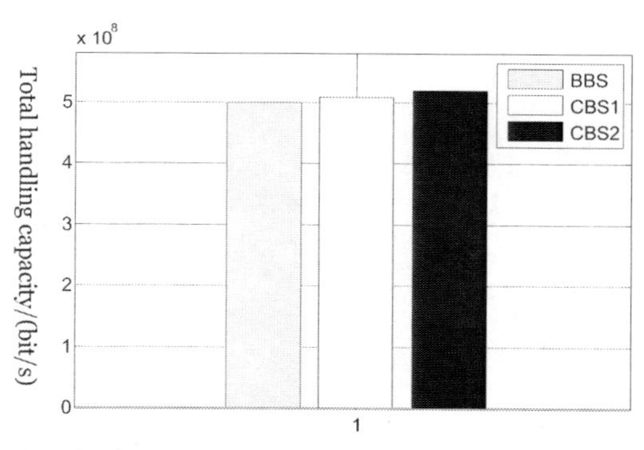

(a) Performance of the total handling capacity

Aiming at analyzing the influence of reducing the computational complexity on MUFPS algorithm, the simulation makes a comparison of three conditions, namely all services of single utility factors buffer queue without reducing the computational complexity (all), services of the forepart of the single utility factors queue that are larger

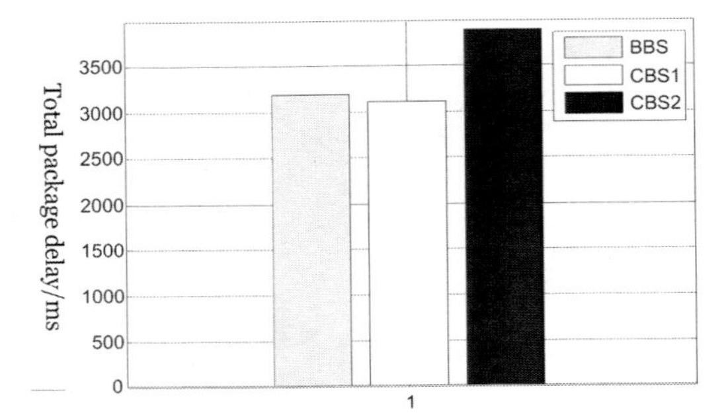

(b) Performance of the total package delay

Figure 4. Performance analyses of parameters of channel condition.

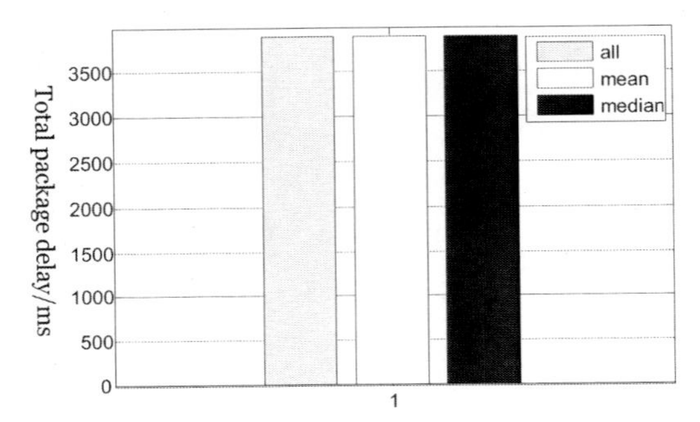

(b) Performance of the total package delay

Figure 5. Performance analyses of reducing the computational complexity.

than the mean of single utility factors (mean) and services of forepart of the single utility factors queue (median). Performances of the total handling capacity and the total package delay are provided in Figure 5(a) and 5(b). It can be seen that performance gains are equal, which means that the influence of reducing the computational complexity on the algorithm is quite small. Although services of the forepart of the single utility factors queue are selected for scheduling, the priority is not influenced and the scheduling of users with higher priority is also guaranteed. Performance gains of the total handling capacity and the total package delay can be guaranteed thusly.

4 CONCLUSION

The joint scheduling of broadcast and unicast that guarantees the QoS of real-time multimedia mixed services is the core problem of wireless resources management in the mode of large and small base stations. The MUFPS scheduling algorithm proposed in this paper is able to guarantee the service quality of broadcast and unicast services, completing the joint scheduling of large and small bases stations effectively and realizing the seamless integration of broadcast and communication in the wireless communication system. The simulation indicates that the algorithm has prominent performance gains in time delay and lower computational complexity with a certain handling capacity as the price. Our next work is to improve the utilization of system resources by energy-saving scheduling under the condition of meeting the requirements of QoS of different services.

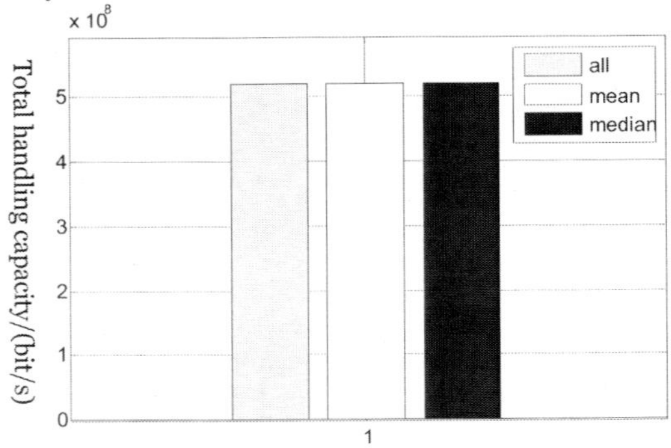

(a) Performance of the total handling capacity

ACKNOWLEDGMENT

This paper is supported by Beijing Natural Science Foundation (4154072).

REFERENCES

[1] Zhen Kong, Yu-Kwong Kwok, Jiangzhou Wang. (2009) A low-complexity QoS-aware proportional fair multicarrier scheduling algorithm for OFDM systems. *IEEE Transactions on Vehicular Technology*, 58(5):2225–2235.

[2] Cecchi M, Fantacci R, Marabissi D, et al. (2008) Adaptive scheduling algorithms for multimedia traffic in wireless OFDMA systems// *GLOBECOM 2008. New Orleans: LA*, pp:1–5.

[3] Yinsheng Xu, Hongkun Yang, Fengyuan Ren, Chuang Lin, Xuemin Shen. (2013) Frequency domain packet scheduling with MIMO for 3GPP LTE downlink. *IEEE Transactions on Wireless Communications*, 12(4):1752–1761.

[4] Guocong Song, Ye Li. (2005) Utility-based resource allocation and scheduling in OFDM-based wireless broadband networks. *IEEE Communications Magazine*, 43(12):127–134.

[5] Zixiong Chen, Kai Xu, Feng Jiang, *et al.* (2008) Utility based scheduling algorithm for multiple services per user in MIMO OFDM system// *ICC 2008. Beijing: China*, pp:4734–4738.

[6] Xuan Wang, Lin Cai. (2014) Stability region of opportunistic scheduling in wireless networks. *IEEE Transactions on Vehicular Technology*, 63(8):4017–4027.

Electric, Electronic and Control Engineering – Shao, Shu & Tian (Eds)
© 2015 Taylor & Francis Group, London, ISBN: 978-1-138-02842-5

A new model for automatically locating the perceptional cues of consonants

Fan Bai
Department of Electronic Engineering, University of Electronic Science and Technology of China, Chengdu, China

ABSTRACT: Previous studies have shown the effectiveness of the 3-Dimensional Deep Search (3DDS) method in searching for the perceptual cue region of consonants in human natural speech. This method was introduced and used in searching for cues of stop consonant and later again for cues of fricative consonants. But the drawback of this method is that it always locates the perceptual cue manually rather than automatically. In this paper, we analyzed how high-pass filter and low-pass filter experiments locate the primary cue regions respectively. Then their data curves are combined together, which creates two curve patterns: overlapping and non-overlapping pattern. We analyzed how these two patterns can locate the primary cues for all consonants. Finally, we created a new model for locating perceptual cues automatically.

1 INTRODUCTION

In December 2009, 3DDS was invented and used to explore the perceptual cues of stop consonants (Li et al., 2010). The paper not only successfully got the stop consonants' cue but also got a very important conclusion: these natural speech syllable sounds often contain conflicting cue regions which are cues for other consonants. So it can lead to confusion after the original target cue region is removed by filtering or masking noise. Based on this, they explained the reason why some consonant is easy to be confused by some other specific consonants. Through the manipulation of original and conflicting cues, one consonant can be morphed into another or a perceptually weak consonant can be converted into a strong one (Li and Allen, 2011; Kapoor and Allen, 2012). Natural fluctuation in the intensity of the plosive consonant cue regions was found to account for the large variations in the AI (Singh and Allen, 2012). Later, the 3DDS method was successfully used to explore the perceptual cues of fricative consonants. (Li, Trevino and Allen, 2012). But the drawback of this method is that it always located the perceptual cue manually rather than automatically.

2 THE PRINCIPLE OF 3DDS

The 3DDS methodology was invented to investigate (evaluate) where the most critical subcomponent of a human natural speech articulation is, in terms of perception in time-frequency domain. Since the speech can be expressed in three dimensions: time, frequency and intensity, we use two independent isolating experiments to locate where the critical subcomponent is in the time - frequency domain, which was called acoustic cue from us. Then we assess the identified cues by noise masking experiment from the perspective of intensive dimensions—the third dimension. Last, we further verify these cues by a special software package designed for the manipulation of acoustic cues (Allen and Li, 2009) based on the short-time Fourier transform (Allen, 1977; Allen and Rabiner, 1977).

In the 3DDS method, we based on the following previous studies and techniques: In time dimension, Furui (1986) used the time-truncation experiments to examine the relationship between dynamic features and the identification of Japanese syllables. In intensity dimension, the masking noise was used to study consonant confusion (Miller and Nicely, 1995; Wang and Bilger, 1973; Allen, 2005) and vowel recognition (Phatak and Allen, 2007). In frequency and intensity dimension, in the 1920s, Fletcher and his colleagues used masking noise along with high- and low-pass filter to investigate the contribution of different frequency bands to speech intelligibility. Almost 10 years work, their studies result in the articulation index (AI) model of speech intelligibility. After integrating (combine) the AI model and Fletcher and Galt's critical-band auditory model, Lobdell and Allen developed a computational model, which was called AI-gram. It can stimulate the effect of noise masking on audibility (Lobdell, 2009; Lobdell et al., 2011) and make audibility visible.

3 ANALYSIS OF HOW FILTER EXPERIMENTS WERE LOCATED THE PRIMARY CUE REGION

For analyzing how filter experiments were located the primary cue region, we take the high-pass filter as an illustration:

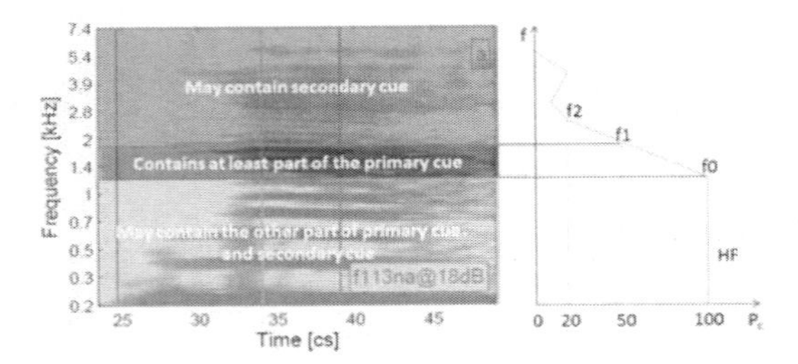

Figure 1. Illustrate how high pass filter experiment data curve locate the primary cue region.

In the figure 1, through the high-pass filter experiment, even though we have deleted the region under frequency f0 (green region), experiment subjects still can 100 percent rightly recognize the target consonant. It means that after we delete the green region, the region above frequency f0 (blue and red region) still contains enough information to make experimental subjects correctly recognize the target consonant. Note that it doesn't mean that the spectral region under the f0 (green region) doesn't contain any important useful information (or feature) for correct recognition of target consonant. After we continue to delete more region (red region) from f0 to f1, the Pc drop to 50%, which means that AFTER deleting the green region, for most people (50%), the red region becomes a necessary information region to be used in correct recognition.

So the red region (f0 to f1) should belong to cue region or at least part of cue region, because it contains critical information and without it most people (50% people) can't correctly recognize the target consonant, after deleting the green region. Again, note that the blue region (above frequency f1) may still contain some information or feature for people or some people to correctly recognize the target consonant. As shown in the figure 1, the blue region may contain secondary cue.

The low-pass filter tells us the same story. It will help us to detect the other part of cue region. It is totally same as what high-pass filter did.

4 TWO REGIONS' RELATIVE POSITION CREATE TWO PATTERNS

According to the above analysis, the perceptual cue is combined by two regions. One region was found by the high-pass filter experiment. The other region was found by low-pass filter. These two regions may have an overlapping region or may not. When they have a common area, we call it as "overlapping pattern. When they have no common area, we call it as "non-overlapping pattern". Note that when we say "overlapping" or "non-overlapping", we are talking about whether these two regions are overlapping or not, rather than their data curves is overlapping or not.

4.1 *Non-overlapping pattern*

As the pattern is shown in the figure 2, the two regions (pink region, which was founded by high-pass filter and red region, which was found by low-pass filtering) have no any overlapping area. It means that people can clearly identify the target consonant by hearing either of these two regions. For correct recognition, these two don't need to be present or heard at the same time. So people can clearly identify the target consonant by only hearing part of the cue region and they even don't need to fully hear any of these two regions for correct recognition. It's understandable that when we try to recognize something, we don't need to see or hear all the features of the target. Someone may recognize them by very few features, and someone else may need more features to recognize them. The number of the features which people need to recognize target are based on how familiar people are with the target and whether there are any similar competing target or not. We may only need to perceive one of the features or part of a feature for correctly identifying them.

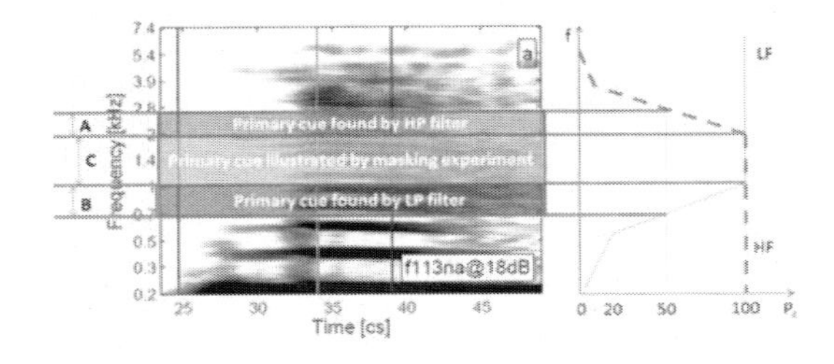

Figure 2. Illustrate the non-overlapping pattern. The red region, which is located by a low pass filter and pink region, which is located by a high pass filter have on any overlapping area.

Based on high pass filter and low pass filter experiment results, we are sure that the combined region of A and B is cue region. But by high and low pass filter experiment result, we can't get any conclusion about if the middle region C (green region, which is between A and B) contain any critical information for correct recognition of target consonant. So we try to test this region by masking experiment in Barren. After we delete the other regions and only leave the green region as shown in the following three examples, we heard the modified sound and found that we still can hear the target consonant very clearly, which means that the green region contains the very important information for correct recognition too. As a result, the green region should belong to the cue region. So we believe that the whole region, which was combined by A, B, C regions, is a cued region of target consonant.

We show following two examples illustrate that middle green region still contains critical information. The symbols in the examples are defined in pervious paper. (Li and Allen, 2011). This green region should belong to cue region:

In the figure 3, example 1 is m118ma. Even though we only leave the middle region(C part, green region), which is around 700 Hz, and delete the other regions, we still can hear a clear /ma/. So this tells us that green region also contains very important information and it should belong to cue region.

In the figure 4, example 2 is m120na. After we only leave the middle region C (green region), which is around 1.4 kHz, and delete the other regions, we still can clearly hear a /Na/, which means that middle part C (green region)contains very important information for correctly identifying the target consonant /na/. So this region should belong to cue region.

This idea that the green region should belong to cue region also successfully works in other consonant besides the nasal consonants.

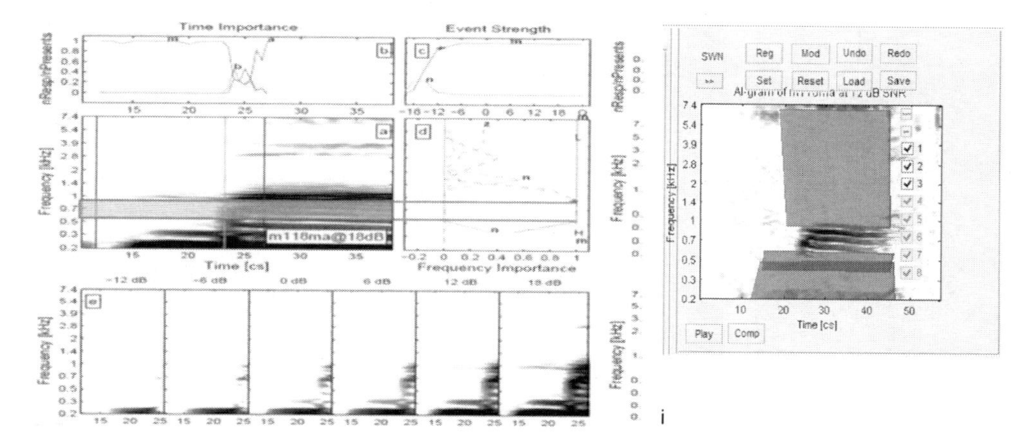

Figure 3. m118ma, we only leave the green region shown in the left panel. We can hear a clear /ma/. Illustrate that the green region should belong to cue region.

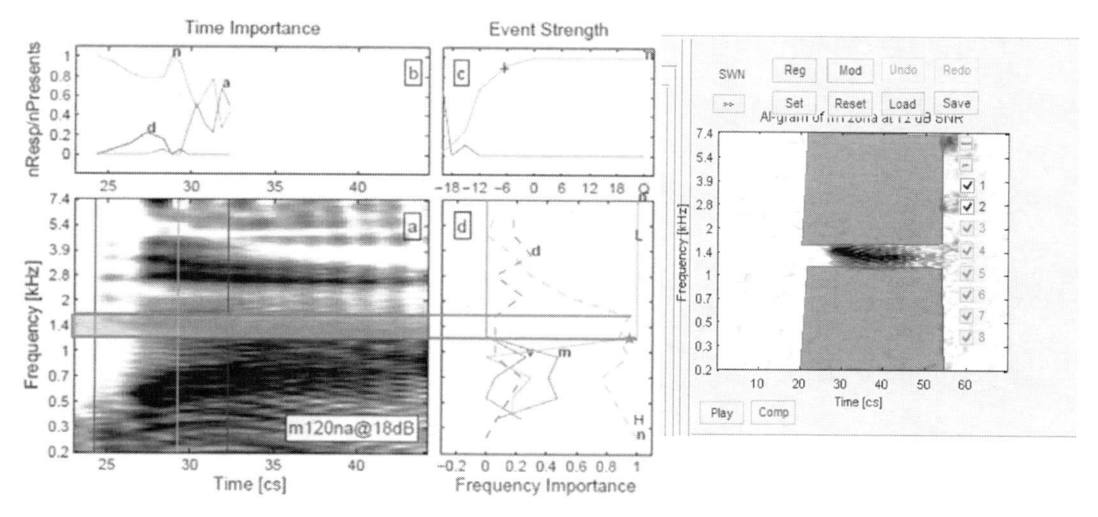

Figure 4. m120na, we only leave the green region shown in the left panel. We can hear a clear /na/. Illustrate that the green region should belong to cue region.

In conclusion, for non-overlapping pattern, the perceptual cue region equal to D=A+B+C. We name the whole region D as a distributed cue region.

There is a critical case that the green region is zero. We call this critical case as "critical nonoverlapping".

4.2 *Over-lapping pattern*

This pattern is shown in the figure 3, the two regions which are found by low and high pass filter have at least part of overlapping region, which means that the low and high pass filter find part of common region or feature (overlapping region). The area B is found by High-pass filter and the area A is found by low-pass filter. So the perceptional cue region is C=A+B.

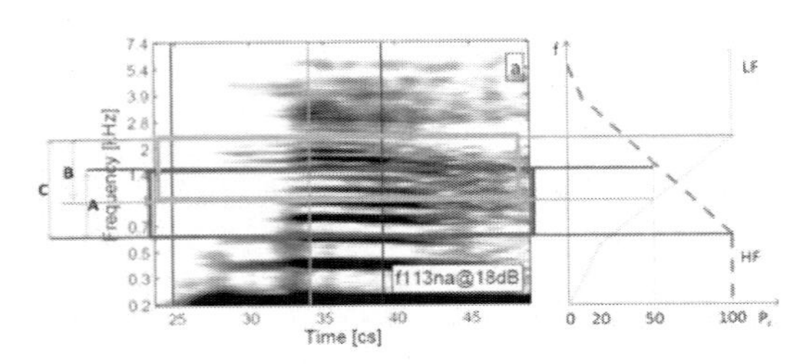

Figure 3. Illustrate the overlapping pattern. The cue region is C=A+B.

There is a critical case that these two regions (green box and red box) totally overlap, which means that the high pass and low pass filter find the totally same region.

Since people can recognize the target consonant only by any one of A and B, no matter in which one of the two patterns, the primary cue region C contains some redundant information for correct recognition. We can't say which region (A or B) is more important than the other one, except above critical case, in which high, low pass filter find the totally same region.

5 SUMMARY

In this study, a new model for automatically locating perceptional cue of consonants in the 3DDS is created, which is combined by two patterns of high-low pass filter experiments data curve. It makes the analysis of huge amount of data from the 3DDS experiments become feasible.

REFERENCES

[1] Allen, J. B. (1977) Short time spectral analysis, synthesis, and modification by discrete Fourier transform, *IEEE Trans. Acoust., Speech, Signal Process*. 25:235–238.

[2] Allen, J. B., and Rabiner L. R. (1977) A unified approach to short-time Fourier analysis and synthesis, *Proc. IEEE*, 65:1558–1564.

[3] Allen, J. B. (2005) Consonant recognition and the articulation index, *J. Acoust. Soc. Am*. 117:2212–2223.

[4] Furui, S. (1986) On the role of spectral transition for speech perception, *J. Acoust. Soc. Am*. 80:1016–1025.

[5] Kapoor, A., and Allen, J. B. (2012) Perceptual effects of plosive feature modification, *J. Acoust. Soc. Am*. 131(1):478–491.

[6] Lobdell B. E. (2009) Models of human phone transcription in noise based on intelligibility predictors, University of Illinois at Urbana-Champaign, Urbana, IL.

[7] Lobdell B. E., Allen J. B., and Hasegawa-Johnson M. A. (2011). Intelligibility predictors and neural representation of speech, Speech Commun. 53(2):185–194.10.1016/j.specom.2010.08.016.

[8] Li F., Menon A., and Allen J. B. (2010) A psychoacoustic method to find the perceptual cues of stop consonants in natural speech, *J. Acoust. Soc. Am*. 127(4):2599–2610.

[9] Li F., and Allen J. B. (2011) Manipulation of consonants in natural speech, *IEEE Trans. Audio, Speech, Lang. Process*. 19(3):496–504.

[10] Li F. (2012) A psychoacoustic method for studying the necessary and sufficient perceptual cues of American English fricative consonants in noise. *J. Acoust.Soc. Am*, 132(4):2663–2675.

[11] Miller, G. A., and Nicely, P. E. (1955) An analysis of perceptual confusions among some English consonants, *J. Acoust. Soc. Am*. 27:338–352.

[12] Phatak, S., and Allen, J. B. (2007) Consonant and vowel confusions in speech-weighted noise, *J. Acoust. Soc. Am*. 121:2312–2326.

[13] Singh, R., and Allen, J. B. (2012) The influence of stop consonants perceptual features on the Articulation Index model, *J. Acoust. Soc. Am*. 131(4):3051–3068.

[14] Wang, M. D., and Bilger, R. C. (1973) Consonant confusions in noise: A study of perceptual features, *J. Acoust. Soc. Am*. 54:1248–1266.

Electric, Electronic and Control Engineering – Shao, Shu & Tian (Eds)
© 2015 Taylor & Francis Group, London, ISBN: 978-1-138-02842-5

A detection method based on image segmentation applied to insulators

Yongjie Zhai, Yang Wu, Di Wang
Department of Automation, North China Electric Power University, Hebei, China

ABSTRACT: To improve the efficiency of insulator fault diagnosis when routing patrolling, a method of recognizing insulators based on image segmentation is put forward. We firstly use Otsu method to realize binary segment after preprocess of the aerial photography. Then morphology for quadratic filter smoothing is used with removing the small divided area. We obtain the image skeleton with processing of image thinning. Finally, according to the regularity of the insulator skeletons and other skeletons which are lack of regularity, we detect insulators by Hough transformation correctly.

KEYWORDS: image thinning; Hough transformation; morphology; insulator; recognition

1 INTRODUCTION

The helicopter is efficient, fast, and reliable, and is not affected by district for patrolling the transmission lines. And it has already taken the place of manual patrol in many developed countries. With rapid economic development, the scale of high voltage transmission lines have expanded gradually and the geographical environment where line corridor pass through is more complex, which bring a lot of troubles to line maintenance. It is crucial to be safe for operating transmission lines. The insulator is a key component of transmission lines, and it is also fragile, which is easy to cause a fault. So it is the production development needs to develop helicopter patrolling. For visible image from the helicopter power-line inspection, using image segmentation technology could provide the basis for subsequent parts of fault diagnosis, which would directly improve the helicopter patrolling intelligence technology.

During the process of helicopter inspection, we capture one or more pictures of the target insulators, detect the insulator status by image processing and analysis, and then we can locate the defects and buried danger of insulators. To achieve this automatic diagnose, the most important part is the recognition and the location of insulator's image in processing of insulator photography. For now, it exists in some preliminary study on the research. A few recognition methods have already been proposed. Paper [3] transformed colored insulator images from RGB space to HIS space, extracted S components, and then segmented the image by calculating its eigen values. Paper [4] realized image segmentation based on maximum entropy thresholding method, applied connecting regional approach to extract and recognize insulator string. Paper [5] adopted adaptive partical swarm optimization, combined with traditional OTSU method, to segment insulators, and then applied mathematical morphology to process the divided images, in order to identify the insulators. Paper [6] put forward a method of recognizing and fault diagnosis by utilizing insulator's infrared thermal images.

This paper proposed an effective method of locating insulators based on image enhancement. The flow chart is shown in Fig.1. First of all, we utilize OTSU algorithm to make binary segmentation of aerial insulator photography after enhancing the image. Then we use morphology to filter the image and eliminate small area to avoid interference. After these steps, we apply thinning processing theory to get the skeleton. At last, we detect lines by Hough transformation. Above all, we can locate and recognize the insulators correctly.

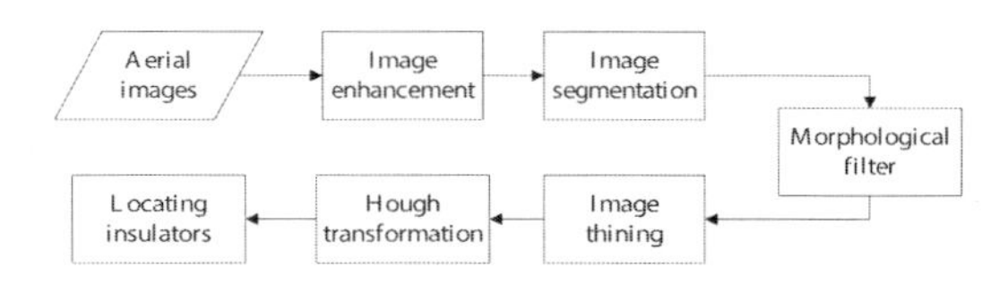

Figure 1. The flow chart of insulators recognition method.

2 RECOGNITION METHOD

2.1 *Image enhancement*

Histogram equalization is also called gray scale equalization. It is the conversion from gray input images to gray out images and the pixels of gray level are the same. The gray levels of the histogram equalization could be distributed equally. So, thin kind of images owns higher contrast and larger dynamic range. So, the overall contrast was improved [18].

Histogram equalization is directly utilized as an enhancing method of image pixels and it can be shown in Eqn. (1)

$$t = T(s) \tag{1}$$

Image enhancement transformation function T should meet two constraint conditions below.

1 T(s) is single-valued and is a monotonic increasing function.
2 For $0 \leq s \leq L-1$, $0 \leq T(s) \leq L-1$. Here, L is the gray levels of the digital images.

Condition (1) guarantees the same arrangement order after the transformer. And it is inverse transformation. Its purpose is to avoid the inverse transformation of the image gray levels after image enhancement. Two conditions above are the processing premises, and CDF (Cumulative Distribution Function) just happens to meet both these two limiting conditions. At the same time, the cumulative histogram of the original images is just the CDF of S, where

$$t_k = T(s_k) = \sum_{i=0}^{k} \frac{n_i}{n} = \sum_{i=0}^{k} p_s(s_i) \tag{2}$$

Here, the value of t_k is in [0,l],and also needs to be expanded to the range [0,L−1].The calculation formula of the gray level t_k of output images are calculated in Eqn. 3.

$$t_k = \text{int}\left[(L-1)t_k + 0.5\right] \tag{3}$$

This paper proposes contrast adaptive histogram equalization method. It means implementing histogram equalization in small area in image and the contrast of it gets enhanced. Then double linear differential method is used to connect small areas, in order to eliminate boundary caused by local area. The preprocessing of insulator is shown as Fig.2.

(a)

(b)

Figure 2. Image enhancing of insulator.

2.2 *Image segmentation*

Image segmentation is one of the toughest tasks of image processing and the general method still does not exist. Otsu method is a segmentation algorithm based on threshold theory, proposed by Japan scholar Dajinzhanzhi (Otsu) in 1979, and also is named as Otsu algorithm. It is proposed based on the least square and it is an automatic unsupervised non-parametric method. Otsu method divides image pixels into two different types and calculate their variance, the bigger the variance is, the larger difference between target and background is and the selecting threshold is much more reasonable [24].

According to the threshold M, the image pixels are divided by gray levels into two types, C0 and C1.The probability, mean, and variance of the two types are calculated respectively. The variance within classσ_W, variance between class σ_K and overall variance σ_r are calculated to evaluate threshold advantages. Here, we introduced some evaluation function.

$$\lambda = \frac{\sigma_K^2}{\sigma_W^2}, \kappa = \frac{\sigma_r^2}{\sigma_K^2}, \eta = \frac{\sigma_K^2}{\sigma_r^2} \tag{4}$$

Our problem is converted into an optimization problem. That means making three evaluation functions above owning the biggest values by calculating the optimal value M. λ, κ, η meets $\sigma_r^2 = \sigma_W^2 + \sigma_K^2$ and σ_r^2 is a constant which doesn't depend on M. So η is the simplest one. Thus, we conclude that the function of deciding classification performance is calculating the variance between two classes, which means the threshold values making σ_K^2 maximum is reasonable. The best threshold M is noted below.

$$\sigma_K^2(M^*) = \max_{1<M<L}\{\sigma_K^2(M)\} \tag{5}$$

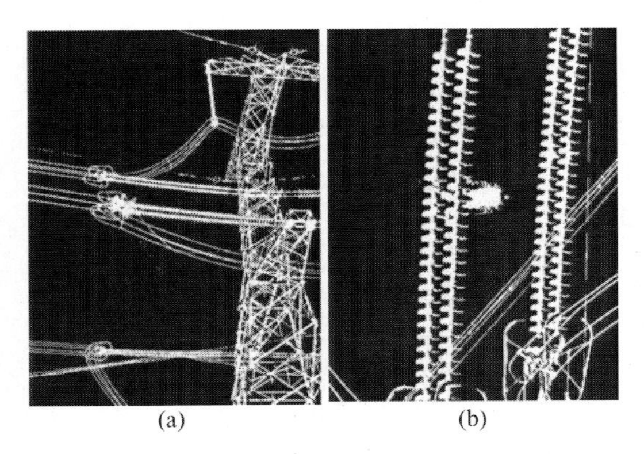

(a) (b)

Figure 3. Segmentation results.

By Otsu's Method, image segmentation results are shown in Fig. 3. Because there are electronic wires and electrical equipment in power patrolling excepting insulators, the gray values among them are not different greatly. It may cause the Otsu's segmentation results less-than-ideal and significant influence of power lines and powers.

2.3 *Morphological filtering*

The basic calculations in mathematical morphological filtering are dilation and erosion. Many algorithms could be derived based on them. The definition of dilation is in Eqn. (6) and the erosion is in Eqn. (7).

$$X\Theta S = \{x \mid S + x \subseteq X\}. \tag{6}$$

$$X \oplus S = \{x \mid S + x \neq \Phi\}. \tag{7}$$

The arithmetic method is called open operation which operates of corrosion on the image first, and then makes the expand operations. The arithmetic method is called close operation which operates of expand on the image first, and then makes the corrosion operations. Generally, open operation can smooth the image contour, weaken the narrow parts, remove the slender, edge burr and isolated spots, adhesion of the disconnected targets, and can keep the target size remains the same. Closed operations are commonly used to populate the target within the small holes and cracks, connect the near disconnected targets, at the same time, remain the basic target size unchanged [10]. This article adopts the combination of morphological open operation and close operation to improve the effect of the OTSU segmentation.

2.4 *Connected domain processing*

In binary images, the area is called connected region which is consisted of the pixels which are adjacent and their value is 1. There are two kinds of common adjacency relations: the four adjacency way and the eight adjacency way. As is shown in figure 3, the four pixels which are up and down or so of the pixel P is called four adjacent way, combined with diagonal four pixels. The total of eight pixels is called the eight adjacent way. This article uses the eight adjacency way to tag on the connected domain. By iterating through the image and make a note of the continuous mass and the tagged equivalences in each row (or column), then re-mark the original image through the equivalence, then calculate the area of each connected domain and remove the small area. The further processed insulator segmentation results which used the method above are shown in figure 4 below.

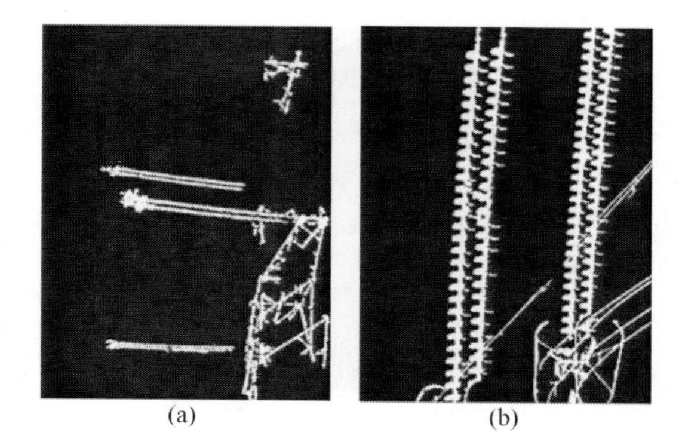

(a) (b)

Figure 4. Connected domain processing.

As can be seen from the diagram above, all the power line and part of the tower were effective segmentation in the aerial image. The background interference is greatly reduced, and the shape of the insulator is highlighted, which provides favorable conditions for subsequent coarse location.

2.5 *Image thinning*

We refine the insulator image processing, which obtain insulator image skeleton. Elaboration is a method which converts the object in the binary image into a set of fine skeleton. In the process of reduction, we constantly delete pixels on the boundary of the object until it becomes single pixel, but we do not allow objects to disconnect. As a result, these fine skeletons still retain the important information of the original object. Based on the characteristics of skeleton extraction and the shape of the insulator, this paper uses refining operations to extract the skeleton in the insulator image, as shown in figure.5.

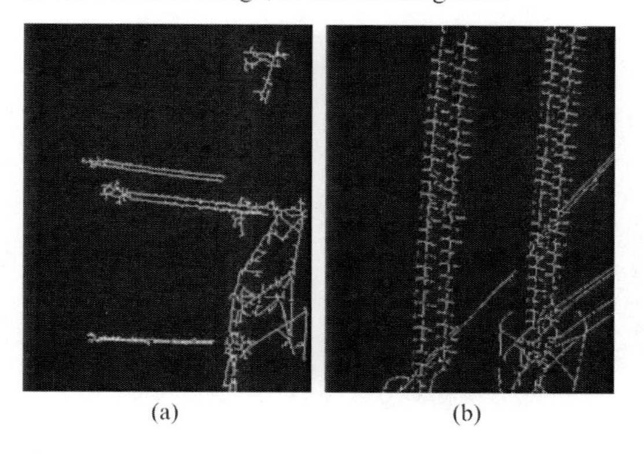

(a) (b)

Figure 5. Insulator image thinning.

2.6 *Hough transform*

The Hough transform of the straight line detection application has become more widely, Hough transform is a mapping relationship from image space to the parameter space, which maps the complicated edge characteristic information of the image space to the clustering detection problem in parameter space. Therefore, the Hough transform method has clear analytical geometry, strong anti-interference ability and easy to implement parallel processing, etc. The biggest advantages of Hough transform is that it's anti-jamming is strong, even if the curve has small disturbance, clearance, even a dotted line, after the Hough transform, it still can form obvious peak point in the parameter space. It is one of the basic methods of image recognition geometry.

The principle of Hough transform can be explained by point - line dual transformation, we assume that the image space has a straight line. The linear equation for the black matrix principle is as follows:

$$y = kx + b$$

Type: k and b is a parameter, the slope and intercept, respectively. Obviously, points(x, y) belong to the straight line in the image space all meet the formula (3–12). Therefore, the line in the parameter space can be expressed as (k, b). As shown in figure 6.

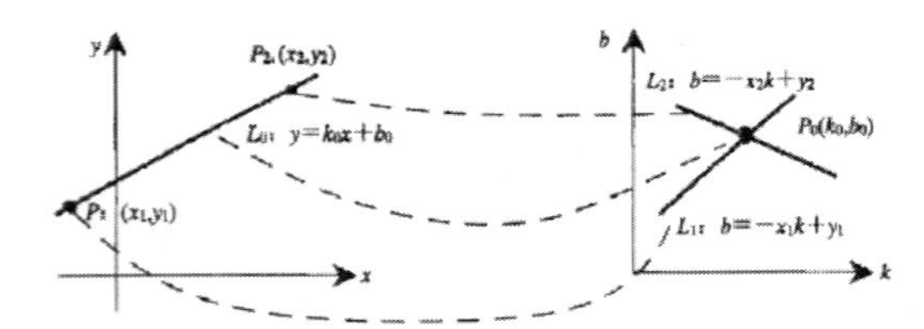

Figure 6. Hough transform.

Considering the linear skeleton of insulator string, this paper uses Hough transform to detect the skeleton extraction after insulator image in a straight line, and function the positioning of the insulator. The results are shown in figure 7 below, visible from the figure, all insulator were detected.

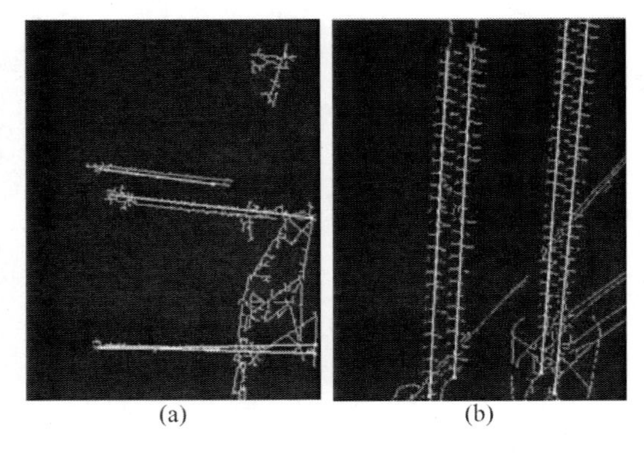

(a) (b)

Figure 7. Insulator positioning.

3 CONCLUSION

In this paper, a kind of effective aerial insulator location method by applying image processing technology is mentioned. The aerial image gray-scale is enhanced in the first place to improve the contrast, and then we use Otsu algorithm to segment image, and use morphological filtering smoothing, then eliminate interference with further filtering based on connected domain. Finally, based on the differences between insulator skeleton characteristics and other transmission equipments, we obtain image skeleton based on image processing. Thus, positioning uses Hough transformation, which detects line insulators. Through the actual test in aerial images, this method has good robustness, accuracy and validity, and reflects its value in practical application, for subsequent transmission line fault detection and repair of laid a foundation.

ACKNOWLEDGMENTS

This research is supported by the Fundamental Research Funds for the Central Universities in China (GN: 2014MS140).

REFERENCES

[1] Weiguo Tong, (2011) *Based on Aerial Image Transmission Line Identification and State Detection Method Research*. Baoding: North China Electric Power University.

[2] Huiling Zhang; Baichuan Lu; Yong Yu, (2008) The method of Otsu used in the video detection of vehicles, *Intelligent Computation Technology and Automation (ICICTA), 2008 International Conference on, 20–22 Oct. 2008*, 2:547 551.

[3] Chunyu Yao; Lijun Jin; Shujia Yan, (2012) The identification of grid image review insulator. *Journal of System Simulation*, 09:1818–1822.

[4] Xiaoning Huang; Zhenliang Zhang; (2010) Helicopter aerial image review insulator image extraction algorithm, *Power System Technology*, 01:194–197.

[5] Jing Cao, (2012) *Aerial Transmission Line Insulator Components Extraction in Image*, Liaoning: Dalian Maritime University.

[6] Zuosheng Li; Wenli Li; Jiangang Yao; Yingjian Yang. (2010) Application of insulator infrared thermal image processing field pollution level detection method. *Proceedings of the CSEE, 2010*, pp:132–138.

[7] Heming Li; Shenghui Wang; Chengfang Lv; Yunpeng Liu, (2010) Based on the polluted insulator discharge characteristics of ultraviolet imaging studies, *Journal Of North China Electric Power University (Natural Science Edition)*, 01:1–6.

[8] Bing Qi; Liangrui Tang; Jing Zhang, (2008) Hydrophobic image detection method research in detail, *Proceedings of the CSEE, 2008*, 31:120–124.

[9] Gonzalez, (2011) *Digital Image Processing*, Beijing: Electronic Industry Press.

Electric, Electronic and Control Engineering – Shao, Shu & Tian (Eds)
 ISBN: 978-1-138-02842-5

Service restoration strategy of active distribution network based on multi-agent system

Hui Ji & Liangwei Ma
Shanghai Second Polytechnic University, Shanghai, China

ABSTRACT: With the development of new energy power generation technology, more and more distribution generators will access the levels of distributed distribution network, and the permeability will increase continuously. When there is a fault occurs in power network, how to make full use of DG for power restoration has important significance for increasing the power supply reliability. The multi-objective mathematical model for service restoration of active distribution network is established in the paper, in which the maximum load restoration and the island operation stability are all considered. Then a multi-agent decision making system is established, and it includes the coordination layer, decision layer and equipment layer. Based on the structure of the distribution network and the capacity, type and control mode of DG, the power supply restoration scheme is given. The load transfer method and the range of each island are all obtained. Finally, an example of active distribution network is simulated, and the optimal restore scheme is given. The feasible and validation of the model and strategy proposed in the paper is verified.

KEYWORDS: active distribution system; service restoration; multi-agent system, island operation.

1 INTRODUCTION

As the core function of smart distribution grid (SDG), the service restoration after the fault is an important guarantee to realize the self-healing of SDG. In recent years, the development of distributed generation (wind, photovoltaic, battery-energy storage station, micro gas turbine and so on) gains lots of attention to the characteristic of being clean, highly efficient and renewable. As a complement to the concentrated generation, distributed generation (DG) connecting to distribution network has been the inevitable trend. Making the best use of generating capacity of DG and restoring service to as many loads as possible are the important researching contents of SDG which improves the active ability and the disaster prevention ability of the distribution operation.

Traditional distribution network fault restoration, which takes into account the interruption customers, network-constrained, feeder residual capacity and power transfer, and so on restores power supply fast by the stand-by tie-lines based on the redundant design of the distribution network. Many mature methods have been proposed to solve the restoration problem mainly including mathematical optimization means, artificial intelligence algorithm, heuristic search method or graph theory[3–7]. DG connecting to distribution network brings new resources to the service restoration. For instance, researchers [8–9] divide the restoration decision process into two steps. The former first makes DG each independent island by DG's power supply radius, and then restores the non-failure power blackout area outside the island with a service restoration scheme based on the binary particle swarm optimization (BPSO). The latter first restores service based on the traditional algorithms ignoring DG, and then makes the DG which is the back-up power of the system switch to the islanding operation mode to restore service further. In the literature [10-13] researches service restoration supposing all the DG have the ability of islanding operation. In Wang et al. [10] does research on service restoration based on the constraint-satisfying-problem model when a large scale blackout occurs in the distribution network within DG. In Li et al. [11] builds a multi-agent system of distribution service restoration by managing each bus with bus agents. In Xu et al. [12] proposes a method based on the genetic algorithm and multi-agent technology to restore service. While in Feng et al. [13] develops the power supply ranges of each restoration area. As can be seen from the researches above, the service restoration based on DG islanding operation gains a widespread attention. However, very little attention has been focused on the types of DG and their control modes. Also, most previous researches haven't taken into account the ability of islanding operation stability and the capacity margin after restoring service. So it is hard to ensure the islanding operation sustainable and stable.

This paper solves the problem of service restoration with synthesizing optimization approaches fully considering stand-by tie-lines and all types of DG inside. Moreover a concept of static stability margin about the ability of islanding operation is given and a multi-objective model for service restoration of active distribution network is established. DG resources connecting to distribution network are proposed to be classified for the DG with the ability of voltage and frequency regulation under the control of V/f defined as main power while the random and uncontrollable DG defined as common power. Then, under the master-slave control mode, this paper develops the optimization scheme of service restoration based on the multi-agent system. Simulation results show that the restoration scheme which fully utilizes various resources this paper develops is effectively feasible to improve the ability of distribution service restoration after faults.

2 ISLANDING MASTER-SLAVE CONTROL MODE AND DEFINITION OF STATIC STABILITY MARGIN

2.1 *Islanding operation modes and category of DG*

Two control modes exist in the grid based on the DG islanding operation, master-slave control mode and peer-to-peer control mode[14-16].

The DG under the master-slave control mode play different roles according to their control mode in the grid which operates on the island. One is supposed to be a main power to stabilize the islanding voltage and frequency in the rated value. While other subordinate powers output some active and reactive power, according to the voltage and frequency the main power supply. Master-slave control is actually the main control mode in the grid operating on the island.

Peer-to-peer control mode which is based on the DG outer characteristic descent method associates frequency with active power and voltage with reactive power to simulate the active-frequency characteristic curve and the reactive-voltage characteristic curve using some control algorithm to realize the function of voltage and frequency auto-adjusting. The relationship of DG under peer-to-peer control mode is not subordinate but equal. This peer-to-peer control mode based on the DG droop characteristic only thinks of primary frequency modulation and gives no considerations to the non-error recovery of voltage and frequency. Also, some key problems about control and application remain to be solved, so that this control mode is in the stage of experimental study at present.

Therefore, this paper based on the more mature control mode of master-slave researches the service restoration scheme of each operating island after the restoration of active distribution network.

Three unit control modes of DG exist in the islanding grid, which is under the master-slave control, *V/f*, *PQ* and *PV* mode. Among them, the DG under *V/f* control, of which the voltage and frequency of the access point remain constant, has the ability to be the main power to establish voltage and frequency of islanding operation. So in this paper, this kind of DG is called main power DG. The DG under *PQ* control generally operates remaining constant power or in a way of maximum power point tracking (MPPT). While the DG under *PV* control, of which the access point voltage remains constant, outputs the given active power. To maximize service restoration and make the best of DG, in this paper, not entirely controllable DG under *PV* and *PQ* control mode are taken account into the service restoration scheme. And this kind of DG is called common DG. Even though lacking the ability of islanding operation, this kind of DG restores more loads following the main power connecting to grid in case that main power under *V/f* control exists nearby.

According to the definition and category, main power generally consists of energy-storage, diesel engine, fuel cell, and so on, which has the ability of independent operation and the voltage and frequency regulation. While lacking the ability of independent operation, common DG operates following the reference voltage and frequency given outside, consisting of the random and uncontrollable DG such as wind power, photovoltaic generation, small hydro power, and so on.

2.2 *Definition of islanding static stability margin*

Traditionally, after the service restoration based on stand-by tie lines of large power grid, with no need to consider the supply capacity of large power grid too much, keeps power supply stable in the capacity scope of the transformer and line. Essentially different from the service restoration based on large power grid, the service restoration based on DG islanding operation fully needs to consider the ability of islanding sustainable and stable operation. So this paper defines the static stability margin to be the constraint and optimizing objective of the mathematical model in active distribution network comprehensive service restoration.

This paper assumes in the islanding distribution network, which is under master-slave control the available power output range of the main power which undertakes loads and DG power fluctuations is $[P^{m}_{min}, P^{m}_{max}]$. The actual power of main power is P^{m}. The islanding total power supply load is P_{L} and the total power of the uncontrollable DG inside the island is P_G. So the islanding static stability margin is defined as follows:

$$K_p = \frac{\min(P^m - P^m_{min}, P^m_{max} - P^m)}{|P_L| + |P_G|} \tag{1}$$

where $P^m - P^m_{min}$ is the reducible capacity of the main power output power and $P^m_{max} - P^m$ is the room for growth of the main power output power. Choosing the smaller capacity $\min(P^m - P^m_{min}, P^m_{max} - P^m)$ acts as the power variation which can be undertaken by the main power. Let the sum of the absolute value of load power P_L and random uncontrollable DG output power P_G be the denominator. So the power fluctuations of load and DG output power are considered together.

The static stability margin index of the traditional large power grid is used to measure if the reserved active power capacity is enough as load increasing. Of which concrete formula is as follows:

$$K_p\% = \frac{P_{max} - P_L}{P_L} \cdot 100\% \tag{2}$$

where the increasable power range in the system is acted as the numerator while the actual load power as the denominator. In formula (1), the islanding static stability margin is similar to it with the increasable or reducible power range being the numerator and the uncontrollable factors that the sum of the absolute value of load power and random DG output power being the denominator.

The larger the K_p is, the higher the islanding storage capacity is, and the stronger the ability of undertaking loads and DG power fluctuations is, also the better the islanding stability is. In large power grid, the output power of most generators is under control, and it is easy to cut generators for the large quantity of generators, so generally active power remains in balance when the load drops. Therefore the ratio of the increasable output power in the system to the actual load in power grid is always used to define the similar static stability margin index, like formula (2). While in DG island, besides the load is uncontrollable, part of the DG power isn't under control and fluctuates randomly like load, and also the active power balance needs to be considered for the island scale being small. As a consequence, according to the difference between an island and large power grid, in the definition of islanding static stability margin index above which is used to measure the islanding static stability, this paper makes the sum of the absolute value of load power and DG output power be the denominator and the increasing or decreasing the lower limit of power range of the main power undertaking the power fluctuations be the numerator.

3 MATHEMATICAL MODEL OF SERVICE RESTORATION IN ACTIVE DISTRIBUTION NETWORK

3.1 *Optimization objective*

Under the emergency blackout in distribution network, it is necessary to find the optimal restoration scheme as soon as possible to restore some load by the stand-by tie lines while some load by making the DG switch to the islanding operation mode. At this time, the network loss isn't so important. So in this paper, the complex flow calculation is avoided efficiently by ignoring the network loss which benefits making the restoration scheme fast. The primary optimal objective F_1 of service restoration is represented as formula (3), which is to maximize the service restoration.

$$\max F_1 = \sum_{i \in Z} x_i P_i \tag{3}$$

Where Z stands for the blackout zone; P_i stands for the active power of the blackout load i; x_i is a 0 and 1 variables to express whether the load i is restored service with 1 standing for restored service while 0 standing for not restored service.

To make sure the islanding stability operation, in this paper, the islanding static stability margin is acted as the second optimization objective of service restoration scheme F_2:

$$\max F_2 = \min(K_{pi}), \quad i = 1, 2, \cdots, N \tag{4}$$

where K_{pi} stands for the static stability margin of island i and N stands for the number of islands after service restoration.

3.2 *Constraints*

General constraints of distribution operation consist of node voltage constrain, branch power constraint, radial constraint and so on. If the service restoration is based on the DG, there's still a DG capacity constraint, islanding power balance constraint, islanding sustainable stability operation constraint and so on, which are as follows:

$$U_{Li} \le U_i \le U_{Ui} \tag{5}$$

$$S_i \le S_{i\max} \tag{6}$$

$$S_{Gi\min} \le S_{Gi} \le S_{Gi\max} \tag{7}$$

$$g_i \in G \tag{8}$$

$$K_p \geq C \tag{9}$$

where U_i, U_{Ui}, U_{Li} respectively stand for the voltage of node i and its upper and lower limits; S_i and $S_{i\max}$ stand for the power and allowable maximum in branch i; S_{Gi}, $S_{Gi\max}$, $S_{Gi\min}$ stand for the actual output power and the maximum or minimum output power of DG i; g_i is the network structure after service restoration; G is the allowable radial network structure; K_p is the islanding static stability margin defined above. C is a constant which stands for the allowable lower limit of islanding static stability margin.

4 DG EQUIVALENCE OF THE STAND-BY TIE LINES IN ACTIVE DISTRIBUTION NETWORK

In order to ensure the distribution network reliability, some stand-by lines are laid during the distribution network construction, or feeders connect with each other and are mutual stand-by each other, so as to transfer the blackout load when faults occur, which is the main method of service restoration after the fault in traditional distribution network. After DG connecting to a distribution network, it's necessary to take into account DG and stand-by lines to make the global synthesis optimization. However it's quite different of the work modes, operation characteristics, the mathematical models and so on from DG and stand-by lines. How to synthetically optimize the two kinds of resources to achieve the best possible optimization is the main problem to be solved in service restoration. In this paper the stand-by tie-line equivalence to DG is presented to simplify the problem, as shown in Fig. 1.

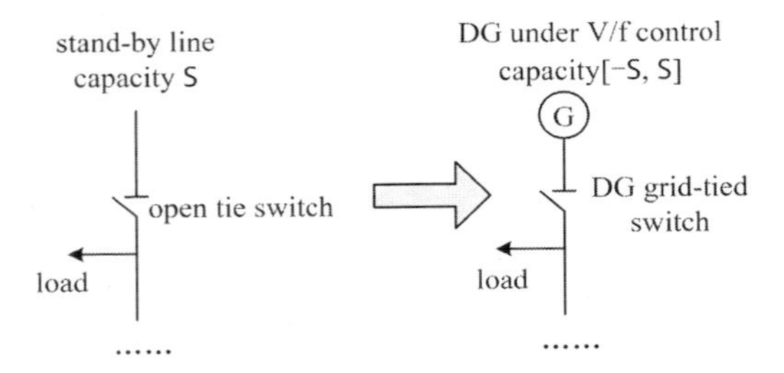

Figure 1. Stand-by tie-line equivalence to DG.

The stand-by tie-line with a certain capacity provides the voltage and frequency of the network outside. Assume that the capacity of the stand-by tie-line is S, and make it equivalent to the controllable DG whose output capacity ranges $[-S,S]$, also the DG outputs stable voltage and frequency which is the voltage and frequency of the network outside with maximum output power S and minimum output power $-S$ like a rechargeable storage power station. Obviously, the virtual DG after equivalence is considered to be a main power, which operates independently.

DG becomes the restoration power of service restoration by making all the stand-by tie-lines the equivalent to DG above in the blackout zone. So the synthesis optimization question of service restoration based on DG and stand-by lines is simplified into the service restoration problem entirely based on the DG.

5 MULTI-AGENT SYSTEM OF SERVICE RESTORATION SCHEME IN ACTIVE DISTRIBUTION NETWORK

5.1 *Multi-agent system structure of service restoration scheme in active distribution network*

Multi-agent system (MAS), by which distributed problem and dynamic problem are conveniently solved, has been applied in many fields of power system, like distributed load forecasting [19], reactive power optimization for distribution network [20], distributed optimal schedule [21] and so on. In MAS, each agent keeps its own independence and autonomy that can asynchronously solve the problems which belongs to its own territory, and meanwhile these agents manage public affairs through competition and coordination or other mechanisms. Service restoration of distribution network consists of multiple zones, multiple stages and the formed island operates independently, which matches the application conditions of MAS well. The fault handling methods become fast and flexible with each agent performing its own duty and processing in parallel. Meanwhile, by the interaction with each other, the global problems are coordinated and the global optimal service restoration scheme is achieved.

This paper constructs a three-layer, multi-agency service restoration system composed of coordination layer, decision layer and the equipment layer in the distribution network, as shown in Figure 2. The coordination layer, which is made up of a service restoration coordinating Agent which is used to manage and coordinate the conflicts and tensions between coordination layer Agents from a global perspective to maximize the global objectives. The decision layer, which is made up of main power Agent, optimizes and develops the islanding ranges, and maximizes load restoration with the satisfaction of the islanding operation constraint. So the main power Agent is also responsible for its optimization of islanding ranges besides managing the operation and control mode of the main power. The equipment layer, which is made up of common DG except main power and the Agent of the power load which remains to be restored, the common DG Agent is used to manage and update its real-time information and give the specific restoration orders, while the Agent of the power load which remains to be restored is used to manage the attributes of this load and give the control commands of load restoration.

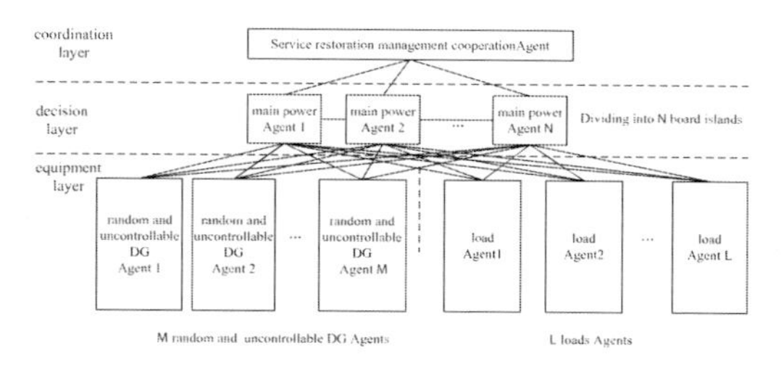

Figure 2. Architecture of multi-agent system for service restoration of distribution system containing DG.

After simplifying the distribution network by the stand-by tie-line equivalence to DG, the system consists of N main powers, so there are N main power agents in the decision layer of the multi-agent system above. The distribution area is divided into no more than N broad islands. It is called the broad island for part of the DG being the stand-by tie-line equivalence to, of which the loads are restored by the virtual DG islands, while actually by the stand-by tie-lines. If this distribution network possesses M random and uncontrolled DG and L loads needing to be restored. So the equipment layer contains M L agents of M common DG agents and L load agents. In this multi-agent system, the whole system optimization objective is the sum of each DG optimization objective meanwhile the complicated operational constraints of the distribution network and DG become the constraints inside the island, which is calculated and examined in the charge of each main power Agent. By which the scale of the problem is reduced and the computational speed accelerates greatly relative to the centralized synthesis optimization, so that the multi-agent technology is well realized.

5.2 *Island range decision of main power agent*

According to the grid structure, each main power Agent in decision layer forms a search tree with the main power being the root node. Of which the end of each branch is either its last load or the load adjacent to other main power. Each main power forms a search tree which consists of a main power located in the root node and some load nodes or common DG nodes downstream. These load nodes and common DG nodes are all the possible resources which may operate in island with the main power. In the certain distribution network, the tree-net structure is for sure and has nothing with the operation modes of power grid and DG. So that the network structure of the main power search tree can be topologically generated once and saved in each main power Agent that only needs to be updated when distribution network upgrades or builds new lines.

After each main power in decision layer forms a search tree, main power Agent based on its capacity limits uses the improved breadth-first search algorithm to searches the feasible islanding ranges, which is the optimal range searched based on the possible load, the capacity of common DG and the islanding static stability margin constraints by the main power Agent. The specific process of islanding ranges search calculations is as shown in Fig. 3, which starts to search from the root node of each island (main power point). Restore service to all the loads in some layer of the search tree if the islanding operation constraints are satisfied and then search the loads in the next layer, otherwise use the 0-1 knapsack optimization algorithm to choose to restore part loads, which restores the most loads around the main power when the islanding ranges which the algorithm gets satisfy the power capacity constraints. After finishing searching, each main power Agent uploads the islanding ranges they develop to the coordination Agent in coordination layer.

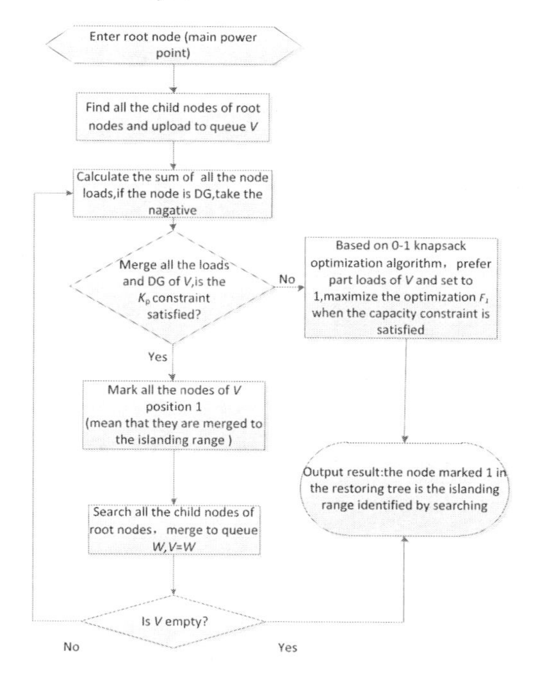

Figure 3. Improved breadth-first search algorithm integrated with 0-1 knapsack optimization.

Compared with the breadth-first search algorithm, the depth-first search algorithm effectively shortens the power supply radius of main power, also connects the loads and common DG to distribution network and restores service to them, at the same time reduces conflicting between decision Agents to some extent that reduces the crossover between each island.

5.3 *Coordination optimal calculation of coordination agent*

Each main power develops the optimal island ranges based on their own optimization objective and ignores the effects on others, so the islanding ranges may cross with each other and there is a situation that the neighbor islands cover a same load or common DG. Therefore, after each decision Agent uploads its islanding range to coordination Agent, coordination Agent judges if there exists conflicts. If there is no conflict, output the optimization results; but if there is conflict that power supply areas cross, coordination Agent makes schemes, based on which main power topologically searches and uploads islanding ranges again, meanwhile coordination Agent calculates the coordination benefits of each scheme and combines the stability margin index of each island to choose the optimal service restoration scheme based on the multi-objective and layered optimization algorithm, specific as follows.

For the objective function F_l, which coordination Agent measures by the coordination benefits, such as there exists conflicts between the islanding ranges which both contain load L that two neighbor main power Agent A and B develop. So the calculation formula of coordination benefits is as follow:

$$E = (P'_A + P'_B) - (P_A + P_B - L) \tag{10}$$

where P_A and P_B respectively stand for the power supply loads of island A and B both containing before coordinating, and L is the loads which island A and B both contain before coordinating, so $(P_A + P_B - L)$ is the total restorable loads of island A and B before coordinating; P'_A and P'_B are the power supply loads after a second optimal search of island A and B according to the coordination schemes, at this time there is no overlapping power supply ranges, for there is no conflict therefore $(P'_A + P'_B)$ is the total power supply. So E is actually the load restoration increases after coordinating, which is defined as coordination benefits. Prefer the coordination scheme of which E is the maximum in coordination process.

If each coordination scheme restores the same loads, which means the coordination benefits are the same, coordinate the power supply ranges of island A and B according to the second optimization objective to improve the stability margin of the two islands, which is:

$$\max F_2 = \min(K_{pA}, K_{pB}) \tag{11}$$

where K_{pA} and K_{pB} respectively stand for the static stability margin of island A and B after coordinating.

This layered optimization coordination scheme embodies the priority of different optimization objectives in the multi-objective service restoration which means maximizing the loads restoration on the premise that the constraints are satisfied, which is the primary objective, and takes into account the second objective to improve the stability margin in the case that the primary objective values are the same.

Conflicts just exist between the neighbor islands in distribution network. If the two preliminarily uploaded neighbor islands contain N loads together, there exist N 1 coordination schemes at most because of the sequentiality of loads restoration that it cannot restore the next-level loads passing over the upper-level loads. Therefore the workload of coordination calculation increases not exponentially but linearly with the growth of distribution network, so this coordination scheme is still practicable even if for large-scale outages in distribution network.

6 CASE STUDY

To validate the feasibility of the model and algorithm in this paper, a simulation analysis is carried on to the service restoration after the fault in the active distribution network which is shown in Fig. 4, where 10 kV feeders are powered by a 110kV step-down substations, Feeder 1 is a multi-sections and multi-connections overhead distribution line, while Feeder 2 and Feeder 3 are cable single ring net. This system contains 2 normally open reserve tie-switches, 15 sectioned switches, 7 DG including battery-energy storage station, fuel cell, diesel generator one each which has the ability to regulate voltage and frequency independently, and the rest 4 DG are random and uncontrollable power containing wind station and photovoltaic station 2 each. There are also 18 load nodes and the fault loads are shown in Tab. 1 which totals 11. 14MW. The maximum and minimum output of controllable DG and the output power during the faults of uncontrollable DG are separately shown in Tab. 2 and Tab. 3.

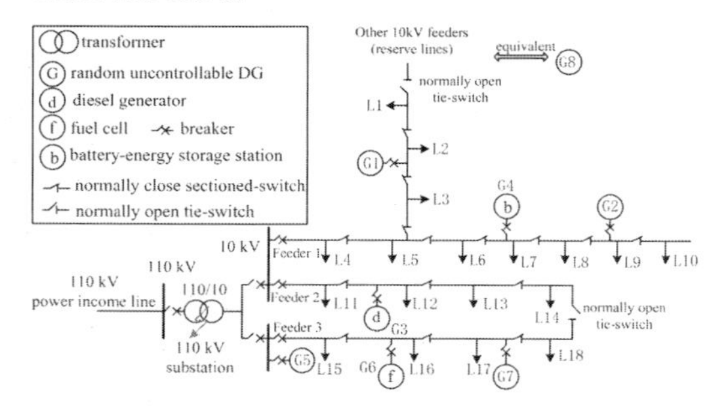

Figure 4. Distribution system containing DGs.

Table 1. Load capacity of the distribution system.

Load number	Load capacity/ MW	Load number	Load capacity/ MW	Load number	Load capacity/ MW
L1	0. 24	L7	0. 94	L13	0. 51
L2	0. 69	L8	0. 68	L14	0. 66
L3	0. 81	L9	0. 13	L15	0. 77
L4	0. 37	L10	0. 69	L16	0. 92
L5	0. 59	L11	0. 26	L17	0. 80
L6	0. 48	L12	0. 39	L18	0. 81

Table 2. Controllable DG capacity.

DG number	Power type	Minimum power/MW	maximum power/MW
DG3	diesel generator	0	3
DG4	battery-energy storage station	-3. 6	3. 6
DG6	fuel cell	0	3
DG8	equivalent DG of tie-lines	-2	2

Table 3. Uncontrollable DG capacity.

DG number	Power type	Output power/MW
DG1	Wind power	1. 58
DG2	Photovoltaic power	0. 84
DG5	Photovoltaic power	1. 38
DG7	Wind power	1. 08

When the failures occurs to the substation in the active distribution network above, the transformer protection acts and the downstream 10kV feeders lose service. So the service restoration scheme is needed to restore the outage load by the stand-by tie-lines and the DG connecting to the grid. As shown in Tab. 2, after the stand-by lines being equivalence to DG, there exist 4 main powers, DG3, DG4, DG6 and DG8. So the outage zone may be divided into 4 independent operation zones around these 4 main powers and these 4 main power Agents forms the search tree with its own root node, as shown in Fig. 5, where the islanding ranges initially developed by the decision Agent according to the improved breadth-first search algorithm is in the dotted box. During the optimal

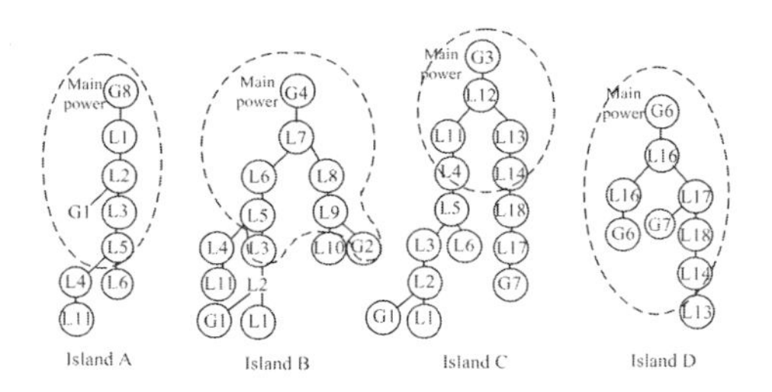

Figure 5. Search trees of each main DG and original island range.

search of this range, the limit value of the islanding static stability margin index K_p is 15%.

There exists a certain contradiction between the islanding ranges initially developed above, like island A and island B both contain load L3 and L5. After coordinating by the coordination Agent with the multi-objective and layered optimization algorithm previously mentioned, the optimal service restoration schemes are as shown in Tab. 4. Island A contains the virtual distributed generation DG8 which is the equivalence of the stand-by lines, so it is a virtual island and the service is restored by this stand-by tie-switch connecting to the grid, totally restoring loads 2.14MW, while island B, C and D are real islands that separately restore loads 3.51, 1.53 and 3.96 MW. Each island has a better static stability margin index, of which island B is the lowest, so it has a better ability of bearing the power fluctuations and ensures the sustainable and stable operation to make the restoration schemes more feasible.

Table 4. Each island range and the optimal restoration result.

Island	Island ranges	Restore loads/ MW	Static stability margin index/%
A	DG8,DG1,L1—L3	2. 14	38. 71
B	DG4,DG2,L5—L10	3. 51	21. 38
C	DG3,L4,L11—L13	1. 53	96. 08
D	DG6,DG5,DG7,L14—L18	3. 96	23. 36

It can be seen from this simulation case that all the 3 feeders in this system satisfy the N-1 principle and have stand-by tie-lines, but without using DG it at most restores 2MW loads by the service restoration based on the traditional stand-by tie-lines when the superior power failures happen and the restoration rate is just 17. 9%. While making the best of DG, all the 100% loads are restored. Therefore, with the large-scale development of DG, the service restoration based on DG has a good feasibility especially for the failures in the large power grid, which benefits reducing dependency of distribution network on stand-by lines, improving utilization of equipment and reducing investment of the grid construction. Certainly it is seen from the scheme process that to improve the reliability of distribution network cannot all the DG inside be the random and uncontrollable DG and a certain number of controllable DG is needed to be the main power to improve the islanding operation ability of the distribution network.

7 CONCLUSION

Making the best of the stand-by tie-lines and DG inside the distribution network to restore service can effectively improve the power supply reliability of distribution network, but because of DG with many types, various control modes and the different tasks undertaken during the restoration, it is difficult to make all the DG be a unified power in the service restoration scheme process. Based on this, in this paper the DG is classified by the types and control modes, also under the master-slave control mode based on the multi-agent system the service restoration scheme is carried out after the fault in the distribution network. Meanwhile, through the equivalence of stand-by tie lines to DG the complexity of comprehensive optimization is effectively simplified. The output power fluctuation of DG is taken into fully account in the restoration scheme process, and a multi-objective optimization mathematical model of the service restoration scheme is established based on the islanding static stability margin index in this paper, so that the ability of islanding sustainable and stable operation is elevated which makes the restoration scheme more feasible.

In the research, it is found that making the best of DG benefits, reducing dependency on distribution network on stand-by lines, so the comprehensive scheme of DG with the distribution network and the rational allocation of all types DG and stand-by tie-lines have important significance that reduces investment of the grid construction, also effectively improves the initiative and self-healing capacity after the fault in an active distribution network. The comprehensive optimization scheme of DG and distribution network will be further researched in the future.

REFERENCES

[1] Li Xun, Gong Qingwu, Hu Yuanchao, *et al.* (2011) Discussion of smart distribution grid system. *Electric Power Automation Equipment*, 31(8):108–111.

[2] Nguyen C P, Flueck A J. (2012) Agent based restoration with distributed energy storage support in smart grids. *IEEE Trans on Smart Grid*, 3(2):1029–1038.

[3] Perez-Guerrero R, Heydt G T, Jack N J. (2008) Optimal restoration of distribution systems using dynamic programming. *IEEE Trans on Power Delivery*, 23(3):1589–1596.

[4] Zang Tianlei, Zhong Jiachen, He Zhengyou, *et al.* (2012) Service restoration of distribution network based on heuristic rules and entropy weight. *Power System Technology*, 36(5):251–257.

[5] Liu Jian, Xu Jingqiu, Cheng Hongli. (2004) Algorithms on fast restoration of lager area breakdown of distribution systems under emergency states. *Proceedings of the CSEE*, 24(12):132–138.

[6] Perez-Guerrero R E, Heydt G T. (2008) Distribution system restoration via subgradient-based Lagrangian relaxation. *IEEE Trans on Power Systems*, 23(3):1162–1169.

[7] Zhang Haibo, Zhang Xiaoyun, Tao Wenwei. (2010) A breadth-first search based service restoration algorithm for distribution network. *Power System Technology*, 34(7):103–108.

[8] Zhao Jingjing, Yang Xiu, Fu Yang. (2001) Smart distribution system service restoration using distributed generation islanding technique. *Power System Protection and Control*, 39(17):45–49.

[9] Lu Zhigang, Dong Yuxiang. (2007) Service restoration strategy for the distribution system with DGs. *Automation of Electric Power Systems*, 31(1):89–92.

[10] Wang Zengping, Zhang Li, Xu Yuqin, *et al.* (2010) Service restoration strategy for blackout of distribution system with distributed generators. *Proceedings of the CSEE*, 30(34):8–14.

[11] Li Hengxuan, Sun Haishun, Wen Jinyu. (2012) A multi-agent system for reconfiguration of distribution systems with distributed generators. *Proceedings of the CSEE*, 32(4):49–56.

[12] Xu Yuqin, Zhang Li, Wang Zengping, et al. (2010) Algorithm of service restoration for large area blackout in distribution network with distributed generators. *Transactions of China Electrotechnical Society*, 25(4):135–141.

[13] Feng Xueping, Liang Ying, Guo Bingqing. (2012) A dynamic programming method based on graph theory for restoration of distribution system with DGs. *Power System Protection and Control*, 40(9):24–29.

[14] Chen Weimin, Chen Guocheng, Cui Kaiyong, *et al.* (2008) Running control of grid-connected dispersed generation systems in islanding situation. *Automation of Electric Power Systems*, 32(9):88–91.

[15] Borrega M, Marroyo L, Gonzalez R, *et al.* (2013) Modeling and control of a master-slave PV inverter with N-paralleled inverters and three-phase three-limb inductors. *IEEE Trans on Power Electronics*, 28(6):2842–2855.

[16] Lu Xiaonan, Sun Kai, Huang Lipei, *et al.* (2013) Improved droop control method in distributed energy storage systems for autonomous operation of AC microgrid. *Automation of Electric Power Systems*, 37(1):180–185.

[17] Tang Yong. (2012) Framework of comprehensive defense architecture for power system security and stability. *Power System Technology*, 36(8):1–5.

[18] Wang Shanshan, Sun Huadong, Yi Jun, *et al.* (2013) Large-scale power grid power system security and stability standard suggestion on editing and revision. *Power System Technology*, 37(11):3144–3150.

[19] Melo J D, Carreno E M, Padilha-Feltrin A. (2012) Multi-agent simulation of urban social dynamics for spatial load forecasting. *IEEE Trans on Power Systems*, 27(4):1870–1878.

[20] Zhang Xiaolian, Yang Wei. (2008) Research on reactive power optimization control of electric power system based on multi-agent system. *Power System Technology*, 32(S2):146–149.

[21] Li Jinghua, Wei Hua. (2011) A multi-agent system to draw up energy-saving power generation dispatching schedule. *Power System Technology*, 35(5):90–96.

[22] Liu Yujuan, Wang Xianghai. (2006) Two kinds of expanding forms of 0–1 knapsack problem and its solution methods. *Application Research of Computers*, 23(1):28–30.

Electric, Electronic and Control Engineering – Shao, Shu & Tian (Eds)
© 2015 Taylor & Francis Group, London, ISBN: 978-1-138-02842-5

A research and implementation of the automatic synchronization strategy under difference frequency power grid based on fuzzy control principle

Huan Hu
XJ Electric Co., Ltd., Xuchang Henan, China

Jiyao Zhou
Province-Ministry Joint Key Laboratory of Electromagnetic Field and Electrical Apparatus Reliability, Hebei University Of Technology, Tianjin, China

Yaguang Fu, Lili Zeng, Yanlong Li, Yapu Yang, & Yafei Yue
XJ Electric Co., Ltd., Xuchang Henan, China

ABSTRACT: Aiming at the synchronization of power system, this paper discusses the principle of the synchronization, establishes the model of control based on the Taylor series. According to the control model, it concretely discusses the synchronization algorithm both of phase and time. Based on this two algorithm, studies the algorithm of the capture of the time of the phase equal to zero when the synchronization of the power system. And using this algorithm, a measure and control bay device was designed. Experimental results show that the algorithm has a good performance and the synchronization operator command action is correct and quickly order under conditions, with which both lines frequency from 47hz to 53hz, and the frequency change rate are 0.02-0.09hz/s.

KEYWORDS: automatic synchronization, slip frequency, Taylor series, algorithm, DFU410.

1 INTRODUCTION

With the development of the power system, the power grid connection is closer than ever, resulting in the application of both closing operations of circuit breaker in the power plant and closing operation of circuit breaker in the transformer substation, more and more relying or depending on the synchronization function [4–6]. Due to the capacity of the power system is bigger and bigger, the voltage is higher and higher, the inappropriate synchronization operation will lead to serious consequences. Improving the accuracy and reliability of grid synchronization operation has a great practical significance for the reliable operation of the power system [8]. It is necessary to study the algorithm of synchronization operation. Literature [3] gives a discussion of the uncertain interconnection principle of slip frequency, but is limited to theoretical discussion; Literature [4] discusses a new automatic synchronization algorithm, it is easy to implement, and its feasibility and effectiveness is proved by simulations, but it is also hard to use in practice. The proposed algorithm has been applied to the automatic synchronization device, and the expected performance is obtained. This article introduces an achieved good effect in practice during the same period closing algorithm.

2 THE WORKING PRINCIPLE OF THE ELECTRIC POWER SYSTEM

As shown in figure1, Ux indicates power grid voltage, Wx is its angle frequency; Ug indicates the voltage of another power grid prepared to synchronization, Wg is its angle frequency; Us is the vector difference between Ug and Ux .

When power grid parameters are fixed, the impact current of the synchronization operation depends on the Us value. To reduce the impact of the breaker, Us value as small as possible when synchronization closed, the biggest impact current shall not exceed the allowable values. Ideally Us is zero, the system parameters satisfy the following relations:

$$\omega g = \omega x \quad \text{or} \quad fg = fx \tag{1}$$

$$Ug = Ux \tag{2}$$

$$\sigma = 0 \tag{3}$$

In this condition, the impact current of the synchronization breaker is zero at this moment, and then the system can run smoothly in the synchronizing state, without any disturbance.

In power grid operation process, the actual conditions of the power grid are very difficult to satisfy the above three types. In real practice, small impact current is only required, it will not do harm to electrical equipment, the small system impact will help the quick synchronization after the connection of the grid.

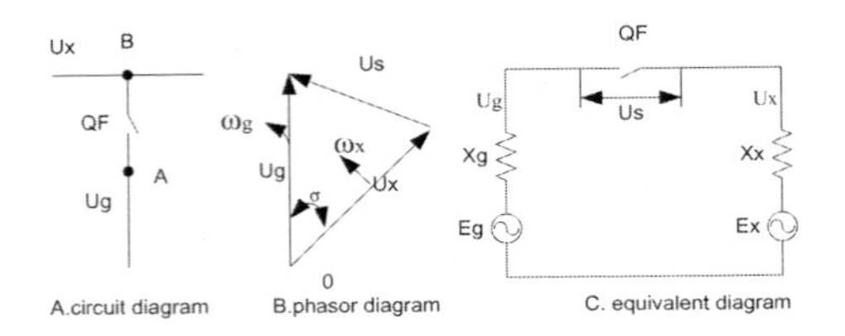

Figure 1. The automation synchronization operator diagram.

The vector diagram of the power grid and generator are shown in figure 1 B, and Ug Ux; fg≠fx or ωg≠ωx, the pulsating voltage Us is the voltage difference between the two sides of the circuit breaker QF

$$U_s = U_g \cos(\omega_g t + \varphi_1) - U x \cos(\omega_x t + \varphi_2) \tag{4}$$

Set the initial phase Angle $\varphi_1 = \varphi_2 = 0$,

$$U_s = U_s \cos(\frac{\omega_g + \omega_x}{2} t) \cos(\frac{\omega_g - \omega_x}{2} t) \tag{5}$$

$\dot{U}_s = 2U_g \sin(\frac{\omega_g - \omega_x}{2} t)$ is the pulsating voltage amplitude, thus

$$U_s = \dot{U}_s \cos(\frac{\omega_g + \omega_x}{2} t) \tag{6}$$

Of which ωs = ωg−ωx, and slip angular frequency, Us waveform is shown in figure 2, thus Ux and Ug phase Angle difference between the two voltage phase as

$$\sigma = \omega_s t \tag{7}$$

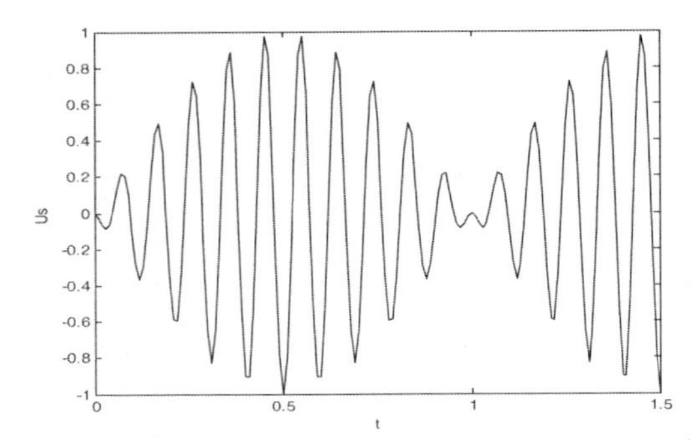

Figure 2. The waveform of Us.

During the synchronization process of difference frequency grid, especially when the generator on the system interconnect, the speed of the generator varies under the influence of the governor, so the frequency offset between generator and power grid system is not constant, but contains a first and second order or higher order derivative. Combined with circuit breaker has an inherent switching time tk at the synchronization time, closing command should be output by the same device before the time tk of the appearance of zero phase difference, to ensure the realization of grid connection when φ 0°. Or contemporaneous device should be in a perspective of φ 0° to impending closing commandone angle φk earlier and the circuit breaker closing time tk , the frequency offset and frequency offset first derivative $\frac{d\omega_s}{dt}$ and second derivative $\frac{d^2\omega_s}{dt^2}$ of frequency offset. Its mathematical expression is:

$$\varphi_k = \omega_s t_k + \frac{1}{2}\frac{d\omega_s}{dt} t_k^2 + \frac{1}{6}\frac{d^2\omega_s}{dt^2} t_k^3 + \cdots\cdots \tag{8}$$

Synchronization device need to solve the differential equation rapidly, to get ideal for current switching Angle φk in the process of synchronization operator. And fast continuously measuring the current coordinate points on either side of the circuit breaker has difference φ, when φ=φk closing command is established, as to realize accurate zero phase difference of the synchronization process.

3 FUZZY CONTROL PRINCIPLE IS APPLIED TO IMPLEMENT SYNCHRONIZATION OPERATION.

3.1 *Fuzzy theories is applied to implement the control of voltage and frequency*

Fuzzy logic is suitable for the description of qualitative, imprecise or uncertain system; it is a combination of fuzzy logic control with input and output variable fuzzy, fuzzy reasoning, fuzzy judgment and decision algorithm and other parts. Usually it takes the deviation and the deviation change rate of the controlled output variables as its input variables, and the charged amount as output variables of the fuzzy controller, reflecting the input and output variables and language control rules of fuzzy quantitative relationships and its algorithm structure. In the process of implementation, the computing machine will implement fuzzy inference and fuzzy decision-making of the collected control information by language control rules of, fuzzy sets, the amount of control is obtained by fuzzy judgment to get an accurate amount of the output control applied to the controlled object, the charged process to achieve the expected control effect.

Generally, we can simplify the principle to express below:

$$u = g(e, de) \tag{9}$$

u - control quantity, e-accused of quantity of the deviation of a given value, de-accused of quantity deviation rate, g - fuzzy control algorithm.

Fuzzy control theory is based on the fuzzy mathematics to get accused of quantity deviation and its rate of change of the fuzzy control decision-making. The fuzzy control reasoning rules table below can describe its essence. Tables deviation e fuzzy values into board to negative big five gears, deviation rate de fuzzy values into board to negative big five files, and their corresponding controller to control the amount of u from fuzzy value is 25, from the board to the negative big five class value. Frequency modulation control, for example, such as controller to measure frequency offset $\omega_S = \omega_F - \omega_X$ (F, X, respectively, to stay and angular frequency of the generator and system) is negative, and frequency difference rate $\frac{d\omega_s}{dt}$ of change is negative, then control the quantity u is zero (lower right corner of the table). This shows that although the generator the system frequency is lower than power grid, but the current generator frequency is changing in the direction of higher direction at high speed, there is no need to control the generator frequency thus the normal value can be restored.

Table1. The command chart of fuzzy control.

de \ u \ e	NB	NS	ZR	PS	PB
NB	PB	PB	PS	PS	ZR
NS	PB	PS	PS	ZR	ZR
ZR	PS	PS	ZR	ZR	NS
PS	PS	ZR	ZR	NS	NS
PB	ZR	ZR	NS	NS	NB

For grid operation in power system, according to formula (1) (2) relating to the adjusting of voltage and frequency, especially for the adjustment of the frequency and the generator grid synchronization, the change of frequency is a random process. With two variables of fuzzy control, according to the current system frequency and flat rate change rate, the frequency range and frequency variation range can be classified into is divided into five ranges of board to plus big to minus big range according to actual situations in accordance with table 1. Figure 3 shows the control flow chart of the system.

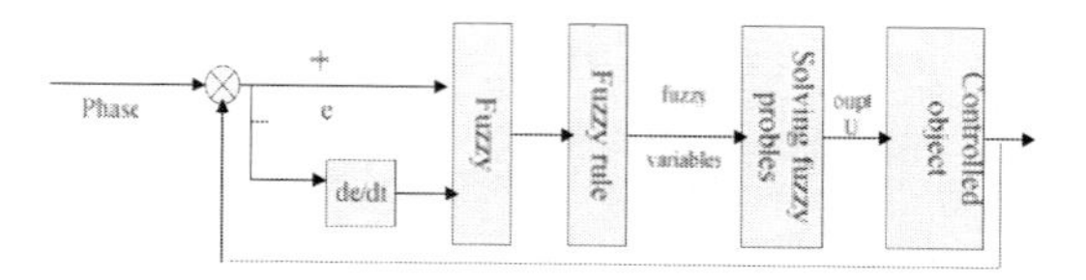

Figure 3. Process chart of fuzzy control.

3.2 *Difference frequency grid algorithm research*

In the process of synchronization operation, if the voltage difference and the frequency difference of generator and power grid voltage are in the allowable range, the synchronization close break time can be calculated by the sample value of the phase difference and the phase change rate.

According to the mathematical model in the second quarter, the phase difference and the slip ω_s are phase Taylor series, in the actual calculation process, higher than the second order items can be omitted in the Taylor series, (8) can be simplified as

$$\varphi_k = \omega_s t_k + \frac{1}{2}\frac{d\omega_s}{dt} t_k^2 \tag{10}$$

Because the frequency of U_g and U_x is a changing process, ω_s it is not a constant. Adjusting the ω_s of the generator according to the sample value is the key to the automatic synchronization device outputs the synchronization close command.

Two synchronization close time calculation method are used, which are leading phase angle calculation method and leading time calculation method.

3.2.1 *Leading phase angle calculation method research*

In the process of the collection of phase difference of the power grid, following assumptions were made: φi indicate the i times phase samples value (set $\varphi_i \leq \varphi_{i-1}$, the phase difference in the decreasing process, the following discussion is based on this condition), Ts is sampling time, then, the i times slip angle velocity as shown in Equation 11:

$$\omega i = \frac{\varphi_i - \varphi_{i-1}}{T_s} \tag{11}$$

t_c is the action time of circuit breaker , t_{QF} is the close time of the circuit breaker, then:

$$t_{DC} = t_{FQ} + t_c \tag{12}$$

As the change of the ω_s between the two calculation point is very small, so the slip acceleration was calculated once after numbers of sample points, the slip acceleration was expressed as shown in equation 13:

$$\frac{d\omega_s}{dt} = \frac{\varphi_i - \varphi_{i-k}}{2t_x k} \tag{13}$$

φ_i , φ_{i-k} indicates the slip velocity of current and the former k times.

According to equation (10), closing leading phase angle difference φ_{yq} can be calculated, this value and the present sampling phase angle difference φ_i , under the condition of

$$|(2\pi - \varphi_i) - \varphi_{yq}| \leq \xi \text{ (} \xi \text{ is permissible threshed value)} \quad (14)$$

Close operation command was acted immediately. Otherwise, the next calculation was needed, until the condition meet necessity.

3.2.2 *Leading time calculation method research*

Like leading angle calculation method, leading time t_{yq} calculation uses the same principle, but calculation output value is leading time. With the former i points, the equation 15 should be met:

$$\varphi_k = \omega_s t_{yq} + \frac{1}{2}\frac{d\omega_s}{dt} t_{yq}^2 \quad (15)$$

According to equation 11, 12, t_{yq} can be obtained. Under the condition of the equation 16

$$|t_{dc} - t_{yq}| \leq \xi_t \quad (16)$$

Close operation command was acted immediately; otherwise the next calculation was need, until the condition meets the need.

3.2.3 *The calculation of phase Angle difference $\varphi = 0$*

Normally, above two kinds of algorithm can achieve good performance. Because of the certain error is allowed with both the leading time and the leading phase angle, the closed operate at the zero angle difference is very rare. How to improve the accuracy of the close operation has important significance to synchronization device.

In the process of calculation, using first method leading phase angle difference φ_{yq} can be calculated. According to the equation 14, the current phase angle difference φ_i and φ_{yq} have error ξ, after time t_{DC}, U and U_x phase Angle difference are not zero, therefore closing phase angle difference φ=0 cannot be captured at current time.

Based on the current calculation of leading phase angle difference φ_{yq} and the leading time algorithm, condition should meet the demands as follow

$$\omega_s t_q + \frac{1}{2}\frac{d\omega_s}{dt} t_q^2 \quad (17)$$

For convenient of the calculation, omit secondary item, we can get:

$$t_q = \frac{|\varphi_i - \varphi_{yq}|}{\omega_s} \quad (18)$$

Above, t_q is time after time t_q from now, at t_q time, phase angle difference $\varphi = 0$, so it is the time to output the synchronization operation.

4 THE IMPLEMENT OF THE SYNCHRONIZATION ALGORITHM AND EXPERIMENTAL VERIFICATION

4.1 *The implement of the synchronization algorithm in the DFU410 distributing measure and control device*

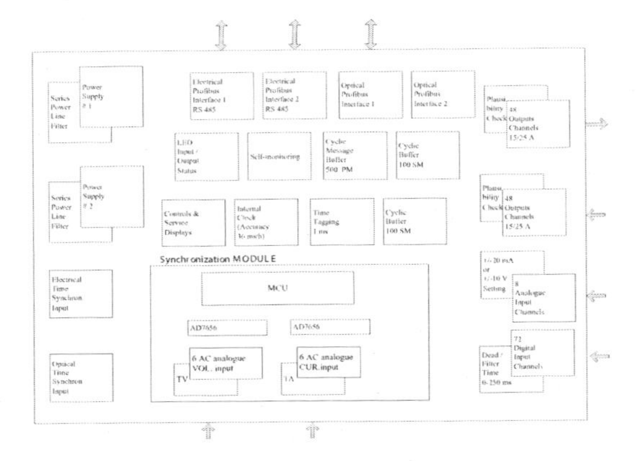

Figure 4. DFU410 function diagram.

DFU410 is developed for data collection, data pre-processing, data transfer, command output and the synchronization function at the field control level in medium-voltage and high-voltage substations, power stations and industrial plants. It is designed to be located in the immediate vicinity of the process with hard-wired parallel interfacing to the various control devices, a distributed intelligence ensures that process data is collected, logged and pre-processed where it is generated. Communications with the central control station is made via a PROFIBUS field bus.

The following functions are implemented:

1. Acquisition and filtering of binary signals, e.g. from switchgear control and protection units
2. Acquisition of analogue values
3. Transmission of process data to the central control station

4 Reception, conversion and output of relay commands from the central control station
5 Time tagging and recording of messages
6 Guaranteed data delivery of messages
7 Self-monitoring / plausibility checking
8 Monitoring of communications links
9 Synchronization operation

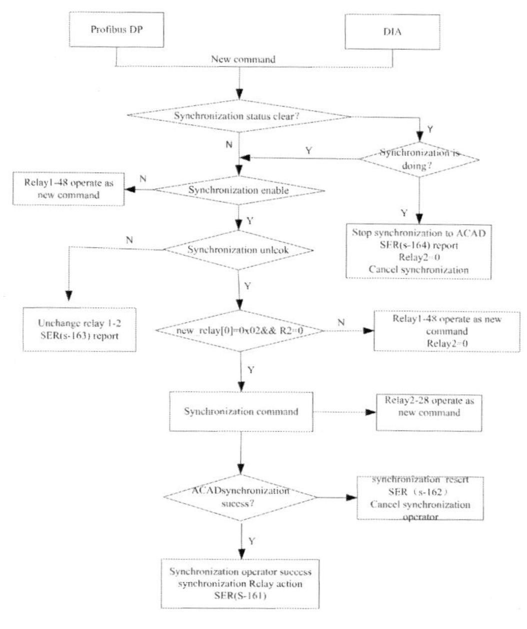

Figure 5. process chart of synchronization control in DFU410.

As shown in figure 5, the synchronization operator was integrated into the main processor board, which is CPU board and ACAD Board, the CPU board receives the command from the DP master linked to a control or protecting device, and the ACAD board is responsible for the synchronization algorithm implementing. To complete the full synchronization operation, some auxiliary function is necessary, such as SER, and reports.

4.2 *Experimental verification*

Application has been discussed about difference frequency of power system interconnection algorithm, in our new generation of measurement and control device to realize the function of difference frequency grid, relay protection and automation equipment quality supervision and inspection center, the Third-party laboratory testing device setting difference frequency of 0.02-0.09hz/s,5 hz frequency offset, the leading time for 300 ms, the device on line a f1 from 47 hz to 53 hz at 0.02 hz/s - 0.09 hz/s change, line 2 f2 from 47hz to 53 hz at 0.02 hz/s - 0.09 hz/s change, start the order during this period, the plant feed motion within a specified time, the OMICRON detection, closing time, system of phase within 3 degrees. And the synchronization success rate is 100% with 20 times operator, as show in table2.

Table 2. The experimental verification under different working conditions.

Line 1 frequency	Line 2 frequency	Frequency change	Success rate in 20 times operator
47-53	50	0.02	100%
50	47-53	0.02	100%
47-53	47-53	0.05	100%
47-53	47-53	0.07	100%
47-53	47-53	0.09	100%

5 CONCLUSION

Difference frequency interconnection of power system, the author of this paper, under the condition of the realization of the function of the same period, established the grid-connection control model at the same time, using the fuzzy control principle, in the process of the discussion for the interconnection of system voltage and frequency adjustment method, according to the grid control model at the same time, based on the given leading phase angle difference algorithm and leading time period, at the same time, further discussed the capture phase angle difference $\varphi = 0$, application of the algorithm, in the same period of measure-control device in the switching function, the type test, the device can correct closing, closing phase Angle difference is small, the algorithm has strong theoretical and practical significance.

REFERENCES

[1] Yang Dongjun, Ding Jianyong, Li Jisheng, et al. (2011) Analysis of power system forced oscillation caused by asynchronous parallelizing of synchronous generator. *Automation of Electric Power Systems*, 35(10):99–103.
[2] Chen Qingxu, Yu Huawu, (2010) Quasi-synchronization device with line-selection function. *Electric Power Automation Equipment*, 30(4):105–108.
[3] Li Xianbin, Chen Shihe et al. (1990) Control principles for microprocessor-Aided synchronization on the uncertainty condition of the slip frequency. *Proceedings of the CSEE, 1990*, 10(3):54–59.

[4] HUANG Chun, HE Yigang, JIANG Yaqun. Novel algorithm automatic synchronization. *Proceedings of the CSEE, 2005*, 25(3):60–64.
[5] Zhang Xiaoying, Dang Cunlu, Wang Shudong. (2007) Design of generator automatic quasi-synchronizing device based on microprocessor and CPLD. *Electric Power Automation Equipment*, 27(8):102–104.
[6] Yang GuanCheng. (2005) *Power System Automation Equipment Principle*. Beijing: China Power Press.
[7] Mo Shaoqing, Li Xiaocong. (2003) An improved high-accuracy algorithm for frequency measurement based on Fourier transform. *Automation of Electric Power Systems*, 27(12):48–49+54.
[8] Bu Guangjin, Dai Jing, Li Ming. (2007) Fixed frequency sampling measurement based on frequency tracing. *Electric Power Automation Equipment*, 27(4):68–70.

Electric, Electronic and Control Engineering – Shao, Shu & Tian (Eds)
© 2015 Taylor & Francis Group, London, ISBN: 978-1-138-02842-5

A method of coal identification based on D-S evidence theory

Tong Wang, Liang Tian & Weining Wang
North China Electric Power University, Baoding Hebei, China

ABSTRACT: Aiming at the problem of coal identification in power plant, we put forward a method of on-line coal identification based on Dempster-Shafer evidence theory. We establish a typical sample database of coal with the historical data. According to the density function we build the reliability function density. Taking coal industry analysis, elemental analysis and low heating values as the evidence, the reliability function density is obtained according to the typical sample database. After establishing the multiple value typical sample distribution according to probability density of logarithmic normal distribution, and normalizing the belief function, we finally get the result of identification using D-S fusion rule. The simulation results show that using this online method of coal identification the result is more accurate and the subjectivity of evidence theory reliability function and the hysteresis are reduced. It can be well applied to control coal blending in power plant.

KEYWORDS: Coal identification; Dempster-Shafer evidence theory; basic probability assignment; Logarithmic distribution

1 INTRODUCTION

In recent years, coal blending has become a major means to improve economic benefit and reduce NO_X for coal-fired power plant[1–3]. However, when the actual coal deviates from the design-coal, the security and economy of the unit is influenced badly. Conventional power plants choose chemical testing method, which can effectively identify the type of the coal. But there is a large hysteresis, and it is non-realtime. Therefore, monitoring the type of the coal in real time has become a critical issue. A lot of researchers have been working on the online coal identification. Dong Shixian[4] proposed a fuzzy neural network for online coal quality identification. Dong Junhua [5] proposed a RBF neural network to identify the type of coal online. However neural networks are often affected by the complication of training samples and the network structure. Once deviation happens, over study or other issues may occur. Tan Cheng [6] put forward a concept of joint density determination, which can reduce the number of samples and can identify unknown coal type. However it requires a lot of real-time monitoring data, which slows online monitoring to a certain extent, and may lead to more errors.

This article puts forward a coal identification technology based on D-S evidence theory, taking industrial analysis, elemental analysis and low calorific value as the evidences. According to online element analyzing method presented by thesis[7] and the online low calorific value identification method proposed by literature [8], we obtain the online evidences. On the basis of these methods, this article presents a method of obtaining basic probability assignment (BPA) using density curve of lognormal. Utilizing D-S evidence theory, we have conducted 20 simulations and measured confidence intervals of coal types. Real-time identification of coal types has reduced the hysteresis and subjectivity, and can be used as an effective means of real-time monitoring and control of power plants.

2 IDENTIFICATION ALGORITHM

The historical operating data that can be updated to establish a representative sample library, and use a representative sample gain the basic probability assignment of evidences and then gain the identification results. Due to unstable operation of the site or the sensor error, the typical sample is often not a particular value, but rather fluctuates within a certain range, and meets a certain distribution. We can get the appropriate distribution by data mining techniques and establish a representative sample letter density function.

3 D-S EVIDENCE THEORY

3.1 *Frames of discernment*

The domain of D-S evidence theory is called frames of discernment, denoted by Q, which contains a limited number of basic propositions denoted by

$\{u_0,u_1,u_2,\ldots,u_n\}$. The corresponding probability of basic events is denoted by primitives. All primitives of Q are antagonistic and mutually exclusive in target recognition modes.

If there is a set of function $m: 2^{\Theta} \rightarrow [0,1]$ that satisfies the following formula:

$$m(\varphi) = 0 \quad (1)$$

$$\sum_{n} m(u_i) = 1, u_i \in \Theta \quad (2)$$

then, the frames of discernment Θ is regarded as basic probability assignment(BPA). If $m(u_i) \neq 0$, u_i is called focal elements of basic probability assignment(BPA). $m(u_i)$ represents the value of the BPA of the evidences of the Focal element u_i, and represents the reliability and level of support to the objectives.

3.2 *Dempster's rule of combination*

According to Dempster's rule of combination, if m_1, m_2 correspond to the same frame of discernment Q and focal elements are $\{u_{11},u_{12},\ldots,u_{p1}\}$, $\{u_{12},u_{22},\ldots,u_{q2}\}$, assuming $\sum_{u_{n1} \cap u_{n2} = \varphi} \{m_1(u_{p1}) m_2(u_{q2})\} < 1$, function is defined as follows:

$$m(u_k) = \sum_{u_{p1} \cap u_{q2} = u_k} \frac{\{m_1(u_{p1}) m_2(u_{q2})\}}{1 - \nabla} \quad (3)$$

If $u_k = \varphi$, then $m(u_k) = 0$ (i、j、k= 1、2、…n),

$$\nabla = \sum_{u_{p1} \cap u_{q2} = \varphi} \{m_1(u_{p1}) m_2(u_{q2})\cdot$$

This function is called Dempster's rule of combination.∇is the sum of the products of the BPA with complete conflict hypothesis u_{p1}、u_{q2}. The so-called complete conflict means u_{p1}、uq2 are incompatible events. The results of D-S evidence theory are not affected by the sequence of evidences[9].

In fault diagnosis, focal elements $\{u_{11},u_{12},\ldots,u_{p1}\}$, $\{u_{12},u_{22},\ldots,u_{q2}\}$ and target modes to be identified $\{u_0,u_1,u_2,\ldots,u_n\}$ correspond to the different modes of the system, including normal mode and various failure modes. m (u_i) indicates the BPA of a particular mode assigned to a evidence.

3.3 *Rules of mode determination*

1 The determined mode should have the greatest PBA.
2 The difference between the BPA of determined type and other types should be greater than a certain value.
3 The BPA of uncertain ones must be less than a certain threshold.
4 The BPA of determined mode should be greater than uncertain[10].

4 TYPICAL SAMPLES AND BASIC PROBABILITY ASSIGNMENT

4.1 *Typical samples*

Due to the fixity of Dempster's rule of combination, the finally fused results are decided by BPA of evidences. BPA represents the degree of support to the objective mode, which is subjective to a certain extent. Different construction methods and subjective judgments will construct different BPA. This paper proposes a method to reduce the subjectivity of PBA.

Assuming a set of evidence $\{x_1,x_2,x_3,\ldots,x_m\}$ $(x_i(i \in m))$ are the output values of a certain sensor. Target mode$\{u_0,u_1,u_2,\ldots,u_n\}$, $u_j(j \in n)$ are to be identified. For any target mode, we need to determine the individual sensor output in this mode of the typical values $\{x_{1j},x_{2j},x_{3j},\ldots,x_{mj}\}$, which are called typical samples.

Typical sample acquisition method is quite difficult. Literature [11,12] mention the concept of a typical sample, but they the methods are simplified too much, and can't be applied to practical production. Because of exist of sensor error, effect of various surroundings and interaction of various parts of equipment during operation, a target mode u_i's eigenvalue is often not fixed, but deviates from the typical value to a certain degree.

4.2 *Distribution of typical sample*

According to the historical data of a typical sample, typical samples obey different distributions with the increase in the number of samples in its confidence interval. Literature [10] shows a typical sample in [0–1] uniform distribution. Literature [11, 12] proposed an acquisition approach which typical samples obey normal distribution. Based on the above findings, we present a utilization of lognormal to fit typical samples when typical samples deviate from the normal distribution.

4.3 *Density function of statistical model*

The eigenvalue denoted by x_{ij} (i 1,2,…,m) of a certain mode is a random variable. x_{ij} can be any value in the confidence interval and should obey a certain probability density function f(x), which satisfies:

$$P = \quad a < x \leq b \quad = \int_{q}^{b} f(x) dx \; 0 \leq x_{ij} \leq +\infty \quad (4)$$

When x_{ij} obeys lognormal distribution, the lognormal probability density function is:

$$f(x;\mu,\sigma)=\frac{1}{x\sigma\sqrt{2\pi}}e\frac{-(\ln x-\mu)^2}{2\sigma^2} \quad (5)$$

In the issue discussed in this article, σ ($\sigma=\sigma_{ij}$) is the standard deviation of the lognormal distribution. Since the introduction of the confidence interval, characteristic variable x_{ij}'s probability density function changes into $\omega_{ij}(x)$:

$$\omega_{ij}(x_i)=\frac{1}{x_i\sigma_{ij}\sqrt{2\pi}}e\frac{-(\ln x_i-x_{ij}^*)^2}{2\sigma_{ij}^2}$$

$$x_{1(\partial)}^*\le x\le x_{2(\partial)}^* \quad (6)$$

4.4 *Statistical characteristics of the lognormal distribution*

From the mathematical statistical results, we found some typical values highly approximate lognormal distribution. When we conduct statistical analysis, we use maximum likelihood estimation as follows:

$$\hat{\mu}=\frac{1}{n}\sum_{i=1}^{n}\ln(x_i) \quad (7)$$

$$\hat{\sigma}^2=\frac{1}{n}\sum_{i=1}^{n}(x_i-\frac{1}{n}\sum_{i=1}^{n}\ln(x_i))^2 \quad (8)$$

Denoting confidence level as α, using t-distribution we obtained:

$$x_1^*=\frac{1}{n}\sum_{i=1}^{n}\ln x_i-t_{\frac{\partial}{2}}(n-1)*\frac{W}{\sqrt{n}} \quad (9)$$

$$x_2^*=\frac{1}{n}\sum_{i=1}^{n}\ln x_i+t_{\frac{\partial}{2}}(n-1)*\frac{W}{\sqrt{n}} \quad (10)$$

$$W^2=\frac{n}{n-1}\hat{\sigma}^2$$

Therefore, the confidence interval of α can be obtained by the corresponding confidence intervals and the probability density function.

4.5 *BPA density function*

BPA is closely related to the eigenvalues, thus BPA density function can be obtained by the probability density function. The $\omega_{ij}(x)$ is defined as the BPA density function in the mode of u_i[7].

$\omega_{ij}(x)$ should satisfy the following formula:

$$\text{I} \quad \omega_{ij}\ge 0 \quad (11)$$

$$\text{II} \quad \int_{x_1^*}^{x_2^*}\omega_{ij}(x)\,dx=1-\alpha \quad (12)$$

Fig.1 is the confidence density function of x_i under u_i

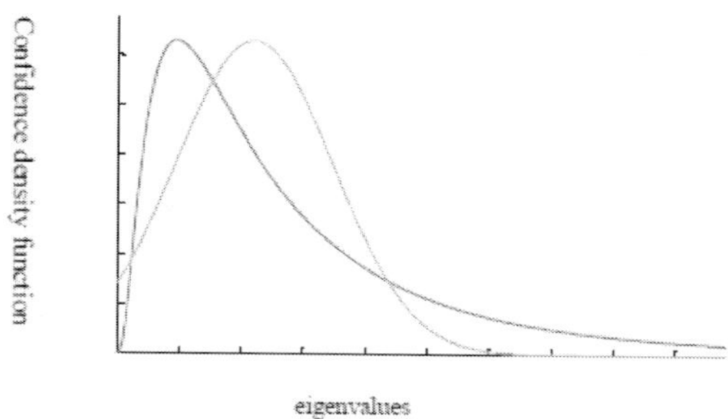

Figure 1. Confidence density function of x_i under u_j.

Figure 1 shows that:

1 Lognormal distribution can highly approximate the data from measuring points. Therefore, the resulting belief function is more objective.
2 By theoretical calculations, we conclude:

$$\text{When } x=e^{\bar{x}-\sigma^2},\ \max[\omega_{ij}(x_i)]=\frac{1}{x_i\sigma_{ij}\sqrt{2\pi}} \quad (13)$$

$$\text{When } x=e^{x_{ij}^*\pm\delta_{ij}},\min[\omega_{ij}(x_i)]=\frac{1}{113x_i\sigma_{ij}\sqrt{2\pi}}\approx 0 \quad (14)$$

In the above formulas, max $[\omega_{ij}(x)]$ is the reliability density of typical empirical value, which represents the greatest degree of support to the objective mode under the historical experience. min $[\omega_{ij}(x)]$ is the confidence interval's edge value, which is approximately equal to zero.

4.6 *Acquisition and normalization of BPA*

In the target recognition, there are *n* target modes denoted by $\{u_0, u_1, u_2, \ldots, u_n\}$. We use *m* sensors to measure *m* fault symptoms of the mode. *m* feature variables denoted by x_{ij} can be obtained. A set of evidences $\{x_{1j},x_{2j},x_{3j},\ldots,x_{mj}\}$ are output. Each evidence x_i (i=1,2,3,…,m) has *n* BPA density values: $\{[\omega_{i1}(x_i)],[\omega_{i2}(x_i)],[\omega_{i3}(x_i)],[\omega_{i4}(x_i)],\ldots,[\omega_{in}(x_i)]\}$. Each BPA density value means the degree of support under certain circumstances.

Assuming that:

$$\omega_{ij}(x_i) = \omega_{ij}(\theta) / \max[\omega_{ij}(x_i)] + \omega_{ij}(\theta) \tag{15}$$

in which $\omega_{ij}(\theta) = \min[\omega_{i1}(x_i), \omega_{i2}(x_i), \ldots \omega_{i2}(x_i)]$,

$\omega_{ij}(\theta)$ means BPA of system's uncertainty density function. The sum of each BPA of evidence x_i (i=1,2, 3,...,m) is called S_i:

$$S_i = \sum_{j=1}^{n} \omega_{ij}(x_i) + \omega_{ij}(\theta) \tag{16}$$

The BPA of each target patterns are normalized in order to make it meet the requirement of BPA. The BPA of evidence x_i under the mode u_i is:

$$m_{ij}(u_{ij}) = \omega_{ij}(x_i) / S_i \tag{17}$$

It can also be obtained that the BPA of system's uncertainty is:

$$m_i(\theta) = \omega_i(\theta) / S_i \tag{18}$$

Though calculating and proving, the above algorithm meets the definition of BPA. In the case of determining the representative sample of the system, we can gain the BPA density function, and then we can get the BPA of each mode u_j(j=1,2,3,...,n).

5 ANALYSIS OF PRACTICAL INSTANCE

5.1 *Statistics and data processing*

Using the above method for obtaining BPA, we process the typical sample data in reference[10]. We use statistical methods to get the expectations m and the RMS s of typical sample data.

Expectations m can be obtained through formula (7) and RMS s can be obtained through formula (8). The results are shown in Table 1.

Table 1. Expectations and RMS of typical sample data.

	Water (M_{ar}%)		Ash (A_{ar}%)		Sulfur (S_{ar}%)		Calorific (Q_{ar}%)	
Coal type	μ	σ	μ	σ	μ	σ	μ	σ
A	6.2	3.29	22.1	6.67	0.54	0.75	23.4	5.07
B	5.6	3.46	21.7	5.20	2.57	3.75	23.7	2.67
C	8.7	16.3	19.1	16.8	1.13	1.76	23.1	8.39
D	23	4.24	23.5	9.33	0.55	1.48	14.0	2.48

The evidences are as follows: base moisture content, ash content, sulfide content, and calorific. In order to facilitate the description, they are defined as follows:

The coal type to identify is denoted by $\{u_1,u_2,u_3,u_4\}$. u_1 represents anthracite. u_2 represents Lean coal. u_3 represents Bituminous coal. u_4 represents Lignite. 4 evidences are x_1 (Base moisture content), x_2 (Ash content), x_3 (Sulfide content), and x_4 (Calorific).

Based on the above discussion, we can get the typical sample matrix:

$$\begin{bmatrix} [6.2\pm1.494] & [5.6\pm1.706] & [8.7\pm5.018] & [23\pm22.452] \\ [22.1\pm0.886] & [21.7\pm0.709] & [19.1\pm2.271] & [23.5\pm1.478] \\ [0.54\pm3.032] & [2.57\pm3.204] & [1.13\pm3.329] & [0.55\pm4.357] \\ [23.4\pm0.643] & [23.7\pm0.337] & [23.1\pm1.056] & [14\pm0.207] \end{bmatrix}$$

5.2 *Acquisition of BPA*

A set of evidences from site date are x_1=5.82, x_2=13.04, x_3=0.52, x_4=25.32. We get the BPA density function values shown in table 2.

Table 2. The BPA density values of evidences towards target mode.

Evidences	ω(u_1)	ω(u_2)	ω(u_3)	ω(u_4)	ω(θ)
Anthracite	0.0701	0.0717	0.0221	5.0259e-007	-
Lean coal	0.1035	0.0402	0.0323	0.0799	-
Bituminous coal	0.3107	0.7180	0.4840	0.0105	-
Lignite	0.0014	2.9690e-007	0.0078	2.1705e-010	-

Then using the above normalization algorithm, we can get the BPA under D-S evidence theory. the results are shown in Table 3.

Table 3. BPA of evidence.

evidences	m(u_1)	w(u_2)	m(u_3)	m(u_4)	m(θ)
Anthracite	0.4278	0.4374	0.1348	3.06E-06	3.06E-06
Lean coal	0.3591	0.1395	0.1121	0.2772	0.1121
Bituminous coal	0.2026	0.4682	0.3156	0.0069	0.0068
Lignite	0.1522	3.22E-05	0.8478	2.36E-08	2.36E-08

Then we can get the result though Dempster's rule of combination. The finally results are shown in table 4.

Table 4. Intermediate results and final results of evidence fusion.

Target mode	u_1	u_2	u_3	u_4	u_θ
The first round of fusion(x_1,x_2)	0.3561	0.1944	0.0534	3.6E-06	0.3962
The first round of fusion(x_3,x_4)	0.07389	3.5E-05	0.6337	1.1E-09	0.2924
Fusion results	0.2453	0.0873	0.4614	1.6E-06	0.2059
final results	0.2452	0.0873	0.4614	0	0.2060

The final result is that the type of coal is u_3 which represents bituminous coal.

From the above conclusion, it is shown that the application of the proposed method is well applied in practical, and the results are more objective.

6 CONCLUSION

This paper proposed a new method to construct BPA with D-S Evidence Theory. It is well applied in power plants' for online coal identification. Concerning that the volatile of combustible is the main impact factor, and coals are classified as anthracite, lean coal, bituminous coal, and lignite according to the content of volatile in coal's dry ash-free radical. The paper uses 4 evidences: base moisture content, ash content, sulfide content and calorific. Through processing and analyzing statistical data, we get the BPA of each evidence. Finally we obtained the result though Dempster's rule of combination. The example shows that the method has explicit theoretical implications which improve the objectivity of BPA. The authors provide a new method of online coal identification. However, the method calls for high quality data of typical sample, which will be the direction of further study.

REFERENCES

[1] Duan Xuenong, Zhu Guangming, Bin Yiyuan. (2010) Application and development of coal-blended combustion technology in boiler in Hunan power plan. *Electric Power*, 43(2):48–51.

[2] Yang Zhongcan, Yao Wei. (2010) Study on burning blending coals in coal-fired boilers of power plants. *Electric Power*, 43(11):42–45.

[3] Wang Wenhuan, Pan Bingchao, Wang Aichen, Pan Weiguo. (2014) Analysis of the economy and environmental protection features based on blending lignite in coal-fired boiler. *Boiler Technology*, 45(2):67–70.

[4] Dong Shixian, Xu Xiangdong. (2004) Application of fuzzy-neural network in identifying information of coal in boilers. *Journal of Tsinghua University (Science and Technology)*, 44(8):1092–1095.

[5] Dong Junhua, Xu Xiangdong. (2006) Coal-information online identification based on RBF neural network. *Journal of Tsinghua University (Science and Technology)*, 46(8):1430–1433.

[6] Tan Cheng, Li Xiaomin, Xu Lijun, Wu Yuting. (2010) Techniques coal type identification based on joint probability density arbiter and neural network. *Journal of Mechanical Engineering*, 46(18):18–23.

[7] Yue Pengcheng, Meng Zhidong, Liang Xiaoyu. (2013) Study on mathematical models for quick-calculating ultimate analysis from proximate analysis. *Journal of China Jiliang University*, 24(3):327–330.

[8] Zhu Daqi, Yang Yongqing, Yu Shengli. (2004) Dempster-Shafer information fusion algorithm of electronic equipment fault diagnosis. *Control Theory & Applications*, 21(4):659–663.

[9] Tian Liang, Chang Taihua, Zwng Deliang, Liu Jizhen. (2005) Fault diagnosis of a boiler milling system on the basis of a typical-swatch data fusion method. *Journal of Engineering for Thermal Energy and Power*, 20(2):163–166.

[10] Zhao Liangyu, Tian Liang, Wang Qi, Liu Jizhen. (2011) Diagnosis technology of coal type based on improved belief function distribution method. *Journal of North China Electric Power University*, 38(4):71–75.

Electric, Electronic and Control Engineering – Shao, Shu & Tian (Eds)
© 2015 Taylor & Francis Group, London, ISBN: 978-1-138-02842-5

Research on Green Spline interpolation algorithm application in optical path computation in aerodynamic flow field

Yonghui Zheng, Huayan Sun & Yanzhong Zhao
Academy of Equipment, Beijing, China

ABSTRACT: Faced with the disadvantage of current interpolation methods for optical path computation in an aerodynamic flow field that are simple and of poor applicability, Green Spline spatial interpolation algorithm application were studied in this paper. Green spline method, which used in the fields of geological modeling were introduced to the interpolation computation of aerodynamic flow field. The feasibility of the interpolation method was tested and verified with calculated data of a subsonic aerodynamic flow field around a hemisphere-cylinder turret, the precision was obtained and compared with current methods. The statistical characteristics of the aero-optical distortion at typical incident angles were computed, and the improvement of the turbulent aero-optical calculation was obtained by using a Green Spline algorithm.

KEYWORDS: Green Spline; aero-optical; optical path computation; spatial interpolation,

1 INTRODUCTION

In the numerical computation of aero-optical effects, optical distortion is obtained by calculating the propagation of light in the aerodynamic flow field grids obtained by computational fluid dynamics (CFD). Since CFD grids are composed of a finite number of discrete nodes, interpolation is required to estimate the air density (or refractive index) at the positions on the light paths. However, no systemic studies have focused on this subject so far. There are three kinds of common interpolation methods as follows.

The first method is the nearest neighbor method (NN for short) which is the simplest and the most widely used interpolation methods. MIT Lincoln Laboratory [1] adopts this method in the integrated aero-optical analysis software so as to improve the computational efficiency. Y. Hsia, W. Lin and H. Loh (The Boeing Company) improved the NN method[2] by employing the nearest node and the gradient in the neighborhood of the point to estimate the air density of the target point. The second interpolation method is the linear interpolation method. In Sutton and Pond's [3] study, the air density of each node is obtained by linear interpolation of four vertices in a quadrangle containing the node. The last method is the second order gradient interpolation method proposed by Kan Wang and Edwin Mathews [4,5] from the University of Notre Dame. The air density of each point was estimated by its second order spatial gradient and it has been applied successfully in the aero-optical computation of the LES aerodynamic flow field of a turret.

The NN method is the simplest one, however, it does not consider the relationship between the interpolation points and the other given data points in the neighborhood. So it is effective only when the CFD grids are dense enough, otherwise the errors are quite large. The linear interpolation method at Boeing Company is improved using gradient, however, the gradient along the direction of the interpolated point and its nearest node is often difficult to estimate with given grid data. The linear interpolation method is only applicable to two-dimensional interpolation, but not applicable to the interpolation of unstructured spatial grid data. The second order gradient interpolation method is not applicable to the interpolation of unstructured spatial grid data either.

In the fields of geological modeling, it is often required to make spatial interpolation estimation using a small number of giving sample points, and Green Spline interpolation method is the most widely used methods. In this paper, Green Spline Method is introduced into the computation of aero-optical spatial interpolation. The classical hemisphere-cylinder turret case is studied with CFD and aero-optical calculation. The performances of the interpolation method in different aerodynamic zones are studied accurate aero-optical distortion analysis is performed.

2 AERODYNAMIC FLOW FIELD COMPUTATION OF THE HEMISPHERE-CYLINDER TURRET AND GREEN SPLINE INTERPOLATION ALGORITHM

The turret structure of the hemisphere-cylinder is simple and has fairly large field range. It also facilitates the beam projection and reception and thus becomes a typical 3D model for aero-optical analysis. In recent years, there have been a lot of studies on the computation of aero-optical distortion simulation and wind tunnel tests related to this model [6–8]. Aero-optical distortion simulation of a hemisphere-cylinder turret is performed in this study.

2.1 *3D model and CFD computation*

The model sizes of the turret and wind tunnel are shown in Figure 1. The turret was installed on one of the tunnel side walls. The width, height and length of the wind tunnel are 36 inches, 36 inches and 195 inches, respectively. The flow velocity is 137m/s, the flying altitude is 2.18km, and the corresponding Mach number is 0.4., the distance between the center point of the cylinder and the entrance of the wind tunnel is 45 inches, and the distances to both side walls of the wind tunnel are 14.5" and 19.5" respectively.

Preprocessor software Gambit was used to generate the computational mesh with unstructured meshes. The results are shown in Figure 2. The target region is meshed in 37 zones with a total of 3.2 million elements. CFD computation is performed using ANSYS Fluent. In order to obtain the characteristics of turbulent flow field accurately, firstly an initial value for transient model is obtained based on k-ω turbulence viscosity model. Secondly, shear-stress transport (SST) transient turbulent model is employed and calculated for 0.2 s with a time step of 10 μs. Finally, detached eddy simulation (DES) model is employed and simulated for 0.16 s. An extra 0.015s is simulated when the statistical characteristics of the flow field get stable, and the flow field data are saved at an interval of 50 μs for turbulent characteristic analysis.

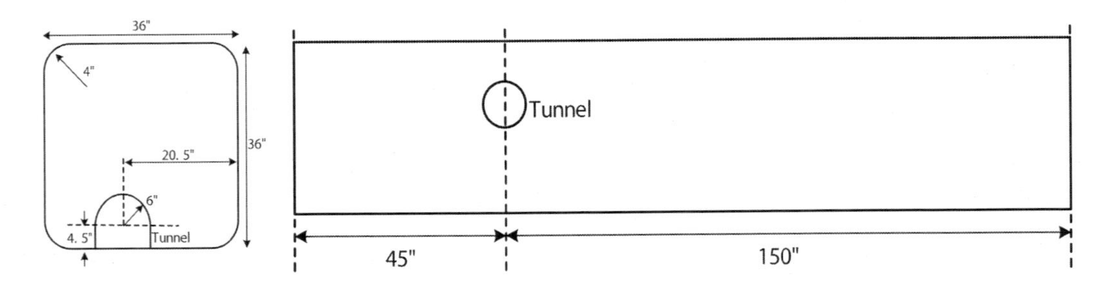

Figure 1. Model sizes of the turret and wind tunnel.

The pressure coefficient of the center line on the turret surface is used to verify the results of CFD computation. Figure 3 shows the comparison between CFD results and the wind tunnel test data[6]. As seen from Figure 3 that the CFD calculated results (especially DES results) are consistent with the results of the wind tunnel test. Therefore, the CFD aerodynamic flow field structure is considered to be consistent with the actual flow field.

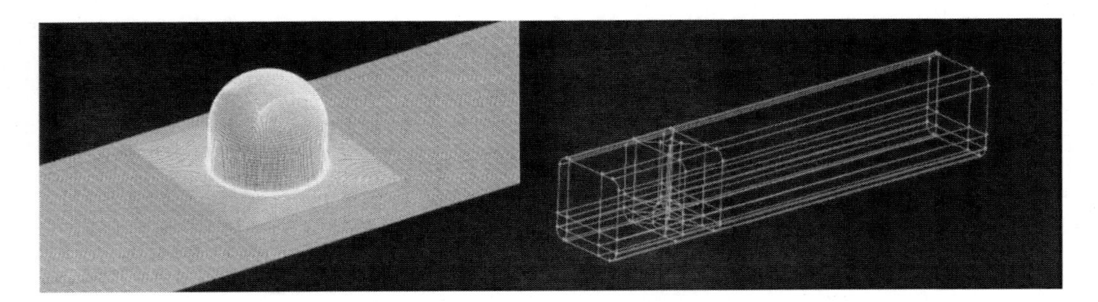

Figure 2. Grid modeling and zones strategy.

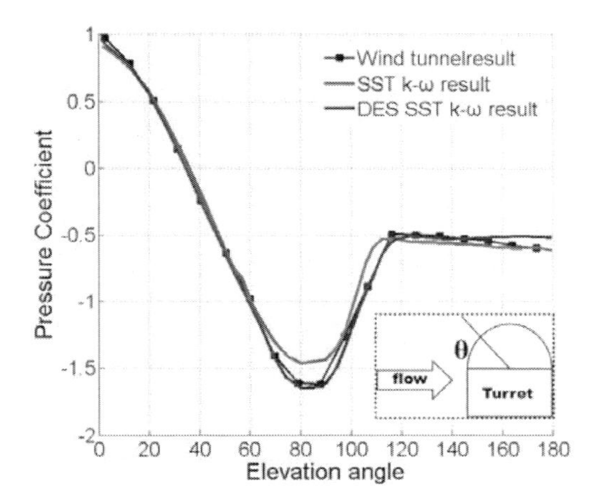

Figure 3. Pressure coefficient of the center line on the turret surface.

2.2 *Green Spline interpolation method*

For Green Spline Interpolation Method, the air density value at the newly interpolated point can be expressed as the weighted sum of n Green functions of the point and each given point in the neighborhood[11]:

$$D(x_0, y_0, z_0) = T(x_0, y_0, z_0) + \sum_{i=1}^{n} \omega_i g\left[(x_0, y_0, z_0), (x_i, y_i, z_i)\right] \quad (1)$$

Where D(x0,y0,z0) denotes the density value of the newly interpolated point (x0,y0,z0); g[,] denotes the Green function; ω_i denotes the unknown weighting coefficient; (xi,yi,zi) denotes the spatial coordinate of the *i*th given point; *T*(x0,y0,z0) denotes the given component which is difficult to express by Green function. The component is usually a constant or special trend in a certain area, and its value is 0 in this study. Substitute the n nodes in the neighborhood into Formula (1), we can get n linear equations for solving ωi. If we selected m additional points with known density gradient and substitute them for the derivative form of Formula (2):

$$\nabla D(x_0, y_0, z_0) = \sum_{i=1}^{n} \omega_i \nabla g\left[(x_0, y_0, z_0), (x_i, y_i, z_i)\right] \quad (2)$$

m additional linear equations will be obtained. Gradient constraints are helpful when the original data have continuous and smoothly varying gradients. The Green function must satisfy the inhomogeneous partial differential equation:

$$\nabla^2\left[\nabla^2 - p^2\right] g\left[(x_0, y_0, z_0), (x', y', z')\right] = \delta\left[(x_0, y_0, z_0) - (x', y', z')\right] \quad (3)$$

where ∇^2 is the Laplacian operator, δ is the Dirac Delta function, and p is the tension efficient.

The expression of the Green function of the 3-D rectangular coordinate system and its gradient are expressed in Formula (4):

$$g = \begin{cases} (1/pr)(e^{-pr} - 1) + 1 & r \neq 0 \\ 0 & r = 0 \end{cases},$$

$$|\nabla g| = \begin{cases} (1/pr^2)\left[1 - e^{-pr} \quad pr + 1\right] & r \neq 0 \\ 0 & r = 0 \end{cases} \quad (4)$$

where, $r = |x - x'|$ denotes the spatial distance between point x and point x', and tension efficient p is usually expressed by another variable t: $p^2 = t/(1-t)$, $(0 \leq t < 1)$. The larger the tension coefficient, the smoother the interpolation data performs, t equals 0.1 in this study.

Moreover, if the interpolation point is very close to a given point during the interpolation process, errors may be generated with Green interpolation method. Therefore, attention should be paid to eliminate the singular interpolation point or use the value of the nearest point instead.

For the purpose of comparison, Average interpolation method and NN method are carried out in this study. Because the method of Sutton and Pond and that of Edwin Mathews and Kan Wang are not applicable to unstructured grid, they are not performed and analyzed in this study.

3 VERIFICATION OF INTERPOLATION METHOD AND AERO-OPTICAL DISTORTION COMPUTATION

3.1 *Data verification*

Three spherical regions, Zone 1, Zone 2 and Zone 3 are selected from the data field obtained by CFD as the interpolation zones. The three zones are shown in Figure 4. The radius of each zone is 5 cm, and the three zones are located at the upstream direction, right above direction and downstream direction respectively, 25 cm away from the center of the hemisphere of the turret. The three zones represent three kinds of different turbulent structures of the compressed attachment zone, the transitional zone and the separation zone respectively. The number of known points in the three zones are 2162, 4708 and 1974 respectively. In order to verify the interpolation precision of the simple average method, NN method and Green Spline Method, interpolation of each known point in the region is carried out. During the interpolation process, suppose the density value of one point is unknown, and then it is estimated using other given points, and then compared to its original data to evaluate the interpolation accuracy. This method is called one-at-a-time cross-validation method[11]. We use the sum of the square of the estimated residuals (SSR) of each point in the zone as the evaluation function for the interpolation precision. Furthermore, the standard deviation σ of the evaluated error for each point in the interpolation region is calculated. Assuming that the evaluated errors is normally distributed, the percentage ξ of evaluated errors falling into the 95% confidence interval (within the interval of $(-1.96\sigma,\ 1.96\sigma)$) is obtained. Interpolation computation for the transient flow field of both the SST and DES results are carried out, the results are as shown in Tables 1 and 2, where the shaded region represents the item with the highest interpolation precision.

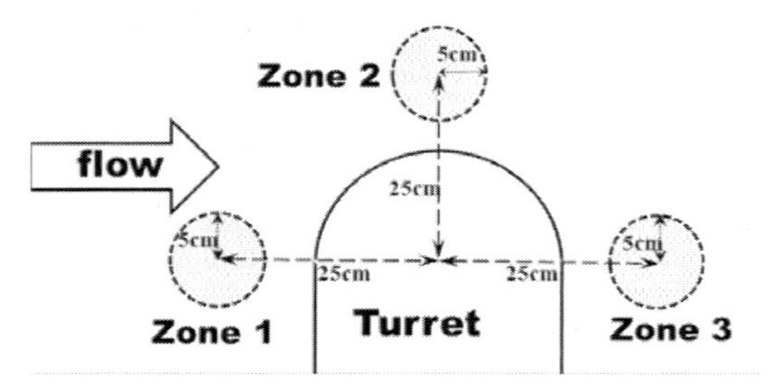

Figure 4. Interpolation zones in the flow filed.

Table 1. Interpolation results of the SST flow field.

	Zone 1		Zone 2		Zone 3	
	SSR	ξ	SSR	ξ	SSR	ξ
Simple average	3.87e-4	92.00	3.31e-6	91.55	0.0036	94.78
Nearest Neighbor	0.0050	1	1.24e-4	92.76	0.0092	95.09
Green spline (n 500)	1.28e-4	92.14	7.85e-6	95.16	1.94e-4	93.47

Table 2. Interpolation results of the DES flow field.

	Zone 1		Zone 2		Zone 3	
	SSR	ξ	SSR	ξ	SSR	ξ
Simple Average	2.25e-4	92	0.0112	91.96	0.0059	92.86
Nearest Neighbor	0.0029	1	0.0264	94.84	0.0096	94.22
Green Spline (n 500)	1.21e-4	93.29	1.65e-4	95.69	5.65e-5	94.93

We can see from Tables 1 and 2 that Green Spline Interpolation Method is the optimal method. In the neighborhood of 500 known points, smallest SSR and highest precision are obtained in the interpolation with the Green Spline method in the three zones of both SST flow field and DES flow field. The interpolation precision of the NN method is lower than that of the other two methods, which is consistent with the anticipated conclusion and this is due to that the relationship between the interpolation points and the other given data points in the neighborhood is not considered.

From the perspective of the percentage (ξ) of interpolation errors falling into the confidence intervals of $(-1.96\sigma,\ 1.96\sigma)$, the distribution characteristics of the interpolation errors obtained by three interpolation methods are not so significantly different. ξ is generally higher than 90%, and some are even higher than 95%.

Generally speaking, Green spline interpolation is the optimal method in which very small SSR can be obtained while no calculation of parameter and settings are needed, and is appropriate of any structure flow field. The standard deviation of the evaluated error is generally one order of magnitude higher than NN method. However, the high accuracy comes from estimations with Green functions for hundreds of given points in the neighborhood of one interpolation point, requiring great computation time.

3.2 *Interpolation computation of aero-optical distortion*

Aero-optical distortion is computed by using the Green Spline Interpolation Method, NN method and Simple Average method. OPD computation is carried out for 300 DES transient aerodynamic flow fields at three directions with the elevation angle of 60°, 90° and 120° respectively. The wavelength is 532nm, and the light beam has a rectangular shape with the size of 12.7cm×12.7cm. Flow field interpolation and optical path computation are carried out for the light beam using 31×31 matrix grid, and Optical Path-Difference (OPD) matrix under each transient flow field condition is obtained therefrom. Root-mean-square (RMS) of OPD is calculated after removing the tilting term, thus 300 OPDrms values are obtained at three directions respectively, which represent the statistical characteristics of the turbulent aero-optical distortions. The distribution of probability density function (PDF) of OPDrms value is as shown in Figure 5. The red, black and blue lines in the figure represent the NN, Simple Average and Green Spline results, respectively.

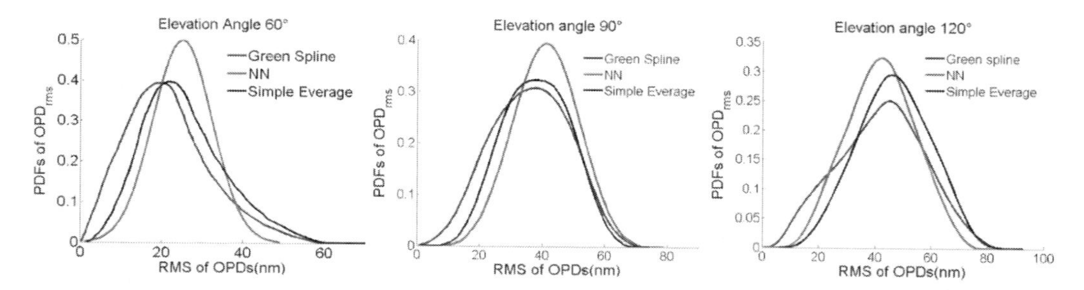

Figure 5. PDF of OPDrms obtained by three interpolation methods at elevation angle 60°(left), 90°(middle) and 120°(right).

As the interpolation accuracy is verified above, difference of OPDrms PDF shown in Figure 5 represents the improvement of aero-optical computation generated by the Green Spline method, Compared with the NN and Simple Average method. At elevation angle 60° and 90°, the Green Spline method obtain the smallest OPDrms, NN method gets the biggest OPDrms, and Simple Average method between them, mean OPDrms of Green Spline is generally 5nm smaller than the NN method. At elevation angle 120°, mean OPDrms obtained by these three methods are close, NN method and the Green Spline method achieve more concentrated distributions of OPDrms, with a smaller variance.

4 CONCLUSION

Green Spline interpolation method, which is used in the field of geological modeling, was introduced into the optical path computation in the aerodynamic flow field to improve the accuracy and applicability of aerodynamic optical distortion computation. Subsonic aerodynamic flow field data around a hemisphere-cylinder turret was calculated with CFD method, and data verification of the interpolation algorithm was performed with the flow field data in three selected zones. Interpolation precision of the Green Spline Method is found to be one order of magnitude higher than generally used NN method. Since the relationship between the interpolation points and other given data points in the neighborhood is not considered, the interpolation precision of the nearest node method is lower than that of the other four methods.

Green Spline method, NN method and Simple Average method are used to calculate the optical distortion statistical effects in the aerodynamic flow field under typical angles, result shows that Green Spline method achieves a weaker turbulent aero-optical distortion at elevation angle 60° and 90°, while the interpolation method only affect the variance of the turbulent aero-optical distortion at elevation angle 120°.

REFERENCES

[1] Mark Bury, Keith Doyle, Thomas Sebastian, et al. (2013) An integrated method for aero-optical analysis. *51st AIAA Aerospace Sciences Meeting including the New Horizons Forum and Aerospace Exposition*, 07 – 10 Jan. 2013, Grapevine, Texas, AIAA 2013–0286, 2013.

[2] Y. Hsia, W. Lin, H. Loh, P. Lin, D. Nahrstedt, and J. Logan, (2005) CFD-based aero-optical performance prediction of a turret. *Laser Source and System Technology for Defense and Security*, edited by Gary L. Wood, Proc. of SPIE (SPIE, Bellingham, WA, 2005) Vol. 5792· doi: 10.1117/12.603001.

[3] G. W. Sutton, J. E. Pond, R. Snow, (1993) Hypersonic interceptor performance evaluation center: aero-optics performance predictions. *2nd Annal AIAA SDIO Interceptor Technology Conference*, June 6–9, 1993, Albuquerque, NM.

[4] Edwin Mathews, Kan Wang, Meng Wang, and Eric Jumper, () LES analysis of hemisphere-on-cylinder turret aero-optics. *52nd Aerospace Sciences Meeting 13–17 January 2014*, National Harbor, Maryland AIAA 2014–0323.

[5] Wang, K. and Wang, M. (2012) Aero-optics of subsonic turbulent boundary layers. *Journal of Fluid Mechanics*, 696:122–151.

[6] Juan M. Ceniceros, David A. Nahrstedt, Y-C Hsia, et al. (2007) Wind tunnel validation of a CFD-based aero-optics model. *38th AIAA Plasmadynamics and Lasers Conference*, 25–28 June 2007, Miami, FLAIAA 2007–4011.

[7] Stanislav Gordeyev and Eric Jumper, (2009) Fluid dynamics and aero-optical environment around turrets. *40th AIAA Plasmadynamics and Lasers Conference*, AIAA-2009-4224,22–25 June, 2009, San-Antonio, TX (Notre Dame).

[8] Stanislav Gordeyev n, EricJumper, (2010) Fluid dynamics and aero-optics of turrets. *Progress in Aerospace Sciences*, 46:388–400.

[9] Paul Wessel, (2009) A General-purpose Green's function-based interpolator. Computers & Geosciences, 35:1247–1254.

[10] John Joseph, Hatim O. Sharif, Thankam Sunil, Hasanat Alamgir,(2013) Application of validation data for assessing spatial interpolation methods for 8-h ozone or other sparsely monitored constituents. *Environmental Pollution*, 178:411–418.

Electric, Electronic and Control Engineering – Shao, Shu & Tian (Eds)
© 2015 Taylor & Francis Group, London, ISBN: 978-1-138-02842-5

A digital image steganography with low modification rate

Xudong Wan & Rener Yang
Faculty of Information Science and Technology, Ningbo University, Ningbo Zhejiang, China

ABSTRACT: In order to embed secret information into images with low modification rate, this paper proposes a digital image steganography based on embedding two bits in the method of the LSB. Firstly, LSB is improved by comparing the divided carrier and the private information before embedding. A flag can be used to control the inversion in every group to reduce the change to the carrier. Experimental results have also demonstrated that the proposed steganography has advantages in embedding efficiency, modification rate, embedding rate, compared with LSB and the new steganography is also faster than the matrix embedding.

KEYWORDS: Steganography; embedding efficiency; modification rate; embedding rate

1 INTRODUCTION

The development of information technology brings people convenience. At the same time, the security of information also calls for people's attention. The steganography algorithm deduces the possibility of disclosing information by hiding it. Traditional information security technology, such as cryptography will encrypt private information. The secret information may not be cracked. But encrypted information has some features that may attract attackers, so it still has some hidden dangers [1]. Digital image steganography embeds information and make it unnoticeable[2]. As a result of it, it won't attract the attackers' attention and increase the security.

LSB algorithm is one of the most classic and simplest digital image steganography[3]. Due to the redundancy of people's eyes when identifying images[4], it can transfer private information into binary and embed into image's every pixel's least significant bit. The advantages of LSB algorithm are simplicity and large embedding rate. Nevertheless, it changes images a lot, that is, its embedding rate isn't high, which enables attackers test information's existence by some means.

In order to improve the LSB algorithm's embedding efficiency, a series of algorithms have been raised. Wu's group raised a digital image steganography algorithm based on PVD[5]. On the basis of it, Chung-Ming Wang raised an improved PVD algorithm based on human's vision system[6], which reduces steganography's influence on image quality to a certain extent. These algorithms can increase images' quality to a certain extent, but it has to be improved in SNR and other performance.

Fridrich J raised an algorithm that uses parity-check matrix embedding based on the structure of linear block codes[7]. On the basis of it, Wang Chao raised a matrix embedding algorithm based on Bit-control[8]. These algorithms all increase embedding efficiency, but solving the matrix makes the calculation more complex and it takes more time in the information process.

This paper's algorithm compares with the secret information and its carrier's two least significant bits by establishing control bits. When the comparison is more than 50%, the secret information is inverted by setting up control bits and then embeds. This algorithm has linear time complexity and a larger processing efficiency than current algorithms. This algorithm improves embedding efficiency and deduces modification rate so that its security increases[9].

2 COMMON STEGANOGRAPHY ALGORITHMS

2.1 *LSB algorithm*

LSB algorithm has easy principles, less complex calculation and large embedding rate. It is a typical digital image steganography. The image pixel's lowest bit has a limited influence in vision, so information can be embedded into the image pixel as a bit stream. In order to implement LSB, we can replace pixels' LSB layer with the secret information directly, which is called LSB replacement[10].

2.2 *Improved PVD algorithm based on people's vision system*

Chung-Ming and Wang's group raised an improved PVD algorithm based on people's vision system[8]. By this way, difference of steganography image and the original image's can't be distinguished from people's eyes.

It uses PVD technology to get the two adjacent pixels' difference, which represents the capacity of hiding information. The larger the difference is, the larger the capacity.

2.3 *Fast matrix embedding based on bit control*

In order to reduce the amount of calculation of matrix solving, Wang Chao's group proposed a fast matrix embedding based on Bit-control[8]. The traditional matrix embedding based on random linear codes needs to traverse the solution space of linear equations to find the minimum weight solution. And the solution space grows linearly with the code dimension.

Wang Chao's group proposed the algorithm that controls bit or carrier block to create new carrier, performing matrix embedding in the new carrier. Then, we modify the model so as to increase the number of control bits at an exponential rate increased by adjusting whether to flip the control bit, but searching computational complexity of the minimum weight modified mode is only increased linearly. Thus, it can be faster to achieve high embedding efficiency.

3 LOW MODIFIED RATE OF STEGANOGRAPHY ALGORITHMS

Although LSB algorithm is simple and embedding rate is excellent, it changes the picture a lot, which makes embedding efficiency low. The algorithm proposed in this paper compares the lowest level grouping pixels and hiding information by preprocessing. And then we determine whether the difference is more than half. If it is more than a half, the flag bit is 0, otherwise is 1. When the flag bit is 0, the method of embedding secret information is in accordance with the LSB method. If the flag bit is1, the secret information is inverted and embedded into the carrier. The detail demonstrated as follows:

Step 1: We set pixels' LSB and sub-LSB respectively as LSB_0 LSB_1 .

$$LSB = (LSB_1, LSB_0) \tag{1}$$

Then the experiment divides information to be embedded into some block of lengths of about $l-1$ ($l-1$ is odd). Together with the control bits, each block is written to the data for the total length l of the carrier.

Step 2: If the length of the image carrier is n, it is possible to embed secret information into the low double bits. So the carrier can be divided into $2n/l$ blocks. We can suppose carrier's block NO.i is:

$$a_i^T = (LSB_{i1}, LSB_{i1}, ..., LSB_{il/2}) = (LSB_{11,1}, LSB_{i1,2}, LSB_{i2,1}, LSB_{i2,2}, ..., LSB_{il/2,1}, LSB_{il/2,2})$$

Step 3: This step is to suppose that the secret information of block No.i is $m_i^T = (m_{i,1}, m_{i,2}, ..., m_{i,1-1})$. Before embedding, it can be compared each carrier and the secret information, and statistic the different number d in correspond place. There are two cases as follows:

1 First case, if d is more than half the length of the block, the flag bit, which is the last bit of each block is 1. Then m_i^T is inverted before embedded, and finally getting the embedded information is:

$$a'^T_i = (\overline{m}_{i,1}, \overline{m}_{i,2}, ..., \overline{m}_{i,l-1}) \| 1 \tag{2}$$

2 Second case, if d is less than $1/2$, the flag bit is set to 0. m_i^T is embedded according to the classical LSB method, and finally getting the embedded information is:

$$a'^T_i = (m_{i,1}, m_{i,2}, ..., m_{i,l-1}) \| 0 \tag{3}$$

After merging the two cases, finally getting the embedded information is:

$$a'^T_i = (m_i^T \oplus flag) \| flag \tag{4}$$

The flow chart of algorithm of the improved algorithm is as shown in Fig. 1.

As can be seen from fig.1, in the first block of example, what the secret information and carriers are having two differences. According to the improve algorithm, we can embed information to carrier's front 7 places, and the flag bit is set to 0. In the second block of cases, there are five differences. So after inverting information, it's embedded in carrier's front 7 places, and finally the flag bit is set to 1.

Compared with the original carrier, the number of images' modification is 3 and 2. Obviously, the improved algorithm can reduce the amount of modification of the information carrier.

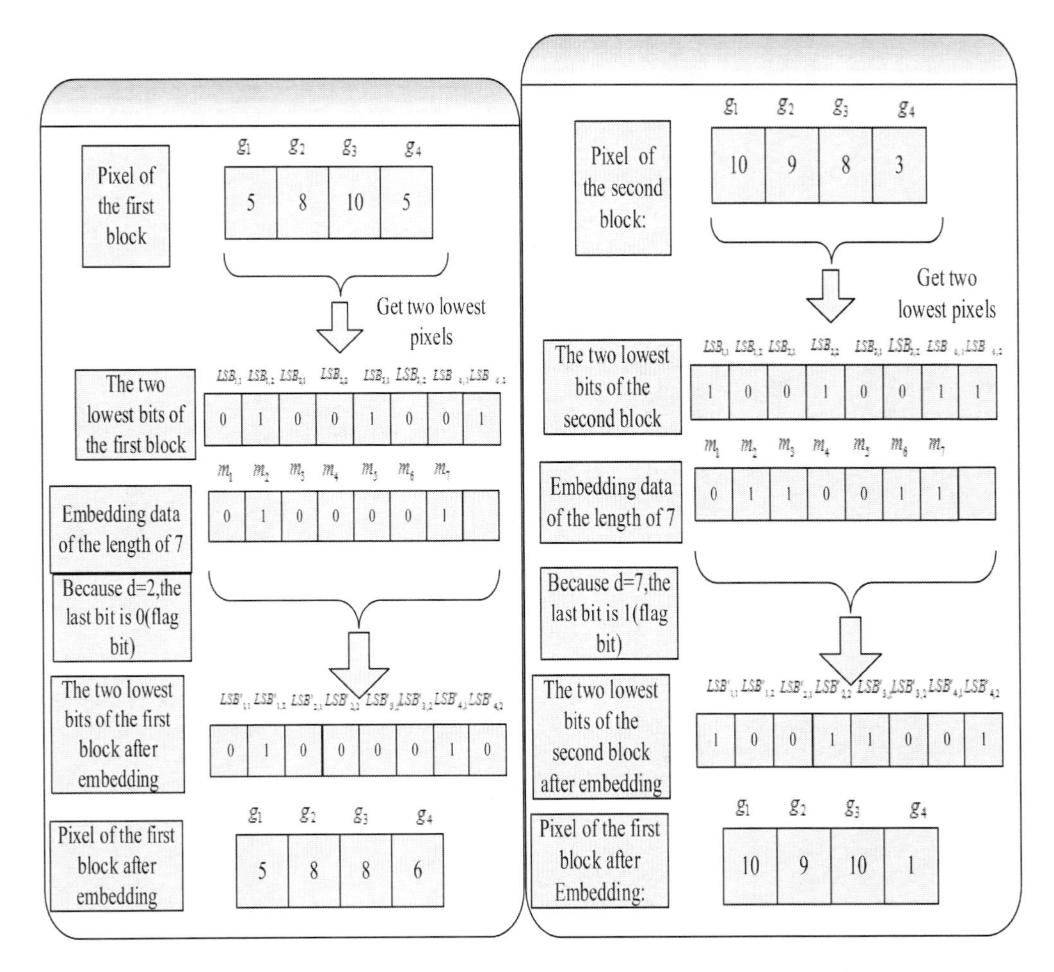

Figure 1. Make an example for steganography with low modification rate. (1) $d < l/2$ (2) $d > l/2$.

4 EXPERIMENT RESULTS AND ANALYSIS

This paper uses 512* 512 gray images as carrier images and test environment was run on a Notebook with Intel Core i3, 2.1GHz CPU and 4GB RAM. Also, the compiler environment is Microsoft Visual C ++ 6.0 under the Windows 7 operating system. Each block's length is 8, and each block contains 7 bits of information and a control bit. Last, we test and get the experimental result of SNR, variation, modification rate and embedding efficiency, comparing with some previously proposed algorithm.

SNR is an important objective indicator in evaluating steganography[11]. The experimental SNR data compared with the literature [6] are shown in Table 1.

Table 1. Comparison of algorithm in literature [6] and improved algorithm with SNR.

Name of image	Lena	Baboon	Peppers
Algorithm in literature [6]	44.1	40.3	43.3
Improved algorithm	45.68	45.69	45.69

Experimental data show that the improved algorithm's SNR is significantly lower than the algorithm proposed in [6]. This shows that the algorithm can greatly improve the image quality, reducing distortion of the image.

The proposed algorithm in this paper performs variation test and the results compare with LSB algorithm. The results are shown in Table 2.

Table 2. Comparison of LSB algorithm and improved algorithm with variation.

The image number	1	2	3	4	5
LSB algorithm	262725	261864	262582	262162	262476
Improved algorithm	190573	190212	190478	190388	190620

Reducing variation to make the hidden information be found possible smaller, and it is of great significance to evaluate an algorithm[9]. As can be seen in Table 2, the variation dealing with improved algorithm is much less than the LSB algorithm.

Table 3. Modification rate and embedding efficiency.

The images number	1	2	3	4	5
modification rate	0.3625	0.3640	0.3644	0.3634	0.3631
Embedding efficiency	2.4136	2.4039	2.4013	2.4077	2.4099

The modification rate and embedding efficiency results are shown in Table 3. Embedding efficiency is the ratio of the modification amount compared with the embedded information amount, which is equivalent to changing the value of one for each bit of information can be embedded approximately 2.4bit.

Next step is to treat with three thousand 512* 512 gray images by the algorithm proposed in this paper, and then we compare embedding speed in this paper's algorithm with it in the literature [8] proposed and the results are shown in Table 4.

Table 4. Comparison of algorithm in literature [8] and improved algorithm with embedding speed.

Algorithm in literature [8]	Improved algorithm
4000kbps	86.56Mbps

It can be seen from the experimental results of embedding speed in this paper is significantly higher than the one in literature [8]. This is due to the relatively simple algorithm proposed in this paper, and is proportional to the amount of embedding data.

5 CONCLUSION

LSB algorithm gets attention by its simple and fast characteristic. This paper proposes a low modification rate steganography algorithm and it can significantly reduce the modification rate in the original algorithm LSB. It also ensures that the algorithm is a linear complexity, which can have fast embedding speed. And when compared to some of the newly proposed algorithm, the proposed algorithm has certain advantages on some indicators. The results also show that the algorithm at the expense of embedding rate, but SNR, modification rate, embedding efficiency and embedding speed have certain advantages compared to some previous algorithms.

ACKNOWLEDGMENT

This work was supported in part by National Science Foundation of China (No.61271399); Zhejiang Science and Technology Innovation Team Fund Project (No.2012R10009-04); Ningbo University Project(XKXL1302); Ningbo Innovation Team Fund Project (No.2011B81002).

REFERENCES

[1] Liang Xiaoping, He Junhui, Li Jianqian. (2004) Steganalysis—principle, actuality and prospect. *Acta Scientiarum Naturalium Universitatis Sunyatseni*, 43(6):93–96.

[2] Li You, Zhang Dinghui. (2010) Information hiding techniques based on steganography. *Information Technology*, 7:119–122.

[3] Huang Qinhua, Ouyang Weiming. (2011) Region selection in LSB embedding of steganography. *Computer Engineering and Applications*, 47(10):151–153.

[4] Huang Haibo, Yang Sen. (2011) Research on image information hiding technology based on LSB. *TECH VOCABULARY*, 3(201):32–34.

[5] Wu, D.C., Tsai, W.H. (2003). A steganographic method for images by pixel-value differencing. *Pattern Recognition Letters*. 24 (6):1613–1626.

[6] Chung-Ming Wang, Nan-I Wu, Chwei-Shyong Tsai, Min-Shiang Hwang. (2008) A high quality steganographic method with pixel-value differencing. *The Journal of Systems and Software*, 81:150–158.

[7] Fridrich J. and Soukal D. (2006) Matrix embedding for large payloads. *IEEE Transactions on Information Security and Forensics*, 1(3):390–394.

[8] Wang Chao, Zhang Weiming, Liu Jiufen. (2011) Fast matrix embedding based on bit-control. *Journal of Electronics & Information Technology*, 33(9):64–69.

[9] Zhang Yan. (2012) *Study on Image Steganalysis Algorithms*. Hangzhou: Hangzhou Dianzi University.

[10] Chen Jiayong. (2012) *Theory and Method for Research on Covert Secrecy Communication Based on Steganography*. PLA Information Engineering University.

[11] Ma Xiuying, Lin Jiajun. (2011) Performance evaluation of information hiding. *Journal of Image and Graphics*, 16(2):209–214.

Electric, Electronic and Control Engineering – Shao, Shu & Tian (Eds)
© 2015 Taylor & Francis Group, London, ISBN: 978-1-138-02842-5

A wireless method for detection of half-wave direct current

Huifen Zhang
School of Electrical Engineering, University of Jinan, Jinan, China

Zhiguang Tian
School of Informatics, Linyi University, Linyi, China

Enping Zhang
School of Electrical Engineering, University of Jinan, Jinan, China

ABSTRACT: Field operation proves that the principle of fault line selection and fault point location based on half-wave current injection method has a high ability of against transition resistance for detecting the single-phase earthed fault in small current grounding system. Magnitude of the half-wave direct current (DC) is the protection criterion. A method is presented for detecting discontinuous 50Hz half-wave DC in a wireless way. According to the principle of electromagnetic induction, when a coil with iron core is placed in the magnetic field generated by the 50 Hz half-wave DC, the induced potential in both ends of the coil is a 100Hz discontinuous periodic signal which will become a continuous 100Hz periodic signal after the analog and digital filtering. Magnitude of the 50Hz half-wave DC can be acquired through the detection of 100Hz periodic signal. The simulation results and the detectors field operation verify the effectiveness of the method.

KEYWORDS: single-phase earthed fault in small current grounding system; line selection and location method with DC injecting; half-wave DC; wireless detecting

1 INTRODUCTION

Medium- and low-voltage distribution networks in China are usually small current-grounding systems. Problems concerning the single-phase grounding fault line selection and location of these systems have not been completely solved on-site for many years. Therefore, traditional sequential switch out and visual line inspection methods are adopted to identify fault line and locate fault point in many cases. This problem has attracted increasing attention from relevant experts and researchers [1,2]. Half-wave direct current (DC) injection is a very effective single-phase grounding fault line selection and location technique [3,4]. This principle is based on the variation of the neutral-to-ground voltage of the distribution network when single-phase grounding fault occurs. Large half-wave DC is injected from the neutral point of the primary side of the system. Fault line selection and location is achieved by detecting the flow path of the injected half-wave DC. Therefore, the injected half-wave DC needs to be detected first.

Most distribution line outlets are equipped with two-phase current transformer (CT), whereas few outlets have zero-sequence CT. Detecting DC by using the existing zero-sequence CT of the distribution network is not feasible. When half-wave DC injection is used to locate the fault, the DC needs to be detected along the fault line, which is especially inconvenient for the CT detection to be applied in overhead lines.

The development of DC supply in various industries has led to unavoidable DC measurement problems for different voltage classes and accuracies. To date, DC measurement methods mainly include shunt, DC transformer, Hall sensor, and DC comparator; these methods play important roles in corresponding measurement and control systems [5–8]. The shunt method needs to be incorporated into detected lines by a series connection. Furthermore, the shunt method cannot be used in the detection of single-phase grounding fault. Although the Hall sensor has a small size, it is easily affected by the nonlinear and temperature characteristics of Hall elements. The DC comparator has higher precision than the other DC measurement methods but has a complex structure that is easily affected by poor working environments.

Several deficiencies and inconveniences exist in measuring the half-wave DC of the distribution network by the existing DC measurement methods. Therefore, this study proposes a simple and practical wireless method for detecting half-wave DC. And the detection method has been authorized by the China invention patent [9]. The proposed method has been used to realize the principle of line selection and location with half-wave DC injection.

2 INTRODUCTION OF THE PRINCIPLE FOR LINE SELECTION AND LOCATION WITH HALF-WAVE DC INJECTION

The principle for line selection and location by using half-wave DC injection [3,4] is shown in Fig.1. The DC generator is composed of a high-voltage silicon stack and current-limiting resistance and is connected to the neutral point of a small current-grounding system. When single-phase grounding fault occurs, the neutral point voltage of the distribution system increases, and the control switch Kd is switched on. The DC generator is connected to the fault system to inject the half-wave DC. The injected half-wave DC only flows in the fault line (Fig. 1, dotted lines). Line selection and location can be realized by tracing the half-wave DC. The DC generator is switched off after line selection and location.

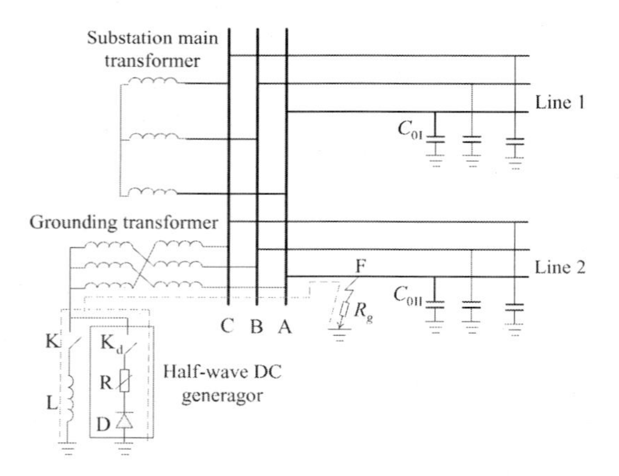

Figure 1. Schematic of the line selection and location principle by using half-wave direct current injection.

The alternating current (AC) grid frequency in China is 50 Hz. When single-phase grounding fault occurs in the small current-grounding system, the neutral point voltage of the grounding transformer is the sinusoidal signal with 50 Hz power frequency. After half-wave rectification by the silicon stack, the 50 Hz sinusoidal signal generates a half-wave DC (also called the pulsating DC) (Fig. 2). The generated DC is injected into the fault system for line selection and location.

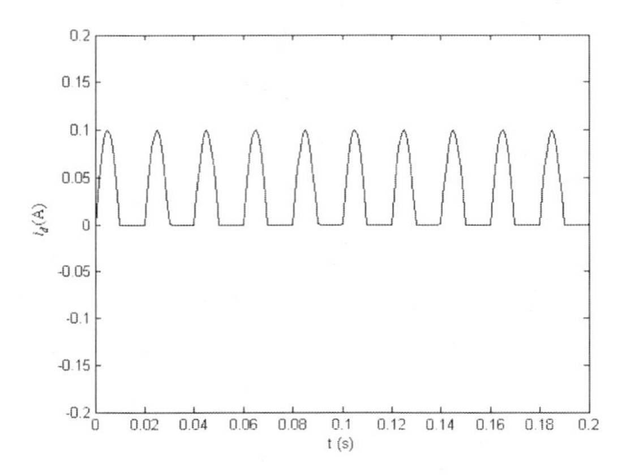

Figure 2. Half-wave DC after silicon stack rectification.

3 DETECTION METHOD OF HALF-WAVE DC

According to the theory of electromagnetic field, the flow of half-wave DC in the fault line produces a magnetic field, which needs to be detected to acquire a half-wave DC. The neutral-to-ground voltage is assumed as the sinusoidal signal.

$$u_0 = \sqrt{2}U_0 \sin \omega t \tag{1}$$

In Eq. (1), the waveform of voltage u_0 is a 50 Hz sinusoidal signal. The pulsating DC current produced by the DC generator is defined as follows:

$$i_d = \sqrt{2}I_d \sin \omega t \qquad (i_d \geq 0) \tag{2}$$

The part of the DC current i_d that flows to the fault feeder is denoted as i_{dc}:

$$i_{dc} = \sqrt{2}I_{dc} \sin \omega t \qquad (i_{dc} \geq 0\ ) \tag{3}$$

where I_{dc} is the effective value of the pulsating DC current in the fault line. In Eqs. (2) and (3), the waveforms of currents i_d and i_{dc} are the half-wave DCs (Fig. 2).

Based on the principle of line selection and location with half-wave DC injection, the half-wave DC needs to be detected at the outlet of each feeder and along the fault feeder. The half-wave DC current flowing through the fault feeder is i_{dc}, which produces a magnetic field. The vertical distance between any point P outside and inside the wire is d (Fig. 3).

Hence, the magnetic induction intensity of point P is expressed as follows:

$$B = \frac{\mu_0 i_{dc}}{2\pi d} \tag{4}$$

The magnetic vector potential of point P is deduced as follows:

$$A = -\frac{\mu_0 i_{dc}}{2\pi} \ln d \tag{5}$$

In the above two Eqs., μ_0 is the magnetic permeability of the vacuum.

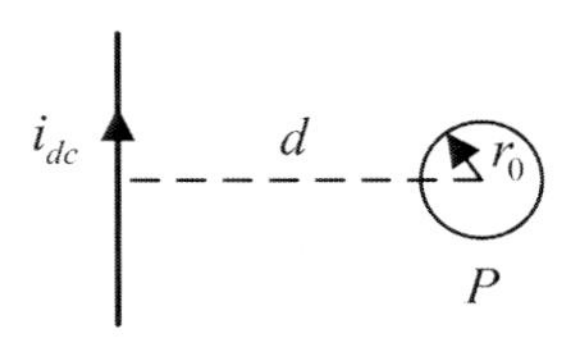

Figure 3. Schematic of the magnetic field calculation of infinitely long straight wire.

The high-permeability material with relative magnetic permeability μ_r is considered as the iron core. The iron core is wounded with N-turn coils as the receiving antenna. The receiving antenna is placed at point P to detect the magnetic fields produced by i_{dc} (Fig. 3). The iron core is supposed to be a column with radius r_0. So, the magnetic flux passing through the iron core of circular cross-section is expressed as follows:

$$\phi = \int_s B\,ds = \int_s \frac{\mu_r i_{dc}}{2\pi d} ds = \mu_r i_{dc}(d - \sqrt{d^2 - r_0^2}) \tag{6}$$

Thus, the induced potential e at both ends of the N-turn coils is denoted by the following:

$$e = -N\frac{d\phi}{dt} = -\sqrt{2}N\mu_r(d - \sqrt{d^2 - r_0^2})I_{dc}\omega\cos\omega t = -\sqrt{2}E\cos\omega t \tag{7}$$

where $E = N\mu_r(d - \sqrt{d^2 - r_0^2})I_{dc}\omega$ is the effective value of the induced potential.

According to Eq. (7), the induced potential e is the cosine signal in the half-wave. This induced potential is also the intermittent and periodic signal of the 100 Hz frequency. The waveform of the induced potential is shown in Fig. 4. By acquiring the signal, the half-wave DC i_{dc} can be obtained.

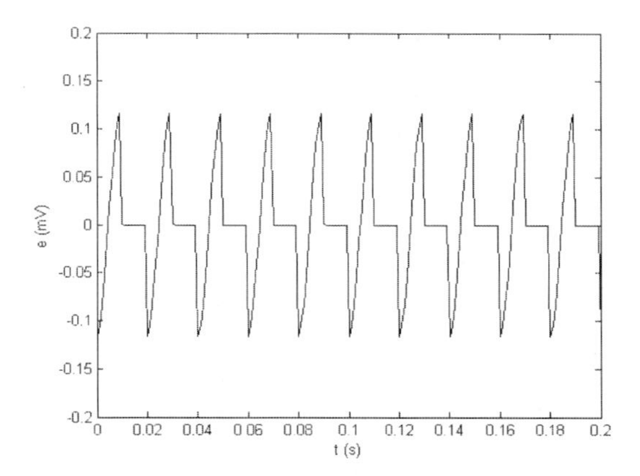

Figure 4. Induced potential at both ends of the receiving antenna.

Suppose that N = 1200, μ_r = 1000, d = 10 m, r_0 = 5 mm, $\omega = 2\pi f$, and f = 50Hz in Eq. (7). The corresponding induced potential E can be determined when I_{dc} changes. The computed results are shown in Table 1. These computed results can be considered the bases in realizing hand-held half-wave DC detectors. Based on the measured distance and precision requirements, the coil turns are adjusted and the suitable iron core material is selected to realize the half-wave DC detectors with different sensitivities.

Table 1. Computed results.

I_{dc}(A)	E(mV)
0.1	0.06
0.2	0.12
0.3	0.18
0.4	0.24
0.5	0.3
0.6	0.36
0.7	0.42
0.8	0.48
0.9	0.54
1.0	0.6
1.5	0.9
2.0	1.2

The half-wave DC can be acquired after the half-wave rectification of the 50 Hz sinusoidal signal. The electromagnetic induction principle is applied to detect the magnetic field produced by the half-wave DC. Finally, the half-wave DC is obtained by detecting the amplitude of the 100 Hz sinusoidal signal.

4 EXPERIMENTAL VERIFICATION

4.1 *Simulation verification*

The 6 kV isolated neutral system of an oil production plant has eight overhead lines with a total length of 98 km but has no overhead ground wires. The overhead lines are LGJ-70 wires that are arranged horizontally and placed 45 cm apart. This system is used as a model in this study. The simulation is carried out by Power Systems Computer Aided Design (PSCAD). The half-wave DC generator is connected to the neutral point of the grounding transformer. At t = 0.1s, fault occurs. The original fault data are obtained from the fault line to verify the proposed detection method of the half-wave DC.

1. The half-wave DC current in a fault line when metallic grounding fault occurs is presented in Fig. 5. The parameters of the receiving antenna in Section 3 are used to obtain the induced potential, which is amplified 200 times (Fig. 6). A 100 Hz digital band-pass filter is used for data filtering (Fig. 7) to obtain the corresponding 100 Hz voltage signal (Fig. 7).
2. By increasing the transition resistance, the amplitude change curves of the 100 Hz AC signal of the fault and non-fault lines are obtained (Fig. 8). The 100Hz AC signal of the fault line is always larger than that of the non-fault line with increasing transition resistance. Therefore, the detection of the 100 Hz AC signal realizes the principle of line selection and location with DC injection.

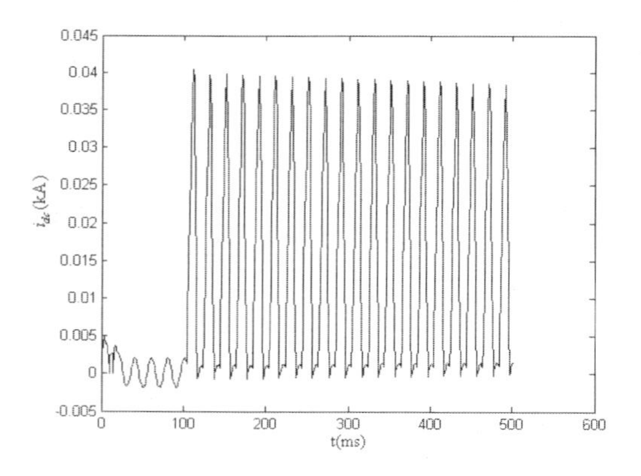

Figure 5. Half-wave DC on the fault line when metallic grounding fault occurs.

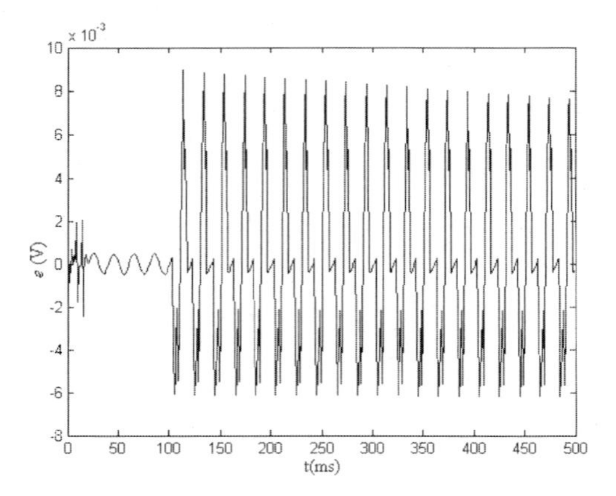

Figure 6. Induced potential amplified 200 times.

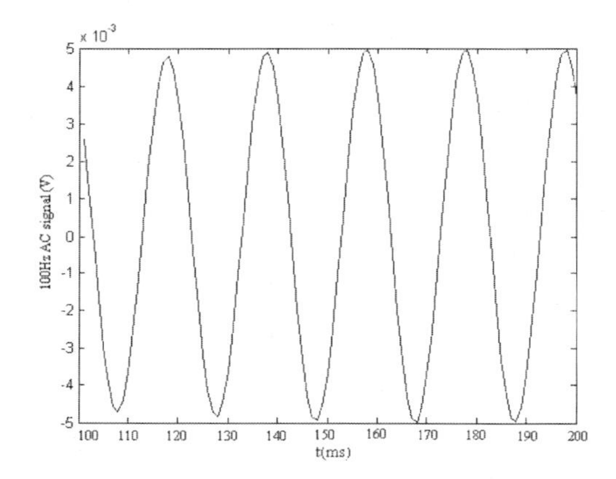

Figure 7. The 100 Hz signal after digital filtering.

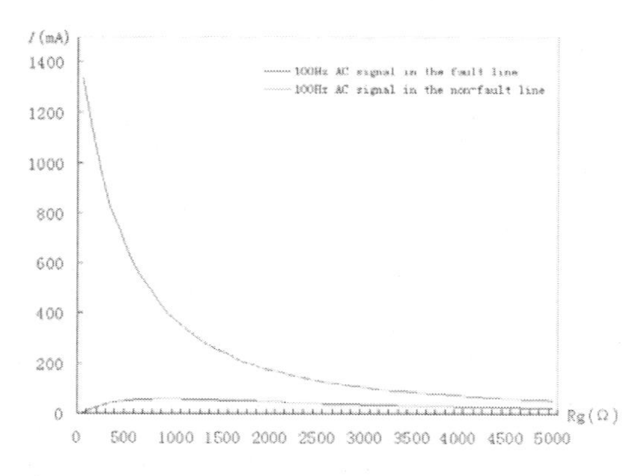

Figure 8. Changes of the 100 Hz signal curve (half-wave DC injection) with transition resistance.

4.2 *Realization and on-site test of half-wave DC detector*

Based on theoretical deduction and application, a wireless half-wave DC detector is developed that consists of a receiving antenna, signal preprocessor, analog-to-digital (A/D) converter, digital signal processor, and display (Fig. 9).

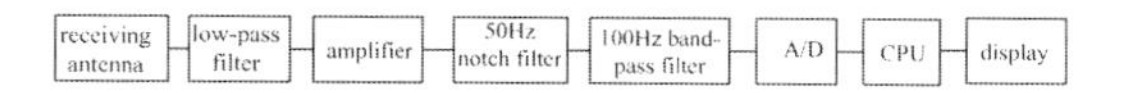

Figure 9. Block diagram of the half-wave DC detector configuration.

The receiving antenna is used to acquire half-wave DC signal. The receiving antenna is composed of a high-permeability iron core that is wounded with coils externally. The 100Hz intermittent and periodic signal can be obtained at both ends of the coils. Through low-pass filtering, amplification, 50 Hz notch, and 100 Hz band-pass filtering, the intermittent and periodic 100 Hz signal is converted to a digital signal by A/D conversion. The collected digital signal is subjected to 100 Hz digital filtering. The amplitude of the 100 Hz signal is calculated by Fourier algorithm. The measured results are presented on liquid crystal display.

According to the above design, three kinds of half-wave DC detectors are developed. Their vertical distances between the receiving antenna and the measured power line are 35cm, 2m and 10m respectively. And they are tested by on-site trial. Their installation conditions are shown in Fig. 10(a, b and c) respectively. Fig. 10(a) shows the wireless line selection detector installed in the relay instrument room of a switch cabinet externally equipped with the receiving antenna. Fig. 10 (b) presents the automatic location detector installed on an overhead line pole. Fig. 10 (c) shows the hand-held location detector internally equipped with the receiving antenna.

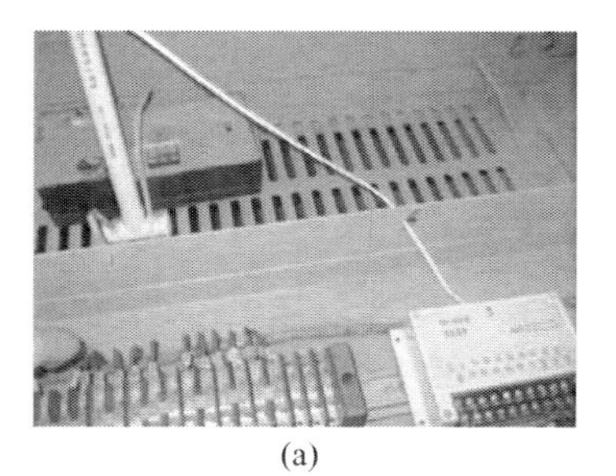

(a)

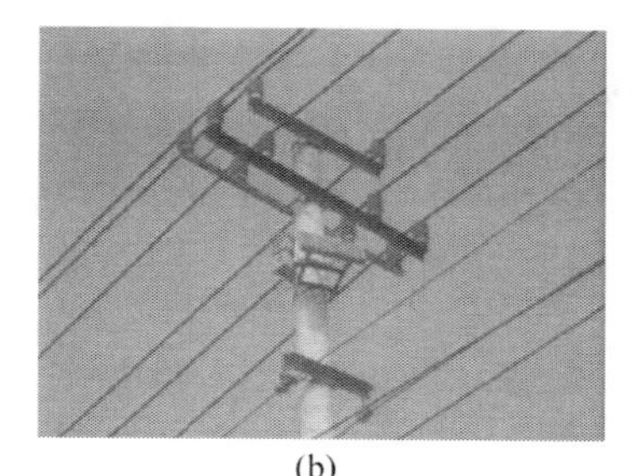

(b)

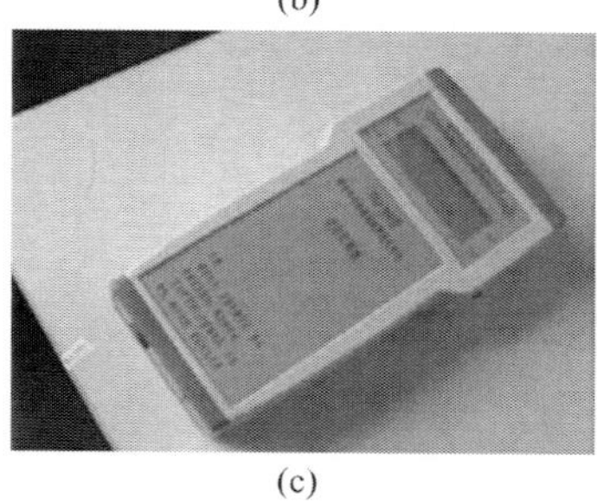

(c)

Figure 10. Photos of the half-wave DC detector
(a) Wireless line selection detector installed in the relay instrument room of a switch cabinet; (b) Automatic location detector installed on overhead line pole; (c) Hand-held location detector

The 6kV isolated neutral system of the oil production plant discussed in Section 4.1 is subjected to artificial high-resistance single-phase grounding test with 3U0 set as 24 V and a transition resistance of 2400Ω. The experimental results are as follows:

1. Among the DC current measured values for each feeder outlet sent by the line selection detector (Fig. 10(a)), the value of the fault line is the greatest among all.
2. The hand-held location detector (Fig. 10 (c)) is subjected to artificial line selection behind a switch cabinet. The read value of the fault line detector is the greatest among all. The hand-held location detector searches for the fault point along the fault line. The reading of the detector on the fault branch is obviously larger than that on the non-fault branch.

5 CONCLUSIONS

The amplitude of the half-wave DC current is the criterion for DC injection line selection and location. This study presents a detailed theoretical analysis and full experimental verification of the proposed wireless detection of half-wave DC. The following conclusions are obtained:

1 Half-wave DC is obtained after the half-wave rectification of the 50 Hz sinusoidal signal. The receiving antenna is used to detect the magnetic fields generated by the half-wave current. Finally, through the detection of the amplitude of the 100 Hz sinusoidal signal, the half-wave DC is acquired.
2 Three half-wave DC detectors with different sensitivities are developed based on the proposed method. These detectors are used in line selection, automatic location, and artificial location and are tested on-site. These detectors are currently in the trial operation stage.
3 The simulation and on-site tests verify the effectiveness of the proposed method. However, this method is vulnerable to external interference. Therefore, future studies should focus on how to resist such interference effectively.

ACKNOWLEDGMENTS

The authors would like to thank the financial support of Natural Science Foundation of Shandong Province (ZR2012EEL19) and the Science and Technology Development Program of Linyi City (201312018).

REFERENCES

[1] Guo Qingtao, Wu Tian, (2010) Survey of the methods to select fault line in neutral point ineffectively grounded power system, *Power System Protection and Control*, 2:146–152.

[2] Ni Guangkui, Bao Hai, Zhang Li, Yang Yihan, (2010) Criterion based on the fault component of zero sequence current for online fault location of single-phase fault in distribution network, *Proceedings of the Chinese Society for Electrical Engineering*, 31:118–122.

[3] Linpan Oil Extraction Plant of Shengli Oilfield Cooperation, University of Jinan, A method of fault line selection and location based on DC injection, China Patent 0014996.6, January 2011.

[4] Gao Zhipeng, Zhang Huifen and Sun Xuna, (2013) A method of fault line selection and fault point location with half-wave DC injection in distribution network, *Power System Protection and Control*, 13:139–145.

[5] Li Qian, (2004) *Study of the Principle of a Novel Sensor for DC Measurement Application*, Huazhong University.

[6] Qian Zheng, Ren Xuerui, Liu Shaoyu, (2008) A DC electronic current transformer based on the giant magneto-resistance effect, *Automation of Electric Power Systems*, 13:71–75.

[7] Jia Xiufang, Li Baoshu, Huang Huifu, (1999) Principle of Holzer magnetic potential balance applied in detecting DC small current, *Journal of North China Electric Power University*, 4:15–18.

[8] Li Weibo, Mao Chengxiong, Lu Jiming, (2005) Principle of a novel DC comparator centered on the structure of saturable reactor, *Automation of Electric Power Systems*, 4:77–81.

[9] Jining Electric Power Corporation, University of Jinan, A method of detecting half wave direct current, China patent 0019617.7, April 2011.

Electric, Electronic and Control Engineering – Shao, Shu & Tian (Eds)
© 2015 Taylor & Francis Group, London, ISBN: 978-1-138-02842-5

Design and implementation of intelligent toolbox based on RFID technology

Minli Wang, Yaning Li, Yongpeng Teng & Jie Tan
Institute of Automation, Chinese Academy of Sciences, Beijing, China

ABSTRACT: Against the problem of traditional artificial style inventory work is hard to meet the needs of the safety management of aviation maintenance tools, we develop an intelligent toolbox system based on RFID and embedded techniques. We carry out specific design and implementation of the hardware and software components of the system and test the performance of the whole RFID intelligent toolbox system. The practical testing results imply that our system can in part solve the reliable label accessing issue in complex electromagnetic environments consisting metal and nonmetal instruments. The intelligent toolbox, which can strengthen the ability of the instrument management and meet the application requirements to some extent in aviation companies, has reached the expectation in this project.

KEYWORDS: RFID; ARM; Embedded Linux; Toolbox.

1 INTRODUCTION

Safety is the eternal theme of aviation enterprises, so maintenance work is an important guarantee for aviation enterprises' safety. With the size of the various aviation enterprises expanding, the type and number of maintenance tools increasing, in the repair work, accidents often occur because of the poor tools management, which causes a serious impact on flight safety. Currently, toolboxes used in aircraft maintenance are generally equipped with a large number and variety of tools, and the maintenance situation is very complex, the traditional inventory methods cannot meet the needs of maintenance tools' safety management, so, there is important practical significance by using modern science and technology to achieve intelligent maintenance tool management.

The technology of radio frequency identification [1] (referred to RFID) is a kind of technology with automatic, accurate and convenient access to get information via using radio frequency signals. RFID technology began to develop in 1940, while with the development of wireless electronic tags, RFID technology adapts to either short-range or long-range identification has been gradually introduced to the aviation industry. Since 2008, each aircraft manufacturer and aviation enterprise around the world has put forward development planning on how to apply RFID technology better, it shows that with the development of RFID technology in the future, its position in the aviation field will be further improved, and its function will be gradually highlighted. While, its application in aviation field is still limited, so it is necessary for the need of aviation industry's development to introduce this technology. This paper focuses on the design and implementation of intelligent toolbox based on RFID technology, the system can realize automatic identification and management of tools in the case, and it can significantly improve the efficiency of the use of tools and the reliability of tools-maintenance, protect repair tools' security and integrity, as well as make the aircraft maintenance toolbox intelligent.

2 HARDWARE DESIGN

We use S3C6410 processor of the SAMSUNG Company as the embedded control module, to the design and implementation of its peripheral circuits. For the Radio frequency identification module, we use ultra-high frequency module SkyeModuleTM M9 [2] of SkyeTek Company. According to the need, we design a multi antenna switching circuit, each antenna switches by the way of time-sharing search to extend the reading range of the system. The design of the overall architecture of this system's hardware platform is shown in Figure 1.

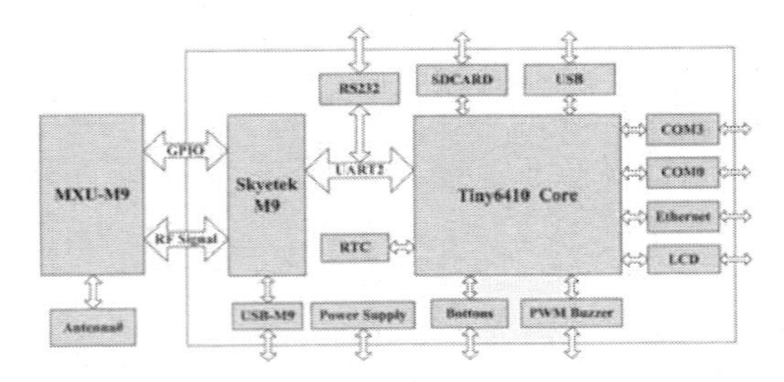

Figure 1. The overall architecture of intelligent toolbox's hardware platform.

According to the function, the system's hardware platform is divided into six parts, including the core control module, radio frequency identification module, test and application interface module, human-computer interaction (HCI) module, power module and multi-antenna switching circuit. In the concrete hardware design, the first five modules are integrated on a circuit board as the main hardware platform system. Among them, the M9 module communicates with the core control module through the UART port, at the same time, converting UART port's TTL level to RS232 level, testing and development of M9 module for PC machine, in order to test and develop M9 module for PC.

Multi antenna switching circuit is individually designed into a piece of circuit board, which communicates with system's main hardware platform via the control signal (GPIO) and the radio frequency signal.

2.1 *The design of system's main hardware platform*

2.1.1 *The core control module*

We choose chip ARM11, S3C6410 processor of the SAMSUNG Company as the core control module. S3C6410 [3] is a 16/32 bit RISC processor based on ARM1176JZF-S core, using 64/32 bit internal bus architecture, the internal integrates rich hardware peripherals, providing a cost-effective, low power and high performance processor application solutions for digital portable equipment. S3C6410 has a high operating frequency and a high electromagnetic compatibility demand for the design of the PCB.

Tiny6410 [4] is an embedded core board with ARM11 chip S3C6410 as the main processor, it has a high density of 6 layer board, integrates of DDR RAM,SLC NAND Flash or MLC NAND Flash memory, realizes various core voltage conversions that CPU required within the board, and leads a variety of common interface resources out.

2.1.2 *The radio frequency identification module*

We use a SkyeModuleTM M9 module of SkyTek Company as the system's RFID module. The M9 module uses a standard CF interface with 50 pins, and the working frequency is 862-955 MHz, which also has USB, UART and SPI host interface.

2.1.3 *The power module*

In this system, the core module, RFID module and LCD etc. all need 5V power supply; the SD card, network chip, TTL power conversion chip etc. use 3.3V power supply. We choose PTH08080 DC voltage conversion module of the TI Company to provide a 5V output voltage, and AMS1086 chip of Austria's Microelectronics Corporation convert the voltage from 5.0V to 3.3V.

2.1.4 *The test and application interface module*

We design SD card, RS232, USB and Ethernet interface for the system, and S3C6410 supports two start-up mode of NAND flash and SD card. The system has 2 SDIO interfaces, while we lead out the SDIO0 port of the core board for ordinary SD boot card, and it provides a great convenience to transplant and development of Linux.

The Tiny6410 core board leads out four serial ports of UART0, 1, 2, 3. Among them, UART1 is five-line-function. We don't use it, and the others are all three serial lines. We set UART0 used as a debug port after RS232 level conversion with UART3, and they are led to COM0, 3. UART2 is connected to the UART of M9 module, to communicate with the host computer. At the same time, after RS232 level conversion, it is led to COM2 as an extension use of serial as the same as COM3.

In order to make the system generality and expansibility, we design USB and net exports. The USB interface is used for connecting the mouse, keyboard, U disk and other peripherals, as the same as ordinary PC machines' USB. NFS service is based on a network interface circuit which must be used in cross compilation. The system adopts DM900AEP network chip, which supports 10/100Mbps bandwidth, and be suitable for products with network function based on ARM and DSP. The network interface circuit belongs to the high frequency circuit. When drawing PCB, you need to consider the problem of electromagnetic compatibility: the RXI +, TXI +pins of DM9000AEP should be as close as possible to 50 ohms, and the connection with RJ45 interface should be parallel, and be as isometric as possible. In addition, the crystal cannot go below the signal line as far as possible, and the protection of the pure crystal oscillator circuit improves the stability of the network port circuit. It is better not to put any signal line below the crystal, in order to protect the purity of crystal oscillator circuit to improve the stability of the network port circuit.

2.1.5 *The HCI module*

The system needs to have a good human-computer interaction interface, so the maintenance personnel check the integrity of tools in the box via the keys and LCD touch screen. If it is abnormal, the alarm buzzer works.

2.2 *The design of multi-antenna switching circuit*

According to the analysis of the toolbox's function and performance, use a single antenna to read cannot meet the needs of the system. We design multi-antenna switching circuit, that is, we expand the single RF interface module from M9 to multi-channel, which can connect a plurality of antennas, and they are respectively arranged at the bottom and the side of the box. As a result, the reading range of the RFID system is expanded. The design principle of the multi antenna switching circuit is shown in Figure 2.

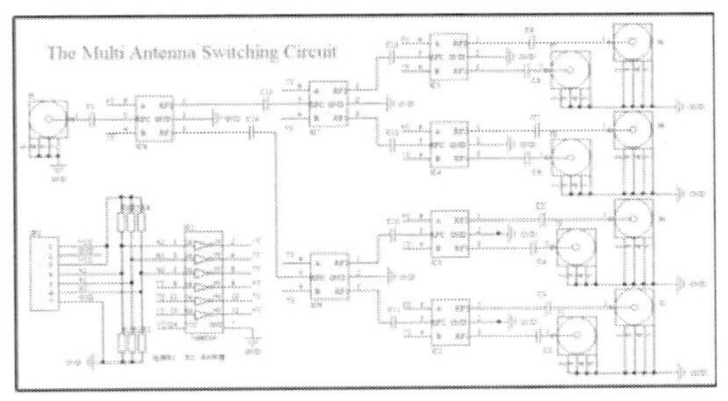

Figure 2. The design principle of the multi antenna switching circuit.

The M9 module's firmware supports multi antenna extension. There are 7 GPIO signals in the CF interface, and they all can be used as the control signal of the antenna switching circuit. In this system, we set GPIO0/1/2 signals as the control signal, while the level of GPIO0/1/2 is determined by the value of the 8 bit register whose address is 0x000A in M9 module. We use inverter 74HC04 of Philips Company and RF switch HMC595E of Microwave Company to realize the multi antenna switching function. GPIO0/1/2 is respectively connected with A0/1/2, via the inverter and three-level RF switch, 8 antenna paths are expanded, which uses the way of time division multiplexing to read tags one by one, that is there is only one antenna working at the same time.

2.3 *The Selection of antenna and tag*

According to the actual application background, the system must have a high reliability for reading and writing that leads to a very high demand for the performance of the antenna and tag. Popular UHF antenna at present includes PCB antenna and ceramic antenna. PCB antenna has the advantages of simple manufacture and low cost, but after testing, this kind of antenna is found that the effective reading range is small, and the effect is not stable, so it cannot meet the needs of the application. After investigation, we choose the 50 ohm UHF ceramic antenna of Jietong Technology Co., Ltd.

UHF tags existed in our lab including self-adhesive passive tags and part of anti-metal tags. The former has no anti metal performance, so it cannot meet the demand. The existing anti metal tags' size is big. After testing, the effective identification distance becomes narrow within 2cm, so it cannot meet the demand well, neither.

3 SYSTEM SOFTWARE DESIGN

3.1 *Communication API design of the RFID module*

In the system, M9 module needs to communicate with the core control module. In the physical layer, the UATR of M9 module is directly connected to the UART2 of the core control module. In the data link layer, the core control module sends the binary data stream formulated by the underlying protocol [5] to M9 module in request frame format. M9 receives the data and phases the data, and then execute related operations according to specific instructions. After the execution, M9 module returns the data stream formulated by underlying protocol to the core control module in response frame format. The core control module receives the data and phases the data, and then informs the user through the graphical user interface.

3.1.1 *Communication protocol of M9 module and its realization*

If we want to execute a write command, firstly, we need to define a request structure, and fill in the objects in the structure corresponding to request framed format according to the requirements of specific instructions. Then, call the Build_request() function. This function completes the construction of request frame based on the type of command and the flags in the request structure and saves the request frame in the meg byte array. Next, call the Write_request() function. This function sends the meg byte stream (the request frame) in the request structure to M9 module. M9 module executes related operations, according to command contents, and return data to the core control module. Lastly, call the Read_response() function. This function saves the response frame to the meg array of the response structure, and phases out related data formulated by protocols according to the answer frame structure and saves to the related members in the answer structure.

The three functions mentioned above are the bases of underlying communication API of M9 module. The following will focus on the design and realization of these three functions.

3.1.2 *Communication API design of M9 module*

1 Build_request() function
STATUS Build_request (*PRO_REQUEST req);

The function receives a pointer of PRO_REQUEST type. Members in the req* structure corresponding with request frame have been registered. The function completes the construction of functional domains according to reg->flags and reg->cmd, and save them to the related location in reg->msg array, then call crc16 check function to do cyclic redundancy checks for request frame. The check results save to a related location in the reg->msg array to complete the build of request frame.

2 Write_request() function
STATUS Write_Request (const char *pathname, *PRO_REQUEST req);

The function is to send the request frame in reg* structure to the serial device that "pathname" represents for.

3 Read_response() function
STATUS Read_Response (const char *pathname,*PRO_REQUEST req, *PRO _ RESPONSE resp, int timeout);

The function is to receive the returned frame from M9 module and save it to the meg array in PRO_RESPONSE structure, and extract functional domains from it.

The definition of the above three functions and related data structures are saved in the file M9_ Protocol.c. They are declared in M9_protocol.h. We can design the communication API interface function of M9 module based on them, such as the Get_SystemParameter function, firstly, load functional domains like the Command for reading system parameters to request structure and call the Write_request function to send request frames to M9 module. Then, call the Read_request function to receive the frame returned from M9 module and extract related information of system parameters. Different parameters represent different functions, so all kinds of functional subprograms can be realized based on this function. Communication API interface functions, such as Set_SystemParameter, Get_DefultSystemParameter, are implemented in the file M9_protocol.c .

3.2 *Application program design*

3.2.1 *Functional subprogram design of M9 module*

Use multiple antenna converters to expend multiple antennas. Antennas switch in polling mode. Only one antenna reads tags at the same time. We design the subprogram Polling_GetTags to complete the function. The specific process of the Polling_GetTags subroutine as shown in Figure 3:

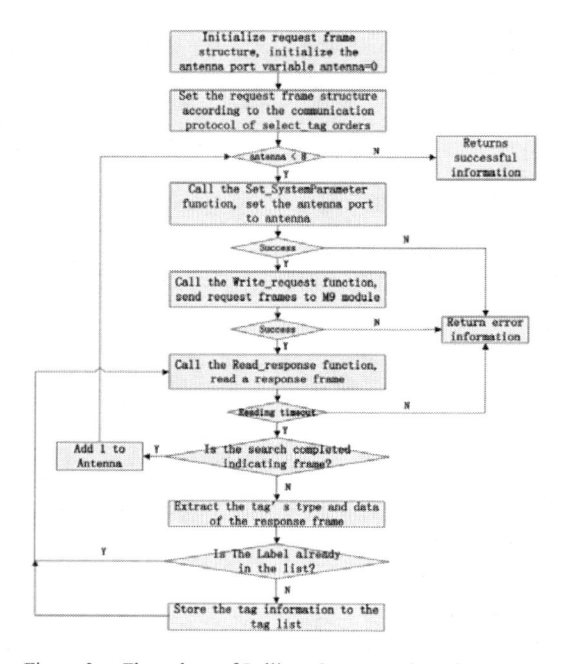

Figure 3. Flow chart of Polling_GetTags subroutine.

3.2.2 *Graphical user interface program design*

We use Qt creator to build the interface and Qt designer integrated in the Qt creator to design the interface. After the interface is designed, we design slot function for the click action of all function buttons. The system can call function subprograms to complete related function in the function, and show the results in each display component.

We design three functional buttons in the interface. After clicking the search button, the core control module sends the SELECT_TAG command to M9 module. M9 module search the tags in the box according to the configuration of each option in the interface, and display the quantity, consumed time, EPC data of the tags have been found and part of the debugging information in the interface. Click the clear button, the information in the display components will be clear. Click the exit button, the graphical interface will be closed. Among them, the design of the slot function of the search button's click signal is the key to realize the system function. The process is shown in Figure 4:

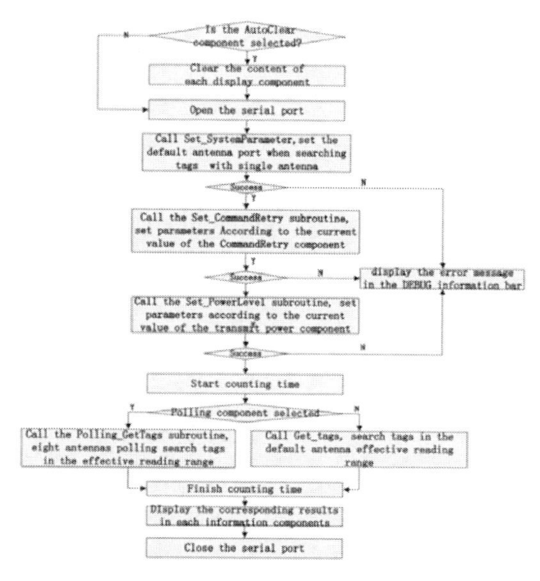

Figure 4. Processes of slot function of the search button's click signal.

The smart toolbox's user interface as shown in Fig. 5

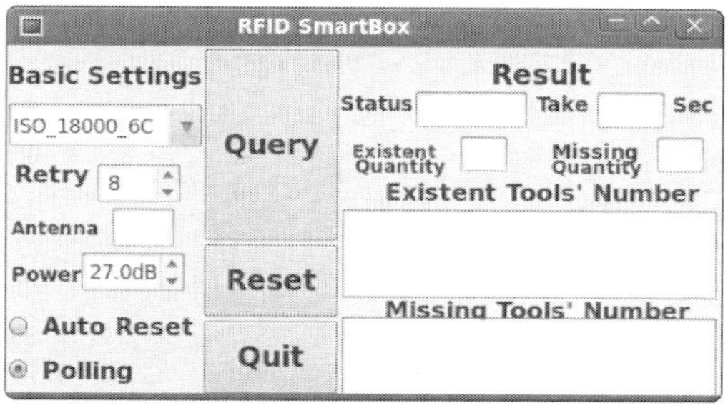

Figure 5. The smart toolbox's user interface.

In order to simulate the actual application, we select 15 representative repair tools, which are numbered from 1 to 15. When the "Search" button is clicked, the system searches tools in the box according to the functional configuration of the interface's left side. When the searching is completed, the status information and time-consuming of the searching, the quantity and number of both the tools in the box and the missing tools will be displayed in the query results area. If there is a lack of tools, the alarm buzzer triggered, prompting the staff to check and replenish the tools. Clicking the "Reset" button to clear results in the display area, clicking "Cancel" to quit.

4 THE EXPERIMENTAL RESULTS

Burn the smart toolbox's user interface program to the flash, and set it to boot in the Linux system, after that, install the main hardware platform of the system, the multi antenna switching circuit, the ultra-high frequency ceramic antenna, the battery, the touch screen and the switch etc.inn the box according to the expected design, composed of RFID intelligent toolkit prototype (as shown in Fig. 6).

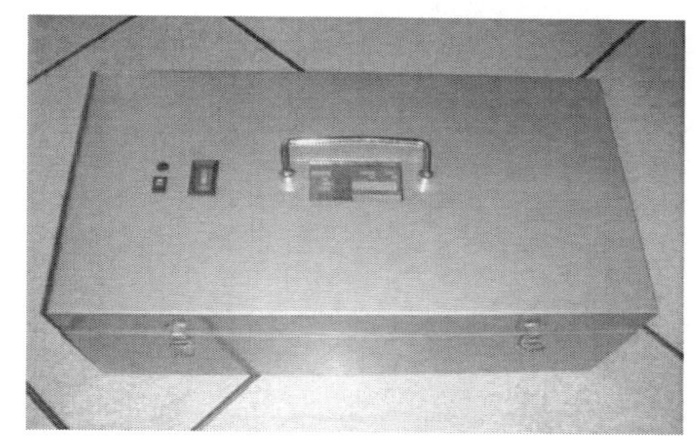

Figure 6. The RFID smart toolbox prototype.

Choose 15 representative repair tools, and then attach electronic tags to the tools in an appropriate manner. After most of the tools have been placed correctly, the attached the labels are on the effective identification antenna direction and an area at the bottom of the box, a few of tools' tags can only be attached to the round face at the bottom of the handle, such as the screwdriver, the flashlight, etc., after placed correctly, the attached labels are in the effectively identify the direction and the region of the internal antennas in the box's side.

Close the lid after finished the arrangement, check the number of tools in the box via an RFID intelligent system of the toolbox. Open the system switch and enter the user operation interface, set the label type as ISO-18000-6C, the command retry as 8 times. When the transmitting power is 27.0dBm and the searching mode is polling in the function setting zone on the left side of the interface, click the search button after the setup has been finished, the system return the result data in about 2 seconds, the results displayed in the touch screen (as shown in Fig.7), we can see all the 15 tools are in the box, testing success.

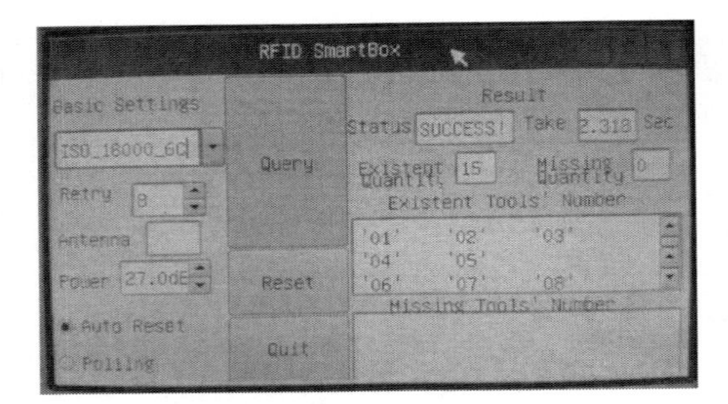

Figure 7. The user interface program found all the tools.

Remove the No.3 and 13 tools from the box, after closing the lid. Click the reset button to clear the last query results, and then click the search button. The system returns the result data in about 2 seconds, the buzzer beeps twice, prompting a lack of tools, the touch screen tool number bar and the deletion tool number bar, respectively, shows the tools and the existing and the missing tool numbers, testing success.

5 CONCLUSION

The practical application and testing results of the RFID smart toolbox prototype show that the system solves expected technical difficulties in a certain extent, achieves each basic function, the operation is stable and reliable, basically achieve the design goals. To a certain extent, the work of this thesis has a certain positive significance in implementing safety, accuracy and real-time automatic identification and management of the repair tools in the toolbox, improving the safety of aircraft maintenance work and protecting the integrity of the repair tools. In addition, the method and the related key technology proposed in this paper are universal, for example, but it has a certain application prospect in valuables intelligent identification management, firearms management and other aspects.

ACKNOWLEDGMENT

This work was financially supported by the National Natural Science Foundation of China (U1201251).

This work was supported by the National Science and Technology Support Program (2014BAF07B04).

REFERENCES

[1] Quanlin Li, Longyan Guo. (2006) Summary of RFID technology and its applications. RFID Technology and Applications, 1:51–62.
[2] SkyeTek, Inc. SkyeModuleTM M9 Reference Guide Version. 2007.
[3] Samsung Electronics, Inc.S3C6410X_USER'S MANUAL_REV 1.10. 2008.
[4] Guangzhou Friendly Arm Computer Technology Co.Ltd.Tiny6410 Hardware Instruction Manual. 2012.
[5] SkyeTek, Inc. SkyeTek Protocol v3 Reference Guide Version. 2008.

Electric, Electronic and Control Engineering – Shao, Shu & Tian (Eds)
© 2015 Taylor & Francis Group, London, ISBN: 978-1-138-02842-5

Design and implementation of the control system in the underwater hydraulic grab

Hui Zhang
School of Naval Architecture and Ocean Engineering, Zhejiang Ocean University, Zhoushan Zhejiang, China

ABSTRACT: In order to simplify the structure to make underwater hydraulic grab bucket better underwater capabilities, a method, namely to design the control system used in the underwater hydraulic grab, was presented in this paper. By analyzing the characteristics of the various types of underwater hydraulic grab bucket, based on the principles of green design, a kind of underwater hydraulic control system is designed. As the single rope hydraulic grab bucket is taken as an example, the overall design is put. The key components of the control system are designed and verified. The design result shows the effectiveness of the application of control system in the underwater hydraulic grab.

KEYWORDS: underwater grab; hydraulic grab bucket; control system; single hydraulic grab

1 INTRODUCTION

In recent years, with increasingly serious water pollution, environmental protection has become increasingly important [1,2,3]. The quality requirements of national inland waterway dredging are more and more stringent and construction companies are also pursuing higher productivity. Thus, underwater hydraulic grab will be toward the low energy consumption, high efficiency, high precision and high frequency response, pollution and other aspects of development. It provides strong protection for future offshore development [3,4,5]. But most of the existing underwater hydraulic grab are complicate, prone to shortcomings such as unfit for ship use, limited the scope of marine transportation, cargo transportation prolonged the duration of increased transportation costs. There is an urgent need to study a simple underwater hydraulic grab with high reliability, energy saving and environmental protection.

2 ANALYSIS OF THE UNDERWATER HYDRAULIC GRAB

2.1 *Types of underwater hydraulic grab bucket*

Underwater hydraulic grab buckets are mainly used for underwater dredge with the crane supporting [7,8]. According to complementary cranes lifting form, it can be divided into the four ropes, the double rope machinery and the hydraulic transmission. Depending on the underwater work conditions is divided into:

1. The double grab for dredging or seizing pebble. A hydraulic grab is described in [9]. It includes the hydraulic grab bucket and the hydraulic tank, weight ratio is 2–4, a cutting angle is 10°–15°. Because of the weight capacity of the grab bucket is quite high, high strength and wear-resistant, it can be used for blasting clearing operations that the flow rate of the water is less than 3M/s.
2. The heavy duty grab for clay gravel and soil mining. A hydraulic grab is introduced in [10]. It includes the grab bucket and hydraulic tank, weight ratio is 1–2, cutting angle of 10°–12°. The grab has a high filling factor. Coupled with advantages hydraulic grab itself, it can greatly improve the efficiency of mining underwater construction. Operating costs can be saved. It applies to dredging silt, clay, clay, loose sand and small grain size of crushed stone.
3. The large stones grab used for fishing and giant fishing grab.

2.2 *Control system of the hydraulic grab bucket with water*

The hydraulic grab bucket is usually connected by a hydraulic cylinder and a jaw, pillars, a lug, earmuffs, bucket teeth and tooth parts. When the hydraulic grab bucket works, motor clockwise turns. The hydraulic oil can be transmitted to pressure P1 through the check valve of the hydraulic pump from the tank. The hydraulic oil flows to a hydraulic cylinder piston side KS by a check valve, relief valve and pipes. When the hydraulic cylinder extends, Oil will return to the tank through pilot operated check valve and the oil return pipe and filter T. Correctly adjusted pilot controlled check valve prevents grab "quiver". Main relief valve unloads under low pressure and ensure that the cylinder pressure of the piston end is always set value, thereby preventing abnormal grab open and also prevent heat up of oil to too quickly. The pressure can be measured through the upper pipe compression joints, pressure testing hose and pressure gauge. When motor counterclockwise rotates, the hydraulic oil is transmitted to pressure P2 by the hydraulic pump from the fuel tank and the check valve. The hydraulic oil is sent to both sides KSS of the hydraulic cylinder piston rod by one-way valves and piping B. When recycling of hydraulic cylinder, oil flows back to the fuel tank by the return pipe A, the pilot operated check valve, the relief valve and filter. At the same time, opened working pressure can be tested through the lower pipeline fittings, connected pressure testing hose and pressure gauge. As shown in Figure 1.

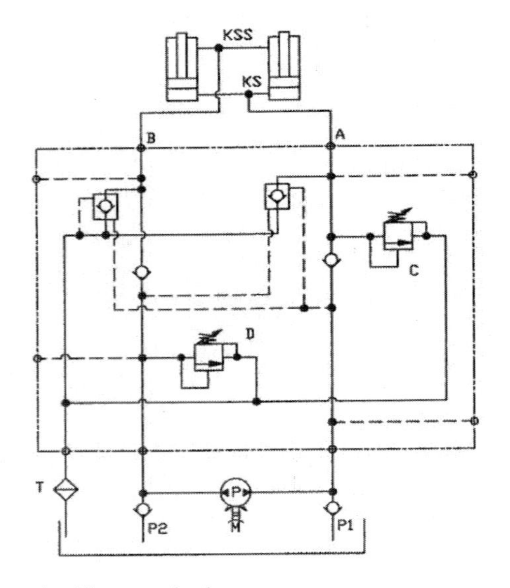

Figure 1. The control valve group.

Because of a simple structure and high reliability of the single rope hydraulic grab bucket, the control system of the single cable hydraulic grab is designed in this paper.

3 THE DESIGN OF CONTROL SYSTEM OF UNDERWATER HYDRAULIC GRAB

3.1 *The overall design of single hydraulic grab*

A single rope hydraulic grab consists of the upper beams, hydraulic cylinders, Pistons, and the grab and other components (as shown in Figure 2). There is a hook above the upper beam, and connected to the crane. The hydraulic grab bucket is sent to grab material underwater. Then the hydraulic cylinder 2 is turned. That the piston 3 upwards. So the gear 5 turns and finally the grab 4 is closed. The hydraulic grab bucket is risen by the crane. The hydraulic cylinder reversal and the piston move downward, so that the gear turns, the grab opens and the material releases. Through the above operations, underwater items are transported on board or on land.

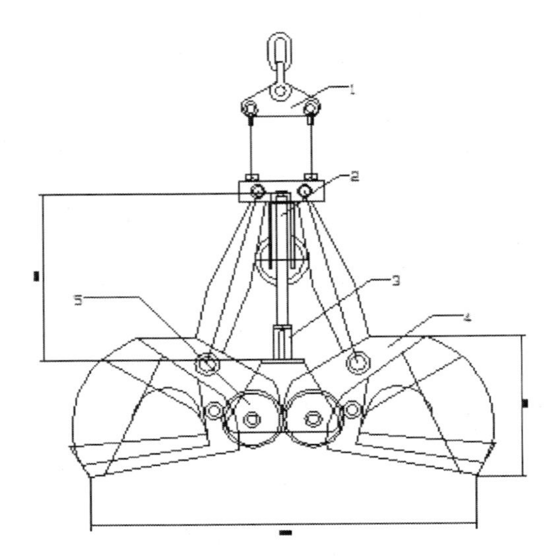

1-the upper beam, 2-the hydraulic cylinder, 3-the piston, 4-the grab, 5-the gear
Figure 2. The single hydraulic grab.

3.2 *Design of the key component of control system*

The hydraulic grab bucket is controlled by the hydraulic cylinder, and then the hydraulic cylinder is the main control element of the hydraulic grab bucket, its structure is shown in Figure 3.

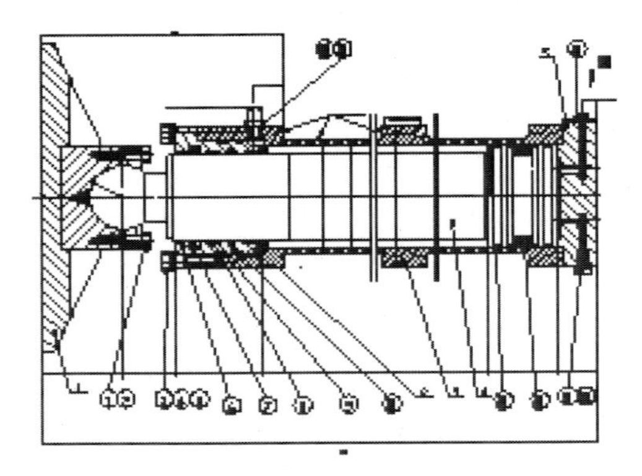

Figure 3. The hydraulic cylinder.

3.2.1 *Selection of the hydraulic cylinder*
There are two forms of the hydraulic cylinder:

- Single-acting hydraulic cylinder. The one of the directions of movement can be realized by the hydraulic pressure and returned by the external force such as gravity or a spring. The one is with only oil and the other end is in contact with air in the two cavities of fuel tanks. It has low cost, simple structure and convenient control but unsafe. The immediate opening and closing control cannot be achieved.
- Double acting hydraulic cylinders. It is a kind of hydraulic cylinder transmitted the pressure oil from both sides of the piston. Both sides of the fuel tank have pressure oil both working stroke and the return trip. It has a high safety coefficient and the immediate opening and closing control can be realized, grab practicality is enhanced. But it has a higher cost and complex structure.

Considered by the designing of a kind of underwater hydraulic grab bucket, single-acting hydraulic cylinder is the end of oil, and exposed to the air at one end. A double-acting hydraulic cylinder is chosen by the safety issues.

3.2.2 *Cylinder calculations*
1 The working pressure in the cylinder
In the case of maximum lifting moment, a single fuel tank maximum thrust:

$$P_{max}=2400KN \tag{1}$$

Total extension length of the cylinder: $L_{max}=7505mm$
The total shrinkage at length of the cylinder: $L_{min}=4165mm$
The cylinder stroke: $L=3340mm$
Cylinder diameter: $D=320mm$
Cylinder diameter: $D_1=356mm$
Piston rod diameter: $d_1=213mm$
Piston rod diameter: $d=250mm$
Guide sleeve length: $L_D=200mm$
Piston length: $L_H=120mm$
The maximum working pressure of the cylinder:

$$p=\frac{4P_{max}}{3.14D^2}=\frac{4\times 245000}{3.14\times 32^2}=304.6kg/cm^2=30.5MPa$$

2 Cylinder checking
Piston rod of moment of inertia:

$$J_1=\frac{3.14(d^4-d_1^4)}{64}=\frac{3.14(25^4-21.3^4)}{64}=9070.88cm^4$$

Piston rod cross-sectional area:

$$A=\frac{3.14(d^2-d_1^2)}{4}=\frac{3.14(25^2-21.3^2)}{4}=134.5cm^2$$

Moment of inertia of the cylinder:

$$J_2=\frac{3.14(D_1^4-D^4)}{64}=\frac{3.14(35.6^4-32^4)}{64}=27372.4cm^4$$

Cross section area of steel tube:

$$A_2=\frac{3.14(D_1^2-D^2)}{4}=\frac{3.14(35.6^2-32^2)}{4}=191.1cm^2$$

Slenderness ratio:

$$\frac{l}{k}=\frac{l}{\frac{1}{4}\sqrt{d^2+d_1^2}}=\frac{750.5}{0.25\sqrt{25^2+21.3^2}}=91.4$$

$$m\sqrt{n}=85\sqrt{1}=85$$

According to Euler's formula, the critical capacity of the hydraulic cylinder:

$$P_K=\frac{n3.14^2EJ}{l^2} \tag{2}$$

N-coefficient of terminal conditions, n=1
Condition (1): cylinder extension, P=160000kg,

$$P_K=\frac{3.14^2nEJ}{l^2}=\frac{3.14^2\times 2.1\times 10^6(25^4-21.3^4)}{750.5^2}=333785.2kg$$

Safety factor:

$$n_K=\frac{P_K}{P_{max}}=\frac{333785.2}{160000}=2.1$$

The fuel tank meets Stability requirements.
Conditions (2): cylinder when not extending fully, when $L=700.5cm$, $P_{max}=245000kg$

$$P_K=\frac{3.14^2nEJ}{l^2}=\frac{3.14^2\times 2.1\times 10^6(25^4-21.3^4)\frac{3.14}{64}}{700.5^2}=383135.2kg$$

Safety factor:

$$n_K=\frac{P_K}{P_{max}}=\frac{383135.2}{245000}=1.56$$

The fuel tank meets Stability requirements.

3 Checking of the piston rod strength

Initial deflection of hydraulic calculation:

$$\delta_0 = \frac{(\Delta_1 + \Delta_2) l_1 l_2}{2al} + \frac{G l_1 l_2}{2pl} Cos\alpha \quad (3)$$

In the formula:

Δ_1- clearance of the piston rod and cylinder, Δ_1=0.0922cm

Δ_2- clearance of the piston and cylinder, Δ_2=0.2289cm

l_1- the middle distance between piston pin hole in the head to guide, l_1=373.5cm

l_2 - the distance between low tail-cylinder pin hole to midpoint of the guiding sleeve, l_2=377cm

l - when the piston rod fully is stretched, the distance between the cylinder pin hole on both ends, l=750.5cm

a - when the piston rod extended, the distance between guide bushings slide from end to end of the sliding surfaces of the piston, a=47cm

G - the weight of the fuel tank, G=1250kg

α - the angle of the hydraulic cylinder with water level, $\alpha = 89°$

P - the maximum thrust of the fuel tank, P=160000kg

$$\delta_0 = \frac{(\Delta_1 + \Delta_2) l_1 l_2}{2al} + \frac{G l_1 l_2}{2pl} Cos\alpha = 0.653\text{cm}$$

Calculation of the maximum deflection of the hydraulic cylinder

According to$J_2 \leq 5J_1, l_1 \neq l_2$, the formula for calculating of maximum deflection value:

$$\delta_{max} = \frac{\delta_0 l}{\left(\frac{K_1}{t_1} + \frac{K_2}{l_2}\right) l_1 l_2} \quad (4)$$

In the formula:

$$K_1 = \sqrt{\frac{P_{max}}{E_1 J_1}} = \sqrt{\frac{160000}{2.1\times10^6 \times 9070.88}} = 2.9\times10^{-3}$$

$$K_2 = \sqrt{\frac{P_{max}}{E_1 J_2}} = \sqrt{\frac{160000}{2.1\times10^6 \times 27372.4}} = 1.67\times10^{-3}$$

$$t_1 = tg(57.3K_1 l_1) = tg(57.3\times2.9\times10^{-3}\times373.5) = 1.89$$

$$t_1 = tg(57.3K_2 l_2) = tg(57.3\times1.67\times10^{-3}\times377) = 0.73$$

$$\frac{K_1}{t_1} = \frac{2.9\times10^{-3}}{1.89} = 1.53\times10^{-3}$$

$$\frac{K_2}{t_2} = \frac{1.67\times10^{-3}}{0.73} = 2.29\times10^{-3}$$

$$\delta_{max} = \frac{\delta_0 l}{\left(\frac{K_1}{t_1} + \frac{K_2}{l_2}\right) l_1 l_2} = 1.21\times10^{-3} cm$$

Strength check of piston rod.

When the piston rod is loaded by the axial compression load and the hydraulic cylinder gravity, the synthesis of stress on maximum deflection ($X=l_1$) section is:

In the formula:

P- maximum thrust, P=122500kg

δ- maximum deflection value of the cylinder, δ=0.324cm

A- the section area of the piston rod, A=87.2cm

ω- the piston rod section modulus

$$\omega = \frac{3.14(D^4 - d^4)}{32D} \frac{3.14(20^4 - 17^4)}{32\times20} = 375.4$$

$$\sigma = \frac{P}{A} + \frac{P\delta_{max}}{\omega} = \frac{122500}{87.2} + \frac{122500\times0.324}{375.4} = 1510 kg/cm^2$$

Piston rod made of 45# steel, $\sigma_s = 3000 kg/cm^2$

Safety coefficient:

$$n = \frac{\sigma_s}{\sigma} = \frac{3000}{1510} = 1.98$$

Meet the requirements.

4 Calculation of the cylinder wall thickness

Cylinder material is 27SiMn, $\sigma_b = 10000 kg/cm^2$, Safety coefficient n=4, stress is:

$$\sigma = \frac{\sigma_b}{n} = \frac{10000}{4} = 2500 kg/cm^2 = 250 MPa$$

The maximum working pressure of the cylinder: $P = 216.5 kg/cm^2 = 21.65 MPa$

$$\delta \geq \frac{PD}{2[\delta]} = \frac{21.65\times30}{2\times250} = 1.299 cm$$

The actual wall thickness of the fuel tanks:

$$\delta = 1.5cm$$

$$\delta > [\delta]$$

Meet the requirements.

4 CONCLUSIONS

The underwater hydraulic grab is a widely used tool. With social development and progress of science and technology, the design requirements are getting higher and higher to the grab. By analyzing various types of underwater hydraulic grab enough, based on design principles of the single cable hydraulic grab bucket, a type of underwater hydraulic control system is designed. With single rope hydraulic grab as an example, the overall design is put. The key components of the control system are designed and checked. The result shows that the design is feasible. Next step work will be to improve the product's performance, make it simple structure, energy saving and environmental protection, better operational capabilities under the water.

REFERENCES

[1] Yuan Yongsheng, Cao Shouying. (2003) Green design technology in mechanical product development. *Machine Tool and Hydraulic Pressure*, 5(1):43–44,137.

[2] Liu Zhifeng, Liu Guangfu. (1999) *Green Design*. Beijing: Mechanical industry publishing.

[3] Li Yan. (2011) Green product design based on sustainable development. *Journal of Art and Design*, pp:219–220.

[4] Chen Huixin. (2006) *Submarine Project*. Beijing: China Electric Power Press.

[5] Wang Jingxiang. (2010) *Dredging Resistance Theory Research and Development of DSG-800 Hydraulic Grab Bucket*. Beijing: China University of Geosciences.

[6] Zhang Chunhua, Wu Feng. (2003) Improved design of large flap hydraulic grab bucket. *Construction Machinery*, pp:64–65.

[7] Wang Yueming, Tong Minhui, Zhang Yuqin, Tong Ruiguo, Gaoxiang. (2011) Deep water salvage 500t grab. *Hoisting and Conveying Machinery*, (8):3–6.

[8] Zhou Zaiyou, Xu Benju, Zhong Shibing. (2008) Applies to jet stream waters of underwater hydraulic grab bucket [P]. CN200810069408.9.

[9] Wang Feng, Zhou Zaiyou, Xu Benju, Zhong Shibing. (2008) Underwater blasting clearing with clam shell hydraulic grab bucket [P]. CN200820097430.

[10] Wang Feng, Zhou Zaiyou, Xu Benju, Zhong Shibing. (2008) Clam shell hydraulic grab bucket for dredging under water [P]. CN200810060318.

Electric, Electronic and Control Engineering – Shao, Shu & Tian (Eds)
© 2015 Taylor & Francis Group, London, ISBN: 978-1-138-02842-5

Analysis on the pier influenced by sea dike causeway bridge

Tao Yue
Sino-steel Ma'anshan Institute of Mining Research Co., Ltd, Ma'anshan, China
State Key Laboratory of Safety and Health for Metal Mine, Ma'anshan, China
Huawei National Engineering Research Center Co., Ltd. China

Dianjun Zuo
Tianjin Research Institute for Water Transport Engineering, Tianjin, China

Quansheng Mao & Jin Zhang
Sino-steel Ma'anshan Institute of Mining Research Co., Ltd, Ma'anshan, China
State Key Laboratory of Safety and Health for Metal Mine, Ma'anshan, China
Huawei National Engineering Research Center Co., Ltd. China

ABSTRACT: According to Zhejiang, a coastal land reclamation projects, using theoretical calculations and numerical simulation method to wear a causeway bridge foundation pile load additional stress caused by the impact on the pier. The results showed that: a causeway pier pile heap load on a certain impact, but the cause of displacement is smaller piers, negligible; vehicle loads imposed a greater impact on the piers, piles negative skin friction drag, thrust and lateral pier displacement value by about 1.4 to 4.8 times the numerical simulation values larger than the theoretical value, the error in the 10% to 15% within the control of construction vehicles, the impact load on the pier is crucial.

KEYWORDS: pile, negative skin friction; lateral pressure; simplified analysis, numerical simulation.

1 INTRODUCTION

With the rapid development of China's marine economy and the large-scale of coastal land reclamation works, the cross of coffer and other traffic buildings is inevitable. Therefore, during the construction and operation, the foundation effect (such as pier pile) of coffer loading on traffic building is crucial for the smooth implementation of the project. The traditional approach is to make the additional stress generated from the coffer loading transfer to deeper side based on the foundation treatment of coffer (such as the use of pile foundation treatment, both sides of the coffer are isolated by sheet pile), but for the coffer with no high loading height, this practice is of high cost and low construction speed.

The effect of coffer loading on bridge pile foundation is mainly manifested in two aspects: one is that the coffer loading causes negative frictional resistance of pier pile; the other is that the coffer loading causes the additional stress in the foundation level to transfer to the pier pile, for the performance of lateral thrust. For the negative friction resistance of building foundation pile caused by surface loading, domestic and foreign scholars have been relatively mature, China's *Construction Specification for Pile Foundation* (JGJ94-2008) [1] has detailed provisions, but for the lateral thrust of pier foundation pile caused by coffer loading, domestic scholars' study is relatively less, the relevant specification has not given a clear answer. In recent years, some scholars calculated the lateral thrust of pier caused by coffer loading[3] through stress solution[2] of the half space elastic body Boussinesq [2] and numerical simulation, but for the pier group pile foundation, calculation model is complex.

In this paper, taking the land reclamation coffer of the eastern part of Haiyan, Jiaxing City across the Hangzhou Bay Bridge as an example, on the basis of previous studies, according to the theory of passive pile bearing capacity calculation method and numerical simulation, study the effect of coffer loading on pier foundation pile. The results have certain reference significance for design and construction of related engineering.

2 PROJECT OVERVIEW

The eastern reclamation project of Haiyan, Jiaxing City, is located in Haiyan coast, the north of Hangzhou Bay, project construction range from Coastal East Road, Wuyuan Town, Haiyan to Hangzhou Bay Beach in Zhengjiadi, the beach part about 10.1091km long, north-south depth of about 0.5~1.1km. The beach area is bounded on the west by the provincial Yuling Stone Pond, Chengbei Road, Wuyuan Town, Haiyan County, current beach height of 1.39 ~ 2.60M, on the east by third-phase dam, Zhapu Port, Jiaxing Port (see Fig. 1). Assume the coffer length crossing the Hangzhou Bay Bridge as 570m, by gravity cast stone embankment, levee crest height of 3.50M, between the 44# and 45# bridge pier, the single span length of pier of 50m. Pier cap adopts a circular pile cap, diameter of 14 meters, thickness of 4 meters. Pier adopts pile column layout, the piers on both sides of dam use 7 Φ2.0m cast-in place piles of single bridge main pier, the piles are evenly arranged according to the circle of diameter 10.4 meters.

Figure 1. Locations of Jiaxing Port and Hangzhou Bay Bridge.

The engineering geological conditions are as follows: 1. 0.8~1.6m: sandy powder soil, yellowish gray, slightly dense, thick layered, medium compressibility; 2. 1.6~13.6m: the silty clay, grey, plastic flow, thick layered to scaly, high compressibility; 3. 13.6~18.6m: clayey silt, gray, slightly dense, thick layered, medium compressibility; 4. 18.6~26.6m: silty clay, grey, plastic flow state, scaly, high compressibility; 5. the 26.60 ~ 31.60m: silty clay, grey to greenish gray, soft plastic state, thick layered, medium high compressibility; 6. 31.60 ~ 34.60m: silty clay, grey, soft plastic to nearly soft plastic, thick layered mainly, medium high compressibility; 7. 34.60 ~ 40.60: silty clay soil, grey, soft plastic to nearly soft plastic, thick layered, medium high compressibility; 8. clay, grey, soft plastic to nearly soft plastic, thick layered, medium high compressibility. The physical and mechanical properties of soil are shown in table 1.

3 THE SIMPLIFIED CALCULATION MODEL OF THEORY

3.1 *Negative friction resistance calculation model of foundation piles*

According to the *Technical Specification for Building Pile Foundation* (JGJ94-2008), the negative friction is the downward friction resistance caused by the soil around the pile due to self-weight consolidation, collapsibility, effect of ground load which is greater than that of settlement of foundation pile on the pile surface. The key point of calculation of negative friction resistance is to determine the depth of the neutral point of foundation pile when the pile settlement is equal to the settlement of soil around the pile. The bearing stratum of bridge pier pile foundation of Hangzhou Bay Cross Sea Bridge is located on the hard plastic clay layer. The operation of the bridge has been a long time, the settlement of pile foundation is mainly the after work settlement which is relatively small, therefore, assume that

Table 1. Dike foundation of basic physical and mechanical properties of soil.

Soils	$\omega/\%$	$\gamma/kN/m^3$	e_0	$\omega_L/\%$	$\omega_P/\%$	C/kPa	$\varphi/^0$	$E_{S0.1\sim0.2}/MPa$
Sandy powder soil	26.7	19.6	0.745	28.4	22.3	10.3	31.7	9.96
Silty clay	37.8	18.3	1.075	40.8	22	21.4	11.8	3.2
Clayey silt	30.1	19	0.851	29.4	21.8	9	28.2	8.41
Silty clay	37	18.2	1.049	33.3	20.4	15	13.2	3.55
Silty clay	25.6	19.7	0.734	27.7	16.7	29	16.5	4.71
Silty clay	33.1	18.8	0.933	36.2	21	29.6	16	5.01
Silty clay	32	18.9	0.905	36.2	21.2	30	19.1	6.38
Clay	34.8	18.7	0.982	40.2	22	29.4	15.5	5.28

the depth of the neutral point of foundation pile is equal to the settlement calculation depth of the soil around the pile when calculating the negative friction resistance.

This paper mainly calculates the negative friction resistance of pile foundation of the bridge pier caused by the loading of bridge section of the dam (hereinafter referred to as the dam), the settlement calculation depth of the soil around the pile can take the calculation depth of dam foundation deformation. According to the *Code for Design of Building Foundations* (GB50007-2011) [4], in the presence of adjacent influence, the calculation depth of foundation deformation shall consider stress ratio (additional foundation stress and gravity stress ratio) and deformation ratio. The former has been studied for many years with considerable experience, but this method does not take into account the structures and properties of soil, paying too much emphasis on the effect of load on the compression layer while paying the insufficient attention to the important factors of the foundation size; the latter calculation of independent foundation and strip foundation, the value is too large, but for the width of foundation B=10~50m, the value is close to the measured data. Therefore, considering the dam with width of 14m, length of 140m, and the ratio of length to width of 10, the dam can be simplified for strip load, the calculation depth of dam foundation deformation adopts stress ratio (namely $\sigma_z/\sigma_c \leq 0.1$), i.e. the depth of neutral point of the negative friction resistance of pile foundation of the bridge pier caused by the dam loading can be determined by the ratio of additional stress of dam loading on the pier foundation pile and the soil gravity.

3.2 *Lateral thrust*

The dam adopts sloping type riprap embankment. In the process of the embankment construction, foundation soil will be lateral sliding, which will produce lateral thrust on the pile foundation of bridge pier, the pier pile group foundation will generate the horizontal displacement in the lateral thrust. According to *Technical Specification for Building Pile Foundation* (JGJ94-2008), the pier of Hangzhou Bay Cross Sea Bridge belongs to the structure sensitive to displacement, horizontal displacement control value of 6mm.

For the lateral thrust of ground loading on pile group foundation, the *Technical Specification for Building Pile Foundation* (JGJ94-2008) have not given the explicit calculation method. There are two ways to calculate the lateral thrust of ground loading on the pile group foundation: one is to calculate the horizontal stress of foundation pile based on the elastic half space Boussinesq stress solution; the other one is to calculate the lateral load of the foundation pile according to the sliding force caused by the sliding of foundation soil of ground loading. The former can calculate the lateral stress of the foundation additional stress caused by the ground loading on any foundation pile of group piles foundation, but it does not consider the effects of group piles and pile cap system stress transfer, the results are relatively large; the latter considers the group piles foundation as a whole, adopts passive pile stress calculation model, assume the distribution mode of the lateral stress of soil around pile, calculate the lateral load of group piles foundation, then calculate the foundation pile load sharing according to the behavior of pile under lateral load. According to the test results of theoretical analysis, numerical simulation, the geotechnical centrifuge model test[5–7] by domestic and foreign scholars, there are four main modes of ultimate soil pressure distribution of the adjacent loading in the soft soil area on pile foundation, as shown in Figure 2. The calculation of this paper uses the third distribution mode.

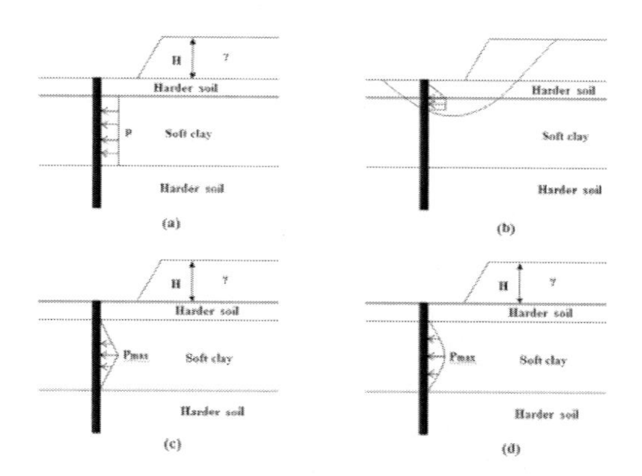

Figure 2. Diagram of heaped generated near the limit on the pile of earth pressure distribution pattern diagram.

4 NUMERICAL SIMULATION

4.1 *Calculation model*

This paper adopts ABAQUS large limited software[8] to carry out calculation analysis of the effect of penetration bridge coffer on the bridge piers. In order to make the numerical analysis process simple and easy to convergence and not to cause great influence on the calculation result, make the following assumption of the numerical simulation in this chapter:

1 The same soil layer is homogeneous and isotropic;
2 The influence of pile side soil seepage and consolidation tin he loading process is not taken into account when analyzing.

In the calculation model, take the horizontal calculation range of 30 times of the pile diameter,

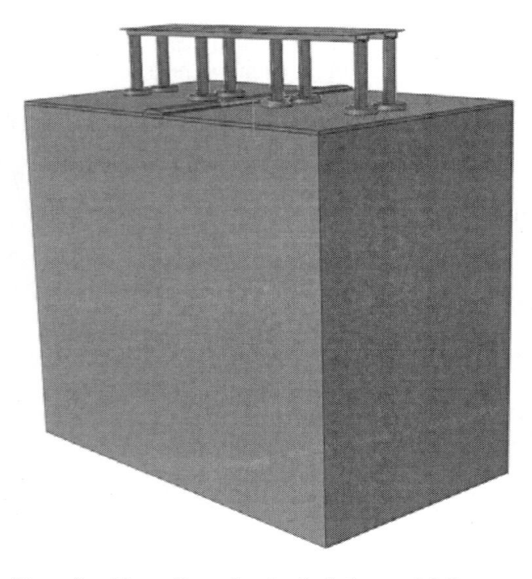

Figure 3. Three-dimensional calculation model diagram.

Figure 4. Dam on the r pile soil he calculation model meshing.

vertical calculation range of 2 times of pile length, build overall modeling of penetration bridge dam loading, bridge, pier and group pile foundation. According to the characteristics of the program and the general finite element theory, bridge, pier, group pile foundation adopt 3D ten-node nonconforming unit (C3D10), while dam, foundation soil adopt 3D ten node coordinating unit (C3D10I). Add specific meshing at the part of loading to be applied at the pile top, the model calculation range and meshing are shown in Figure 3~4.

4.2 *Calculation parameters and constitutive model*

For the bridge, pier, group pile foundation, the unit weight and elastic modulus and Poisson are relatively fixed, which can be obtained by test data; for soil, use modified Cambridge model, and the material parameters of soil can be conversed through the plasticity index, the selection of the material parameters of the finite element calculation model as shown in table 2.

The parameters of modified Cambridge model include Poisson ν, initial void ratio e_0, compression curve slope λ, rebound curve slope κ, the slope of critical state line M, constant β, K, among which K is in the range of [0.778,1], $\beta < 1$, ν , $e_0$can be determined by conventional method, other parameters are determined by formula (1)~(3).

$$M = \frac{6\sin\phi'}{1-\sin\phi'} \tag{1}$$

$$\lambda = \frac{C_c}{\ln 10} \approx \frac{C_c}{2.303} \tag{2}$$

$$\kappa = \frac{C_s}{\ln 10} \approx \frac{C_s}{2.303} \tag{3}$$

Table 2. Soil properties in K32+840 section.

Soil layer	Constitutive model	Model parameters					
		ν	λ	κ	M	Constant β	Constant K
①	cam-clay	0.40	0.04	0.01	36.96	0.10	0.778
②	cam-clay	0.35	0.11	0.02	31.49	0.10	0.778
③	cam-clay	0.25	0.01	0.01	41.02	0.10	0.778
④	cam-clay	0.35	0.14	0.03	28.83	0.10	0.778
⑤	cam-clay	0.30	0.02	0.00	39.06	0.10	0.778
⑥	cam-clay	0.30	0.12	0.02	30.37	0.10	0.778
⑦	cam-clay	0.30	0.09	0.01	33.17	0.10	0.778
⑧	cam-clay	0.30	0.11	0.02	31.49	0.10	0.778

5 THE CONTRASTIVE ANALYSIS OF THE CALCULATED RESULTS

Table 3 is the calculation results table for negative friction resistance and lateral thrust of foundation pile. Seen from the table and figure, the negative friction resistance value of 44 # and 45 # pier foundation pile caused by coffer loading is 14.3 kPa, the maximum value of negative friction resistance along the depth distribution is 6.7 kPa, the maximum value of lateral thrust is 10.4 kN, so that the vertical displacement increment of piers is 1.3 mm, the maximum horizontal displacement increment is 1.0 mm. Vehicle construction load has a certain influence on the foundation pile of bridge pier, when the vehicle construction load is 30 kPa, neutral point calculation depth of foundation pile is 6.3 m, which is 1.5 times than that not considering the vehicle construction load, the value of negative friction resistance increases accordingly to 19.5 kPa, the maximum value along depth distribution of 8.3 kPa, which is 1.4 times than that not considering vehicle construction load, the vertical displacement increment of piers reaches 2.0 mm, horizontal displacement increment value of pier reaches 1.2 mm; When the vehicle construction load is 40 kPa, neutral point calculation depth of foundation pile is 11.0 m, which is 2.6 times than that not considering the vehicle construction load, the value of negative friction resistance increases accordingly to 32.0 kPa, the maximum value along depth distribution of 9.5 kPa, the vertical displacement increment of piers reaches 3.0 mm, horizontal displacement increment value of pier reaches 1.5 mm; When the vehicle construction load is 60 kPa, neutral point calculation depth of foundation pile is 18.0 m, which is 4.5 times than that not considering the vehicle construction load, the value of negative friction resistance increases accordingly to 60.1 kPa, the maximum value along depth distribution of 16.8 kPa, the vertical displacement increment of piers reaches 4.0 mm, horizontal displacement increment value of pier reaches 2.2 mm;

Numerical results are consistent with the theoretical calculation results, and the calculated value is larger than the theoretical calculation results. The calculated value of negative friction resistance is about 10% ~ 15% larger, The calculated value of lateral thrust is about 8% ~ 12% larger.

6 CONCLUSIONS

1 The results of theoretical calculation and numerical simulation show that the effect of penetration bridge riprap coffer loading on the pier is limited, the negative friction of foundation pile is 16.3kPa, and the maximum negative friction resistance is only 8.5kPa which has little impact on the bridge pile foundation. The theoretical calculation and numerical analysis show that the maximum settlement values of bridge pier pile top are 1.3mm and 3.2mm respectively which are very small, this also shows that the negative friction resistance caused by dam loading on the bridge pier piles is negligible.
2 The degree of influence of dam load on the 44# and 45# pier is very small, and the maximum lateral thrust is only 21.4kN, so the impact on the bridge pile foundation is little. The theoretical calculation and numerical analysis show that the horizontal displacement values of bridge pier pile top are 1.0mm and 1.4 mm respectively which are very small, this also shows that the lateral thrust caused by dam loading on the bridge pier piles is negligible.
3 By elastic theory calculation, on the basis of dam loading, in consideration of vehicle construction load respectively by 30kPa, 40kPa and 60kPa, the effect scope of vehicle construction load and dam load on pier foundation pile increases, so that the 43# and 46# pier will be affected, but the influence degree is small, which can be neglected, the effect on 44# and 45# pier increases.

Table 3. Soil properties in K32+840 section.

Analysis method	Pile		Pier	
	Negative friction resistance /kPa	Lateral thrust /kPa	Vertical displacement /mm	Horizontal displacement /mm
Theoretical	6.7	10.4	1.3	1.0
	8.3	14.3	2.0	1.2
	9.5	28.8	3.0	1.5
	16.8	48.3	4.0	2.2
Numerical	8.5	21.4	3.2	1.4
	9.3	30.7	3.5	1.8
	10.1	35.2	4.0	2.2
	21.6	41.3	5.8	4.6

REFERENCES

[1] Technical code for building pile foundation [S]. Beijing: China Architecture and Building Press, 2008.

[2] Poulos H G, Davis E H. (1973) Elastic Solutions for Soil and Rock Mechanics. New York: John Wiley & Sons.

[3] Bitang Zhu, Xinglong Wu, Min Yang, *et al.* (2007) Effect of piled embankments on adjacent piles for bridge abutments. *Chinese Journal of Geotechnical Engineering*, 19(7):1009–1017.

[4] Technical code for building pile foundation[S]. Beijing: China Architecture and Building Press, 2011.

[5] DeBeer, 1977, Piles Subjected to Static Lateral Loads. Proc9 ICSMFE, SpSess. No10.

[6] Tschebotarioff, G.P. (1973) *Foundations, Retaining, and Earth Structures*. New York: McGraw-Hill.

[7] Spring man, S. M (1989) *Lateral Loading of Piles due to Simulated Embankment Construction*. London: University of Cambridge.

[8] Xiangrong Zhu, Jinchang Wang. (2004) Introduction to partly soil models in ABAQUS Software and their application to the geotechnical engineering. *Rock and Soil Mechanics*, 24(s):145–148.

Electric, Electronic and Control Engineering – Shao, Shu & Tian (Eds)
 ISBN: 978-1-138-02842-5

Study on influence of new-built sea dike on submarine optical fibre cable

Yongjun Luo & Deyi Tang
Zhejiang Guangchuan Engineering Consulting Co., Ltd., Hangzhou Zhejiang, China

ABSTRACT: Based on ANSYS FEM software, a new-built sea dike across the existing submarine optical fibre cable was studied. The results showed that: in a reasonable arrangement of the construction process conditions, dike construction has little effect on the fiber optic cable; combination in extremely adverse conditions, when completed cable suffered a causeway maximum axial force of 5.9t, the maximum tensile strain 0.44%, were within the normal operating range of the cable; considering the sensitivity of the foundation soil parameters, the elastic modulus of the foundation soil 10% reduction in the calculation, the results show a causeway suffered when completed cable maximum axial force is 6.86t , the maximum tensile strain 0.51%, still less than the allowable value cable work. Calculation result of the analysis has a certain reference value for the coastal causeway across the submarine cable and pipeline protection design.

KEYWORDS: new-built sea dike; submarine optical fibre cable; numerical simulation; foundation deformation

1 INTRODUCTION

At present, the communication traffic volume of the submarine optical cable takes up about 90% of the international communication traffic volume, and it has become the main force of the contemporary transcontinental communication [1–4]. In order to ensure the smooth operation of the information and meet various needs of people for information, it is rather important to keep the submarine optical cable away from damage. Researches show that: 95% of submarine optical cable damages are caused by human fishery and shipping activities [5–7]. In view of this, measures to protect submarine optical cable have been put forward at home and abroad, such as, setting up closure of fishing areas, strictly controlling the numbers of the fishery workboat, setting up international laws and regulations for protecting submarine optical cable and the like [8, 9]. Studies on stress property of the submarine optical cable can provide references for preventing the damage of the seawall optical cable, and various scholars have carried out numerical simulation and theoretical analysis of the seawall optical cable under external stress. Such as, Xue-Jun Zhou and others [10–11] have reasonably simulated the deformation process of the submarine optical cable under stress on the basis of flattening experiment and finite element method for the submarine optical cable. And the simulation process can be used to study the damage factors and damage mechanism of the submarine optical cable. Lu Y G and others [12] have carried out parametric modeling on the lateral pressure resistance and hook resistance models of the submarine optical cable via the finite element method, the studies of the Zeng Shaolei and other [13] have shown that there is a long submarine optical cable being damaged once the hook tension of the submarine optical cable exceeds the damage limit of the submarine optical cable.

In recent years, with the increasing of the offshore engineering construction in our country, such as the coastal highway[14], seawall engineering[15], reclamation engineering[16], artificial hydraulic reclamation [17] and the like, the ratio that the offshore engineering construction crosses the built submarine optical cable has greatly increased. At present, there is less report on the influences of the offshore engineering construction on seawall and optical cable at home and abroad. Dong Xiuchun and others [12] have reported the influences of the ocean routing survey on seawall optical cable, and Miller [19] has introduced the working conditions that the submarine optical cable causes damage to the Milwaukee port. However, there is no similar engineering report on the influences of the seawall or reclamation project construction on the built seawall optical cable at home and abroad, and in the *Design Code of Seawall Engineering* (SL 435-2008) [20] , the engineering measures demanded for the seawall or reclamation engineering construction across the submarine optical cable are still not mentioned.

In view of the above, based on the construction state of a certain newly-built seawall while crossing the submarine optical cable, this paper studies the mechanical properties of optical cable and analyzes the stress-strain relationship of the optical cable via test methods. Three-dimensional numerical model is adopted to simulate the construction process of the seawall, research the foundation deformation caused by seawall construction and influences on optical cable, and to calculate and analyze the deformation and tension distribution of the optical cable, which is capable of providing certain references for similar engineering construction and optical cable protection.

2 PROJECT OVERVIEW AND GEOLOGICAL CONDITIONS

2.1 *Project overview*

A certain reclamation project is located in northeast Ningbo and southbank marine outfall of the Hangzhou Bay. The project is border on the Hui Beiyang ocean to the east, close to the Xin Hongkou reclamation engineering to the south, near to Ni Luoshan mountain reclamation engineering to the west and arrives the Cixi boundary to the north. Figure 1 is the design profile of the east reclamation seawall, and there is seabed communication optical cable in the reclamation region, and the buried depth is about 4.8m.

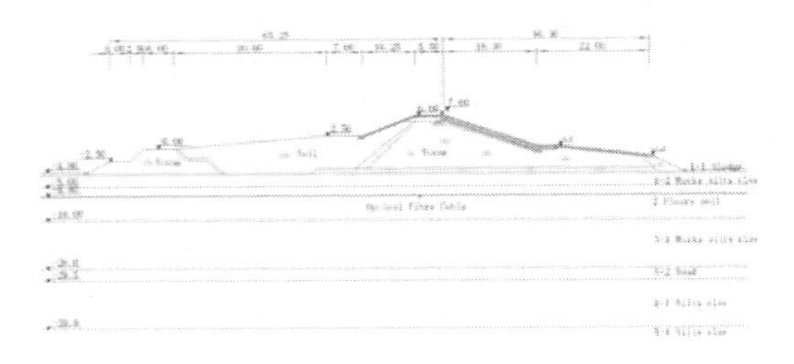

Figure 1. Soil profile of east sea dike.

2.2 *Engineering geological condition*

The geological soil layer of the project mainly includes: (1) 1–1 sludge: the sludge on the surface of the tidal marsh, physical and mechanical property is very poor; (2) 1–2 mucky silty clay: flowing model, slight bedding, it is the newly-deposited mollisol with high water content, low strength and high compressibility, physical and mechanical property is very poor; (3) 2 floury soil: loose to slightly dense, the partial contains muddy silty clay soft rock strata, relatively uniform soil texture, and physical and mechanical property is better; (4) 3–1 mucky silty clay: it is the mollisol with high water content, low strength and high compressibility, physical and mechanical property is poor; (5) 3–2 fine sand: slightly dense to intermediately dense, the soil thickness is uneven; (6) 4–1 silty clay: plastic to hard plastic state, soil texture is more uniform, and physical and mechanical property is better; (7) 5–1 silty clay: soft drifting plastic model, the physical and mechanical property is poorer. The physico-mechanical index for each soil layer of foundation is summed up in the Table 1.

3 TENSILE TEST AND RESULT ANALYSIS OF THE SUBMARINE OPTICAL CABLE

3.1 *Test overview*

The researched optical cable is the GYTA33 military communication optical cable, and the buried time of the optical cable has been 31 years so far. In order to study the mechanical properties of seabed communication optical cable, determine the mechanical parameters of the optical cable and to research the influences of the cofferdam construction on the optical cable, this paper carries out the tensile test to the optical cable through a universal testing machine, as shown in the figure 2. The sample length is 1m, 6 test-pieces in total. The 6 test-pieces were tested under different final tension, and then the final tension is respectively as follows: 10t, 20t, 30t and 40t. The test instrument is a 2000 KN universal testing machine. The loading rate of the test is 2 KN/s. The data acquisition frequency is 10 times/s.

Table 1. Physical and mechanical indices of in-situ soils.

Number	Name	H(m)	w (%)	ρ(g/cm³)	e	k(cm/s)	C(kPa)	φ(°)	E (MPa)
1–1	Sludge	0.6	/	1.80	/	/	0.6	12.0	1.76
1–2	Silty clay	2.8	42.4	1.69	1.174	5.5×10-6	13.5	11.1	2.48
2	Floury soil	8.0	30.0	1.89	0.852	1.7×10-5	14.0	23.0	6.25
3–1	Silty clay	10.0	44.2	1.75	1.273	1.3×10-6	13.0	12.5	1.80
3–2	Sand	3.5	25.6	1.74	0.754	1.1×10-4	0.0	34.9	9.68
4–1	silty clay	10.4	30.4	1.77	0.852	5.9×10-7	17.8	15.6	6.11
5–1	silty clay	15.0	33.4	1.79	0.961	7.9×10-7	14.8	27.0	4.90

(a)

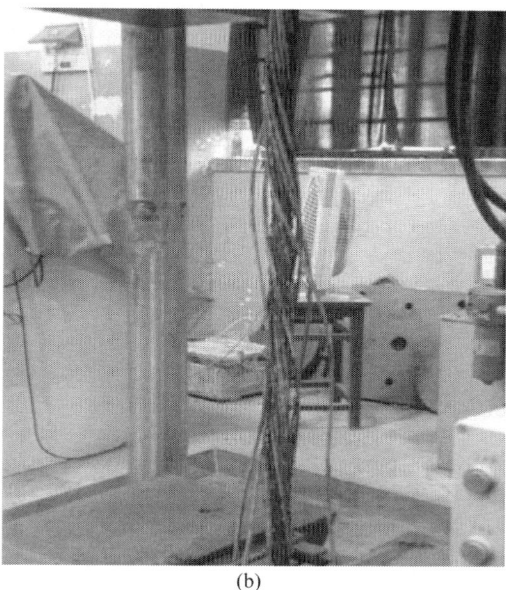

(b)

Figure 2. Diagram of loaded samples: (a) sample during loading; (b) sample after destruction.

3.2 *Result analysis*

Figure 3 is the graph of the optical cable stress and strain relation, it can be seen that after the sample experienced the physical deformation in initial period (OA period), the stress-strain relation of AB section is basically elasticity, collect the test data of AB section to obtain the curve relation of the optical-cable elasticity modulus with the change of the strain, as shown in the Figure 4, according to the test result, it can be seen that the average elasticity modulus of the optical cable is about 14 GPa, and its Poisson's ratio is 0.3 according to the material characteristics of the optical cable.

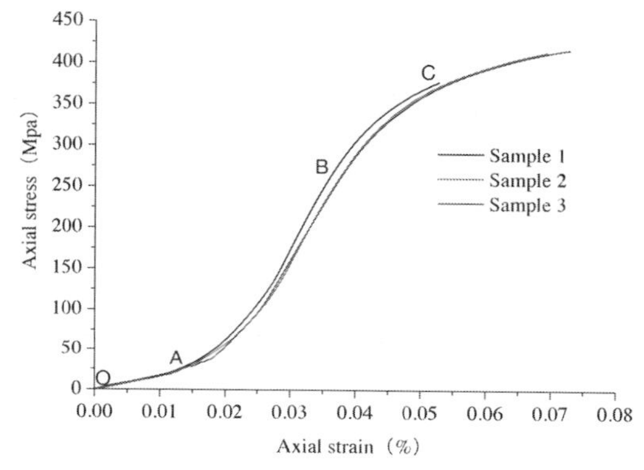

Figure 3. Stress-strain curve of samples.

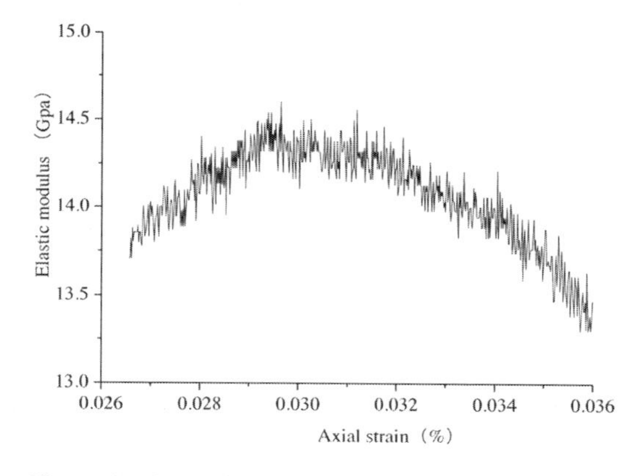

Figure 4. Curve between elastic modulus and strain of optical fibre cable.

4 NUMERICAL MODELING AND RESULT ANALYSIS

4.1 *Calculation model*

Numerical simulation adopts the large-scale commercial software ANSYS, it takes -40m elevation in the foundation depth direction (z direction), the length is 220m in the vertical-seawall axis direction (x direction), the foot of the seawall extends 50m respectively towards landside and sea side, and it takes 150m along with the axis of the dike (y direction).

Figure 6 is the established three-dimensional finite element model, there are 17728 entity units generated in total, the optical cable simulation is implemented via 57 inhaul cable units, and the simulation of the cofferdam and foundation is implemented through hexahedron entity unit, the optical cable adopts the link10 three-dimensional bar unit simulation provided by the ANSYS software, the model bottom is applied three-way constraint, and the surrounding boundary of the model is applied normal constraint.

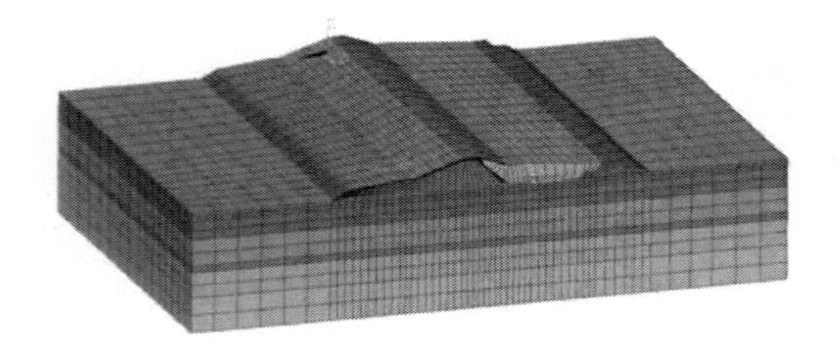

Figure 5 3-D. finite element model.

4.2 *Computational process*

Calculation is to simulate the seawall construction process via gradation loading, and it is divided into 10 levels in total: (1) Simulate the foundation and the initial ground stress balance; (2)Seawall filling to average low water (namely -2.50 elevation); (3~7): Seawall filling, 1.50m for each level, and meanwhile, carry out the construction of the outside surface structure; (8): Seawall leakage stopping and earthwork filling; (9): Seawall construction broken stone hardcore and levee crown roads; (10): Seawall construction wave wall structure.

Calculation process is as follows: Calculate the initial ground stress of the foundation before construction, serve the initial ground stress of the foundation as the initial conditions, carry out zero cleaning calculation to the initial displacement of the foundation (balancing process of the ground stress), and the simulate the construction process of the seawall.

The external optical cable is wrapped with steel strand, the rigidity is far more than the rigidity of the surrounding soil mass, during the construction process of the seawall, the soil mass of the foundation sediments, relative displacement would occur between the optical cable and soil mass. In axial direction, it would appear the relative sliding between the optical cable and soil mass, this relative sliding would balance the axial stress of the optical cable. Meanwhile in the settlement process of the soil mass, it would produce "earth cutting" due to the larger rigidity of the optical cable, namely, the soil mass appears "streaming around" relative to the optical cable, which makes the vertical displacement of the optical cable less than the settlement of the soil mass. In this way, the axial force of the optical cable is less than the optical cable axial force while the optical cable sediments together with the soil mass. From the angel of security, this calculation considers the most extreme case, instead of the relative deformation between optical cable and the surrounding soil mass, it is regarded that the deformation between the two is consistent.

4.3 *Results*

Figure 6 and Figure 7 show the settlement contour map and horizontal displacement contour map when the seawall filling is completed. Calculation results show that under seawall load effect, the maximum sediment of the mollisol foundation is 161.7 cm, appearing in the foundation surface near the seawall axis. The maximum horizontal displacement in the foundation pointing the sea direction is about 25cm, and about 20cm pointing the land side, and the maximum horizontal displacement in the foundation is located in the 3–1 mucky silty clay. Figure 9 is the vertical displacement distribution of the fiber cable, because of not considering the relative displacement between optical cable and soil mass, the maximum sediment of the fiber cable is larger, the sediment distribution law is similar to the sediment distribution law of the foundation face. Figure 10 is the axial-force distribution curve of the optical cable. The maximum axial tension is

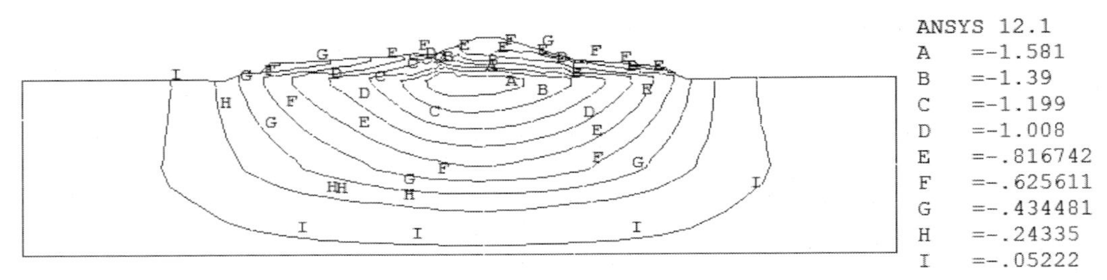

Figure 6. Cloud chart and contours of settlement distribution after embankment filling completed.

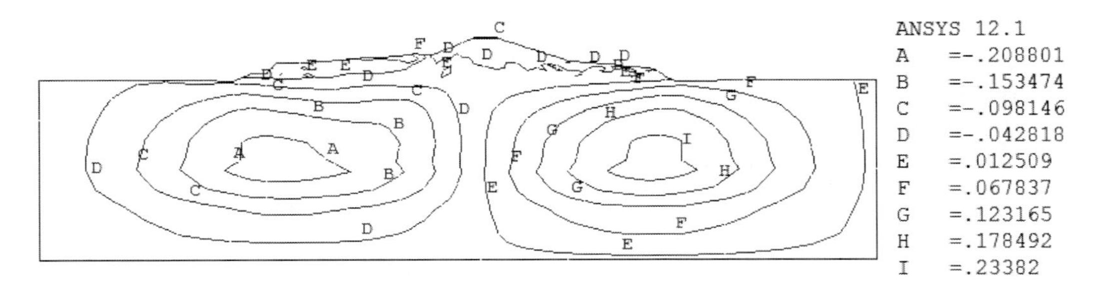

Figure 7. Cloud chart and contours of horizontal displacement distribution and after embankment filling completed.

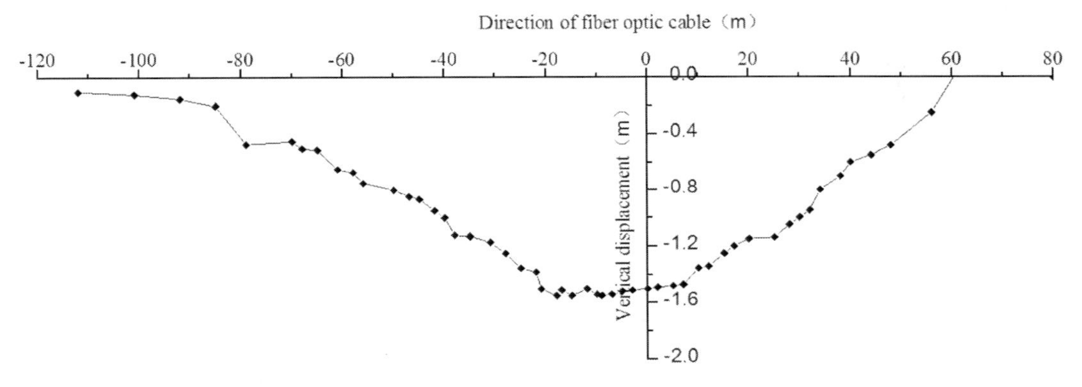

Figure 8. Vertical displacement of submarine optical fibre cable after embankment filling completed.

located near the seawall axis, the maximum tension is 58.75 KN (about 5.9t), the value is less than the allowable load of the optical cable within the safety range. Figure 11 is the axial tensile strain distribution curve of the optical cable. The calculation result shows that under the effect of seawall load, the maximum tensile strain of the optical cable is 0.436%, less than the 4% of the optical-core failure strain in optical cable, accordingly, it can be seen that the optical cable still can work well under the most extreme case.

5 CONCLUSIONS

1 The tensile test results of the optical cable sample show that with the protection of steel strand, the tensile strength of the optical fiber is higher, the tensile failure resistance exceeds 40t. But when tension exceeds 30t, the optical cable strain is larger and maybe exceeds the allowable tensile strain of the optical fiber. If the tension on the optical cable is less than 10t, the strain is very small, and when tension is up to 20t, the strain is between 1% and 2% within the allowable range basically.
2 The three-dimensional finite element is used to simulate the deformation of the foundation soil under the effect of seawall load and the stress-strain of the optical cable, the calculation results show that: if there is no such worst condition, namely the relative displacement between optical cable and surrounding soil mass, at the completion of the seawall, the maximum axial force on the optical cable is 5.9t, the maximum tensile strain is 0.44%, belonging to the normal working scope of the optical cable.
3 Considering the uncertainty of soil mass parameters, carry out the further study on the sensitivity of the foundation parameters, and calculate it after decreasing 10% of the elasticity modulus of the foundation soil layer, results show that at the completion of the seawall, the maximum axial force on the optical cable is 6.86t, the maximum tensile strain is 0.51%, being still less than the allowable value of the optical cable under normal work.

REFERENCES

[1] Worzyk T. (2009) Submarine power cables design, installation, repair, environmental aspects. *Springer Series: Power Systems*.

[2] Ye Yincan. (2006) Development of submarine optic cable engineering in the past twenty years. *Journal of Marine Sciences*, 24(3):1–10.

[3] Williams D O. (1997) An oversimplified overview of undersea cable systems [EB/OL]. Available: http:// www.cern.ch/davidw/public/SubCables.html, 1997-05-13.

[4] Amano K. (1990) Optical fiber submarine cable systems. *Journal of Light wave Technology*, 8(4):595–609.

[5] Lu Xiaoling, Ye Yinshan, Li Dong. (2009) Analysis of the factors affecting submarine optical cable safety in the East China Sea. *The Ocean Engineering*, 27(4):121–125.

[6] Cao Huojiang. (2006) Protection and management of electric (optic) submarine cables. *Electric Wire & Cable*, (3):34–38.

[7] Qiu Shengmei. (2005) Localization and testing for submarine cable fault. *The Ocean Engineering*, 23(3): 94–98.

[8] Zhang Hanjia. (2001) A proposal to guard against cable man-made damaging on sea bottom. *Modern Fisheries Information*, 16(6):11–13.

[9] Wagner E. (1995) Submarine cables and protections provided by the law of the sea. *Marine Policy*, 19(2): 127–136.

[10] Liu Jia, Zhou Xuejun, Wang Jingjuan. (2008) Study on the fault location technique for submarine optic fiber cables based on the electric field method. *Electric Wire & Cable*, (6):32–33(36).

[11] Wang Hongxia, Zhou Xuejun, Wang Ping. (2006) Fault location and repair of submarine optical communication cables. *Electric Wire & Cable*, (1):29–31(36).

[12] Lu Y G, Zhang X P, Dong Y M, et al. (2007) Optical cable fault locating using billion optical time domain reflect meter and cable localized heating method. *Journal of Physics: Conference Series*, 48(1):1387–1394.

[13] Zeng Zhaolei, Zhou Xuejun. (2009) A mathematical modeling of towing anchor for salvaging deep sea submarine fiber optic cables. *Optical Fiber & Electric Cable*, (1):41–43.

[14] Zhou Dongsheng. (2003) The application of plastic drainage board in the foundation consolidation of motorway embankment. *Journal of Railway Engineering Society*, (4):124–127.

[15] Huang M, Liu J. (2009) Monitoring and analysis of Shanghai Pu dong seawall performance. *Journal of Performance of Constructed Facilities, ASCE*, 23(6):399–405.

[16] Chen Pan, Li Yonghe, Wang Jili, *et al.* (2012) Effect of squeezing silt by blasting on compression characteristics of marine soft clays. *Rock and Soil Mechanics*, 33(S1):49–55.

[17] Dong Zhiliang, Zhang Gongxin, Zhou Qi, *et al.* (2011) Research and application of improvement technology of shallow ultra-soft soil formed by hydraulic reclamation in Tianjin Binhai New Area. *Chinese Journal of Rock Mechanics and Engineering*, 30(5):1073–1080.

[18] Dong Xiuchun, Dong Xianghua. (2007) Analysis of effect of marine route survey on submarine fiber-optic cable engineering. *Coastal Engineering*, 26(1):11–14.

[19] Miller S J. (2010) Submarine cables cause dock wall failure at the port of Milwaukee. *12th Triennial International Conference on Ports 2010, Building on the Past, Respecting the Future, ASCE, Jacksonville, Florida, United States*, pp:434–441.

[20] The Ministry of Water Resources of the People's Republic of China. Code for design of sea dike project (SL 435-2008) [S]. Beijing: China Water Power Press, 2008.

Electric, Electronic and Control Engineering – Shao, Shu & Tian (Eds)

Study on overall stability of sand-filling bag embankment by centrifuge model test

Yi Meng, Xiaoyu An & Yue Zhao
Tianjin Research Institute for Water Transport Engineering, Tianjin, China

ABSTRACT: The main influencing factors on stability of embankment filled with sand tube bag are engineering geology, foundation treatment measures, resistant to impact, durability of the geo-membrane bag and so on. Centrifuge modeling test was adopted to study the seawall engineering stability basing on the sea wall moving outwards practical engineering in Hangu district Tianjin city. The modeling technology, modeling method and the operate method were presented in details in the centrifuge modeling test which can provide a certain reference. Test results show that in the sea outside the block works scour depth 2m conditions, the sea dike construction and toe gear shift is small, one year after the vertical displacement were 58mm and 28mm, display scour depth of the sea outside the safe and stable block works less impact on the region sea gear engineering design has a certain reference value.

KEYWORDS: Sand tube bag; Centrifuge model test; Slide stability

1 INTRODUCTION

Engineering practices show that the factors influencing the stability and safety of the dike body of the sand-filling pipe/bag embankment mainly include the engineering geological condition, ground treatment measure, scouring resistance of the surface cover structure, durability of geomembrane bag and other factors. Scholars have studied the stability of the soft-foundation seawall engineering for many times. Chen Xiaoping and others [1] have established the soft-foundation creep model and analytic model of the coupling stress field under seepage effect and verified it according to the strain aging characteristic of the mollisol, coupling characteristic of the self-weight stress field and rheolytic field. Zhao Yukun and others [1] have studied dam slope stability of the embankment downstream of the Yellow River under the working condition of rapid drawdown. Gao Feng [3] thought that it is necessary to consider the influences of the pore air pressure and hydraulic pressure on the stability of the embankment in soft soil area under construction period, at the same time, to consider the influence of the improved foundation consolidation degree on the stability of the embankment. Zhao Shougang and others [1] have put forward the suggestions on the seepage failure standard of the Yellow River embankment, and have discussed the seepage stability of the embankment. Azm and others [1] have put forward the calculation software (PTDSSA) to calculate the 3-dimentional slope stability of the dam. Hasani and others [1] have used Geo-studio software to carry out slope and seepage analysis aiming at the on-site practical engineering.

Based on the practical engineering of the coastal levee moving-outwards in Hangu District of Tianjin, this paper studies the security and stability of the seawall engineering through the centrifuge model test. It specifically introduces the simulation ways and modes adopted in the process of the model test, obtains the deformation physical map and displacement field distribution of the dike before and after souring, the test result has certain reference significances to the coastal levee engineering design within this area.

2 BACKGROUND

The design elevation of the top seawall of the Coastal levee moving-outside engineering is 6.0m, the crest level of the breast wall is 7.5m. The width of the dike top is 8.0m, clear width of the top road is 6.0m. The dike body is filled with the 50cm through-length sand filling pipe/bag. The mud-filling pipe/bag is sewed by 200g/m^2 polypropylene knited mesh (or woven microplush). The sectional view of the sea-intercepting engineering is as shown in Figure 1.

Figure 1. Sectional view of Hangu sea block relocation project.

3 CENTRIFUGE MODEL TEST

The soil engineering centrifuge in Hong Kong University of Science and Technology is used in the test, the centrifuge was built up in 2001, there are complete testing/monitoring equipments, and it is equipped with four-way manipulator capable of achieving multi-function simulation [6-7]. Centrifuge contains the major design index below: capacity is 400g-t, radius of gyration is 4.2m, maximum acceleration is 150g (as shown in Figure 2). Figure 3 is the two-dimensional model box adopted in the test, and the size is 1.25m (length) x 0.35m (width) x 0.85m (height).

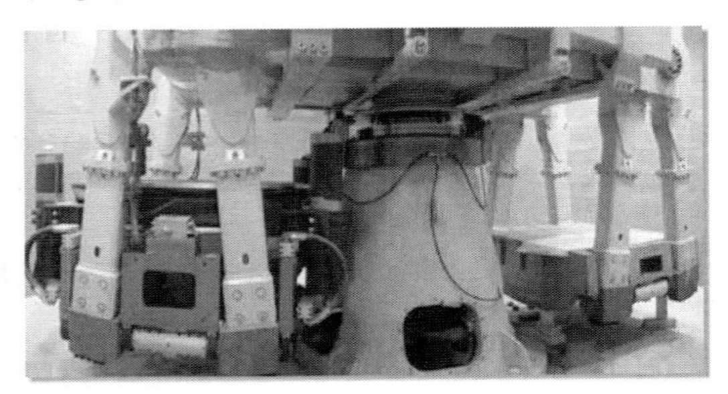

Figure 2. Centrifuge of HKUST (400g-t).

The purpose of the test is to study the stability of the dike, therefore, the stability and displacement of the dike, and the settlement deformation of the lower dike soil mass are the major test objects. The model box is equipped with two sets of Digital Video (pixel: 1024 x 1024), two sets of high-definition digital camera (pixel: 2560 x 1920), and two LVDT vertical displacement sensors for monitoring the deformation in the test. Digital Video is used for recording the whole process of the test, one is dead against the side of the dike, and the other one is installed right above the dike. The high-definition digital camera is used for recording the video of the test process, when releasing heavy-fluid is to simulate the stress relief effect, the interval of the camera shooting is set one piece per 30 seconds, which is used for carrying out the displacement field analysis of the soil mass with PIV software after the test. Meanwhile, LVDT vertical displacement sensor is respectively installed in the dike top and slope toe to monitor the vertical displacement of the soil mass.

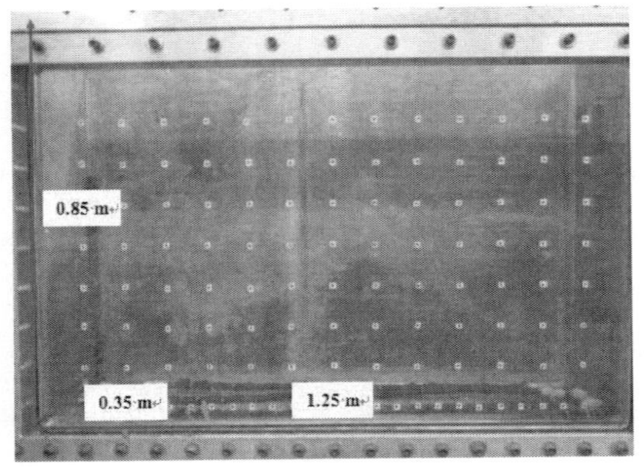

Figure 3. Two dimensional model box of centrifuge.

1. Model ratio: The physical quantity proportional relation of the centrifuge model test is listed in the table 1, there is the rated scale relation between the result obtained in the centrifuge model test and the measured one. 1:100 (model: prototype) centrifugal test model is adopted in this test.
2. Simulation method: Simulation method adopted in the model test mainly includes the following: (1) to simulate the sludge scouring via the method of releasing heavy-fluid ($ZnCl_2$) [8–10]; (2) inside alluvial sludge is replaced by the Toyoura sand bed with same load; (3) put the Toyoura sand in the closed soil engineering knitting bag for simulating the sand-filling pipe/bag, the Toyoura sand in the pipe has same density with the soil mass on site; (4) the rockfill material with particle diameter range obtained from calculation is used to simulate the surface cover structure for building.

Figure 4 is the centrifuge test model design drawing of the standard section in Coastal levee engineering. The left side of the model box is preserved with trapezoid groove to simulate the scour area, compared with the field actual situation, obtain the depth, bottom width and upper width being 20mm, 260mm and 300mm respectively. The dike is located in the intermediate model box, the width of the upper part is 80mm, the width of the lower part is 400mm, height is 80mm,

Table 1. The main physical ratio relationship.

Physical quantities	Scale relationship (Model/ prototype) when centrifugal acceleration is ng
Acceleration a	n
Linear scale L	$1/n$
Area A	$1/n^2$
Volume V	$1/n^3$
Time(consolidation/ seepage) t	$1/n^2$
Seepage velocity v	n
Coefficient of Permeability k	n
Settlement s	n
Stress σ	1
Strain ε	1
Force F	$1/n^2$
Density ρ	1
Mass m	$1/n^3$
Bending rigidity EI	$1/n^4$

the nearest souring distance away from the dike foot is 175mm. the sand bed thickness of the external dike is 10mm, simulating the on-site siltation layer with 1m thickness. The test earth materials mainly adopted in the model test include: silty clay, rockfill material and Toyoura sand bed. The next bed of the silty clay is a layer of 50mm Toyoura sand bed, it is used for providing drainage path, and the upper part is the sand-filling pipe/bag and mud sedimentation bed.

Model test setting: Figure 5 is the set-up diagram of the model test, in which figure 5 (a) is the reference point and control point of the photo processing technic (PIV) and two-dimensional model box adopted in this test, the distance between the reference point and control point is 80mm. After the test, according to the reference point and control point, carry out calibration to obtain the absolute displacement of the soil mass. Figure 5 (b) is the drainage system adopted in the soil mass consolidation, it mainly consists of four strips of PVC pipes, multiple weep holes are preserved in each tube, the external is wrapped with soil engineering knitting cloth. In the figure, the left two water pipes are connected to the bottom drainage system for controlling water level. Figure (c) is the preparation of the Toyoura sand bed, in order to obtain uniform sand bed, Toyoura sand is scattered from the top of the model box in a way of sandy rain, and the thickness is 50mm. Waterhead saturation is adopted to exclude the air. Figure5 (d) is the preparation of the silty clay bed, in the test, the foundation soil is the existing silty clay of the Hong Kong University of Science and Technology, it is divided into two layers (10.5cm and 11.5cm) for sample preparation. The dry density same with the site is regarded as the control parameters of the sample preparation. The target dry density of the mucky silty clay (1-a) and floury soil (1-b) is 1.24 g/cm^3 and 1.66 g/cm^3, in which, water content is 16%. The total soil thickness is 220mm, it is divided into 8 lays for sample preparation, and the thickness of each soil layer is 25 to 30mm. During the process of sample preparation, the water content in each layer has different degrees loss. Therefore, after preparing every two soil samples, and carry out the water content test. After completing each soil mass, spread black sand on the soil mass of the organic glass (used for analyzing the displacement field of the soil mass), and scratch marks on the surface of each soil mass. Figure 5 (e) is the front view and vertical view of the sample after preparation. Figure 5 (f) is the rigidity-free rubber film bag for containing heavy-fluid, the size of the rubber film bag bottom is consistent with the preserved digging chute. After installing water bag, carrying out test, no water leakage, after removing the fresh water, pouring the heavy-fluid configured in advance into the rubber membrane.

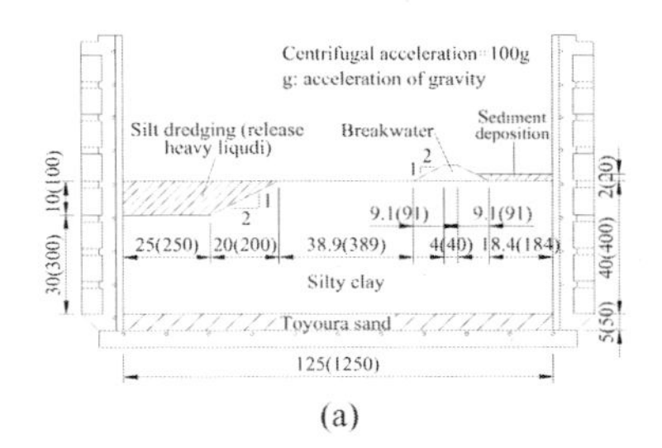

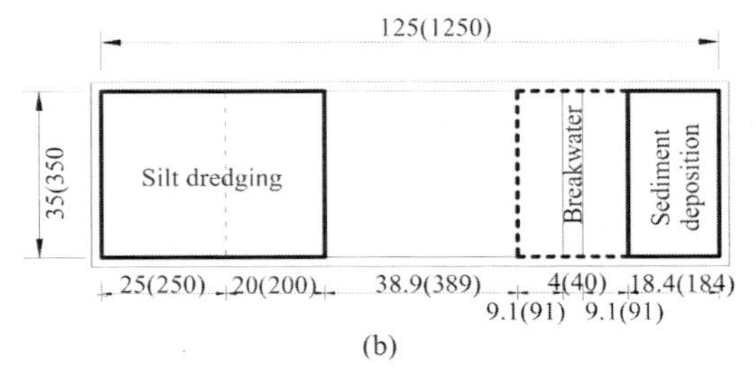

Figure 4. Diagram of centrifuge test model.

3 Test flow: In this test, carry out the secondary consolidation to the soil mass, and simulate the excavation effect under high-power acceleration. Specific steps are as follows: (1) Rise to 100g,

carry out the first solidification, the target acceleration of the centrifuge is 100 times of the gravitational acceleration (100g), the whole process of rising g is 10min. According to the method to judge the consolidation degree put forward by Asaka (1978) [11], when the soil mass consolidation measured by displacement sensor is up to more than 95% of the final settlement, it is regarded that the primary consolidation of the soil mass is completed. (2) Reduce to 1g, pile up the sand-filling pipe/bag, armour rock and sand bed, after completing the first consolidation, reduce the gravitational acceleration to 1g within about 60min, and pipe up the sand-filling pipe/bag, armour rock and sand bed. Figure 6 is the diagrammatic figure of the sand-filling pipe/bag and sand bed after packing. After the completion of packing, carry out the secondary high-power acceleration consolidation. (3) Rise to 100g, carry out the secondary consolidation, at the end of packing, the centrifuge acceleration re-rises to 100 times of the gravitational acceleration (100g), the whole process takes 3h. (4) To simulate scouring is based on the method to judge the consolidation degree put forward by Asaoka(1978)[11], when measured soil sediment reduces to above 95% of the final settlement again, carry out scouring depth simulation. Open the valve controlling excavation, slowly discharge the $ZnCl_2$ solution which is placed in the rubber bag of the excavation region in advance to be used for simulating the stress release effect produced by sludge scouring. Maintain the centrifuge carry out high-speed revolution

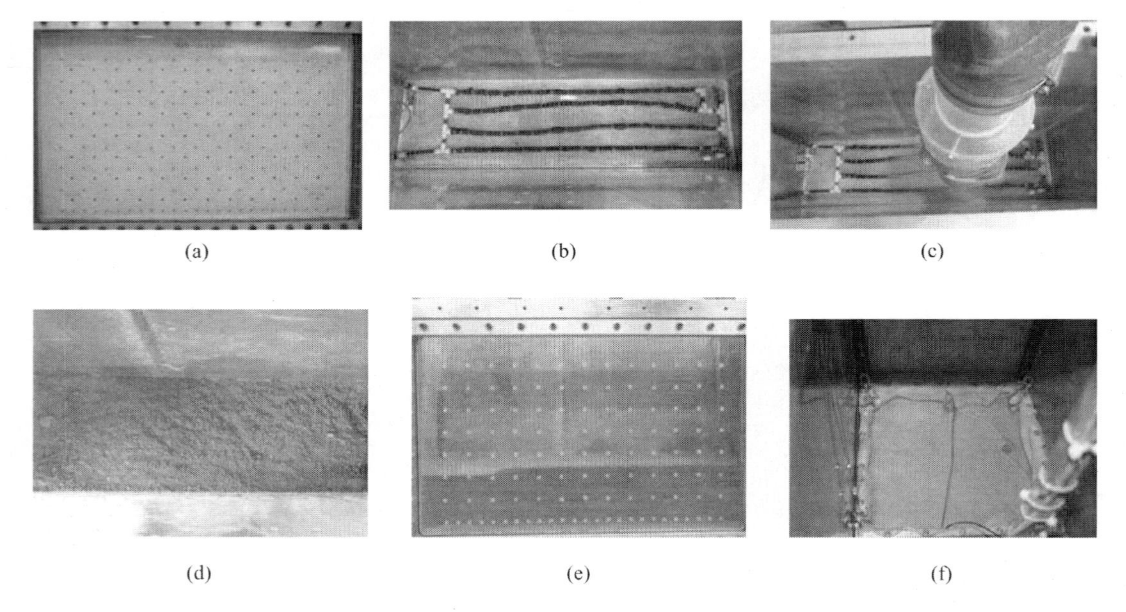

Figure 5. Model test setup.

(100g) for 1h, and obtain the foundation soil and coastal levee response through one-year sludge scouring.

4 RESULTS

Figure 6 and Figure 7 are the front view and vertical view of the dike section before scouring and after one-year outside 2m-depth scouring, the depth of the sludge scouring is 2m, the soil mass displacement caused by scouring is smaller, and has less influence on the coastal levee engineering. Figure 8 is the test result diagram of the soil mass displacement field, in the test, two displacement sensors are installed in the dike top and slope toe to measure the vertical displacement of soil mass. Three months later, the vertical displacement in the dike top and slope toe is 28mm and 14mm respectively. Considering the outside scouring with 2m scouring depth for one year, the vertical displacement in the dike top and slope toe is 58mm and 28mm respectively.

According to the scouring displacement field in Figure 8, it can be seen that the displacement of the sand-filling bag/pipe foundation soil appears the extends of horizontal slip and vertical upheaval, indicating that the sand-filling bag/pipe banking exists slide instability risk, at the same time, from the point of displacement field distribution area, it forms obvious slip band outside of the dike body.

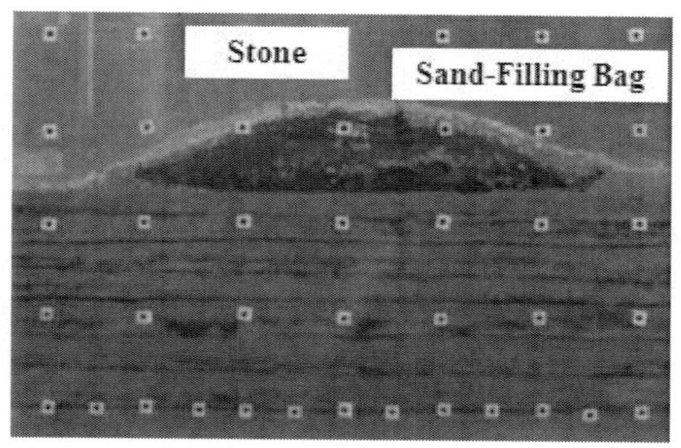

(a) before scouring

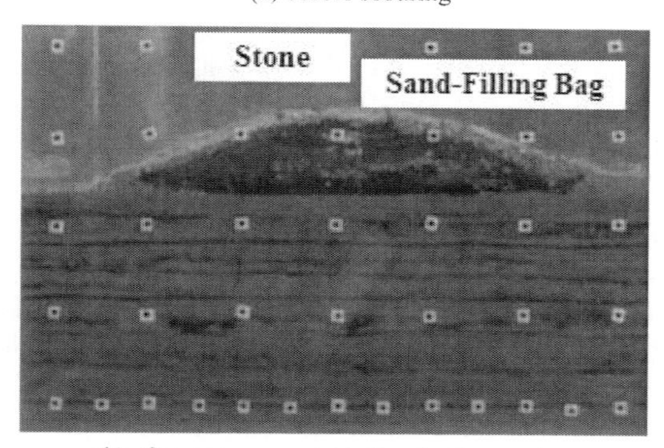

(b) after one-year outside 2m-depth scouring

Figure 6. Front view of the dam.

(a) before scouring

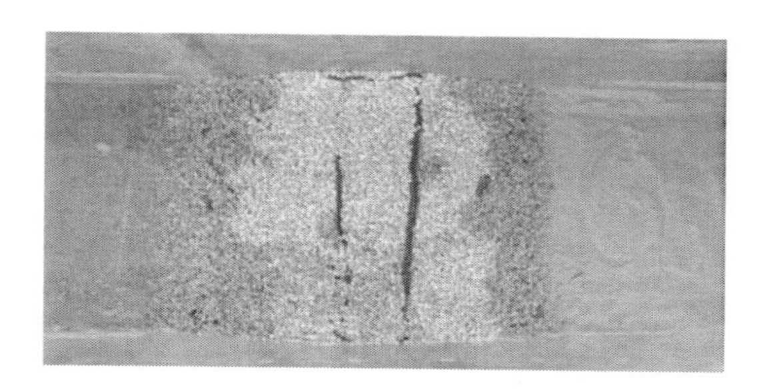

(b) after one-year outside 2m-depth scouring

Figure 7. Top view of the dam.

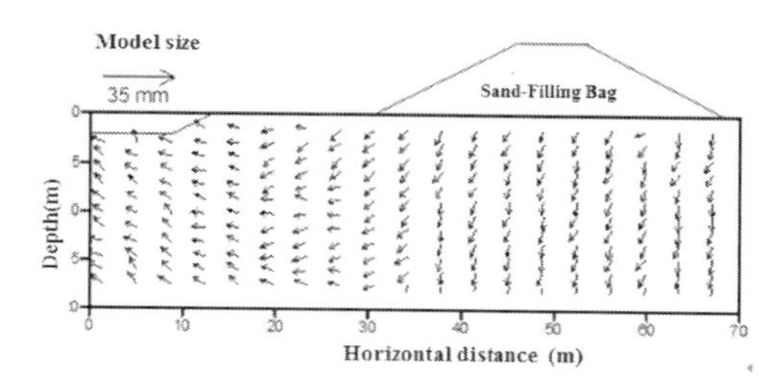

Figure 8. Field displacement of soil erosion after one year.

Therefore, it can be seen that geotechnical centrifuge test can accurately simulate security and stability of the sand-filling bag/pipe banking.

5 CONCLUSIONS

Aiming at the Coastal levee moving-outside engineering in Hangu District of Tianjin, this paper studies the security and stability of the Coastal levee project via the centrifuge model test, and the research results are as follows:

1 In the process of the centrifuge model test, firstly, simulate the crustal stress balance situation of sand-filling bag/pipe banking with the technological means of pre-consolidation, and then, the pressure pumping is adopted to replace the outside scouring condition of the sand-filling bag/pipe banking, the process of the model test has a certain inventiveness.
2 The situation that the scouring depth of the outside beach face of the Coastal levee engineering is 2m has less influence on the security and stability of the Coastal levee engineering, after three months, the vertical displacement at the dike slope toe and top is 28mm and 14mm respectively, and a year later, the vertical displacement at the dike slope toe and top is 58mm and28 mm respectively.

3 The results of the model test displacement field show that sand-filling bag/pipe banking easily develops the slip band under the condition of external scouring, indicating that the study on security and reliability of the Coastal levee engineering via centrifuge model test has certain reference significances to the design of the Coastal levee engineering within this region.

REFERENCES

[1] Xiaoping Chen, Jingwu Huang, Liming Zhang, *et al.* (2007) Stability study for coastal levee on soft foundation. *Rock and Soil Mechanics,* (12):2495–2500.

[2] Yukun Zhao, Handong Liu, Qingan Li. (2011) Slope stability analysis of lower reaches' dikes of Yellow River under flood immersion and water level rapid drawdown. *Rock and Soil Mechanics,* (05):1495–1499.

[3] Feng Gao. (2009) An analysis method of stability of embankment during construction period. *Rock and Soil Mechanics,* (S2):158–162.

[4] Shougang Zhao, Xiangqian Chang, Pan Shu. (2007) Reliability analysis of seepage stability on standard dyke of Yellow River. *Chinese Journal of Geotechnical Engineering,* (05):684–689.

[5] Al-Homoud A S, Tanash N. (2001) Monitoring and analysis of settlement and stability of an embankment dam constructed in stages on soft ground. *Bulletin of Engineering Geology and the Environment,* 59(4):259–284.

[6] Hasani H, Mamizadeh J, Karimi H. (2013) Stability of slope and seepage analysis in earth fills dams using numerical models (Case Study: Ilam DAM-Iran). *World Applied Sciences Journal,* 21(9):1398–1402.

[7] Garnier J, Gaudin C, Spring man S M, *et al.* (2007) Catalogue of scaling laws and similitude questions in geotechnical centrifuge modeling. International *Journal of Physical Modeling in Geotechnics,* (3):1–23.

[8] G Zheng, S Y Peng, Y Diao, *et al.* (2010) In-flight investigation of excavation effects on smooth single piles[C].

[9] G Zheng, S Y Peng, Ng C W, et al. (2012) Excavation effects on pile behaviour and capacity. *Canadian Geotechnical Journal,* 49(12):1347–1356.

[10] Hong Y, Ng C W W. (2011) In-flight centrifuge modeling of multi-stage excavation in soft clay with an underlying aquifer [Z].

[11] Asaoka A. (1978) Observational procedure of settlement prediction. *Soils and Foundations,* 18(4):87–101.

Electric, Electronic and Control Engineering – Shao, Shu & Tian (Eds)
© 2015 Taylor & Francis Group, London, ISBN: 978-1-138-02842-5

The technology research about vibration reduction of vehicle borne laser communication stabilized turntable

Hanyuan Hu
Department of Mechanical Engineering Changchun, Changchun University of Science and Technology, Changchun, China

Lizhong Zhang
Department of Mechanical Engineering Changchun, Changchun University of Science and Technology, Changchun, China
Department of Optics and Electronics of Space, Changchun University of Science and Technology, Changchun, China

Jin Hong & Yangyang Bai
Department of Mechanical Engineering Changchun, Changchun University of Science and Technology, Changchun, China

ABSTRACT: Aiming at the problem of vibration characteristics of vehicle platforms on the turntable axis stabilization and tracking precision influence, firstly a reasonable scheme for passive vibration reduction is putting forward based on the Characteristic analysis of vehicles platform's vibration and derivation of damping vibration; secondly, it has the vibration simulation analysis on the model of turntable's vibration system on the base of ADAMS/Vibration and gives the optimum damping ratio; at last, it has a vibration simulation test of the damping system of turntable. Simulation result shows that the damping system could better isolate high-frequency vibration of vehicle platform and ensures the stability and tracking accuracy of the turntable axis.

KEYWORDS: laser communication; vehicle-borne stabilized sighting turntable; damping system; vibration test.

1 INTRODUCTION

With the development of laser communication technology, space laser communication system demands of higher precision of dynamic tracking and stability in the communication. Though the vehicular-stabilized turntable and laser communication payload mostly are connected rigidly, the vibration of vehicle platform will transmit to the photoelectric equipment directly and cause sight axis sharking. Therefore, vehicle borne laser communication system is very sensitive to the vibration characteristics of the platform, so it's necessary to suppress or isolated vibration effect. The low frequency disturbance and high frequency vibration of vehicle platform which is an important interference source tracking servo system. It also has the maximum influence on the tracking accuracy[1]. Therefore, this paper introduces a principle which could eliminate the low frequency disturbance and restrain the high frequency vibration of vehicle sight stabilized turntable.

2 ANALYSIS ABOUT VIBRATION CHARACTERISTICS OF PLATFORM VEHICLE

Vehicle platform which is a multi-DOF and a complex quality - stiffness damping vibration is a system. It is composed of multiple intercoupling vibration-subsystems, Such as vibrations of the movement of vehicular body, of the engine torsional, and of the shaft system inside the gearbox, etc. The resonance frequency of the subsystems is not only an important index of vehicle platform vibration performance but the important design basis of on-board equipment as well. Figure 1 shows the power spectrum curve in the yaw angle and roll angle direction measuring in the vehicle platform moving. Figure 2 shows the total power spectrum curve measured in the moving vehicular body.

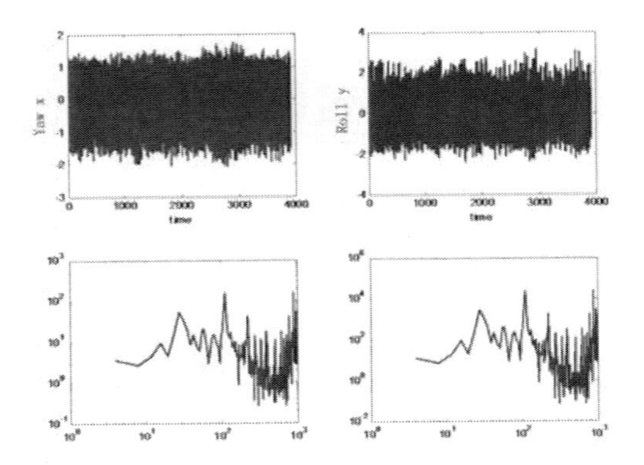

Figure 1. The power spectrum curve in the yaw angle and roll angle direction measuring in the vehicle platform moving.

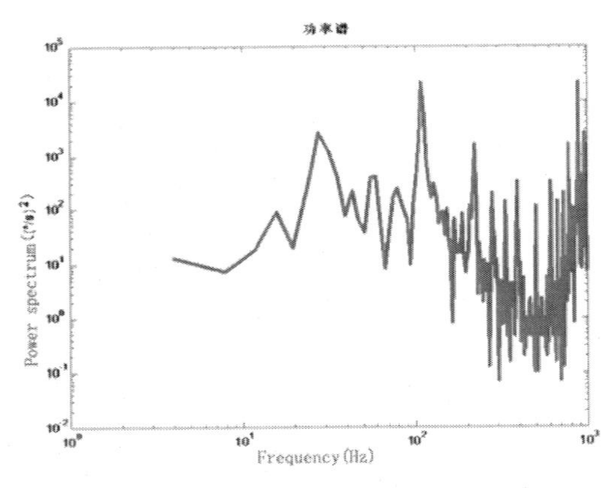

Figure 2. The total power spectrum curve measured in the moving vehicular body.

From Figures 1 and 2, we could see that the vehicle platform's mechanical resonance has been mainly reflected in two ranges: 0 ~ 10 Hz and 10 ~ 1000 Hz; These vibrations could make the visual axis of turntable which produced slight or larger chatter and showed the outburst characteristic of which the frequency is opposite with the amplitude [2]. Therefore, it's necessary to fix laser communication turntable in order to obtain high stability and tracking accuracy.

3 DESIGN OF THE STRUCTURE OF THE SIGHT STABILIZING TURNTABLE

For vehicular laser communication, it usually adopts dynamic long distance or static short distance communication due to the serious effect of which blocked out by the ground building and communicating distance is usually 20 km ~ 50 km vehicular turntable could realize the azimuth and the pitching rotation range, which is ±175° and -90°~60° (the scope of work is ±45°), the tracking accuracy is better than 0.15 mrad.

According to the requirements of turntable design, this vehicle tracking servo turntable of the laser communication system using the spherical turret appearance of diaxon and two frame structure, the reasons as following: (1) High frequency vibration of vehicle platform is not very outburst, so it wouldn't need designing into diaxon and four frame structure. For poor environment of the platform vibration, a passive vibration isolation system could be designed for turntable; (2) Technology of diaxon and two frame structure system is relatively mature, the system designing is simple, it has such a good stiffness that it's easy to achieve a higher resonance frequency and bandwidth control; (3) The ball turret shape structure has good aerodynamic performance and the wind disturbance torque is small; it contains the collective temperature control function, therefore it could meet all kinds of demand which means all kinds of adaptability of space environment [3]. The vehicular turntable structure as shown in Figure 3.

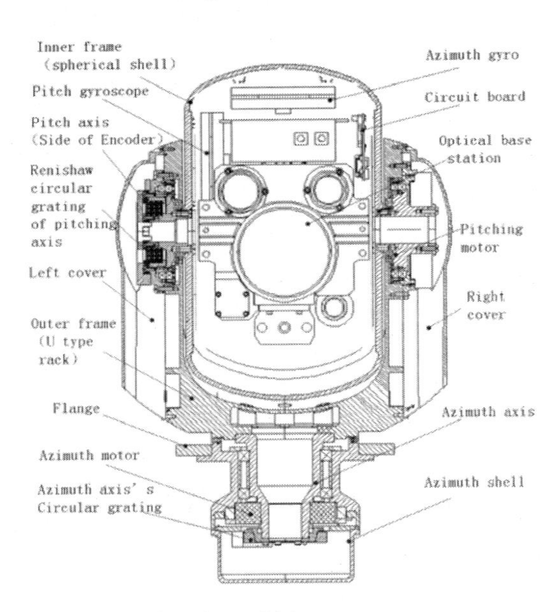

Figure 3. On-board turntable's structure.

As the on-board stabilized turntable is a high precision optoelectronic tracking equipment which is based on the gyro stabilized platform technology and stabilized technology, it could effectively restrain the low frequency disturbance caused by the vehicle while working with the servo system, and track moving targets in real-time. Though due to the limits which of turntable gyroscope bandwidth, of the intrinsic parameters of the control element and of the vehicular turntable structure, it couldn't suppress the influence of carrier frequency disturbance by increasing the servo bandwidth.

4 PRINCIPLE OF VIBRATION REDUCTION

In order to keep the tracing ability and stability of the accuracy of on-board turntable's visual axis precision, for low frequency disturbance stability it could be used high broadband rate gyro for inertial stable in order to keep the tracking servo system of the residual vibration reduction in a stable range; For high frequency vibration, due to shock absorber could effectively isolate high frequency vibration from carrier, therefore we could consider using passive vibration reduction measures.

4.1 *The principle of controlling low frequency disturbance*

As shown in Figure 4, the schematic diagram of the turntable system's control. The turret tracking system structure is composed of inner and outer frame: in the outer frame, there are two fiber optic gyros installed in a way which mutually perpendicular to each other, which in order to measure the inner framework and outer frame shaft's angular velocity and constitute a speed loop in servo system; There are circular grating which have been equipped within the inner and outer shaft frame for measuring two shaft angle and constitute a position loop of the servo control.

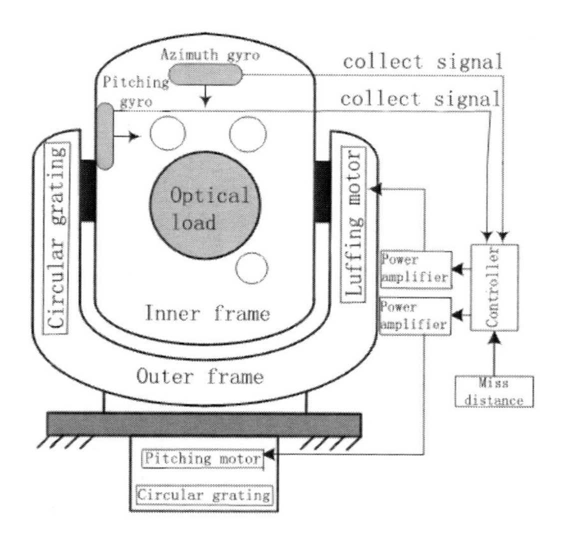

Figure 4. The image of stabilization system's control.

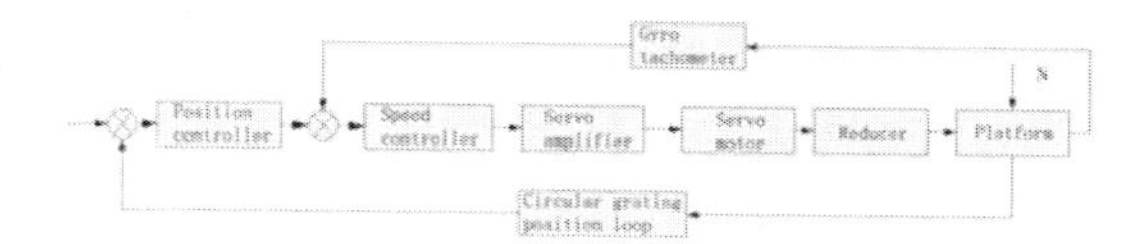

Figure 5. The control system of stabilized platform schematic diagram.

As shown in Figure 5 is the working principle of the two axis stabilized turntable in one direction.

The principle of the control system's work: first, rate gyroscope is used to measure angular velocity signals of the stabilized turntable, and feedback the processed signal to the corresponding servo motor to drive turntable movement to keep it stable in the desired location; At the same time, using the position sensor's grating measures angle position information of the turntable which has turned round and comparing the information with the angle of the feedback signal gyro after processed by gyro gets the miss distance as input values after calculating by the controller to control turntable's movement to keep its stability in the target position [4].

4.2 *Passive vibration reduction system*

1 The principle of passive vibration reduction

It usually used elastic supports as the vibration isolator in order to prevent carrying equipment effected by the influence of external vibration sources and reduce or stop the transmission of vibration impact so as to achieve the goal of reducing vibration, which is called a passive vibration reduction measure. Generally, we control vibration energy transmission for using spring and damping, so the stiffness of spring and damping of the system is the two mainly parameters in the passive controlled system. The passive vibration isolation system has the advantage such as its structure is simple and reliable, it's easy to implement, it's not expensive and it's effectively isolates high frequency vibration of the vehicle, but the system is not good at restraining the low frequency disturbance reduction's effect, usually a passive vibration reduction could suppress the vibration frequency of more than 10 Hz. It is shown by the vibration characteristics of vehicle platform that the passion vibration reduction scheme is the best choice to suppress the high frequency vibration. As shown in Figure 6, the mechanical model which has single DOF and passive vibration, referred to as k - m - c system [5].

As shown in Figure 6, the quality of the content block is m; Spring stiffness is k; Damping is c;

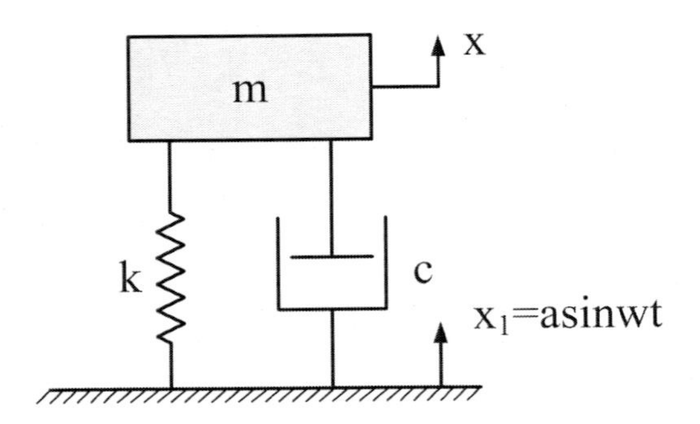

Figure 6. Single degree of freedom of vibration principle diagram.

The system moves as the simple harmonic motion $x_1 = a\sin wt$

Draw a content block dynamic equation is:

$$m\ddot{x} + c\dot{x} + kx = ka\sin wt + cwa\cos wt \tag{1}$$

Set type (1) under the steady state vibration of special solution for:

$$x = b\sin(wt - \varepsilon) \tag{2}$$

Plug in type (2) into type (1):

$$b = a\sqrt{\frac{k^2 + c^2w^2}{(k - mw^2)^2 + c^2w^2}} \tag{3}$$

The transmissibility for the system is η; The damping ratio for system is ξ; Based excitation frequency is w; System natural frequency is w_n; The ratio of frequency is λ.

By type (3) getting the stable amplitude -as we called-the transmissibility:

$$\eta = \frac{b}{a} = \sqrt{\frac{k^2 + c^2w^2}{(k - mw^2)^2 + c^2w^2}} = \sqrt{\frac{1 + 4\xi^2\lambda^2}{(1 - \lambda^2)^2 + 4\xi^2\lambda^2}},$$

$$\lambda = \frac{w}{\sqrt{k/m}} = \frac{w}{w_n}, \quad w_n = \sqrt{\frac{k}{m}}, \quad \xi = \frac{c}{2\sqrt{k/m}}$$

In a word, the curve among the relations of λ, ξ, η is shown in Figure 7.

As shown from Figure 7:

1 when $\lambda = 1$, η is much greater than 1; η tends to infinity, $\eta = 0$, and the resonance has occurred;
2 when $0 \prec \lambda \prec \sqrt{2}$, $\eta > 1$; vibration isolation system amplifies ;

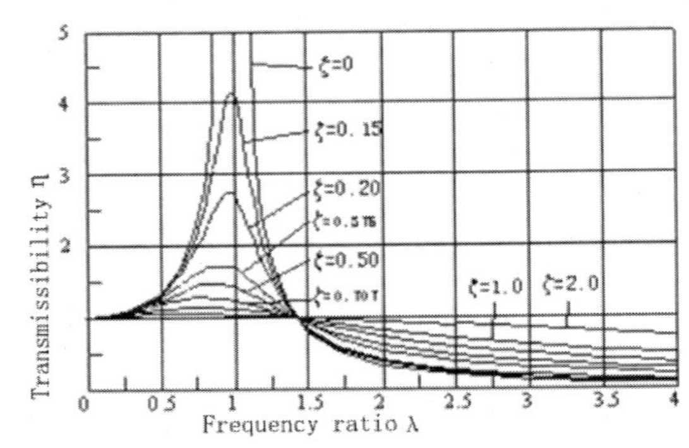

Figure 7. The curve among the relations-η and λ, ξ.

3 when $\lambda > \sqrt{2}$, $\eta < 1$; , the value of η is the increasing numbers of smaller with the increasing of λ. vibration isolation system have the effect on vibration isolation and the more the value of λ increasing, the better of the results are. But the larger value λ could make the device stable worse;
4 when the $\lambda < \sqrt{2}$, η increases with the decreasing of the λ, while the source of vibration frequency close to resonance frequency, the result will produce larger vibration and leads to a decline of vibration damping performance.

In a word which based on the analysis, we could establish system transfer function which according to the laws of the change of rigidity and damping, by comparing the system transfer function under different parameters, we could get the optimized design parameters. When we are designing vibration isolation system, on the one hand, we shall ensure η is not too large while the system resonating; on the other hand, it should achieve a good vibration isolation function.

4.3 *Design vibration system*

There are two kinds of shock absorber installation way: one is an overall vibration reduction; the other is a partial vibration reduction. So-called overall vibration reduction is that the shock absorber installed on foundation surface between the carrier and the turntable; and partial vibration reduction method is to add shock absorber between the inside and outside of the frame, in order to avoid the center of gravity of the inner frame and the shock absorber spring heart don't coincidence, and install the shock absorbers in both sides of the inner frame. According to the fact that characteristic analysis of vehicle platform vibration and the turntable structural framework that both inside and outside for internal structure are rigidly connected, therefore, only could we choose the overall layout of vibration reduction due to it couldn't design vibration reduction measures inside. We choose shock absorbers which have equal stiffness and no angular displacement on three directions, using 8 shock absorbers in parallel. The layout of the shock absorber is shown in Figure 8 below.

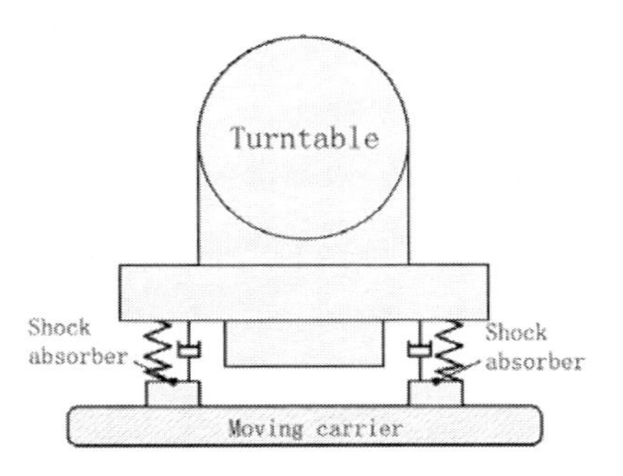

Figure 8. Diagram overall layout of vibration reduction.

The overall vibration way which is layout way, has an advantage of simple structure, flexible operation, easy to implement, good at vibration damping, etc. But the vibration reduction plane is far away from the shock absorber damping center, when the foundation is motivated by the larger vibration, it's easy to swing and instability.

5 VIBRATION SIMULATION AND EXPERIMENT

5.1 *The analysis of vibration simulation*

Using ADAMS/Vibration module for Vibration system to analysis, analysis steps [6] as shown in Figure 9. According to the below vibration analysis steps: in the vertical direction (Y direction) set the input channel parameters, the input conditions: in physical Parameter column selects Swept going; in Force

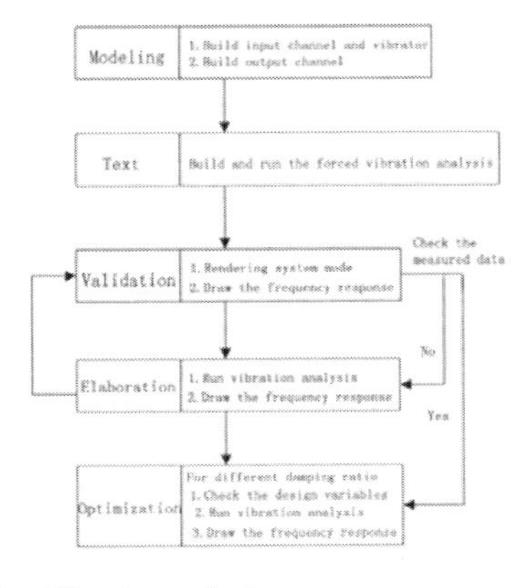

Figure 9. Vibration analysis steps.

Magnitude column type 1, the largest amplitude of Force is 1; in Phase Angle (deg) column input 0, namely the Phase Angle is 0. Simulation results that the vertical direction affects the vibration of two frequency, amplitude response and phase frequency response curve of the maximum of two points, is shown in Figure 10.

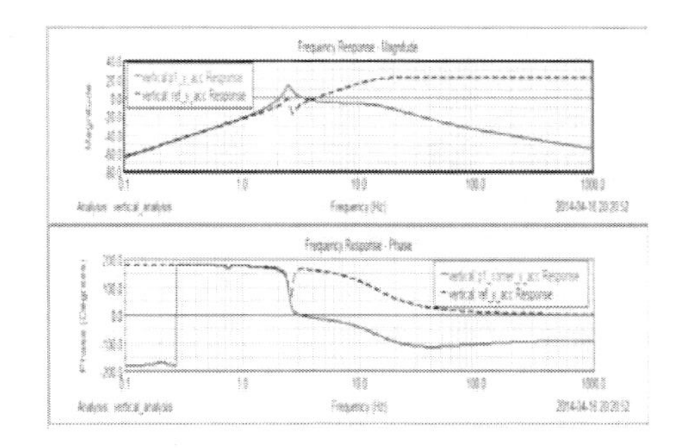

Figure 10. Frequency response curve.

We could know from Figure 10 that: within the scope of the 10 Hz ~ 1000 Hz, vibration frequency response amplitude in the vehicle platform's vertical direction is less than 30μrad, shown that vibration reduction system could isolate high frequency vibration from carrier better. In vibration forces by the sine sweep cases, by changing the size of the vibration situation of the simulation turntable's the damping ratio, as shown in Figure 11. Setting parameters: the stiffness of the initial value is 10 N/mm, range:

5 N/mm ~ 20 N/mm, damping the initial value is 0.25 N. the SEC/mm, range: 0.1 ~ 0.5 N sec/mm.

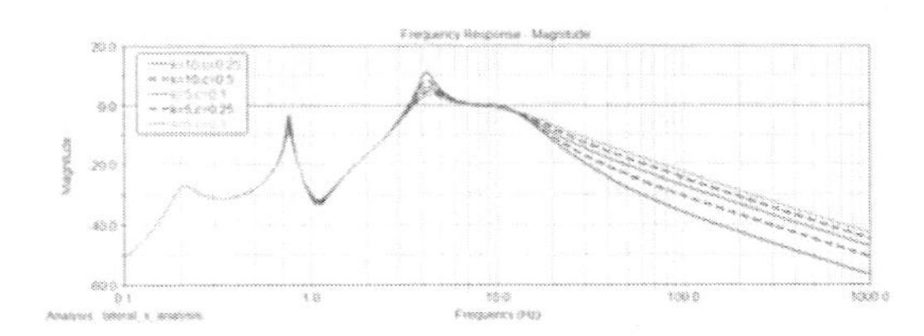

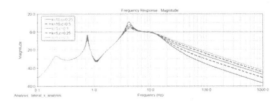

Figure 11. Vertical response curves under different rigidity and damping.

Known from the analysis of Figure 11:

1 If we won't change stiffness, the larger damping, the smaller the response of the system, the more decreasing gently; In the case of damping won't change, the smaller the stiffness, the lower the natural frequency of the system, the less the amplitude frequency response.
2 For vehicle platform's high frequency vibration, vibration damping system has a good damping

effect, and for high frequency vibration (> 10 Hz), the optimum design parameters is k = 10, c = 0.25.

5.2 *The vibration test*

This paper designs a passive vibration reduction measures of vehicle platform's high frequency vibration suppresses the effect of stabilized turntable, it verifies the inhibition effect of vibration isolation system through vibration test.

5.2.1 *Composition of the test*

The vibration test is mainly by the testing equipment: electrodynamic vibration test rig, vibration table control cabinet, several sensor cables, the data transmission module, the vibration reduction system structure (such as three-way rigidity shock absorber and the connection plate etc), stabilized turntable, the turntable control system, PC, power supply, etc. As shown in Figure 12, the vibration table which used for vibration test is produced in Sunan vibration experiment instrument factory typed DC-600-6-c/S0808. Main components of the vibration test table are: vertical platform, horizontal sliding table, the acceleration sensor, vibration control meter, the data transmission module, PC and switch power amplifier and cooling fan, etc. DC-600-6-c/S0808 electric vibration table control system's working principle diagram is shown in Figure 13.

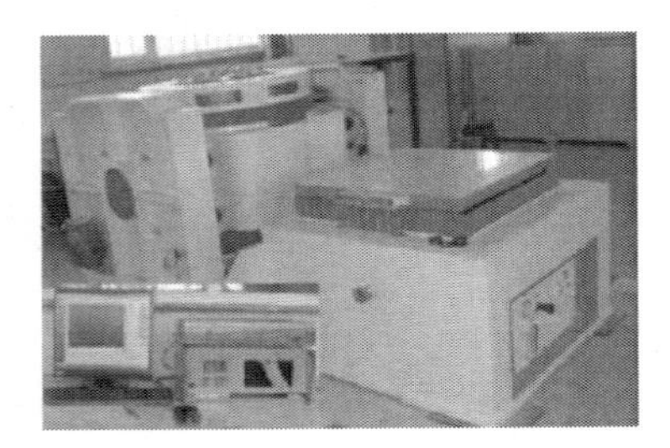

Figure 12. DC-600-6-c/S0808 electric vibration table.

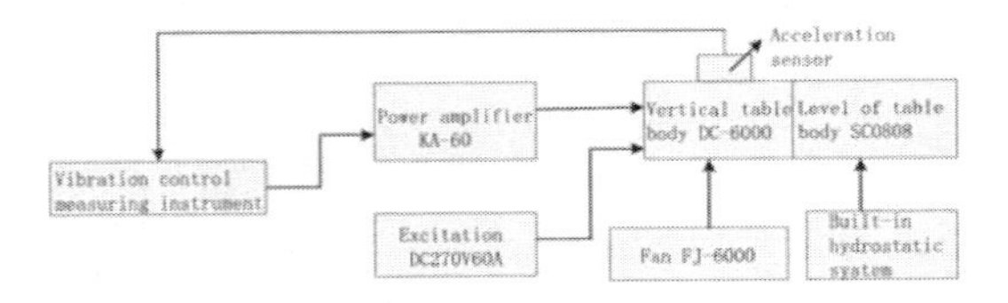

Figure 13. Vibration table control system schematic diagram.

5.2.2 *The principle of test*

Stabilized turntable vibration test principle diagram is shown in Figure 14. The principle of vibration test: Firstly, power on stabilized turntable in order to prevent the coupling influence from the internal structure of turntable movement to effect test results, that is eliminated the impact of exercise from the internal structure's moving in vibration test; Secondly, fasten vibration isolation system and sensor installation, a random vibration signals (such as sine sweep frequency curve, the PSD curves or user input vibration curve, etc.) is sent by a vibration table control system of vibration table to the table; Vibration signals is passed to the vibration table by data lines, vibration table outputs corresponding vibration simulation motion signal according to the parameters of the input signal; stabilized turntable is fastened by the angular displacement of shock absorbers, fittings and vibration table, this is the experiment of passive vibration isolation system, so analogs the output signal of the vibration is passed on to the turntable by the damping system, make the turntable indirectly affect by forced vibration; By installing sensors on test turntable to get the output of the vibration data of the turntable in all directions (X, Y, Z), the data are sent to the computer to carry on the analysis after the signal processing system, finally the power spectrum curve get of the direction.

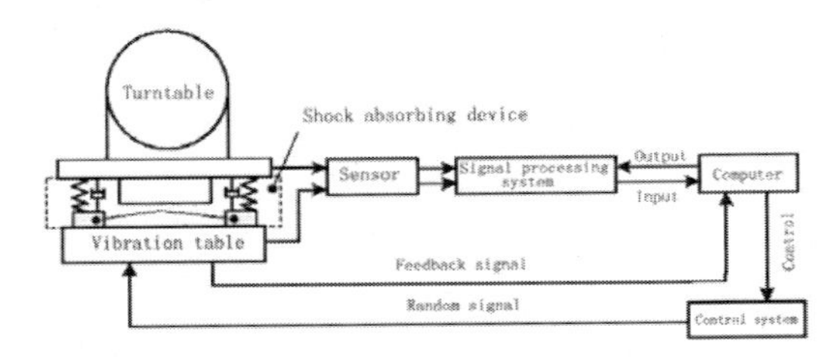

Figure 14. Stabilized turntable vibration test principle diagram.

5.2.3 *The random vibration test*

In order to detect the damping effect of the vibration reduction system, random vibration produced by the vibration table is respectively applied 3 directions excitation for sight stabilizing turntable. Then we test frequency response characteristics of the turntable framework respectively, roughly evaluates the damping effect of the vibration reduction system. Setting sensor installation position and direction: set the positive direction which is pointing to the optical light of the base stations of the exit portal and the direction is the X direction that the horizontal direction is parallel and perpendicular to the plane internal frame (that is perpendicular to the horizontal plane) on the orientation to Y direction, Z direction according to the right-hand rule. Through a vibration table control system, respectively do the random vibration frequency sweep test for a vibration table in the sensor of X, Y, Z three directions, measure random vibration frequency sweep test output curve by the sensor in the X, Y, Z three directions, as shown in Figure 15.

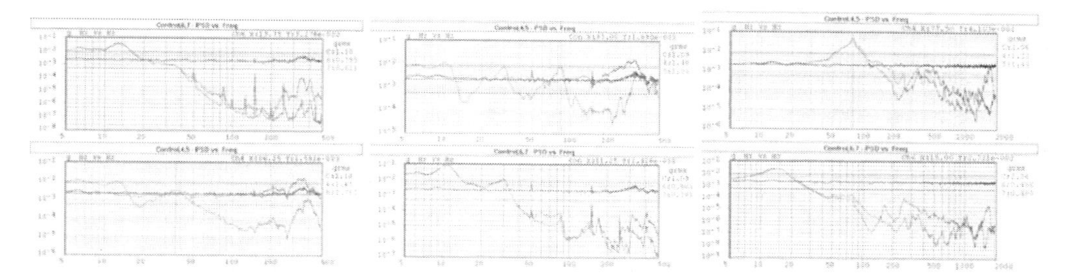

Figure 15. Vibration power spectrum in the X direction, Y direction, Z direction.

Known from Figure 15: in the condition of random vibration, the maximum stable precision of the inside and outside framework is 17.86μrad; between 0 ~ 20 Hz frequency for vibration, the damping system has the function of amplifying; for more than 20 Hz high frequency vibration, the system has a obvious attenuation effect and good inhibition effect.

6 CONCLUSION

Through the analysis about the characteristics of the vehicle vibration platform, the vibration causes relatively strong by vehicle platform's using the environment and the movement of the body. In order to ensure the tracing ability and stability precision accuracy of vehicle laser communication stabilized turntable's the visual axis. The vibration isolation system is designed on the basis of the principle of passive Vibration, and we optimize shock absorbers parameters by using the ADAMS/Vibration module and get the optimal stiffness and damping parameters is that $k = 10$, $c = 0.25$. Experiments verified that the vibration reduction system has a better inhibition effect of high frequency vibration to ensure the stable turntable accuracy is better than that of 30μrad and meet the requirements of space laser communication.

REFERENCES

[1] Li Yan, Fan Dapeng, (2007) Error analysis of three-axis turntable aimed at assembling based on multi-system kinematics theory, *Acta Armamenri*, 28(8):981–988.

[2] Li Long, (2014) Research on Laser Communication Stabilized Sighting Turntable for Vehicle Platform, *Changchun University of Science and Technology*.

[3] Jiang Huilin, Liu Zhigang, Tong Shoufeng, (2007) Analysis for the environmental adaptation and key technologies of airborne laser communication system, *Infrared and Laser Engineering*, (S1):299–302.

[4] Lu Zhufei, Hu Zhigang, (2008) Design of two-axis stable platform controller based on P89V51, *Computer Measurement&Control*, 12:1863–1864+1875.

[5] Yan Lijun, (2013) Optimization technology research on the stable structure of aviation pods, Changchun University of Science and Technology.

[6] Ja Changzhi, Yin Zhijun, Xue Wenxin, (2010) *MD ADAMS Virtual Prototype from Entry to the Master*. Beijing: Mechanical Industry Press.

Electric, Electronic and Control Engineering – Shao, Shu & Tian (Eds)
 ISBN: 978-1-138-02842-5

The study of micro-molecule polymer product engineering based on decision model

Jimei Song
School of Chemical Engineering & Environment, Weifang University of Science and Technology, Shouguang Shandong, China

Su'e Li
Shandong Houzhen Central Hospital, Shouguang Shandong, China

ABSTRACT: High polymer material is developing rapidly in recent years, and the process of polymerization product structure is complex. The paper makes a deep description about the new chemical product. With the rapid development of macro-molecule polymer materials, and many high performance polymer products are produced. Based on the present developing situation of polymer and macro-molecule materials, the paper makes a detailed analysis. The paper adopts mathematical methods to make a quantitative research about the macro-molecule polymer chemical engineering. In the developing of polymer, we need to pay more attention about the basic theory of the polymer reaction.

KEYWORDS: macro-molecule material; polymer product; chemical engineering; level analysis fuzzy evaluation.

1 INTRODUCTION

Polymer means the micro-molecule chemical compound, and the simple molecule can form a super polymer under the complex condition. The common polymers include plastic, fiber, synthetic rubber, and so on, which belongs to the chemical product, however the structure is very complicated and various, and the producing condition needs strict requirement. The properties of polymer is the stronger part to be decisive than the structure and level of the polymer itself.

The national economy development depends on chemical industries. The chemistry provides various materials, which makes an extremely important role in industry, agriculture, chemistry. With the development of oil industry, the chemical industry is improving day by day. The chemical industry provides a condition for the mass manufacturing of the national economy and society. Therefore, in the new period, the branch of chemical industry, polymer molecule material is rising rapidly, studying polymer products has a strategic significance in the future of China.

2 POLYMER PRODUCTS

The internal structure and morphology with the micro shape is very important in the process of producing polymer production. The technology of polymerization usually needs to be finished under the corresponding catalyst and the certain conditions. With the progress of the chemical technology, the polymerization is improved to a more simple, low cost, and more significant to protect the environment. And many scientists try various ways to improve the polymerization technology. To choose a better technology to improve the output is more important. By polymerization, the polymer structure can be controlled in a precise standard, and the property of polymer can also be improved. After a deep analysis of dynamic model from polymerization nature, we can know the output from the view of dynamics, and then adopt the simple producing technology to improve the production efficiency, meanwhile, make sure the blending process of molecules after a micro and macro research. Besides, the polymerization reactor is also a essential part in the process. Some scientists have made a special research for the alloying of the polymerization reactor.

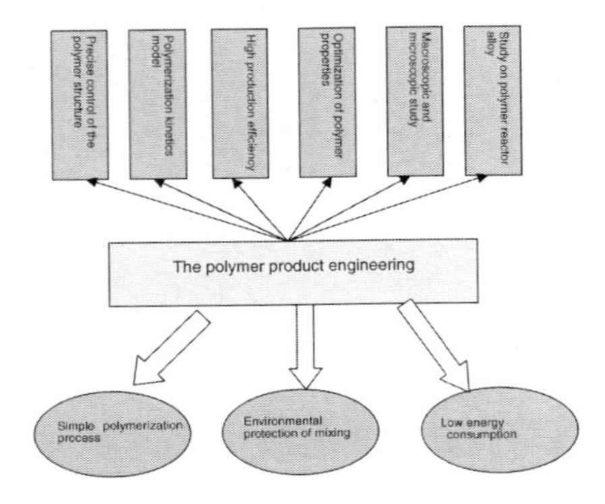

Figure 1. The hierarchical relationship between the polymer product engineering and the macro-molecule material.

3 FUZZY HIERARCHICAL AND MULTI-DECISION-MAKING MATHEMATICAL MODEL

The fuzzy means we can't differ with the critical state or transient state, and there are no clear differences between them. Not only in the reality but also in the engineering study, our problems are not the precise state, but the fuzzy state. Based on the fuzzy number phenomenon, Prof L.A. Zadeh promoted fuzzy math concept in 1965, fuzzy math became a new branch of math field. The fuzzy comprehensive evaluation is a method based on many elements or standards of the system evaluated by the application of the grade of the system to the fuzzy relationship synthetic membership, making a final result.

3.1 *The steps of fuzzy evaluation.*

Step 1: The factor domain of establishing the evaluation objects U, $U=(u_1, u_2 \quad, u_n)$:u_1 stands for pushing the polymer product, u_2 stands for strengthening the relationship between the polymer and macro-molecule, u_3 stands for improving the polymer product property, $u_4$stands for meeting the requirement of polymer product from the society.

Step 2: To determine the grade of qualitative evaluation in the factor domain, the words often can be described as: good or bad, high or low, importance, etc. In the paper, the evaluation is set to be (most effective, more effective, effective, less effective, least effective), expressed as v_n $(n=1,2,3,4,5,)$.

Step 3: To determine the domain F of values, there is $F=(f_1, f_2, f_3, f_4, f_5, f_6)$, which stand for precisely control the polymer structure, polymer reactive dynamic model, effective production, optimize polymer property, micro and macro blending study, polymer reactor alloying study. Establishing the fuzzy relationship matrix based on the degree of the relative evaluation, the specialists give them scores as follow:

Table 1. Degree numerical table.

	Most effective v_1	More effective v_2	Effective v_3	Less effective v_4	Least effective v_5
u_1	0.16	0.42	0.23	0.12	0.07
u_2	0.54	0.21	0.16	0.09	0
u_3	0.53	0.23	0.14	0.10	0
u_4	0.13	0.19	0.44	0.18	0.06

From the above, we can determine the functional formula expressed as follow:

$$R_1=\begin{pmatrix} 0.16 & 0.42 & 0.23 & 0.12 & 0.07 \\ 0.54 & 0.21 & 0.16 & 0.09 & 0 \\ 0.56 & 0.23 & 0.14 & 0.10 & 0 \\ 0.13 & 0.19 & 0.44 & 0.18 & 0.06 \end{pmatrix}$$

r_{ij} stands for the degree of the factor to the evaluation comments. The second matrix are fuzzy matrix. By the same theory, we can get the other value as follow:

$$R_2=\begin{pmatrix} 0.46 & 0.25 & 0.18 & 0.11 & 0 \\ 0.14 & 0.39 & 0.28 & 0.19 & 0 \\ 0.57 & 0.23 & 0.10 & 0.08 & 0.02 \\ 0.15 & 0.15 & 0.48 & 0.12 & 0 \end{pmatrix}$$

$$R_3=\begin{pmatrix} 0.08 & 0.18 & 0.12 & 0.39 & 0.23 \\ 0.13 & 0.23 & 0.28 & 0.19 & 17 \\ 0.11 & 0.13 & 0.21 & 0.28 & 0.26 \\ 0 & 0 & 0.13 & 0.54 & 0.33 \end{pmatrix}$$

$$R_4=\begin{pmatrix} 0.18 & 0.22 & 0.25 & 0.19 & 0.16 \\ 0.09 & 0.13 & 0.28 & 0.47 & 0.03 \\ 0.16 & 0.38 & 0.25 & 0.18 & 0.03 \\ 0 & 0 & 0.13 & 0.67 & 0.20 \end{pmatrix}$$

$$R_5=\begin{pmatrix} 0 & 0.12 & 0.33 & 0.28 & 0.27 \\ 0.63 & 0.21 & 0.15 & 0 & 0 \\ 0.23 & 0.18 & 0.31 & 0.28 & 0 \\ 0.58 & 0.27 & 0.15 & 0 & 0 \end{pmatrix}$$

$$R_6=\begin{pmatrix} 0.31 & 0.23 & 0.18 & 0.15 & 0 \\ 0.21 & 0.56 & 0.12 & 0.09 & 0.01 \\ 0.30 & 0.41 & 0.0.15 & 0.12 & 0.02 \\ 0.07 & 0.13 & 0.18 & 0.46 & 0.16 \end{pmatrix}$$

4 To determine the weight of factor domain, there is a weight vector W, $W=(w_1, w_2, \cdots w_n)$, a_i stands

for the relationship of the evaluation system, which is the distribution of the importance of every factors in the evaluation system. There are two ways of determination, one is the direct assignment by the specialists interview. The other is calculated from compare the established weight vector by the degree analysis.

Concrete steps:

1 Constructing the comparison matrix

All factors will be compared in a pairwise comparison, and then expressed the results as follow, such as comparing u_i, u_j, the result is shown as a_{ij}, then after comparing all factors, the comparison matrix will be shown as follow:

$$A = \begin{pmatrix} a_{11} & a_{12} & \cdots & a_{1j} \\ a_{21} & a_{22} & \cdots & a_{2j} \\ \vdots & \vdots & \ddots & \vdots \\ a_{i1} & a_{i2} & \cdots & a_{ij} \end{pmatrix}$$

The value a_{ij} can be expressed by the number of 1~9 and its reciprocal, Saaty thinks the comparison way conforms to the people's judgment ability. The number1–9 will stand for something as follow:

The explanation of 1~9 scale

scale	The explanation
1	Two factors are equal important to the aim
3	The former is much more important than the later
5	The former is more important than the later
7	The former is a little more important than the later
9	The former is extra more important than the later
Even number	Means the importance is between two odd number
Reciprocal	Means the order of factors compared positive to the negative.

2 the test of variance consistency and the calculation of weight vector

The definition of consistent matrix: for the matrix $A = \left(a_{ij}\right)_{n*n}$, if the factor of matrix conform to $a_{ij}a_{jk} = a_{ik}$, the matrix will be a direct matrix. $a_{ij} > 0$, $a_{ij} = 1/a_{ji}$. To calculate the weight vector of factors, the inconsistent of matrix will be accepted. But the complicated problem is that we can't calculate all factors when we are in a pairwise comparison, leading to an inconsistent situation.

The consistent standard of judging matrix CI, with the ratio CR, the computation is expressed as follow:

$$CI = \frac{\lambda_{\max} - n}{n - 1}$$

n stands for the order of matrix, which are the number of the comparison factors .

$$CR = \frac{CI}{RI}$$

RI stands for Random Consistency Index, as follow:

RI Index table

n	1	2	3	4	5	6	7	8	9	10	11
RI	0	0	0.58	0.90	1.12	1.24	1.32	1.41	1.45	1.49	1.51

As $CR \geq 0.1$, we should adjust the matrix if the matrix is inconsistent. As $CR < 0.1$, if the judgment matrix is in the acceptable scope, we can go on. The next step is testing the calculation of the general order and the compounding consistence.

Calculating the weight vector, there are many ways, such as definition calculation, computer interaction, power principle and sum formula, the sum formula is the simplest method. If the weight vectors have An factors:

$$W = (w_1, w_2, w_3 \quad w_n)$$

The judgment matrix is :

$$A = \begin{pmatrix} w_1/w_1 & w_1/w_2 & \cdots & w_1/w_n \\ w_2/w_1 & w_2/w_2 & \cdots & w_2/w_n \\ \vdots & \vdots & \ddots & \vdots \\ w_n/w_1 & w_n/w_2 & \cdots & w_n/w_n \end{pmatrix}$$

Based on the above matrix property, we can normal all weight vectors and get matrix D

$$D = \begin{pmatrix} a_{11} & a_{12} & \cdots & a_{1n} \\ a_{21} & a_{22} & \cdots & a_{2n} \\ \vdots & \vdots & \ddots & \vdots \\ a_{n1} & a_{n2} & \cdots & a_{nn} \end{pmatrix} \bullet \begin{pmatrix} 1/\sum_{i=1}^{n} a_{i1} & 0 & \cdots & 0 \\ 0 & 1/\sum_{i=1}^{n} a_{i2} & \cdots & 0 \\ 0 & \vdots & \ddots & \vdots \\ 0 & 0 & \cdots & 1/\sum_{i=1}^{n} a_{in} \end{pmatrix}$$

After calculating the sum of normalization, the sum of every line is matrix E

$$E = D \bullet \begin{pmatrix} 1 & 1 & \cdots & 1 \end{pmatrix}^T_{1\times n}$$

$$E = \begin{pmatrix} e_{11} & e_{12} & \cdots & e_{1n} \end{pmatrix}^T$$

After normalizing to matrix E, the weight vector will be:

$$W = \begin{pmatrix} w_1 & w_2 & & w_n \end{pmatrix} = \begin{pmatrix} e_{11} / \sum_{i=1}^{n} e_{i1} & e_{12} / \sum_{i=1}^{n} e_{i1} & & e_{1n} / \sum_{i=1}^{n} e_{i1} \end{pmatrix}$$

Calculating the maximum eigenvector

$$\lambda_{max} = \frac{1}{n} \sum_{i=1}^{n} \frac{(AW)_i}{w_i}$$

3 the result of weight vector

The judgment matrix is:

$$P = \begin{pmatrix} 1 & 2 & 1/3 & 4 \\ 1/2 & 1 & 1/4 & 5 \\ 3 & 4 & 1 & 7 \\ 1/5 & 1/3 & 1/7 & 1 \end{pmatrix}$$

After calculating, the weight vector will be:

$$W = \begin{pmatrix} 0.25 & 0.14 & 0.55 & 0.06 \end{pmatrix}$$

The maximum eigenvector:

$$\lambda_{max} = 4.068$$

The consistent test ratio:

$$CR = 0.025 < 0.1$$

the computation result passed the test.

4 CALCULATE AND ANALYZE THE RESULT

4.1 *Scheme evaluation computation*

We can get the weight vector W, fuzzy matrix. Using $M(\bullet, \oplus)$ in A and R to get $B = (b_1, b_2, \quad b_m)$. According to the evaluation model, we can compare every scheme, and get the final proper scheme.

$$B_1 = WOR_1 = \begin{pmatrix} 0.25 & 0.14 & 0.55 & 0.06 \end{pmatrix} O \begin{pmatrix} 0.16 & 0.42 & 0.23 & 0.12 & 0.07 \\ 0.54 & 0.21 & 0.16 & 0.09 & 0 \\ 0.56 & 0.23 & 0.14 & 0.10 & 0 \\ 0.13 & 0.19 & 0.44 & 0.18 & 0.06 \end{pmatrix}$$

Then,

$$B_1 = \begin{pmatrix} 0.43 & 0.27 & 0.18 & 0.11 & 0.02 \end{pmatrix}$$

The results calculated by the same principle:

$$B_2 = \begin{pmatrix} 0.46 & 0.58 & 0.32 & 0.25 & 0.14 \end{pmatrix}$$

$$B_3 = \begin{pmatrix} 0.10 & 0.04 & 0.06 & 0.07 & 0.09 \end{pmatrix}$$

$$B_4 = \begin{pmatrix} 0.15 & 0.07 & 0.11 & 0.11 & 0.11 \end{pmatrix}$$

$$B_5 = \begin{pmatrix} 0.25 & 0.79 & 0.33 & 0.25 & 0.06 \end{pmatrix}$$

$$B_6 = \begin{pmatrix} 0.28 & 0.24 & 0.32 & 0.14 & 0.13 \end{pmatrix}$$

4.2 *Priority computation for every scheme*

Calculating the priority order, we can determine the of the comprehensive practice process input proportion of the scheme.

Quantify the fuzzy value of evaluation, as

$$V = \begin{pmatrix} 0.5 & 0.4 & 0.3 & 0.2 & 0.1 \end{pmatrix}$$

The priority computation is

$$N_k = B_k V^T, (k = 1, 2, 3, 4, 5, 6)$$

We can get the matrix B by 4.1:

$$B = \begin{pmatrix} 0.43 & 0.27 & 0.18 & 0.11 & 0.02 \\ 0.46 & 0.58 & 0.32 & 0.25 & 0.14 \\ 0.10 & 0.04 & 0.06 & 0.07 & 0.09 \\ 0.15 & 0.07 & 0.11 & 0.11 & 0.11 \\ 0.25 & 0.79 & 0.33 & 0.25 & 0.06 \\ 0.28 & 0.24 & 0.32 & 0.14 & 0.13 \end{pmatrix}$$

The calculation result is:

$$N = \begin{pmatrix} 0.4 & 0.6 & 0.11 & 0.17 & 0.6 & 0.37 \end{pmatrix}^T$$

Normalizing the above results:

5 THE RESULT ANALYSIS AND THE CONCLUSION

5.1 *The result analysis*

Based on the analysis of polymer, we can build multi-target hierarchy structure. After calculating the multi-target policy, we can precisely control the weight vector of every scheme: precisely controlling the polymer structure0.18, polymer reactor dynamic model 0.27, effective productivity0.05, priority property of polymer0.07, blending micro and macro study 0.26, polymer reactor alloying study 0.16. All those can be proved to improve that it is the effective way of producing polymer productions.

Therefore, in the background of high-molecule material chemical industry, the future of polymer is full of hope, so is the polymer production. In the initial time, we should pay more attention to the polymer reactor dynamic model, blending micro and macro study, precisely control the polymer structure, and make a good theoretical basis for the developing polymer industry.

REFERENCES

[1] Ke Yangchuan. (2012) The new field of high performance molecule material—heavy oil polymer material. *High Molecule Science and Engineer*, 6.

[2] Ge Chunli. (2010) *The Continuing Lighting Polymer Reactor Set with High Property of High Molecule Material*. Beijing: Beijing University of Chemical Technology.

[3] Wang Huashan, Zhao Zinia, Li Ruining, Gao Yuzhuo. The Exploration and Thinking based on the excellent engineering to the high molecule material producing process engineering major polymer model education.

[4] Dong Qi. (2008) *Performance Optimization of Structural Load Type Titanium Catalyst Catalyzed Ethylenepropylene Copolymerization and PP/EPR Reactor Alloy*. Zhejiang: Zhejiang University.

[5] Wang Qi, Li Li, Chen Ning, Liu Yuan, Bai Shi Bing. (2011) Preparation of high performance polymer materials by the method of molecular composite supramolecular system. *Journal of Polymer Science in September*.

[6] Li Bogeng. (2005) *Product engineering chemical reaction engineering new development progress of. Chemical*. 24(4).

[7] Wang Qi. (2010) *Preparation of High Performance Polymer Materials (D) through the Method of Molecular Composite Supramolecular System*. Zhejiang: Zhejiang University.

Electric, Electronic and Control Engineering – Shao, Shu & Tian (Eds)
© 2015 Taylor & Francis Group, London, ISBN: 978-1-138-02842-5

Brief analysis about the application and prospect of bentonite and modified bentonite products in chemical engineering

Chunxia Li
School of Chemical Engineering & Environment, Weifang University of Science and Technology, Shouguang Shandong, China

ABSTRACT: As one of the abundant nonmetallic mineral resources in China, bentonite contains very high value in use. From preservatives to industrial production and treatment of the three wastes, it is easy to find the application of bentonite in various aspects of social activities.

Starting from the theoretical basis of bentonite, this paper analyzes the structure, chemical composition, apparent color and correlated characteristics of it. It can thus be seen that the extensive application of bentonite is closely bound up with its chemical composition, structure and characteristics. Next, this paper gives an introduction to the application background and the practical application domains of bentonite. It builds mathematical models, and makes vertical analysis and horizontal analysis on the application value and development prospect of natural bentonite soil, organic bentonite and inorganic modified bentonite with the conclusion that compared with inorganic modified bentonite and natural bentonite soil, organic bentonite has greater development prospect and higher superiority in light industries such as chemical engineering, textile industry, and papermaking. Organic bentonite shall be set as the main direction for study.

KEYWORDS: bentonite; modified bentonite; chemical engineering; mathematical models

1 INTRODUCTION

As science and technology are in continuous advancement along with the development of various industries, people are becoming more and more interested in nonmetallic materials. Bentonite is one of the most important contemporary nonmetallic minerals. People have started to extract and purify those minerals for more satisfying nonmetallic materials.

During his study on the application and prospect of contemporary bentonite and modified bentonite in chemical engineering, Jie Yang took the current social structure of China into account and looked up plenty of document literature. His finding is as follows: the extraction on contemporary bentonite minerals is lagging in technology, and there're so many resources waiting to be made full use of. This paper offers concrete research on the application of bentonite and modified bentonite products in chemical industry. Many experimental materials for chemical engineering need to be extracted and purified from bentonite. The extraction process is relatively simple while the extracted components are stable, and thus bentonite and modified bentonite have been widely used in chemical engineering. Besides, this paper also proposes the conception about the important role that bentonite will play in the future of chemical engineering.

In his *The Application of Bentonite and Modified Products in Environmental Engineering*, Jun Qiu analyzed the characteristics of the impact that bentonite and its modified products have left on contemporary environment in recent years. His finding is as follows: bentonite is non-toxic with no chemical pollution. It does minor harm to the environment. As it contains good absorbability and lubricity, bentonite has been widely used in environmental engineering and chemical engineering industries. With the increasing attention paid to social environment recently and the extensive application of bentonite in environmental engineering, the importance of bentonite and its modified products has been manifested.

During her *The Modification of Bentonite and Its Application in Printing and Dyeing*, Hualing He made specific study on the application of bentonite in printing and dyeing in China. She pointed out that compared with those of more developed countries, our printing and dyeing industry is far more behind. One of the reasons lies in technology while the other is because we are lack of various raw materials. As the raw materials for burning can be refined from bentonite and the impact on the environment is little, the finding of bentonite fills up our gap in raw materials. This is the future development direction for our printing and dyeing industry.

While studying on bentonite, Xuefeng Hu made deep analysis on bentonite preparation and came to study its adsorption property. The author found bentonite contains strong adsorption property which can be further developed into adsorbent and thus broaden the new market of bentonite.

Firstly, this paper gives a brief introduction to the theoretical knowledge of bentonite, and then makes analysis based on the fundamental information about the application of bentonite and its modified products in various industries in China. At last, the prospect for the development of bentonite industry in each area of China for the following five years is also given in this paper.

2 MODELING

2.1 *Theoretical analysis*

Bentonite is aqueous clay vanadium mineral mainly composed of montmorillonite. The main part of theoretical analysis will study the basic properties of bentonite, including its structure, chemical composition, color, characteristics, etc.

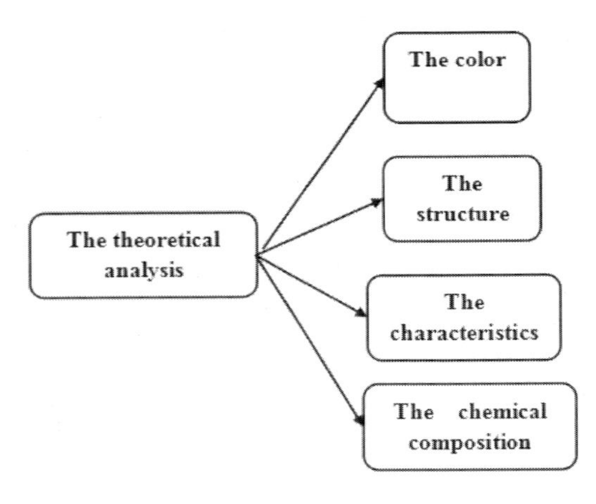

Figure 1. Analysis of bentonite theory.

The study process diagram for the theoretical analysis of bentonite is shown above. Definite introduction to the above points specific to bentonite is as follows.

2.1.1 *Characteristics of bentonite*

Among the multiple characteristics contained by bentonite, the outstanding ones are its absorbability, viscidity and cation exchangeability.

1 Expansibility and hygroscopicity are the outstanding characteristics of bentonite. To measure the amount of absorbability, the adsorbed water quantity can be used for comparison. Average bentonite can adsorb 8 to 15 times of water quantity compared to the original water quantity contained in itself.
2 Submerged into water, bentonite can disperse into colloidal substances. The formed final solution contains lubricity and certain glutinousness.
3 According to the variety of substances, bentonite has different degrees of absorbability. For example, there's big difference between the absorptive power for gas and liquid while that for organic substances can be as much as 5 times.
4 Bentonite can be mixed with water, mud or fine sands, and the formed final admixture contains plasticity and cohesiveness.
5 Moreover, there's one kind of bleaching clay that has surface activity. It presents acidity and can adsorb colored ions.

Thus it can be seen that each characteristic of bentonite can decide its domain and application area. The application of bentonite has a wide range. To develop the actual application value of bentonite according to each characteristic of it found in the study is the key for the utilization of bentonite at present stage.

2.1.2 *Structure of bentonite*

Bentonite is montmorillonite. Its basic structure is a 2:1 type crystal structure consisted of two Si & 0 tetrahedrons with one octahedron formed by a layer of Al & 0 placed in the middle.

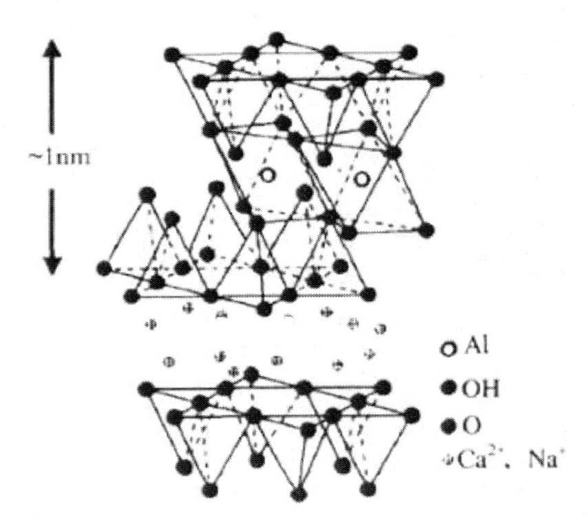

Figure 2. The structure of montmorillonite.

There're many positive ions in the structure mentioned above, such as ***Cu, Mg, and Na***. The relation between montmorillonite and the positive ions corresponding to these elements is not stable. As the positive ions can be easily exchanged by others, the ion exchange is strong.

2.1.3 *Apparent color of bentonite*

Some bentonite can be as loose as soil while some can be very hard with compact structures. Average bentonite presents faint yellow while some can be white. Resulting from the iron content in it, the color of bentonite can also vary into light grey or light green. Some bentonite can even present ash black.

2.1.4 *Chemical composition of bentonite*

Bentonite is also called montmorillonite. Its main elements are *Si* and *Al* while the main molecular formula is $Al_2 \bullet Mg_3)[SiO_{10}][OH]_2 nH_2O$.

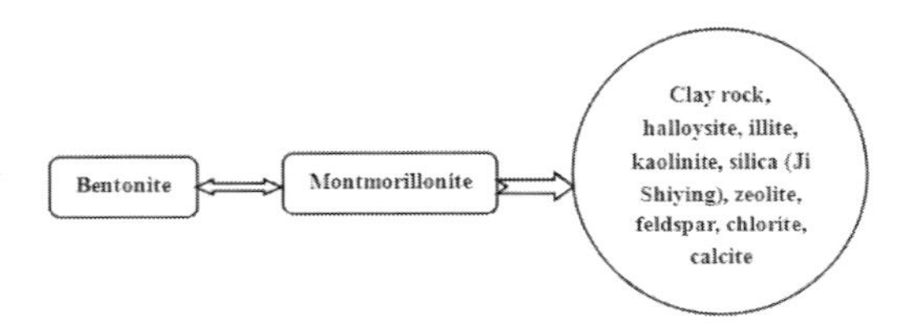

Figure 3. Chemical composition of bentonite.

The main chemical components of bentonite are SiO_2, Al_2O_3 and H_2O, and it also contains *Fe*, *Mg* and *Ca*. The *NaO* content and *CaO* content have great influence on the relative physic-chemical properties of bentonite and its technical performance.

2.2 *Application background introduction*

It can be seen from the introduction to bentonite theory that bentonite is a product with broad design range and high utilization value. In foreign countries, bentonite has already been applied in industrial & agricultural production among which more than 20 domains have had it come into service and over 300 products have been generated.

In China, the original bentonite is used as detergent with no wide application range. Furthermore, some even have no idea about what bentonite is.

Although the bentonite resource in our country is abundant, our utilization on it is less than that in foreign countries. Some of our products are on the lower side and our technologies need to be improved. Therefore, at the present stage, the study on bentonite and modified bentonite is the key to break through the bottleneck of our bentonite development.

2.3 *Practical application*

As bentonite has good physical and chemical properties and a wide range of application, it can be used in steel casting, petroleum exploitation, metallurgical pellet, chemical products, engineering industry and agricultural industry. Among the application domains, bentonite is mainly used as various chemical reagents, such as catalyst and stabilizer. See Figure 4 below for details.

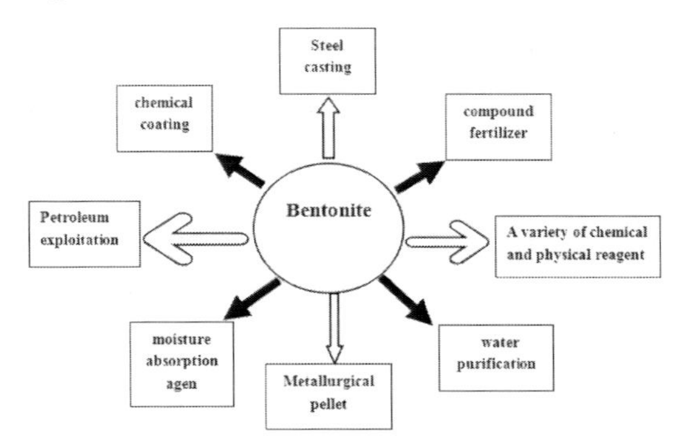

Figure 4. Application of bentonite.

In addition, bentonite can also be used in daily life and industrial production. For example, it can be used to remove the trace amount of toxin in cooking oil; to deal with the waste water generated from industrial production; and to purify the impurities in gasoline and kerosene.

Besides, light industry belongs to the application area of bentonite as well. In paper manufactory industry, bentonite can be used as padding to optimize coating. In textile industry, bentonite can be used as yarn sizing. And starch is within the application domain of bentonite.

In short, bentonite is a mineral material with a broad range of application. It can be widely used in engineering, agricultural industry and light industry, such as paper-making industry and textile industry; and it can also be applied in daily-use chemical industry.

2.4 *Modified bentonite*

By certain means (solution intercalation or copolymerization mixture), modified bentonite is made by mixing calcium bentonite with sodium carbonate or sodium bicarbonate and the dodecyl trimethyl ammonium bromide, cetyl trimethyl ammonium bromide and octadecyl trimethyl ammonium bromide in the positive ion long chain.

In most areas of our country, the bentonite is calcium based. As there're many defects in calcium-based bentonite, such as weak dispersity and easiness in deposition, calcium-based bentonite needs to be modified in order to broaden its application range.

2.5 *Study on the prospect of bentonite and modified bentonite*

By looking up a flood of document literature, natural bentonite soil, inorganic modified bentonite and organic bentonite have been set as the three main directions for study on the prospect of bentonite and modified bentonite, among which:

1 Natural bentonite soil: preservatives, desiccant, synthetic molecular sieve, fire extinguisher configuration model and cat litter configuration model.
2 Inorganic modified bentonite: inorganic pillared bentonite and activated clay.
3 Organic bentonite: coating chemical, petrochemical, textile and paper industries, such as high temperature lubricant and daily-use chemical industry.

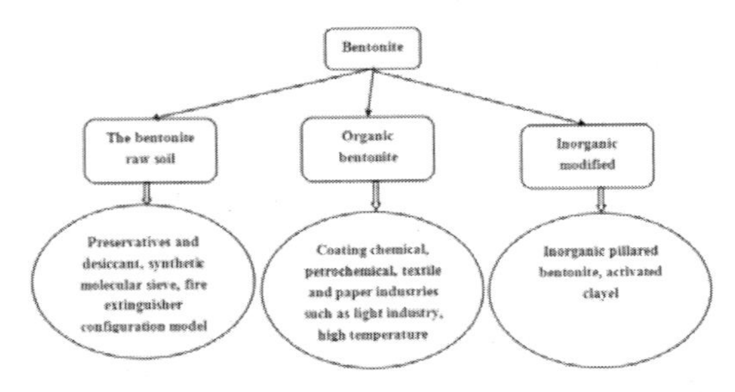

Figure 5. Bentonite and modified bentonite used classification.

Make vertical analysis on the development prospect of natural bentonite soil, inorganic modified bentonite and organic bentonite; and confirm the main direction for the future development of bentonite. Make horizontal analysis on the specific application of all bentonite; and further confirm the application domain with the most value of each kind.

2.5.1 *Vertical analysis*

For the analysis on the prospect of bentonite, we shall firstly make vertical analysis of natural bentonite soil, inorganic modified bentonite and organic bentonite, and build vertical analysis model.

From document and literature in quantity, it can be known that three aspects should be considered for the assessment of bentonite——production cost, use value, and market sales volume. According to:

$$U=\{u_1,u_2,\cdots,u_m\},m=1,2,3,4$$

While applying expert evaluation method to assess bentonite, the basis is as follows:

$$V=\{v_1,v_2,\cdots,v_n\},n=1,2,3,4$$

Evaluation grade set= {very good, good, normal, bad}.

The main representing method of weight function is given below:

$$w=\{\mu_1,\mu_2,\cdots,\mu_m\},m=1,2,3,4$$

Among which: $\sum_{m=1}^{6}\mu_m=1$

AHP (analytic hierarchy process) and normalization method are mainly used to confirm the weight of evaluation index. Normalization formula is as follows:

$$w_i=\frac{\frac{C_i}{S_i}}{\sum_{i=1}^{n}\frac{C_i}{S_i}},(i=1,2,\cdots,m)$$

w_i refers to the monitoring value of evaluation index i and $\bar{S}_i$ refers to the arithmetic mean value of m-level standard. The weight set is given below:

$$w=\{w_1,w_2,\cdots,w_m\}。$$

Normalization formula shall be used to calculate the weight here. The result is as follows:

$$w=\{0.30\quad 0.29\quad 0.28\quad 0.13\}$$

Membership principle can be used to obtain fuzzy relation matrix R, which is:

$$R=\begin{pmatrix}0.1 & 0.2 & 0.1 & 0.4\\ 0.3 & 0.4 & 0.4 & 0.1\\ 0.5 & 0.3 & 0.2 & 0.1\\ 0.1 & 0.1 & 0.3 & 0.4\end{pmatrix}$$

As $W=(\mu_j)_{1\times m}$ and $R=(r_{ji})_{m\times n}$ are given, according to

Fuzzy evaluation set S can be obtained, among

$$S=w\circ R=(\mu_1,\mu_2,\cdots,\mu_m)\circ\begin{pmatrix}r_{11} & r_{12} & \cdots & r_{1n}\\ r_{21} & r_{22} & \cdots & r_{2n}\\ \vdots & \vdots & \vdots & \vdots\\ r_{m1} & r_{m2} & \cdots & r_{mn}\end{pmatrix}=(s_1,s_2,\cdots,s_n)$$

which "∘" refers to the fuzzy composition operator. Take $M(\cdot,\oplus)$ operator as the fuzzy operator here, which is:

$$s_k=\min\left(1,\sum_{j=1}^{m}\mu_j r_{jk}\right),k=1,2,\cdots,n$$

The following set can be obtained by bringing the above calculation result into the above equation:

$$S=\begin{pmatrix} 0.298 & 0.312 & 0.390 \end{pmatrix}$$

According to the maximum membership principle, if fuzzy evaluation sets $S=(S_1,S_2,\ \ ,S_n)$ (among which S_i refers to the membership degree of level v_i in the fuzzy evaluation set) and $M=\max(S_1,S_2,\ \ ,S_n)$ are given, the elements corresponding to M are the evaluation results of the comprehensive assessment.

It can be known from $S=\begin{pmatrix} 0.298 & 0.312 & 0.390 \end{pmatrix}$ that:

$$M=\max(S_1,S_2,\ \ ,S_n)=0.356。$$

Therefore, from the results of the above model calculation, we can confirm that organic bentonite should be set as the main direction for study instead of natural bentonite soil or inorganic modified bentonite.

2.5.2 *Horizontal analysis*

As natural bentonite soil is bentonite without modification, it basically contains the original form and characteristics of bentonite. On this basis, the reasonable utilization made on natural bentonite soil is mainly accomplished by processing it into preservatives and desiccant; or applying it to make synthetic molecular sieve; or using it as fire extinguisher configuration model and cat little configuration model.

As natural bentonite soil has unique cell unit structure and can be made into preservatives and desiccant, it possesses important status in food preservative and waterproofing areas. With a structure similar to that of molecular sieve and high acid resistance, thermal stability and non-corrosiveness, natural bentonite soil can be used in the operation of synthesizing molecular sieve. To make use of the non-reactivity, non-flammability and high hygroscopicity of natural bentonite soil, it can be applied in the configuration of a new fire extinguishing material which means great usefulness in the study on fire extinguishing agent at this stage.

Inorganic modified bentonite results from inorganic modification on bentonite. Its main application lies in making inorganic pillared modified bentonite and activated clay. Compared with natural bentonite soil, activated clay has wider application range in chemical industry and possesses certain development prospect.

From the above vertical analysis model, we are able to make respective analysis on natural bentonite soil, inorganic modified bentonite and organic bentonite, and thus can find out the proper application for these three kinds of bentonite.

1 Natural Bentonite Soil

On the basis of:

$$U=\{u_1,u_2,\ \ ,u_m\},m=1,2,3,4$$

The following equation can be used while assessing natural bentonite soil by expert evaluation method:

$$V=\{v_1,v_2,\ \ ,v_n\},n=1,2,3,4$$

Evaluation grade set= {very good, good, normal, bad}.

The main representing method for weight is given below:

$$w=\{\mu_1,\mu_2,\ \ ,\mu_m\},m=1,2,3,4$$

Among which: $\sum_{m=1}^{6}\mu_m=1$

Normalization formula shall be used to calculate the weight here. The result is as follows:

$$w=\begin{Bmatrix} 0.30 & 0.29 & 0.28 & 0.13 \end{Bmatrix}$$

Membership principle can be used to obtain fuzzy relation matrix R, which is:

$$R=\begin{pmatrix} 0.1 & 0.2 & 0.1 & 0.4 \\ 0.3 & 0.4 & 0.4 & 0.1 \\ 0.5 & 0.3 & 0.2 & 0.1 \\ 0.1 & 0.1 & 0.3 & 0.4 \end{pmatrix}$$

And thus S can be obtained:

$$S=\begin{pmatrix} 0.356 & 0.312 & 0.256 & 0.231 \end{pmatrix}$$

Which is: $M=\max(S_1,S_2,\ \ ,S_n)=0.356$

Thus, it can be concluded that the application of natural bentonite soil in synthesizing molecular sieve is the direction for future study while the making of preservatives and desiccant has high use value as well.

2 Inorganic Modified Bentonite

By means of the above method, we can confirm that activated clay is the main direction for future study on inorganic modified bentonite, and it is also the application with the highest use value for inorganic modified bentonite.

3 Organic bentonite

The use value of organic bentonite in petrochemical engineering and coating chemical industry is high. In addition, organic bentonite has good development prospect in daily-use chemical industry.

3 CONCLUSIONS

Bentonite and modified bentonite are the main aspects of bentonite study at the present stage. This paper mainly starts from study on bentonite theory and its application in industrial industry, agricultural industry and chemical industry, and thus analyzes the development prospect of bentonite and modified bentonite.

Firstly, this paper focuses on analyzing the theories related to bentonite, including its structure, chemical composition, apparent color and related characteristics. It can be seen from the analysis that the widespread application of bentonite is closely bound up with its chemical composition, structure and characteristics.

Secondly, based on theoretical study, this paper analyzes the application background of bentonite and gives an introduction to the practical application domains of it. At last, by building mathematical models, this paper makes vertical analysis and horizontal analysis on the application value and development prospect of natural bentonite soil, organic bentonite and inorganic modified bentonite with the conclusion that compared with inorganic modified bentonite and natural bentonite soil, organic bentonite has greater development prospect, thus it shall be set as the main direction for study.

REFERENCES

[1] Yang J. (2008) The Application and Prospect of Bentonite and Modified Bentonite in Chemical Engineering. *Chemical Engineering and Equipment*.

[2] Qiu J. (2013) The Application of Bentonite and Modified Products in Environmental Engineering. *China Non-metallic Mining Industry Herald*.

[3] Hu X.F. (2008) *Preparation of Modified Bentonite and Study on Its Adsorptive Property*. Wuhan University of Technology.

[4] Chen S.P (2007) *Preparation of Modified Bentonite and Study on Its Capacity in Absorbing Indoor Polluting Gases*. Northwest Normal University.

[5] Xin F.C. (2007) *Study on Modification of Bentonite and Its Application in Wastewater Treatment*. Northwest Normal University.

[6] He H.L. (2009) *Modification of Bentonite and Its Application in Printing and Dyeing*. Hebei University of Science and Technology.

[7] Luo C.Y. (2006) *Modification of Bentonite and Study on Its Absorbing Capacity in Heavy Metal Ion of Wastewater*. Anhui Agricultural University.

[8] Yang M. (2006) *Preparation of Modified Bentonite Adsorbent and Study on Nitrogen and Phosphorus Removal Capacity of Modified Bentonite*. Kunming University of Science and Technology.

Electric, Electronic and Control Engineering – Shao, Shu & Tian (Eds)
© 2015 Taylor & Francis Group, London, ISBN: 978-1-138-02842-5

Preliminary study on the application of humic acid chemistry & chemical engineering materials in biological bacteria preparations

Bin Peng
School of Chemical Engineering & Environment, Weifang University of Science and Technology, Shouguang Shandong, China

Qingjun Peng
Hualong Junior Middle School, Shouguang Shandong, China

ABSTRACT: From detailed agricultural planting work, it has been discovered that single variety of bacteria preparations has very limited antagonistic effect. It is especially true in the case where the advantages of biological population cannot be effectively manifested after the bacteria preparations are sprayed into the crop fields. As a result, the inhibitory effect on targeted bacteria is weak. Some study suggested the combination of the single variety of bacteria preparations with chemistry and chemical engineering can realize our goal of improving microorganic environment, and maintain the vigor of the bacteria preparations in a stable and ideal state. Hence, the application of biological bacteria preparations in eliminating pathogenic bacteria will result in a satisfying consequence. In combination with the actual situation of the diseased bananas in some certain district, this paper will make preliminary analysis and study on the application of humic acid chemistry and chemical engineering materials in biological bacteria preparations.

KEYWORDS: biological bacteria preparations; chemistry and chemical engineering materials; bacillus licheniformis; antagonistic bacteria; experiment

As a critical component of microorganism system, bacillus licheniformis is widespread, and it has very important influence on the performance of soil structure system. [1] Related studies [2-4] suggested that besides their effect on disease prevention, some bacillus licheniformis can help boost yield, adjust and improve rhizospheric environment for plants, and promote the benign growth of crops. From the study result, we can see the potential of conducting deeper study based on the previous one. The special effect and value owned by bacillus licheniformis can be applied to develop new microorganic preparations. If the development is achieved, the new preparations can even replace traditional chemical pesticide to a certain degree. In order to deepen the study on this issue, the following analyses will be given from two aspects in this paper:

1 ANALYSIS ON BACILLUS LICHENIFORMIS RESEARCH PROGRESS

1.1 *Character analysis on bacillus licheniformis*

It has already been pointed out in report that bacillus licheniformis can be used to control the pathogenic bacteria in plants suffering from blight. The mass propagation of pathogenic bacteria can be completely eradicated by antagonism treatment, and thus the goals of improving plant survival level and enhancing plant yield can be realized. The character indexes for this kind of bacillus licheniformis are mainly as follows: 1) thallus length is around 1.75um with width of about 0.7um; 2) the main insertion method of thallus endospore is growing in the central; 3) the main form of thallus endospore is short rod-like. Meanwhile, bacillus licheniformis should also be in accordance with the following utilization features of chemical substances: positive expression in citrate, glucose acid production, arabinose acid production, wood sugar acid production, mannose acid production, amylolysis, casein hydrolyzation, methyl-red reaction, glucose anaerobic fermentation and catalase reaction.

1.2 *Strain cultivation of bacillus licheniformis*

There are certain differences of nutritional requirement in various microorganisms. They need continuous nourishment supply from the surroundings, and energy can be obtained by this means while new cellular materials can be synthesized. However, it shall be paid special attention [5] to that for culture media of different nutritional ingredients, the synthesis of cellular materials can leave very important impact on cultivation results. Some study has already applied starch, cane sugar and glucose with mixing proportion of 2:1:1 as carbon source [6] for the composite design and analysis on strain cultivation. Analysis has been made on the optimal culture medium quality matching solution according to analysis data, and the reached conclusion is that the matching ratios of total sugar, bean pulp, yeast extract, MgSO4 and K2HPO4 should be 2.4%, 3.0%, 0.3%, 0.06% and 0.2% respectively. Unit productivity has been effectively improved by adjusting the culture media based on the matching ratio solution mentioned above.

1.3 *Antagonistic activity of bacillus licheniformis*

According to related research data, the reason for the disease prevention effect of antagonistic bacteria results from the active substance secreted during the regrowth phase of antagonistic bacteria. Other than cell, the active substance has strong effect on antagonism specialization, and is able to definitely inhibit the growth of cells from same bacterium species but different strains or from relative species. In this regard, the antagonistic activity of bacillus licheniformis mainly functions in two parts: one is that antagonistic bacteria can make direct damage to the generation of pathogenic bacteria: after the treatment on pathogenic bacteria realized by antagonistic bacteria, certain distortion will appear on the hyphae of pathogenic bacteria, and thus lead to thinner color. In some cases, hyphae will have swelling reaction with significant substance decrease in the cells while obvious fracture signs can be found on the appearance of hyphae [7]; the other part is that antagonistic bacteria can effectively inhibit the growth of pathogenic bacteria hyphae: previous studies have suggested that antagonistic bacteria transplantation in different time can have great influence on the inhibition effect of pathogenic bacteria growth. The best bacteriostatic action comes from transplantation in advance. Moreover, the second best bacteriostatic actions comes from simultaneous transplantation of antagonistic bacteria and pathogenic bacteria. Delayed antagonistic bacteria transplantation will significantly weaken the antibacterial effect.

1.4 *Biological prevention and control for bacillus licheniformis*

The approaches to realize biological prevention and control for bacillus licheniformis are as follows: firstly, certain antifungal protein and antibiotics can be generated during the biological metabolism reaction process of antagonistic bacteria. They can have specific inhibitory effect on the growth and reproduction of pathogenic bacteria. (Enzyme degradation made by cells can also inhibit the growth of pathogenic bacteria for sure.) Secondly, antagonistic bacteria can obtain sufficient nutrition and water through the rapid growth and reproduction of antagonistic bacteria, and some pathogens sharing the same biological environment with antagonistic bacteria can be eliminated in mutual competition. Thirdly, plenty of sexual structures and branches can appear after part antagonistic bacteria enter the main hyphae of pathogenic bacteria. The occurrence of a series of symptoms evoked by pathogenic bacteria can be avoided in this way. Lastly, as anti-pathogenic bacteria, bacillus licheniformis can also promote plant growth to a certain degree. They can not only control the appearance of pathogenic bacteria, but also improve seed germination rate, root length, sprout length, and plant vigor [8–9].

2 ANALYSIS ON BIOLOGICAL BACTERIOSTASIS AND ANTAGONISM EXPERIMENT

Take XX district as an example. Banana planting is the main agricultural business in this district. In recent years, the increasing occurrence frequency of banana chlorotic disorder has affected a broad range. Banana yield of this district has been seriously damaged due to low control efficiency. Based on the biological prevention and control effect of biological bacteria preparations, this paper will try to conduct antagonism experiment with the chlorotic disorder pathogenic bacteria in this district and the bacillus licheniformis existing in surrounding soil, and analyze the data of this antagonism experiment, in order to demonstrate the feasibility of biological bacteriostatic effect and antagonistic effect [10]. Specifically, the key operating steps can be summarized into the following parts:

2.1 *Pathogenic bacteria cultivation method*

In the strains coming from the root system of the diseased or infected bananas in XX district, 11 species of pathogenic bacteria strains have been extracted. In order to get purified pathogenic bacteria and antagonistic bacteria cultures, and make further analysis on their characters, reasonable pathogenic bacteria cultivation method must be set: in phase one, put each formula into different test tubes and 5 to 8 triangular flasks. Reserve them for future use after high-pressure sterilization treatment. In phase two, take the junction issues from diseased pathogenic bacteria, and cut them into small pieces. Immerse the pieces into ethanol water with concentration of 70% for 60.0s, and then put them into mercury bichloride solution with concentration of 0.1% for 60.0s. Use sterile water to wash the pieces over and over again for 3 times, and then place them on a PDA (PSA) plate. Keep the cultivation under temperature-constant state for 7 days until white cotton-shaped hyphae appear. Pick relatively dispersive bacterial colonies and use inoculating loops to inoculate them onto test tube slant. Cultivate the pathogenic funguses under temperature from 24.0°Cto 26.0°C.

2.2 *Antagonism experiment method*

Use the pathogenic bacteria identified by the method mentioned above and the bacillus licheniformis separated from the soil neighboring diseased or infected root segments to conduct antagonism experiment. PSA should be applied as the culture medium, inoculation environmental temperature should be kept under 25.0°C, and the growth time of pathogenic bacteria should be controlled more than 5.0 days. Observe and analyze the results of antagonism experiment.

See the table below (Table 1) for the schematic table of cingula variation caused by different cultivation days for 20# and 21# antagonistic bacteria. For 20# and 21# antagonistic bacteria, the perfect effect can be obtained by inoculating antagonistic bacteria in the meantime after inoculating antagonistic bacteria one day earlier. The data suggested at the same time that under the condition with varied treatment days, ageing process occured in the performance of antagonistic bacteria to some extent. Antagonistic activity level was significant lower, basically consistent with the result of the research report.

Table 1. Schematic table of cingula variation caused by different cultivation days for 20# and 21# antagonistic bacteria.

Treatment time of antagonistic bacteria	20#			21#		
	5d	6d	7d	5d	6d	7d
2 days in advance	0.6	0.4	0.1	0.5	0.3	0.2
1 day in advance	1.0	0.9	0.5	1.2	1.1	0.6
Synchronized	0.8	0.5	0.1	1.3	1.1	0.4
1 day in delay	0.4	0	0	0.7	0.3	0.4
2 days in delay	0.1	0	0	0.4	0.2	0

Special attention needs to be paid that during this antagonism experiment, pathogenic bacteria are required to be cultivated for 5 days. However, it is too hard to observe the antagonistic activity of the antagonistic bacteria before those 5 days. Use PSA as the culture medium in this experiment with temperature of 25°C, and then the antagonistic bacteria are able to grow in the pathogenic bacteria during the best period. If obvious antagonistic activity can be manifested, this kind of antagonistic bacteria can be used to inhibit the occurrence of banana chlorotic disorder.

3 ANALYSIS ON BIOLOGICAL BACTERIA PREPARATIONS FIELD EXPERIMENT

The main idea for the analysis in this phase is as follows: by conducting separated cultivation of antagonistic bacteria with the analysis on indoor antagonistic experiment data, strains with ideal bacteriostatic effect can be selected for further fermental cultivation. Reproduction can be expanded based on the above procedures, and biological bacteria preparations can be made by mixing the bacterial liquid from fermental cultivation and chemical materials. At last, demonstrate the practical effect of the biological bacteria preparations in field experiment. The detailed procedures are covered by the following parts:

3.1 *Experimental method*

Select 18#, 20#, 21# and 26# strains containing definite bacteriostatic effect in indoor laboratory experiment for further field experiment. Base on the previous procedure, use 21# strain to make three formulas for further observation. In the meantime, according to local banana planting farmers' experience, sterilizing effect can be achieved by supplying the soil under high-temperature state at noon with water flooding for continuous 7 days. Thus, the farmers' method can be set as the method for control group. Compare these two methods in disease control effect and per unit area yield.

3.2 *Experimental result*

See the table below (Table 2) for the schematic table of bacteriostatic effect on pathogenic bacteria in different bacterial strain field experiments. According to related data in Table 2, the bacteriostatic effect of 20# and 21# bacterial strains are very precise with low morbidity, low severity but high prevention and cure efficiency.

Table 2. Schematic table of bacteriostatic effect on pathogenic bacteria in different bacterial strain field experiments.

Bacterial strains	Total strain quantity(unit)	Diseased strain quantity(unit)	Morbidity (%)	Severity (%)
18#	125	31	24.8	19.3
20#	147	27	18.4	10.5
21#	133	11	8.3	5.7
26#	142	45	31.7	23.9

See the table below (Table 3) for the schematic table of impact effect left on banana yield by antagonistic bacteria. According to related data in Table 3, formula II and formula III have ideal effect on inhibiting pathogenic bacteria and improving planting yield. Water flooding treatment can only have short-term effect while the impact that pathogenic bacteria have on bananas cannot be significantly controlled. The yield increase efficiency of treatment method II and III can 36.28% and 43.51% respectively. Although water flooding method can lower morbidity, its efficiency on disease control is very limited, basically the same with that of the control group.

Table 3. Schematic table of impact effect left on banana yield by antagonistic bacteria.

Antagonistic bacteria	Total strain quantity(unit)	Diseased strain quantity (unit)	Morbidity (%)	Severity (%)
Treatment I(empty)	127	22	17.3	13.6
Treatment II(formula)	125	27	21.6	11.3
Treatment III(formula)	133	10	7.5	4.0

4 CONCLUSION

Although biological pesticide has definite disease control efficiency in the crop planting process at present, its vitality can be impacted and interfered by natural conditions. There're abundant types of chemistry and chemical engineering materials with diverse functions. Full use of these materials can create favorable conditions for the development and spread of biological bacteria preparations, and will consequently promote its integrated application effect.

REFERENCES

[1] Zhao T.T., Zhang Y.R. & Zhang L.J. et al. (2012) Root-position Intensification on Methane Oxidation Accomplished by Amphitrophic Bacteria from Mineralized Refuse. *CIESC Journal*, 63(1):266–271.

[2] Yang Y., Li X. & Han F.H. et al. (2012) Analysis on the Metabolome in the Industrial Fermentation Process of Pleurotus Mutilus. *CIESC Journal*, 63(4):1182–1188.

[3] Effect of a preparation containing lactic fermentation bacteria on the hygienic status and aerobic stability of silages. *Polish journal of veterinary sciences*, 2008, 11(1):35–39.

[4] Zhou Deming, Li He, Li Rong et al. (2012) Research on the preparation of fir bacterial fertilizer using biological material. *Mechatronic Systems and Materials Application*. pp:100–105.

[5] H.Y. Moustafa. (2006) Preparation and characterisation of grafted polysaccharides based on sulphadiazine. *Pigment &Amp; Resin Technology*, 35(2):71–75.

[6] Hyun-Woo Shim, Ji-Hye Lee, Bo-Yeol Kim et al. (2009) Facile Preparation of Biopatternable Surface for Selective Immobilization from Bacteria to Mammalian Cells. *Journal of Nanoscience and Nanotechnology*, 9(2):1204–1209.

[7] A. Sivasamy, M.Krishnaveni, P.G. Rao, et al. (2001) Preparation, Characterization, and Surface and Biological Properties of N-Stearoyl Amino Acids. *Journal of the American Oil Chemists' Society*, 78(9):897–902.

[8] Nyczepir, A. P., Kluepfel, D. A., Waldrop, V. et al. (2012) Soil solarization and biological control for managing Mesocriconema xenoplax and short life in a newly established peach orchard. *Plant Disease*, 96(9):1309–1314.

[9] Li J.T., Li Y.N. & Yuan Y.J. et al. (2006) HIV Protease Inhibitor—Molecular Docking Study of Streptolydigin. *CIESC Journal*, 64(24):2491–2495.

[10] Ma Y.L., Liu N.P. & Zhang Z.Y. et al. (2011) Application of Tylosin Dreg in Complex Enzyme Preparations Production and Related Effect Research. *Environmental Science & Technology*, 34(10):134–137.

Electric, Electronic and Control Engineering – Shao, Shu & Tian (Eds)
© 2015 Taylor & Francis Group, London, ISBN: 978-1-138-02842-5

Preparation and electrochemical performance study of polyvinylidene fluoride microporous polymer electrolyte

Jinxiu Liu
School of Chemical Engineering & Environment, Weifang University of Science and Technology, Shouguang Shandong, China

ABSTRACT: This paper will describe the following procedures: mutually mix ionic liquid and organic electrolyte; decide the best mix proportion of ionic liquid and organic electrolyte by conducting inflammability experiment and electric conductivity experiment; and then prepare and form polyvinylidene fluoride microporous polymer dielectric through phase inversion. At last, give comprehensive analysis and demonstration about the electrochemical performance of polymer electrolyte according to the analysis of the void ratio, imbibition volume and conductive mechanism of fire-resistant electrolyte polyvinylidene fluoride-polyethylene glycol polymer electrolyte.

KEYWORDS: polyvinylidene fluoride; polymer electrolyte; preparation; electrochemical performance

1 INTRODUCTION

Polymer electrolyte is not only applicable to the preparation of lithium ion battery, but can also be flexible with a solar cell. According to the volume and leakage variation problems, possibly existing in the current batter preparation process, polymer electrolyte system [1] can be introduced to improve battery working life under normal service conditions. In the meantime, the application of polymer electrolyte can make it easier to form hull cells, so as to reach our goal of improving the utility rate of space [2]. From the vast existing studies, we can see that the huge application potential in polymer electrolyte to be used for electrochemical device preparation, such as a battery, is catching great attention from all parties. How to accomplish better development has become technicians' important research issue in this field. In combination with existing research, polymer electrolyte has three main types: the first is a solid polymer electrolyte, the second is a gel polymer electrolyte, and the third is a microporous polymer electrolyte. Looked from industrial application area, microporous polymer electrolyte contains overall application advantages over the other two among these three polymer electrolyte types. Therefore, the application range of microporous polymer electrolyte is widespread in current lithium secondary battery or other electrochemical devices. And there're numerous types of plasticizers being used in mircroporous polymer electrolyte applications, mainly including diethyl carbonate and DMC (dimethyl carbonate). After introducing the plasticizers mentioned above, the charge-discharge cycle times of electrolyte has been improved with less chances of having capacity loss. However, damage appears on battery flammability and volatility, which can lead to serious adverse influence on battery use safety.

Previous research suggests that ionic liquid is composed of organic cations and inorganic anions or organic anions. The salt element presenting in liquid form under indoor temperature or neighboring temperature contains zero vapor pressure and non-volatility. Meanwhile, ionic liquid has ideal heat-conducting property and electrical conductivity. It possesses high thermal energy storage density, high thermal capacity and good thermal stability. Based on the above advantages, ionic liquid has become the solid liquid with special functions and features in industrial chemistry area. It has high feasibility in constituting various multifunctional materials and media due to the combination its features. It has also been pointed out in some research reports that ionic liquid may have the possibility of replacing traditional electrolytes [3–6] in lithium battery because of its high electrical conductivity, low vapor pressure, reliable fire resistance and wide electrochemical window. Certainly, in order to improve the liability of battery electrochemical performance, interaction with organic solvent can be applied based on ionic liquid. And thus, mixed electrolyte can be formed, so as to improve battery application safety.

According to the results of the research mentioned above, the basic research idea of this paper is as follows: mutually mix ionic liquid and organic electrolyte; decide the best mix proportion of ionic liquid and organic electrolyte by conducting inflammability experiment and electric conductivity experiment; and then prepare and form polyvinylidene fluoride microporous polymer dielectric through phase inversion. At last, give comprehensive analysis and demonstration about the electrochemical performance of polymer electrolyte according to the analysis of the void ratio, imbibition volume and conductive mechanism of fire-resistant electrolyte polyvinylidene fluoride-polyethylene glycol polymer electrolyte.

2 EXPERIMENTAL METHOD ANALYSIS

2.1 *Raw materials for experiment and pretreatment*

The raw materials for experiment are as follows: 1) the methylimidazole provided by XX Chemical Plant. Reduced pressure distillation should be applied in advance. 2) The polyvinylidene fluoride provided by the XX Material Company. Vacuum drying treatment under 100.0°C should be given for continuous 48.0 hours in advanced. 3) The polyethylene glycol provided by XX Company. 4) The ES-002 electrolyte solution provided by XX Company. Arrangement standard for electrolyte solution is ethylene carbonate (1): DMC (dimethyl carbonate) (1): diethyl carbonate (1). The electric conductivity under room temperature is 9.9mS cm-1. 5) Bromoethane provided by the XX Research Institution. It works as chemically pure analytical reagent. 6) 1-ethyl-3-methylimidazole tetrafluoroborate ionic liquid made by the laboratory.

2.2 *Preparation of mixed electrolyte*

Mix the 1-ethyl-3-methylimidazole tetrafluoroborate ionic liquid made by the above standard and ES-002 electrolyte with certain volume ratio standard. The whole operation should be completed in a simple glove box. After the mixing procedure is done, the entire mixed electrolyte will present as a liquid. Based on the above procedure and results, further test analysis needs to be made in order to ensure the safety of the mixed electrolyte as much as possible. Detailed operation procedures are as follows: immerse glass fibers into the mixed electrolyte for 5.0 minutes. Use a nipper to take the glass fibers out and remove the electrolyte exposed on the surface. Ignite the alcohol lamp and wait until the flame stays stable. Place the glass fibers right above the alcohol lamp flame with a distance of 10.0cm [7–5] and maintain it for 10.0 seconds. If the glass fibers do not burn during the process, or they stop burning after the alcohol lamp is removed, it can be certain that the mixed electrolyte is inflaming retarding.

2.3 *Preparation method of polyvinylidene fluoride-polyethylene glycol polymer film*

Existing research has suggested that phase transfer can be used in the preparation process of polyvinylidene fluoride-polyethylene glycol polymer film to obtain the ideal preparation effect. Detailed operation procedures are as follows: blend 1.0g polyvinylidene fluoride material and 1.0g polyethylene glycol material in the mixed solution of DMF (dimethyl formamide) and glycerin. Sufficiently stir the mixed solution under 80.0°C temperature for continuous 12.0 hours. After transparent mucus appears, pour the mixed mucus on a glass slide to help it achieve well-proportioned curtain coating. And then place the mixed mucus in a vacuum drying oven for the next treatment. Keep the environment temperature at 100.0°C and continue the drying treatment for 36.0 hours. Milk white microporous polymer film with certain mechanical strength can be obtained by the above procedures.

2.4 *Characterization analysis method*

1 Imbibition volume analysis method: cut the microporous polyvinylidene fluoride-polyethylene glycol polymer film obtained by the above treatment into a 16.0mm-diameter wafer. Immerse the wafer into ES-002 electrolyte for continuous 48.0 hours. Use filter paper to absorb the electrolyte exposed on the surface of microporous film and measure the quality of the film before and after it is immersed into the ES-002 electrolyte. And then, calculate the imbibition volume, according to the equation given below:

Imbibition volume(unit:%)=100*(quality after immersion−quality before immersion)/quality before immersion;

2 Electric conductivity analysis method: use the TH2818 LCR self-motion component analyzer manufactured by XX Limited Company to make the dried samples into wafers with diameter of 13mm. Scan the wafers with a frequency range from 20HZ to 300KHZ and measure impedance ($|Z|$) and phase angle (θr). The sampling number is 100. Take $Zre = Z' = |Z| \cos\theta$ as x-coordinate and $-Zim = -Z'' = -|Z| \sin\theta$ as y-coordinate to draw Nyquist chart. Polymer body resistance can be achieved by making alternating-current impedance, and thus the electric conductivity can be obtained [9–10].

3 EXPERIMENTAL RESULT ANALYSIS

3.1 *Analysis of imbibition volume experimental result*

See the figure below (Figure 1) for the schematic diagram of the corresponding relation between microporous polymer film imbibition and temperature. Based on the change characteristics of the corresponding relation between microporous polymer film imbibition and temperature in Figure 1, it can be observed that the absorption curve corresponding to the microporous polymer electrolyte gradually ascends with the continuous extension of ES-002 electrolyte immersion operation time until it reaches a balanced state. It can also be seen from Figure 1 that the imbibition volume of the microporous polymer electrolyte reaches the maximum value when the Es-002 electrolyte immersion operation time is 72.0 hours. The reason for why the maximum value got in this way is higher than that of the traditional organic electrolyte is as follows: the viscosity of the mixed electrolyte is around 17.2mPa·s which is obviously higher than that of the traditional organic electrolyte (which is around 0.98mPa·s in most cases).

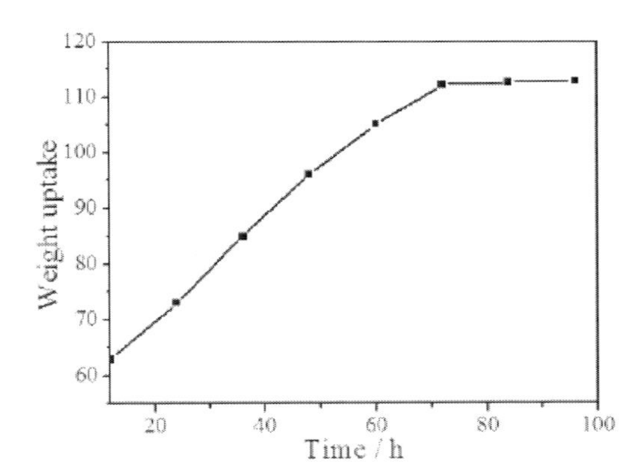

Figure 1. Schematic diagram of the corresponding relation between microporous polymer film imbibition and temperature.

3.2 *Analysis on electric conductivity experiment result*

See the figure below (Figure 2) for the schematic diagram of the corresponding relation between the ionic conduction of polyvinylidene fluoride-polyethylene glycol polymer electrolyte with mixed electrolyte and temperature. Based on the change characteristics of the corresponding relation between the ionic conduction of polyvinylidene fluoride-polyethylene glycol polymer electrolyte with mixed electrolyte and temperature in Figure 2, it can be observed that the electrical conductivity corresponding to the microporous polymer electrolyte has certain trend to rise. In the meantime, the electrolyte ionic conduction value of the microporous polymer is mainly affected by the connected cellular structure and the electrolyte in gel. Moreover, the fact that ionic conduction conforms to VTF equation means the movement of positive ions and negative ions is related to electrolyte viscosity. Hence, the apparent ion transport activation energy of the polymer electrolyte can be reasonably calculated by standard Arrhenius equation. According to this standard calculation method, the apparent ion transport activation energy value corresponding to the microporous polymer electrolyte in this experiment is within 8.0kJ/mol. From this perspective, polyvinylidene fluoride microporous polymer electrolyte still has an obvious influence on ion transport even if it is in microporous form.

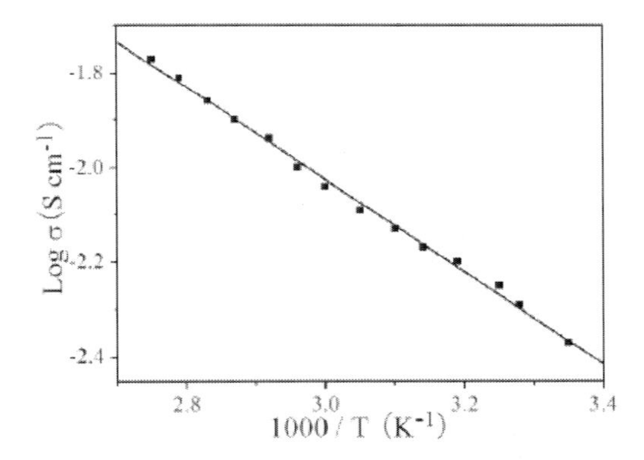

Figure 2. Schematic diagram of the corresponding relation between the ionic conduction of polyvinylidene fluoride-polyethylene glycol polymer electrolyte with mixed electrolyte and temperature.

4 CONCLUSION

In the research described in this paper, the following procedures have been given: mutually mix ionic liquid and organic electrolyte; decide the best mix proportion of ionic liquid and organic electrolyte by conducting inflammability experiment and electric conductivity experiment; and then prepare and form polyvinylidene fluoride microporous polymer dielectric through phase inversion. At last, analyze the void ratio, imbibition volume and conductive mechanism of fire-resistant electrolyte polyvinylidene fluoride-polyethylene glycol polymer electrolyte. Research data suggest that 1-ethyl-3-methylimidazole tetrafluoroborate ionic liquid has high feasibility in organic electrolyte flame retardant additive application. The safety of battery use can be improved by the application of polyvinylidene fluoride micro-porous polymer electrolyte.

REFERENCES

[1] Han K.M., Lu D.R. & Na M. (2008) Preparation and Electrochemical Performance of SiO2/LiClO4/PVDF-HFP Plural Gel Polymer Electrolyte. *Composite Material Journal*, 25(3):57–62.

[2] Xu J.M., Jiang Y.X. & Zhuang Q.C. et al. (2007) The Application of FTIR Spectrum Emission Technology in Polymer Electrolyte Conductive Mechanism Research. *Optical Spectroscopy and Spectral Analysis*, 27(2):247–249.

[3] Zhao S.J., Fan L.Z. & Lin Y.H. et al. (2005) PEO-LiClO4-SiO2 Composite Polymer Electrolyte. Rare Metal Materials and Engineering, 34(z2):1122–1125.

[4] Guan H.Y., Lian F. & Wen Y. et al. (2014) Thermal Stability of New-type Polyvinyl Formal Gel Polymer Electrolyte. *Chemical Journal of Chinese Universities*, (1):80–84.

[5] Mohamed Abudakka, Daniel S. Decker, Logan T. Sutherlin et al. (2014) Ceramic/polymer interpenetrating networks exhibiting increased ionic conductivity with temperature control of ion conduction for thermal run-away protection. *International Journal of Hydrogen Energy*, 39(6):2988–2996.

[6] Wei Zhai, Huajun Zhu, Long Wang et al. (2014) Study of PVDF-HFP/PMMA blended micro-porous gel polymer electrolyte incorporating ionic liquid [BMIM] BF_4 for Lithium ion batteries. *Electrochimica Acta*, 133:623–630.

[7] Ji-Ae Choi, Ji-Hyun Yoo, Woo Young Yoon et al. (2014) Cycling characteristics of lithium powder polymer cells assembled with cross-linked gel polymer electrolyte. *Electrochimica Acta*, 132:1–6.

[8] Hiromori Tsutsumi, Asami Suzuki. (2014) Cross-linked poly (oxetane) matrix for polymer electrolyte containing lithium ions. *Solid State Ionics*, 262:761–764.

[9] Nair, J.R., Destro, M., Gerbaldi, C. et al. (2013) Novel multiphase electrode/electrolyte composites for next generation of flexible polymeric Li-ion cells. *Journal of Applied Electrochemistry*, 43(2):137–145.

[10] Jae-Kwang Kim, Jou-Hyeon Ahn, Per Jacobsson et al. (2014) Influence of temperature on ionic liquid-based gel polymer electrolyte prepared by electrospun fibrous membrane. *Electrochimica Acta*, 116:321–325.

Electric, Electronic and Control Engineering – Shao, Shu & Tian (Eds)
© 2015 Taylor & Francis Group, London, ISBN: 978-1-138-02842-5

Technical economy analysis on the comprehensive utilization of high-carbon natural gas chemical industry

Peng Hu
School of Chemical Engineering & Environment, Weifang University of Science and Technology, Shouguang Shandong, China

ABSTRACT: With the rapid development of the society, energy problem has become a key factor that imposes restrictions on social advancement. Resulting from oil shortage and coal shortage, natural gas has been merged into our life as an indispensable emerging energy resource. Gradually, high-carbon natural gas has developed as one of the important parts in chemical engineering application.
This paper analyzes the technological processes of MTO, MTP and DMTO. Based on fuzzy comprehensive evaluation model, a comprehensive assessment was given on these three technologies from five aspects: production scale, equipment cost, raw material cost, conversion rate and revenue. The assessment results show that DMTO is better than MTO and MTP in the technical economy analysis on the comprehensive utilization of high-carbon natural gas chemical industry.

KEYWORDS: high-carbon natural gas; chemical engineering utilization; fuzzy comprehensive evaluation; DMTO technology

1 INTRODUCTION

Currently, as oil resources are in shortage, the development of new-type energy is extremely urgent. As a new-type energy resource, natural gas has gradually stepped into our life. A series of practical applications, such as natural gas heating and natural gas vehicles, have manifested that the need for natural gas in energy market will be far beyond that for petroleum. With the increasing use of natural gas, more and more urban gas will be needed while natural gas used in chemical engineering will be reduced. Thus, it is very necessary to develop and utilize high-carbon natural gas.

In 2005, Ming Ma explained low-carbon olefin preparation technology, sulfur recovery technology and aromatic hydrocarbon preparation from methane anaerobic aromatization technology which are all based on natural gas in his *The Current Status of Natural Gas Resource Development and Utilization in Our Country*. The chemical engineering utilization of natural gas includes indirect conversion and direct conversion of methane. However, the key to prepare ethylene from methane is to study catalyst. Research findings suggest that that high-quality catalysts include alkali metal, alkaline earth metal oxide, transition metal oxide and rare-earth oxide. The methods to prepare ethylene from methane cover the catalytic conversion under aerobic condition and anaerobic condition. The fundamental purpose of sulfur recovery from natural gas is for sulfuric acid preparation.

Dan Luo points out the method to utilize high-carbon natural gas in chemical engineering industry in his *Technological Analysis on the Utilization of High-carbon Natural Gas in Chemical Engineering Industry*, which is to remove carbon dioxide from high-carbon natural gas and apply the remaining natural gas as fuel or process it into chemical engineering products. Moreover, with the precondition that technical capacity is strong, direct utilization of high-carbon natural gas can be obtained. While using high-carbon natural gas, we must select the strategic thought that is in accordance with national development policy. Take local resources as the basis for development. Based on overall planning, we shall target at improving our own competitive strength and identify the core of the program. Value engineering is an important means to analyze and improve the utilization of high-carbon natural gas in chemical engineering. The significance of value engineering lies in the equilibrium point between cost and function. It is able to acquire high-quality functions with low cost to the greatest extent and result in obtaining the optimum point between cost and revenue.

In 2010, Jianwen Zhang suggested three ways to prepare low-carbon olefin from high-carbon-dioxide natural gas in his *Comprehensive Utilization of High-carbon-dioxide Natural Gas in Chemical*

Engineering. One is the MTO technology process route from UOP/Hydro Corporation. There're three kinds of raw materials in this company: naphtha, natural gas and ethane. One is the MTP technology process route from Lurgi Corporation. Lurgi uses MTP technology as the key means to produce propylene and polypropylene based on methanol. The last one is the DMTO technology from DICP (Dalian Institute of Chemical Physics). It takes coal or natural gas as the replacement of petroleum. This technology uses coal and natural gas to produce ethylene and propylene, making a breakthrough in shattering some restrictions of thermodynamic.

This paper will take comprehensive consideration of high-carbon natural gas utilization program in chemical engineering industry from several aspects, such as production cost, production revenue and rate of investment return.

2 COMPREHENSIVE UTILIZATION OF HIGH- CARBON NATURAL GAS IN CHEMICAL ENGINEERING INDUSTRY

The basis for utilization of high-carbon natural gas is the process of methanol reaction. The matching of hydrogen, carbon dioxide and carbon monoxide shall conform to the matching of reaction equation.

$$CO + 2H_2 = CH_3OH \tag{1}$$

$$CO_2 + 3H_2 = CH_3OH \tag{2}$$

From reaction equation (1) and (2), we can see that the molar ratio of hydrogen and carbon monoxide is 2 while the molar ratio of methanol formed with carbon dioxide is 3. If carbon monoxide and carbon dioxide coexist, the expression of H-N ratio f is shown in equation (3).

$$f = \frac{n_{H_2} - n_{CO_2}}{n_{CO} + n_{CO_2}} = 2.10 \sim 2.15 \tag{3}$$

In general, natural gas with high CO_2 content can satisfy the need for CO_2 in the reaction. It suits the environmental requirement of contemporary society better to apply natural gas as the raw material for low-carbon olefin preparation. In this means, the CO_2 in natural gas can be fully used and the goal of zero release may be realized.

2.1 *MTO process route*

the chemical reactions involved in the process route are three main chemical reactions and three side chemical reactions. Methanol turns into dimethyl ether after the first dehydration, and continues to turn into low-carbon olefin such as ethylene and propylene after following dehydration operations. A small part of low-carbon olefin can generate hydrocarbon through a series of reactions, such as polycondensation, dehydrogenation and hydrogen transfer.

$$2CH_3OH \rightarrow CH_3OCH_3 + H_2O \tag{4}$$

$$CH_3OH \rightarrow C_2H_4 + C_3H_6 + C_4H_8 \tag{5}$$

$$CH_3OCH_3 \rightarrow C_2H_4 + C_3H_6 + C_4H_8 \tag{6}$$

$$CH_3OH \rightarrow CO + 2H_2 \tag{7}$$

$$CH_3OH \rightarrow CH_2O + H_2 \tag{8}$$

$$CH_3OCH_3 \rightarrow CH_4 + CO + H_2 \tag{9}$$

Reactions (4), (5) and (6) are main reactions and reactions (7), (8) and (9) are side reactions. The technological process of the reaction is shown in Figure 1.

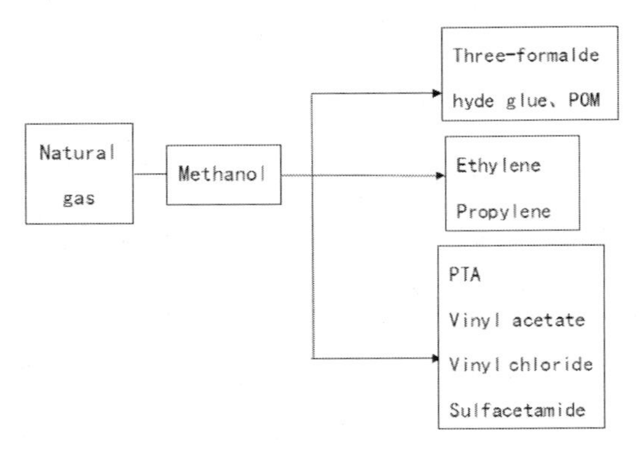

Figure 1. MTO technical process flow diagram.

See Table 1 for the productive rate of this technology.

Table 1. Productive rate of MTO industrial demonstration plant.

Component	Product volume	Productive rate of carbon base/%
Ethylene	52.75	48.0
Propylene	36.50	33.0
Butylene	10.50	9.6
C_5^+	2.63	2.4
$H_2,C_1 \sim C_4$	3.90	3.5
CO_x	0.53	0.5
Coke	0.53	3.0
H_2O	142.66	
Total discharge	250.00	100

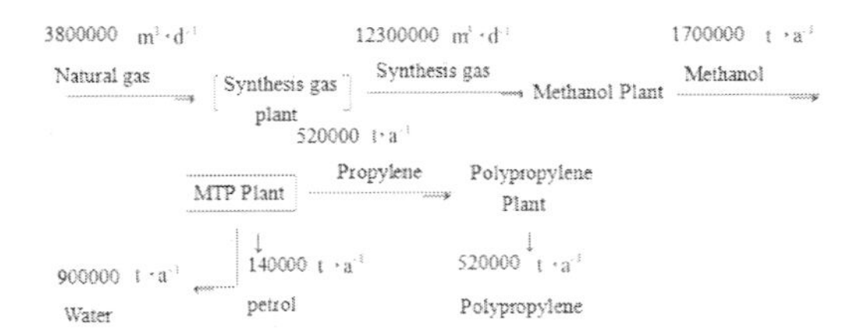

Figure 2. Material balance chart.

2.2 *MTP process route*

MTP technology is an important means to use the methanol processed by the synthesis gas to produce polypropylene after turning natural gas into synthesis gas. See Figure 2 for the material balance chart using MTP technology to produce polypropylene.

From Figure 2, we can see that natural gas can generate synthesis gas after passing through synthesis gas plant. While synthesis gas passes through methanol plant, methanol can be generated using synthesis gas as the raw material. Then methanol can generate propylene through MTP technology while water and gasoline appear as the accessories in the meantime. In the last phase, propylene leads to polypropylene through the processing of polypropylene plant.

Table 2. Rate of MTP investment return.

Item	High propylene reaction yield	High gasoline reaction yield
Investment fee	7.30	7.30
Owner's fee	1.46	1.46
Production fee	1.543	1.507
Income	3.566	3.172
Gasoline	0.186	0.316
Polypropylene	3.38	2.856
Rate of investment return/%	23.1	19.0
Internal rate of return before tax/%	25.1	20.6
Internal rate of capital stock return /%	36.8	30.3

See Table 2 for the rate of investment return of this technology.

2.3 *DMTO process route*

DMTO technology is a technical means to produce alkene materials by natural gas or coal instead of petroleum. This technology covers two parts which can be utilized jointly or individually, as shown in Figure 3. Hence, DMTO technology contains strong flexibility.

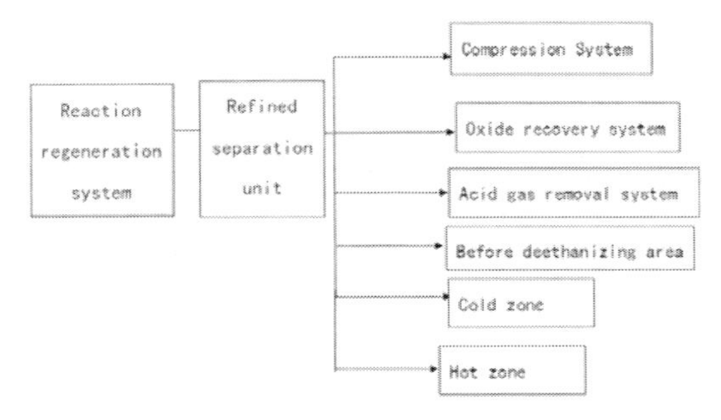

Figure 3. DMTO process diagram.

See Table 3 for the catalyst comparison results of MTO technology and DMTO technology. From Table 3, we can see that there are differences existing between DMTO technology and MTO technology.

Table 3. Catalyst comparison results of MTO technology and DMTO technology.

Item	MTO technology	DMTO technology
Scale of pilot plant test	0.75	0.06~0.1
Raw material	methanol	Dimethyl ether
Type of zeolite	SAPO-34	SAPO-34
Type of reactor	Fluidized bed	Fluidized bed
Price of catalyst	High	Low
Trade mark of catalyst	MTO-100	DO-123
Raw material consumption	2.659	1.845(equals to 2.567 methanol)
Mass fraction of olefin		
Ethylene	34~46	49
Ethylene + propylene	76~79	>79
Ethylene + Propylene + butylene	85~90	About 87

3 FUZZY COMPREHENSIVE EVALUATION

In general conditions, three numbers are involved in fuzzy comprehensive evaluation. Suppose there are n factors relevant to the evaluation object, denoting as $U=\{u_1,u_2,\cdots,u_n\}$. We call U the factor set. And then, suppose there are m comments in total that may appear, denoting as $V=\{v_1,v_2,\cdots,v_m\}$. We call V the judgment set. Resulting from the various statuses owned by each factor with different functions, weight appears, working as yard stick and denoting as $A=\{a_1,a_2,\cdots,a_n\}$.

3.1 *Procedures of comprehensive evaluation*

the procedures of fuzzy comprehensive evaluation are as follows:

1 Set up factor set $U=\{u_1,u_2,\cdots,u_n\}$.
2 Set up judgment set $V=\{v_1,v_2,\cdots,v_m\}$.
3 Get $r_i=\{v_{i1},v_{i2},\cdots,v_{im}\}$ from evaluation of single factor.
4 Construct comprehensive judgment matrix:

$$R=\begin{bmatrix} r_{11} & r_{12} & \cdots & r_{1m} \\ r_{21} & r_{22} & \cdots & r_{2m} \\ \vdots & \vdots & & \vdots \\ r_{n1} & r_{n2} & \cdots & r_{nm} \end{bmatrix} \quad (10)$$

5 Comprehensive assessment: calculate $B=A\circ R$ for weight $A=\{a_1,a_2,\cdots,a_n\}$, and make assessment according to maximum membership principle.

3.2 *Definition of operator*∘

While making comprehensive judgment, there are different model resulting from different definitions of operator∘.

1 Model I: $M(\wedge,\vee)$——principal divisor determining type
The computing method is

$$b_j=\max\{(a_i\wedge r_{ij}),i=1,2,\cdots,n\}(j=1,2,\cdots,m) \quad (11)$$

The result of this judgment model is determined by the factor working as the main function of the total evaluation while other factors have no influence on the judgment. Relatively speaking, this model is suitable for the case in which comprehensive judgment is considered as the best if the judgment for single factor is thought as the best.

2 Model II: $M(\bullet,\vee)$——principal divisor highlighting type
The computing method is

$$b_j=\max\{(a_i\bullet r_{ij}),i=1,2,\cdots,n\}(j=1,2,\cdots,m) \quad (12)$$

This model shares some similarity with Model I, however, it is more detailed. Model II not only highlights the principal factors, but also manifests other factors. It is applicable for the range that is not suitable for Model I, meaning that it can be used in the situation in which various factors cannot be separated but need to be made detailed.

3 Model III: $M(\bullet,+)$——weighted average type
The computing method is

$$b_j=\sum_{i=1}^{n}a_i\bullet r_{ij}(j=1,2,\cdots,m) \quad (13)$$

This model takes all influencing factors into consideration according to the importance of each factor. Comparatively speaking, Model III is applicable for the situation in which the best comprehensive judgment is required.

4 Model IV: $M(\wedge,\oplus)$——sum of the minimum upper bound type
The computing method is

$$b_j=\min\left\{\left(1,\sum_{i=1}^{n}(a_i\wedge r_{ij})\right)\right\}(j=1,2,\cdots,m) \quad (14)$$

While using this model, special attention should be paid: each a_i cannot be taken too big, or it might result in each b_j equaling to 1; and each a_i cannot be taken too small, or it might results in each b_j equaling to the sum of a_i, leading to the loss of information relative to evaluation of single factor.

5 Model V: $M(\wedge,+)$——balanced average type

The computing method is

$$b_j = \sum_{i=1}^{n}\left(a_i \wedge \frac{r_{ij}}{r_0}\right)(j=1,2, \quad ,m) \tag{15}$$

In which, $r_0 = \sum_{k=1}^{n} r_{kj}$. This model is applicable for the situation in which the element of comprehensive judgment matrix R is too big or too small.

The models established in this paper take operator of principal divisor determining type.

Table 4. Evaluation grading results.

Evaluated aspect	MTO technology	MTP technology	DMTO technology
Production scale	0.4	0.5	1.0
Equipment cost	0.2	0.4	0.8
Raw material cost	0.6	0.5	0.9
Conversion rate	0.3	0.8	0.7
Revenue	0.5	0.4	0.6

4 ESTABLISHMENT PROCEDURE

Conduct technical economy evaluation on the three comprehensive utilization methods for high-carbon natural gas. The evaluation will be processed according to five aspects: production scale, equipment cost, raw material cost, conversion rate and revenue. First of all, set up grading for the evaluation on each aspect. Grading ranks are in direct proportion to evaluation results (grading ranks will be higher if the evaluation results show "better"). See Table 4 for the results for evaluation grading.

Establish judgment set according to the data in Table 4. Take production scale, equipment cost, raw material, conversion rate and revenue as u_1, u_2, u_3, u_4 and u_5 respectively.

Weight set $A = (0.12, 0.11, 0.20, 0.26, 0.32)$

Set up comprehensive judgment matrix R according to the data in Table 4:

$$R = \begin{bmatrix} 0.4 & 0.5 & 1.0 \\ 0.2 & 0.4 & 0.8 \\ 0.6 & 0.5 & 0.9 \\ 0.3 & 0.8 & 0.7 \\ 0.5 & 0.4 & 0.6 \end{bmatrix}$$

Respectively calculate those two weights based on Model I——$M(\wedge,\vee)$, resulting in:

$$B = A \circ R = (0.22, 0.29, 0.49)$$

Thus, we can see that DMTO technology is better than MTO technology and MTP technology according to the technical economy analysis on the comprehensive utilization of high-carbon natural gas in chemical engineering industry.

5 CONCLUSION

The fuzzy comprehensive evaluation method used in this paper generally can solve the problem of a permutation or selection. The key for the overall process is to establish fuzzy comprehensive judgment matrix. The constituents of the matrix are the evaluation results for each factor. Readers can set up the weight according to their own experience or the data in other references. However, improper weight can lead to wrong computing process and influence the results.

Currently, the utilization of high-carbon natural gas in chemical engineering industry is the hotspot of study. Till now, there are three methods to utilize high-carbon natural gas: MTO technology, MTP technology and DMTO technology. Based on fuzzy comprehensive evaluation model, this paper makes comprehensive assessment on these three technologies from five aspects: production scale, equipment cost, raw material cost, conversion rate and revenue. The results suggest that DMTO technology is the best one.

REFERENCES

[1] Wang H.S. (2010) Technical Economy Analysis on the Comprehensive Utilization of High-carbon Natural Gas in Chemical Engineering Industry. *Master's Thesis from Beijing University of Chemical Technology*.

[2] Zhang J.W. & Wang H.S. (2010) Comprehensive Utilization of Natural Gas with High Carbon Dioxide Content in Chemical Engineering Industry. *Chemical Engineer*, (2):34–36.

[3] Zhang G.L., Jiang W. & Zhoug Q.X. (2009) Discussion on Carbon Filling Method for Methanol Production Based on Degassing of Mixing Hydrogen Preparation from Natural Gas. [J] *Natural Gas Chemical Industry*, (4).

[4] Ma M. et al. (2005) Current Situation of the Natural Gas Resource Development and Utilization in Our Country. *Petrochemical Engineering*, (4):394–395.

[5] Zhuo J.W. et al. (2010) *Application of Matlab in Mathematical Modeling*. Beijing: Beijing University of Aeronautics and Astronautics Press.

[6] Zhou Y.Z. et al. (2010) *Mathematical Modeling*. Shanghai: Tongji University Press.
[7] Wang X.H. (2007) *Probability and Mathematical Statistics*. Beijing: Science Press.
[8] Wang X.Y. et al. (2010) *Mathematical Modeling and Mathematical Experiment*. Beijing: Science Press.
[9] Lei Y., Liu Q. & Tang W.Q. (2011) The Role of Energy Conservation and Emission Reduction Implemented by Natural Gas Processing Plant in Low-carbon Economy. *Petroleum and Natural Gas Chemical Engineering*, (4).

Electric, Electronic and Control Engineering – Shao, Shu & Tian (Eds)
© 2015 Taylor & Francis Group, London, ISBN: 978-1-138-02842-5

The Swarm intelligence optimization method and its application in chemistry and chemical engineering

Yuzhen Dong
School of Chemical Engineering & Environment, Weifang University of Science and Technology, Shouguang Shandong, China

ABSTRACT: With the development of economy, competitions in various fields are becoming more and more intensive in recent years. In order to be outstanding in a certain field, it is obliged to strengthen the enterprise management platform so that each link can be optimal. A great many excellent methods have been found by people as for these optimal problems of combining science and engineering. As a new technology, the swarm intelligence optimization algorithm has been successfully applied in various fields for the solution of difficult and complicated problems. It has aroused wide concern of scholars in relevant fields. The structural system of chemical enterprises has undergone tremendous changes with the increasingly intensified economic globalization. Optimizations are conducted in different levels, ranging from products production to supply management, so that multidirectional concepts of optimization model are integrated in traditional chemical and engineering disciplines. A variety of characteristic functions, like simple linear functions and nonlinear and dynamic functions, have made optimization problems more and more complicated, which are difficult to be achieved by conventional algorithms. Thus, it is particularly important to make use of the swarm intelligence optimization method. Studies on operating steps of the swarm intelligence optimization algorithm in chemical engineering are quite less. Based on this, the paper conducts a bionic optimization calculation by simulating the foraging behavior of ant colony with the swarm intelligence optimization method. Therefore, this paper discusses the algorithm in detail, establishes a swarm intelligence optimization model, points out implementation steps of the optimization process, and expounds such aspects of chemistry and chemical engineering as studies on optimized metrology, applications in development, applications in chemical metrology, applications in extraction process, and applications in chemical process optimization. The prospect of the algorithm is also discussed in the end.

KEYWORDS: swarm intelligence optimization method; optimization; ant colony algorithm

1 INTRODUCTION

In the past few decades, competitions of products have become globalized, environmental pollution has aggravated and energy prices have risen as well. Facing these pressures, enterprises have to reduce cost for development. Technology optimization is a key indicator for enterprises to improve economic performance, involving all aspects from product design to supply management. Applications of technology optimization in the chemistry and chemical engineering field are relatively hot in recent years. Material energies are dispersed in the nonlinear form because of storage difficulties in the process of transformation. The classical optimization theory is not suitable for the development of chemistry and chemical engineering field in the face of numerous difficulties. For this reason, technology optimization for intelligent algorithms becomes increasingly urgent.

Random optimization algorithms show strong ability in solving practical problems, such as the particle swarm optimization, the simulated annealing algorithm, the genetic algorithm, taboo search, etc. They are widely used in fields like biology, thermodynamics and manual work. Methods of this kind is called intelligent optimization algorithm. The swarm intelligence optimization method features global optimization, distributed form and rapidity in the calculation of complicated optimization problems. This method is not only a mathematical analysis method but also a machine learning method without tutor supervision. It belongs to difficult problems of NP and is widely applied in various fields. Predecessors have made a great deal of efforts in studies on this algorithm with certain achievements. For example, the swarm intelligence optimization method has been applied in solving TSP problems with good effects since 1996 when the improvement theory was proposed by M.Dorigo. As for applications of random algorithm in chemical engineering, Biao Cheng proposed a nonlinear dynamic driving model established by the artificial network and applied it in chemical engineering. This paper proposes that cyclic subspace regression model of RBF should be modified so as to optimize indicators in line with the prediction performance of the model. The hybrid optimization genetic algorithm is also proposed in this paper as the training model, which is applied in chemical engineering with good results. As for problems of optimization constraints that cannot be solved by the particle swarm optimization, the author puts forward a linear constraint particle swarm model that is improved and applies it in linear constraints of chemical engineering, such as conservation of matter, conservation of atom and conservation of mass. It is a highly efficient constraint algorithm that maintains swarm feasible space [2]. As for applications of the swarm intelligence optimization method in chemistry and chemical engineering, Yijun He indicated that the current swarm intelligence optimization method lacks an efficient mechanism. According to deficiencies of the swarm intelligence optimization method, the author presents a feature optimization model with multiple constraints, multiple dynamic conditions and multiple objectives and applies it in chemistry and chemical engineering. The final result indicates that the improved swarm intelligence optimization method has a better performance in terms of overall situation and adaptability [3].

Based on previous studies, this paper proposes a swarm intelligence optimization method for studies on chemistry and chemical engineering problems, providing a theoretical basis for the future development.

2 THE SWARM INTELLIGENCE OPTIMIZATION METHOD AND OPTIMIZATION IMPROVEMENT

2.1 *The idea of swarm intelligence optimization*

The swarm intelligence optimization method provides an effective means for the solution of complicated problems. It includes the particle swarm algorithm and the ant colony algorithm. This paper mainly carries out an analysis and a study on the ant colony algorithm. Through the study on the foraging process of ant colony, it is not difficult to find that cooperation in ant colony is realized by simple follow and message release. The most complex behavior is the occupational behavior of ant colony, which refers to the shortest path of ants between food and an ant cave.

The schematic diagram of ant cave, food and path is shown in Figure 1.

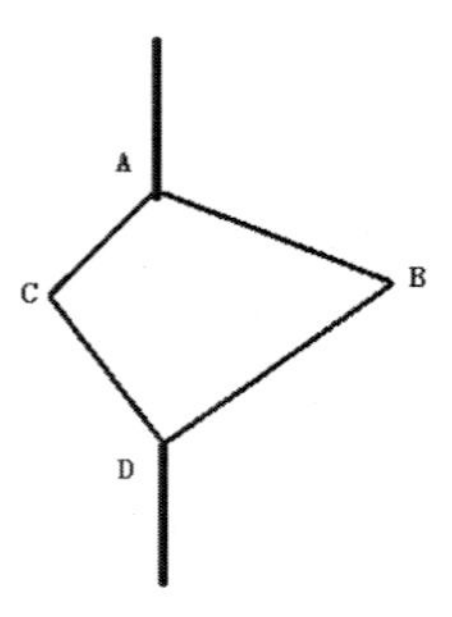

Figure 1. Ant colony foraging schematic.

If there are n ants from A to D, the duration time is Δt. The speed of these ants are constant and in the same unit. Pheromone left by these ants in the process of crawling is in the same unit as well. If $t = 0$, the pheromone left by ant on the path $A\text{–}B\text{–}D$ and the path $A\text{–}C\text{–}D$ is 0. Therefore, probabilities of choosing paths from A to D are equal. Half of the ants choose the path $A\text{–}B\text{–}D$ while half of the ants choose the path $A\text{–}C\text{–}D$

Under the above-mentioned circumstances, Δt is taken as the time period. Starting from $t = 0$, $\frac{n}{2}$ ants choosing the path A-C-D reach D point in advance. In the meantime, pheromone of $\frac{n}{2}$ is left on the path $A\text{–}C\text{–}D$.Then, these ants go back the same way after the time period of Δt, leaving $\frac{n}{2}$ pheromone on the path$A\text{–}C\text{–}D$.However, the duration time of $\frac{n}{2}$ ants passing from A to D along the path $A\text{–}B\text{–}D$ is $2\Delta t$. So, the pheromone left on the path $A\text{–}B\text{–}D$ is$\frac{n}{2}$. Foraging in line with different pheromones starts from the moment when $t = 2\Delta t$. If there are n ants are leaving for D point, there $\frac{2n}{3}$ ants choose the path A-B-D for foraging. There will be a vast majority of ants choose the relatively shortest path if this process is repeated.

It can be concluded from the above principle that there are multiple choices of foraging paths for ants and the relatively shortest path will be selected by ant colonies according to the concentration of pheromone until all the ant choose the path.

2.2 *Swarm intelligence optimization model*

Ant colony algorithm includes three kinds of relatively typical swarm intelligence optimization algorithms, namely the max-min ant system, the ant colony system and the ant system. There are differences among the three algorithms in the update process of ant information contents. This paper selects the ant system for analysis in order to fit applications in chemistry and chemical engineering fields.

The ant system was first proposed by *Dorigo* in 1990s. He updated in accordance with different information contents, including the number of ants, the cycle of ants and the density of ants. The cycle of ants means that information will be updated when the construction of all the ants is completed. The number of ants and the density of ants are conducted between two cities without interference of distance, but the performance is relatively poor. So, scholars usually choose the cycle version of the ant system.

If M ants are put randomly on N points, the probability that the kth ant on the ith point choosing point j for crawling is shown in Formula (1):

$$p^k(i,j)=\begin{cases}\dfrac{[\tau(i,j)]^{\alpha}\bullet[\eta(i,j)]^{\beta}}{\sum\limits_{s\notin tabu_k}[\tau(i,s)]^{\alpha}\bullet[\eta(i,s)]^{\beta}}, & if \quad j\notin tabu_k \\ 0 & otherwise\end{cases} \quad (1)$$

In Formula (1), $\tau(i,j)$ stands for the concentration of pheromone on (i, j), $\eta(i,j)=\dfrac{1}{d(i,j)}$ stands for the heuristic information. Here, $d(i,j)$ stands for the length of (i, j), α,β respectively reflects the importance of pheromone and heuristic information, $tabu_k$ is the list of points through which the ant k has passed.

The update of pheromone after all point having been passed through is shown in Formula (2):

$$\tau_{ij}(t+n)=\rho\cdot\tau_{ij}(t)+\Delta\tau_{ij} \quad (2)$$

In Formula (2), the global volatilization rate of pheromone is expressed with ρ, which is smaller than 1. In$\Delta\tau_{ij}=\sum\limits_{k=1}^{M}\Delta\tau_{ij}^{k}$, the information content left by the kth ant when passing through the path (i, j) is shown in Formula (3):

$$\Delta\tau_{ij}^{k}=\begin{cases}\dfrac{Q}{L_k} & ij\in l_k \\ 0 & otherwise\end{cases} \quad (3)$$

In Formula (3), Q stands for a constant, the path of the kth ant is expressed with l_k, and the path length is L_k.

The design flow of the prototype system of the swarm intelligence optimization algorithm is provided in Figure 2.

Steps of the iteration process designed according to Figure 1 are shown below:

1 Initialization. Establish a tabu list randomly and put initial nodes in the list;
2 Iteration
  $k=1$
  while $k\leq$ *It Count do* (execute iteration)
    for $i=1$ *to M do* (circulation of M ants)
      for $j=1$ *to* $N-1$ *do* (circulation of N points)

Put node j in the tabu list and ants are transferred to node j as the initial node. Repeat the above steps

end for

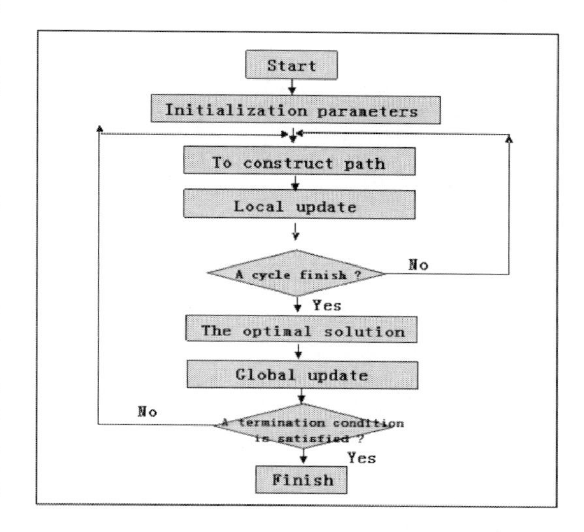

Figure 1. Ant colony algorithm prototype system design flow.

end for

Information on all paths of ants can be updated in line with the above-mentioned steps;

3 Output results and end the algorithm.

2.3 *The improvement idea of the swarm intelligence optimization*

Information contents of the environment of several practical ant colonies are determined by moving directions. The movement on a plane is called artificial ant. The process of foraging of ant colonies can be abstracted as the track between node and phase that can be perceived by ants on each node. From the initial point, select the final goal node according to the transition probability of specific state so as to obtain the results. The swarm intelligence optimization mainly sets the selection of system parameters and the changing pattern and then carries out variation improvement of parameters by introducing the variation mechanism. The swarm intelligence optimization method has the advantage of being easily combined with other intelligent algorithms in terms of the information contact mode of subsystems, the structural mode and the overall structure. Therefore, improvements can be realized by the combination in such aspects as the effect of advantage complementation among algorithms.

2.4 *The improvement model of the swarm intelligence optimization*

The possibility of ants choosing k cities is that the provided heuristic information and the residual information content on the path from the city of ant to the target city are in the relation of a certain function. It is a complete cyclic process after all the ants having passing through all n cities in order to prevent excessive submerging of information. So, the information concentration on the path ij at (t+1) is:

$$\tau_{ij}(t+1) = (1-p).\tau_{ij}(t) + \sum_{k=1}^{m} \Delta T_{ij}(t) \qquad (4)$$

Here, the path of k ants at the moment t is represented by $T_k(t)$, the length of which is $L_k(t)$. Q is the related adjustment parameter and Q>0. The optimization path of TSP can be thusly obtained, corresponding codes of which are shown below [5]:

```
%Ant system for solving the traveling salesman problem;
<Input of matrix D of distances>
<Initialization of the parameters of the algorithm: α,β,Ρ,Q, and τo>
m=n                 %Number of ants is equal to the number of towns
For i=1: n                  % for each edge
For j=l: n
If i=j
η(i.j)=1/D (i.j)            % Visibility
τ(i,j)= τo              % pheromone
Elaeτ(i,j)=0
End lf
End For
End For
For k=l: m
<Allocate ant k in randomly chosen town >
End
<select a conditionally shortest route T + and cclulate its length L+>;
%Main; loop
For t=1: tmax                  % tmax-number of iterations
Fork=1: m                  % for each ant:
<Build a routeTk(t) aceording to rule(l)>
<Calculate Lk(t),the length of the routeTk(t)>
End
If"Is the best solution found? ="<U [dateT+and L+>End
For i=l: n
For j=l: n %for each edge
<Upate pheromone trails according to rule (2)>
End
Eod
End
<Output the shortest route T+its length L +>
```

3 APPLICATIONS OF THE SWARM INTELLIGENCE OPTIMIZATION ALGORITHM IN CHEMISTRY AND CHEMICAL ENGINEERING

Applications of optimization problems are widespread in the chemical engineering field. An optimization algorithm with excellent performance is always the pursuit in the chemical engineering field. The swarm intelligence optimization algorithm is highly efficient globally, the seeking ability of which is stronger than that of other algorithms. The swarm intelligence algorithm has attracted much attention due to this feature. Applications of the swarm intelligence optimization algorithm in chemistry and chemical engineering are introduced simply as follows.

3.1 *The application of the swarm intelligence optimization method in the exploitation of catalyst*

The method of logical operation is the core content of the computer-aided catalyst design. But this method has such deficiencies as low convergent speed and low optimization efficiency that impede the rapid development of computer-aided catalyst design. The swarm intelligence optimization method has certain advantages in complex optimization problems so that studies on catalyst design are hot topics. There are still lots of work needed to be done.

3.2 *The application of the swarm intelligence optimization method in chemical metrology*

P. S. Shelokar[6-7] compared the ant colony algorithm with the genetic algorithm and the taboo search in terms of results. The ant colony algorithm is able to forecast data collection and reveal the internal law, quality and processing time of which are relatively reasonable and practical.

Yaping Ding[8-9] et al. proposed the chemical ant colony algorithm through the spectrum analysis with the swarm intelligence method. It can be known from the comparison that the convergent speed is 40% faster than that of the genetic algorithm. Besides, this method can be applied in systems of multiple groups like black, white and gray.

3.3 *The application of the swarm intelligence optimization method in the extraction process*

A. I. Papadopoulos et al. composited heptanes with the genetic algorithm, the simulated annealing algorithm and the swarm intelligence optimization algorithm, and extracted the extraction agent composed by water, methylbenzene, dimethylbenzene, acetic acid, chloroform, acetone, etc. CPU provides the operating time, which can be regarded as the evaluation standard. It can be finally concluded that the advantage of the swarm intelligence optimization algorithm is relatively prominent. However, the simulated annealing algorithm is very reliable in terms of robustness. Standard deviations of all the objective functions are the smallest.

3.4 *The application of the swarm intelligence optimization method in chemical engineering analysis*

It was first put forward by V. K. Jayaraman that the swarm intelligence optimization algorithm can be applied to solving chemistry and chemical engineering problems. They proposed a swarm intelligence algorithm in a paper, the global optimization design of which is an intermittent chemical engineering process with intermittent changes. It is difficult to handle nonlinear mathematical models according to previous experience. V. K. Jayaraman et al. maintained the relatively stable diversity of ant colonies by taking advantage of variation characteristics and appropriate hybridization and solved the model of combined optimization problems, the production constraints of a single product, continuous function optimization model of intermittent production of multiple products, as well as combined optimization problems of intermittent production scheduling. The optimization result with relatively fierce competitions is 0.573, which is extremely close to the iteration result 0.5735. The yield of intermediate products is 0.03% smaller than the optimization result of the least square method in terms of error.

Thus, Yijun He et al. constructed a continuous ant colony optimization system (CACS) for constraint problems through the method of searching food so as to obtain the ant colony system of continuous optimization problems. If the system model is applied in the preparation of butenoic alkane, the optimal solution of CACS is more optimized than that of αBB (α-*based branch and bound*). Now, σ=5*10exp (-4) and the constraint violation is much smaller than that of αBB. However, when this method is inferior toα BB, σ=5*10exp (-6) orσ=0. This process is not in violation of the constraints, but the final solution of α BB is in severe violation of the constraints.

4 CONCLUSION AND PROSPECT

1 The swarm intelligence algorithm is a new bionic algorithm, establishing an approximate model in line with the behavior of ant foraging for spatial problems. In the meantime, heuristic operators are introduced to improve the practicability. The algorithm has a relatively low requirement for hardware, which is convenient for application and promotion.
2 The swarm intelligence algorithm has the same effect with the genetic algorithm and the annealing algorithm in terms of applications in the chemical engineering field. However, this algorithm features high robustness in chemistry and chemical engineering processes and provides new solutions for studies on chemical process optimization and chemical metrology.
3 Because of the late starting, the swarm intelligence optimization algorithm is still in a primary stage compared to the simulated annealing algorithm and the genetic algorithm and still lacks strict mathematical proof. It is a major research direction of applications of the swarm intelligence optimization method in chemistry and chemical engineering that most parameter can be acquired from experiments and experiences.
4 The swarm intelligence optimization method is still in the exploratory phase in terms of application in chemistry and chemical engineering. With the solution of a series of problems, the algorithm plays an important role not only in chemical engineering and chemical clustering but also in chemical difficulties.

REFERENCES

[1] Wei, P. & Xiong, W.Q. (2002) An ant colony algorithm for function optimization, *Computer Science*, 29(9):227–2291.

[2] Cheng, B. (2007) *The Improvement of Two Random Optimization Algorithm and Applications in Chemical Engineering*, Zhejiang University, 6.7:5–11.

[3] He, Y.J. (2008) *The Swarm Intelligence Optimization Method and Its Application in Chemistry and Chemical Engineering*, Zhejiang University, 1.1:4–10.

[4] Yan, C.Y., Zhang, Y.P. & Xiong, W.Q. (2007) A new information content update strategy of the ant colony optimization algorithm, *Application Research Of Computers*, 24(7):86–88.

[5] Du, L.F. & Niu, Y.J. (2011) The implementation of the ant colony algorithm, Information Technology, 6.

[6] Wu, Q.H, Zhang, J.H. & Xu, X.H. (1999) The ant colony algorithm with variation features, *Computer Research and Development*, 36(10):1240–1245.

[7] Shtovba SD. (2005) Ant algorithms: theory and applications, *Programming and Computer Software*, 31(4):165–176.
[8] Mitra K, DebK, GuPta S K. (1998) Multi-objective dynamic optimization of an industrial nylon 6 semi-batch reactor using genetic algorithm, *Journal of Applied Polymer Science*, 69:67–85.
[9] Ravi G GuPta S K, Ray M B. (2002) Multi-objective Optimization of Cyclone Separators Using Genetic Algorithm, *Industrial & Engineering Chemistry Research*, 39:4270–4285.
[10] Srinivas N, Deb K. (1995) Multi-objective function optimization using non-dominated sorting genetic algorithms, *Evolutionary Computation*, 2(3):222–245.
[11] Zitzler E, Thiele L. (1999) Multi-objective evolutionary algorithms: a comparative ease study and the strength pareto approach, IEEE Transaction on Evolutionary Computation, 3(4):256–170.
[12] Wang, Z., Liu, G.Q. & Chen, E.H. (2009). The K. means algorithm for the optimization of the initial center, *Pattern Recognition and Artificial Intelligence*, 22(2):298–305.
[13] Song, L., Li, M.Y. & Li, X.Y. (2008) The K. means algorithm based on particle swarm optimization and its applications, *Computer Engineering*, 34(16): 201–203.
[14] Wu, Y.W. & Hu, X.G. (2007) An optimization scheme of the K. means algorithm, *Journal of Chaohu University*, 9(6): 21+24.

Electric, Electronic and Control Engineering – Shao, Shu & Tian (Eds)
 ISBN: 978-1-138-02842-5

The research of the long-distance runner's physical index based on data analysis

Cheng Liu
Pingdingshan Institute of Education, Pingdingshan Henan, China

ABSTRACT: With the development of information technology, database technology has been widely used in the database management in the last two years. The technology of data mining is the most advanced technology which has been widely used in the large-scale companies in the area of communication, bank, transportation, insurance, and so on. The paper analyzes the physical indexes of the long-distance runners and the theory of data mining in order to solve the problems. If we provide the right data to decision-makers, a better plan will be made in advance.

KEYWORDS: data mining; athlete; arithmetic; physical index.

With the rapid development of science and technology, great changes have been taken in the field of sports training, and scientific training emerged as an important tool of the sports training, which had a positive impact on the athlete's good performance. Physical test of the athlete seems to be indispensable, so we can find much important information and regular pattern by analyzing the data of the test. The coach could develop a scientific training program for long-distance runners according to this data. Above all, data mining plays an important role in this process.

1 DATA MINING TECHNOLOGY

1.1 *The meaning and task of data mining*

Data mining is a method of data analysis. Finding out the useful data seems like extracting the iron from the ironstone, so we should analyze the large amount of data. Data mining is a technology to extract the most important things from the original data, which seem to be valued, unknown and implicated. The type of knowledge is called data mining tasks, and its main part is to summarize the rule of mining, classify the data and so on.

1.2 *The methods and techniques of data mining*

Different data mining tasks require different methods and means. On the whole, it can be divided into the following categories: statistical analysis, data mining, decision trees and rough sets. Statistical analysis is to analysis the data by using the traditional statistics, which mainly contain traditional statistical parts and systems of subjective guide. The intelligence of data mining can be achieved based on the genetic algorithms and neural network techniques, and self-organizing networks and former spy network have been widely applied in the neural network. In the field of engineering, decision tree is a simple method to express things which will be gradually separated into different areas and represent different categories. The sub-branch will help to establish a decision tree.

1.3 *Classification and application of data mining*

The classification of data mining are not all the same, it is based on different type of mining, and it can be classified on the basis of knowledge such as the character rules and deviated rules. It can also be classified by the usage, such as data driven mining, spontaneous knowledge mining, etc.. Based on the diversity, there exist two kinds of database: deductive databases and media databases. The mining technology had been widely used in the market via data collection. It will confirm the consumers' habits in a particular way, which will predict the next step of them. We will gain more money by analyzing the information on them.

2

2.1 *The process of data mining*

The process of the data mining refers to the way we get the information from the big data, which contains lots of aspects such as data collection, the assessment of the mode, etc.. In this paper, we take the mode of

John as an example, which has brought some simple introductions of this process. The model attaches more importance to personnel as well as experts of data mining during the whole process, so that experts can describe it clearly because of sufficient background knowledge, and the scholar will pay more attention to practical techniques and problems.

2.2 *Functional analysis of the TENNIS-DAMS*

TENNIS-DAMS system can solve the problems of the long-distance runners by analyzing the indexes. The management issues consist of three parts, like management & analysis of data test. It will analyze and classify the associated functions, and the data mining process can be roughly divided into stages such as data collection and management, data preprocessing, data mining (pattern discovery) and pattern processing and evaluation phase. The original database has been shaped by the collection of data, and this stage is separated into input processing and maintenance module of data. The stage of data preprocessing includes the comparative analysis and query module. The goal of data mining in this system is to discover the relationship between the physical indicators and the physical state of the athletes, which may be the most important part in this article.

2.3 *Logical structure of system and the storage of database*

In the system, the database must be established, for which the database is the central function, and it is also the interface among the functional modules. The module of the system has linked to each other by using the database, so that the respective functions could be completed (see Figure 1). In the process of mining, you can create the target database and schema libraries by yourself according to the reality, and it will be cost savings, easy management, query and analysis.

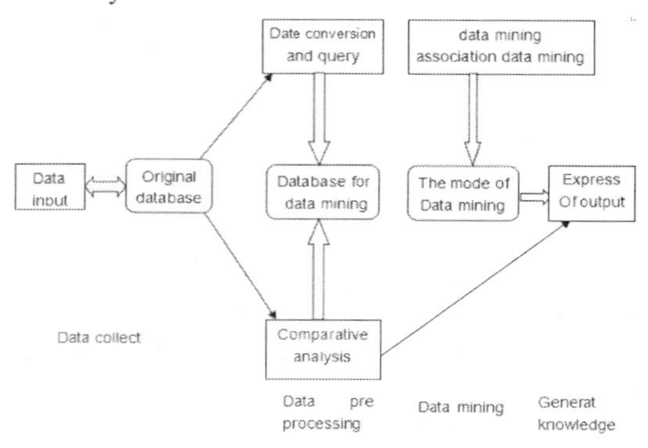

Figure 1. The module of the system.

3 MAIN RULES AND ALGORITHMS OF DATA MINING

3.1 *The association rule mining kernel*

The concept of association rules is to make I = (i ,, i:. , I¨ . . , i} as a set of attributes of possible values, which is called the set of data items, and i- ($1 \le k \le n$) is called data item to record a property's value in database. The number of elements in the database was called the length of the data item, while the length of the data item is called the n-dimensional. Attribute association rules can be described by confidence, support, hope and confidence. Suppose D contains C% and Y, C is considered as the confidence of the association rule $X \ge Y$, as C%=The number of Transactions($X \cup Y$)/ The number of Transactions(X); Suppose D contains s% of X and Y, we will consider S% as the supporter of the association rule $X \ge Y$. as S%=the number of Transactions $X \cup Y$/ The number of Transactions(D); Suppose D contains e% of Y, we will consider e% as the connection of the association rule $X \ge Y$. It describes the appearance probability of Y without effect.

3.2 *The step of data mining association rules*

In order to discover the meaningful association rules, the association rule mining needs to develop minimum support and confidence level. Problems of the association rule can be divided into two parts, and the degree of the support is larger than the minimum support degree. Because people pay great attention to the association rule, data mining algorithm has always been used as follows: we need to process the data and seek out all data items to meet the minimum degree of support; it will match the minimum confidence level and then explain R output it.

3.3 *The usage of algorithm in the system*

The main objective of TENNIS-DAMS system association rule mining is to analyze the potential association in physical fitness test indicators. The core algorithm of the system finds out that we should use the classical association rule mining such as Apriori algorithm and DHP algorithm. For example, Apriori is a width-first algorithm, which could be achieved by scanning the database D, and we should only take the same length k (projects contained in each scan trip number) into consideration. The algorithm uses the hashing technique DHP to find out the next project, such as Hash function

$$h([x, y]) = ((order\ of\ x) * 10 + (oder\ of\ y)) mod\ 7 \quad (1)$$

Hash function considered the first candidate item to generate C1 = ((A), (B), (C), (D), (E)). And then scan the database to get the statistics of the support, thus generate a 1- frequent set L1 = ((A), (B), (C), (D), (E)); In the meanwhile, the statistics can be used to construct the project candidate and so on.

3.4 *Mining tasks and model evaluation*

In the system of TENNIS-DAMS, the discovery of the mode can be performed by a combination of the two algorithms. In the stage of data preprocessing, we need to divide the attribute of each pattern based on the actual value of the physical indicators, and define the minimum degree of the support and confidence level, by using a DHP algorithm to calculate the data mining. And each line of it is an association rules. Iilj, the first column is the degree of support, and the second is the confidence level. We can get some confirming and enlightening rules of mining from the analysis.

3.5 *Multi-valued association rule mining*

Multi-valued association rules and the basic problems have some differences. The collection of data items has changed into Ir = I × P × P, i.e., Ir = {(x, l, u) I x∈I, l∈P, u∈P, l≤x≤u), (x, l, u) ∈Ir x represents the attribute values between l and u, I is the set of attributes, P is the set of positive integers. The algorithm of the multi-value association rules is MAQA, Boolean association rules problem had changed from multi-value association rules. Obtaining valuable rules require the use of Boolean association rules and so on.

4 EMPIRICAL ANALYSIS

4.1 *Overview of neural network technology*

The neural network has obtained from the biological inspiration, connected with others by using a large number of simple processing units and constitute an information system in some way. Nonlinear processing, association's ability and massively parallel computing are all its advantages. It can be divided into five parts according to the principles. Due to the way of learning, we can study the network by ourselves or learn from the teachers. The Hopfield network and BP network of the neural network are widespread use. The information processing of it has been divided into the phases of execution and the phases of learning, and the execution stage is to process the neural network such as:

$$U_i(t+1) = \sum_{j=1}^{n} W_{ij}(t+1)X_j(t) - \theta_i(t+1) \tag{2}$$

$$Xi(t+1) = f_i[u_i(t+1)] \tag{3}$$

where w is the weight coefficient of j and f is the nonlinear function of neuron. The learning phase of the neural network is a self-improvement, and we apply the following mathematical formula to represent learning formula:

$$W_{ij}^{ii} = \phi[W_{jj}^{n-i}(t), \eta_{ij}^{i}, \frac{\partial E(t)}{\partial W_{ij}(t)}] \tag{4}$$

In the light of some learning rule, we usually modified the link between nerve cell weight coefficients w in order to achieve a minimum error function E. The learning rule of the neural network is usually divided into the rules related learning, correction rules learning and unsupervised learning.

4.2 *Back-propagation training method*

In the multi-layer network, because of the hidden layer, the output of the network is unable to continue. The appearance of back-propagation training will correct the weights the layers. It is so called BP algorithm, which had been widely used in networked learning, and it is also used for multi-layer feed forward network learning. The main process of it was the back-propagation learning process and forward propagation process. The input sample X was added to the layer, and represented the sum of the k, and the function of the neuron excitation is f. We can use the following mathematical formula to express the relationship between the variable:

$$X_i^k = f(U_i^k) \qquad U_i^k = \sum_j W_{ij} X_j^{ki} \tag{5}$$

The essence is to calculate the minimum value of the error function. The BP algorithm is applied to the multilayer feed forward networks. Sigmoid function is used as the excitation function, and we can use the following steps to strike the weights

$$U_i^k = \sum_{j=1}^{n+1} W_{ij} X_j^{k1}, X_{n+1}^{k-1} = 1, W_{i,n+1} = -\theta \quad X_i^k = f(U_i^k) \tag{6}$$

The error of each layer d_i^k for the output layer k = m, there are other layers

$$d_i^k = X_i^k(1 - X_i^k) \sum_l W_{li} d_l^{k-1} \tag{7}$$

Right correction factor $^{W_{ij}}$ Threshold $^{\theta}$

$$\Delta W_{ij}(t+1)=W_{ij}(t)-\eta \cdot d_i^k \cdot X_j^{k-1} \tag{8}$$

By (5–25)

$$\Delta W_{ij}(t+1)=-\eta d_i^k \cdot X_j^{k-1}+\alpha\Delta W_{ij}(t) \tag{9}$$

Include

$$\Delta W_{ij}(t)=-\eta d_i^k \cdot X_j^{k-1}+\alpha\Delta W_{ij}(t-1)=\Delta W_{ij}(t)-\Delta W_{ij}(t-1) \tag{10}$$

In this process, the output sample X and Y must be performed. We should know all the position of the inputs and outputs, the process of the BP algorithm program is shown in Figure 2.

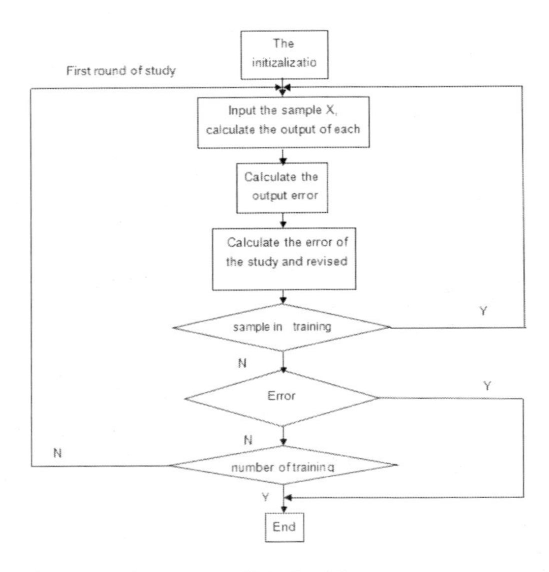

Figure 2. The process of BP algorithm program.

4.3 *Evaluation of long-distance runners' physical condition*

4.3.1 *Problem statement*

In the long-distance running training, the coach needs to analyze the athlete's physical fitness index data carefully in order to evaluate the physical condition. On the basis of the data, they divided the physical state into poor, fair and good levels. However, the evaluation process is not a simple add, instead of it, it is difficult for coaches to evaluate the process according to their own experiences.

4.3.2 *Network model*

We select 18 items to design the layers of input and output, and evaluate the physical state of the long-distance runners. The athlete's gender, age, height, weight and other 14 indicators of physical fitness will be replaced with nodes. The coach usually divided the physical condition of athletes into three states, which is good, generally and poor, and therefore we need to set three different output layers to represent the three different states. When we make the choice for hidden layers, it will be extracted the character of the input layer. The ability of the neural network could be improved with the increase of the hidden layer, and it will increase the time and date of training. By considering the scale and limits on sample data, we could solve the problem by hidden layers, such as BP network model. The layers of BP neural network had a great impact on the functional performance of the network, so that the inner layer nodes need to make the right choice. With the purpose of evaluating the physical condition, the design of the BP network should be matched with the input nodes and output nodes, training, and the number of samples. Since the network is 18 inputs and three outputs, it can be regarded as a characteristic compression process. The intermediate layer number N matches 18: N = N: 3, thus available N is approximately equal to 7. 4. After several experiments, we finalized that when the hidden layer nodes is 8, the network performance is relatively stable.

4.4 *Final results and analysis.*

Table 1. The result of physical test

Study sample	Test sample	Node in hidden layer	Rate of average accuracy
50	100	5	54.3%
50	100	7	58.24%
50	100	12	56.1%
50	100	8	59.12%
75	75	8	59.4%
100	50	8	61.28%
125	25	8	66.66%

Three months is the cycle of the long-distance runner test, so we can predict the period of physical indicators of long-distance runners by adjusting the input samples above network model.

4.5 *Problems and improvements*

When we talk about the data entry, there is no denying the fact that problems exist in the data due to the facticity of the data. One reason is the small quantity of the data. In terms of the network model, we only take

Table 2. The result of physical test.

Study sample	Test sample	Node in hidden layer	Rate of average accuracy
50	70	7	48.2%
100	20	7	50.1%
50	70	8	51.3%
100	20	8	55.82%
100	20	9	50.1%

the common factors of the distance runners' physical state into consideration, but ignored their personality factors such as injuries, health conditions and others. These factors play an important role in the test, and we should learn to make full use of the information to improve ourselves. In the meanwhile, we need to increase the number of the training input samples and filter the data if necessary. The sample used in the training can be described as follows: the distribution of three different physical state data is more reasonable, so we need to learn from the coach and personal factors should be considered in the process of input nodes, and we can improve the efficiency of study by accelerating algorithm.

5 CONCLUSION

The athletes' all-round abilities include physical ability and psychology ability which is supplemented for each other. The ultimate goal of long-distance runners' physical training is to enhance their overall quality. By using the techniques and methods in this paper, the management of physical index and the capability of analysis will be achieved. Meanwhile, the relationship between physical fitness and skills testing can be discovered by data mining techniques, and it has provided a scientific and effective method to enhance the ability of athletes.

REFERENCES

[1] Yan Qi, Ren Manying. (2012) The training of physical function. *Journal of the coach in China.* (01):16–18.

[2] Wang Dongyue. (2010) The study of the athlete of synchronised swimming. *Journal of Zhejiang Physical Science,* (02):33–36.

[3] Isaacs L.S. (1998) Comparison of the Vertical and Just Jump system for measuring height of vertical jump for young children. *Perceptual and Motor Skills,* pp:66–77.

[4] Huang Tao. (2011) The characteristic of the track and field competition. *Journal of The Technology Information.* (16):657–658.

[5] Teford et al. (1989) A simple method for the assessment of general fitness: the tri-level profile. Australian *Journal of Science and Medicine in Sport,* pp:45–65.

[6] Jeffreys I. (2002) Developing a progressive core stability program. *Strength Cond J.,* pp:11–13.

[7] Zhou Xingwang. (2004) The teaching method of the track and field competition. *Journal Of Wuhan Physical Institute,* (04):125–126.

[8] Li Jianguo, Li Jiantao. (2011) The research of the core competition for track and field competition. *Journal of Gunagzhou Physical Institute,* (03):78–83.

Electric, Electronic and Control Engineering – Shao, Shu & Tian (Eds)
© 2015 Taylor & Francis Group, London, ISBN: 978-1-138-02842-5

An analysis on the maintenance system of electric power equipment based on data mining technology

Kangning Wang, Laifu Zhang, Min Jiang, Yun Tian & Tianzheng Wang
Electric Power Research Institute, State Grid Shanxi Electric Power Company, Taiyuan Shanxi, China

ABSTRACT: The correct evaluation on electric power equipment status has become a key point concerned by related researchers. How to extract the rules from multifarious data is also what needs to be solved by data mining technology. So there is a corresponding point between the two. On the basis of analyzing the application status and process of data mining technology, a status evaluation frame model and a SOA condition based maintenance system of electric power equipment are designed in this paper so as to provide theoretical basis for scientific service of electric power equipment.

KEYWORDS: electric power equipment; data mining; frame model; SOA; condition based maintenance.

1 INTRODUCTION

Zhang Yuan (2008) points out that methods of constructing disaggregated models of data mining now include decision tree, statistics, machine learning, neural network, analogical learning, genetic algorithm, rough set, inference based on cases, etc[1]. Data mining is a new technology of data analysis, through which valuable potential information can be extracted from a large amount of data. Yang Guoqing, *et al.* (2012) indicates that, with the continued trend of explosive growth of data in electric power system database, it is imperative to introduce data mining technology to the on-line monitoring system of electric power equipment[2]. In condition based maintenance system of electric power equipment, quantity of state in the evaluation module is determined through the analysis of basic data based on former experience, which is wasting time and energy. When quantity of state is inadequate, reasonable data need to be added according to former experience so as to ensure the accuracy of the final evaluation results. Wang Shishuang, *et al.* (2012) proposes that a reasonable method of data processing is urgently needed for making up the insufficiency of existing methods, thereby acquiring more accurate status evaluation information and improving the reliability of condition based maintenance[3].

Therefore, in this paper, data mining technology is introduced to the condition based maintenance system of electric power equipment so as to lay a foundation for making good strategies for condition based maintenance.

2 THE ANALYSIS ON THE CURRENT SITUATION OF DATA MINING'S APPLICATION

Chen Chaojin (2009) points out that the expansion of database has brought large amounts of data with the rapid development of information technology, yet information supporting decision-making cannot be identified by people and the need of information mining cannot be met by traditional searching and reporting tools[4]. Gao Ningning (2008) puts forward the concept of data mining, namely the technology of finding potential laws from large amounts of data and extracting useful knowledge[5]. Data mining technology has been widely applied in recent years, mainly in such aspects as Telecom customer analysis, agricultural industry data prediction, sales forecasting of retail, web page customization of weblog, fraud of bank customer, biological gene analysis, classification of celestial body, analysis of electric power equipment maintenance, etc[6].

The main research content of this paper is the application of data mining in electric power system. There is still a great potential for application of electric power system, the one with extremely multifarious data information. The application situation of data mining technology in electric power system is shown in Figure 1.

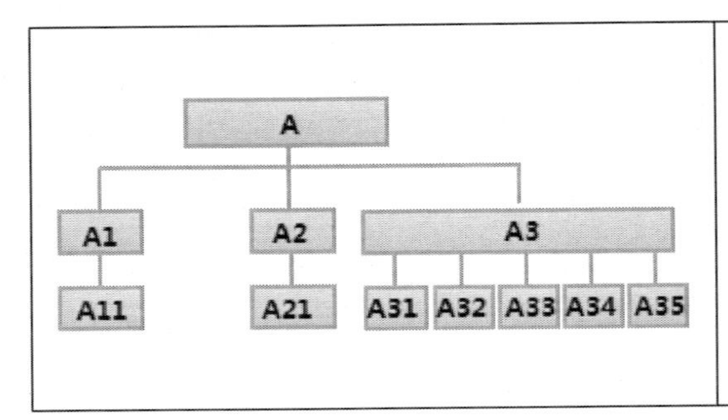

A. Application of data mining technology
A1. Management informatization
A2. Dispatch informatization
A3. Dispatch automation
A11. Feature extraction of power customer
A21. Power system planning
A31. Load forecasting
A32. Monitoring of equipment operation
A33. Feature extraction of power system fault.
A34. Online safety assessment
A35. Statistical analysis on system stability

Figure 1. The schematic diagram of application situation of data mining technology in electric power system.

3 DESCRIPTION OF DATA MINING PROCESS

The process of data mining is generally composed of five links, namely data selection, data pre-processing, data conversion, data mining and data explanation. Besides, such four conditions as effectiveness, innovativeness, serviceability and simplicity need to be satisfied in mining process. Data mining process and conditions required are summarized in Table 1.

Table 1. Links of data mining process and requirements.

Links	Description	Requirements	Description
Data selection	Generally use the Internet to select relevant data with problems needed to be solved.	Effectiveness	This requirement implies the importance of laws and knowledge and indicates whether they are applicable to unknown data.
Data pre-processing	The process of filtering information		
Data conversion	Convert qualitative data into quantitative data, namely feature extraction.	Innovativeness	Instead of being closely related with prior knowledge, it is an important new discovery in the practice process.
Data mining	Search for important models hidden in the database, namely the discovery process of data law.	Serviceability	Being useful and interesting for customers.
Data explanation	Evaluate and explain the results of data mining, namely the acquisition of knowledge.	Simplicity	Laws shall be as simple as possible and be able to create and explain complex data.

He Youquan (2004) holds that data mining is now being rarely used in electric power departments. Applications of data mining in electric power system include status evaluation of electric power equipment, system load forecasting and classification, classification of system operational mode, operation status and equipment monitoring, equipment fault diagnosis, power dispatch optimizing and power system modeling, etc.[7].

The main content of this paper is the application of data mining in status evaluation system of electric power equipment, which has a great impact on the condition based maintenance of equipment. An excellent evaluation system provides a guarantee for service life and maintenance effect of equipment. The flow chart of management on condition based maintenance of electric power equipment and trigger processing of dominant and hidden problems is shown in Figure 2.

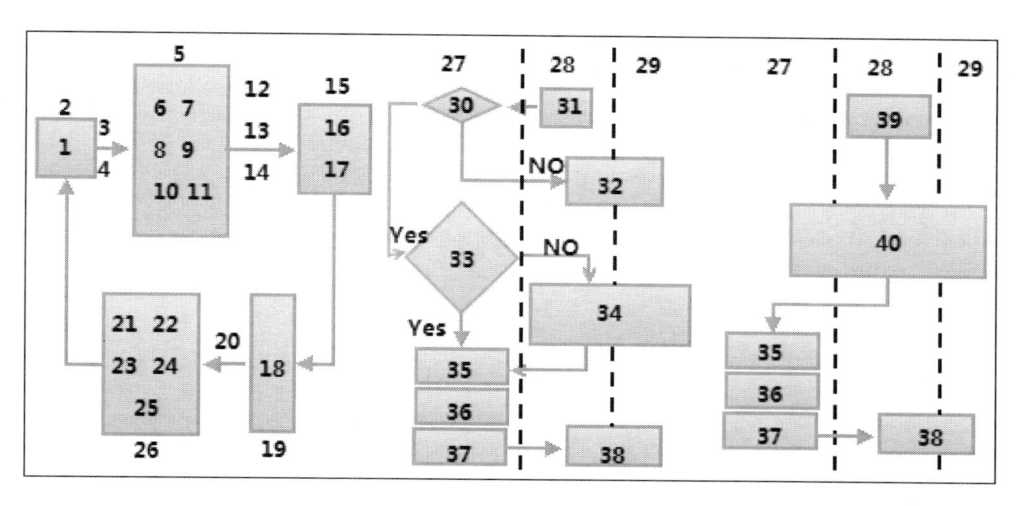

Figure 2. Flow chart of management on condition based maintenance of electric power equipment and trigger processing of dominant and hidden problems.

The left part of the chart is the management process of condition based maintenance of electric power equipment, the center part is the trigger processing of dominant problems, and the right part is the trigger processing of hidden problems. Symbolic meanings of 1–40 are presented in Table 2.

Table 2. Symbolic meanings in the flow chart of management on condition based maintenance of electric power equipment and trigger processing of dominant and hidden problems.

Number	Meaning	Number	Meaning	Number	Meaning	Number	Meaning
1	Electric power equipment	11	Testing of agency	21	Maintenance	31	Alarming
2	Assets management	12	data transmission	22	Replacement	32	False alarm recording
3	Online monitoring	13	communication network	23	Monitoring measures	33	Comprehensive confirmation
4	Energized monitoring	14	Manual download	24	Maintenance measures	34	Expert diagnosis
5	Monitoring	15	Data management	25	Review of design	35	Defect elimination
6	Online partial discharge	16	Database	26	Implementing & improving	36	Overhauling
7	Temperature measurement	17	Human-computer interface	27	Production & operation	37	Technical improvement
8	Oil pressure & air pressure	18	Trend	28	Condition monitoring	38	Releasing
9	Dissolved gas in oil	19	Expert diagnosis	29	Technical section	39	Diagnostic report
10	Gas mass	20	Decision -making	30	Defect confirmation	40	Defect confirmation

4 EVALUATION MODEL AND MAINTENANCE SYSTEM FRAMEWORK OF ELECTRIC POWER EQUIPMENT STATUS BASED ON DATA MINING TECHNOLOGY

4.1 *Evaluation model of electric power equipment status*

Condition based maintenance actually means analyzing results in accordance with the operational status of equipment and determining the health condition of equipment so as to decide the maintenance plan. The evaluation model of electric power equipment status based on data mining technology, the frame of which is shown in Figure 3, can perfectly analyze the operational status of equipment, aiming at providing a reference for the planning of condition based maintenance.

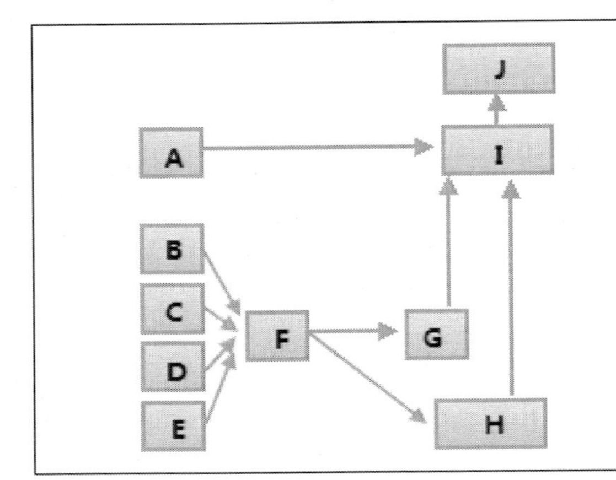

Figure 3. Diagram of status evaluation frame model of electric power equipment.

The vibration alarm record of a steam turbine in a power plant is shown in Table 3. With the record as original data, the fault phenomena record can be obtained by mining algorithm of association rules.

Table 3. Vibration alarm record of a steam turbine and fault phenomena obtained by mining algorithm of association rules

Alarming time	Event record	Event symbol	TID	Events
2013-02-01 01:13:23	A vibration alarm	A	1	A、C
2013-02-01 01:33:02	C vibration alarm	C		
2013-03-12 08:40:30	A vibration alarm	A	2	A、D、C
2013-03-12 08:42:16	D vibration alarm	D	3	B、A、C
2013-03-12 10:00:38	C vibration alarm	C	4	A、B
2013-05-28 19:35:12	B vibration alarm	B		
2013-05-28 19:49:52	A vibration alarm	A		
2013-05-28 20:01:06	C vibration alarm	C		
2013-06-12 05:12:52	A vibration alarm	A		
2013-06-12 05:20:06	B vibration alarm	B		

Note: The minimum support is 20%, and the minimum confidence coefficient is also 20%. T: the set of part items; I: the set of all items; D: the set of all objects. T is included in I, and each object can be represented with a unique identifying symbol TID.

It can be inferred from Table 3 that A=>C and both support degree and confidence coefficient are 0.75. Thus it can be predicated that C would start vibration alarm right after the vibration alarm of A.

4.2 *Maintenance system framework of equipment status*

Liu Jinsong (2010) points out it is ultimately a problem of application system integration that how to obtain data from other application systems for the maintenance system of electric transmission and transformation equipments. In order to solve the current and future data acquisition problems of the condition based maintenance system, it is required to establish a basic data service platform for the maintenance system, which shall be able to adapt to interfaces of various heterogeneous data sources and to provide data information for upper applications, such as condition monitoring, life evaluation, forecast evaluation, aid decision-making, etc [8].

The IT framework of maintenance system of electric transmission and transformation equipments shall be able to take ever-changing challenges of business requirements, because electric power applications exist in a changeful environment. A service infrastructure based on SOA is proposed in this paper, which is an effective way for maintenance system of electric transmission and transformation equipments to meet the requirements of varied businesses. In a service-oriented framework, not only the six functions of CBM but also GDA, HSDA, TSDA and GES of basic data service platform can release Web Services. A condition based maintenance system of electric power equipment based on SOA is shown in Figure 4.

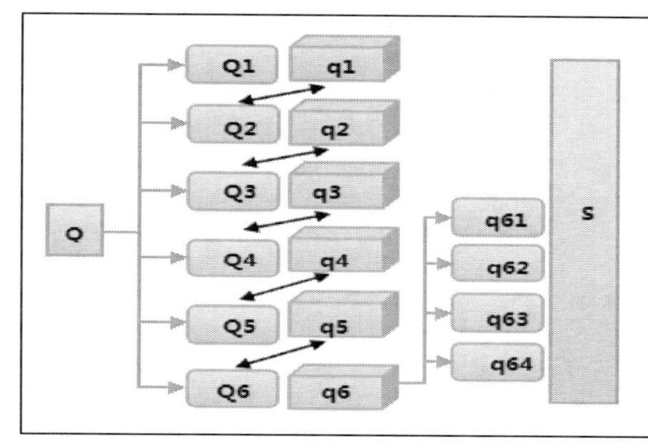

Q. Application
Q1、Q2、Q3、Q4、Q5 和 Q6 表示 OSA-CBM Web Service
Q1, Q2, Q3, Q4, Q5 and Q6 stand for OSA-CBM Web Service
q1is aid decision-making，q2 is forecast evaluation
q3 is life evaluation，q4 is condition monitoring
q5 is data processing，q6 is data acquisition
q61 is GES Web Service
q62 is GDA Web Service
q63 is HSDA Web Service
q64 is TSDA Web Service

Figure 4. A condition based maintenance system of electric power equipment based on SOA.

An agile strain capacity in business is the primary goal of condition based maintenance system of electric power equipment based on SOA, as shown in Figure 4. It is an effective way for condition based maintenance system of electric power equipment to meet the requirements of varied businesses that the IT framework featuring loose coupling and reusability to be matched with businesses.

5 CONCLUSION

The key of making condition based maintenance strategies of electric power equipment lies in scientific and accurate evaluation on equipment condition, and the key of evaluation on equipment condition lies in data mining of equipment condition. Evaluation on equipment condition can be conducted correctly through data mining technology. Based on the analysis of the application status and process of data mining technology, a status evaluation frame model of electric power equipment is constructed in this paper. A condition based maintenance system framework of electric power equipment based SOA is also designed in this paper, in order to realize visualization and operability of the evaluation model and provide a reference for condition based maintenance of electric power equipment in China.

REFERENCES

[1] Zhang Yuan, et al. (2008) An analysis on sorting algorithm of data mining and the quantitative research. *Journal of Northwestern Polytechnical University*, 26(6):718–722.
[2] Yang Guoqing, et al. (2012) The application of data mining technology in condition based maintenance of electric power equipment. *Journal of Shanghai University of Electric Power*, 28(2):176–181.
[3] Wang Shishuang, et al. (2012) A feasibility analysis report of data mining technology in condition based maintenance of electric power equipment. *Computer Science & Application*, (2):251–254.
[4] Chen Chaojin. (2009) A research overview of condition based maintenance technology based on data mining. *Guangdong Electric Power*, 22(9):21–24.
[5] Gao Ningning. (2008) The Modeling of Data Mining based on Excel. *Journal of the Central University for Nationalities*, (1):49–52.
[6] Teng Guangqing, et al. (2005) An analysis on data mining application and development abroad. *Statistical Research*, (12):68–70.
[7] He Youquan. (2004) *An Analysis on Data Mining and its Application in Fault Diagnosis of Electric Power System*. Chengdu: Southwest Jiaotong University.
[8] Liu Jinsong, et al. (2010) An analysis on basic data service platform of condition based maintenance system of electric power equipment. *East China Electric Power*, 38(2):216–219.

Electric, Electronic and Control Engineering – Shao, Shu & Tian (Eds)
 ISBN: 978-1-138-02842-5

The control strategy of offshore wind power participating in system frequency modulation through flexible DC transmission

Gangui Yan, Xin Liu & Tingting Cai
Electrical Engineering College, Northeast Dianli University, Jilin Jilin, China

ABSTRACT: In order to improve the capacity of wind farm participating in system frequency regulation, this paper proposes a frequency control strategy of wind turbine generator unit and DC converter with drooping characteristics. According to the principle that the maximum wind power capture is abandoned by the doubly-fed fan, this paper also proposes a load shedding control scheme of wind turbine generator unit and designs the control strategy of rotor sided converter and stator sided converter of wind power generator. Through the simulation analysis, it is verified that wind farm is able to reduce the change rate of frequency in system fluctuation with its controllable capacity of frequency modulation under the proposed control strategy.

KEYWORDS: wind farm; drooping characteristics; maximum power capture; load shedding; frequency modulation.

1 INTRODUCTION

New problems and challenges are brought to the operation of power grid due to large-scale development and utilization of wind energy, especially problems of wind power consumption and operational reliability. The primary cause is that the intermittent characteristic of wind power decides the volatility of wind power. Therefore, it has become an urgent key problem of the development and utilization of wind power that how to restrain the influence of wind power volatility on power grid furthest and improve the acceptance of grid for wind power. The development of offshore wind power starts to become a new focus to which attentions are paid gradually with more and more saturated development of onshore wind resources. Offshore wind resources have a relatively high development value and involve no land expropriation due to the high density and stability. In the European market, for example, the planned installed capacity of wind power in 2030 will reaches 300GW, of which 150GW comes from offshore wind farms (mainly in the north sea). According to the planning of renewable energy in China, however, the installed gross capacity of offshore wind power will reach 5GW by 2015. The newly added installed capacity of offshore wind power in the UK in 2010 is about 700,000kw with an accumulative installed capacity of 1,200,000kw, ranking the first of the world.

In order to improve the acceptance of grid for wind power, it is obliged to enhance grid's ability of regulation and control on the one hand and increase the regulation ability of wind power on the other hand. Wind farm participates in frequency modulation of power grid with its own capability of frequency modulation, which is one of the important characteristics of "grid-friendly" wind farm. As the mainstream type of wind farm [1–3], the variable-speed wind turbine generator unit is able to adjust active and reactive power flexibly, track the maximum power point, increase wind power utilization and reduce harmonic pollution on grid brought by wind power networking operation through power electronic converters. But there is no longer a coupling relationship between the rotational speed of wind power generator and power grid frequency in order to realize the maximum wind power capture, so that the wind turbine generator unit fails to respond grid frequency changes and lose the normal ability of inertial response and frequency modulation [4–7]. Therefore, large-scale wind farms will inevitably lead to the reduction of system inertia and incompetence of frequency modulation.

At present, theoretical studies focus mainly on rotor speed control and pitch angle control of wind turbine generator unit in accordance with control strategies of wind farms participating in system frequency modulation. With the deepening of studies on frequency control technology of variable-speed wind turbine generator unit, the proportional plus derivative controller is adopted currently, which combines derivative control and active/frequency droop control that are superposed on the maximum power tracking control together [8–13].

With the aim of improving the frequency modulation ability of wind farms, this paper proposes a load shedding control strategy of wind turbine generator unit through the analysis on drooping characteristics and primary frequency modulation ability of wind turbine generator unit and VSC converter, reserving required spare capacity for system frequency modulation.

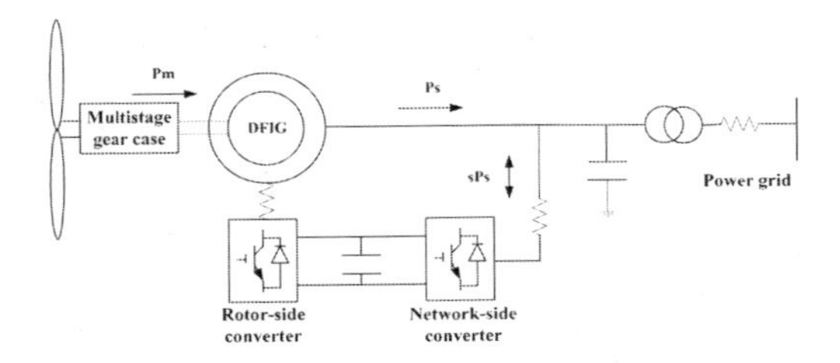

Figure 1. The structure of a double-fed wind turbine generator unit.

2 THE MAXIMUM WIND POWER TRACKING MECHANISM OF DOUBLE-FED WIND TURBINE GENERATOR UNIT

A double-fed induction generator (DFIG) is adopted in this system. The stator is directly connected to grid while the rotor realizes AC excitation through the three-phase AC-DC-AC converter. Exchanges between electric power and power grid are realized through the dual-channel of stator and rotor, which is shown in Figure 1. The rotor excitation converter adopts the structure of a three-phase two-level voltage PWM converter in a back-to-back form. Two PWM converters can be respectively called grid-side converter and rotor-side converter according to their positions.

The relation between output power P_0 and rotational speed ω of wind turbine controlled by the fixed pitch under different wind speeds is shown in the figure. It can be seen that output powers of different units with the same wind speed are different. There is an optimal rotational speed in the same wind speed, under which the wind turbine is able to reach the optimal tip speed ratio and capture the maximum energy and output power under this wind speed. Therefore, the maximum wind power tracking control of wind turbine with fixed pitch is closely related to the rotational speed control of wind turbine.

As shown in Figure 2, a curve of the optimal power P_{out} can be formed by connecting relevant maximum power points of the optimal rotational speeds under different wind speeds. The expression of P_{out} is:

$$P_{out}=K_w\omega_w^3 \quad (1)$$

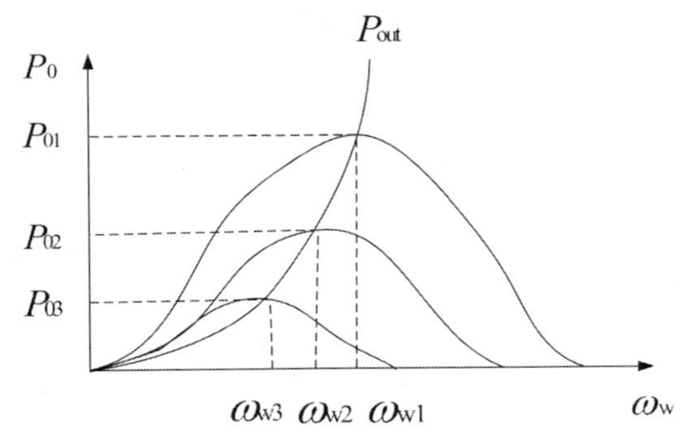

Figure 2. The relation between output power and rotational speed of wind turbine under different wind speeds.

The diagram of the maximum wind power tracking control of double-fed induction generator is shown in Figure 3. In this figure, P_v is the output wind power of wind turbine and P_s^* is the reference value of active power (power instruction) of DFIG. The key of power control is the correct calculation of power reference value.

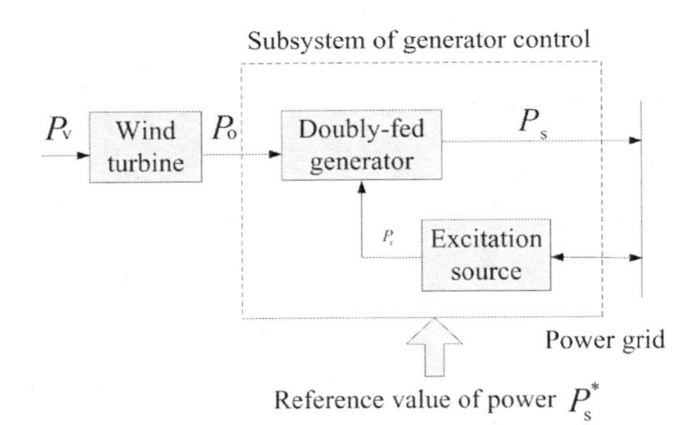

Figure 3. The diagram of operational control of the maximum wind power tracking.

With the above-mentioned traditional power control strategy, the double-fed wind turbine generator unit outputs power to grid by following the maximum wind power tracking instruction. The double-fed wind turbine generator unit is able to give up the moving trajectory of the maximum wind power tracking and modulate system frequency fluctuation when system frequency increases with the decrease of active load. But the double-fed wind turbine generator unit fails to provide the system with active support and frequency modulation when the system frequency reduces with increase of active load. Besides, there is no coupling relationship between the rotational speed of wind power generator and power grid frequency as well as inertial response that inhibits system frequency changes. So, the double-fed wind turbine generator unit should have the capability of upward and downward frequency modulation.

3 FREQUENCY CHARACTERISTICS AND PRIMARY FREQUENCY MODULATION CONTROL OF WIND TURBINE GENERATOR UNIT

3.1 *Load shedding control of wind turbine generator unit*

System frequency is normally modulated as more than 20s. The wind turbine generator unit should participate in frequency modulation according to static frequency characteristics required by power grid and modulate the mechanical power captured by wind turbine continuously so as to completing the task of primary frequency modulation. Necessary spare capacity should be provided for system frequency modulation because the wind turbine generator unit fails to operate with full load.

It is able to provide system frequency modulation with necessary spare capacity by altering the maximum power tracking curve of the wind turbine generator unit. Under different wind conditions, the wind power utilization coefficient can be changed by enlarging the pitch angle so as to reserve certain spare capacity for primary frequency modulation.

According to the aerodynamic knowledge, the output power of wind turbine can be expressed as:

$$P_v = \frac{1}{2}\rho S_w v^3 \tag{2}$$

In the above formula, ρ is the air density; S_w is the windward swept area of wind turbine blades; v is the velocity of air before entering the swept surface of wind turbine (namely undisturbed wind speed).

Because wind energy passing through the rotary surface of wind wheel cannot be completely absorbed by wind turbine, the wind power utilization coefficient C_p can be defined to represent the capability of capturing wind energy. Thus, the output mechanical power of wind turbine is:

$$P_0 = C_p P_v \tag{3}$$

The wind power utilization coefficient C_p is an important parameter representing the operational efficiency of wind turbine. It is closely related to wind speed, rotational speed of blades, rotor diameter and pitch angle. In order to discuss the characteristics of C_p more conveniently, another important parameter of wind turbine is defined——tip speed ratio λ, namely the ratio of wind speed and linear speed on blade tip.

Characteristics of wind turbine with variable pitches are frequently expressed by the wind power utilization coefficient C_p, which is shown in Figure 4. The wind power utilization coefficient C_p is the comprehensive function involving tip speed ratio λ and pitch angle β, namely $C_p(\lambda, \beta)$. It can be seen that the curve C_p decreases sharply when pitch angle β increases.

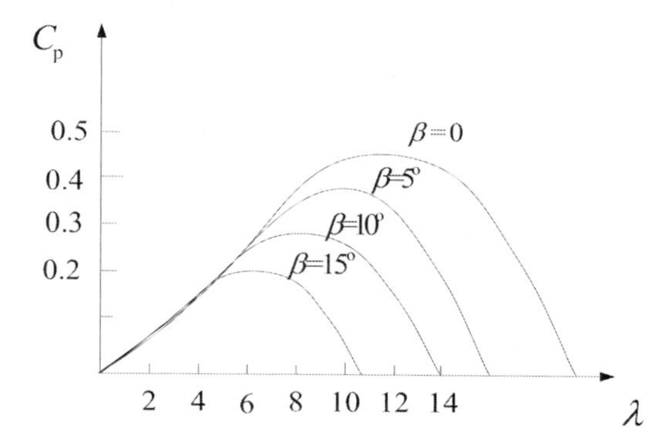

Figure 4. Performance curve of wind turbine with variable pitches.

Therefore, the double-fed wind turbine generator unit is able to change the wind power utilization coefficient in line with the relation between λ and β in different regions of the wind turbine generator unit so as to realize load shedding control of the wind turbine generator unit.

In actual operations, the wind turbine generator unit judges the current operational state in accordance with wind speed. With the C_p-λ-β characteristic of wind turbine, the load shedding level can be set by adjusting pitch angle so as to provide solutions for primary frequency modulation control.

3.2 *Characteristic analysis of primary frequency modulation*

Static frequency characteristics of power system reflect the capability of primary frequency modulation. Imbalanced power caused by system frequency changes can be effectively shared by the participation of double-fed wind turbine generator unit in primary frequency modulation. But reasonable static frequency characteristics should be adjusted.

Static difference modulation coefficient is of great significance for maintaining the stability of system frequency. The smaller the static difference modulation coefficient is, the stronger the frequency modulation capability of generator unit is. Thus, it is easier to guarantee the stability of frequency. In actual operations, however, the excessive small difference modulation coefficient will certainly lead to unreasonable load distribution of generator units in the system as well as unstable operation of the speed regulation system. Therefore, the static difference modulation coefficient should be adjusted reasonably by the primary frequency modulation.

4 SIMULATION ANALYSIS AND RESULTS

4.1 *System introduction*

In this paper, the AC excitation double-fed induction wind turbine system with variable speed and constant frequency is applied to verifying the effectiveness of the control strategy. As shown in the figure, the simulation system is established with the PSCAD simulation software, including a heat-engine plant of 600MW and a wind farm (15MW·66 double-fed wind turbine generator units). The load is active load with the capacity of 560MW.

4.2 *Simulation analysis of VSC converter frequency modulation for sudden load increase*

When the wind speed is 8m/s and the initial operation of wind turbine generator unit is in the condition of the maximum power capture, the load increases suddenly by 30MW, leading to the reduction of system frequency. Simulation results are shown in the figure.

Dynamic response processes of frequency modulations with VSC converter and without VSC converter are compared in Figure 5. The system frequency change is 0.0842HZ when frequency is modulated without VSC; the system frequency change is 0.0841HZ when frequency is modulated with VSC. It can be known from the figure that VSC converter cannot provide the system with effective frequency support. So the effect of VSC converter in frequency adjustment can be neglected. The diagram of drooping characteristics is shown in Figure 6.

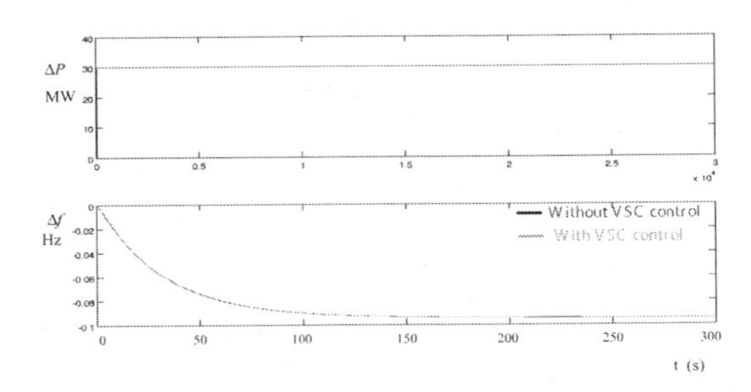

Figure 5. Comparison of system dynamic responses when load increases suddenly.

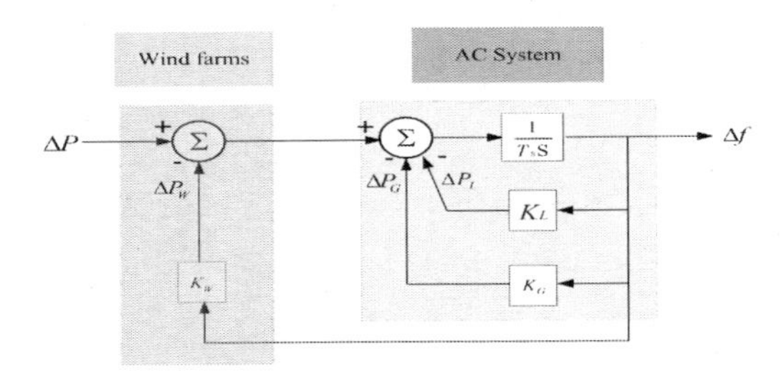

Figure 6. The diagram of drooping characteristics of wind farm participating in system frequency modulation.

4.3 *Wind farm frequency modulation of different control methods*

The wave pattern of grid frequency in 50 minutes is shown in Figure 7. As shown in Figure 8, four kinds of different control methods are introduced in accordance with the fluctuation of grid frequency, that are respectively wind farm with no frequency modulation, wind farm frequency modulation with full load, wind farm frequency modulation with load shedding of 5% and wind farm frequency modulation with load shedding of 10%.

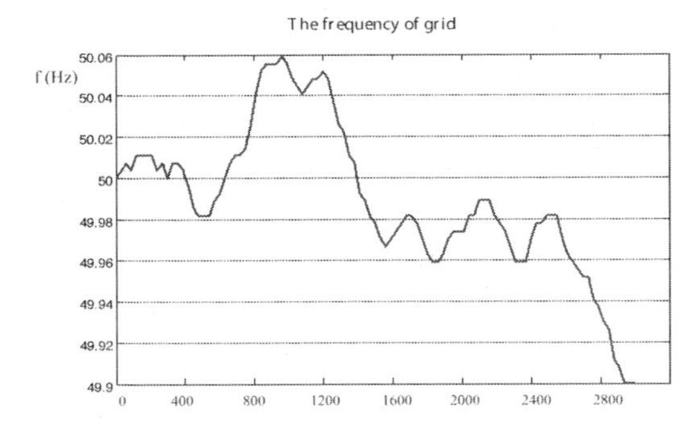

Figure 7. Dynamic wave pattern of grid frequency.

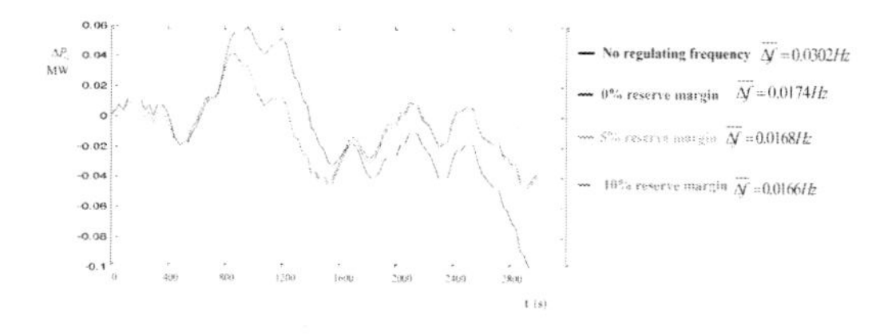

Figure 8. Increased or reduced power of wind turbines in four kinds of control methods.

It can be concluded from the comparative analysis that the larger the load shedding of wind farm is, the better the system frequency modulation quality is. But there is not much difference between 5% load shedding and 10% load shedding. Besides, the wind farm with 5% load shedding is more economic than the wind farm with 10% load shedding, consuming less electric power.

5 CONCLUSIONS

Large-scale wind power integration relatively reduces the rotational inertia of system and brings challenges for the frequency stability. According to the requirements of relevant guides of wind power integration in different countries, wind power participating in system frequency modulation will become an inevitable requirement for the future development of power grid. Based on the load shedding control scheme of variable-speed wind turbine generator unit, this paper further not only proposes a control scheme that combines VSC converter with primary frequency modulation organically so that the wind turbine generator unit is able to conduct the primary frequency modulation according to static frequency characteristics, but also improves the ability of wind farm of supporting system frequency. The following conclusions are drawn from the theoretical and simulative analysis:

1. During the process of dynamic response of system frequency, VSC converter participates in system frequency modulation through absorbing or releasing electric power. As for the control method of no participation in system frequency modulation, the change rate of system frequency is so small that the function of frequency change cannot be improved. Therefore, the function of VSC converter in system frequency modulation can be neglected.
2. The wind turbine generator unit should operate with load shedding in order to respond to system frequency drop and reserve part of active spare capacity for primary frequency modulation. According to different wind conditions and CP characteristics of wind turbine, the load shedding level of the wind turbine generator unit can be controlled by adjusting the pitch angle. It can be known from the comparative analysis that the more the reserved active spare capacity is, the better the ability of system frequency modulation can be. But there is not much difference between 5% reservation and 10% reservation in terms of the effect of system frequency modulation. It is the most appropriate to reserve 5% if the economic aspect is taken into account.

ACKNOWLEDGMENT

This paper is supported by Funding project of Natural Science Foundation of China (512111038).

REFERENCES

[1] Guan, H.L., Chi, Y.N., Wang, W.S., *et al.* (2007) Simulation on frequency control of doubly fed induction generator based wind turbine. *Automation of Electric Power Systems*, 31(7):61–65.

[2] Lin, C.W., Wang, F.X., Yao, X.J. (2003) Study on excitation control of VSCF doubly fed wind power generator. *Proceedings of CSEE2003*, 23(11):122–125.

[3] Li, S.H., Haskew. T.A., Xu, L. (2010) Conventional and novel control designs for direct driven PMSG wind turbines. *Electric Power Systems Research*, (80):328–338.

[4] Ekanayake J, Jenkins N. (2004) Comparison of the response of doubly fed and fixed-speed induction generator wind turbines to changes in network frequency. *IEEE Transactions on Energy Conversion*, 19(4):800–802.

[5] Holdsworth L, Ekanayake J B, Jenkins N. (2004) Power system frequency response from fixed speed and doubly fed induction generator-based wind turbines. *Wind Energy*, 7(1):21–35.

[6] Morren J, Haan S, Kling W L, *et al.* (2006) Wind turbines emulating inertia and supporting primary frequency control. *IEEE Transactions on Power Systems*, 21(1):433–434.

[7] Xue, Y.C., Tai, N.L. (2011) Review of contribution to frequency control through variable speed wind turbine. *Renewable Energy*, (25):1671–1677.

[8] Gautam D, Goel L, Ayyanar R, *et al.* (2011) Control strategy to mitigate the impact of reduced inertia due to doubly fed induction generators on large power systems. *IEEE Transactions on Power Systems*, 26(1):214–224.

[9] Lalor G, Mullane A, O'Malley M. (2005) Frequency control and wind turbine technologies. *IEEE Transactions on Power Systems*, 20(4):1905–1913.

[10] Mauricio J M, Marano A, Expósito A G, *et al.* (2009) Frequency regulation contribution through variable-speed wind energy conversion systems. *IEEE Transactions on Power Systems*, 24(1):173–180.

[11] Cao, J., Wang, H.F., Qiu, J.J. (2009) Frequency control strategy of variable-speed constant-frequency doubly-fed induction generator wind turbine. *Automation of Electric Power Systems*, 33(13):78–82.
[12] Li, J.J., Wu, Z.Q. (2011) Small signal stability analysis of wind power generation participating in primary frequency regulation. *Proceedings of CSEE 2011*, 31(13):1–9.
[13] Chien L R C, Lin, WT., Yin, Y.C. (2011) Enhancing frequency response control by DFIGs in the high wind penetrated power systems. *IEEE Transactions on Power Systems*, 26(2):710–718.

Electric, Electronic and Control Engineering – Shao, Shu & Tian (Eds)
© 2015 Taylor & Francis Group, London, ISBN: 978-1-138-02842-5

Application of the new portable injection pump in Guhanshan coal mine

Hucheng Wu & Deyu Wei
Henan College of Industry and Information Technology, Jiaozuo Henan, China

Jianzhou Guo
Jiaozuo Coal Group Company, Henan Energy and Chemical Industry Group, Jiaozuo Henan, China

ABSTRACT: For solving the problem of the low concentration of gas extraction in Guhanshan coal mine, we develop an efficient portable four-ram injection pump based on analyzing sealing process, transporting two sealing chemical seriflux with 1:1 mixture. The pump has the high output pressure, can compact the fracture in bore with the high pressure seriflux and block the air leakage passage. Comparing the old one, the pump is efficient, high quality and can get the high concentration of gas extraction in the sealing borehole. It is a Safe and reliable and efficient mechanical equipment for hole sealing

KEYWORDS: the gas extraction; borehole sealing; four-ram injection pump; application analysis

1 INTRODUCTION

As a coal and gas outburst mine, the designed production capacity of Guhanshan coal mine is 1.5 million t/a. There're two coal seams are for mining. No. 15 mining area and No. 16 mining area are under coal mining process at present. The gas content of these two mining areas is 1.23~13.15 m^3/t. The gas in the extraction regions of both mining areas is pre-extracted by a method combining transverse drilling and concordant drilling. The gas extraction process is as follows: firstly, use drilling machines to make extraction boreholes; secondly, seal the orifices of the boreholes with hole-sealing machine; at last, connect the gas extraction pipes of each borehole to start gas extraction [1]. The sealing quality of boreholes has great impact on the concentration and flow of gas extraction. Currently, the reason for why the gas extraction concentration in Guhanshan coal mine is low (the average concentration is less than 20%) is given as follows: the applied hole-sealing machines and tools fail to satisfy the complicated geological conditions of the working face underground. There's air leakage and fracture resulting from loose sealing on the borehole orifices. According to the geological conditions of this mine, we have developed a new type of hole-sealing injection pump [2] that can keep the hole sealing tight on gas extraction boreholes. With good filed service effect, this hole-sealing injection pump has great popularization and application value.

2 CURRENT SITUATION AND EXISTING PROBLEMS OF BOREHOLE SEALING

The original hole-sealing drilling machine applied for gas extraction in Guhanshan coal mine is Gu Tejie single-plunger pump. This pump has one air cylinder in one direction, driving two single-plunger pumps in the meantime. Those two plunger pumps carry two kinds of grout respectively. The grout generates hybrid reaction at the delivery outlet and then are sent to the hole mouth for sealing [3]. This pump is simple in structure and easy to operate with low price. However, there're some problems existing in the sealing process: one is that the measurement is not accurate while the pumps deliver those two kinds of grout. There's no guarantee that the grout can be prepared by 1:1 equal proportion. Insufficient reaction can lead to under mixing chemical grout and cause the resultant nature to be unstable. After solidification, the hardness and density of the resultant decreases while the crystalline compaction is bad. As a result, air leakage appears in the passage and the concentration of gas extraction becomes lower [4]. The second problem is the failure rate of this pump is high. During the eight working hours in one shift, half of the time needs to be spent in pump maintenance, resulting in low working efficiency and high labor intensity. And the third problem is the waste of grout in sealing operation. The chemical grout running off in the workplace under the shaft can cause environmental pollution. Furthermore, this pump is big in volume and heavy in weight, making it difficult for workers to move it (two are needed to move it in most cases).

3 COMPOSITION AND WORKING PRINCIPLE OF THE NEW-TYPE PORTABLE INJECTION PUMP

3.1 *Composition of the new-type pump*

The newly developed four-plunger pump is shown in Figure 1. The whole injection pump is consisted of six main parts——pressure system, lubrication system, hydraulic system, cleaning system, operating system and auxiliary components. The pressure system is formed by air filter, control valve, pressure gage, reversing valve, double-acting four-piston rod cylinder (the cylinder of traditional injection pump is double-acting single-piston rod one) and silencer. The hydraulic system is mainly composed of (4) plunger pumps, (8) one-way valve, switching valve, mixing agitator and injection pistol.

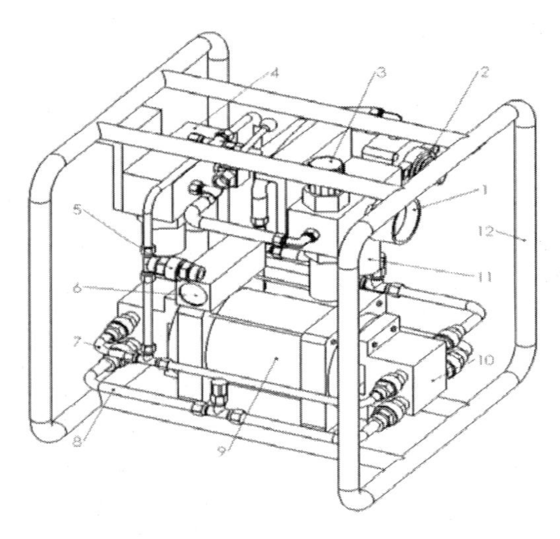

Figure 1. Structure of the new-type injection pump.
1. Gas-pressure meter; 2. Manual operating valve; 3. Pressure regulating valve; 4. Air filter; 5. One-way cleaning valve; 6. Gas reversing valve; 7. Bleeder tube; 8. Pick-up tube; 9. Air cylinder; 10. Hydraulic cylinder; 11. Priming can; 12. Rack

The lubrication system is mainly composed of priming can, reversing valve and working cylinder. The cleaning system is consisted of air filter, control valve, and cleaning tube. The operating system is mainly made up of various control valves while the auxiliary components cover rack, dedicated grout bucket, various pipeline and accessories.

3.2 *Working principle of the new-type pump*

1 Pressure System and Lubrication System
See Figure 2 for the working principle of pressure system.

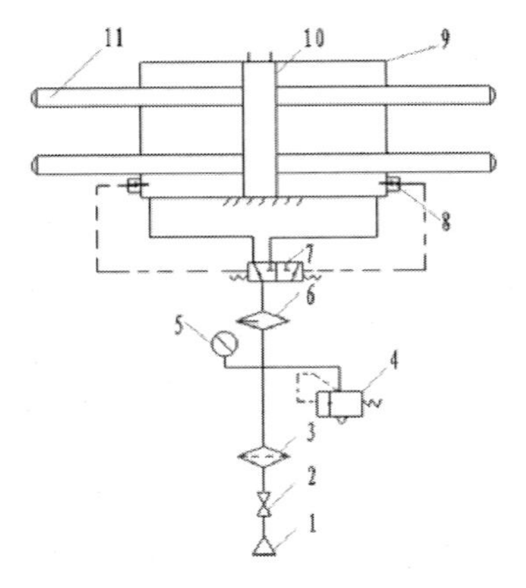

Figure 2. Pressure system and lubrication system.
1. High pressure air source; 2. Stop valve; 3. Air filter; 4. Pressure control valve; 5. Gas-pressure meter; 6. Atomized lubricator; 7.Two-position-three-link pneumatic operated reversing valve; 8. Travel switch; 9. Cylinder block; 10. Piston; 11. Piston rods (4)

High pressure gas in the pit passes through the manual switching valve→air filter 3→pressure control valve 4→pressure gage 5→priming can 6→pneumatic operated reversing valve 7→left & right bodies of the cylinder drive the cylinder piston to move on the left and right.

The high-frequency left and right movement of the piston inside the cylinder drives the piston rod above to stretch on the left and right. Meanwhile, the piston rod also works as the power element of the plunger pump in hydraulic system.

The left and right movement of the piston inside the cylinder is controlled by the automatic reversing of (pneumatic) two-position-three-way reversing valve 6. Connected as a whole, the reversing valve and the cylinder form an automatic reversing gear together with piston, travel switch, one-way valve, reversing valve element and silencer. This automatic reversing gear is a patented technical product.

The lubrication system is attached to high-pressure gas circuit[5], and lubricating oil is mainly provided by the priming can 6 in the gas circuit. When high-pressure gas passes through, according to the oil absorption and atomization principle under negative pressure, the lubricating oil will be atomized into the mixture of oil gas and high-pressure gas and be sent to the automatic reversing valve 7. After lubricating the valve element and valve body, the mixture gas continues to enter the cylinder and lubricates the piston and the inner wall of the cylinder body. All components and parts in high-frequency movement will be lubricated.

2 Hydraulic System

The working principle of hydraulic system is shown in Figure 3.

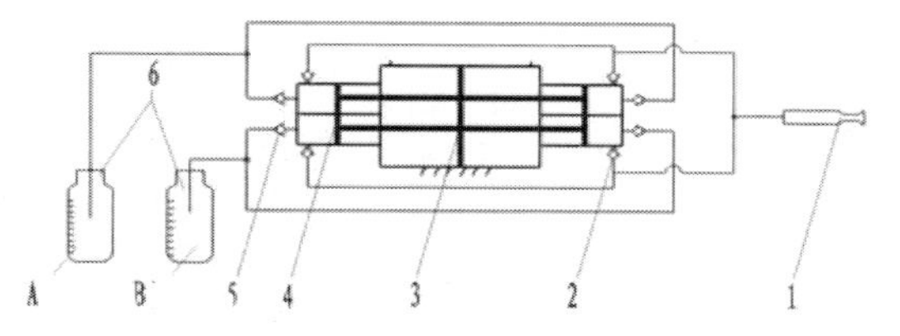

Figure 3. Hydraulic System.
1. Injection pistol; 2. Grout drainage valve; 3. Cylinder piston; 4. Piston of injection cylinder plunger pump; 5. Grout absorbing valve; 6. Liquid container (with scale)

The liquid absorbing and drainage is operated by the left and right movement of the main piston inside the cylinder. The piston movement drives the plunger 2 of the hydraulic cylinder to move, so as to manage the liquid absorbing and drainage operation. Major pressure difference is generated from the area difference of the big and small pistons, and thus high-pressure liquid can be exported externally.

For the structural design of the main engine, pressure can be increased according to the mechanics equation of $P_1 \times S_1 = P_2 \times S_2$, and thus high-pressure seriflux can be exported. The principle can be represented by:

$$P_2 = \frac{S_1}{S_2} \times P_1$$

Among which:

S_1——active area of the piston inside the cylinder, m^2;

S_2——area of the piston inside the hydraulic cylinder (grout cylinder), m^2;

P_1——bleed pressure provided in the pit, MPa. It is usually a fixed value (0.2–0.6 MPa).

P_2——pressure of the grout exported by the hydraulic cylinder (grout cylinder), MPa.

As S_1 is much bigger than S_2 in this structural design, and thus P_2 is much bigger than P_1, which means the exported liquid pressure P_2 is much higher than the bleed pressure P_1. The grout pressure exported by this pump station can reach 10MPa. Higher pressure can bring more compacted injection and the fraction around the boreholes can be more effectively filled. As a result, air leakage on the orifices can be prevented and gas extraction concentration can be improved.

In order to improve mechanical transmission efficiency, the actuating cylinder of the pressure system is combined with the plunger pump of the hydraulic system as a whole and connected by plunger rods. In the meantime, plunger rods are the primary members of the plunger hydraulic pump. To ensure the 1:1 mixture of chemical grout A and chemical grout B in equal volume and equal quantity, four plunger pumps are symmetrically designed with two on each side for synchronized operation. The plunger pumps on the same side absorb or discharge the two kinds of liquid (chemical grout A and chemical grout B) at the same time. This design can ensure that chemical grout A and chemical grout B can be "continuously and efficiently "injected into gas extraction drilling process "with equal quantity and in the meantime".

After these two kinds of chemical grout are mixed in the hydraulic system, a liquid mixing agitator is designed in front of the injection pistol in order to guarantee sufficient and uniform mixing. There's a plastic static screw rod placed inside the agitator. When high-pressure mixing grout passes through the screw leaves, the screw rod will be driven to rotate and thus stir the mixing grout. In this phase, these two kinds of chemical grout will have chemical reaction. Hole-sealing raw materials can be saved to the maximum extent in this way. And the compactness of the solidified grout can bring the best hole-sealing effect.

3 Self-cleaning System

As the solidification resulting from the mixing of chemical grout A and chemical grout B or from the low-temperature of chemical grout can block the valve channel or pipelines of the hydraulic system, all chemical grout must be cleaned up after each shift of injection. No chemical grout is allowed in the hydraulic system. At present, the applied injection pump station is not complete in functions. The system is always blocked by the grout crystalline remained inside. In some cases, vales can be even damaged. Thus, the related maintenance workload is great while working efficiency is low.

See Figure 4 for the self-cleaning system specially designed in this injection pump. When each shift ends, the self-cleaning system can connect high-pressure air into the hydraulic system through the valves for exclusive use. All remained grout in the main engine cylinder and grout delivery pipelines can be discharged, so as to realize self-cleaning. This design can ensure that no grout will be left in the hydraulic system when the pump station is stopped.

Cleaning functions include the cleaning of the mixing agitator and the cleaning of the whole hydraulic system.

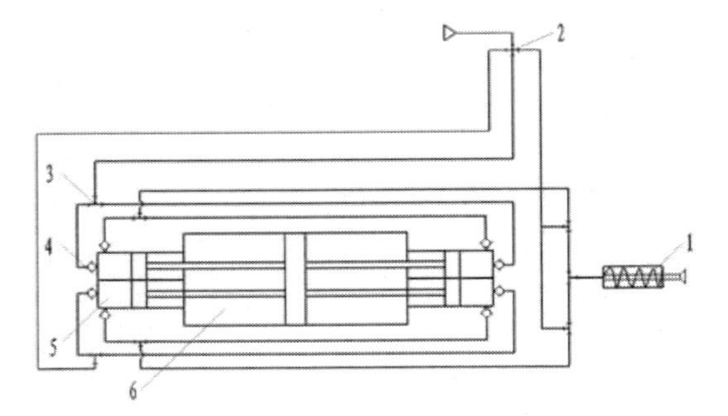

Figure 4. Cleaning system.
1. Stirrer; 3. Four-way valve; 3. Three-way; 4. One-way valve; 5. Grout (liquid) cylinder; 6. Air cylinder

4 Operating System
The whole pump station system is consisted of a manual reversing valve and four stop valves. It is easy to control and convenient to operate.

5 Auxiliary Components
Auxiliary components mainly include the liquid buckets (with scale marking) containing the two kinds of chemical grout [6] and rack, etc.

4 PERFORMANCE FEATURES OF THE NEW-TYPE PUMP

1 There is flow control valve exclusive for this pump which strictly controls he absorbing volume of the chemical grout and thus can ensure the chemical grout will be 1:1 mixed with equal quantity. Spiral agitator is designed for the mixing process. It can help the chemical grout mix uniformly and have sufficient chemical reaction. By this means, the resultant features resulting from reaction can meet the borehole strength requirement.
2 The output pressure of this pump is higher than that of the old pump. As the output chemical grout pressure of this pump can reach 10MPa, the grout can be pressed into the fraction of the borehole wall. By this process, the stability of extraction borehole wall can be reinforced while the air leakage passage can be sealed.
3 This pump is designed with a self-cleaning system. When each shift of work ends, the self-cleaning system will blow gas into the hydraulic system for cleaning. No chemical grout will be left in the system, and thus the orifices and pipelines of the valves are free of being blocked by the solidification crystal formed by remaining grout. The self-cleaning system can decrease the failure rate of the pump and reduce the workload for maintenance.
4 Low hole-sealing cost[7]. As the pump is able to control the grout usage volume and its failure rate is low, the chemical grout is greatly prevented from running off or being wasted. The average material cost of the pump for sealing a borehole is 230RMB which is 160RMB less than that of the old pump. Besides, no pollution will be brought by the new-type pump to the working place.
5 The new-type pump is simple in structure, light in weight (25kg), and small in volume (430×400×380mm). Thus, it is convenient to move or carry the pump in the confined space under the shaft.

5 HOLE-SEALING TEST AND EFFECT ANALYSIS

5.1 *Hole-sealing test for concordant coal seam drilling*

Since August, 2014, the new-type portable injection pump and the original Gu Tejie pump (the old pump) had been simultaneously used in No 16021 air return way of Guhanshan coal mine. In No. 16021 air return way, the new pump and the old pump were alternatively used for hole sealing with certain intervals (the pump was replaced by the other after sealing 3 to 4 boreholes), and the test records of gas extraction concentration were made at once. On September 1st, 2014, a test was made on 30 boreholes. 16 boreholes were tested for the hole sealing of the new pump while 14 boreholes were tested for the hole sealing of the old pump. On September 8th,a test was made on 32 boreholes. 16 boreholes were respectively tested for the hole sealing of the new pump and the old pump. By the same means, 36 boreholes were tested respectively on September 10th, September 15th and September 22nd. See Table 1 for the data of the tested 36 boreholes on September 10th.

See Figure 5 for the gas extraction concentration comparison of the new injection pump and the old injection pump resulting from the summarization of the data for the five tests within a month.

Analysis result: from the comparison of the test made on the 36 boreholes on September 10th in

Table 1. Hole-sealing gas extraction concentration test data of the new pump and the old pump in no. 16021 return airway.

Number of borehole	Hole-sealing concentration of the new pump %	Hole-sealing concentration of the old pump %	Number of borehole	Hole-sealing concentration of the new pump %	Hole-sealing concentration of the old pump %
1	31		19	39	
2	12		20	55	
3	19		21	26	
4		22	22		15
5		24	23		3
6		23	24		18
7	14		25		21
8	10		26	2	
9	37		27	14	
10		21	28	25	
11		19	29	36	
12		12	30	35	
13	47		31		13
14	44		32		11
15	32		33		4
16		25	34	18	
17		20	35	16	
18		17	36	12	

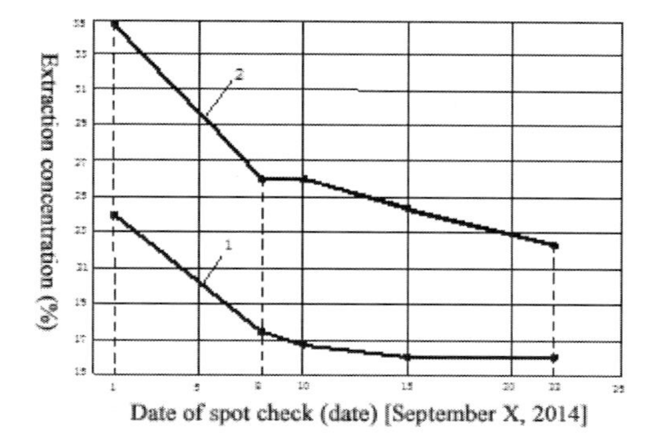

Figure 5. Comparison of the hole-sealing gas extraction concentration data of the new pump and the old pump.
1. Hole-sealing spot check curve of the old pump; 2. Hole-sealing spot check curve of the new pump

Table 1, we can see that the new pump hole sealing tested 20 holes while the old one tested 16 holes. It can be seen from the tested gas extraction concentration that the average of the old pump hole sealing was 16.75% while that of the new pump was 26.2%. The average gas extraction concentration was improved by 56.4%. From the extraction concentration comparison within a month in Figure 2, we can see that the average concentration of the old pump hole sealing was 18.1% while that of the new pump was 28.8%. The average concentration was improved by 59.1%.

5.2 *Transverse drilling hole-sealing test*

50 new boreholes were made among the original boreholes in the No. 1602 end-located drainage roadway. The new-type injection pump was used for hole sealing. In October, 2014, 20 boreholes respectively sealed by the new and old pumps were selected for gas extraction concentration test each day. The average concentration was measured and recorded in every period (10 days). The recorded data is shown in Table 2.

See Figure 6 for the histogram of the average extraction concentration value of the 20 holes in each ten-day period. It can be seen from the data that the hole-sealing extraction concentration of the new pump was higher than that of the old pump by 81%, 97% and 61% respectively in the first, second and third ten-day periods. The monthly average gas extraction concentration was improved by 79.7%. Thus, it can be concluded that the hole-sealing

Table 2. Test data of the hole-sealing gas extraction concentration of the old and new pumps in no. 1602 end-located drainage roadway.

Extraction concentration after the hole sealing of the new pump (%)			
Number of hole sealing	The first ten days	The second ten days	The third ten days
S21	11	9	15
S22	64	51	44
S23	79	76	37
S24	49	46	40
S25	30	22	20
S26	8	14	20
S27	14	20	12
S28	51	49	36
S29	38	13	15
S30	35	27	22
S31	5	11	12
S32	23	19	22
S33	75	71	70
S34	18	14	16
S35	34	30	37
S36	60	57	52
S37	51	49	44
S38	21	27	20
S39	11	25	22
S40	13	17	16
Average concentration (%)	34.5	32.5	28.5

Extraction concentration after the hole sealing of the old pump (%)			
Number of hole sealing	The first ten days	The second ten days	The third ten days
1251	9	8	7
1252	7	6	12
1253	6	5	9
1254	5	8	16
1255	17	5	6
1256	25	19	22
1257	19	6	10
1258	34	26	22
1259	49	37	31
1260	6	17	14
1261	17	19	13
1262	5	15	11
1263	19	17	21
1264	31	25	20
1265	18	21	29
1266	38	25	20
1267	34	28	26
1268	16	15	18
1269	20	18	22
1270	7	11	17
Average concentration (%)	19.1	16.5	17.7

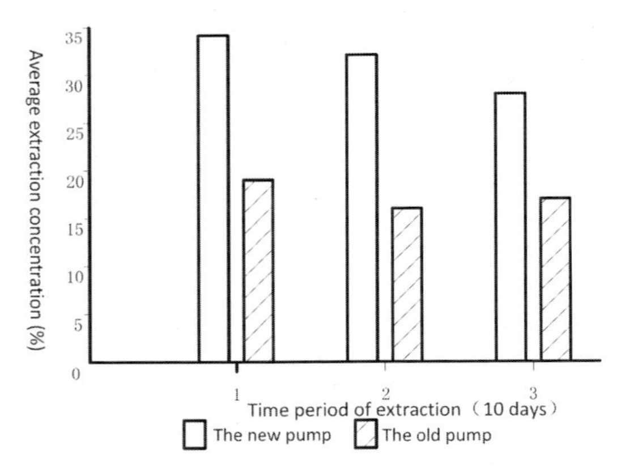

Figure 6. Comparison of the hole-sealing gas extraction concentration of the old and new pumps in no. 1602 end-located drainage roadway.

effect for transverse drilling gas extraction of the new pump in the end-located drainage roadway is better.

6 CONCLUSION

1 The new-type portable injection pump was given hole sealing test in No. 16021 return air course and No. 1602 end-located drainage roadway of Guhanshan coal mine. Gas extraction concentration text was continuously conducted for three months. The test results showed that the extraction concentration in No. 16021 return air course was higher than that of old-fashioned injection pump by 59.1% on average while the extraction concentration in No. 1602 end-located drainage roadway was higher than that of old-fashioned injection

pump by 79.7% on average. They can explain that the new-type pump is free of air leakage; it has great hole-sealing compaction and good gas extraction effect.

2 During the three-month service, failure only appeared once in the new-type portable injection pump. The failure was an operation error: details were not clearly handed over during shift changes, as a result, the worker put the pipet for chemical grout A into the bucket of chemical grout B and put the pipet for chemical grout B into the bucket of chemical grout A. These two kinds of liquid had reactive crystallization in the hydraulic system and the resultant solidification blocked the one-way drainage valve. The test proves that this new-type pump has low failure rate, reliable working ability and durable service quality.

3 The new-type pump can take full advantage of the compressed gas in the pit as its power source [8] while no chemical grout is wasted and no pollution is generated in between. Thus, it is able to realize working environment optimization for the workers and keep the hole sealing cost at a low level. In this way, energy can be saved while the environment can be protected in the meantime.

4 The new-type portable injection pump is simple in operation and convenient to carry. It can greatly ease workers' labor intensity.

REFERENCES

[1] Sang P., Yang. L., Han W., et al. (2014) Study on "two-sealing-and-one-injection" concordant borehole sealing technology and its application. *China Coal*, 40(3):98–105.

[2] Guo J.Z. & Guo J.Z. (2013) *Double-liquid Pneumatic Injection Pump*: China, ZL201320550966.3 [P]. 2013-9-6.

[3] He J. & Dong K.Y. (2014) Deep analysis on the reasonable sealing for gas extraction borehole in the coal roadway of large cross section. *China Coal*, 40(2): 101–104.

[4] Liu X., Lin H.X., Zhang S.B., et al. (2014) The repair mechanism of gas extraction borehole and study on its application, *China Coal*, 40(1):102–105.

[5] Yan L.Q., Pan F., Lv B.M., et al. (2014) Existing problems of the gas extraction in XinAn Coal Mine and improvement measures. *Zhongzhou Coal*, 2014(9): 48–49.

[6] Yan B.Y., Wang F. (2014) Study on the hole sealing technology for central bypass capsule-bag type injection and its application. *Zhonogzhou Coal*, 2014(11): 61–63.

[7] Sun Y.Y., Zhang B. & Chen J.L. (2010) Experimental analysis on the one-way valve shear pin for the hole sealing operation of capsule-bag type injection. *Coal Mine Machinery*, (3):62–64.

[8] Ren Q.S., Ai D.C., Chen X.C., et al. (2014) Application of high-water expanding material grouting process in the hole sealing for gas extraction borehole. *Coal Technology*, 33(8):233–236.

Electric, Electronic and Control Engineering – Shao, Shu & Tian (Eds)
© 2015 Taylor & Francis Group, London, ISBN: 978-1-138-02842-5

Economical optimal operation of the CCHP micro-grid system based on the improved artificial fish swarm algorithm

Xiaomeng Yu, Peng Li & Haijiao Wang
State Key Laboratory of Alternate Electrical Power System with Renewable Energy Sources, North China Electric Power University, Baoding Hebei, China

Haijiao Wang
Renewable Energy Department China Electric Power Research Institute, Beijing, China

ABSTRACT: Under the consideration of generating costs of various distributed resources, environmental costs and equipment maintenance costs of the micro-grid and on the basis of satisfying the operational constraints of the micro-grid, the total operation costs of the system can be the lowest by optimizing the power outputs of different distributed resources and energy storage systems in the micro-grid. A micro-grid, which uses Combined Cooling, Heating and Power (CCHP), can effectively improve its distributed power efficiency, thereby economy improves. Finally, improved Artificial Fish Swarm Algorithm (AFSA) is adopted to verify the economy of a CCHP micro-grid mode through the Matlab numerical calculation tool.

KEYWORDS: micro-grid, economical operation, Combined Cooling Heating and Power (CCHP), distributed resources, Artificial Fish Swarm Algorithm (AFSA).

1 INTRODUCTION

With the popularization of smart grid, the micro-grid, which is a system of combing the distributed resources and the load, gets increasing attention. The distributed resources are composed of wind power generation, photovoltaic power generation, fuel cells, micro-turbines, and so on. A micro-grid is closed to the load side, which can reduce power losses during the power transmission and construction costs of the transmission lines. CCHP system is established on the basis of the concept of energy cascade utilization, which uniformly solves the supply problem of the power energy and the (cold) thermal energy and improves the utilization efficiency of various distributed resources. As a result, it has a favorite economy. A CCHP micro network is economical and environment-friendly with good social and economic benefits, attracting a wide attention domestically and on abroad[1–3].

An economic operating mathematical model of the CCHP micro-grid connecting to an external power grid is established and the cost optimization is achieved by a formula. The fuel consumption cost of each power unit and the system operation and maintenance costs are considered in the cost function. Optimization is a concept that minimizes the operation cost of a system on the premise of satisfying the local demand of heating and electricity load. The CHP micro-grid is a complex energy system which contains a balance among different kinds of energies. In the premise of meeting the needs of the user loads, it is a quite consequent content to make a system operation scheme for a further period, according to the micro-source configuration so as to acquire a better economic benefit for the whole system [4]. By far, the research in domestic is still constricted to the level of power micro-grid[5], and does not involve many contents of the CCHP, while researches on CCHP are already underway in some foreign countries.

This paper establishes a CCHP micro-grid system containing renewable energies. The CCHP system can achieve the aim of high efficient energy utilization, which leads to an economic operation and the lowest costs.

2 MATHEMATICAL MODEL OF DISTRIBUTED RESOURCES

2.1 *Wind power generation model*

The output model of Wind turbines is as follows [6]:

$$P_{WG} = \begin{cases} 0, & v < v_{ci} \\ av^2 + bv + c, & v_{ci} \le v < v_r \\ P_{r\_WG} & v_r < v < v_{co} \\ 0, & v \ge v_{co} \end{cases} \quad (1)$$

Where: P_{WG} is the output power of wind turbines, P_{r_WG} is the rated power of wind turbines, v_{ci}, v_{∞}, v_r, v are the cut-in wind speed, cut-out wind speed, the rated wind speed and the actual wind speed respectively. Coefficients a, b, c can be acquired by Matlab fitting of wind turbine output power curve provided by manufacturers.

2.2 *Photovoltaic power generation model*

The power output of PV modules generally is corrected with the system output, which is restricted to the standard test condition (the solar radiation intensity is 1000W/m2, and the cell temperature is 25°C). It can be expressed[7] specifically as follows:

$$P_{PV} = P_{STC}\frac{G_{ING}}{G_{STC}}[1 + k_{PV}(T_c - T_r)] \tag{2}$$

Where: P_{PV} is the actual power output of PV modules, G_{ING} is the real radiation intensify of the sun, G_{STC} is the solar radiation the photovoltaic cell assemblyrd test condition, P_{STC} is the maximum power output of photovoltaic cell assembly under standard test condition, k_{PV} is the power temperature coefficient, T_C is the temperature of the battery.

2.3 *The model of fuel cells*

Fuel cells are widely recognized as efficient, convenient and environmentally-friendly green energy devices. They convert energies discharged by the chemical reaction of materials to electricity directly and need a continuous supply of active materials (fuels and the oxidants) during their work. The reason why calling them fuel batteries is that they convert energies released by material chemistry reaction to power output. On the surface, they have positive and negative electrodes and electrolytes, just like storage batteries, but in fact they cannot store energy but are power generating devices. Its model is as follows [8]:

$$C_{f\_FC}(P_{FC}) = C_{fuel}\sum_{t}\frac{P_{FC}}{\eta_{FC}} \tag{3}$$

where: C_{f_FC} is the consuming fuel cost (yuan) during the operation of fuel cells, C_{fuel} is the fuel price (yuan/m^3), P_{FC} is the output power(kW) of fuel cells, η_{PC} is the power generation efficiency of fuel cells.

2.4 *Micro-turbines*

The micro-turbine (MT), as a new type of gas turbines, has received great concerns and has an increasing development. Its power is generally not more than 1000kW and it suits micro-grid system with CCHP ideally[9–11] due to its high power generation efficiency, high fuel adaptability, low fuel consumption rate, low pollution emission, high reliability, the ability of remote operation, low noise, long life, low vibration, the self-diagnosis function and other advanced technical features. The mathematical model for the CCHP micro-turbine is as follows[12]:

$$\begin{cases} Q_{MT} = \dfrac{P_e(1-\eta_e-\eta_l)}{\eta_e} \\ Q_{he} = Q_{MT} \times COP_{he} \\ Q_{co} = Q_{MT} \times COP_{he} \\ V_{MT} = \dfrac{\sum P_e t_1}{\eta_e \times LHV_{NG}} \end{cases} \tag{4}$$

where: Q_{MT} is the exhaust waste heat of turbines, Q_{he} heat capacity of gas turbines providing, Q_{co} is the cooling capacity of gas turbines providing, η_e is the heat loss coefficient, COP_{co} is the refrigeration coefficient, COP_{he} is the heating coefficient, is the running time of gas turbines, V_{MT} is the amount of gas consumed within the operating time gas turbines, LHV_{NG} is the calorific value of the natural gas, generally 9.7kwh/m^3.

3 A MATHEMATICAL MODEL OF THE ECONOMIC OPERATION OF MICRO-GRID

3.1 *The objective functions*

Objective:

1 The lowest operating costs [13].

$$\min C_{OPE}(P_t) = \sum_{i=1}^{N}\left[C_f(P_{it}) + C_{OM}(P_{it})\right] + C_{PE}(P_{gridt}) + I_{SE}(P_{gridt}) \tag{5}$$

$$C_{PE}(P_{gridt}) = \begin{cases} 0 & P_{gridt} < 0 \\ C_{pt}P_{gridt}\Delta t & P_{gridt} \geq 0 \end{cases}$$

$$I_{SE}(P_{gridt}) = \begin{cases} C_{st}P_{gridt}\Delta t & P_{gridt} < 0 \\ 0 & P_{gridt} \geq 0 \end{cases} \tag{6}$$

where: *COPE* is the operating cost (yuan) of the micro-grid system in an optimizing cycle, is the number of DG units in the micro-grid system, P_t is the active power of all DG units, is the running time of the micro-grid system, C_f is the energy consumption costs (yuan) of DG units, P_{it} is the active power(KW) of DG unit numbering i, C_{OM} is the operating maintenance costs (yuan) of DG units, C_{PEt} is the expenses (yuan) when micro-grid system purchases electricity from the main grid system, P_{gridt} is the interaction power(kW) between the micro-grid power system and the main grid system, I_{set} is the revenue (yuan) when the micro-grid system sells electricity to the main grid system, K_{OM} is the operation and management coefficient of a micro-grid system, DG units (Yuan/kWh), Δt is the interval (h) of optimizing operation, C_{st} is the electricity price (yuan/kWh) of micro-grid system selling electricity to the main grid system, and C_{Pt} is the electricity price (yuan/kWh) when the situation above is converted.

2 The lowest pollutant processing cost of the micro-grid system.

$$\min C_{ENV}(P_t) = \sum_{k=1}^{M} 10^{-3} \beta_k (\sum_{i=1}^{N} \alpha_{ik} P_{it} + \alpha_{gridk} P_{gridt}) \quad (7)$$

where: C_{ENV} is the emission abatement cost of micro-grid system (yuan), M is the amount of all types of emitted pollutants of the system, β_k is the cost coefficient of managing contaminants consumed (yuan/kg), α_{ik} is the emitting coefficient of various pollutants according to different distributed resources (g/kWh).

3 The highest overall efficiency of micro-grid, or a lowest total cost.

$$\min C(P_t) = C_{OPE}(P_t) + C_{ENV}(P_t) \quad (8)$$

Where: C is the total cost of an operating micro-grid.

3.2 *Constraints*

1 Power supply and demand balance:

$$\sum_{i=1}^{N_d} P_{it} = P_{Li} - \sum_{j=1}^{N_{nd}} P_{jt} \quad (9)$$

$$\mathrm{N_d} + N_{nd} = N + 1 \quad (10)$$

Where: $P_{jt}, P_{it}, P_{lt}, N_d, N_{nd}$ are power outputs of DG units that cannot be dispatched in the Micro-grid system, power output of GD units that can be dispatched in the micro-grid system, demanding power of loads (kW), and the number of DG units that cannot be dispatched in a micro-grid system respectively.

2 The limiting values of the power output of DG units

$$P_i^{\min} \leq P_{it} \leq P_i^{\max} \quad (11)$$

$$P_{Li} = \sum_{i=1}^{N_d} P_{it} + \sum_{j=1}^{N_{nd}} P_{jt} \quad (12)$$

Where: $P_i \min, P_i \max$ DG are the lower and superior limit output of DG units in a micro-grid system respectively.

3 The maximum capacity constraint that can be interacted between the micro-grid system and the main grid system.

$$P_{line}^{\min} \leq P_{linet} \leq P_{line}^{\max} \quad (13)$$

Where: $P_{linet}, P_{line}^{\min}, P_{line}^{\max}$ are the interacting powers for real-time between the main power grid system and the micro-grid system, and the lower and superior limit interacting power capacity that can be allowed between the micro-grid system and the main grid system respectively.

4 Ramp constraints of generating Units

$$\left| P_{FC}^{t} - P_{FC}^{t-1} \right| \leq P_{FC}^{\max} \quad (14)$$

$$\left| P_{MT}^{t} - P_{MT}^{t-1} \right| \leq P_{MT}^{\max} \quad (15)$$

Where: P_{FC}^{t} and P_{FC}^{t-1} are the power outputs of fuel cells at time t and time t-1 respectively; P_{MT}^{t} and P_{MT}^{t-1} are the power outputs of the micro-turbine at time t and t-1 time respectively; P_{FC}^{t} and P_{MT}^{t} are the superior limits of the ramp constraints of fuel cells and micro-turbines respectively.

4 A IMPROVED AFSA FOR MODEL RESOLVING

AFSA is a kind of swarm intelligence optimization algorithm based on animal behaviors, which was proposed by LI Xiao-lei of Zhejiang University in 2002. The algorithm searches the optimal solution through simulating the behaviors of fish foraging, bunching and so on, which is a specific application representing the group smart thinking [14].

As shown in Figure 1, the current position of a virtual artificial fish is X, and its view scale is the Visual. Location X_v is its view position at a certain time. On the one hand, if the food intensifies of position X_v is higher than that of the current pmoving,

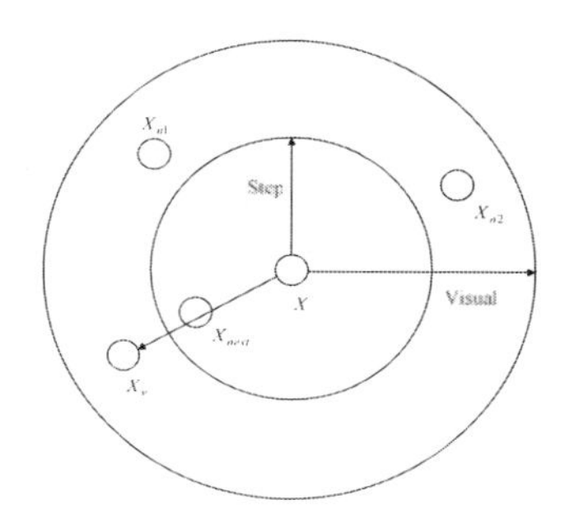

Figure 1. Artificial fish vision and moving step.

it will consider to move a step toward position X_v, which would reach X_{next}; On the other hand [the fish would continue to search other positions within the field of vision[15].

In figure 1, position$X = (x_1, x_2, ..., x_n)$, position $X_v = (x_1^v, x_2^v, ..., x_n^v)$, the process can be represented as follows:

$$x_i^v = xi + visual \bullet r \qquad i = 1, 2, ..., n \tag{16}$$

$$X_{next} = \frac{X_v - X}{\|X_v - X\|} \bullet Step \bullet r \tag{17}$$

Where: r is a random number in [-1,1] interval; Step is the step length. Because the number of companions in a particular environment is limited, the method of adjusting one's position according to the positions of peers within the field of vision is similar to the formulas above.

This paper applies AFSA to the optimization operational problems of CCHP micro-grid system, and also improves the algorithm.

1 The optimizing operation of the CCHP micro-grid system is a nonlinear, complex multi-constrained optimization problem[16].In this paper, we consider the total capacity of all kinds of micro powers as state vectors of artificial fish $X_i = (x_1^i, x_2^i, ..., x_5^i)$ .And there are five kinds of state vectors, which are photovoltaic power, wind power, fuel cells, micro-turbines and energy storage devices respectively, including fuel cells and micro-turbines providing Cooling and Heating respectively. Then the difference Y between the benefit from selling electricity and the sum of operation cost of various micro-sources and the processing cost of pollutants is taken as the adaptation value of the food intensity. That is

$$Y = C_{benefit} - C_{OPE}(P_t) - C_{ENV}(P_t)$$

2 The basic AFSA algorithm considers only the range of all variables instead of the processing of various constraints, however, this paper fully considers a number of constraints such as the ramp constraints of turbines, power balance and so on. Penalty factor C_{PF} is introduced to the objective function in the ramp constraints, therefore the function of food intensify adaptation value can be rectified as follows:

$$\begin{aligned} Y' = {} & C_{benefit} - C_{OPE}(P_t) - C_{ENV}(P_t) \\ & - \mathrm{C_{PF}} \ (\left|P_{FC}^t - P_{FC}^{t-1}\right| - P_{FC}^{\max}) \\ & - \mathrm{C_{PF}} \ (\left|P_{MT}^t - P_{MT}^{t-1}\right| - P_{MT}^{\max}) \end{aligned} \tag{18}$$

3 In the swarming behavior and foraging behavior of the artificial fish, the initial convergence speed is faster due to the values of the vision and step length are fixed, however, it is not easy to converge at the end and the optimal solution is also not precise enough. Therefore, it ought to measure the distance between each artificial fish before iteration. According to the distance, the vision should be adjusted itself based on the nearest and the optimal artificial fish adaptation principles, and the moving direction should also be determined by comparing various food intensifies. Therefore, the improved moving direction should be:

$$X_i^{t+1} = X_i^t + \frac{X_V - X_i^t}{\|X_V - X_i^t\|} \ Step \left|1 - \frac{Y}{Y'}\right| Rand() \tag{19}$$

In the AFSA of the paper, the total number of artificial fish is 100 and the state vector is two dimensions. The attempting times are 300 and the iterating times are 50.The congestion level is 0.618.

5 EXAMPLE ANALYSIS

5.1 *The example system*

The micro-grid system model established in this chapter is shown in Figure 2.A distributed power is composed of photovoltaic power device PV (10kW), micro-turbines MT (65kW), wind turbine WT (20kW), fuel cell FC (45kW) and energy storage device (100kW).

The micro-grid system established in this paper will provide the demand of a particular residential district heating and cooling loads during different seasons. Figure 3 and Figure 4 are the typical-day-load curves in different seasons respectively.

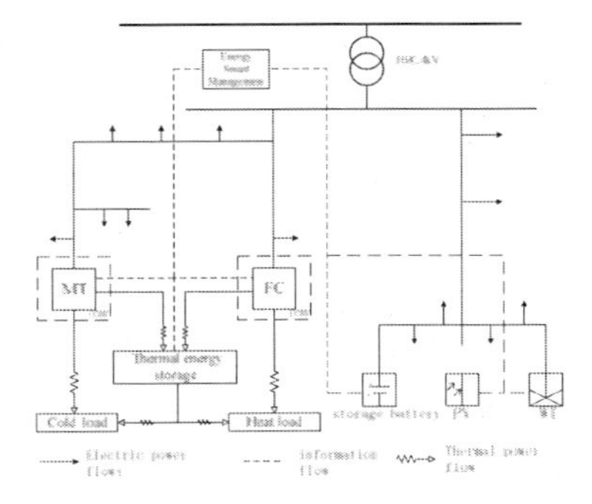

Figure 2. Model of micro-grid system.

It can be seen from Figure 3 and 4 that the main loads are the electrical load and the thermal load in winter because of the large use of heating equipment.

And on the contrary, because of the extensive use of refrigeration equipment in the summer, the main loads are the electrical load and the cooling load. The system should firstly satisfy the demands of cooling in summer and heating in winter and the power is provided partly by micro-turbines and the remaining should be satisfied by other distributed power. If the demand is still unable to be satisfied, it can purchase electricity from the main distribution grid; and if the electricity generated is surplus, it can sell electricity to the external grid [17].

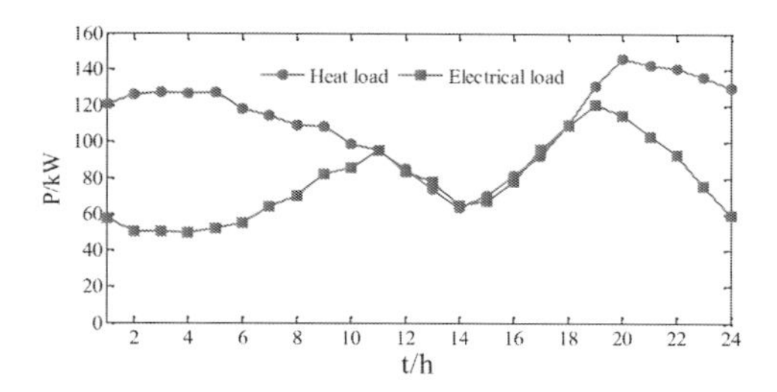

Figure 3. Heat and electrical load in winter day.

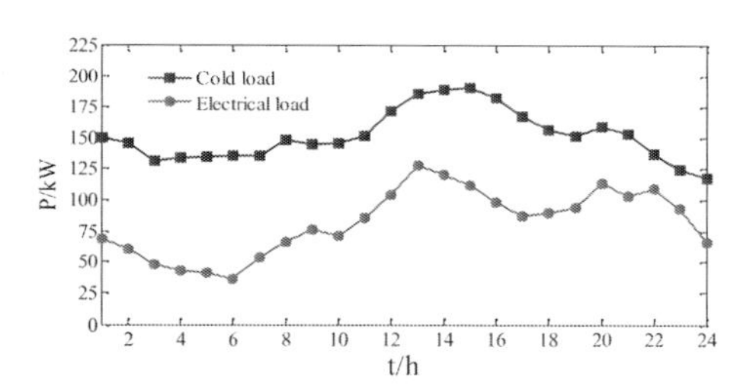

Figure 4. Cold and electrical load in summer day.

However, such behavior must be considered in the economic view. That is, purchasing electricity is considered only when the electricity generating cost is higher than the high electricity price and also selling electricity is considered only on the contrary situation.

The peak periods are at table: 10: 00 ~ 15: 00, 18: 00 ~ 21:00.The average periods are: 7:00 to 10: 00, 15: 00 ~ 18: 00,21: 00 ~ 23:00. Valley Period is: 23:00 - 7:00.

Table 1. Local electricity price.

Purchase/sale of electricity periods	Price (yuan/kwh)
Purchase of peak power segment	0.83
Purchase level segment	0.49
Power Purchase Valley section	0.17
Peak electricity sales segment	0.65
Sale-level segment	0.38
Valley section of the sale of electricity	0.13

5.2 *Calculation results*

According to the optimal model of the CCHP micro-grid system established previously, power outputs of different micro-resources on the typical days of different seasons and selling and purchasing conditions of the CCHP micro grid system can be concluded by analysis and calculation.

The main load is the thermal load in winter. Various outputs of micro-resources and the condition of system purchasing and selling electricity are as follows:

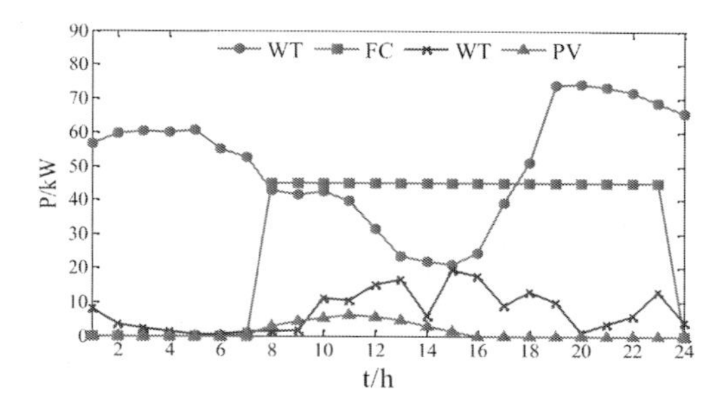

Figure 5. Micro power source output in winter day.

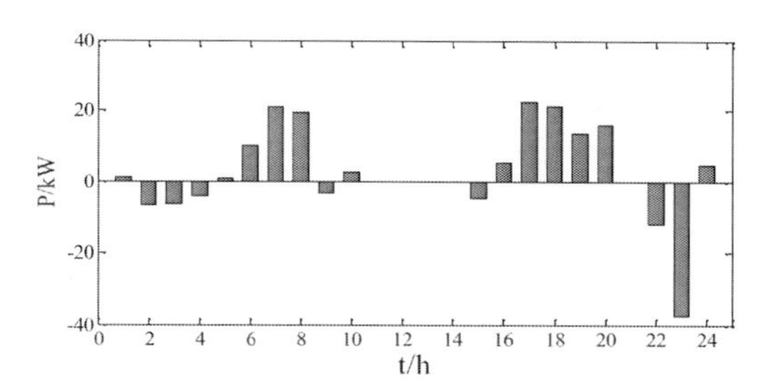

Figure 6. The date case of purchase and sale electricity in winter.

The main load is cooling loads in summer. Various outputs of micro-resources and the condition of system purchasing and selling electricity are as follows:

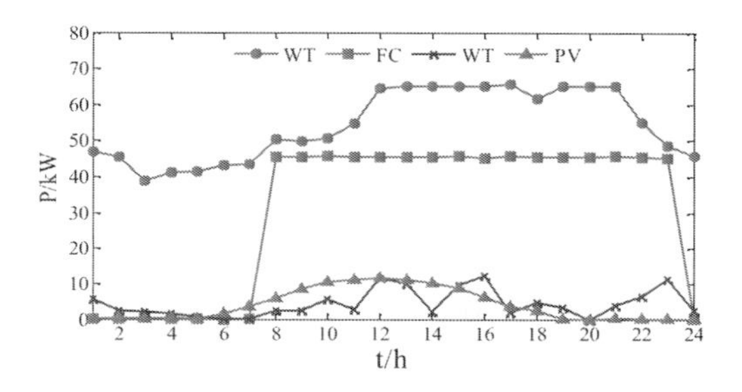

Figure 7. Micro power source output in summer day.

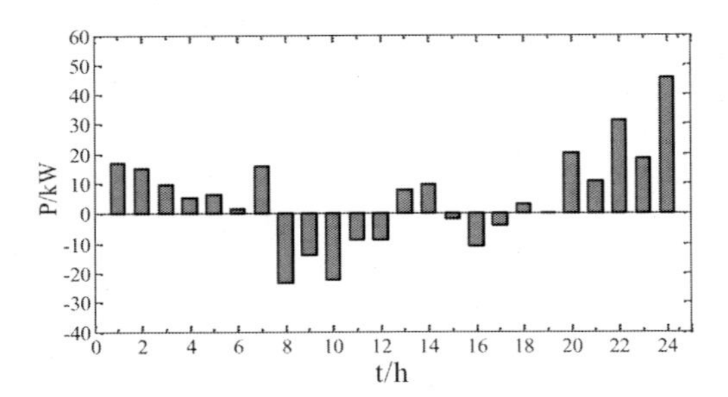

Figure 8. The date case of purchase and sale electricity in summer.

6 CONCLUSION

This paper establishes a mathematical model of economic operation in the state of the CCHP micro-grid system connected with the external network. And the cost function proposed considers the fuel-consuming costs of various generators, pollutant-processing costs of the micro-grid system and costs of selling and purchasing electricity and so on. The power outputs of various micro-resources and the electricity transaction between the micro-grid and the main grid are concluded according to the heating and cooling loads of a particular residential area on the typical days in different seasons. Also the calculating result is analyzed. It indicates that the CCHP micro-grid system has a high energy-consuming efficiency and the pollution to the environment is also alleviated. It can also be able to achieve the energy cascade usage, which plays a significant role in energy conservation and emission reduction and improving the economy of the micro-network.

ACKNOWLEDGMENT

This work was supported by National High Technology Research and Development Program of China (863 Program) (2015AA050104); National Natural Science Foundation of China (50977029); the Natural Science Foundation of Hebei Province (E2013502074).

REFERENCES

[1] Jablko, R., *et al.* (2005) Technical and economical comparison of micro CHP systems. *Future Power Systems, 2005 International Conference on. IEEE, 2005.*

[2] Aki, Hirohisa. (2007) The penetration of micro CHP in residential dwellings in Japan. *Power Engineering Society General Meeting, IEEE, 2007.*

[3] Quanxue, Dong, Tian Kuo, Zhang Yuanmin, and Zeng Ming. (2009) The situation and problems of CHP industry in China. *In Management and Service Science, 2009. MASS'09.* International Conference on, pp. 1–4. IEEE, 2009.

[4] Wang Rui, Gu Wei, Wu Zhi. (2011) Economic and optimal operation of a combined heat and power micro-grid with renewable energy resources. *Automation of Electric Power System*, 35(8):22–27.

[5] Djing Ming, Zhang Yingyuan, Mao Meiqing, *et al.* (2009) Steady model and operation optimization for micro-grids under centralized control. Automation of Electric Power System, 33(24):78–82.

[6] Wu Jiandong, (2009) Innovation development of smart power grid and interaction smart grid in China. Power System and Clean Energy, 25(04):5–8.

[7] Mohamed, Faisal A., and Heikki N. Koivo. (2007) Online management of micro-grid with battery storage using multi-objective optimization. *Power Engineering*, Energy and Electrical Drives, 2007. POWERENG 2007. International Conference on. IEEE, 2007.

[8] Mohamed F A, Koivo H N. (2007) Online management of micro-grid with battery storage using multi-objective optimization. *Power Engineering*, Energy and Electrical Drives, 2007. POWERENG 2007. International Conference on. IEEE, 2007: 231–236.

[9] Du Jianyi, Wang Yun, Xu Jianzhong. (2004) Development and application of distributed energy system and micro-turbine. *Journal of Engineering Thermophysics*, 25(5):786–788.

[10] Chen Yuehua. (2007) *Study of Modeling and Control for Molten Carbonate Fuel Cell/Micro Gas Turbine Combined Generation System Based on Intelligent Strategy.* Shanghai: Shanghai Jiao Tong University.

[11] Guo Li, Wang Chengshan, Wang Shouxiang, *et al.* (2009) The grid-connection schemes comparison of two types of micro turbines with dual mode operation. *Automation of Electric Power System*, 33(8):84–88.

[12] Wei Bing, Wang Zhi-Wei, Li Li, *et al.* (2007) Analysis of economic efficiency for cold, heat, and electricity triple co-generation system with miniature gas turbine. *Thermal Power Generation*, 36(9):1–5.

[13] Chen Jing, Li Yu-Wei, Xi Peng, *et al.* (2012) Optimization of micro-grid economic operation. *East China Electric Power*, 02:167–172.

[14] Yang Li, Liu Gaofeng, Yang Zhijie, *et al.* (2011) Cavity detection based on artificial fish swarm algorithm. *Computer Engineering and Application*, 47(11):30–33.

[15] Li Xiaolei. (2003) *A new Intelligent Optimization Method-Artificial Fish School Algorithm.* Zhejiang: Zhejiang University.

[16] Li Xiaoyu, Li Tao, Li Peng, *et al.* (2013) Adaptive artificial fish swarm algorithm based optimization configuration of gas turbine unit in large-scale photovoltaic power station. *Shaanxi Electric Power*, 41(8):15–20.

[17] Guo Jiahuan, Sheng Hong, Huang Wei. (2009) Research on economical operation of micro-grid. *Power System and Clean Energy*, 25(10):21–24.

Electric, Electronic and Control Engineering – Shao, Shu & Tian (Eds)
© 2015 Taylor & Francis Group, London, ISBN: 978-1-138-02842-5

Short-term load forecasting of power system based on wavelet analysis improved neural network model

Yulong Wang, Peng Li & Kai Zhang
State Key Laboratory of Alternate Electrical Power System with Renewable Energy Sources, North China Electric Power University, Baoding Hebei, China

Weiping Zhu
Jiangsu Electric Power Research Institute, Nanjing Jiangsu, China

ABSTRACT: This paper, firstly, establishes a fuzzy decision-making model to analyze the relevancy of the given historical load data, and to extract useful data for forecasting by removing "bad data", which means accomplishing the original data preprocessing. Secondly, the article establishes a basic neural network model. Then combining the advantages of wavelet analysis and neural network, this paper establishes a structural model of the improved wavelet neural network. The improved model predicts the electrical load values at 96 time points which need to be measured. Accurate analysis of the model application verifies its rationality. Furthermore, the proposed model of load forecasting was applied in a certain area and the region's load characteristics and regularity were discussed.

KEYWORDS: power system, wavelet analysis, neural network, load forecasting.

1 INTRODUCTION

The regularity of load change is mainly subject to the laws of production and the lives of people, and influenced by meteorological factors. The total load of a given region is the sum of innumerable individual load, so inevitably there is a random variation component in the load. The cyclical and stochastic load change is a contradiction. Growth and decline between them determines the load predictability, and are important factors affecting the accuracy of load forecasting [1]. Improving the accuracy of forecasting, is the goal of all researchers engaged in forecasting, but in fact, historical load data for modeling, the error of the model itself, and the error between the established forecast model and the forecast model, will seriously increase the difficulty of the load forecasting.

There are many conventional methods to forecast such as fuzzy linear regression method [2], regression analysis [3], gray model method [4], optimal combination forecasting method [5], and so on. However, with the development and improvement of Nonlinear Science and artificial intelligence, neural network with strong nonlinearity and generalization ability is widely applied in the field of load forecast [6, 7]. Comparing with the classic algorithm of neural networks, these algorithms improved in terms of convergence speed and precision. But due to the excessive dependence on initial value, the over-learning phenomenon, and the existence of local minimum problems in the training process, such methods' convergence speed is relatively slow.

Based on the above analysis of the neural network, the main contribution of this paper is to present a new neural network model based on wavelet analysis, effectively combining the time domain and frequency domain characteristics of wavelet transform and the self-learning function of neural networks. Forming the neural network of the wavelet network after regarding the wavelet function as the basic function can reduce errors, at the same time, quantity of states influencing the predicting error most will be extracted and is taken as the input of neural network.

2 ESTABLISHMENT AND SOLUTION OF THE MODEL

According to the known conditions, we can get the matrix $\tilde{F}$ and $\tilde{w}$ by means of translating the given historical load data into triangular fuzzy numbers. At the same time, we insist that all of the indicators' weight coefficients are equal. In the matrix $\tilde{F}$, there are n fuzzy indexes marked as $\tilde{x}_i = (a_i, b_i, c_i)$, $(i = 1, 2, \cdots, n)$. After normalization, $\tilde{x}_i$ can be expressed as:

$$\tilde{y}_i = \left(\frac{a_i}{\max(c_i)}, \frac{b_i}{\max(b_i)}, \frac{c_i}{\max(a_i)} \wedge 1 \right) \quad (1)$$

We set matrix $R = (y_{ij})_{m\times n}$ as the fuzzy indicator matrix after normalization. Then a fuzzy decision matrix $D = (r_{ij})_{m\times n}$ is obtained through weighted approach of matrix R. Where r_{ij} is a fuzzy decision indicator, and can be written as:

$$\tilde{r}_{ij} = \tilde{w} \cdot \tilde{y}_{ij} \quad (2)$$

Then we supposed that $M = (M_1\%, M_1\%, L\%, M_j\%)$, (where $\tilde{M}_j^+$ is the maximum fuzzy set corresponding to Fuzzy parameter values in a fuzzy decision matrix $\tilde{D}$ whose subscript is j, which can be expressed as $\tilde{M}_j^+ = \max\{\tilde{r}_{1j}, \tilde{r}_{2j}, \cdots, \tilde{r}_{nj}\}$; on the contrary, M_j^- is the minimal fuzzy set, which can be expressed as $\tilde{M}_j^- = \min\{\tilde{r}_{1j}, \tilde{r}_{2j}, \cdots, \tilde{r}_{nj}\}$. Both of them are solved by using MATLAB software. Determine the distance functions between evaluation objects i and $\tilde{M}^+ \cdot \tilde{M}^-$, which are defined as d_i^+, d_i^- and can be respectively expressed as:

$$d_i^+ = \sqrt{\sum_{j=1}^{5} [d(\tilde{r}_{ijL}, \tilde{M}_{jL}^+) + d(\tilde{r}_{ijR}, \tilde{M}_{jR}^+)]^2} \quad (3)$$

$$d_i^- = \sqrt{\sum_{j=1}^{5} [d(\tilde{r}_{ijL}, \tilde{M}_{jL}^-) + d(\tilde{r}_{ijR}, \tilde{M}_{jR}^-)]^2} \quad (4)$$

In the fuzzy optimization, decision-making process, the evaluation object i is subject to the fuzzy positive ideal in the form of membership degree μ_i, which can be expressed as:

$$\mu_i = \frac{(d_i^-)^2}{(d_i^+)^2 + (d_i^-)^2} \quad (5)$$

Membership degree μ_i is listed in descending order. The bigger the membership degree μ_i is, the better the evaluation object i will be. That is to say, the larger the correlation of load data is, the more useful it is for predicting.

According to the results of d_i^+, d_i^-, μ_i given by MATLAB software, all the membership degrees of historical load samples can be achieved. At the same time, the curve of membership degree of historical load samples will be made, as shown in Figure 1.

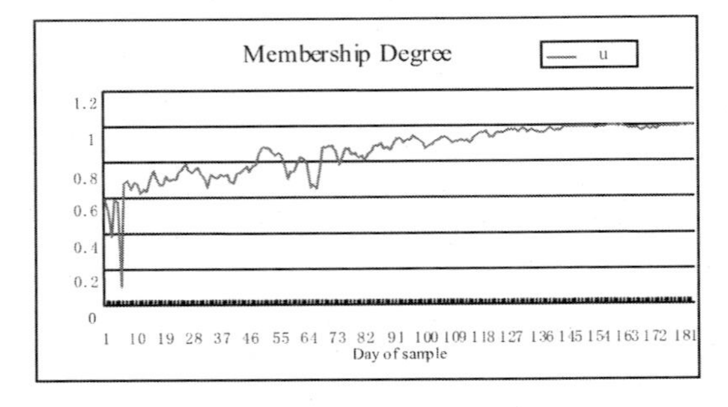

Figure 1. Membership degree change curve.

According to the analysis above, the pretreatment of the original sample is shown, as follows:

a. Get rid of the historical sample data whose membership degree is less than 0.65. After this treatment, the sample can be substantially eliminated, including the date of omission and the data of large mutations.
b. If there is a greater mutation of membership degree occurred in samples closed to prediction day, we select its similar day samples with a similar membership degree instead of excluding them directly. According to the weighted arithmetic in Equation (6), we can substitute for original data.

$$p_t^- = \frac{\sum_{i=1}^{n} p_i u_i}{\sum_{i=1}^{n} u_i} \quad (6)$$

3 IMPROVE NEURAL NETWORK BASED ON WAVELET ANALYSIS

3.1 *Traditional neural network model*

After data preprocessing, by selecting the appropriate neural network model, the regularity of change can be found out according to the historical data of the power load. And the prediction model about the power load neural network will be established. In order to predict the demand for next period, according to the electric power demand of the previous several hours in a row, the prediction model can use neural network structure with multiple inputs and multiple outputs.

In addition, any continuous function with the future, can be accurately approached by using a three layer feed-forward neural network with a structure of $n \times (2n+1) \times m$. Based on the existing data, the three layer structure of the BP network can meet the needs of the power load forecasting, the structure is shown as follows:

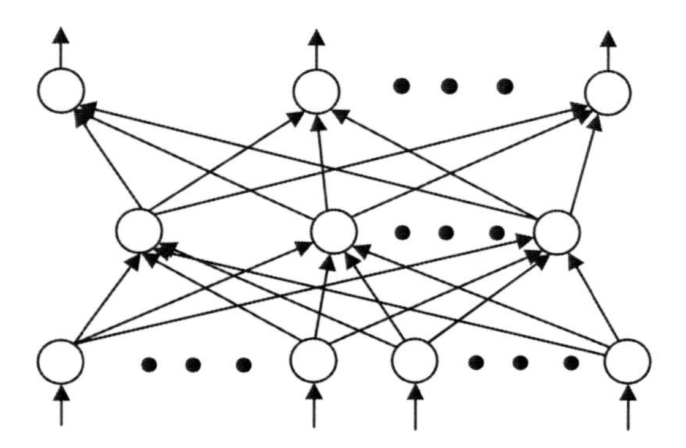

Figure 2. Three-tier structure of BP network.

3.2 *Wavelet analysis*

Evolved from a windowed Fourier transform (WFT), Wavelet transform is an integral transformation, attempting to analyze the local characteristics of nonlinear and non-stationary signal. We can analyze the signal's local characteristics of each moment and each range through comparing the signal to be analyzed with the known basic function after pan and zoom [8].

Wavelet packet decomposition is able to provide a more precise signal analysis, used in load forecasting for improving the prediction accuracy. Wavelet function's double scale equation can be defined as [9]:

$$\begin{cases} u_{2n}(x) = \sum_{k\in Z} h_k u_n(2x-k) \\ u_{2n+1}(x) = \sum_{k\in Z} g_k u_n(2x-k) \end{cases} \tag{7}$$

Where h_k, g_k are conjugate quadrature filter coefficients, $g_k = (-1)^k h_{1-k}$; $u_0(x)$ is the scaling function; $u_1(x)$ is the wavelet function.

For the existing frequency aliasing phenomenon of wavelet packet, the main contribution of this paper is to present an improved algorithm based on wavelet packet decomposition and single-node reconstruction. This algorithm can be characterized as:

$$\begin{cases} X_k = \sum_{n=0}^{N_j} (\frac{1}{N_j} \sum_{n=0}^{N_j} X_k W^{-kn}) W^{kn}, \ 0 \le k \le \frac{N_j}{4} \\ X_k = 0, \quad \frac{N_j}{4} < k \le N_j \end{cases} \tag{8}$$

Where N_j indicates the data of length; $k = 0,1,\cdots N_j$; $n = 0,1,\cdots N_j$. Actually, the improvement of Wavelet packet decomposition and single-node reconstruction is to apply Fast Fourier transform (FFT) to analyze the result of the convolution, and set the spectral value of redundant frequency components in the frequency spectrum to zero. After applying Fourier inverse transformation on the zeroed frequency spectrum, the result of the fast Fourier inverse transformation instead of the result of the convolution of wavelet filter is selected to continue wavelet packet decomposition and reconstruction.

In the process of wavelet transform, the signal denoising based on threshold value quantization of high frequency coefficients of wavelet decomposition is substantially to suppress the unwanted portion of the signal. And threshold denoising is to quantify the thresholds of high frequency coefficients under different decomposition scales [10,11]. This thresholding method is carried out through the processing of coefficients greater or less than a certain threshold after wavelet decomposition. With the aid of flourier inverse transforms, wavelet coefficients with treatment are used to reconstruct the denoised signal. How to select the threshold value and ascertain the appropriate threshold estimation principle are shown in [12].

The model of noise signal can be expressed as:

$$q(t) = S(t) - f(t) \tag{9}$$

Where, $S(t)$ is the model with noise signal, $f(t)$is the model of useful signal, $q(t)$is the model of noise signal.

Threshold processing can adopt the soft-thresholding and hard-thresholding method, which are respectively shown in (10)–(11).

$$w = \begin{cases} w-\lambda & w \ge \lambda \\ w+\lambda & -w \ge \lambda \\ 0 & |w| < \lambda \end{cases} \tag{10}$$

$$w = \begin{cases} w & |w| \ge \lambda \\ 0 & |w| < \lambda \end{cases} \tag{11}$$

The handle method of soft threshold and hard threshold is shown in Figure 3 [13]:

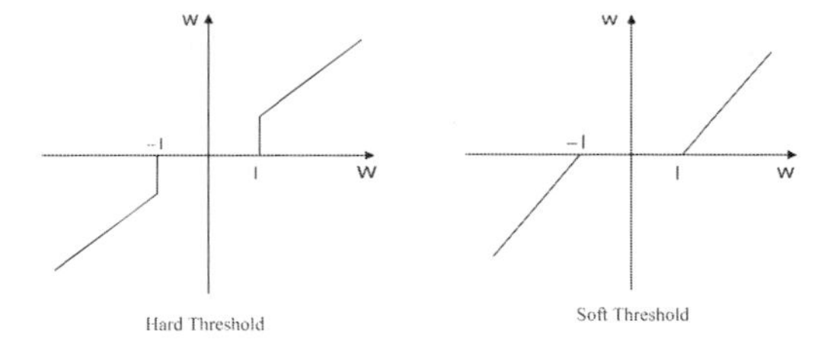

Figure 3. Threshold handle method.

3.3 *The procedure of load forecasting*

1 Consider the design of input layer and output layer first. According to the power load forecasting model, 100 input units (each period load, the maximum and minimum daily load, the average daily load and daily load factor) and 96 output unit (96 power load) are set up. And the rolling forecast method is adopted. In order to improve the accuracy of prediction, two improvements are organized as follows:

a. Set the number of pretreated sample is n. The training will be organized based on the input matrix and $(n-1)\times 96$ output matrix, which are respectively comprised of the front n-1 samples and the latter n-1 samples. The $(n-1)\times 96$ output matrix named A is obtained after applying the training results to the $(n-1)\times 100$ matrix comprised of the latter 100 samples. The last line of matrix A is the forecast load value of the first day;
b. In order to improve the accuracy of prediction, the predicted results are added to each sample set amount for the next forecast.

2 Design the hidden layer. The number of hidden layer nodes can refer to the formula: $num = \sqrt{n+m} + \alpha$.
Where num is the number of hidden units, n is the number of input units, m is the number of output units, α is a constant between 1 and 10. According to the previous analysis, n is 100, and m is 96, so the number of hidden units (num) is 150.

3 With the purpose of improving the training efficiency of network [14], pretreatment of data needs to normalize the raw data, and the formula of normalization can be expressed as: $\overline{x_i} = \dfrac{x_i - x_{\min}}{x_{\max} - x_{\min}}$.
Where $\overline{x_i}$ is the normalized value of the i th element of the input data column x; $x_{\min}$ is the minimum value in all elements of input data column x; $x_{\max}$ is the minimum value in all elements of the input data column x. After the normalization, data distributes in the interval [0, 1].

4 Determine transfer functions of input node, output node and hidden node. The transfer function of input node adopts a linear function; the transfer function of hidden layer and output node uses hyperbolic tangent function with Sigmoid type, such as $f(x) = \dfrac{1}{1+e^{x}}$; the error back-propagation need to use the transfer function $f(x) = \dfrac{1}{(1+e^{x})^2}$.

5 In the *BP* network algorithm, for any given function $f(x_1, x_2, ..., x_n)$ and error precision $\varepsilon > 0$, there is always existing a total input-output relationship $Y = \overline{f}(x_1, x_2, ..., x_n)$, which can approximate any given function, such as $f(x_1, x_2, ..., x_n)$, with specified accuracy.

In this case, the existing power load sequences and the other factors, such as maximum load and minimum load, are defined as the input data $x_1, x_2, ..., x_n$. After repeated training, a network will be obtained to seek out the relationship, $F = \overline{f}(x_1, x_2, ..., x_n)$, between the total input and output. The power load sequence of $f(x_1, x_2, \cdots, x_n)$ can be predicted within prescribed accuracy range ε. The operation process expressed in the mathematical model is shown in Equation (12)–(14).

$$x_{ij}^{[s]} = f[\sum W_{ij}^{[s]} x_{ij}^{[s-1]} - Q_{ij}] \qquad (12)$$

Where $x_{ij}^{[s]}$ is output forecasting load of the sth layer of the network at j o'clock of the i th day; $W_{ij}^{[s]}$ is the related influencing factor of the sth layer of the network at j o'clock of the i th day; $x_{ij}^{[s-1]}$ is forecasting load of the $(s-1)$ th layer of the network at j o'clock of the ith day; Q_{ij} is actual system loss value of each layer of the network at j o'clock of the i th day.

$$O_l = \overline{f}[\sum T_{ij}^{s+1} x_{ij}^{s-1} - Q_{ij}] \qquad (13)$$

Where O_l is the final forecasting power load of the network; T_{ij}^{s+1} is output forecasting power load correction of the network at j o'clock of the i th day;

$$C = \sum_{k=1}^{p} c(k) < \varepsilon \qquad (14)$$

where $c_{(k)}$ is system error of the kth sample; $c(k) = \sum_{l=1}^{n} \left| t_l^{(k)} - O_l^{(k)} \right|$; C is final output system error of the network; p is the number of actually tested load sample; ε is allowable value of system error; $t_l^{(k)}$ is expected value of final output of the network; $O_l^{(k)}$ is final load forecasting value of the network. n is the number of nodes in the network .

4 EXPERIMENTAL VALIDATION

According to the above discuss, a period between February 1st and February 7th of a region for experimental validation was implemented. And the load distribution is shown in Figure 4. Comparing the predicted values and the original data, a graphical comparison of the two models is obtained as shown in Figure 5. After error analysis processing, the absolute error is shown in Figure 6, and the relative error is shown in Figure 7.

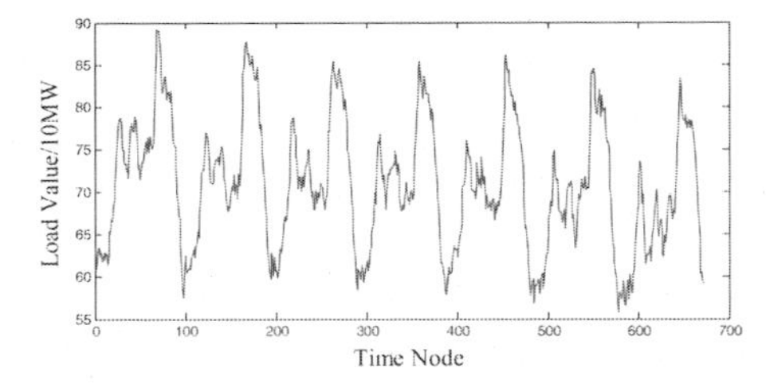

Figure 4. Actual value of the load.

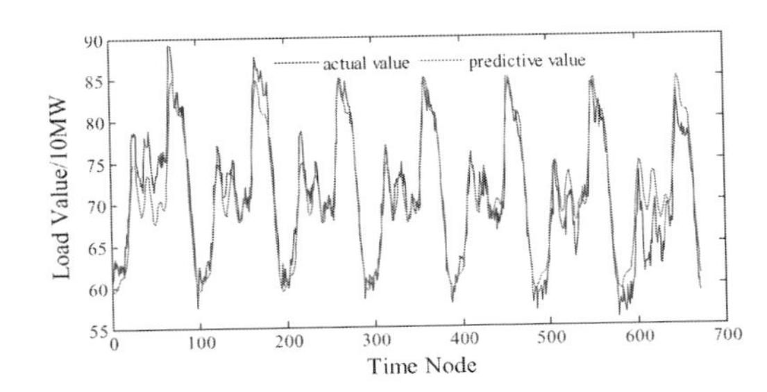

Figure 5. Contrast of the forecast results.

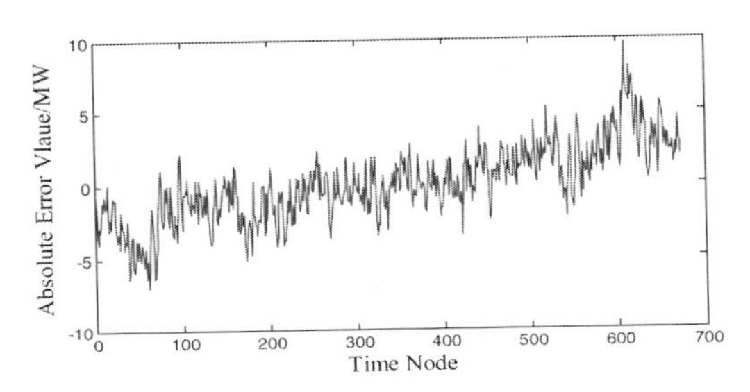

Figure 6. Absolute error curve.

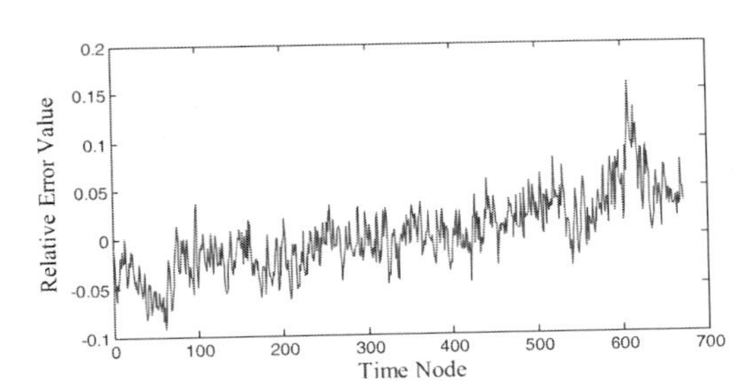

Figure 7. Relative error curve.

4.1 *Accuracy analysis*

This paper makes an accuracy analysis with the data source from February 1st to 7th in a region, putting forward the following analysis methods:

1 Mean Absolute Percentage Error

$$MP = \frac{1}{N}\sum_{i=1}^{N}\left|\frac{Y_i - X_i}{X_i}\right| \tag{15}$$

It should be noted that the smaller MP shows the higher precision, and the greater MP indicates the lower precision.

2 Determination Coefficient

$$RSS = \sum_{i=1}^{n}(y(i) - y(\hat{y}))^2 \tag{16}$$

$$TSS = \sum_{i=1}^{n}(y(i) - \overline{y})^2 \tag{17}$$

$$\overline{y} = \frac{1}{n}\sum_{i=1}^{n} y(i) \tag{18}$$

Then the determination coefficient can be written as:

$$R^2 = 1 - \frac{RSS}{TSS} \tag{19}$$

Therefore, the predictive effect of this proposed method can be obtained through analyzing the determination coefficient and comparing the ratio of residual sum of squares and the sum of deviation squares of actual value.

3 Comparison of the Predicted Results

Table 1. Contrast of forecast precision

	Neural network model	Proposed model
Mean absolute percentage error (MP)	5.28%	4.32%
Determination coefficient (R^2)	0.785	0.823

Table 1 shows that the improved neural network model with wavelet analysis has certain enhancement than the basic neural network model, both in terms of mean absolute percentage error and determination coefficient. It is clear that the improved model is more superior.

4.2 *Discussion on the error of forecasting*

Short-term load forecasting is equivalent to a future value estimating problem of a time series at math area, consequently cannot be completely accurate. The forecasting result relates to not only the rationality of the adopted model algorithm and certain approximate of hypothesis sum, but also the inherent randomness of the system limits the accuracy of the prediction.

1 Mathematical model approximation. Only the historical load data were considered, which cannot fully reflect the complex load changes.
2 Quality problem of the data. Vast and high accuracy data are the basic requirement of load prediction. Bigger error in load prediction often exists due to incomplete and inadequate data. Historical data given in this paper are only a fraction of the

total months of the previous year, so regularity that load varies from the seasonal cycle cannot be discovered within historical data.

3 Grid incidents result in the decrease of prediction accuracy, which is inevitable.

5 CONCLUSION

The application of neural network plays an important role in power system short-term load forecasting. In this paper, on the basis of current technology, in order to make a more accurate load forecasting and lower its dependence on historical data, the wavelet analysis is combined. This article selects 672 data groups from February 1st to 7th in the Baishan region as research object, establishing a short-term load forecasting model of the power system. In addition, the error analysis and comprehensive evaluation of the established model is made to verify its rationality.

ACKNOWLEDGMENT

This work was supported by the National Natural Science Foundation of China (50977029), the Natural Science Foundation of Hebei Province (E2013502074), and Science and Technology Project of Jiangsu Electric Power Research Institute (5210EF14001A).

REFERENCES

[1] Mu Gang, Hou Kaiyuan, Yang Youhong, *et al.* (2001) Load forecasting method based on intrinsic error evaluation. *Automation of Electric Power System*, 22:37–40.

[2] Geng Guangfei, Guo Xiqing. (2002) Application of fuzzy linear regression to load forecasting. *Power System Technology*, 26(4):19–21.

[3] He Jing, Wei Gang, Xiong Lingling. (2003) Fuzzy improvement of linear regression analysis for load forecasting. *East China Electric Power*, 11:21–23.

[4] Ding Xining, Niu Chenglin, Li Jianqiang. (2005) Electric load forecast based on decision tree and expert system. *Journal of North China Electric Power University*, 05:59–63.

[5] Chu Jinsheng, Chen Yuping. (2013) Application of an optimization combined forecasting method in short-term load forecast. *Electric Switcher*, 2:86-87+90.

[6] Shao Nenglin, Hou Zhijia. (2006) New short-term load forecasting principle with the wavelet transform fuzzy neural network for the power system. *Proceedings of the CSEE 2006*, 26(19):41–46.

[7] Mohan Saini L, Soni M K. (2002) Artificial neural network-based peak load forecasting using conjugate gradient methods. *Power Systems, IEEE Transactions on*, 17(3):907–912.

[8] Wang Lijie, Dong Lei, Liao Xiaozhong, *et al.* (2006) Short-term power prediction of a wind farm based on wavelet analysis. *Proceedings of the CSEE 2006*, 26(19):41–46.

[9] Jiang Tao, Yuan Shengfa. (2014) Rolling bearing faults diagnosis based on the improved wavelet neural network. *Journal of Huazhong Agricultural University*, 33(1):131–136.

[10] Leng Jianwei, Fu Xianglian. (2014) Application of Wavelet Analysis in Data Processing of Low-voltage Load Forecasting. *Journal of Power Supply*, 01:110–113.

[11] Gelal E, Jakllari G, Krishnamurthy S V. (2006) Exploiting diversity gain in MIMO equipped ad hoc networks[C]. *Signals, Systems and Computers, 2006. ACSSC06. Fortieth Asilomar Conference on.2006*, pp:117–121.

[12] Liu Weifeng. (2013) Research on Channel Estimation for OFDM System with Wavelet De-noising. Harbin: Harbin Institute of Technology.

[13] WU Xuehua. (2014) A wavelet based method for monthly electricity demand forecasting. *Jiangsu Electrical Engineering*, 33(2):8–11.

Electric, Electronic and Control Engineering – Shao, Shu & Tian (Eds)
© 2015 Taylor & Francis Group, London, ISBN: 978-1-138-02842-5

Application of energy storage system in islanded microgrid

Kai Zhang, Peng Li & Xiaomeng Yu
State Key Laboratory of Alternate Electrical Power System with Renewable Energy Sources, North China Electric Power University, Baoding Hebei, China.

Bo Zhao
Electric Power Research Institute of State Grid Zhejiang Electric Power Company, Hangzhou Zhejiang, China

ABSTRACT: Energy storage system is a device with the ability of rapid response. It could increase the flexibility of microgrid control system, and improve the security and economic benefits of microgrid. The energy storage system plays a key role in the microgrid operation. In this paper, the benefits of islanded microgrid using energy storage system are investigated. Microgrid optimal scheduling scheme is used to solve a long-term unit commitment problem, and is used in the storage system utility model. Probabilistic reliability is used in this paper to calculate the cost of reliability in the microgrid, Example simulation verifies the effectiveness of this method.

KEYWORDS: energy storage system; islanded microgrid; optimization; reliability.

1 INTRODUCTION

With the increasingly urgent issue of energy, new energy and renewable energy distributed generation (DG) technologies have got the attention of the world[1,2]. Microgrid technology can effectively integrate the advantages of the new energy and renewable energy distributed generation, providing a new efficient technological way for large-scale application of new energy and renewable energy grid-connected generation[2–5].

The energy storage system can store energy when it is excess and export it when the market price is high or there is not sufficient energy. Microgrid storage system is one of the most important ingredients in microgrid and plays a significant role in the short-term and long-term operational reliability of microgrid. Distributed energy and controllable load can achieve flexible and efficient operation for microgrid in the industry of production and consumption[6]. To install an energy storage system in microgrid can enhance its flexibility, the fast response ability of energy storage systems will increase the reliability and economy of microgrid. Additionally, it will reduce frequent and rapid power changes of new energy. Consequently, energy storage system can enhance reliability of microgrid by solving volatility and intermittent problem of new energy. Also, energy storage system can be used in tracking changes in load, voltage and frequency stability, peak load management, improving power quality and delay upgrading equipment investment, etc. [7–8]. A coordinate optimization model is made in [9], combined with microgrid reliability assessment, it achieves microgrid reliability assessment considering the influence of coordinative optimization of loads and storage devices. In [10], a coordinated control method is proposed between diesel generator and battery storage in an isolated AC microgrid.

Similar to the grid company, the purpose of micogrid is also providing power to the electricity customers, which need to have sufficient generating capacity to meet the load demand at any moment. On this basis, define two operation modes of microgrid, which are named grid-connected operation mode and islanded operation mode. In grid-connected mode, microgrid is a part of the main grid. In this mode, microgrid can either sell its excess electricity to the grid (as a power generate system of main grid), or buy electricity from the grid (as a load of main grid). Compared to islanded mode, grid-connected mode has certain advantages. In grid operation mode, the microgrid can purchase electricity from the grid at low price periods, sell excess electricity to main grid at high price periods, so as to enhance economic efficiency of microgrid. When the energy storage system is existent in microgrid, the aforementioned benefits will become more apparent as the amount of electricity it purchases at low price periods and sells at high price periods may increase to raise economic benefits significantly. In addition, because it is connected to a larger system, the frequency of the microgrid is more stable in grid-connected mode. However, it is inevitable for the microgrid to operate in islanded mode under conditions of power grid failure. In that case, the microgrid will switch to islanded mode, in order to prevent the failure expanding from the grid to the microgrid. Also, it can be that the main grid and micro-grid systems is not connected successfully, so the microgrid will be islanded operation. In islanded mode, microgrid operation will be independent of the main grid. In this mode, the power consumption demand of local microgrid customer should be met by local distributed generation equipment. In a typical microgrid system, distributed generation equipment includes wind turbine, PV panels, micro-turbines, fuel cells and energy storage system. As disconnection to the main grid, the reliability of microgrid operating in islanded mode is essential.

In this article, we focus on the study of energy storage applied in islanded mode microgrid, and observe its feasibility in islanded mode, and furthermore, study the contribution of the energy storage system to increase economy and reliability of microgrid. This paper develops a operation plan of microgrid generation units and storage systems by a long-term economic optimization scheduling problem, which is used to solve one year reliability assessment, and applied to the energy storage system in an accurate model. We define energy storage system microgrid installs is the optimal size. If the energy storage system is too small, it may not provide sufficient economic benefits, desired flexibility and expected reliability standard of microgrid. Vice versa, the large energy storage system requires higher investment costs and operation and maintenance costs, which is not conducive to economic of microgrid. Thus, to calculate its optimal size, is an important way to prove its economic feasibility, and can avoid waste caused by excessive capacity and achieve the use purpose unsuccessfully caused by insufficient capacity.

2 MODEL DESCRIPTION

A microgrid optimization model is made to study the problem of energy storage system in islanded microgrid. The superiority of energy storage system in islanded microgrid is learned by solving the optimization model. The objective of an optimization problem is minimization of operating costs and reliability cost of microgrid. Operating cost includes the cost of power generation unit start-stop and fuel cost. Microgrid reliability is related to the operating mode of microgrid, reliability assessment is an important issue with the long-running term of microgrid. Reliability satisfaction ensures that microgrid has sufficient capacity and reserve capacity to meet its load changes requirements during long-term operation. In this paper, the microgrid reliability standard is represented by expected energy not served (EENS). EENS expected electricity shortage, whose value is obtained according to the system load plan. This probability reliability criterion can accurately assess the reliability of the power system. Reliability cost can be calculated by multiplying EENS with load loss value. Lower reliability cost means more reliable unit, which also means a reduction of the amount of load shedding.

Reliability can be assessed by direct analysis or simulation. Monte Carlo simulation (MCS) method is adopted to calculate the reliability of the microgrid in this paper. Monte Carlo simulation is a completely random process. In this method, fault time and repair time of microgrid component are generated by appropriate probability distribution, and the generated random number is used to determine the status of each member. In the Monte Carlo simulation, the system operation state will be a very large number, and generate different topology structures and load rating to simulate the system operational status in a particular state. When electricity demand can be met by the simulation component status and load levels, the operating status of the system will be determined. Using this method, the economy and reliability energy storage devices in the microgrid system can be reflected its advantages.

3 OPTIMIZATION PROBLEM DESCRIPTION

Long-term economic optimization problem determines that the generator scheduling plan can reduce operating cost and reliability cost under the condition of meeting current constraints, the objective function is as follows:

$$Min\{\sum_{i=1}^{NG}\sum_{t=1}^{NT}[F_{ci}(P_{it})I_{it}+|SU_{it}+SD_{it}|]+\lambda\sum_{t=1}^{NT}v_t\} \quad (1)$$

Where, F_{ci} is the cost function of generating unit, I_{it}^{s} the star-stop state of unit i at time t, NG is the quantity of generating unit, P_{it} is the output power of unit i at time t, SU_{it} and SD_{it} are the start-up cost and shut-down cost of unit i at time t respectively, λ is the coefficient of load loss, v_t is the load shedding value.

The objective function consists of two aspects, the first part is the cost of the internal microgrid system, including fuel cost as well as start-stop cost of each unit, renewable energy, such as wind power, photovoltaic, running cost of which are negligible. The second part is the reliability cost, which is obtained by the load loss value and load shedding value. Since it is assumed that the microgrid is under islanded operation mode, disconnected to the main power grid, so the cost of purchase electricity and the profit of sale electricity are not considered. The operation of microgrid system should follow the constraints of actual unit and system conditions, these constraints, including the power balance constraints, unit capacity constraints, climbing constraints, unit start-stop time constraints.

Power balance constraints:

$$\sum_{i=1}^{NG} P_{it} I_{it} + P_{S,t} + v_t = P_{D,t} \tag{2}$$

unit capacity constraints:

$$P_{i,\min} I_{it} \le P_{it} \le P_{i,\max} I_{it} \tag{3}$$

climbing constraints:

$$P_{it} - P_{i(t-1)} \le UR_i[1 - I_{it}(1 - I_{i(t-1)})] + P_{i,\min}[I_{it}(1 - I_{i(t-1)})] \tag{4}$$

$$P_{i(t-1)} - P_{it} \le DR_i[1 - I_{i(t-1)}(1 - I_{it})] + P_{i,\min}[I_{i(t-1)}(1 - I_{it})] \tag{5}$$

$$[X_{i(t-1)}^{on} - T_i^{on}][I_{i(t-1)} - I_{it}] \ge 0 \tag{6}$$

$$[X_{i(t-1)}^{off} - T_i^{off}][I_{it} - I_{i(t-1)}] \ge 0 \tag{7}$$

Where, $P_{S,t}$ is the power of energy storage system, negative when charge, positive when discharge; $P_{D,t}$ is load power, $P_i^{\max}$ and $P_i^{\min}$ are the maximum and minimum output power of unit i respectively, UR_i and DR_i are the constraint of rate of power increase and power decline respectively, T_i^{on} and T_i^{off} are shortest start time and stop time of unit i respectively, X_i^{on} and X_i^{off} is the actual start time and stop time.

The power balance equation can ensure that the local distributed power and energy storage systems adequately supply the local load. In addition, in this equation additional variable is added to simulate load shedding. When the power supply cannot meet the load electricity consumption demand, the load shedding phenomenon occurs. A unit capacity constraint is a physical characteristic of the power generating unit itself. Unit climbing constraint ensures that the amount of change of power generating unit between two successive times cannot exceed its allowable change rate value. To start-stop time constraint, the unit can neither stop after the start time period, nor start again in the post-closure period. Reserve capacity of the system is implicitly taken into account in the model. There are enough power generating units to meet the reserve capacity needs in the model. This method is more efficient than conventional deterministic methods, for example, in large capacity unit systems. Here are classical mathematical models of unit composite constraints, mixed integer programming method of these constraints can be found in [11], more detail unit composite solutions and constraints are mentioned in [12].

Table 1 shows the operation mode state of energy storage system, the mentioned charging and discharging of the storage devices take place at a constant power level.

Table 1. Operation state of energy storage system.

Operation mode	Binary variable	
	u_t	w_t
Charge	0	1
Discharge	1	0
Idle	0	0

In addition to power generation unit and system constraints, the energy storage system constraints are also added to the model. The constraints are as follows:

$$u_t + w_t \le 1 \tag{8}$$

$$-P_S^R v_t \le P_{S,t} \le k P_S^R w_t \tag{9}$$

$$SOC_t = SOC_{t-1} - P_{S,t}\Delta t \tag{10}$$

$$0 \le SOC_t \le SOC_{\max} \tag{11}$$

Where, u_t and w_t are the charge signal and discharge signal respectively, the value is 0 or 1. P_S^R is the rate power of energy storage system. $P_{S,t}$ is the charge and discharge power of energy storage system. k is the depth of discharge. SOC_t is the electricity quantity of energy storage system at time t. $SOC_{\max}$ is the maximum capacity of energy storage system. Here interval is set to 1 hour, so $\Delta t = 1$.

4 EXAMPLE ANALYSIS

4.1 *Microgrid structure and parameters*

The islanded microgrid structure is adopted in this paper to analyze and explain this method, as shown in Figure 1. The purpose of this method is to solve long-term economic optimization problems and check the influence energy storage made on it. The peak and valley electricity price and peak and valley time in one day is shown in Figure 2. Characteristics of the power generating units are shown in Table 2. The defined log-term is one year, fault rate is 4%, the load loss coefficient is defined as 180 yuan/kW, peak load of microgrid is 13.58MW, the total energy of load demand is 52700MWh.

According to the characteristics of energy storage system, three operation strategies are formulated in this paper:

1 Microgrid system without energy storage devices;
2 Install 1MW energy storage device in islanded microgrid;
3 Install optimal capacity and power energy storage device in islanded microgrid after optimization calculation.

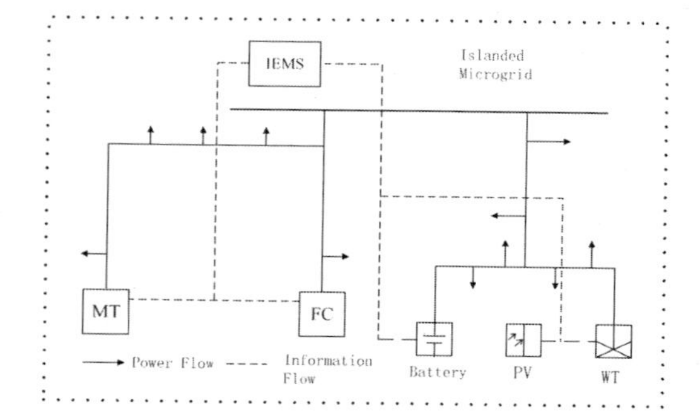

Figure 1. Schematic diagram of the microgrid system model.

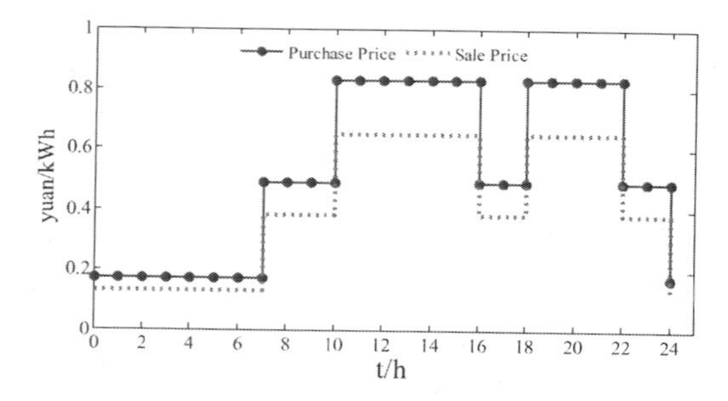

Figure 2. Schematic diagram of local peak and valley price.

4.2 *Optimized calculation*

Strategy 1: Microgrid system without energy storage devices. In this case, the load has priority to meet the demand for electrical energy by renewable energy within the microgrid. When renewable energy like wind power and photovoltaic cannot meet the demand, it is necessary to use thermal units to meet the load electricity consumption. By then, expected energy not served (EENS) value is 46.61MWh. In peak load hours, condition that system generating power cannot meet load demand easily occurs, so there is need for load shedding under that condition.

Table 2. DG parameters.

DG type	Life/ year	Power lower limit (MW)	Power superior limit (MW)	Start-stop cost (thousand yuan /MW)
MT	10	1	4	0.9
FC	10	0.5	2.5	1.16
PV	10	0	0.5	0
WT	15	0	0.5	0

DG type	Shortest start time(h)	Shortest stop time(h)	Power increase rate (MW/h)	Power decline rate (MW/h)
MT	3	3	2	2
FC	3	3	1.2	1.2
PV	1	1	0.5	0.5
WT	1	1	0.5	0.5

Strategy 2: Microgrid system installs 5MWh capacity, 1MW power energy storage device, which can get to superior limit of capacity at maximum charging state. The device storage energy at the off-peak load time and release energy at the peak load time to compensate for the lack of power generated by generating units. So the condition of active load shedding during peak load hours should be avoided, and EENS will be zero. Under this strategy, the total cost of microgrid is 954 million yuan, of which contains only operating cost, because the reliability cost is zero in that case.

Strategy 3: The rated power of the energy storage system is considered as a variation of the problem, so there must be an optimum size of the storage system. The optimum energy storage device power of microgrid is 3MW after optimized calculation, and the optimum capacity is 15MW. EENS is 0 in this case, there will be no active load shedding happen, so microgrid is completely reliable. Under this strategy, the total cost of microgrid is 9.96 million yuan, slightly higher than the second strategy, as smaller energy storage system will reduce the amount of operating cost, but smaller energy storage system will also reduce the cost of investment. In summary, the application of the optimal capacity and power energy storage device will improve the economic efficiency of microgrid.

The relationship between EENS and the power of energy storage system is shown in Figure 3. The relationship between microgrid operation cost, reliability cost and the size of energy storage device is shown in Figure 4.

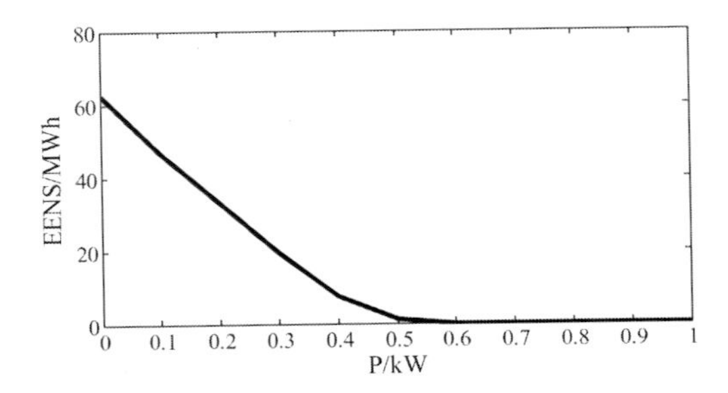

Figure 3. Relationship between EENS and install power.

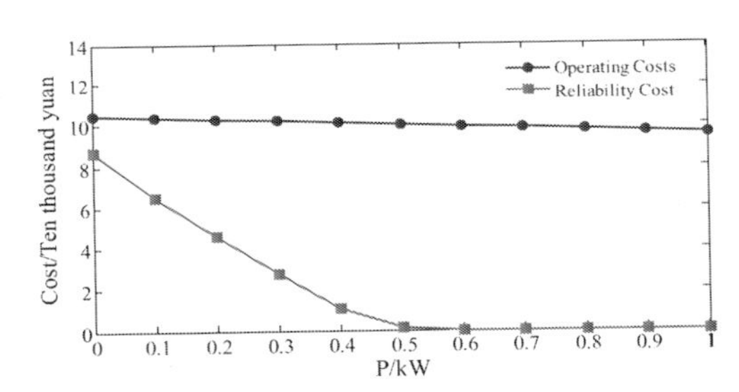

Figure 4. Relationship between cost and install power.

5 CONCLUSION

The application of energy storage system in islanded microgrid is studied in this paper and an accurate energy storage system is in islanded microgrid. The proposed method analyzes the influence energy storage system made on islanded microgrid at the terms of economic and reliability with a long-term optimization model. Reliability index is denoted with expected energy not served (EENS). Case analysis verifies that application of energy storage system is helpful of the economic and reliability of islanded mictogrid.

ACKNOWLEDGMENT

This work was supported by National High Technology Research and Development Program of China (863 Program) (2015AA050104); National Natural Science Foundation of China (50977029) and the Natural Science Foundation of Hebei Province (E2013502074).

REFERENCES

[1] Wang Chengshan, Wu Zhen, Li Peng. (2014) Research on key technologies of microgrid. *Transaction of China Electrotechnical Society*, 29(2):1–12.

[2] Ding Ming, Zhang Yingyuan, Mao Meiqin, *et al.* (2011) Economic operation optimization for microgrids including Na/S battery storage. *Proceedings of the CSEE 2011*, 31(4):7–14.

[3] Wang Chengshan, Xiao Zhaoxia, Wang Shouxiang. (2009) Multiple feedback loop control scheme for inverters of the micro source in microgrids. *Transactions of China Electrotechnical Society*, (2):100–107.

[4] Changhee Cho, Jin-Hong Jeon, Jong-Yul Kim. (2011) Active synchronizing control of a microgrid. *IEEE Transactions on Power Electronics*, 26(12):3707–3719.

[5] Li Peng, Zhang Ling, Wang Wei, *et al.* (2009) Application and analysis of microgrid. *Automation of Electric Power Systems*, 33(20):109–115.

[6] Shahidehpour M. (2006) Role of smart microgrid in a perfect power system[C]//*Power and Energy Society General Meeting, 2010* IEEE. IEEE, 2010: 1–1.

[7] Joseph A, Shahidehpour M. (2006) Battery storage systems in electric power systems[C]//*Power Engineering Society General Meeting, 2006*. IEEE. IEEE, 2006: 8.

[8] Marwali M K C, Haili M, Shahidehpour S M, *et al.* (1998) Short term generation scheduling in photovoltaic-utility grid with battery storage. *Power Systems, IEEE Transactions on*, 13(3):1057–1062.

[9] Bie Chaohong, Li Gengfeng, Xie Haipeng. (2014) Reliability evaluation of microgrids considering coordinative optimization of loads and storage devices. *Transaction of China Electrotechnical Society*, 29(2):64–73.

[10] Guo Li, Fu Xiaopeng, Li Xialin, *et al.* (2012) Coordinated control of battery storage and diesel generators in Isolated AC Microgrid Systems. *Proceedings of the CSEE 2012*, 32(25):70–78.

[11] Carrión M, Arroyo J M. (2006) A computationally efficient mixed-integer linear formulation for the thermal unit commitment problem. Power Systems, IEEE Transactions on, 21(3):1371–1378.

[12] Fu Y, Shahidehpour M, Li Z. (2005) Security-constrained unit commitment with AC constraints. *Power Systems, IEEE Transactions on*, 20(2):1001–1013.

Electric, Electronic and Control Engineering – Shao, Shu & Tian (Eds)
© 2015 Taylor & Francis Group, London, ISBN: 978-1-138-02842-5

The research of seawater chemical oxygen demand measurement technology with ozone oxidation method

Hechao Zang, Lei Li & Zhonghai Zou
Institute of Oceanographic Instrumentation of Shandong Academy of Sciences, Qingdao, Shandong, China
National Engineering and Technological Research Center of Marine Monitoring Equipment, Qingdao, Shandong, China

ABSTRACT: Seawater COD measurement technology with ozone oxidation method is an advanced technology in marine water quality monitoring field, which has characteristics of short cycle measurement, high accuracy and less frequent replacement of oxidizing agents. These characteristics make it unattended working on the case of long-term, real-time, continuous water-quality monitoring. Pollution happens in major marines could be early warned in first-time, to prevent serious pollution disasters happen. For comprehensively evaluating the accuracy, stability and applicability of this technology, it needs to do the authoritative third-party experiment, the authoritative third-party is the National Marine Environmental Monitoring Center of China.

KEYWORDS: seawater; COD measurement technology; ozone oxidation method.

1 INTRODUCTION

Chemical Oxygen Demand (COD) is a composite indicator to measure the levels of organic pollution in seawater. Domestic measurement of the chemical oxygen demand (COD) is generally used in laboratory by the methods of the permanganate analysis or dichromate oxidation titration analysis. Although the method is the GB method, due to the complexity of the analysis process and the long experiment cycle, it is not qualified the job of real-time continuous pollution monitoring, the ability of China's ocean monitoring is largely restricted.

Seawater chemical oxygen demand analyzer with an ozone oxidation method developed by the ozone oxidation chemical theory with seawater COD measurement technology with ozone oxidation method, which is used to detect the content of seawater COD indicators. Comparing with similar instrument, it is better with short measurement cycle and high accuracy. It uses self-generated ozone as the oxidizing agent, without frequent replacement of oxidizing agents, no secondary pollution to environment, and be able to work reliably in the harsh marine environment. It's suitable for ship and marine stations, offshore oil platforms and ocean buoy stations, be able to detect seawater COD indicators of near-shore coastal and marine economic zone with on-site, real-time, continuous monitoring.

2 EXPERIMENTAL METHODS AND PROCEDURES

To verify the authenticity of the instrument to be, November 2011, the third-party environmental experiment of the instrument had been done at the National Marine Environmental Monitoring Center for a period of 10 days, through a large number of comparison experiments on the GB method and instrument method.

2.1 *Methods and principles*

The experiment was the comparison between the determination of seawater chemical oxygen demand analyzer with ozone oxidation method (instrument method) and «GB 17378.4 -2007 marine monitoring Part 4: Seawater analysis» chemical oxygen demand (COD) in a predetermined measurement method (GB method), the technical roadmap is Figure 1.

Seawater chemical oxygen demand analyzer with ozone oxidation method contains an ozone generator which produces O_3. By a vacuum pump, O_3 air mixtures are pumped from the ozone generator to the reaction chamber when water sample is pumped into the reaction chamber by vacuum pump (liquid). Gas and water samples are thoroughly mixed with organic matter in the water sample, meantime chemical light signal resulted in oxidation reduction is detected by a photomultiplier tube and calculated

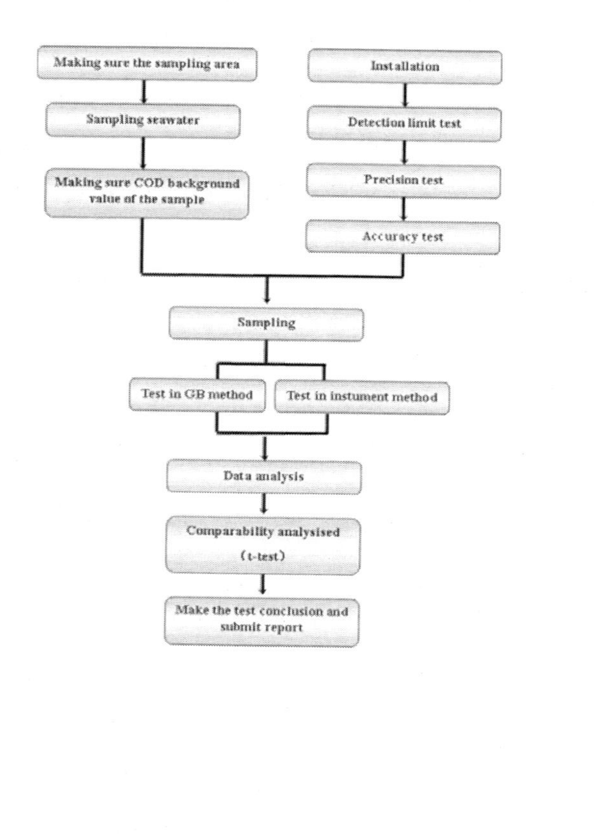

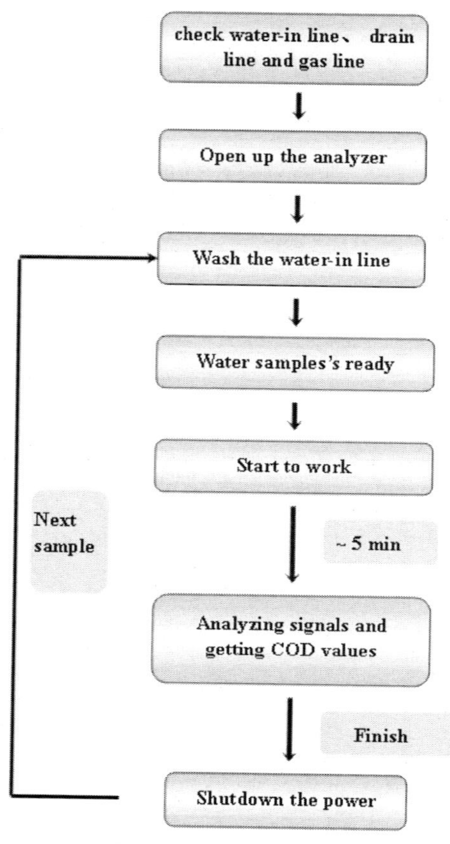

Figure 1. Roadmap of the comparative experiment (left) and instrument working steps (right).

by the microprocessor according to the algorithm between the optical signal and COD value. After being installed and debugged, it passed through a comparative experiment of 10 days.

2.2 *Seawater samples collected*

Through data collection and pre-experiment, we chose the place near the sea area called Heishijiao as sampling site and found different sampling sites with high, medium and low COD concentrations, collected surface seawater samples and carried back to the laboratory for analysis immediately.

Sampling device, sampling cable and other equipments, cleaning programs, preservation of the sample bottle, sampling operation at sampling sites, storage and transportation of samples and other related requirements were executed by the relevant requirements of GB17378.3.2007.

3 EXPERIMENTAL DATA ANALYSIS

We got 24 sample groups (n = 5 sub-samples) of data with Instruments method and GB method in the comparative experiment.The data had been analyzed in different ways.

3.1 *Linear responses of the instrument*

With artificial seawater, the high concentration seawater was diluted to samples that were different multiples ($V_{\text{seawater samples}}$: V_{total}), and then measured the samples with the instrument method, the experimental results are Table 1.

"Seawater chemical oxygen demand analyzer with ozone oxidation method" linear response experimental results are Figure 2. The abscissa is the dilution multiples ($V_{\text{seawater samples}}$: V_{total}), the ordinate presents COD value. In the experimental concentration range, the correlation coefficient (R) of the

Table 1. Concentration gradient experiment results Unit: mg / L

Dilution ratio	Measured value with instrument method					Mean value of instrument method	Mean value of GB method
1/4	1.74	1.74	1.73	1.75	1.74	1.74	2.23
1/3	2.01	2.10	2.05	1.98	2.01	2.03	2.43
2/3	2.79	2.67	2.79	2.73	2.67	2.73	2.91
3/2	3.88	3.84	3.72	3.82	3.78	3.81	4.16

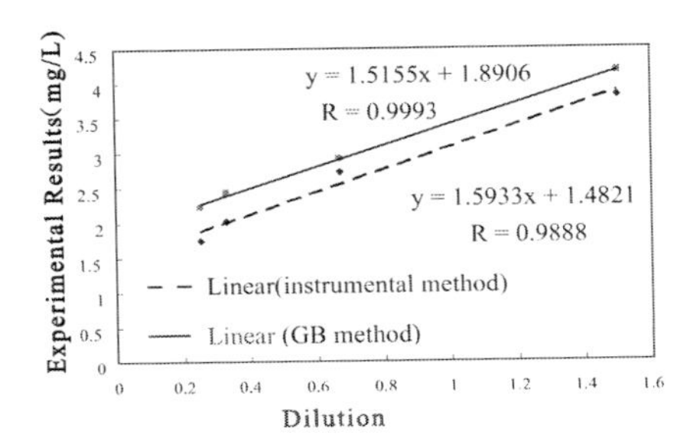

Figure 2. Linear relationship between the instrument measurement result and the different dilution multiples of the high concentration seawater.

instrumental measurement results and dilution multiples is 0.99, indicating that the "seawater chemical oxygen demand analyzer with ozone oxidation method" shows good linear responses with different concentration seawater samples. The instrumental measurement results all below the GB measurement results.

3.2 *Sensitivity of the instrument*

It's listed as instrumental method sensitivity experimental data (Table 2), we found the COD detection limit of the instrument was 0.013mg/ L, it was lower than the COD detection limit of GB method, that means the sensitivity of instrumental method is higher.

3.3 *Precision of the instrument*

Because the reference material of the instrument was not found yet, we used seawater samples to evaluate the precision of the instrument, and there were 24 groups of data to be measured with the GB method and instrumental methods. The relative standard deviation of the analysis results is Table 3, parallel five

Table 2. The experimental data of instrumental method detection limit. Unit: mg / L.

COD value	0.02	0.02	0.02	0.03	0.03	0.02	0.02	0.02	0.02	0.02	0.02
Standard deviation	0.004										
The detection limit	0.013										
The quantitation limit	0.04										
GB method detection limit	0.15										

Table 3. The precision of instrumental method.

Method \ Range	RSD (%) 0%~5%	5%~10%	10%~ 15%	> 15%
instrumental method	21	3	0	0
GB method	23	1	0	0

times experimental data of the instrument which RSD is below 5% are 87.5% of the total data numbers, that number of GB method is 95.8%, it indicates the precision of instrumental method is nearby of the GB method, fulfilling the requirements of the seawater sample analysis.

3.4 *The comparison of instrumental method and GB method average values*

After having the experimental data of instrumental method and GB method, we removed abnormal one of these data with Dixon method. Calculated the average value of parallel five times experimental data, and evaluated separately the average value of these two methods with the t-experiment, the experiment results are Table 4.

Table 4. Comparing results of seawater samples.

	Instrumental method	GB method
The amount of data	120	120
The amount of abnormal data	8	2
The proportion of insignificant different data(P = 0.95)	16.7%	—

Comparing the average values of instrumental method and GB method experimental data shown in Figure 3, the linear relation of above two methods is good match.

4 SUMMARY

The conclusions of the third-party report issued by the National Marine Environmental Monitoring Center:

1 The instrument performs well in these aspects, stably running, highly automated, easy operation and fast measurement (5 minutes a sample), no warm-up, no need chemical reagents and other advantages.
2 The operator panel of the instrument is user-friendly, shown the COD value of samples directly without data processing, compared with the GB method, it good, at shorting the time-consuming of running, after further improvement, it could be used in the sea environment protection field.
3 The linear relation of instrumental method and GB method results is good, correlation coefficient R = 0.97, the results of these two methods show certain coherency.
4 The instrument performs good precision, low detection limit, after further improvement, it can be gradually used for the analysis of seawater samples.

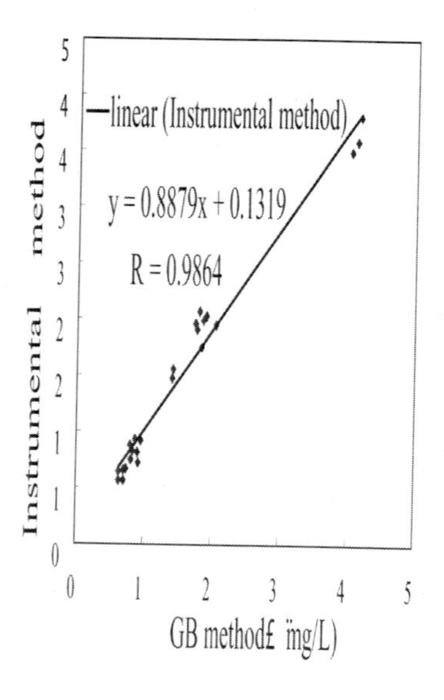

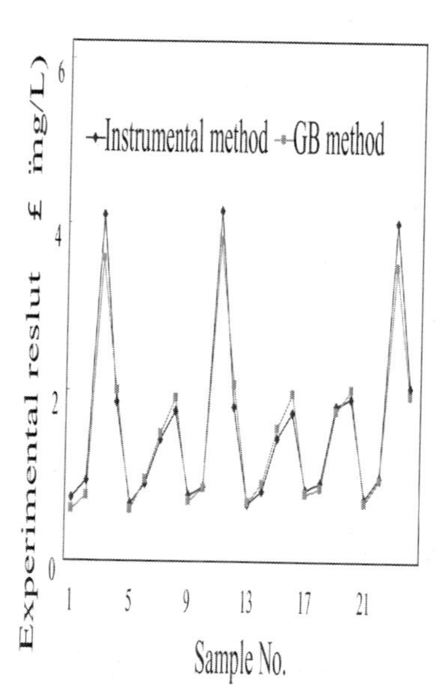

Figure 3. Comparing the average values of instrumental method and GB method experimental data. Linear relation analyzed (left) Average value comparison (right)

ACKNOWLEDGMENT

The project is supported by the Marine Commonweal Project (201005025-1).

REFERENCES

[1] The people's Republic of China State Bureau of Technical Supervision (1998). GB17378.2-1998. *The specification for marine monitoring: The second part: Data processing and quality control of analysis.* Beijing, Chinese Standards Press.

[2] The people's Republic of China State Bureau of Technical Supervision (1998). GB17378.4-1998. *The specification for marine monitoring: The fourth part: Seawater analysis*. Beijing, Chinese Standards Press.

[3] P.S. Epstein & M.S.Pleasst. (1950) On the stability of gas bubbles in liquid-gas solution. *Chem. Phys*, 18:1505–1509.

[4] L.Somoza, V. Diaz-del-Riob & R. Leon. (2003) Seabed morphology and hydrocarbon seepage in the Gulf of Cadiz mud volcano area. *Marine Geology*, 195:153–176.

[5] A.J. Guwy, L.A. Farley, P. Cunnhy, F.R. Hawkes, D.L. Hawkes, M. Chase & H. Buckland. (1999) An automated instrument for monitoring oxygen demand inpolluted waters. *Water Research*, 33(14): 3142–3148.

Electric, Electronic and Control Engineering – Shao, Shu & Tian (Eds)
 ISBN: 978-1-138-02842-5

Application of improved Canny operator in medical cell image edge detection

Mingbin Luan, Gongwen Xu & Qinghua Xu
School of Computer Science and Technology, Shandong Jianzhu University, Jinan, China

Xiaoyan Wang
School of Computer Science and Technology, Shandong Xiehe College, Jinan, China

Honglaun Wang
School of Computer Science and Technology, Shandong Jianzhu University, Jinan, China

ABSTRACT: In computer vision and image processing, image edge detection is a classical problem. The traditional Canny edge detection operator is analyzed in theory and experiment in the beginning of the paper. With the analysis of the traditional Canny algorithm limitations, the contradiction between the smoothness and approximation is carried out. Then a new filtering method is proposed, which uses improved mathematical morphology and can solve the contradiction effectively. With the help complex mathematical morphology filtering, the precise image contour can be obtained, and the noise can be effectively restrained. In the end, the edge detection experiment was carried out on the Lenna image and the medical cell image with noise. The experiment results show that the improved algorithm can detect the edge of image and remove the noise more effectively than the traditional Canny Operator method.

KEYWORDS: Canny operator; edge detection; mathematical morphology; cell image.

1 INTRODUCTION

The main research contents of image processing mainly include edge detection, image segmentation, pattern recognition, etc.. Digital image processing means turning image signal into digital signal and processed by computer program. With the advent of the digital age, digital image processing and analysis research work is more and more important. Digital image processing is widely applied in industry, agriculture, transportation, finance, geology, marine, meteorological, biomedical, military, public security, electronic commerce, satellite remote sensing, robot vision, object tracking, autonomous vehicle navigation, multimedia information network communications and other fields. It has achieved remarkable social and economic benefits[1].

Edge detection is a basic topic for image understanding, analyzing and image recognition. It provides the basis and method for image segmentation[2].

The edge of the image contains most information of the image, and a large number of the information is not continuous because of the step difference of the pixel gray value[3]. The edge is defined as area where there is larger step in the image gray value and it widely exists between the object and the object, or the object and the background. In the image edge detection, the image is preprocessed firstly, and then the edge is extracted and judged[4].

Edge detection was put forward in 1959 for the first time. In 1965, Roberts and some other people began to study the detection technology systematically. From the nineteen-seventies, the edge detection began to attract people's attention[5]. Because the image edge detection is a key technique in image processing and it occupies an important position in the image processing and computer vision, it has been paid more attention. Now people have put forward many types edge detection algorithm. For the image edge detection, there generally are the following requirements in the detection process:

- The image edge can be correctly detected.
- The edge can be positioned accurately.
- The edge detection response should be as less as possible. The best result is that there is only one response.
- The process has strong adaptability, and the miss rate and mistake rate should be lower.
- It should have good anti-noise capacity[6].

2 CANNY EDGE DETECTION

2.1 *Canny operator*

In 1986, Canny put forward the Canny edge detection operator, and Canny also carried out the "optimal operator" edge detection criterion. Canny edge detection operator is based on the following three standard criterions[7].

- Signal to noise ratio

The criterion for judging the signal-to-noise ratio is that there are lower real edge loss rate and lower fake edge rate. The bigger of the signal to noise ratio is the better of detection effect will be. The mathematical expression for the SNR is showed below.

$$SNR = \frac{\left|\int_{-\infty}^{\infty} E(x) f(x) d_x\right|}{n_0 \sqrt{\int_{-\infty}^{\infty} f^2(x) d_x}} \tag{1}$$

$f(x)$ is the filter impulse response in the field of $[-\omega, +\omega]$, $E(x)$is the edge, n_0 is the Gauss edge RMS(root mean square).

- The positioning accuracy

The positioning refers to the relationship between the detected edge and the actual edge, and the ideal effect is the edge points locate in the actual edge point center. The mathematical expression is showed below.

$$LOC = \frac{\left|\int_{-\infty}^{\infty} E'(x) f'(x) d_x\right|}{n_0 \sqrt{\int_{-\infty}^{\infty} f'^2(x) d_x}} \tag{2}$$

$E'(x)$ and $f'(x)$ are the first derivative of $E(x)$ and $f(x)$. n_0 is the same as above. The positioning precision is high if the requirements are met in the expression.

- The unilateral response

In the detecting process, there may be more than one false edge responding to a real edge. The unilateral response criterion aims to reduce the probability of multiple false edges. The false edge response should be pressed dramatically. When f responses to the noise, the two largest adjacent noise response values should be twice as the average distance of the zero crossing point. That is:

$$x_{max}(f) = 2x_{zerocross}(f) \tag{3}$$

$$x_{zerocross}(f) = \pi \left[\frac{\int_{-\infty}^{+\infty} f'^2(x) d_x}{\sqrt{\int_{-\infty}^{+\infty} f''(x) d_x}} \right]^2 \tag{4}$$

If this criterion is satisfied, it is guaranteed that there is only one edge response.

Canny edge detection operator has optimal edge detection effect; it has better detection effect and denoising effect for both weak edges and strong edges. But the calculation amount of the operator is larger than other operators[8-9].

2.2 *Analysis of the limitations of Canny operator*

Step 1: Gauss filter smooths the image.

$$I' = I * G \tag{5}$$

$G=h(x, y, \delta)$ is the 2-dimetions Gauss function. $I'=g(x, y)$ is the smoothing image. $I=f(x, y)$ is the initial image.

Step 2: Calculating the gradient magnitude and direction by Gauss first-order partial derivatives of the finite difference.

Firstly, the image is carried out the convolution operation with two differential convolution templates respectively, and then the result of the convolution is squared and added up. After extraction of square root, the gradient magnitude is obtained. The gradient direction is obtained by reverse triangle operation.

A 2X2 first-order differential template is designed to calculate the partial derivatives ($f_x'(x, y), f_y'(x, y)$) of x and y approximately.

$$f_x'(x,y) \approx [f(x+1,y) - f(x,y) + f(x+1,y+1) - f(x,y+1)]/2 \tag{6}$$

$$f_y'(x,y) \approx [f(x,y+1) - f(x,y) + f(x+1,y+1) - f(x+1,y)]/2 \tag{7}$$

The gradient magnitude and direction are showed below respectively:

$$M[x,y] = \sqrt{f_x'(x,y)^2 + f_y'(x,y)^2} \tag{8}$$

$$\theta[x,y] = \arctan(f_x'(x,y) / f_y'(x,y)) \tag{9}$$

Step 3: Using non-maximum value to suppress the gradient magnitude.

With a 3 × 3 window, in each center pixel, the domain center pixel is compared with the two pixels along the gradient direction. If the value of this pixel is no more than two adjacent points, then it will be set to 0.

Step 4: The edge detection is connected using dual-threshold method.

The following is the analysis of Canny algorithm[10].

Firstly, two definitions are given.

def 1: The smoothness of the function: if a function is derivative, and its derivative function is continuous, then it is a smooth function.

def 2: The degree of approximation of the function: any two quadratically integrable functions ($f(x)$, $g(x)$) in the definition $[a,b]$. The approximation degree can be specified as the following formula.

$$A = \|f - g\|_2 = \left[\int_a^b [f(x) - g(x)]^2 \, dx\right]^{1/2} \tag{10}$$

Smoothness and approximation: let Gauss function be $G(x) \in L(R)$, and $g(x) = f(x) * G(x)$. Where $g(x), f(x), *$ are respectively on behalf of the processed image, the original image, convolution. According to the differential properties of convolution, we can obtain the following expression.

$$\frac{d^n g(x)}{dx^n} = \frac{d^n}{dx^n}[f(x) * G(x)] = f(x) * \frac{d^n G(x)}{dx^n} \tag{11}$$

Based on the above formula, we put forward a proposition: setting $a(x) = b(x) * c(x)$, where $c(x)$ is n-order smooth function, then the following can be got.

$$M = \int_{-\infty}^{+\infty} |a(x) - b(x)|^2 \, dx \tag{12}$$

The value of the above expression increases with the increase of n.

The Canny Operator has better detection effect than other operators, but there still are defaults lying in it [11].

- The scale of Gauss filter is single. Gauss filter has better removal effect to Gauss noise, but for other noise, such as salt and pepper noise, it does not work effectively. The experiments show that the Canny operator filtering salt and pepper noise is inferior to the Gauss noise image.
- The gradient magnitude value is calculated via the finite difference mean value of the image which is filtered in the 2x2 neighbor domain. This method cannot effectively filter noise and eliminating the false edge.
- The high and low threshold is not reasonable. If the threshold is set too high, some information we needed may be missed. If the threshold is too low, the false edge may be emerged, which will obscure the image edge[12].

3 MATHEMATICAL MORPHOLOGY EDGE DETECTION

From the above-mentioned content we can conclude that the smoother the function is, the worse the approximation degree is. In order to solve the imbalance problem between smoothness and approximation, a kind of improved mathematical morphology filter method was proposed. The noise can be suppressed effectively with the use of complex mathematical morphology filtering, and the precise image contour can be acquired.

Mathematical morphology is a nonlinear filtering method, which is mainly used in noise suppression, pattern recognition, image segmentation, texture analysis, and other image information processing methods.

Mathematical morphology is suitable for mastering image specific information, and mainly scouts the pending goals using different kinds of structural shapes as a "probe". All the operation includes corrosion, expansion, open and close operations. And the image processing results will be collected as a set, which is the result of morphological edge detection.

Mathematical morphology uses the set theory to analyze the geometry problems, and it breaks the traditional numerical modeling method, which is one kind of nonlinear signal analysis theory. When different structure elements are used in image signal detection, the different image analysis is obtained. This is the reason that the structural element size and shape are affected by the information image structure. Mathematical morphology method can remove irrelevant information of the images and still maintains the basic shape characteristics of the image. In this way, the image data is simplified. So the mathematical morphology is widely used in filling, denoising, skeletonization and other image processing methods.

Let images I and S be suitable structure elements. Under the help of complex mathematical morphology detection and antinoise morphological edge detection, further improvement was carried out in this paper. Let image F be the original image, "open" is the first operation, then the "corrosion" processing, F1 image is obtained. Now the original image is operated by "close", and then expansion processing, F2 image is obtained. F can be expressed as a linear and operation of F1 and F2.The formula is showed below.

$$F = \frac{1}{2}[(I \bullet S) \oplus S + (I \circ S) \Theta S] \qquad (13)$$

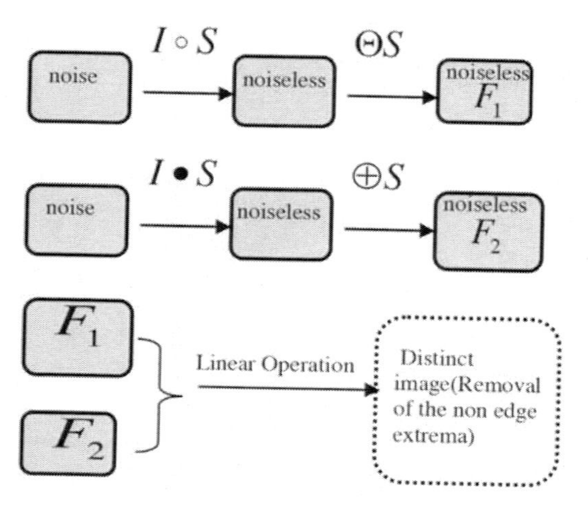

Figure 1. The operation of F, F1 and F2.

Mathematical morphology is an important nonlinear filter developed in recent years. From the beginning of the two-valued filter, to the gray-scale morphological filter, using the geometric characteristics and algebraic its properties, they are widely used in the shape recognition, edge detection, texture analysis, image restoration and enhancement and other fields. The filter operation is mainly put forward by the open and close operators. The open operation can remove isolated dots, burr and bridge, while the total position and shape keep unchanged. It is a filter based on geometric operation. Close operation can fill holes, bridging the small cracks, while the shape and position keep same, filtering image by filling the image concave.

When the filter consisting of mathematical morphology open operation and corrosion makes linear operation with the original image, the bur noise points and bridge noise points can be restrained. At the same time, the smooth boundary point of image edge can be effectively gotten. When the filter consisting of mathematical morphology close operation and expansion filter makes linear operation with the original image, the holes noise points and small joint noise points can be restrained. At the same time, the smooth outer boundary points of image edge can be effectively gotten. When the complex filter applying to the Canny operator, the maximum noise point in the target region can be removed. The ridge belt extremum points set near the image edge can be more precise. The non-maxima value suppression aims to obtain the local maximum of amplitude image array. In order to determine the edge, the ridge belt in the amplitude image should be refined to retain the points that amplitude value changes most greatly. So that the edge is refined and the process provides the conditions getting accurate edge. In addition, the complex approach above can get rid of all noise specific to the structural elements, in order to maintain the image structures are not blunted.

From the above analysis, the improved algorithm can be shown in the following.

Step 1: Smoothing the image with the improved morphological filter proposed above.

Step 2: Calculating the gradient magnitude and direction by Gauss first-order partial derivatives of the finite difference.

Step 3: Suppressing the gradient magnitude using non-maximum values.

Step 4: Detecting the connecting edge using dual-threshold method.

4 THE EXPERIMENTAL ANALYSIS

4.1 *The effects of lenna image with noise*

To further verify the effect of the improved edge detection algorithm, the traditional Lena image was used in the experiment.

In order to verify the feasibility and effectiveness of the algorithm respectively, the image with salt and pepper noise of 0.02 was used, which is shown in figure 2. Now the edge detection effects are shown in figure 3 and 4.

Figure 2. The original image with noise.

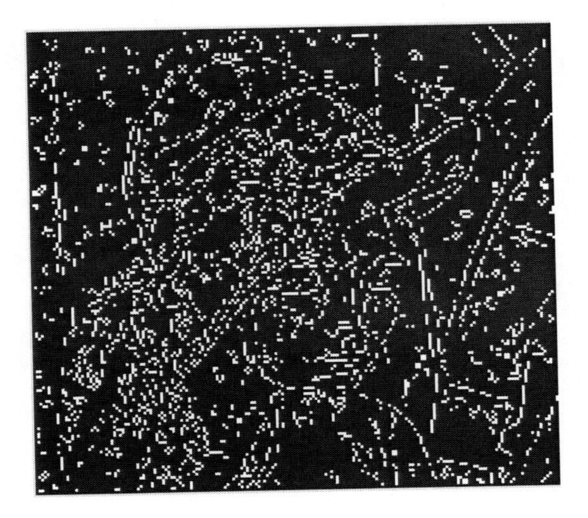

Figure 3. Traditional Canny operator edge detection.

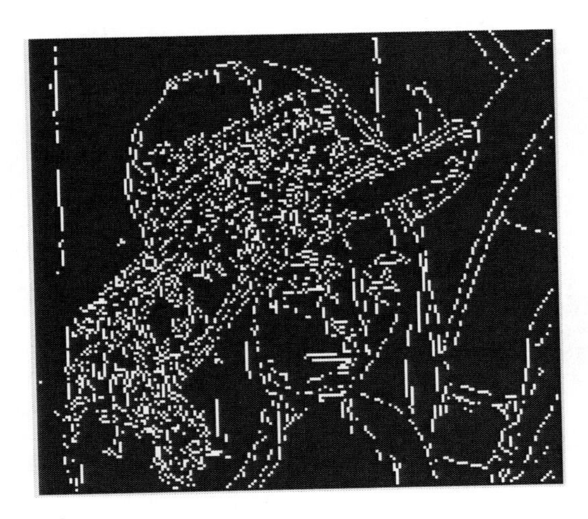

Figure 4. The improved algorithm edge detection.

Compared with the traditional Canny algorithm, when completing salt and pepper noise image edge detection, the algorithm proposed in this paper can obtain the continuous edge. What's more, the image edge detail information is more abundant, the contour is clearer, and most of the noise has been filtered. The improved algorithm has better effect on the image with high salt and pepper noise, and a more clear and detailed edge can be achieved.

4.2 *The effects of medical cell image with noise*

Edge detection is an essential step in medical image processing, as the effective edge detection is helpful in improving the doctors' diagnostic level and alleviating the suffering and medical expenses of the patients.

The original image figure 5 is a 338 * 350 pixel medical cell image with noise. Figure 6, 7 are the detection results of two images obtained by using the traditional Canny Operator algorithm and the Improved Algorithm.

Comparison of the two edge detection results of figure 6 and 7, we can find that the improved algorithm has improved the detection template; the new algorithm has better results than traditional methods. The improved algorithm has better search ability and more complete edge. As the medical cell image is mixed with noise, the experiment results show that the improved algorithm also has the ability of anti-noise ability.

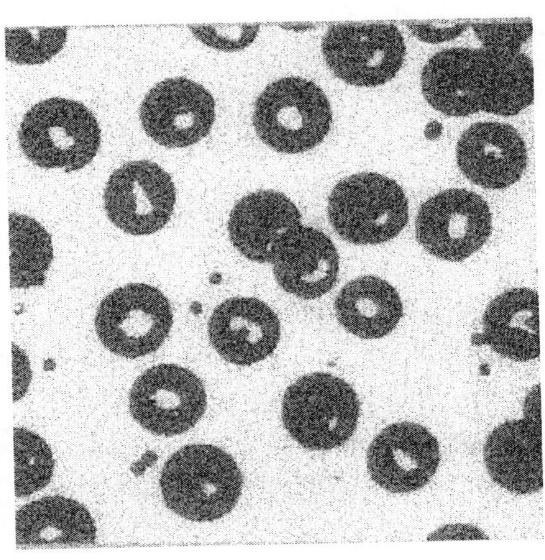

Figure 5. The initial medical cell image.

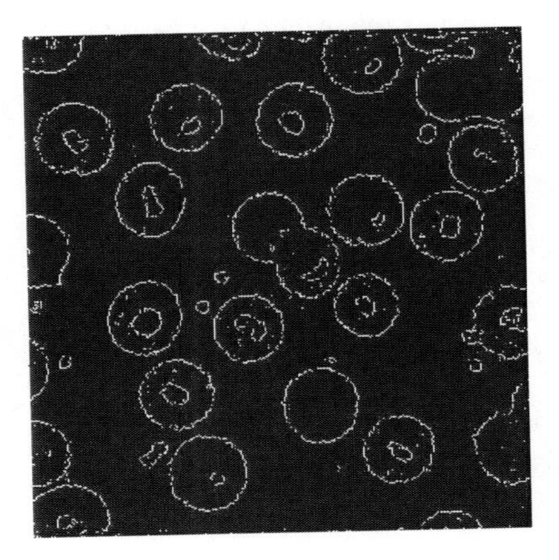

Figure 6. Traditional Canny operator edge detection.

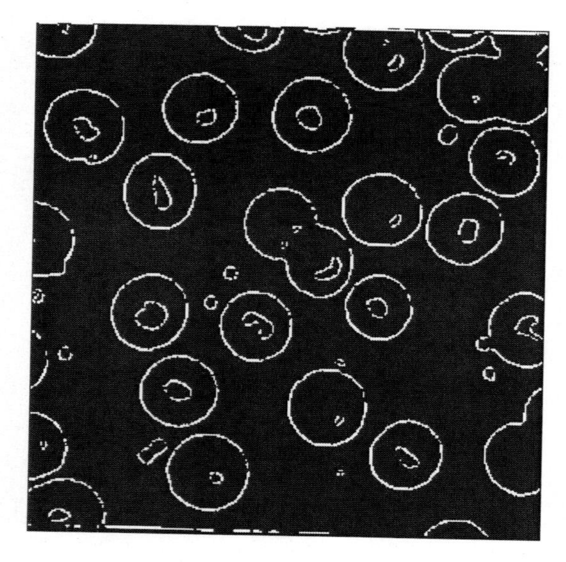

Figure 7. The improved algorithm edge detection.

5 CONCLUSIONS

In the field of image processing and computer vision, the image edge detection is one of the basic and the most important research area, which has been widely applied in all areas of life and production.

The image may be interfered by the unwanted details in the scene, so the noise suppression and edge determination cannot be met at the same time. The Gauss function filter has good effect in noise suppression, but there are over-smoothed problem, which will increase the edge uncertainty. In the image recognition and understanding of medical cell, the morphological parameters of image are needed. If the edge is inaccurate, the shape of medical cell image will change, resulting in misidentification. An improved algorithm was put forward in this paper, and it is verified by Lenna images and the medical cell image edge detection. After comparison, it is obvious that the improved algorithm has better ability of denoising and edge detection.

From the experiment we can see that the improved algorithm proposed in this paper has better detection effect than the traditional Canny operator detection, this algorithm also solves contradiction between the smooth and the approximation in the Gauss smoothing process.

ACKNOWLEDGMENTS

The work is supported by the School Fund project of Shandong Xiehe College (XHXY201431) and Jinan Higher Educational Innovation Plan (201401214, 201303001), the Project of Shandong Province Higher Educational Science and Technology Program (J14LN59, J13LN11, and J12LN31), the natural science foundation of Shandong Province (ZR2012GQ010).

REFERENCES

[1] Xun Wang and Jianqiu Jin. (2007) An edge detection algorithm based on Canny operator. *Intelligent Systems Design and Applications. ISDA 2007.* Oct. 20–24, 2007, pp:623–628.

[2] Zou Qijun, Li Zhongke. (2008) Image edge detection algorithm contrastive analysis. *Computer Application*, 6(28):215–216.

[3] Canny J. (1986) A computational approach to edge detection. *IEEE Trans on PAMI*, 8(6):679–698.

[4] Zhang Jie, Tan Jieqing. (2009) Based on anisotropic diffusion equation Canny edge detection algorithm. Computer Application, 8(8):2049–2051.

[5] Zhang Yuwei,Wang Yaoming. (2006) Based on Wavelet Transform Canny operator edge detection. *Shanghai Normal University (Natural Science Edition)*, 6(3):35–38.

[6] Li Mu, Yan Jihong, Li Ge, Zhao Jie. (2007) Adaptive Canny operator edge detection technology. *Journal of Harbin Engineering University*, 9(9):1002–1007.

[7] Xu Gongwen, Zhang Zhijun, Yuan Weihua, Xu Li'na. (2014) On medical image segmentation based on Wavelet Transform. *ISDEA 2014. 15–16 June 2014*. pp:671–674.

[8] Chen Yanlong,Zhu Hucheng. (2008) Improvement algorithm based on Canny operator edge detection. *Computer Application and Software*, 8(8):51–53.

[9] Jansen M, Malfait M, Bultheel A. (2009) Generalized cross validation for wavelet thresholding. *Signal Processing*, 56(1):33–44.

[10] M. Wen and C. Zhong, (2008) Application of Sobel algorithm in edge detection of images. *China High-tech Enterprise*. (6):57–62.

[11] J. Li, X. K. Tang, and Y. J. Jiang, (2007) Comparing study of some edge detection algorithms. *Information Technology*. 9(38):106–108.

[12] X. L. Ci and G. G. Chen. (2008) Analysis and research of image edge detection methods. *Journal of Infrared*, (7):20–23.

Electric, Electronic and Control Engineering – Shao, Shu & Tian (Eds)
 ISBN: 978-1-138-02842-5

Design and realization of wireless communication module over ZigBee

Jun Zhang
College of Electronic and Information Engineering, SIAS International University, Zhengzhou Henan, China

ABSTRACT: In the context of wireless communication, Beige wireless communication modules are compared with Bluetooth and IEEE802.11b in respect of energy continuation, time to respond, effective range and data transmission. Separate studies are conducted on Beige wireless communication for its application in house building, medical apparatus and industrial control. The design method for ZigBee wireless modules in different areas and the coverage of Beige wireless communication based on cellular model are proposed.

KEYWORDS: ZigBee technology; wireless communication; GPRS network.

1 INTRODUCTION

ZigBee is a low-cost, low-energy wireless communication protocol that transmits low data within a short distance. At the same time, it is a low-energy technology that can be used for the long time. When used to transmit data, ZigBee has proven to be highly reliable with large network capacity and good security behavior.

In wireless communication, ZigBee is composed of two parts: a full function device (FFD) that can be connected to any of the transmission devices capable of data acquisition to serve as a coordinator, and a reduced function device (RFD) that locally collects and transfers data to FFD. The wireless monitoring system of a thermal power company, for example, consists typically of three parts: a management layer, a control layer and a local instrument layer. The local instrument layer contains a master ZigBee nodes and ZigBee sub-nodes. By connecting ZigBee to GPRS network, data is collected to the master node from individual sub-nodes and transmitted to the control layer. Daily data are collected and collated and transmitted steadily using ZigBee technology to allow automatic control of the thermal network so that the entire heat supply system supplies heat uniformly with low loss and high energy efficiency.

2 CELLULAR MODEL

2.1 *The radiation center theory*

When the radiation center theory is used to discuss the spatial layout of the signal transmitter and router, central place model has to be used first.

Here, the following assumptions are made:

1. The radiation center model discusses the data transfer services provided by the signal transmitters and the routers and is established at any center densely covered by nodes;
2. A low-level radiation center can reduce the number of signal transmitters and routers; and
3. There are fewer high-level radiation centers and these centers receive more node data and cover a wider area.

Based on the radiation center theory, the positions with dense nodes are selected to save resources.

2.2 *Model construction*

ZigBee wireless modules include a ZigBee network coordinator, a ZigBee repeater and ZigBee terminal devices connected into a network structure. Fig.1 shows the network connection of these devices.

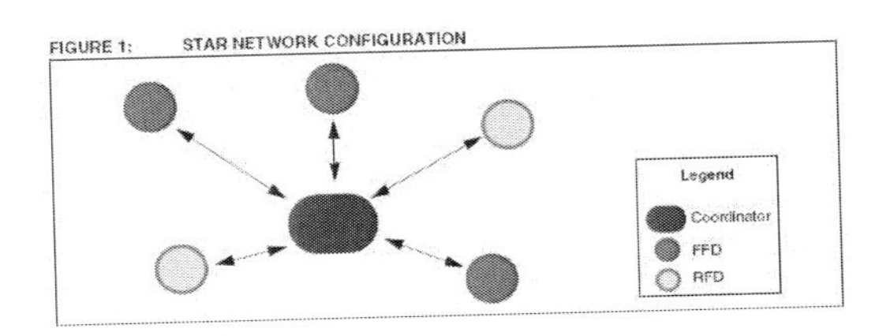

Figure 1. Single coordinator.

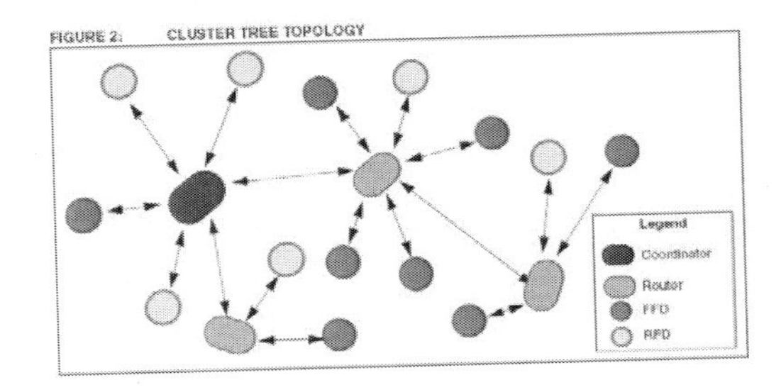

Figure 2. Router combination.

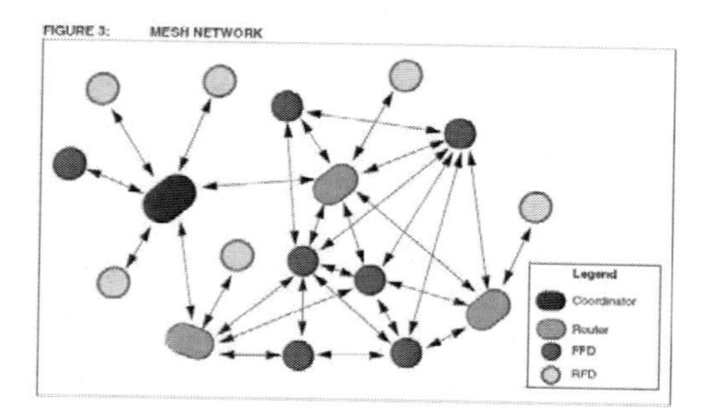

Figure 3. Multilayer combination.

In our study, the signal areas diverged from the router are least overlapped to remove other factors. From the number of routers connected by the ZigBee nodes, as the signal coverage of the router is circular, given that the radiation radius r remains the same, the geometric values of three shapes in terms of overlap area are obtained as given in Tab.1.

Table 1. Comparison of three shapes.

Cell Shape	Regular triangle	Square	Regular hexagon
Distance from adjacent cell	r	$\sqrt{2}r$	$\sqrt{3}r$
Cell area	$1.3r^2a$	$2r^2$	$2.6r^2$
Overlap width	r	$0.59r$	$0.27r$
Overlap area	$1.2\pi r^2$	$0.73\pi r^2$	$0.35\pi r^2$

From this table, the regular hexagon is the most suitable since it has the best similarity to circular in terms of shape and can effectively accommodate the coverage. Hence, the center of the regular hexagon is used as the cellular structure and extended outwards as shown in the chart below.

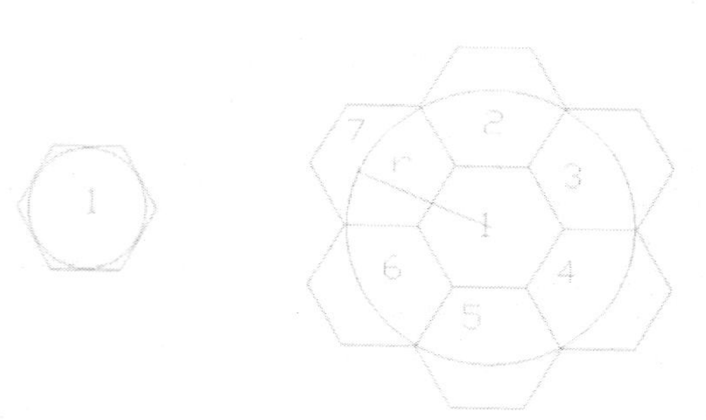

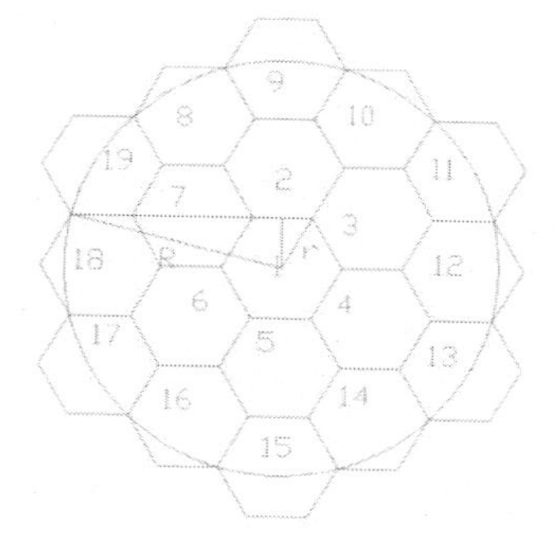

From this we can find the relation between the diameter d and the number N:

Figure 4. As shown in Fig.4, the boundary of the circular area is at the center of the outermost hexagon. According to the law, we have

$$N = 12n^2 + 30n + 19$$

Here: N is the number of hexagons needed to cover the circular area; n is

$$n = \frac{D}{d}$$

Here: D is the diameter of the circular area; d is the diameter of the incircle of the hexagon.

The comparison of wireless communication module of Zigbee, Bluetooth and IEEE802.11b is shown as Table 2.

Assume the smallest number of routers is used to allow data transfer at the nodes within a 40-mile radius circular area, where each of the routers contains a fixed range of coverage that varies with the density level of the node, to obtain the optimum coverage, the overlap area of coverage between two adjacent routers must be the smallest. Assume the regional area is indefinitely large, the coverage areas needed for different number of routers are calculated and given in Tab.3.

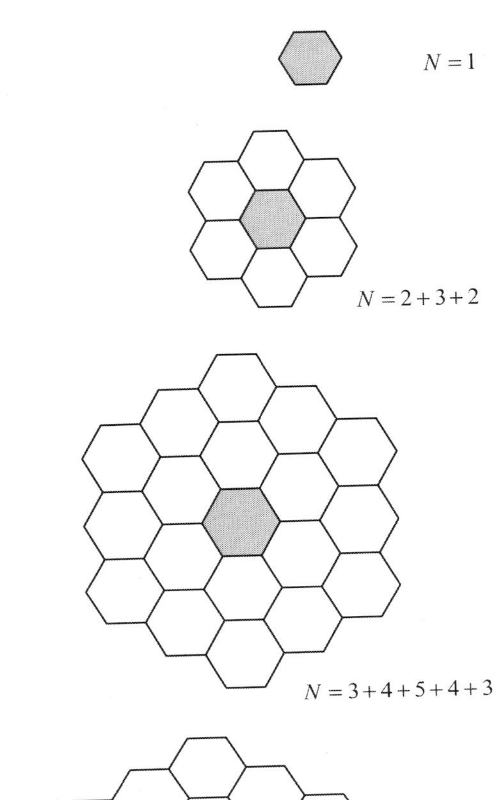

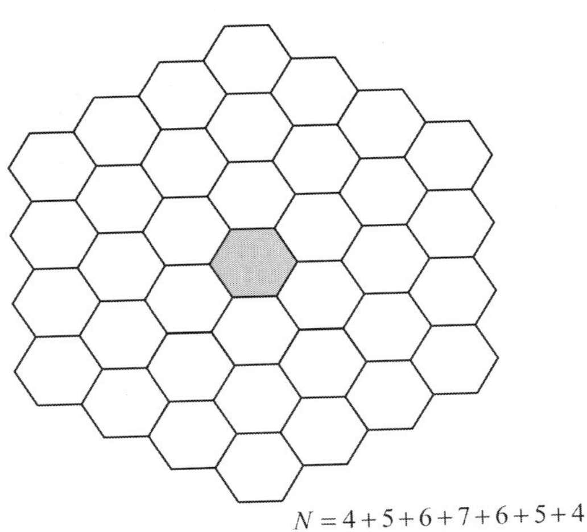

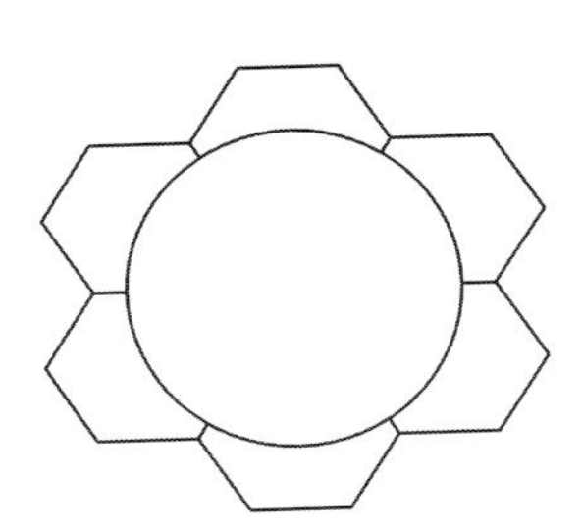

Fig 4.

Table 2. Comparison of three communication protocols.

Feature	ZigBee	Bluetooth	IEEE802.11b
Energy continuation	1	6d	A few hours
Complexity level	Low	Very high	High
Time to respond	35mm	8sec	3.5sec
Effective range	75m	15m	100m
Extendability	Yes	No	Yes
Data transmission	250Kbps	1Mbps	11Mbps
Number of nodes	64000	7	32

Table 3. Summary.

Number of Routers N	Coverage Radius r	Number of Routers N	Coverage Radius r
1	40	261	1.859
7	20	387	1.818
19	11.09	519	1.701
37	8	657	1.667
61	7.184	801	1.568
91	6.667	951	1.538
27	5.298	210	1.454
69	5	226	1.428
17	4.193	243	1.355
71	4	261	1.333
31	3.468	279	1.269
97	3.333	297	1.25
69	2.957	316	1.194
57	2.857	336	1.176
61	2.577	357	1.126
71	2.5	378	1.111
87	2.283	399	1.066
19	2.222	421	1.053
17	2.049	444	1.012
141	2	468	1

Tab.4 gives the signal intensities of different wireless communication modules from different distances during short-distance point-to-point data transmission.

Tab.5 gives the signal intensities of different wireless communication modules from different distances during long-distance point-to-point data transmission.

Table 4. Signal intensity test.

	0.2m	0.5m	1m	1.5m	2m
$M00 \rightarrow M00$	46.4 dB	44.1 dB	35.9 dB	31.8 dB	25.9 dB
$M00 \rightarrow M00$	55.2 dB	48.6 dB	41.2 dB	36.2 dB	35.8 dB
$M00 \rightarrow M00$	64.1 dB	62.3 dB	59.2 dB	56.2 dB	52.8 dB
$M00 \rightarrow M00$	38.9 dB	61.8 dB	60.1 dB	61.3 dB	59.6 dB

Table 5. Communication distance and loss rate test.

	5m	10m	15m	20m	25m
$M00 \rightarrow M00$	8.32%	-	-	-	-
$M00 \rightarrow M00$	0	0	1.39%		
$M00 \rightarrow M00$	0.87%	-	-	-	-
$M00 \rightarrow M00$	0	0.56%	0.87%	2.60%	3.61%

3 DESIGN OF A ZIGBEE WIRELESS NETWORK SYSTEM

Network sensors are set to the specific environment of each area according to different environmental parameters.

It applies to the hierarchical allocation of the wireless network system and consists of a monitoring host computer, node sensors, the master network coordinator and wireless routers.

1 Design and realization of network coordinator

The node design of the master network coordinator includes the selection of a master module compatible with CC2430 chips, an antenna, power supply and interface model, powered by AC power supply. Displays and keyboards should also be tested and set. As the network coordinator is a device that serves as a critical hub, it is connected between the monitoring host computer and the routers and node sensors. When connected to the host computer, RS485 series bus can be used. The signal received is transferred to the display of the host computer.

2 Router settings

A router node is generally composed of a master module consisting of the display, power module, antenna and chips. The node is maintained by the routing table and neighbor table. Between the sensor nodes and the coordinator, data is repeated and transferred by the data transfer nodes. If necessary, the environment information can be acquired and the sensor module can be configured according to the specific conditions.

If the sensor and router nodes are far apart from each other and the greenhouse area is too large, an additional CC2591 module will add to the signal transmission performance to establish a large ZigBee wireless network.

3 Wireless sensor nodes

The network terminal node consists of CC2430 as the master module. Other elements include the temperature and humidity sensors and their interface circuits, the debugging interface, power supply, display and antenna as detailed in the diagram below.

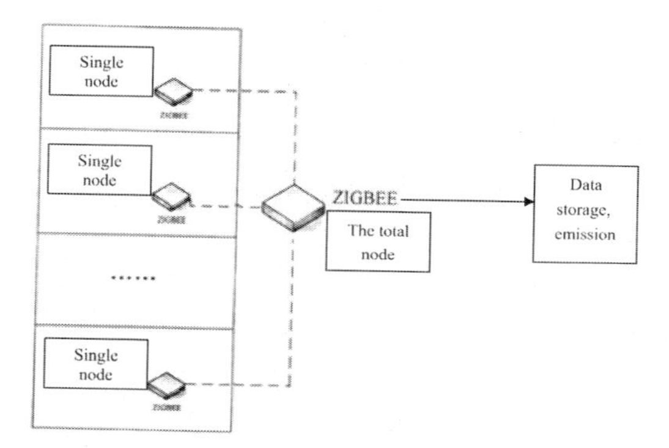

Figure 5. Transmission from wireless sensor node to regulator.

The sensor nodes are used to acquire temperature and humidity data. As the first terminal of the channel, data obtained from the main sources and basis for the automatic environment control of the control device. At present, LCD display is mostly used to receive data collected from sensor nodes to minimize energy consumption and facilitate observation. The energy for the sensor is mainly supplied from the battery module to allow mobility. If more current is needed by the sensor, solar cell can be used to allow convenient and energy-efficient operation.

4 SIMULATION APPLICATION OF WIRELESS MODULES IN ZIGBEE

4.1 *Greenhouse control*

Greenhouses are widely used in agricultural design. With its very flexible node arrangement and low energy consumption, ZigBee allows effective, safe and convenient monitoring of greenhouses including temperature, humidity and the operation of automatic agricultural machines in the greenhouses.

Fig.6 shows the application of ZigBee based on wireless modules in greenhouse control.

By combining the measuring and control instruments in the greenhouse with the wireless modules of ZigBee, the resulting wireless communication modules form part of the ZigBee wireless network connecting the nodes as well as an upper decision-maker for the monitoring and control.

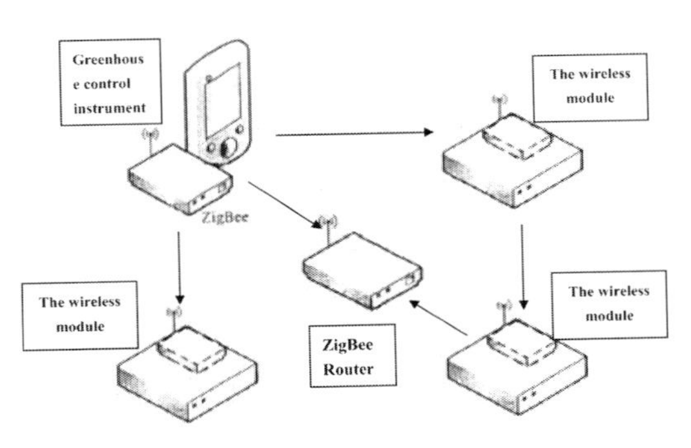

Figure 6. The application of ZigBee based on wireless modules in greenhouse control.

In our design, the greenhouse sensor is SHT11. The interface is the sensor interface circuit. Fig.7 shows the design principle of the greenhouse sensor.

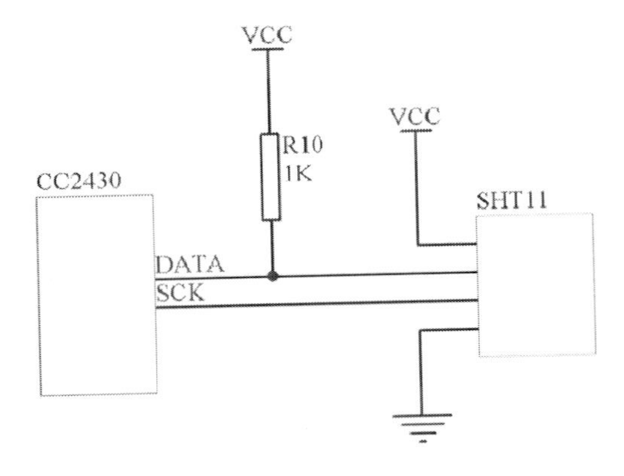

Figure 7. Interface circuit of SHT11 sensor.

Operation command steps of SHT11: Start Command, transform, send text, complete, read verified number of bytes, read low bytes, *RH* humidity measurement time sequence. Fig.8 shows the program operation flowchart.

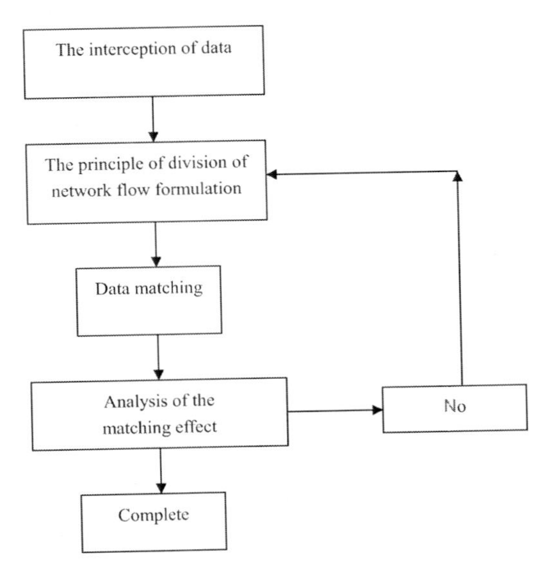

Figure 8. Program operation flowchart.

4.2 *Medical apparatus*

The application of ZigBee in medical apparatuses is typically concentrated in remote medical technology, where computer is used as the medium to transfer medical information and allow remote monitoring by the medical workers. As wireless modules in ZigBee operate under working or non-working state, during the former of which the ZigBee transmission efficiency is very low and during the latter, of which the ZigBee nodes are asleep, energy consumption is very low. Besides, ZigBee also provides reliable data transmission and large capacity.

1 Design of the wireless monitoring system

Wireless monitoring system has very good application prospect in the medical area and has proven good performances in medical monitoring and remote treatment. ZigBee can be connected to the ECG monitoring system so that the ECG monitoring is connected to the remote monitoring system into a remote ECG monitoring system.

2 Body area medical design

Within the human body, to transmit some body information to the medical apparatus, some low-energy devices are needed to perform data transmission, where sensing devices are placed in some part or nodes of the human body to deliver information of the nodes to the local receiving system and the GPRS is connected to the medical monitoring system as shown in Fig.2.

The entire wireless monitoring system of a thermal power company, for example, consists typically of three parts: a management layer, a control layer and a local instrument layer. The local instrument layer contains a

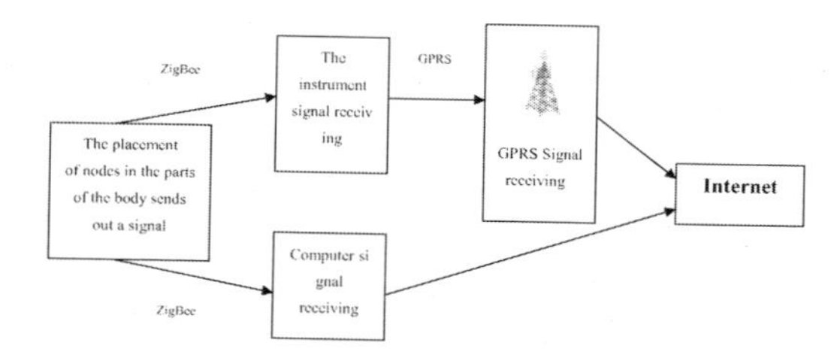

Figure 9. Body area medical data transmission.

master ZigBee nodes and ZigBee sub-nodes. By connecting ZigBee to GPRS network, data is collected to the master node from individual sub-nodes and transmitted to the control layer. Daily data are collected and collated and transmitted steadily using ZigBee technology to allow automatic control of the thermal network so that the entire heat supply system supplies heat uniformly with low loss and high energy efficiency.

5 CONCLUSIONS

The application and design of wireless communication modules contained in ZigBee are studied, the merits and applicability of the low-cost, low-energy ZigBee wireless communication protocol over its counterparts, particularly bluetooth wireless transmission, are discussed. Cellular model is used to examine the ZigBee wireless coverage. The results indicate that the wireless area radius covered by each router varies with the number of routers used.

ACKNOWLEDGMENT

This paper is supported by Hebei Province Technology Support Project "Synthesis of sulfonic group functionalized ionic liquid and its applications in the synthesis of isoamyl acetate" (GN:13211409).

REFERENCES

[1] Wang, T.T. (2007) *Remote Monitoring System of Heat-supply Network Based on Zigbee.* Nanjing: Nanjing University of Science & Technology.
[2] Zhou, M.G. (2007) *Study of Smart Greenhouse Controller Based on IEEE1451.2 Standard and ZigBee Protocol.* Hangzhou: Zhejiang University.
[3] Xu, Z.M. (2008) *Study on ECG Monitoring System Based on Wireless Communication.* Shanghai: Fudan University.
[4] Wan, L., Wang, P. (2010) Design and realization of temperature and humidity monitoring network based on ZigBee. *Low Voltage Apparatus*, 14(20): 32–36.
[5] Ao, C.B. *Design of Temperature Data Acquisition and Detection System Based on ZigBee.* Jilin: Jilin University, pp:13–16.
[6] Qin, T.G., Dou, X.Q., Huang, W.B. (2007) Application vof ZigBee technology in wireless sensor network. *Instrumentation Technology*, 12(1):57–59.
[7] Sun, C., Zhang, S.Q., Zhang, X.L., Yang, J. (2006) Application of wireless sensor network in greenhouse environment monitoring. *Journal of Agricultural Mechanization Research*, (9):194–195.

Electric, Electronic and Control Engineering – Shao, Shu & Tian (Eds)
© 2015 Taylor & Francis Group, London, ISBN: 978-1-138-02842-5

Design and implementation of USB-AKA protocol

Xuesi Zhang
Zhengzhou Information Science and Technology Institute, Zhengzhou, China

Cui Li
Unit of PLA, Xi'an, China

Jianjun Li
Information Assurance Laboratory, Beijing, China

Bin Yu
Zhengzhou Information Science and Technology Institute, Zhengzhou, China

ABSTRACT: In the USB protocol, the lack of security mechanism causes some security risks during the data transmission and storage. Based on the USB communication features, combined with the idea of the encrypted key exchange protocol (EKE), a USB authentication and key agreement protocol (USB-AKA) is promoted in this paper. Meanwhile, the validity of the protocol is verified via the Strand Space Model. And the performance of the designed protocol is tested on FPGA. The results show that the designed protocol is propitious for the USB system, and guarantees the legality of all the users, hosts and USB devices in a communication group, which can resist various attacks, such as the counterfeit attack, the replay attack, the bus monitoring attack and so on. After the execution of the designed protocol, the session key for the bus transmission can be consulted.

KEYWORDS: Mutual authentication; Authentication and Key Agreement; Strand Space Model; USB Transmission.

1 INTRODUCTION

Because of their outstanding advantages such as low-cost, convenience, hot-plug and so on, USB devices are widely used in today's various information exchange and storage situations. However, due to the lack of security mechanism, the data security problems are coming out one after another. In order to increase the transmission channel safety, the authentication for USB devices is studied.

The studies of USB authentication converge on the user's identification at the beginning. By identifying the legality of the USB device's user, the safety is protected. Wang Wei[1], Sun-Ho Lee[2] et al. identify users based on what they know. Yi Qingsong[3] add security memory chip into device's hardware to store users' passwords and implement the user's identification. Hu Jianyong[4] promotes a two-factor authentication based on what users know and what they have by utilizing an ID card. All the approaches above can identify the users uniquely, but the legality of the host as well as the device is not considered.

Aiming at the host and device authentication, Hou Xingchao[5] fulfills a one-way authentication between host and storage device. Since the authentication is unidirectional, the device can't identify the host. Therefore, via illegal host, attackers can get confidential information from legal device and the protection becomes in vain. In order to solve this problem, Helena Handschuh[6], Yin Wenhao[7] et al. design a mutual authentication protocol to identify both host and device. Meanwhile, the user's legality is also identified by verifying what they know. But, there still exists a safety risk, that is, the USB bus may be monitored [8]. The most efficient way to resist bus monitoring is to create a session key shared by the two transmission parts. By encrypting with the shared session key, the bus data can be protected. Practically, there are two ways to create session keys, key distribution and key agreement. The reason why the key agreement is generally adopted is that the two transmission parts must supply information equally and isochronously, so that the freshness of the keys can be insured. Combined with key agreement, Yang Xianwen[9] designs a USB authentication protocol UAKA, but two parts' IDs are not included in any transmission information. As a result, the device can only be used in one host, but not in a communication group. Besides, two parts have to store a large quantity of public keys and private key, which is difficult to be managed.

Borrowing from the encrypted key exchange (EKE) protocol, the defects above can be made up. The first EKE protocol is Diffie-Hellman encrypted key exchange protocol (DHEKE)[10] put forward in 1976. Later, Li Li [11] simplifies it by proposing a SEKE protocol with higher safety and lower cost. Lo Naiwei[12] demonstrates a simple three-party password authenticated protocol to resist an undetectable on-line dictionary attack based on the idea of EKE. However, the above schemes are designed for network, not appropriate for USB devices.

Ellipse curve Cryptography (ECC) [13] is a new kind of cryptography system with shorter keys and smaller storage space. And the ECC is always used in limited computing power and space environment such as electronic devices. So, in order to make the EKE suit for USB devices, the ECC can be brought into design.

In this paper, based on the idea of EKE and ECC, we will design a USB-AKA protocol through improving the SEKE. The scheme design will be introduced in chapter 2. The validation of the protocol will be proofed in chapter 3 via the Strand Space Model [14]. And the performance will be tested on FPGA, presented in chapter 4. At last, the work is going to be summarized in chapter 5.

2 USB-AKA PROTOCOL DESIGN

The USB-AKA design contains two parts, the mutual authentication and key agreement between host and device, and the user's authentication. The main symbols are illustrated in Table 1.

Table 1. Main symbols.

Symbol	Illustration	Symbol	Illustration
A	Host	B	Device
G	Base point of the elliptic curve	Q	Elliptic curve point order
A	Temporary private key randomly produced by A	B	Temporary private key randomly produced by B
N_A	Fresh number randomly produced by A	N_B	Fresh number randomly produced by B
K_{AB}	Symmetric key shared by host and device	$H(\ )$	Hash function
$USER$	User name	PW	User password
SK_A	Session key calculated by A	SK_B	Session key calculated by B
SK	Agreed session key	IDx	ID information of x

In the system building stage, host, device and users all need to be registered in the trusted center. Host and device must supply their IDs to the center to apply for the registration first. Then, the center will distribute the shared symmetric key to both host and device. Users must supply their names and passwords to the center. And the center will make the device storage hash values of every user's name and password separately. Then, the user's registration is finished.

2.1 *Mutual authentication and key agreement*

The protocol contains the following steps.

Step 1: A chooses N_A, and sends the massage (ID_A, N_A) to B. The massage is the connection of A's ID with N_A.

Step 2: B produces the temporary private key $b(1 < b < q)$, calculates the temporary public key bG, and produces N_B. Then, B encrypts (bG, N_A, N_B) with the shared symmetric key K_{AB}. The result is connected with B's ID and sent to A. The sent massage is $(bG, N_A, N_B)_{K_{AB}}, ID_B$.

Step 3: A produces the temporary private key $a(1 < a < q)$, and calculates the temporary public key aG. At the same time, A checks out the shared symmetric key according to B's ID. Then, A decrypts the received massage with the checked K_{AB}. From the decryption, A can get an N_A. Compared it with the produced N_A in step 1, if they are equal, A authenticates B successfully; else, the authentication fails. After the successful authentication, A encrypts N_B, aG with K_{AB}, the result $(N_B, aG)_{K_{AB}}$ is sent to B。

Step 4: B decrypts the massage and gets N_B and aG. Compare N_B with the produced N_B in step 2, if they are equal, B authenticates A successfully; else, authentication fails. After successful authentication, B calculates the hash value ofaG, N_B and ID_B, then $H(aG, N_B, ID_B)$ is sent to A.

Step 5: A calculates the hash value of$aG, N_B$and ID_B, and compares it with the received $H(aG, N_B, ID_B)$. If they are the same, A can confirm that the temporary public key sent by B in step 3 is received successfully. This step ensures that A and B can both receive each other's temporary public key and they all have the ability to calculate the session key.

After the steps above are executed, A calculates the session key $SK_A = H[a(bG), N_A, N_B]$, B calculates the session key $SK_B = H[b(aG), N_A, N_B]$, $SK_A = SK_B = SK$. The session key is used to

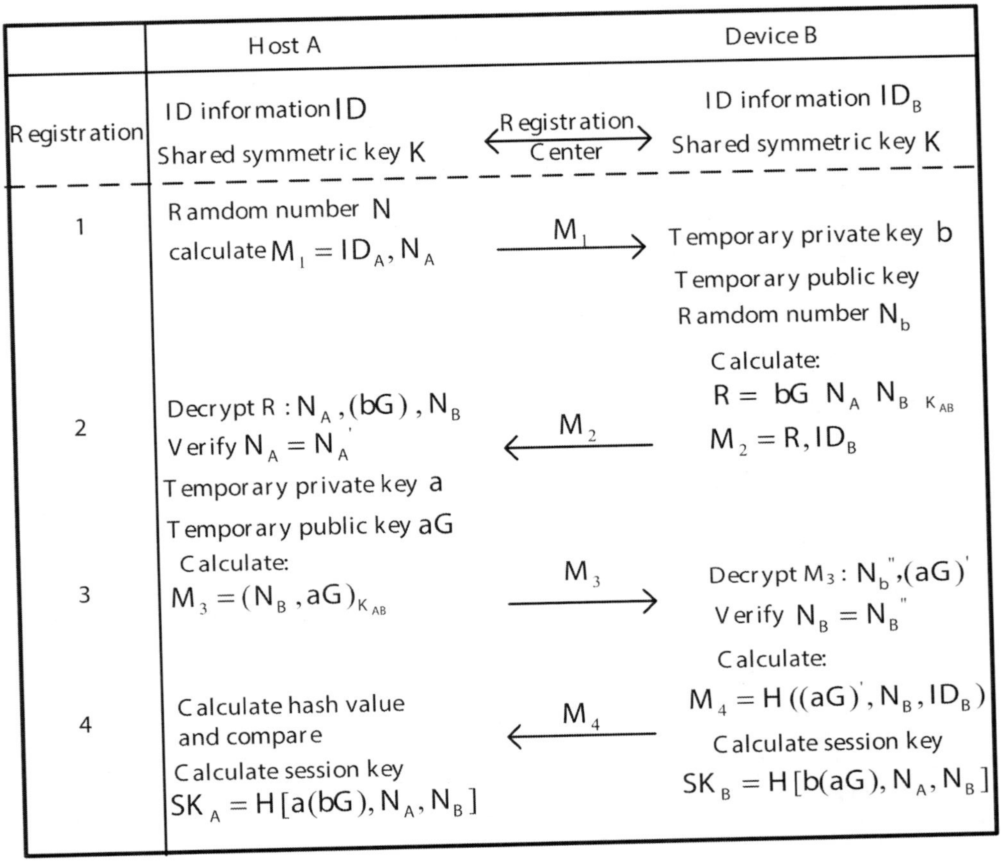

	Host A		Device B
Registration	ID information ID Shared symmetric key K	⟷ Registration Center	ID information ID_B Shared symmetric key K
1	Ramdom number N calculate $M_1 = ID_A, N_A$	M_1 ⟶	Temporary private key b Temporary public key Ramdom number N_b
2	Decrypt R : N_A ,(bG) , N_B Verify $N_A = N_A'$ Temporary private key a Temporary public key aG	⟵ M_2	Calculate: $R = bG \; N_A \; N_B \; K_{AB}$ $M_2 = R, ID_B$
3	Calculate: $M_3 = (N_B , aG)_{K_{AB}}$	M_3 ⟶	Decrypt M_3 : N_b'',$(aG)'$ Verify $N_B = N_B''$ Calculate:
4	Calculate hash value and compare Calculate session key $SK_A = H[a(bG), N_A, N_B]$	⟵ M_4	$M_4 = H((aG)', N_B, ID_B)$ Calculate session key $SK_B = H[b(aG), N_A, N_B]$

Figure 1. Mutual authentication and key agreement.

protect the user's authentication information and other transmission data between host and device. The process of this protocol is shown in Figure 1.

2.2 *User's authentication*

The user's authentication is as follow.

Step 1: When user uses the device, the user name and password are inputted through the software in host. USER and PW are sent to device after they are encrypted.

Step 2: Device decrypts the received massage and calculates the hash value of the USER and PW. The value is compared with the device stored information inputted in the registration stage. If the values are the same, the user is considered legal. Else, the user is considered as an attacker, not able to use the device.

The process is shown in Figure 2.

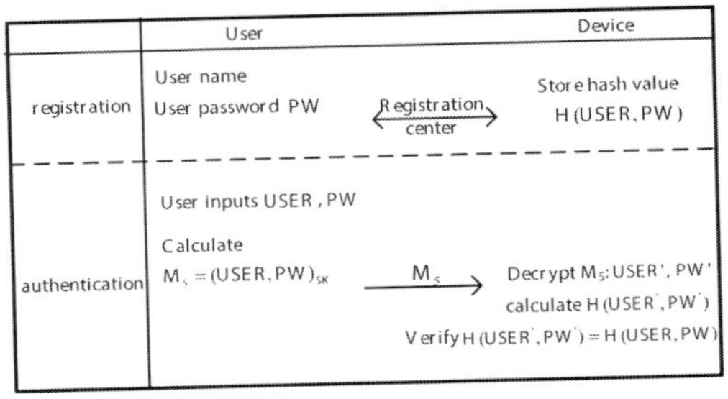

	User		Device
registration	User name User password PW	⟷ Registration center	Store hash value H (USER, PW)
authentication	User inputs USER , PW Calculate $M_5 = (USER, PW)_{SK}$	M_5 ⟶	Decrypt M_5: USER', PW' calculate H (USER', PW') Verify H (USER', PW') = H (USER, PW)

Figure2. User's authentication.

3 VALIDATION

The validation is proved via the Strand Space Model.

3.1 *Basic concepts*

In protocol, all massages sent or received by subjects comprise the set *M*. The random numbers and subject names in massages comprise the set T. T_{name} is subject name set and $T_{name} \subseteq T$. A signed term can be represented by a binary group $<\sigma, a>$. Among which, a is the element of M, representing a protocol massage; σ is "+" or "-", representing the massage operation receiving or sending.

The strand is events sequences of protocol subjects, including sending and receiving sequences. The set of strands comprise the strand space. When s is a strand in a strand space, the trail of s is composed of all the finite sequences with signed items in s, marked as $tr(s)$. And $len(tr(s))$ represents the length of the finite sequences. The node is an ordered pair $<s,i>$. i is an integer and $1 \le i \le len(tr(s))$. Every node belongs to a unique strand and if the node n satisfies $n = <s,i>$, we always call that the node n is in the strand s. $term(n)$ represents the massages sent or received by node n. Nodes and their edges comprise a directed graph. If C is a sub-graph, C is called a bundle. That the node n ($n = <s,i>$) is in the C can be marked as $n \in C$. The maximum i in strand s that satisfies $<s,i> \in C$ is called the C height of strand, represented as $C - height(s)$.

3.2 *Protocol description*

According to the basic concepts and the designed protocol, the formal description is shown in Figure 3.

The infiltrate strand space of the designed protocol can be marked as (Σ, P), then Σ is defined as the composition of the following three types of strands.

1. Intruder strand $p \in P$. P is the set of attackers.
2. Initiator strand $S_{init} \in Init[ID_A, ID_B, N_A, N_B,, aG, bG]$ having the trail $< +(ID_A, N_A)$, $-[(bG, N_A, N_B)_{K_{AB}}, ID_B]$, $+(N_B, aG)_{K_{AB}}$, $-H(aG, N_B, ID_B)>$. All the strands having this kind of trail comprise a set, represented by $Init[ID_A, ID_B, N_A, . N_B, aG, bG]$ The related subject is A.
3. Responder strand $S_{resp} \in Resp[ID_A, ID_B, N_A, N_B, aG, bG]$, having the trail $< -(ID_A, N_A)$, $+[(bG, N_A, N_B)_{K_{AB}}, ID_B]$, $-(N_B, aG)_{K_{AB}}$, $+H(aG, N_B, ID_B)>$. All the strands having this kind of trail comprise a set, represented by $Resp[ID_A, ID_B, N_A, N_B, aG, bG]$. The related subject is B.

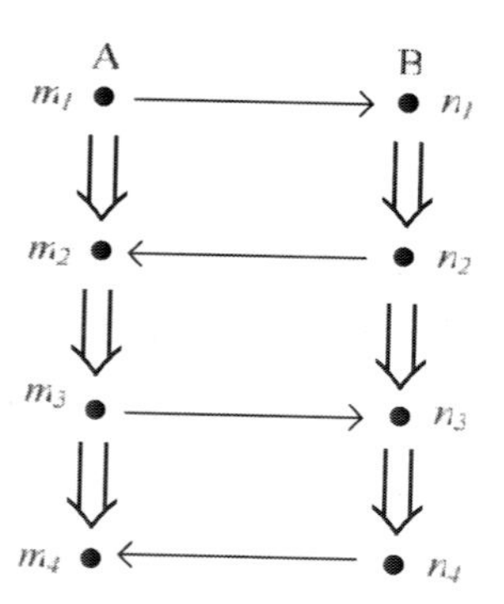

Figure 3. Protocol description.

$ID_A, ID_B \in T_{name}$, $N_A, N_B \in T$. According to definitions above, for any strand $s \in \Sigma$, the strand type can be confirmed on the basis of its trail's form uniquely.

3.3 *Authentication proof*

The authentication proof contains two parts. The one hand is to prove that the host identifies the device; the other hand is to prove the device identifies the host. As the initiator, A thinks that the certain data item is used to complete a protocol round by A itself and subject B. Thus, for subject B, there must be an only protocol round corresponding to subject A, and B uses the same data item as a responder.

According to the authentication test theory of the Strand Space Model, the device authentication can be transformed into proving proposition 1, and the host authentication can be transformed into proving proposition 2.The proof is based on the free hypothesis of the Strand Space Model.

Proposition 1: C is a bundle of this protocol's strand space. $ID_A \ne ID_B$. aG and N_A are only originated in C. $aG \ne bG$. $K_{AB} \notin K_p$. K_p is the set of intruders' keys. If s is a initiator strand, $s \in Init[ID_A, ID_B, N_A, N_B, aG, bG]$ and C-$height(s)$ =4, then there must be a normal responder strand $t \in Resp[ID_A, ID_B, N_A, N_B, aG, bG]$, and C-$height$ (t) =4.

From the protocol description, N_A is only originated in m_1. If $t_1 = (bG, N_A, N_B)_{K_{AB}}$, $t_1 \not\subset term(m_1)$, $N_A \subset t_1$, $t_1 \subset term(m_2)$, $K_{AB} \notin K_p$, then according to the definitions of input authentication test, the edge $m_1 \Rightarrow^+ m_2$ is t_1's input test edge about N_A.

According to input authentication test rules, there are normal nodes m and m' in bundle C, making $m \Rightarrow^+ m'$ become the transforming edge of N_A, and $t_1 \subset term(m')$. Since m' is a positive node, from the bundle massage form, m' can only be a node in a certain responder strand t, $t \in Resp[ID'_A, ID'_B, N'_A, N'_B, (aG)', (bG)']$. Meanwhile, $m = <t,i>$ (i is an integer and $1 \le i \le len(tr(t))$). $t_1 = (bG, N_A, N_B)_{K_{AB}} \subset term(<t,i>)$. Comparing $(bG, N_A, N_B)_{K_{AB}}$ with the contents in responder strand, since $K_{AB} \notin K_p$, and $(bG, N_A, N_B)_{K_{AB}}$ can only be produced in $<S_{resp}, 2>$, from the encryption free axiom, the following can be got: $ID'_A = ID_A$, $ID'_B = ID_B$, $N'_A = N_A$, $N'_B = N_B$, $(bG)' = bG$. That is, $m \Rightarrow^+ m'$ is actually $<S_{resp}, 1> \Rightarrow^+ <S_{resp}, 2>$, and also is $n_1 \Rightarrow^+ n_2$.

aG is only originated in m_3. If $t_2 = (N_B, aG)_{K_{AB}}$, $t_2 \subset term(m_3)$, $aG \subset t_2$, $t_2 \not\subset term(m_4)$, $K_{AB} \notin K_p$, according to the definitions of output authentication test, the edge $m_3 \Rightarrow^+ m_4$ is t_2's output test edge about aG. According to output authentication test rules, there are normal nodes n, n' in bundle C, making $n \Rightarrow^+ n'$ become the transforming edge of aG and $t_2 \subset term(n)$, Since n is a negative node, from the t_2's massage form, can only be a node in responder strand t, $t \in Resp[ID_A, ID_B, N_A, N_B, (aG)', bG]$. Meanwhile, $n = < t, i >$ (i is an integer and $1 \leq i \leq len(tr(t))$). Comparing $(N_B, aG)_{K_{AB}}$ with the contents in responder n strand, since $K_{AB} \notin K_p$, and $(N_B, aG)_{K_{AB}}$ can only be produced in $< S_{resp}, 3 >$, so $(aG)' = aG$. $n \Rightarrow^+ n'$ is $< S_{resp}, 3 > \Rightarrow^+ < S_{resp}, 4 >$, that is, $n_3 \Rightarrow^+ n_4$, and $C\text{-}height(t)=4$.

In a word, $t = S_{resp}$, so the responder strand $t \in Resp[ID_A, ID_B, N_A, N_B, aG, bG]$ is contained in bundle C, and $C\text{-}height(t)=4$.

Proposition 2: C is a bundle of this protocol's strand space. $ID_A \neq ID_B$. N_B is only originated in C. $K_{AB} \notin K_p$. K_p is the set of intruders' keys. If t is a responder strand, $t \in Resp[ID_A, ID_B, N_A, N_B, aG, bG]$ and $C\text{-}height(t) = 4$, then there must be a normal initiator strand $s \in Init[ID_A, ID_B, N_A, N_B, aG, bG]$ and $C\text{-}height(s) = 3$.

From the protocol description, N_B is only originated in n_2. If $t_3 = (bG, N_A, N_B)_{K_{AB}}$, $t_3 \subset term(n_2)$, $t_3 \not\subset term(n_3)$, $N_B \subset t_3$, then according to the definitions of output authentication test, the edge $n_2 \Rightarrow^+ n_3$ is t_3's output test edge about N_B.

According to output authentication test rules, there are normal nodes m and m' in bundle C, making $m \Rightarrow^+ m'$ become the transforming edge of N_B, and $t_3 \subset term(m)$. m' is positive.

m is a negative node. From the bundle massage form, m can only be a node in a certain initiator strand s, $s \in Init[ID'_A, ID'_B, N'_A, N'_B, (aG)', (bG)']$. Meanwhile, $m = < s, i >$ (i is an integer and $1 \leq i \leq len(tr(t))$), and $t_3 = (bG, N_A, N_B)_{K_{AB}} \subset term(< s, i >)$. Comparing t_3 with contents in initiator strand, since $K_{AB} \notin K_p$, and m must be $< S_{init}, 2 >$, so it can be inferred that $ID'_A = ID_A$, $ID'_B = ID_B$, $N'_A = N_A$, $N'_B = N_B$, $(bG)' = bG$.

If $t_4 = (N_B, (aG)')_{K'_{AB}}$ is a test component of node m', $K_{AB} \notin K_p$, and t_4 is not sub-item of any normal test component, according to output authentication test rules, since $m'' = < S_{resp}, 3 >$, $(N_B, (aG)')_{K'_{AB}} = term(< S_{resp}, 3 >) = (N_B, aG)_{K_{AB}}$, with the help of the free hypothesis of the Strand Space Model, it can be inferred that $(aG)' = aG$. Then, $m \Rightarrow^+ m'$ is $< S_{init}, 2 > \Rightarrow^+ < S_{init}, 3 >$, that is $n_2 \Rightarrow^+ n_3$, so $C\text{-}height(s)=3$.

In a word, $s = S_{init}$, so the initiator strand $s \in Init[ID_A, ID_B, N_A, N_B, aG, bG]$ is contained in bundle C, and $C\text{-}height(s)=3$.

From the above proof, we can get a conclusion, that is, the designed protocol can realize the mutual authentication.

4 PERFORMANCE

4.1 *Safety analysis*

The designed protocol can authenticate all the entities in communication, so the protocol is able to resist the counterfeit attack. The protocol is executed in the way of challenge-response; the fresh factors confirm that the massages in channel can't be remade. And the massages are design not symmetrical in form deliberately, so the replay attack can be resist. Besides, the session key is agreed in the protocol, as a result, the data in channel can be protected by the encryption, so the bus monitoring attack can also be resisted. Compared with other schemes, the results are listed in Table 2, showing that our scheme has a higher safety.

Table 2. Safety comparison.

Schemes	bus monitoring attack resistance	counterfeit attack resistance	Group usage	replay attack resistance
Reference [7]	×	√	√	√
Reference [9]	√	√	×	√
Reference [12]	√	√	√	√
Our Scheme	√	√	√	√

Note: √ able; × unable

4.2 *Experiment test*

In host, as a man-machine interactive interface, the software is designed to input users' names and passwords, as well as display the system working state. Meanwhile, the USB driver is modified. Through adding the process of handling with the self-defined authentication requests in the driver dispatch, the host can send authentication request to the device. Besides, the encryption/decryption process for bus data is also added. The data will be encrypted before sent out, while decrypted before received.

In device, a simple USB device is constructed based on the USB device controller Intellectual Property (IP) core. Through adding the safety function to the core, the authentication request can be responded. The device runs on an FPGA. The FPGA we chose is EP2C20Q240C8, which is the second generation of Cyclone series products made by Altera. The physical transceiver we chose is USB3280 made by SMSC. A computer is connected with FPGA through the serial port to show the device state. Synthesizing in Quartus II, the device costs are 2490 Logic Elements and 1449 Registers in all.

The host and device work together to realize the designed protocol. Thus, the protocol performance can be tested. The test environment is shown in Figure 4.

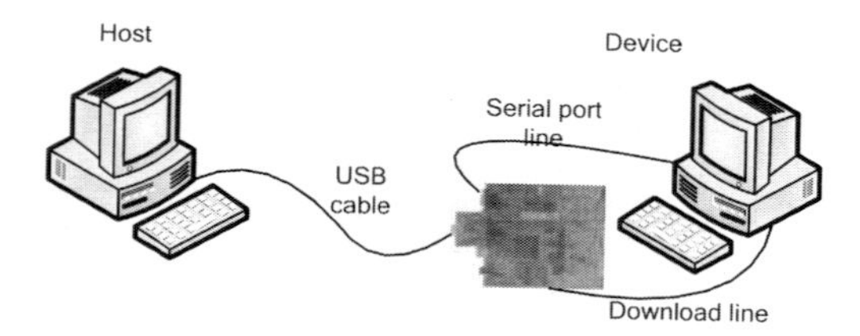

Figure 4. Test environment.

Protocol performance is usually measured through the numbers of communication rounds and the loads of calculation. Since reference [12] doesn't contain the specific process of user's authentication, the test only aims at the mutual authentication process. The comparison results are listed in Table 3. Among the table, the point multiplication means the sum number of point multiplication operated by the two communication parts in one protocol execution process. The symmetrical encryption and hash mean the sum numbers of the relevant operations too.

From the table 3, with the hybrid cryptosystem, our scheme decreases the cost and increases the safety. The protocol efficiency is improved.

5 SUMMARY

We have concluded the authentication schemes and problems of USB devices. Based on which, we propose a novel protocol relying on the ECC and EKE. Through the design of mutual authentication and user's authentication, the communication entities can all be identified. We use the authentication test theory of the Strand Space Model to prove the protocol validation. And then, we modify the USB IP core to realize the protocol. The results show that our designed protocol has a good performance that several kinds of attacks can be resisted and the protocol is more efficient. More management can be implemented based on the protocol, so that the data safety can be protected thoroughly.

ACKNOWLEDGMENT

This paper is subsidized by the Foundation of Science and Technology on Information Assurance Laboratory (No. KJ-14-103) and the Science and Technology Key Problems Tackling Project of Henan Province (No. 132102210003)

REFERENCES

[1] Wang Wei. (2010) *Research and Implementation of the U_Disk with Identity Authentication and Data Encryption*. Shenyang: Shenyang Institute of Aeronautical Engineering.

[2] Sun-Ho Lee, Jin Kwak, Im-Yeong Lee. (2009) The study on the security solutions of USB memory. *2009 International Conference on ubiquitous Information Technologies & Applications, ICUT2009, Fukuoka, Japan*, pp:470–473.

[3] Yi Qingsong, Su Jinhai, Yue Yuntian, Dai Zibin. (2007) Hardware design of secret USB disk based on CY7C68013. *Computer Engineering and Design*, 28(6):1297–1302.

[4] Hu Jianyong, Su Jinhai. (2007) Design and implementation of a secret carrier USB disk based on FPGA. *Electronic Technology*, 36(11):95–99.

[5] Hou Xinchao, (2007) *Research and Design on Host Information Leakage Prevention System Based on the Management of Mobile Storage Device*. PLA Information Engineering University.

[6] Handschuh Helena, Trichina E. (2008) Securing flash technology: how does it look from inside? *Information Security Solutions Europe Conference, 2008, Madrid (ES)*, pp:380–389.

[7] Yin Wenhao, (2011) *Design of Security U-Disk and Application in Management of Files with Multi-Security Levels*. PLA Information Engineering University.

[8] Wang An, (2011) *Research on Key Technologies of Cryptographic System on a Chip*. Shandong University.

[9] Yang Xianwen, Li Zheng, Wang An, Zhang Yu. (2010) System design of cryptographic security USB device controller IP core. *J. HuaZhong Univ. of Sci. & Tech. (Natural Science Edition)*, 38(9): 59–62.

[10] Diffie W, Hellman M E. (1976) New directions in cryptography, *IEEE Transactions on Information Theory*, 22(6):644–654.

[11] Li Li, Xuerui, Zhang Huanguo, Feng Dengguo, Wang Lina. (2005) Security analysis of authenticated key exchange protocol based on password. *Acta Electronica Sinica*, 33(1):166–170.

[12] Luo Naiwei, Ye Kuohui. (2011) Simple three-party password authenticated key exchange protocol.

Table 3. The comparison of scheme costs.

Schemes	Communication Rounds	Point Multiplication	Hash	Symmetrical Encryption	Matrix operation
Reference [12]	3	10	4	2	2
Our scheme	4	4	4	4	—

— not exist

Journal of Shanghai Jiaotong University (Science), 16(5):600–603.
[13] Cao Tianjie, Lei Hong. (2008) Privacy-enhancing authenticated key agreement protocols based on elliptic curve cryptosystem. Acta Electronica Sinica, 36 (2):397–401.
[14] Yue Leixiao, Yu minwang, Liao junpang. (2012) Verification of trusted network access protocols in the strand space model. *IEICE Transactions on Fundamentals of Electronics, Communications & Computer Sciences*, E95-A (3):665–668.

Electric, Electronic and Control Engineering – Shao, Shu & Tian (Eds)

Author index